C000077784

PARSONS 2003

Organising Committee

R. D. Conroy (Chairman)	Siemens Power Generation
W. M. Banks (Vice-Chairman)	University of Strathclyde
M. Blackler	Howmet Ltd
J. Leggett	Rolls-Royce plc
G. McColvin	Demag Delaval Industrial Turbomachinery Ltd
J. Monaghan	Trinity College Dublin
G. Scaife	Trinity College Dublin
S. Simpson	Power Technology, PowerGen
M. Smith	Mitsui Babcock (US) LLC
F. Starr	European Technology Development Ltd
A. Strang	University of Leicester
R. W. Vanstone	ALSTOM Power

International Liaison Committee

H. -O. Andren	Chalmers University of Technology, Sweden
P. Auekari	VTT, Finland
R. Blum	ELSAN Energy, Denmark
H. Cerjak	Technical University of Graz, Austria
A. Czyrska-Filemonowicz	University of Mining and Metallurgy, Poland
P. J. Ennis	Research Centre Juelich, Germany
V. Foldyna	JINPO Plus, Czech Republic
J. Hald	Technical University of Denmark
T. -U. Kern	Siemens Power Generation
T. Khan	ONERA, France
K. Kimura	NIMS, Japan
V. Sklenicka	IPM, Czech Republic
T. M. Theret	SNECMA, France
M. C. Thomas	Rolls-Royce Allison, USA
V. R. Viswanathan	EPRI, USA

Acknowledgements

The Conference Organising Committee and The Institute of Materials, Minerals and Mining would like to thank the following organisations for their generous support of the conference.

- ALSTOM Power
- Demag Delaval Industrial Turbomachinery Ltd
- Mitsui Babcock
- Power Technology, Powergen
- Rolls-Royce plc
- Siemens Power Generation

PARSONS 2003
Engineering Issues in Turbine Machinery, Power Plant and Renewables

Proceedings of the Sixth International
Charles Parsons Turbine Conference
16–18 September 2003
Trinity College Dublin, Ireland

Edited by
A. Strang, R. D. Conroy, W. M. Banks
M. Blackler, J. Leggett, G. M. McColvin
S. Simpson, M. Smith
F. Starr and R.W. Vanstone

MANEY
FOR THE INSTITUTE OF MATERIALS, MINERALS AND MINING

B0800
First published in 2003 for
The Institute of Materials, Minerals and Mining by
Maney Publishing
1 Carlton House Terrace
London SW1Y 5DB

© IOM[3] 2003
All rights reserved

ISBN 1-904350-20-8

Typeset in India by Emptek Inc.
Printed and bound in the UK by
The Charlesworth Group, Huddersfield

Contents

Operation and Maintenance

Materials, Processing and Microstructures

Modelling and Prediction of Creep Properties

Gas Turbine Materials for Blades and Discs

Environmental Impact

Novel Materials

Repair Technologies

Renewables 1007

Alternative Cycles and Regenerative Systems

Renewables

Welcome to Trinity College Dublin – The Parsons Connection

It is with great pleasure that we welcome delegates and distinguished guests to Trinity College Dublin for this the 6th International Charles Parsons Turbine Conference.

The University of Dublin has one college, Trinity College and is Ireland's oldest university having being established over 400 years ago in 1592 by Queen Elizabeth 1 of England. As the oldest university in Ireland, and one of the oldest in the Western Europe, Trinity is very conscious of its teaching and research heritage and the efforts of both students and staff in continuing to foster intellectual drive.

Engineering has been taught in Trinity College for just over 160 years, with the first professor, of engineering appointed in 1842. From its foundation as one the first engineering schools in the world, engineering in Trinity has been in the forefront of technological progress. This continues today through teaching and research on contemporary topics such as aerospace applications, bioengineering advanced manufacturing technology, information technology, microchip fabrication, urban transportation, and much more.

The connection between the Parsons family and Trinity goes back to William Parsons the Third Earl of Rosse who was an astronomer and was responsible for building the giant six foot reflecting telescope at Birr Castle in Co Offaly. William was elected President of the Royal Society in 1848, and became chancellor of the University of Dublin in 1862. However it is generally through the efforts of Charles Parsons, Williams's youngest son, that the link between Trinity College and the Parsons family is better recognised. Charles was educated at Birr by tutors employed as astronomical assistants and following his early education he, like his three older brothers, entered Trinity as a student in 1871. In 1873 he transferred to Cambridge University where he graduated in 1877 with a degree in mathematics. On graduation, Charles served as a premium apprentice at Sir W. G. Armstrong's Elswick plant at Newcastle upon Tyne. In 1884 he became a partner of Clarke Chapman where he built and patented the world's first practical steam turbine generator. He opened his own manufacturing plant at Heaton in Newcastle in 1889 and was joined there by Gerald Stoney a brilliant engineering graduate of Trinity College Dublin who acted as Parsons' right hand man until 1912.

The association between Charles Parsons, Stoney and Trinity was further strengthened by the presentation of the number 5 turbine to College in 1911. Other links between Trinity and the Parsons companies included Alex Law, an engineering graduate of Trinity, who joined Parsons in 1900 and who eventually became chief electrical designer and a director of the company. Another connection was through Charles' elder brother Laurence, a Trinity science graduate, Chancellor of Dublin University and chairman of Parsons marine turbine companies.

In 1911 Charles was knighted in recognition of his contribution to British industry. Sadly however until quite recently his achievements were not honoured in his native country. However the Royal Irish Academy, the Institution of Engineers of Ireland, and the Republic of Ireland branch of the Institution of Mechanical Engineers helped rectify this omission through their sponsorship of the First Parsons Turbine Conference held at Trinity in 1984. Parsons' achievements have recently been further recognised by the Royal Irish Academy through the inauguration of its own Parsons Medal, which is to be presented this year at the PARSONS 2003. It is therefore fitting to see the 6th International Charles Parsons Turbine Conference returning again to Trinity nearly two decades later. It is also gratifying to see that the conference has once again received support from old friends in both the UK and Ireland such as ALSTOM Power, Mitsui Babcock, Rolls-Royce, Siemens Power Generation, the Royal Society, the Institution of Mechanical Engineers, the Institute of Materials, Minerals and Mining, the Royal Irish Academy, the Institution of Engineers of Ireland and the Republic of Ireland Branch of the Institution of Mechanical Engineers and the Electricity Supply Board.

We hope you enjoy your visit to Trinity College Dublin and that while you are here you will take the opportunity to explore the College and spend some time among buildings that would have been very familiar to Charles Parsons and his family over 130 years ago.

John Monaghan & Garrett Scaife

Foreword

PARSONS 2003, the Sixth International Charles Parsons Turbine Conference has arrived back in Dublin after an absence of 19 years. The conference series which started in Dublin in June 1984, celebrating the centenary of Charles Parsons' steam turbine and generator patents, has proved and continues to be a lasting forum for the presentation and discussion of design and materials engineering technologies for aero and industrial gas turbines and power generation plant.

Both industries have been severely affected by the dramatic global market changes that have taken place in the past two years. However, despite these changes, the need for innovative improvement and change in terms of the development and economic manufacture of efficient environmentally friendly turbines and power plant remains paramount. This is clearly shown in the diversity of conference papers presented on gas turbines, steam plant technology, advanced energy conversion systems, as well as the rapidly expanding development of renewable energy resources for power generation.

The Conference opening plenary session contains major keynote papers on these four main topic areas and ends with a final plenary discussion on the challenge of moving towards the 'holy grail' of zero emission power generation. In this latter respect continued recognition of the fast growing importance of renewable energy generation together with regenerative and alterative generation cycles is shown by their prominence in the Conference. The drive for improved efficiency low emission fossil fired power generation also continues to feature strongly through papers on the design and manufacture of advanced steam power plant. The materials and design technologies developed and introduced over the past two decades for steam power plant operating up to 620°C are now fairly well established with thermal efficiencies approaching 50% now being achieved. Further development in the properties of advanced creep resistant martensitic steels may enable steam temperatures to be raised to 650°C with consequent improvements in thermal efficiency. However for operating temperatures above this, with potential thermal efficiencies of 55% and greater, the use of nickel base alloys will be required and continuing development to determine these long-term microstructural and property changes and the developments for even higher temperature operation are addressed. These developments are being driven by the need to reduce emissions and operating costs and are also mirrored by the technology development for aero and industrial gas turbines.

A major development during the past decade has been the growth of plant upgrading, inspection and life extension, which reduces operator's costs by improving performance and plant life. Many of the papers presented include descriptions of new technologies that can be applied to both new and existing plant which together with property and material microstructural assessment and inspection can be applied for the safe and improved performance of gas turbines and steam plant.

The 'Parsons' conferences have always been a good opportunity for delegate networking and no better tribute can be made to the genius of Sir Charles Parsons than the voice of stimulating discussion on the continuing development of the technologies he invented in 1884. There is no doubt that were Parsons alive today he would be astounded by the developments in materials, design and manufacturing which have taken place in gas turbines and steam generating plant and the fact that the legacy of his discoveries of more than a century ago still provide the principal source of power in today's world.

It is particularly pleasing that Trinity College Dublin has taken a very active role in the organisation of this Conference, as indeed they did for the first held at Trinity College, Dublin in 1984. It is also pleasing that Professor Garrett Scaife, who chaired the Organising Committee for the first Parsons Conference and who is a leading authority on the life and works of Charles Parsons, presents the PARSONS 2003 Conference celebrity lecture entitled 'Charles A Parsons - The Substance of a Great Man'.

The Organising Committee and the Institute of Materials, Minerals and Mining are also very pleased and honoured to host the Royal Society's annual Charles Parsons Memorial Lecture. This year's lecture, 'Industrial gas turbines – a challenge for the future', will be given by Mr Frank Carchedi, technical director of Demag Delaval Industrial Turbines Ltd. It is also fitting that this year's lecture should given during the PARSONS 2003 Conference in Ireland, the land of Charles Parsons' family, childhood and youth.

Finally thanks must be given to all of the Conference Organising Committee members for their considerable efforts in planning PARSONS 2003 and to the Institute of Materials, Minerals and Mining Conference and Events Department and our hosts at Trinity College Dublin for all of the detailed arrangements.

Roger Conroy, Conference Chairman
Andrew Strang, Proceedings Editor

INTRODUCTION

Royal Society Charles Parsons Memorial Lecture

Industrial Gas Turbines – Challenges for the Future

F. CARCHEDI

*Demag Delaval Industrial Turbomachinery Ltd**

P.O. Box 1, Waterside South, Lincoln LN5 7FD, UK

ABSTRACT

The Industrial Gas Turbine for power generation and mechanical drive applications has been the subject of continuous development over the past fifty years with dramatic improvements in fuel efficiency, power density and exhaust emissions.

The evolution of engine design parameters in the Demag Delaval (formerly ALSTOM) range of industrial gas turbines illustrates the continuous development that is needed to maintain competitive performance. Engine performance improvement relies mainly on advances in turbomachinery technology and advances in this area have much in common with aircraft engine technology.

Significantly different operating conditions, engine environments and fuel types are evident for aircraft propulsion gas turbines, large combined cycle gas turbines for base-load power generation and smaller gas turbines for combined heat and power or oil and gas industry applications. These differences are reflected in both the engine design and materials choice.

The challenge of meeting legislation to reduce noxious emissions from industrial gas turbines to very low levels has been successfully met by the introduction of lean burn premixed combustion systems. Initially introduced for natural gas fuel, these systems are now being applied to a wider range of liquid and gas fuels.

While electronic engine control systems have been used for the past thirty years, modern computing and telecommunications systems are now allowing sophisticated remote control and monitoring of engine operation. The development of continuous health monitoring and trending, fault monitoring and life assessment will provide enhanced plant reliability and allow 'on condition' maintenance in the future.

INTRODUCTION

The gas turbine engine has now become the first choice for power generation in addition to it's dominance of the aerospace propulsion market. Over the past twenty years, the base-load electricity generation market has seen the emergence of the natural gas fired combined

* On 30 April 2003 ALSTOM sold its Industrial Gas Turbines businesses in Lincoln, Spain and Singapore to Siemens. The registered name of the business is Demag Delaval Industrial Turbomachinery Ltd. and is a wholly owned Siemens company.

cycle gas turbine (CCGT) as the most efficient, cost effective and clean means of generating power. The gas turbine continues to be the power plant of choice for the oil and gas industry and is a major contributor to efficient energy usage in combined heat and power applications. This position has only been achieved by intensive development work to enhance performance while reducing noxious emissions and minimising cost of ownership.

There has also been a widening of the range of powers offered by industrial gas turbines to meet evolving markets which now range from micro turbines at 30 kW to large 50 Hz engines at 260 MW.

For the future, all major gas turbine manufacturers are looking to target further improvements in performance, but in addition, new technologies such as gas turbine fuel cell hybrids and zero emissions technologies offer exciting challenges for the future.

Performance improvement will continue to drive turbine inlet temperatures higher, but the need to burn alternative fuels with a wide range of calorific values, as the availability of premium fuels such as natural gas declines, will place increasing demands on high temperature materials and coating technology.

The technology synergies between the aircraft gas turbine and the industrial gas turbine have allowed considerable benefits to be derived from technology transfer and collaborative component technology projects.

Engine Development Trends

For the simple cycle gas turbine, while there is a strong inter-relationship between firing temperature and pressure ratio for optimum performance, in broad terms, firing temperature will increase engine specific power and pressure ratio will increase thermal efficiency. For civil aircraft engines where fuel efficiency is all important, the emphasis has been on high pressure ratios, which can reach up to 40 in the latest multi-spool high bypass ratio engines.

On engines designed for industrial duty, exhaust heat availability for cogeneration applications is an important issue and in CCGT plants efficiency is more dependent on operation at high firing temperatures. In purpose designed industrial gas turbines, lower pressure ratios are chosen to provide a more optimum mix of fuel efficiency and exhaust heat for industrial applications.

This is demonstrated by the trends for the Demag Delaval range of small industrial gas turbines generating less than 15 MW shown in Figure 1. This indicates that over the past 25 years firing temperatures have increased by almost 20°C per year and over the same time period engine pressure ratio has increased from 7 to 17.

Prior to the start of this period, increases in firing temperature were achieved as a result of the development of improved, higher temperature materials. However, the introduction of cooled turbine blading to industrial gas turbines from around 1980 onwards has resulted in an increased rate of firing temperature increase, relying both on high temperature materials development and the development of more complex cooling systems.

The impact of the increases in firing temperature and pressure ratio on engine performance is shown in Figure 2 and indicates the dramatic improvements that have been obtained over the past 25 years. Specific Power has increased by almost 100%, whilst Specific Fuel Consumption has been reduced by over 30% in the same period.

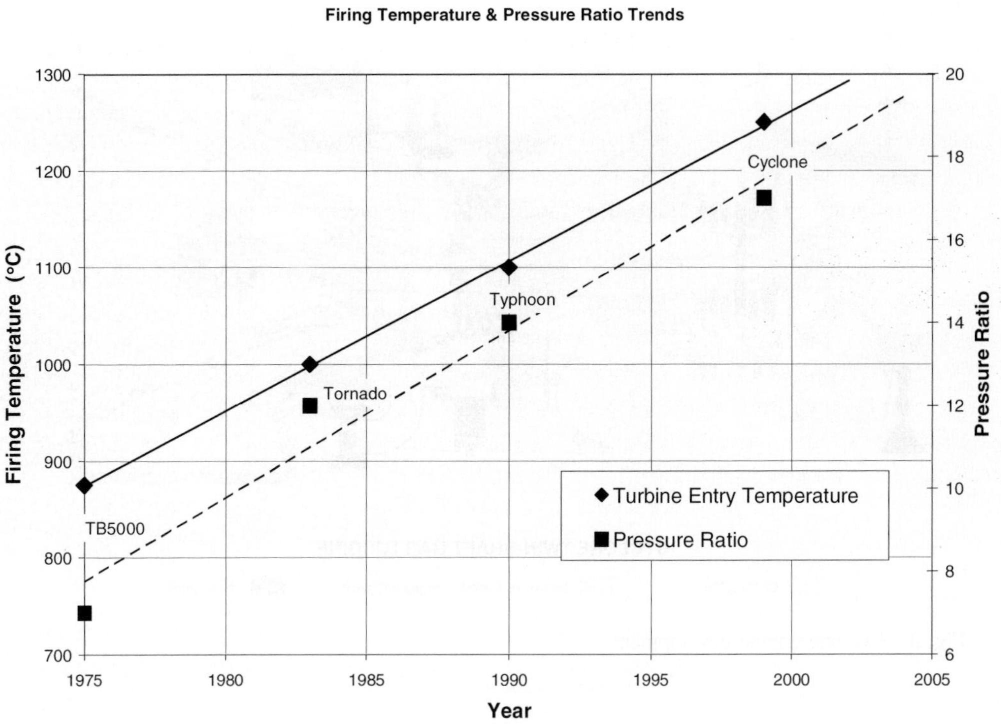

Fig. 1 Increase in firing temperature and pressure ratio over time.

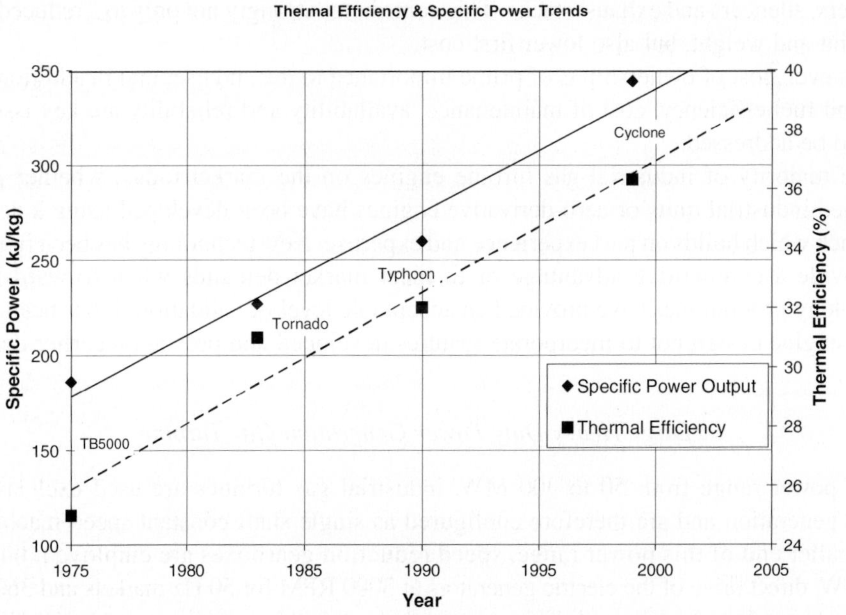

Fig. 2 Increase in thermal efficiency and specific power output over time.

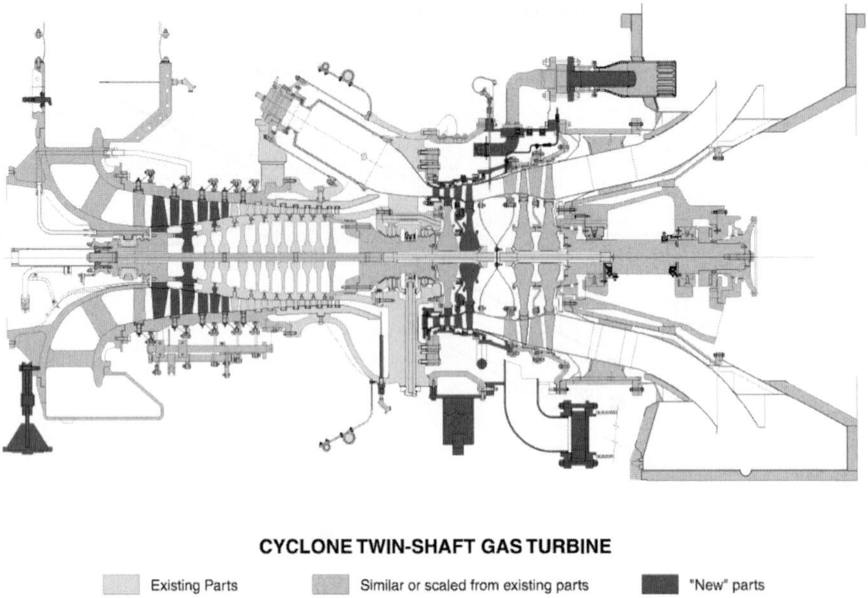

Fig. 3 Cyclone engine development.

Improvements in specific power reduce the airflow requirement for a given power output and has resulted in smaller engine sizes, and smaller sizes for ancillary equipment such as air filters, silencers and exhaust stacks. This contributes strongly not only to a reduced engine footprint and weight, but also lower first cost.

However, cost of ownership is of prime importance to the end user and in addition to first cost and fuel efficiency, cost of maintenance, availability and reliability are key issues that need to be addressed.

The majority of industrial gas turbine engines on the market today, whether purpose designed industrial units or aero-derivative engines have been developed using a derivative approach which builds on past experience and expertise. New technology has been introduced to provide a competitive advantage or to meet market demands when forward-looking technology programmes have provided an acceptable level of validation. It has been rare for a new engine design not to incorporate features developed and proven on earlier designs.[1]

Large Heavy Duty Power Generation Gas Turbine

In the power range from 50 to 300 MW, industrial gas turbines are used exclusively for power generation and are therefore configured as single shaft constant speed machines. At the smaller end of this power range, speed reduction gearboxes are employed, but above 100 MW, direct drive of the electric generators at 3000 RPM for 50 Hz markets and 3600 RPM for 60 Hz markets is universal. This means, that other things being equal, a 50 Hz unit is 44% larger in airflow and power than a 60 Hz unit.

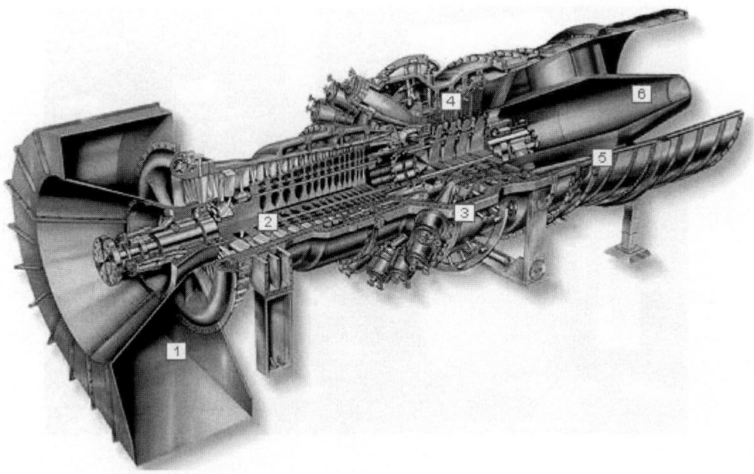

Fig. 4 Siemens W501G 253 MW industrial gas turbine.

The introduction of large gas fired combined cycle gas turbines increased power generation efficiencies from 40% for large coal fired steam turbines to over 50% for CCGT. Although in terms of operating economics this was offset to some extent by the higher gas fuel cost, capital costs for CCGT were lower and there were large benefits in reduced emissions. SO_2 emissions are largely eliminated, CO_2 greenhouse gases were reduced in line with the improved fuel efficiency and the introduction of lean pre-mix combustion technology reduced NOx emissions by over an order of magnitude.[2]

For the CCGT cycle, efficiency and unit power output are both maximised by increased turbine firing temperature and with conventional turbine cooling technology firing temperature has been increased to over 1300°C with efficiencies of over 57%. However, the competing demands on air usage for cooling air in turbines and combustion air in lean burn low emissions combustors is setting a limit on the firing temperatures that can be achieved with air cooled blading in available superalloy materials. Alternative technologies are being pursued to overcome this. Use of steam rather than air for cooling and sequential combustion to apply reheat within the turbine stages are features of the latest large base-load power generation gas turbines. Steam cooling has allowed firing temperatures to exceed 1400°C and an efficiency level of 60% to be achieved.

Small & Mid Sized Industrial Gas Turbines

In the power range from 1 to 50 MW, the main applications for industrial gas turbines are for industrial cogeneration, power generation in peaking applications and oil and gas industry applications. The latter requires engines both for power generation and for mechanical drive of pumps and compressors. Mechanical drive requires variable speed operation with a twin

Fig. 5 Cyclone heavy duty GT.

shaft engine having a free power turbine, an engine configuration quite different to that required purely for power generation, but one found in many aircraft gas turbines.[3]

While twin shaft engines can be used to drive electric generators, frequency control and transient response to load changes are worse than what is achievable on a single shaft engine, particularly at smaller sizes. It is therefore common to find that below 10 MW, both single shaft and twin shaft variants of an engine are produced to maximise market acceptability for all applications.

The oil and gas industry has demanded rugged and reliable products and in simple cycle applications, fuel efficiency is governed more by cycle pressure ratio than by firing temperature. As a result, there has been less of a driver to push towards the high firing temperatures typical of the larger CCGT cycles and firing temperatures on recently introduced engines are of the order of 1250°C. However, the smaller industrial GT will experience a significantly more severe cyclic operation than a large baseload GT in terms of both starts and load changes, and will experience an increased level of thermo-mechanical fatigue damage.

Heavy duty industrial gas turbines in this power range can provide cycle efficiencies of up to 37% and in cogeneration applications, an overall thermal utilisation of around 90%. These engines compete with aero-derivative industrial gas turbines, which when derived from the latest high bypass ratio civil engines, can offer cycle efficiencies of up to 42% but with lower exhaust temperatures that are less suitable for cogeneration applications.

NEW DEVELOPMENTS

Today's rapidly changing energy markets and emerging legislation have stimulated interest in novel gas turbine technologies which require significant departures from previous practice

Fig. 6 RLM5000 aeroderivative GT.

and these are being supported by research programmes initiated by government agencies in Europe, Japan and the USA.

Distributed generation to provide power close to the point of use and minimise reliance on transmission grids has stimulated interest in complex cycles for power generation at smaller sizes. Technologies such as exhaust recuperation, intercooled compressors and humidified or steam injected cycles are currently being re-evaluated. The first commercial products to emerge as a result of this have been recuperated microturbines in sizes up to 100 kW with thermal efficiencies of around 30%. At the present time, these have had a much smaller impact on the power generation market than had been predicted at the time of their introduction. Mid-size complex cycle GT concepts at up to 50 MW with an efficiency of around 50% have the potential to offer generation costs competitive with large CCGT and are currently the focus of US DOE programmes.[4]

Gas turbine/fuel cell hybrid systems utilising high temperature fuel cells such as the Solid Oxide Fuel Cell (SOFC) or Molten Carbonate Fuel Cell (MCFC) have the highest potential efficiency levels in the 65 to 70% range. Development effort in this field is mainly aimed at scale up, life enhancement and cost reduction of the high temperature fuel cell technology.

Reductions in CO_2 greenhouse emissions are stimulating developments in biomass fuelled gas turbines burning a fuel derived from energy crops or from waste which is CO_2 neutral.[5] An alternative approach to eliminate greenhouse gases is CO_2 sequestration where the CO_2 is captured within the engine cycle and stored either underground or on the seabed under hydrostatic pressure.

Pilot plants to develop and demonstrate these technologies are in operation or development.[6,7] Market forces and increasing legislation on emissions and energy efficiency are likely to see commercial development of the most promising of these new technologies in coming years.

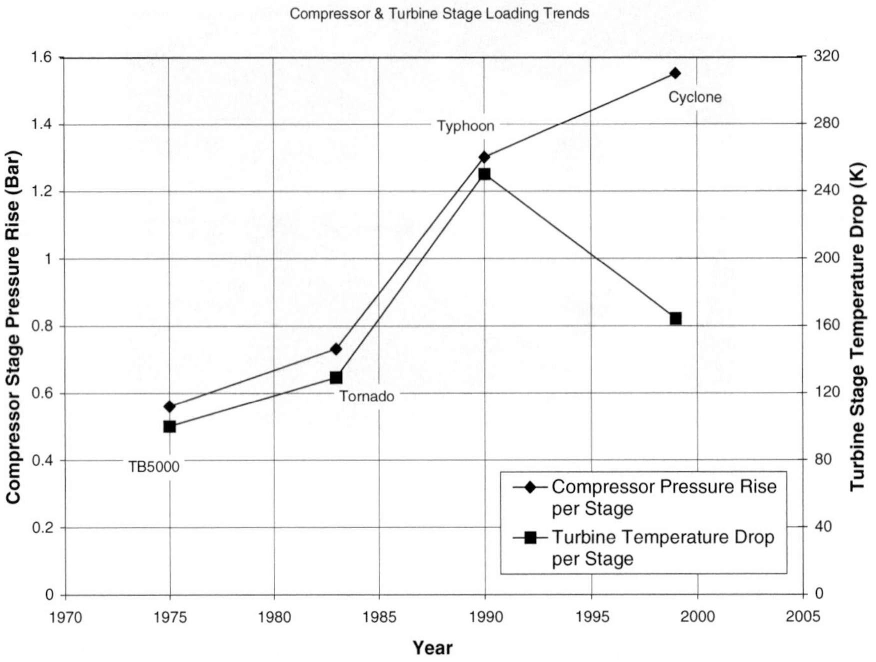

Fig. 7 Development in compressor and turbine stage loadings for small industrial gas turbine engines.

TURBOMACHINERY TECHNOLOGY

In parallel with improvements in the engine cycle parameters, advances in component loadings, both aerodynamic and mechanical, have been made which allow increases in engine pressure ratio to be achieved with little increase or even a reduction in the number of stages in the compressor and turbine. Stage Loading is defined by the loading parameter $\Delta H/U^2$, where ΔH is the work per stage and U is the peripheral blade speed. Increases in both the loading parameter itself and blade speeds have enabled advances in compressor pressure rise per stage and turbine work output per stage as demonstrated in Figure 7.

The benefits of the improved design and test facilities are illustrated by a comparison of the TB 5000 compressor, which employs 12 compressor stages to achieve a pressure ratio of 7, with that for the Cyclone, which achieves a pressure ratio of 17 with only 11 stages.

These improvements have resulted in transonic flow conditions within both the compressor and turbine sections and the levels of performance achieved have only been made possible by the application of computational fluid dynamics (CFD) analysis methods in conjunction with extensive validation testing of new designs in aerodynamic rigs.[8,9] Despite the increase in stage loadings, improved understanding of aerodynamic loss mechanisms and their control has resulted in increased component efficiency levels as shown in Figure 8 below.

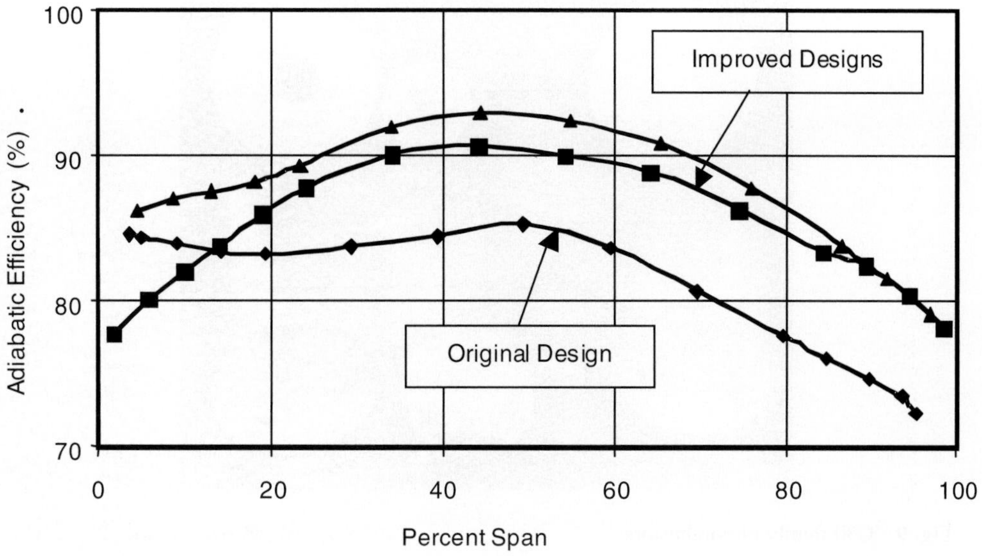

Fig. 8 Efficiency improvement abtained from new 3D high load turbine design.

Increased blade speeds increase the centrifugal loads and stress levels on rotating components and to ensure that adequate component lives are achieved, improved materials in conjunction with more rigorous stress and life prediction analyses have to be employed.

Recent developments in 3D unsteady CFD analysis coupled to finite element analysis should with further work provide an analytical tool to avoid flow induced vibrations and high cycle fatigue failure. Such methods have only become possible by utilising the latest developments in computing power.[10]

COMBUSTION AND EMISSIONS

Legislation to control NO_x emissions has driven the gas turbine industry to introduce Dry Low Emissions (DLE) technology in order to meet emissions targets. The technology chosen to achieve this is Lean Pre-Mix combustion to yield NO_x emissions of less than 10 ppm although development of catalytic systems for further emissions reduction is being pursued.

Design objectives for the combustor encompass the following areas

- Reduced costs and reduced emissions (NO_x, CO, CO_2 and SO_X).
- Improved turndown operation to low loads.
- Increased lifetime and integrity.
- Ability to meet the demands for new innovative cycles.
- Ability to burn a wide range of fuels.

Fig. 9 G30 family of combustors.

Increased levels of combustor dynamic pressure oscillations have caused reliability problems for some manufacturers and it has been essential to develop a clear understanding of the combustion process and the influence of such factors as flow turbulence and thermoacoustic oscillations on mechanical integrity. Testing and modelling methods, backed up by field experience, have been developed to help optimise combustor design and operation.[11] Figure 9 shows a series of combustor cans developed for intermediate sized engines.

Combustor exit temperatures range from 1525 to 1650 K, depending on the engine size and duty cycle. The targets set are for dual-fuel operation producing less than 10 ppm NO_x for natural gas and less than 35 ppm for distillate liquid fuel with a life in excess of 24,000 hours. These targets are being met in service as shown in Figure 10.

Further reductions in emissions levels in lean pre-mix systems are being pursued and catalytic combustion systems have the potential to offer even lower levels but this may only be possible by accepting some limitations on operability.

The other major area of combustion development is fuel flexibility. While, the first choice fuel for an industrial gas turbine is natural gas with liquid distillate as a back up, these fuels are not universally available and increasingly operators are looking to use alternative fuels such as :-

- Oilfield Wellhead Gases (depleted wells have increasing H_2S content in the gas),
- Liquified Petroleum Gas (LPG),
- Naptha,
- High Hydrogen Refinery Tail Gases,
- Medium Calorific Value Gases (Landfill, Gasified Waste, Coal Bed Methane),
- Low Calorific Value Gases (Biomass derived) and
- Syngas from coal gasification plant.

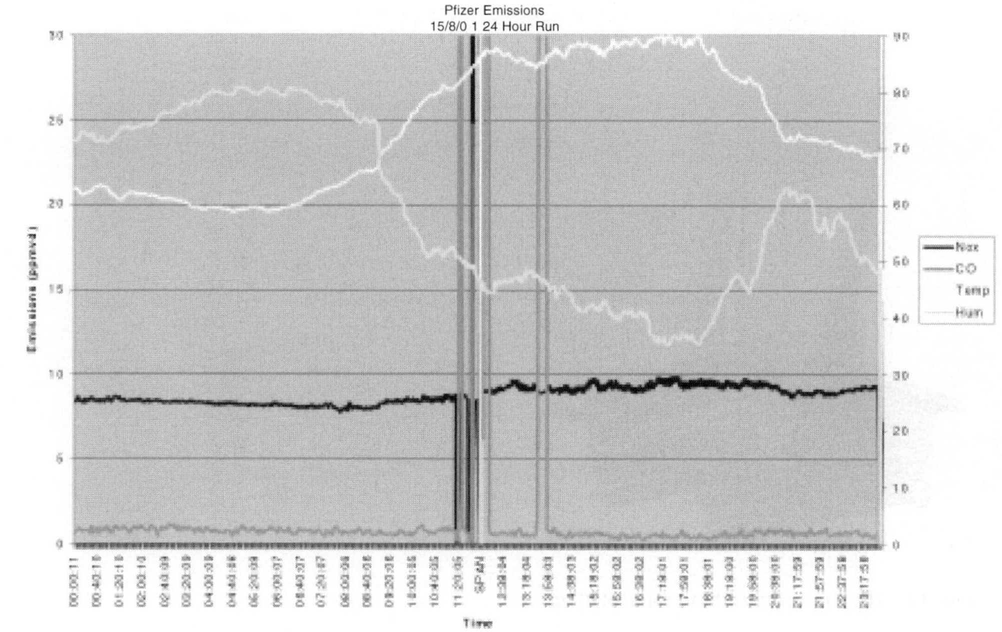

Fig. 10 NO_x emissions from a dual fuel 8 MW engine in field service.

Each of these fuel types present a particular range of component operation and materials problems for the design engineer and developing low emissions combustion systems tailored to these fuels has significantly increased the workload in the combustion area.

TURBINE COOLING

As Turbine entry temperatures have increased, the amount of air used for cooling the combustion system and the turbine blading has increased. In recent engines, the first four rows of turbine blading are internally air cooled with up to 20% of the compressor air flow being used for turbine cooling. Cooling systems in use feature impingement cooing, surface films, complex multipass passages and turbulators to enhance internal heat transfer. Research programmes are in progress to enhance internal cooling geometries and improve heat transfer predictions.[12]

Such a blade may have to operate in a 1200°C gas temperature, subject to high stress levels due to rotation and reliably achieve a life of 4 years continuous operation. Even with today's highest strength single crystal superalloys, this requires the blades to be cooled by 300 to 400°C and extensive analysis and validation testing are essential to assure satisfactory operation.

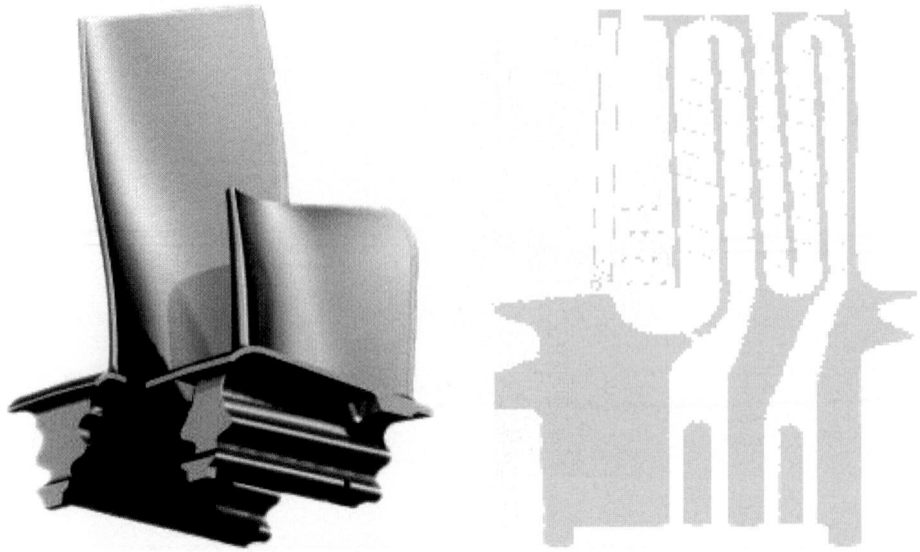

Fig. 11 High pressure turbine rotor blading.

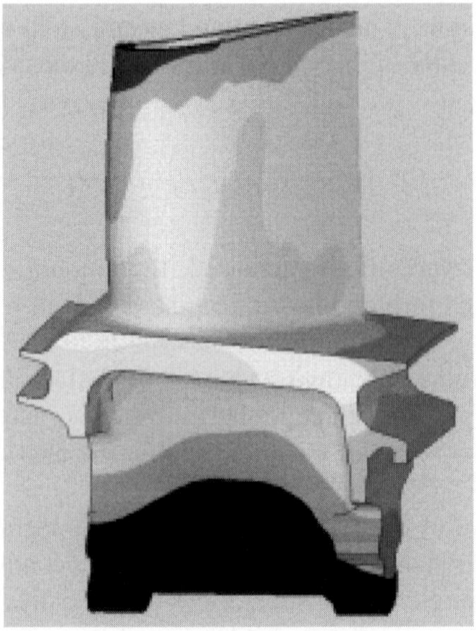

Fig. 12 3D thermal analysis.

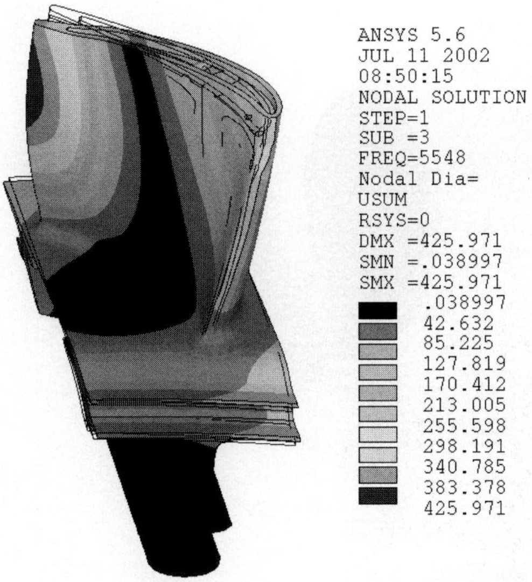

Fig. 13 Finite element vibration analysis.

MATERIALS TECHNOLOGY

Compressor

The temperature loading through the compressor ranges from –50 to 500ºC and a series of low alloy and ferritic stainless steels (825 M 40 W, FV535) has provided adequate materials properties capable of meeting these conditions. Materials selection for extreme low temperature environments as well as the need for improved corrosion resistance of compressor blading (e.g., use of martensitic 17/4 PH steel) remain key issues to be addressed. To meet the need for increased temperatures, pressure ratios and rotor speeds, higher strength steels are likely to be needed, as well as aeroderivative technology such as titanium alloys, nickel alloys and composites. This would, however, present a significant increase in cost and manufacturing complexity (forgings, machining, joining, component lifing) as well as operational difficulties (component handling, overhaul, repair, cleaning) and may introduce additional problems associated with thermal mismatch and fretting fatigue. The use of welded compressor rotor drums is a feature of several recent designs and weldability is then an additional materials issue that needs to be addressed.

Combustor

The materials used presently are generally wrought, sheet-formed nickel-based superalloys, such as Haynes 230. These provide excellent thermomechanical fatigue, creep and oxidation

Fig. 14 CMC combustor liner.

resistance for static parts and are formable to fairly complex shapes such as combustor barrels and transition ducts. Equally of importance is their weldability, enabling design flexibility and the potential for successive repair and overhaul operations, which is crucial to reducing life cycle costs.

The high thermal loading and need to minimise cooling air consumption has led to the widespread use of ceramic thermal barrier coating (TBC) systems to protect combustion hardware. Current TBC technology for combustor applications is based exclusively on multi-layered systems comprising of an MCrAlY bondcoat and a ceramic topcoat applied using plasma spray deposition techniques. For industrial gas turbines application of this technology generally aims to limit peak metal temperatures to 900 to 950°C. Future developments are aimed at applying thicker TBCs to enable higher flame temperatures and/or reduce metal temperatures further. Other programmes are aimed at increasing the phase stability and resistance to sintering of the ceramic topcoat at temperatures above 1250°C. Increased operating temperatures and corrosive/erosive fuels will require the further development of coating technologies.[13]

In the longer term, technology acquisition programmes for combustor designs are aimed at replacement of conventional nickel-based products with either oxide dispersion strengthened (ODS) metallic systems or ceramic matrix composites (CMCs) based on either SiC-SiC or oxide-oxide matrix-fibre combinations and capable of running at up to 1600 K. Candidate ODS and CMC materials have been evaluated and demonstration hardware designed and manufactured (see Figure 14) and, in the case of CMC components, engine tested. Much work remains to be done to improve lifetime prediction methods and to develop coatings to provide thermal and environmental protection of the liner at temperatures in excess of 950°C.

Turbine

For the hot section of the engine, Nickel base superalloys continue to be the materials of choice for both turbine blading and turbine discs.[13] Research is continuing into the use of structural ceramic materials but it is evident that many problems remain to be solved before these can be used for highly stressed rotating components.

For small and medium sized gas turbines the high strength Nickel/Iron alloy IN718 is widely used as a disc material with disc rim temperatures up to 550°C. Nickel base alloys such as Waspalloy and Udimet 520 are available to provide increased temperature capability. A number of materials and component design activities are aimed at further increasing both the temperature and loading carrying capabilities of the turbine discs, for example powder nickel alloy forgings and dual alloy. In the past, manufacturing issues limited large heavy duty gas turbines to the use of steel turbine discs with disc rim temperatures limited to below 450°C. However, materials and forging process developments now enable large gas turbine disc forgings to be produced in a high strength nickel alloy such as IN706 with significant benefits in reduced cooling air consumption.

To meet the demand for increased turbine temperatures, more advanced blading materials have been introduced into the hot gas path of industrial gas turbines. For vanes and blades there has been a gradual move away from conventionally cast nickel-based superalloys, such as IN738 towards directionally solidified (DS) alloys, such as Mar-M247DS, and single crystal (SC) alloys such as CMSX-4. The introduction of these alloys, manufactured using near-net shape investment casting has provided significant benefits in terms of much improved creep and thermal fatigue properties. Significant advances in both design and casting process development have now allowed these materials to be used for the largest heavy duty baseload gas turbine engines.

However, the increased cost of manufacture, due to high alloying levels and parts rejection, needs to be controlled by the use of revert materials and control of the casting conditions, and offset against the improved component lifetime and more efficient running capability. Successively higher levels of alloying additions (Al, Ti, Ta, Re and W) have been used to increase creep resistance and have been key to the success of the aero gas turbine industry and increasingly, that of the land-based sector. However, as the level of alloying has increased the chromium (Cr) content has had to reduce significantly to avoid phase instability problems and the lower Cr addition has reduced the corrosion resistance of the alloys. The need for industrial gas turbines to operate on a wide range of fuels, often with increased potential for corrosion due to higher levels of sodium, sulphur etc. and in industrial and offshore atmospheres, has led to concerns over hot corrosion attack and significantly reduced material fatigue properties. To combat this, a series of protective coating systems (aluminides and overlays) have been developed and are applied both externally and internally on cooled components, to meet the range of fuel types used by various operators. These coatings are applied to provide increased component lifetimes, but can exhibit low strain to failure properties that impact upon the TMF endurance.[14]

Figure 15 illustrates the diverging requirements of the aero and industrial sectors for SC blade alloy technology. The need for increased strength for aero-engine applications took precedence with alloy developers/component suppliers, there has been a deleterious effect on the stability of the alloys during long term exposure which has been attributed to the high precipitate volume fractions (>70%) needed to achieve high temperature creep strength.

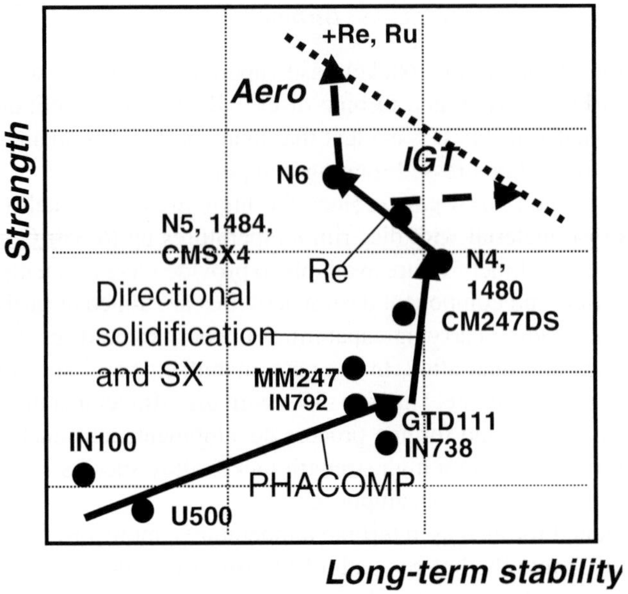

Fig. 15 Strength-stability diagram for nickel-based superalloys.

This has prompted development of industrially specific alloys having improved castability, higher corrosion resistance and reduced heat treatment times such as MK4 and SCA425. Further work is needed in this field and it will remain a key issue for future IGT turbine designers, impacting directly on the performance and RAM capabilities that can be offered.

As on combustors, the application of Thermal Barrier Coatings (TBC) to cooled turbine blading offers significant benefits in increased temperature capability or reduced cooling air requirements. However the demanding TMF environment for turbine blading can lead to spallation of the TBC and local overheating. Considerable effort is being expended to develop robust material systems combining the high strength superalloy with a protective bondcoat and ceramic TBC using improved coating deposition techniques such as Electron Beam Physical Vapour Deposition (EBPVD).

The interaction of coatings with superalloys and their impact on material properties are areas where significant work is ongoing to understand the behaviour of these high temperature material systems.[15] The figure below indicates the effect that coating application, environment and temperature can have on the fatigue strength of a directionally solidified superalloy.

PACKAGING AND SYSTEMS

In addition to the core gas turbine engine, a gas turbine power generation set will include the following systems and equipment to produce a fully functioning unit:-

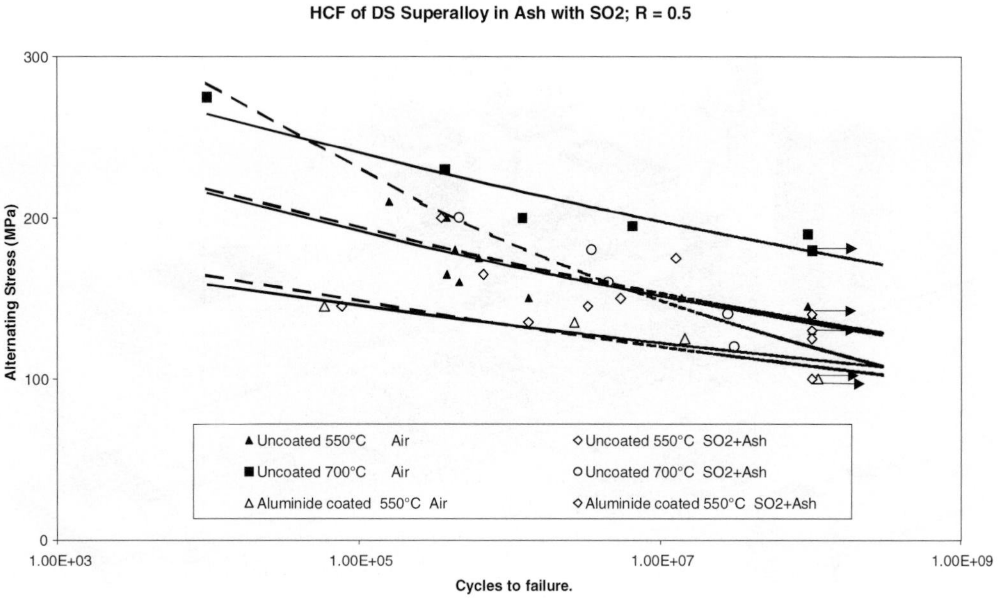

Fig. 16 Impact of coating and environment on fatigue life.

- Underbase and Acoustic Enclosure,
- Reduction Gearbox,
- Alternator,
- Electronic Control System,
- Lubricating Oil System,
- Fuel System,
- Fuel Gas Compressor (if required),
- Starting System,
- Intake Air Filtration System,
- Intake and Exhaust Silencers & Ducts,
- Enclosure Ventilation System and
- Fire & Gas Safety Systems.

The cost of the package and systems is typically 60% of the total price of a gas turbine genset with the gas turbine itself accounting for the remaining 40%. It is therefore, equally important to pay attention to good design practice on the package. Considerable effort has recently been applied to develop a generic package layout to increase modularity of the systems and improve access to both the core gas turbine and systems in order to minimise downtime during maintenance.

Where gas turbines are used in the oil and gas industry either onshore or on offshore platforms, increased attention has to be paid to ensure that systems meet all applicable health and safety regulations when operating in potentially hazardous environments. A typical package layout for a 13 MW(e) genset is shown in Figure 17.

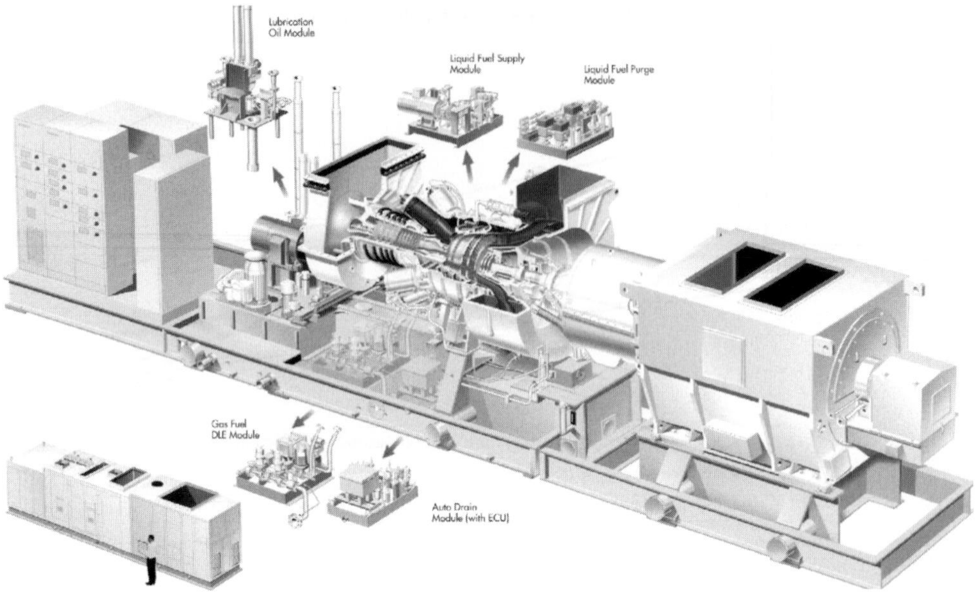

Fig. 17 Cyclone 13 MW gas turbine power generation package.

AVAILABILITY AND RELIABILITY

The economics of a gas turbine engine applications are increasingly being judged in terms of the overall life cycle costs covering operation and maintenance costs in addition to capital cost and fuel cost.

Whilst the demand for increased fuel efficiency and reduced specific costs has driven engine temperature and pressure levels higher, there has been no relaxation in the need to minimise maintenance costs, by achieving component design lives and maximising engine availability. This is particularly true in the Oil and Gas industry, where the value of the product pumped in a single day can exceed the cost of the gas turbine installation; avoidance of unscheduled downtime is a key factor and engines with a proven reliability record are preferred.

To minimise uncertainties in this area, manufacturers are increasingly offering maintenance contracts that provide guaranteed levels of support for both routine and unscheduled maintenance operations in order to assure the desired availability level.

Reduction of the cost of maintenance and overhaul is therefore of interest to both manufacturers and operators, alike. With more sophisticated designs for today's high temperature components, component repair is becoming more cost effective. This is placing increasing importance on component remnant life assessment procedures and repair and overhaul practices, including strip and re-coating, rejuvenation and blade tip re-profiling using laser deposition techniques.

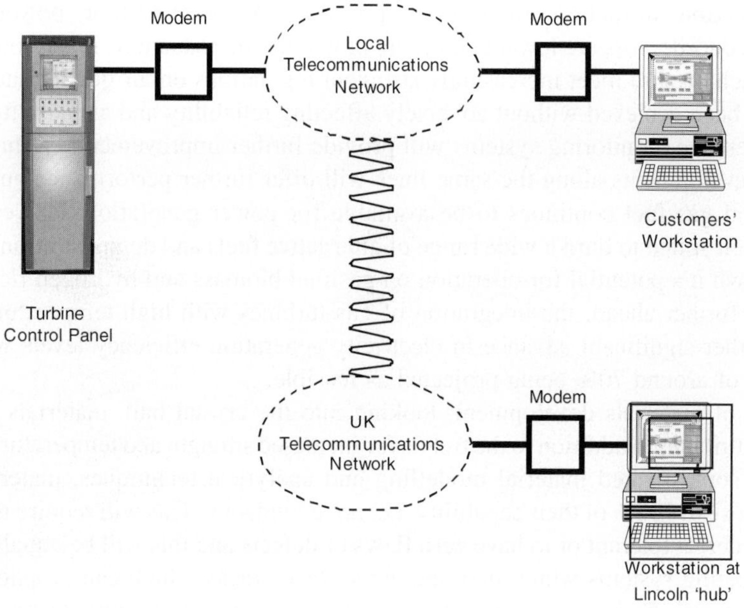

Fig. 18 Electronic data exchange network.

While electronic control systems have been in use for over twenty years, the computing power available today, and the communications capabilities offered by satellite systems and the world wide web will revolutionise the way engines are controlled and operated in the future.

The company has introduced the Electronic Data Exchange Network (EDEN™) system, which currently offers remote monitoring of all engine sensors and control parameters, trending of parameters for health monitoring purposes and the ability to adjust control system settings remotely from the factory or service centres.

At present EDEN, is available as an option for all engines with a suitable standard of control system, and is fitted to all new engines delivered during the initial year's warranty period and is proving invaluable for setting up control systems during engine commissioning and for fault detection.

For the future, the ability to continually monitor engine operation, opens up the potential for further improvements in health monitoring, accurate component life usage monitoring and the opportunity to offer maintenance "on condition" rather than at fixed service intervals.

CONCLUSIONS

Over the past 25 years, whether in large base-load combined cycle power stations or in smaller cogeneration and oilfield applications, industrial gas turbines have been developed to offer reductions in fuel consumption of up to 30%. At the same time, power plant sizes have been reduced, costs are lower and orders of magnitude reductions in noxious emissions have been achieved to meet increasingly stringent regulations on air quality and pollution.

This has been achieved without adversely affecting reliability and availability levels and web based engine monitoring systems will provide further improvements in this area.

Future developments along the same lines will offer further performance enhancements while natural gas fuel continues to be available for power generation. However, the gas turbine is well suited to burn a wide range of alternative fuels and demonstration plants have already shown it's potential for operation on gasified biomass and hydrogen rich fuels.

Looking further ahead, the integration of gas turbines with high temperature fuel cells offers a further significant advance in electricity generation efficiency levels with thermal efficiencies of around 70% being projected as feasible.

In terms of materials development, looking into the crystal ball, materials will need a number of attributes in addition to the trends for increased strength and temperature capability. With more sophisticated material modelling and analytical techniques, materials will be exploited close to 100% of their capability. i.e. no redundancy. This will require the materials either to be defect tolerant or to have zero flaws or defects and this will be equally critical in regard to coating systems which may initiate surface cracks which can propagate into the substrate material. To assure improved material quality, non-destructive testing techniques will need to keep pace with materials developments.

With a desire to extend component lives to up to 50,000 hours in the high temperature areas of the gas turbine and 150,000 hours in lower temperature areas, retention of material stability and properties throughout component life is increasingly important.

The general economic trend of reducing cost/kW for gas turbines in real terms can be expected to continue, as it has since their introduction in the 1950's. For materials, this implies two things, more reliable processing to increase component yields or using lower cost processes more effectively. For example, using air melt to produce materials, but achieving today's vacuum melt properties. Other improvements may come from chemical and process modelling development whereby a material with specific property requirements can be 'custom engineered' to order! Alternatively, a 'discontinuity' technology or breakthrough is required. This was predicted in the past but has failed to materialise with ceramics.

Without a doubt, the technical advances needed for the next generation of gas turbines will be as challenging and interesting as the achievements made over the past 25 years.

REFERENCES

1. Brian M Igoe and Martin McGurry: 'Design, Development and Operational Experience of ALSTOM's 13.4 MW Cyclone Gas Turbine' ASME GT-2002-30254.
2. D. Eckardt and P. Rufli: 'Advanced Gas Turbine Technology: ABB/BBC Historical Firsts', *ASME Journal of Engineering for Gas Turbines and Power,* **124**(3), 2002, 542–549.
3. Anders Hellberg, Georg Norden and Sergey Shukin: 'GT10C: 30 MW Gas Turbine for Mechanical Drive and Power Generation' ASME GT-2002-30253.

4. Kristin Jordal, Mikael Jonsson, Jens Fridh and Uif Linder: 'New Possibilities for Combined Cycles Through Advanced Steam Technology' ASME GT-2002-30151.
5. M. F. Cannon and M. J. Welch: 'Gas Turbines and Renewable Fuels: The Technologies, the Risks and the Rewards', Institute of Mechanical Engineers, London, 2002.
6. Torbjörn Lindquist, Per Rosen and Tord Torisson: ' Evaporative Gas Turbine Cycle – A Description of a Pilot Plant and Operating Experience', *ASME International Mechanical Engineering Congress & Exposition 2000.*
7. S. E. Veyo, L. A. Shockling, J. T. Dederer, J. E. Gillett and W. L. Lundberg: 'Tubular Solid Oxide Fuel Cell/Gas Turbine Hybrid Cycle Power Systems: Status', *ASME Journal of Engineering for Gas Turbines and Power*, **124**(4), 2002, 845–849.
8. Y. S. Li and R. G. Wells: 'The Three-Dimensional Aerodynamic Design and Test of A Three-Stage Transonic Compressor', ASME 99-GT-68.
9. Brian Haller and James Anderson: 'Development of New High Load/High Lift Transonic Shrouded HP Gas Turbine Stage Design – A New Approach for Turbomachinery', ASME GT-2002-30363.
10. W. Ning, Y. S. Li and R. G. Wells: 'Predicting Bladerow Interactions Using A Multistage Timelinearized Navier-Stokes Solver', ASME GT-2002-30309.
11. D. J. Cramb and R. McMillan: 'Tempest Dual Fuel DLE Development and Commercial Operating Experience and Ultra Low NO_x Gas Operation', *PowerGen Europe*, 2001.
12. D. A. Rowbury, S. Parneix, D. Chanteloup and A. Lees: 'Research into the Influence of Rotation on the Internal Cooling of Turbine Blades', *Proceedings of the NATO Advanced Vehicle Technology Panel Symposium on Heat Transfer and Cooling in Propulsion and Power Systems*, RTA Report RTO-MP-069(SYB-22), 2001.
13. J. Hannis, M. B. Henderson and G. McColvin: 'Materials Issues for the Design of Industrial Gas Turbines', *UEF Advanced Materials and Processes for Gas Turbines Conference, Copper Mountain*, 2002.
14. J. R. Nicholls: 'Design of Oxidation-Resistant Coatings', *Journal of Materials*, 2000, 28–35.
15. M. Nazmy, J. Denk, R. Baumann and A. Kunzler: 'Environmental Effects on the Tensile and Low Cycle Fatigue Behaviour of a Single Crystal Nickel Base Superalloy', *COST522 Liege Conference*, 2002.

PARSONS 2003 Conference Celebrity Lecture

Charles A. Parsons – The Substance of a Great Man

W. GARRETT SCAIFE
Trinity College Dublin

When the cortege bearing the human remains of Sir Charles Algernon Parsons entered the parish Church at Kirkwhelpington in Northumberland on a bitterly cold February day in 1931, the coffin was draped with the White Ensign of the Royal Navy. It was a rare honour for a civilian, and it was a mark of the high esteem in which he was held by Naval men throughout the world. His widow, Katharine Lady Parsons accompanied by her daughter sat beneath the memorial to her son, who had been killed in the first World War. At her request there were neither hymns nor anthems, there was nothing ostentatious about Katharine or her deceased husband. Elsewhere in St Nicholas' Cathedral in Newcastle upon Tyne and at Westminster Abbey in London more elaborate services commemorated someone who had been a public figure for much of his life. During his lifetime he had been honoured by Universities and by Engineering and Scientific Societies in Britain, America and Europe. He had served as a Vice-President of the Royal Society, and it was under the aegis of the Society that a plan was developed for a memorial to his life and work. Each year in turn, one of eight Institutions and Societies with which he had been associated, undertook to present a Parsons Memorial Lecture. This tradition has continued with barely any pause from 1936 until the present day. That in itself is perhaps the best testimony to Parsons' achievements.

CHILDHOOD INHERITANCE

As time has passed it has become easier to discern more clearly the influences which combined with his genetic inheritance to mould a truly great engineer. The first place to begin must be with his family circumstances. Charles was the youngest of the four boys who reached manhood from among the children who were born to Mary the wife of William Parsons, third Earl of Rosse. Seven other siblings died before becoming adults, so in that sense he was already privileged. When Charles was born in 1854 his father was 54 years of age, and he died only thirteen years later. Despite the brevity of these years, the father left a clear imprint on his son. The Parsons family had long valued education. Charles' grandfather, Laurence, was a graduate of Trinity College and a founder member of the Royal Irish Academy in 1785. Charles' father William had established a reputation for himself as a scientist with considerable engineering talents and had served as President of the Royal Society.

When William married in 1836 he decided to repeat what had been his own experience and not to send his children to public schools. One advantage of this was that they did not imbibe the attitude of disdain which was exhibited among the wealthy in Britain, for those who might get their hands dirty by engaging in industry. But this carried a disadvantage in

[1] Assistants were employed to make observations with the telescopes and among them were three men who became university professors, Robert Ball, Johnstone Stoney and J.F. Purser and another, Bindon Blood Stoney who became chief engineer for the Port of Dublin.

that they missed an experience which would perhaps have 'knocked the corners off them'. He was able to provide his children with the benefit of being educated by a series of exceptionally gifted private tutors at their home in Birr Castle before attending university.[1] All the children, girl cousins included, were encouraged to make use of the workshop machinery. This was not free of dangers. Charles made himself an airgun and when, in the process, a sliver of steel became embedded in the white of his eye, he ran to his mother who coolly drew it out. On another occasion while experimenting with gunpowder he succeeded in burning off his eyebrows. He certainly was fearless. In later years he recalled as a ten year old

> making contrivances with strings, pins, wires, wood, sealing wax and rubber bands as motive power, making little cars, toy boats and a submarine... When working in my father's workshop, dealing with metals became more attractive, and amongst other things I made a spring trap, and I well remember the delight of catching my first rabbit with it.

In later years he was to devote much of his effort to 'boats' and while designing *Turbinia* he began with models equipped with '*rubber bands as motive power*'.

His father was an altogether unusual aristocrat. Working from scratch, the young Lord Oxmantown as he then was, had developed the workshops which would produce the largest telescopes in the world in Birr, a town which was distant from any industrial centre. We have quite a clear picture of William Parsons' way of working because he filled his early papers in Brewster's Journal and later in the Transactions of the Royal Society with much practical detail setting out the difficulties which he had encountered in building the giant telescopes. He described the means by which he tracked down the causes and solutions which he had devised, with a view to easing the path of others who might wish to emulate his work. He wrote concerning his machine for grinding and polishing concave mirrors,

> The results obtained by machinery are very nearly uniform. Where a uniform combination of motions produces a defect, that defect will uniformly recur and may therefore, with great facility, be traced to its source and corrected...

It is fair to say that this expresses very well his son's method of working too, when he grew to manhood. But of the engineering features, the construction of the massive masonry walls and the design of the counterweights which made the huge telescope tube manageable, there is not a word.

William Parsons was remarkable in another respect which no doubt influenced his youngest son. He got on very well with those he employed to build his telescopes, he wrote about an early experience when he was learning the craft of casting brass,

> All my workmen were trained in my own laboratory without the assistance of any professional person, and none of them had seen any process in the mechanic arts; and I was not myself then acquainted with the precautions necessary to insure the production of an alloy of zinc and copper in the due proportions.

After his death a letter was printed in the Bristol Times headed 'A noble working man' which paints a vivid picture,

> About ten years ago I was in Parsonstown close to which is the demesne of the noble working man, now no more…(being near) I availed of the opportunity (to see the giant telescope). I saw not only the great telescope but I saw the Earl, the telescope maker himself - not in state, with his coronet and ermine robe on, but in his shirt sleeves, with

his brawny arms bare. He had just quitted the vice at which he had been working and, powdered with steel filings, was washing his hands and face in a coarse ware basin placed on the block of an anvil, while a couple of smiths sledging away on a blazing bar on another, were sending a shower of sparks about his lordship which he as little regarded as though he were a 'Fire King'. This was in a spacious, rude, smithy which almost occupies one side of the court yard of the castle and in which not only were swing bridges and force pumps, and tackle for scientific instruments constructed, but common and everyday articles in the shape of agricultural gates, sub-soil ploughs etc. for use of his farms.

I alighted on the noble Vulcan by accident, and feared I might be thought intrusive, but it was not so; for he spoke in an easy friendly way to the gentleman with whom I was, and whom he knew by name, giving permission to see everything about the house and demesne.

As he drew on his coat… the Earl looked an intelligent foreman… a man with his "head (as the phrase is) screwed in the right place".

Although most of the ground breaking work in the workshops at Birr had been completed a decade before Charles was born, the men who had been trained by William Parsons in the skills of casting metal, grinding, cutting and polishing bronze alloy, still worked to keep the telescopes in service. Whether in conversation with his father, or with his tutors who were employed as assistants in the Observatory, or with the mechanics who had built and operated the machinery in the workshops, he learned much about the scientific approach of his father.

SEAFARERS

As children, the family of Lord Rosse were quite used to travel. Because he was a member of the House of Lords and President of the Royal Society, he would frequently bring his family to London where he rented a house for the Season. This involved crossing the Irish Sea, but they were seasoned sailors anyway, because Lord Rosse owned an iron hulled yacht manned by a crew of 15. Although he came to sailing when past middle age, Lord Rosse sailed with the family, including his wife Mary Parsons, on quite lengthy journeys from Dublin. His second son Randal in his Reminiscences described one occasion when they rounded Cape Wrath in the north of Scotland in stormy conditions to reach the Isle of Wight, and other trips when they visited Antwerp, in 1862 when Charles was only eight years of age, as well as Coruna, in Spain in 1865. Scott Russell, the naval architect and collaborator with the great I.K. Brunel was a regular visitor to their house, and his advice was sought when they were competing in the Royal Cup Race around the Isle of Wight. For much of the year their father was a busy man, but on such sailing expeditions he was close to the boys, sharing with them his enthusiasm for visiting shipyards and seeing over the new steam powered warships of the Royal Navy. This familiarity with the sea was of great importance for both Laurence and Charles Parsons as their careers evolved in later years. Their mother had acquired considerable skill in the newly evolving technology of photography and she brought her equipment with her on holiday expeditions. Having lost so many of her children in infancy and early childhood, she kept a close eye on her boys. Since they did not attend Public Schools the boys lived a rather sheltered life, lacking in an element of rough and tumble. Moreover they were accustomed to deference from most acquaintances. It gave them little chance to develop social skills such as the ability to 'read' another person's reactions, and so what may well have been a family tendency anyway, became a real drawback for many of them in adult life.

Laurence, the eldest son studied science at university and succeeded his father as the fourth Earl. He took over the running of the astronomical observatory which his father had established. His infrared measurements of the temperature of the Moon broke new ground, and though he chaired the Board of Charles' marine turbine company, he showed no great inventiveness or engineering aptitude. He was shy and withdrawn, traits which were no doubt accentuated by a sheltered childhood. After William's death in 1867 it fell to his widow Mary to oversee the completion of the boys education. She was of course from an English family and this no doubt steered her sons across the Irish Sea. One of them, Randal, had avoided any contact with technology as a youngster preferring gardening as a hobby, and after he graduated from Trinity College Dublin he was ordained for the Church of Ireland. Richard Clere who was next to Charles in age, studied engineering, winning high honours as an undergraduate at Trinity College Dublin. Very unusually for a son of a nobleman, he went on to complete his education as an apprentice to Easton and Anderson in their works at Erith in Kent. While there he studied the performance of shrouded propellers for ships as well as centrifugal pumps, and for his paper on the latter he was awarded the Miller Scholarship of the Institution of Civil Engineers in 1876. Clere held a considerable number of patents, and during the time that he managed Sir James Kitson's Airedale foundry in Leeds, he collaborated with his younger brother helping to develop the latter's rotary steam engine which was being manufactured by Kitsons. Afterwards he joined George Westinghouse as a director when the latter was building his huge Trafford Park Works in Manchester. In the end he developed a successful practice as a consulting engineer.

On the face of it Clere was best placed to become a great inventor, so why did this not happen? The answer to this throws light on some of the gifts which were given to Charles and were denied to Clere. In 1880 while Charles was nearing the end of his apprenticeship at the Elswick Works of Sir William Armstrong in Newcastle upon Tyne, he filed a patent on behalf of Clere who was abroad at the time in Calcutta. This was for a pneumatically powered mechanism to be used to operate 'punkhas'. Punkhas were a form of fan used in hot climates, and consisted of a cloth screen which was hung from the ceiling, and which had a horizontal rod located on its lower edge. A rope attached to this rod allowed the screen to be swung to and fro by a servant, to cool those beneath. Clere proposed to use a steam engine to provide compressed air for powering his machinery which would make the 'punkha wallahs' unnecessary. An annual financial saving which he estimated as £75,000, could be achieved in hospitals and other Government premises by adopting his system. At that time India was in the grip of a famine, and even then its population was huge. One might say that the case for labour saving in such circumstances was not compelling!

FUNDING INVENTIVE ACTIVITIES

Normally one does not think of Charles Parsons as a businessman but in fact he did make a financial success of his enterprises, and he relied on this income to allow him to pursue his primary interest, research and invention. Throughout his life Charles remained focused on costs and market opportunities.

While still at university he developed a quite unique steam engine, one in which the four cylinders were located in a casting which rotated at 600 r.p.m. After that he worked for a couple of years on an alternative to Whitehead's locomotive torpedo. The idea was to replace the engine which was driven by compressed air, with a turbine powered by the gases from a rocket casing filled with gunpowder. He funded these ventures by a joint venture agreement with Sir James Kitson. Although he had considerable technical success, when it became clear that neither of these ventures was likely to be a commercial success he did not hesitate to abandon them. He observed that

> in my business I... have made many more failures than successes in the effort to progress, ...yet the failures... have been soon discerned and force concentrated on the successes.

Using some £24,000, •4–5 million in today's money, which he had inherited, he entered into a partnership with Clarke Chapman in Newcastle upon Tyne in 1884, and before the year was out he had designed and constructed the world's first turbo-generator. Throughout his career from this time onwards he succeeded in always generating sufficient profits to pay for his research and development costs. There was one notable exception and that involved the application of the steam turbine to marine propulsion. Here the sums involved were too great for him to manage unaided, and he needed to move forward with speed if he was to retain the advantage he had already gained over his rivals by being first in the field. He needed funds to pay for the building of his demonstration vessel the *Turbinia* which would show the advantages of steam turbines for marine propulsion, so in 1894 he raised £15,000 of cash from shareholders, wealthy friends and influential figures who were keen to profit from the exploitation of his key patent 394 of 1894. When he put *Turbinia* through its paces at the Naval Review in 1897, held to celebrate Queen Victoria's Diamond Jubilee, the shy inventor showed that he had an appreciation for the value of publicity. The demonstration aroused so much interest that he was easily able to find much greater amounts of capital. This was needed for the building and equipping of another factory at Wallsend-on-Tyne, as well as for working capital to cover the cost of speculative ventures which were aimed at securing the interest of private ship-owners and the Royal Navy.[2] The Parsons Marine Steam Turbine Company Limited was set up in 1897 and it took over the earlier company, while the capital was increased to £208,000.

It was as an electrical engineer that Parsons had been employed by Clarke Chapman in 1884, to take charge of the manufacture of electric lighting equipment, but naturally he sought to push sales of his turbo-generators. Eventually when a conflict of priorities caused the partnership to be dissolved in 1889, Charles was able to regain his original investment but, impetuously, he refused to pay the price asked for the return of his key 1884 patents because he thought it too high. Undeterred by this setback he crossed to the north bank of the Tyne where he set up his own private company, C.A.Parsons and Co., Electrical Engineers, in his works at Heaton close to Newcastle. A period of intense development followed as he evolved a series of completely new patents which circumvented all the features of the 1884 patents. He was able to call on the help of wealthy friends to avoid seeking funds from

---

[2] For example in 1898 the Admiralty required a guarantee that it would be paid £100,000 should the destroyer HMS Viper fail to meet its specified performance.

public shareholders. Eventually in *1913 'for personal and family reasons'* he did turn this business into a private Limited Company with 3,000 shares. Its capital value had risen by now to an estimated £421,207.

From the early years Parsons followed a very successful course of action which ensured a flow of income from his *intellectual* property, his patents, while at the same time it protected him in large measure from wasteful and costly legal challenges of the kind that had robbed many an inventor of the fruits of their efforts. He offered other manufacturers a deal. In return for a relatively modest payment, he would give a license to use his patents in certain market regions, together with an undertaking to make available technical support. For their part the licensees agreed to make available to Parsons any improvements which the licensees might develop themselves, and in addition they undertook not to make a legal challenge to the validity of Parsons' patents during the period of the license. Typical licenses would cover several patents, and would extend over a period of several years. These were carefully thought out legal documents.[3] To manage licenses in other countries C.A.Parsons and Company and the Parsons Marine Steam Turbine Company jointly established the 'Parsons Foreign Patents Company', PFPCo, in 1899. One of the first European licensees of PFPCo was the youthful Swiss firm of Brown Boveri.

Parsons took a hard line in regard to any infringement. When Mark Robinson of Willans and Robinson, was negotiating a license, he suggested that Parsons' patent 8698 of 1896 would not stand up in law, and if necessary he would challenge it, adding that *'of course there were no hard feelings'*. Parsons' response was sharp,

8th October 1903

> Dear Mr Robinson,
> I must ask you further to dismiss from your mind that I shall not consider your further action as personal, for it will give me and my colleagues a great deal of trouble and take up our time which should be devoted to furthering the introduction of the steam turbine, to which we have devoted most of our lives.
> ...If it becomes necessary, we shall not hesitate to carry through any law cases that may arise.
> *Charles A. Parsons*

From an episode which occurred after his death we can form some idea of how potentially damaging legal challenges could have been. In 1924 an action had been taken in the American Courts against the United States Government for having manufactured and used steam turbines for marine propulsion without a license from the owner of the patents. This action was not pursued at the time, but after Parsons' death one of the executors of his will decided to reopen the matter. The United States Court of Claims ruled that the plaintiffs were not entitled to recover damages, and with their judgement issued an extensive review of the patents, vigorously challenging the validity of many of them. Had such a judgement been delivered during his lifetime it would have struck a crippling blow against his lucrative licensing business. Some impression of the importance of license fees and royalties to profitability can be gleaned from data published by the Parsons Marine Steam Turbine Co. In the 14 years up to 1911 income from these amounted to £173,427. For the same period the profits

[3] For example one License for the manufacture of steam turbines authorised Willans and Robinson Ltd. to build turbines in the United Kingdom only, not for marine propulsion, and for sale in a specified list of countries. It listed 21 patents and was valid for ten years from February 1905.

distributed as dividends and bonuses amounted to only £179,016. In other words manufacturing contributed virtually nothing, though of course without an active manufacturing program further technical development would have been impossible.

After his death in 1931, Parsons' estate was valued at £1,214,355. Clearly during his lifetime Charles Parsons had successfully watched over his resources, without ever becoming obsessed with money-making. So although one is inclined to think of him primarily as an inventor and designer, one should appreciate that he was also an effective businessman. Just where this talent for business came from is unclear, unless it was a manifestation in this branch of the Parsons family which had been evident for well over two centuries in Ireland of an ability to husband its wealth, helped where occasion offered with the aid of marriage dowries.

MATHEMATICS

After his brother Clere graduated from TCD his mother returned to London and Charles Parsons transferred from TCD to the University of Cambridge. During his lifetime much was made of his success in the Cambridge Tripos examinations. These were largely devoted to applied mathematics with a small quantity of what would be described today as engineering science. Certainly to have achieved the ranking of 11th 'Wrangler' in a class of 36 was no mean feat, but Parsons in later life made very little use of mathematics. His notebooks are filled chiefly with arithmetic and shrewd estimates relating to his experimental results, with not an integral in sight! In 1914 he recalled,

> After that (schooling at Birr) were interposed five years of pure and applied mathematics including the Cambridge Tripos; I recall that the strain was more severe than anything I have experienced in business life. Luckily for me boat racing interfered with reading.

But it would be a mistake to dismiss the importance of his university experience for it allowed him to hear lectures by the Scotsman, James Stuart, the Professor of Mechanism and Applied Science, and he became a pupil of the Canadian born E.J. Routh who published papers on the stability of motion, including the behaviour of governors. In later life, when Parsons broke entirely new ground by building rotating machinery which ran at unprecedented speeds, he displayed a wonderful instinct for devising solutions to the accompanying vibration problems. Charles' childhood and early education had prepared him well for later life, giving him a particular mind set which caused him to analyse engineering problems in a very logical and scientific fashion, as well as providing opportunities to acquire considerable skill with his hands. Although he wrote many papers, these were largely filled with figures for the number and size of his products. They gave little insight into the thought processes which led to his achievements. After he read the first paper on his invention in 1887 one member of the audience commented that he could not

> see from Mr Parsons' paper that this machine could in its present form, attain to the efficiency which was there alleged. It might be so theoretically. But it was not shown, for instance, that the blades and guides had the proper curvatures to get the maximum effect from the steam, which acted on the blades by impact. That was one point that struck him as being omitted. The difference of angle of the blades on the various wheels was not

mentioned in the paper at all, and he first became aware of it by noticing it on the machine exhibited on the table.

Parsons replied

Mr Weighton had drawn attention to some of the most important points, which enabled him to explain more fully some portions which he regretted had not been made clearer.

Judging by the report of the discussion which followed the paper, he doesn't seem to have kept this promise. Of course turbo machinery was a completely new area at this time and Parsons faced design problems in two areas, both of which involved developing an understanding of the flow of fluids over moving surfaces, what has since become known as fluid mechanics.

INVENTION

(a) Inspiration

If Charles Parsons is remembered today it is primarily as an inventor, so it is worth looking more closely at the way he went about his work. It is possible to distinguish two distinct aspects. First let us consider some examples of inspiration, sudden flashes of insight. These were evident during 1884 when he was creating a practical steam turbine in less than one year. He directed all his effort to solving problems which <u>had</u> to be overcome, leaving matters like efficiency to be dealt with at a later date. His machine ran at 18,000 r.p.m.. Had he not mastered the control of shaft vibrations encountered at such speeds he would have fallen at the first fence. He did this by evolving bearings which had some elasticity and significant damping, and by dividing the shaft into two with a flexible coupling joining the turbine to the generator. Of equal importance were his designs for a novel governor which achieved unparalleled steadiness in shaft speed. But perhaps the key feature was the construction of a dynamo in which the conductors were secured safely against centrifugal forces which reached a magnitude of 11,000 times that due to gravity (each kg of wire experienced a force of 11 tonnes). This electric dynamo allowed him to extract power from his turbine while avoiding gearing or belt drives. It was precisely the problem of extracting power in a useable form that had baffled Professor Osborne Reynolds when he succeeded in building his own steam turbine, in 1875. It ran successfully at 12,000 r.p.m., but Reynolds could see no way of extracting power from a device at such speeds, and he abandoned his project.

(b) Hard Slog

The early machines did have special advantages which made them attractive but they were very inefficient - commentators called them 'steam-eaters'. The cure for the difficulty lay in the design of turbine blades and his approach to this illustrates a second aspect of the inventor's work, which called for many short steps, carefully reviewed and refined. Parsons' chief competitor in regard to steam turbines was the Swedish engineer Carl Gustaf Patrik de Laval whose approach would be to convert the thermal energy of steam into kinetic energy in one step using his newly devised convergent divergent nozzle to accommodate the pressure drop.[4]

There was no pressure drop at all across the turbine blades which were given a straightforward symmetrical shape. Parsons on the other hand allowed pressure to fall across both the fixed components and the moving blades. The drop of pressure which accelerated the steam as it passed through the passages between the moving blades contributed a 'reaction' effect. In its final form, Parsons' blade has a shape which is far from intuitive and suggests the cross section of an aircraft wing.

His first experiments were with a turbine, powered by the gases from a rocket casing and were part of a novel design of torpedo. The little we know of this suggests that the blade surfaces were flat. In his prototype steam turbine, the blade passages, both fixed and moving, were simply straight sided slots, cut in the rim of discs which were carried on the shaft. Thanks to the French research on waterwheels and to the practical experiments of J.B. Francis the English engineer living in the United States, it was well understood that the passages in a turbine must be shaped and oriented in an appropriate fashion to accommodate the direction and velocity of fluid flow.[5] Although Parsons says he knew of Francis' work he made little attempt to use these ideas in his first machine. What he did do was to chose the number of stages in his machine so that the steam velocity developed by the pressure drop across each stage was equal to twice the blade speed. Also by adjusting the angle at which the passages were orientated and by increasing their height, he sought to accommodate the increasing volume of flow as the pressure fell. This is shown in the drawings of the 1884 patents but it is difficult to see. His paper in 1888 described how he had arranged the moving blades at a successively larger radius in three stages of expansion. This paper also shows, without comment, that in the final row, the blades have been undercut to allow the leading edge to be bent using a pliers. This achieved better compatibility with the steam leaving the preceding fixed blades, and incidentally it narrowed the steam passages as the steam approached the exit so causing the acceleration needed for a reaction effect. The original 1884 patent 6735 had foreseen that blades could also be made from sheet metal and so could be shaped more accurately, but this was only done for the first time when he was developing his radial flow machines. When an agreement was reached with Clarke Chapman in 1893 which returned the 1884 patents to Parsons, these were used with success in *Turbinia* and successive vessels, but the greatly increased steam consumption of these more powerful machines called for much longer blades. These were made from drawn brass strip, but they had to be thickened at a point just ahead of midway to improve their strength in bending. The resulting profile allowed the angle at inlet and exit to be chosen to suit design conditions, and it narrowed the passage as the steam approached the exit. The benefit of all these changes was checked by careful measurements of steam consumption, and in 1891 this meticulous work was rewarded when his 100 kW turbo-alternator for Cambridge Corporation equalled the efficiency of the best steam engine.

---

[4] De Laval's patent for the convergent divergent nozzle, number 7143, was granted in 1889, five years after Parsons' key patents.

[5] Francis' book contains a description of 'the path described by a particle of water in passing through the (rotor) wheel'. Charles Parsons would also have been familiar with his brother Clere's (1879) paper 'On the loss of power in the screw propeller and the means of improving its efficiency' which also directs attention to the path of water through a propeller.

A PLACE FOR MODELS

Surprisingly, models played little part in the evolution of blade shapes, and it was several years later that a form of water tunnel was devised to allow the visualisation of the flow through rows of blades.[6] It seems to have been initiated in response to the adoption by Brown Boveri of a proposal to combine a set of de Laval nozzles and blades at the inlet to the turbine, with Parsons' stages for the remainder. This idea had been first proposed by Gerald Stoney several years earlier in 1897, but was rejected by Parsons. It meant that his rotors were much longer than those of competitors, especially disadvantageous on board ship where space is costly. It was a bad judgement, a case of *'not made here'*. The episode illustrates the fact that Parsons was for the most part a loner. Although he gathered an impressive array of able men around him to help in research and development and the running of his factories, he placed little reliance on their views and opinions and he could be quite stubborn. It was a weakness, and it meant that he did not leave behind him a team capable of embodying a tradition which expressed his unique mixture of invention, adventure and courage.

Another decision which was made without ever receiving any comment, was to make both moving blades and fixed blades of identical shape, in modern parlance, to create equal enthalpy drops. Remember that in an impulse turbine this is not so. In passing, it may be noted that there was never any mention in his papers of the thermodynamics of the reaction stage which Parsons alone had chosen to develop. In the discussion which followed his paper to the Institution of Civil Engineers in 1905 the only mathematical analysis was that in the written submission by the Hungarian Professor Donat Banki who quoted from *Steam Turbines* by Professor A. Stodola of the Zurich Polytechnicum, already in its third edition.

The use of models as an aid to design was apparent for the first time when Parsons began work on a demonstration vessel, the *Turbinia*. He began by constructing models to help optimise the shape of the hull. From the force required to tow his models he was able to estimate the power likely to be needed to reach his target speed of 30 knots. He had evidently fully absorbed the procedure which allowed the results of model tests to be used to predict full scale behaviour, and which had only recently been mastered by Froude. He was also able to explore the effect of changing details such as the shape of the stern. After such painstaking work with the most modest facilities he settled on a design for the hull.

Meanwhile he pressed forward the design of his alternative, radial flow turbine till it reached the then unparalleled output of 1650 h.p. at 1600 r.p.m. But do what he might, he just could not exceed speeds of 20 knots. When raising money for this venture Parsons had made his target crystal clear to his friends, so this failure was a hard blow. At this point Parsons had the good fortune to read a paper newly published by Thornycroft. This drew attention to the latter's observation that there seemed to be a limit on the power which could be handled by a single screw propeller, and he gave an estimate for this. When Parsons revisited his design he realised that he had been attempting to develop too much power with his single propeller. Meanwhile he had recovered ownership of his 1884 patents in 1893, so he was able to revert to a turbine of the original, axial flow kind. This allowed him to split up

[6] Sections of blades cut from wood were arranged in a row and sandwiched between glass sheets. A second group was mounted so that it could move parallel to the first, fixed, group so simulating fixed and moving blades. When water was caused to flow through the space between the glass plates, the behaviour of the flow could be made visible by injecting jets of dye. Parsons placed great reliance on having a visual image of the phenomena that he was studying and he showed a flair for achieving such demonstrations.

the total output among three propeller shafts and it solved his problem. In 1897 he reached a record 32.76 knots on the measured mile.

This was fine for a vessel 100 ft long and displacing 40 tons, but it provided no basis for designing propellers for larger vessels, where incidentally huge sums of money were at hazard. Ten years on, in 1907, the *Mauretania* was completed, displacing 36,000 tons and achieving a record speed for such a vessel of 26 knots. To prepare himself to meet such challenges Parsons needed a better understanding of the behaviour of propellers, so he built a small and primitive water tunnel. It was shaped like a race track and had windows in the walls which gave a view of a model propeller as it rotated in the stream of water. Froude had speculated that the limitation which Thornycroft had observed was due to the appearance of voids around the propeller blades as the water 'boiled' locally, a phenomenon which he described as 'cavitation'. Rankine's theory of propeller action had predicted that the water pressure at the propeller blades would fall, but there was a disparity between what theory predicted and what tests revealed. Parsons had to get to the bottom of this, so he developed ways of varying the temperature and pressure in the water at the location of the propeller. In this way he could induce or suppress 'cavitation' at will. He took a further step by illuminating the propeller with an intense beam of light from an arc lamp. This could be interrupted by a spinning shutter synchronised with the propeller speed, which allowed for strobing the action to 'freeze' it. In this way he could observe and even photograph the onset and gradual development of cavitation until the propeller completely lost its ability to develop any thrust. Such observations were of critical economic significance. If cavitation developed beyond a certain point, a fraction of the turbine power would be wasted as useless heat with corresponding waste of fuel. Furthermore in another piece of research Parsons established that it was this cavitation which caused the pitting of the propeller blades, frequently destroying them in a matter of months. Eventually a large scale water tunnel was built at Wallsend and it provided a valuable service to ship owners, but interestingly it was never described in the literature during his lifetime.

PIONEERING OF A NEW CONCEPT

Inevitably the efficiency of the first turbines which had a very small power output, was low. It was only possible to market them at all by concentrating on niche applications. One such was electric lighting for use below decks on board ship. The steadiness of the dynamo output of his turbo-generators greatly enhanced the life of the newly invented, and very costly, incandescent lamps. Also the lubrication arrangements for his machines were greatly superior to those of conventional steam engines and there was no need to employ an attendant to keep an eye out for overheating or bearing seizure.

Parsons' character and temperament were at once both sources of strength and of weakness. This can be readily seen on the occasion when he lost his rights to the key, 1884, turbo-generator patents. He knew that the efficiency of his motors could only be raised if their output could be greatly increased. When a group of businessmen decided to set up the Newcastle upon Tyne Electric Supply Company (NESCO), they could not be persuaded to adopt the new style of prime mover. No doubt with his urging, a rival, the Newcastle and District Electric Lighting Company (DISCO), came into being also at this time, and Parsons himself took on the responsibilities of managing director. He was able to persuade the Board of DISCO to purchase two of his 75 kW turbo alternators. His partners Clarke Chapman felt that he could not continue to work as a partner while at the same time acting as managing director of another company, but Parsons was not willing to back off. The partnership was dissolved in an orderly way but Parsons would not pay the asking price for the return of his patents which Clarke Chapman demanded. He resorted to law and had two eminent expert witnesses to speak for him, Dugald Clerk and Professor J.A.Ewing, while Clarke Chapman engaged Sir William Thomson (later Lord Kelvin) and Gisbert Kapp. There was an element of farce in the outcome for he ultimately failed to convince the arbitrator Gainsford Bruce Q.C. that his precious patents *were really not worth the price* demanded by Clarke Chapman!

The loss of his patents was a devastating blow and a fiery streak in Parsons' makeup certainly played some part in the way matters developed. By 1890 there were two credible challengers; in Sweden C.G.P. de Laval was making progress with his 'impulse' design, while in the U.S.A. C.G.Curtis was also experimenting. Undaunted, Parsons set to work and systematically sought, and in just a couple of years found, alternatives to each of the key features of his earlier patents, as well of course of those of his competitors. It was an achievement which required enormous determination and confidence in his own capability, fully equalling that of creating his first turbo-generator. He was fortunate in that although in the 1884 *provisional* specification he had described in general terms the principles of a radial flow machine as an alternative to the axial flow layout, these did not form part of the final specification, so he was free to pursue this concept. Within three years Parsons had sold one of his new radial flow designs rated at 100 kW to the Cambridge Electric Lighting Co., of which by the way, he was also Managing Director! More to the point, for the first time his turbine exhausted to a condenser, so greatly improving its efficiency. This was of enormous long term significance because the turbine is significantly better suited to taking advantage of low vacuum than steam engines. Just for good measure during this time he also checked out a number of other designs which had been suggested by others and dismissed them as practical alternatives. During a period in which he was bereft of his key patents and which might have been disastrous for his ambitions, he actually advanced the efficiency and power output of the turbine. It is important to bear in mind that during the 15 years after 1884, Parsons was virtually alone in promoting the steam turbine. By the time competitors appeared on the scene he had firmly established its suitability both for generating electricity as well as for propelling fast ships. It was no longer a risky venture.

ASPECTS OF CHARACTER

One trait of Parsons' personality was a fieriness which often wrought havoc on those around him. We have seen one example of this in the episode where the partnership with Clarke Chapman broke up. There were others. C.Turnbull recalled that

> Few people who met Parsons casually, thought of the explosive energy behind his introspective appearance, but it could come out when something happened. The first time I encountered this was when the accident happened to the blades of the axial flow turbine for the Turbinia. The short blades of ordinary turbines were skimmed up in the lathe, but

> the Turbinia L.P. rotor had blades of unprecedented length, and when the turner tried to machine the tips, the blades were bent right over. Dick Williams, the Works Manager, stopped the lathe instantly, and when Parsons came to look at the mess, the man disappeared in a hurry.

It appears that the unfortunate turner was sacked on the spot. On another occasion

> a workman whom he (Parsons) personally wanted early in the morning was not forthcoming. Parsons became more and more impatient, and when the man ultimately arrived half an hour late he said 'You're fired' and would not hear a word in mitigation or excuse. However a member of the engineering staff insisted on stating the man's case, which was that he had been sitting up all night with a sick wife. Parsons was filled with remorse, decreed that the man's wages were to be increased five shillings a week, and personally called at his home with presents of grapes and other delicacies suitable to an invalid.

Parsons had close and devoted friends, people like Gerald Stoney his Irish chief engineer who knew him so well that they allowed for his failings. Others found it harder to deal with him. Mark Robinson for example observed perceptively that it seemed to be in Parsons' nature to suspect the good faith of those with whom he did business. And yet when he gave his word on a transaction, he never failed to keep it. Yarrow the naval architect and shipbuilder crossed swords with Parsons over a patent dispute. Speaking after his death in 1931 Yarrow said,

> Parsons was a very jealous man, and he wrote me a very rude letter, practically saying, 'I don't want to have anything to do with you. So we took up with another kind (*meaning Rateau's design of turbine*).

In 1905 when Parsons was courting Sir William White to have him join the Board of PMSTCo, White refused unless Parsons first settled his quarrel with Yarrow. Yarrow continued,

> Then Parsons turned up at the factory one day and said 'You do not seem to be inclined to work with me' and I replied 'You have written me such a rude letter I cannot continue dealings with you.' Parsons said he did not remember the letter… and I showed it to him. Parsons read it and confessed 'I don't know how I came to write such a letter, but it was just at the time the Cobra went down. (1901)' I thereupon forgave him and asked 'why didn't you add to your letter that you were not in a fit state? We parted friends and remained so ever after.

Writing to Sir William White in 1904, Yarrow had remarked,

> It is universally recognised that Parsons is a man who is difficult to get along with; therefore when any of his staff come to me and say they could not get on with Parsons, my natural inclination is to believe them and attribute their leaving to Parsons' peculiar temperament.

The *Cobra* referred to above was one of the first two turbine powered destroyers for the Royal Navy, and had been built in Armstrong's Elswick yard. On 18th September 1901 it floundered off Flamborough Head on its way south from Newcastle; only 12 of the 79 men aboard, naval personnel and civilians, survived. Robert Barnard who had often served as helmsman on *Turbinia* trials, and who was manager of PMSTCo. was among those drowned. Parsons was deeply affected by this accident and the loss of life. When the news reached him at Heaton the workmen were sent home for the rest of the week, while he retreated to his office that day and remained there in silence with his thoughts for several hours. For the rest of his life he showed concern for the dependants of those drowned. No doubt the thought must have struck him that just such a disaster could well have cost him his own life, or that of other senior staff during the numerous trials in which they had participated, often in foul weather in the North Sea during the last seven years.

The ferocious concentration on work which characterised Parsons behaviour affected his private life. Recalling their courting days, Katharine Bethell, his future wife, remarked that he was known '*as a strange and weird young man*'. He was absentminded and on occasion would forget to collect her after leaving work. On one occasion '*he blew up*' she said, and '*I walked away and did not own him*'. Their two children were born within the first couple of years of marriage, but he did not emulate his parents who had 11 children. While Katharine gave him lifelong support and loyalty, the tensions had clearly impacted their marriage bed. One may speculate that his sexual impulse became sublimated, and reinforced his drive towards success. His relations with his own children were little different, for he simply didn't have the time to devote to them. In the long term neither followed his footsteps in business.

As a child living at Birr Castle in Ireland, Parsons enjoyed out door activities. The artificial lake which his father had made was just right for trying the boats and a submarine which he had made in the workshops. He also fashioned his own spring trap for catching rabbits and an air gun for shooting game. When he moved to Newcastle he became a keen cyclist, but most of all for recreation he loved to fish and when possible to shoot game. He certainly carried no surplus weight even in later years.

MANAGERIAL STYLE

As Turnbull's story and the remarks of Yarrow which have been quoted, make clear, there were difficulties in working in close proximity to Parsons. As has been indicated his factories were not run primarily to make money, but to facilitate technical development on full size machines. The only manufacturing experience Parsons had been exposed to was at Armstrong's Elswick works and at Clarke Chapman's in Gateshead, neither of which offered role models for a man-manager. As a consequence works organisation had a low priority in his scheme of things. Parsons spent much of his time at the Heaton works and there was a saying among the staff, that the closer you got to the top, the nearer you were to the gate, (and exit). Many very able men, Fullagar, Gerald Stoney and Alex Law among them, found it impossible to work in such an environment. Even his own children who spent time as directors of the Heaton Works quit the scene. One of his nephews, Arthur, who was a son of Clere had travelled to Pittsburgh to serve an apprenticeship with Westinghouse after studying engineering at Cambridge. An impression of his ideas on factory management can be seen in a paper in which he wrote in 1916 on production planning where he described documentation for the control of turbine manufacture, with the caveat that,

> care must be taken to guard against the tendency of building up a system of forms only…and of the whole system developing into so much waste paper, and the work being completed before the system designed to assist... can come into use.

Arthur was just what C.A. Parsons and Co. needed, but he was unable to work for long in close proximity to his uncle.

Parsons simply could not 'read' other people's reactions. This was probably a consequence of his personality, but it was not helped by the deference that his social station demanded in Victorian society nor by his sheltered upbringing. Only those who could bend with the wind were able to function in such a stormy environment. It is perhaps noteworthy that his last senior manager was the exception which proved the rule. He was a self assured young Australian, Claude Dixon Gibb (1898-1959) an airman who had graduated from Adelaide University after the end of the War with a degree in engineering. He came to Heaton in 1924 as an apprentice when Parsons was 70 years of age. In the following year he was brought to Parsons' attention by the resourcefulness which he had exhibited in dealing with a problem encountered during the commissioning of a turbine at the Barking station in London. He was immediately put in charge of the Test House, then moved to the Design Offices with a commission to overhaul the organisation there. In 1929, he was called to Parsons' office to discuss some incident when, out of the blue, "*Sir Charles said quite suddenly and unexpectedly, 'I am making you a Director, come and have a cup of tea'*". It seems that Parsons acted without reference to other members of the Board in appointing him a Director and Chief Engineer.

SCIENTIST, ARC LAMPS AND DIAMONDS

All the while that Parsons was running his factories and heading up a program of development, he retained an interest in science. Take just one example. While he was working with Clarke Chapman he became interested in compressing carbon dust into electrode rods for arc lamps, and his first scientific paper arose from this and was communicated in 1888 to the Royal Society by his brother, now fourth Earl of Rosse. Just then diamond mines were being developed in South Africa while at the same time reports (erroneous as it transpired) were being published of successful attempts at synthesising diamond from graphite. Parsons had the technical resources to achieve high pressures as well as such powerful sources of electric current as would create high temperatures when passed through the graphite samples. Using these he tried to convert graphite dust into diamond crystals. Although this effort was a side line to his main business, he pursued it with just as much dogged determination and inspired inventiveness. He carried out hundreds of tests and thousands of chemical analyses. He also carried out a great many experiments designed to reach high transient pressures, an approach which is now exploited commercially. The details of this program were set down in meticulous detail in his paper to the Royal Society in 1918. His style recalls his father's scientific publications setting out the sequence of events which led to success in creating 6 foot diameter telescope mirror, having overcome many difficulties on the way.

Just one example of his ingenuity may be quoted. When he was searching for a means of achieving very high pressures he took out patent 169274 of 1921. This described a method of imploding a hollow lead sphere with a layer of explosive covering the outside surface. Less than twenty five years after this, this technique was developed by a team of scientists at Los Alamos as one of two alternatives for detonating the first nuclear bombs.

While developing his high pressure equipment and calibrating it, Parsons arranged for some careful measurements of compressibility of graphite, and of liquids like ethyl ether, water and paraffin oil. These results were reported in 1911 to the Royal Society of which he had been elected a Fellow in 1898. Having expended between £20,000 and £30,000 of his money he thought for many years that he had achieved success, and he claimed as much in the Bakerian lecture to the Royal Society in 1918, but eventually he was forced to admit that he had been deceived in this. In 1955 the American General Electric Company finally did succeed and it turned out that while Parsons had reached sufficiently high temperatures, his maximum sustained pressure, of 15,000 atmospheres, was perhaps a quarter of what is required. Despite his lack of success he had been working on the right track.

In the middle of this program which ran in parallel with his turbine development, Parsons said to Gerald Stoney by way of excuse, "*We have now made a bit of money and deserve to have some fun*". Evidently some of the driving force which sustained him was the pleasure and satisfaction which his work gave him. In the case of Charles' father, engineering skills were a by-product of a program for developing a scientific instrument, the giant telescope. In his own case scientific studies were a by-product of his engineering ventures. In both cases science and engineering were seamlessly joined. It was fitting that it was for his '*application of the Steam Turbine to industrial purposes, and for its recent extension to navigation*' that the Royal Society awarded him its Rumford Medal in 1902.

A MODEST GENIUS

In 1933 Lord Rayleigh recalled that '*Parsons was a singularly modest man, and did not seem at all to realise his own standing in the world. His lack of self assertion was at times almost comic*'. In 1922 he travelled to a conference of the International Astronomical Union at Rome. Arriving at his hotel he was piqued not to have received the customary invitations to various receptions. '*He seemed to think that this was a sign that he was not appreciated and was much hurt*'. Instead of protesting or making enquiries, he stayed silent, only to find on his return home that the missing invitations awaited him there. J.H. Bruce an engineer with Parsons Foreign Patents Company remembered the way engineers of other nations regarded him, '*Frequently, one heard during meetings of his Continental friends and colleagues the question, 'Was sagte der Parsons dazu' or 'wie denkt der Parsons daruber' - always 'der Parsons' - that retiring genius*'.

CONCLUSION

Any account like this must just skim the surface of the subject. I have not attempted to fill the text with references for the reason that it is drawn from a book in which there is space to deal in greater depth, and which for that reason I recommend to those who may be interested in a better understanding of one of the greatest engineers.[7]

[7] W. G. Scaife, From Galaxies to Turbines - science, technology and the Parsons family 2000 (Bristol).

OPENING PLENARY SESSION

Advances in Aeroengine Materials

M. A. Hicks
Rolls-Royce
Derby, UK

M. C. Thomas
Rolls-Royce
Indianapolis, USA

INTRODUCTION

The history of the aeroengine has been intrinsically linked with the development of advanced materials technologies. The last decade has seen the launch of a large number of new engine programmes which have provided the opportunity to introduce many new materials. Over this period the emphasis has changed from being dominated by the need to improve engine performance to one in which cost is the prime requirement. Consequently, new technologies have to show a clear business case in order to 'buy their way' onto an engine. The commercial aerospace industry has also been going through a very difficult period since the events of September 11th. Now, as well as the terrorist threat, we are experiencing a further downturn because of the SARs outbreak in SE Asia. As a consequence, there will be fewer new aircraft launched, less research funding available to develop new technology and there will be an even greater emphasis on cost and supporting the aftermarket.

In terms of materials technology this means that we will need to achieve bigger changes in capability when an engine opportunity presents itself, to focus on fewer technologies with the biggest business impact and to do things faster and smarter.

This paper reviews the development of aeroengine materials and discusses emerging technologies for future engines against this changing business scene. As well as reviewing developments in current titanium and nickel alloys, advances in ceramics and composites will be covered. Finally, emerging technologies in smart materials and nano materials are discussed.

TITANIUM ALLOYS

Titanium Discs

Titanium alloys are very attractive materials for aerospace applications, offering high strength at intermediate temperatures at a density almost half that of steel or nickel based alloys. At

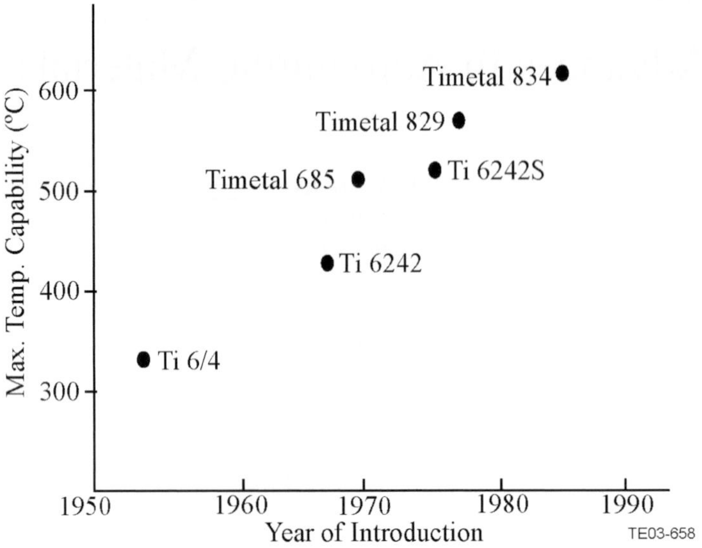

Fig. 1 Titanium alloy temperature capability has doubled.

present they typically account for approximately 25% of the weight of large commercial engines, finding application in discs, blades and casings.

The most common alloy in current use is Ti-6Al-4V which has been applied for almost 50 years. Subsequent developments have mainly been directed at higher and higher temperature applications (Figure 1). The latest alloy in the series, Timetal 834, has a temperature capability of approximately 630°C, close to 300°C above that of Ti-6Al-4V. This increase has been achieved through the careful optimisation of the balance of primary alpha and transformed beta to give the best combination of high tensile strength, fatigue resistance and creep behaviour.

Operating at temperatures above 650°C for long periods of time is seen as a barrier for future developments of conventional titanium alloys. Under these conditions the problem is not restricted to high temperature creep. Long term surface and metallurgical stability effects can also lead to degradation in low temperature properties – in particular tensile and low cycle fatigue strength. In addition, there are concerns about the effects of contaminants such as salt and the increased risk of a titanium fire following a blade rub or foreign object damage.

Current interest in titanium alloys is focused on weight and cost savings. This focus has resulted in the recent introduction of integrally bladed discs (blisks) in which up to 50% weight savings have been achieved by the elimination of blade fixing features. For small engines, blisks tend to be machined from a single forging but larger engines require the use of linear friction welding to attach the aerofoils to a separate disc forging. The use of integral blading will also raise issues of repairability when aerofoils sustain foreign object damage. Here again linear friction welding will be a key technology. Further weight saving may be achieved through the introduction of titanium metal matrix composite (TiMMC) technology. The high performance of the continuous, silicon carbide fibres means that when they are embedded in a Ti-6Al-4V matrix, the overall system can yield a 50% increase in strength and a twofold improvement in stiffness compared with the parent material. There are a number of potential applications for TiMMCs but the greatest benefit would arise in the compressor as a natural development beyond the blisk. It would be feasible to completely eliminate the bore section of the component as the hoop stresses can be born by the fibre reinforced ring or bling. (Figure 2) Demonstration components have been manufactured and tested by Rolls-Royce and the first TiMMC bling was engine tested in 1991. The most promising manufacturing route for these critical components is metal coated fibre processing, in which the metal coating is applied directly to the fibre using electron beam, physical vapour deposition. Consolidation is achieved by HiPping. This yields a near perfect fibre distribution and eliminates 'touching fibre' defects which can cause a fatigue penalty (Figure 3). Another key technology is that of fibre coatings. These are required in order to prevent an interaction between the fibre and the matrix resulting in a low strength brittle interface. Consequently, all commercially available fibres have a protective surface coating generally based on layers of carbon.

The first TiMMC bling application is likely to be in a military vehicle where weight is critical. However, cost still remains an issue and there are a number of projects underway looking at reducing fibre, processing and consolidation costs.

Titanium Blades

Titanium blade alloys have largely derived from the disc alloy developments. The major 'novelty' has been in the advanced processing techniques such as the superplastically formed and diffusion bonded Ti-6Al-4V hollow fan blade.

Titanium aluminides have long been recognised as offering potential for cost competitive improvements in performance through significant density reductions. The problem, in common with all intermetallics, has always been their low temperature brittleness. Selective additives coupled with microstructural control have gone some way toward addressing this issue and ductilities of up to 3% have been observed in some current alloys. Manufacturing routes have been established and specimen and component test results have been very promising. The first application of these materials is likely to be in static and rotating compressor aerofoils. Once engineering confidence is gained in running this class of materials in production engines, it is likely that other applications will emerge. The biggest prize would be a low pressure blade application where savings of over 250 lb could be achieved for large engine applications.

Again, cost is a factor for components manufactured from titanium aluminide but cost is likely to come down significantly as volumes increase and scrap costs are reduced. The initial components will be forged but recent developments in casting technologies have opened up further opportunities for cost reduction.

Another recent development is that of burn resistant titanium alloys. One of the current restrictions placed on how far back in the compressor titanium can be used, particularly for aerofoils, is the risk of a titanium fire. As a result, a family of alloys has now been developed

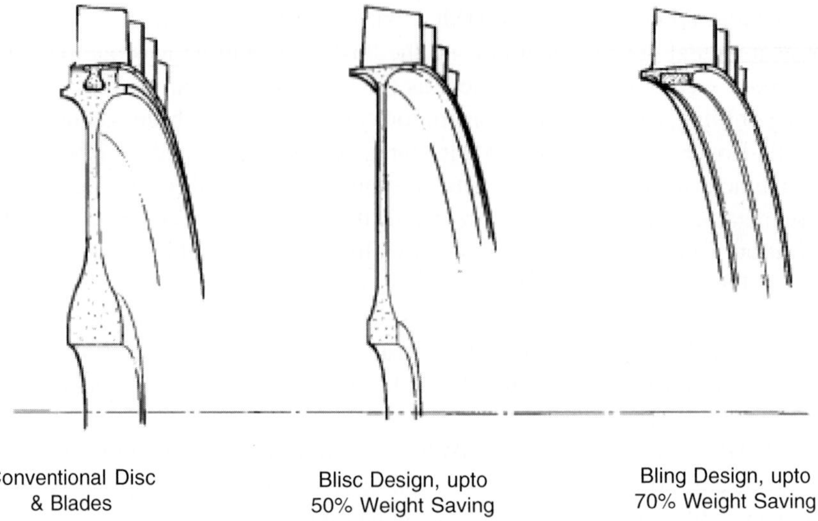

Fig. 2 The TiMMC bling offers significant weight savings.

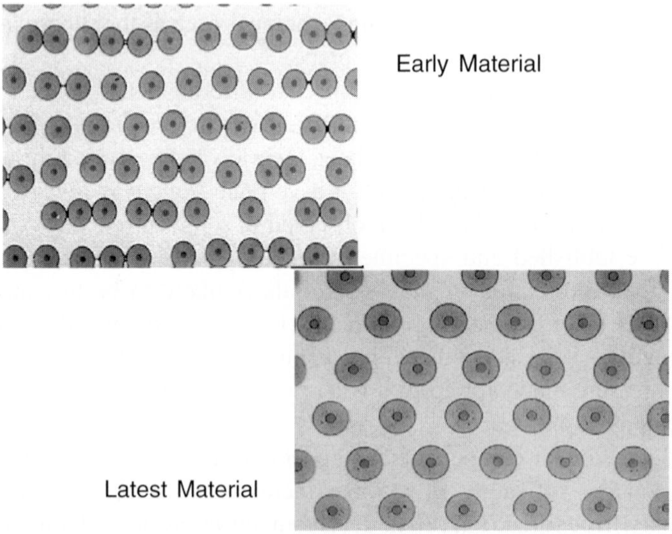

Fig. 3 TiMMC process development has improved fibre position control.

which have performed very well in fire tests. These burn resistant titanium blades are undergoing demonstrator and engine tests.

ORGANIC COMPOSITES

Polymer matrix composites (PMCs) currently account for almost all the composite materials used in aeroengines despite being limited to relatively low temperatures. Most of the European applications are in components such as nacelles or bypass ducts. Further expansion of these materials into the core of the engine will require a greater maturity from both the materials and the associated design methodologies. The higher temperatures experienced will demand the development of a high temperature matrix system capable of being formed economically into complex geometries.

Rolls-Royce is pursuing the goal of higher temperature capability on the Joint Strike Fighter programme, where PMCs are being applied in a number of static aerofoil components. Twenty-five percent or greater weight savings have been achieved together with significant cost savings. A key requirement in keeping costs down is the ability to use resin transfer moulding as the primary manufacturing route.

One recent advance in temperature capability is the Superimide® polyimide resin capable of use up to 375°C.[1] These materials are composed of mixed phenylenediamines (PDAs) with a biphenyldianhydride diester acid (BPDE) and a nadic ester acid (NE) as a cross linking end group. The high temperature capability and good chemical resistance is primarily attributable to the very high aromatic character based on the use of BPDE and PDA in combination.

These materials can be formulated to meet specific performance criteria of thermal oxidative stability, glass transition temperature and strength.

Rolls-Royce currently uses low cost compression moulded bushings with chopped fibre fillers made from these materials in the AE 3007 engine for the Embraer regional jet, and as the range of 'tailored' properties expands, further applications will emerge.

One other area of composite usage is the GE 90 composite fan blade. GE has developed a wide chord composite fan blade as their solution to the high performance light weight issue. Rolls-Royce uses a hollow titanium blade as their solution. The GE blade is a toughened epoxy graphite with a polyurethane coating for wear resistance and a titanium leading edge guard for ingestion protection. Materials, manufacturing and computing advances have provided the necessary technology to enable successful utilisation of this concept.

NICKEL ALLOYS

Nickel Discs

The developments in nickel disc alloy capability have been one of the major success stories of metallurgy over the last half century. The initial development of gamma prime strengthening was a leap forward, giving much increased temperature capability and thermal stability from the previous austenitic and ferritic materials. As demands for higher temperature capability continued, alloy developers generated increasingly complex chemistries that combined solid solution strengthening, carbide precipitation, grain boundary strengthening and ever higher volume fractions of gamma prime.

These alloys became harder to melt and cast without serious segregation and deleterious intermetallic phase precipitation. Thus, processing was refined with the introduction of triple melting using sophisticated control systems to hit ever narrowing melt and solidification rate windows.

Ultimately the use of inert gas atomisation powder metallurgy became necessary for the successful manufacture of the most complex and highly alloyed materials.

Current work has focused on the area of control of forging, heat treatment and residual stress management in order to squeeze the last performance capability from these materials. Sophisticated modelling is used to define and control chemistry, melting, forging, heat treatment and precipitate morphology. In these alloys, not only do chemistry and melting have to be controlled extremely tightly, but heat treatment must be selected to be close to or above the gamma prime solvus to optimise both grain size and strengthening phase precipitation during aging. This requires accurate gamma prime solvus control and determination.

Since these alloys are designed to have high strength at high temperatures, fast cooling from solution treatment is required to prevent excessive precipitate coarsening prior to ageing. Consequently, the discs can retain significant residual stresses following cooling from solution treatment. Thus controlled cooling at the optimum rate independent of cross section is important. The advent of computer simulation and computer control of cooling rates offers a cost effective way of balancing these conflicting requirements, enabling these alloys to achieve their potential. The Supercooler at Ladish Company is a prime example of this technology.

Further temperature capability and weight optimisation can be achieved using dual structure disc concepts, whereby the hottest areas of the disc are selectively grain coarsened to enhance resistance to creep and dwell crack growth. Dual alloy discs and blisks may be employed in selective future applications to further enhance temperature capability, wherein creep resistant cast alloys are bonded to the disc material.

It is however becoming clear that as temperatures continue to climb at the back of the compressor and in the high pressure turbine that either sophisticated thermal management systems featuring active disc cooling designs or as yet undefined materials systems will need to be developed.

Nickel Blades

Nickel based alloys have come to dominate the high temperature stages of aeroengines. Blades are required to withstand high centrifugal loads while operating in a gas stream at temperatures in excess of the melting point of the alloy. Operation in this environment makes severe demands on both the mechanical properties and environmental stability of the blade system and success is only possible through the close interaction of design, materials and manufacturing technologies. Major advances in cooling efficiency bring complexity to the manufacturing technology and require significant effort to control costs.

All engine companies use single crystal alloys in the most demanding high pressure turbine stages. They were first introduced in the early 1980s and the continued drive for more temperature capability has resulted in successive generations of alloy with ever more exotic alloying additions. The current fourth generation alloys employ ruthenium as well as rhenium, which was introduced in increasing amounts in the second and third generations.

With increasing blade operating temperatures, the intrinsic resistance of the metal to environmental attack is no longer sufficient. Protective surface coatings are then required to provide the necessary oxidation and corrosion resistance. These are commonly aluminide coatings and may include the addition of platinum to further enhance oxidation resistance. MCrAlY overlay coatings have also been employed, where M represents Ni or Co.

The latest blades also use thermal barrier coatings (TBCs). These coatings are low thermal conductivity ceramics which prevent the flow of heat from the gas stream to the underlying blade. This maximises the benefit obtained from the complex blade cooling and offers a potential increase in operating temperature of over 100ºC. The majority of TBCs are based on yttria stabilised zirconia, with the most recent developments containing additional elements to further reduce thermal conductivity. Coatings are applied by electron beam physical vapour deposition to develop the columnar grain microstructure necessary to resist thermal and mechanical strains. The coating system must be very durable since any breach would expose the blade to gas stream temperatures which could not be endured by the metal alone. Bond coats are therefore an essential technology and the subject of much current research. Likewise, component life prediction and assessment are critical in the safe and economical use of these systems.

The total effect of all the advances in blade cooling technology coupled with alloy and coating development has been to increase metal temperatures by around 300°C since the original wrought alloys.

The big unanswered question is what comes after nickel blades. Ceramics have come in and out of favour many times but it is unlikely that they will be considered for rotating components for many years (see Ceramics section). Recent interest in the United States has focused on niobium and molybdenum silicides and a major development programme is underway supported by the U.S. Air Force Materials Directorate. Niobium silicides comprise a two phase system of niobium and titanium silicides in a niobium matrix. They contain Laves phase for creep and oxidation resistance. Figure 4 shows a typical microstructure and property comparison. However there are still questions over oxidation resistance, fatigue crack growth and low temperature brittleness of these materials.

The most pressing near term driver for blade materials, as with all components, is cost reduction. Some of the advanced blade alloys have long and complex heat treatment cycles which are very expensive. Some recent work has show that significant reductions can be achieved without impairing mechanical properties. Other approaches are addressing manufacturing yield. Clearly with such complex castings and coating systems, the possibility of generating scrap is quite high. Process modelling, process control and quality acceptance standards are all being addressed so as to maximise the number of 'good' blades produced from each casting assembly. Grain boundary strengthening additions are also being added to

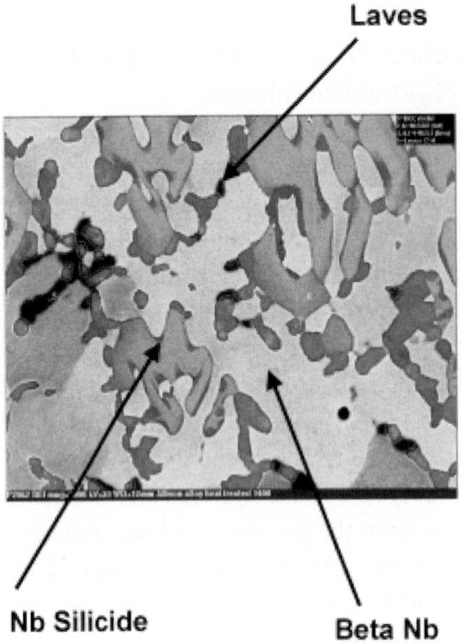

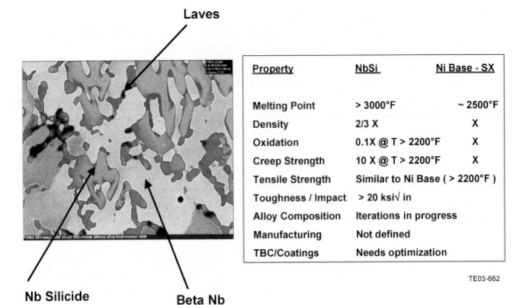

Property	NbSi	Ni Base - SX
Melting Point	> 3000°F	~ 2500°F
Density	2/3 X	X
Oxidation	0.1X @ T > 2200°F	X
Creep Strength	10 X @ T > 2200°F	X
Tensile Strength	Similar to Ni Base (> 2200°F)	
Toughness / Impact	> 20 ksi√ in	
Alloy Composition	Iterations in progress	
Manufacturing	Not defined	
TBC/Coatings	Needs optimization	

TE03-662

Fig. 4 Structure, properties and challenges of niobium silicides.

single crystal alloys to mitigate the risk of blades being scrapped for certain grain boundary features.

MORE ELECTRIC ENGINE MATERIALS

Specialist electrical and magnetic materials will eventually permit the development of technologies which lead to more efficient and lighter engines and better integration with airframe systems. For example, there is likely to be greater use of embedded electrical machines and electromagnetic bearings, (Figure 5) with electricity becoming the primary power source for many applications, replacing the current hydraulic/pneumatic equipment and mechanical accessories.

Real progress will depend on the ability to manufacture electric motors/generators and associated power electronics with sufficient power density and tolerance to operate reliably in very harsh environments.

Several projects developing materials for these machines/devices are underway, including:

- Theoretical limits of magnetic materials.
- Optimising the properties of cobalt-iron alloys.
- Reliable compact capacitors.
- High temperature conductors.
- Electrical characterisation of high temperature electrical insulation.

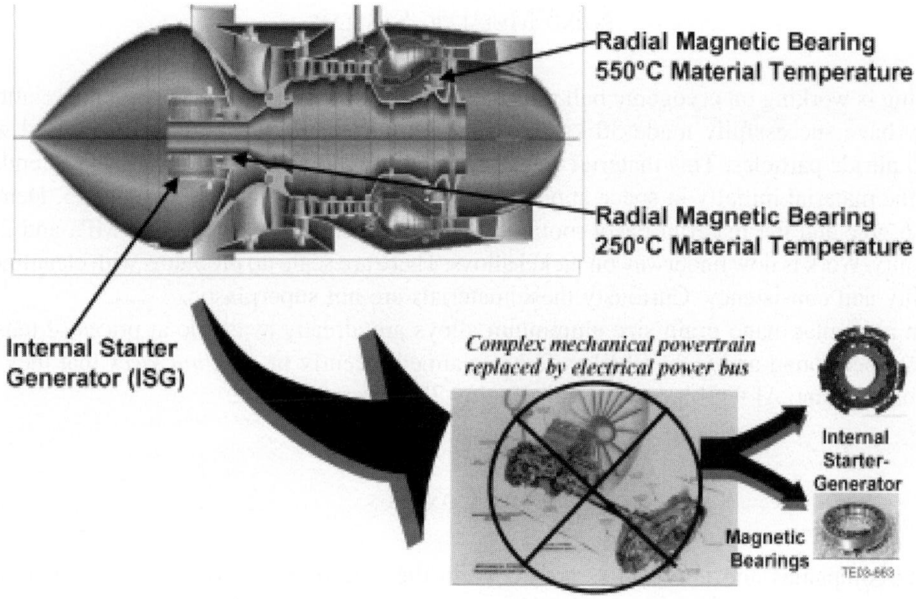

Fig. 5 The more electric engine (MEE) has significant materials challenges.

These technologies will be inserted on aeroengines in a progressive manner over the next 10 years.

Current efforts include oxide dispersion strengthening (ODS) of Fe-50 Co for applications up to 550°C. There is also interest in nano crystalline Fe Co alloys and amorphous alloys.

NANO MATERIALS

Nano technology is the generic name given to characterisation, manipulation and manufacture at the length scale of 1–100 nanometers. These technologies are finding application in the areas of materials science and engineering, electronics, computing, biology, chemistry and physics. There is currently a U.S. National Nanotechnology Initiative funded at 605M$ and many similar, but smaller, Japanese and European programmes.

NANO POLYMERS

There is current work in polyhedral oligomeric silsesquioxanes (POSS) and polyethylene terephlalate (PET) at nano sizes as polymer reinforcements and these show significant promise, as additions to polymers to reduce diffusion rates, and hence enhance stability.

Nano Metallic Systems

Boeing is working on cryogenic ball milling (in liquid nitrogen) to create nano phase alloys. They have successfully made 30 kg lots of aluminium nano phase alloys, stabilised with nano nitride particles. This material can be HIPped, forged and extruded. Boeing intends to use the material initially in space applications to replace titanium and Inconel 625. Density is 2.6 g/cc and the material has a room temperature strength of around 500 MPa and 20% ductility. Work is now underway on nickel alloys. There are scale up problems with cleanliness, quality and consistency. Curiously these materials are not superplastic.

In particular, nano grain size aluminium alloys are already available at prices of tens of dollars per pound and General Electric has claimed recently in *Aviation Week* that they are evaluating nano Al fan blades for their Boeing 7E7 aircraft engine.

Nano Coatings

Many companies and universities are active in the area of nano coatings. Rolls-Royce is involved with a programme run by the Ohio Aerospace Institute and University of Cincinnati. This programme is developing an epoxy polyimide based system reinforced with clay nano composites of chemically formed conducting polymer (CCP) coating. These coating are intended for use as primers for environmental resistance.

Nano Materials Summary

There are many issues to address with nanosize materials such as stability, passivation (decreasing the particle size by 100× increases surface energy factor by 1000×), productionisation and inevitably, cost. However with the level of effort being expended in Japan, U.S. and Europe in particular, major advances are expected by the end of the decade.

SMART MATERIALS

There is increasing interest and funding for the development of 'intelligent engine' technology. This is a focus area for the VAATE program, (Versatile Affordable Advanced Turbine Engine) a key U.S. Department of Defence initiative. A major part of this work involves the application of sensors which monitor performance and permit control of the engine.

The materials contribution to this capability involves the development and application of smart materials. These are materials that respond to their environment and could include shape memory alloys, piezoelectric materials and magnetostrictive materials as well as coatings which react to the ambient conditions. This technology is in its infancy, but holds much promise.

One example of current work in smart materials is the nanocomposite tribological coating work of Voevodin et al.[2] They have developed nanocomposite coatings made of yttria stabilised

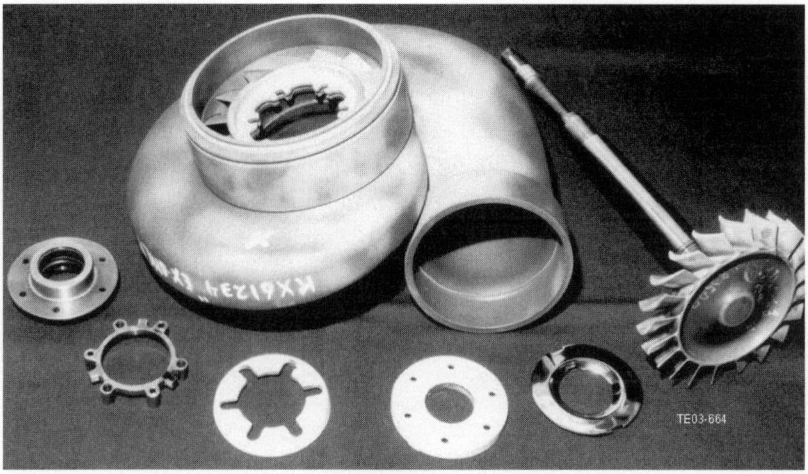

Fig. 6 Rolls-Royce ceramic engine components.

zirconia (YSZ) in a gold matrix with encapsulated nanosized reservoirs of MoS_2 and diamond-like carbon. Depending on the environment, the coating friction surface changes its chemistry and structure between graphitic carbon for sliding in humid air; MoS_2 for sliding in dry nitrogen or vacuum and metallic gold for sliding in air at 500°C. Friction coefficients change from 0.1 in humid air to 0.007 in dry nitrogen.

CERAMICS

Rolls-Royce has been involved in the design, development, manufacture, inspection and testing of ceramic component for gas turbines since the early 1970s.[3]

Components have been successfully manufactured and tested for over 5100 hours in an all ceramic rig (Figure 6) Silicon nitride rotors ran successfully for over 1000 hours at 1370ºC and silicon nitride vanes have run in industrial engines for 2300 hours. (Figure. 7).

As well as these monolithic ceramics, ceramics matrix composites (CMCs) have been under development since the 1980s. Two systems have emerged as having significant potential, SiC/SiC and oxide/oxide.

Recent development in CMCs include several important steps.

1. Sufficient toughness to assure notch insensitive behaviour and to avert foreign object damage degradation has been demonstrated. The concept uses thin 'debond' interlayers between the fibres and the matrix that mechanically detach when locally stressed.
2. SiC/SiC fibres that demonstrate adequate stability and creep strength have been developed in Japan and the U.S.

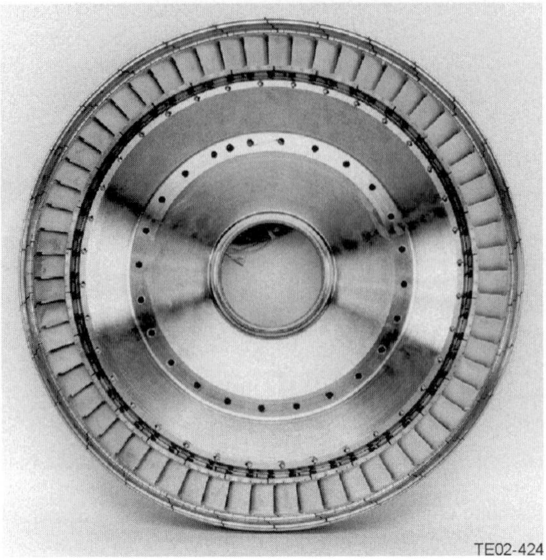

Fig. 7 Rolls-Royce 501-K industrial engine ceramic vane assembly.

3. A melt infiltration process based on reaction bonded SiC that significantly reduces cost and increases through-thickness conductivity has been devised. These materials are capable of durable operation at temperatures of 1300°C.
4. SiC/SiC composites degrade at high temperatures and cannot be used in aeroturbines without an environmental barrier coating (EBC). Developments in the last 5 years have resulted in EBCs that are 'impervious' to water vapour and oxygen. The system comprises a glass ceramic matched to the thermal expansion coefficient of the composite and of similar thermodynamic stability. Such EBCs have performed well in engine tests at both Solar and GE.

There are still issues of fibre availability and system cost, but these will surely be addressed and SiC/SiC components will find application in aero gas turbines in the near future. First applications are likely to be in combustor liners and static vanes.

Oxide/Oxide system have undergone parallel development primarily at 3 M and Rockwell Scientific. Fibres with high creep strength were developed and marketed by 3 M in 1998 (Nextel 720). Rockwell developed a 'debond' interlayer-matrix technology, based on monazite, that assures durable composite performance. The temperature capability is now limited by the creep strength of the fibres. Recent Rockwell data indicate good overall performance at temperatures up to about 1200ºC. These materials have low through-thickness conductivity, which adversely affects their use in cooled configurations. However, a capability for the design and manufacture of thin sections, based on 3D weaving, has the potential to obviate the conductivity limitation. Cost remains to be assessed.

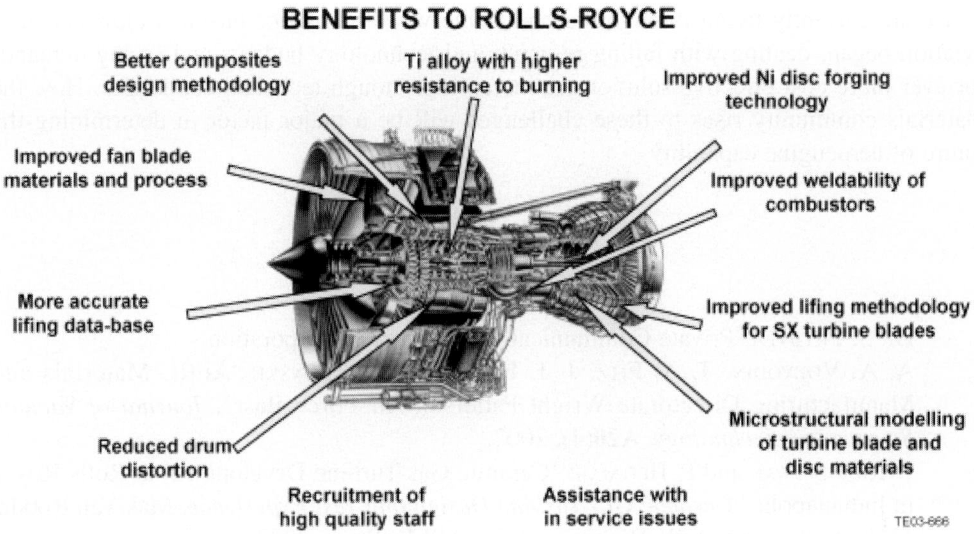

Fig. 8 Materials technology from University technology centres.

GENERAL REMARKS

Rolls-Royce has used its University Technology Centres (UTCs) to enable the development and introduction of new materials capabilities into the product line. The UTC concept has been very successful in focusing the research funding onto collaborative programs that are aimed at real world issues. Improved hollow fan blade materials and processing, burn resistant titanium alloys, more accurate lifing databases and many other benefits have accrued from these partnerships (Figure. 8).

Since the inception of the gas turbine, materials have been at the forefront of enabling technologies, and this is not likely to change in the foreseeable future. Until now, metallic systems have dominated gas turbines, and even the next generation U.S. fighter engine the JSF is 80% metallic. However, operating conditions, particularly temperatures, are compelling materials scientists to seriously consider other systems such as composites (polymer, metallic, ceramic and hybrids) as well as novel concepts such as nano technology and smart materials.

The issues facing the materials technologist are increasingly economic. This emphasis is evidenced by the current focus on repair technology and the United States Air Force Metals Affordability Initiative (MAI). The aim of repair technology is to create methods to permit the reuse of components after overhaul by rejuvenation, strip and recoat and additive techniques to restore dimensional conformance. MAI is a significant U.S. program involving engine and airframe manufacturers and many in the supply base to reduce the cost of metallic components. It supports activities from melting, casting, forging and subsequent processing.

We are currently living through the one of the worst aerospace business climates since aviation began, dealing with falling research and technology budgets and facing demands for ever more cost effective solutions to increasingly tough technical challenges. How the materials community rises to these challenges will be a major factor in determining the future of aeroengine capability.

REFERENCES

1. Dr. S. Prybyla: Private Communications, Goodrich Corporation.
2. A. A. Voevodin, T. A. Fitz, J. J. Hu and J. S. Zabinski: 'AFRL Materials and Manufacturing Directorate Wright Patterson Air Force Base', *Journal of Vacuum Science and Technology*, **A20**(4), 2002.
3. P. Khandelwal and P. Heitman: 'Ceramic Gas Turbine Development at Rolls-Royce in Indianapolis,' *Ceramic Gas Turbine Design and Test Experience,* Mak Van Roode, M. K. Ferber and D. W. Richerson, eds., ASME Press, 2002.

Advances in PF Plant, 1990 to 2010

SVEN KJAER
Tech-wise A/S
D-7000 Fredericia, Germany

INTRODUCTION

Some 40% of power generation worldwide is based on coal, which makes coal the backbone of power generation; moreover a major part of the capacity is based on well-proven pulverised coal-fired (PF) technology.

This paper will review the recent 10–15 years of improvements of PF technology and highlight some of the remarkable results which have been achieved concerning plant performance and lower cost of electricity. Focus will be on European developments although the Japanese results are remarkable too and their belief in the potential of steel for power generation should be recognised.

Section 2 will review steam parameters and plant efficiency, and recent improvements of PF technology will be looked at in section 3. An analysis of cost for fuel, capital, operation and maintenance and for electricity will be presented in section 4, and finally section 5 presents an outlook to 2010 on the joint European AD700 project.

Throughout the paper the abbreviation PF stands for pulverised coal-fired, SC for super critical, CoE for cost of electricity and O & M for operation and maintenance.

STEAM PARAMETERS

The 1950s was a period with considerable improvements of steam parameters and both single and double reheat cycles were introduced. Steam temperatures rose to a range of 520–530°C. In general steam pressures stayed sub-critical, but several attempts were made to introduce very advanced super critical (SC) steam parameters in power plant design based on austenitic steels for the thick-walled sections. Most famous is Philadelphia P & L's 300 MW Eddystone plant which started with 350 bar and 650°C, but soon had to reduce steam parameters to the range of 300 bar and 600°C due to concerns on thick-walled austenitic steam lines and headers. However, it has operated on these conditions ever since, similarly to a number of smaller utility-owned base load installations in the USA (~125 MW) and power generating installations at chemical factories in Germany, which were operated at steam temperatures beyond 620°C for many years.

A considerable number of large SC units were built in the USA during the 1960s and early 1970s. Typical steam parameters were 250 bar and 540°C (1000°F) and these units are still generating electric power at very low prices. In the same period European steam

Table 1 Danish super critical coal-fired installations.

Unit	Comm. Year	Net Output MW	$p_{main\ Steam}$ bar	$t_{Main\ Steam}$ °C	$t_{Reheat\ Steam}$ °C	Net Eff. %
Studstrup 3 & 4	1984/85	350	250	540	540	42.0
Amager 3	1989	250	245	545	545	42.4
Avedøre 1	1990	250	245	545	545	42.4
Fyn 7	1991	385	250	540	540	43.5
Esbjerg 3	1992	385	250	560	560	45.0
Nordjylland 3	1998	385	290	582	580/580	47.0

Table 2 European super critical coal-fired installations outside Denmark.

Unit	Comm. Year	Net Output MW	$p_{Main\ Steam}$ bar	$t_{Main\ Steam}$ °C	$t_{Reheat\ Steam}$ °C	Net Eff. %
Staudinger 5	1992	500	262	545	562	43.0
Amer 9	1992	600	230	540	568	41.3
Meri Pori	1993	560	250	540	560	43.5
Rostock	1994	500	262	545	562	43.0
Hemweg 8	1994	630	260	540	568	42.6

parameters stayed sub-critical based on drum boilers or once-through boilers, and most new installations were fuelled by heavy fuel oil. Oil was extremely cheap and combined with a large programme for nuclear capacity, coal became less and less important for power generation.

Two energy crises in the 1970s forced the power generators to focus on fuel prices and re-introduction of coal for power generation. Therefore, development programmes for new high-temperature materials were started, and during the 1980s important results started to show up for all hot sections of the water/steam cycle.

The situation also became very difficult for Danish power generators in late 1970s as they were not allowed to build nuclear capacity, and oil became more and more expensive. Therefore, Elsam became among the first utilities in Europe to re-introduce coal and base its conversion to electricity on the super critical technology, which showed a good balance between capital cost and fuel savings. Table 1 lists the Danish super critical coal-fired installations currently in operation.

The result of the Danish interest in super critical technology soon spread to the neighbouring countries, and all new european power projects changed to super critical steam parameters as illustrated in Table 2. Table 2 also clearly shows that no new projects have been started for more than 10 years.

Table 3 German super critical lignite-fired installations.

Unit	Comm. Year	Net Output MW	$p_{Main\ Steam}$ bar	$t_{Main\ Steam}$ °C	$t_{Reheat\ Steam}$ °C	Net Eff. %
Schkopau	1995 + 1996	450	285	545	560	40.0
Schwarze Pumpe	1997 + 1997	820	268	547	565	40.6
Boxberg	2000	910	266	545	583	41.7
Lippendorf	1999/2000	940	267	554	583	42.3
Niederaussem	2003	940	269	580	600	45.3

Lignite is the only major coal resource in Europe still being mined at competitive cost, and a new generation of very advanced boilers has been developed and demonstrated successfully in Germany. Table 3 shows the remarkable improvements of steam parameters and net efficiencies. Modest increases in capital cost by moving towards more and more advanced steam parameters are also indicated, as additional capital costs have to be paid by savings of a relatively cheap fuel like lignite.

It may be concluded that after several failed attempts to establish super critical high-temperature technology during the 1950s and early 1960s this technology has now finally succeeded in europe. Main and reheat steam temperatures rose from the range of 530–540°C to 580–600°C meaning a jump in steam temperatures of 50°C with steel still in use for all high-temperature sections. This kind of development was hardly foreseen before 1990 but it is similar to what has happened in Japan.

In other coal importing countries like Korea, SC technology was introduced in the 1980s, and now it is being introduced also in countries with very large coal reserves like Australia, Canada, China and India, which indicates a worldwide break-through for super critical technology.

TECHNOLOGY

The success of PF technology is based on its enormous flexibility: it can be designed to handle and burn a wide range of coal qualities, PF installations can be build at any output ranging from less than 100 MW to more than 1000 MW, it can be optimised for low or high specific coal prices and for many different site locations, etc. In particular Europe, which imports many relatively expensive bituminous coals, has a long tradition of being able to manufacture the most advanced boilers and turbines for PF technology.

Due to the international, parallel development of steam turbines and high-voltage grids throughout the 20th century, PF technology complies excellently with the characteristics of the high-voltage grid and the demands of national and international load dispatchers and power pools. Further, despite more and more advanced steam parameters, operating flexibility of super critical power stations has remained constant.

New martensitic-ferritic 9–10 CrMoV(W, Nb, N)-steels for thick-walled sections like outlet headers, steam piping, valve bodies, turbine casings and rotors were basic to all improvements of steam parameters and plant economy at constant thermal flexibility since the mid-1990s. The latest development named P92 allows for 600°C and has been applied in the Danish power generator, Energy E2's new #2 at the Avedøre power station. P92 is an excellent steel which has doubled the creep strength at 600°C compared with the old high-temperature steel X20CrMoV 121, and it seems difficult to make further substantial improvements of the creep strength of ferritic-martensitic steel.

The progress in PF technology becomes more remarkable if the extremely difficult situation for equipment manufacturers with very few new projects and corresponding economic problems in the period considered is taken into account. Unfortunately, the ongoing deregulation of the European power market and the reluctant attitude of the European establishment towards coal do not contribute to reduce the problems of the whole industry or generate new projects. Hopefully, the emerging awareness of considering imported bituminous coal as a fuel of strategic importance to security of European energy supply and price stability may help to change this situation.

Boilers

Boiler design is an extremely difficult discipline, which in particular needs to consider and make compromises between the many different kinds of coal that are being mined worldwide.

Pulverised coal-fired boiler technology was introduced some 80 years ago to boost boiler output beyond the performance of grate-fired boilers. Basic to supercritical PF technology are the relatively simple once-through type boilers, which allow single-phased water/steam simply to be pumped through the boiler walls and superheaters. Feed water flow is being controlled by an enthalpy measurement of the steam shortly after passing the furnace, and at low loads, a circulating system through the furnace walls is being superimposed on the main cycle to guarantee a constant minimum flow for cooling.

During the 1990s the tower boiler got a well-deserved breakthrough in Europe based on the very homogenous and symmetrical flow conditions for flue gas throughout the boiler and water/steam inside the furnace wall tubes. A drawback of the tower boiler is the increased height, which enhances the visual impact of the boiler in heavily populated areas and might increase boiler construction cost in more isolated areas.

At part load once-through type boilers allow a perfect co-ordination of heat uptake in furnace and super heaters by letting the main steam pressure slide with boiler load. The advantage of sliding pressure control of the boiler is enhanced through:

- Introduction of effective speed-controlled motors as feed pump drives.
- More simple design of the HP turbine, to be addressed in the section on turbines.
- Life of thick-walled components like headers, main steam lines and HP-turbine is enhanced through reduced creep during part load operation with reduced pressure and stresses.
- Main steam temperature can typically be held constant in a very broad load range between rated load and some 35–40%, which improves the thermal flexibility of the SC power plant at part load operation.
- Reheat temperatures can typically be held constant between rated load and some 50%.

Table 4 Environmental performance of PF boilers (6% O_2).

	1988	2003	As Measured
SO_2	400	200	2-35
NO_x	650	200	150
Particles	50	30	7-25

Boosting main steam temperature beyond 540°C typically means that the final super and reheaters need to be designed with austenitic materials due to concerns on high-temperature corrosion, oxidation and creep strength. However, austenitic materials also mean problems with the growth of ferritic magnetite and hematite layers on the tube inside surfaces, which will scale off during intransient operation caused by the different thermal movements of ferrite and austenite. It seems that the problem of scaling can be contained within acceptable limits through the introduction of a Japanese austenitic steel named TP347HFG which combines high creep strength with high-corrosion resistance thanks to a fine-grained structure obtained through additional heat treatments. Furthermore, optimised design and plant operation will help to reduce the problem.

In the once-through type boilers the same water circulates all the time and therefore, control of water/steam quality is crucial to SC technology, which meant a worldwide breakthrough for the combined water chemistry during the 1990s.

Since the 1970s, flue gas cleaning has been improved and improvements are still ongoing. Most remarkable improvements have been achieved through the introduction of:

- Flue gas desulphurisation, most of the installations were of the wet scrubber type.
- DeNOx installations mostly of the SCR type.
- Improved combustion, typically based on improved burner technology, with staged combustion, over burner air and over fire air.
- Re-use of by-products like fly-ash (for road stabilisation, cement and concrete etc) and gypsum (for wall boards).

Table 4 shows the improvements since 1988 in EU legislation concerning flue gas emission (in ppm) from new coal-fired capacity. Additionally, typical 'as measured' values from the Nordjylland installation are shown.

Turbines

Since the introductions of the impulse and reaction types of steam turbines by Gustav de Laval and Charles Parsons more than 100 years ago, this technology has been improved continuously and improvements are still ongoing.

De Laval's and Parson's different approach to turbine design meant that many different turbine designs were started based on either the impulse or the reaction type of blading. However, the number has been reduced dramatically throughout the 1990s, and after many mergers and take-overs, only two major manufacturers of utility-size turbines remain: Alstom and Siemens. Fortunately it can be said that through the company mergers many excellent designs have also been merged, so that the present technology represents a selection of the best solutions from the merged companies.

Therefore it seems logical that in many areas the solutions of Alstom and Siemens also look similar. The sections, where the largest differences still exist, seem to be the HP turbine where Alstom offers a double casing design with the inner casing kept together by shrink rings, and Siemens offers an HP turbine of the barrel type. Also rotor designs are different with Alstom offering welded rotors and Siemens offering mono-block rotors.

Output from steam turbines has increased tremendously, so that now more than 1000 MW can be generated in single-shaft arrangements running at 3,000 rpm. Separate HP and IP casings instead of combined casings are typical for European turbine design, which guarantees a good thermal flexibility of the machine with short start and stop times and fast load change capability. Only one bearing between the casings is another topic of European turbine design, which improves the vibrational behaviour of the turbines and reduces O & M cost.

HP-turbine, turbine blading and last stage of LP turbines are three major topics from the development of European turbine design during the 1990s, which need to be addressed in more detail.

The impact of sliding pressure operation on HP turbine design has been strong and positive as it led to a more simple design. Both manufacturers reduced the number of inlet valves from four to two, which are fully open during once-through operation. Furthermore, the inlet valve bodies are now bolted directly to the HP turbine outer casing and the turbine inlet sections are designed with full arc admission and no control stage. Finally, sliding pressure operation means constant steam and metal temperatures at all loads, which improves the thermal flexibility of the HP turbine.

The reaction type of turbine blade with its 50% reaction requires more stages with longer blades arranged on a smaller diameter than the impulse type, and by the end of the many European mergers, the reaction principle now forms basis for the blade design of new utility turbines. However, as blading and rotor design look very different for the two types of blade design, it is difficult to find any major indication of which principle is most advantageous concerning cost and efficiency. For retrofit business both types of blades are being offered depending on the customer's needs.

Modern computerised fluid dynamics – or CFD calculations – and modern manufacturing technology by 5-axis milling machines meant large improvements of blade design so that three dimensional (3D) blades could be produced allowing stage efficiencies to be improved from the low 90s to more than 95%. Alstom has achieved further improvements through the introduction of the inlet spiral, which improves the inlet conditions of the steam flow to the first row of stationary blades on HP, IP and LP turbines.

The modern, extremely flexible way of manufacturing turbine blades have enhanced the freedom of blade designers additionally so that 3D blade profiles can be optimised for each stage and with a reaction in the range of 40–50%.

CFD calculations also meant substantial improvements in the understanding of the flow through the last stages of the LP turbines. In particular the last stage can now be designed with very long moving blades in the range of 1200 mm in steel or Titanium, and Siemens is also offering 1400 mm Titanium blades. This allows for reduction in the number of LP turbines or – if the number of LP turbines remains constant – improvement in condenser pressure and/or reduction of outlet losses.

Table 5 Specific fuel cost.

Year		1990	2003
Coal Price	Euro/GJ	1.75	1.40
Average Net Efficiency	%	42	48
Specific Fuel Cost	Euro/MWh	15.00	10.50

COST OF ELECTRICITY

The major drive behind coal-based power generation is the large-scale production which has guaranteed low electricity prices and stable electricity supply. This section will investigate how the three major issues influencing the specific cost of electricity, namely fuel cost, capital cost, and operation and maintenance cost have changed during the 1990s. All calculations are based on zero inflation.

Europe is now the largest importer of coal in the world; the coal typically being brought into Europe by large vessels from all coal exporting countries like South Africa, Columbia, the USA and Australia. During the 1990s prices for the internationally traded bituminous coal with a lower heating value around 25 MJ/kg were relatively stable but with a falling tendency from some 1.75 Euro/GJ to some 1.5 Euro/GJ, based on a very effective and still improving mining industry. The recent large increases in oil and gas prices, which have nearly tripled, also meant short increases in coal prices up to 2 Euro/GJ, but coal prices have resumed the previous price level again contrary to oil and gas prices.

The effect of the fall in coal prices is increased by the improvements of plant efficiency, which in general have improved by 20% (relatively) as indicated in Tables 1 through 3 so fuel cost decreased by some 30% during the 1990s as demonstrated in Table 5.

Coal-fired power stations are expensive to build and the impact from capital cost on CoE is nearly as large as the impact from fuel prices. Therefore many efforts have been spent investigating how to reduce capital cost, and the outcome seems to be positive, as specific investment cost of coal-fired capacity has come down from former more than 1,000 Euro per kW to a range of 800 to 900 Euro per kW. This reduction is partly based on cost savings through less redundancy in systems design, which will also lower the needs for inspections and overhauls.

Table 6 illustrates how the specific capital costs for a power plant with 600 MW net output have changed since 1990. The calculations are based on an annuity of 9% for 1990 where capital was more expensive to borrow than the present level in the range of 8%. In both cases the load factor is set to 80% (or 7000 equivalent full-load operating hours), which corresponds to base load operation needed to guarantee the economic viability of coal-fired power generation. Interest rates during construction are not included.

Although costs for operation and maintenance (O & M) of the power plant are only a minor share of CoE, experiences from the 1990s demonstrated that it was still possible to demonstrate relatively large savings on O&M, and some topics will be addressed shortly.

Equipment monitoring in modern power stations has reached a high quality level based on modern control systems. Furthermore, design and manufacturing of major parts of the

Table 6 Specific capital cost.

Year		1990	2003
Load factor	%	80	80
Annuity	%	9.50	8.00
Specific investment cost	Euro/kW	1000	850
Specific capital cost	Euro/MWh	13.55	9.70

Table 7 Change in cost of electricity 1990-2003.

		1990	2003
Fuel Cost	Euro/MWh	15.00	10.50
Capital Cost	Euro/MWh	13.55	9.70
O&M	Euro/MWh	3.00	2.70
CoE	Euro/MWh	31.55	22.90

rotating equipment has improved so they can operate for very long periods (more than ten years) without being opened for inspections. Therefore, maintenance principles are now changing from periodic overhauls to overhauls based on indications from the monitoring systems, and with a target of more than halving the annual outage time for planned overhaul.

Costs for O & M were also strongly impacted by the ongoing opening of the electricity markets in Europe starting in the late 1990s where monopolistic structures had to be reshaped into competing shareholding companies. The transformation was typically combined with rationalisation, outsourcing and reductions in the number of employees of more than 20% to reduce costs for O & M, and in total annual O & M costs were reduced by some 10%.

In total, CoE based on coal-fired SC technology was lowered by 25–30% during the 1990s as illustrated in Table 7.

AD700

The AD 700 Project

In 1994 a large group of European power generators and equipment manufacturers started a joint R & D project named 'The Advanced 700°C Pulverised Coal-Fired Power Plant' or shortly: the AD700 project. The project is based on the idea that the period 2000 to 2010, where little new capacity is planned, shall be used to develop a competitive, innovative and high-efficient, pulverised coal-fired technology with maximum steam temperatures beyond 700°C.

Nickel-based materials are foreseen for the high-temperature sections of boiler piping and turbine as they seem well adapted for the temperature range 700–800°C. However, considerations on risk control limited the first step to an introduction of nickel-based materials into the cycle combined with a major jump in steam temperatures into the 700°C range. After successful demonstration of the technology at 700°C, the full potential of the nickel-based materials can be fully exploited in further stepwise developments, typically for pulverised coal-fired technology.

A programme for development of materials for the AD700 technology and demonstration of their properties is listed below:

- New nickel-based superalloys for long-term operation at steam temperatures in the range of 700–720°C. Superalloys will be developed for thin-walled super and reheater tubes, thick-walled outlet headers and steam piping and castings and forgings for the turbine. Targets of the 10^5 hours creep strength are 100 MPa at 750°C, which seem to be reached.
- New austenitics for boiler tubes operating in the temperature range 600–700°C to minimise the use of expensive super-alloys. Targets of the 10^5 hours creep strength are 100 MPa at 700°C, which seem to reached.
- New ferritic-martensitic pipe materials for boiler headers operating in the temperature range 600–650°C to minimise the use of expensive austenites. Targets of the 10^5 hours creep strength are 100 MPa at 650°C and development is still ongoing.
- New boiler and plant structures – shortly named 'Compact Design' – will be developed, which will allow steam lines between boiler and turbine to be shortened substantially thereby reducing investment cost.

The AD700 project is staged into six phases and follows an overall time schedule as shown in Figure 1. Phase 1 started in 1998 based on 40% funding from the Commission's THERMIE programme (DG TREN) under the 4th Framework Programme. Phase 2 started in 2002 granted through a contract with DG RTD also under the Commission's 5th Framework Programme but with 50% funding. Table 8 shows that Phase 3 should start in 2004 but unfortunately and incomprehensibly, the 6th Framework Programmes does not contain any support for fossil fuel-based projects. Therefore, great efforts are ongoing to find other ways to support the AD700 project.

A number of 40 respectively 36 industrial partners from all over Europe have joined the Phase 1 and 2 works and in both phases the project organisation is divided into a boiler, a process and a turbine group.

Phase 1 results indicate net efficiencies in the range of 50-51% for a power plant with a single reheat cycle cooled by a wet cooling tower, and 53-54% for a double reheat cycle cooled by seawater. However, development needs of the double reheat cycle are larger than for the single reheat cycle. Figure 2 shows the planned water/steam cycle for an inland location.

Design studies from phase 1 indicate that this will all be accomplished without jeopardising the simplicity, plant economy and high reliability characteristics of the components in a more traditional power station. The AD700 technology will be mature shortly after 2010 and long-term targets after 2020 are net efficiencies above 55% based on maximum steam temperatures in the range of 800°C.

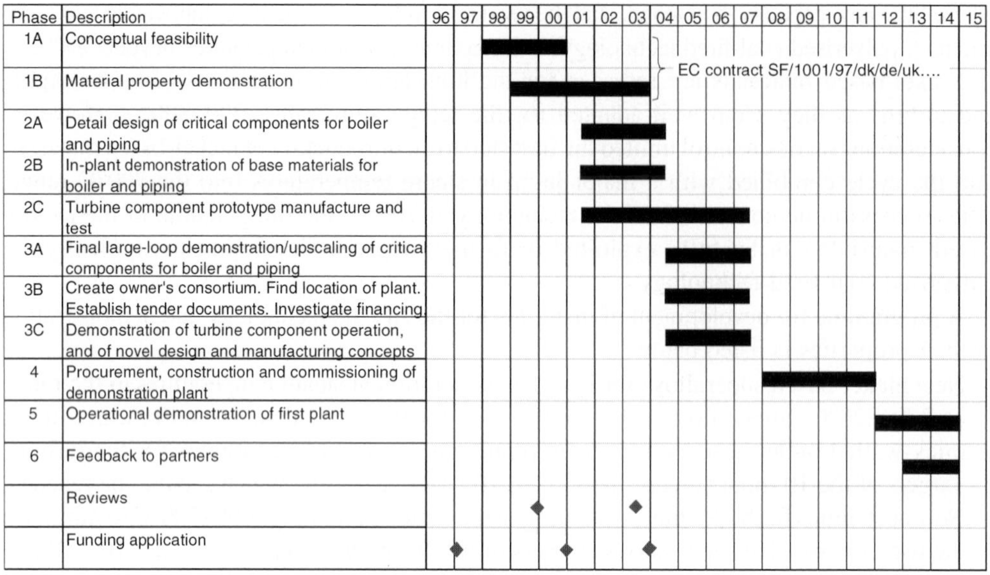
Phase
Description
96
97
98
99
00
01
02
03
04
05
06
07
08
09
10
11
12
13
14
15
1A
Conceptual feasibility
EC contract SF/1001/97/dk/de/uk....
1B
Material property demonstration
2A
Detail design of critical components for boiler and piping
2B
In-plant demonstration of base materials for boiler and piping
2C
Turbine component prototype manufacture and test
3A
Final large-loop demonstration/upscaling of critical components for boiler and piping
3B
Create owner's consortium. Find location of plant. Establish tender documents. Investigate financing
3C
Demonstration of turbine component operation, and of novel design and manufacturing concepts
4
Procurement, construction and commissioning of demonstration plant
5
Operational demonstration of first plant
6
Feedback to partners
Reviews
Funding application

Fig. 1 Time schedule for development of the AD700 technology.

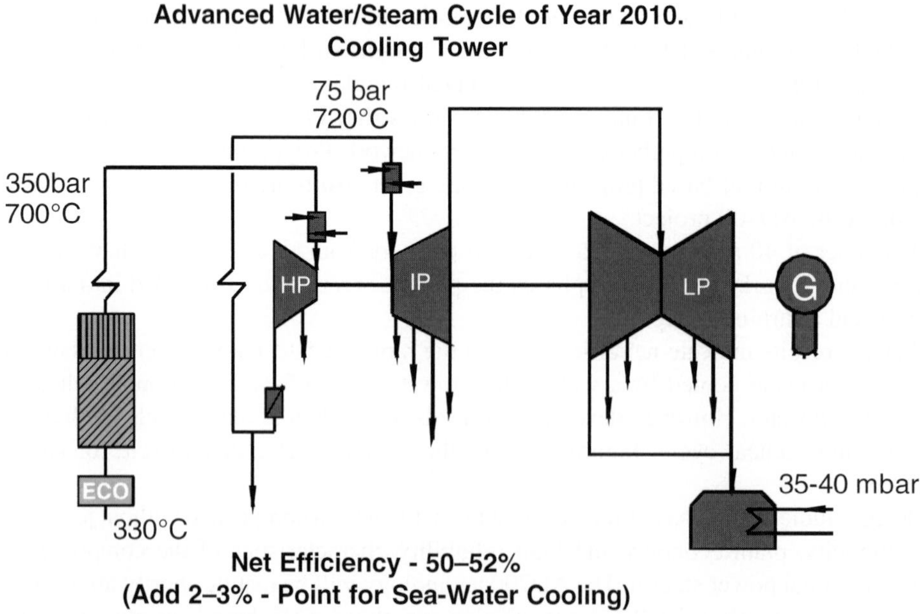

Fig. 2 Water/steam cycle of an AD700 plant with wet cooling tower.

CONCLUSIONS

This paper demonstrates the viability of well proven super critical PF technology which has achieved remarkable improvements concerning cost of electricity and environmental performance during the recent 10 to 15 years. It may be concluded that on basis of the present coal prices, capital cost, and financing conditions, new super critical capacity seems very competitive. Furthermore, the paper shows the potential of future improvements, which however needs political and economical support to be deployed. The situation for European power manufacturing industry is severe so, hopefully, ongoing efforts to find a solution with the European Commission and the European Parliament will succeed.

Cyclic Operation of Advanced Energy Conversion Systems

FRED STARR
European Technology Development Ltd.
Ashtead Surrey, UK

ABSTRACT

A variety of turbine based advanced energy conversion systems, which are likely to come into service after 2010, are evaluated in terms of their likely operability and damage to key components. Current problems with pulverised fuel fired steam and, CCGT, plant are briefly described. Some of these difficulties are due to failings of older materials of construction and to shortcomings in plant design and operation. Hopefully, component problems with advanced versions of these systems will be restricted to genuine engineering issues, which, in steam plant, will include thermal fatigue of austenitic and nickel based superheater and turbine materials. As some of these alloys will be supplied in the precipitation hardened condition, there may be problem with the thermal fatigue of weldments, particularly in steam environments. With advanced CCGT plants, most problems will be with gas turbine components. Here, high rates of heat transfer, resulting from higher turbine inlet temperatures and innovative methods of blade cooling, will require increased work on the life assessment of, TBCs, and single crystal blades.

In evaluating competitors to these systems, the paper focuses on developments of turbine based systems that are in commercial use, or have reached the major pilot plant scale. Entrained bed, IGCC, plants will not be used in a cycling mode, and cycling will not be significant. Circulating Fluidised Bed and Pressurised Fluidised Bed plants have good load following capability, but will be likely to suffer increased thermal fatigue and corrosion problems, because of the need to run these at increased temperatures. Potential cycling problems with recuperators in advanced gas turbines, at the kilowatt and megawatt sizes, are also reviewed. Of nuclear systems, the high temperature gas reactor, , should have excellent load following capability and be virtually free from thermal fatigue problems.

INTRODUCTION

Plant cycling has always been a major factor in power station operation. As this is a conference, at which we celebrate the introduction of the Parsons steam turbine, it is worth reflecting that cycling, for early power generation plant, was about as bad as it can get. Power plants were very small, producing less than a megawatt, with most of the power going to domestic lighting or small-scale industry. The over-night load was negligible, with most plants having to be equipped with banks of lead acid batteries. These were run down in the day, to help meet the load, and recharged during the night when the demand had fallen away.

The ideal power unit would have been able to maintain good efficiency at the lowest loads, dispensing with the need for battery storage. It was here that the Parsons steam turbine began to score over the reciprocating steam engine. In addition to its many advantages, such as high speed of rotation, small size and ability to make full use of low condenser pressures, the steam turbine could maintain its efficiency advantage down to relatively low loads. As early as 1892, Ewing commented that:[1]

> The efficiency under comparatively small fractions of the full load is probably better than in any steam engine, and is a feature of special interest in relation to the use of the turbine in electric lighting from central stations.

By the first decade of the 20th Century, the steam turbine had ousted its older rival, but it is around this time that we see the first hint of the mechanical and metallurgical problems that have now come to haunt us, when we have to get power plant to two shift or load follow. In his 1921 book on power station equipment, Snell points out that two shortcomings of the Parsons turbine are the very tight clearances at the HP end and, in addition, the relatively long rotor, both features characteristic of reaction turbines. Snell claimed that any bowing of the rotor, due to temperature differences, or differential thermal movement between the rotor and the casing, was likely to lead to stripping of the HP blading.[2]

Any power plant will have problems of this type, and it is necessary to decide whether these are endemic or whether design changes or better operating procedures will eliminate these difficulties. For example, many of the cycling problems in CCGTs come from the need to purge the plant after a shut down, to clear any inflammable gases from the system. Hence, the high temperature sections of the gas turbine and in the heat recovery steam generator, HRSG, are subject to a blast of cold air. If we could eliminate this purge step, we could improve CCGT plant life significantly and, at the same time, improve its startup response. The intention of this paper is to point out potential problems in the cycling of advanced energy conversion plants, and, where these do seem to be a feature of the design, to point out what materials technology can offer.

PLANT CYCLING IN THE FUTURE

Concern about global warming is a major factor in the drive to improve power plant efficiencies, which can only be realised through operation at the highest possible temperatures and pressures. Materials must be pushed to the limit; whether these limits are determined by mechanical properties or by resistance to high temperature corrosion. With development, we should be able achieve electrical efficiencies of about 50% (HHV) when using coal as fuel, and around 65–70% (HHV) when using natural gas.[3–5] This would reduce the emissions of CO_2 by 10–15% compared to the best plants currently operating.

Meeting these efficiency targets will be difficult enough in itself. Just as important is that, in coming decades, power plants will need to cycle much more frequently than today. Indeed, in 10–15 years time, the probability is that newly commissioned plants will have to two shift and load follow from their first day of operation. This is in complete contrast to the present, where a plant is not expected to start cycling until 5–10 years after commissioning. Concern over cycling has led the industry to commission reports on improved operating practices.[6–8]

A UK conference was held on this issue in 2001, with the discussion being subsequently reported in Materials at High Temperature.[9, 10]

Fossil fuel plants will need to cycle much more in the future. The principal cause is the growth of renewable energy, the majority of which comes from wind, wave and solar sources. If renewables are to have to have a significant effect on CO_2 levels, they will need to take up as much of the base load as possible. Most CHP (Combined Heat and Power) and nuclear plants will also be of the base load type, with the net result that even high performance fossil fuel plants will be pushed down the merit order. But when the wind doesn't blow, or the sun doesn't shine, fossil plants will have to make up the difference, often at short notice. All of this will have an adverse effect on fossil fuel plant cycling. Long term German predictions indicate that, in the summer of 2050, fossil plants will be off line for days at a time![11] One bright spot is that new designs of nuclear reactor, such as the pebble bed type, seem to be more able to load follow than existing units of the PWR and AGR type.

One new entrant in the field of the generation of electricity is 'distributed power', which is promoted on the basis of better energy usage. Here too, there has been very little mention of the effects plant cycling on operability, part load efficiency, or likely damage to key components. Distributed power, however, means different things to different people. To some it may signify stand-alone gas turbine systems of around 100 MW, whose main advantage is to reduce line losses and plug gaps in the network. Others take it to mean the large-scale use of CHP or cogen systems for apartment blocks, offices or factories. Here power outputs are typically in the range 100 kW to 10 MW. In the UK, and other parts of Northern Europe, distributed power might imply electricity generation on a house by house basis, with units offering as little as a kilowatt.

CONVENTIONAL CENTRAL PLANT DEVELOPMENTS

Advanced CCGT Systems

For the foreseeable future, until natural gas becomes uneconomic, the CCGT will be the energy conversion system of choice. Even after natural gas begins to decline, it will be possible to run CCGTs on fuel gas manufactured from coal or biomass, or from hydrogen produced from the electrolysis of water, as these gases can be piped over long distances. In practice, many CCGT plants may be built as part of an IGCC (Integrated Gasification Combined Cycle) facility.

Gas Turbine Issues

Up to recently, designers took the best of aero engine technology and used it to improve the efficiencies of the gas turbine in CCGT plants. Advanced blade cooling techniques have been used, along with oxidation resistant and thermal barrier coatings and directionally solidified blades. Turbine inlet temperatures are now around 1,400°C but, metallurgically, CCGTs are still a few years behind the aerospace sector, with only one manufacturer, General Electric, using single crystal blades. Manufacturers are not keen to broadcast problems with their machines, but it is clear that blade and coating cracking, due to two shifting, is becoming a major issue. Some of the problems reported, during a recent survey by ETD Ltd, included:[8]

- Blade cracking during load following, possibly accelerated by the use of more oxidation resistant, but less ductile coatings,
- Increased blade seal wear and loss of TBCs from cans,
- Cracking from cooling holes on rotor and stator blades,
- Transition duct cracking,
- Loss of TBC from the leading edge of blades and
- Cracking of oxidation resistant coatings.

However, design techniques are well in advance of aircraft type gas turbines. Steam cooling of the transition duct has already been introduced on the Westinghouse 701 G, with the intention of reducing the parasitic extraction of air from the main gas turbine flow, thereby increasing turbine output and efficiency. The GE Frame 7 and 9H machines, which use steam cooling of the first and second stage HP stator and rotor blades, are even more advanced in this respect. Here, since there is no flow of fluid to the outside of the blade, the aerodynamics are improved. With this, other things being equal, local heat transfer rates will be reduced and perforation of the blades with stress raising cooling holes is avoided.

The downside of steam cooling is the need to get steam, free from condensate, into the gas turbine, and then back into the HRSG. Westinghouse does this by giving the transition duct steam a mild degree of superheat. GE takes the 'blading steam', from the HP steam turbine exhaust. How feasible is this, on a day-to-day basis, must be a moot point.

An alternative technique is to pre-cool the air needed for blade cooling. This will also reduce the amount of air abstracted from the compressor, but, here again, there may be difficulties in ensuring that the pre-cooled air is free from condensed moisture. Air-drying equipment will have to be reliable. Quite different to all of these advanced cooling techniques is the reheat gas turbine of the GT24/26 type, currently marketed by Alstom. There is nothing in the aviation sector like this, where, after partial expansion along the HP turbine spool, extra fuel is burnt before the remainder of the expansion occurs in the LP spool. This procedure raises the specific output of the gas turbine and, by increasing the inlet duct temperature to the HRSG, gives the potential for raising steam temperatures to around 600°C.

Sophisticated cooling techniques only become profitable when combined with thermal barrier coatings (TBCs). There is great concern about TBC reliability under cycling conditions. Workers at Siemens have distinguished three different mechanisms involved in coating failure.[12] These are:

- High Temperature Sintering,
- Cyclic Fatigue and
- Bond Coating Oxidation.

The actual failure mechanism depends on the turbine inlet temperature, heat transfer rate from the combustion products into the blade, and the amount of blade cooling. For example, if the turbine inlet temperature and heat transfer rates are high, but the blade cooling is good, the likely result will be that outer surface of the TBC will be in the sintering range. The TBC will then tend to fail, relatively quickly, by delamination of the outer surface. Conversely, if the inlet temperature and heat transfer rates are low, but the cooling rate is modest, the risk is that the bond coat will be at too high a temperature. Failure will now tend to occur along the line of the bond coat oxide. Cyclic fatigue failures tend to occur within the depth of the TBC, at intermediate conditions to those described.

Bond coat oxidation is relatively slow, compared to sintering. It is currently viewed as being the chief TBC failure mechanism in industrial machines, where inlet turbine temperatures are about 300°C less than in military aircraft engines. In this respect, a recent paper from FA Jülich indicates that the recent introduction of columnar structures has improved the strain tolerance of the TBC itself.[13] Hence, much of the current R&D is aimed at improving the oxidation and mechanical properties of the bond coat, and modelling bond coat-oxide and TBC interactions. The use of enhanced cooling techniques, combined with higher turbine inlet temperatures, must push up the surface temperature of TBCs, indicating that efforts to improve fatigue resistance and reduce sintering must not be neglected.

Enhanced cooling techniques will also tend to increase blade thermal gradients. Viswanathan and Scheirer quote Pratt and Whitney work which indicates that the use of directional solidified and single crystal material increases the thermal fatigue resistance by factors of 6 and 9 respectively, compared to equiaxed material.[14] In contrast, QinetiQ emphasises that, as single crystal structures are anisotropic, this will result increase in local stresses. In tests an isotropic model overestimated the cyclic life a test bar with a cooling hole by some 5,000 times, whereas the anisotropic model came within a factor of two of the actual life.[15]

Heat Recovery Steam Generators

The other part of the CCGT is the HRSG, in which temperatures are now at the same levels as those in steam plants. The highest HRSG outlet temperature is now 565°C, but it is moot point whether HRSGs will ever need to run at temperatures much over 600°C, even in plants which are using an Alstom/ABB reheat type gas turbine. CCGT plant manufacturers have many options available to them to take full advantage of any increase in gas turbine inlet temperatures. Pressure ratios need to be increased, and this will have the effect of holding down the outlet temperature.

There is a tendency to regard the HRSG as being just another variety of steam boiler but the differences are profound. In contrast to steam plant boilers, where flue gas temperatures in the evaporator region are around 2000°C, and exit furnace temperatures are around 1,450°C, the temperatures in CCGT plant are much more modest. Exit temperatures from the gas turbine are in the 600°C region, and the temperature in the flue gas drops more or less uniformly all the way along the duct. The driving force for heat transfer is limited at all points and, as the author has argued elsewhere, heat transfer considerations have tended to outweigh the requirements for good mechanical design.[16]

In two shift duty, older forms of HRSGs have experienced many problems, even when steam temperatures have been quite modest. The main concern is the HP superheaters, where creep fatigue is the issue. Some manufacturers are now supply fatigue life models, which take into account the final operating temperature and the ramp rate.[17] Curves of this type are, it would appear, only applicable to units that are started from cold, or where, if they are two shifting, condensate drainage is good. Pearson has shown that the real problem is the tendency of the HRSG to build up condensate during the purge period.[18] This is most severe in horizontal HRSGs, where the condensate accumulates in the bottom header. On restart, condensate

Table 1 Likely attainable strength levels.

Temperature ° C	650°	700°	750°	800°	850°
Likely Attainable Rupture Strength (MPa)	219.9	146.2	97.2	64.8	43.0

will tend to block or waterlog tubes, leading to local overheating and differential expansion of individual tubes. More serious is the situation where the condensate starts to boil off and flashes over into the header, leading to quench cracking. Here relatively trivial design changes, to prevent condensate build up, can make a significant improvement to HRSG life. In addition, higher strength construction materials will offer a better performance under cycling conditions, whether or not condensate quenching is an issue.

Steam Plant

Superheaters and Connecting Pipework

If current projections are to be believed, by 2015 steam temperatures could be around 700°C, with pressures in the 350–400 bar range, and superheater metal temperatures up to 740–780°C. Much of the connecting steam pipework will be in high strength austenitic alloys, with the hottest section of superheaters and steam turbines in nickel based materials. How are these types of construction likely to fare in a two shift situation?

Ideally steam plant tubing and pipework should be supplied in the 'solution treated' wrought condition, with the precipitation of strengthening phases occurring soon after entering service. This avoids the need for an ageing treatment, and reduces the risk that the weld and HAZ regions will be significantly weaker than the parent metal. Whether it will be practical to obtain an acceptable strength level with materials that will not require a precipitation treatment is arguable. Starr and Shibli have suggested, using a correlation based on accepted ideas about dislocation climb, that the likely maximum strength level for highly developed solution treated alloys, at 750°C, will be just under 100 MPa[19] see Table 1. Given that superheater pressures will be over 350 bar, this implies very high values of wall thickness for both headers and superheater tubing. This impinges on the susceptibility of tubing and headers to thermal fatigue. Current wrought alloys, which rely on in-service precipitation, such as Tempaloy AA-1, Haynes 230 and Alloy 617, fall somewhat below the 'likely attainable strength levels' postulated in Table 1.

The alternatives are materials, which after solution treatment at around 1,100°C, are age hardened between 700–850°C. This induces the formation of stable precipitates, which would not form at the much lower service temperatures. Materials of this type include older alloys such as IN718, Alloy 263, and more recent developments such as Alloy 740 and improved versions of Inco 617. Even with these, it seems difficult to get much more than 100 MPa at 750°C, and there must be concern about weldment properties.

During startup, the rapid changes in plant temperatures will lead to significant through wall gradients. One estimate of relative susceptibility to thermal stress is given by the function $\alpha \cdot E \cdot k^{-1}$ where:
α = Coefficient of Thermal Expansion (10^{-6} K^{-1}),
E = Elastic Modulus (GPa) and
k = Thermal Conductivity (W.m.K^{-1}).

Table 2 shows values of this thermal stress parameter at 600°C. The lower the value the more resistant is the alloy, indicating that the austenitic and nickel based alloys are much poorer than the so-called ferritics.

A review by Skeleton, of thermal fatigue, has drawn attention to a modification of this somewhat simple thermal stress parameter, which takes into account the short term yield strength of the material. The argument is that high temperature yield properties are the ruling factor during start up, since the stronger the material is, the less likely is the risk of local plasticity, leading to crack initiation.[20] This newer function is termed the 'R' or the 'merit order', and is given by an equation of the type:

$$R = k \cdot \sigma_y \cdot \alpha^{-1} \cdot E^{-1}$$

where σ_y is the yield strength at temperature.

Table 1 also shows values of the *R* function, where high *R* values indicate good resistance to thermal fatigue. It can also be argued that a high yield stress will be important if condensate quenching was to occur. Fortunately, this should be less likely in advanced supercritical plants, as such units do not have a steam drum, and drainage arrangements should be better.

Temperature changes in advanced supercritical plants will be about 150–200°C more than in today's units, giving a strain range about 25% higher than that for which we currently design. Discounting this, how do the advanced nickel and austenitic alloys compare with those now used? The message of Table 2 is that we should hope that the *R* value is more important than the older thermal stress parameter, since, as one might expect, the latter shows that the ferritic type alloys are much better than the austenitics, but in terms of R values, the two nickel based solution treated alloys, Haynes 230 and IN 617 are quite reasonable. The values of *R* for the precipitation hardened alloys, IN 718 and 263, are even better, as a result of their excellent yield strengths. Here, there must be a word of caution, particularly in the case of IN 718. The ductility value of this material is under 20%, and may be deemed unacceptable by plant designers. In contrast, Inco 740 has quite a high *R* value but appears to have acceptable ductility. Both solution treated and aged alloys are likely to overage in service, due to temperature escalations, reducing their yield strength. In-service deterioration with T22 alloys, leading to a drop in the *R* values, was noted by Skelton.[20] An estimated *R* value is also given for the advanced Japanese austenitic Tempaloy AA-1, which, although somewhat better than Type 316, does not do as well as the ferritics. This is probably the case with any normal stainless in which the nickel content is in the 25–45% range.

Steam Turbines

The main concern of operators has been temperature mismatch that can occur when steam is first brought onto a hot rotor, after an overnight start up. The quenching effect of cold steam has a serious effect on the life of HP rotors.[23] Improved control systems and critical attention

Table 2 Relative susceptibility of candidate superheater alloys to thermal stress.

Alloy	Conductivity (W.m.K^{-1})	Coefficient of Thermal Expansion (10^{6}.K)	Young's Modulus (GPa)	Thermal Stress Parameter $\alpha^{-1}.E^{-1}.k$	Yield Strength At Temperature	Merit Order "R"
T22	33	14.6	167	73	175MPa@500°C	2300
P91	30	12.7	175	74	220(min)MPa@600°C	3000
P92	c.27	c. 12	c.180	c.80	c.300MPa@600°C	c.3750
12Cr (X20)	26	12.3	198	93	250MPa@550°C	2700
Type 316	22.5	18.0	150	120	98MPa@600°C	800
Tempaloy-AA1	c.22	c.18	c.155	c.120	c.140@ 700°C	1100
Haynes 230	20.4	14.5	177	125	280MPa@700°C	2200
IN 617	22.5	14.0	173	108	250MPa@700°C	2300
IN 718	20.6	15.5	185	139	850MPa@700°C	6100
IN 263	16.3	14.7	185	167	500MPa@700°C	3000
Inco 740	20.2	15.0	178	132	c.650MPa@700°C	4900

to superheater drainage should eliminate this problem in the future. The metallurgical challenge in the reheat turbine rotors, for very advanced "700°C type" plant, is the most serious question, particularly where double reheat is employed as with the more advanced Thermie and Vision 21 concepts.

With Thermie the proposed reheat inlet temperatures, at 720°C, are 20°C higher than that of the HP turbine. Reheat turbine outlet temperatures will be in the range 400–450°C, again significantly above current designs. Superheat and reheat inlet temperatures in Vision 21 concepts are higher still at 760°C. The second reheat stage is particularly critical, due to the low inlet pressure of 30 bar and high steam temperature, which will result in a large rotor size, compared to most current single reheat designs. The addition of an extra stage of reheat only gives a fuel saving of about 3%, but has the advantage of increasing the specific power, which will reduce the size of plant and components for a given output. This may help bring the second reheat turbine down to a more manageable size.

These temperatures would suggest that some of the high nickel alloys, mentioned for superheater duty, could be suitable. Assuming existing criteria are applicable, a room temperature yield strength of over 600 MPa and a creep rupture strength, at 720–760°C, in excess of 100 MPa will be needed.[22, 23] Heat treatment of rotor materials can be a significant issue, since, unlike superheater tubing, the stresses on a rotor are so high that maximum strength is needed at the outset. In contrast, with superheater tubing, one can allow an appreciable period to pass before strong precipitates begin to form. Fortunately, only the inlet to the turbine is subject to the highest temperatures, and because of the short blade length at this point, stresses are relatively low. One possible option is to build the front end in a nickel based alloy, with the back end in a martensitic based alloy, the two pieces being welded together. If viable, this points to the use of one of the simpler nickel based alloys as the weldment area, in a precipitation hardened alloy, is certain to be significantly weaker than the parent material. Efforts to recoup these properties by heat treatment would have a detrimental effect on the martensitic alloys.

Rösler et al. have made a preliminary analysis of the types of material that will be required for 700°C steam turbine rotors.[24] Their provisional conclusions are that although, in principle, steam turbine rotors can be made, none of the alloys under test were ideal due to a variety of problems, including freckle formation in the castings, difficulties in forging, through-section heat treatment problems, and microstructural stability at operating temperature. The standard heat treatment resulted in very poor properties with high strength alloys of the γ'/γ'' type. The basic cause of this is air induced SAGBO (Stress Assisted Grain Boundary Oxidation). This can be countered by inducing a limited amount of grain boundary precipitation. With IN706, eta phase is produced at the grain boundaries, through a precipitation treatment at 820–850°C. Similar treatments to the solid solution/carbide strengthened alloys such as IN 617 and the γ' strengthened types like Waspaloy will also induce the formation of grain boundary precipitates. However, it would seem to the writer, that a major issue that needs to be addressed is the possibility of SAGBO type cracking occurring in steam, since it is now recognised that steam is much more aggressive than air. Although to date there have been no reported SAGBO problems with steam cooled turbine blades, as in the GE 7H and 9H machines, this does not necessarily suggest that there is absolutely no cause for concern. These particular turbine alloys are somewhat different in composition to the rotor materials and more importantly do not contain grain boundaries.

ADVANCED CENTRALISED COAL FUELLED CONCEPTS

IGCC Systems

Coal gasification, using steam and oxygen (or air) as reactants, can be used to produce a fuel gas containing H_2, CO and, depending on how low is the reaction temperature, CH_4 and higher hydrocarbons. This 'synthesis gas' can be used instead of natural gas in a CCGT plant. Certain of the gasification processes were intended to produce a substitute natural gas, in these some of the gasification reactions occur at temperatures under 1,000°C, to synthesise as much methane as possible. Examples are HyGas, Lurgi, and the British Gas-Lurgi Slagging Gasifier. With such processes the gasification stream can operate independently of the CCGT plant. The comparatively small amount of waste heat produced in these processes is utilised to bring the reactants up to temperature and to raise steam for the oxygen plant. Gasifiers of this type can operate independently of the CCGT , putting surplus gas into the gas-grid during the overnight period. In addition "fixed bed" type processes, such as the Slagging Gasifier, can be put into a hot hold mode for long periods, during which no gas is produced. Gasifier outlet temperatures are around 600°C, and the outgoing gas is quenched with recirculated tar, such that downstream problems, of the thermal fatigue type, are absent and a light coating of tar protects heat exchangers.

The British Coal process makes use of a low temperature fluidised bed containing coal and limestone to absorb sulphur. As with all such processes, only the more volatile constituents in the coal can be gasified. The ungasified char is routed through a transfer pipe to a fluidised bed combustor where it is used to raise steam for gasification reactions, and for power production. Here too, it is possible to envisage the gasifier and combustor being shut down at night, as the two fuel beds will retain a considerable amount of heat. Some thermal fatigue problems might be expected in the transfer pipe and in the steam raising section of the plant.

Most systems that are currently in operation are of the high temperature entrained bed IGCC (Integrated Gasification Combined Cycle) type, as exemplified by the Prenflow, Shell processes. Gasification takes place at around 30 bar, using oxygen plus steam or water to produce complete gasification of pulverised coal. The mixture of reactants carries the particles along at high velocity. Temperatures are in the 1,400–1,600°C range, with the mineral matter in the coal forming slag droplets. These are solidified by ‘quenching’ the temperature down to about 800–900°C, by mixing the stream with cold recycled gas, before entry to the ‘syngas cooler’. Here 15% of the thermal energy in the plant is converted into HP steam, which is given very little superheat, implying metal temperatures around 400°C. Temperatures are too high in the gasifier for limestone to absorb sulphur, and the syngas cooler unit is the most limiting section of the plant, being subject to sulphidation, carburisation and chloridation. In the downtime period tubing will suffer from acid condensate and stress corrosion. Fe-31Ni-28Cr-3Mo alloy, Sanicro 28 or Incoloy 825 have the necessary resistance to stress and pitting corrosion plus reasonable resistance to sulphidation.[25]

The efficiency of entrained gasification plants, at base load, is broadly comparable to the best coal based European plants. 43% has been quoted to the author for one system but, at 80% load, efficiency fell to the low thirties. This is not unexpected. At the Buggenum Shell based plant, coal input corresponds to 585 MW, but the ‘own electrical consumption’ of this plant and heat losses total 111 MW, presumably because of the complexity of the gas treatment processes and the need to recirculate massive amounts of quench gas. Although these plants can operate down to 50% load, entrained bed IGCC systems are not suited to cycling, and the perception is that they must be integrated into some other type of energy conversion system. See Figure 1. At present this would be a refinery complex, but in future IGCCs could be used to produce pure hydrogen, with the CO_2 being easily sequestered.

Efficiencies in entrained bed systems are constrained by the syngas cooler, because of the poor resistance of stainless type alloys to sulphidation. Most, if not all, of the superheating has to be done in the high temperature section of the HRSG of the CCGT, suggesting that such units are best linked into reheat type gas turbines. Karg et al., using an IGCC plant at Puertallano in Spain as a base case, suggest that the efficiency could be improved from the present level of 47.5 to 52% (net) fairly easily.[26] Most of this improvement comes through the use of better gas turbines, rather than through enhanced syngas conditions, although the plant become even more complex in its usage of heat.

Ideally one would wish to match the steam temperatures and pressures from the syngas cooler with those from the gas turbine HRSG, suggesting that the syngas cooler would need to run at about 570–600°C. The types of coatings and alloys, which are likely to have the required resistance, are high in silicon or aluminium and are of low ductility, even if not totally brittle.[27, 28] Even with more ductile materials, Guttmann et al found that cracking in gasification environments, at 600°C, was likely to occur at more than 1% strain.[29]

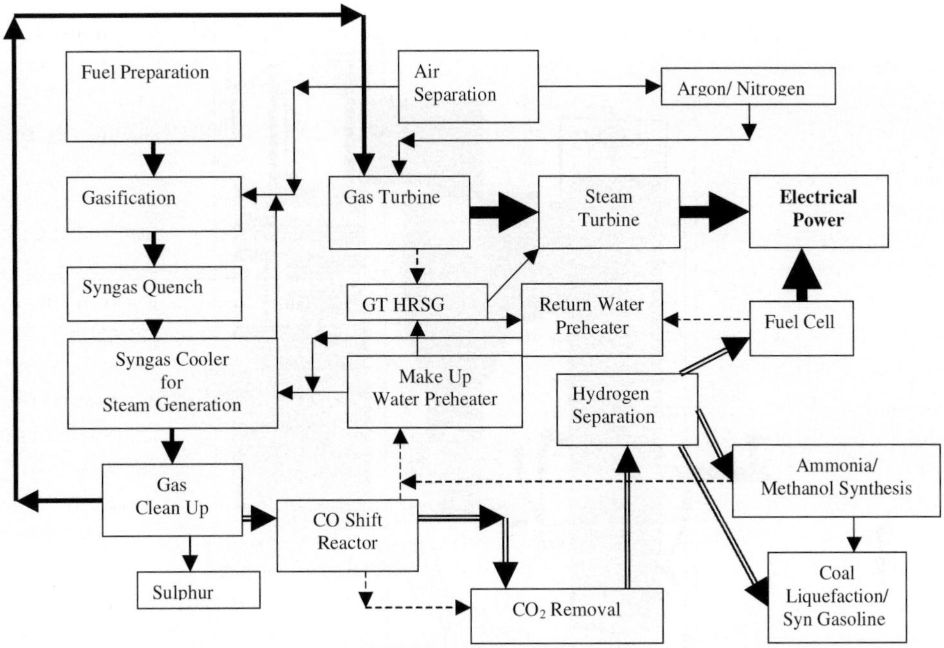

Fig. 1 Typical IGCC plant layout for advanced petrochemical/refinery complex. (Large black arrows power, medium black arrows-synthesis gas for GT, light arrows-steam, oxygen and nitrogen, lined arrows syngas and hydrogen for process, dotted arrows-heat for steam raising).

Circulating Fluidised Bed Boilers

Although subject to the same Rankine cycle considerations, the, CFB (Circulating Fluidised Bed) boiler, is the biggest competitor to conventional pulverised coal fired boilers. Low reaction temperatures minimise NOx formation, and much of the sulphur can be captured by the addition of powdered limestone to the bed. Among the claimed attractions of the CFB are excellent load following capability, good turn down to 33% load, and ease of start up from the hot condition since the slumped bed retains much of its heat.[30]

In some respects the CFB resembles a two pass pulverised fuel boiler Steam is generated in water wall tubes, which line the CFB furnace and in many plants the superheater is located in a downwards flowing 'back pass' see Figure 2. However, because the furnace exit temperature is relatively low, some units have a superheater section located part way down the water wall, and Foster Wheeler have introduced the so-called INTREX superheater, which is located in a bubbling type fluidised bed adjacent to the CFB furnace proper.[31] The bottom section of the furnace is the hottest region, partly because the volatiles in the coal burn off at this point, but also because the bottom section of the furnace is protected with a layer of refractory.

In the CFB the fluidised velocity is sufficiently high (c.4–6 m/s), even at ambient pressure, to transport the individual particles of coal and ash up to the top of shaft where they are de-entrained using separators, and allowed to fall back down the 'solids return line'. Against

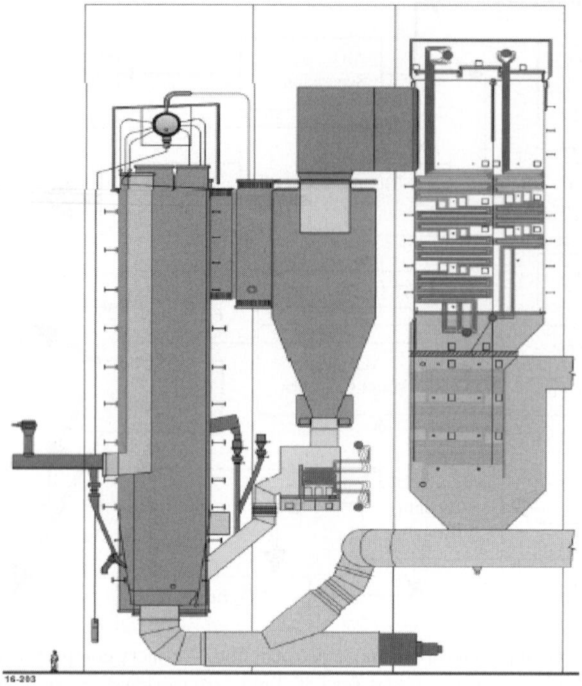

Fig. 2 Schematic of a CFB Boiler. The furnace is on the left, with the solid return system in the centre. Superheaters and economisers are on the right.[31]

this net flow in the CFB furnace, a stream of particles, adjacent to the water walls, slide back downwards, as the fluidising airflow is lower at the sides. This results in erosion-corrosion particularly where there is a change in section, for example, where the refractory terminates, or where the flow direction is changing.[32] However, in CFBs erosion appears to be much less of a problem than in bubbling bed type combustor systems. This could be due to the reasons, evinced by Natesan, who pointed out that the partial pressure of sulphur is much higher in bubbling bed systems.[33]

A recent review by Johnk Engineering gives an insider's view of the cycling issues in CFBs. The view is expressed that a turn down to about 50% load is fairly easy.[34] As the load drops, however, furnace temperatures begin to sag, since although the fuel input is cut, the amount of air, needed for fluidisation, cannot be cut below a certain level. To get down to the 33% level, and run comfortably, usually means modifications to the air circulation and distribution arrangements. In many cases it is best to shut down at low loads, as the boiler can usually be started up without the use of auxiliary fuel, if this is done within an eight-hour period.

The main materials problems caused by cycling are the usually caused by thermal and differential expansion of the boiler parts. It should be possible to use arrangements other than that of the pendant design of superheater, such that condensate quenching may be less of a risk in CFBs. An immediate concern is the effect of cycling on refractory linings, which can cause them to crack and break off leading to excessive tube wear.

These are, however, all current problems. The first supercritical CFBs are under construction, but what are the prospects for CFB power plants with steam temperatures in the 700°C range, and will they be likely to experience more or different problems during cycling? No conclusive answer can be made at this stage, but it would seem that, if high steam temperatures are to be attained, the final superheater will need to be located either within the bottom section of the CFB or, alternatively, within a bubbling bed INTREX unit. Work at Stoke Orchard, by Michener and others, suggests that, when limestone is added to the bed, the corrosion rates are beginning to be unacceptable above 650°C.[35] The cause is the formation of deposits of calcium sulphate on in-bed tubing, which induces sulphidation of the metal. It would seem, therefore, that there could be serious difficulties in raising steam temperatures in CFB plants as much as we would wish.

Another potential problem with in-bed tubing, where a CFB has to two shift, is the possibility of in-bed conditions becoming reducing, through the presence of unburnt char. Conditions will switch back to oxidising when the plant is brought back on line. This is perhaps analogous to the two shift, water wall corrosion, involving FeS formation with low NO_x burners, as recently described by Bakker.[36]

Pressurised Fluidised Bed Boilers

Whereas the CFB utilises an entrained bed, the PFBC (Pressurised Fluidised Bed Combustor) makes use of a coal, ash and limestone bed, that is kept in suspension by a relatively slow moving current of pressurised air and combustion products. The PFBC bed has the appearance of a liquid in a gently boiling condition, having a definite level taking up the shape of the pressure vessel, which it fills. The rate of combustion and the heat transfer rate are quite high, as the bed is pressurised to between 10–20 bar. This overcomes a major shortcoming of earlier versions of this technology, in which the beds ran at atmospheric pressure. Tubes for the evaporator, superheater and reheater tubing are located within the bed, although some of the tubing can be above the bed when the PFBC is running at part load.

The average bed temperature is around 850°C, and the combustion products leave the bed at this temperature. After most of the entrained ash is removed, the combustion products are used to drive a gas turbine, which provides pressurised air for the combustion process as well as producing electricity. The design is a two stage compression unit, with intercooling, each compressor being driven by separate turbines the HP set being linked to an alternator.[37, 38] However, more than two thirds of the power comes from the steam turbines. These derive heat from a variety of sources including the fluidised bed itself, in which steam is raised and superheated, an economiser attached to the gas turbine, the gas turbine intercooler, and steam turbine feedheaters of the common type. This complexity enables the PFBC to about equal, in efficiency terms, the best pulverised-fuel boilers see Figure 3. Despite this complexity, the load following ability of the PFBC is quite good, at 4%/minute. This stems from the use of a variable speed LP gas turbine set, and the ability to add and discharge material from the PFBC fuel bed. What could be critical at such times is maintenance of bed temperature. If this rises, sintering of the bed can occur.[39]

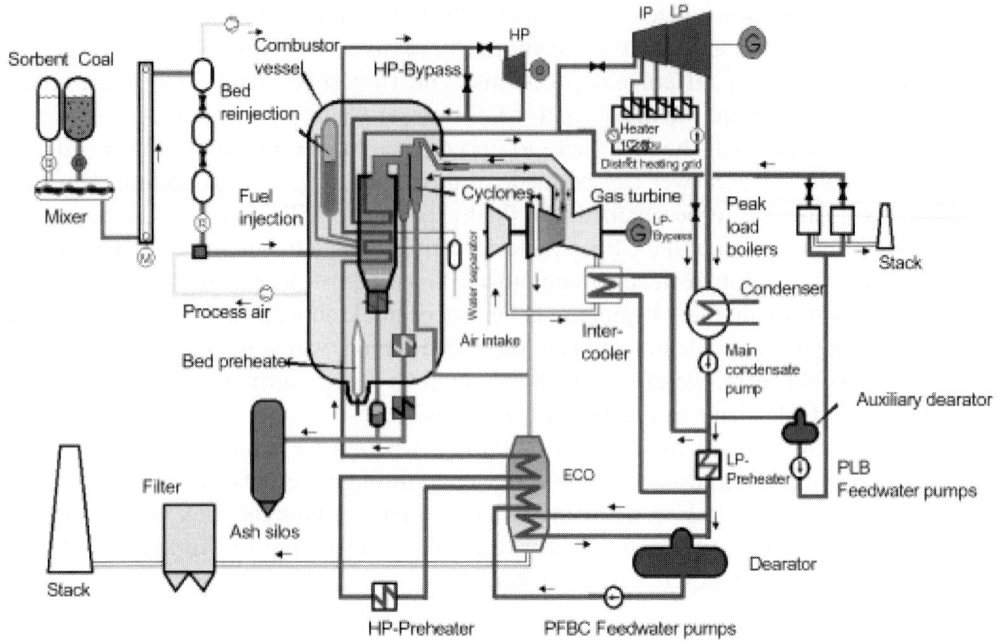

Fig. 3 Simplified flow diagram of the Cottbus CHP plant.

Bed sintering imposes an efficiency limit on the PFBC, such that it is impossible to increase the temperature of the bed combustion products to boost gas turbine output. A large increase in steam temperatures may also be problematic, because of the risk of high temperature corrosion, as mentioned in the previous section.[35] Efforts to increase efficiency have focused on the burning of additional fuel at the entrance to the gas turbine. The fuel could be natural gas, or could come from the gasification of coal char. Unfortunately, with this increase in temperature, any ash, which is carried over, will melt and stick to the blading; so that the development of a near perfect, low pressure drop filtration system is critical to future progress.

With the advanced PFBC, two shift operation or severe load following may cause some problems with the gas turbine itself. Much depends on the design of the turbine. Because of the possibility of ash carry over blades will probably be of the solid type, limiting inlet temperature to 1,100°C. Even solid blades are subject to harmful thermal gradients effects.[40, 41] If blade cooling is used, the cooling flow will probably be of the internal type, in which, according to Suciu, high blade temperatures and thermal stress could still be problematic.[40]

RECUPERATIVE GAS TURBINES FOR DISTRIBUTED POWER

The recuperative gas turbine is one solution to the need for distributed power; that is the provision of electricity close to the point of demand, avoiding the need for long distance transmission systems. The simple-cycle gas turbine has been used for this duty, but a major shortcoming is poor part load efficiency. Turbine inlet temperatures of such units drop quite markedly as power output falls. The recuperative gas turbine, to a large extent, avoids this difficulty. Inlet temperatures are maintained even at low loads, through the use of the recuperator, so there is only a modest drop off in efficiency. By its nature the recuperative gas turbine is intended to load follow, and would be expected to two shift, as well as being expected to start up and shut down very rapidly. Both the recuperator and the hot section components need to be designed for cycling.

The recuperator consists of a highly compact sheet metal heat exchanger, which preheats the combustion air from the compressor, using the heat in the turbine exhaust. There are distinct differences between the recuperators that are used in micro gas turbines and those used in bigger multi-megawatt sized machines. The two sizes of machine are also likely to suffer rather different metallurgical challenges to the turbines when operated in a cycling mode.

Micro Gas Turbines

Smaller recuperative machines have to be linked into CHP or cogen schemes to make them economic. They are much less efficient in terms of power production, because of their reliance on radial compressors and turbines. It is difficult to use blade cooling with radials, and inlet temperatures are under 1,000°C. Typical power outputs are between 0.05 and 1 MW. Recuperators of these smaller machines operate over even harsher temperature regimes than the big axial flow types, as turbine outlet or exhaust temperatures tend to be higher.

The need for better recuperators is likely to become imperative, if the micro turbine cooling problem is overcome, or fifty years of effort to introduce ceramics into radial turbines suddenly bears fruit. Figure 4 shows the effect of increasing the turbine inlet temperature on the exhaust temperature. Assuming that the pressure ratio is constant at 4.5, which is, for a variety of reasons, typical of recuperative radial type gas turbines, the turbine outlet temperature increases linearly with turbine inlet temperature, such that, at an inlet temperature of 950°C, the outlet temperature is 630°C. At an inlet temperature of 1,250°C, the outlet temperature is over 850°C. With recuperator metal temperature being only about 80°C lower than this, it is an extremely punishing regime for what are paper-thin exchanger surfaces. With simple cycle gas turbines, pressure ratios change quite markedly with inlet temperatures, which have the effect of keeping outlet temperatures more or less constant at well under 500°C, as shown by Figure 4.

In micro turbine recuperators, pressure differences are modest and volumetric flows between the high and low-pressure sides are similar, implying that recuperators for these machines will be of the 'prime surface type'. Prime surface designs consist of highly convoluted thin sheets of material, in which the heat passes directly through the wall.[42] See Figure 5. Sheet thickness is well under a millimetre, with the passage size being about a millimetre across. These dimensions reduce the stresses caused by pressure differences to a very low order, but maximise heat transfer rate, although the flows are laminar.

A typical example of a primary surface recuperator is that in the Solar Turbine Inc, Mercury 50 gas turbine. The heat exchange surfaces consist of thin sheets of Type 347 stainless steel 0.1 mm thick, which are of the cross wavy type. When the engine is running, the convolutions

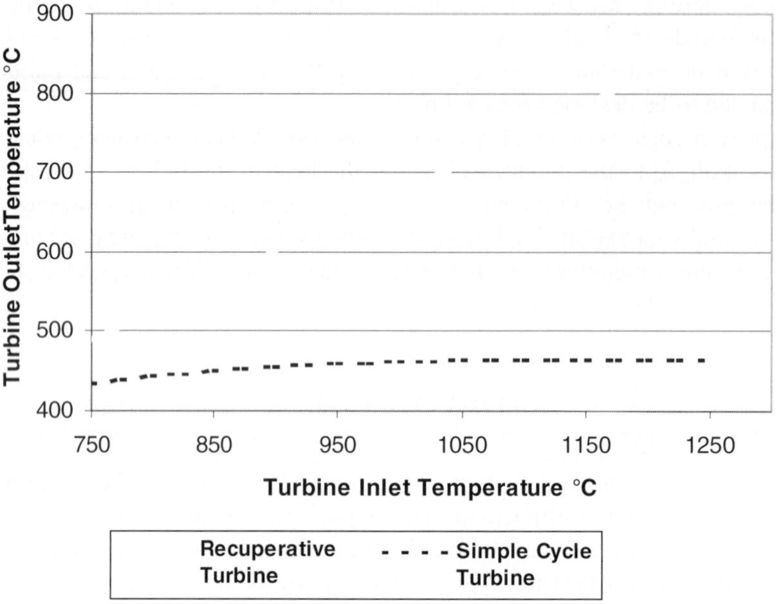

Fig. 4 Turbine outlet temperatures for recuperative and simple cycle gas turbines.

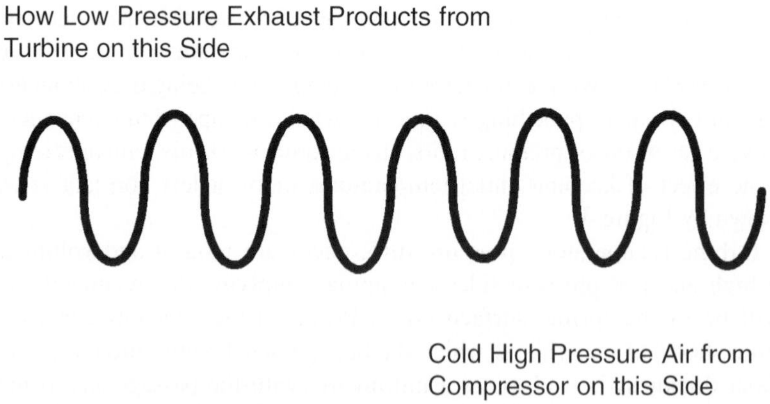

Fig. 5 Schematic of one row of heat exchanger passages in primary surface recuperator. Heat flow is through the thickness of the convolutions.

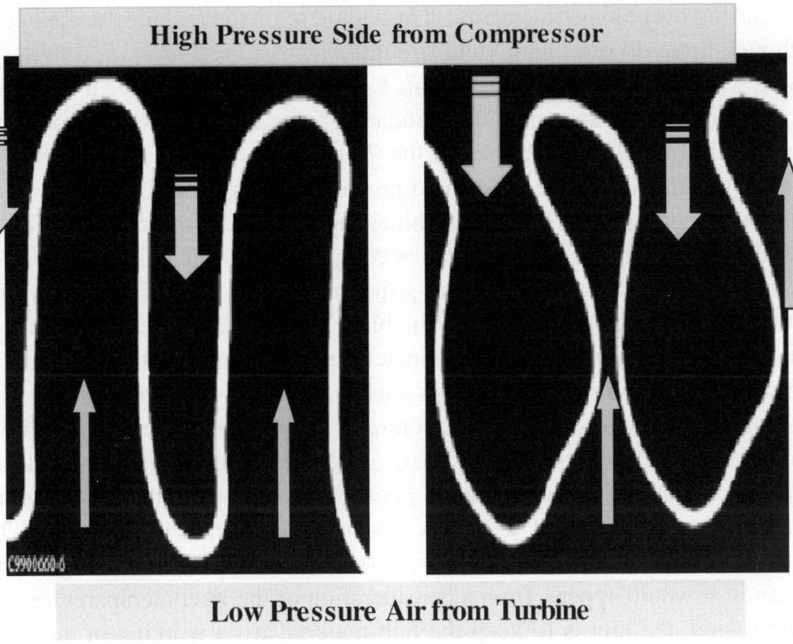

Fig. 6 Collapse of cells of a primary surface recuperator.

between opposing pairs of sheets move slightly, because of the pressure difference, and so give support to one another.[43]

One failure mechanism of the recuperator is the inwards collapse of the low pressure, or turbine exit side, channels. This will have a serious effect on the performance of the recuperator, as it is the turbine exit side that has the largest pressure drop. Any further reduction in channel size will increase pressure drops still further. Figure 6 is based on pictures forwarded to the author from Oak Ridge National Laboratory. The pictures have been manipulated to show more clearly the collapse of the cells. For cycling duties, a material with high yield strength at temperature is needed, as this is probably more critical than creep rupture strength. It is arguable whether a very high elongation to failure is a critical parameter, in view of the extreme flexibility of the 'foil type' heat transfer surfaces.

This is a personal view of the author, but seems to be supported by the fact that designers seem to be unwilling to push existing recuperator materials, namely the type 300 stainless steels, up into the true creep range. Some authorities have argued strongly for the use of ceramics, which have virtually no ductility or crack resistance compared to metals. Good oxidation resistance in cycling conditions is vital, as the foils are so thin that oxide /metal differential expansion effects cannot be ruled out, particularly if a weak denuded zone were to form. The author has also come across a form of cracking in highly worked stainless steels, in the 550°C range, in which the cause appeared to be selective oxidation of slip planes, occurring over along time period. This would seem to preclude the use of cold worked foils, although these do offer high yield strength.

Masiasz, who is presenting a paper at this Conference, has identified a range of materials that are considerably cheaper than the standard grades of alloy, which would be used for recuperators with metal temperatures in the 750°C range. Some of these have been heat treated, and as such would give good yield properties as suggested above.[44]

The other key component in microturbines susceptible to the effects of cycling is the radial turbine itself. This is a highly complex component, in which the hub, disc and blading are cast as one unit, and whose thermo-structural design is difficult. There is a severe temperature drop of around 300°C from the blade tips, which run a little below the turbine inlet temperature, down to the hub. Fortunately blade tip temperatures are 50–100°C less than turbine inlet temperatures. Tip stresses are low, but damage from oxidation is significant. In contrast, the centrifugal stresses at the centre of the hub are of the order of 300 MPa. This stress temperature combination requires the use of complex age hardened nickel based alloys. The thermal conductivity of these is quite poor, exacerbating the thermal stress problem.

The thermal stress is, indeed, a serious design issue, even in existing units. Early work showed the necessity of scalloping the disc between the blades, even though it spoils blade aerodynamics. It would appear, from a consideration of the likely temperatures and stresses in the hub region, the aim is to keep the hub material from working in the creep regime. Unfortunately, the blade thickness, in micro turbines, is only a few millimetres at most, and the introduction of cooling passages is difficult. Made properly cooling holes, with air passing through them, would reduce temperature gradients in the blades as well as reducing metal temperatures, as shown in an early paper by Suciu.[40] If the micro turbine is really to make an impact, effort is needed on the fabrication of cooled radial turbines.

Large Recuperative Turbines

The larger type of recuperative machine is likely to use aero engine components, and will use either directionally solidified or single crystal blades for the HP turbine section. Outputs are above 10 MW. Full load efficiencies, for big recuperative machines using axial flow compressors, axial turbines and intercooling, are above 40%, dropping to about 35 % at half load. Turbine inlet temperatures are in excess of 1,200°C, and the hot section will be prone to the problems previously described . An advantage of recuperative designs is that, when tripped, the thermal mass of the recuperator prevents a sudden temperature drop in hot gas path components and turbine discs, suggesting that some of the criteria for assessing the effect of cycling, might be too harsh. For example, General Electric states that, for the MS7001 and 9001 simple cycle machines, a trip from full load is equal to 9 normal shut downs.[45]

Recuperation, because of the increased pressure drops, tends to pull down output, so to restore power, intercooling can be used between the compressor stages, reducing compressor work. Intercooling also helps reduce recuperator temperatures, both directly and indirectly. Outlet temperatures from the compressor are lower and, for thermodynamic reasons, intercooled turbines maintain efficiency at higher pressure ratios, and as we have seen, this results in low turbine outlet temperatures.

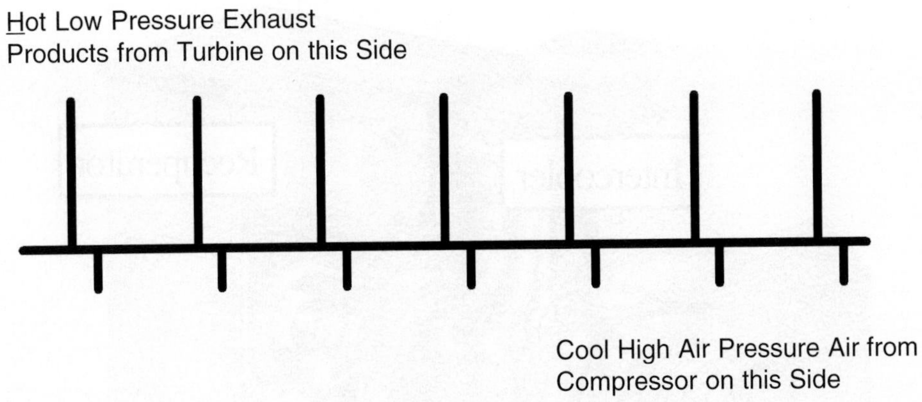

Fig. 7 Schematic of one row of heat exchanger passages in a secondary surface recuperator. Much of the heat flow is along the fins.

The downside of a more sophisticated approach is that pressure differences between the outlet compressor and outlet turbine are in the 10–20 bar range. Stresses are higher and the recuperator enclosure must withstand higher loads. Furthermore, if the channel widths and geometry are the same on both sides, as with primary surface designs, the pressure drop on the turbine outlet side becomes quite unacceptable. Hence, secondary surface types tend to be used for this duty. Figure 7 shows such a design, sometimes described as a plate-fin exchanger. Note that the sheet, through which the heat must flow, is set with fins, which are longer on the low pressure side as the heat transfer coefficient is poor.

The secondary surface design is more complex because the fins must be brazed or welded into place. The fins do, however, help to strengthen the sheet against pressure loads, and, within limits, heights can be adjusted to even out the disparity in heat transfer coefficients. Much of the heat needs to flow down the fins, and materials with good conductivity are needed, which also helps to reduce thermal stresses. In reviewing this type of issue, McDonald has emphasised the fact that thermal stress is proportional to the heat transfer rate, and that the wrought nickel based alloys are superior to the austenitics.[46] Here again, the argument needs to be based on high temperature yield rather than creep properties. Figure 8 shows a cutaway drawing of the WR-21, a subject of a paper at this conference.[47] The recuperators are situated in the tanks in the upper right hand corner. Peak metal temperatures in this type of design are unlikely to exceed 500°C. It is understood that consideration was given to the use of a precipitation hardened martensitic stainless steel, as the yield strengths of these alloys are extremely high.[48] However, the idea was dropped because of the inferior conductivity of this class of alloys.

Improved efficiency can be obtained by increasing the turbine inlet temperature. This needs to be done without increasing the amount of cooling air to the turbine, as this will unbalance the upward and downward flows through the recuperator still further. Any difference between these flows will reduce the amount of waste heat that can be recuperated, and

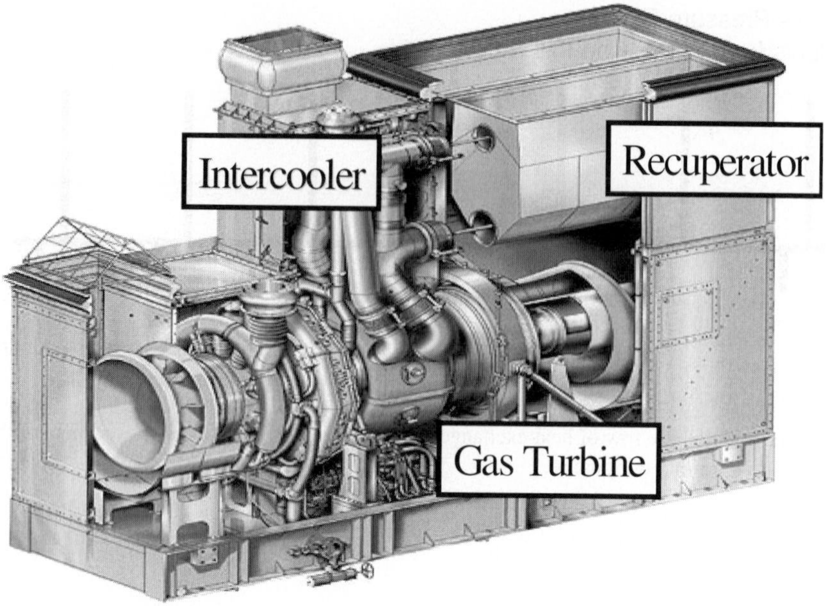

Fig. 8 Cutaway of the WR-21 westinghouse rolls recuperative gas turbine.

thermal efficiency drops. For this reason, in advanced recuperative designs, we are likely to see the use of blading of the sealed convection type, in which a high pressure fluid circulates inside the blade, before being taken off to an external cooling arrangement. As with advanced CCGT designs, thermal stresses will be higher and it is likely that the blades will be of the single crystal, thermal barrier coated type. Recuperator temperatures are also likely to be higher in future.

HIGH TEMPERATURE NUCLEAR SYSTEMS

In the introductory section it was emphasised that a shortcoming of current nuclear plants is the reluctance of operators to cycle them. This can be due legislative obstacles, problems with reactor control at low power levels, concern about the effects of changing temperatures on components, plus the commercial fact that it is impractical to run these units at anything other than full output. The one major exception to the rule is EDF, but even this company has had to revert to base load operation because of problems with control rods.

It is a truism that, if nuclear power is to have a future, capital costs should be cut to those of fossil power plants, efficiency be improved from the present meager level of 32% for the PWR, and the inventory of highly radioactive materials be reduced, to minimise clean up and storage costs. Water type reactors are clearly limited in efficiency improvements, and

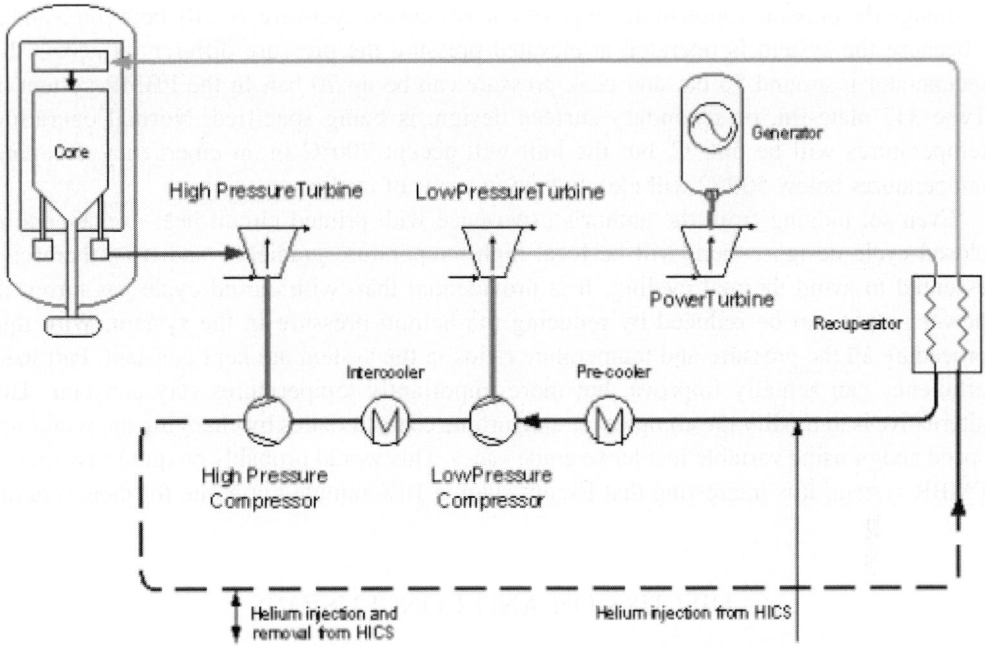

Fig. 9 Schematic of helium loop and turbine arrangement for the escom pebbled HTGR.

the High Temperature Gas Reactor or HTGR is coming back into vogue. The HTGR is based upon a recuperative gas turbine cycle, in which the combustor is replaced by a reactor containing uranium or plutonium fuel, running at about 900°C. The fuel consists of oxides of these elements, formed into pebbles that are encased in a silicon carbide shell, eliminating the thermal stress problems to which conventional fuel rod elements or pins are subject.

The gas turbine is of the closed cycle type, in which helium is the working fluid. Multistage compression is used, with intercooling to reduce negative work and improve power output and efficiency. After the final stage of compression, the helium flows through a recuperator, picking up heat from the turbine exhaust, before entering the reactor. Here the helium absorbs heat from the fuel and also cools the graphite reflector, which surrounds the fuel mass. The helium, now at a temperature of about 850°C, enters the turbine, which is used to produce electric power and drive the compressors. After leaving the turbine the helium enters the low pressure side of the recuperator. Any remaining heat in the helium is rejected to cooling water, before the helium passes around the circuit once again. The actual turbine arrangement is varied the PBMR unit promoted by Escom Enterprises using three separate turbines to drive, respectively, each compressor and the alternator.[49] See Figure 9. That of General Atomics has but one turbine driving two sets of compressors and the alternator.

The most critical unit on this system, in terms of cycling duty, is the recuperator, some sections of which are expected to be mechanical stressed, due to the high operating pressures.

Although the pressure ratios in this type of nuclear closed cycle are usually between 3 and 5 , because the system is operated at elevated pressure the pressure differential across the recuperator is around 50 bar and peak pressure can be up 70 bar. In the PBMR system, a Type 347 plate-fin, or secondary surface design, is being specified. Normal operating temperatures will be 500°C, but the unit will accept 700°C in an emergency. Keeping temperatures below 500°C will clearly help in terms of cycling.

Even so, judging from the author's experience with printed circuit heat exchangers in closed cycle designs, there will be local high temperature gradients, and it is, therefore, essential to avoid thermal cycling. It is providential that, with closed cycle gas turbines, power output can be reduced by reducing the helium pressure in the system. With this procedure all the pressure and temperature ratios in the system are kept constant. Part load efficiency can actually improve, but more importantly temperatures stay constant. The alternative is to modify the compressor and turbine characteristics by changing the rotational speed and/or using variable incidence guide vanes. This would probably be quite easy on the PMBR system. It is interesting that Escom claim a 10%/minute ramp rate for their system.

DISCUSSION AND CONCLUSIONS

Tables 3 and 4 summarise the main points made in this short review. These tables indicate the likely ability of a plant to two shift and load follow, as well as identifying which components are likely to suffer as a result of cycling. In drawing up the tables, an effort has been made to forecast those cycling problems that could be eliminated, through more thoughtful design and better operating procedures.

Table 3 summarises the problems of current pulverised coal and CCGT plants and what might expected of very advanced developments of these. It will be seen that although, hopefully, the condensate quench problem can be eliminated, thermal fatigue of superheaters will still be a big issue on advanced 700°C steam plant.

Table 4 summarises the problems of novel fossil fuel fired and HTGR nuclear systems. One feature of some of the coal based systems is the highly integrated nature of these plants, which could suggest problems in two shifting and even in load following. From what one gathers about entrained bed IGCC systems, load following will not be a good idea because of the fall in plant efficiency. Genuine two shifting is not possible, because of the explosion risk, which is always present during the start up of gas making plants. Wide scale IGCC is, perhaps, best suited to a time when the hydrogen economy is fully with us. For this reason, although there are IGCC components that could be liable to the effects of thermal fatigue, in practice no such failures will occur since these plants will not cycle. Hence, the relevant sections of the table are marked as 'Not Relevant'. However, there are other gas making processes, which can load cycle with ease, and can be made to produce a synthetic or substitute natural gas, which could go into the existing gas network. Plant complexity is, however, not necessarily a problem. It all depends on the system. The PFBC seems to load follow quite easily despite the multiplicity of feed heaters, economisers and evaporators that are needed. The issue here, as with the CFB, is whether they can beat the coal fired 700°C steam plant in efficiency terms.

Table 3 Likely operability, maintenance and major component problems in current and highly advanced pulverised fuel steam and CCGT plant.

Plant	Operability and Maintenance			Critical Components and Failure Mechanism					
	Ease of Two Shifting	Ease of Load Following	Cycling Damage	Economiser, Evapoator and Feed Heaters	Superheater and Reheater	Steam Turbines	Combustion or Gas Turbine	HRSG	Recuperator
Current Pulverised Fuel Subcritical	Good	Good to Excellent	Significant	Corrosion and Thermal Fatigue Furnace Wall Attack	Thermal Fatigue and Shock	Thermal Fatigue and wear	NA	NA	NA
Current Pulverised Fuel Supercritical	Good to Excellent	Excellent	Significant	Corrosion and Thermal Fatigue Furnace Wall Attack	Thermal Fatigue and Shock	Thermal Fatigue and Wear	NA	NA	NA
Current CCGT	Good to Excellent	Poor to Good	Significant	NA	NA	Thermal Fatigue and Back End Erosion	Increased Oxidation and Hot Section Cracking	Thermal Fatigue and Shock	NA
700°C Pulverised Fuel	Good	Good to Excellent	Significant	Thermal Fatigue	Severe Thermal Fatigue and Shock Fireside Corrosion	Severe Thermal Fatigue and Wear	NA	NA	NA
Advanced CCGT	Good to Excellent	Good to Excellent	Significant	NA	NA	NA	Increased Thermal Fatigue TBC Damage	Thermal Fatigue	NA

Table 4 Likely operability, maintenance and major component problems in 2015 fossil fuel and nuclear based systems.

Plant	Operability and Maintenance			Critical Components and Failure Mechanism					
	Ease of Two Shifting	Ease of Load Following	Cycling Damage	Economiser, Evaporator and Feed Heaters	Superheater and Reheater	Steam Turbines	Combustion or Gas Turbine	HRSG	Recuperator
IGCC	Non-Existent	Poor	Not Relevant	NA	NA	Not Relevant	Not Relevant	Not Relevant	NA
Advanced CFBC	Excellent	Excellent	Significant	Thermal Fatigue	Thermal Fatigue and In-Bed Corrosion	Thermal Fatigue and Wear	NA	NA	NA
Advanced PFBC	Good to Excellent	Fair to Good	Significant	Reduced Thermal Fatigue	Reduced Thermal Fatigue	Reduced Thermal Fatigue	Thermal Fatigue and Wear	No Problems	NA
Advanced Micro and Mini Gas Turbines	Excellent	Good to Excellent	Moderate	NA	NA	NA	Increased Thermal Fatigue TBC Damage	NA	Some Thermal Fatigue
Advanced Intercooled Recuperative Gas Turbines	Good to Excellent	Excellent	Moderate to Severet	NA	NA	NA	Increased Thermal Fatigue TBC Damage	NA	Modeate
High Temperature Gas Reactor	Poor	Good to Excellent	Low	NA	NA	NA	Low	NA	Low

This review has shown that, not surprisingly, the overriding failure mechanism is, of course, one of thermal fatigue, although the R&D will need to be directed to some new areas. These include.

- Thermal fatigue of austenitic and nickel based alloys in steam,
- Performance of transition welds in steam pipework and steam turbines,
- Behaviour of weldments in precipitation hardened alloys,
- Life assessment of single crystal and directional solidified gas turbine blades,
- Behaviour of TBCs and bond coatings under very high heat flux conditions and
- Failure mechanisms of recuperators due to thermal fatigue.

It is also likely that plant cycling will exacerbate corrosion type phenomena, such as steam side oxide spalling, fireside attack and sulphidation in bubbling bed combustors. Here, much can be done through improved design and the correct choice of materials.

Table 4 also shows that HTGR nuclear systems are, more or less, immune to the effects of load following. This is a direct result of the design being based on a closed cycle gas turbine, whereby power can be reduced, by taking helium out of the system, so that all temperatures, including those of the reactor, turbines and recuperator stay the same. This feature of the closed cycle was one reason why a good deal of work was done on a natural gas fired closed cycle by the author's team at British Gas during the early nineties. Efficiencies of around 50% appeared possible, with excellent part load turndown, with little thermal damage.

The author and his colleagues, at ETD Ltd, are currently working on microturbine developments. For this type of unit to be successful, improved versions will be needed, some running at higher temperatures. This type of small scale unit, with an output of 50 kW or less, would, in a sense, be returning to the type of boiler room, ancillary power unit, that Parsons manufactured to get the steam turbine off the ground. His first units had efficiencies of around 20%, being categorised as "steam eaters", and we will need a lot better than that. If small-scale power did become widespread, microturbines could be used to lop off the peaks that bedevil large scale power generation.

Peaks could be reduced still further, making use of power conditioners, which are a necessary adjunct of micro turbines, as rotational speeds are far above the 50 and 60 Hz standard. This technology, which can turn any voltage and frequency in that demanded by the Grid, was not available to Parsons, his turbines having to be geared down or limited to supplying direct current. As such power conditioners linked to high-speed micro flywheel/ generator systems, could store surplus electrical energy from those times when renewables are producing an excess. In this way, whilst retaining highly efficient centralised power generation, which hopefully would have to do only a mild amount of load following, we could link this into a big renewables sector, utilising the widespread use of micro-CHP to ensure that power and heat is being used in the most effective manner.

ACKNOWLEDGEMENTS

To ETD Ltd for permission to publish this work. However all opinions expressed are those of the author and do not necessarily represent the views of ETD Ltd.

REFERENCES

1. R. H. Parsons: (quotation) 'Early Days of the Power Station Industry' Cambridge University Press, 1939, 174–175.
2. J. F. C. Snell: 'Power House Design' *Review of Reaction Turbines*, Longmans, Green and Co., 1921, 199–200.
3. Q. Chen and G. Scheffknecht: 'Boiler Design and Materials Aspects for Advanced Steam Plants', *Proceedings of 2002 Liege Conference on Materials for Advanced Power Engineering,* Lecomte-Beckers, Carton, Schubert and Ennis, eds., FA Juelich, Germany, 2002, 1019–1073.
4. R. Viswanathan, R. Purget and U. Rao: 'Materials for Ultra-Supercritical Coal-Fired Power Plant Boilers', *Proceedings of 2002 Liege Conference on Materials for Advanced Power Engineering*, Lecomte-Beckers, Carton, Schubert and Ennis, eds., FA Juelich, Germany, 2002, 1109–1129.
5. Vision 21 Home Page www.fe.doe.gov/coal power/vision21/index.shtml.
6. F. Starr, J. Gostling and A. Shibli: 'Damage to Power Plant Due to Cycling', ETD Report 1002-HP-1001 c/o European Technology Development Ltd 2000, www.etd1.co.uk
7. HRL Technology and Rohan Fernando: 'Non Base Load Operation of Coal Fired Power Plant', c/o IEA Coal Research www.iea-coal.org.uk
8. I. A. Shibli and F. Starr: 'Damage to Combined Cycle Gas Turbines Due to Cyclic Operation', ETD Report 1012-iip-09, c/o European Technology Development Ltd ., 2002, www.etd1.co.uk
9. I. A. Shibli, Starr, Viswanathan and Gray: eds., *Cyclic Operation of Power Plant-Technical Operation and Cost Issues*, Science Reviews, London, 2001.
10. R. P. Skelton and F. Starr: 'Introduction (to Special Issue on Plant Cycling), *Materials at High Temperature*, **18**(4), 2001, 193–209.
11. Trieb and Dûrrschmidt: eds., *Concentrating Solar Power Now*, German Federal Ministry for Environment, Nature Conservation and Nuclear Safety, 2002.
12. M. Oechner, J. Goedjen, W. Stamm and R. Subramanian: 'Life Prediction for Thermal Barrier Coating Systems for Land Based Gas Turbines' *EPRI Conference on Advances in Life Assessment and Optimisation of Fossil Plants*, Orlando, Fla, USA., 2002.
13. L. Singheiser, R. Steinbrech, W. J. Quadakkers and R. Herzog: *Materials at High Temperature*, **18**(4), 2001, 249–259.
14. R. Viswanathan and S. T. Scheirer: 'Materials Technology for Advanced Land Based Gas Turbines', KL-1-2 Proc Creep 7 JSME Tsukuba, Japan, 2001.
15. D. P. Shepherd, A. Wisbey, G. F. Harrison, T. J. Ward and B. Vermeulen: *Materials at High Temperature*, **18**(4), 2001, 231–247.
16. F. Starr: 'Background to the Design of HRSG Systems and Implications for CCGT Plant Cycling', OMMI, 2003.
17. J. Briggs: 'Repowering Florida' *Modern Power Systems*, 2001, 57–61.
18. R. A. Anderson and M. Pearson: 'Reliability and Durability from Large HRSGs', *CCGT Plant Components: Development and Reliability*, Professional Engineering Publishing Ltd., 1999, 21–46.

19. F. Starr and A. Shibli: 'Microstructural Issues in the Design of Austenitic and Nickel Based Materials for Superheater Systems in700°C Steam Plant', *Proceedings of 2002 Liege Conference on Materials for Advanced Power Engineering*, Lecomte-Beckers, Carton, Schubert and Ennis, eds., FA Juelich, Germany, 2002, 1233–1240.
20. P. Skelton and B. E. Beckett: 'Thermal Fatigue Properties of Candidate Materials for Advanced Steam Plant', *Advances in Material Technology for Fossil Power Plants*, ASM, Ohio, 1987, 359–366.
21. J. L. Bolton and J. Muscroft: 'The Two Shifting Capability of Large Steam Turbines'.
22. M. E. Staubli, K-H. Mayer, T. U. Kern and R. W. Vanstone: 'COST 501/COST 502–The European Collaboration in Advanced Steam Turbine Materials for Ultra Efficient Low Emissions Power Plant', *Parsons 2000*, Strang, Banks, Conroy, McColvin, Neal and Simpson, eds., IOM Communications Ltd., 2000, 98–128.
23. A. Feldmûller and T. U. Kern: 'Design and Materials for Modern Steam Power Plants an Actual Concept', *Parsons 2000*, Strang, Banks, Conroy, McColvin, Neal and Simpson, eds., IOM Communications Ltd., 2000, 143–156.
24. J. Rösler, Böttger, M. Wolske, H. J. Penkalla and C. Berger: 'Wrought Nickel Based Alloys for Advanced Gas Turbine and USC Steam Turbine Rotor Applications', *Proceedings of 2002 Liege Conference on Materials for Advanced Power Engineering,* Lecomte-Beckers, Carton, F. Schubert and Ennis, eds., FA Juelich, Germany, 2002, 89–106.
25. J. Van Liere and W. T. Bakker: 'Present Status of Advanced Coal Fired Plants' *Corrosion in Advanced Power Plants*, Bakker, Norton and Wright, eds., *Science and Technology Newsletters 1997*, (Special Issue Materials at High Temperature), **14**(2/3), 1997, 4–9 and 7–12.
26. J. Karg, G. Haut and G. Zimmermann: 'Optimised IGCC Cycles for Future Applications', *Conference 2000 Gasification Technologies*, San Francisco, 2000.
27. J. F. Norton, M. Maier and W. T. Bakker: 'Corrosion of 12% Cr Alloys with Varying Si Contents in a Simulated Dry-Feed Entrained Slagging Gasifier', 81–91, ibid.
28. K. Natesan: 'Corrosion Performance of Alumina Scales in Coal Gasification Environments', 71–79, ibid.
29. V. Guttmann, K. Stein and W. T. Bakker: 'Deformation-Corrosion Interactions in Selected Advanced High Temperature Alloys', 61–70, ibid.
30. P. Basu, C. Kefa and L. Jestin: 'Fluidised Bed Boilers' *Boilers and Burner-Design and Theory*, Springer, 2000, 302–345.
31. S. J. Goldich and R. G. Lundqvist: 'The Utility CFB Boiler- Present Status, Short and Long Term Future with Supercritical and Ultra-Supercritical Steam Parameters', PowerGen, Europe, Milan, 2002.
32. I. G. Wright: 'A Review of Experience of Wastage in Fluidised Bed Boilers', *Corrosion in Advanced Power Plants*, Bakker, Norton and Wright, eds., *Science and Technology Newsletters*, 1997, (Special Issue Materials at High Temperature, **14**, 2/3), 207–218.
33. K. Natesan: 'Assessment of Corrosion in FBC Systems', *Proceedings of Corrosion-Erosion Wear of Materials at Elevated Temperature*, Levy, ed., NACE, 1990, 3–1–3-30.
34. C. Johns: 'Cycling Operations of Circulating Fluidized Bed Boilers', cbjohnk@inetworld.net, 2001.
35. A. J. Michener, E. A. Rogers and J. Stringer: 'The Behaviour of Iron, Nickel and

Cobalt Based Alloys in Fluidised Bed Combustion Systems, *Behaviour of High Temperature Alloys in Aggressive Environments*, The Metals Society, 1979, 29–44.

36. W. Bakker: 'Complex Fireside Corrosion Mechanisms in Boilers Using Staged Combustion Systems', *Proceedings of 2002 Liege Conference on Materials for Advanced Power Engineering,* Lecomte-Beckers, Carton, Schubert and Ennis, eds., FA Juelich, Germany, 2002, 815–832.
37. H. Komatsu, M. Maeda and M. Muramatsu: 'A Large Capacity Pressurized Fluidised Bed Combustion Boiler Combined Cycle Plant', *Hitachi Review*, **50**(3), 2001, 105–109.
38. Karita PFBC Internet Plant Description www.neeco.co.jp/pfbc/english/6-1.htm
39. P. S. Weitzel, D. K. McDonald, S. A. Whitney and N. Oda: 'Directions and Trends for PFBCs and Hot Gas Clean Up', *Pittsburgh Coal Conference*, 1996.
40. S. N. Suciu: 'High Temperature Design Considerations', *AGARD Conference*, Florence, 1977.
41. J. K. Hepworth: 'The Effect of Cyclic Operation on the Life of Stage 3 and 4 Gas Turbine Blading S5-3-1 to S5-3-10', *Cyclic Operation of Power Plant- Technical Operation and Cost Issues*, Shibli, Starr, Viswanathan and Gray, eds., London, Science Reviews, 2001.
42. E. Ultainen and B. Sundén: 'Evaluation of the Cross Corrugated and Some Other Heat Transfer Surfaces for Microturbine Recuperators', *Journal of Engineering for Gas Turbines and Power Trans ASME*, **124**, 2002, 550–560.
43. 'Solar Mercury 50- A Fresh Approach to Efficiency', Turbomachinery International, 1997.
44. P. Maziaz and R. W. Swindemann: 'Selecting and Developing Advanced Alloys for Creep Resistance for Microturbine Recuperator Applications', *Journal of Engineering for Gas Turbines and Power Trans ASME*, **125**, 2003, 310–315.
45. R. Hoeft and E. Gebhardt: 'Heavy Duty Gas Turbine Operating and Maintenance Considerations', c/o GE Power Systems, 2000.
46. C. F. McDonald: 'Gas Turbine Recuperator Advancements', *Materials Issues in Heat Exchangers in Boilers Conference*, F. Starr and B. Meadowcroft, eds., IOM, 1997, 337–369.
47. E. R. Watson, M. L. Parker and D. A. Branch: 'Development and Testing of the WR-21: An Intercooled and Recuperative Gas Turbine', Paper 19 INEC 96, Institute of Marine Engineers, 1996.
48. F. Starr: 'Advanced Materials for Advanced Heat Exchangers', *Materials for High Temperature Power Generation and Process Plant Applications*, A. Strang, ed., IOM Communications, 2000, 79–151.
49. J. A. De Beer and Z. Olsha: *Proceedings of 2002 Liege Conference on Materials for Advanced Power Engineering,* Lecomte-Beckers, Carton, Schubert and Ennis, eds., FA Juelich, Germany, 2002, 1801–1808.

STEAM PLANT TECHNOLOGY

Supercritical Boiler Technology for Future Market Conditions

JOACHIM FRANKE and RUDOLF KRAL
Siemens AG, PGW7, P.O. Box 3220
D-91050 Erlangen, Germany

INTRODUCTION

The requirements for environmental protection and operating economy in future steam power plants make high efficiency levels and operating flexibility a matter of course not only in the EU but also in increasing measure around the world. Existing technologies have currently enabled fulfillment of these requirements by pulverized-coal-fired power plants and in part also by power plants with circulating fluidized bed (CFB) combustion systems.

Higher efficiencies can be achieved only along the path of higher steam temperatures and pressures.

STATE OF THE ART

Power plants operating at supercritical pressure and high steam temperatures were already being constructed in the 1950s (Figure 1). The 1960s saw a series of supercritical plants in the U.S. (such as those equipped with the universal pressure boiler) and in the last twenty years supercritical plants were used exclusively in Germany and Japan. The latter were designed for sliding-pressure operation and thus also fulfill the requirements for high operating flexibility and high plant efficiencies at part load. (Figure 2).

To date, CFB power plants have been used especially for smaller power output levels, generally with drum boilers. Plants up to 350 MW are in the meantime already in operation and several plants equipped with Benson[1] boilers have also been constructed. Supercritical plants for ratings above 400 MW are planned.

Power plants operating at supercritical steam pressure have already demonstrated their operational capabilities and high availability over decades. The transition to steam temperatures of 600°C and higher is now a further major development step, which decisively affects many aspects of the design of the power plant, especially of the boiler. Whether the transition to these high steam temperatures is economical also depends not only on the choice of main steam pressure, reheat pressure and feedwater temperature, but also on the range of fuel.

[1]Benson is a registered trademark of Siemens AG.

1924	Siemens buys the „BENSON Patent" from Mark Benson
1926 to 1929	Siemens manufactures three BENSON boilers (30 t/h to 125 t/h)
1933	Siemens introduces variable-pressure operation
1933	Siemens discontinues it's own production; awarding licences instead to several boiler manufacturers
1949	The world's first once-through boiler with high steam conditions (175 bar/610°C)
1954	The first BENSON boiler with supercritical pressure (300bar/605°C)
1963	The world's first spiral-tubed water walls in membrane design
1999	First once-through HRSG with vertical tubing and natural circulation flow characteristic
2000	1000 BENSON boilers sold with > 700.000 t/h in total
2002	First once-through boiler with vertical tubed water walls in low mass flux design (natural circulation flow characteristic)
2003	Order for first supercritical CFB boiler (Siemens low mass flux design)

Fig. 1 Milestones in the field of BENSON boilers.

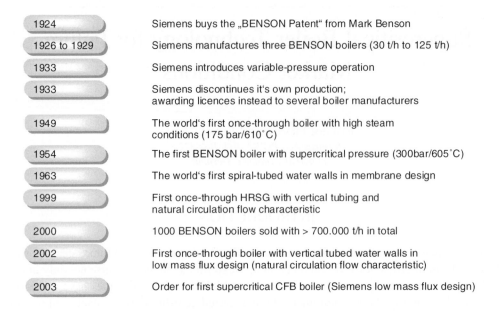

Power Plant	Output [MW]	Design Pressure *) [bar]	Steam Temperature Boiler Outlet [°C]	Year of Commisioning	
Avedorevaerket 2	415	332	582 / 600	2001	
Boxberg	915	285	545 / 580	2000	
Skaerbaekvaerket	410	310	582 / 580 / 580	1999	
Lippendorf	2 x 930	285	554 / 583	1999	
Nordjyllandsvaerket	410	310	582 / 580 / 580	1998	
Aghios Dimitrios	350	242	540 / 540	1998	
Schkopau	2 x480	285	545 / 562	1997	Europe
Neckar 2	340	285	545 / 568	1997	
Rostock	550	285	545 / 562	1994	
Hemweg	660	261	540 / 540	1994	
Meri Pori	550	240	540 / 540	1994	
Staudinger 5	550	285	545 / 562	1992	
Fynsvaerket	430	275	540 / 540	1992	
Tachibanawan	1050	285	605 / 613	2001	
Tachibanawan 1	700	275	570 / 595	2000	
Haramachi 2	1000	280	604 / 602	1998	
Matsuura 2	1000	275	598 / 596	1997	
Nanao Ota	500	275	570 / 595	1995	Japan
Shinchi	1000	275	542 / 567	1994	
Noshiro	600	293	542 / 567	1993	
Hekinan 2	700	275	543 / 569	1992	
Shin Miyazu	450	270	541 / 569	1991	

*) max. allowable working pressure at boiler outlet

Fig. 2 Large supercritical BENSON boilers in Europe and Japan - references.

To date, the focus was on material development for the superheaters and the thick-walled components for high steam temperatures. However, investigations indicate that the wall heating surfaces can become the limiting components for further increases in steam parameters. One reason for this is the increasing fraction of superheater heat to be transferred with increasing steam parameters.

EFFECT ON DESIGN

Size of Heat Exchange Surfaces

Higher steam temperatures automatically diminish the temperature differences between the flue gas and steam, with relatively large superheater and reheater heating surfaces as a consequence. As higher tube wall temperatures also mean an increased tendency to fouling, corresponding heating surface reserves must be provided.

Feedwater temperature has a large effect on the size of the heating surfaces in the cooler flue-gas path. Values of 290 to 300°C or higher are necessary for high-efficiency plants. As on the one hand the flue-gas temperature downstream of the economizer is set in the design case at roughly 400°C – the temperature window for $DeNO_x$ and on the other hand the water outlet temperature from the economizer is limited to avoid steaming, the upstream superheaters must absorb more heat with increasing feedwater temperature. At higher steam conditions, especially at increasing reheat pressures, the exhaust steam temperatures from the HP section of the turbine and thus the reheat inlet temperatures also increase. While these temperatures are still approx. 320°C at a design main steam temperature of 540°C, they already increase to over 350°C in a 600°C main steam temperature design and even up to over 420°C in a 700°C design. This considerably decreases the temperature difference to the flue gas, with the consequence of still larger heating surfaces in the reheaters.

Under consideration of a cost-effective heating surface design, feedwater temperatures should not exceed 300°C, and HP exhaust steam pressures should lie in the range of 60 bar.

Separator Configuration

The location of the separator determines the location of the end of the evaporator on startup and at low load in recirculation mode. Usually the separator is configured such that its temperature is slightly superheated at the lowest once-through load point. Design of the boiler for high steam temperatures and pressures leads to this being already the case in lower areas of the furnace walls instead of as from the outlet first pass or in the boiler roof. The reason for this is the increasing degree of superheat and correspondingly decreasing fraction of evaporation in the heat input to the HP section with increasing steam parameters. At a load of 40%, the degree of superheat in a 540°C boiler is approximately 27%, and this increases to 39%, for example, in a design for 700°C main steam temperature (Figure 3). As the highly loaded heating surface area must lie upstream of the separators for reasons of evaporator cooling and the separator thus cannot be moved arbitrarily toward the burners, a

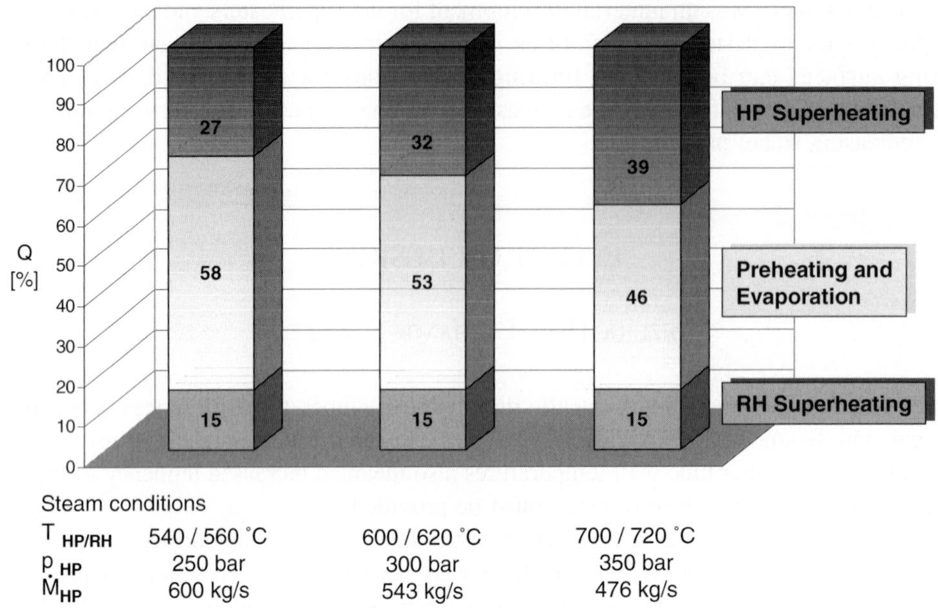

Fig. 3 Heatflow distribution in variable pressure operation at 40% load.

significantly larger degree of superheat will result at the lowest once-through operating point (Figure 4). This considerably increases the downward step of the steam temperatures on the transition to recirculation mode. In order to extensively prevent this temperature change, the transition from once-through to recirculation mode must be placed at a very low load point, requiring recirculation mode only for startup. An evaporator design based on the 'Benson Low Mass Flux' design with vertical rifled tubes enables loads to below 20% in once-through operation.

Water Walls

The water walls in boilers for subcritical steam conditions are generally configured as evaporators. At increasing steam temperatures and pressures, the fraction of evaporator heating surfaces decreases, with the result that parts of the water walls must also be configured as superheaters, i.e. downstream of the separator.

In the highly loaded furnace area, spiral-wound evaporator tubing is usually used with smooth tubes and high mass fluxes – approximately 2000–2500 kg/m^3s. As spiral-wound furnace tubing of this type is not self-supporting, it is reinforced with support straps which are welded to the tube wall with support blocks.

High steam parameters also lead to higher material loading in the evaporator. The previously existing design reserves are no longer available, with the result that a detailed stress analysis

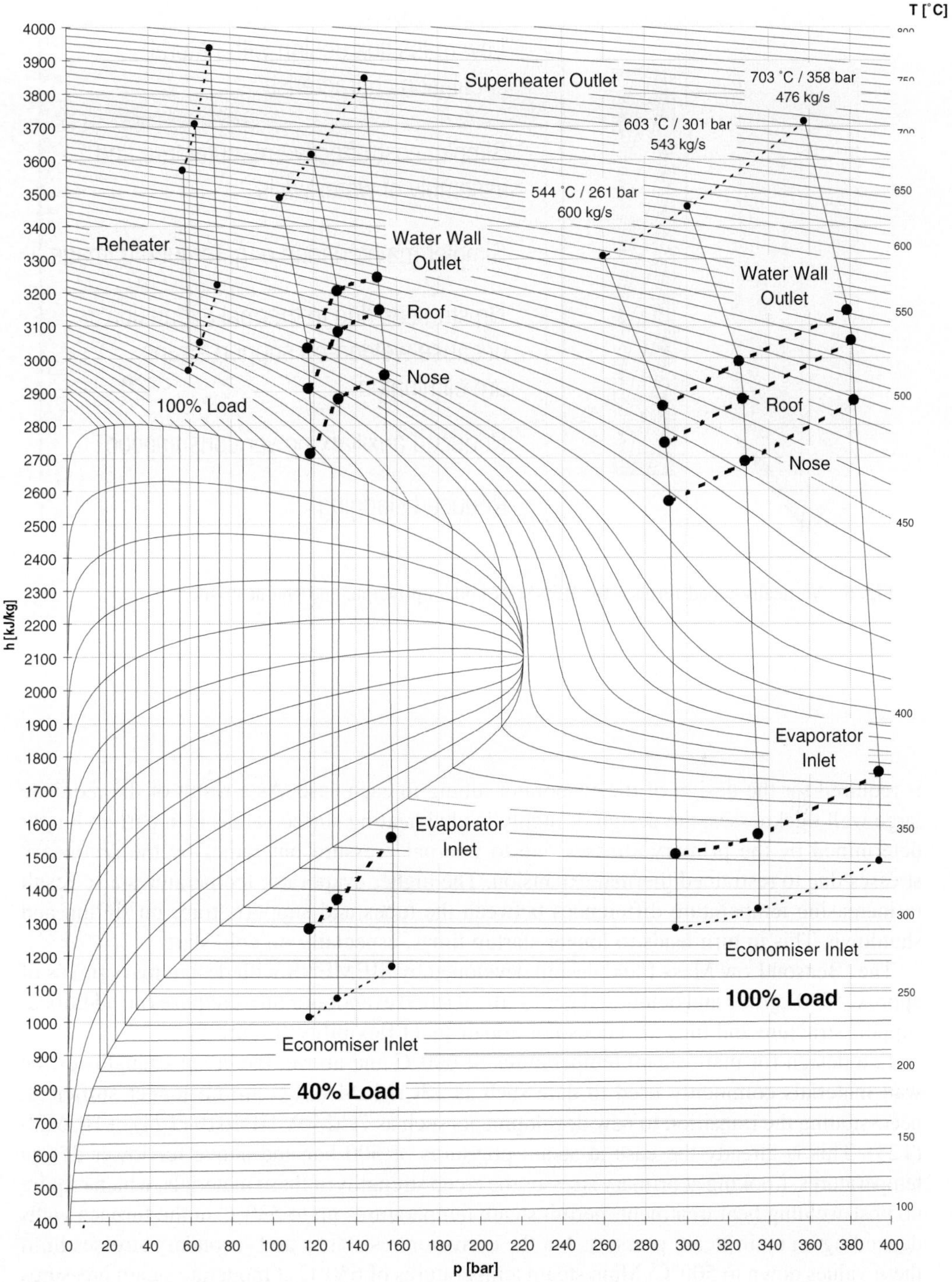

Fig. 4 Water and steam temperatures in the h-p diagram.

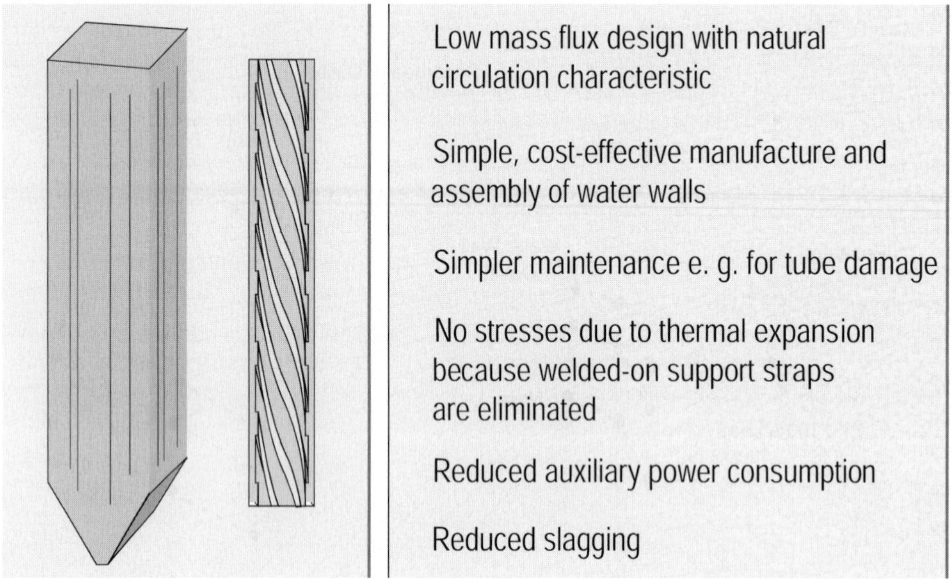
Low mass flux design with natural circulation characteristic
Simple, cost-effective manufacture and assembly of water walls
Simpler maintenance e. g. for tube damage
No stresses due to thermal expansion because welded-on support straps are eliminated
Reduced auxiliary power consumption
Reduced slagging

Fig. 5 Vertically-tubed furnace for BENSON boilers principle and characteristics.

is required for the design of the evaporator tubing in each case. As a result of the requisite large wall thicknesses, the design of highly loaded heating surface areas is in part no longer determined by the primary stresses due to internal pressure but rather by the secondary stresses due to restrained thermal expansion. The higher evaporator temperatures also result in increasing temperature differences between the tubes and support straps on startup and shutdown. This in turn leads to longer startup times, especially on cold start.

The 'Benson Low Mass Flux' design developed by SIEMENS with design mass fluxes of approx. 1000 kg/m²s and below and with vertical rifled evaporator tubes requires no additional support structure and thus also does not impair plant flexibility (Figure 5).

In a design for main steam temperatures of 600°C and above, the creep strengths of the wall materials commonly used to date such as 13CrMo44 (T12) are no longer sufficient, necessitating the transition to new developments such as 7CrMoVTiB1010 (T24) or HCM2S (T23). This is already the case at steam pressures of 300 bar and above for lower design temperatures. Looking at primary stresses the creep strengths of these materials, which require no post-welding heat treatment, permit steam temperatures up to 530°C in the furnace walls depending on main steam pressure, but the corrosion resistance and secondary stresses limit these values down to 500°C. Main steam temperatures of 630°C at moderate steam pressures are thus achievable as regards the walls.

At higher steam temperatures, materials such as HCM12 or T92 are required which must be heat-treated after welding. In order to minimize the manufacturing expenditure in such a design, the erection welds on evaporator tubes must be reduced to the absolute minimum possible. This is currently feasible only with vertical tubing. The relatively complex welds in the corners for spiral-wound furnace tubing are eliminated and the individual wall segments are welded together only at the fins. Welding of tubes may become necessary only in the horizontal plane. Solutions are also available for this which minimize expenditure on heat treatment on erection.

In all cases, it can be stated that the problems in the design of the water walls increase disproportionately with increasing steam pressures. A reduction of main steam pressure from 350 to 250 bar reduces the efficiency of a 700°C plant by 0.7% points but it also reduces the wall outlet temperature from 540 to 500°C and makes a design with materials without post weld heat treatment possible. Main steam pressures far above 250 bar should therefore be avoided, also in plants with high steam temperatures.

Evaporator/Superheater Dividing Point

At high steam parameters the water walls can no longer be designed entirely as an evaporator. The transition from evaporator walls to superheater walls then lies above the furnace. This transition must be designed so as to minimize the temperature differences between the evaporator and superheater sections of the walls which automatically result on water filling after shutdown, especially on water filling after an outage. Values of up to 80 K represent no cause for concern. For higher values such as can occur at very high steam conditions as well as in large furnaces, a flexible connection, not necessarily welded gastight, should also be taken into consideration for this transition.

Superheater Heating Surfaces

For steam temperatures up to approx. 550°C, all heating surfaces can be constructed of ferritic or martensitic materials, while at 600°C austenitic materials are necessary for the final superheater heating surfaces for both the HP section of the boiler as well as the reheater. In addition to the strength parameters, corrosion behaviour on the flue-gas and oxidation behaviour on the steam sides is especially determinative for material selection. Figure 6, Superheater materials for high temperatures, shows a selection of available materials. With regard to strength parameters, construction of superheater heating surfaces for steam temperatures up to 650°C is currently already feasible with austenitic steel materials. The corrosion resistance of the available materials however reduces the design limits to about 630°C.

Thick-Walled Components

In the first steam generators with very high steam temperatures, austenitic materials were used for the hot headers and connecting lines. However, the poor thermoelastic behaviour

	Maximum HP Steam Temperature limited by		Approved by
	Creep Rupture Strength*	Corrosion	
X3CrNiMoN1713	595	580	EN
AC66	605	620	VdTÜV
Esshete	615	580	VdTÜV / BS
TP 347 H (FG)	620	600	VdTÜV / ASME MITI
Super 304H (FG)	635	600	ASME / MITI
NF 709	645	620	MITI
HR 3C	630	630	VdTÜV / ASME MITI
Save 25	655	630	**under develop-ment / MITI**
Alloy 617 A130	685	720	**under development**

* 100 MPa at Steam Temperature +35K

Fig. 6 Available superheater tube materials.

low thermal conductivity, high thermal expansion render these materials unsuitable for boilers which are implemented in power plants with a large number of load changes and minimum startup times.

The development of chromium steels such as P91, P92 or E911 has enabled steam temperatures up to 620°C without the use of austenitic materials for thick-walled components. More recent developments such as NF12 and Save 12 could extend the limits of implementation at moderate main steam pressures up to 650°C in the near future.

With regard to the thick-walled components, especially for the main steam headers, it proves that the main steam pressures should more likely lie below 300 bar for optimum component utilization (Figure 7).

Effect on Operation

Power plants which are designed for fast load changes and short and frequent starts must necessarily be operated in sliding-pressure mode. Only then does the material loading of the turbine remain acceptable: in sliding-pressure operation usually between full load and 40% load - the temperature curve in the turbine remains nearly constant over the entire load

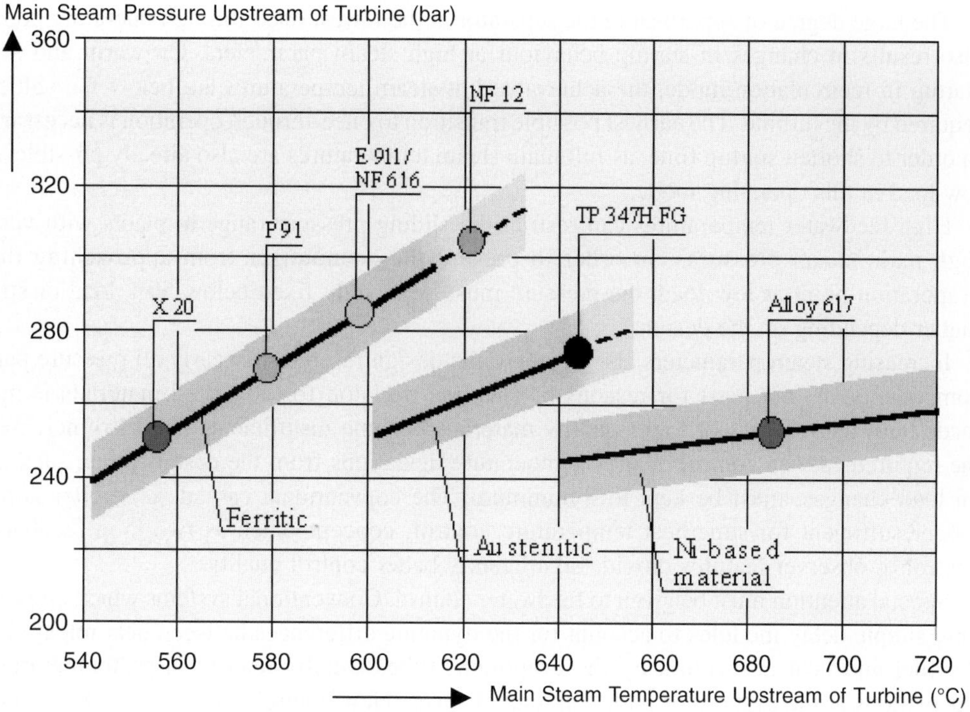

Fig. 7 Optimum main steam conditions with given main steam header dimensions.

range. These advantages for the turbine contrast with disadvantages for the boiler. For example, the temperatures in the water walls decrease from full load to part load by approx. 100 K. Due to their magnitude and the ordinarily larger wall thicknesses at the elevated steam parameters, the temperature changes during start up and load variations place increased requirements on the design of the thick-walled components such as multiple parallel passes, but also on the design of the tube walls, such as vertical tubing, in order to achieve similar startup times and load change rates to those in plants with conventional steam parameters.

With increasing steam parameters, the degree of superheat at the outlet of the evaporator sections of the water walls at the lowest once-through load point also increases. A high degree of superheat leads to a temperature reduction at the evaporator end and in the superheaters in the transition to recirculation mode. The separators are therefore moved as far as possible toward the burner zone. Operating measures to reduce the degree of superheat are increased excess air, flue-gas recirculation and use of the uppermost burner levels. The higher the steam temperatures and pressures become, the more important is the lowest possible load point in once-through operation, so that the once-through/recirculation mode transition need be traversed only on startup.

The large degree of superheat in the separator at the lowest once-through operating point also results in changes in startup behaviour at high steam parameters. On warm and hot startup in recirculation mode, the achievable hot steam temperatures are below the values required by the turbine. The earliest possible transition to once-through operation is necessary in order to shorten startup time, as full main steam temperatures are also already possible at low load in this operating mode.

High feedwater temperatures can restrict the sliding-pressure range in plants with very high main steam pressures. In order to prevent the economizer from approaching the evaporation point at low load, the pressure must be already fixed below 50% load or still higher depending on the design.

Increasing steam parameters also decrease the design reserves of nearly all pressure part components, as, not least for reasons of cost, the decision for advanced materials is not made until the reserves of lower quality materials become insufficient. This also increases the requirements on control quality: temperature deviations from the design value, such as on load changes, must be kept to a minimum. The conventional cascade controller is no longer sufficient for superheat temperature control; concepts such as two-loop feedback control or observer features provide significantly better control quality.

Special attention must be given to feedwater control. Conventional systems which employ only simple delay modules to account for the dynamic differences between heat release by the fuel and heat absorption by the evaporator tubes usually lead to large temperature fluctuations at the evaporator outlet on load changes. New control concepts which account for effects such as those of changes in the evaporator inlet temperature or the thermal storage capacity of the tube wall in the form feed forward control (Figure 8) increase control quality decisively and thus minimize the use of more expensive, higher-quality materials.

For high degrees of superheat at the lowest once-through load point, the transition from recirculation mode to once-through operation and back can no longer take place without delay due to the relatively large temperature change; the control must be adapted accordingly for a sliding transition.

OTHER EFFECTS

Design of the tube walls in particular is impeded by the high steam temperatures and pressures. The design parameters should be selected as best as possible so as not to necessitate the use of materials for which heat treatment must be performed after welding. A significant aspect for this is selection of the fuel. Coals with low ash deformation temperatures require large furnaces, associated with high heat input to the walls. A 100 K lower ash deformation temperature leads in a comparable boiler concept to a temperature increase at the wall outlet of about 25 K. Because of this for the currently available wall materials without post-welding heat treatment, the ash deformation temperature for a 600°C boiler may not be much lower than 1200° (Figure 9).

The implementation of flue-gas recirculation extraction of the flue gases if possible upstream of the air heater in order to reduce the negative effect on exhaust-gas temperature can shift the limits to higher steam parameters.

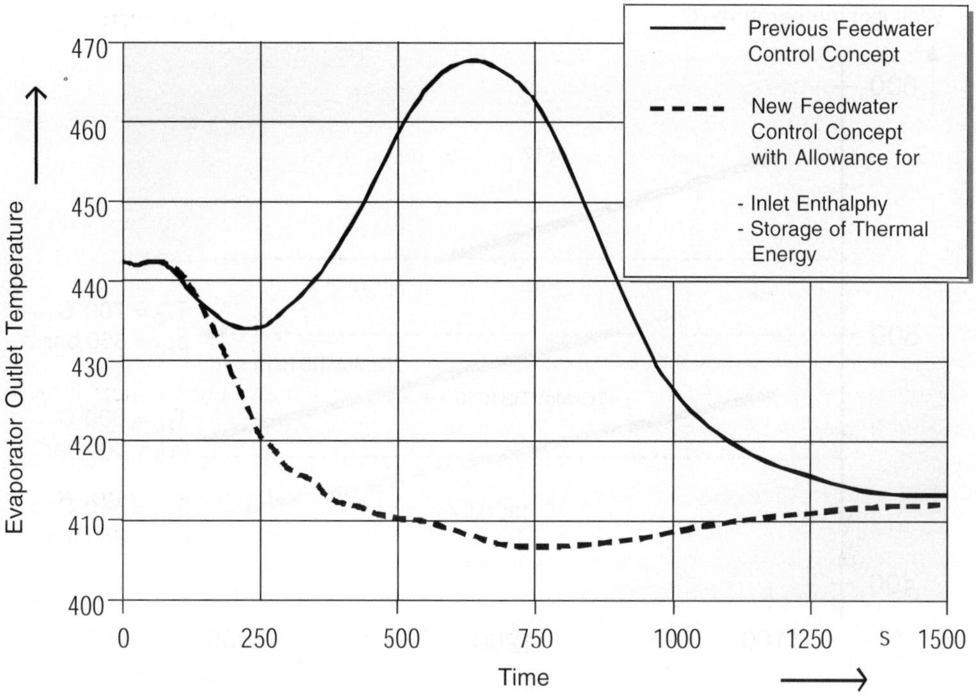

Fig. 8 Comparison of feed water control concepts load reduction from 100 to 50%.

Steam generators for power plants with high steam parameters and hence high plant efficiencies are consequently also designed for high boiler efficiencies. The lowest possible exhaust-gas temperatures 115 to 110°C can be achieved depending on the coal and lower excess air are prerequisites for this. Both of these factors lead to an increased heat input to the evaporator and thus impede the design of the wall heating surfaces.

The high tube wall temperatures of the superheater heating surfaces as well as lower excess air and low-NO_x firing systems increase the corrosion problem. For the selection of superheater materials the resistance to scale formation from the flue-gas atmosphere and steam is therefore just as important as creep resistance.

SPECIAL ASPECTS FOR CFB

The advantages of CFB technology are uncontested for low-grade fuels or for fuels with widely fluctuating quality as well as for low exhaust-gas emissions without post-combustion control measures. CFB plants up to capacities of 350 MWe are currently in operation. However, only once-through operation with high steam conditions render CFB technology

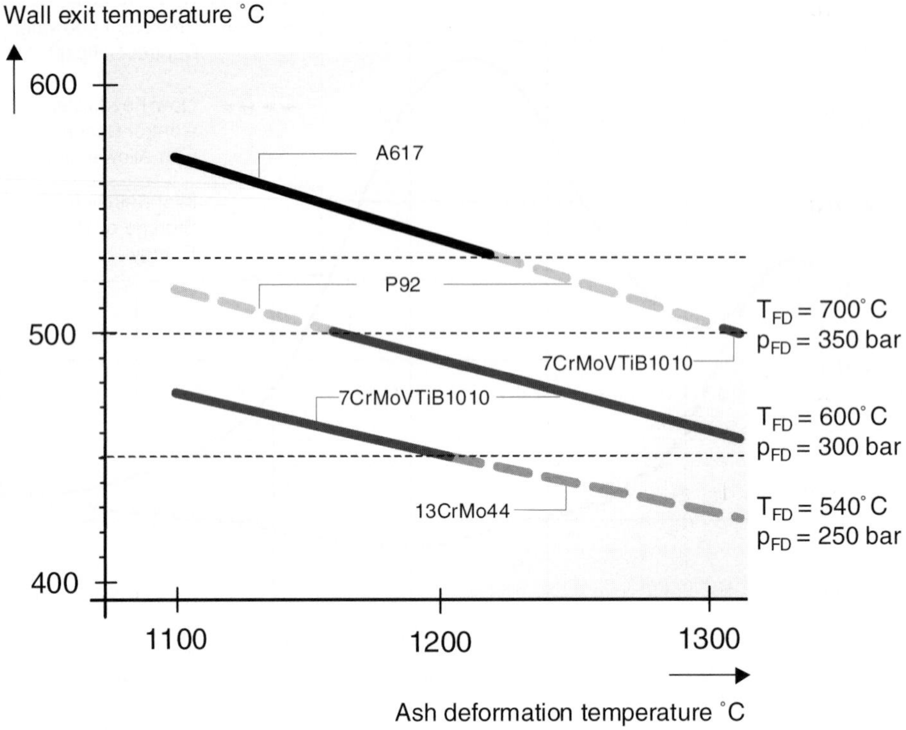

Fig. 9 Design limits for water wall materials.

serious competition for pulverized-coal firing. A plant for approx. 460 MWe with steam parameters of 560/580°C and 265 bar was developed in an EU research program. The BENSON 'Low Mass Flux' design was selected as the evaporator concept. It fulfills the requirements of a fluidized bed to a special degree: the tube orientation parallel to the flue gas/ash flow ensures low susceptibility to erosion, and temperature variations between the evaporator tubes are extensively prevented, as non-uniform heat inputs are evened out by the natural circulation flow characteristic of the low mass flux design. It also features an especially simple construction, as flow through all of the tubes in a single pass is parallel, thus eliminating the need for elaborate water/steam distribution.

The suitability of this evaporator system for sliding-pressure operation also fulfills all requirements for a power plant with regard to operating flexibility.

COMBINED-CYCLE PLANTS

Heat-recovery steam generators downstream of gas turbines are usually designed as drum boilers. Increasing exhaust-gas temperatures downstream of gas turbines as well as the increasing requirements on flexibility of a combined-cycle plant with frequent starts also make the use of once-through systems interesting here. Elimination of the drum on the one hand increases operating flexibility and on the other hand is a noticeable cost aspect. In the Cottam combined-cycle plant, a heat-recovery steam generator with a once-through evaporator based on the Benson "Low Mass Flux" design was constructed for the first time and runs successfully in commercial operation since Sept. 1999. This evaporator concept is characterized by extremely low mass fluxes which still lie far below those of fired boilers.

SUMMARY AND OUTLOOK

Steam temperatures of 600 to 620°C are currently possible as a result of efforts in materials development. However, not only are new materials necessary for higher temperature ranges, but further development was also necessary for the wall materials. On further temperature increases, previous design concepts can no longer be adopted without modifications. New designs are necessary for the evaporator in particular in order to give boilers for high-temperature plants similar flexibility to that of previous once-through boilers.

The Low Mass Flux Design provides an evaporator concept which meets the new requirements and which permits further development to higher steam parameters for pulverized-coal-fired boilers and for boilers with circulating fluidized bed firing as well as for heat-recovery steam generators downstream of gas turbines.

A further increase in steam temperatures appears possible in the next years with continuous materials development, but without using nickel based materials not more than 10 to 20 K. From the current standpoint, the jump to 700°C will not take place until the next decade. However, from an economic perspective, the high steam temperatures will only be selected given correspondingly competitive materials prices and if, among other things, the appropriate main steam and reheat pressures are selected and the fuel ranges are limited.

Design and Material Aspects for Advanced Boilers

GÜNTER SCHEFFKNECHT, QIURONG CHEN and GERHARD WEISSINGER
ALSTOM Power Boiler GmbH
Stuttgart
Germany

ABSTRACT

Since the 1950s the technology of coal-fired steam power plants has experienced significant developments. The main improvements have been focused on higher efficiencies, lower generation costs, higher reliabilities and lower emissions. In order to increase the efficiency and reduce the emissions, supercritical and ultra-supercritical technology with continuously higher steam parameters are used which will represent also the major development trend in the near future. The supercritical cycle is realised by means of the well-known and since the 1960s commercially used once-through technology. This avoids the introduction of a totally new clean coal technology which may have some substantial risks. Only material selection and dimensioning of the pressure parts have to be adapted to the new requirements, and the design and construction have to consider the limitations on the material side.

The increase of the steam parameters is limited on the boiler side by water walls, tubing of the final superheaters and reheaters as well as the thick-walled components, mainly the high-pressure outlet headers and the piping to the turbine. Several boiler materials with improved mechanical properties have been developed in the last years. Some new materials are still in the stage of development. Besides the aspects of mechanical properties and workability of the materials, attention must be paid to the corrosion and oxidation behaviour at high temperatures. In order to investigate the suitability of the various tube materials in high-temperature areas, several test installations are now in operation to forecast the corrosion and oxidation properties at high temperatures.

The paper shows the design and material aspects for advanced boilers with ultra-supercritical conditions. It reviews the available boiler materials with respect to mechanical and corrosion properties and gives limitations on the maximum achievable steam conditions. Finally, a forecast for further increase of the steam conditions considering pending and future material developments will be given.

INTRODUCTION

An improvement in the thermal efficiency of the power plant does not only reduce the fuel costs but also reduces the release of SO_2, NO_x and CO_2 emissions. Despite its long history, coal-fired power plant technology is still evolving, and efficiency has been improved significantly since the late 1980s and will be improved still further. The increase of steam parameters to ultra-supercritical conditions is the main development trend to improve the power plant efficiency. Figure 1 shows the evolution of the efficiency improvement of the

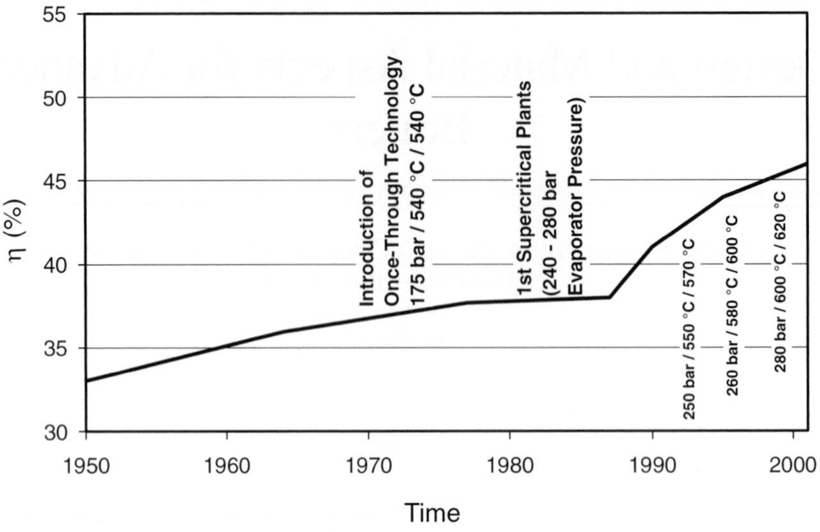

Fig. 1 Efficiency development of power plants in europe.

power plants. In order to realise the supercritical conditions, the well-known and since the 1960s commercially used once-through technology has to be applied to the boiler design. As the components of the power plants being operated with this technology are well-proven, higher steam conditions do not lead to a new boiler design. Only material selection and dimensioning of the pressure parts have to be adapted to the new requirements. Figure 2 shows a typical coal-fired once-through boiler with supercritical conditions and a power output of 900 MW.

Most coal-fired units which were built in the 1980s were characterised by moderate steam conditions. The steam parameters of the power plants which are currently realised are most supercritical with much higher steam temperatures. In Germany, a lignite-fired power plant with steam parameters of 275 bar/580/600°C started successfully its commercial operation.[1] In Denmark, two power plants with steam parameters of 290 bar/582/580/580°C (double reheat system) have been built.[2] In Japan, some new power plants with steam parameters up to 250 bar/600/610°C have been commissioned in recent years.[3] For the next years, different development programs in Europe and Japan will lead to the planning and erection of ultra-supercritical power plants with steam pressures of about 300 bar and steam temperatures above 600°C.

In an on-going research project within the framework 5 (former THERMIE) programme which is funded by the European Commission, investigations are performed for an advanced power plant with the maximum steam parameters of 375 bar/700/720°C which are significantly higher in comparison with the state of the art power plants. The efficiency of the advanced power plants will be increased from the recent range of 43–47% to above 50%.

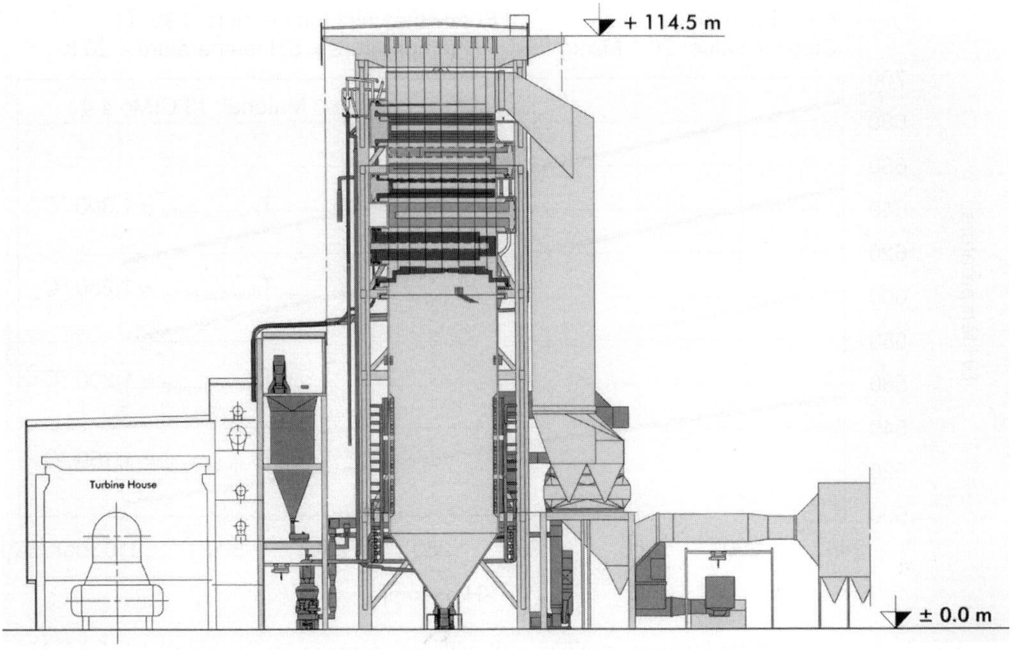

Fig. 2 A coal-fired once-through boiler.

Due to the higher thermal efficiency of the power plants, the fuel consumption and emissions will be reduced by almost 15% compared with the best plant currently available.

By increasing the steam parameters, the design margin for some boiler components will be more and more exhausted. A further increase of steam conditions is limited especially by water walls, the tubing of the final superheaters and reheaters as well as the thick-walled components, mainly the high-pressure outlet headers and the piping to the turbine.

Several boiler materials with improved properties have been developed in the last years. The use of the materials for the boiler with advanced steam parameters will be first limited by the relevant creep strength specifications. The range of application of the materials for water walls and thick-walled components is additionally restricted by the manufacturing aspects. For superheater and reheater heating surfaces, further restriction results from the risk of flue gas side high temperature corrosion and steam-side oxidation. For advanced steam conditions, new materials with higher strength, higher corrosion resistance and good weldability and formability are required.[4]

WATER WALLS

The water wall as flue gastight enclosure of the furnace and the heat exchanger section is a very important component of the boiler. The conventional water wall materials used so far

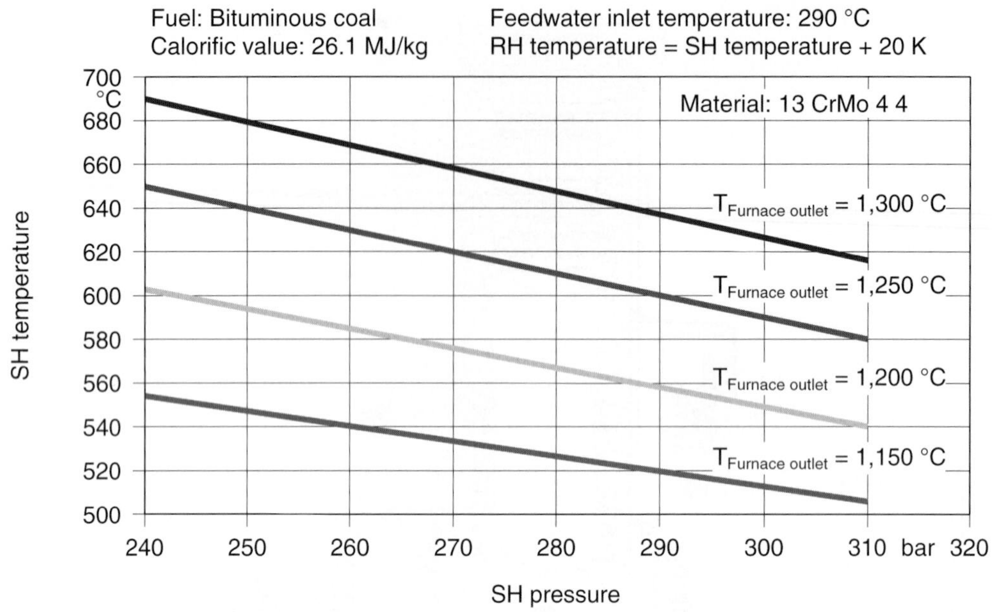

Fig. 3 Steam parameter limits at specified wall outlet temperature for the material 13CrMo44 (pressure and temperature refer to operating conditions at boiler outlet).

are low-alloy steels like 15Mo3 and 13CrMo44 (T12) which are characterised by an excellent workability for the manufacturing of the membrane walls.

Figure 3 shows the application limit of the material 13CrMo44. Based on a furnace outlet temperature of 1250°C and certain fuel properties, the steam conditions up to 290 bar/600°C (SH outlet) can be realised. This corresponds to a water wall outlet temperature of about 460°C which is shown in Figure 4.

From a design point of view, two aspects can be considered for higher steam conditions:

1. A possibility to reduce the steam temperature on the water wall is the increase of the furnace outlet temperature. If the furnace outlet temperature is increased to 1300°C, the limit of steam parameters is increased to 300 bar/620°C (SH outlet). As a consequence, however, the range of the applicable coals would have to be limited.
2. Another possibility to shift the heat absorption from the water wall to the convective heating surfaces is the application of flue gas recirculation. A cold flue gas recirculation of 10% leads to a reduction of the wall outlet temperature of approximately 20 K. This method leads, however, to very large air preheaters or it may be even impossible to reach the desired low exhaust gas temperature.

On the material side, new water wall materials with much higher creep strengths can be used for higher steam conditions. With regard to the manufacturing and erection aspects, materials which do not need any post-weld heat treatment should be preferred. In this case,

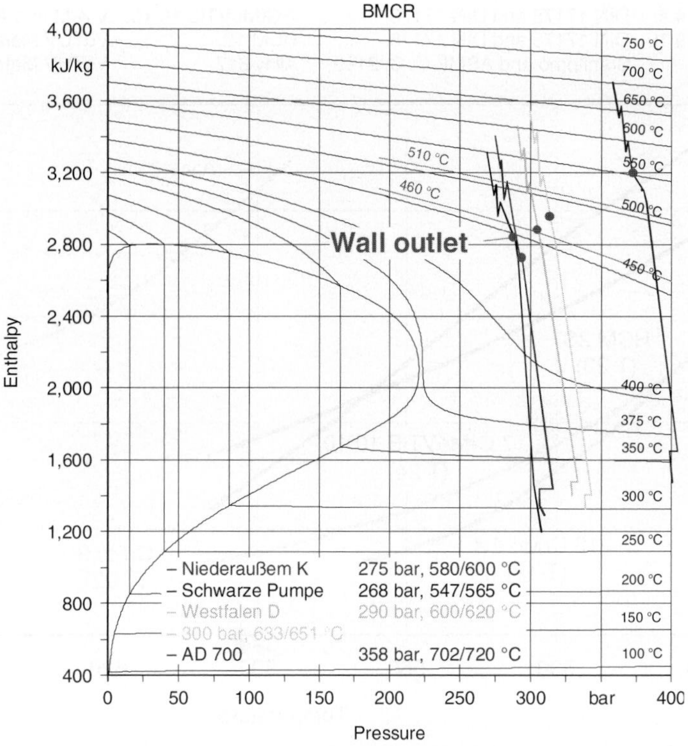

Fig. 4 h, p-diagram of different boilers.

the maximum hardness levels in the as-welded condition have to be limited to avoid any risk of hydro-gen-induced stress-corrosion cracking. Table 1 shows the chemical composition of the candidate water wall materials in comparison with conventional ones. The 100,000 h creep rupture strengths of the materials are shown in Figure 5.

With the material 10CrMo910 (T22), the steam parameters cannot be increased considerably. A substantial improvement of the creep strength with limited hardness levels in as-welded conditions can be reached by the newly developed 2–2.5% chromium steels. The typical examples of this group of steels are the material HCM2S (T23) and 7CrMoVTiB1010 (T24) which have been intensively investigated by different German research programs. By use of the T24 or T23 steels, the steam temperature limit at the water wall can be increased by approximately 50 K in comparison with the 13CrMo44 steel (see Figure 4). For a furnace outlet temperature of 1250°C, the steam conditions of 300 bar/ 640°C (SH outlet) can be realised.

For very high steam conditions such as 375 bar and 700°C at boiler outlet with maximum metal temperatures of above 600°C in the water wall, the 2–2.5% chromium steels are not adequate due to insufficient creep strength and lower corrosion and oxidation resistance. In order to fulfil the strength and corrosion requirements, higher alloyed materials are needed.

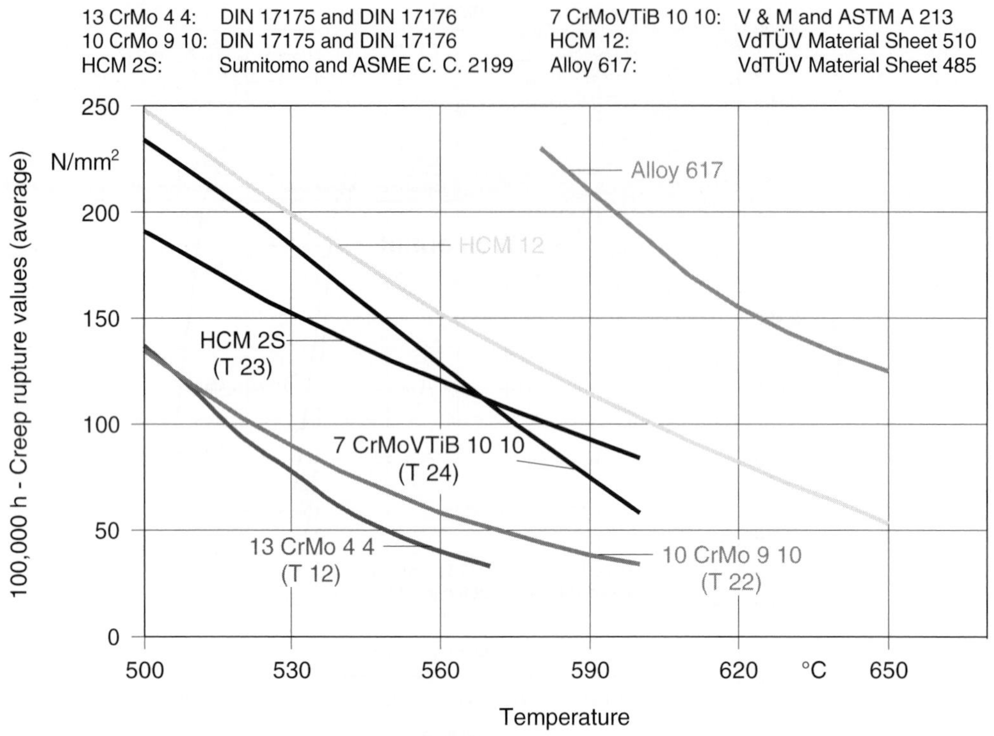

Fig. 5 100,000 hours creep rupture strength for water wall materials.

Table 1 Chemical composition of water wall materials.

Material	C	Cr	Mo	W	Ti	Co	Others
1% Cr-Steels:							
13CrMo44 (T12)	0.10–0.18	0.70–1.10	0.45–0.65	-	-	-	-
2–2.5% Cr-Steels:							
10CrMo910 (T22)	0.08–0.15	2.00–2.50	0.90–1.20	-	-	-	-
T23	0.04–0.10	1.90–2.60	Max. 0.30	1.45–1.75	-	-	V, Nb, N, B
T24	0.05–0.10	2.20–2.60	0.90–1.10	-	0.05–1.10	-	V, N, B
9–12% Cr-Steels:							
HCM12	Max. 0.14	11.0–13.0	0.80–1.20	0.80–1.20	-	-	V, Nb
Ni-Based Alloys:							
Alloy 617	0.05–0.10	20.0–23.0	8.0–10.0	0.80–1.20	0.20–0.50	10.0–13.0	Ni, Al

The austenitic steels cannot be used as candidate water wall materials because of their large thermal expansion coefficient which inhibits a welded joint between austenitic and ferritic water wall sections. The candidate materials can be first selected from the group of modified 9–12% chromium steels like T91, T92 and HCM12. Under the European COST program, it was demonstrated that the material HCM 12 is more promising in that it has a sufficient creep strength and higher corrosion and oxidation resistance, and the hardness value in as-welded condition is the lowest among the 9–12% chromium steels. However, it has to be investigated in more detail whether a post-weld heat treatment is needed for the manufacturing of the water walls. Another group of candidate materials for the water wall are the nickel-based alloys for which a post-weld heat treatment is not necessary. A well-examined nickel alloy is the material Alloy 617 which was originally developed for gas turbine applications. The alloy is used in the solution-treated condition and has the advantage of being relatively easy to weld without any requirement for complex post-weld heat treatment. Other advantages of this material are very high creep rupture strengths, high corrosion and oxidation resistance and similar heat expansion coefficients as martensitic steels. However, the Ni-based alloys are expensive. In the new framework 5 programme, an evaporator panel made of HCM12 and Alloy 617 will be manufactured and installed in an existing boiler to gather the manufacturing and operating experience.

SUPERHEATER AND REHEATER TUBES

For superheater and reheater tubes, the creep strength of the used materials should be high enough at the relevant pressure and temperature range. In addition to this requirement for high strength, the corrosion resistance of the materials both on the flue gas side and on the steam side has to be considered. The oxide layer on the inside of the tube leads to higher material temperatures and could cause creep damage. External high-temperature corrosion on the flue gas side reduces the effective wall thickness of the tubes.

In the design of boilers with very high temperatures, it should be considered that the reheater outlet group is more critical regarding the corrosion and oxidation risk due to the often higher steam temperatures and the lower heat transfer coefficients. Therefore, it should be arranged on the flue gas side downstream of the final SH heating surfaces.

On the material side, the chemical composition, especially the chromium content has a large influence on the flue gas side corrosion and oxidation behaviour . The resistance against corrosion and oxidation increases with increasing chromium content. Table 2 gives the chemical composition of some typical superheater materials. The 100,000 hours creep rupture values of the materials which are taken from the relevant standards or based on manufacturer's specifications are shown in Figure 6.

Both due to the requirement for creep strength and corrosion resistance, the martensitic steels can be only used up to steam temperatures of approximately 560–570°C. For higher steam temperatures, higher-alloyed austenitic steels have to be used.

In the group of 18% chromium steels, the steels TP347HFG, Tempaloy A-1 and Super304H are new developments. All three steels have higher creep strengths than the conventional 18Cr-8Ni steels. The steam oxidation resistance at the tube inside is also increased due to a fine-grained structure. For even higher steam temperatures, the 18% chromium steels could

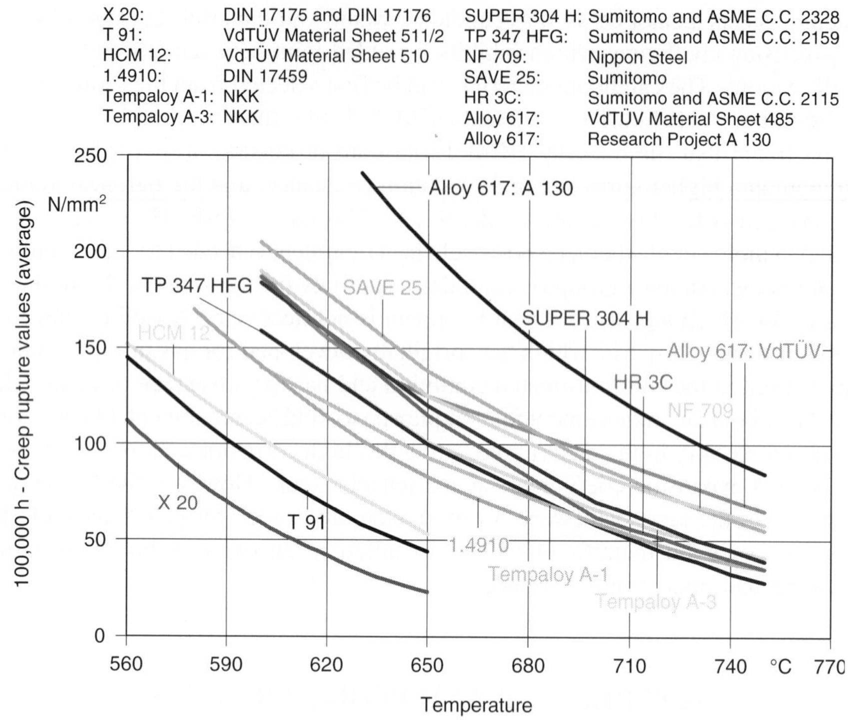

Fig. 6 100,000 hours creep rupture strength for superheater and reheater materials.

Table 2 Chemical composition of superheater and reheater materials.

Steel	Cr	Ni	Mo	Nb	Ti	Co	Others
Martensitic Steels:							
X20CrMoV121	10.0–12.5	0.30–0.80	0.80–1.20	-	-	-	V
T91	8.0–9.5	Max. 0.40	0.85–1.05	0.06–0.10	-	-	V, N
HCM12	11.0–13.0	-	0.80–1.20	Max. 0.20	-	-	V
Austenitic Steels:							
X3CrNiMoN1713	16.0–18.0	12.0–14.0	2.0–2.8	-	-	-	N
AC66	26.0–28.0	31.0–33.0	-	0.60–1.00	-	-	Ce, Al
Super304H	17.0–19.0	7.5–10.5	-	0.30–0.60	-	-	Cu, N
Tempaloy A-1	17.5–19.5	9.0–12.0	-	Max. 0.40	Max. 0.20	-	-
TP347HFG	17.0–20.0	9.0–13.0	-	Max. 1.0	-	-	-
NF709	18.0–22.0	22.0–28.0	-	0.10–0.40	0.02–0.20	-	N, B
Tempaloy A-3	21.0–23.0	14.5–16.5	-	0.50–0.80	-	-	N, B
SAVE25	21.0–24.0	15.0–22.0	-	0.30–0.60	-	-	W, Cu, N
HR3C	24.0–26.0	17.0–23.0	-	0.20–0.60	-	-	N
Ni-based Alloy:							
Alloy 617	20.0–23.0	Rest	8.0–10.0	-	0.20–0.50	10.0– 13.0	Al

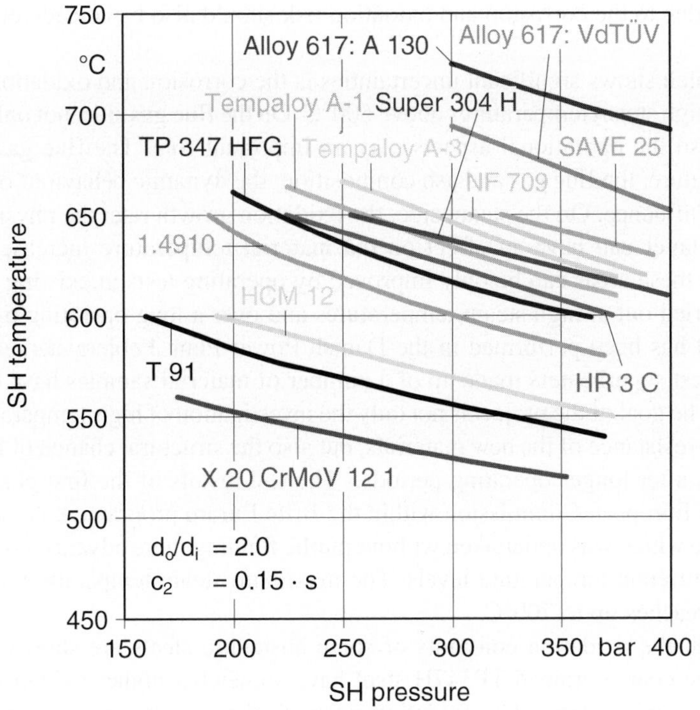

Fig. 7 Steam parameter limits for different superheater materials from the strength point of view (pressure and temperature refer to operating conditions at boiler outlet, design temperature = operating temperature + 35 K).

be insufficient in view of strength and corrosion resistance. In this case, higher-alloyed materials with chromium contents above 20% for which typical examples of the newly developed steels are HR3C, Tempaloy A-3, NF709 and SAVE 25 should be used.

For very high steam conditions such as 375 bar and 700 °C at boiler outlet the austenitic steels are not adequate due to insufficient creep strengths. Here, nickel-based alloys are needed. The well examined Alloy 617 can be used. In an on-going German research project, this material should be further optimised in the chemical composition to achieve higher creep strength values. Parallel to this project, a new nickel based alloy with the name Alloy 740 will be developed in the on-going European framework 5 project on the basis of the material Alloy C263 which must be aged after solution treatment to produce the strengthening dispersion and thus needs more complex post-weld heat treatment. The aim is to have an average 100,000 hours creep strength value of 100 MPa at 750°C.

The limit of steam conditions which can be realised with the materials from the point of view of creep strength is given in Figure 7. In this diagram, a corrosion allowance of 15% of the wall thickness has been considered. Besides the creep strengths, the application limits of the materials due to the corrosion and oxidation risk should also be considered in the design calculation.

A topic which shows significant uncertainties is the corrosion and oxidation mechanism, especially at high steam temperatures above 600°C. On the flue gas side, not only the material aspect, but also the operational aspects play an important role. The flue gas temperature, steam temperature, the flue gas and ash composition, the dynamic behaviour of the plant are all factors of influence. On the steam side, the oxidation growth rate, the physical properties of the oxide layer can have an effect on the material temperature increase. The lack of knowledge in these areas can be only improved by operating tests in existing boilers which should be carried out at high steam temperatures and over a long operating time.

Such a test has been performed in the Danish Power Plant Esbjergvaerket Unit 3 since 1995. Many test superheaters made up of a number of material samples have been installed in the boiler. The goal of the project is not only the investigation of high-temperature corrosion and oxidation resistance of the new materials, but also the structural change of base materials and the welds after longer operating periods.[5] The test results of the first phase which was funded by the European Commission within the Brite Euram programme are published.6 In a further phase which was undertaken without public funding more advanced boiler materials are tested at different temperature levels. The maximum steam temperature at the outlet of the test loop reaches up to 700°C.

In Figure 8, the oxidation constants of some austenitic steels are shown. The material 1.4988 and the coarse-grained TP347H steel have somewhat higher oxidation rate among investigated austenitic steels. The oxidation rates of other austenitic steels are significantly lower than expected. A considerable difference between the fine-grained steels TP347HFG and Super304H as well as the higher-alloyed material NF709 cannot be found.

HIGH-PRESSURE OUTLET HEADERS AND PIPING

By increasing steam parameters, the wall thickness of pressure parts will increase. The large wall thickness of thick-walled components limits the allowable rate of temperature change. This leads to disadvantages in the operation of the power plants.

A possibility to reduce the wall thickness of high-pressure outlet headers and piping is the increase of parallel lines of the headers to reduce the inside diameter.

From the material point of view, the wall thickness can be reduced by use of materials with higher creep strengths. As candidate materials for thick-walled components, main attention has been paid up to now to the improvement of the modified 9 – 12% chromium steels. The main reason is to avoid using austenitic steels for steam parameters expected in the near future. Table 3 gives the chemical composition of the new materials in comparison with conventional ones. The 100,000 hours creep rupture values are shown in Figure 9.

In Europe, the conventional steel for thick-walled components has been for a long time X20CrMoV121. Extensive operating experience is available in the steam temperature range up to 560°C. With an increase of the steam parameters, the limit of the steel X20CrMoV121 is quickly reached. The next development led to the application of the material P91 which can be meanwhile regarded as well-examined material. In a further development, the creep rupture values of the new steels are further improved through the addition of tungsten. The

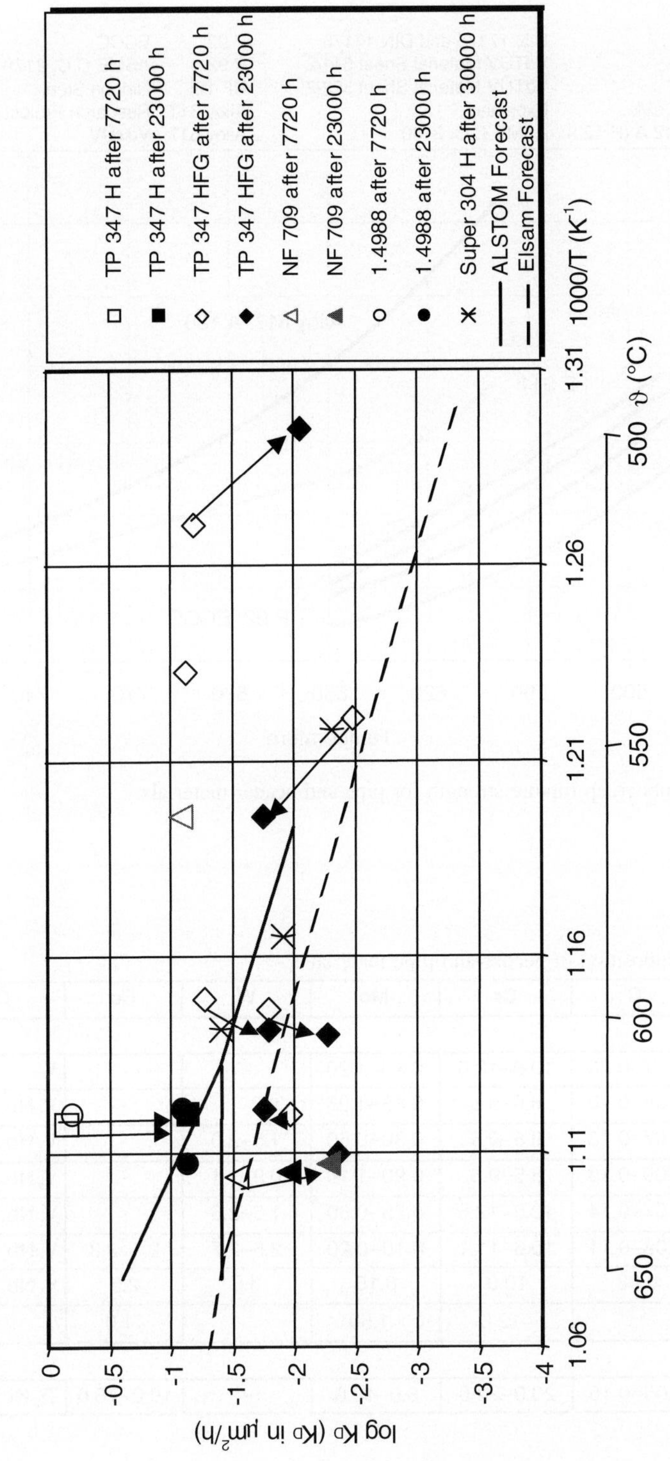

Fig. 8 Oxidation constants for austenitic materials.

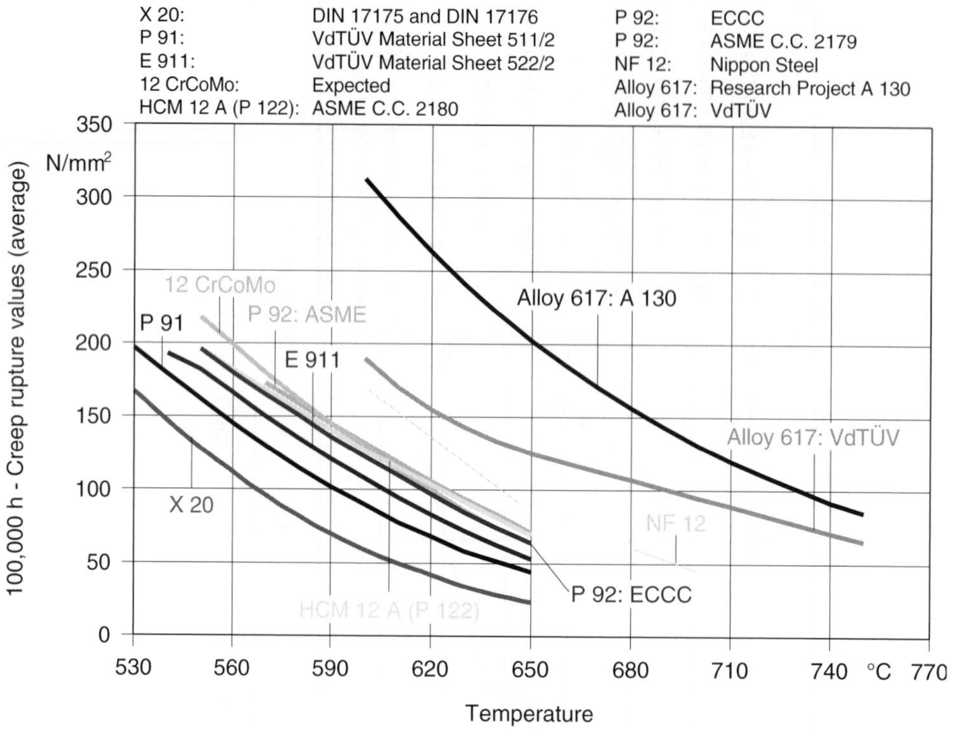

Fig. 9 100,000 hours creep rupture strength for pipe and header materials.

Table 3 Chemical composition of header and pipe materials.

Steel	C	Cr	Mo	W	Co	Others
Martensitic steels:						
X20CrMoV121	0.17–0.23	10.0–12.5	0.80–1.20	-	-	V
P91	0.08–0.12	8.0–9.5	0.85–1.05	-	-	V, Nb, N
P92 (NF616)	0.07–0.13	8.5–9.5	0.30–0.60	1.5–2.0	-	V, Nb, N, B
E911	0.09–0.13	8.5–9.5	0.90–1.10	0.9–1.1	-	V, Nb, N, B
P122 (HCM12A)	0.07–0.14	10.0–12.5	0.25–0.60	1.5–2.5	-	V, Nb, N, B, Cu
NF12	0.08–0.11	10.6–11.1	0.10–0.20	2.5–2.7	2.4–2.8	V, Nb, N, B
SAVE12	0.08	10.0	0.15	3.0	2.5	V, Nb, N, B, Cu
12CrCoMo		~12.0	~1.50		~1.0	
Ni-Based Alloys:						
Alloy 617	0.05–0.10	20.0–23.0	8.0–10.0	-	10.0–13.0	Ti, Ni

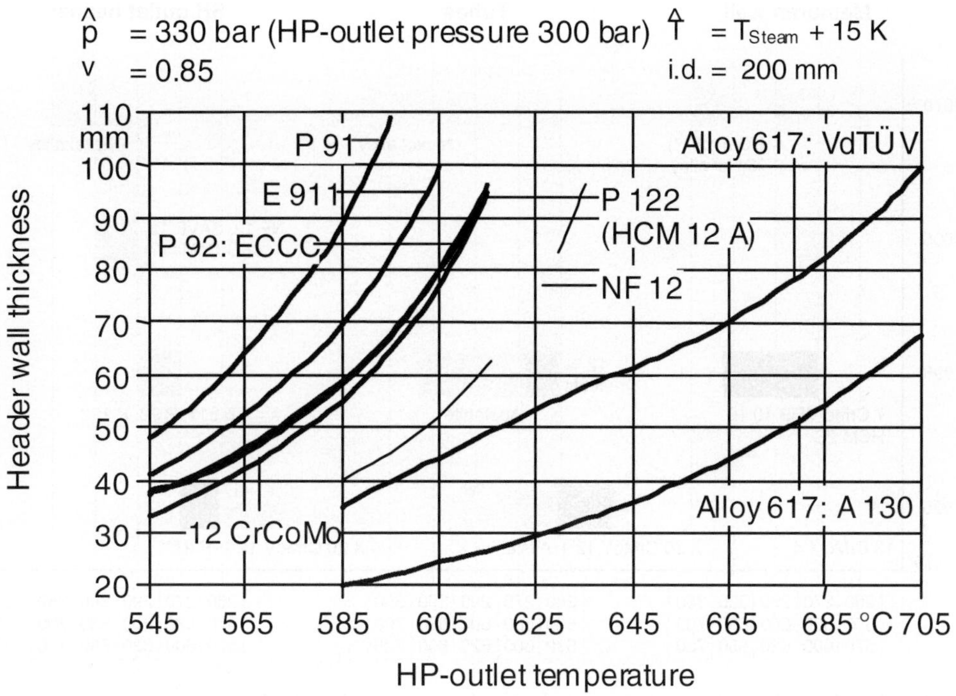

Fig. 10 Wall thickness of header materials for different steam conditions (pressure and temperature refer to operating conditions at boiler outlet).

typical materials from this new development are the tungsten-alloyed chromium steels E911, P92 and P122 (HCM12A). The steel E911 developed during the European research project COST is standardised in a VdTÜV material sheet. For other steels P92 and P122, ASME code has been approved.

The recent developments on the martensitic steels are the 12Cr-Co steels NF12 , SAVE12 and the 12CrCoMo steel which is being developed by COST 522 programme. All materials are in a very early development stage. For all materials, even higher strength values than the above mentioned 9–12% Cr-steels can be expected. But no long-term strength values are available up to now.

For the very high steam conditions of 375 bar/700°C, nickel-based alloys like Alloy 617 and Alloy 740 are needed. For such high steam parameters, the load change rate of the thick-walled component is very low. The power plants should be operated preferably as base-load units.

Figure 10 shows the wall thickness of the final SH outlet header for different materials at a pressure of 300 bar. If the wall thickness is limited by 100 mm which represents an outer diameter to inner diameter ratio of 2, the steam parameter limit lies approximately at

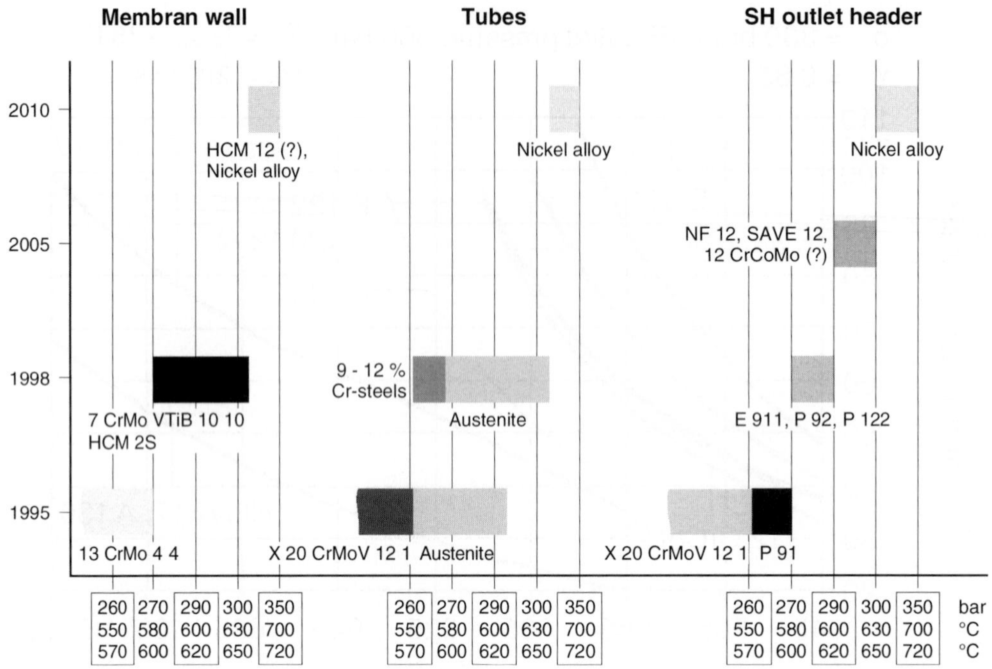

Fig. 11 Material development stages and related steam parameter limits.

300 bar/610°C (SH outlet) by using the material P92 or P122 and at 300 bar/700°C by using Alloy 617 (based on the current VdTÜV values). If the wall thickness should be smaller, the maximum steam conditions are accordingly lower. After upgrading of the Ni-based alloy 617, the wall thickness of the headers can be reduced by approximately 30% at the same steam parameters.

SUMMARY

In order to increase the power plant efficiency, once-through technology with advanced steam parameters can be used for pulverised coal-fired boilers. The critical components at high steam conditions are the water walls, the final stages of superheater and reheater tubes as well as high-pressure outlet headers and piping.

Figure 11 shows the different material development stages and related steam parameter limits. By use of the newly developed materials which have been qualified in the 90s, coal-fired power plants with the steam parameters 290 bar/600/620°C can be realised. Various on-going R & D projects like COST 522 etc. will push the steam conditions to higher than 600°C.

Table 4 Research projects for boiler materials in Germany.

Research Projects	Time Schedule	Materials	Boiler Components
A77 A129	01.1994–12.1998 01.1998–12.2001	7CrMoVTiB1010, E911	Waterwall, Piping, Header
A109/A170 (MARCKO DE1)	06.1996–06.2000 01.2001–12.2002	T91, E911, 1.4910	Header, Tubing, Dissimilar Welds
A130 (MARCKO DE2)	04.1999–12.2004	Inconel 617, Alloy 617	Header, Tubing
A180	04.2002–03.2006	12CrCoMo	Waterwall, Header, Tubing
KOMET 650	01.1998–12.2004	Austenites, Ni-Based Alloys	Header, Piping, Tubing
Collab. Com. f. Heat-Resisting Steels and High-Temp. Materials	01.1998–12.2004	Heat-Resisting Materials, Austenites, Ni-base Alloys	Waterwall, Header, Piping, Tubing, Valve
A152	07.2000–06.2003	P91, E911	Piping
FDBR/VGB Joint Project	09.2000–06.2002	Bends	Piping
A176	01.2002–12.2004	Internally Coated Tubes	SH/RH Tubes
A196	07.2002–06.2005	Longitudinally Welded Tubes, P91 and E911	Header, Piping
VGB-FDBR-AiF Project	03.2003–02.2006	Cold Forming of Austenites a. Ni-Based Alloys	SH/RH Tubes
MARCKO DE2	03.2003–02.2007	7CrMoVTiB1010, P92, 12CrCoMo, Alloy 617	Waterwall, Header, Piping

In Table 4, an overview of research projects which are performed in Germany to qualify the materials for the application in boilers is listed.

It should be mentioned that for the steam parameters 300 bar/630/650°C currently no candidate materials can be regarded as promising materials for the SH outlet headers and piping system. The 12% Cr-steels NF12, SAVE 12 and 12CoCrMo could have sufficient creep rupture strength (no long-term material properties are up to now available), but the oxidation resistance of these materials may not be high enough at such high temperature level.

A completely new perspective is opened with the application of Ni-based alloys which marks the next development step for advanced coal-fired power plants.7 A major efficiency increase can be reached through a significant rise of steam parameters from the currently available 290 bar/600/620°C to 350 bar/700/720°C. The realisation of such an advanced 700 °C power plant can be expected in the next 10 years.

REFERENCES

1. T. Tippkötter, M. Schütz and G. Scheffknecht: 'Start-up and Operational Experience with the 1000 MW Ultra-Supercritical Boiler Niederaussem in Germany', *Power-Gen Europe 2003*, Düsseldorf, Germany, 2003.

2. ELSAM: 'Highest Supercriticality for Skaerbaek and Nordjylland', *Modern Power Systems*, 1995.
3. K. MURAMATSU: 'Development of Ultra Supercritical Plant in Japan', *Advanced Heat Resistant Steels for Power Generation*, R. Viswanathan and J. W. Nutting, eds., IOM Communications Ltd., London, 1999, 543–559.
4. G. SCHEFFKNECHT, Q. CHEN and A. KATHER: 'Dampferzeuger Mit Fortschrittlichen Dampfparametern', *VDI-GET Conference Entwicklungslinien der Energie- und Kraftwerkstechnik*, Siegen/Germany, 1996.
5. G. SCHEFFKNECHT, R. BLUM, Q. CHEN and A. VANDERSCHAEGHE: 'Ein Vorhaben zur Ermittlung der Hochtemperaturkorrosionseigenschaften Verschiedener Dampferzeugerwerkstoffe unter Betriebsbedingungen und Dampftemperaturen bis 620°C', *VGB Kraftwerkstechnik*, **76**, 1996, 856–862.
6. R. BLUM, Q. CHEN, C. COUSSEMENT, J. GABREL, C. TESTANI and L. VERELST: 'Betriebsversuche von Überhitzerwerkstoffen bei hohen Dampftemperaturen in einem steinkohlegefeuerten Dampferzeuger', *VGB-Conference Research for Power Station Technology 2000*, Düsseldorf/Germany, 2000.
7. S. KJAER, F. KLAUKE, R. VANSTONE, A. ZEIJSEINK, G. WEISSINGER, P. KRISTENSEN, J. MEIER, R. BLUM and K. WIEGHARDT: 'The Advanced Supercritical 700°C Pulverised Coalfired Power Plant', *Power-Gen Europe 2001*, Brussels/Belgium, 2001.

Yaomeng — The First Full Scale Application of Optimised Rifled Tubing in a Once-Through Boiler

M. J. Smith, J. E. Jesson, C. -H. Chen and D. -M. Finch
Mitsui Babcock

BACKGROUND

Yaomeng Power Plant is situated in Henang Province near the city of Pingdingshan in the People's Republic of China. It is operated by Yaomeng Power Generation, Ltd. (YPGL). Unit 1 was the first Chinese designed 300 MWe coal-fired unit and, as such, is well-known and holds a special place in the history of power generation in China. The boiler is a subcritical once-through type. It was designed with a twin-cell furnace with division wall for tangential indirect firing and top-supported from steel beams set on a concrete structure – a feature significantly affecting the final solution.

In a bid to control tube metal temperatures, the original furnace was constructed using a high fluid mass in vertical tubes. In this design, the fluid made two passes through the unit, starting with the division wall panels and ending with the outside walls to cover the periphery of the furnace cells. Each circuit had flow balancing valves that enabled the operators to manually pre-set water flow rates. The design, however, proved intolerant to load changes and made start-up difficult due to flow instability in some tubes. This resulted in tube overheating and numerous failures. Also, the high temperature differentials between adjacent tubes created undue stress in headers which led to reduced life and failure. The boiler was designed for steam temperatures of 570°C, but due to incorrect material selection in the superheater and reheater the steam temperatures were limited to 545°C, which meant that the unit could only reach 270 MWe output.

In pre-conversion tests, furnace tube metal temperatures were monitored at different loads. Differential temperatures between some adjacent tubes were up to 70 K. At loads below

230 MWe, overheating of the tubes was observed. At 200 MWe, metal temperatures reached 450°C — significantly above the design limit of 400°C.

Thermo-hydraulic analysis of the existing boiler concluded that the main reason for the poor performance was due to the high fluid mass flux employed in the design. It showed that dynamic losses accounted for a very large proportion of the total pressure losses. This created a negative flow characteristic; i.e., an increased heat flux to the tube resulted in a lower fluid flow and with the lower flow, the metal temperatures became elevated.

The operational restrictions imposed by the above and the resulting poor availability led YPGL to solicit proposals for the design and supply of a new furnace concept. Several options were available for the furnace retrofit:

Natural Circulation: Conversion to a natural circulation unit was quickly ruled out because the existing support structure would need major modifications to accommodate the additional weight of a steam drum, the drum operating pressure would have been above generally accepted limits and superheater modifications would have been required.

Spiral Furnace Once-Through: Most modern once-through boiler units use a spiral wound furnace design in the high heat flux zone to avoid multi-passes and this would have been a simple solution were it not for the complication of the division wall panels. Their removal would result in major modifications to the boiler heating surface to recover the lost heat in a lower temperature zone and have caused an increased potential for slag formation on superheaters and increased NO_x.

Positive Flow Response Once-Through: In recent years, Mitsui Babcock has been offering a new technology utilising vertical, ribbed bore tubes with a low fluid mass flux. It was chosen to provide the best technical and cost-effective solution for Unit 1. As this would be a world first, full scale commercial application, special care would be needed at all stages of design, manufacture and commissioning to ensure total success.

INTRODUCTION

In 1986 Mitsui Babcock foresaw the advantages of using low water mass fluxes in internally ribbed tubing for the construction of once through boiler furnace walls for steam generation at pressures both above and below the critical pressure. The internal helical ribbing of boiler tubes had been recognised since the 1950s to sustain nucleate boiling at higher pressures, higher steam fractions and lower mass fluxes than plain bore tubes and had found applications in high pressure sub-critical drum type boilers.

To date the majority of existing once-through boilers have furnace walls built with plain bore tubes having water mass velocity in the region of 2500 $kg/m^2/s$ at full boiler output. Other units, including a significant number of domestically-designed boilers in the PRC, have internally ribbed tubes with water mass fluxes between 1800 to 2000 $kg/m^2/s$ at full load. While assuring heat transfer, these high mass flow rates, create disadvantages such as intricate construction, and high feed water pump power consumption.

This paper will show by reference to Boiler #1 at Yaomeng that boilers designed or retrofitted to operate with low mass fluxes using optimised internally ribbed tubes are freed from these effects, retain adequate cooling of furnace walls as well as offering a number of operating cost advantages to their owners.

FURNACE TUBE LAYOUT

Benson once-through boilers, which were intended for operation at supercritical steam pressures, were first built some 80 years ago, and numerous furnace designs have been used in their construction throughout that period. The most up-to-date and cost effective method of furnace construction uses a fully welded steel membrane tube wall forming a gas-tight combustion enclosure with a minimum of air infiltration. The tubes carry water and steam at sufficiently high mass flux to provide cooling to the tubes and membrane. To limit its metal temperature, the membrane strip is made narrow enough to be well cooled by conduction to the adjacent tubes as well as being partly shielded.

In a once-through boiler furnace, there is no re-circulation to increase the total water flow. Using plain bore tubes that require a mass flux of 2000 to 2500 kg/m²/s, and with the water flow set by the boiler output, the number of furnace tubes is insufficient to clothe the furnace periphery in a single vertical pass (the simplest for support and construction). Thus, many furnace designs have emerged involving bands of tubes arranged horizontally or inclined. Some designs use multiple water passes that bring problems with phase separation at re-entry to the furnace. This design limits its application to supercritical pressures, and denies boiler operators the full gain in station efficiency through sliding pressure at part load.

The essence of all good furnace wall design dictates that the layout of the tubes forming the furnace walls and their water supply system, should result in a water flow in each individual tube that closely matches the aggregate heat absorption to that tube. Failure to abide by this guideline leads to variations in outlet steam quality, uneconomic selection of the tube material and unacceptable tube metal temperatures.

The majority of once-through boilers constructed in the last 30 years use walls formed by an inclined helical winding, generally known as a spiral winding, of the band of tubes. This has generally been successful in achieving a near balanced heat absorption since all tubes pass through the different walls and zones of heat flux. On the water-side, the tube resistances were made as similar as possible by careful layout so that the water flows are also well balanced. This arrangement, however, creates a number mechanical design issues related to supporting tubes, furnace framing, burners, ash loads etc., principally to prevent unacceptable bending stresses in the tube wall without imposing undue start-up restrictions due to thermal stresses.

In a furnace design with vertical tubes on the other hand, the support problems are greatly reduced, but the aggregate heat absorptions are by no means balanced. Tubes in the furnace corners are subject to much lower heat inputs than at the centres of walls.

For natural circulation boilers, water flows match the heat input by relying on the density differences caused by differences in heat input. (See Figure 1). In a once-through boiler, the same physical phenomena that drive natural circulation systems are available. Although flow distribution can be changed, the total water flow cannot. Figure 1 also shows, contrary to popular misconception, that the necessary density differences are just as strongly present even at supercritical pressures. However, if density differences are to be effective in regulating furnace tube flows, mass flux must be reduced well below 2000 kg/m²/s so that dynamic losses become a much smaller component of the total pressure drop.

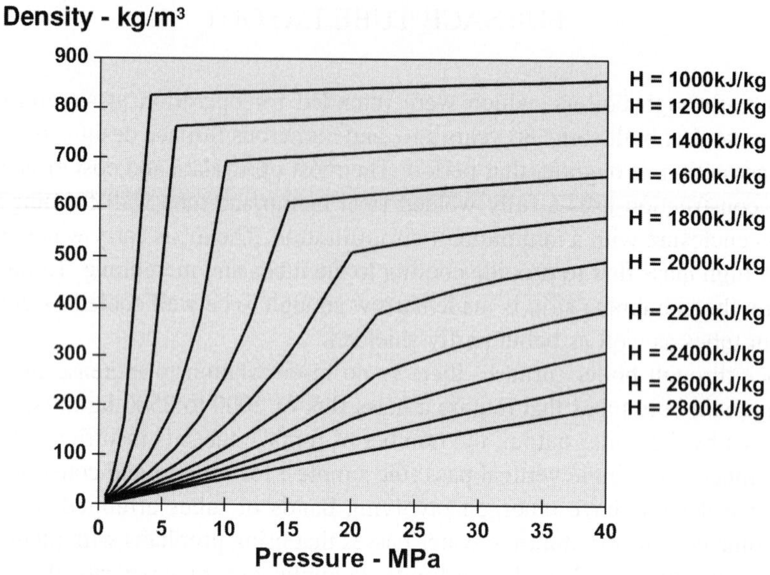

Fig. 1 Density of steam/water.

FURNACE TUBE FLOWS

The concept of the new furnace design at Yaomeng is based on a low water mass flux. This allows the static head changes in response to heat supply to increase water flow in individual tubes and gives a positive gradient to the curve of water flow versus heat input. This is particularly important in coal-fired boilers where unpredictable ash deposition can take place, a characteristic not suited to furnaces with preset flow distributors. Not only does flow change positively in tubes getting more heat and requiring more cooling, but overall pressure losses are cut dramatically. Thus, full advantage can be gained during sliding pressure operation where the boiler feed pump power requirement is reduced. The improved efficiency is also evident at part loads.

Figures 2 and 3 show the static and dynamic pressure losses calculated for a design with high mass flux and for another with a low mass flux. In both cases, at a given flow, any increase in the heat supply to an individual tube indicates reduction of the static pressure loss. In the high mass flux design, this static pressure loss reduction is overwhelmed by the increase in the dynamic losses – indicating that pressure drop increases with heat supply, but in the low mass flux design, the change in dynamic loss is small so that the total pressure drop reduces with an increase in the heat supply. Since the tubes in both designs are arranged between common headers, the flow in the tube with above average heat will change until the pressure loss matches the average figure. To attain this, the tube flow in the high mass flux

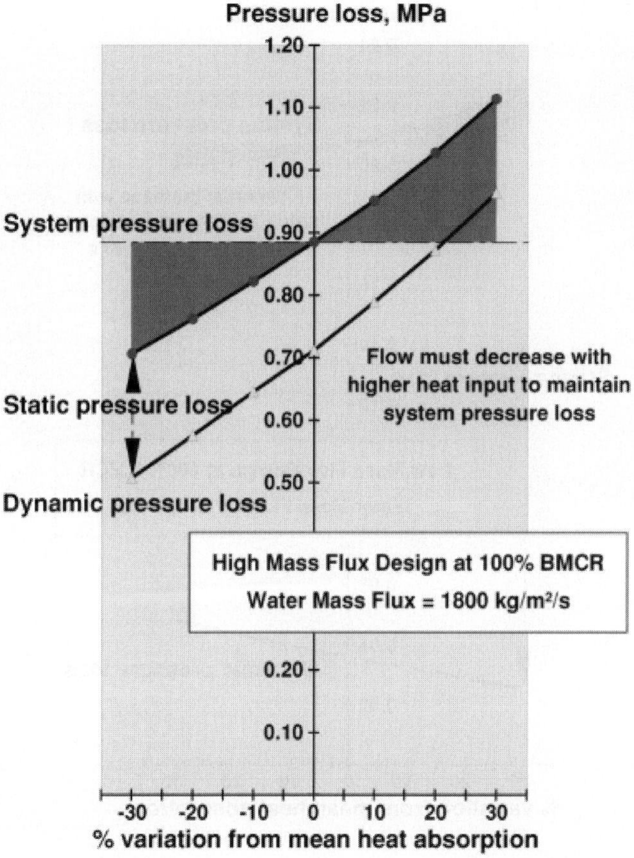

Fig. 2 High mass flux response.

design must be reduced, a negative response, and in the low mass flux design the flow must be increased - a desirable positive flow response.

Site data from Yaomeng following the retrofit has confirmed the mean pressure losses as calculated. Incident heat fluxes and individual tube flows have also been measured at Yaomeng and confirmed the design expectations. Hence we can conclude that a positive flow response has been achieved at Yaomeng.

FLOW INSTABILITY

Two forms of flow instability can give rise to problems in operation, namely static instability and dynamic instability. Static instability is a well-understood phenomenon in which two

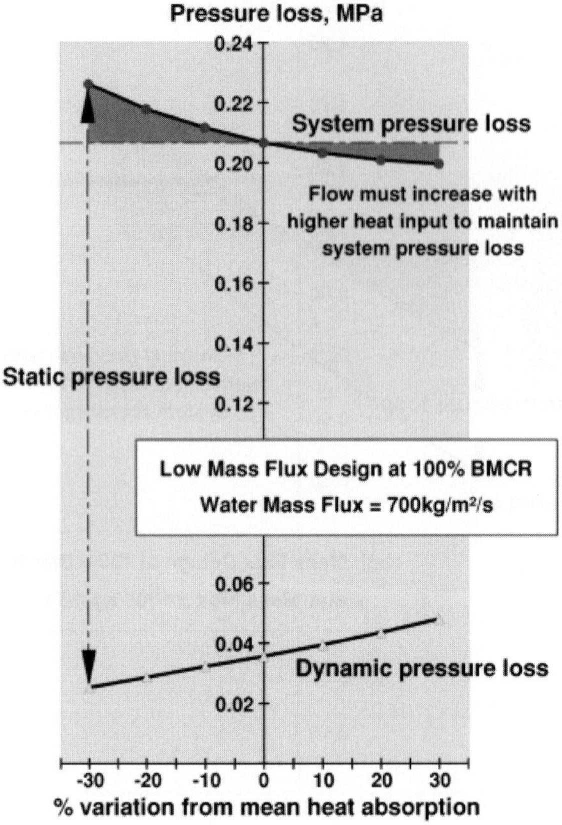

Fig. 3 Low mass flux response.

separated flows can give rise to the same pressure drops. This was not found during the design of Yaomeng.

Parallel channel instability, another unstable furnace thermo-hydraulic characteristic was predicted in some of the preliminary designs for the Yaomeng furnace. It became a complex process to eliminate this phenomenon working with variables such as, mass flux, tube selection, mix points and operational limits. Figure 4, illustrates what may occur if the furnace is either operated outside the recommended limit envelope or is not designed properly.

If the furnace were operated at 3 MPa (30 bar), ie below the recommended pressure, calculations showed that the flow pulsation, also called dynamic instability, would occur over a cycle of 10 – 30 seconds. Susceptible circuits may start to oscillate following a heat input, pressure or flow change, all of which are common events in furnace wall tubes. Although the input and output pressures for every tube do not change, intermediate pressures are changing significantly. The phenomenon may be understood by first considering that part of the cycle when the intermediate pressure is rising. This is due to an increased rate of steam

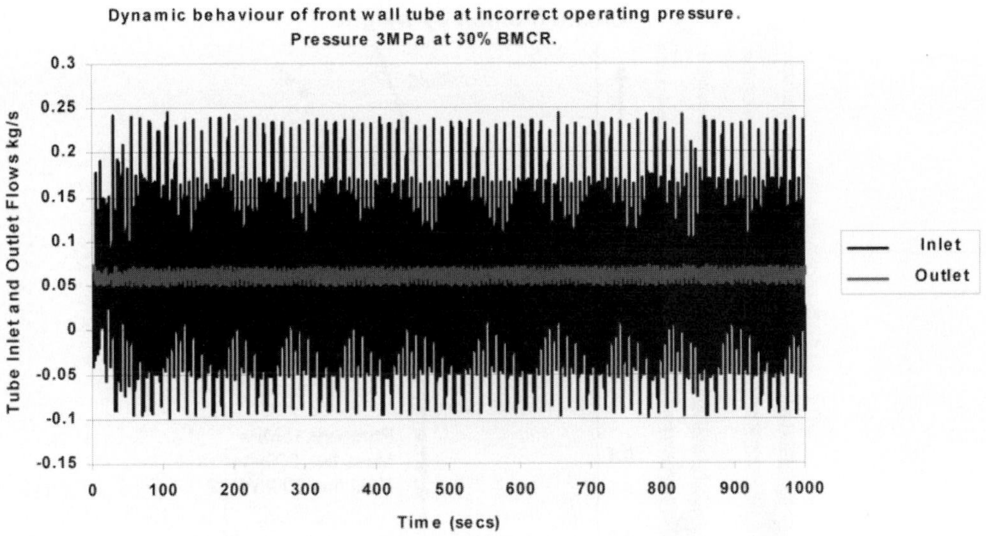

Fig. 4 Unsafe operation below minimum pressure.

production locally. The rising pressure naturally increases the tube exit flow and reduces the tube inlet flow. Because dynamic losses in the water-cooled portion of the tube are much lower, the inlet flow change is much greater and in some cases inlet water flows may reverse.

The high rate of evaporation can only be sustained whilst the boiling region of the tube contains water. Eventually the pressure falls and the water flow and steam flows tend towards their mean values. However, steam generation does not start so quickly from the sub-cooled water now flowing into the tube, and the pressure falls below the mean value, causing water in-flow to rise above the mean and the steam outflow to drop below. Thus the cycles continue. The predictions of unstable behaviour are complex, they show that the inlet flow does not behave in a simple periodic way but that a number of frequencies are involved.

The cause for concern is fatigue damage from repeated thermal stress cycles caused by temperature and heat transfer fluctuations. With a cycle time of 20 seconds, and perhaps 27 million cycles in a plant lifetime, the allowable thermal stress range and permissible temperature differentials are small.

All of the Yaomeng circuits have been designed to be free of this type of instability and none has been detected by post retrofit flow measurements at site.

HEAT TRANSFER

As identified above, water-side heat transfer coefficients would be insufficient in plain tubes at such low mass fluxes. To address this issue, Mitsui Babcock's boiler designers advanced the concept of a low mass flux vertical tube design with internally ribbed tubing. Ribbing

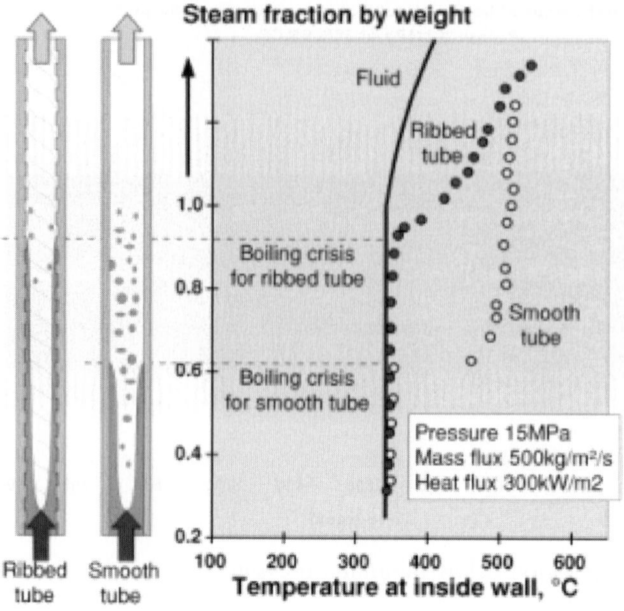

Fig. 5 Boiler tube test loop data.

enhances water side heat transfer to a remarkable extent and, as shown in Figure 5, is able to sustain nucleate boiling by keeping a water film in contact with the tube wall at low mass fluxes and high steam qualities. This is exactly what is required for a vertical tube once through furnace wall with low mass flux.

To prove the validity of the concept, heat transfer and pressure loss tests on a range of tube sizes and rib configurations were commenced. Tests were commissioned by Mitsui Babcock and performed in the UK and continued by Siemens Power Generation in Germany. An extensive database has been built up covering a wide range of rib and tube parameters, water mass fluxes and external heat fluxes for a wide range from sub- to supercritical pressures. The resultant optimised rib has a modified lead angle and a taller rib than standard ribbed tubes with their modest rib heights and 30º lead angle.

THE CONTRACT

It was decided that conceptual engineering would be undertaken in the UK and the detail engineering and project management would be undertaken in the PRC. The contract, as well as involving leading edge technology, had a number of commercial and organisational features of note. The contract language and law was Chinese and all documentation to and from YPGL had to be in Chinese. To maximise local content, the majority of the hardware was sourced in the PRC through Mitsui Babcock's local office.

SUPPLY AND FABRICATION

The advanced ribbed tubing for the furnace and division walls was to be manufactured for the first time on a commercial scale. The desired geometry of the internal ribs and the fine tolerances required, presented a considerable challenge both in respect of tool design and in achieving consistent results.

Tube and draw bench preparation were also discovered to be critical in achieving production at an economic rate. Several trial pulls for each tube size were undertaken before any full-scale production took place. There was full participation of Mitsui Babcock, the UK tube supplier and the toolmaker to establish the optimum drawing parameters. This ensured that the tubes produced were of consistently of high quality and within the dimensional tolerances allowed to achieve the essential pressure loss behaviour and heat transfer performance. Apart from the tubing, there were other offshore suppliers for the start-up circulating pump, safety valves, control valves and the steel-plate gilled economiser tubing.

To fulfil the intention of maximising the work content in the PRC, as mentioned earlier, major areas of detailed design were undertaken by design institutes and subcontractors in the PRC. The two major orders subcontracted were for the manufacture and fabrication of the pressure parts, and the supply of the eight burner assemblies. Orders were also placed in the PRC for framing, slings, pipework and supports, flame monitors, actuators, valves and casings, sootblowers, galleries and ladders.

The pressure part conceptual and detail design were undertaken by Mitsui Babcock in the UK. A PRC boiler contractor then completed the manufacturing detailing to suit their in-house processes. The ribbed tubing for the membrane walls was supplied by Mitsui Babcock, while the remaining pressure parts (start-up system, partial supply of superheater and reheater) were supplied by the local boiler contractor. The majority of the pressure part manufacturing activities took under five months to complete.

Although the original furnace and the replacement furnace are both tangentially-fired, the burners were extensively redesigned to ensure that the required NO_x and combustion efficiencies would be attained. This redesign caused revisions of the membrane wall design and the supporting structures. Significant design changes had to be made to the boiler slings as the weight distribution had changed and, because of deterioration, the lower level of slings was completely replaced. The boiler framing was also entirely redesigned to meet new requirements. In both of these areas, the conceptual design was completed in the UK and the detailing and supply were done by subcontractors in the PRC.

The first deliveries of equipment took place in August 2001 and by the end of 2001 the majority of major equipment had been delivered to the site.

CONSTRUCTION AND COMMISSIONING

Dismantling began in September 2001. YPGL dismantled the furnace, the superheater, reheater and economiser sections. They moved the induced draught fans, removed the grit collector system to provide space for the erection of an electrostatic precipitator and cleared the deaerator area to accommodate a major refurbishment. The control room and control

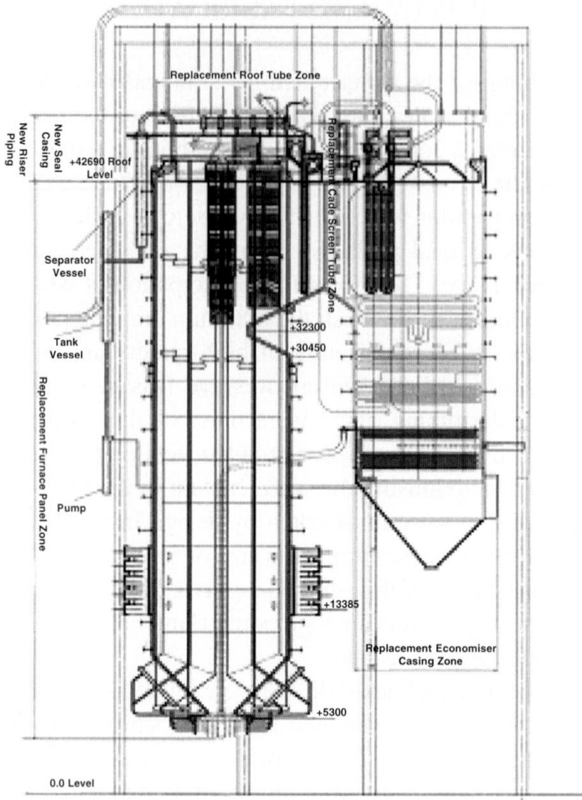

Fig. 6 Side elevation.

equipment room were also stripped out to enable the new DCS system to be installed. At the same time, work to replace the turbine rotor and refurbish the feed water systems went underway. As soon as dismantling had progressed sufficiently to clear the dead load, work commenced on welding of reinforcing plates to the existing members.

Erection of the new pressure parts began in the middle of November 2001, and continued through both the Christmas and Chinese New Year periods. The finished boiler side elevation is illustrated in Figure 6, while the 3D model, from which this 2D 'drop-off' was derived is shown in Figure 7. The hydraulic testing of the boiler took place successfully by the middle of March 2002, and the pre-service acid cleaning of the boiler was completed a week later. Cold commissioning of the boiler was started whilst the installation of the insulation, turbine, electrostatic precipitator, control systems, galleries and ladders was still continuing.

Substantial testing of the air systems and airflows in the furnace, burners and flue & ducts were completed during this period. Most significantly the firing of the boiler took place in the middle of April 2002 and the steam purging of all new systems was completed before the end of April. Figure 8 shows steam discharges during steam purging. To prevent dynamic instability at high steam flows and low pressures, something which is not called for during normal service conditions, steam purging was conducted by releasing steam accumulated in

Fig. 7 3D model.

the boiler after firing had been stopped. This enabled more than adequate disturbance factors to be achieved to ensure cleanliness.

After re-instatement of the boiler and other systems, safety valves were set and the generator was synchronised and commenced electricity generation during the first week of May 2002.

Senior engineers from Mitsui Babcock and Siemens were based at site whilst the unit underwent extensive optimisation of the control systems to ensure that all the control and logic functions were correctly implemented.

OPERATING EXPERIENCE

Furnace outlet steam and water separation with a circulating pump to provide a circulation during start-up, and for low load operation have not been featured in the majority of once through boilers built in PRC. However their use, which has been almost universal in Benson once-through boilers for the last 30 years, has been very successfully applied to Yaomeng. It proved to be a considerable improvement over the original equipment.

Fig. 8 Steam purging the superheater.

A high degree of instrumentation was fitted to assess the success of the low mass flux design and to assist the boiler designer on future installations. However, commercial considerations prevent placing all of this accumulated information into the public domain.

Monitoring of furnace tube outlet temperature differentials revealed one area of concern. During operation at low loads, with wet steam at the lower furnace circuit outlets, there was no difference between tube outlet temperatures. As load increased, some deviations appeared as expected according to heat absorption differences. The most significant temperature differences developed at the outlet of the furnace division wall. Improvements in the balancing of pulverised fuel supplies to the furnace reduced the temperature variations significantly. Higher than expected temperatures remained in tubes that had extra length and consequently extra heat absorption in the upper sections of the wall. These had been formed with rectangular loops to control tube alignment. Detailed analysis showed that additional heat absorption, particularly in this location, produced a negative flow response. Modifications were therefore devised to correct this and maintain tube alignment.

Individual tube flows and associated heat flux measurements at Yaomeng confirmed the physical principles discussed in this paper.

Operation of Unit 1 has continued ever since its full-load 168 hours reliability run on May 26, 2002, delivering peak output at 327 MWe and meeting all generation requirements.

The low mass flux furnace at Yaomeng has a significantly lower water/steam side pressure loss enabling the maximum continuous boiler output to be increased from 855 to 950 te/h with a peak output of 1000 te/hr to meet generation demands. As the steam turbine now operates with a sliding pressure regime, cycle efficiency has been increased and feed pump power reduced. Perhaps the most important gain for YPGL is the absence of any restrictions in the boiler to follow daily load demands. This is a major improvement over the original plant, which would otherwise have been permanently shut down.

SUMMARY OF GAINS FOR YPGL

The vertical tube low mass flux furnace retrofit at Yaomeng Unit #1 has provided the owners the following advantages:

- Lower pressure loss that translates to reduced feed pump power consumption.
- Lower heat rate through sliding pressure operation and power saving.
- No valves between the furnace and superheater.
- Simple, self-supporting furnace walls without start-up restrictions.
- Short site construction time.
- Positive flow characteristic to suit variable heat fluxes.
- Reduced furnace tube temperatures.
- Smaller tube to tube temperature differences and lower thermal stresses.
- No flow-balancing valves.
- Increased boiler continuous output.
- New peak load capability.
- Faster start-ups
- Improved availability

There were other gains from the upgrade, such as improved boiler efficiency and lower emissions due, only in part, to the furnace upgrade.

FURTHER APPLICATIONS

This demonstration of the low mass flux furnace technology has finally come to reality more than 15 years after Mitsui Babcock put forward the concept. It has opened the way for the upgrading of a number of identical and similar boilers in PRC that are suffering from the same operational restrictions.

With the completion of the project, a panel of experts that convened in Beijing to evaluate the reports on the project, have endorsed the boiler design and its performance, and confirmed its success.

The principles demonstrated at Yaomeng are equally applicable at supercritical pressures and have been demonstrated in full-scale test installations. Many utilities with supercritical boilers of the UP (Universal Pressure) type which have limited flexibility to change load or cannot take full advantage of sliding pressure operation, will be able to benefit from the successful application of this technology. Any new supercritical plant proposed should seriously consider this technology as there is no "downside" when compared to existing technologies.

ACKNOWLEDGEMENTS

B. C. Mackintosh	:	for the concept
J. E. Jesson	:	for the commitment
Siemens AG	:	for the research and support
Xu Tao	:	for the sales drive
Yaomeng, Power Generation Ltd	:	for the opportunity
M. B. Yaomeng, Project Team	:	for the reality

REFERENCES

1. J. E. Jesson and M. J. Smith: 'The World's First Vertical Tube Low Mass Flux Furnace for Once-Through Utility Boilers', *MES Technical Review # 178,* 2003.
2. J. E. Jesson, M. J. Smith and Xue Yimin: 'Upgrade and Refurbishment of a 300 MW Unit at Yaomeng Power Plant. The Worlds First Benson Boiler with Vertical Ribbed Tubes and a Low Water Mass Flux', Not Published.
3. J. Franke, R. Cossmann and H. Huschaur: 'Benson Steam Generator with Vertically Tubed Furnace', *VGB Power Technology*, **75**, 1995, 353–359.
4. M. J. Smith, D-M. Finch and C-H. Chen: 'Low Mass Flux Vertical Tube Furnace Retrofit at Yaomeng in the Peoples Republic of China', *Power Station Retrofitting, Optimising Power Plant Performance,* Professional Engineering Publishing, 2002.
5. R. Brundle: 'World Firsts for Yaomeng with Vertical-tube, Low-mass-flow Benson Unit', *Modern Power Systems,* Wilmington Publishing, 2002.

The Exploitation of Advanced Blading Technologies for the Design of Highly Efficient Steam Turbines

MATHIAS DECKERS and ERNST WILHELM PFITZINGER
Siemens AG, Power Generation (PG)
Muelheim an der Ruhr
Germany

ABSTRACT

In this paper, Siemens' latest improvements in steam turbine blading and bladepath design tools for high pressure (HP) and intermediate pressure (IP) turbine modules are presented. A description of the key characteristics of these blading technologies and tools is given and typical design applications are discussed.

It is shown that this new blading generation is characterised by both an extremely modular and flexible construction as well as a fully three-dimensional airfoil shape allowing the exploitation of all primary blading design parameters (e.g. stage reaction and stage loading) for highest efficiencies. This, in combination with the latest design tools, yields considerable increases in turbine efficiency and reduces drastically the bladepath design cycle time.

As a result, turbine cylinder efficiency levels of up to 96% can be obtained nowadays in modern high power rating steam turbine modules and, in parallel, design cycle time reductions of up to 80% can be achieved.

INTRODUCTION

In a world of limited primary energy resources and increasing awareness of environmental pollution, the aim of designing steam turbine power plants with as high an efficiency as possible will continue to be of primary importance. Since the overall cycle efficiency depends strongly on the steam turbine performance, continuous improvements are sought to increase the turbine efficiency. These efforts are directed primarily towards improvements in the blading as the key-component of the turbine.

Modern steam turbines are customized products aimed at accurately meeting all the individual customer needs.[1] As a result, the turbine bladepath is to be designed individually in most cases in order to meet the guaranteed efficiency levels. In addition to that, ever shorter steam turbine delivery times are demanded nowadays. So, these individual bladepath designs must not only lead to very high levels of efficiency but they must also be carried out in a very short period of time. This requires both a new type of steam turbine blading technology featuring a high degree of flexibility and a wide range of application as well as advanced design tools ensuring a high level of automation and standardization in the design process without sacrificing the flexibility to meet the individual customer's needs. This is not only important for new machines but even more so in service upgrade applications due to the inherent design constraints that are typically present for the latter.

Recognising that traditional blading technologies and design methods were a limiting factor in meeting these requirements, considerable effort has been spent during the past decade within the steam turbine blading development group of Siemens Power Generation (PG). As a result, a highly standardised but flexible blading technology has emerged. This is essentially based upon the latest generation of fully three-dimensional blading with compound lean (3DS™) and variable stage reaction (3DV™). However, since not only the technology but also the quality and speed of the design process decide, whether or not the overall performance and lead time requirements are met, the entire bladepath design process has been automated within a very powerful design system.

In this paper, Siemens' latest improvements in steam turbine blading and blading design tools are presented. The blading technology is referred to as 'HP/IP blading technology' and is applied to all drum-type stages in the HP and IP turbines as well as the front stages of LP turbines featuring integrally shrouded blades. In the paper, an overview of the blading technology is given, their main characteristics are highlighted and examples of typical design applications are discussed.

HP/IP BLADING TECHNOLOGY

Blade Construction and Installation

The drum-type HP/IP blade construction feature a fully integrally shrouded blade design concept, Figure 1. The key characteristics of this concept are based upon using both rhombic T-roots and rhombic integral shrouds. The integral shrouds provide two basic functions: Firstly, they form a circumferential steam path boundary allowing efficient seal designs to be employed; and secondly, they provide individual blade tip support between neighbouring blades in each blade row. The latter is accomplished by pre-stressing elastically the rhombic shrouds when the blades are installed in the circumferential rotor T-grooves. The primary advantage of these pre-stressed blades, which are installed carefully under controlled conditions, is the fact that they display an excellent damping behaviour enabling to absorb dynamic stresses under all operating conditions without endangering their long-term reliable life time.[2] Siemens PG has now accomplished more than forty years of successful experience with these fully integrally shrouded blade designs.

The backbone of the HP/IP blading technology is a strictly modular concept of bladepath construction from standard and proven elements (e.g. airfoils, roots, grooves, shrouds, extractions, locking devices, ...). As an example, the composition of a single blade from

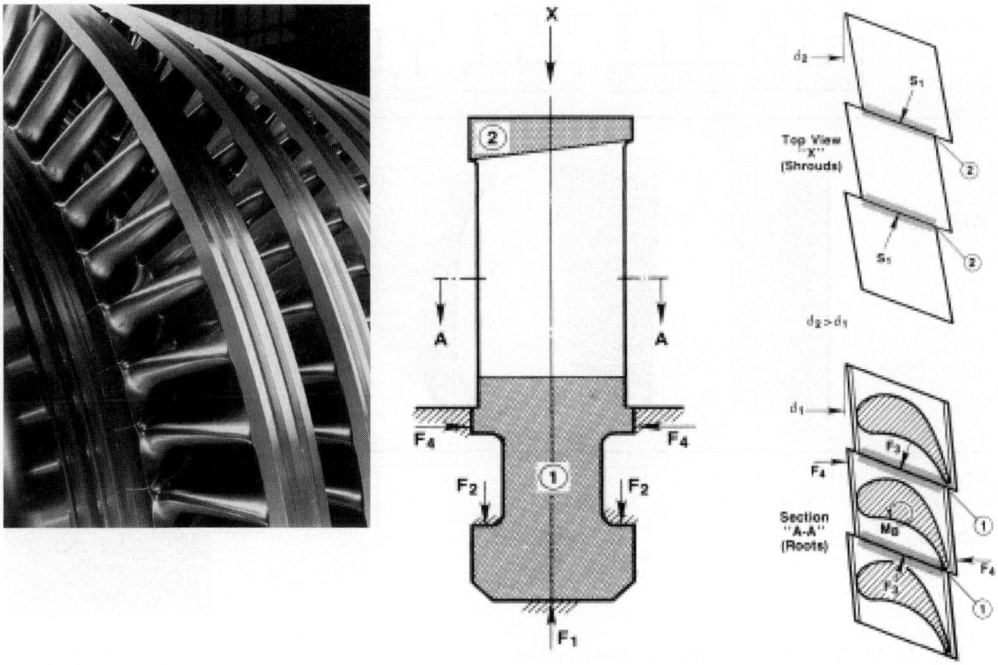

Fig. 1 Integrally shrouded blade design.

root, shroud and airfoil is shown in Figure 2. For each element different types exist for the various applications, each type having its own advantages with respect to performance, mechanics and costs. Within the modular concept, all these different types may be combined arbitrarily together to give an optimum blade for the specific design boundary conditions such as aerodynamics, forces, materials and temperatures. Hence, cylindrical, twisted or bowed airfoils can be assembled with any of the available roots or shrouds.

Advances in Blade Airfoil Designs

Steam turbine blade airfoil design is inherently associated with the identification of loss origins and an understanding of the mechanisms producing loss. The overall aerodynamic loss of a turbine blade can be roughly divided into profile loss, annulus loss, secondary flow loss and leakage loss and their contribution to the overall loss depends primarily upon blade geometry, blade loading and flow turning.

The profile loss is generally defined as the loss due to boundary layer growth along the blade surface, to mixing in the blade wake and, in transonic flow, to dissipation in shock waves. The mechanisms involved are considered to be basically two-dimensional and so

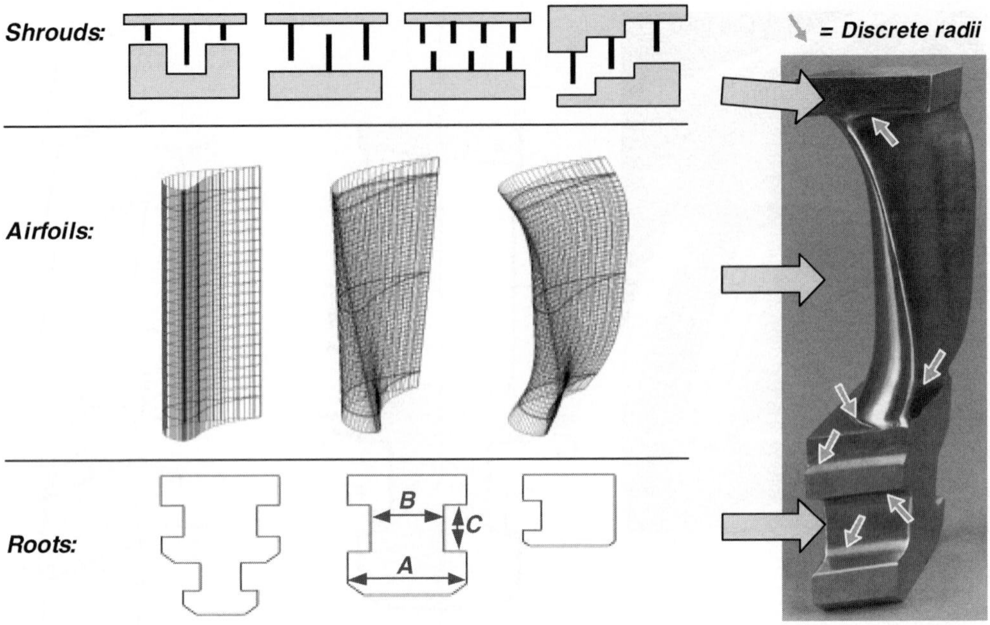

Fig. 2 Modular concept of blade construction.

unaffected by the viscous (secondary) flows near the end walls. The annulus loss is considered here as the loss due to boundary layer growth along the hub and casing whilst the secondary flow loss is the loss caused by so-called secondary flows which result from the turning of inlet vorticity, i.e. the turning of the inlet boundary layer. In practice, it is very difficult to separate the latter two loss sources and so the annulus loss is often included in the definition of the secondary flow loss. The leakage loss is caused by the fluid flow that passes through the unavoidable gap between rotating and stationary turbine parts. In doing so, the leakage mass flow does not contribute to the useful work output. In addition, the leakage causes mixing losses when it is re-injected into the main flow and influences the radial exit flow distribution which, in turn, may lead to further incidence and secondary flow losses in the downstream blade row.

The relative flow losses at various locations in the high pressure (HP) and intermediate pressure (IP) turbine cylinders, respectively, are given in Figure 3. The blade profile loss can be seen to be the largest single source of loss in the whole turbine. This clearly indicates that the overall loss can be most easily reduced by reducing the profile loss throughout the entire turbine. The secondary flow loss is of significance for those stages that are characterised by a low aspect ratio (ratio of blade span to blade chord), i.e. the front stages of the HP and IP turbine component. Leakage losses are relatively high in the admission sections to the HP and IP turbines.

In general, the flow field in steam turbines is, at least to some extent, characterised by fully three-dimensional effects. This is particularly true for low aspect ratio turbines where

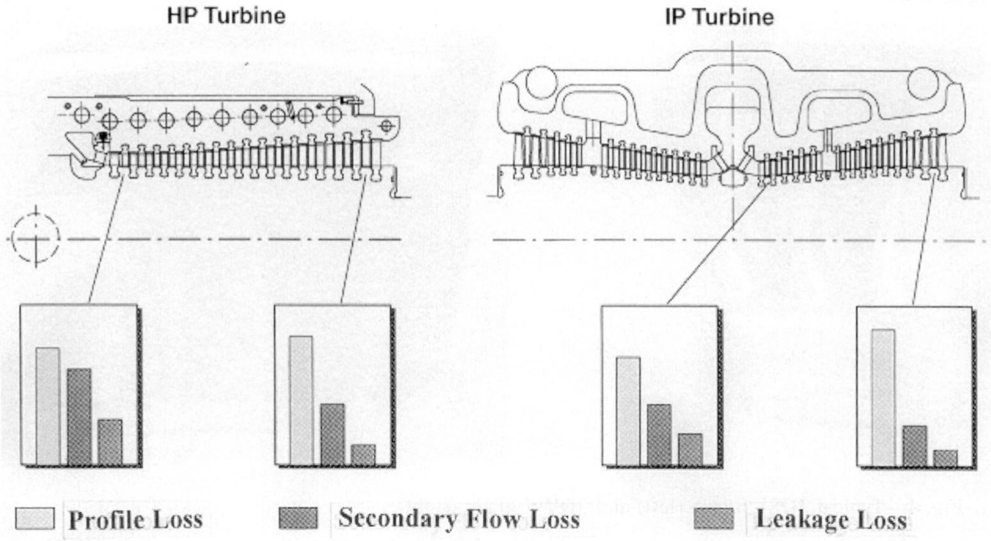

Fig. 3 Relative flow losses in HP and IP turbine components.

the secondary flows generated close to the end walls may influence the entire flow field. As a result, the loss caused by these secondary flows can contribute significantly to the overall aerodynamic loss, particularly in the front stages of the HP and IP turbine component, Figure 3. In order to achieve the high efficiencies that are required these days, secondary flow losses must be addressed and this is the point at which three-dimensional blade design comes into its own.

In recent years, an intensive research and development programme was carried out at Siemens PG aimed at developing a completely new generation of fully three-dimensional steam turbine blades for highest efficiencies. In a first step, the so-called 3DS™ blade was developed. This blade was designed specially for use in front stages of HP and IP turbine modules and features substantially reduced secondary flow losses. A typical 3DS™ blade is presented in Figure 4. Detailed experimental investigations performed on a 4-stage close-loop model turbine under representative Mach and Reynolds numbers show that a stage efficiency improvement of up to 2% points can be achieved with these 3DS™ blades in comparison to conventional blades.[3]

This benefit of employing 3DS™ blades has since been confirmed for a multitude of real steam turbine designs in new apparatus and major service upgrade applications. The most prominent example certainly is the steam turbine power plant Boxberg (907 MW) which is regarded as the world's most modern lignite-fired power plant with a record-breaking gross-efficiency of 48.5%.[4] The turbine features fully three-dimensional bladepath designs using the 3DS™ blading technology throughout the HP and IP flows which, in turn, contributed

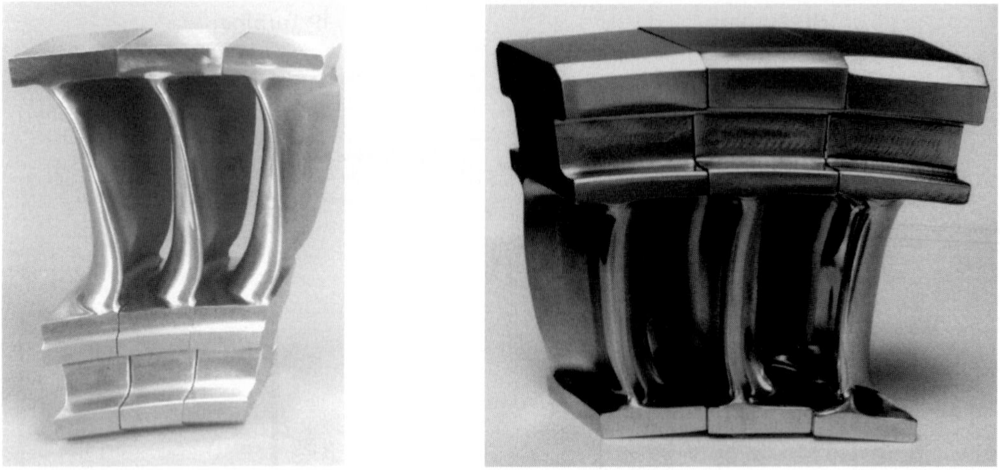

Fig. 4 Typical 3DS™ blade (left) and 3DV™ blade (right).

considerably to the measured cylinder efficiencies of 94.2% and 96.1% for the HP turbine and IP turbine, respectively.

It should be noted that a major pre-requisite for the development of the 3DS™ blades was the introduction of advanced CFD methods[5] and modern blade manufacturing facilities. The latter enables the cost-effective production of individually designed fully three-dimensional steam turbine blades. Furthermore, the availability of these advanced design and manufacturing methods enabled even more design parameters to be varied in order to obtain the highest possible efficiency levels. This, in a second major development step, has led to the development of the 3DV™ blading technology at Siemens PG. The 3DV™ blade also features a fully three-dimensional airfoil shape but it's main characteristic is the possibility to vary both stage reaction and stage loading, Figure 4. The exploitation of these additional free parameters yields further efficiency improvements by up to 1% point in cylinder efficiency.[6] The 3DV™ blading technology was first applied to the world's largest super-critical lignite-fired power station Niederaußem[7] with a nominal output of approx. 1000 MW. Since then, it has been employed in all new machines and the majority of service upgrades.

HP/IP BLADEPATH DESIGN SYSTEM

Overview

The exploitation of the benefits of the aforementioned fully flexible three-dimensional blading technology requires the quality-assured automation within a bladepath design tool. But

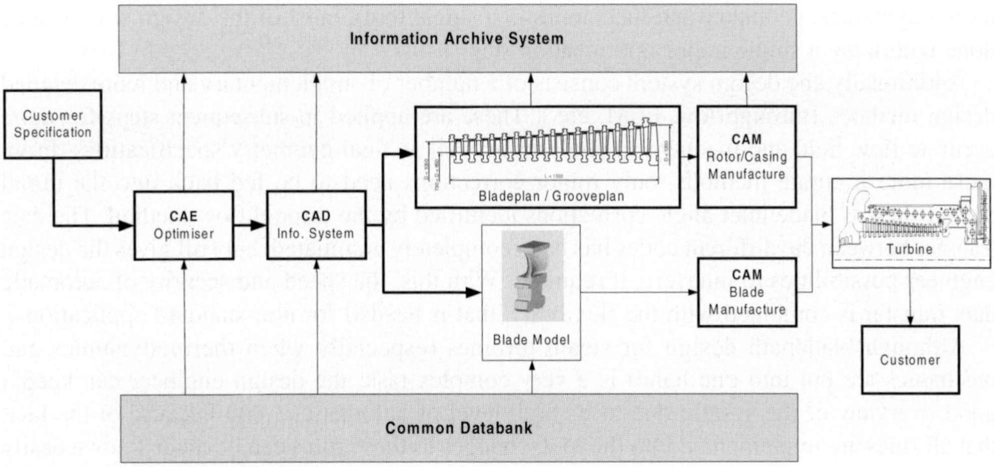

Fig. 5 Schematic overview of the design system.

automating a very complex process as bladepath design is a delicate task. The sensitive balance between the degree of automation and remaining possibilities for manual intervention will decide ultimately about performance and flexibility of the automated design process. Therefore, much effort has been spent also at Siemens PG in recent years to develop a modern and very sophisticated bladepath design system taking full advantage of the modular concept of the flexible blading technology using generic models and design rules.

A schematic overview of the design system is presented in Figure 5. The system essentially consists of the CAE-Tool, the CAD-Tool and the CAM-Tool. These tools are linked through clearly defined interfaces and are supported by a common databank ensuring single-sourcing of data. The design system covers the entire design process from the initial bladepath design up to the final generation of all manufacturing data and drawings.[8] In the following, the key characteristics are summarised for the sake of completeness.

The backbone of the design system is an extended meanline code within the CAE-Tool, that also covers partially 2 and 3 dimensional aspects of the design. Although the main thermodynamic design is performed on a meanline basis, all primary influences from the second and third dimensions are also taken into account by means of appropriate models. Typical examples are the spanwise distributions of inlet and exit angles, the exact geometry of roots, grooves and shrouds as well as numerous mechanical constraints on roots, claws and airfoils. For the fully 3 dimensional blades with compound lean, both the aerodynamic benefits and the mechanical impact are modelled within the tool.

In order to speed up the complex design task with several hundreds of variables, a gradient based optimisation method has been integrated into the extended meanline code. And since this advanced method has the full view of all relevant aspects of the design (i.e. aerodynamics, thermodynamics, geometry and mechanics in a single tool), most of the design work can be done within the a single major optimisation step.

Additionally, the design system consists of a number of supplementary and more detailed design methods (throughflow, FEM, etc.). These are applied in subsequent steps for more accurate flow field and mechanical analyses as well as final geometry specifications. From these more accurate methods, only minor corrections need to be fed back into the initial design such as blade inlet angle corrections identified by the throughflow method. The data transfer between the different codes has been completely automated, but still gives the design engineer possibilities to interfere, if required. With this, the speed and security of automatic data transfer is combined with the flexibility, that is needed for non-standard applications.

Although bladepath design for steam turbines (especially when thermodynamics and mechanics are put into one hand) is a very complex task, the design engineer can keep a good overview of the system due to its high level of automation. And because of the fact, that all rules are implemented into the tool, changes to these rules can be made known easily to all design engineers by means of a new release of the tool. It is therefore fair to say, that the tool incorporates the rules and to a wide extend the design process itself. Experience tells us, that its application leads to a very high level of product quality, while at the same time the overall effort and the number of errors are reduced substantially.

Typical Applications

The new system can be employed for drum-type bladepath designs for all turbine cylinders. It was implemented into the company's steam turbine design environment in 1999 and was first applied to the world's largest super-critical lignite-fired power station Niederaußem. Since then, the system has been used for all new apparatus steam turbine units and a number of major service upgrades at the companies primary design centres in Europe and the US. Hence, designers throughout the company work with a uniform set of very sophisticated design tools. In order to demonstrate some of its main strengths and benefits, a few typical examples of the application of the system will be given in the following.

Mechanical Impact on Efficiency

The first example underlines the importance of an integrated design method, which incorporates aero-/thermodynamic and mechanical aspects in a single system. Within a given limited time frame (dictated by market requirements), a sequential design process will not achieve an equally good balance between the different disciplines and hence will deliver a design with generally larger mechanical safety margins than required. This results in a negative impact on bladepath performance especially for turbines with limited axial length. To demonstrate this effect, a 60 Hz combined cycle IP bladepath has been optimised for different bladepath lengths. Thereby, all mechanical safety factors have been increased by 10% and 20% from their standard values. The resulting deficit in bladepath efficiency is given in Figure 6. The results clearly show, that for restricted bladepath lengths considerable performance losses are to be taken into account, if the design process does not ensure a design close to the allowable mechanical limits of the blades.

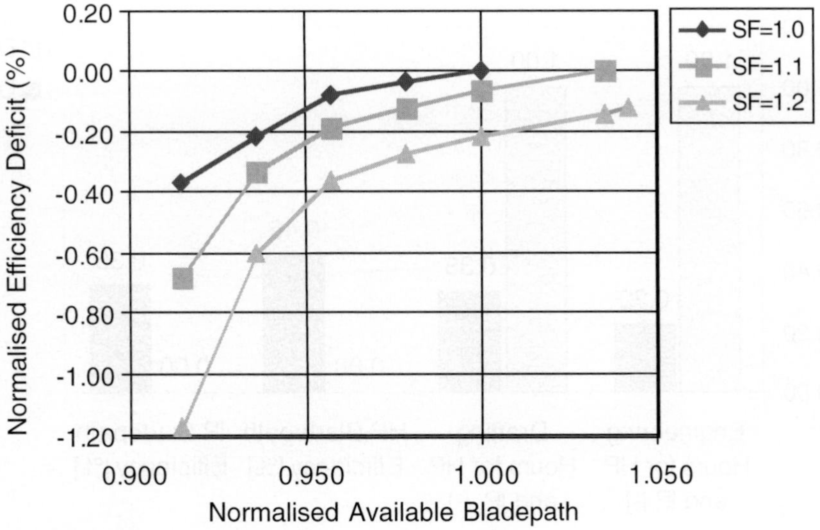

Fig. 6 Mechanical impact on efficiency with restricted bladepath length.

HP/IP-Redesign for Supercritical Unit

In the second example, the results of two different design systems are compared. The basis for the comparison are the HP and IP bladepath designs of a supercritical steam turbine unit. Design I has been performed with a previous, (partly) sequential and less automated set of tools, whereas Design II is a redesign from a recent study with the new system. Both designs are based on similar technology and identical boundary conditions.

The amount of man-hours for both engineering and drafting tasks could be reduced dramatically for Design II, Figure 7. This is due to the high degree of automation of the new consistent program chain that completely avoids all manual data transfer within the process. In addition to the reduction in design cycle time, the bladepath efficiencies could be improved considerably by 0.43 and 0.35%, respectively.

Sketches of the resulting IP bladepath layouts are compared in Figure 8. Although the axial bladepath length was reduced in Design II, all blade heights and yet the blade aspect ratios could be increased. In addition to that, a modified hub contour design was chosen by the system and adjusted the flow path to the steam expansion more homogeneously.

A more detailed comparison of the two designs showed, that the integrated design methodology and the application of a numerical optimisation method allowed a better balance between the different disciplines. In Design I, the blade heights were limited by single blades with stresses close to their allowances. But for Design II, the design system unloaded these limiting blades by carefully varying aerodynamic load, degree of reaction, hub diameters, root dimensions, etc. In consequence, a more uniform exploitation of the mechanical properties of all blades could be achieved, in which the safety margins on average are closer to their allowable limits.

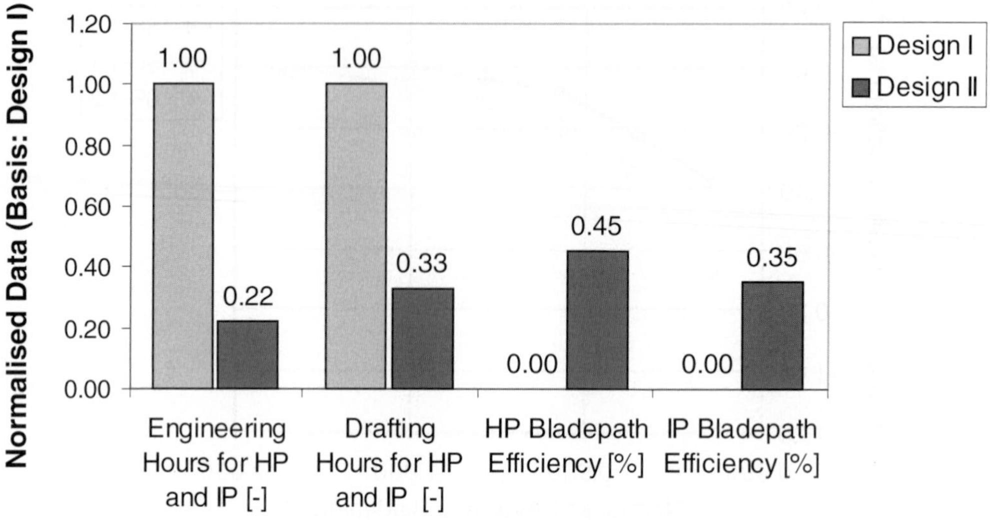

Fig. 7 Increase in design process efficiency.

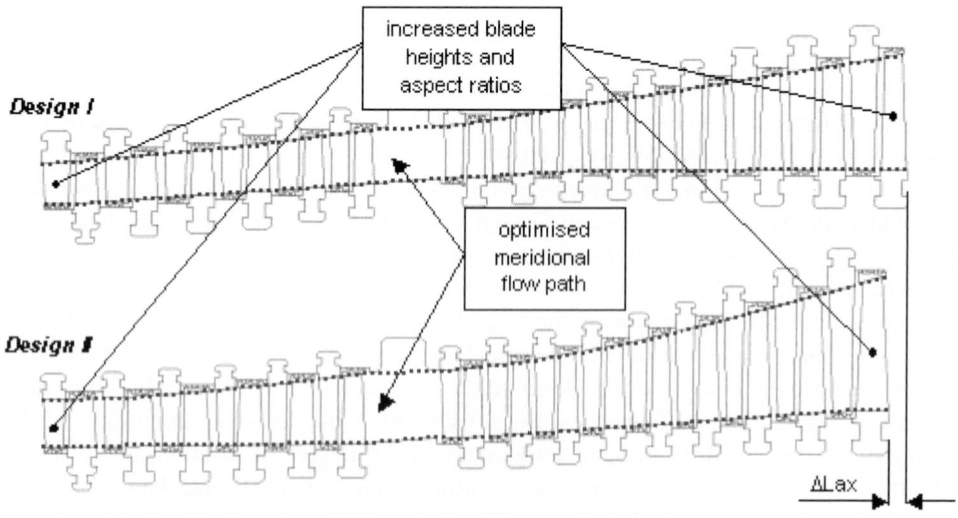

Fig. 8 Comparison of bladepath designs (IP turbine).

SUMMARY AND CONCLUSIONS

In this paper, Siemens' latest improvements in steam turbine blading and blading design tools are presented. This technology offers improved performance and highest efficiencies for a wide range of steam turbine applications.

It is shown that this new blading technology is characterised by both an extremely modular and flexible construction as well as a fully three-dimensional airfoil shape allowing the exploitation of all primary bladepath design parameters such as stage reaction and stage loading. This, in combination with the latest design tools, yields highest cylinder efficiencies of up to 96% and reduces drastically the bladepath design cycle time by up to 80%.

The backbone of the blading technology are a strictly modular construction of the entire bladepath from generic models, design rules derived from extensive parametric studies and highly automated tools. The latter incorporates a numerical optimisation method based on the design rules to give a highly efficient design system. The core idea is to standardise the *way to the product*, i.e. the design process, instead of the product itself in order to combine all benefits from standardisation with a flexibility that enables contract specific designs.

The design system allows to consider all design aspects in parallel, instead of having subsequent design steps for thermodynamics and mechanics, leading to larger improvements per design cycle. Fewer design cycles are required to achieve the same level of 'goodness' of the overall bladepath design. In combination with the automation, which enables more design cycles in a given time, this leads to a much more efficient design process reaching a better balance between mechanical limits and performance in a shorter period of time. The system enables the design engineers to design a fully three-dimensional bladepath to the customers specific needs in a minimum amount of time and combines all benefits of standardisation and flexibility.

The blading technology and tools can be applied to turbines of any power rating for new apparatus installations as well as for service upgrade applications.

ACKNOWLEDGEMENTS

The authors would like to thank Siemens Power Generation (PG) and, particularly, Dr. W. Ulm and Dr. R. Bell for the encouragement and permission to publish the paper. Special thanks are also due to all the members of staff in the Steam Turbine Blading Development group at Siemens Power Generation (PG) for their contributions.

REFERENCES

1. H. Oeynhausen, A. Drosdziok and M. Deckers: *Steam Turbines for the New Generation of Power Plants*, VGB Kraftwerkstechnik 76, (12), 1996.
2. K. Neumann, G. Stannowski and H. Termuehlen: 'Thirty Years Experience with Integrally Shrouded Blades', *Proceedings of the Joint Power Generation Conference*, Dallas, Texas, USA, 1999.

3. M. Jansen and W. Ulm: 'Modern Blade Design for Improving Steam Turbine Efficiency', *Proceedings of 1st European Conference on Turbomachinery, Fluid Dynamic and Thermodynamic Aspects*, Universität Nürnberg-Erlangen, 1995.
4. U. Hoffstadt: *Boxberg – Ein neuer Benchmark für moderne Kraftwerkstechnologie*, BWK Bd. 53, Nr. 3, 2002.
5. T. Thiemann, A. de Lazzer and M. Deckers: 'The Application of Advanced CFD-Methods to the Design of Highly Efficient Steam Turbines', *Proceedings of the 3rd International Conference on Engineering Computational Technology*, B. H. V. Topping and Z. Bittnar, eds., Civil-Comp Press, Stirling, Scotland, 2002.
6. V. Simon, I. Stephan, R. M. Bell, U. Capelle, M. Deckers, J. Schnaus and M. Simkine: 'Axial Steam Turbines with Variable Reaction Blading', *Advances in Turbine Material, Design and Manufacturing, Proceedings of the 4th International Charles Parsons Conference*, UK., 1997, 46–60.
7. R. J. Heitmüller, H. Fischer, J. Sigg, R. M. Bell and N. Hartlieb: 'Lignite-Fired Niederaußem K Aims for Efficiency of 45 per cent and more', *Modern Power Systems*, 1999.
8. M. Deckers, S. G. C. Hadden, E. W. Pfitzinger and V. Simon: 'A Novel Bladepath Design System for Advanced Steam Turbines', *Proceedings of the 4th European Conference on Turbomachinery – Fluid Dynamics and Thermodynamics 2001*, Firenze, Italy, 2001.

Brush Seals in Steam Turbine Power Plant

SIMON I. HOGG
ALSTOM Power
Newbold Road, Rugby
Warwickshire
CV21 2NH, UK

ABSTRACT

Brush seals are a new type of advanced aerodynamic seal for turbomachinery applications. They have found increasing industrial use in recent decades, principally in gas turbine aeroengines and their derivatives. Brush seals essentially consist of a pack of bristles (usually Haynes 25 – a cobalt based alloy) which form a physical barrier to flow between stationary and rotating components. The pack of bristles has sufficient compliance to accommodate rotor excursions resulting from vibration and thermal growth during normal operation. Industry experience of this technology is increasing rapidly and brush seals have now operated successfully for over 25,000 hours in steam turbines and 40,000 hours in industrial gas turbines.

The technology has progressed to the point that brush seals are becoming standard features of some steam turbine designs. This paper discusses the engineering and economic constraints that govern the development of brush seal technology for applications in steam turbines. The performance benefit to be gained from improved sealing at various locations along the steam path is considered. This allows a ranking to be established to show the most cost effective areas for improved cylinder performance (and hence reduced emissions) through the use of brush seals. The engineering design and operational issues that influence the application of brush seals are also described and the need for further base technology development is identified. This allows a second ranking to be defined, based on technical difficulty/risk of applying brush seals in various areas under consideration. Finally, some current examples of brush seal applications in ALSTOM steam turbines are given.

INTRODUCTION

Brush seals are a new type of advanced aerodynamic seal for turbomachinery applications. They have found increasing industrial use over the last decade, principally in gas turbine aeroengines and their derivatives. Brush seals essentially consist of a pack of super-alloy (usually Haynes 25 a cobalt based alloy - nickel based Haynes 214 or Inconel X750 are options for nuclear steam turbine application to avoid the use of cobalt) bristles which form a physical barrier to flow between stationary and rotating components. The pack of bristles has sufficient compliance to accommodate rotor excursions resulting from vibration and thermal growth during normal operation. Bristle wear is clearly a key issue governing application of the seals. Development of this technology has now progressed to the point that seal lives of the order of 10,000's hours are being claimed by gas turbines manufacturers. Interest in brush seals in the steam turbine industry is growing as a result of this experience and the increased drive for greater efficiencies. Some steam turbine manufacturers are offering brush seals within their product range and a large number of steam turbines are operating worldwide with brush seals. A number of these units have now been in successful operation for several years.

COMPARISON OF BRUSH AND LABYRINTH SEALS

Conventional Labyrinth Seals

Labyrinth type seals have been used to reduce leakage flows between rotating and stationary steam turbine components since the very earliest designs. This type of seal is non-contacting during normal operation and consists of a series of sharp-edged baffle fin restrictions. The baffle fins can be mounted or machined directly on to both stationary and rotating components. The clearance between the baffle fin and the opposing sealing surface is set to accommodate the largest expected rotor excursions during operation, including allowances for the effects of manufacturing and assembly tolerances. The clearance is typically 1.0 mm or less for HP's rising to over 2.0 mm for some LP turbine stage sealing applications. The radial height of the baffle fins are an order of magnitude larger than the fin clearance. The axial spacing between successive fins is similar to their height. With this arrangement, any steam driven through the labyrinth by the pressure drop across it, will be accelerated as it passes under the baffle fins so that it gains kinetic energy at the expense of static pressure. The kinetic energy will then be dissipated as the jet expands in an uncontrolled fashion into the relatively large cavity immediately downstream of the baffle. This process is then repeated at the next baffle fin. The labyrinth creates a very torturous path for flow due to this series of high loss generating rapid expansions, resulting in low leakage flow. The radial clearances between the rotor and stationary components, at blade tips and shaft glands, are sized to ensure optimum cylinder thermal performance is maintained whilst avoiding radial contact over a wide range of operating conditions.

Conventional labyrinth seal designs are often mounted on spring backed glands in order to maintain relatively small baffle fin clearances in steady state operation, whilst at the same time accommodating the relatively large radial rotor movements that can occur under low load conditions as the machine is brought up to speed through criticals. The clearance requirements are calculated based on rotor seal diameter, rotor speed, casing geometry, and the position of the stationary component within the cylinder. Allowance is made for any radial thermal differential expansions, casing distortion, key clearances, movement of the rotor within the bearings, and rotor dynamic effects. The values of these allowances vary according to the different operating conditions i.e. barring, run-up to speed, early loading, full load operation etc. The limiting condition needing the largest requirement is used to determine the clearance provided. When labyrinth seals do come in contact during operation the most immediate effect is that the sharp knife edging on the seal tips is lost. In severe cases the rubbing can lead to vibration problems and can cause localised heating leading to rotor vibration. However even in mild cases although the loss of the knife edge is not a significant mechanical issue, both clearance and the discharge coefficient for the seal increase, making the seal far less effective.

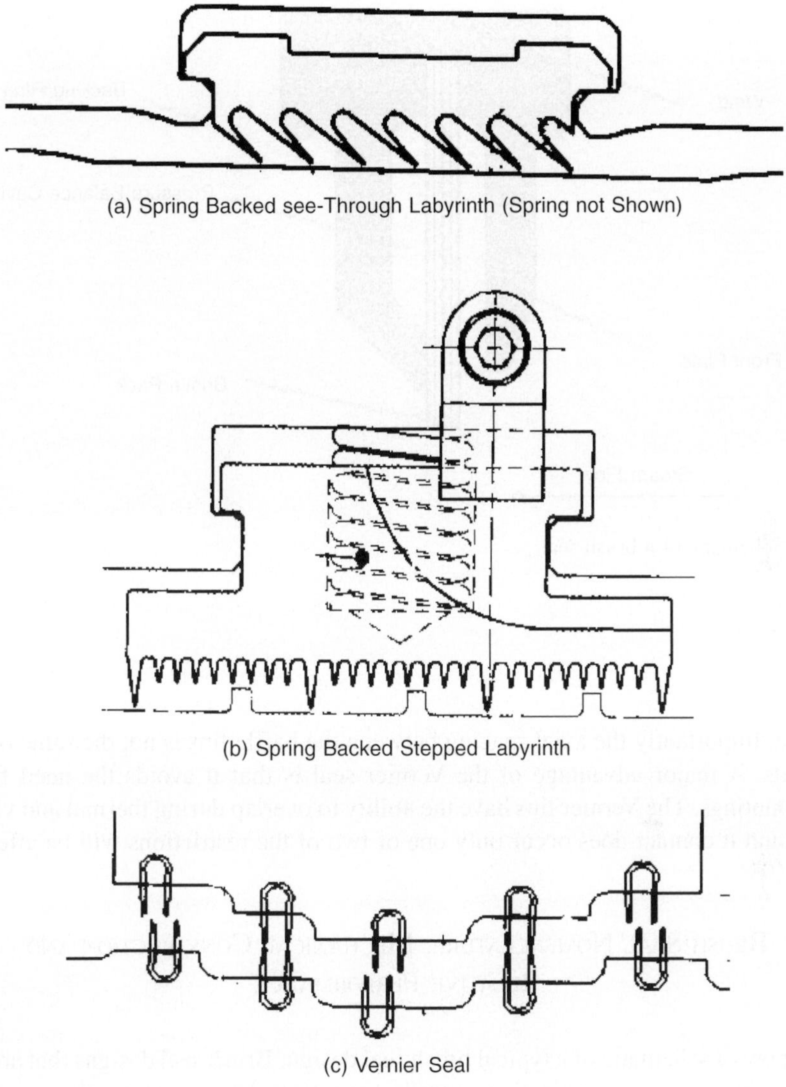

(a) Spring Backed see-Through Labyrinth (Spring not Shown)
(b) Spring Backed Stepped Labyrinth
(c) Vernier Seal

Fig. 1 Different labyrinth seal types.

Figure 1 shows three typical types of modern labyrinth seal design. The simplest design is the see-through labyrinth shown in Figure 1a. The stepped labyrinth (Figure 1b) avoids 'carry-over' (resulting in increased leakage) of undissipated kinetic energy, as the leakage jet from one restriction impinges onto the next. The leakage through a see-through seal design is typically 50% higher than that through an equivalent stepped labyrinth seal. The third type of labyrinth seal shown (Figure 1c) is the Vernier seal. The stepped Vernier design, as shown in the figure, has a similar leakage performance to the conventional stepped labyrinth

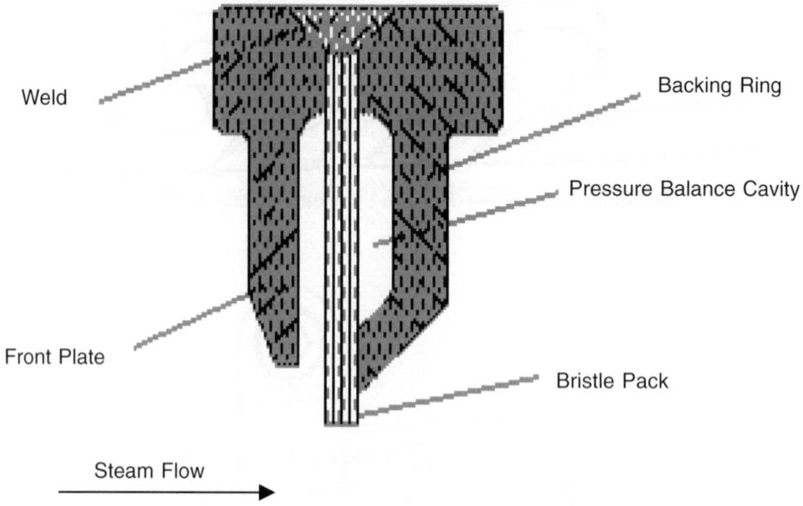

Fig. 2 Schematic of a brush seal.

seal design. Importantly the axial spacing between the baffle fins is not the same on the two components. A major advantage of the Vernier seal is that it avoids the need for spring backed mountings. The Vernier fins have the ability to overlap during thermal and vibrational transients and it contact does occur only one or two of the restrictions will be affected.

Brush Seal Nomenclature, Mechanical Construction and Relative Performance

Figure 2 shows a schematic of a typical brush seal design. Brush seal designs that are suitable for steam turbines sealing applications usually have 10 to 15 rows of bristles axially with a wire diameter in the range 0.1 to 0.15 mm. The bristles are inclined radially, often by 45 degrees, in order to prevent them from picking up on the rotor. The bristles are supported by a backing ring on the downstream side of the pack. Compared to labyrinth seals, brush seals offer a dramatic improvement (between 50 and 90% reduction) in leakage performance. The superior performance of the brush seal comes from the blockage between the backing ring and the rotor from the flexible bristle pack. The brush seal is often located so that a clearance exists between the nominal bristle tip location and the rotor under steady operating conditions, in order to minimize wear. During operation, the aerodynamic forces on the bristles due to the leakage flow, will cause them to move down and close up the bristle tip clearance further reducing leakage flow. This is the well know bristle '*blow-down*' effect.

A front plate is positioned on the upstream side of the bristle pack. This plate helps to protect the bristle pack from the effects of inlet swirl, flow instabilities, particle impact damage, and from handling damage during installation. The bristles are held in place by a circumferential weld between the front plate and the backing ring running around the periphery of the brush seal.

The backing ring design shown in Figure 2 has a pressure balancing cavity immediately behind the bristles. The aim of this feature is to reduce friction between the bristle pack and the backing ring, enabling the bristles to follow rotor excursions more readily. The velocity of the flow between the bristles in the vicinity of the cavity is quite low and so the pressure in the cavity is quite close to the pressure upstream of the seal. Tests by several companies have shown that the pressure balancing feature significantly improves the mechanical operating characteristics of the seal without having a large impact on leakage flow.

The brush seal design described above is a welded construction. Most suppliers use welded construction techniques to manufacture brush seals. One supplier however, uses a novel crimped construction method that avoids the need for welding completely. In order to decide which brush seal supplier to use, ALSTOM has carried out back-to-back endurance rig tests in steam on brush seals.

The backing ring of the brush seal shown in Figure 2 is thinned down towards the inner radius, so that the consequence of a rub with the rotor is similar to that for a conventional labyrinth strip. i.e. significant localised heating can occur causing rotor vibration and the leakage performance of the seal is reduced. This is typical of brush seal designs for turbomachinery applications. It is normally necessary to mount brush seals in spring-backed or retractable glands in steam turbine applications where these would be used with the equivalent conventional labyrinth glands. Unfortunately, for steam turbines in general, opening up the backing ring clearance to avoid this would result in overloading causing the bristles to deform plastically due to the higher bending loads caused by the pressure drop across the seal.

In addition to contact with the backing ring, shaft heating from the rubbing contact with the bristles is also an important effect that influences brush seal design. This is discussed in more detail later in the paper.

EVALUATION OF BRUSH SEALS IN STEAM TURBINES

The author's company offers both reaction and impulse technology steam turbine products.[1] Figure 3 shows the HP cylinder of a typical reaction machine. Four sealing locations are identified in the figure where the company is currently active in developing new sealing designs using brush seals. It is also intended to apply the new seal designs using brush seals at corresponding locations in IP and LP cylinders, when the technology evaluates. The Company is also developing brush seals designs for application on fixed blade diaphragm glands, for it's impulse technology cylinders.

For blading flows, the contribution of leakage loss to the total stage loss decreases with increasing blade height. At the present time, brush seals are high cost items compared to conventional labyrinth seals. Nevertheless, they usually evaluate in HP and IP cylinders due to the dramatic reduction in leakage flow. Sometimes, brush seals do not evaluate in LP cylinders and often they only evaluate on the early dry stages.

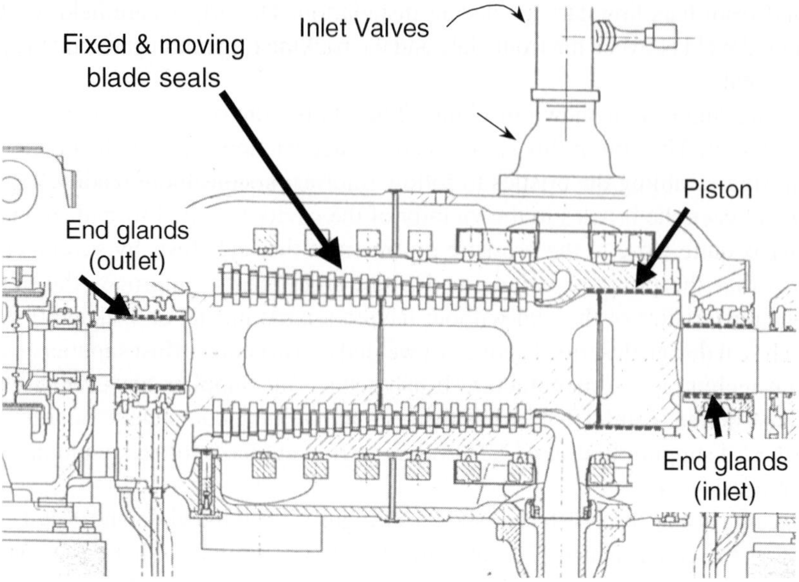

Fig. 3 Principal steam seal locations in a typical HP/IP cylinder.

Table 1 shows the power loss due to various leakages for a typical large (500 to 600 MW) and medium sized (200 to 300 MW) impulse steam turbine. The values are intended to give a feel for the relative magnitudes of the various losses. Clearly the absolute values will vary for different designs, but overall the general picture is representative. Balance gland and shaft end gland leakage losses are similar for reaction machines. Fixed and moving blade leakage losses differ in that they are essentially identical in reaction cylinders and larger than the corresponding losses for impulse machines (reaction stages are penalised by higher leakage loss but benefit for lower blade profile losses compared to impulse stages).

Table 1 indicates that applying brush seals on the fixed and moving blade of short height (HP) stages will produce the greatest benefit in terms of increased shaft power. As discussed earlier, the benefit is less for the blades in IP and LP cylinders because stage leakage losses are less important for longer blades. Significant power increases can also be achieved by reducing leakage on balance glands and shaft end glands. The potential gains, though generally lower than for short height blading, are still quite large.

Brush seal component costs are high (typically \$75 per circumferential inch of brush seal). Sales volumes to the turbomachinery industry are not yet large enough for significant cost reductions to be achieved. For blading seals (particularly moving blade tips), large seal diameters are required for a significant number of stages (over 20 stages for some reaction cylinders). The total cost of installing brush seals on blades is considerable and even though the power gains are large, the evaluation rate (\$ per kW) is often not that high (particularly on small machines). Brush seals tend to evaluate much better on balance glands and shaft end glands. Pressure drops on these applications are higher than for blade seals (usually over

Table 1 Distribution of leakage losses for typical impulse technology cylinders.

Sealing Location	Power Loss Due to Leakage Flow (kW)	
	Large Machine (500 to 600 MW)	Medium Machine (200 to 300 MW)
HP Fixed Blades (all stages)	1000 kW	600 kW
IP Fixed Blades (all stages)	300 kW	100 kW
LP Fixed Blades (all stages)	500 kW	200 kW
HP Moving Blades (all stages)	1700 kW	1300 kW
IP Moving Blades (all stages)	500 kW	400 kW
LP Moving Blades (all stages)	900 kW	300 kW
HP Balance Gland	800 kW	900 kW (Centre HP/IP)
IP Balance Gland	700 kW	-
HP Shaft End Gland (per end)	700 KW	500 kW
IP Shaft End Gland (per end)	100 kW	50 kW
LP Shaft End Gland (per end)	300 kW	150 kW

100 bar for HP balance glands compared to under 10 bar for most blading seals). A number of brush seals in series are required for these high pressure drop applications. Nevertheless, the total number of seals needed is less than for sealing on blades and seal diameters are lower. This combination generally results in much more favourable evaluation rates for brush seals on balance glands and shaft end glands compared to fixed and moving blade seals.

Although the economic evaluation described above shows that the best gains can be made from developing brush seal systems for balance glands and shaft end glands, there are a number of brush seal operational issues (in addition to the high pressure drops already mentioned) that must be taken into account in order to correctly prioritise development efforts for the various potential sealing applications. These are discussed in the next sections.

OPERATION OF BRUSH SEALS

There are a number of phenomena that need to be understood, over and above the basic mechanical design of the brush seal to avoid over-pressurizing the bristles etc., in order to apply brush seals successfully in turbomachinery applications. These are discussed below.

Heat Dissipation from Rubbing Contact

During operation of a brush seal, the aerodynamic loading on the bristles exerts a force on them towards the rotor. The aerodynamic forces increase the torque exerted on the rotor by

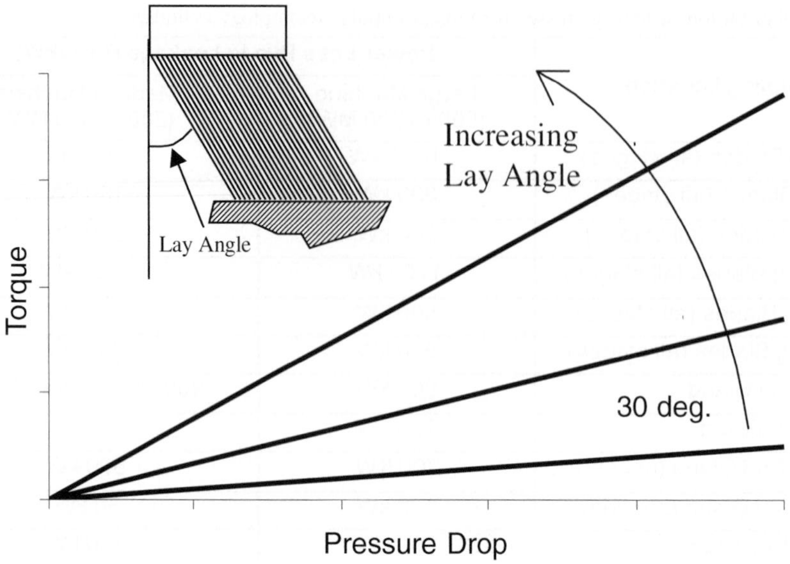

Fig. 4 Effect of bristle angle on rotor torque.

the brush seal due to friction. The aerodynamic loading on the bristles is a strong function of bristle angle. Figure 4 below shows how bristle lay angle (defined as the bristle angle to the radial direction) influences torque in a typical brush seal test. The aerodynamic stiffening of the bristle pack dominates the mechanical stiffness of the bristles and results in much greater brush seal torques as the bristle lay angle is increased.

In steam turbines, unlike gas turbines, seals are often required to run against solid shafts. It is well-known that rubs from conventional labyrinth gland seals can generate sufficient localised heating of solid rotors to bend rotors unacceptably. Similar problems can arise with brush seals unless the heat dissipated by the frictional contact between the bristles and the rotor is carefully managed. Fortunately, the bristle pack itself acts as a very effective heat exchanger, so, if the brush seal is operating with a reasonable pressure drop across it, most of the heat generated is convected away by the leakage flow through the bristle pack. However, during start-up and shut down when the machine is unloaded and under low steam flow conditions, any rubbing can result in significant non-uniform circumferential heating of the rotor, in particular when the rotor passes through critical speeds. One method for avoiding this situation is to mount the brush seal on retractable packing so that there is no contact between the bristles and the shaft under low load conditions. An alternative approach is to carefully design the brush seal and its rubbing surface, so that bristle heating is reduced to acceptable levels. ALSTOM Power are currently developing a number of solutions that avoid unacceptable shaft thermal distortion when running brush seals against solid rotors in several different applications.

Bristle Stability and Rotordynamic Stability when Running Multiple Brush Seals in Series

There are two stability issues that must be taken into account when running multiple brush seals in series. The first of these concerns the aeroelastic stability of the bristle pack itself. Great care needs to be taken when designing brush seal configurations with more than one seal in series, to ensure that the leakage jets from the seals do not impinge directly onto the bristles of the next seal downstream. If this happens, the combination of bristle excitation from the impinging jet and the radially outward flow up the front face of the bristle pack due to the recirculating flow in the inter-seal cavity, can destabilise the bristles to the extent that they experience very large oscillations.[2] This results in rapid wear and bristle failure.

The second stability issue associated with multiple brush seals in series or brush seals embedded in labyrinth seals, concerns the stability of the rotor. Conventional multi-fin labyrinth glands can generate destabilising aerodynamic forces on the rotor , if swirl is present in the inlet flow to the gland and the gland clearance is small. As the seal clearance is reduced, the potential for generating significant circumferential variation in static pressure in the cavity trapped between adjacent restrictions when the rotor moves off line increases. The pressure variation essentially produces a cross-couple stiffness force on the rotor which can destabilise the rotor. In this context, brush seals can be viewed as conventional labyrinth sealing strips with low clearance. Destabilising forces can not be generated by a single seal , but if more than one seal is used in series the effect on the stability of the rotor needs to be accounted for in the design phase. Based on test results ALSTOM has modified it's standard labyrinth seal methods to account for brush seals.

Effect of Disturbances in the Upstream Flow and Disturbances on the Brush Seal Running Surface

It is important to avoid significant disturbances in the flow immediately upstream and downstream of the brush seal. High inlet swirl can be detrimental to brush seal performance producing a bristle failure mode similar to that noted in the previous section concerning brush seals in series. Large localised disturbances can also result in damage over just the section of the bristle pack that is immediately downstream of the disturbance. In addition, discontinuities in the rotor surface (bumps between integral shrouded blades for example) can exert dynamic forces on the bristles at frequencies that are sufficiently close to critical values for the bristles resulting in a fatigue type failure. All of these factors need to be taken into account when designing brush seal systems.

Brush Seal Product Introduction

Many of the phenomena described above are particularly relevant to brush seal applications on balance glands and shaft end glands , in particular the potential for shaft thermal distortion from running brush seal directly onto solid shafts and destabilising rotordynamic forces due to multiple brush seals in series. These are not relevent to brush seal operating on moving blade tips (heat input to the moving blade shroud can not thermally distort the shaft and only a single seal is required on each stage to accommodate the pressure drop. It is for these reasons that ALSTOM's earliest experience of brush seal developments for steam turbines applications has been on moving blade tips. A number of units are now operating in the field with brush seals installed at this location. More recently ALSTOM has registered patents for new brush seal systems for balance pistons and shaft end glands. Prototype designs have now been produced for these applications and they are currently being evaluated in test rig and site trials. Some details of ALSTOM's current site experience of brush seals in steam turbines are given in the next section.

FIELD EXPERIENCE

At present ALSTOM Power has brush seals operating on the HP and IP moving blade seals of a number of retrofit units. They are also operating on the balance piston of one of the Company's HP cylinders. In total, ALSTOM Power has accumulated several years of steam turbine operating experience, on these units.

Brush seal performance has been monitored over many thousands of hours operation in a test facility (using high pressure steam as the working fluid) and on a trial application on the end-glands of a boiler feed pump turbine. No significant degradation in leakage performance over time has been observed in the tests. Figure 5 shows one of the brush seals installed on the feed pump turbine after approximately 6 months operation. The seal is viewed from the downstream side in the picture. The bristles are visible between the backing ring and the rotor and they do not seem to have suffered any significant degradation during operation. When this photograph was taken the rotor was axially displaced from it's running position. The apparently rough rotor surface is not where the seal runs. The seals have now been operating successfully for several years without experiencing any significant drop in performance.

There is sufficient industry experience to indicate that long-term operation of brush seals over the life of large steam turbine power plant can be achieved. As indicated previously, many of ALSTOM Power's brush seal applications in steam turbines to date are on the moving blade tips of short height stages. Figure 6 shows brush seals installed in one of the Company's impulse technology steam turbine retrofits. The brush seals are mounted directly into the diaphragm extension rings and they are backed up by additional conventional labyrinth sealing strips.

Brush seals are also being developed for the moving blade tips of ALSTOM's reaction technology steam turbines. Figure 7 shows brush seals being installed in the inner casing of a HP cylinder reaction retrofit. The brush seal is mounted in a groove machined into the inner casing of the cylinder. The casing grooves to locate the fixed blades are also visible in Figure 7. In this example the brush seal is backed up by two conventional labyrinth seal strips. The performance benefit from brush seals operating in tip seal applications similar to these examples is predicted by ALSTOM's brush seal leakage models to be approximately 0.5% on cylinder efficiency.

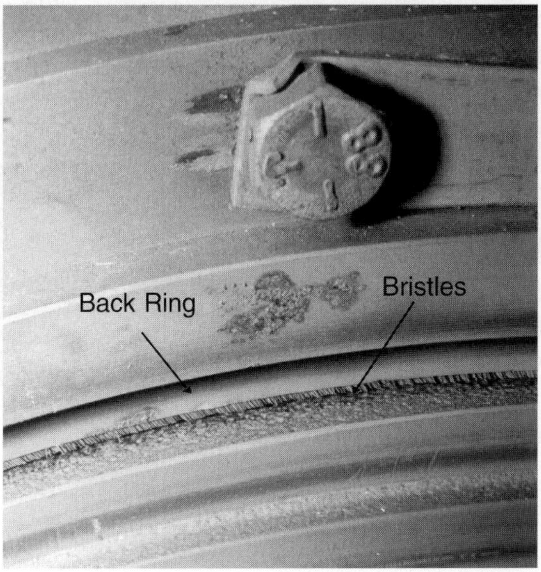

Fig. 5 Brush seals installed on a boiler feed pump turbine after 6 months operation.

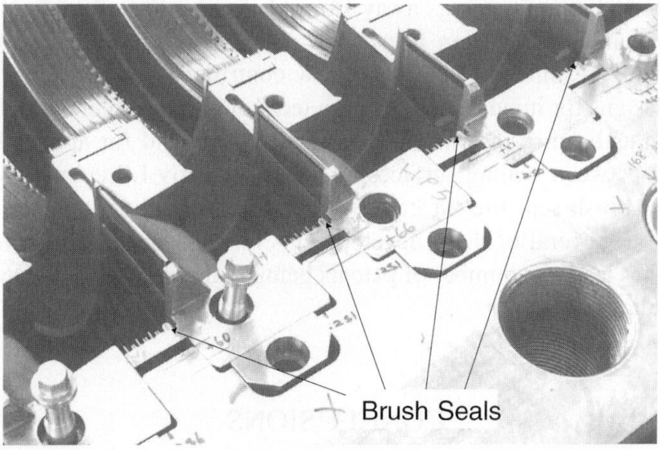

Fig. 6 Brush seals installed in an ALSTOM impulse retrofit steam turbine.

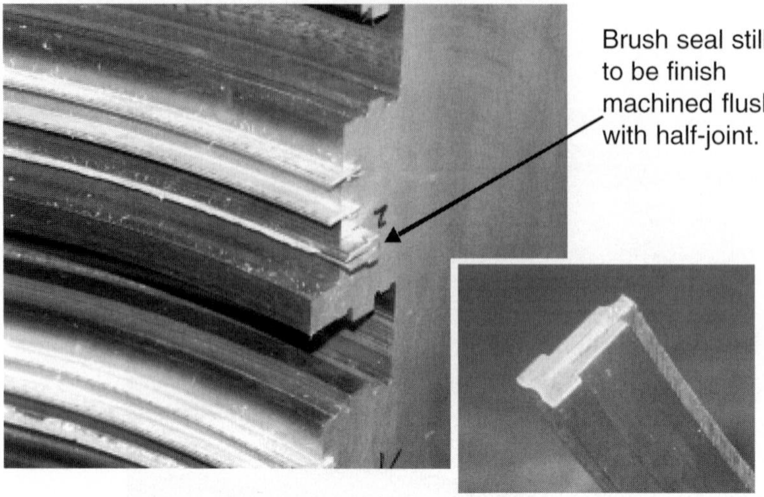
Brush seal still
to be finish
machined flush
with half-joint.

Fig. 7 Brush seals installed in an ALSTOM reaction steam turbine.

Brush seal applications on moving blade tips avoid some of the operational issues described in the previous section, e.g. any heat generated by bristle friction at the blade tips can not bend the shaft. Although, there are other aspects of this application which make this a challenging environment.

ALSTOM currently has brush seal development programs underway aimed at developing the technology for other applications away from blade tips. For example, the mechanisms associated with shaft heating from brush seal rubbing have been subjected to detailed investigation.[3] These results and others are now contributing to the development of new sealing system designs for impulse and reaction technology fixed blades, high pressure drop applications such as balance pistons and shaft end glands and for applications with large radial movement between sealing surfaces i.e. seals for early LP cylinder stages. Systems for safe-guarding brush seal life on balance pistons and end-gland applications are also under development. Several of these developments have now reached the trial application stage. The work has lead to a number of patents being registered by ALSTOM in the general area of brush sealing.

CONCLUSIONS

ALSTOM Power has identified brush seals as the new sealing technology that has the potential to bring about a significant reduction in leakage losses in the Companies steam turbines products. It has had a major development programme aimed at applying brush seals in all of the rotating steam sealing applications where the technology evaluates. The technology is being developed for application in the Company's impulse and reaction technology products. An economic evaluation of the potential gains from brush seals indicates that applications on balance glands and shaft end glands have the greatest benefit. However, consideration of the operational issues surrounding the successful application of brush seals have shown that these locations are also the most challenging for applying the new technology. For this reason ALSTOM has initially focussed on applying brush seals on the moving blade tips of short height stages. The Company has already accumulated several years operating experience for this application. Developments are currently underway to extend the range of applications within the turbine, with several new designs for balance glands and shaft end glands in the trial application phase.

REFERENCES

1. J. A. Hesketh, S. I. Hogg and D. Stephen: 'A Stage Efficiency Prediction Method and Related Performance Aspects of Retrofits on Disc/Diaphragm Steam Turbines', *International Joint Power Gen. Conference*, Atlanta, Paper No. IJPGC2003-40145.
2. A. T. O'Neill, S. I. Hogg, P. A. Withers, M. T. Turner and T. V. Jones: 'Multiple Brush Seals in Series', *ASME Turbo Expo.*, Orlando, Paper No. 97-GT-194, 1997.
3. A. K. Owen, T. V. Jones, S. M. Guo and S. I. Hogg: 'An Experimental and Theoretical Study of Brush Seal and Shaft Thermal Interaction', *ASME Turbo Expo.*, Atlanta, Paper No. GT2003-38276, 2003.

Shrunk on Disk Technology in Large Nuclear Power Plants — The Benchmark Against Stress Corrosion Cracking

WALTER DAVID, ANDREAS FELDMÜLLER and HEINRICH OEYNHAUSEN
Siemens Power Generation

INTRODUCTION

The modernisation of steam turbines to improve plant efficiency plays an important role in reaching the targets set at Kyoto.

Especially blades, developed in the last years, combined with new sealing technologies and turbine casings with optimised flow paths[1] are key factors for ecological, as well as economical, improvements of a power plant.

Regarding the modernisation of power plants, along with efficiency is the issue of reliability of the equipment a driving force.

In the last decades, many turbines, especially in nuclear plants, were affected by stress corrosion cracking (SCC), mainly in the large LP rotors.[2]

Repair work is often time consuming and neither supports the economics nor solves general design deficiencies. Therefore newly designed equipment, applying latest technology to improve reliability, as well as efficiency, is often a more advantageous option.

As large half speed (1500/1800 RPM) LP turbines were first developed in the late 60s, qualified monoblock rotors were not available in the required sizes. Therefore most of the turbine suppliers started to develop LP turbines with shrunk on disk designs.

Although the basic principle was the same, the design details of the suppliers were extremely different varying varied widely from excellent in their ability to avoid SCC, to poor, i.e., strongly affected by SCC. Figure 1 shows typical problems with stress corrosion cracking found on shrunk on disk rotors in the past.[2] All major manufacturers worldwide, except one, had to improve their design due to a large numbers of reported failures which led to the shrunk on disk generally receiving a poor design reputation. Because of this, most suppliers converted to using monoblock rotors as their design principle, based on improvements in large forgings, as well as welded rotors.

This paper will show that in contrast to this, a shrunk on disk rotor is not an design, but rather <u>the benchmark for all other nuclear LP rotor designs</u>.

With a closer examination of the background of the SCC phenomenon, it will be clear to all that the design details are the distinguishing factors and not the design principle itself. This notion is supported by the fact that the turbines built with the technology described in

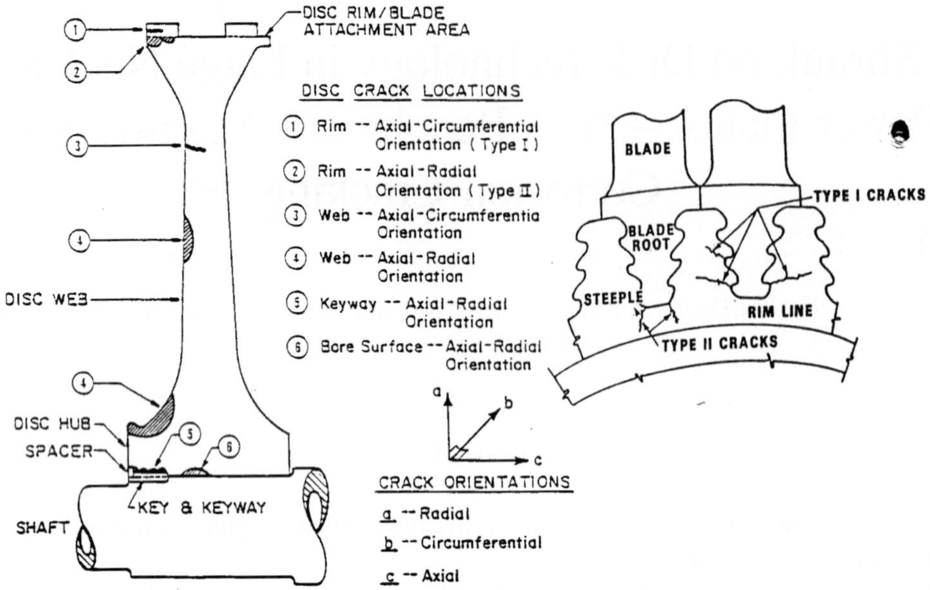

Fig. 1 SCC problems published in the past.[2]

this paper did not experience an SCC problem. This technology was chosen to replace LP turbine rotors built by GE and Alstom, including all varieties of design principles, such as:

- Shrunk on disk [3]
- Monoblock [4]
- Welded disk

The reason for the replacement of these non OEM rotors was the gain in efficiency and power, but often a reliability problem (e.g., SCC in the blade attachments) was an additional reason for the modernisation.

All these different designs which can be seen in Figure 2 were replaced by the advanced shrunk on disk rotor design developed by Siemens PG.

DESIGN HISTORY OF THE ADVANCED SHRUNK ON DISK ROTORS

The original shrunk on disk design for large nuclear LP turbines was based on ten disks per rotor (see Figure 3). With the advanced shrunk on disk rotor designs the number of disks per flow was reduced as this was found to be beneficial for efficiency, as well as SCC prevention. As an example, the larger first disk, allows for an increased number of blade stages to reduce stage loading and increase blade efficiency. Moreover the compressive stresses applied during heat treatment (see below) benefit from this design modification.

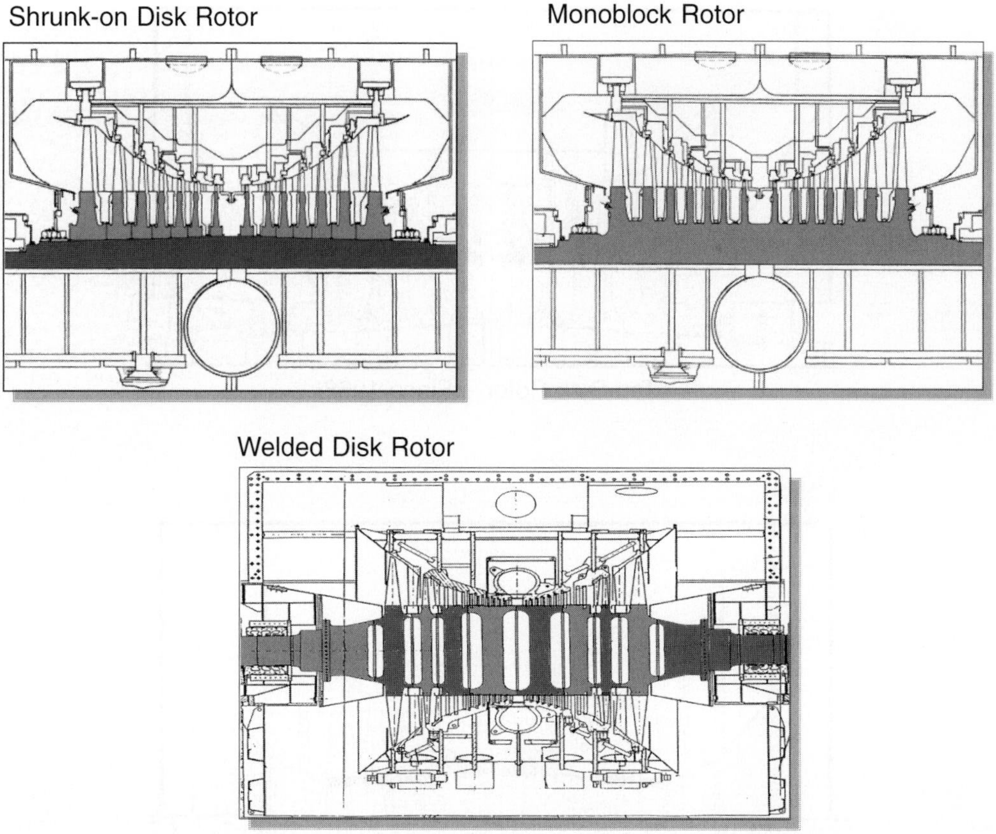

Fig. 2 Advanced disk rotors replace various rotor designs.

On top of the reduced number of disks, it was found that in the last two disks per flow, anti-rotation keys were not required if a sufficient shrink fit was applied.

The following portion of this paper describes the SCC phenomenon and the preventative measures to avoid SCC attacks on the rotor components.

STRESS CORROSION CRACKING OF TURBINE DISKS

In Figure 4, the factors influencing the stress corrosion cracking behaviour (SCC) and the Siemens measures to avoid SCC are summarized. Although Siemens low-pressure turbines were not affected by stress corrosion cracking, Siemens PG began extensive studies into the causes and prevention of stress corrosion cracking as a precautionary measure in the early 80's.

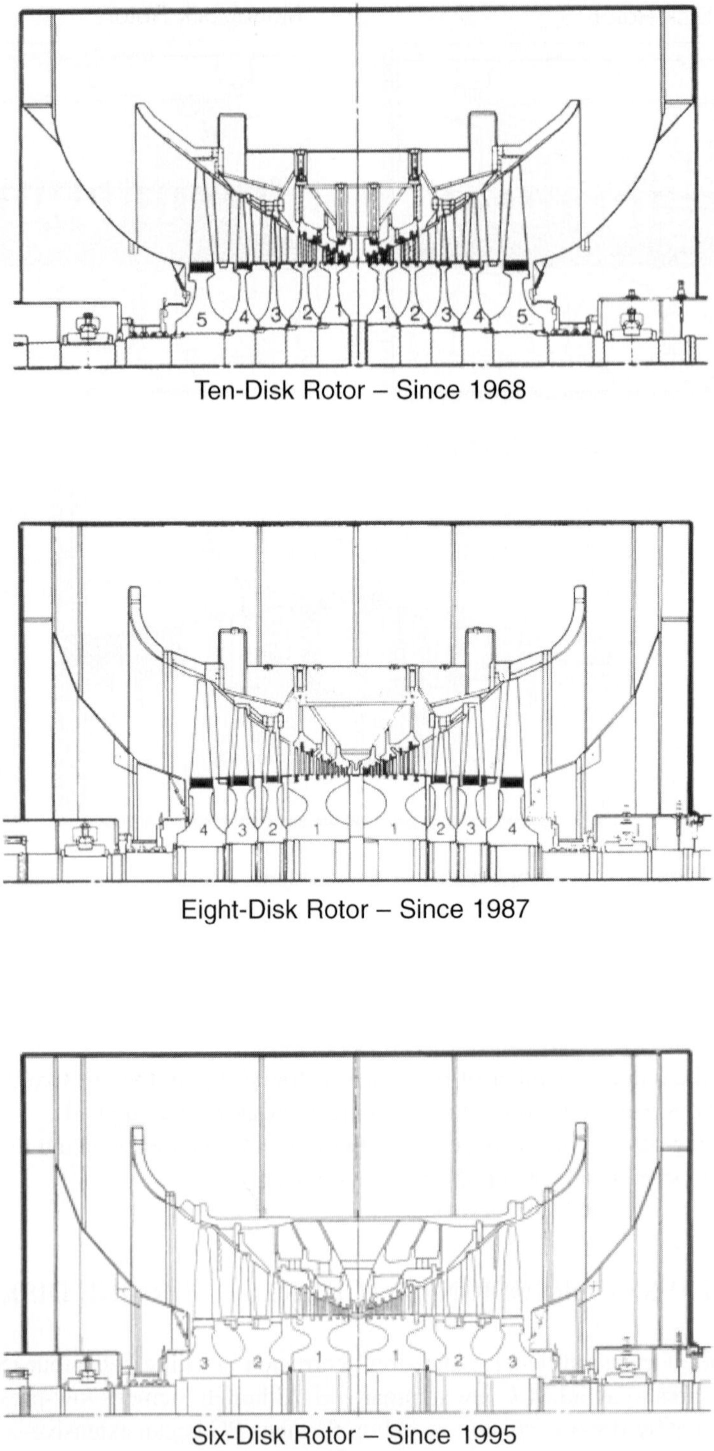

Fig. 3 Development of disk rotors.

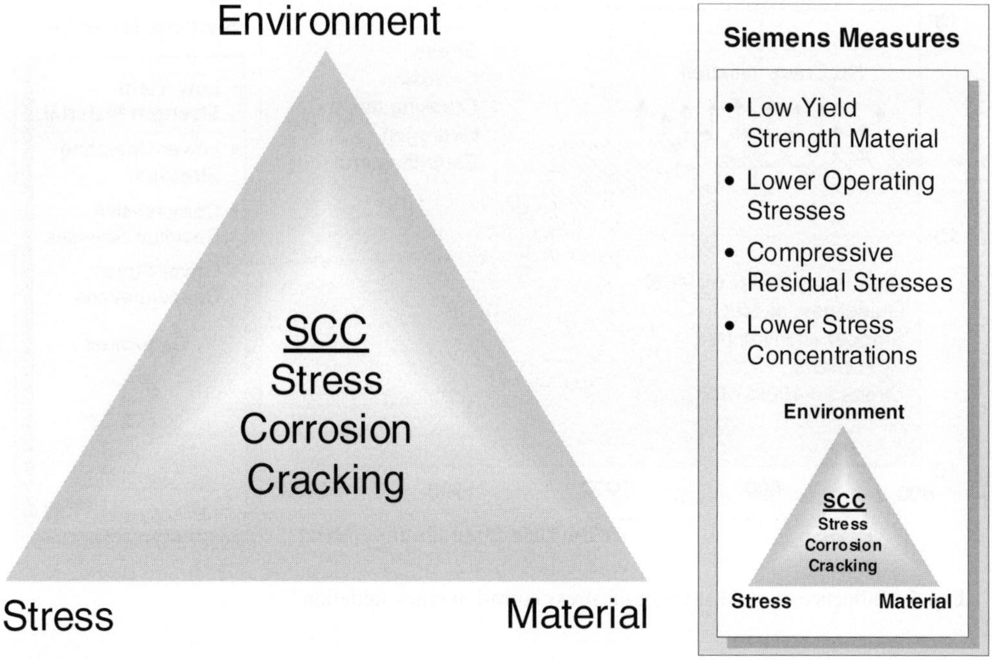

Fig. 4 Stress corrosion cracking (SCC) and prevention.

MATERIAL PARAMETERS

The quenched and tempered alloy steels that were investigated had different amounts of nickel, chromium, molybdenum and vanadium. Also considered were various degrees of purity including steels with segregations and elevated inclusion content and a 'super-clean' variant. Some steels were tempered at different temperatures to achieve 0.2% yield strength levels between 700 and 1250 MPa.

LABORATORY STUDIES OF STRESS CORROSION CRACKING

Siemens PG studied the stress corrosion cracking in steam turbine disk steels as a function of water chemistry and various material parameters, such as strength, chemical composition, steel purity, melting process, etc. High-purity water with a conductivity of < 0.2 μS/cm, both deoxygenated (oxygen < 10 ppb) and oxygen saturated, at temperatures of up to 150°C, was used as the corrosive medium. To simulate irregularities in the water steam cycle, gaseous and other types of impurities were added to increase conductivity.

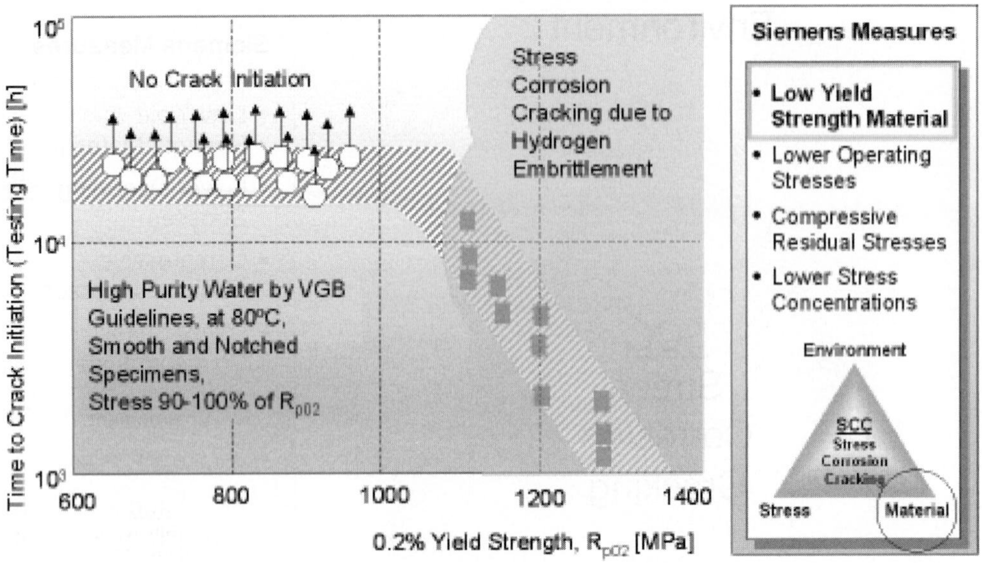

Fig. 5 Influence of yield strength on stress corrosion crack initiation.

STRESS CORROSION CRACK INITIATION

In high-purity water with a conductivity of < 0.2 μS/cm, only the quenched-and-tempered condition, i.e., the hardness of the steel, influences initiation of stress corrosion cracking. Up to a 0.2% yield strength of 970 MPa, stress corrosion cracking did not initiate cracking in any of the quenched-and-tempered conditions studied. Nor did it cause cracking in smooth or notched test pieces at stresses above the 0.2% yield strength (Figure 5).[5–7] This is independent of the purity of the steel. In high-purity water, those steels conventionally melted 25 years ago perform just as well as high-purity steels made using the ESR process.[8] Under high purity water conditions, even non-metallic inclusions (such as Al_2O_3, MnS, etc.) in the surface of the material do not act as crack initiators and have no effect on stress corrosion cracking resistance. This holds true in both pure oxygen and oxygen-saturated water. No localised, anodic dissolution on the material surface (pit formation) was observed under these high-purity corrosion conditions.

Under quenched-and-tempered conditions the steels with a 0.2% yield strength > 1085 MPa were damaged by hydrogen-induced stress corrosion cracking (Figure 5).

STRESS CORROSION CRACK PROPAGATION

The two phases, crack initiation and crack propagation, must be looked at separately with stress corrosion cracking. As stated above, stress corrosion cracking does not initiate cracking

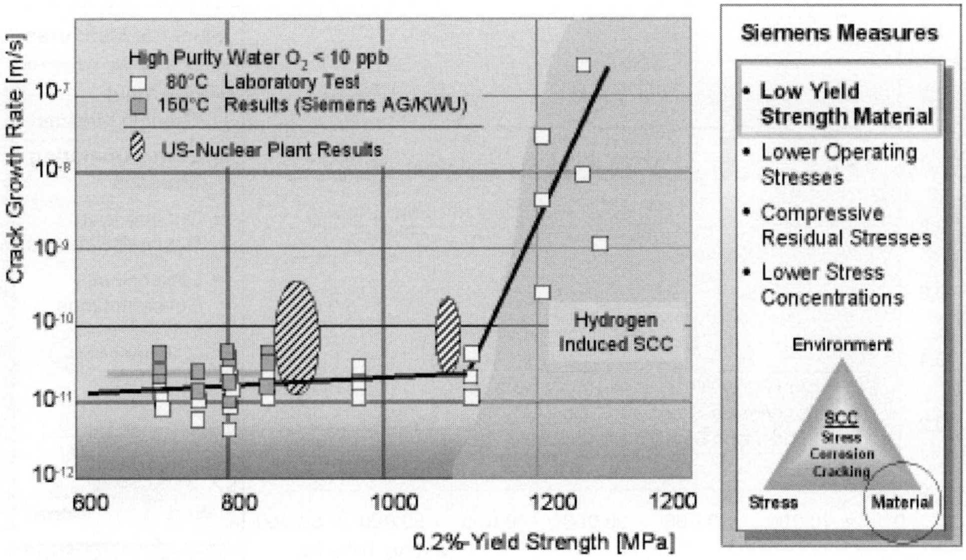

Fig. 6 Influence of yield strength on stress corrosion crack growth rate.

in either smooth or notched specimens in high-purity water. However, stress corrosion cracking does result in crack propagation in statically loaded fracture mechanics test specimens with very sharp, deep cracks in high-purity water.

Crack propagation results are shown in Figure 6. As with crack initiation, there is a 0.2% yield strength threshold between 1000 and 1100 MPa above which the mechanism, and subsequently the crack propagation rate, drastically changes. With 0.2% yield strength between 650 and 1085 MPa, propagation rates for stress corrosion cracking are almost entirely independent of strength. At a service temperature of 80°C, a crack would, depending on the material's strength, propagate by 0.2 to 0.8 mm per year due to stress corrosion cracking.

Higher aggressiveness in the media only has a slight influence on crack propagation since the electrolyte in the crack is almost decoupled from the surrounding medium and nearly constant electrolyte conditions are established at the crack tip.[6, 7]

When the 0.2% yield strength exceeds roughly 1100 MPa, the crack propagation rate increases drastically by several powers of ten. In this strength range, hydrogen-induced cracking plays a decisive role during crack propagation.

In summary, whereas crack initiation does not occur in high-purity water at 0.2% yield strength below roughly 1000 MPa, existing cracks will propagate also under high purity water conditions due to stress corrosion cracking. Cracks which have initiated in media with elevated conductivity can therefore continue to propagate even if high-purity water is present outside the crack.

Therefore the maximum allowable service stresses, to avoid stress corrosion crack initiation, under different steam and water conditions must be defined.

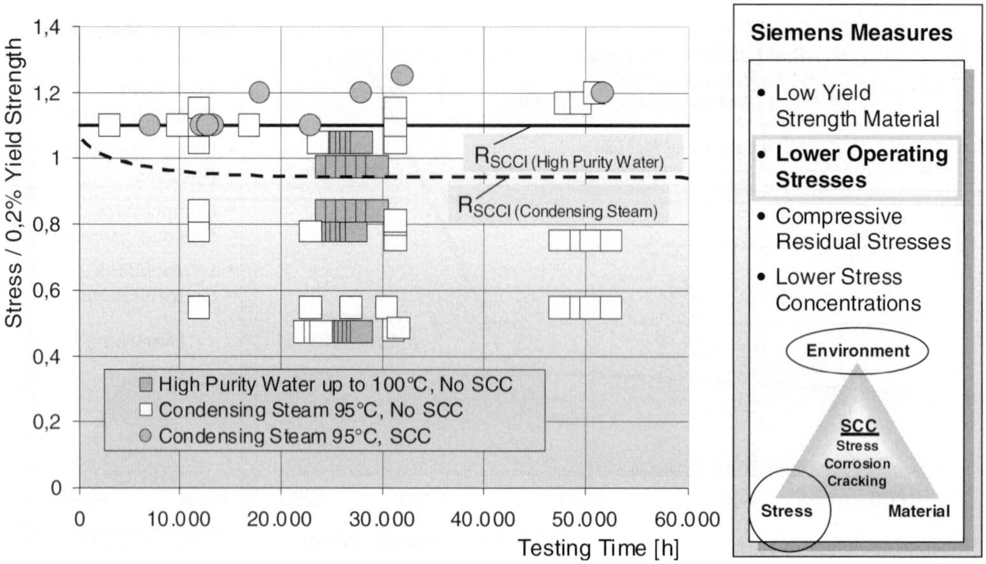

Fig. 7 Stress corrosion crack initiation tests on turbine disk steels, 0.2%-yield strength < 1000 MPa.

THRESHOLD VALUES TO PREVENT SCC INITIATION

Our own results and data from the literature regarding studies of stress corrosion cracking initiation in low-alloy quenched-and-tempered steels for steam turbine disks with 0.2% yield strength <1000 MPa were summarized and disk used below. The goal was to define stress limits for the prevention of stress corrosion crack initiation as a function of the corrosive medium.

The results in high purity water are summarized in Figure 7.[5–8, 10, 11] Since no stress corrosion cracking occurred, the limit stress for the prevention of crack initiation was defined as

R_{SCCI} (high-purity water) = 1.1

(R_{SCCI} [service stress/0.2% yield strength] = threshold value for preventing stress corrosion crack initiation in rotor steels 0.2% yield strength <1000 MPa).

Figure 7 also includes results of stress corrosion crack initiation studies in condensing steam.[11–13] These studies simulate conditions as can occur during operation in the region of the Wilson line. In this region, the condensate first forms on the surface of the material. Because of the coefficients of distribution, enriched concentrations of water-soluble impurities in the water/steam cycle can occur here and cause pitting and crack initiation.

R_{SCCI} (condensing steam) = 0.9

Similar threshold curves as shown in Figure 7 were developed with data under more aggressive corrosion conditions, such as

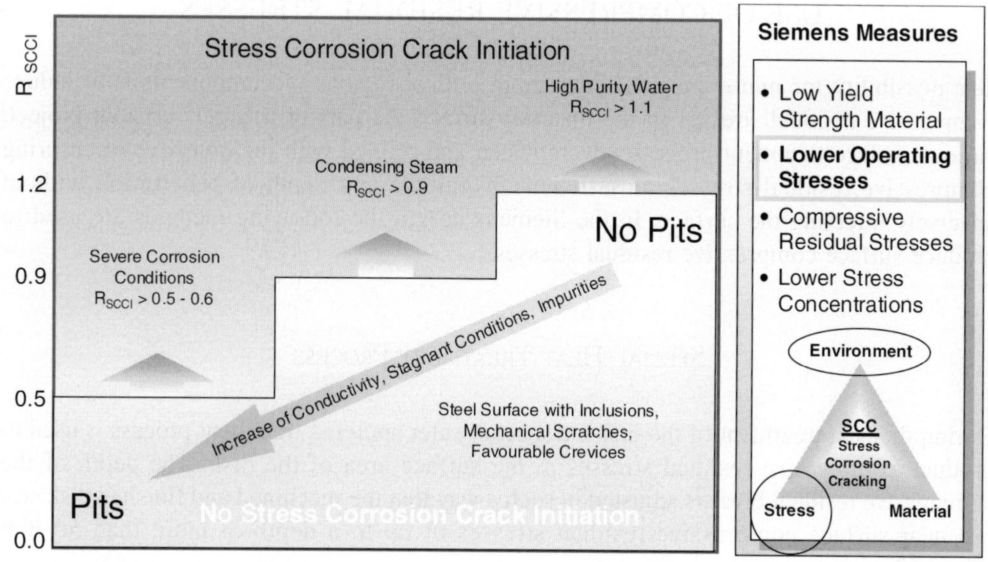

Fig. 8 Stress Corrosion Crack Initiation Threshold Value R_{SCCI} for Turbine Disk Steels, 0.2%-Yield Strength < 1000 MPa.

- condensing steam with crevice conditions[11]
- water with gaseous impurities[5, 6, 8, 10, 14]
- stagnant water without refreshing[6–8, 10, 14]

This simulates strong increase in conductivity, which however is not typical for normal operating conditions. In these relatively aggressive media for low-alloy steel, high stresses are likewise required to produce stress corrosion cracking. The majority of test specimens exhibited no stress corrosion cracking with stress/0.2% yield strength ≥ 0.9.

Even in 30% NaOH solutions at 100–120°C, the SCC-initiation time is still clearly dependent on stress.[15–17] Crack initiation can therefore be precluded even during short disturbances if the associated service stresses remain below 50 to 60% of the 0.2% yield strength:

R_{SCCI} (aggressive medium conditions) = 0.5 – 0.6

Figure 8 provides a summary of which stress limits must be exceeded to initiate stress corrosion cracks in low-alloy quenched-and-tempered steels for steam turbine rotors with a 0.2% proof stress < 1000 MPa. These stress limits are independent of the material parameters chemical composition, steel purity, surface inclusions or similar metallurgical factors as can occur with heats produced using conventional industrial techniques. This includes surface imperfections, such as mechanical notches, surface scratches and comparable surface damage as can occur in singular cases despite careful processing and stringent quality assurance measures.

USE OF COMPRESSIVE RESIDUAL STRESSES

One possibility for minimizing the stresses in critical regions of components is to induce compressive residual stresses in the material surface. As part of a larger research project, various surface treatment processes were tested and refined with the objective of ensuring compressive residual stresses of sufficient magnitude and depth of penetration without adversely affecting the surface. In the Siemens design the following methods are used to produce surface compressive residual stresses:

Special Heat Treatment Process

During the heat treatment of the disks, a special water spraying treatment process is used to produce compressive residual stresses in the surface area of the disk. The depth of the compressive residual layer is adjusted in such a way that the machined and finished disk still has near surface compressive residual stresses of up to a depth of more than 50 mm (Figure 9).

Machining and Shot Peening

To reduce the potential for stress corrosion cracking, the machining of the grooves and the entire rim section of the disks have been optimised to minimize residual tensile surface stresses at any particular location. In nearly all cases after machining, a thin layer of compressive residual stresses exists. In the rotor sections which could possibly be exposed to stress corrosion, the disk surfaces, and certain sections of the blade attachment zone, are additionally shot peened to produce deeper compressive surface residual stresses (Figure 10).

Rolling of the Keyways

After the shrinking of the disks, the keyway bores are drilled, followed by a rolling and honing procedure. This honing process guarantees the maximum level of compressive residual stresses directly at the surface (Figure 11).

COMPRESSIVE RESIDUAL STRESSES AND STRESS CORROSION CRACK INITIATION

Comparative studies of stress corrosion cracking were performed on test specimens with rolled and non-rolled bores. Tests were performed in sodium hydroxide to study the mechanism of anodic crack tip dissolution while cathodic polarization was used to simulate

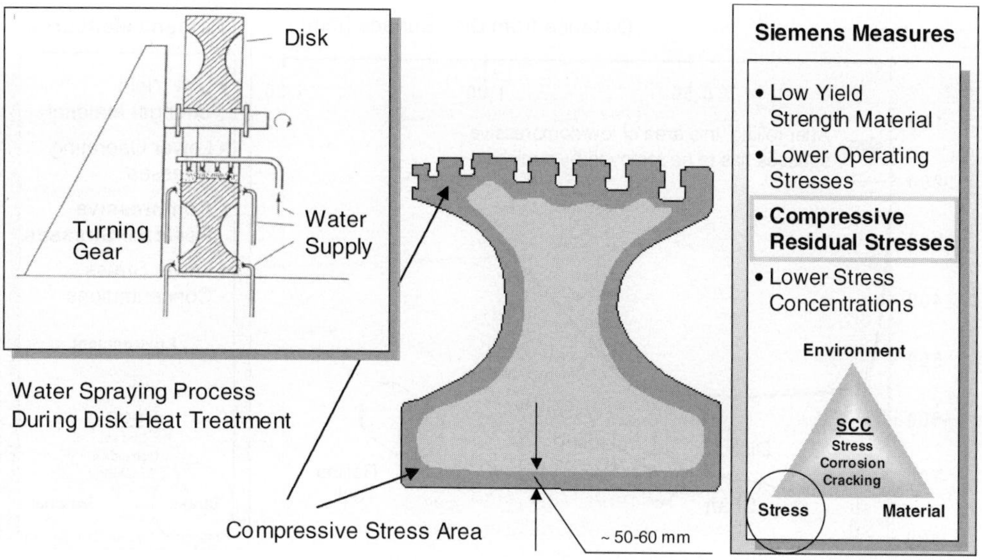

Fig. 9 Compression residual stresses produced during heat treatment process.

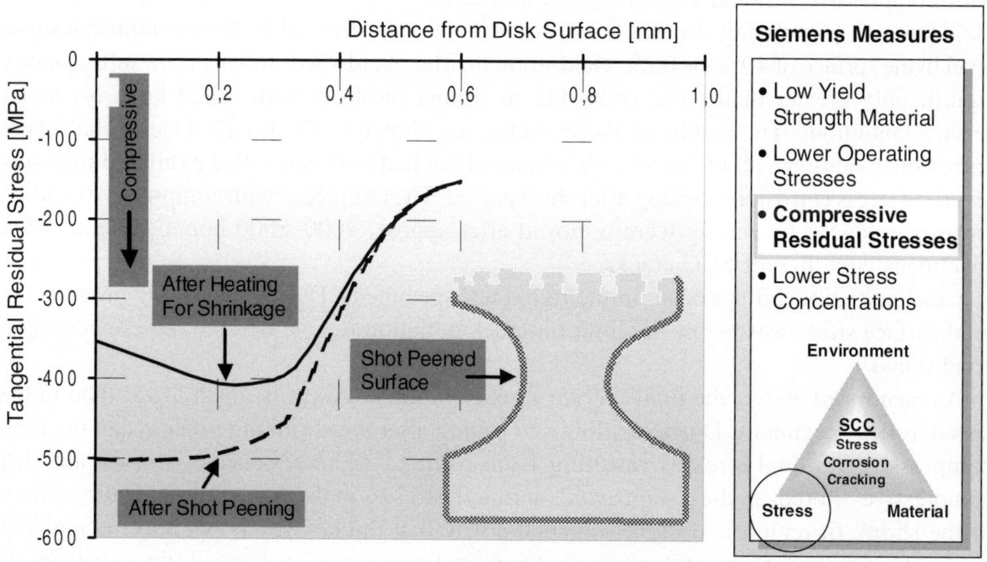

Fig. 10 Compression residual stresses produced by shot peening process.

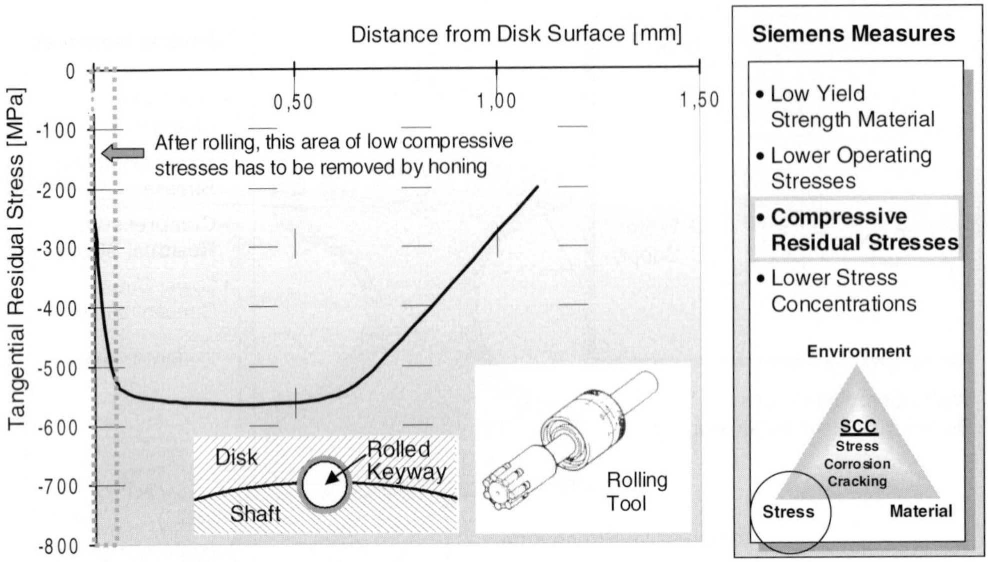

Fig. 11 Compressive residual stresses produced by rolling process.

the mechanism of hydrogen-induced cracking. In the non-rolled test specimens, the stresses in the bore were on the same order of magnitude as the 0.2% yield strength. For the tests with compressive residual stresses, the test specimens were first pre-stressed to the 0.2% yield strength (simulating shrink fit), cold-rolled and then stressed again to establish a stress level in the surface of 40% the 0.2% yield strength. The established stresses were subsequently significantly greater than those occurring in shrunk on disks with rolled keyways under service conditions. The results of these studies are shown in Figure 12. Due to the severe corrosion conditions, all of the test specimens which had not been rolled exhibited more-or-less deep stress corrosion cracking after short period. The test pieces with compressive residual stresses produced by rolling were removed after approx. 4000–8000 hours. All the rolled specimens were free of SCC-cracks.

Crack initiation did not occur in the rolled test specimens. Due to the rolling process, the total surface stress was below the limit for crack initiation R_{SCCI} = 0.5–0.6 (severe corrosion conditions).

As mentioned above, the total service stresses in the keyways is much lower than in the tested, rolled specimens. Due to drilling and rolling after the shrinking process and the high compressive residual stresses resulting from rolling, the total service stresses are still compressive. Moreover, the design avoids shrink fit stresses in the keyway area. The geometry of the shrink fit region is designed in such a way that the keyway is open to the ambient steam. That means steam or condensed steam continuously is exchanged, thus stagnant or crevice conditions with higher conductivity are avoided (Figure 13).

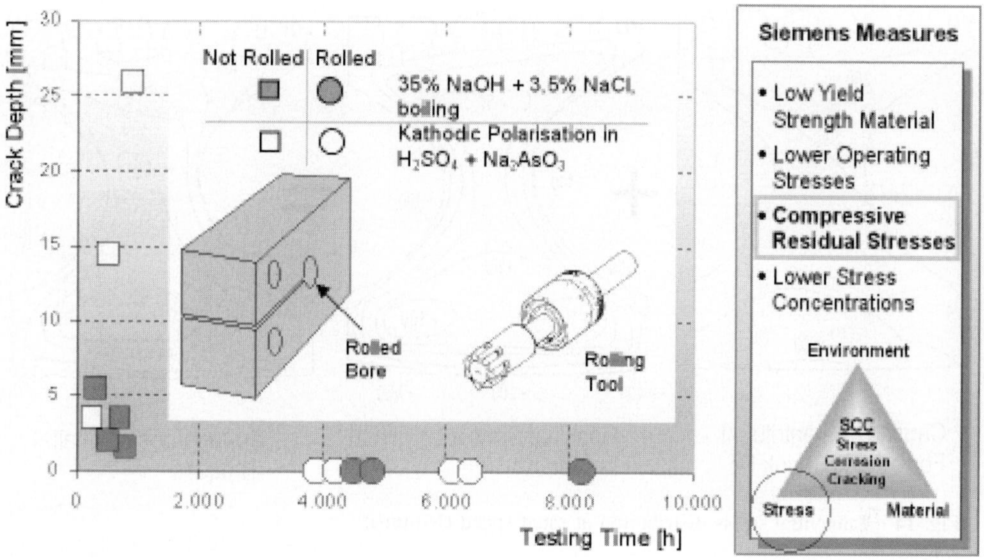

Fig. 12 Stress corrosion crack initiation - modified CT-specimens with residual compression stresses (rolled bore).

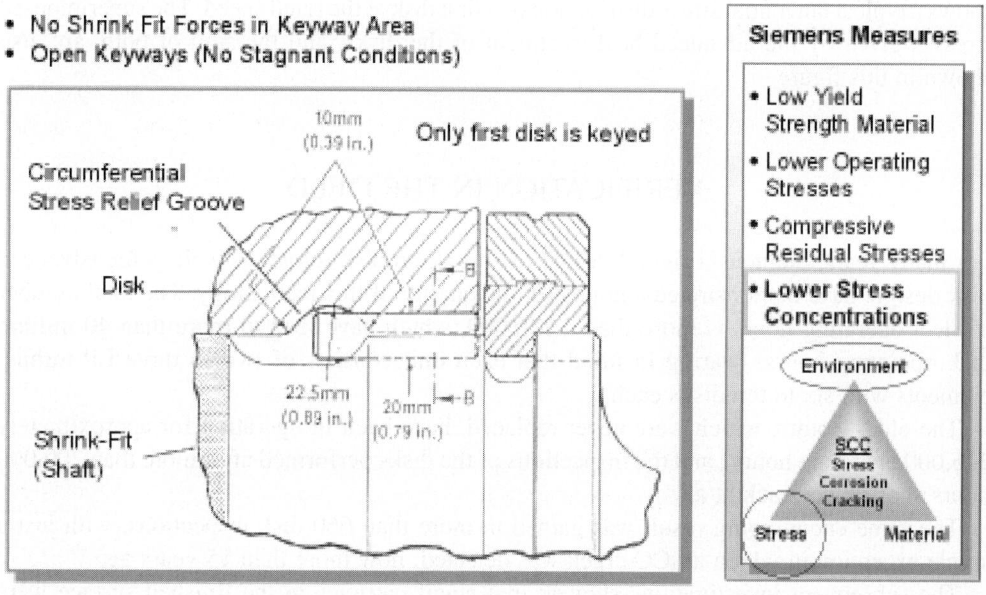

Fig. 13 Design details from shrink fit and keyway area.

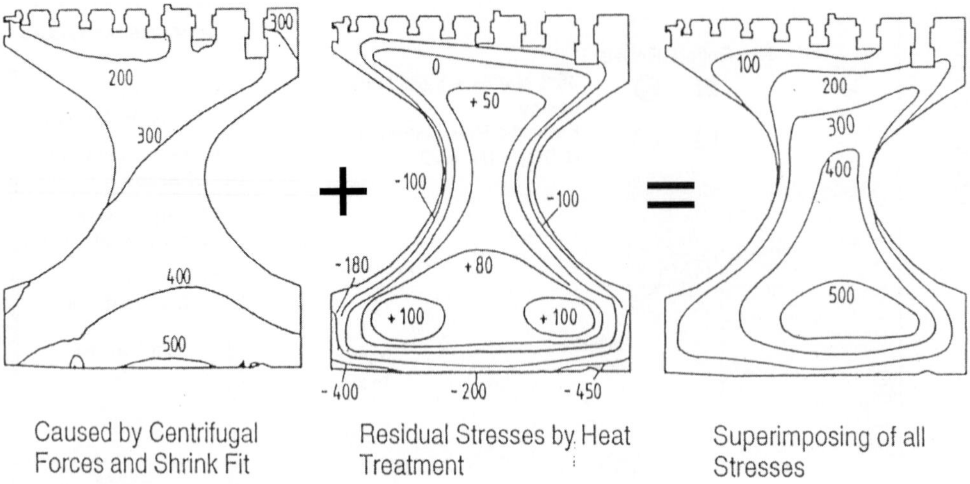

Fig. 14 Tangential stress distribution at rated speed (N/mm²).

REDUCED OPERATING STRESSES

In addition to the locally optimised design in the keyway area, the global geometry of the disks has been designed to reduce all tensile surface stresses to the target level. Figure 14 shows a typical tangential stress distribution of a first disk at the rated speed. The superimposed stresses given by the advanced heat treatment of the disks, and the sum of both, are also shown in this figure.

VERIFICATION IN THE FIELD

For many years, Siemens' original design of shrunk on disk rotors, as well as the advanced disk design, have demonstrated and proven the quality of this technology. The total number of fleet operating hours is more than 2,750,000, which have lead to more than 40 million disk operating hours, bearing in mind that each unit consists of two to three LP turbine elements with six to ten disks each.

The oldest rotors, which were never replaced, have been in operation for approximately 225,000 operating hours , and the inspections of the disks performed after more than 200,000 hours showed no crack at all.

The same encouraging result was gained in more than 660 disk inspections, with just a single exception in which a SCC crack was detected, now more than 15 years ago.[18]

The subsequent investigations showed that small particles in the finished surface were responsible for the crack initiation. This disk was manufactured more than 25 years ago, at

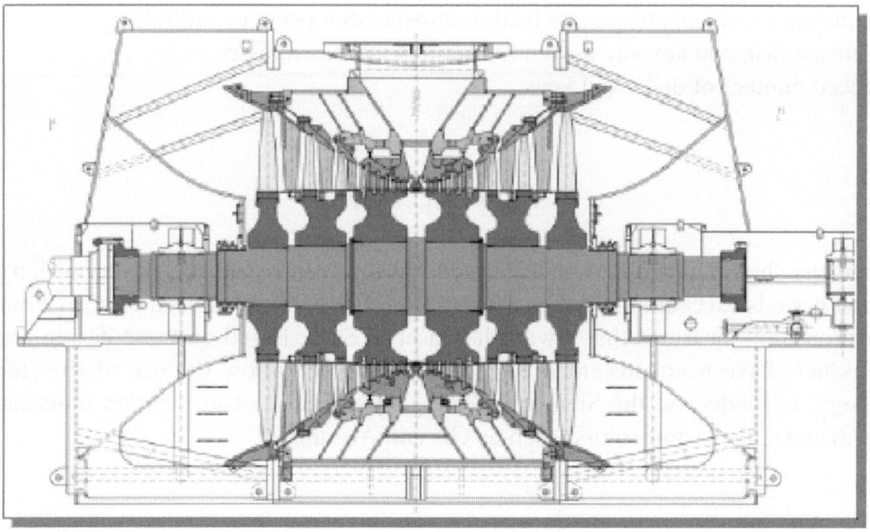

Fig. 15 Modernisation of a Westinghouse BB81/281 LP turbine with advanced disk rotors.

a time when the reasons for SCC were not known in detail. The optimisation of the quality controlled manufacturing process and the developed knowledge of SCC ensured the avoidance of similar events.

Due to this impressive operational record and the availability of new blade and seal designs for improving the turbine efficiency, turbine modernisations were performed in order to optimise the cost effectiveness of the plants with advanced disk technology.

Excellent results were obtained in the German 1400 MW nuclear power plant, Emsland, in May 2000.[19] In only 16 days, two LP inner casings and rotors, including re-work of the outer casing, were able to be fully completed. Additional 32 MW showed the success of this modernisation by the fact that the guarantees were fully met. It has to be noted that this improvement took place despite the fact that the size of the last stage blades remained constant.

NEW PRODUCTS

Many of the turbines originally designed by Siemens have been modernised in order to improve efficiency. Furthermore, products have been made available to improve non OEM turbines like in Figure 2 with the advanced disk design.

Figure 15 shows the advanced shrunk on disk LP rotor, designed for Westinghouse LP turbines BB81 and BB281, including all major design features described in this paper. These include the following factors:

- Compressive residual stresses by heat treatment, shot peening and rolling.
- Optimised disk and keyway design to minimise tensile stresses.
- Reduced number of disks and keys.

CONCLUSION

The Siemens shrunk on disk rotor technology has proven its reliability during long-term operation periods exceeding 30 years. In addition, the advanced disk rotor design employs further protection measures that thwart the onset of SCC. Based on these design features, new products have been presented to the market which allow the use of this superior technology, to modernise the Siemens Westinghouse fleet, not to mention units built by turbine manufacturer competitors, such as GE and Alstom.

REFERENCES

1. M. Deckers and D. Doerwald: 'Steam Turbine Flow Path Optimisations for Improved Efficiency', *Power-Gen Asia '97*, Singapore, 1997.
2. EPRI Report NP-2429, 'Steam Turbine Disk Cracking Experience', Volume 1 through 7, Research Project 1398–5, 1982.
3. M. W. Smiarowski and R. C. Green: 'Turbine Steam Path Retrofit Project Completed at Peco Nuclear's Limerick Generating Station Units 1 and 2', PowerGen International, New Orleans, 1999.
4. J. A. Bartos, M. S. Gorski, L. E Olah and M. W. Smiarowski: 'Susquehanna Steam Electric Station Turbine Retrofit/Generation Uprate: Decision Factors for Long Term Reliability and Improved Performance' PowerGen International, Las Vegas, 2001.
5. W. Engelke, K. Schleithoff, H. A. Jestrich and H. Termuehlen: 'Design, Operating and Inspection Considerations to Control Stress Corrosion of LP Turbine Disks', *American Power Conference*, Chicago, Illinois, 1983.
6. W. David, K. Schleithoff and F. Schmitz: 'Spannungsrißkorrosion in Hochreinem Wasser Von 3 3, 5% NiCrMoV Vergütungsstählen für Dampfturbinen Scheiben und Wellen', Teil 1 und 2, Mat. wiss. u. Werkstofftech. 19, 1988, 43 50, 95 104.
7. H. Oeynhausen, G. Röttger, J. Ewald, K. Schleithoff and H. Termuehlen: 'Reliable Disk Type Rotors for Nuclear Power Plants', *American Power Conference*, Chicago, 1987.
8. W. David, K. Schleithoff, F. Schmitz and J. Ewald: 'Stress Corrosion Cracking Behaviour of Turbine Rotor Disk Materials, Crack Initiation and Propagation, Measures to Prevent SCC', *High Temperature Materials for Power Engineering 1990*, Liege Bel-gium, 1990, 577–588.
9. W. David, G. Röttger, K. Schleithoff, H. Hamel and H. Termühlen: 'Disk-Type LP Turbine Rotor Experience', *International Joint Power Conference*, Kansas City, Missouri, ASME Power Division, 1993.
10. H. Termuehlen, K. Schleithoff and K. Neumann: 'Advanced Disk Type LP Turbine

Rotors', *EPRI Workshop, Stress Corrosion Cracking in Steam Turbines*, Charlotte, North Carolina, 1990.

11. B. W. Roberts and P. Greenfield: 'Stress Corrosion of Steam Turbine Disk and Rotor Steels', *Corrosion-NACE*, **35**(9), 1979, 402–409.
12. S.R. Holdsworth, G. Burnell and C. Smith: 'Factors Influen-Cing Stress Corrosion Crack Initiation in Super Clean 3.5NiCrMoV Rotor Steel', *Superclean Rotor Steels, Work. Proc.*, Sapporo, Japan, 1989, 299–319.
13. S. R. Holdsworth and G. Burnell: 'Stress Corrosion Crack In-itiation in LP Turbine Rotor Steels', *High Temperature Materials for Power Engeneering*, Liege, Belgium, 1990, 555–566.
14. J. Tavast: 'Initiation of Stress Corrosion cracking in LP Rotor Materials', COST 505, Final Report Project Sl, 20.01.89.
15. M. O. Speidel: 'Stress Corrosion Cracking in Steam Tur-bine Rotors, Effects of Materials, Environments and Design', *EPRI Seminar on Low Pressure Turbine Disk Integrity*, San Antonio, Texas, 1983.
16. J. E. Bertilsson and B. Scarlin: 'Betriebssicherheit von Niederdruck Dampfturbinenwellen unter Spannungsriß-Korrosionsbedingungen', Brown Boveri Mitlt. 3/4 84, 169–174.
17. R. B. Scarlin and J. Denk: 'Stress Corrosion Cracking Behaviour of a Clean 3.5~ NiCrMoV Steel in Comparison toConventionell Steels', *Superclean Rotor Steels, Work.Proc.*, Sapporo, Japan, 1989, 263–284.
18. G. Jacobsen, H. Oeynhausen and H. Termuehlen: 'Advanced LP Turbine Installation at 1300 MW Nuclear Power Station Unterweser', *American Power Conference*, Chicago, 1991.
19. H. P. Claßen, H. Oeynhausen and J. Riehl: 'Upgrading of the Low-Pressure Steam Turbines of Nuclear Power Plants', *Siemens Power Journal*, 2001, 26–29.

RISKWISE™:
A User Oriented Risk-Based Maintenance Planning Product

BRIAN J. CANE
TWI Ltd., Granta Park
Great Abington, Cambridge
CB1 6AL, UK

ABSTRACT

Risk-based approaches are an integral part of holistic asset management aimed at all aspects of improved safety and business performance. The benefits of risk-based methods for inspection and repair or replacement optimisation are now recognised by the power and process industry sectors. This paper briefly reviews the development of risk-based inspection & maintenance (RBM) methods including reference to best-practice guidance and industry surveys conducted by TWI in close contact with regulators and operators across several industry sectors.
The requirement to retain application efficiency and user friendliness is an important issue in RBM software tools. The concept of Remaining Life Indicator (RLI) incorporated in TWI's RISKWISE™ software demonstrates that the single semi-quantitative approach can be sufficient for plant-wide inspection planning as well as establishing risk management measures. Incorporation of fully quantitative tools in risk-based methods in most cases should be considered only as a refinement in terms of inspection planning although such approaches have their place where the remaining life is critical.
The paper identifies the key issues for effective implementation of RBM tools and these are illustrated by example application using RISKWISE™.

INTRODUCTION

In recent years privatisation and deregulation of the electricity supply industry has led to a fierce competitive climate for operators striving to maximise profitability whilst maintaining security of supply. Both planned and unplanned outages are now missed opportunities. Maintenance must accordingly be time and cost efficient.

He process of plant asset management is thus increasingly incorporating risk assessment followed by identification of optimum inspection and maintenance measures to selectively mitigate risks to levels consistent with target maintenance outage plans. This process is the essence of risk-based inspection & maintenance (RBM). Products which address RBM are

essentially management support tools which formalise the methodology and practice of inspection and maintenance planning.
The aim of RBM is thus to:

- Offer an integrated methodology for risk management,
- Focus resource on items of greatest risk such that plant safety, inspection and maintenance targets can be achieved in a cost effective manner and
- Increased inspection and maintenance intervals and reduced outage costs.

REGULATIONS AND GUIDELINES

It is essential that risk-based tools recognise safety requirements as set out by the regulators. In the UK, the Health and Safety Executive (HSE) is responsible for drawing up guidelines on tolerability limits for risks at work. These guidelines have provided the basis of safety rules which have been adopted in several other countries. Owing to the upsurge of interest in risk-based approached TWI has worked closely with the HSE and plant insurers to develop an industry best practice guide on risk-based inspection. TWI's RBM software, RISKWISE has been developed to be compliant with such best practice.

A review of regulations and guidelines leading up to risk-based methods has been provided in a recent publication.[1] ASME and API approaches to risk-based inspection are compared as well as methods for determining inspection frequency.

INDUSTRY FEEDBACK SURVEY

In a recent joint industry project, TWI carried out a questionnaire-based survey to gain a better understanding of the needs of plant operators. Approximately ninety of the questionnaires that were distributed worldwide, were completed and returned to TWI.

The majority of respondents were based in the oil & gas refining, fossil power, chemical and petrochemical industries and classified themselves as producers, operators or manufacturers and engineering service contractors. Engineering insurers, equipment manufacturers or suppliers and safety regulating authorities were under-represented in the survey. A summary of the indications and directions resulting from the survey is attached in the ppt format but briefly described below.[2]

- Most of the respondents' companies (69%) had 'Previously implemented' RBM, or were 'Currently implementing' RBM. Several respondents, in all sectors, indicated that they were currently implementing RBM.
- Approximately 98% of all respondents (whose company's had previously undertaken a RBM assessment), indicated that the results of their RBM program had 'Met expectations' or 'Exceeded the expectations' of their company.
- General pressure vessels, piping and heat exchangers are the types of equipment to which RBM is more often applied.
- Approximately 20% of all respondents indicated that their company had established and documented a uniform RBM policy or guidance for application throughout their company.

- Approximately 60% of all respondents whose company is currently using, or intends to use RBM software, indicated that their software was not linked to any other electronic data management system, or other software system.
- Approximately 24% of all respondents indicated that both the input variables and output results of their RBM software were linked to another data management or software system.
- Respondents indicated that there is no substantial increase in the accuracy of semi-quantitative compared to qualitative (subjective) RBM methods.
- Approximately 57% of all respondents indicated that their Safety Regulating Authority accepted RBM as an alternative basis for determining inspection and maintenance intervals.
- For those respondents that had previously undertaken RBM assessments, the two most important reasons for implementing a RBM program were:
  (a) improving the overall safety of critical plant and
  (b) reducing the duration of inspection or maintenance outages.
- Similarly, for those respondents that were currently undertaking a RBM assessments the two most important reasons for implementing its program were:
  (a) improving the overall safety of critical plant and
  (b) extending the interval between inspection or maintenance outages.
- For those respondents that had previously undertaken RBM assessments, the most critical success factors for its implementation programs were:
  (a) having the 'right' people in the assessment team,
  (b) appointing a suitable assessment team leader and
  (c) reliability of the RBM analysis methodology.
- For those respondents that were currently undertaking RBM assessments, the most critical success factors were:
  (a) having the 'right' people in the assessment team,
  (b) reliability of the RBM analysis methodology, and
  (c) appointing a suitable assessment team leader.
- For those respondents that currently use and intend to use RBM software, the most important attributes of the assessment software were:
  (a) overall user-friendliness of the software system; and
  (b) based on well-known or published model or methodology.
- For those respondents that have not and do not intend to use RBM software, the most important attributes were:
  (a) based on well-known or published model or methodology; and
  (b) tracking system for actions.

EFFECTIVE RBM IMPLEMENTATION

The process of RBM should form part of an integrated strategy for managing the integrity of all assets and systems throughout the plant or facility. From the 'Best Practice Guideline' developed by **TWI** in association with the UK Health & Safety Executive the major steps within the process are as follows.[3]

- Establish requirements and clear statement of objectives,
- Define systems, system boundaries and equipment to be addressed,

- Specify the RBM management team and responsibilities,
- Assemble plant database,
- Evaluate failure scenarios, damage mechanisms and uncertainties,
- Perform risk audit based on event probability/likelihood and consequence analyses,
- Review risk management measures and develop risk-focused inspection plan,
- Implement inspection plan and any associated operational/maintenance measures,
- Assessment of inspection findings in terms of remaining life and fitness-for-service,
- Update and feedback to plant database, risk audit and inspection plan on a continuous basis.

Some of the main requirements and critical issues for the effective implementation of RBM which should be recognised by suppliers of RBM programmes and software products, are summarised below.[4]

- User-friendly Software.
- Incorporation of all Damage Mechanisms.
- Audit Team Approach, comprising of relevant experts.
- Holistic Risk Management Measures.
- Formal Link to Inspection Frequency taking Remaining Life into account.

RISKWISE™ SOFTWARE

RISKWISE™, TWI's RBM software for optimising plant maintenance planning, has been developed and continuously updated to meet the foregoing requirements of both users and regulators.[2, 4] Specific outputs are:

- Unit-wide risk audit enables repair and maintenance resources to be risk-focused.
- Safe inspection periods are formerly obtained based on an implicit time dimension of risk.
- Risk mitigation measures are signalled and selected to meet maintenance frequency targets.

RISKWISE™ Has the Following Functionality

- Assesses the likelihood and consequence of failure for equipment items and produces an itemised risk and remaining life profile for each plant unit.
- Is convenient to use and can be easily learned as the risk model allows qualitative as well as quantitative input.
- Includes a regularly updated database with all relevant damage mechanisms, as well as guidance on formulating the likelihood or probability and the consequence of failure.
- Allows the user to appraise or focus/defocus the probability and consequence attributes for each component, e.g. the level of inspection and maintenance and thus mitigate the risk of loss or optimise the current inspection programme.
- Identifies the most likely damage locations in each component and allows inspection to be properly targeted.
- Determines the risk of failure with time which provides a formal basis for assessing remaining life and hence establishing the maximum period between major inspection outages. Figure 1 illustrates this for the case of boiler plant.

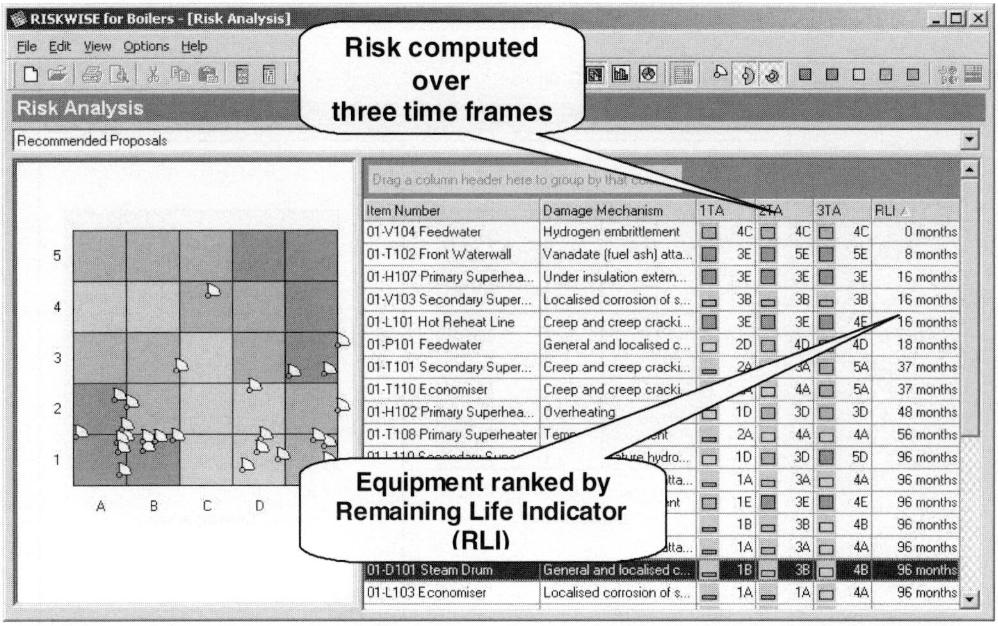

Fig. 1 RISKWISE™ summary screen.

Key Features of RISKWISE™ Include

- User friendly software, fully transparent – ensures buy in by users.
- Uses an audit team approach – accommodates plant experience.
- Readily interfaces with computerised maintenance management systems.
- Incorporates a time-based risk auditing module enabling equipment to be ranked by risk and remaining life.
- Determines inspection/maintenance frequency based on a practical treatment of formal reliability rules - remaining life indicator (RLI) module.
- A risk-management or focus/defocus module – facilitates selection of optimum risk mitigation measures (see Figure 2).
- Fully auditable output – acceptability to insurers/regulators.
- Suite of products includes application to power generation, pipelines, storage facilities and process plants.

Application areas in the energy utility industry include power stations, gas transmission and distribution pipelines, LNG and storage facilities, wind farms, power distribution systems, etc.

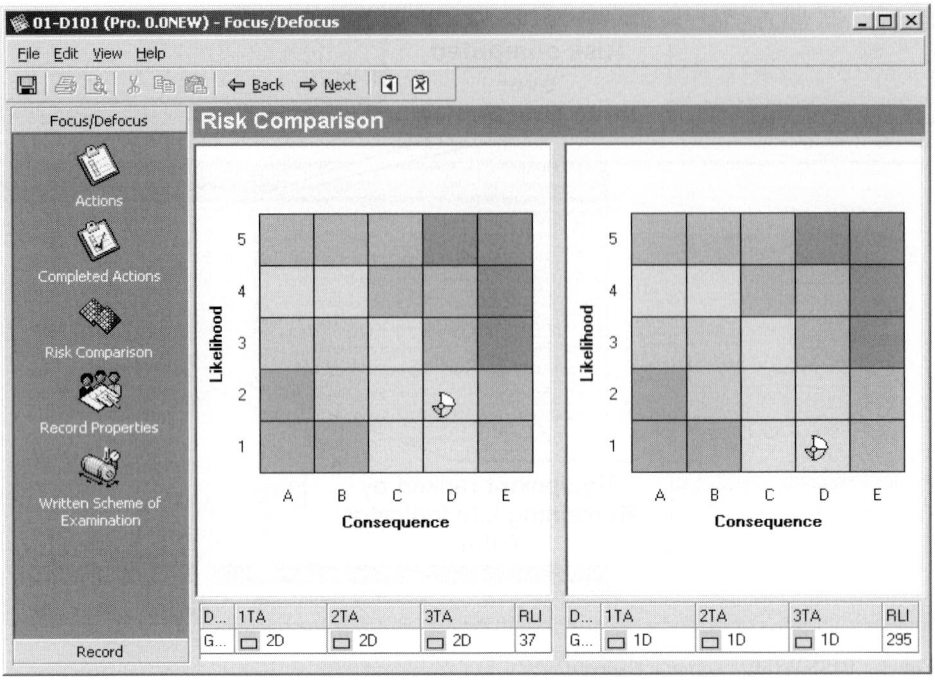

Fig. 2 Comparison of risk matrix and RLI (months) before (LHS) and after (RHS) risk mitigation measure.

SUMMARY

The use of risk-based approaches to plant inspection & maintenance planning are important for survival in deregulated energy markets. User friendly tools such as RISKWISE™ which provide requisite levels of precision in setting safe inspection and maintenance intervals ensure uptake by plant personnel as well as meeting the needs of regulators.

REFERENCES

1. B. J. Cane: 'Risk-Based Inspection and Maintenance – Industry Feedback and User Needs', *International Seminar on Risk-Based Management of Power Plant Equipment*, Institute of Materials, London, 2002.
2. J. B. Speck and A. T. M Iravani: 'Industry Survey of Risk-Based Life Management Practices', *ASME PV & P Conference*, 2002.
3. J. B. Wintle, B. W. Kenzie, G. Amphlett and S. Smalley: 'Best Practice for Plant Integrity Management by Risk-Based Inspection', TWI Public Report No. 12289/1/01, Produced for the UK Health and Safety Executive (HSE), 2001.
4. B. J. Cane: 'User Oriented Risk-Based Integrity Management Tools', *WTIA Conference*, Australia, 2001.

Fractographic and Remnant Life Details of Growing Cracks in LP Turbine Blades

J. H. BULLOCH and P. J. BERNARD
Power Generation
ESB, Head Office
Dublin 2, Ireland

ABSTRACT

The present paper describes details concerning the fractographic nature and predicted remnant life assessment of a series of pin hole root cracks found in the LP blades of a 250 MW steam turbine during a routine inspection. It was shown that the sub-critical crack growth was fatigue in nature which had been initiated from surface corrosion pits that were around 0.2 mm deep.

The fractographic details suggested that near threshold fatigue crack growth conditions were prevalent since no evidence of ductile striations was observed. Also, it was established that fatigue cracks growth was almost identical in different LP blades and that the last stages of growth corresponded to an active fatigue stress range of around 50 MPa.

The fatigue crack growth details of these pin hole root cracks, known as corner cracks, were assessed using

(a) an estimation of the critical crack depth at which fast failure ensued,
(b) the average value of a range of published data which indicated that crack growth rates along the bore and flat surfaces were different and
(c) applying constant probability fatigue crack growth curves to a range of fatigue crack growth data for ferritic steels in water at 120°C.

For the existing LP blade cracks remnant life estimates for the deepest crack were around 3 years and 9 months for respective probability of failure, POF, values of 0.5 and 0.05.

Finally, remnant life estimates were calculated for a range of initial crack depths from around 1 to 3.55 mm.

INTRODUCTION

Martensitic stainless steels, typically containing around 12% chromium, are commonly used as the blading material for both turbine driven compressors and the LP stages of the steam turbines. The condensation process in a turbine can cause local concentrations of impurities at locations where wetting and drying are repeated. As a consequence the blading material can be exposed to fairly aggressive environments; indeed Rottoli and Sigon[1] have reported chloride concentrations of up to 22% on LP blades in service which would greatly accelerate any localised corrosion process. This process would cause corrosion pits to form which will be prime nucleation sites for fatigue cracks. Fatigue cracking is one of the primary causes of

Fig. 1 Details of the corner crack in blade 62.

damage in steam turbine blades with most fatigue crack growth failures occurring in the low pressure, LP, blades.

Recently a series of cracks were found at the pin hole location on LP blades from a 250 MW steam turbine which had around 106,000 hours of operation. The present paper deals with the following,

a. the assessment of these cracks in terms of their fractographic details in an effort to reveal the actual crack initiation and subsequent crack growth mechanisms under real service conditions and
b. using this information to assess some realistic remnant life for these cracked blades.

FRACTOGRAPHIC DETAILS

Two of the blades, 62 and 74, exhibited significant corner cracks at the surface hole location and both blade fracture surfaces exhibited similar features. A general view of the entire corner crack found in blade 62, see Figure 1, showed that the crack shape was likened to that of a quarter ellipse with the crack lengths at the bore and flat surfaces being 3.55 and 1.79 mm respectively. Numerous crack arrest lines, or beach marks, were readily observed. It was evident that this crack had been initiated at a surface corrosion pit which was around 0.64 mm in surface length by some 0.2 mm deep, see Figure 2. Also from Figure 1 a dark

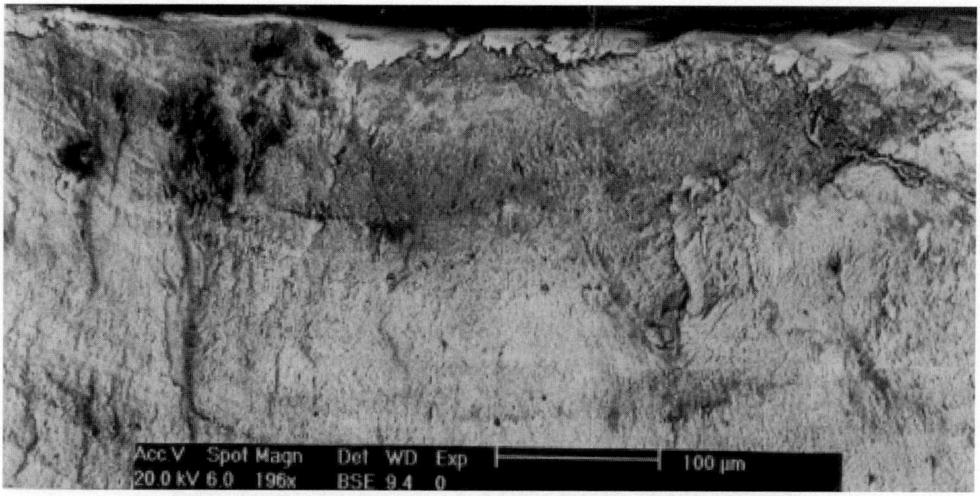

Fig. 2 Details of the surface corrosion pit.

coloured ribbon of some 0.2 mm in size adjacent to the final fast fracture region was evident. The start of this feature coincided with a prominent beach mark and its actual length in the bore, 45 degree and flat locations were assessed at 0.246, 0.194 and 0.117 mm respectively. A detailed view at the 45 degree location, see Figure 3, indicated a series of fine beach marks within the dark ribbon of final sub-critical crack growth. Indeed at the bore location eleven fine beach marks were counted within the 0.246 mm ribbon.

The fracture surface was typical of fatigue crack growth and a general view is illustrated in Figure 4. Where the fracture surface was faceted in nature and was indicative of crack growth in the near threshold fatigue region. Isolated areas of intergranular failure, which ranged from 3 to 6%, were recorded only within the final 0.2 mm of crack growth, see Figure 5. The intergranular facets were around 10 µm in size which coincided with the average microstructural grain size of the present 12%Cr blading steel. Also there was no evidence of ductile striations on the fracture surfaces of the two blades subjected to a detailed examination.

REMNANT LIFE ASSESSMENT

A R6 analysis indicated that the critical crack depths were 13 and 9 mm for the bore and flat face locations respectively. As such the remnant life of the cracked LP blades was taken as the average time to extend the 3.55 mm bore crack to 13 mm.

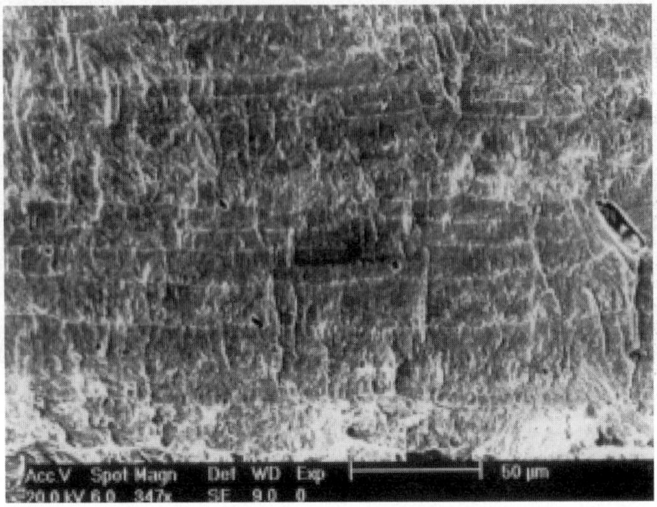

Fig. 3 Details of a series of fine be beach marks over the last 0.2 mm of crack growth.

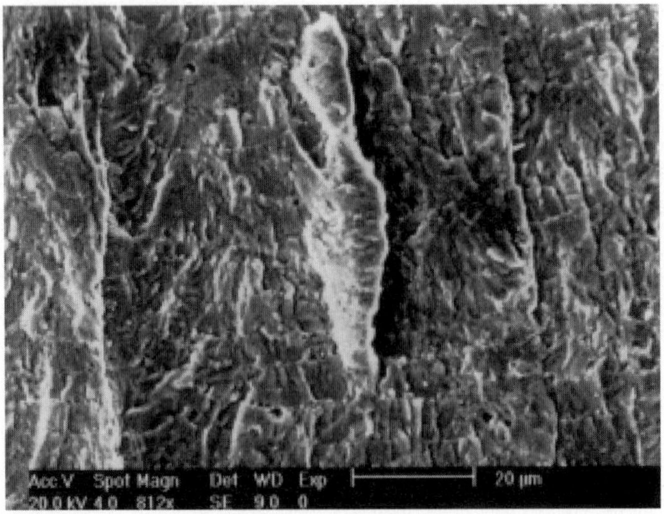

Fig. 4 General view showing ductile fatigue crack growth.

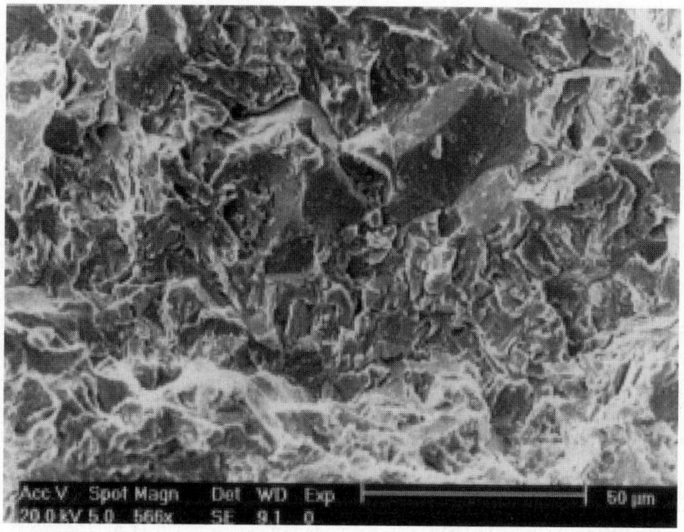

Fig. 5 Deails of isolated intergranular facets within the final 0.2 mm of crack growth.

Since the temperature of the cracked LP blades was estimated to be 120°C an average crack growth law, obtained from a wealth of low alloy ferritic steel ASME high temperature water data at the above temperature, was adopted.[2–4] Also a constant probability fatigue crack growth approach[5] was taken and the fatigue crack growth law was expressed as follows,

$$da/dt \text{ (mm/second)} = C \times (\Delta K)^{4.19}(-\ln P)^{-\beta} \quad (1)$$

where ΔK is the stress intensity range, P is the probability of failure and C and β are scaling constants of value 8.48×10^{-12} and 0.53 respectively.

The fatigue cracks in the LP blades are called corner cracks and it is well known that fatigue crack growth rates along the bore and flat surfaces are different as a result of the differing values of the geometric Y-Factor in the expression,

$$\Delta K \text{ (MPa m)} = Y \times \Delta\sigma\ \pi \times a \quad (2)$$

where $\Delta\sigma$ is the stress range and a is the crack depth.

In order to carry out a fatigue crack growth assessment an estimate of the fatigue cyclic stress, $\Delta\sigma$, which drives the fatigue event can be obtained from both fatigue limit and fatigue threshold data for 12%Cr steels. From the data reported by Gabetta and Torri,[6] Ishii et al.[7] and Usami and Shida[8] it was evident that the stress range required to initiate a fatigue crack from a small corrosion pit at the pin hole location surface varied from 35 to 80 MPa. Other data in water at 80°C[9] indicated a $\Delta\sigma$ crack initiating value of 60 MPa while Congleton and Wilks[10] reported a value of 50 MPa in a steam environment. In the present study the onset of

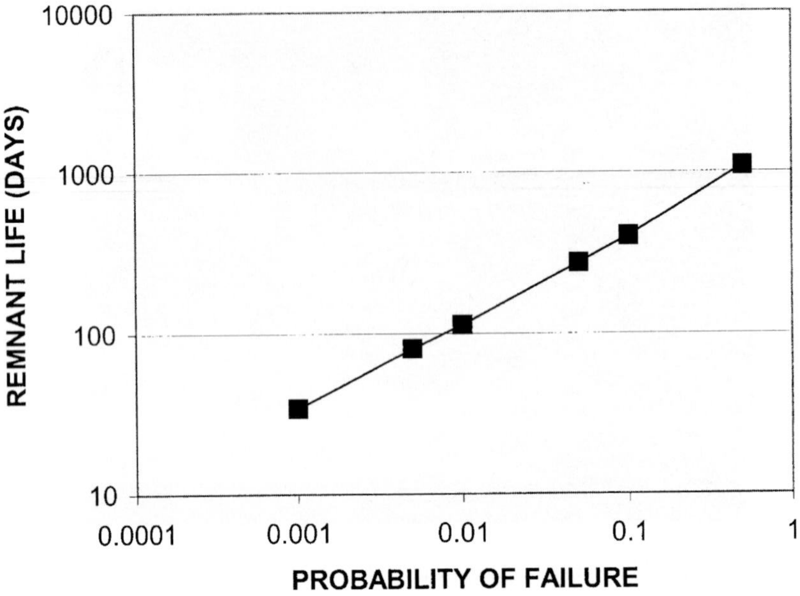

Fig. 6 The influence of POF level on the remnant life on an initial 3.55 mm deef bore face crack.

intergranular facets within the last 0.2 mm of sub-critical fatigue crack growth exhibited good agreement with another study concerning a 13%Cr steel[7] which showed that intergranular facets appeared when the reversed plastic fatigue zone attained the same size as the microstructural grain size. Interestingly this occurred at a stress range of some 50 MPa. Also recently Ebara[11] established that ductile striated fatigue crack growth in 13% Cr steels started at a ΔK value of around 12 MPa m which effectively indicated that the stress range driving the present LP blade fatigue crack growth was less than around 70 MPa. From all this information it was estimated that any remnant life assessment should consider stress ranges from 40 to 60 MPa and that the most likely stress range value driving the present LP blade cracks was some 50 MPa. Interestingly the present fatigue crack initiating conditions observed in blade 62 using a stress range of 50 MPa indicated a threshold stress intensity range value of 1.5 MPa m which would correspond to actual lower bound threshold value for 12%Cr ferritic steels fatigued under high mean stress conditions.

In the present study the various Y-Factor values for corner cracks reported by Bell et al.,[12] Schijve,[13] Heath and Grandt,[14] Shin[15] and Gandt[16] were used to calculate the DK level which were employed to assess the remnant life and the predicted remnant life values for cracked blade 62 using a stress range of 50 MPa are listed in Table 1 where it was noted that the Bell et al. and Schijve approaches gave the longer remnant life values, averaging some 1440 days, while those from the Heath and Grandt and Shin approaches gave the same remnant life value which was around half of the above value. The average remnant life value , where the POF value was 0.5, from all the individual values for the bore location was assessed at some

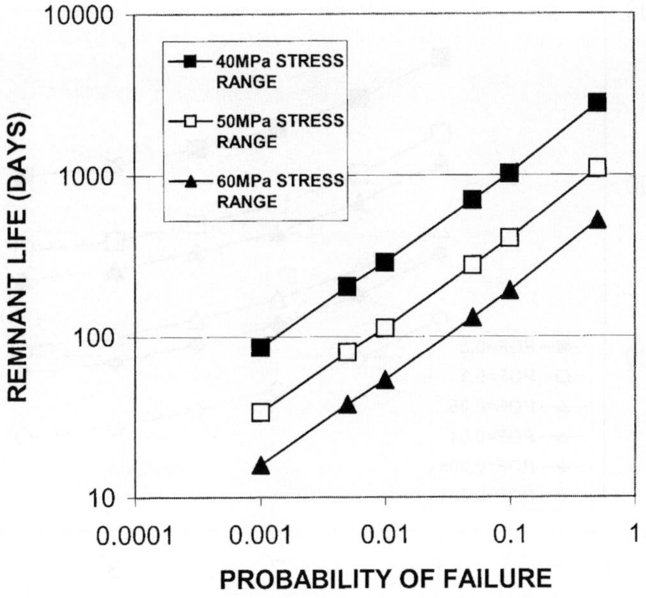

Fig. 7 The effect of POF level on the remnant life of an initial 3.55 mm deep bore face crack for a series of fatigue stress ranges.

Table 1 Individual remnant life values for the 3.55 mm bore and 1.79 mm flat cracks in blade 62.

Approach	Remnant Life, (Days)	
	Bore	Flat
Ref. 12	1548	2490
Ref. 13	1338	1343
Ref. 14	740	–
Ref. 15	740	817
Ref. 16	–	817

1092 days, or 3 years and a plot of remnant life with POF value , see Figure 6, indicated that the remnant life dramatically decreased to only around 4 months at a POF of 0.01. Essentially a POF value of 0.01 means that there is a one in 100 chance that the real remnant life of the cracked LP blades was less than the predicted remnant life value.

The influence of stress range , from 40 to 60 MPa, and probability of failure, POF, value on the predicted remnant life values of the LP cracked blade 62 are shown in Figure 7. From this figure it can be seen that both increasing stress range and reducing POF value range

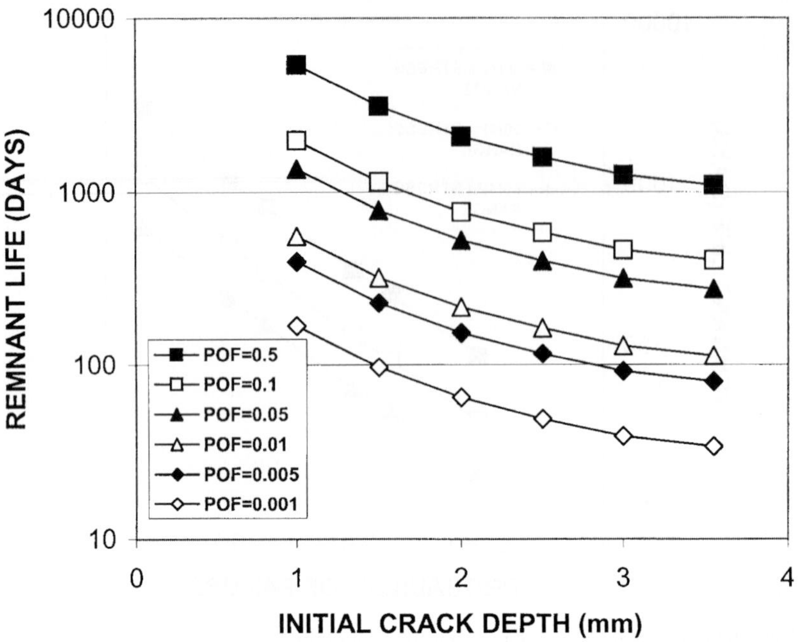

Fig. 8 The effect of initial bore face crack depth on the remnant life values for a range of POF levels.

from 40 to 60 MPa reduced the remnant life of the 3.55 mm deep crack by over 5 times while reducing the POF from 0.5 to 0.01 decreased the remnant life almost an order of magnitude.

The predicted remnant life values for a range of initial crack depths from 1 to 3.55 mm are shown in Figure 8 as a function of initial crack depth. Note that at a POF value of 0.05 the remnant life of a initial 1 mm deep crack was 5.4 years while that of a 3 mm initial crack depth was only 1.3 years.

CLOSING REMARKS

It has been shown that the remnant life of the deepest crack within the LP blades had an average remnant life, POF of 0.5, of some 3 years. Since power constraints were critical the unit was immediately returned to service with the cracked LP blades and a POF value of 0.05 was thought a reasonable value to adopt since any blade failures in the last few rows would not cause extensive damage to the turbine. The predicted remnant life value for this condition was some 9 months and thus there was a one chance in twenty of blade failure occurring within this service period. As such it was planned to run the unit over the winter period and conduct a blade replacement exercise during the summer of the following year.

REFERENCES

1. M. Rottoli and F. Sigon: 'Deposits on Turbine Blades', *Proceedings of Forty Ninth International Water Chemistry Conference*, Pittsburgh, USA., 1988, 450–458.
2. EPRI Report 'Corrosion Fatigue of Water-Touched Pressure Retaining Components in Power Plants', Final Report TR-106696, 1997, 4-45–4-60.
3. J. H. Bulloch: 'Deaerator Feedwater Storage Vessel Weldment Cracking, Some Fractographic and Crack Extension Details', *International Journal of Pressure Vessels and Piping*, **68**,1996, 81–98.
4. J. H. Bulloch: 'Some Field Observations Involving Deaerator Vessel Cracking 2', *Proceedings of International Conference Assessment and Extension of Residual Lifetime of Fossil-Fired Power Plants*, Moscow, Russia, **3**, 1994, 101–113.
5. J. H. Bulloch: 'Influence of Mean Stress and Frequency on the Fatigue Crack Growth Behaviour of A508 Steel in a Low Temperature PWR Environment', *International Journal of Pressure Vessels and Piping*, **49**, 1992, 139–151.
6. G. Gabetta and L. Torri: 'Corrosion Fatigue in 13%Cr Steels', *Fatigue Fract. Engng. Mats. & Structures*, **15**(2), 1992, 1191–1211.
7. H. Ishii, Y. Sakakibara and R. Ebara: 'Fatigue Crack Growth Rates in 12% Chromium Steels', *Metall. Trans. 13A*, 1982, 1521–1529.
8. S. Usami and S. Shida: 'Fatigue Crack Growth in 12%Cr LP Blading Steels', *Fatigue Fract. Engng. Mats. and Structures*, **1**, 1979, 471–481.
9. EPRI Report 'Corrosion Fatigue of Steam Turbine-Blading Alloys in Operational Environments', Report No. CS-2932, 1984, 220–246.
10. A. Congleton and C. Wilks: 'Corrosion Fatigue of a 12%Cr Steel in Water and Steam Environments', *Fatigue Fract. Engng. Mats. & Structures*, **11**, 1988, 139–148.
11. R. Erara: 'Corrosion Fatigue Fracture Surfaces of Structural Materials in Various Aggressive Environments', *Proceedings of International Symposium on Case Histories on Integrity and Failures in Industry*, Chifi, Milan, Italy, 1999, 187–196.
12. R. Bell, I. A. Pagotto and J. Kirkhope: 'Evaluation of Stress Intensity Factors for Corner Cracked Turbine Discs Under Arbitrary Loading Using Finite Element Methods', *Engng. Fracture Mechanics*, **32**(1), 1989, 65–72.
13. J. Schijve: 'Comparison Between Empirical and Calculated Stress Intensity Factors at Hole Edge Cracks', ibid., **22**(1), 1985, 49–60.
14. B. J. Heath and A. F. Gandt: 'Stress Intensity Factors for Coalescing and Single Corner Flaws Along a Hole Bore in a Plate', ibid., **19**(4), 1984, 665–674.
15. C. S. Shin: 'The Stress Intensity of Corner Cracks Emanating from Holes', ibid., **37**(2), 1990, 424–436.
16. A. F. Gandt: 'Stress Intensity Factors for Some Through-Cracked Fastener Holes', *International Journal of Fracture*, **11**(2), 1975, 283–293.

Blade Root / Blade Attachment Inspection by Advanced UT and Phased Array Technique

Michael F. Opheys, Hans Rauschenbach and Michael Siegel
Siemens AG Power Generation
D-45466 Mülheim, Germany

Graham Goode
Siemens Power Generation
Newcastle, UK

Detlev Heinrich
Cegelec AT GmbH & Co KG
Nürnberg, Germany

INTRODUCTION

International competition in the field of power generation is increasing and customers are demanding economic and efficient power plants. In the long term, continuous power plant availability can only be guaranteed through an effective mode of operation in conjunction with a systematic maintenance and inspection concept.

Apart from boiler, steam piping and valves, the rotating components of the turbine/generator (turbine and generator rotor) also belong to the most highly stressed components in a power plant. Loads result for example from operating parameters, the mode of operation of the machinery, start-up processes, thermal stresses, prestressing, residual stresses from the manufacturing process, as well as loading from the centrifugal forces acting on the rotating components. During scheduled outages, highly-stressed components are subjected to non-destructive testing designed to reliably detect any possible service-induced damage (e.g. cracking) before this can lead to failure of a component and severe consequential damage. For example, damage to a blade in the low-pressure turbine of a South African power plant (600 MW) in January 2003 resulted in the entire turbine generator unit being destroyed. Quite apart from the risk to personal health, such damage can lead to unscheduled outages and plant downtime, as well as unplanned costs for expensive repair and maintenance work on the turbine/generator. In comparison to these risks, the cost of inspecting such highly-stressed components is easily justified, as is the need for reliable and qualified techniques in the field of non-destructive testing.

The following describes two examples for non-destructive techniques used on turbine blade roots and blade attachment grooves.

ULTRASONIC TECHNIQUE FOR INSPECTING TURBINE BLADE ROOTS IN SITU

Description of Inspection Problem

The blades in a steam turbine belong to the most-highly stressed components in a turbine/generator. The high turbine speed (3000 rpm) and the dead weight of the blades means that the last-stage blades in a steam turbine are subjected to enormous centrifugal forces during plant operation. The roots on such blades are designed and calculated using the most up-to-date methods to allow them to accommodate these high loads. Particularly during transient loading conditions (start-up and shutdown processes) certain areas of the blade roots and blade attachment grooves are subjected to high stressing. Under unfavourable conditions unusual events occurring during operation of a turbine (e.g. loss of vacuum, overspeed) can result in damage to blading, with possible crack initiation in the highly-stressed areas of the blade root and subsequent service-induced crack propagation. In addition steam purity is also an important criterion regarding the susceptibility of a turbine blade to corrosion. If the steam is polluted with chlorides this is one of the basic prerequisites for the occurrence of corrosion fatigue in turbine blades, blade roots and blade attachment areas.

In the light of such influences on safe turbine blade operation, the necessity for non-destructive testing becomes particularly apparent.[1–3] Turbine blades and their roots should be examined non-destructively at predetermined intervals to allow timely detection of any damage and the replacement of affected blades.

The task faced here was to develop an ultrasonic testing technique for a special type of blade root to allow inspection of the roots of the last-stage blades in the rotor of a low-pressure steam turbine. When installed in the rotor the most highly-stressed areas of the blade root are not accessible for standard crack testing techniques. The objective was therefore to develop a technique which allowed these highly-stressed areas of the blade root to be inspected in situ, i.e. without removing the blade. The examination system had to provide reliable and reproducible results while remaining cost effective.

Theoretical Investigations

Extensive theoretical investigations had to be performed before any decisions could be made regarding selection of the ultrasonic examination technique. The blade under investigation was a last-stage blade from an LP turbine rotor. Figures 1 and 2 show the root for such a blade. Performance of the inspection on the blade roots of the dual-flow turbine rotor required that a calibration block be fabricated for the right and left side.

Reference reflectors (grooves, 6 mm long, 2 mm deep) were introduced into these calibration blocks at the most-highly-stressed areas. These areas can be found on the pressure side in the vicinity of the leading/trailing edge of the blade root in the first serration of the fir-tree root as well as in the middle of the first serration on the suction side of the blade root.

The theoretical investigations showed that it is indeed practicable to select scanning positions at the blade root which allow reliable detection of the reference reflectors (see

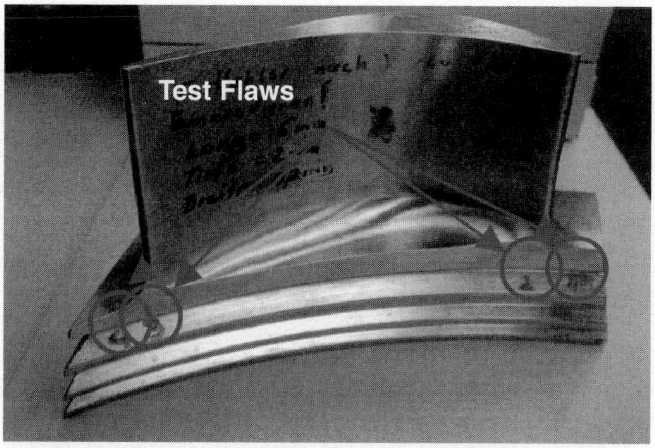

Fig. 1 Blade root for LP rotor with reference reflectors, pressure side.

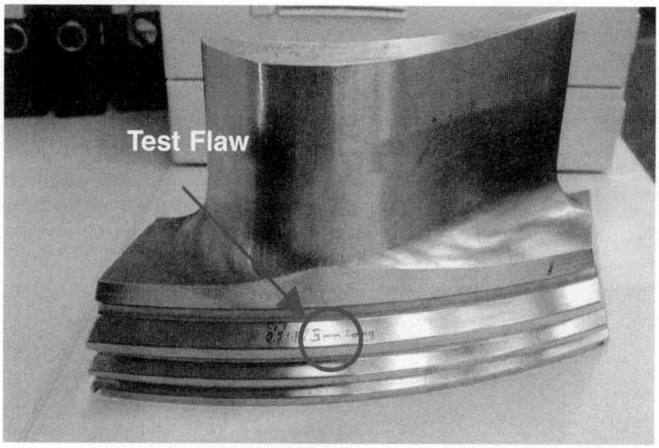

Fig. 2 Blade root for LP rotor with reference reflectors, suction side.

Figures 3 and 4). Along the complex geometry of this blade root these scanning positions were also situated in radii and on other curved surfaces which required a customized inspection solution for the component in question. For this reason, it was decided to fabricate specially-fitted pieces for each area to be scanned, which would allow exact positioning of the ultrasonic

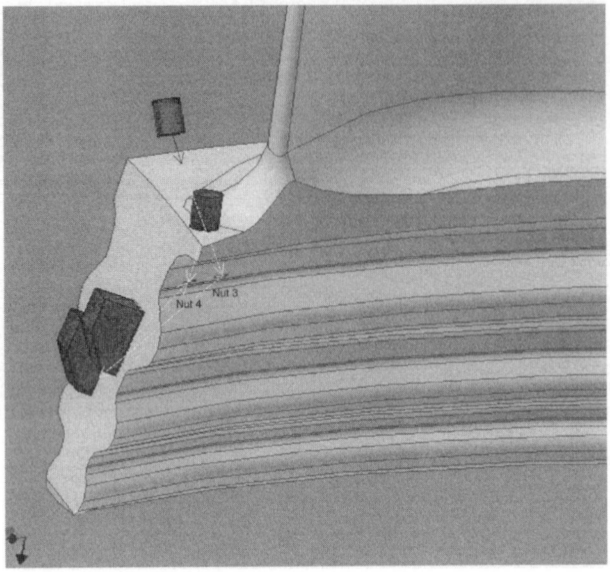

Fig. 3 Graphic for determining possible scanning positions.

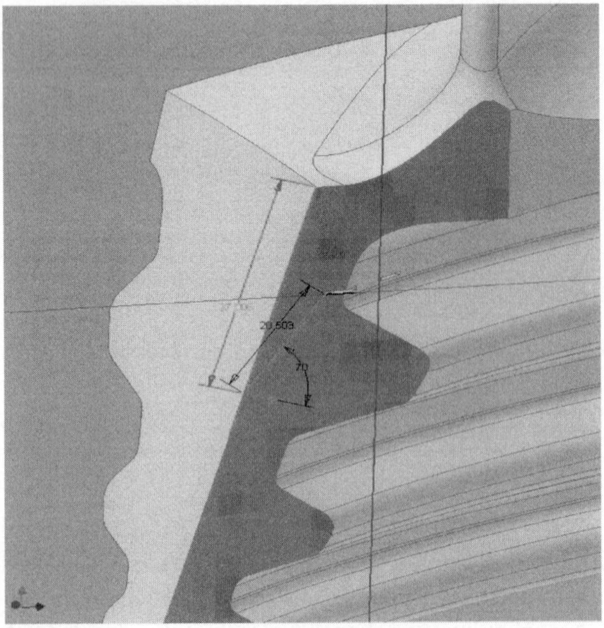

Fig. 4 Graphic for determining possible scanning positions.

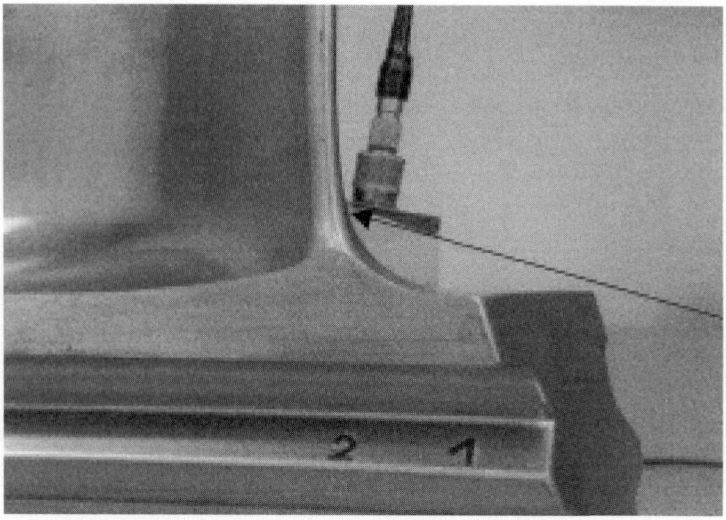

Fig. 5 Scanning positions at transition between root platform and airfoil for inspecting zone #2.

search units. This inspection technique seemed to be the right solution for the inspection problem, providing a suitable tool for power plant inspection services.

Practical Investigations on the Calibration Block

Once various suitable scanning parameters had been determined, practical tests were able to begin on the calibration blocks. The results of investigations with phased-array search units at various frequencies (3, 7 and 11 MHz) indicated only limited reliability. For the customised solution (using contoured probe holders with integrated search units) it was decided to use 5 MHz longitudinal wave search units with a transducer diameter of 6.3 mm as well as 60°C shear wave search units. These search units were equipped with Plexiglas wedges. It was then possible to contour the Plexiglas wedges so that they could be coupled to the surface at the determined scanning positions. Figure 5 is an example showing the scanning position for a search unit at the transition between the blade root platform and the airfoil.

On the basis of the theoretical investigations, at least two scanning positions were determined for each reference reflector which seemed suitable for detecting the reflector. All the scanning positions calculated during the theoretical investigations were checked during the practical tests on the calibration block. This confirmed that all reference reflectors could be reliably detected, in all cases. Owing to the fact that the different reference reflectors were able to be found using various scanning positions and beam angles, it was decided to make use of this during the actual performance of inspections.

Fig. 6 Probe holder #1 for inspecting zone #1 in the blade root.

Development and Qualification of a Practical Inspection System

Once the investigations on the calibration block had confirmed the suitability of the selected inspection technique for the problem in hand, the contoured probe holders mentioned above were fabricated. The probe holders are matched to the blade root contour to guarantee exact positioning of the ultrasonic search units. Three contoured probe holders were fabricated, each containing several ultrasonic search units. Their use in conjunction with a 4-channel ultrasonic instrument (μTomoscan) guarantees an effective inspection. Once the corresponding probe holder has been brought into position, this instrument allows the results from all the integrated search units to be evaluated at a glance. Three contoured probe holders were made for the blade root under investigation (see Figures 6 and 7). When testing the inspection system, investigations were performed using several different blades of the same type, to verify that existing manufacturing tolerances for these large LP blades do not have any influence on the results of the examination. It was shown that the dimensional differences existing between the blades inspected were able to be compensated for using a gel-type couplant and can therefore be neglected. The couplant bridges the gap between the contoured probe holder and the blade root. Manufacturing tolerances were not found to have any effect on test sensitivity/defect detectability. Figure 8 shows the A-scans of the blade root with artifical notches. For comparision Figure 9 shows the result of a blade root without any notches.

SUMMARY

The manual ultrasonic inspection system described above was developed to provide a reliable and cost-effective method of inspecting turbine blade roots. The main considerations during development work were:

Fig. 7 Probe holder #1 on blade root.

- Blades must be able to be inspected in situ.
- Simple handling and operation (no complex manipulators, etc.).
- Reliable and meaningful test results.
- Possibility of verifying indications by using 2 scanning positions.
- Fast test method.

All these requirements are met by the inspection system. The configuration with the contoured probe holder and a systematic inspection procedure means that only a short introduction to the equipment is required before testing can begin.

With respect to the reliability of testing, it proved to be a considerable advantage that each zone for examination was able to scanned from at least 2 scanning positions, thus providing the possibility of verifying the presence of any indications detected, by scanning from a second position.

ULTRASONIC EXAMINATION OF BLADE ATTACHMENT GROOVES OF LP TURBINE SHAFTS

DESCRIPTION OF THE PROBLEM

Due to the world wide SCC issue there is an increasing demand for a non destructive examination of blade attachments of steam turbine rotors. In December 2001, Siemens Power Generation's NDE laboratory received a request to perform a non-destructive examination on blade attachment grooves of a non-OEM turbine (European nuclear power plant).[7] The problem is discussed in greater detail below.

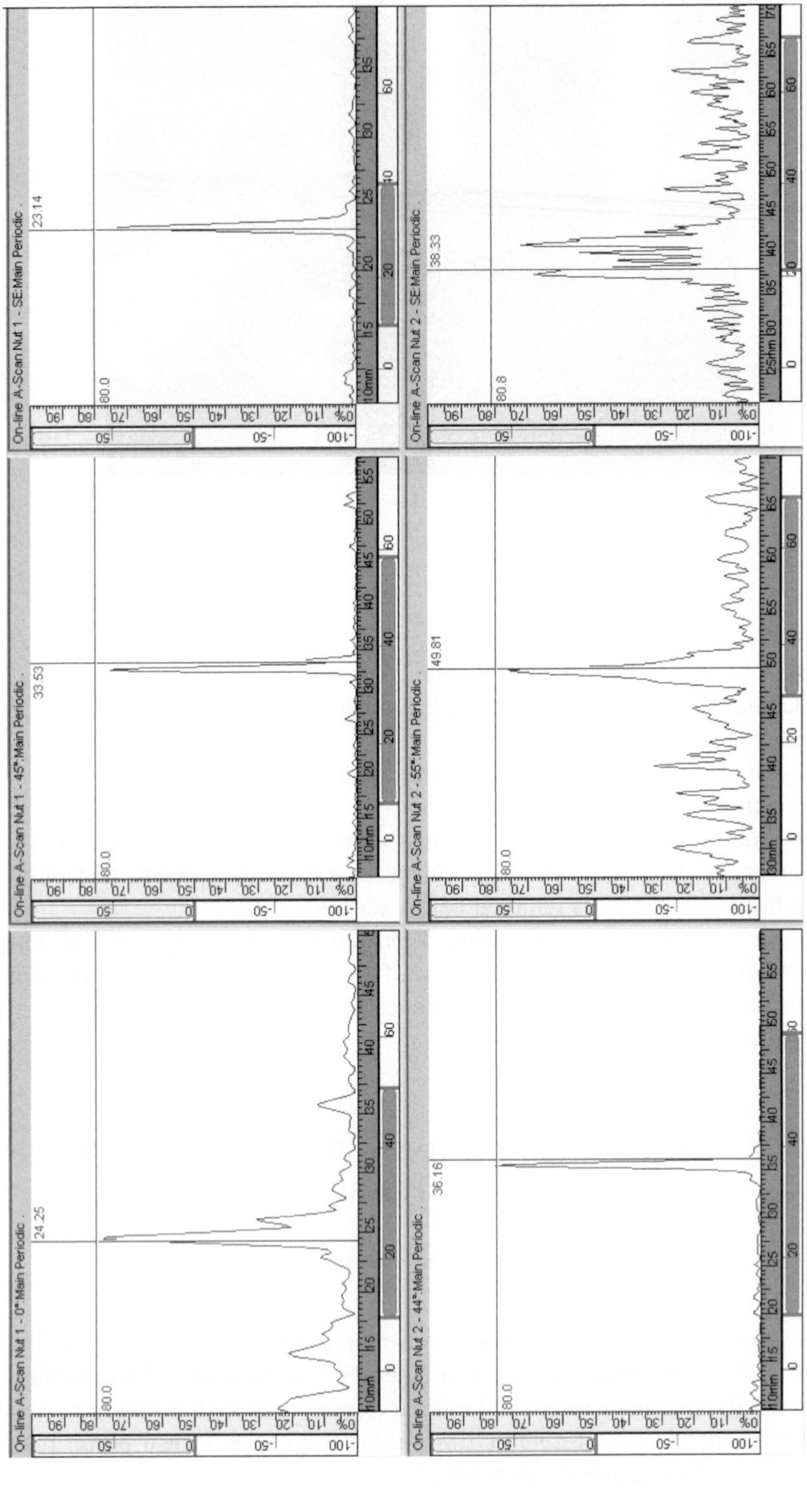

Fig. 8 Test results when inspecting Zone #1 using probe holder #1on a blade with reference reflectors.

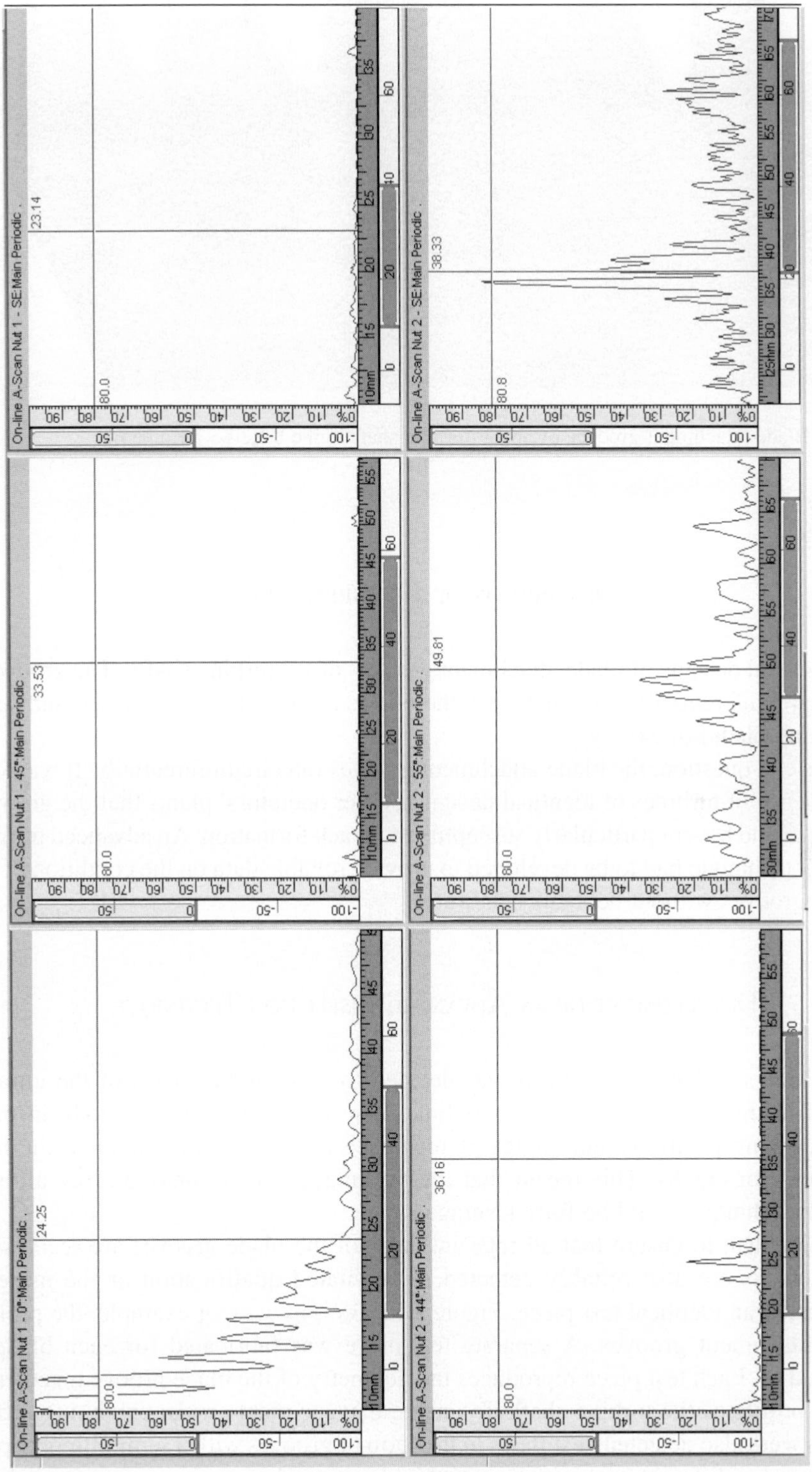

Fig. 9 Test results when inspecting Zone #1 using probe holder #1 on a blade without reflectors.

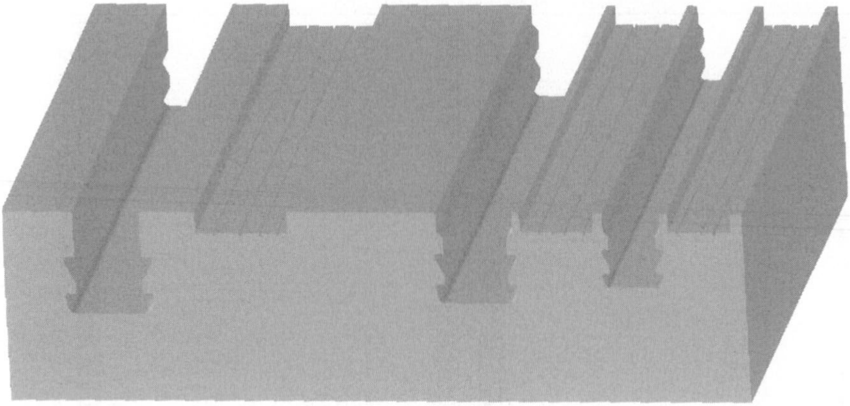

Fig. 10 Blade attachment grooves of an LP turbine shaft Reihe 8 Reihe 7 Reihe 6.

Description of Requirements

There are several designs of blade attachment grooves of LP turbine shafts. The grooves can run either circumferentially (in which case the blades are inserted in sequence and secured with a locking blade) or axially.

In the case in question, the blade attachment grooves ran circumferentially. It was known from experience of turbines of identical design at other operators' plants that the grooves of blade row 6, 7 and 8 were particularly susceptible to crack formation. An advanced ultrasonic examination technique had to be developed to provide reliable data on the condition of blade attachment grooves without deblading the rotor.

Development of an Advanced Inspection Technique

Following analysis of the problem, it was decided to solve it by means of the ultrasonic phased-array technique. Given the different dimensions of the blade grooves to be inspected, different scanning positions and angles of incidence are required to examine the highly stressed areas for cracks. This meant that the advantages of the phased-array ultrasonic examination technique could be fully leveraged.

The requirement to ensure that all relevant areas of the blade grooves are scanned, and that small cracks are also reliably detected, necessitated qualification of the inspection technique using an identical test piece. Figure 10 shows, by way of example, the profile of the blade attachment grooves. A separate test piece was fabricated for each blade row (row 6, 7 and 8). Each test piece reproduces the geometry of the blade groove together with the outside profile of the turbine shaft. To ensure that incipient cracks exhibiting different orientations were also detected, test flaws in the form of grooves with a semi-elliptical profile were introduced at different angles in the most highly stressed areas (inspection zones 1 and 2

Fig. 11 Qualification of the inspection technique on a test block.

in Figure 11). To ensure detection of even the smallest incipient cracks, the dimensions of the semi-elliptical grooves used were as follows (length in mm × depth in mm): 2 × 1, 4 × 1, 4 × 2, 8 × 2. The grooves were positioned in inspection zones 1 and 2.

Qualification of the Inspection Technique using Test Pieces

Following fabrication of the test pieces, qualification of the technique was carried out using a 45 EL3 phased-array search unit (natural angle of incidence: 45°C, search unit frequency 3 MHz, 16 array elements). The tests were carried out using a triple-axis manipulator in conjunction with the SAPHIR+ phased-array system.

Due to the complex geometry of the blade grooves and the associated geometric indications, a vertical scan was programmed in a range from 30 to 85°C in steps of 1°C. Figure 12 shows the result of the vertical scanning using a sector scan presentation. To identify the form echoes more easily the CAD drawing of the inspected blade attachment is overlaid.

Assessment of the TD images of all scans enabled the optimum angle of incidence to be rapidly determined. Figure 13 shows, by way of example, the TD image of a scan in inspection zone 1 (row 8).

Results of Qualification and Conclusions

By scanning the test pieces it was possible to demonstrate that the deployed phased-array ultrasonic inspection technique is suitable for use in field service to examine blade grooves

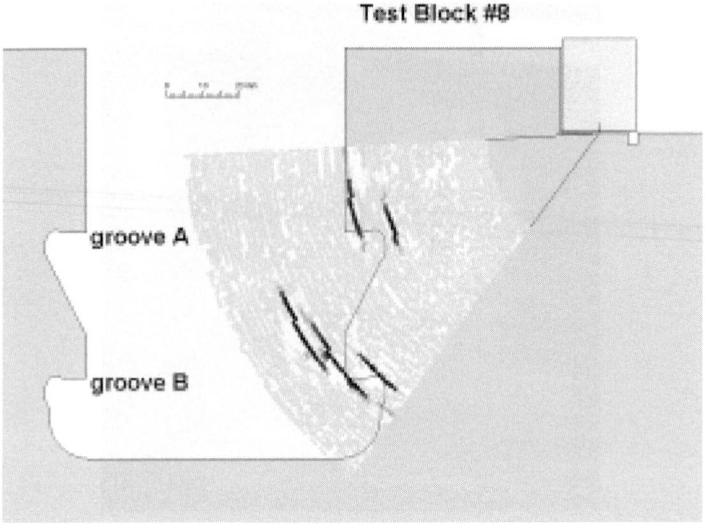

Fig. 12 Overlay of the sector scan representation with the CAD drawing of the blade attachment (row 8).

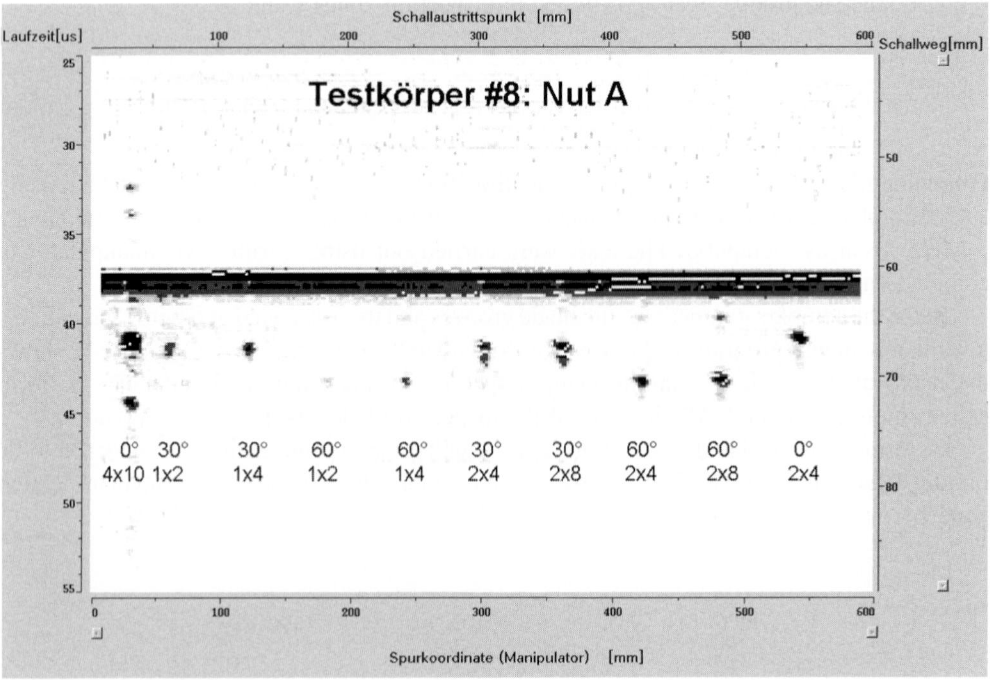

Fig. 13 Scan results in form of a TD image showing all test flaws. (inspection zone1, row 8).

of LP turbine shafts for incipient cracking in highly stressed areas in the assembled condition (in other words without removing the blades).

All of the test flaws in inspection zone 1 and 2 of all three test pieces were detected (the smallest test flaw was a semi-elliptical groove 2 × 1 mm).

These two examples of advanced inspection techniques demonstrate that direct customer benefits can be delivered through the use of problem-focused techniques. Key examples include time savings on component disassembly and reassembly, required with conventional crack inspection techniques, but eliminated when advanced techniques are used. Given the requirement for virtually non-stop power plant availability and the associated reduction in plant downtimes, these kinds of in-situ service techniques are playing an increasingly important role in the planning and performance of plant outages.[5–7]

REFERENCES

1. V. Viswanathan and David Gandy(EPRI): 'Rim Attachment Cracking Promts Developement of Life Assessment Tools', *Fourth International EPRI Conference on Welding & Repair Technology for Power Plants, 2000*, 50.
2. Darryl A. Rosario, Peter C. Riccardella and S. S. Tang: 'Developement of an LP Rotor Rim – Attachment Cracking Life Assessment Code (LPRimLife)', *Power Engineering*, (Structural Integrity Associates San Jose, CA, USA), Marco Island, Florida, USA, 2000.
3. Carlos Arrietta, Francisco Godinez, Marta Alvaro and Andres Garcia: 'Blade Attachment UT Inspection Using Array', *7th EPRI Steam/Turbine Generator Workshop*, (Technatom, SA), Baltimore, MD, 2001.
4. Hans Rauschenbach, Michael Opheys and Uwe Mann: 'Siemens Power Generation Jürgen Achtzehn, IntelligeNDT Framatom: Advanced NDE Inspection Methods for Field Service at Power Plants', *8th European Conference on Non-Destructive Testing,* Barcelona, 2002.
5. Richard Fredenberg: 'Dovetail Blade Attachment Experience using Phased Array Ultrasonic Test Techniques', *7th EPRI Steam/Turbine Generator Workshop*, (Wes Dyne International), Baltimore, MD, 2001.
6. Petru Ciorau, et al.: 'In Situ Examination of ABB L-0 Blade Roots and Rotor Steeple of Low-Pressure Steam Turbine, Using Phased Array Technology', *15th WCNDT*, (Ontario Power Generation Inc.), Rome, 2000.
7. A. Lamarre, N. Dube, P. Ciorau and P. Bevins: 'Feasibility Study of Ultrasonic Inspection Using Paced Array of Turbine Blade Root – Part 1', EPRI Workshop, RDTech, Canada, Notario Power Generation Inc., 1997.

Pitting and Cracking of Steam Turbine Disc Steel

A. Turnbull and S. Zhou
Materials Centre
National Physical Laboratory
Teddington, TW11 0LW, UK

ABSTRACT

Stress corrosion cracking of steam turbine discs continues to be an area of concern for the power industry because of the uncertain impact of environmental variables, the consequences of failure and the need for extended life. Here, the focus is on the role of solution chemistry variables. A brief overview is given of present knowledge and understanding of the characteristics of the liquid film condensate formed on turbine discs and blades. On-load, the anion (e.g. sulphate and chloride) concentration under normal operation conditions would be typically about 300 ppb but could rise to 1.5 ppm for poorly controlled conditions. However, the condensate film will be essentially free of dissolved gases except during start-up or following a significant air-leakage into the turbine chamber. Off-load, the condensed liquid will be near-pure water but will be aerated, unless nitrogen blanketing is used.

The effect of different levels of aeration and of chloride content of the solution on the pitting and cracking of self-loaded 3% NiCrMoV disc steel immersed at 90°C was evaluated. Tests were carried out in deaerated pure water, aerated water, and aerated 1.5 ppm chloride water, the latter representing a major system upset. Pitting was observed in all cases with the growth rate being greatest for the aerated chloride water. No cracking was observed after 22 months in deaerated pure water but cracks initiated in aerated water between 13 and 19 months and in less than 7 months in aerated 1.5 ppm Cl solution. The probability of a crack initiating from a pit of specific depth in aerated solution seems unaffected by chloride or exposure time.

INTRODUCTION

The power industry has experienced failures from environment assisted cracking (EAC) of steam turbine blades, discs and rotors. Despite worldwide effort, occasional problems still arise. The major challenge is to predict more reliably the conditions under which cracking is likely and, for those conditions, the evolution of crack size with time so that non-destructive evaluation may be used in a focused manner and informed decisions made about inspection intervals and remnant life. The difficulty is the complexity and transient nature of service conditions, and constraints in their detailed characterisation, e.g. the chemistry of the condensate formed on the turbine steel surface.

Stress corrosion cracking (SCC) of steam turbine disc and blade steels, especially for long cracks, has been studied for more than 30 years. However, the data generated can show

Table 1 The temperatures at which first condensation occurs in the LP, IP and HP turbines for various stations.

Type of Station	Saturation Temperature (°C)		
	LP	IP	HP
Fossil-Fired	< 100	No Condensation	No Condensation
AGR	< 90	No Condensation	No Condensation
PWR	< 78	N/A*	250 – 280
Magnox	70 – 120	N/A*	~ 170

*N/A - Not Applicable

inconsistency for often nominally the same conditions. This is in part due to a lack of control of test conditions but it is also the case that the scientific framework for interpretation of the data has not been adequate in many applications and apparently similar test conditions have subtle differences that are overlooked.

A programme of research is presently underway at NPL to establish best understanding of the condensate chemistry in service,[1] to predict the evolution of pits and the transition to cracking, to provide data for the rate of crack growth in the short and long crack domain, and to put these results in a scientific and engineering context to assist in the transfer from laboratory-based measurements to service prediction. The focus of the present paper is on disc steel and the impact of solution chemistry on pitting and the transition from pits to cracks.

CHARACTERISATION OF SERVICE CONDITIONS

Temperature for the First Condensation and Thickness of Condensate Layer

The distribution of temperature and pressure in the turbines is complex and varies from station to station. Therefore, it is not possible to make general statements on the distribution of steam moisture content in the turbine and the exact location where condensation starts to occur. The temperatures at which the first condensation occurs in the low pressure (LP), intermediate pressure (IP) and high pressure (HP) turbines for various stations in the UK are summarised in Table 1. However, it should be noted that there are variations in the distribution of temperature and pressure in the turbines during operation, especially during start-up and shutdown period. Hence, there are locations that will experience cyclic dry/wet steam conditions. Furthermore, early condensate could form above the saturation temperature due to local supercooling of the steam flow and the presence of impurities.[2] Stress corrosion cracking occurs only when there is wetting and most commonly in the LP turbines because of the higher stresses.

The thickness and continuity of the liquid film depend on the steam moisture content, chemical impurities, wetability and rotation speed of the surface where the film is located.[2–4] Film thicknesses of up to 120 μm have been measured in model turbines at MEI.[3] An EPRI[4] study revealed that at steam moisture contents less than 1%, no continuous liquid film forms on the turbine components in the steam flow path. However, early condensates form at moisture levels of less than 1%. The key issues are the solution chemistry of the thin liquid film and the dissolved gas content.

Impurities in the Steam and Liquid Film

The boiler and feed-water chemistry is tightly controlled under normal operating conditions. However, high levels of impurity in the feedwater and boiler water may arise during chemical excursions due to:

a. the deficiency of water treatment (e.g. demineralisation and deaeration equipment failure),
b. leakage from the condenser,
c. transients in water chemistry during start-up and shutdown period,
d. contamination of corrosion products in the boiler, feed-water tubing, etc.

Impurities can be transported into the steam in different ways. When the steam enters the turbine, the impurities from the steam can be transported onto the metal surface by deposition (vapour to solid) and condensation (vapour to liquid). The deposits can be formed on the LP turbine, as well as on the HP and IP turbines, especially during major chemical excursions. The concentration of impurities in the water phase can be calculated using thermodynamic data[5,6] assuming that the impurities in the vapour (steam) and liquid (condensate) phase are close to equilibrium.[7] The calculated chloride content in the water phase at various values of steam conductivity and steam wetness is shown in Figure 1.

Under normal operating conditions the cation conductivity of the steam is about 0.2 μS/cm. If this were all in the form of chloride, the chloride concentration in the liquid phase at 1% of moisture would be about 1.6 ppm and could be as high as 10 ppm during a chemistry excursion (cation conductivity equal to 1 μS/cm). However, the water chemistry guidelines[5] indicate that a steam cation conductivity of 0.2 μS/cm should not contain so much chloride and the limiting value has been set to 3 ppb. At this concentration, the condensate chloride content is predicted to be 300 ppb and 100 ppb for the steam wetness of 1 and 3% respectively. Higher chloride concentration may exist in the regions where the steam is dry or where wet/dry cycles occur. Similar calculations can be done for sulphate anions. For example, if the sulphate concentration in the steam were controlled at less than 3 ppb under normal operating conditions, the maximum concentration of sulphate ion in the liquid phase would be 300 ppb. Organic species, such as acetates and formates, could lead in principle to some acidification of the condensate layer but reliable data are absent.

The calculated oxygen content in the water phase as a function of temperature at various oxygen concentrations in the inlet steam is shown in Figure 2.

It can be seen that the oxygen concentration in the water phase is less than 1 ppb even when the inlet steam contains 8 ppm oxygen. The calculation is in agreement with the experimental data measured in the condensate.[2] However, high levels of oxygen may exist in the water phase if excessive air leakage were to occur. For carbon dioxide in the steam

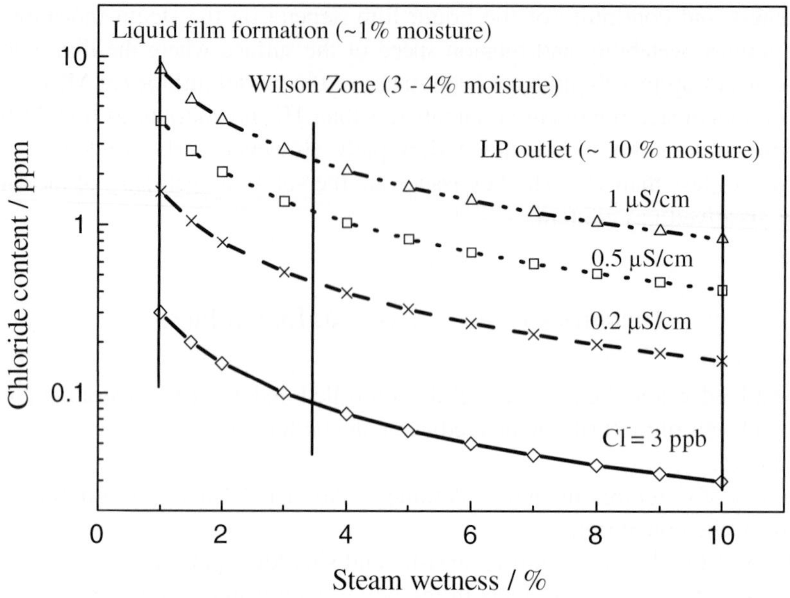

Fig. 1 Chloride content in the water phase as a function of steam wetness at various cation conductivities of the inlet steam.

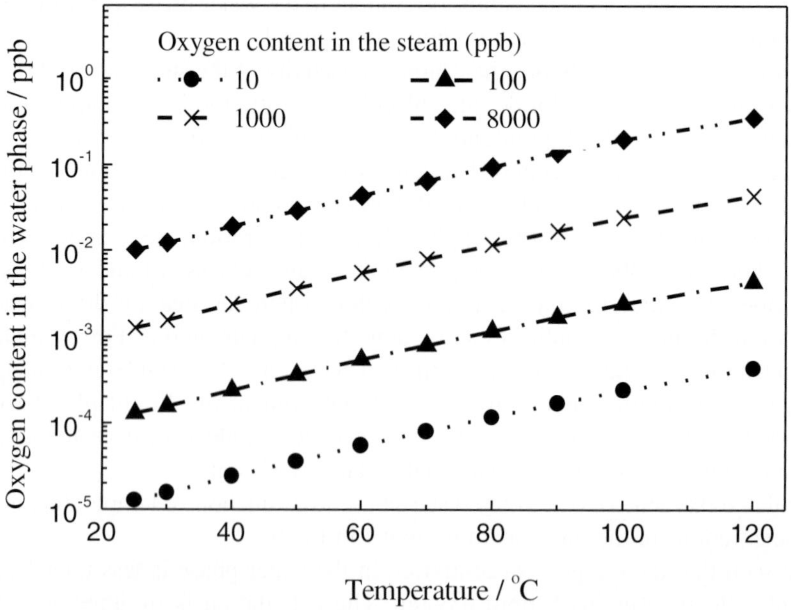

Fig. 2 The oxygen content in the water phase as a function of temperature at various oxygen concentrations in the inlet steam.

phase, a similar prediction is made, i.e. there is generally negligible gas dissolved in the liquid film.

Summary of Service Conditions

Under normal operating conditions on load, the liquid film will have about 300 ppb of anionic impurity (sulphate and chloride) and will be free of dissolved gases. The temperature of relevance would be about 90°C for the LP turbines. Off-load, the condensed layer will be essentially aerated pure water (unless there is nitrogen blanketing) and a temperature ranging from about 70°C to near ambient depending on the shutdown period.

IMPACT OF SOLUTION CHEMISTRY ON PITTING AND THE ONSET OF CRACKING

Stress corrosion cracking is often initiated from corrosion pits, certainly for the low-to-medium strength disc steels. Pit initiation is usually rapid so the important steps in the overall process of crack development are pit growth, the transition from pit to a crack, short crack growth, and long crack growth. At all stages, there are limitations in our knowledge base, particularly in relation to the impact of solution chemistry on the kinetics of pit and short crack growth. Accordingly, testing was initiated using self-loaded cylindrical specimens of a disc steel (3% NiCrMoV), cut from an ex-service steam turbine disc supplied by Powergen with a composition in mass% given by: C: 0.30; Si: 0.28; Mn: 0.45; P: 0.017; S: 0.013; Cr: 0.69; Mo: 0.27; Ni: 2.89; V 0.091; N: 0.21; Fe: bal. The full details of the experimental method are described elsewhere.[9] The overall specimen length was 100.0 mm, the shoulder diameter 16.0 mm, the gauge length 25.4 mm and the diameter 6.4 mm. All specimens, after final grinding, were stress-relieved in vacuum for 2 hours at 625°C and then cooled in a furnace under vacuum. The subsequent values of $\sigma_{0.2}$ and UTS at the test temperature of 90°C were 705 MPa and 827 MPa respectively. To avoid galvanic corrosion, the loading frames and the nuts were also made of the disc steel. The environments chosen for this study are shown in Table 2.

Prior to the detailed review and analysis of Reference 1 in which the chloride content of the liquid film on-load was calculated, the literature on stress corrosion cracking indicated that on-load conditions were best represented by laboratory tests in high purity water. This was the basis for this environment when initiating these long-term tests. It is now realised that 300 ppb of anions is likely in the liquid film whilst on-load. Nevertheless, the tests in pure water would reflect a high steam wetness situation.

For the tests conducted in deaerated high purity water a once-through system was used giving an oxygen level of about 1 ppb and a conductivity of 0.06 μS/cm. For tests in aerated high purity water the inlet water was aerated using an air pump; the conductivity was higher, about 1.1 μS/cm, due to dissolved CO_2. The solution circulation loop for aerated water containing 1.5 ppm Cl^- was slightly different. In this case the aerated feed-in solution was re-circulated from a 40 litre reservoir, which was refreshed weekly. A magnetic stirrer was

Table 2 The environments and test temperatures for pitting and SCC tests.

Environment	Conductivity/ $\mu S\ cm^{-1}$	pH	Test Temperature (°C)	Simulation
Deaerated High Purity Water	0.056	Assume Neutral	90	Liquid Film Chemistry with High Steam Wetness/High Purity
Aerated High Purity Water	1.1	5.8	90	Liquid Film Chemistry During Start-Up/air Leak
Aerated Water + 1.5 ppm Cl^-	6.8	5.9	90	Liquid Film Chemistry During Major System Upset

used in all test cells to improve local mixing. The corrosion potential was measured using an external 0.01 N Ag/AgCl reference electrode designed specially with an extremely low chloride diffusion rate. The reference electrode chamber and the test cell were separated with a glass valve, which was closed except when a measurement was taken. The potentials quoted are with reference to the saturated calomel electrode (SCE) at 25°C. All specimens were loaded to 634.5 MPa, i.e. 90% of the $\sigma_{0.2}$ at the test temperature (90°C). For each environment, the specimens were distributed in two or three different test cells of 1 litre capacity so that specimens could be removed and examined at different test intervals without affecting the remaining specimens.

Testing was carried out for periods of up to about 22 months. The variation of potential with time for the three test conditions is shown in Figure 3.

A low potential for deaerated solution is attained although from related studies we believe that film formation (corrosion product deposition) is a factor in raising this above the value of about –0.7 V SCE or less that might be anticipated for a film-free surface in deaerated solution. The unexpected result is the slow positive drift in potential for the 1.5 ppm chloride solution since a high density of pitting was observed. In conducting solutions, the development of pits often leads to a decrease in potential because the pit current introduced, which would be net anodic, has to be supported by an increased cathodic current. By implication, an increase in potential in a controlled environment situation usually reflects an increased protectiveness of the surface film, yet pits are growing. It is likely that the film is simply corrosion deposit, retarding metal dissolution.

The density of pits and their depths were measured for each specimen in each environment and the extent to which the pits had cracks was also evaluated. For pragmatic reasons, associated with the high number and difficulties in reliable depth estimation, pits with a depth smaller than about 25 μm were not measured. The distribution of pit depths for the three environments at two exposure times is shown in Figures 4-6. The aspect ratio (depth/mouth opening ratio) of the pits in the chloride-free solutions was about a factor of two greater than for 1.5 ppm chloride solution. The most likely explanation is that the tighter pit is favoured because it enhances retention of the local chemistry associated with MnS inclusion

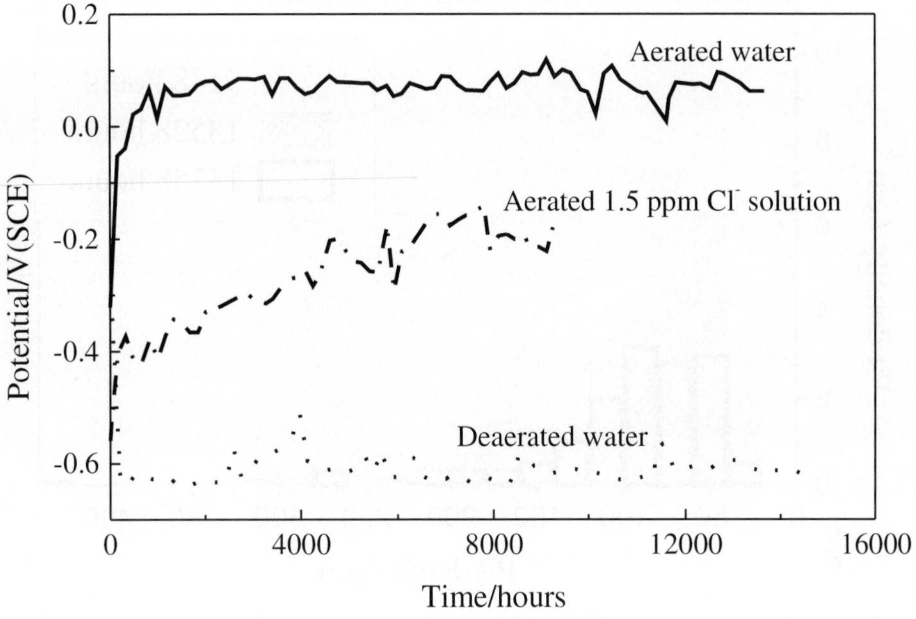

Fig. 3 Variation of the corrosion potential with time for self-loaded disc specimens.

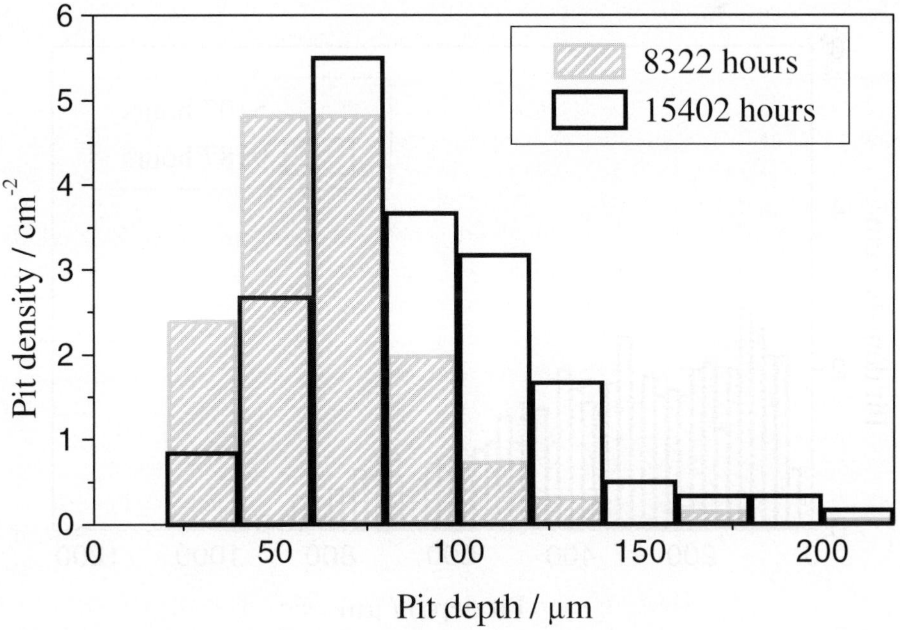

Fig. 4 Pit depth distribution for disc steel in deaerated pure water.

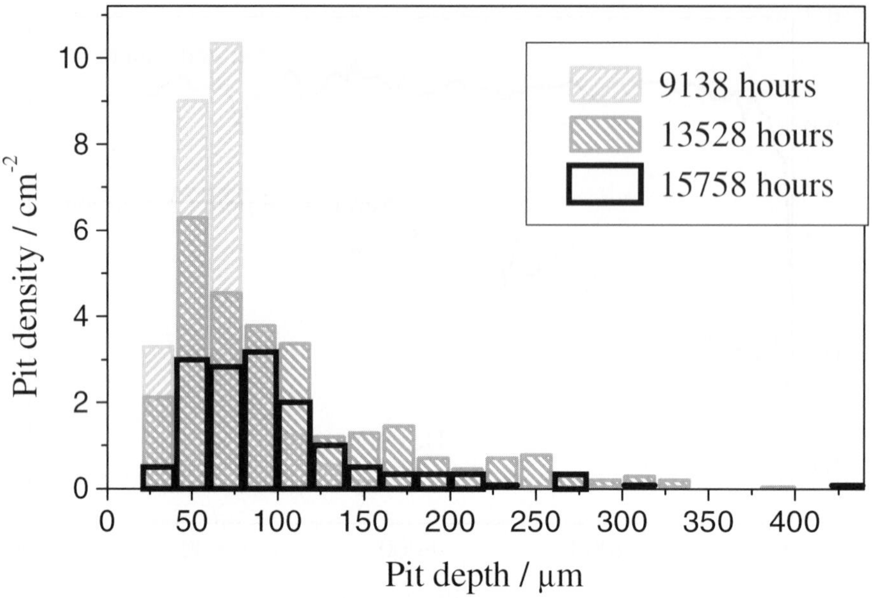

Fig. 5 Pit depth distribution for disc steel in aerated water (The dark border line represents the pit density distribution at 15758 hours).

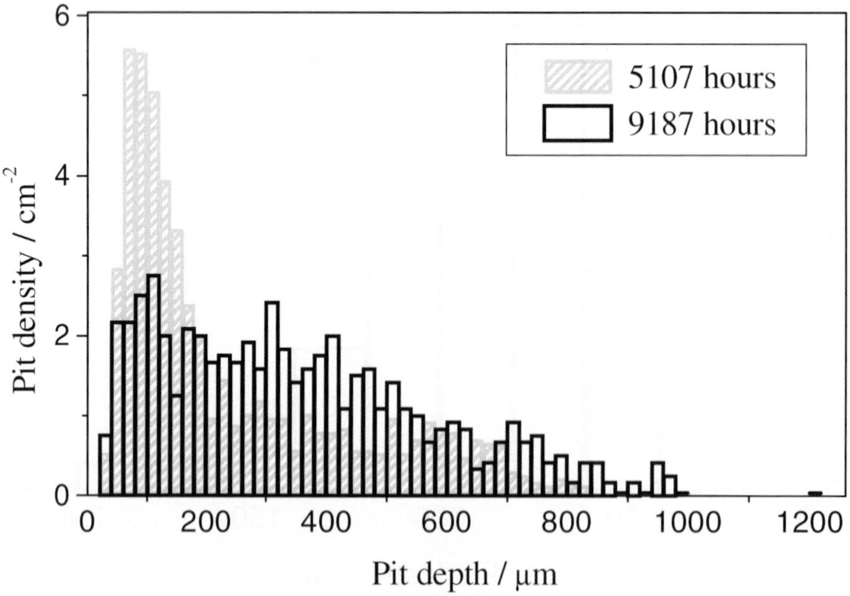

Fig. 6 Pit depth distribution for disc steel in aerated 1.5 ppm Cl- solution.

dissolution, which is likely to determine growth in this relatively benign environment. This is an old steel with a measured inclusion density of $2.2 \times 10^4/cm^2$. The greater width of the mouth of the chloride-containing pits may reflect the more acidic environment anticipated (albeit expected to be between pH 5 and 6). This would reduce the stability of the film adjacent to the mouth of the pit as the lower pH solution mixes with the bulk. In combination with the more noble potential compared to the base of the pit this would encourage some lateral dissolution.

In deaerated pure water, the maximum pit depth after about 22 months was just over 200 μm. The variability in the data from one exposure time to the other put constraints on the conclusions about the evolution of the pit density. For the deaerated water, the indication is that the change with exposure time in the density of pits greater than 25 μm was modest, less than 20%; the pits simply got deeper. This was rather similar for the aerated water (Figure 5), with the exception of the 15758 hours exposure, but the depth of pits achieved, compared with the deaerated water was generally greater, with some pit depths approaching 350 μm. The combination of oxygen and an increased conductivity would seem sufficient to give an increased driving force for pit growth The result at 15758 hours, based on only one specimen (compared with two for the other exposure times) does not conform to expectation and would seem to be an anomaly.

The important feature of the measurements in the aerated water was the observation of cracks on specimens removed at 13528 hours (about 19 months) but not on specimens removed at 9138 hours (about 13 months). By contrast, no cracking was observed in deaerated pure water, although long crack growth does occur in such an environment.[1, 11–12]

The pit density in the 1.5 ppm chloride solution exceeded that for the aerated water by about a factor of two. Data are shown only up to 9187 hours in this case as the test was terminated because of the high level of damage to the specimens. Cracking was very extensive.

The extent of cracking in this environment and in the aerated environment is represented in Figure 7. There are two key features. No cracks are observed below a critical pit depth of about 60 μm. Also, there is a remarkable similarity in the percentage of pits with cracks at specific pit depths despite changes in exposure time and the presence or absence of chloride. Of course, the density of deep pits is greater in the 1.5 ppm solution and this will increase with exposure time. Thus, overall, cracking will be more extensive and the cracks larger in the 1.5 ppm solution. Crack coalescence will also be more significant. The absence of apparent cracking in aerated water at 9138 hours most likely reflects the small number of pits of significant depth, with the deepest about 175 μm and only about 5 or so pits with depths greater than 125 μm. For the deaerated solution, there are some deep pits at long times (rather similar to the aerated data at 15758 hours) and this explanation is not wholly satisfactory.

From a phenomenological perspective the criteria for the transition from a pit to a crack is based on the concept of a minimum pit size plus a requirement that the crack growth rate exceeds the pit growth rate. In the context of Figure 7, the similar probability of cracking for the same pit depth might be presumed to imply that the growth rate at that depth is similar despite the different times of exposure and chloride content. The independence of exposure time is surprising because the cathodic current available for individual pits would be expected to diminish as the pits develop and the total pitting (net anodic) current increases Thus, exposure time might have been expected to have influenced the results. However, the corrosion

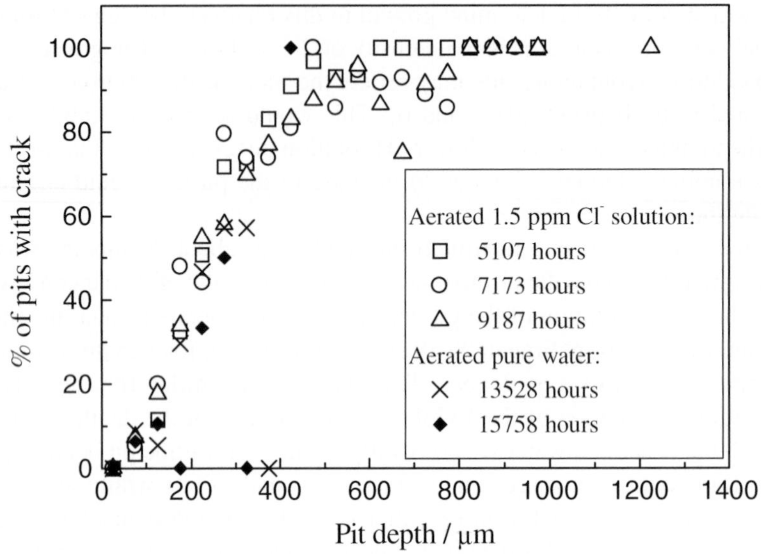

Fig. 7 Likelihood of a crack developing from a pit of specific depth.

potential is not reflecting this behaviour and it is feasible that pits at a specific depth are growing at relatively similar rates.

The important finding is that there is a distinct effect of the solution composition and degree of aeration on the density and growth rate of pits and thus on the extent of cracking. Further work is ongoing involving layer removal and 3-D profiling of the pits and cracks at different exposure times to provide insight with respect to the location of origin of the cracks, how the crack develops around the pit, and also an indication of average short crack growth rates. Preliminary results indicate that the depth of some cracks was smaller than that of the corresponding pit. This suggests the possibility that the crack did not originate from the base of the pit in all cases. The alternate explanation that the pit subsequently grew faster seems less likely because corrosion blunting might have been anticipated. There were many cracks with a depth greater than that of the corresponding pit, but it is difficult intrinsically to ascertain whether the cracks initiated at the bottom of the pits or the initiated crack progressed beyond the depth of the pits. The possibility that the crack does not initiate from the base of the pit has implications for the prediction of the pit to crack transition using fracture mechanics concepts as the stress intensity factor is based usually on treating the pit as an equivalent crack of the same depth.

Further analysis is ongoing to establish what information can be gleaned concerning the short crack growth rate for comparison with long crack growth rates. The difficulty is the uncertainty in time of crack initiation. From a more practical viewpoint, the distinction may be academic since it is the growth rate of the flaw (pit plus crack) that is important.

It should be emphasised that these measurements of pit and crack growth rate are based on continuous exposure and in the case of the 1.5 ppm chloride under very severe conditions. More work is required to translate these measurements into likely growth rates in service, for which the exposure conditions will be transient and in the case of the severe conditions, short-lived.

CONCLUSIONS

- Pitting was observed under all exposure conditions with the pit density and growth rates increasing generally in the order: deaerated pure water; aerated pure water, and aerated 1.5 ppm chloride-containing water.
- Pits formed in the pure water tended to be 'tighter' with a higher depth to mouth opening ratio than for the chloride-containing water. This is rationalised on the basis of retention of local pit solution chemistry influenced by MnS for the pure water bulk solution and a combination of mildly acidic pit environments and potential drop considerations in the chloride-containing solution.
- Extensive stress corrosion cracking from pits had initiated in the chloride-containing solution at exposure time less than 7 months. In aerated water, cracking was observed at 19 months but not at 13 months. A minimum pit size of 50–70 μm for initiation of cracks was identified. However, no cracking was observed in deaerated pure water after 22 months despite pits depths of up to 200 μm.
- The probability of a crack initiating from a pit of specific depth in aerated solution seemed little affected by chloride content or exposure time.

ACKNOWLEDGEMENTS

This work was conducted as part of the 'Life Performance of Materials' programme, a joint venture between the United Kingdom Department of Trade and Industry and an Industrial Group comprising Alstom, BNFL Magnox, British Energy, Innogy, Powergen, and Siemens. Angela Mensah and Linda Orkney provided experimental support.

REFERENCES

1. S. Zhou and A. Turnbull: 'Steam Turbine Operating Conditions and Environment Assisted Cracking - An Overview', NPL Report, MATC (A) 95, 2002.
2. 'Turbine Steam, Chemistry, and Corrosion, Experimental Turbine Tests', Palo Alto, CA, Electric Power Research Institute, 1997, TR-108185.
3. 'Corrosion of Steam Turbine Blading and Discs in the Phase Transition Zone', Palo Alto, CA, Electric Power Research Institute, 1998, TR-111340.
4. O. Povarov, V. Semenov and A. Troitsky: 'Generation of Liquid Films and Corrosive

Solutions on Blades and Discs of Turbine Stages', *Specialist Workshop on Corrosion of Steam Turbine LP Blades and Discs*, EPRI, 1998.

5. *Handbook of Chemistry and Physics*, 74th Edition, CRC Press, London, UK., 1993-1994.
6. MTDATA, Version 4.72, NPL, 1999.
7. M. Nougaret: Alstom, Private Communication, 2001.
8. *Steam Turbines - Steam Purity*, IEC, 2000.
9. A. Turnbull and S. Zhou: 'Effect of Oxygen and Chloride on Pitting and Stress Corrosion Cracking of a Steam Turbine Disc Steel', NPL Report MATC (A) 133, 2003.
10. A. Turnbull and S. Zhou: 'Effect of Oxygen and Chloride on Pitting and Stress Corrosion Cracking of a Steam Turbine Disc Steel', NPL Report MATC (A) 133, 2003.
11. M. O. Speidel and R. Magdowski: *2nd International Symposium Environmental Effects on Degradation in Nuclear Power Plants*, Monterey, 1985, 267–275.
12. M. O. Speidel, J. Denk, and B. Scarlin: 'Stress Corrosion Cracking and Corrosion Fatigue of Steam-Turbine Rotor and Blade Materials', Cost Project, EUR 13186 EN, 1991.

The Effect of Manufacturing Route on the Fracture Properties of an Austenitic Steel Valve Spindle in a Steam Environment

P. J. BERNARD and J. H. BULLOCH
Power Generation, Esb, Head Office
Dublin 2
Ireland

ABSTRACT

A throttle valve spindle in a large steam turbine failed prematurely after only some 3000 hours of service. The spindle, which was fabricated from a A286 stainless steel, failed at the head to spindle transition radius inside the valve casing during operation by intergranular stress corrosion cracking, IGSCC. The failure exhibited multiple crack initiation sites and progressed in an intergranular manner before changing to a transgranular mode of crack growth with both stages exhibiting virtually zero ductility. Consideration of service stresses suggested that the stress source was residual in nature and was produced during some fabrication process.

INTRODUCTION

Recently a throttle valve on a 260 MW unit failed after only around 3000 hours service. The valve spindle was found to have broken at the head to shaft transition location with the fracture being approximately normal to the axis of the spindle, see Figure 1. This valve spindle was around 1 m in length, the shaft was 46 mm in diameter and the head was 120 mm in diameter and was fabricated from the austenitic stainless steel AFNOR Z6CNT2515 which is a French version of A286 stainless steel.

The present paper describes a detailed failure analysis and the effects that the manufacturing route had in dictating the mechanical properties of the valve spindle.

FAILURE ANALYSIS

A general view of the spindle fracture surface are shown in Figure 2 where three distinct regions were identified, viz.,

a. a series of about five discrete semi-circular facets radiating from the spindle surface which was wholly intergranular in nature, see Figure 3,
b. a central portion, labelled A, where again intergranular crack growth prevailed, see Figure 4,

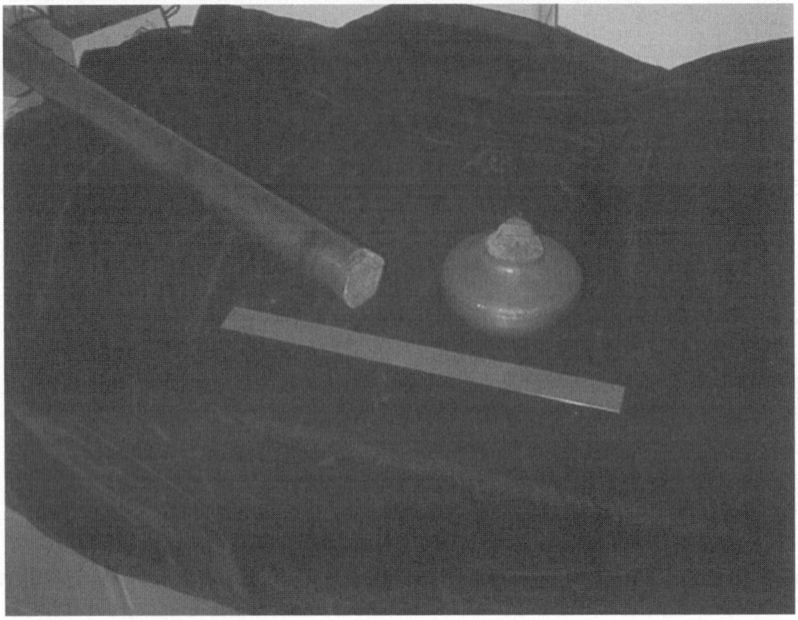

Fig. 1 Details of the broken spindle.

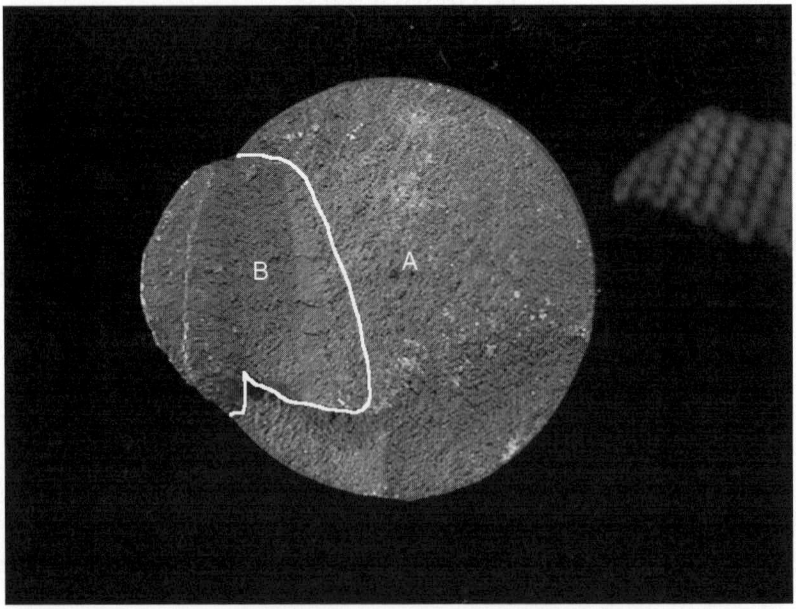

Fig. 2 Macro view of fracture face.

Fig. 3 Intergranular failure initiation site.

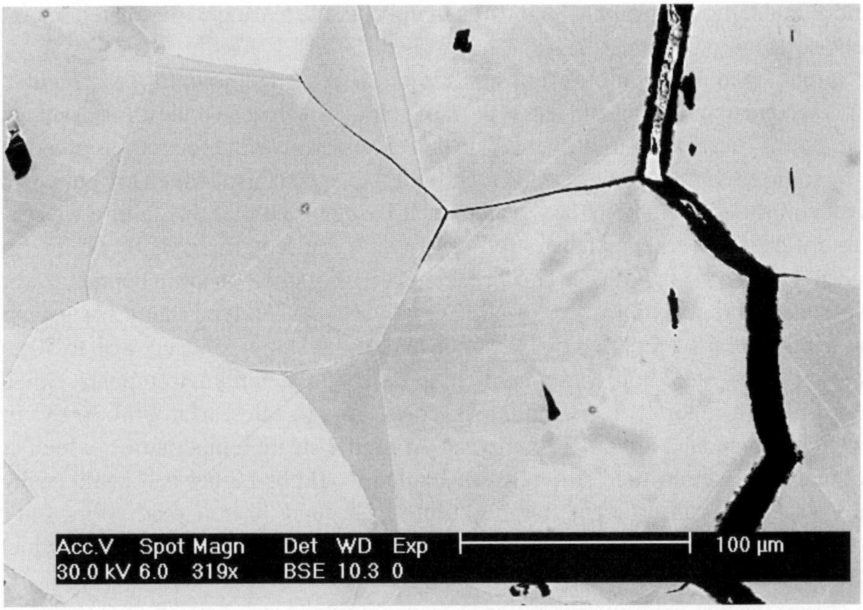

Fig. 4 Profile of intergranular cracking.

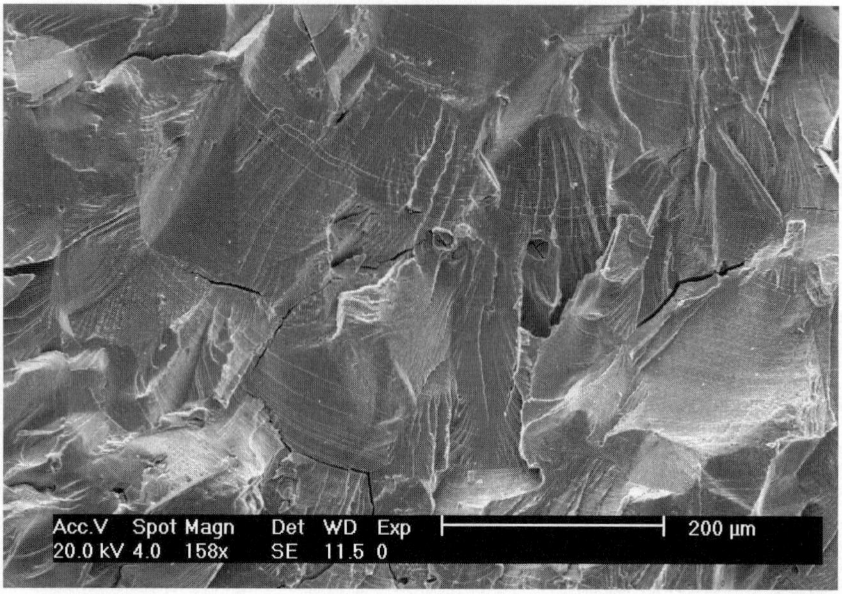

Fig. 5 General view of transgranular cracking.

c. a left of centre portion, labelled B, which was totally transgranular in nature, see Figures 5 and 6, and a final left side portion which was a mainly ductile tearing with isolated intergranular facets, see Figure 7.

The initial fracture initiation event and region A exhibited smooth intergranular facets and much secondary cracking and was typical of crack growth in an embrittled grain boundary microstructure. These facts strongly indicate that the crack growth process was predominantly driven by some environmental assisted cracking, EAC, mechanism which are known to thrive within an embrittled grain boundary structure. It is suggested that this failure was the result of a sub-critical EAC crack extension process which can only realistically be intergranular stress corrosion cracking, IGSCC. This process is very common on grain boundary weakened microstructures and many instances of IGSCC failures in A286 type stainless steels in steam environments have been reported by Hanninen and Aho-Mantila;[1] indeed, a 30 to 50% failure rate of some A286 steel bolt components in a variety of steam environments. Spiedel and Magdowski[2] have reported IGSCC plateau crack growth rates at around 300°C in A286 steels which approached around 30 mm year. At higher steam temperatures, which apply to the present case, in embrittled structures this value would be higher and could easily attain the rates observed in the present spindle failure of around 30 mm year. Other sub-critical crack growth processes, such as fatigue, could not produce crack velocities which even approached such values.

The abrupt change from intergranular to transgranular crack growth is quite common in stainless steels and Ni-based alloys and is the result of a change in the stress state. At this

Fig. 6 Detailed view of transgranular cracking.

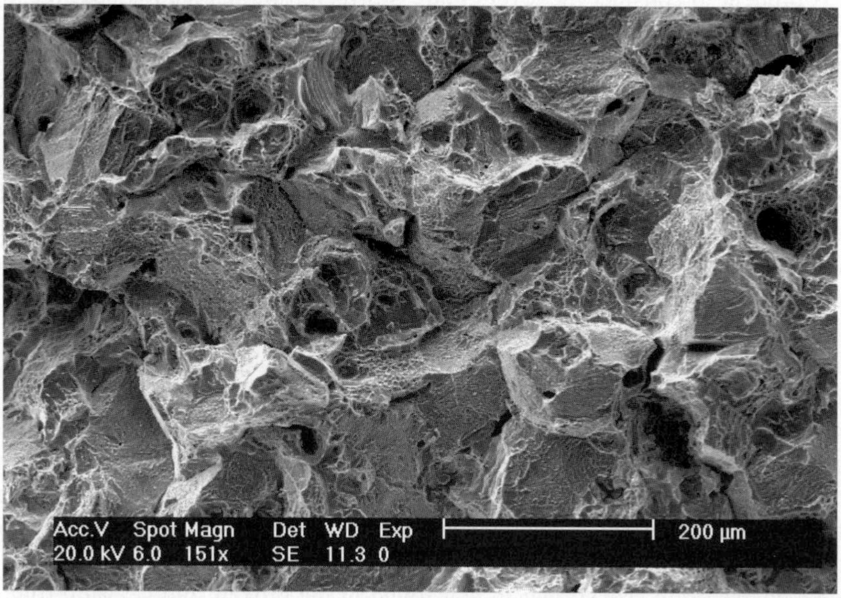

Fig. 7 General view of overload fracture.

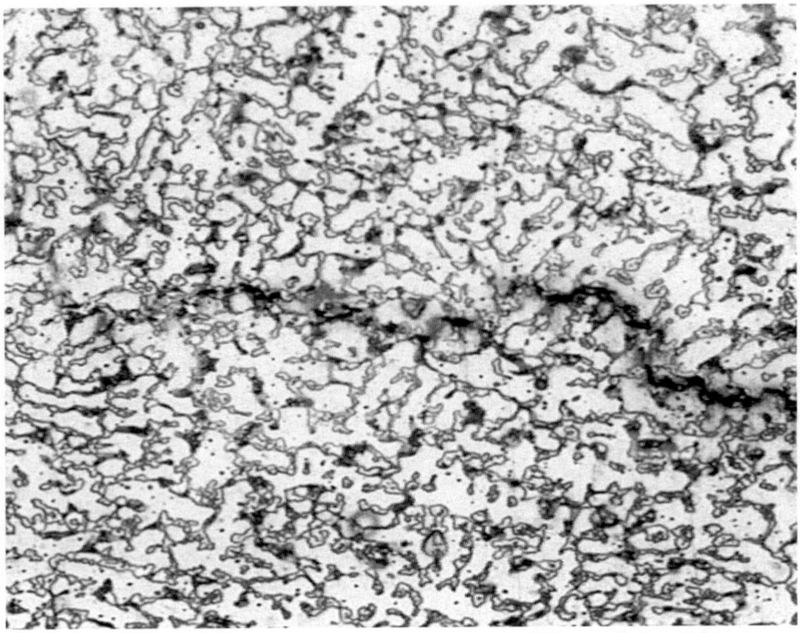

Fig. 8 Details of liquation cracking.

point it is prudent to consider the level of stress involved in this failure and it is suggested that any stress could be caused by the following:

a. Valve back-seating,

b. Steam loading and

c. Residual material stress.

The first option is possible although clearances and the lack of surface markings suggested that the possibilities were remote. Also, calculated steam loading stresses were very low and approached zero. Thus any stresses must have come from residual stresses induced during some fabrication process, such as forging or satellite coating, which could well be significant at the fracture or shoulder location.

The final stages of failure where ductile tearing predominated were caused by simple overloading of the remaining material ligament.

Another identical spindle which had been in service for the same duration as the broken one was subjected to detailed visual and dye-penetrant examinations. A sharp totally circumferential machining groove was observed at the head to shaft transition region and two circumferential cracks of 5 and 7.5 mm long were revealed close to the heat affected zone of a satellite coating run. A detailed view of one of these crack showed them to typically liquation cracking, see Figure 8. Also, a series of small intergranular surface cracks, which indicated a coarse grain size similar to that found in the broken spindle, were found in the parent material adjacent to the satellite coating.

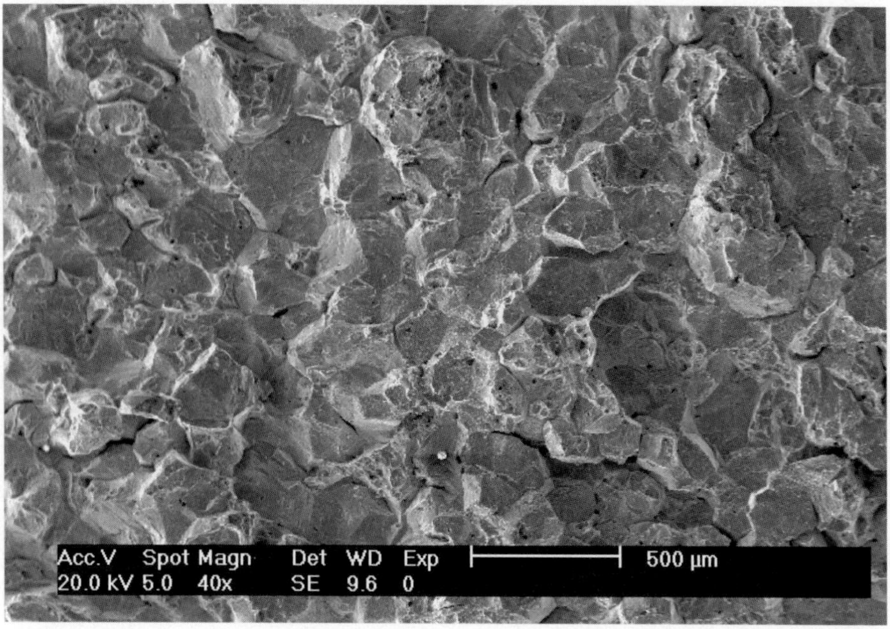

Fig. 9 Intergranular fracture in charpy test.

At this point room temperature Charpy type tests were carried out on the coarse grained head region of the fracture location and the finer grained threaded portion of the spindle and the resulting fracture surfaces were subjected to a detailed fractographic examination.

In the past spindle failures have occurred in the threaded part of the spindle which was at the opposite end to the present failure location. The fracture details of the spindle the head location, see Figure 9, showed that around 85% was intergranular in nature while those of the threaded location were very different with transgranular fracture predominating with only 17% some intergranular fracture, see Figure 10. The transgranular fracture was extremely ductile in character, see Figure 11.

The grain boundary, or bond percolation theory[3] indicated that there is some critical proportion of embrittled, or active, grain boundaries, which will result in total intergranular failure during fracture in an air environment. Also, a significant decrease in resistance to IGSCC resistance was observed when greater than 23% of the grain boundaries were embrittled during constant extension rate tests. Such an abrupt decrease in IGSCC resistance has been interpreted within the context of bond percolation theory. This critical value corresponds to a one-dimensional bond percolation threshold for a three-dimensional array of tetrakaidecahedra and confirmed the earlier results of Wells et al.[4] Effectively, this means that when more than 23% of the grain boundaries are embrittled a connected path of embrittled grains can exist through a steel matrix which causes extensive intergranular failure.

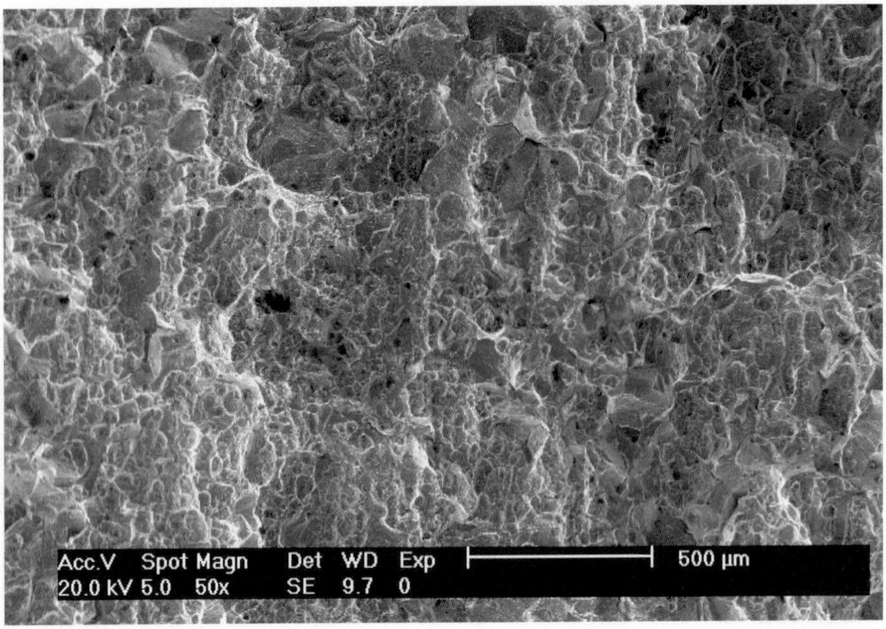

Fig. 10 Transgranular fracture in charpy test.

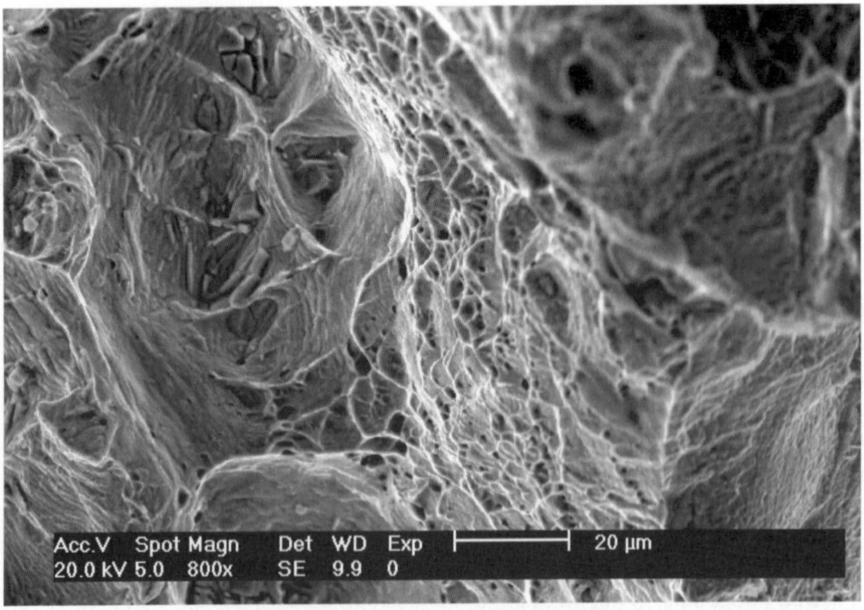

Fig. 11 Detailed view of ductile tearing in transgranular fracture region.

From the fracture details of the head and threaded locations it was clear that the head had suffered extensive 85% grain boundary embrittlement while only a small amount, viz. 17%, were embrittled in the threaded location. Furthermore, Wells et al[4] and more recently Gaudett and Scully[5] have conclusively demonstrated that in an environment an abrupt decrease in IGSCC resistance occurred once the level of embrittled grain boundaries in stainless steels went above around 23%. These fractographic and threshold IGSSC facts readily explain the present IGSCC spindle failure at the head location and also the lack of significant intergranular fracture in a series of previous spindle failures[5] which occurred in the threaded locations, where the extent of embrittled grain boundaries was below 23%. Indeed, the grain sizes of the previous spindle failures were much finer at around 40 to 50 μm and the sub-critical crack growth process was identified as fatigue.

The cracks found in the stellite coating in the unbroken spindle were liquation cracks which were formed by the fusion of low melting point constituents at the interdendritic boundaries of the solidified satellite deposit by the application of a final circumferential run of satellite. It is not known at present whether these cracks exist only within the satellite layer or have extended into the base material. However, since this type of crack growth from a cracked coating into the substrate is well documented, these cracks might well extent into the base material. Also since these cracks were only found under laboratory conditions and a highly polished surface condition it is highly probable that they could well have been missed under normal factory inspection conditions

In conclusion, this spindle suffered a low energy type failure due to the degradation of the material properties at the head location during either the upset forging or heat treatment operations.

PROCESS VARIABLES

The fine grain size, 40 μm, consistently found in these spindles, in regions remote from the head, is consistent with the normal manufacturing route of the material. In contrast, the grain size revealed at the head of the spindle is coarse, an average of 160 μm, with individual grains being as large as 400 μm.

Such a microstructure yields a substantially smaller grain boundary surface area than that of the smaller microstructure, thus leading to the accentuation of any grain boundary embrittlement or sensitisation processes.

A286 stainless steel is known to be sensitive to microstructure and heat treatment in terms of embrittlement, and clearly the parallel and head portions are quite different.

These spindles are conventionally manufactured by a three stage upset forging process, cumulative forging ratios at each stage being 1.5, 2.0 and finally, 4.3 at the finish. This is normally accomplished by heating to 1200°C, and carrying out the three deformation stages without any reheating, with a finishing temperature of 900°C. This results in the recrystallisation of the microstructure on plastic deformation, and allows no opportunity for excessive grain growth.

The forging process for the current spindles differed, in that the starting forging temperature was 1100°C, and the three forging stages were carried out individually, with reheating after each stage. This ensured that the beneficial effects of recrystallisation of the first two stages was eliminated by grain growth at the subsequent heating stage, resulting in a large finished grain size.

CLOSING REMARKS

The spindle failed by subcritical IGSCC, which was caused by degraded microstructure at the spindle head location. The degraded microstructure was developed during the forging process, which had been changed from the original process by the addition of inter stage reheating, allowing excessive grain growth to occur.

REFERENCES

1. H. Hanninen and I. Aho-Mantilla: 'Environment-Sensitive Cracking of Reactor Internals' *Proceedings of Environmental Degradation of Materials in Nuclear Power Station - Water Reactors*, Travers City, USA., 1987, 77–94.
2. M. O. Spiedel and R. Magdowski: 'Stress Corrosion Cracking of Nickel Based Alloys in High Temperature Water' ibid; San Diego, USA., 1993, 361–371.
3. D. Staffer: 'Introduction to Percolation Theory', PA, USA., 1985.
4. D. B. Wells J. Stewart, A. W. Herbert, P. M. Scott and P. E. Williams: 'IGSCC of Stainless Steel Tubes in High Temperature Water' *Corrosion*, **45**, 1989, 649–660.
5. M. A. Gaudett and J. R. Scully: 'Applicability of Bond Percolation Theory to Intergranular Stress Corrosion Cracking of Sensitised AISI 304 Stainless Steel' *Metallurgical Transactions*, **25A**, 1994, 775–787.
6. A. G. Callaghy and J. H. Bulloch: 'Detailed Fracture Analysis of a Failed Main Steam Turbine Throttle Valve Spindle' *Proceedings of International Conference in Advances in Materials and Processing Technologies*, AMPT 93, Dublin, Ireland, **2**, 1993, 885–896.

A Detailed History of Circumferential Cracking in Superheater Tubes in a Small Boiler

J. H. Bulloch and P. J. Bernard
Power Generation, Esb
Head Office, Dublin 2
Ireland

ABSTRACT

The present paper attempts to describe three separate instances of circumferential cracking failures in superheater tubes from a small 30 MW boiler which have occurred over the past thirty years or so. In each case the failures were associated with a crack initiation mechanism which was triggered by corrosion processes which occurred within a low pH fused salt deposit which was identified as potassium sulphate. Effectively, when the fused layer with sulphurous gases a liquid pyrosulphate phase was formed which caused active electrochemical corrosion of the steel tube surface.
The sub-critical crack growth of the circumferential cracks was driven by a sulphidisation corrosion assisted thermal fatigue mechanism. Crack extension occurred in a wholly intergranular fashion.
Experience has shown that the service time required to cause the onset of a spate of tube failures was around 36,000 to 40,000 hours. Also a detailed inspection of a series of replacement superheater tubes after only some 8000 hours service indicated that

(a) the damaging fused salt layer had already been formed and
(b) incipient circumferential crack formations were commonly observed across the boiler tube surfaces.

Finally, it was found that the crack growth times to tube failure events were such that circumferential crack growth rates ranged from 0.7 to 0.9 mm year.

INTRODUCTION

When a particular unit of any steam raising plant is subjected to a forced outage, which resulted from the failure of some critical engineering component, the revenue lost can be significant. Examples of boiler failures are essentially varied in nature and include corrosion, welding imperfections, fabrication defects and high temperature related processes such as microstructural degradation, oxidation and decarburisation. The manner in which different people survey boiler failures is markedly dependent upon the individualist specialist discipline, viz., a water chemist would develop a different perspective from those of a welding engineer or metallurgist. Such differing perspectives, if merged together, can yield a powerful method of understanding and solving boiler failures.

Today, boiler failures, and in particular boiler tube failures, represent the primary limiting problem which confronts steam raising utilities. Although many of the mechanisms which

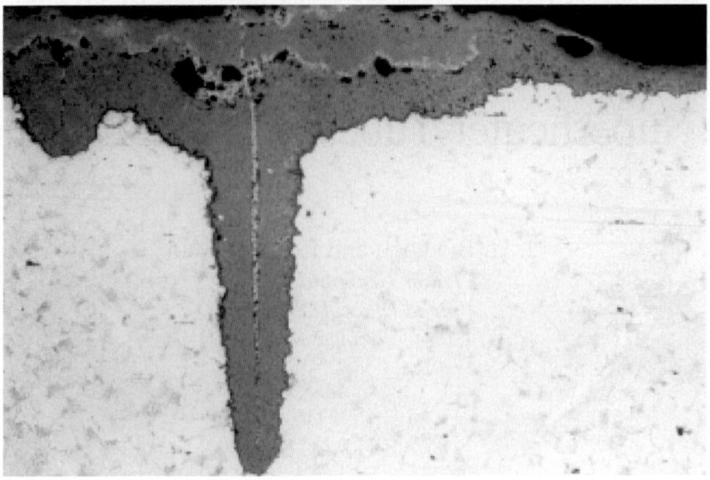

Fig. 1 Details of a typical circumferential crack (×100).

promote the various tube failures have been successfully identified one type of tube failure which is still not completely understood mechanistically is circumferential cracking. It is, however, thought that this type of cracking resulted from a certain mixture of some corrosion dominated process and thermal stressing.

Circumferential cracking, which is being detected in an increasing extent in water filled tubes, can initiate at both the outer and inner diameter locations and can be driven by fatigue or stress corrosion cracking.[1] Indeed, one study on a number of North American utilities has indicated that around half have been affected by circumferential cracking in water filled tubes.[2]

The present paper attempts to describe the history of circumferential cracking in superheater tubes in a small 30 MW peat-fired boiler which has been ongoing over a period of some thirty years or so.

HISTORY AND FAILURE ANALYSIS DETAILS

This unit was commissioned in 1966 and the first spate of tube failures were reported in 1973 after a service time of around 40,000 hours. The tubes were 3.5 mm thick and fabricated from a 0.5 Mo steel. Details of a typical circumferential crack is shown in Figure 1 where it was suggested that the failure was the result of a combination of vibration fatigue failure which was assisted by molten ash deposit corrosion.

The historical details after these spate of failures in 1973 was not recorded in any great detail and the next documented spate of failures in this unit was reported in 1991 after at total service life of some 155,000 hours. From a visual inspection it was clear that a thick

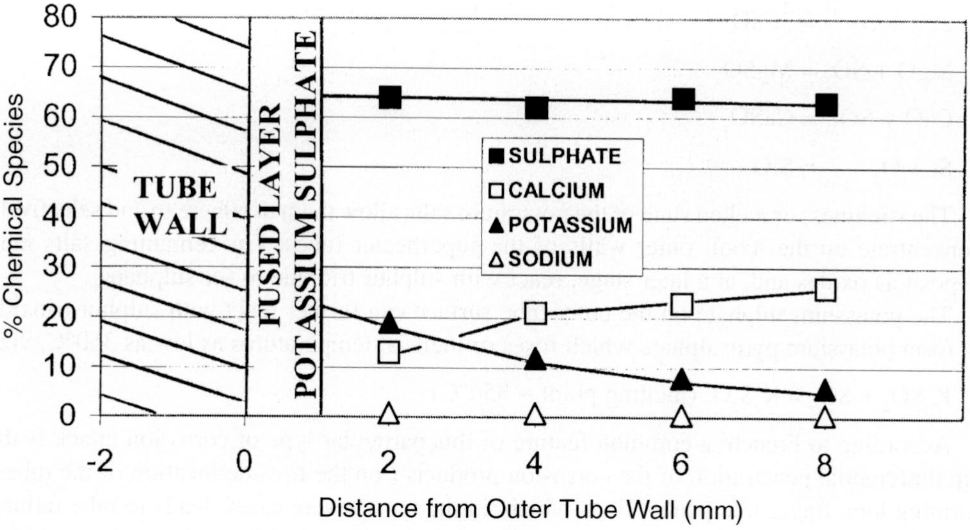

Fig. 2 Chemical composition of the outer tube deposit.

adherent fused slag deposit some 1 mm thick covered the 3.5 mm thick tubes which were fabricated from a 2.25Cr1Mo steel. Also an outer friable deposit of some 8 mm thick was evident. This deposit was chemically analysed at 2 mm intervals and the results are shown in Figure 2 where it was evident that significant metallic sulphates of calcium and potassium were present with the latter increasing markedly in the outer tube wall location. Indeed, the potassium levels at this location were about an order of magnitude higher than those found in a typical peat ash deposit. The innermost fused layer of the deposit was almost pure potassium sulphate which had a pH value of 3.5.

The only reasonable explanation for this selective separation and concentration of potassium sulphate on the outer tube wall surface was that the furnace operation temperatures were at times significantly high enough to melt the potassium sulphate, i.e., greater than 1,070°C, without fusing or melting the rest of the ash constituents. Sodium based salts, which fuse at even lower temperatures, volatilise in the temperature range 1,000 to 1,100°C and thus these were not recorded in the outer wall deposits at this particular boiler location. The chemical reactions that occur in the furnace tube wall locations may be represented by the following expressions,

$$2Na + O_2 = 2NaO \text{ (volatile)}$$

$$4K + O_2 = 2K_2O$$

$$K_2O + SO_3 = K_2SO_4 \text{ (melts or fuses)}$$

$$2Mg + O_2 = 2MgO$$

$2Ca + O_2 = 2CaO$

$MgO + SO_3 = MgSO_4$

$CaO + SO_3 = CaSO_4$

$Si + O_2 = SiO_2$

The stickiness or melted state of the potassium salts allow them to adhere to and selectively concentrate on the 'cool' outer walls of the superheater tubes. The remaining salts may deposit as oxides and, at a later stage, react with sulphur trioxide to for sulphates.

The potassium sulphate on the outer tube surface can further react with sulphur trioxide to form potassium pyrosulphate which fuses or melts at temperatures as low as 350°C, viz.,

$K_2SO_4 + SO_3 = K_2S_2O_7$ (melting point ~ 350°C)

According to French[3] a common feature of this particular type of corrosion attack is the circumferential penetration of the corrosion products (on the fireside location of the tubes) forming long finger-like penetrations which, in the most severe cases, leads to tube failure.

Details of a typical circumferential crack are given in Figure 3 and significant concentration of potassium and sulphur were found running along the entire length of the crack. Figure 4 illustrates the presence of significant sulphur concentrations at the crack tip location. Sub-critical crack growth occurred in a wholly intergranular manner as shown in Figure 5.

It is proposed that the defect initiation process was intimately associated with the fused potassium sulphate outer layer and that subsequent crack growth was the result of a sulphidation corrosion assisted thermal fatigue mechanism. The sulphidation reaction[4] occurs when iron is heated in an environment containing oxygen and sulphur and iron sulphide is formed which grow along grain boundaries causing crack growth. Continued crack growth is thought to occur by a combination of sulphidation and an acting stress cycle which was thermal in nature.

The superheater tubes were replaced in 1992 and the authors inspected these tubes in late 1993 after they had experienced only around 8,100 hours of service. Surprisingly enough, it was found that a 1 mm thick fused layer of potassium sulphate had already been formed and that the outer deposits were around 4 mm thick in total. Chemical analysis indicated similar composition trends to those found in 1991, see Figure 6. Also the presence of 'incipient' circumferential cracks, or indications, were found in large numbers, see Figure 7, and the deepest was only around 160 mm.

In 1997 the first spate of tube failures were reported after a service time of some 36,000 hours from the 1992 tube replacement exercise. Another series of tube failures were reported in 1998 after some 42,000 hours of service. The failure characteristics of the 1997/8 failures were identical to those recorded during the 1991 failure study.

CLOSING REMARKS

It was clear from the previous section that crack growth through the superheater tube was

a. Initially associated with high concentrations of potassium and sulphur due to the presence of the fused layer on the outer tube wall,

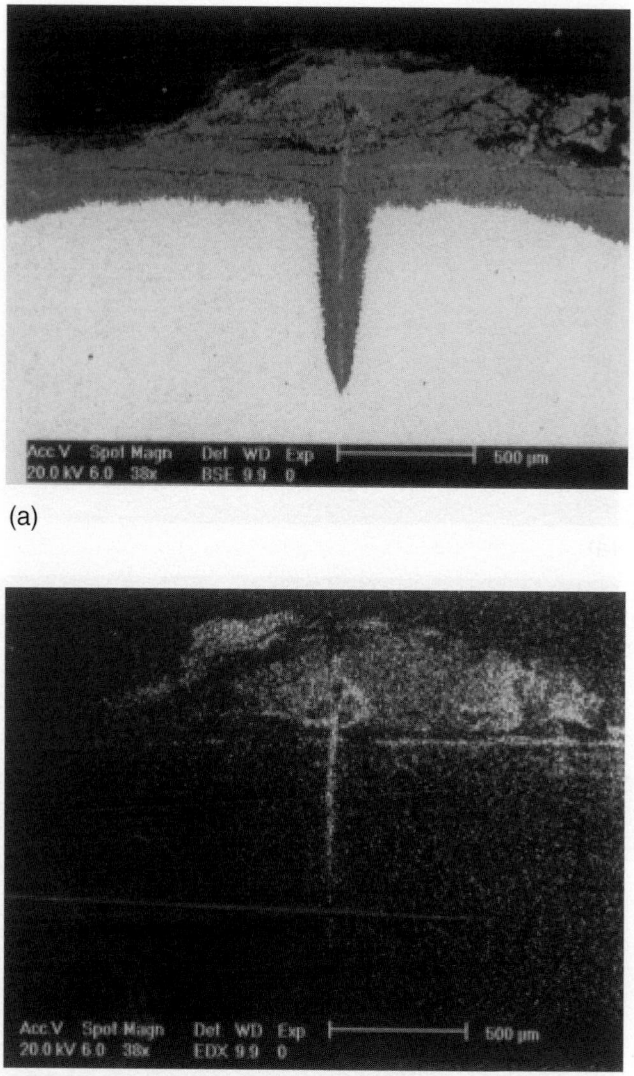

(a)

(b)

Fig. 3 Details of the circumferential cracks (a) metallographic details and (b) sulphur concentration dot map.

b. Seen to be associated with high sulphur concentrations over the entire defect length and even at the crack tip location and
c. Wholly intergranular in nature.

It is proposed that circumferential cracking was initiated by the fused low pH potassium sulphate layer with subsequent sub-critical crack growth being the result of a sulphidisation

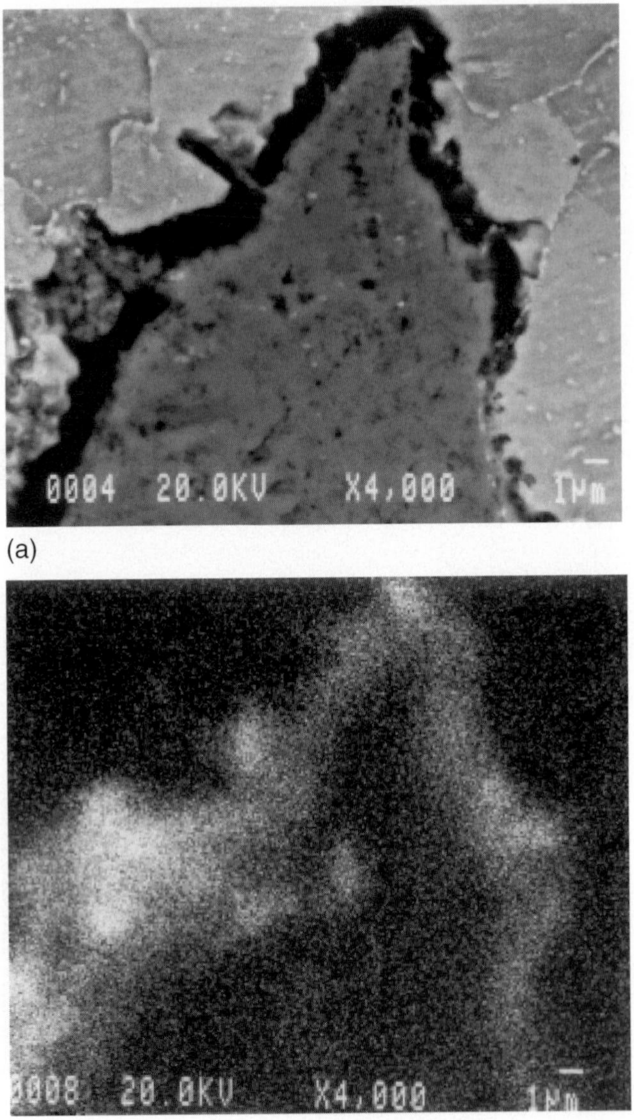

(a)

(b)

Fig. 4 Crack tip details (a) metallographic details and (b) sulphur concentration dot map.

corrosion assisted thermal fatigue process. The source of the sulphur was thought to come from copper sulphate additions to the peat fuel which kept the peat ash friable. Ideally copper oxy-chloride should have been added but copper sulphate was used since it was much cheaper. Also this particular boiler was not subjected to tube washing which would have been beneficial since the sulphate layer was around 60% soluble.

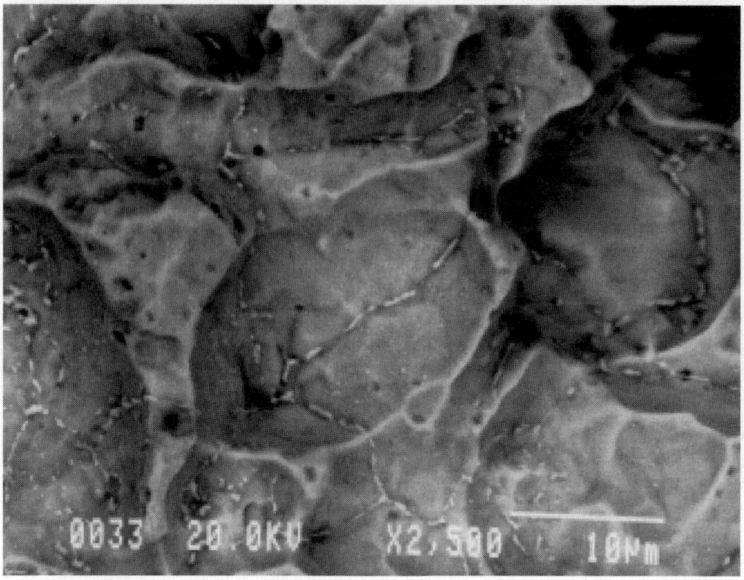

Fig. 5 Details of intergranular crack extension.

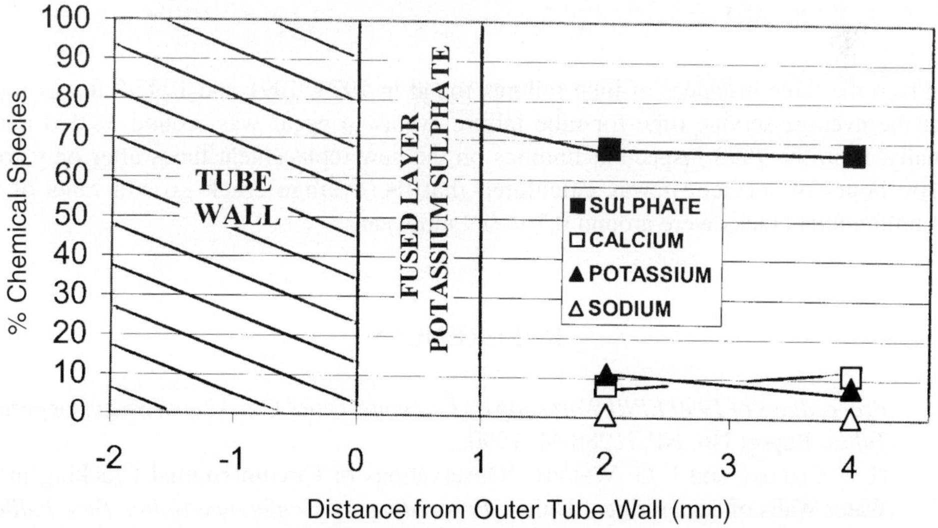

Fig. 6 Chemical composition of the outer tube deposition on the replacement tubes after 8,100 hours service.

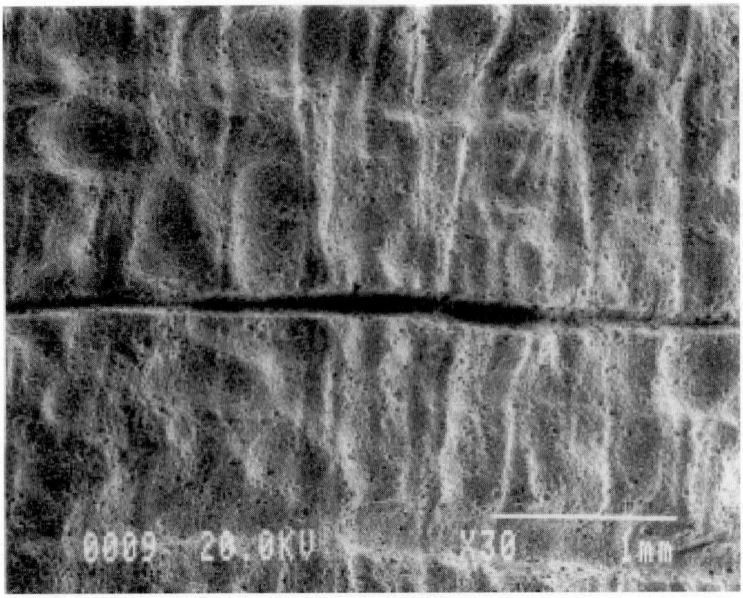

Fig. 7 Details of incipient circumferential cracking after only 8,100 hours service.

From the three instances of tube failures found in 1973, 1991 and 1997/8 it was found that the average service time for tube failure events to occur was around 38,000 hours. Finally, from the 1993 inspection findings on the new replacement tubes after only some 8,100 hours of service, it was calculated that the average crack growth rates of the circumferential cracks were around 0.7 to 0.9 mm/year.

REFERENCES

1. *Proceedings of 1990 EPRI Workshop on Circumferential Cracking of Steam Generator Tubes*, Report No. NP-71988-M, 1990.
2. H. J. Cialone and I. G. Wright: 'Observations of Circumferential Cracking in the Water Walls of Supercritical Units', *Proceedings of Conference Boiler Tube Failures in Fossil Plants*, EPRI, Atlanta, USA., 1987.
3. D. N. French: *Metallurgical Failures in Fossil Fired Boilers*, John Wiley & Sons, New York, 1983.
4. J. Stringer: 'The Sulphidation Reaction in Steels', *High Temperature and Technology*, **3**, 1985, 119–141.

Estimation of Creep Damage for 10Cr-1Mo-1W-VNbN Steel Forging

R. Ishii, Y. Tsuda, K. Fujiyama and K. Kimura
Toshiba Corporation
Power & Industrial Systems
R&D Centre

K. Saito
Toshiba Corporation
Keihin Product Operations

ABSTRACT

In order to develop life assessment techniques for aged components made of 10Cr-1Mo-1W-VNbN steel forgings, specimens were artificially deteriorated by creep and creep interrupted tests at various conditions. Microstructural observations were carried out focused on the martensitic laths. The martensitic lath of crept specimens grew compared with those of as-tempered specimen. The migration and annihilation of the dislocations during creep resulted in the growth of laths. The mean lath width increased with creep time and the increasing trend corresponded to both the creep strain and softening. The theoretical formation of softening was designed based on the kinetics of the dislocation structure. The formula gave a quantitative correlation among the stress, temperature, time and hardness. The creep damage at any given condition was estimated by this formula, and the estimated results were in good agreement with the experimental value of the hardness at any creep time fraction t/tr. A practical creep damage curve reflecting the initial hardness is proposed as one of the life assessment techniques for high temperature steam turbine components. The technique makes it possible to estimate the accurate creep damage even if data are limited.

INTRODUCTION

In order to improve the thermal efficiency of fossil-fired power plant, the steam condition has been raised every year. Several high chromium heat resistant steels with superior creep strength were developed[1, 2] in this process and applied to commercial power plants. Several steels have been operated over 80,000 hours, however, the degradation is anticipated in hot section components. Damage estimation methods for widely used low alloy steels such as CrMoV were established based on huge data.[3, 4] And active investigations on modified 9Cr-1Mo steel have also been carried out in recent years.[5–8] Sawada et al.[5, 6] carried out the quantitative analysis on the martensitic lath structure of deteriorated modified

Table 1 Chemical composition of steels studied (wt.%).

	C	Si	Mn	Ni	Cr	Mo	V	W	Nb	N
Steel A1	0.14	0.03	0.52	0.73	10.4	1.05	0.21	1.06	0.07	0.04
Steel A2	0.15	0.03	0.64	0.69	10.0	0.99	0.19	1.01	0.05	0.04

Table 2 Heat treatment condition of steels studied.

	Normalising	1st Tempering	2nd Tempering
Steel A1	1323 K × 33 hours → O.Q	843 K × 33 hours → F.C	933 K × 49 hours → F.C
Steel A2	1323 K × 5 hours → O.Q	843 K × 14 hours → F.C	923 K × 17.5 hours → F.C

9Cr-1Mo steel. Kimura et al.[7, 8] proposed a useful damage estimation method based on the kinetics of the dislocation structure. In these investigations, the damage estimation was related only to the recovery of the martensitic lath. In the heat resisting steels with high creep rupture strength, lath interfaces are covered with carbides and intermetallic compounds composed by supersaturated solid solution elements during creep.[9] This morphology is also possible to suppress the migration of dislocation and to delay the recovery of martensitic lath. Therefore, in order to establish damage estimation methods for high chromium heat resisting steels, it is important to understand the change of the martensitic lath structure. On the other hand, since the estimation methods need to be useful, the simplified method supported by microstructural change is indispensable. This paper describes the investigation results on martensitic lath structure, hardness and the estimated results of softening on a 10Cr–1Mo–1W–VNbN steel forging during creep.

EXPERIMENTAL PROCEDURE

The chemical composition and heat treatment condition of the steels studied are shown in Table 1 and Table 2, respectively. A 10Cr-1Mo-1W-VNbN steel forging called TOS107[2] was developed by Toshiba in 1980's for high temperature steam turbine rotor. 9 rotors have been favourably operated in the commercial power plants. Specimens were machined from prolong of the commercial rotors. Creep tests were carried out at temperature range from 839 to 903 K, stress range at from 117.7 to 294 MPa. Creep interrupted specimens were prepared for Steel A2 tested at 839 K-245 MPa, 873 K-196 MPa and 903 K-127.5 MPa. Creep interrupted time

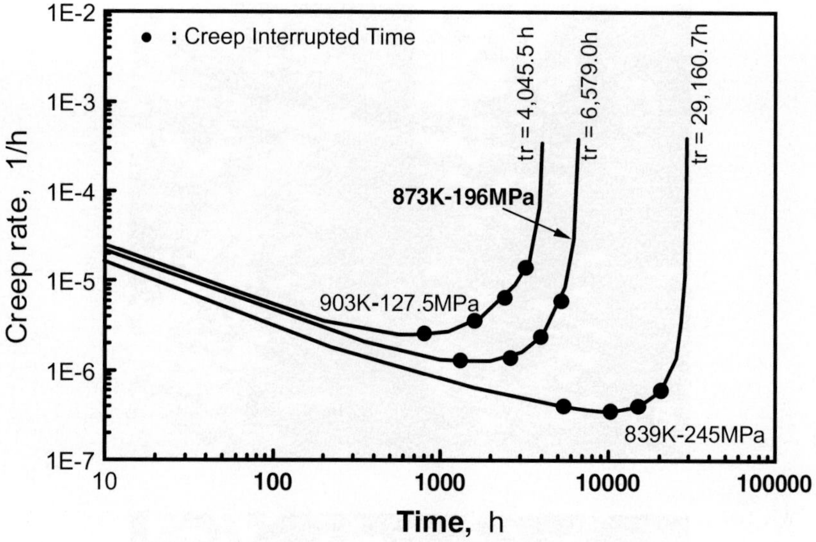

Fig. 1 Creep rate vs. time curves of steel A2.

was approximately 20, 40, 60 and 80% of creep rupture time of each creep testing condition. The normalised creep rupture time (t/t_r) is called creep damage. TEM observations were carried out on as-tempered and crept specimens. The martensitic lath width were measured on more than 100 laths based on the photographs of each specimen. Carbon extracted replicas were taken from a part of the specimen, and the density of fine precipitates in the martensitic lath were also measured. Hardness measurement was carried out on all specimens.

RESULTS AND DISCUSSION

Change of the Dislocation Structure During Creep

Creep rate versus time curves of Steel A2 tested at 839 K-245 MPa, 873 K-196 MPa and 903 K-127.5 MPa are shown in Figure 1. Regardless of the testing condition, the creep interrupted time was in the tertiary creep region except for 20% creep damage specimen. Dislocation structure of as tempered Steel A2 is shown in Figure 2a. The grain was composed of martensitic lath with high dislocation density. The lath was shaped linearly and had small width. Dislocation structure of crept Steel A2 (873 K-196 MPa, tr = 6,579.0 hours) is shown in Figure 2b. It revealed the deformation of lath interfaces, the growth of lath and the decrease of dislocation density in comparison with the as-tempered specimen.

The quantitative analysis performed focused on the lath width. The measurement results for the as-tempered specimen and crept specimen tested at 873 K-196 MPa are shown in

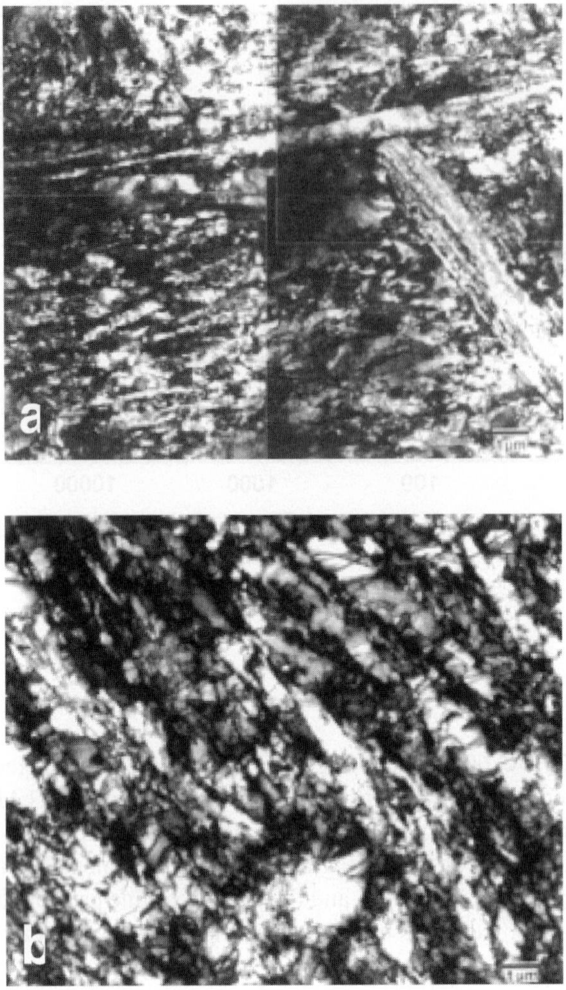

Fig. 2 TEM thin-foil micrograph of Steel A2.

Figures 3a and 3b, respectively. The lath width of as tempered specimen was ranged from less than 0.1 to 0.8 μm, and the mean lath width was 0.33 μm. In the crept specimen the lath width ranged from 0.1 to 1.1 μm and the average lath width was 0.51 μm.

The change of the mean lath width is shown in Figure 4 together with the creep curves for each testing condition. The lath width increased with creep time. The mean lath width of near grain boundaries was larger than that of grain interior. The preferential recovery at near grain boundaries was confirmed in 10Cr-1Mo-1W-VNbN steel as well as in CrMoV steel,[10] 12CrMoVW steel[11] and 9Cr-1Mo-VNbN steel.[12] From these results, it was anticipated the creep deformation was controlled by recovery of the martensitic lath.

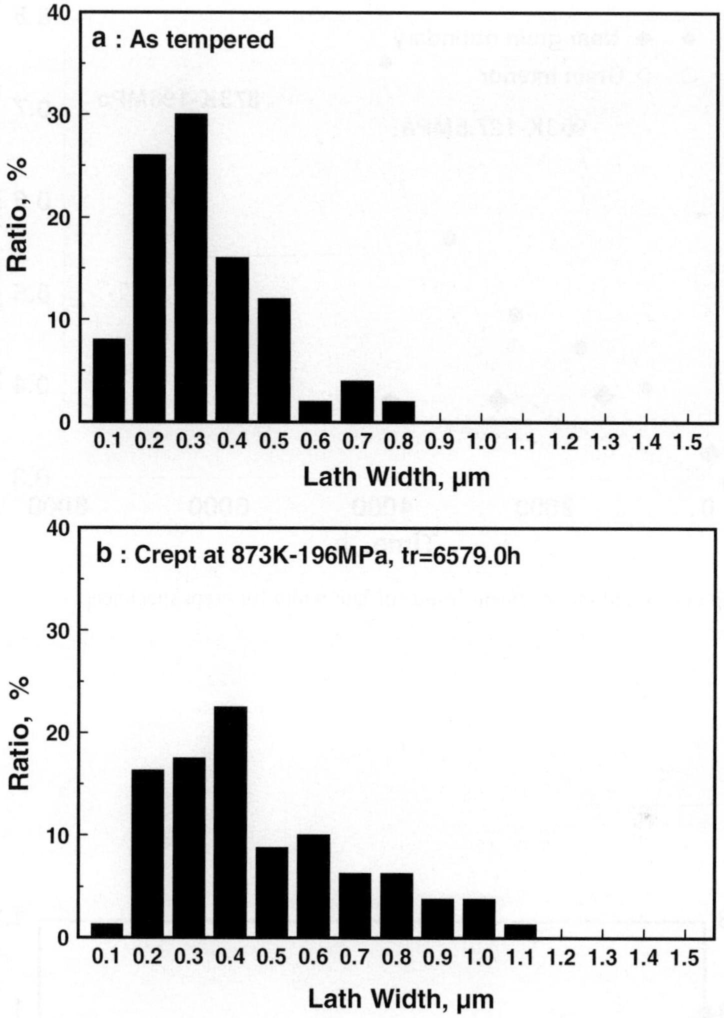

Fig. 3 Measurement result for lath width of as tempered and crept specimen.

Softening During Creep

In the preceding section, good correlation between the increase of lath width and creep strain was confirmed. Watanabe et al.[11] reported that the softening of 12CrMoVW steel occurred by decrease of the dislocation density during creep. Since the migration and annihilation of dislocations resulted in the growth of laths, the correlation between softening, lath width and creep strain was investigated in this section.

Figure 5 shows the results of Vickers hardness measurements on crept Steel A2 tested at 839 K - 245 MPa, 873 K - 196 MPa and 903 K - 127.5 MPa. Creep curves for each testing

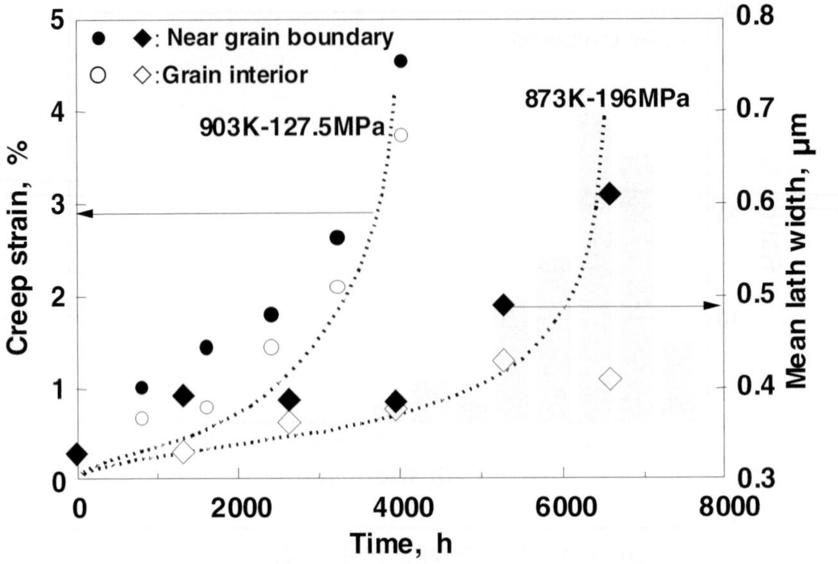

Fig. 4 Creep curves and measurement results of lath width for crept specimens.

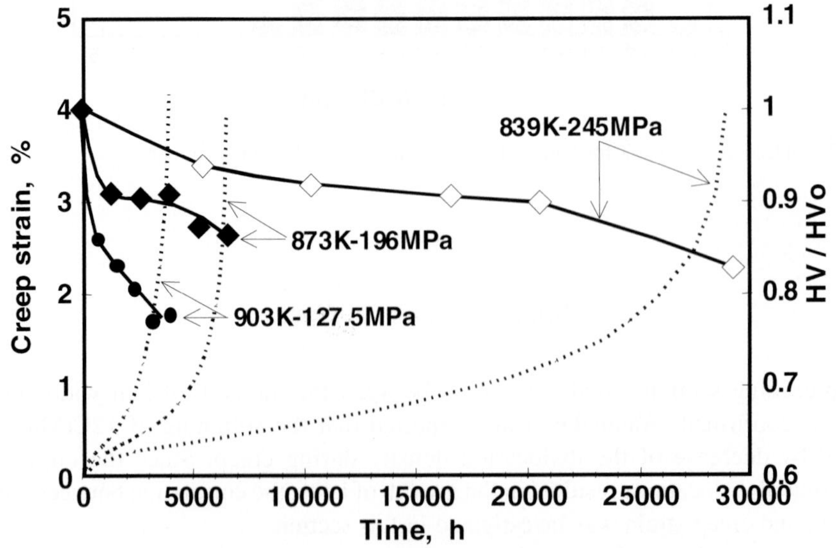

Fig. 5 Creep curves and measurement results of hardness for crept specimens.

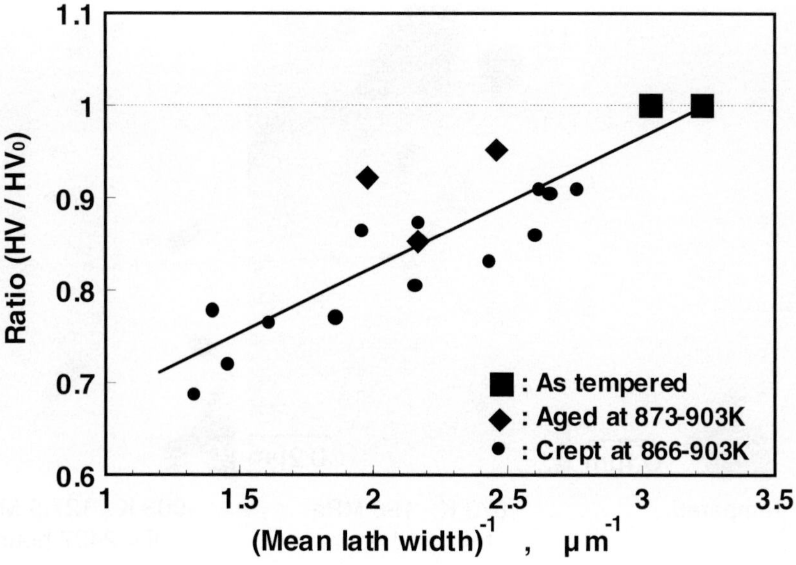

Fig. 6 Relation between mean lath width and hardness.

condition are also drawn in the figure. Regardless of the testing condition, remarkable softening was confirmed after loading. In the tertiary creep region, the softening trend became gradual. This trend depended on both temperature and stress, and corresponded to the creep strain and growth of the lath width. Sawada et al.[13] and Ishii et al.[14] found a similar correspondence in 11Cr-2.6W-3Co-MoVNbNB steel and 9Cr-1Mo-VNbN steel, respectively. Moreover Abe et al.[15] and Sawada et al.[5] found a linear correspondence between hardness and lath width in 9Cr-2W steel and 9Cr-1Mo-VNbN steel, respectively. These relations were independent of both temperature and stress. Figure 6 shows the results for Steel A1 and Steel A2 arranged by the same method. The linear correspondence was confirmed in the 10Cr-1Mo-1W-VNbN steel as well as other steels. The increase of creep strain in the steels revealed martensitic structure corresponded to both the increasing of the lath width and the softening regardless of the chemical composition. Therefore, the softening is regarded as an indirect description of the creep deformation. From these results, it was clear the measurement of softening was a reasonable way to estimate creep damage.

Precipitation Morphology During Creep

The effect of precipitates was not taken into account in the preceding sections. Williams et al.[16] and Bolton et al.,[17] however, reported that the coarsening of precipitates caused degradation in CrMoV steel. And many investigations were reported on the suppression of the dislocation

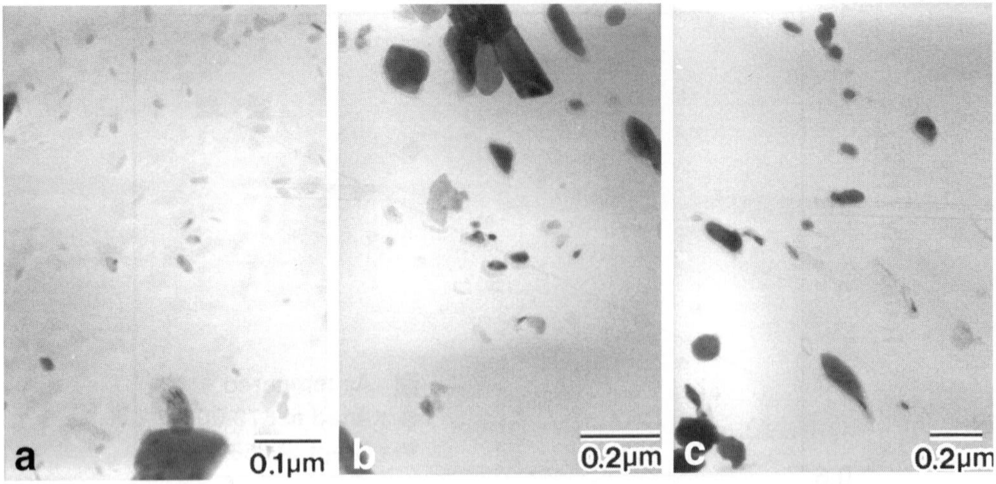

As Tempered

873 K - 196 MPa, ti = 1316 hours

903 K - 127.5 MPa, ti = 2427 hours

Fig. 7 Carbon extracted replicas of as tempered specimen and crept specimen.

migration by fine precipitates within lath. In addition, the effect of the precipitates at lath interfaces on the suppression of the recovery is also shown qualitatively.[9] The morphology at lath interfaces, however, governs directly the increase and the suppression of the lath width. Therefore, the investigation was carried out focused on the fine precipitates within lath.

Figure 7 shows carbon-extracted replicas of as tempered and crept Steel A2. Based on these photographs, the number of the fine particles located in the lath was measured. The fine precipitates identified in the lath were mainly M_2X and MX.[18] The measurement results of particle number fraction (n/n_0) are shown in Figure 8. The particle number fraction observed within lath became 1/5 or less by 1000 hours compared with the as-tempered specimen, and it hardly changed after 1000 hour regardless of the creep testing condition. This trend did not correspond with the increase of the lath width and with the softening. Therefore it is reasonable to estimate the creep damage without taking account of the behaviour of fine precipitates. The effect of fine particles was implicit in the increase of lath width.

Estimation of Softening During Creep

From the results of the preceding sections, the migration and annihilation of dislocation resulted in both the softening and the growth of lath included the change of the precipitation morphology. Accordingly, in order to express the relation between creep damage and hardness, the theoretical formation of softening is designed based on the kinetics of the dislocation structure.

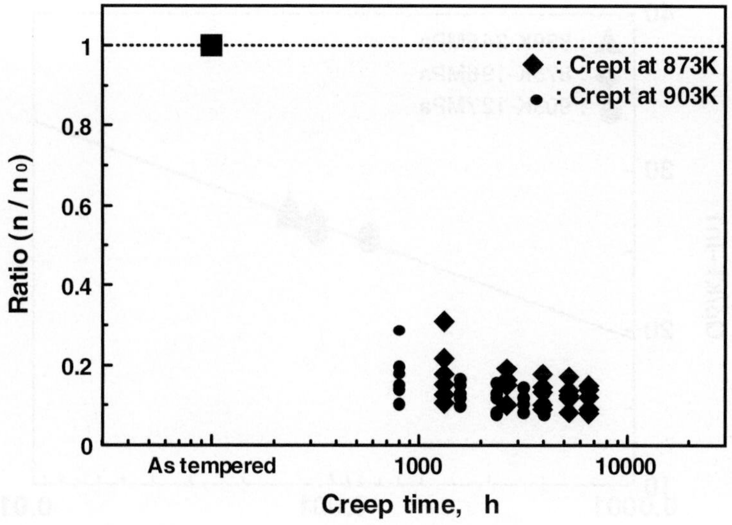

Fig. 8 Change of the number of fine precipitates in lath during creep.

Sawada et al.[5] showed the experimental equation between hardness *HV* and dislocation density ρ in modified 9Cr-1Mo steel. From the correlation between hardness and growth of lath shown in the section 'Softening during creep', eqn (1) is available regardless of the chemical composition.

$$HV = R_0 + \alpha \cdot \ln \rho \tag{1}$$

It was observed that softening due to creep accompanies significant change in dislocation substructure. Therefore, kinetic equation for dislocation density is used.[19]

$$d\rho/dt = A_1\rho - A_2\rho^2 \tag{2}$$

where, ρ is dislocation density, and A_1, A_2 are coefficients.

Assuming the recovery term in eqn (2) governs at high dislocation density, eqn (2) can be rewritten as follows.[20]

$$\frac{d\rho}{dt} = -\beta\left(\frac{\sigma}{E}\right)^{\nu} \exp\left(\frac{-Q_\rho}{RT}\right)\rho^2 \tag{3}$$

where, β and ν are unknown constants and Q_ρ is activation energy.

By integrating eqn (3) and substituting into eqn (1), we obtain the following equation.

$$HV = R_0 - \alpha \cdot \ln\left\{\exp\left(\frac{(R_0 - HV_0)}{\alpha}\right) + \beta\left(\frac{\sigma}{E}\right)^{\nu} \exp\left(\frac{-Q_\rho}{RT}\right) \cdot t\right\} \tag{4}$$

where, E is the Young's modulus, R is the gas constant, and HV_0 is the initial Vickers hardness. eqn. (4) shows the general relation between creep time *t* and hardness *HV* at any temperature

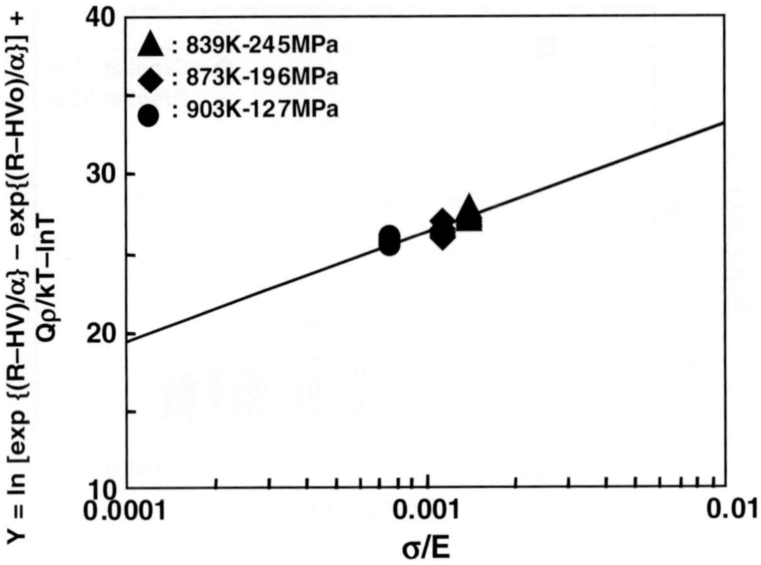

Fig. 9 Result of regression analysis.

and stress. Rewriting eqn. (4) in the linear form for regression analysis to determine unknown ν and β.

$$Y = \nu \ln\left(\frac{\sigma}{E}\right) + \ln \beta \tag{5}$$

where,

$$Y = \ln\left\{\exp\left(\left(\frac{R_0 - HV}{\alpha}\right)\right) - \exp\left(\left(\frac{R_0 - HV_0}{\alpha}\right)\right)\right\} + \left(\frac{Q_p}{RT}\right) - \ln t \tag{6}$$

Eqn (5) shows the linear relationship between Y and normalised σ. The unknown ν and β were determined by using the data obtained from crept specimens. Figure 9 shows the regression analysis result of eqn (5) used the crept data at 839 K - 245 MPa, 873 K - 196 MPa and 903 K - 127.5 MPa. Figure 10 shows the calculated creep damage estimation curves. The curves obtained by eqn (4) roughly approximated the experimental data. From these results, it was confirmed that the estimation of creep damage was possible even in case of the 10Cr-1Mo-1W-VNbN steel revealed complex precipitation morphology.

Based on eqn (4), it is possible to estimate the relation between hardness and creep damage at any conditions of temperature, stress and initial hardness.

Several calculated results for 10Cr-1Mo-1W-VNbN steel are shown in Figures 11, 12 and 13. Each figure shows the stress dependence (Figure 11), temperature dependence (Figure 12) and initial Vickers hardness dependence (Figure 13) on estimation curve, respectively.

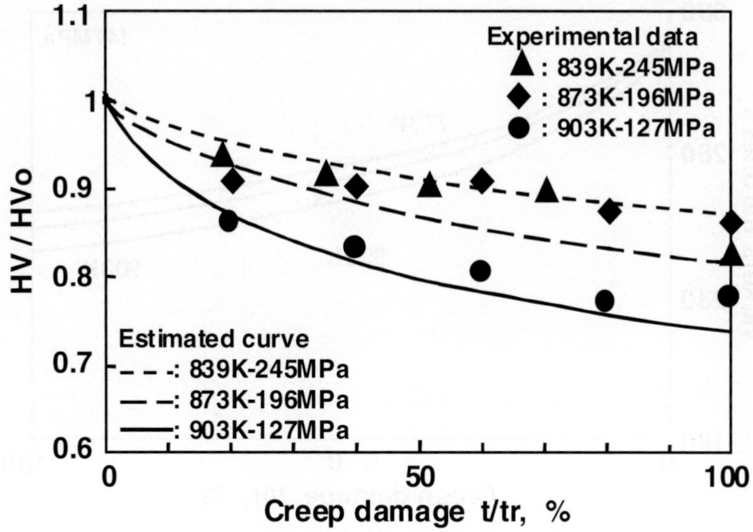

Fig. 10 Comparison between creep damage estimation curves and test results.

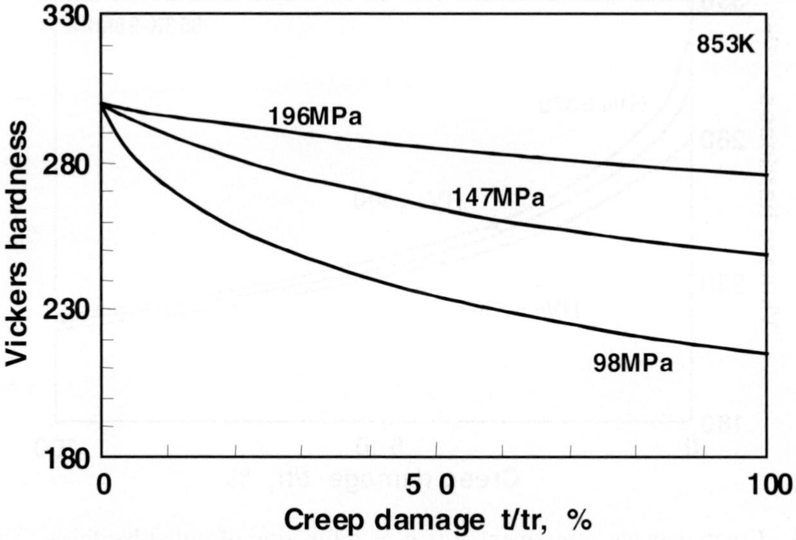

Fig. 11 Creep damage assessment curves as a function of stress.

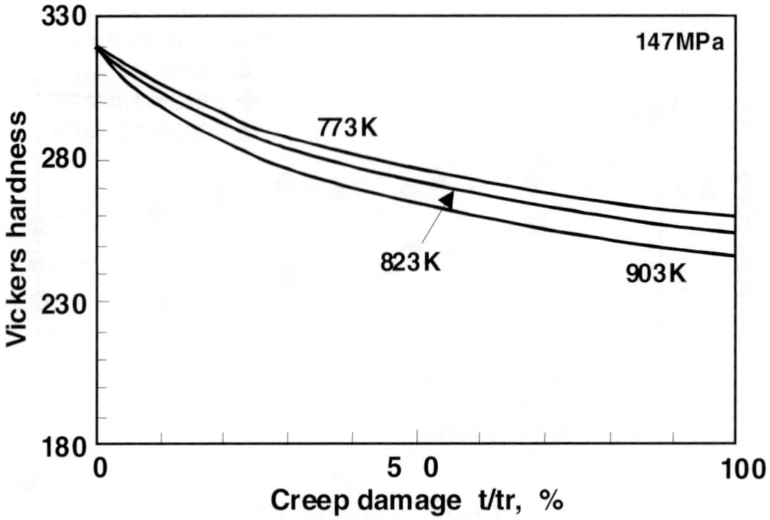

Fig. 12 Creep damage assessment curves as a function of temperature.

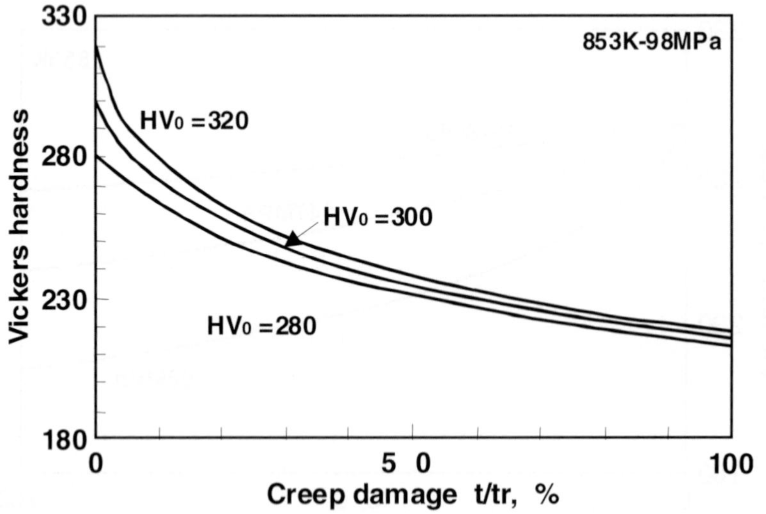

Fig. 13 Creep damage assessment curves as a function of initial hardness.

Although the validity of calculation results needs to be checked by actual data and needs much more data for more accurate estimation, this method is one of the beneficial estimation methods for creep damage easily. This estimation method is also useful for other high strength steels revealed martensitic structure.

SUMMARY

Microstructural observations were carried out on 10Cr-1Mo-1W-VNbN steel deteriorated by creep at various conditions. The theoretical formation of softening was designed based on the kinetics of the dislocation structure. The formula gave quantitative correlation between the creep stress, temperature, time and hardness. The main results obtained through this investigation are summarised as follows.

1. Good correlation was confirmed between the growth of the martensitic lath, softening and the increase of creep strain.
2. The remarkable decrease of the density of fine precipitates within lath was observed in the early stage in the creep. The effect of fine particles was implicit in the increase of lath width.
3. As a result of designing a softening model on the basis of correlation between dislocation density and hardness, the softening during creep was calculated at any temperature, stress and initial hardness without taking the influence of precipitates into consideration seemingly.
4. Although the improvement of the accuracy of the damage estimation curve was required, basic information for establishment of simple and useful damage estimation method was obtained for high chromium heat resisting steel.

REFERENCES

1. T. Fujita: 'Development of High Chromium Ferritic Steels for Ultra Super Critical Power Plant', *Proceedings of the 3rd International Charles Parsons Turbine Conference*, R. D. Conloy, M. J. Goulette and A. Strang, eds., 1995, 493–516.
2. Y. Tsuda, M. Yamada, R. Ishii and O. Watanabe: 'Development of High-Temperature Materials for Steam Turbine', *Proceedings of the 4th International Charles Parsons Turbine Conference*, A. Strang, W. M. Banks, R. D. Conloy and M. J. Goulette, eds., 1997, 283–295.
3. K. Fujiyama, K. Kimura, M. Muramatsu, E. Tsunoda and H. Aoki: 'Procedure for Life Assessment of Degraded Steam Turbine Components and Life Diagnosis System', *Journal of Society of Materials Science*, **37**, 1988, 315–321. (in Japanese).
4. T. Ookoda: 'Remaining Life Evaluation of Thermal Power Plant', *The Thermal and Nuclear Power*, **46**, 1995, 362–368. (in Japanese).
5. K. Sawada, K. Maruyama, R. Komine and Y. Nagae: 'Microstructural Changes During Creep and Life Assessment of Mod. 9Cr-1Mo Steel', *Tetsu-to-Hagane*, **83**, 1997, 466–471. (in Japanese).

6. K. SAWADA, M. TAKEDA, K. MARUYAMA, R. KOMINE and Y. NAGAE: 'Residual Creep Life Assessment by Change of Martensitic Lath Structure in Modified 9Cr-1Mo-Steels', *Tetsu-to-Hagane*, **84**, 1998, 580–585. (in Japanese).
7. H. OKAMURA, R. OHTANI, K. SAITO, K. KIMURA, R. ISHII, K. FUJIYAMA, S. HONGO, T. ISEKI and H. UCHIDA: 'Basic Investigation for Life Assessment Technology of Modified 9Cr-1Mo Steel', *Nuclear Engineering and Design*, **193**, 1999, 243–248.
8. K. KIMURA, K. FUJIYAMA, R. ISHII and K. SAITO: 'Estimation of Creep Damage for the Components of Mod. 9Cr-1Mo Steel', *Transactions of JSME*, **66A**, 2000, 1404–1410. (in Japanese).
9. R. ISHII, Y. TSUDA and M. YAMADA: 'High Strength 12%Cr Heat Resisting Steel for High Temperature Steam Turbine Blade', *Steel Forgings*, E. G. Nisbett and A. S. Melilli, eds., ASTM STP 1259, 1997, **2**, 317–329.
10. K. KIMURA, T. MATSUO, M. KIKUCHI and R. TANAKA, 'The Effect of Stress of Degradation of 1Cr-1Mo-1/4V Steel at Elevated Temperatures', *Tetsu-to-Hagane*, **72**, 1986, 474–481. (in Japanese).
11. T. WATANABE, Y. MONMA, T. MATSUO and M. KIKUCHI: 'Degradation of 12Cr Type Heat Resisting Steels Due to High Temperature Creep', *123rd Committee on Heat Resisting Metals and Alloys Rep.*, **30**, 1989, 1–16. (in Japanese).
12. H. KUSHIMA, K. KIMURA and F. ABE: 'Degradation of Mod.9Cr-1Mo Steel during Long-Term Creep Deformation', *Tetsu-to-Hagane*, **85**, 1999, 841–847. (in Japanese).
13. K. SAWADA, M. TAKEDA, K. MARUYAMA, R. ISHII and M. YAMADA: 'Dislocation Substructure Degradation During Creep of Martensitic Heat-Resisting Steels with and without W', *Proceedings of the 6th Liege Conference Part I, Materials for Advanced Power Engineering 1998*, J. L. Beckers, F. Schbert and P. J. Ennis, eds., 1998, 575–583.
14. R. ISHII, K. KIMURA, K. FUJIYAMA, S. HONGO and K. SAITO: 'Evaluation of Damage for Modified 9Cr-1Mo Steel', *Journal of High Pressure Institute of Japan*, 2000, **38**, 4–11. (in Japanese)
15. F. ABE, S. NAKAZAWA, H. ARAKI and T. NODA: 'The Role of Microstructural Instability on Creep Behaviour of a Martensitic 9Cr-2E Steel', *Metallurgical Transactions A*, **23A**, 1992, 469–477.
16. K. R. WILLIAMS and B. J. CANE: 'Creep Behavior of 1/2Cr-1/2Mo-1/4V steel at Engineering Stresses', *Materials Science and Engineering*, **38**, 1979, 199–210.
17. C. J. BOLTON, B. F. DYSON and K. R. WILLIAMS: 'Metallographic Methods of Determining Residual Creep Life', *Materials Science and Engineering*, **46**, 1980, 231–239.
18. R. ISHII, Y. TSUDA, M. YAMADA and K. KIMURA: 'Fine Precipitates in High Chromium Heat Resisting Steels', *Tetsu-to-Hagane*, **88**, 2002, 36–43. (in Japanese).
19. F. DYSON and M. MCLEAN: 'Creep Deformation of Engineering Alloys : Developments from Physical Modelling', *ISIJ International*, **30**, 1990, 802–812.
20. K. FUJIYAMA, T. ISEKI, A. KOMATSU and N. OKABE: 'Creep Life Assessment of 2.25Cr-1Mo Piping Steel and of Its Simulated HAZ Material', *Material Science Research International*, **3**, 1997, 237–243.

Performance of P91 Steel Under Steady and Cyclic Loading Conditions – Research and Power Plant Experience

I. A. Shibli
European Technology Development
2 Warwick Gardens, Ashtead
Surrey KT21 2HR, UK

ABSTRACT

This Paper is based on a recent review carried out by European Technology Development (ETD) on the use of 9Cr martensitic steels in power plant both as thick section components (headers, steam pipes etc.) and thin section tubing in superheater plant sections.

P91 or 9CrMoVNb steel was originally developed for use in ultra supercritical power plant to operate at temperatures up to about 600°C. However, it is now also being used at much lower temperatures (540–568°C) as a replacement material for the headers being made redundant from some of the ageing plant, and in new heat recovery steam generators (HRSGs) in combined cycle gas turbines (CCGTs). The introduction of this steel in high temperature plant was originally based on very high rupture strength of the base metal compared with that of the traditional low alloy ferritic steels. However, new research work on 9Cr martensitic steels shows that the weldments of this type of steel may show large stress reduction factors. In addition one of the advantages of this steel was envisaged to be the smaller wall thickness required for thick section components such as headers and steam pipes thus reducing through wall temperature gradients and therefore damage due to fatigue. However, recent research work has shown that the weldments of this steel, the weakest link in the chain, show lower ductility than the traditional ferritic steel weldments and this appears to reduce the resistance of P91 welded components to creep-fatigue cracking. Earlier than expected Type IV plant failures experienced in some of the thick section components appear to support these findings. With regards to thin section components, some utilities which use tube steel T91 for high temperature superheater tubing have reported high steam side oxidation rates and earlier failures. R&D work has shown that in Cr steels resistance to steam oxidation only improves above a Cr level of about 11.5%. These aspects are discussed in this paper.

BACKGROUND AND INTRODUCTION

By 1987 it was apparent that P91 did not suffer from reheat cracking and had enough creep ductility to resist weld creep cracking. However, it was recognised by this time that P91, like all ferritic steels, is potentially susceptible to Type IV cracking, e.g.[1]

The basis for the present review by ETD was that P91 steel has been in use, in a limited number of plants, since the late eighties/early nineties, and that although steam temperatures in most of these older units were between 540–565°C (below the maximum design temperature of 600°C for this steel), there have been some disturbing reports, admittedly few in number, about P91 failures. All of these failures were of the Type IV variety in which cracking took place in the fine grain section of the HAZ. These failures are a matter of concern especially because they occurred at much lower temperatures than the maximum design temperature of 600°C for such steels and the present operating temperature of up to 590°C in some of the new higher efficiency European and Japanese plant where 9Cr martensitic steels have been introduced more recently.

Market forces and competition are now dictating the power plants to be cycled on daily basis, to follow load demands and maximise profits. This 'cycling' is driving a redesign of existing and new plants. One redesign solution has been to use higher strength P91 steel for pressure vessel construction which allows pressure-containing components to be made with thinner sections. Thinner components require less time to reach thermal equilibrium and have fewer tendencies towards thermo-mechanical damage mechanisms. In addition, the reduced section thickness increases pipework flexibility. Because of this potential benefit a number of operators/owners of the existing plants are substituting these new higher strength steels for the older materials, especially when a plant is moved from base load to cyclic operation. In fact in the case of new HRSGs it is understood that some manufacturers offer P91 as a standard material for thick section components. However, recent limited R&D work throws some doubt on the validity of the claim that the relatively thinner section components of P91 may be less susceptible to fatigue cracking.

In terms of thin section components some utilities, notably outside Europe, have started suing T91 tube version of the steel for superheater tubing to replace ageing T22 (2.25Cr1Mo) tubing in the hope of avoiding frequent failures in this new high strength steel. However, our review has shown that this experience has not been very successful due to a number of factors.

A prime aim of the ETD review was to forecast whether the failures in thick wall components could presage a series of mid-life problems in units using this material. Furthermore, assuming that operators may experience a number of failures in future the review outlined potential repair weld procedures, although the weld repair aspect is not covered in this paper. The review also covered the performance of dissimilar metal welds but again this aspect is not discussed here.

THICK SECTION COMPONENTS

Recent Failures

Published Failures

Five or six known incidents of cracking and failures have occurred in the UK which has the longest experience of service (installation of P91 thick section headers in the UK started in 1989). All of these failures occurred in the West Burton power plant. As usual, after a few

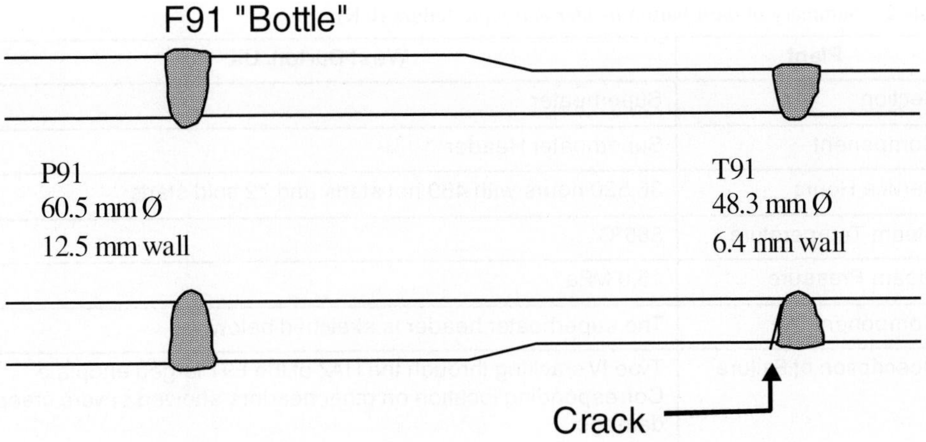

Fig. 1 Transition bottle failures.

Table 1 Summary of West Burton *transition bottle* failures (UK).

Plant	West Burton Power Plant
Section	Superheater
Component	Superheater T91 Transition Bottle
Service Hours	20,000 hours
Steam Temperature	565ºC
Steam Pressure	15.0 MPa
Description of Component	A 75 mm diameter F91 forged rod was bored through to an ID of 35.5 mm and turned to an OD of 60.5 mm. A portion of the length was then further machined to an OD of 48.3 mm with a smooth taper between OD's creating a transition piece dubbed a superheater "bottle." The large diameter end of the fitting was butt-welded to 60.5 mm diameter × 12.5 mm wall T91 tube, while the smaller diameter end was butt-welded to 48.3 mm diameter by 6.35 mm wall T91 tubing.
Description of Failure	*Four* such transition pieces failed, the earliest at 20,000 hours, by cracking through approximately half the circumference of the smaller diameter end of the transition piece in the Type IV band of the heat affected zone.
Failure Analysis Results	There was no evidence of poor welding or installation procedure, but there was evidence that there may have been a deficiency in the heat treatment of the forged stock material.

Table 2 Summary of west burton *header end plate* failure (UK).

Plant	West Burton, UK
Section	Superheater
Component	Superheater Header
Service Hours	36,520 hours with 469 hot starts and 72 cold starts
Steam Temperature	565ºC
Steam Pressure	15.0 MPa
Component	The superheater header is sketched below.
Description of Failure	Type IV cracking through the HAZ of the F91 forged endplate. Corresponding location on other headers showed severe creep damage.
Failure Analysis Results	Zone IV of the HAZ in the F91 material was found to be approximately 20 HV units softer than Zone IV of the P91, and stress analysis showed that the stress at the origin of the crack was 212 MPa and was as high as 110 MPa at a depth of 5–10mm into the end-plate forging. It was concluded that the cause of failure was Type IV Creep.

years of confidentiality, information on these failures has now been published.[2–5] Four of these failures occurred in 'bottle' type joints, and one in a header end cap with another header end cap showing cracking during later investigations/ inspections. The first bottle failure occurred after a service of only 20,000 hours while the end cap failure occurred after a service of mere 36,000 hours. In all cases the nominal operating temperature was 565°C. It should be emphasised that these were the longest serving thick wall components in the UK and perhaps elsewhere, as the West Burton Power Station was the first to receive replacement components of P91 when this programme started in the UK in the 1980s.

As[2, 3] show the failure was attributed to the overtempering of the base material. This was inferred from the low hardness of the failed casts compared with the most of the other P91 material but was not traced back to the heat treatment records presumably due to their non-availability. High aluminium to nitrogen ratio, albeit within the Standards specifications, was also reported to be the potential culprit as this is thought to result in the formation of coarse aluminium nitrides leaving less nitrogen for other creep strengthening fine nitride precipitates. However, no direct proof in terms of finding higher amount of aluminium nitride precipitates is known to have been found. In the case of the end cap failures/cracking design involving sharp change of section near the weld section was also blamed although such design has been successfully used in P22 vessels for a long time and was in accordance with the British Standards design specifications.

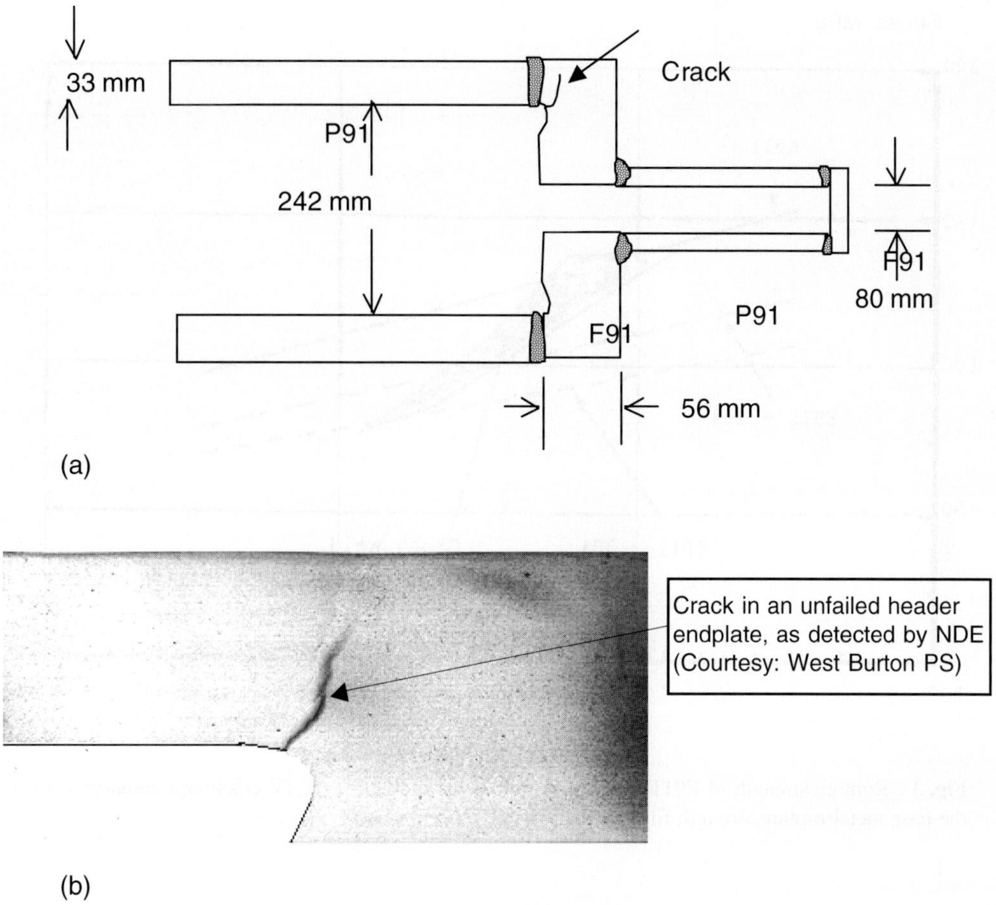

Fig. 2 (a-b) Header end-cap cracking.

Unpublished Failures

Reportedly more recently there have been at least three other P91 Type IV failures in power plant in Europe but these, as happened in the case of the West Burton plant, are not expected to become public for a few more years due to the plant confidentiality.

Creep Rupture and Type IV Failures

As is well known, the introduction of P91 to the power plant industry was based on the rupture strength of the parent metal. Only short-term weld metal data were available when this steel was first introduced in plant use. However, this situation is now changing. Some of

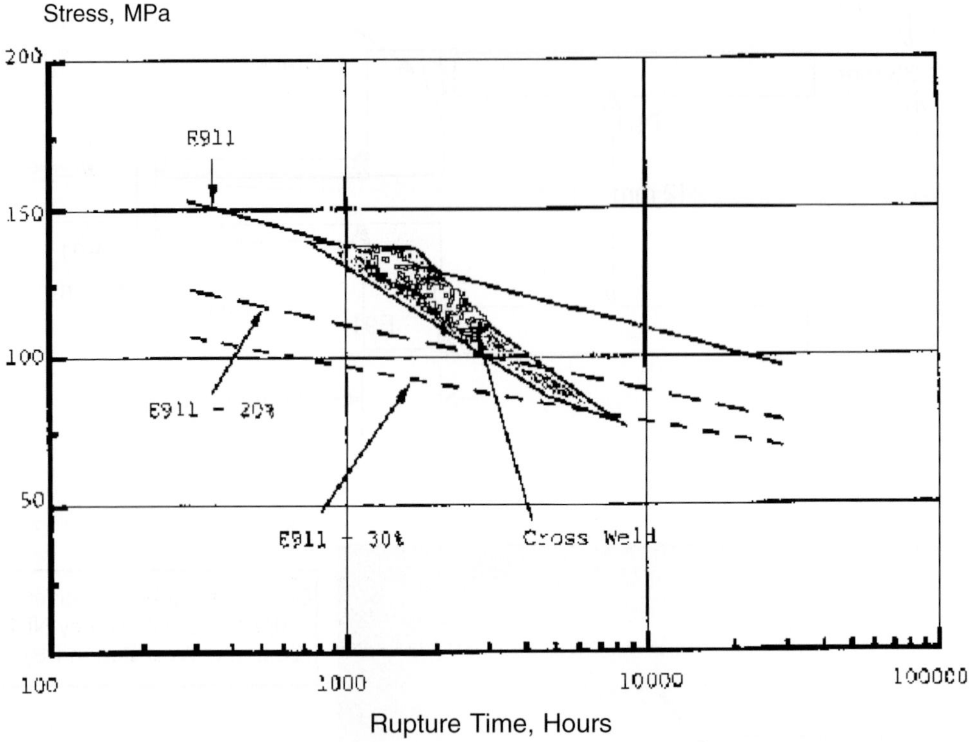

Fig. 3 Rupture strength of E911 cross weld specimens showing Type IV cracking compared with the base metal rupture strength for E911.[7]

the work carried out even as early as late 1980s showed that the cross weld rupture strength of this steel fell much below that of the base metal.[1] Recently a number of other publications on 9Cr Martensitic steels have confirmed this view, e.g.[6,7] This is shown in Figure 3 for the European 9Cr martensitic steel,

Allen et al. through their work on E911 (European version of 9Cr martensitic steel) have shown that the weld strength reduction factor for this steel could be as high as 40% when the failure is in Type IV position.[6] They extrapolated it to 50% for failure to 100,000 hours.[6] This compares unfavourably with 20% strength reduction factor for P22 steel at 550°C – its usual temperature of operation.[8] The criticism of the above work on E911 can be that tests carried out on cross weld specimens were conducted at 625°C, higher than the service temperatures. However, some work on P91 at 600°C (maximum design temperature for this type of steel) carried out in the European Commission supported project SOTA[9] also showed that P91 specimens tested in creep crack growth mode were more vulnerable to Type IV cracking than P22 specimens tested at 550°C.

The UK analysis of experience with the traditional low alloy ferritic steels used in power plants is shown in Figure 4. This shows that although Type IV failures started appearing

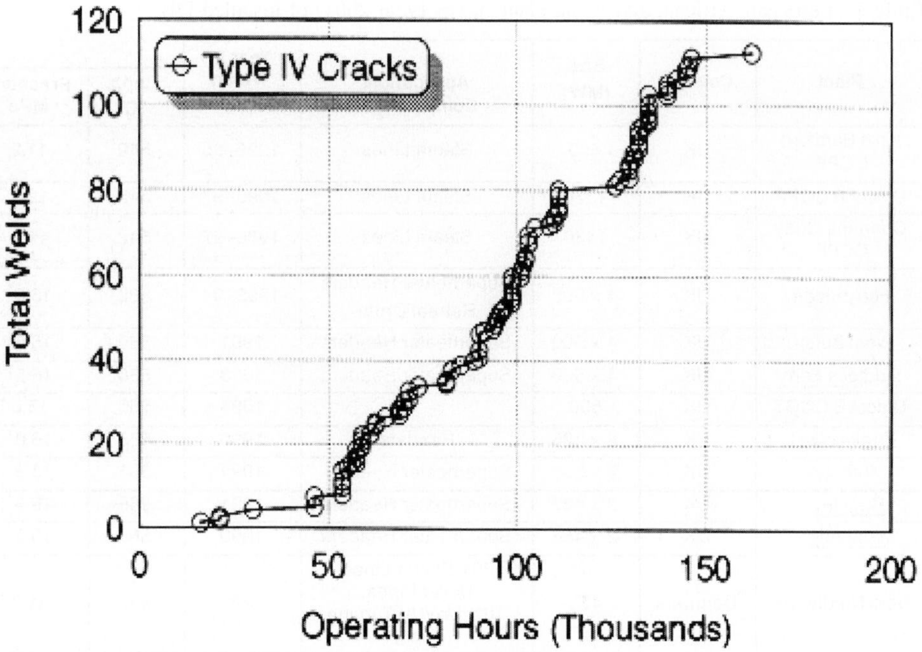

Fig. 4 Type IV weld failures in conventional ferritic steels.[10]

after ~ 20,000 hours most of the failures in these steels appeared after 50,000 hours of service i.e. the Type IV failures were a medium to long term service phenomenon. In the UK where experience with the use of thick section P91 has been the longest (about 45,000 hours to date), even ignoring unpublished failures, at least five failures and one incidence of cracking has manifested itself. This in spite of the fact that the service temperature was only modest (568°C) for this steel and the frequency of use is also very low compared with the conventional ferritic steels, as shown in our Plant survey for UK, Europe, Asia and Africa – see Table 3 which summarises responses by power generating stations. This Table shows station capacity, the year(s) in which P91 was installed, and its nominal steam temperature and pressure. The worrying aspect is that in spite of the low frequency of use the failures in P91 have started occurring after about 20,000 hours of service.

Creep-Fatigue / High Temperature Fatigue Issues

Many power plant components go through some form of fatigue cycling due to normal shutdown and re-start, load following, etc. However, because of privatisation and competition in the electricity industry power plants are now more frequently started up to meet peak

Table 3 European, African, and Asian plant survey (year 2001) of installed P91.

Plant	Country	Size (MW)	Application/ Components*	Year In Which Installed	Temp. ºC	Pressure MPa
Little Bardford CCPP	UK	680	Steam Lines	1995–96	540	11.2
Didcot B CCPP	UK	1370	Steam Lines	1996–97	545	12.5
Qonnahs Quay CCPP	UK	1430	Steam Lines	1996–97	542	11.4
Ferrybridge	UK	4 × 500	Superheater Headers Reheat Drum	1992–94	568	16.6[1]
West Burton	UK	4 × 500	Superheater Headers	1991	565	16.5[1]
Fiddler's Ferry	UK	4 × 500	Superheater Headers	1993	568	16.5[1]
Didcot B CCGT	UK	500		1994	565	15.0
Drakelow C	UK	2 × 325	Headers	1990	566	16.0[1]
Tilbury[2]	UK	2 × 350	Superheater Headers	1994	565	15.9[1]
Rugeley B	UK	2 × 487	Superheater Headers	1991	566	15.9[1]
Ironbridge	UK	2 × 486	Superheater Headers	1990	566	15.9[1]
Nefo Nordjylland	Denmark	425	P91-Steam Lines, Boiler Pipes, 10CrMoVNb Turbine Rotor	1998	587	31.0
Skaerbaek	Denmark	425	P91-Steam Lines and Boiler Pipes. 10CrMoVNb -Turbine	1997	587	31.0
Avedore	Denmark	530	P92-Steam Lines and Boiler Pipes. 10CrMoWVNb Turbine Rotor	2000	580	30.0
Schkopau	Germany	2 × 450	Headers, Pipes and Fittings	1995	550	28.5
Schwarze Pumpe	Germany	2 × 800	Headers, Pipes and Fittings	1997–98	552	25.7
Boxberg	Germany	2 × 800	Headers, Pipes and Fittings	1997–98	550	28.5
Lippendorf	Germany	2 × 930	Headers, Pipes and Fittings	1999-00	559	28.5
La Spezia	Italy	1 unit x 600 MW	Steamlines	2001	568	25.3
Poolbeg	Ireland	2 × 350 MW	Header + Steam Pipe	1997	530	8.0
Ringsend	Ireland	HRSGs	Steam Pipe	2001	530	8.0
Moneypoint	Ireland	3 x 305 MW	RHO Header + RH Tubing (Replacement)	1999–00	~600	4.0
Kawagoe	Japan			1989	565	30.0
Black Point	Hong Kong	6 × 312 HRSGs	Sec Superheater Tubes, Headers, Branch Pipes, Manifold T-Pieces and Bi-Furcations for Main Steam Pipes of P22	1995–98	540	12.8 (set) 10.5 Normal
Matla	South Africa	2 units 600 MW	s/h, r/h Headers, Interconnecting Piping	1994–95	540	18.3

[1] This is the drum safety valve set pressure = design pressure

[2] Two units decommissioned/mothballed

demand and shut down or partially cooled down during the less demanding times. This has enhanced the risk of cracking due to fatigue.

In terms of creep damage, since creep is time, stress and temperature dependent, plant cycling (also known as 'two shifting') and low load operation would be expected to reduce damage due to long term creep. However, during a unit start, or at periods of low load running, there may be some circumstances when temperature overshooting or localised overheating may occur. The cumulative effect of repeated overheating during thermal and load cycling can give rise to extended periods of operation above the design temperature which may result in accumulation of creep damage or its acceleration.

A creep related phenomena in 'two shifting', when using conventional low alloy ferritic steels has been degradation in the microstructure and the concomitant reduction in material properties which has already occurred during the previous term of base load operation. These microstructural changes will have occurred simply as a result of exposure to temperature. The most obvious signs of such degradation are the onset of spheroidisation in carbon-manganese and 2.25Cr1Mo steels. The implications of this are usually reduced ductility, compared to virgin material, and reduced resistance to creep and/or fatigue cracking. Similar phenomena can occur in 9Cr martensitic steels where the precipitates can coarsen or reprecipitation can occur and the matrix can loose creep strengthening elements.

However, by far the most common problem experienced as a result of plant cycling is thermal fatigue damage. This can manifest either in the form of cracking of an individual component or by the mechanical failure of structures. Cracking of a component is attributed to severe thermal gradients arising from excessive steam to metal and through wall temperature differences associated with rapid rates of change of steam temperatures as generally observed during start up, shut down and load changes. The principal components at risk typically comprise any thick walled sections such as boiler superheater headers, steam pipework, valves, HP and IP steam chests and turbine inlet belts. HP heaters and economiser inlet headers are also frequently exposed to similar effects due to rapid cooling by cold feed water. On a wider scale, structures such as boiler framework and tube attachments, boiler supports and pipework support systems are also vulnerable to thermal cycling. Thin walled sections such as boiler tubes, reheater headers etc. are less prone to the problem. A well known example of thermal fatigue cracking in a header ligament is shown in Figure 5.

Creep-Fatigue Capabilities of High Temperature Alloys

It is well known that creep-fatigue interaction of high temperature alloys can reduce life in a non-linear manner. Figure 6 (due to ASME) shows that interaction for P22 steel is quite severe compared to some of the other alloys. The ASME curve in Figure 6 is for base metal but the same rule is expected to apply to the welded components.

As stated earlier, the use of P91 thick section components as a replacement material and in new high efficiency advanced power plant is now considered beneficial from fatigue damage viewpoint in the belief that thinner section components made from high strength P91 will be less prone to fatigue cracking during the cyclic operation due to smaller temperature gradients across the wall thickness. However, recent creep and high temperature fatigue crack growth work in the European Commission funded project HIDA[11] on welded P91 and P22 components has suggested that the creep-fatigue interaction in P91 thick section

Fig. 5 An example of a header ligament cracking.

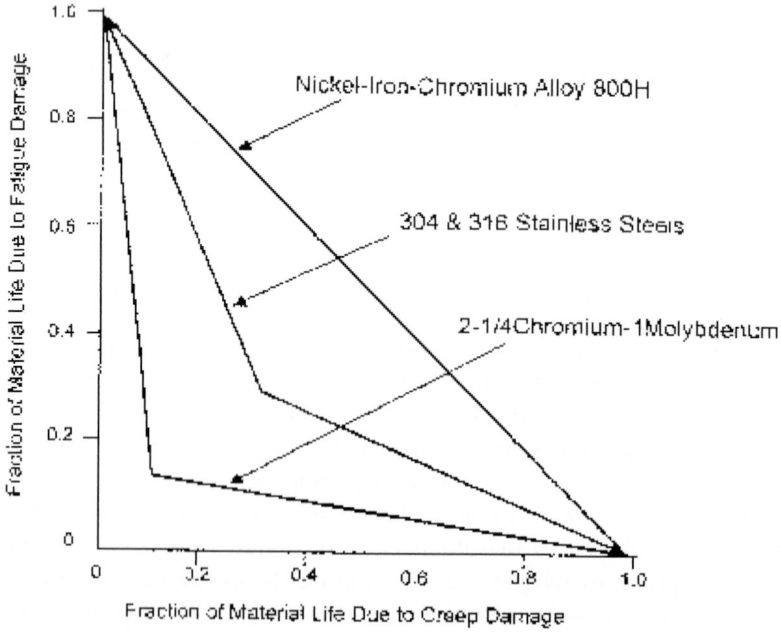

Fig. 6 Creep fatigue interaction of some of the high temperature alloys (ASME).

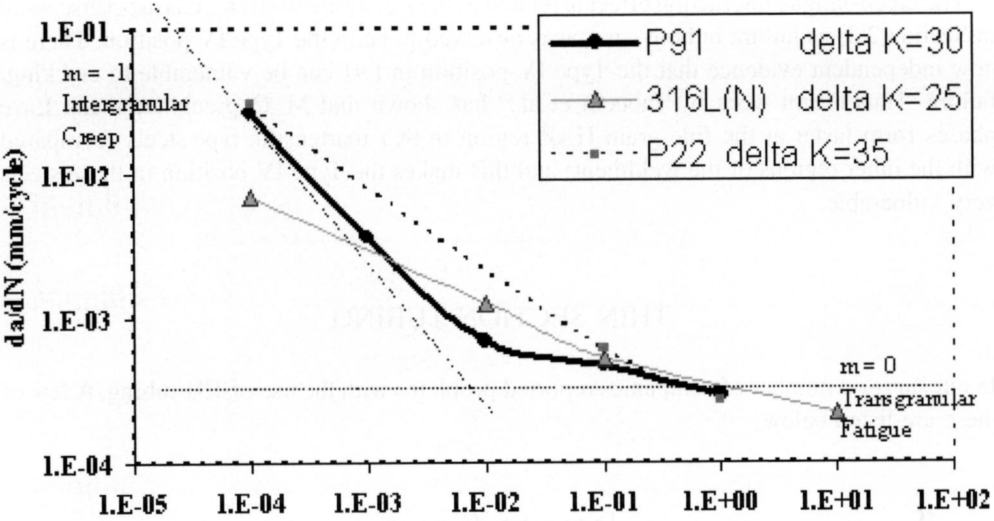

Fig. 7 Creep-fatigue crack growth curves for some of the high-temperature alloys (including P91).

pipes containing welds could be even more severe than P22.[12, 13] These tests were conducted at constant temperature but the load was cycled. Creep-fatigue crack growth tests on standard laboratory specimens were also carried out within the project HIDA, and these also showed more severe effect on 9Cr material, Figure 7.

It must, however, be emphasised at this stage that, although fairly detailed, this work has been carried out on only one cast and at one test temperature of 625°C. Further work on this material is now planned.

The above work involving feature specimen / pipe tests showed that even very low cycling (0.0001 Hz - three cycles a day with about seven hours hold time in each cycle) of large welded pipes adversely affected the crack growth rate in P91 HAZ. The adverse effect of fatigue is not surprising as work on the P91 cross-weld rupture specimens tested in the HIDA programme has shown that when fracture occurs in the Type IV position the ductility of these specimens could be as low as 1 to 2%.

Mohrmann et al.[14] carried out low cycle fatigue (LCF) tests on welded E911 specimens with and without hold times at room temperature and 600°C. These tests showed a decrease in lifetime by a factor of about two compared with the base material. The hold times further reduced the lifetime. Similarly Bicego et al. carried out LCF tests at 600°C on cross weld specimens of P91, P92 and E911 steels[15] – all 9Cr martensitic steel variants - and showed similarly a life reduction of about 2 compared with the base metal.

In summary, the above work shows that: (a) creep fatigue interaction effect on P91 weldments can be more severe than on, say, P22 weldments and (b) the creep-fatigue interaction effect can be more severe on the weldments than on parent metal.

The creep-fatigue interaction effect is related to the steady creep strength issues (discussed in Section 2.2) as failure in both cases was observed to be in the Type IV position. There is now independent evidence that the Type IV position in P91 can be vulnerable to cracking/failure. Thus recent work by Tabuchi et al.[16] has shown that $M_{23}C_6$ precipitates and Lave phases form faster at the fine grain HAZ region in 9Cr martensitic type steels (compared with the other regions of the weldment) and this makes the Type IV position in these steels very vulnerable.

THIN SECTION TUBING

In our survey a number of companies reported problems with the use of T91 tubing. A few of these are listed below.

Published Failure

Secondary Superheater Tube Failures in Hawaiian Electric Company (USA)

Hawaiian Electric Company (HECO) replaced its secondary superheater (SSH) tubes in 1993 after a succession of T22 (2.25Cr1Mo) tube failures which were reported to be mainly due to overheating and creep and were attributed to long term service of the plant i.e. about 14 years (120,000 hours) of plant operation.[17] It was therefore decided by HECO to replace these tubes with T91 to obtain 'trouble free' performance for remaining life of the plant. This replacement was completed and the unit K5 was returned to service in July, 1993.

The unit was taken off line for its 1995 planned maintenance on April 9, 1995. Approximately 14,808 operating hours had expired since the T91 was first placed in service. The unit pre-overhaul hydrostatic test was performed before the boiler wash. The test pressure could not be maintained indicating the presence of a leak somewhere in the system. As a result of further investigations cracking and failures in a number of tubes was discovered. Typical appearance of all damaged areas was swelling on one side of the tube in an area 50 to 75 mm in diameter with longitudinal linear indications or cracks. Similar to the original T22 SSH tube failures, most of the T91 tube failure locations were in the second row tubes. *Reportedly radiant heat of the furnace appeared to have contributed to the overheating.*

Possible Causes of Failures

After investigations the following conclusions were reported by HECO:

- The tube ruptures were characteristic of long-term (subcritical) creep-rupture failures.
- Nothing unusual was observed regarding the chemistry or microstructure of the T91 material away from the ruptures.
- *Near the ruptures, the tubing exhibited very thick (> 0.76 mm) steamside oxide scales.*
- The steamside scales had the characteristics of indigenous (locally grown, rather than deposited) oxide scale.

- Based on the tubing hardness, microstructure, and steamside scale, it was estimated that the tube metal temperature near the ruptures was in excess of 649°C for a *service exposure of 15,000 hours.*
- Away from the ruptures, the tube microstructure and hardness indicated that the tubing had not experienced extensive service degradation.

As Reported by HECO, videoprobe inspection of the damaged elements was performed after the damaged sections were cut out. There were no blockages nor anything unusual sighted. The SSH inlet header, the manifold headers and connecting tubes were also inspected by videoprobe. Again there were no blockages nor anything unusual sighted.

According to HECO it can be safely said that the tube bulges and ruptures were due in part to *localized overheating*. The wall thickness of these tubes was about 6 mm and was therefore not much different from that of T22 tubes.

Unpublished Failures

T91 Tube Failures Due to Soot Blower Problems (USA)

Although no specific information was available on some of the USA tube failures, nevertheless the survey showed that at least one utility had replaced its T91 secondary superheater tubes with 304H stainless steel because of the T91 tube failures due to soot blower erosion.

Reverting to T22 Use Due to Weld Repair Problems with T91 Tubes (USA)

The survey also reported that one USA utility had replaced its ageing superheater T22 tubes with T91 to solve the tube failure problem. However, like HECO, only after about two years it started experiencing multiple failures in T91 tubes. The utility has not specified the root cause of the failure. It also realised that welding of T91 involved post weld heat treatment (PWHT) which meant longer outage periods and higher costs. In consequence, this utility reverted to the use of T22 tubing.

T91 Tubing Failure in Japan

Following is a summary of a premature failure of the superheater tubes made from T91. The boiler concerned generates steam using combustion products from a gas turbine and supplementary burners. The rupture took place at the first row tubes, facing burner flame, in the secondary superheater section. The primary and secondary superheater tubes are installed vertically. The operating time to failure was 63,000 hours. The design and operating conditions for the secondary superheater tubes were as follows:

- Design (metal) temperature: 610°C,
- Design pressure: 9.18 MPa; Operating pressure: 8.2 MPa,
- Nominal outer diameter and thickness of tube was 38.1 and 5.2 mm respectively and
- The steam temperature at the outlet was 540°C.

The Cause of the Failure

This was identified as creep associated with *overheating of tubes and attributed to the inner oxide scale*. The utility reported no problem in the water treatment. Chemical composition of the failed tubes was found to be in the specified range by ASTM. No external corrosion was found with naked eye.

The thickness of the inner scale at the failed position was approximately 0.06 mm. The steam oxide scale was composed of two layers, external one (on the steam side) and inner layer which is denser than that facing steam and containing alloying elements. Studies using optical microscope showed that the scale at the failed position lacked the external layer. Therefore, it can be presumed that the external layer was pulled off at the rupture and *the original thickness was larger than 1 mm*. Usually, the thickness of both layers was roughly the same.

To prevent the recurrence, the first and second row tubes in the secondary superheater section were replaced with Type 347 stainless steel tubes.

In *summary*, the plant survey with the use of T91 superheater tube replacement showed that this experience was not necessarily very successful. Usually the aim behind the replacement of T22 by T91 superheater tubing was to reduce the wall thickness taking advantage of the higher creep strength of T91. It was considered that the thinner wall will help improve the heat transfer. Although, it is known that the oxide scale formed in steam on T91 is smaller than that on T22 due to its higher Cr content, the overheating and premature failure of the replacement T91 tubing experienced by some of the utilities was probably due to the thinner wall (this can be 2 to 3 mm less than T22 tubes). Steam side oxidation has now been shown by many of the researchers and by plant experience to be high for steels with Cr contents below about 11.5%.[18] ETD calculations have shown that in such cases a temperature increase of well over 100°C can result,[19] as confirmed by plant experience - HECO and Japan failures referred to above. The effect of oxidation on thinner wall tube life is such that 'thinner the wall, higher the strength reduction'.[20] Thus the combination of high oxidation rates both in flue gases and steam and resulting metal temperature rise and the use of thinner wall might together be responsible for the unfortunate experience in some of the utilities. Imprecise original heat treatment of the affected tubes cannot be ruled out as a cause. Reportedly some of the tubing in North America had been given a tempering heat treatment of only 704°C (as against the European standard recommendation of 750–760°C) and this of course could have contributed to the failure of the tubing in some of the North American utilities.

A problem with 9Cr tubing can be the extra time required to carry out in-situ PWHT on repair welded thin wall tubes which is not required in the case of low alloy ferritic thin wall tubular components. In the case of larger volume of failures such PWHT can add to the outage period thus making the use of such tubing costly. In our survey of USA utilities one plant quoted the higher outage costs due to PWHT of weld repairs in T91 tubing as the reason for reverting to T22.[19]

When interviewed, one of the more successful European utility in using of the new 9Cr martensitic steels told that they will be reluctant to use 9Cr martensitic type steels for superheater tubing. Instead they use austenitic or 12Cr (X20) steel tubing when temperatures of 600°C or above are required. The reason for this was cited as the lower steam and combustion gas oxidation resistance of 9Cr steels. One of the successful UK utilities told that they only use T91 in reheater and superheater tubing in the deadspace of the boiler. They are firmly of the opinion that the use of T91 tubing without precise heat treatment can be problematic even in the short run. Through interviews it is also known that the use of T91 in Germany in high heat flux areas has proved to be problematic.[19]

There is the extra disadvantage due to the vulnerability of dissimilar metal welds (T91 to T22 and T91 to austenitic stainless steel) and these also need to be taken into consideration.

Some researchers believe that T91 oxide might be spalling off in larger chucks compared with the traditional T22 tubing and thus could be more effective in tube blockage. This is an interesting aspect which needs further investigation.

DISCUSSION

Thick section P91 components have only seen limited service durations, the longest being that in the UK. From this experience and limited research work it appears that thick wall components may be subject to Type IV failures even at lower service temperatures than those experienced in the new higher performance ultra supercritical power plant where 9Cr martensitic steels were introduced only a few years ago. The thick wall UK failures of header end caps, at 36,000 hours duration, were at first attributed to end cap design which put the weld close to the high stress level position of the change in section (Figures 2a and b), although P22 steel has successfully experienced the same British Standards design. However, four other failures of 'transition bottles' even at shorter service durations of 20,000 hours, where the weld was situated away from high stress position, made the plant operators look for other possible reasons. A comparison of the failed and some of the unfailed components finally made them conclude that a hardness difference of about 10 VPN (Vickers pyramid hardness number, also designated as VH) could have been responsible for such failures, the failed components showing a hardness of 10 to 20 VPN below 200 VPN. This lower hardness was attributed to the possible over-tempering heat treatment of the failed casts. The other possible reason was cited to be the low N/Al ratio, albeit still within the Standards specifications, of the failed casts compared with the unfailed casts. This has been attributed to the observation that aluminium nitride creates large size precipitates often situated on grain boundaries, which impair creep ductility significantly. The AlN precipitation can reduce the fine VN particle precipitation which contributes to the creep strength of this steel. Thus, for example, Vitkovice Research Institute in Czech Republic has shown that the lowering of Al content from 0.03 to 0.003% resulted in an increase in creep rupture strength of about 15%.[21] Indicating that coarse AlN particles do not contribute to precipitation strengthening. The lower rupture strength of low N/Al level steels is also supported by other research findings which have shown higher creep strength for high N level 9Cr steels,[18] although, this work has pointed out that this effect is only seen in short term tests.

Although a large amount of test work has been carried out on P91 base metal and some on cross weld specimens little work is known to have been carried out on thick wall welded feature specimens. The exception to this appears to be the HIDA work where tests on both butt and seam welded pipes and on standard laboratory specimens showed similar findings.[12, 13] These were that: (a) P91 appears to be more vulnerable to Type IV cracking than P22 and

(b) that even low cycle fatigue results in the enhancement of crack growth in P91 fine grain HAZ (i.e. Type IV position) compared with the steady state crack growth.[13] The limitation or criticism of this work can be that it was carried out on one cast and at one slightly accelerated test temperature (625°C) while the maximum envisaged temperature for the use of P91 is 600°C. However, in spite of the limited nature of the work the uniqueness of the work means that its findings cannot be ignored until results at service temperatures from any future work on large wall feature test specimens become available. There is of course some support of the HIDA work from[14, 15] as discussed earlier, which shows that even at 600°C low cycle fatigue cycling reduces the life of cross weld specimens by a factor of 2 compared with the base metal. In the project HIDA, the shorter fatigue life was attributed to the very low cross weld rupture ductility of the 9Cr steels. The finding by Tabuchi et al.[16] is of interest where it has been shown that during the creep process $M_{23}C_6$ precipitates and Lave phases form faster at Type IV position which will make this position more vulnerable.

Indeed it is possible that West Burton failures occurred because of the heavy duty cycling that this power plant (see Table 2) has been undergoing, like many other UK plant. The lower creep rupture strength of the weaker casts due to overtempering or low N/AL ratio or both might have accelerated this process. The question therefore that needs to be asked is: ***Is it possible that stronger casts may also start showing Type IV cracking at longer service durations or at higher service temperatures?***

It is well known that 9Cr martensitic steels, like the 12Cr martensitic steels with which some European countries (notably Germany and Denmark) have a much longer experience, are very sensitive to tempering and welding heat treatments. According to V & M Tubes the tempering and PWHT temperatures for 9Cr martensitic steels must be in the correct regime (750–760°C) to give the desired rupture strength.[22] At higher tempering temperatures overtempering will occur resulting in lower rupture strength, and at lower temperatures ductility and other aspects will not be satisfactory. In the case of PWHT the lower PWHT temperature will mean that the hardness difference between Type IV position and weld metal/coarse grain HAZ will be very high and will make Type IV position more vulnerable.[22] The strict observation of the heat treatment regime of this steel during steel production and the welding cycle is so important that Elsam in Denmark, as one of the more successful users of this steel as thick section components, have developed their own strict criteria of material quality check and control. This is outlined in the original ETD P91 Survey Report.[19] In summary, Elsam ensure compliance of all heat treatment of the base and weld metals to European standards and checking of all heat treatment records in detail including the accuracy of the thermocouples used by the material supplier. This is followed by random checks on hardness and dimensions of the supplied material. Where the heat treatment of any material supply appears to be suspect Elsam carry out metallographic checks including transmission electron microscopy to observe precipitates and microstructure in general and compare it with the normal material.

With regards to the thin wall tubing it is clear that the manufacturers in Europe do not have confidence in the use of T91 as superheater tubing mainly because of the high rate of oxidation in steam. This was initially based on research findings but more recent USA experience also supports this. The requirement of PWHT for this steel even for thin section components means longer outage times and loss of revenue. This makes T91 as the unlikely candidate for superheater tubing. It, however, appears somewhat surprising that many utilities in countries outside Europe are still buying T91 tubing for use in superheaters.

CONCLUSIONS

Both plant experience and research work show that it is possible that P91 may start showing mid-life crisis at least in the case of the some of the more vulnerable casts being used around the world, especially in plant subjected to cyclic operation. It is thus important that further work is undertaken to better understand the behaviour of the welded thick section components of P91 and to develop repair and replacement strategies.[23]

So far P91 has been considered to be advantageous steel as thinner wall and smaller component size reduce thermal gradients and hence the adverse effect of fatigue cracking - which in the past has been experienced in thick section components made from the traditional low alloy ferritic steels. However, preliminary R & D studies have shown that P91 welded components may be equally, or even more, prone to Type IV cracking compared with the conventional low alloy ferritic steels, both under creep and creep-fatigue conditions, possibly due to the weakness of this zone and low cross-weld creep ductility of the welded material. It is thus important that preparations should now be made for appropriate inspection and repair strategies of plant using this steel.

Limited experience shows that the use of these steels in superheater tubing may be unwise at present both from the viewpoints of high oxidation in flue gases and in steam of very thin wall tubing and PWHT required for repaired welds which can extend the outage period thus increasing costs.

REFERENCES

1. C. Middleton and E. Metcalfe: 'A Review of Laboratory Type IV Cracking Data in High Chromium Ferritic Steels', Paper C386/027, *IMechE Proceedings*, London, UK., 1990.
2. S. J. Brett, D. J. Allen and J. Pacey: 'Failure of a Modified 9Cr Header Endplate', *Proceedings of Conference on Case Histories in Failure Investigation*, Milan, 1999, 873–884.
3. D. J. Allen and S. J. Brett: 'Premature Failure of a P91 Header Endcap Weld: Minimising the Risks of Additional Failures', *Proceedings of Conference on Case Histories in Failure Investigation*, Milan, 1999, 133–143.
4. S. J. Brett: 'Identification of Weak Thick Section Modified 9Cr Forgings in Service', *Published in the CD version of the Proceedings of the Swansea Creep Conference*, Organised by the University of Swansea and EPRI, Swansea, UK, 2001.
5. S. J. Brett: 'The Creep Strength of Weak Thick Section Modified 9Cr Forgings', *Proceedings of Baltica V*, **1**,2001.
6. D. J. Allen and A. Fleming: 'Creep Performance of Similar and Dissimilar E911 Steel Weldments for Advanced High Temperature Plant', *Proceedings of the 5th Charles

Parsons 2000 Conference on 'Advance Materials for 21st Century Turbines and Power Plant', Churchill College, Cambridge, UK., 276–290.

7. J. Orr, L. W. Buchanan and H. Everson: 'The Commercial Development and Evaluation of E911, A Strong 9% CrM0NbVWN Steel for Boiler Tubes and Headers', *Proceedings of the Conference 'Advanced Heat Resistant Steel for Power Generation'*, R. Viswanathan and J. Nutting, eds., San Sabastian, Spain, IOM Publications, London, UK., 1998.
8. C. F. Etienne and J. H. Heerings: 'Evaluation of the Influence of Welding on Creep Resistance (Strength Reduction Factor and Lifetime Reduction Factor), IIW doc.IX-1725-93, TNO, Appledoorn, The Netherlands.
9. I. A. Shibli: 'Creep and Fatigue Crack Growth in P91 Weldments', *Proceedings of the Swansea Creep Conference*, Organised by the University of Swansea and EPRI, Swansea, UK., 2001.
10. J. D. Parker: 'The Creep and Fracture Behaviour of Thick Section, Multi-Pass Weldments', *Conference Proceedings on 'Integrity of high temperature welds'*, Professional Engineering Publishing, Nottingham, UK., 1998.
11. I. A. Shibli: 'Overview of the HIDA Project', *Proceedings of the 2nd International HIDA Conference on 'Advances in Defect Assessment in High Temperature Plant'*, MPA, Stuttgart, Germany, 2000.
12. I. A. Shibli, N. Le Mat Hamata, U. Gampe and K. Nikbin: 'The Effect of Low Frequency Cycling on Creep Crack Growth in Welded P22 and P91 Pipe Tests', Paper S4-4, *Proceedings of the 2nd International HIDA Conference on 'Advances in Defect Assessment in High Temperature Plant'*, MPA, Stuttgart, Germany, 2000.
13. N. Le Mat Hamata and I. A. Shibli: 'Creep Crack Growth of Seam-welded P22 and P91 Pipes with Artificial- Part 2': *Proceedings of the 2nd International HIDA Conference on 'Advances in Defect Assessment in High Temperature Plant'*, MPA, Stuttgart, Germany, 2000.
14. R. Mohrmann, T. Hollstein and R. Westerheide: 'Modelling of Low-Cycle Fatigue Behaviour of the Steel E911', *Materials for Advanced Power Engineering 1998, Proceedings of the 6th Liege Conference*, **5**(1), 1998.
15. V. Bicego, P. Bontempi, R. Mariani and N. Taylor: 'Fatigue Behaviour of Modified 9Cr Steels, Base and Welds', *Materials for Advanced Power Engineering 1998, Proceedings of the 6th Liege Conference*, **5**(1), 1998.
16. M. Tabuchi, T. Watanabe, K. Kubo, M. Matsui, J. Kinugawa and F. Abe: 'Creep Crack Growth Behaviour in HAZ of Weldments for W containing High Cr Steel', *Proceedings of the 2nd International HIDA Conference on 'Advances in Defect Assessment in High Temperature Plant'*, MPA, Stuttgart, Germany, 2000.
17. M. Chang, W. K. Edmund, N. Clark and M. Robert: 'Service Experience in Operating Fossil Power Plants, T91 Secondary Superheater Tube Failures Investigation, Hawaiian Electric Company, Inc., Honolulu, Hawaii 96840. *ASME PVP Conference*, **335**, 1996.
18. P. J. Ennis and W. J. Quadakkers: '9–12% Cr Steels: Application Limits and Potential for Further Development', *Proceedings of the Fifth Charles Parsons Conference on 'Advanced Materials for 21st Century Turbines and Power Plant'*, Institute of Materials, Churchill College, Cambridge, UK., 2000.

19. ETD Report on 'Performance Review of P91 and Other 9Cr Martensitic Steels: Plant and Research Experience', ETD Report No.1009-IIP-1002.
20. W. J. Quadakkers and P. J. Ennis: 'The Oxidation Behaviour of Ferritic and Austenitic Steels in Simulated Power Plant Service Environments', *Materials for Advanced Power Engineering 1998, Proceedings of the 6th Liege Conference*, **5**(1), 1998, 123.
21. Z. Kubon, V. Foldyna and Vodarek: 'Optimised Chemical Composition of 9–12%Cr Steels with Respect to Maximum Creep Resistance', *Materials for Advanced Power Engineering 1998', Proceedings of the 6th Liege Conference held in 1998*, **5**(1), 1998, 375.
22. Vallourec and Mannessmann Tubes, The T91/P91 Book, 1999.
23. I. A. Shibli: 'Planned Future Work for P91 Investigations on Feature Testing, Weld Repairs and the Behaviour of Dissimilar Metal Welds, Three ETD proposals for Groups Sponsorship, enquiries@etd1.co.uk.

Creep Performance of Dissimilar P91 to Low Alloy Steel Weldments

D. J. Allen
Power Technology
Powergen, UK

ABSTRACT

Modified 9Cr (P91) steel has been widely used in the UK to replace power plant low alloy steel headers. Its superior creep strength permits reduced thickness and hence better resistance to thermal fatigue. Dissimilar welds are required to connect to the CrMoV steel pipework. Alternative weld designs use high or low alloy weld metals and make the joint with or without a 2¼CrMo steel insert. Service performance should clearly determine which design is chosen.

Cross-weld creep rupture and creep crack growth testing has been carried out to assess the alternative welding procedures. Pre-aging heat treatments were applied to investigate carbon migration at interfaces. Comparisons between test types showed how different factors influence crack initiation and growth and determine overall performance. Results are discussed with reference to the different factors governing laboratory creep tests and pipe butt weld performance, and the significance of ductility in cracking of real plant components.

INTRODUCTION

Dissimilar welds in high temperature plant can suffer premature failure. In welds between high and low alloy ferritic steels, carbon diffuses toward the higher chromium material. Hence, weak decarburised zones can form – in the low alloy steel HAZ when a high alloy weld metal is used, but in the weld metal itself when a low alloy steel weld metal is instead selected.

The welding procedure developed for P91 header installation at Drakelow Power Station[1] used a joint between P91 and 2¼CrMo steel (P22). The P22 insert was judged less likely to cause problems than a direct weld to the CrMoV steel pipework. To minimize concerns with PWHT temperature, a non-modified 9CrMo (P9) weld metal was used in a complex double heat treatment procedure involving buttering the P91 and applying high temperature PWHT, then making the joint and applying a second PWHT at a lower temperature appropriate to P22.

Stress rupture testing[1] produced failures in the inner decarburised zone of the P22 HAZ, with creep strength values close to the parent P22 lower bound. However, like-to-like ferritic steel welds often show comparable reductions in rupture strength due to Type IV cracking at the outer edge of the HAZ. Thus, although the P9 filler metal had clearly produced a new zone of weakness in the HAZ, it was not clear whether weld creep performance had been impaired.

A weld with P22 filler and a single intermediate temperature PWHT would however be simpler and cheaper. A direct P91 to CrMoV weld could further reduce costs. However, such initial savings would be small compared with the lifetime costs of in-service inspection and repair that might arise if a weld design with poor service performance were selected.

This project was carried out to investigate the comparative high temperature performance of these alternative dissimilar weld designs. Cross-weld creep rupture[2] and creep crack growth[3] data were obtained on P91 to P22 welds made with P9 filler (weld W17) and P22 filler (weld W18). This paper also provides comparison results on direct P91 to CrMoV welds made to similar procedures with the same P9 filler (weld W19) and P22 filler (weld W20).

Decarburised Zones in Dissimilar Welds

A concern is that decarburisation in service could cause progressive weakening. Creep tests on as-manufactured welds could thus be misleadingly optimistic. To minimise this risk, welds were heat treated to simulate service aging prior to testing. Alternative treatments at different temperatures were applied with the aim of distinguishing the specific effects of carbon migration at interfaces (interstitial diffusion, activation energy Q circa 30 kcal/mole) from general microstructural coarsening (substitutional diffusion, Q 60–80 kcal/mole, hence much more strongly accelerated by higher temperature aging).

EXPERIMENTAL

Weld Manufacture, Heat Treatment, and Simulated Service Aging

Four circumferential butt welds were made between P91 and low alloy pipe sections machined to matching dimensions of 270 mm ID, 57.5 mm wall thickness for the P22 pipe and 245 mm ID, 57.5 mm wall thickness for the CrMoV pipe. The latter had been withdrawn from service after 175,000 hours at 568°C and, while free from detectable creep damage, showed substantial microstructural degradation. Downhand (rotated) manual metal arc welding with 4 mm diameter electrodes was used to fill V preparations with principal sidewall angle 15°. Basic coated P9 consumables were used to reproduce the Drakelow procedure with PWHT at 765°C for 3 hours (buttering) and 720°C for 3 hours (weld). Basic coated P22 consumables were used for the alternative procedure with PWHT at 730°C for 3 hours. Three different aging treatments were then applied to different sections of each weld:

1. 730°C for 500 hours, representing an extreme overaged condition,
2. 650°C for 1,000 hours, approximately simulating long term service at 565°C and
3. 600°C for 10,000 hours, fairly similar to (2) but causing more carbon migration.

Stress Rupture Testing

Long uniaxial cross-weld plain bar stress rupture specimens of 9.06 mm diameter, with gauge lengths extending at least 12 mm beyond each fusion boundary, were used to enable failure anywhere within the weld. To estimate weld creep life at the service temperature of 565°C and a nominal 'worst case' axial stress of 40 MPa, two parallel forms of test acceleration were applied. Isostress tests at 65.5 MPa were carried out at temperatures from 625°C down to 580°C, while isothermal tests at 625°C were carried out at stresses down to 39 MPa.

Waisted specimens from weld W20 were also tested. This geometry was designed to suppress failure in bulk P22 weld metal and hence obtain data on the decarburised P22 weld metal adjacent to the P91. The gauge diameter in the CrMoV and bulk weld metal was increased to 10.1 mm, while a waist was machined to reduce diameter to 9.06 mm within the P91 HAZ and an adjacent short (7 mm) length of the P22 weld metal.

Creep Crack Growth Testing

Creep crack growth tests on specimens from P91 to P22 welds were notched and sidegrooved to grow cracks in specific microstructural zones.[3] Comparative data were thus obtained on cracks in decarburised and non-decarburised P22 HAZs obtained with P9 and P22 fillers respectively. Further comparative data were obtained on cracks in the decarburised P22 weld metal close to the P91 fusion boundary and on cracks growing in the bulk P22 weld metal.

RESULTS

Weld HAZ Microstructure

The P22 HAZ in heat treated P9 filler welds showed a fully decarburised white-etching band adjacent to the fusion boundary. This was about 0.1–0.2 mm width after the 720°C PWHT, growing to about 0.4–0.5 mm width after 1,000 hours aging at either 600 or 650°C. Thus, carbon migration rate is only weakly dependent on temperature, and more migration takes place with aging treatment (3) above than with treatment (2).

The CrMoV HAZ in P9 filler welds showed no white-etching band and much less visible decarburisation. Kozeschnik et al.[4] have shown that whilst $M_{23}C_6$ carbides in a P22 HAZ dissolve readily, the more stable VC carbides in CrMoV steel do not. Hence, in dissimilar welds with high chromium fillers, a P22 HAZ undergoes short-range gross decarburisation, but a CrMoV HAZ shows less local carbide dissolution and more long-range carbon migration.

Cross-Weld Stress Rupture Testing

Test data and failure locations identified by metallography are given in Table 1. Specimens preaged at 730°C showed gross degradation, with different welds failing in different locations

Table 1 Stress rupture test data.

Test Temp. °C	Test Stress MPa	Aging Condition	Rupture Life (h) and Failure Location (All in Low Alloy Steel Except N). W = Weld Metal, H = HAZ, T4 = Type IV, P = Parent Metal. **Bold D = Decarburised Zone Failure.** Bracketed = Unfailed. **N** = Necking (WM & P91 HAZ), C = Conjoint (P22 HAZ Neck)				
			W17 P91–P22 P9 Filler	**W18** P91–P22 P22 Filler	**W19** P91–CMV P9 Filler	**W20** P91–CMV P22 Filler	**W20** Waisted Specimens
625	65.5	(1) 730°C 500 h	53 T4		44 T4	130 W	
610			212 T4		304 T4	249 W	
595			414 T4		587 T4	687 W	
580			1395 T4		1331 T4	1228 W	
625	65.5	(2) 650°C 1000 h	**492 HD**	369 W	682 W	589 W	**609 WD**
610			**1028 HD**	720 WC	1738 W	**1357 WDN**	**1153 WD**
595			**2179 HD**	2107 WC	4359 T4/P	**2627 WDN**	**3133 WD**
580			**5090 HD**	3967 W	(9644)	8592 W	**9154 WD**
625	65.5	(3) 600°C 10000 h	**674 HD**	453 WC	800 W	957 W	
610			**1150 HD**	1063 WC	1694 W	2051 W	
595			**2703 HD**	2342 WC	3742 W	5465 W	
580			**5921 HD**	**4976 WD**	8189 W	(9776)	
625	65.5*	(3) 600°C 10000 h	**674 HD***	453 WC*	800 W*	957 W*	(* - data Copied from Series Above)
	55		**1471 HD**	1550 H	2043 W	**1643 WD**	
	46		**2578 HD**	3016 WC	**2589 HD**	**3551 WD**	
	39		**4412 HD**	4627 WC	**4394 HD**	5784 T4	

but with similar very short rupture lives. This suggests that gross thermal aging can eliminate microstructural strengthening factors and reduce all weld types toward a common denominator 'intrinsic strength' level. Hence, these data are not analysed further. Specimens preaged at lower temperatures showed more realistic behaviour. The shorter 650°C aging treatment proved slightly more damaging than the longer 600°C aging treatment for all four weld types.

P91 to P22 Welds (Aged at 650 or 600°C)

All specimens from the P9 filler weld failed in the decarburised P22 HAZ with low rupture elongations of around 3–4%. Most specimens from the P22 filler weld showed a more complex 'conjoint' failure mechanism with deformation in both the P22 HAZ and adjacent P22 weld metal, leading to failure in the weld metal with higher rupture elongations of typically 16%. The P9 filler welds had slightly longer rupture lives when tested at the accelerated stress level of 65.5 MPa, but slightly shorter rupture lives at lower stress levels appropriate to plant conditions.

P91 to CrMoV Welds – High Alloy (P9) Filler

For welds with the P9 filler, tests at 65.5 MPa showed substantially longer lives in weld W19 to parent CrMoV than in corresponding weld W17 to parent P22. At this stress level, therefore, the decarburised CrMoV HAZ is stronger than the decarburised P22 HAZ. Failure in weld W19 occurred after a longer life and was usually displaced into the P9 weld metal.

However, the two tests conducted at the lowest stress levels showed quite different behaviour. The failure location in weld W19 then reverted to the decarburised low alloy steel HAZ. Further, comparison of the W19 and W17 rupture lives at these stress levels (46 and 39 MPa) shows remarkable similarities. Although the P22 and CrMoV materials differ in composition, the recorded HAZ rupture lives at each stress level are almost identical. Thus, while W19 outperforms W17 at high stress, it shows no advantages under realistic plant stress conditions.

P91 to CrMoV Welds – Low Alloy (P22) Filler

For welds with the P22 filler, weld W20 to CrMoV shows consistently better performance than weld W18 to P22. However, a number of different failure mechanisms were observed.

Failure Mechanisms

Decarburised HAZ Failure

Figure 1 shows a typical P22 HAZ decarburised zone failure. The crack initiated near the fusion boundary and ran parallel to it at distances of 0.1–0.5 mm into the HAZ. Final failure took place by cracking across the bulk HAZ with ductile deformation. A grossly decarburised

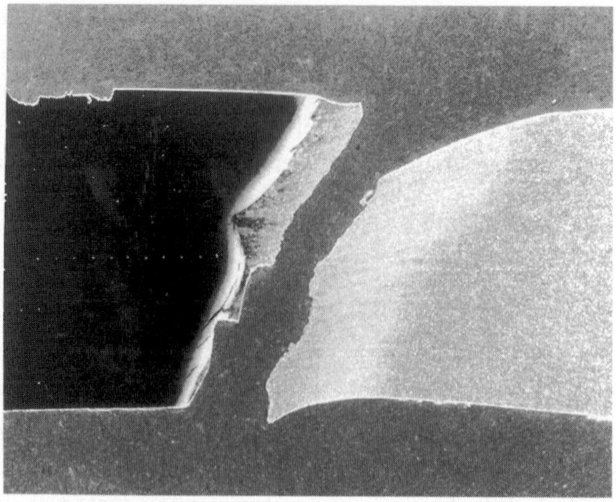

Fig. 1 P22 decarburised HAZ failure.

HAZ zone close to the dark-etching weld metal appears carbide-free and largely undamaged, Figure 2. However, the partially decarburised HAZ slightly further from the weld metal retains some carbides and a fine grain structure, and shows intergranular creep damage and cracking.

Figure 3 shows a corresponding failure in the decarburised CrMoV HAZ. Once again, initial brittle creep cracking occurred in the decarburised zone while final ductile creep failure took place in the bulk non-decarburised outer HAZ. However, in CrMoV the initial crack and creep damage region are much closer to the fusion boundary, Figure 4, there being no zone of gross decarburisation. Hence, cracking again takes place in a partially decarburised HAZ zone.

Conjoint Failure – P22 Weld Metal and P22 HAZ

Most tests on weld W18 failed in weld metal, but most of these failed close to the P22 fusion boundary. The failure region thus included the P22 HAZ, which showed particularly sharp necking, Figure 5, and hence played a role in the failure. A "conjoint" mechanism is identified[2], whereby necking starts in a weak, creep ductile P22 HAZ, but failure then occurs in adjacent, somewhat stronger but less strain tolerant, P22 weld metal. The weak P22 HAZ therefore acts to accelerate ductile creep failure in P22 weld metal. By contrast, conjoint failure does not occur in weld W20, where the CrMoV HAZ is evidently stronger. Hence, the rupture life for P22 weld metal failure is typically a factor of two higher in weld W20 than in weld W18.

Decarburised Weld Metal Failure

The weakest region was expected to be the decarburised weld metal adjacent to the P91. However, only three plain bar specimens from welds W18 and W20 showed a crack growth

Fig. 2 Detail – P22 decarburised HAZ.

Fig. 3 CrMoV decarburised HAZ failure.

mode of failure in this zone, with elongation around 8%, limited local ductility in weld metal, and no deformation in the hard P91 HAZ, Figure 6. In weld W18, early 'conjoint' failure may have intervened to prevent decarburised zone failure. However, the W20 tests also showed more failures in bulk P22 weld metal than in the decarburised zone. Two further

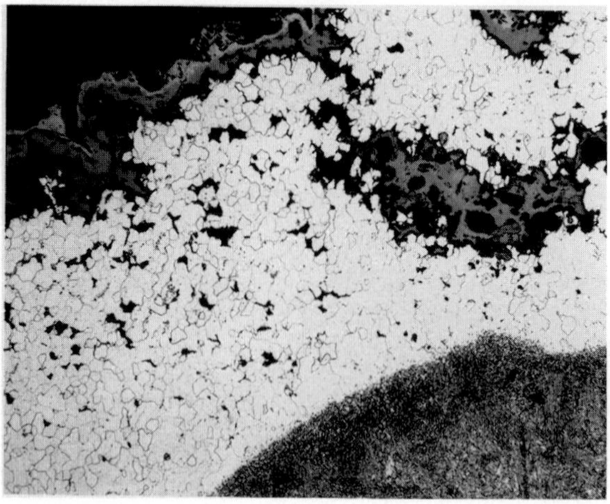

Fig. 4 Detail – CrMoV decarburised HAZ.

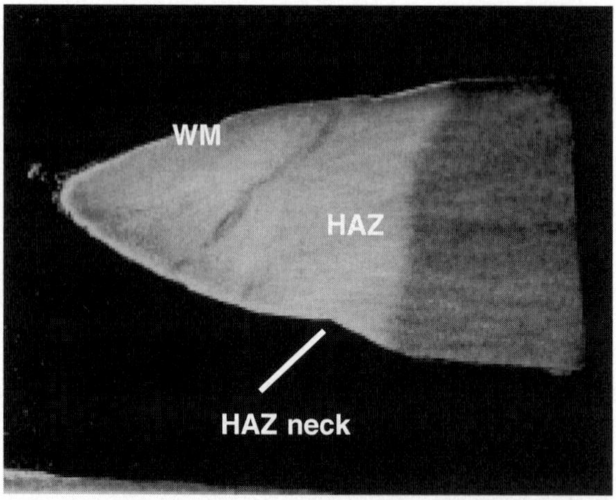

Fig. 5 'Conjoint' P22 weld/HAZ failure.

W20 specimens ('WDN' in Table 1) showed an unusual necking failure mode centred on the P22 weld / P91 parent fusion boundary. Here, both the P22 weld metal and P91 showed severe ductile shear,[2] with P91 flow toward the CrMoV end at the specimen surfaces balanced by P22 weld metal flow toward the P91 end at the specimen core. This peculiar failure

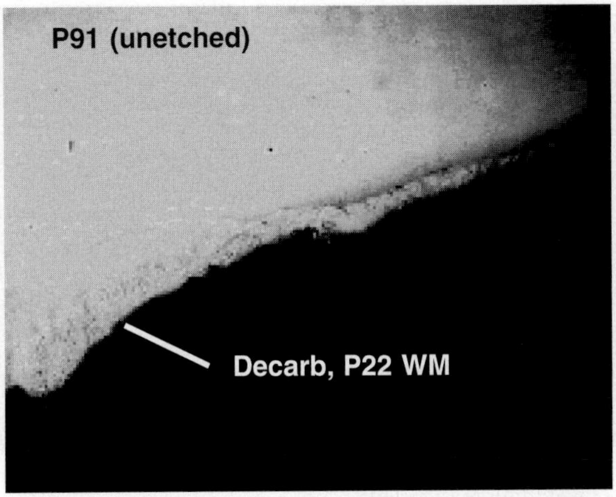

Fig. 6 Decarburised P22 weld failure.

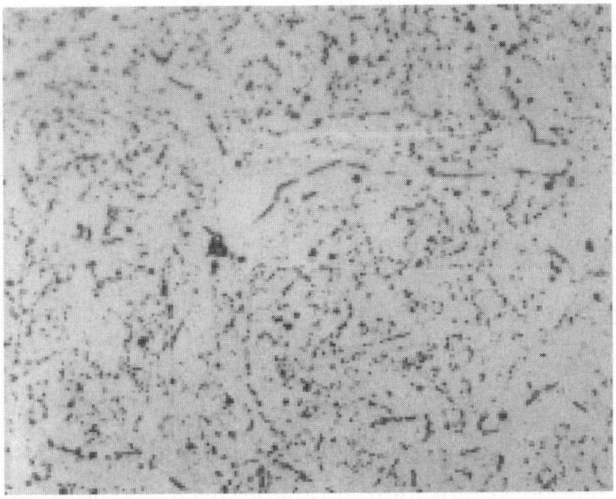

Fig. 7 Degraded P91 parent.

process caused gross strain softening of P91 parent to a hardness as low as 120 VPN, Figure 7. The waisted specimen design suppressed high ductility failure and produced 'typical' decarburised zone failures, Figure 8, but with rupture lives similar to non-waisted specimens, Table 1.

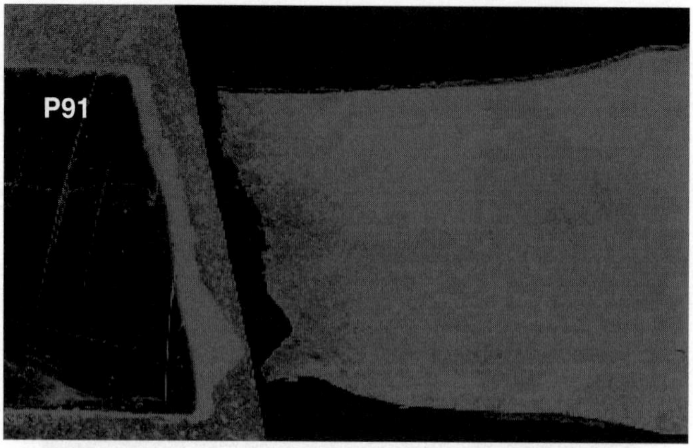

Fig. 8 Decarb. P22 weld failure (waisted).

Table 2 Activation energy for creep rupture.

Filler	Aging Temp °C	Q (Kcal/mole)	Filler	Aging Temp °C	Q
P9	600	75	P22	600	81
P9	650	79	P22	650	83

Rupture Data Analysis

A ratio R = {life after long term 600°C aging[3]/life after short term 650°C aging[2]} gives a measure of the effect of decarburisation. If progressive decarburisation causes weakening, R should be lower for decarburised zone failures. However, while the average R for weld W18 (no decarburisation) is 1.27, that for weld W17 (decarburised P22 HAZ) is 1.22, only marginally lower.[2] Thus, preaging increases decarburised zone width, but this has little effect on life.

Each isostress test series may also be analysed to determine the effective activation energy for creep rupture[2]. Results for P91 to P22 welds, Table 2, are consistent with substitutional diffusion control, again indicating that the low Q decarburisation process has no more than a marginal influence on creep performance of the P9 filler weld W17. It is conceivable that this influence, insignificant at 650°C but perhaps just detectable at 600°C, could become more important in service at 565°C. Longer term tests for ECCC, now under way, should clarify this.

For P91 to P22 welds, extrapolation[2] to 565°C with a 'worst case' 40 MPa axial stress predicts creep lives after preaging treatment (3) of 70,000 hours for P9 filler and 80,000 hours for P22 filler welds. Thus, rupture data alone do not clearly show which type of weld should be used. For comparison, a life estimate for like-to-like CrMoV steampipe welds under similar conditions is about 140,000 hours, in line with plant experience (i.e. only welds with highest system loads crack in this timescale). The predicted shortfall in dissimilar weld life is therefore not severe.

Creep Crack Growth

HAZ and weld metal decarburised zones both showed high crack growth rates as a function of C*. However, analysis of creep crack growth (CCG) rate and creep deformation rate as a function of crack length for specific load values provided more useful information, detailed elsewhere.[3] When a crack grows in a thin weak zone, creep deformation is highly localised, and the overall deformation rate used to determine C* can be low. Hence, an observed high decarburised zone crack growth rate as a function of C* can have more than one cause. It can arise because CCG is genuinely faster, as proved to be the case for the HAZ specimens, where decarburisation appeared roughly to double CCG rate. However, it can also arise simply because deformation is slower. This was the case for the weld metal decarburised zone tests, where the adjacent strong P91 acted to reduce overall creep deformation rate, whilst there was little evidence of any increased CCG rate compared with bulk weld metal data.

DISCUSSION

Creep Failure in Uniaxial Testing

Creep failure in uniaxial cross-weld tests involves complex interactions between zones. Parker[5] has shown that adjacent strong material can constrain deformation and retard failure in a weak zone which is thin compared to the specimen diameter. Experimental confirmation includes data showing that a strong nickel-based weld metal can retard cross-weld Type IV failure (albeit only in high stress tests) in high alloy steel.[6] In the present work, thin low alloy steel zones may be weakened by decarburisation, but the composite test specimen is also strengthened by adjacent high alloy steel. The net result therefore requires careful analysis.

For welds to P22 parent, the use of non-matching P9 filler seems to have no marked net effect on either overall deformation rate in CCG testing,[3] or overall rupture life in uniaxial testing. HAZ decarburisation does, however, cause faster creep crack growth. Hence, it seems likely that the strong P9 weld metal initially exerts a protective constraint effect in uniaxial testing, retarding crack initiation. However, once a crack forms in the weak decarburised zone, its growth to cause uniaxial test failure may be rapid.

For welds to CrMoV parent, the stronger and more decarburisation – resistant high temperature HAZ yields better performance under high stress, when the constraint effect retarding failure in the thinner zone of weakness may be most marked. However, this has little practical importance because of the poor performance at more realistic stresses. A model of cracking in partially decarburised material – weak enough to deform, but with sufficient retained grain boundaries and carbides to nucleate creep cavities – can then account for some otherwise puzzling results. Thus, CrMoV and P22 decarburised HAZ failures show similar rupture lives because both materials must produce a similar worst-case balance of weakening and carbide retention at some location in their HAZs. It may not be important that this worst-case location is closer to the fusion boundary in CrMoV than in P22, or that the worst-case location moves further away from the weld as decarburisation proceeds. Thus, the model explains the result that while a decarburised HAZ zone is weak, widening may not make it weaker.

There is less evidence that the lower carbon P22 weld metal is significantly weakened by decarburisation. Crack growth failure evidently can occur in the decarburised zone, but ductile necking by the conjoint mechanism (W18) or in bulk weld metal (W20) often 'wins the race' to uniaxial test failure. There is some evidence (for W20) of greater tendency toward low ductility thin zone failure at lower stress, but the lowest stress W20 failure shows a further shift into the CrMoV Type IV location, where a like-to-like weld is also expected to fail. It might be surmised that whilst P91/2CrMo weld metal/CrMoV joint performance can be somewhat inferior to that of a CrMoV/2CrMo weld metal/CrMoV joint at high stress, the differential is eliminated at more realistic stresses. Insofar as uniaxial testing provides a realistic guide to plant weld performance, therefore, decarburised weld metal failure does not seem a very serious threat.

Creep Failure in Plant Welds

Many of the observed weld uniaxial test failures are, however, highly artificial. The P22 weld / P22 HAZ conjoint failure mechanism by ductile strain interaction and necking is clearly a peculiarity of the plain bar specimen geometry. The protective behaviour of strong zones such as P9 weld metal can explain why a decarburised HAZ (W17) may outperform a non-decarburised HAZ (W18), but the protective effect could clearly be less effective under non-uniaxial loading. Finally, the behaviour of the P91 / P22 weld metal interface is notable. Commonly, the P22 weld deforms while the much stronger P91 HAZ does not, producing high mismatch stresses at the fusion boundary, Figure 6. It may be that such stresses can precipitate a sudden transition to ductile shear in the P91 HAZ, with strain softening causing formation of a local plasticised zone in 'WDN' (Table 1) type failure. This might perhaps occur late in life, so that rupture life is similar to that of the corresponding waisted specimen failing by crack growth. However, this peculiar behaviour highlights the uncertainties in applying uniaxial data to plant.

Real circumferential pipe butt welds differ from uniaxial test bars in constraint, loading conditions, and stress state. Thin weak zones are highly constrained in both uniaxial test bars and real welds. Hence, uniaxial thin zone test failures should be quite representative of real welds. However, thick weak zones such as P22 weld metal are unconstrained in test bars, but quite constrained in real welds with high wall thickness/weld width ratio and circumferential geometry. Hence, creep deformation in thick weak weld zones may proceed only slowly in real welds, but can cause rapid failure in uniaxial test bars. It follows that while low ductility uniaxial rupture data may predict real weld failure quite accurately, high ductility uniaxial data are liable to underpredict life in real welds. A separate consideration which reinforces these conclusions is the effect of loading condition. Uniaxial test load is unaffected by creep strain, but plant system loading may be relaxed by creep deformation. A thick weak zone is thus likely to absorb strain and relax its loading, but a thin weak zone will show low macroscopic ductility, negligible system stress relaxation, and hence potential early creep failure.

Choice of Welding Procedure

A high alloy filler should therefore not now be used for this dissimilar joint, as it creates a thin weak decarburised HAZ zone with demonstrably impaired properties in either P22 or CrMoV steel. Some shortfall in creep life is probable for joints already in service with a P9 filler, but this may not be a practical problem if these joints are not in high system stress locations.

For the low alloy filler, data which involve ductile uniaxial failure in the weld metal and/ or adjacent material should be largely disregarded, because they relate to mechanisms which will not operate in real circumferential welds. There remains some risk of early failure in decarburised weld metal adjacent to P91, but any shortfall in life compared with like-to-like welds is likely to be fairly small. A stronger low alloy filler could be preferable to P22 weld metal. Insofar as a dissimilar joint is at any greater risk than a like-to-like joint with P22 filler, the risk relates to the interface between P91 and P22 weld metal. Hence, a direct P91 to CrMoV weld should be acceptable, and there is no need to use a P22 insert.

CONCLUSIONS

1. Uniaxial cross-weld creep tests often fail by artificial mechanisms involving complex creep strain interactions between adjacent weld zones. Prediction of real weld behaviour requires careful assessment of differences between constraint factors in laboratory tests and plant components.
2. Dissimilar welds with a high alloy filler show similar uniaxial rupture lives to those with a low alloy filer because of the countervailing effects of a weaker decarburised HAZ and stronger weld metal. However, low alloy filler welds are at lesser risk because constraint and system loading factors make their high ductility failure modes inoperative in real pipe butt weldments.
3. Aging widens the decarburised HAZ, but this does not cause further weakening. Hence, extended aging in service should not act to increase the relative risk of decarburised zone failure, compared with alternative mechanisms such as Type IV cracking.
4. Dissimilar welds with a P9 filler show somewhat poorer high temperature performance than like-to-like welds, but mid-life problems are fairly unlikely.

5. Dissimilar welds with a P22 filler are at low risk of cracking in long term service, but more data are needed on the performance of decarburised weld metal adjacent to high alloy steel.
6. A direct P91 to CrMoV weld, without a P22 insert, shows acceptable creep performance.

ACKNOWLEDGEMENTS

This paper is published by permission of Powergen UK plc.

REFERENCES

1. W. M. Ham and B. S. Greenwell: 'The Application of P91 Steel to Boiler Components in the UK', *The Manufacture and Properties of Steel 91 for the Power Plant and Process Industries*, ECSC, Dusseldorf, 1992.
2. D. J. Allen: 'High Temperature Performance of Dissimilar P91 To P22 Steel Weldments', *Proceedings of 5th EPRI Conference on Welding and Repair Technology for Power Plants*, Point Clear, Alabama, 2002.
3. D. J. Allen: 'Creep Crack Growth in Weldments – Why the C* Correlation is Not Enough', *Proceedings of 3rd International HIDA and INTEGRITY Conference*, Lisbon, 2002, 355–363.
4. E. Kozeschnik, et al.: 'Dissimilar 2·25Cr/9Cr and 2Cr/0·5CrMoV Steel Welds', *Science and Technology of Welding and Joining*, **7**(2), 2002, 63–68(6) and 69–76(8).
5. J. D. Parker: 'The Creep and Fracture Behaviour of Thick-Section Multi-Pass Weldments', *Proceedings of Conference on Integrity of High Temperature Welds*, I. Mech. E., London, UK, 1998, 143–152.
6. D. J. Allen and A. Fleming: 'Creep Performance of Similar and Dissimilar E911 Steel Weldments for Advanced High Temperature Plant', *Proceedings of 5th International Charles Parsons Turbine Conference,* A. Strang, et al., eds., Cambridge, UK, 2000, 276–290.

Steam Oxidation of 9–12%Cr Martensitic Steels: The Influence of Laboratory Practice

S. OSGERBY and A. T. FRY
National Physical Laboratory
NPL Materials Centre, National Physical Laboratory
Teddington, Middlesex
TW11 0LW, UK

ABSTRACT

Three martensitic steels, P92, alloy 122 and X19, have been exposed to steam at 600 and 650°C using a range of laboratory procedures viz flowing argon/50% H_2O at atmospheric pressure, flowing 'pure' H_2O at atmospheric pressure, and static H_2O at 50 bar. The growth kinetics of scales for individual alloys are relatively independent of exposure procedure but the microstructure and spallation behaviour of the scales show significant differences. The difference in oxidation behaviour between the alloys is greater than that due to the laboratory procedures and cannot be explained simply in terms of the chromium content of the alloy.

INTRODUCTION

During service turbine components are exposed to steam at high pressure, flowing at high rates. The precise conditions are difficult if not impossible to reproduce in the laboratory and some degree of simplification is necessary, if only to reduce the cost of laboratory testing. However data are required, both to aid material selection and for detailed design of components.

Several exposure procedures are in common use in the laboratory – in ascending order of complexity (and cost) these are:

- Water vapour in a carrier gas (usually argon),
- Flowing steam at atmospheric pressure,
- Static steam at high pressure and
- Flowing steam at high pressure.

None of these laboratory procedures include heat flux. In order to assess the influence of this parameter exposure in pilot plant is normally required. This is usually the final stage in any material assessment.

A recent review[1] has highlighted the scatter in data arising from the use of different exposure procedures. The purpose of the current work is to identify any differences in oxidation kinetics and scale microstructure arising from the exposure procedure that is used. The materials investigated were the 9–12Cr martensitic steels, P92, X19 and alloy 122.

Table 1 Alloy composition.

Alloy	Composition, wt.%						
	C	Si	S	Cr	Mo	Ni	Al
P92	0.12	nr	nr	8.85	0.42		
X19	0.215	0.33	0.005	10.87	0.66	0.41	0.007
Alloy 122	0.12	0.20	0.68	11.07	0.35	0.31	0.005
	B	N	Nb	V	Cu	Co	W
P92	0.042	nr	0.07	0.20			1.85
X19	0.0015	0.055	0.41	0.22			
Alloy 122	0.002	0.055	0.054	0.22	1.01	0.011	1.94

nr – not reported

EXPERIMENTAL

Material was supplied in the form of tube for P92, forged bar for X19 and a section from a large forging of alloy 122. The composition of each material is given in Table 1.

Specimens were machined in the form of rectangular blocks and the surface was prepared by abrasion using 600 grit SiC paper. All the specimens were cleaned using industrial alcohol prior to exposure.

The specimens were exposed to steam at 600 and 650°C for durations up to 1000 hours using three procedures.

1. Gaseous argon and liquid water were passed into the furnace at rates of 100 ml min^{-1} and 0.0736 ml min^{-1} respectively. The water is gasified on entry to the furnace to give the desired argon/50% H_2O by volume.
2. Water in a reservoir is boiled outside the furnace and the steam flows naturally through the furnace.
3. Steam is generated in a sealed pressure vessel. A nominal pressure of 50 bar was chosen for these tests; excess pressure is released through an outlet valve.

Control of water chemistry during exposure was limited to ensuring that the water was de ionised and de aerated prior to conversion to steam. For the exposures under flowing steam and the argon/50% H_2O mixture, water was taken from a de-ionised supply and distilled, de-aerated by bubbling nitrogen through the reservoir before final conversion to steam by boiling. In the autoclave the water was de-aerated during leak testing of the system under high-pressure nitrogen prior to heating.

After exposure the specimens were plated with nickel to retain the oxide, sectioned and mechanically polished. Oxide scale thickness was measured at 24 positions, evenly spaced along the specimen surface, using a calibrated eyepiece graticule in the optical microscope. The resolution of these measurements is 1 μm. Some specimens were coated with a thin layer of iron oxide to provide interference contrast between the various oxides that were formed during exposure. With the substrate taking a deep red colour, haematite appears orange, magnetite appears olive green and spinel has a lime green hue.

RESULTS AND DISCUSSION

In general the oxide scale consisted of 4 components: haematite; magnetite; iron-chromium spinel and an internal chromium rich phase. X-ray diffraction confirmed the presence of both haematite and magnetite.

However the detailed microstructure of the scale differed according to the exposure procedure used as illustrated in Figure 1 for scales formed during exposure of P92 at 600°C.

The oxide scale that forms under an argon/50% H_2O atmosphere (Figure 1a) consists of two continuous layers an outer haematite and inner spinel. Haematite appears occasionally at the surface and decorating pores and cracks in the magnetite. This is evidence of a substructure within the spinel and an additional corrosion product that occurs at the inner edge of the spinel.

Under flowing steam at atmospheric pressure, Figure 1b, the haematite appears as a continuous layer at the outer edge of the scale. Over a significant distance the haematite and magnetite exist as a mixture before disappearing to leave a continuous layer of magnetite. The spinel appears to have an internal structure, similar to that formed during the exposure to argon/50% H_2O. The unidentified corrosion product at the inner edge of the spinel is also present after this exposure.

The scale that forms during high-pressure exposure to static steam in the autoclave does not have an outer haematite layer, Figure 1c. The detailed structure of the spinel is also different from that which develops using the other exposure techniques, in that a dark distributed phase is prevalent after the high pressure exposure.

The magnetite layer that forms during all three exposure procedures contains defects. Individual defects are larger for scales grown under argon/50% H_2O but more numerous in the magnetite that forms under flowing steam at atmospheric pressure. The magnetite layer that forms in the autoclave is the most compact.

Further differences in scale microstructure are observed when the exposures are carried out at 650°C (Figure 2).

For scales grown under argon/50% H_2O (Figures 1a and 2a) the increase in temperature to 650°C does not have a massive effect on the scale microstructure. Haematite is again present in patches at the surface and around some pores in the magnetite layer. The magnetite layer appears to be more compact after the 650°C exposure than after exposure at 600°C. The largest observable difference between the two exposure temperatures is in the spinel layer where the internal structure after the 650°C exposure is more distinct than after exposure at 600°C.

The increase in temperature from 600 to 650°C for exposures to flowing steam at atmospheric pressure causes the biggest change in scale microstructure. From a continuous layer at 600°C (Figure 1b) the haematite becomes less prevalent and, at 650°C, only occurs in isolated patches at the surface and around pores in the magnetite (Figure 2b). The magnetite is much more porous after the 650°C exposure and the spinel has developed an internal structure similar to that formed during exposure to argon/50% H_2O at 650°C.

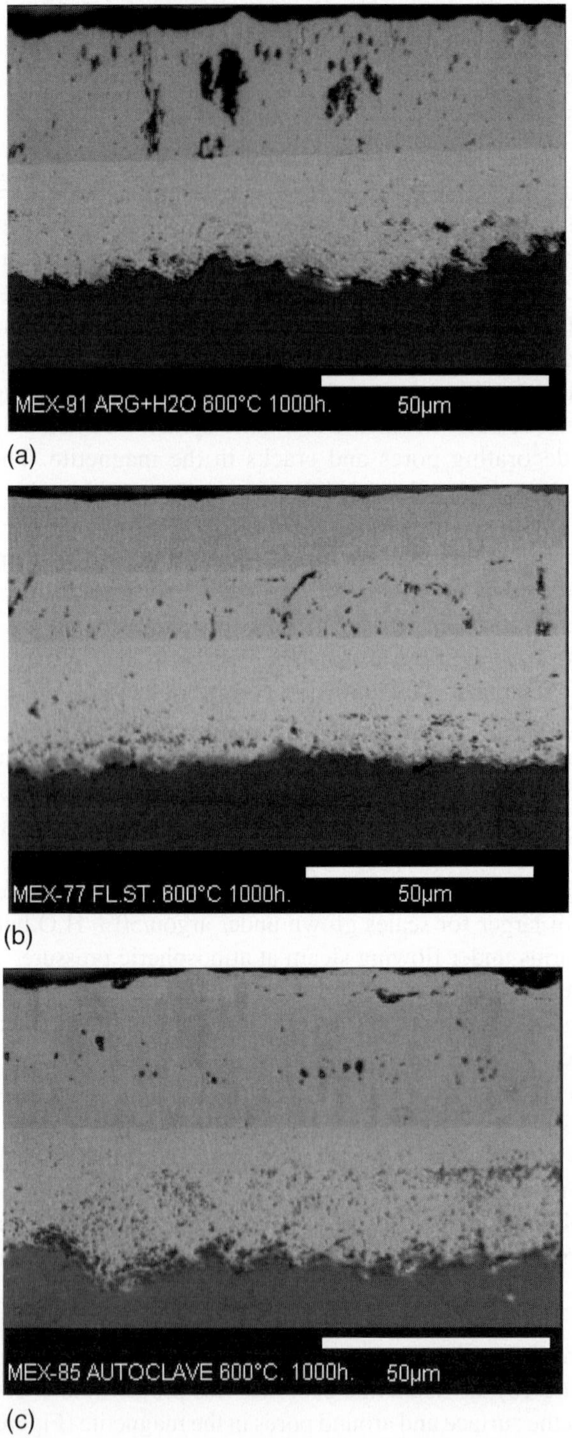

Fig. 1 Oxide scales grown on P92 after 1,000 hours at 600°C (a) Argon/50% H_2O, (b) Flowing steam at atmospheric pressure and (c) 50 bar steam in autoclave.

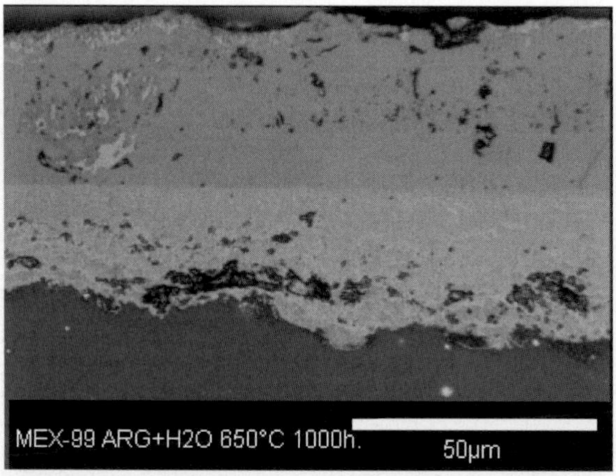

(a)

(b)

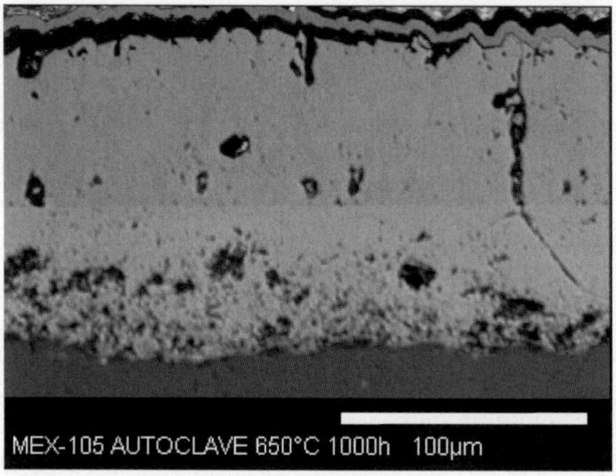

(c)

Fig. 2 Oxide scales grown on P92 after 1,000 hours at 650°C (a) Argon/50% H_2O, (b) Flowing steam at atmospheric pressure and (c) 50 bar steam in autoclave.

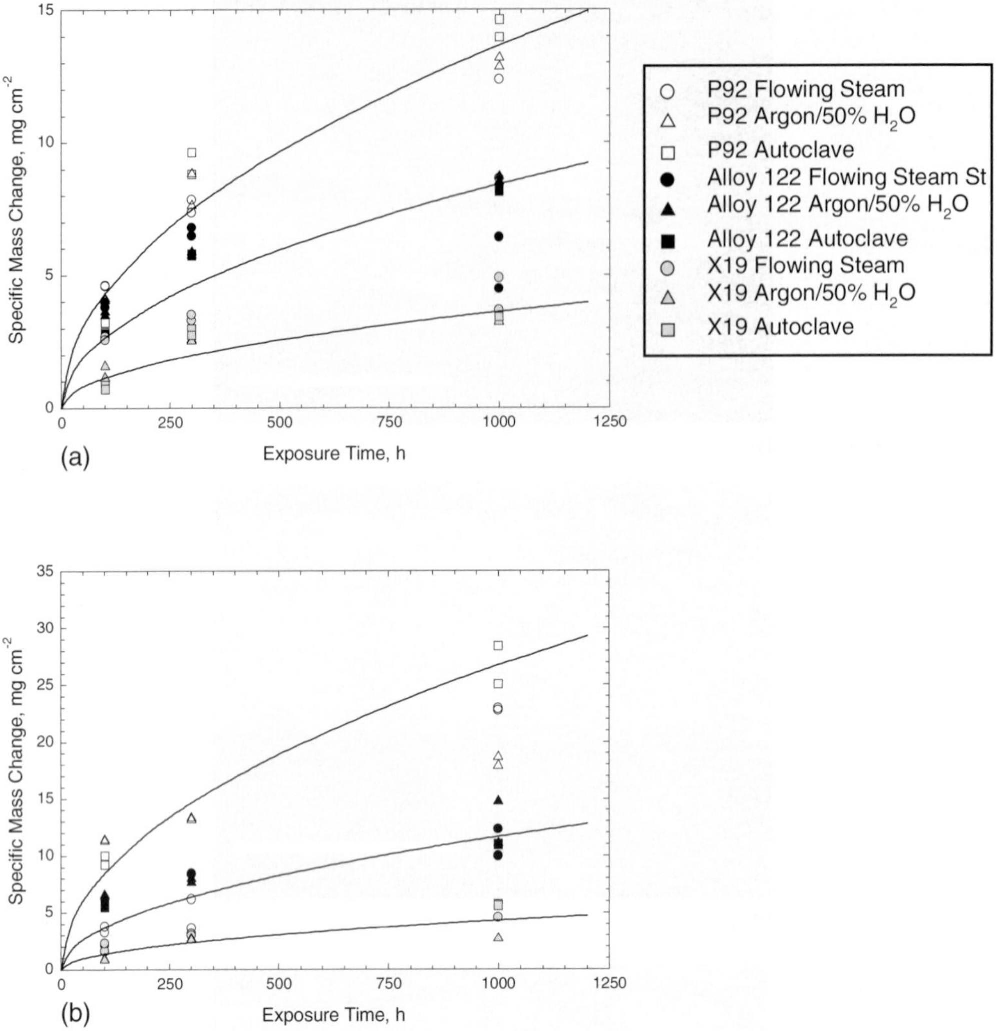

Fig. 3 Mass change as a function of time, exposure procedure and temperature for the alloys P92, Alloy 122 and X19 (a) 600°C and (b) 650°C.

There is little difference in the microstructures of the oxides grown in the autoclave at 600 and 650°C (Figures 1c and 2c respectively). The main observable difference is a higher concentration of pores in both the magnetite and spinel layers after the exposure at 650°C.

The specific mass change as a function of time and exposure procedure for the three alloys under investigation is shown in Figure 3.

At both temperatures the differences in mass change behaviour between the alloys is greater than that arising from the different exposure procedures. In all cases X19 oxidised less rapidly than alloy 122, which in turn, was more resistant to steam than P92. These results confirm the excellent oxidation resistance of X19, which exceeds that of alloy 122 even though alloy 122 has a higher chromium content. This observation suggests that one of the minor elements affects oxidation resistance. Examination of Table 1 suggests that either higher contents of Mo or Nb are beneficial or that the presence of W is detrimental to oxidation performance. Recent work[1] on experimental martensitic steels also indicates a possible increase in oxidation rate in steels where 0.4%W has replaced Mo, compared with W-free steels of otherwise similar composition.

The three alloys show differences in the temperature dependency of their respective oxidation rates. The increase in mass change kinetics on increasing the temperature from 600 to 650°C is nearly a factor of 2 for P92 but only ~1.5 for alloy 122 and ~1.3 for X19.

Spallation of the oxide is usually initiated by through thickness cracks in the magnetite followed by separation at the magnetite/spinel interface.[3] Examination of the defect structure in the magnetite layer in Figures 1 and 2 indicate that fewer defects are observed in the scales that were grown under static steam in the autoclave. This observation, coupled with the low cooling rate inherent to the thermal cycle in the autoclave – due to he high thermal mass of the system – implies that, for scales of equivalent thickness, spallation is less likely in the specimens exposed in the autoclave than exposed using the other laboratory procedures. In addition the magnetite scale that forms under flowing steam at 650°C shows the highest concentration of defects, implying that these scales are the most likely to spall.

This behaviour is confirmed in Figure 4 where the mass change of the specimen is correlated with the scale thickness, measured on polished cross sections. A line of slope 9.65 has previously been shown to correlate these parameters when the scale remains intact[4] and this line has been superimposed on the data. At 600°C (Figure 4a) the data fall the line indicating that the scale remains intact under all conditions. One data point, the longest exposure of P92 using flowing steam does fall below the line possibly indicating the onset of spallation in this alloy under these conditions. At 650°C data for P92 from both the 'flowing steam' and 'argon/50% H_2O' exposures, fall beneath the line implying that spallation has occurred, thereby supporting the hypothesis regarding the influence of exposure procedure on spallation behaviour through changes in the defect structure within the magnetite.

CONCLUSIONS

The martensitic steels, P92, alloy 122 and X19, have been exposed to steam at 600 and 650°C using a range of laboratory procedures viz flowing argon/50% H_2O at atmospheric pressure, flowing 'pure' H_2O at atmospheric pressure, and static H_2O at 50 bar. The growth kinetics of scales for individual alloys have been shown to be relatively independent of exposure procedure, thereby giving confidence that data from different laboratories are reproducible. However the microstructure and spallation behaviour of the scales show significant differences between the laboratory procedures.

The difference in oxidation behaviour between the alloys is greater than that due to the laboratory procedures. These differences cannot be explained simply in terms of the chromium content of the alloy, but may demonstrate an influence of tungsten on oxidation performance.

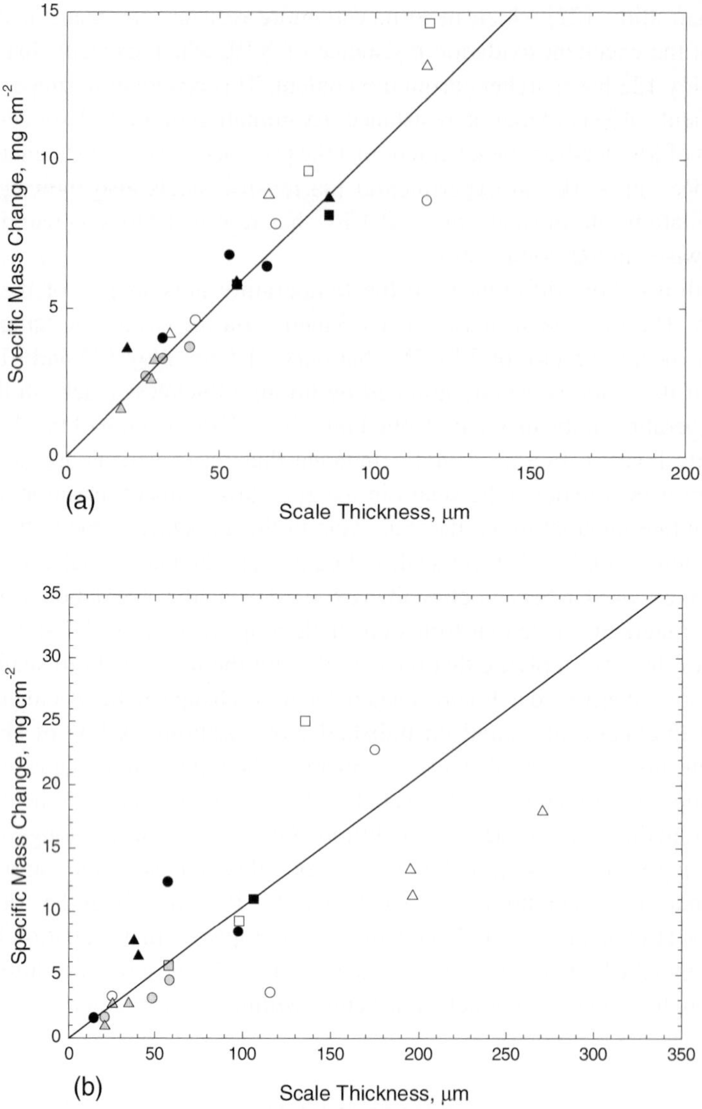

Fig. 4 Correlation of specific mass change with scale thickness for martensitic steels exposed to steam using a range of procedures (a) 600°C and (b) 650°C. (key as in Figure 3).

ACKNOWLEDGEMENTS

This work was carried out under the Life Performance of Materials Programme, a programme of underpinning research funded by the United Kingdom Department of Trade and Industry.

REFERENCES

1. A. Fry, S. Osgerby and M. Wright: *Oxidation of Alloys in Steam Environments – A Review*, NPL Report MATC(A)90, National Physical Laboratory, UK., 2002.
2. P. J. Ennis and W. J. Quadakkers: *Materials for Advanced Power Engineering 2002*, J. Lecompte-Beckers, M. Carton, F. Schubert and P. J. Ennis, eds., Forschungszentrum, Julich, 2002, 1131–1142.
3. S. Osgerby and L. N. McCartney: *Materials for Advanced Power Engineering 2002*, J. Lecompte-Beckers, M. Carton, F. Schubert and P. J. Ennis, eds., Forschungszentrum, Julich, 2002, 1613–1620.
4. R. Knoedler: Private Communication.

Materials for Advanced Steam Power Plants: The European COST522 Action

MARC STAUBLI and BRENDON SCARLIN
Alstom Switzerland, Baden
Switzerland

KARL-HEINZ MAYER
Alstom Power, Nürnberg
Germany

TORSTEN-ULF KERN
Siemens Power Generation, Mülheim
Germany

WALTER BENDICK
Mannesmann Forschungsinstitut, Duisburg
Germany

PETER MORRIS
Corus, Rotherham
UK

AUGUSTO DIGIANFRANCESCO
CSM, Roma
Italy

HORST CERJAK
Technical University, Graz
Austria

ABSTRACT

Today's state of the art large fossil-fired steam turbines comprises live steam conditions of up to 610°C/300 bar and re-heat temperatures of up to 630°C. These ultra super critical (USC) steam parameters significantly increase the plant efficiency and reduce the fuel consumption hereby reducing the emissions of CO_2.

In order to maintain high operational flexibility of large USC plants, the thick-walled components should be manufactured from ferritic materials. A further increase in steam parameters up to the limitations of the ferritic steels will additionally contribute to the reduction of greenhouse effects.

Improved high temperature steels therefore are needed for the critical components such as rotors, casings, bolts, tubes/pipes and waterwalls. The 9–12%Cr class of steels offers the highest potential to meet the required property level. Therefore a significant effort to increase the application

temperature of this class of steels was focused within the European COST501 programme. This effort led to improved materials for forged and cast components and pipework. These materials for 600°C applications are already successfully applied in a number of advanced European power plants. Within the current European programme COST522 (1998–2003) the main aim of working group 'Advanced Steam Power Plant' is to develop and evaluate ferritic steels for applications under steam conditions up to 650°C. The 9–12%Cr steels still offer some further potential regarding their creep strength. Nevertheless the oxidation resistance must be carefully taken into consideration when temperatures up to 650°C are targeted.
A large number of new ferritic-martensitic compositions have been designed on the basis of the positive experience from the previous activity as well as the results of advanced thermodynamic calculation tools. Several full size cast and forged components have been successfully manufactured from the most promising compositions. Extended mechanical investigations are ongoing on these components. Oxidation resistance, microstructure and weldability of these new materials are specifically addressed within this COST activity.

INTRODUCTION

Readily available energy at an economical price is a major factor affecting the success of manufacturing industry, upon which the general well-being and the standard of living of the population depend. Energy production on the other hand is faced with the introduction of increasingly stringent emission regulations. This is to safeguard health and preserve the environment for future generations. Coal, oil and natural gas remain the fuel basis for the next twenty to thirty years.

But the reserves of oil and natural gas are unlikely to be sufficient to fully satisfy the projected increased power demands. Therefore coal may be the only fuel available in substantial quantities by the middle of the 21st century. This fact drives the utilities to increase the efficiency of coal fired power plants. In Europe, coal or lignite will also play an important role in the foreseeable future in spite of the limited requirements for new plants. Hence there will be a substantial market for the upgrading or replacement of existing plants.

The thermal efficiency is influenced by several means, but the adoption of supercritical conditions by increasing steam temperatures and pressures plays a key role. The application of steam parameters of 600 to 650°C/300 bar offers improvements of 8 to 10% and a corresponding reduction in CO_2 compared to 540°C/180 bar.[1] These advanced steam parameters require materials with adequate creep strength and resistance to oxidation. Experience with austenitic materials was unsatisfactory showing considerably restrictions in the operational flexibility of the plants. Hence the 9–12%Cr class of steels offers the highest potential to meet the required property level for the critical components in steam power plants.

INTERNATIONAL RESEARCH PROGRAMMES AND COST

Since the late 1970s the research efforts have been focused world wide on the further development of ferritic-martensitic 9–12%Cr steels. Strong activities are ongoing in Japan, U.S.A. and Europe:

JAPAN

The research programmes initiated and supported by EPDC (Electrical Power Development Company) in Japan led to a number of new steels with improved creep properties at 600°C and even at higher temperatures. EPDC has ordered several power plants in which new materials are used.[2] The steam temperature in the power plants ordered recently is 610°C while material test programmes for 625°C applications are ongoing. The Japanese scenario for coal-fired power plants foresees as a last stage the use of ferritic materials for thick-walled boiler components at 650°C and 350 bar.

U.S.A.

In the United States, EPRI launched as early as 1978 a study for more economic coal-fired plants. The study was followed in 1986 by a widely supported extended programme in which not only American but also European and Japanese companies took part. This RP1403 programme was focused on the development of materials for thick-walled boiler components and the steels NF616, HCM12A and TB12M were validated. ASME approved the first two steels, bearing the designations P92 (Nf616) and P122 (HCM12A). Components manufactured from these materials were subsequently applied in several advanced European power plants. However the programme RP1403 led also to the introduction of an advanced cast steel of the P91 type.

EUROPE

In Europe the development activities were concentrated in the COST programmes. **COST** (**Co**-operation in the field of **S**cience and **T**echnology) is a long established European programme aimed at co-ordinating pre-competitive research activities in numerous areas of science and technology. The current action is titled 'COST 522 – Energy generation in the 21st century; ultra-efficient, low emission plants' (1998–2003). It is founded on the success of earlier related COST actions, specifically COST 501 (1986–1997), which have established strong trans-European networks in this field.

The effectiveness of the past COST actions is exemplified by the introduction of newly developed ferritic steels in forgings, castings and pipework. These improved steels are commercially in operation in advanced European power stations. They have made it possible to increase the operating steam temperatures from 530–565°C to 580–600°C with a corresponding increase in the thermal efficiency. The close co-ordination of activities related to turbine and boiler components within COST led to the successful development of advanced materials for applications in the turbine as well as in the boiler.

The target of the current COST 522 action is focused on the development of suitable materials, coatings and surface treatments for

- Steam power plants with inlet temperatures of up to 650°C and
- Gas turbines with combustion temperature of up to 1450°C and NO_x emissions <10 ppm.

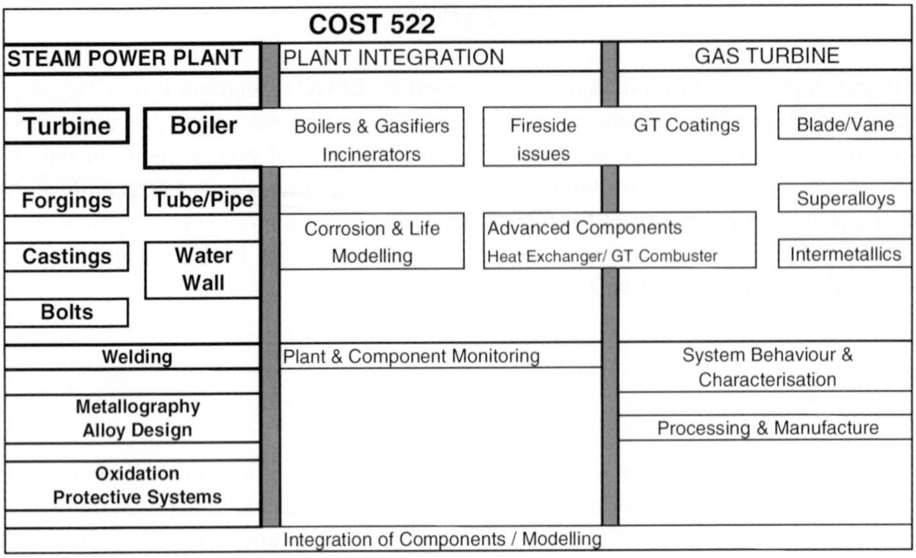

Fig. 1 Organigramme of the COST522 action.

Further objectives are the improvement of the lifetime prediction methods, the preparation of material models to characterise the creep and LCF behaviour, and the improvement of plant simulation techniques and the operating condition monitoring. Accordingly, the entire action is divided into different sub-projects as shown in Figure 1. It involves 16 European countries and co-ordinates well over 100 research projects involving over 70 different organisations including all of the main utilities, manufacturers, materials suppliers and research establishments.

The steam power plant (SPP) working group is aiming for materials for highly loaded components at steam inlet temperature of up to 650°C. Corresponding efficiencies of approximately 50% may be achieved by the use of ferritic-martensitic steels. This working group is sub-divided into 'Turbine' and 'Boiler' groups. A close collaboration between these two groups ensures the use of synergies whenever possible.

MATERIALS DEVELOPMENT FOR 600°C APPLICATION / PROPERTIES & EXPERIENCE

Within COST 501 a series of advanced steels for forgings, castings and pipe/tube application as given in Table 1 was qualified. As compared to traditionally used materials the increased creep strength of these new materials offers a significant gain in application temperature as shown in Table 2.

The improved creep rupture strength does not, however, prejudice other design-relevant material properties, such as fracture toughness or low cycle fatigue behaviour. Additionally,

Table 1 Compositions of improved ferritic steels developed in COST 501.

	C	Cr	Mo	W	V	Ni	Nb	N	B	Status
Forged Steels										
Type F	0.1	10	1		0.2	0.7	0.05	0.05		- Applied in operating
Type E	0.1	10	1.5	1	0.2	0.6	0.05	0.05		power stations
Type B	0.2	9	1.5		0.2	0.1	0.05	0.02	0.01	- Trial rotor manufactured
Cast Steels										
10CrMoWVNbN	0.1	10	1	1	0.2	0.8	0.07	0.05		- Applied in operating
10CrMoVNbN	0.1	10	1		0.2	0.8	0.07	0.05		power stations
Tube/Pipe Steel										- Applied in operating plants
E911	0.1	9	1	1	0.2	0.3	0.05	0.07		appl. for ASME/ASTM

Table 2 Temperature for creep strength 100 MPa/100,000 hours of conventional and improved ferritic steels.

Forged Steels	100 MPa 100,000 h	**Cast Steels**	100 MPa 100,000 h
28CrMoV4 9 (1CrMoV)	550°C	G17CrMoV5 10	538°C
X20CrMoV12 1 (12CrMoV)	570°C	GX23CrMoV12 1	562°C
12CrMoWVNbN10 11 (E-type) **12CrMoVNbN10 1 (F-type)**	**597°C**	**GX12CrMoWVNbN10 1 1** **GX10CrMoVNbN9 1**	**592°C**
18CrMoVB9 1 (B-type)	**620°C**		

Table 3 Advanced steam power plants using European technology.

Steam Plant	Fuel	Output MW	Live Steam Bar/°C	Re-heat Steam °C	Thermal Efficiency %
Skaerbaek	Gas	400	290/582	580/580	49*
Nordiyland	Coal	400	290/582	580/580	48*
Avedore	Biomass, coal	530	300/582	600	47*
Schkopau	Lignite	450	285/545	560	40
Schw. Pumpe	Lignite	800	264/542	560	40.6
Boxberg	Lignite	900	260/540	580	41.7
Lippendorf	Lignite	900	268/554	583	42.3
Niederaussem	Lignite	1050	265/576	600	> 43
Isogo	Coal	600	251/600	610	> 43
Westfalen (design study)	Coal	350	283/600	620	> 43

the new materials provide benefits in medium-load and peak-load power plant operation. This is due to their favourable physical properties, such as reduced thermal expansion as compared to low alloyed steels and consequently shorter start-up times.[3]

The development of these new materials and their properties and application are reported in[4–6] for the tube/pipe steel E911 and in[7–10] for forged and cast steels. It's noteworthy that creep tests of >85,000 hours on rotor steels type 'E' and 'F' are showing very similar long-term properties independent of the W addition, the heat treatment and the production route i.e. ESR vs. VCD. These new cast and forged COST steels are currently applied in several advanced power plants at elevated steam parameters using European technology (Table 3).

The experience of the qualified manufacturers of these plant components has been positive throughout. The acceptance of the new rotor alloys of the 1W-1Mo and the 1.5Mo types has been demonstrated by the number of manufactured and installed power plant components e.g. steam & gas turbine rotors, discs & shaft ends and various other forged components. Six foundries have successfully produced some 350 castings with the weight ranging from 1 to 60 tonnes.[11]

In order to confirm the long-term properties of the improved steel grades for 600°C application and to establish a comprehensive database for these materials, a programme was launched with the support of VGB in Germany. This programme involves the investigation of commercial components e.g. 24 heavy components manufactured from the new cast and forged steels (W-bearing and W-free). These components were produced by different suppliers including the use of different production routes.[12]

Creep tests, tensile and LCF tests at different temperatures have been performed. The creep tests with durations > 40,000 hours show a good agreement of these production components with the trial components tested within COST as shown in Figures 2a and b.

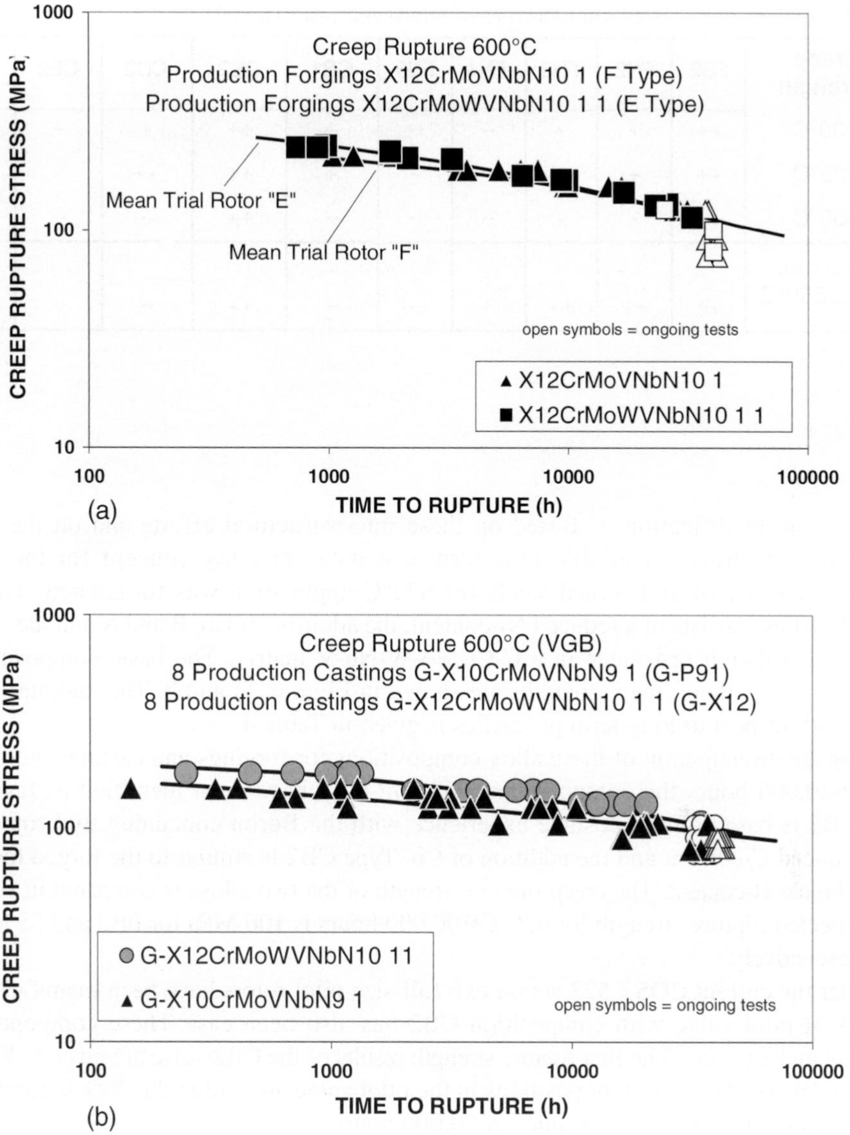

Fig. 2 (a-b) Long-term creep tests of production forgings and castings / VGB Progr.

NEW TURBINE MATERIALS/FORGINGS & CASTINGS

Extensive microstructural investigations have been carried out in parallel to the mechanical tests within the metallography and alloy design working group of COST. This was in order to ascertain the causes for the differences in properties observed and to provide support for

Table 4 Ranking of the long-term properties of forged and cast trial melts.

Creep Strength	FB2	FN2	FN3	FN4	FN5	CB1	CB2	CD2	CE2	CF2
600°C	++	+	+	--	--	+	++	+	--	--
625°C	++	+	+	--	--	+	++	++	--	--
650°C	++	+	+	--	--	--	++	--	--	--
Exposure 600 & 650°C	++	++	++	+	--	--	++	--	+	+

further alloy modifications.[13] Based on these microstructural efforts and on the careful evaluation of already available long-term test data, an alloy concept for the further modification of cast and forged steels for 620°C application was formulated. The main characteristics consists of a reduced Ni-content, the addition of Co, B and N and the increase of Cr for oxidation resistance in a C-Cr-Mo-W-Nb-V matrix. The basic compositions of these alloys and their creep strengths are shown in Figures 3a and b. The ranking of these alloys with respect to long-term properties is given in Table 4.

After the investigation of these alloy compositions for forgings and castings with testing times >30,000 hours the compositions FB2 and CB2 have been identified as favourites. Type FB2 is based on the positive experience with the Boron containing trial rotor B but with reduced C-content and the addition of Co. Type CB2 is similar to the forged melt FB2 with a higher B-content. The creep rupture strength of the two alloys is compared in Figure 4. The expected rupture strength for 625°C/100,000 hours is 100 MPa for FB2 and 85 MPa for CB2 respectively.

Under the current COST 522 action two full-size pilot rotors have been manufactured in FB2. A 5t pilot valve with composition CB2 has also been cast. These components are currently under testing. The first rupture strength results of the CB2 valve are given in Figure 5. The test data of three different positions in the pilot valve are within the 20% scatterband of the trial melt and have reached times of 20,000 hours.

FB2 and CB2 offer a significant improvement over the E & F type steels and the cast steels GX12CrMo(W)VNbN10 1(1) and are suitable for 625°C application. However for a further increase in steam parameters up to 650°C, an improvement of these ferritic-martensitic grades is necessary with respect to creep strength taking into consideration the increased oxidation occurring at this temperature level.

On the basis of the promising compositions FB2 and CB2 modifications were made in the C, Cr and Co-contents. The well known contradictory effect of Cr on the creep strength and the oxidation resistance is limiting these modifications. The adverse effect of increasing the Cr-contents on the creep rupture strength is given in Figure 6 for different 9-12%Cr steels.

It was considered that the relevant alloying aspects with respect to creep strength may be valid for both, forged and cast materials. Therefore the new compositions for forged and cast

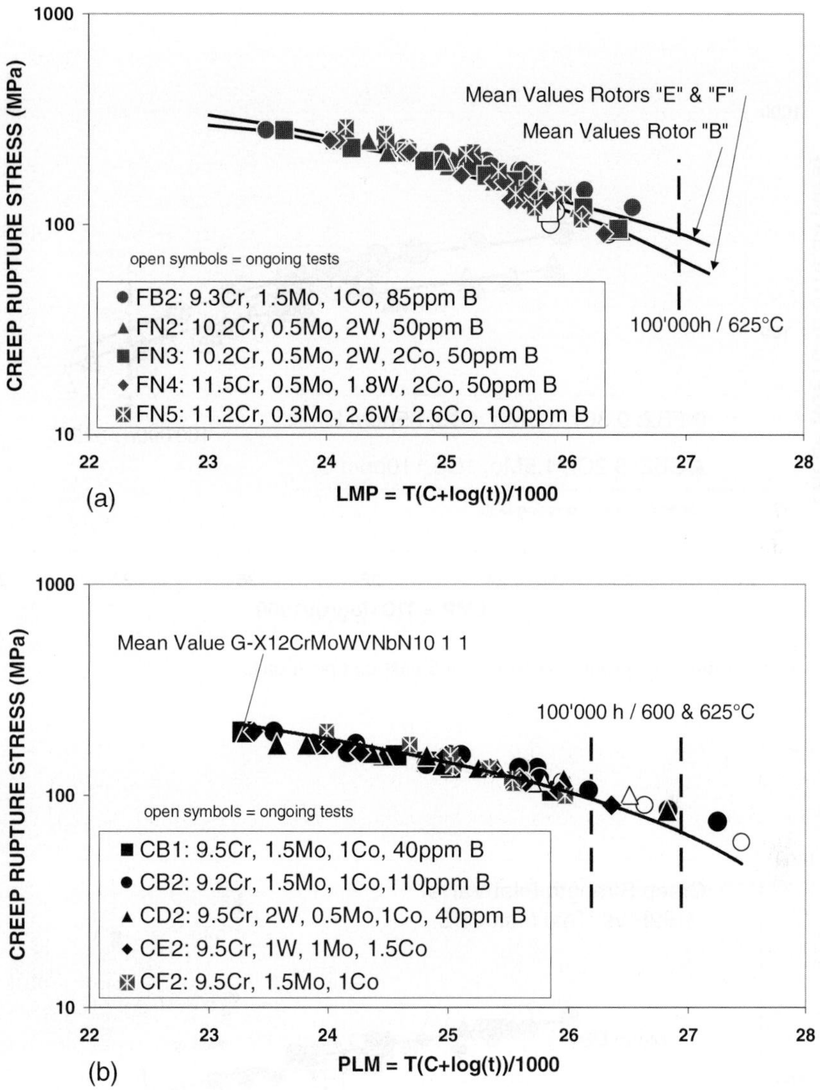

Fig. 3 (a-b) Creep rupture strength of forged and cast trial melts for 625°C application.

trial melts show a strong similarity. Seven forged trial melts of approximately 150 kg, were forged into bars and heat treated to simulate the core position of a heavy forging. In addition six cast trial melts have been fabricated as plates of 100 mm thickness and were given a common heat treatment.

For the forged and cast trial melts an extended test programme has been started including metallographic investigations, short-term and long-term mechanical properties, oxidation

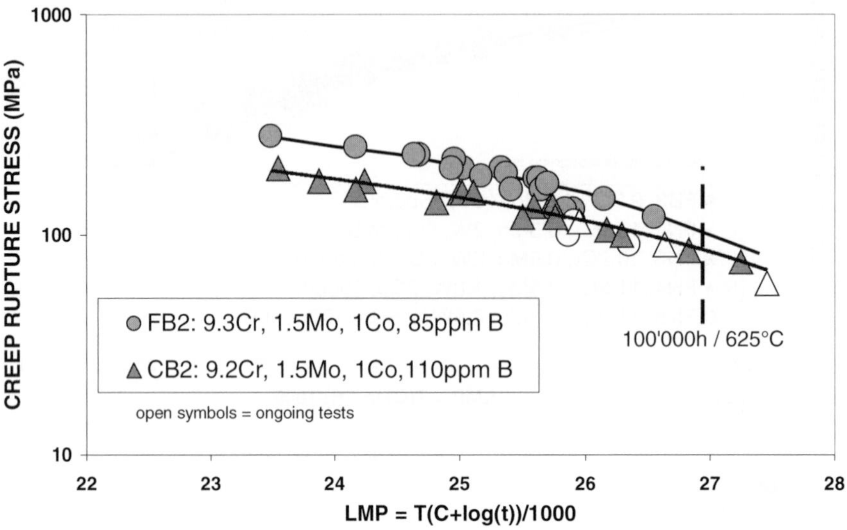

Fig. 4 Creep rupture of promising forged and cast compositions.

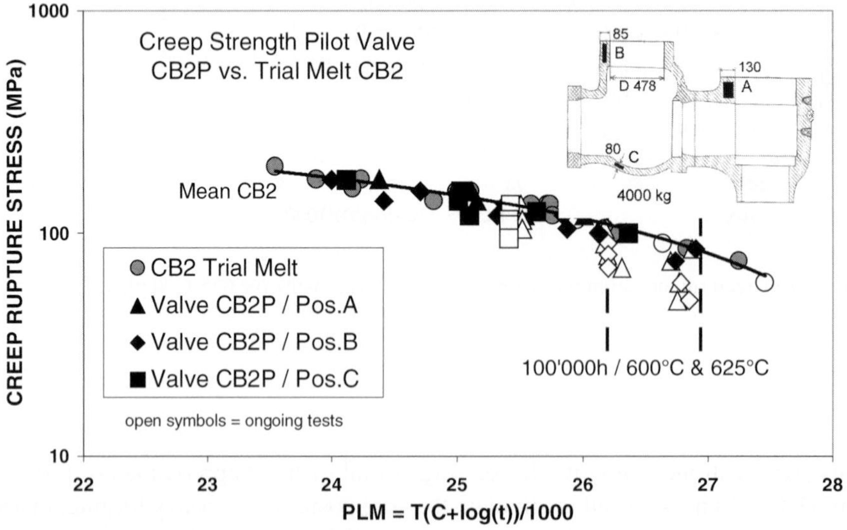

Fig. 5 Creep strength of pilot valve CB2.

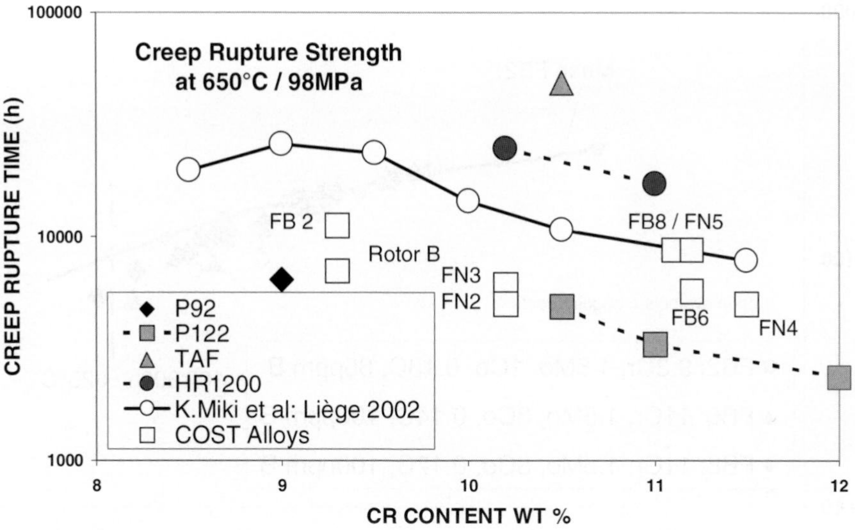

Fig. 6 Effect of Cr content on creep strength.

tests, evaluation and qualification of filler metals for the new cast steels and the investigation of production welds on the components. The compositions FB5 and FB6 have also been produced as 1t industrial heats. Two large melts have been cast by two foundries in alloy CB6.[14] Investigations have also been started on these industrial melts.

As shown in Figure 7 for some forged compositions and in Figure 8 for the cast compositions, there is no candidate alloy with a clear advantage over FB2 and CB2 respectively. In particular there is no increased creep strength at 650°C. However these alloys may offer the potential for a higher oxidation resistance due to their increased Cr content. A qualitative summary of creep results is given in Table 5.

BOILER MATERIALS

The following components are considered critical in the boiler for increased steam parameters:

- Waterwall,
- Final superheater and reheater tubing and
- Thick walled components such as HP headers and main steam pipes.

Materials for the fabrication of these components should mainly offer an appropriate creep strength. However there are additional restrictions such as weldability without post weld heat treatment (PWHT) for waterwalls to reduce the hardness in the as welded condition and avoid the risk of hydrogen induced stress corrosion cracking. The resistance to high temperature corrosion and steam oxidation must be sufficiently high to avoid thick insulating

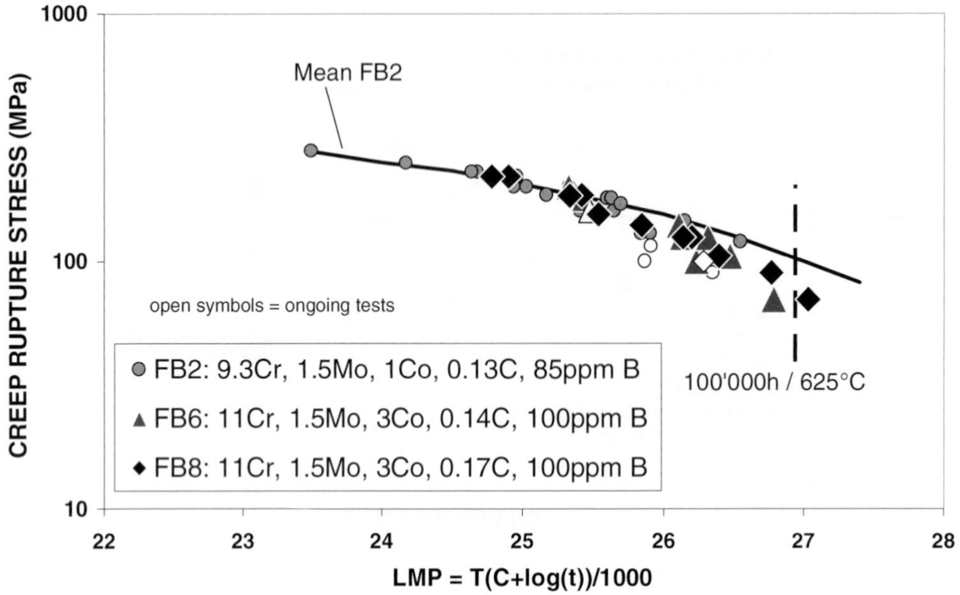

Fig. 7 Creep strength of selected CrMoCoB forged trial melts vs. FB2 results.

Table 5 Creep rupture strength of forged and cast trial melts.

Forged Steels Compared to FB2 Cast Steels Compared to CB2	Duration max. (h)	600° C	625° C	650° C
FB5: 10.1Cr, 1.5Mo, 3Co, 0.13C, 100ppm B	21,000	+/--	+	--
FB6: 11.2Cr, 1.5Mo, 3Co, 0.14C, 100ppm B	16,000	+/--	--	--
FB7: 10.2Cr, 1.5Mo,3Co, 0.16C, 90ppm B	15,000	+/--	--	--
FB8: 11.1Cr, 1.5Mo, 3Co, 0.17C, 100ppm B	19,000	+	+	+/--
FB9: 11Cr, 1.5Mo, 6Co, 0.18C, 90ppm B	21,000	--	+/--	+/--
FB10: 10.3Cr, 0.9Mo, 0.4W, 3Co, 0.12C, 70ppm B	13,000	+	+	--
FB11: 11.2Cr, 0.9Mo, 0.4W, 3Co, 0.19C, 70ppm B	13,000	+/--	+	--
CB5: 10.1Cr,1.5Mo, 3Co, 0.13C, 110ppm B	21,000	+	+/--	+/--
CB6: 10.9Cr, 1.5Mo, 3Co, 0.13C, 110ppm B	24,000	+/--	--	--
CB8: 10.9Cr, 1.5Mo, 3Co, 0.17C, 110ppm B	24,000	+/--	+/--	+/--
CB9: 10.9Cr, 1.5Mo, 6Co, 0.17C, 110ppm B	21,000	+/--	--	--
CB10: 10.2Cr, 0.9Mo, 0.4W, 3Co, 0.13C, 80ppm B	12,000	--	--	--
CB11: 10.7Cr, 0.9Mo, 0.4W, 3Co, 0.17C, 120ppm B	16,000	--	--	--

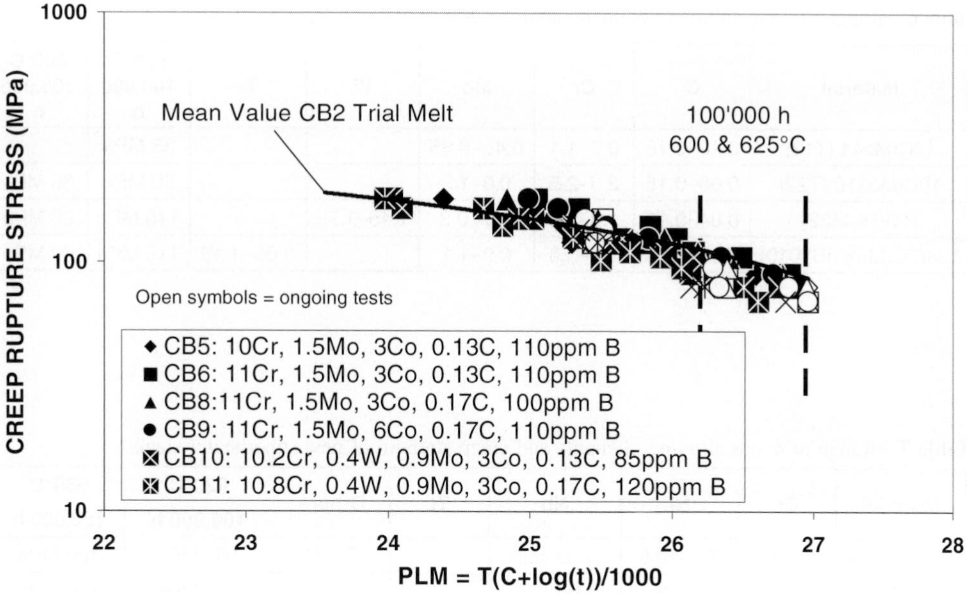

Fig. 8 Creep strength of cast CrMoCoB trial melts vs. CB2 results.

layers on the surface and subsequently increased creep load to the component. A recent overview on materials requirements for advanced boilers is given by Chen and Scheffknecht.[15]

WATERWALL

For waterwall application the newly developed 2-2.5Cr steels offer a significant increase in temperature as compared to low alloyed steels such as 13CrMo44 (T12) or 10CrMo910 (T22) used in the past. T23(HCM2S) and T24(7CrMoVTiB1010) are typical alloys in this group. The composition ranges and typical creep properties are given in Table 6. These steels are suitable for superheater parameters up to 300 bar/640°C. With respect to manufacturing and erection aspects, it has been demonstrated, that PWHT is not mandatory for these new steels. No development for new waterwall material was made within COST522 however various investigations of similar welds in T23 and T24 and dissimilar welds with 9%Cr steels were performed.

For even higher steam parameters T23/T24 have to be replaced by 9–12%Cr steels such as P91, E911, P92 or HCM12 with superior creep strength. Within earlier COST activities it has been shown that P91 may be used for the fabrication of waterwall panels. However these panels needed a PWHT. The high hardness in the as welded condition is common for all materials of this group i.e. P91, E911. P92 and HCM12 they all need a PWHT thus fabrication

Table 6 Range of major alloying elements and creep strength of waterwall steels.

Material	C	Cr	Mo	W	Ti	570°C 100,000 h	600°C 100,000 h
13CrMo44 (T12)	0.10–0.18	0.7–1.1	0.45–0.65			35 MPa	
10CrMo910 (T22)	0.08–0.15	2.0–2.5	0.9–1.2			50 MPa	35 MPa
T23(HCM2S)	0.04–0.10	1.9–2.6	Max.0.3	1.45–1.75		115 MPa	80 MPa
T24(7CrMoVTiB1010)	0.05–0.10	2.2–2.6	0.9–1.1		0.05–1.10	115 MPa	60 MPa

Table 7 Range of major alloying elements and creep strength of new superheater steels.

Material	Cr	Ni	Nb	Ti	Others	620°C 100,000 h	650°C 100,000 h
Super304H	17–19	7.5–10.5	0.30–060		Cu, N	160 MPa	120 MPa
TP347HFG	17–20	9–13	Max.1.0			134 MPa	100 MPa
NF709	18–22	22–28	0.1–0.4	0.02–0.2	N, B	165 MPa	125 MPa
SAVE25	21–24	15–22	0.3–0.6		W, Cu, N	180 MPa	140 MPa
HR3C	24–26	17–23	0.2–0.6		N	160 MPa	115 MPa

and erection of waterwalls is restricted. In addition all but HCM12 are basically 9%Cr bearing and therefore no significant improvement over T23 and T24in steam oxidation or corrosion may be expected.

Superheater and Reheater Tubing

Ferritic materials are no longer suitable for the steam parameters aimed at in COST522 due to their creep strength and corrosion resistance. Higher alloyed austenitic steels have to be used instead. Austenitic steels with increased creep strength have been developed in the past years, some of them having an increased oxidation resistance due to their fine grain structure.

The ranges of chemical compositions and some creep properties of selected modern austenitic steels are given in Table 7.

On the basis of the well known Esshete1250 further developments of austenitic steels were performed within the COST522 working group. Trial melts of these experimental alloys as given in Table 8 were produced and tested. Short term creep rupture tests at 650 and 700°C (Figure 9) show for some alloys, rupture strength to be close to NF709.

Steam oxidation tests at 650°C with a duration of 1000 hours exhibit a clearly reduced weight gain for the new experimental alloys as compared to Esshete1250 (Figure 10).

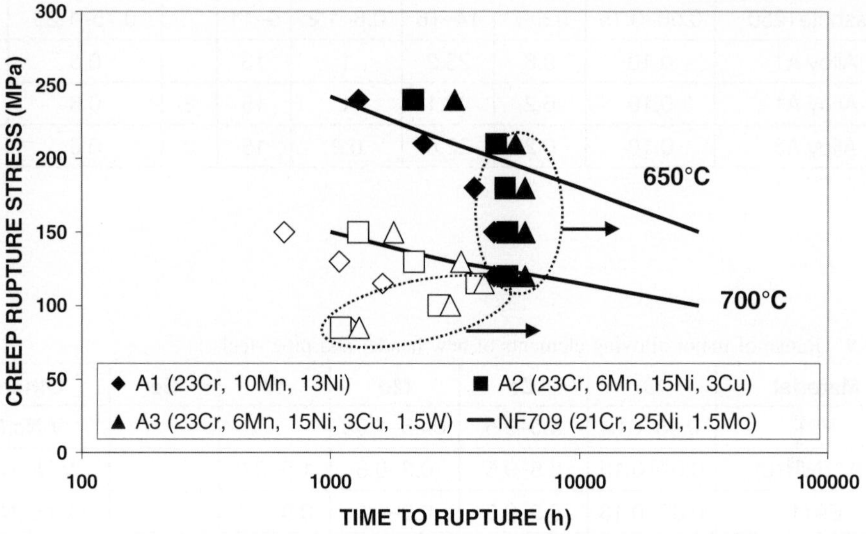

Fig. 9 Creep strength of austenitic alloys developed in COST.

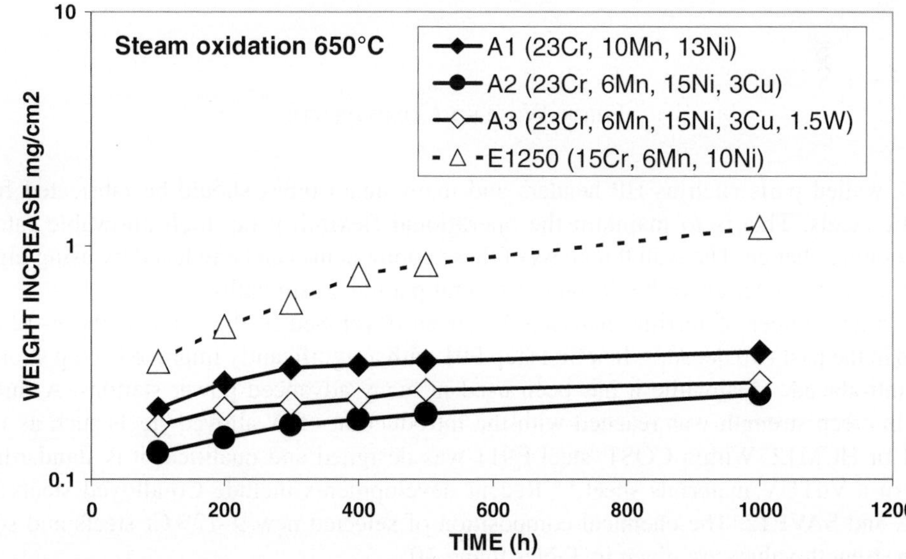

Fig. 10 Steam oxidation at 650°C of austenitic alloys developed in COST vs. Esshete 1250.

Table 8 Major alloying elements of new experimental austenitic steels / COST522.

Material COST	C	Mn	Cr	Mo	Ni	Cu	Nb	W
Esshete1250	0.06–0.15	5.5–7	14–16	0.8–1.2	9–11		0.75–1.25	
Alloy A1	0.10	9.8	23.2	1	13		0.6	
Alloy A1	0.10	6.2	23.1	1	15	3	0.6	
Alloy A3	0.10	6.2	22.8	0.2	15	3	0.6	1.4

Table 9 Range of major alloying elements of new header and pipe steels.

Material	C	Cr	Mo	W	Co	Others
P91	0.08–012	8.0–9.5	0.85–1.05			V,Nb,N
P92 (Nf616)	0.07–0.13	8.5–9.5	0.3–0.6	1.5–2.0		V,Nb,N,B
E911	0.09–0.13	8.5–9.5	0.9–1.1	0.9–1.1		V,Nb,N,B
P122 (HCM 12A)	0.07–0.14	10.0–12.5	0.25–0.60	1.5–2.5		V,Nb,N,B,Cu
NF 12	0.08–0.11	10.6–11.1	0.1–0.2	2.5–2.7	2.4–2.8	V,Nb,N,B
SAVE 12	0.08	10	0.15	3.0	2.5	V,Nb,N,B,Cu

Thick Walled Components

Thick walled parts such as HP headers and main steam pipes should be fabricated from ferritic steels. This is to maintain the operational flexibility i.e. high allowable rate of temperature change. The wall thickness of these components can be reduced by using higher creep strength materials or by designing several paralle steam paths.

A large number of ferritic materials has been developed in the class of the 9–12%Cr steels in the past two decades. In a first step P91 with a significantly improved creep strength was introduced. Meanwhile it has been used in many advanced power stations. A further step in creep strength was reached with the introduction of W-alloyed steels such as P92, E911 or HCM12. Within COST, steel E911 was designed and qualified, it is standardised now in a VdTÜV materials sheet.[4–6] Recent developments include Co-alloyed steels like NF12 and SAVE12. The chemical composition of selected new 9–12%Cr steels and some creep strength values are given in Tables 9 and 10.

In the framework of COST522 a large number of Co-bearing new alloys were designed (Table 11). Experimental heats were produced and tested; promising compositions were also produced as 20 t industrial melts for the fabrication of tubes and tickwalled pipes. The Cr content is increased to 11–11.5% for all melts and the resulting steam oxidation resistance

Table 10 100,000 hours creep rupture strength of new header and pipe steels.

Material	600°C	625°C	650°C	Source
P91	90 MPa	58 MPa	48 MPa	VdTÜV material sheet
P92 (NF 616)	131 MPa	101 MPa	72 MPa	ASME SA213/335
E911	108 MPa	76 MPa	53 MPa	VdTÜV material sheet
P122 (HCM 12A)	127 MPa	100 MPa	70 MPa	ASME CC 2180
NF 12	170 MPa	135 MPa	80 MPa	Nippon Steel

Table 11 New Co bearing tube/pipe steels / COST522.

Material	Cr	W	Mo	Co	Si	Nb
P92	9	1.8	0.5			
Alloy F1	11	0.3	1	2	0.15	0.07
Alloy F2	11	0.3	1	2	0.3	0.07
Alloy F3	11	2	0.5	2	0.3	0.07
Alloy F4	11	0.3	1	2	0.3	0.03
Alloy F5	11.5	1.5	0.3	1.5	0.5	
Alloy F6	11.5	1.3	1.3	1.5	0.3	

has been significantly improved over P92 with 8.5–9.5%Cr only. The measured difference in weight gain is almost two orders of magnitude as shown in Figure 11.

Creep tests at 625–650°C have reached ca. 10,000 hours. The rupture strength is within the scatterband for P92 (Figure 12). Test will be continued with special emphasis on material from large industrial melts and products, respectively.

SUMMARY

- The 9-12% Cr steels have been improved so far in the last 15-20 years that today thermal power plants with steam parameters of 610°C/300 bar and re-heat temperatures of up to 630°C can be built.
- By concentrating the European efforts in the COST 501/ 522 programmes, in which about 70 organisations from 16 countries take part, materials with improved properties could be successfully manufactured and tested.

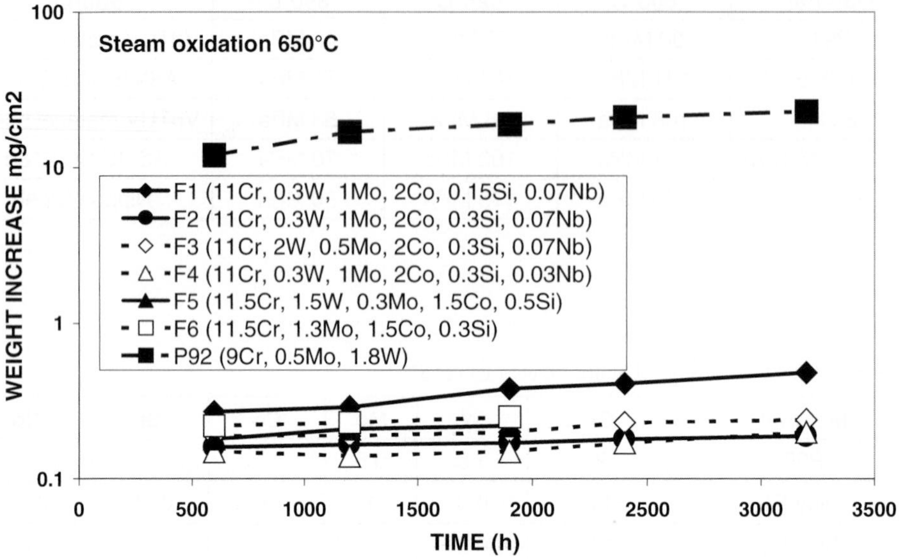

Fig. 13 Steam oxidation at 650°C of ferritic alloys developed in COST vs. P92.

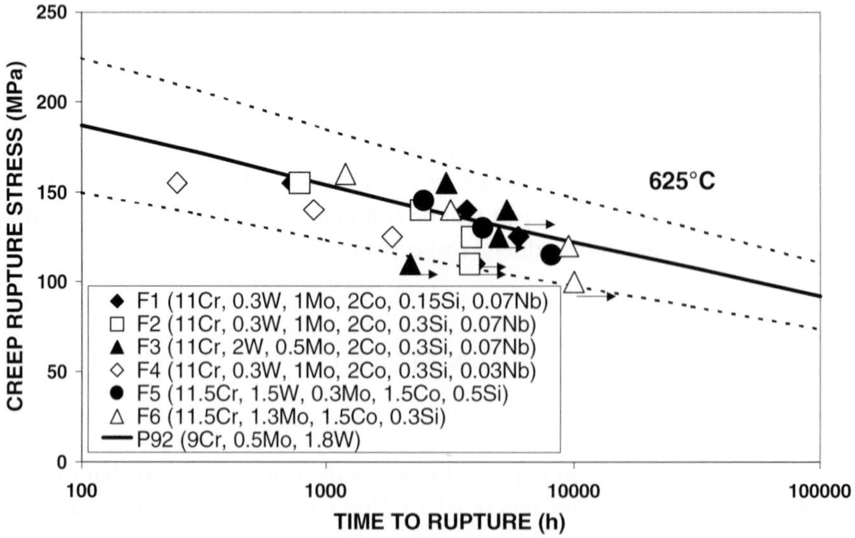

Fig. 14 Creep strength of ferritic pipe steel developed in COST vs. P92 at 650°C.

Table 12 Review on steam parameter limits and available boiler materials.[15]

Component	Materials and Steam Parameter Limits[15]			
	260 bar 550°C 570°C	270 bar 580°C 600°C	290 bar 600°C 620°C	300 bar 630°C 650°C
Waterwall	13CrMo44	T23/T24	T23/T24	T23/T24
Tubes	X20CrMoV12 1	Austenite 9–12%Cr Steels	Austenite Austenite	Austenite Austenite
SH Outlet Header	X20CrMoV12 1	P91	E911, P92 P122	NF12?, Save12? New 12CrCoMo Steels?

- The recently developed materials with improved properties are already applied in several European power plants and will be used in power plants currently under construction.
- New alloy compositions have been identified with the potential to improve the oxidation behaviour at T > 600°C. Further effort is necessary to gain the targeted creep rupture strength of 100 MPa for 100,000 hours at 650°C.
- The new highly alloyed materials are different with respect to the more complex microstructures, the stability and the applicable heat treatment procedures. However also the fabrication of large parts, their weldability and NDT properties have to be considered.
- For critical boiler applications the available materials and the related steam parameter limits are shown in Table 12.[15] With the introduction of the newly developed and qualified steels, plants steam parameters of 300 bar/600°C/620°C can be realised. For even more advanced parameters the materials for thick walled components are most challenging. The oxidation resistance as well as the creep strength of the new 12CrCoMo type materials, have to be confirmed. The development of protective coating systems should be pushed in parallel to get an alternative solution for very high temperatures.

ACKNOWLEDGEMENT

The authors wish to thank their partners in the entire COST 522 collaboration for their contributions and the discussions during the course of the work. Thanks are also extended to the COST management committee and to the national funding bodies for financial support of the individual, national projects.

REFERENCES

1. K. H. Mayer, W. Bendick, R. U. Husemann, T.-U. Kern and R. B. Scarlin: 'New Materials for Improving the Efficiency of Fossil-Fired Thermal Power Stations', *VGB*

Power Technology, 1/98.

2. K. Miyashita: 'Advanced Power Plant', *I. Mech. E. Conference Transaction*, London, 1997, 17–30.
3. T.-U. Kern, R. B. Scarlin, R. W. Vanstone and K. H. Mayer: 'High Temperature Forged Components for Advanced Steam Power Plants', *6th COST Conference*, Liège, Belgium, 1998.
4. M. Staubli, W. Bendick, J. Orr, F. Deshayes and C. Henry: 'European Collaborative Evaluation of Advanced Boiler Materials', *6th COST Conference*, Liège, Belgium, 1998.
5. H. Cerjak, E. Letofsky and M. Staubli: 'The Role of Welding for Components Made from Advanced 9-12%Cr Steels', *6th COST Conference*, Liège, Belgium, 1998.
6. W. Bendick, K. Haarmann and M. Zschau: 'E911- Ein neuer Werkstoff für Dampfleitungen im Kraftwerksbau', *VGB Conference, Materials and Welding Technology in Power Plants 1998*, Hannover, 1998.
7. M. Staubli, K. H. Mayer, T.-U. Kern and R. Vanstone: 'COST 501/COST 522-The European Collaboration in Advanced Steam Turbine Materials for Ultra Efficient', *Low Emission Steam Power Plant, 5th International Charles Parsons Conference*, Cambridge, UK., 2000.
8. T.-U. Kern, M. Staubli, K. H. Mayer, K. Escher and G. Zeiler: 'The European Effort in Development of New High Temperature Rotor Materials up to 650°C – COST522', *7th Liège COST Conference*, Liège, Belgium, 2002.
9. M. Staubli, K. H. Mayer, W. Giselbrecht, J. Stief, A. DiGianfrancesco and T.-U. Kern: 'Development of Creep Resistant Cast Steels within COST522', *7th Liège COST Conference*, Liège, Belgium, 2002.
10. M. Staubli, K. H. Mayer, T.-U. Kern, R. Vanstone, R. Hanuns, J. Stief and K.-H. Schönfelf: 'COST522-Power Generation into the 21st Century; Advanced Steam Power Plant', *9th International Conference on Creep & Fracture of Engineering Materials & Structures*, Swansea, 2001.
11. R. Hanus and K. H. Schönfeld: 'Transformation of Knowledge and Technology from R & D to the Commercial Production of Heavy Steel Castings and Forgings for Power Engineering Made of Advanced Creep Resistant Steels', *Proceedings of 5th International Parsons Conference*, Cambridge, UK., 2000.
12. K. H. Mayer, R. Blum, P. Hillenbrand, T.-U. Kern and M. Staubli: 'Development Steps of New Steels for Advanced Steam Power Plants', *7th Liège COST Conference*, Liège, Belgium, 2002.
13. R. W. Vanstone, H. Cerjak, V. Foldyna, J. Hald and K. Spiradek: 'Microstructural Development in Advanced 9–12%Cr Creep Resistant Steels - A Collaborative Investigation in COST 501/3 WP11', *5th Liège COST Conference*, Liège, Belgium, 1994.
14. A. Contessi, D. Delvecchio, A. Ghidini, S. Valenti, A. Carosi, A. DiGianfrancesco and F. M. Ielpo: 'The First Industrial Cast of CrMoCoB Advanced Steel for Cast Turbine Components', *7th COST Conference*, Liège, Belgium, 2002.
15. Q. Chen and G. Scheffknecht: 'New Boiler and Piping Materials, Design Considerations for Advanced Cycle Conditions', *VGB Conference*, Köln, Germany, 2003.

Continuing Evaluation of a New 11%Cr Steel for Steam Chests

R. W. VANSTONE
Alstom Power, Rugby
UK

S. OSGERBY
NPL Materials Centre, Teddington
UK

P. MULVIHILL
Powergen Technology Centre
Ratcliffe-on-Soar, UK

P. BATES
Sheffield Forgemasters Engineering Ltd
Sheffield, UK

ABSTRACT

The creep, LCF and oxidation characteristics of alloy 122, which had been manufactured as pilot steam chests by casting and forging routes, have been assessed. Creep and oxidation data show that alloy 122 is superior to modified 9Cr1Mo materials that are used existing turbines. Room temperature tensile and impact properties show little change after prolonged ageing. These results demonstrate the continued confidence in this material for potential use for steam chests in the next generation of turbines.

INTRODUCTION

Of all the components in a steam turbine, the steam chest is one of the most demanding in terms of materials property requirements. The steam chest is exposed to the highest operating temperatures and pressures with no practical scope for cooling. Current state-of-the-art turbines operate at temperatures of around 600°C which represents the practical upper limit for 9%Cr steels. However, the next generation of turbines will require materials with adequate creep and steam oxidation resistance at higher temperatures. A 11%Cr alloy, designated alloy 122, is a candidate material for this application and has already been demonstrated to have superior properties to 9%Cr materials in tubing and pipework.

Table 1 Room temperature mechanical properties after simulated PWHT.

Material	Simulated PWHT	$R_p0.2$ MPa	R_m MPa	Elongation %	R A %	Cv J
Forging	10 hours at 760°C	511	697	24	51	20/33/38
Casting (80 mm section)	16 hours at 760°C	523	712	23	54	17/19/22
Casting (120 mm section)	16 hours at 760°C	506	686	22	51	16/17/24

Table 2 Room temperature mechanical properties after thermal exposure.

Material	Thermal Exposure		$R_p0.2$ MPa	R_m MPa	Elongation %	R A %	Cv J
	Time h	Temperature °C					
Forging	3,000	650	519	691	9.4	49.9	26/24/22
Casting	3,000	600	548	699	15.0	20.2	21/20/19
		650	510	674	21.0	46.2	18/18/20
	10,000	600	527	687	13.5	23.6	14/16/20
		650	531	684	18.2	36.2	19/14/19

Full size castings and forgings have been fabricated successfully in alloy 122 and initial integrity and microstructural evaluation showed that the components were sound and homogeneous.[1–2]

MECHANICAL PROPERTIES

Tensile and Impact

Room temperature tensile and impact properties of the two materials in the conditions corresponding to the quality heat treatment and a simulated post-weld heat treatment are given in Table 1.

Material has been exposed for 3,000 and 10,000 hours at 600 and 650°C. Tensile and impact data after these exposures is given in Table 2. Generally, the data are similar for impact and tensile properties pre and post-exposure, confirming the long-term stability of the material.

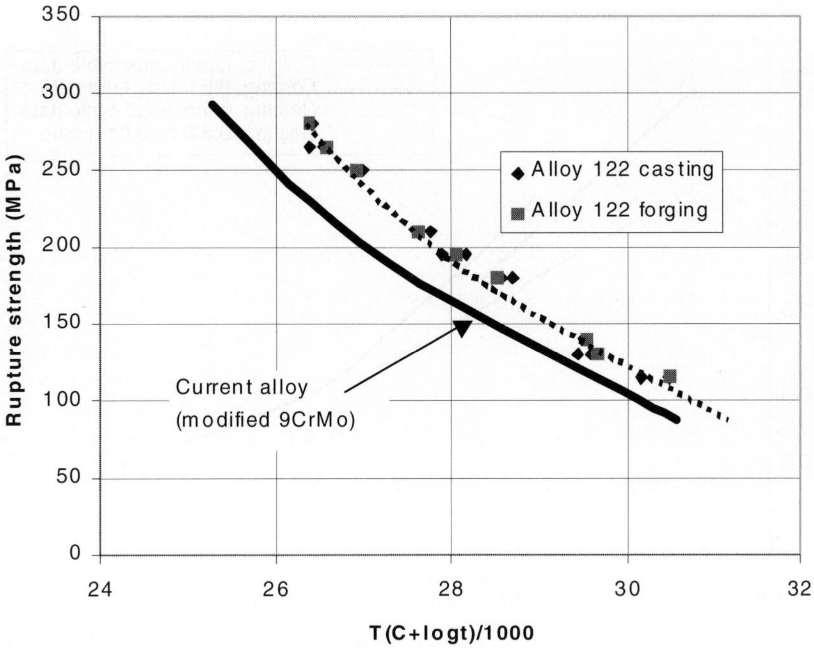

Fig. 1 Summarised creep properties of cast and forged material.

Creep

Creep tests have been carried out in the temperature range 550–650°C for durations up to 10,000 hours. Rupture data are presented as a Larson-Miller plot in Figure 1. The properties of the cast and forged material are similar and show a temperature advantage of ~20°C over modified 9Cr1Mo material - the alloy currently applied to high temperature steam chests. Testing to extend the dataset to 30,000 hours durations is in progress.

Some creep tests were performed using notched test pieces to determine the sensitivity of the material to stress concentrations. For rupture times up to 10,000 hours no notched test pieces failed prior to the rupture duration of the unnotched test under equivalent stress and temperature conditions. These results indicate that the material is not overlay susceptible to stress concentrations.

LCF

Cast and forged materials were tested at 600°C under continuous cycling and with cycles that included a 0.5 hour hold period at maximum tensile strain. The forged material shows

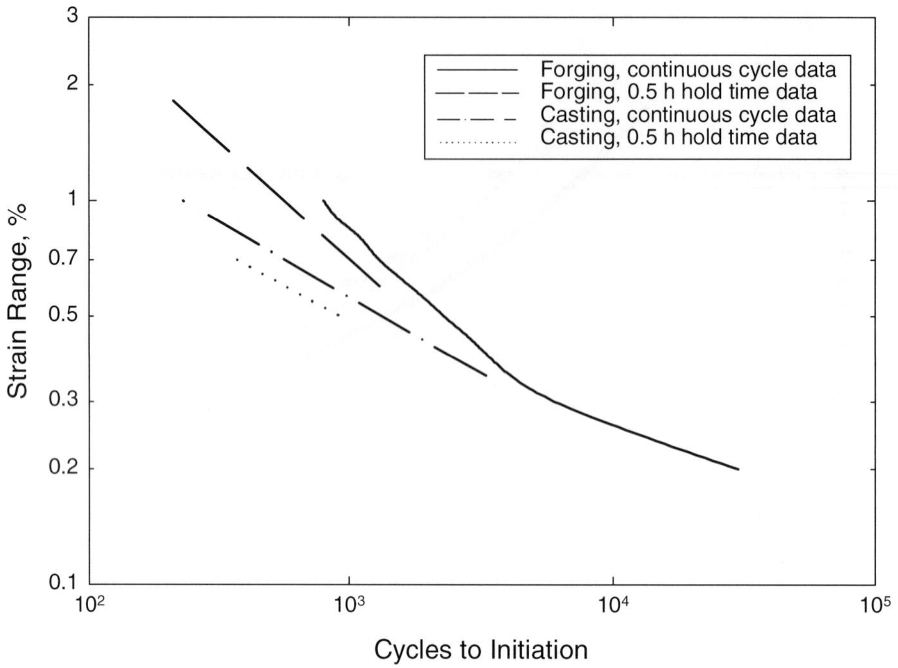

Fig. 2 Low cycle fatigue properties.

superior LCF resistance to the cast material at high strain ranges but at lower strain ranges their behaviour is similar. The introduction of tensile hold periods produces a small decrease in cycles to crack initiation in both materials (Figure 2). The decrease in cycles to crack initiation caused by the introduction of hold periods is similar for both production routes.

OXIDATION

Specimens of cast and wrought alloy 122 were exposed, together with P92 samples for comparison, to flowing steam at 600 and 650°C for durations up to 5,000 hours. The scale growth kinetics were determined by mass change and by direct measurement of scale thickness on cross-sections.

The oxidation behaviour of the materials is shown in Figure 3. At both temperatures there was no significant difference between the behaviour of cast and forged alloy 122 materials, both being superior to P92.

The expected trend for parabolic growth kinetics is shown on both graphs in Figure 3. It is apparent that at long times the P92 material deviates from parabolic growth behaviour. This is due to spallation of the oxide, which has been observed in specimens of P92 that were withdrawn from the furnace after 5,000 hours exposure. The temperature dependency

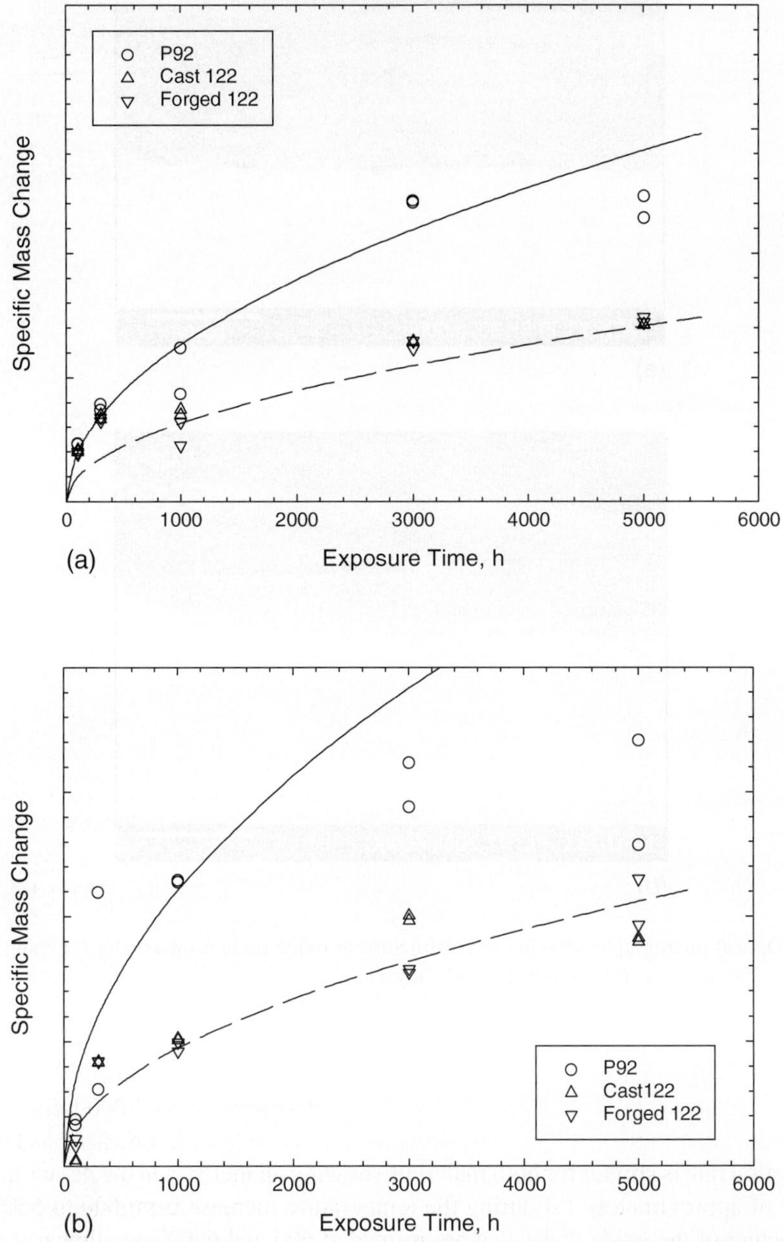

Fig. 3 Specific mass change of Alloy 122 and P92 in flowing steam as a function of time at (a) 600°C and (b) 650°C.

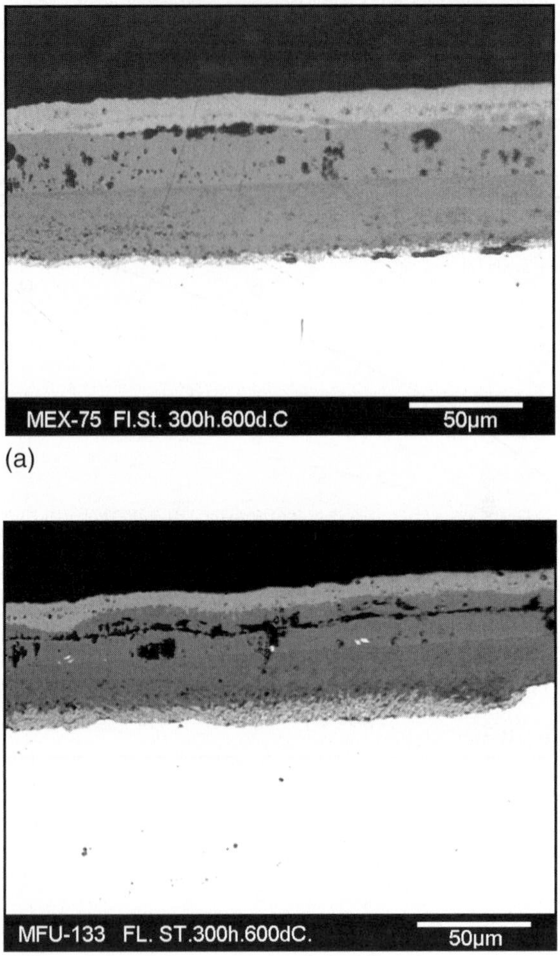

(a)

(b)

Fig. 4 Optical micrographs showing microstructure of oxide scale grown under flowing steam at 600°C for 300 hours (a) P92 and (b) Forged alloy 122.

of the oxidation rate is similar for both materials showing an increase in the parabolic constant by a factor of approximately 1.3 during the temperature increase from 600 to 650°C.

The structure of the oxide scales that are formed at 600 and 650°C are shown in Figures 4 and 5 respectively.

At 600°C the oxidation product that is formed on both materials consists of 4 layers. From the outside these are haematite, magnetite, and iron-chromium spinel scales and an internal chromium rich precipitate respectively. The haematite and spinel scales contain little porosity compared to the magnetite layer.

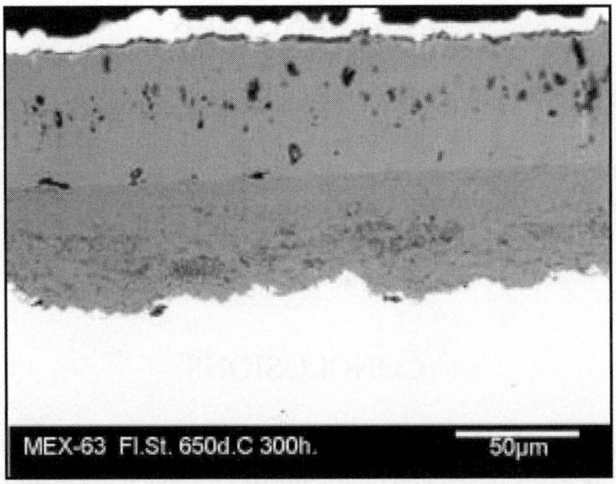

(a)

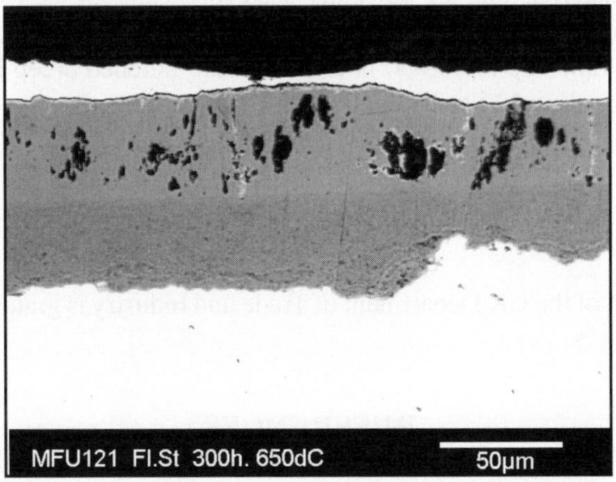

(b)

Fig. 5 Optical micrographs showing microstructure of oxide scale grown under flowing steam at 650°C for 300 hours (a) P92 and (b) Forged alloy 122.

At 650°C the scale structure is somewhat different. The haematite layer is very thin and the internal precipitates are no longer present. There is some indication that the pores in the magnetite are interlinked and open to the surface as some of these are decorated with haematite.

The pores in the magnetite layer that forms on alloy 122 are significantly larger than those that in the equivalent layer that forms on P92. This may indicate a potential susceptibility to scale spallation, although no evidence of this has been observed.

FURTHER WORK

The test programme is continuing and includes:

- Creep tests to 30,000 hours duration,
- Oxidation in steam environments to 10,000 hours duration,
- Creep crack growth on as-received and thermally aged material and
- Creep performance of welded joints.

CONCLUSIONS

Test specimens of alloy 122 taken from pilot steam chests that had been manufactured by forging and casting processes have been characterised for mechanical and oxidation behaviour. The creep and oxidation data indicate that alloy 122 performs significantly better than the modified 9Cr1Mo materials that are currently used for steam chests. No significant change in room temperature tensile and impact properties was observed after thermal exposure at 600 and 650 °C for times up to 10,000 hours. LCF and notched creep behaviour shows no cause for concern for alloy 122 to be used in this application.

ACKNOWLEDGEMENTS

The financial support of the UK Department of Trade and Industry is gratefully acknowledged.

REFERENCES

1. P. Bates, I. Nicholls, D. V. Thornton and R. W. Vanstone: 'Forged and Cast Steam Chests in New 11%Cr Steel', *Parsons 2000: Advanced Materials for 21st Century Turbines and Power Plant*, A. Strang, W. M. Banks, R. D. Conroy, G. M. McColvin, J. C. Neal and S. Simpson, eds., IOM Communications, 2000, 157–170.
2. P. Bates, R. W. Vanstone, S. Osgerby and P. Mulvihill: *Materials for Advanced Power Generation 2002*, J. Lecompte-Beckers, M. Carton, F. Schubert and P. J. Ennis, eds., Forschungszentrum Julich GmbH, 2002, 1261–1268.

Factors Influencing the Creep Resistance of Martensitic Alloys for Advanced Power Plant Applications

P. D. Clarke and P. F. Morris
Corus R, D and T
Rotherham
England

N. Cardinal and M. J. Worrall
Corus Engineering Steels
Sheffield
England

ABSTRACT

Over the last 50 years the rising steam temperatures and pressures needed to improve generating efficiency in fossil fuelled stations have placed increasing demands on the creep and oxidation resistance of boiler and turbine materials. These have been met by the development of martensitic steels with higher chromium levels together with additions such as molybdenum and tungsten resulting in alloys such as Type 91 (9%Cr, 1%Mo) and Type 92 (9%Cr, 2%Mo, 0.5%W). Further proposed increases in operating temperature to 650°C and in the longer term to 700°C mean that alloys with superior performance to these are required.

Greater resistance to oxidation can most easily be obtained by further increases in the chromium content. At levels above about 10% however it becomes ever more difficult to obtain a fully martensitic microstructure due to the formation of delta ferrite. This results in loss of strength and reduces the hot ductility which can give problems during hot working operations such as tube making. It can also influence weldability. The higher chromium levels must thus be balanced by austenite stabilising additions such as copper, nickel and cobalt.

As part of the COST 522 programme, work has been carried out to attempt to improve the creep properties of martensitic alloys. The paper reviews briefly the design of creep resistant 9–12%Cr steels and presents initial results on the mechanical properties and creep rupture properties of a martensitic steel composition designed to have creep properties superior to those of Steel 92. For short term creep data, currently for durations up to 8000 hours, rupture lives in excess of Steel 92 are being achieved for testing at 600–650°C. Testing is continuing and data for durations out to 10,000 hours should be available by the time of the conference.

INTRODUCTION AND BACKGROUND

The move towards higher efficiency power plant has required the development of materials suitable for use under more demanding operating conditions. Over recent years this has led to several international work programmes aimed at developing materials capable of operating at higher steam temperatures and pressures for which improvement of creep strength and oxidation resistance are recognised as essential.

One family of steels that has received considerable attention is the 9–12%Cr martensitic types. These offer greater flexibility of operation for thick section boiler components over more creep/oxidation resistant austenitic steels due to their lower coefficients of thermal expansion and better thermal conductivity. They are also cost effective as a result of the lower alloy content, particularly of expensive additions such as nickel.

Recent developments have included grades such as Steel 92 containing 9%Cr which gives a 100 MPa/10^5 hours rupture life at temperatures approaching 620°C, the highest among commercially available ferritic grades. To achieve enhanced steam oxidation resistance chromium contents above 9% are required. A project in the COST 522 collaborative programme has sought to establish a martensitic 11–12%Cr base composition with better creep strength than Steel 92. The aim of this paper is to briefly describe the design of such an alloy and to present interim test results.

DESIGN OF DEVELOPMENT COMPOSITION

The base composition of 9%Cr-1%Mo has since the 1930's been widely used in coal-fired power plant. This produces a fully martensitic microstructure in air cooled thick walled sections after a conventional austenitising and tempering treatment. It also provides higher creep strength and better oxidation resistance than lower chromium steels, such as the 2¼%Cr-1%Mo grade.

Extensive work in the USA during the mid-1980's led to the development of Steel 91. This involved the addition of Nb and V to the 9%Cr-1%Mo base together with increases in nitrogen and boron producing small, stable precipitates which are important for good creep strength.

More recently the Japanese and Europeans have developed Steel 92 (NF616) and E911 respectively. Both these alloys rely on the addition of tungsten (and reduction of Mo in the case of Steel 92) in relation to the Steel 91 composition to improve creep strength. A further Japanese development was Steel 122 (HCM12A). This contains 11%Cr for superior corrosion resistance and a copper addition for the suppression of delta ferrite which can give toughness and hot workability problems. The compositions of these alloys are shown in Table 1.[1–3]

The temperature for a 100 MPa/10^5 hour creep life for a selection of these alloys is shown in Figure 1.[4] Steel 92 demonstrates the best performance in terms of creep strength and for this reason is recognised as the best of the 9–12%Cr grades currently available.

In order to improve the creep and oxidation performance offered by Steel 92 a development programme was initiated through COST 522. An 11%Cr base was chosen for enhanced oxidation resistance. The following factors were then considered:

- Avoid suppression of the α/γ transition temperature to permit the use of the maximum tempering temperature.[5]
- Obtain microstructural stability through use of low levels of residual elements to reduce coarsening rates of precipitates.[6, 7]

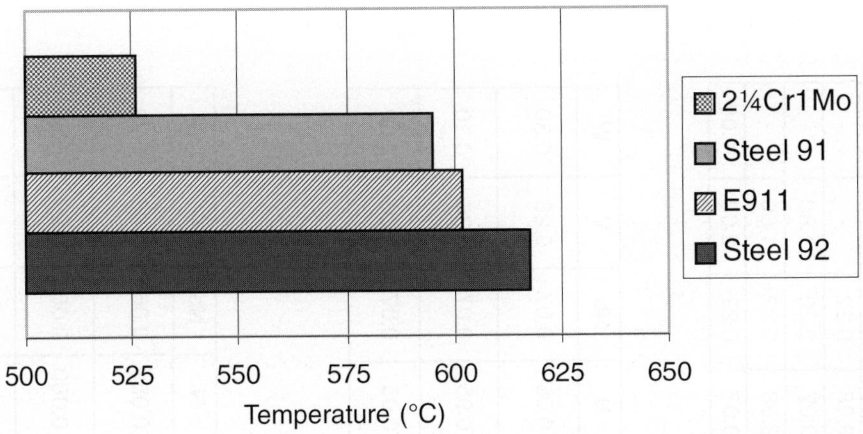

Fig. 1 Maximum temperature in °C for 100,000 hours creep rupture strength of 100 MPa.

- Use Mo and W to provide rupture strength by solid solution strengthening and Laves phase precipitation.
- Use high W to Mo ratio to optimise the long term stability of Laves phase precipitates.[8]

This led to the design of three development alloys. The first alloy contained lower residual levels (Si, Mn, Ni, Al) than found in conventional 9–12%Cr grades. The level of Mo was set at 1%, W at 0.30% and Co at 2.0% the latter to help suppress delta ferrite. The second alloy is of the same base composition but with Si, Mn, Ni and Al levels more typical of those which can be readily achieved in large scale electric arc steelmaking. The third variant contained higher W (2.0%) and lower Mo (0.50%) levels than the other two.

The nominal compositions of the development alloys are shown in Table 2.

PRODUCTION OF MATERIAL AND ASSESSMENT PROGRAMME

Experimental Manufacture

A 50 kg air induction melt was produced for each of the development alloys with the aim compositions given in Table 2. Product analyses are shown in Table 3.

Chemical analyses for the alloys were generally close to the respective aims but the levels of Si, Mn, Ni and Al were slightly above the aim in the low residual material. The levels of Si, Mn and Ni were, nevertheless, much lower than found in normal production material and so were considered suitable to study the effect of low residual contents on high temperature properties.

The ingots were forged into 50 mm square bar and then rolled to 19 mm diameter round bar for the production of test samples. The final product, both before and after heat treatment,

Table 1 Compositions of 9–12%Cr martensitic alloys for use in advanced power plant.

Alloy	C	Si	Mn	Cr	Mo	Ni	Al	B	N	Nb	V	W	Cu
Steel 91	0.10	0.35	0.45	9.0	1.00	0.20	0.020	-	0.050	0.08	0.22	-	-
Steel 92	0.10	0.25	0.45	9.0	0.50	0.20	0.020	0.004	0.050	0.07	0.20	1.80	-
E911	0.11	0.30	0.45	9.0	1.00	0.20	0.010	0.003	0.070	0.08	0.22	1.00	-
Steel 122	0.11	0.25	0.35	11.0	0.40	0.25	0.020	0.003	0.070	0.07	0.22	2.00	1.00

Table 2 Nominal compositions of development alloys (wt.%).

Alloy	C	Si	Mn	Cr	Mo	Ni	Al	B	Co	N	Nb	V	W
Low Residual	0.17	0.10	0.05	11.0	1.00	0.09	0.002	0.01	2.00	0.05	0.07	0.26	0.30
Normal Residual	0.15	0.30	0.20	11.0	1.00	0.40	0.005	0.01	2.00	0.05	0.07	0.26	0.30
High W + Low Mo	0.15	0.30	0.20	11.0	0.50	0.40	0.005	0.01	2.00	0.05	0.07	0.26	2.00

Table 3 Chemical analysis of experimental materials (wt.%).

Alloy	C	Si	Mn	Cr	Mo	Ni	Al	B	Co	N	Nb	V	W
Low Residual	0.16	0.15	0.09	11.0	1.04	0.14	0.008	0.009	2.12	0.059	0.065	0.26	0.32
Normal Residual	0.16	0.30	0.22	11.1	1.04	0.43	0.007	0.011	2.16	0.060	0.067	0.27	0.32
High W + Low Mo	0.17	0.35	0.22	11.2	0.50	0.44	0.007	0.009	2.15	0.063	0.068	0.26	2.10

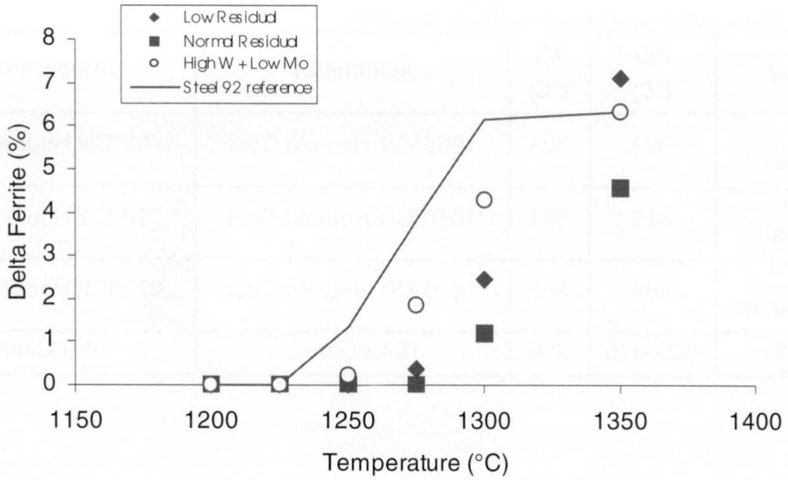

Fig. 2 Delta ferrite as a function of temperature.

was subjected to ultrasonic and alternating current potential drop test techniques to ensure that only sound material was used for the manufacture of test specimens.

The tendency to form delta ferrite on heating was determined metallographically from 40 mm long bar samples held at selected test temperatures between 1200–1350°C for 30 minutes, followed by water quenching. Volume fractions of delta ferrite were then established by point counting. Reference values for Steel 92 were also evaluated using the same technique for an experimental cast made to the composition range shown in Table 1. The area fractions of delta ferrite observed were less than 8% in all cases. Significant levels (>1%) of delta ferrite were observed in the Steel 92 at 1250°C and above, Figure 2. For the low and normal residual materials >1% delta ferrite was observed at 1300°C and above and for the Mo/W alloy at 1275°C and above. Except at 1350°C, which is well above reheating temperatures for hot working, where delta ferrite was observed, the levels in the experimental alloys were below those found in the Steel 92 for any given temperature.

Heat Treatment

A solution treatment temperature of 1060°C was selected, being typical of that used for 9–12%Cr martensitic steels.[3, 9] Phase transformation temperatures determined using dilatometry are shown in Table 4. For determination of α/γ transition temperatures (Ac_1's), samples were heated at a rate of 5°Cmin^{-1}. These were used to allow the maximum tempering temperature to be used without reaustenitisation. The measured Ac_1 temperatures reflected the alloy content ranging from 823°C for the low residual material to 809°C for the Mo/W alloy. Tempering temperatures about 30°C below these were used in subsequent heat treatments.

Table 4 Phase transformation data and heat treatments.

Alloy	Ac_1 (°C)	M_s (°C)	Austenitise	Temperature
Low Residual	823	324	1060°C/1 Hour/Air Cool	790°C/2 Hour/Air Cool
Normal Residual	815	301	1060°C/1 Hour/Air Cool	780°C/2 Hour/Air Cool
High W + Low Mo	809	285	1060°C/1 Hour/Air Cool	770°C/2 Hour/Air Cool
Steel 92	830–870	400	1040°C min.	730°C min

Table 5 Room temperature mechanical properties.

Alloy	Position	$Rp_{0.2}$ (MPa)	Rm (MPa)	A (%)	Z (%)	Hardness (HV)	Impact Energy (J)
Low Residual	Long.	528	756	22	67	235	163
Normal Residual	Long.	568	786	24	67	258	144
High W + Low Mo	Long.	620	837	22	63	266	96
Steel 91	-	415 min.	585 min.	20 min.	-	265 max.	-
Steel 92	-	440 min.	620 min.	20 min.	-	265 max.	-
Steel 122	-	400 min.	620 min.	20 min.	-	265 max.	-
E911	-	440 min.	620-850	20 min.	-	-	-

For M_s determinations, samples were heated at 200°C min^{-1} to 1060°C, held for 5 minutes and then cooled at 20°C min^{-1}. These again reflected the alloy content of the steels ranging from 324°C for the low residual alloy to 285°C for the Mo/W alloy. Data for Steel 92 is included in the table for comparison.[9]

Room Temperature Mechanical Properties

Room temperature tensile, hardness and Charpy impact data for samples heat treated according to the cycles in Table 4 are shown in Table 5. As would be expected, the strength increased with alloy content and the Charpy impact toughness decreased. There was no significant difference in ductility between the three development alloys.

Reference data included in the table show that test values for the development alloys exceeded minimum strength and ductility requirements for current 9–12%Cr martensitic grades.[1–3]

Stress Rupture Testing

Stress rupture tests were carried out at 600, 625 and 650°C at stresses designed to give aim lives of 1000, 3000, 10,000 and 30,000 hours based on data for Steel 92.[4] At each temperature, tests with aim lives of 1,000 and 3,000 hours have been completed whilst those with projected lives of 10,000 and 30,000 hours are still in progress. The status of the test programme as of March 2003 is shown in Figure 3.

At 600°C and aim durations up to 3,000 hours, the performance of the development alloys was superior to Steel 92 with the Mo/W alloy giving lives up to 4 times those expected for Steel 92. Two tests on each alloy with aim durations up to 30,000 hours had been running for approximately 3,220 hours by the beginning of March 2003.

At 625°C the low and normal residual alloys gave failure lives below those expected for Steel 92. The Mo/W alloy again gave lives well in excess of those expected apart from a test at 125 MPa with an aim life of 10,000 hours which failed after 7,749 hours. Close examination of the failure showed that it was almost certainly premature as it had occurred at a machining defect in the sample. The result has been excluded from Figure 3.

At 650°C the rupture lives given by the development alloys fell below the mean line for Steel 92 although the 1000 hours aim test for the Mo/W alloy failed after 1418 hours.

The data clearly demonstrate the beneficial effects of the higher tungsten in the Mo/W alloy. There is also evidence that the use of low residuals does produce an improvement in creep rupture life compared with the higher residual material however, the effect is much smaller than that produced by the tungsten. The limited data are not totally consistent. At 650°C where a direct comparison can only be made for the 110MPa tests the low residual material failed after 1,244 hours compared with 2,068 hours for the higher residual material.

Steam Oxidation Testing

Steam oxidation tests were carried out on samples from the development alloys together with a sample of commercial Steel 92 in the same furnace. Samples of dimensions 20 × 10 × 3 mm were prepared by spark erosion and tested at 650°C for times up to 5000 hours. Samples were removed from the furnace and the weight change measured at several points during the test.

Results of the steam oxidation tests are shown in Figure 4. The development alloys with chromium levels of 11% all showed a noticeably greater oxidation resistance than commercial P92 with 9%Cr. After around 5000 hours, weight gains for the development alloys were 0.17 mg cm^{-2} (normal residual), 0.20 mg cm^{-2} (Mo/W) and 0.62 mg cm^{-2} (low residual) which compared to 25.8 mg cm^{-2} for P92.

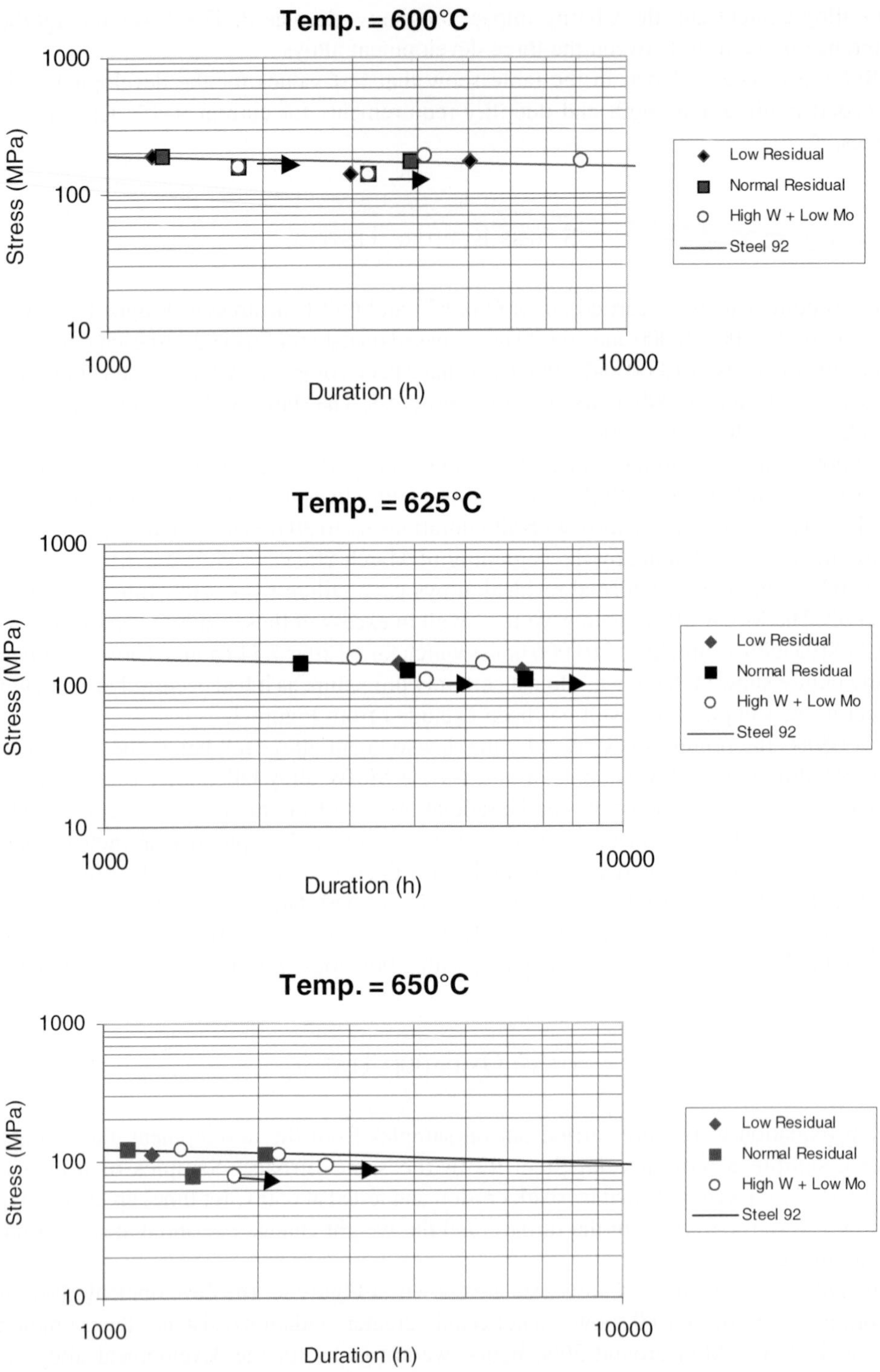

Fig. 3 Creep rupture test values compared with data for Steel 92.

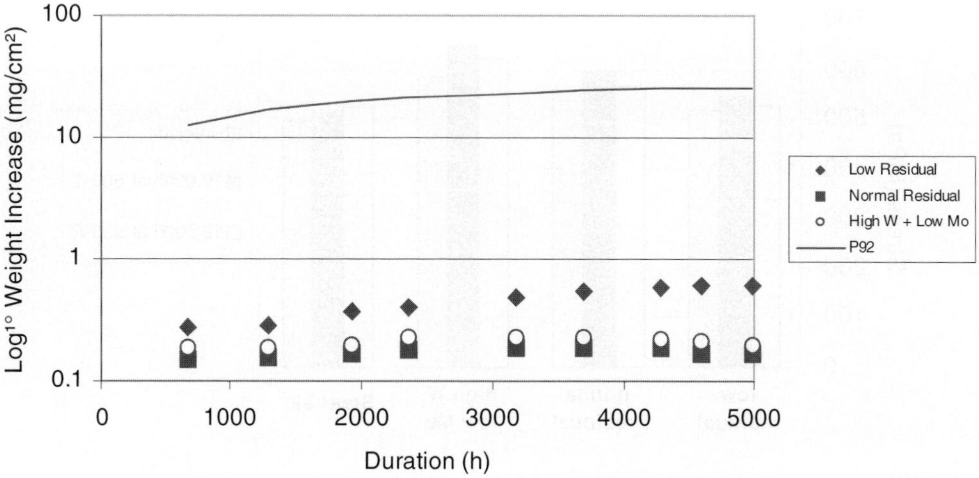

Fig. 4 Steam oxidation test data at 650°C for development alloys and P92.

Long Term Exposure Testing

Tensile and Charpy impact tests were carried out on materials aged for up to 10000 hours at 600 and 650°C covering the likely service range for these alloys.

$Rp_{0.2}$ and Charpy impact data at room temperature are shown in Figures 5a and b together with comparative data for Steel 92.[10] For the three development alloys there was a small increase in the $Rp_{0.2}$ value after aging for 10000 hours at 600°C. After aging for the same time at 650°C there was a small drop of the $Rp_{0.2}$ value compared with the unaged condition. Data for Steel 92 showed little change in $Rp_{0.2}$ on aging at 600°C, and a small drop after aging at 650°C. In all cases the changes were small, at 35 MPa or below.

All three development alloys and the Steel 92 showed a rapid fall in room temperature Charpy impact energy on aging. The fall was greater after aging at 650°C than at 600°C. After 10000 hours at 650°C the absorbed energies of the Steel 92 and the development alloys were similar, in the range 32 to 35J. The absorbed energy of the low residual material was slightly higher, at 46J.

DISCUSSION

The purpose of the work was to develop a martensitic alloy containing 11–12%Cr with improved oxidation resistance and a creep rupture strength equal to or better than Steel 92 at temperatures up to 625°C. A base alloy with 11%Cr, 1.0Mo, 0.3%W, 2.0%Co was studied. Two approaches were taken to improve rupture strength, the use of low residual levels and

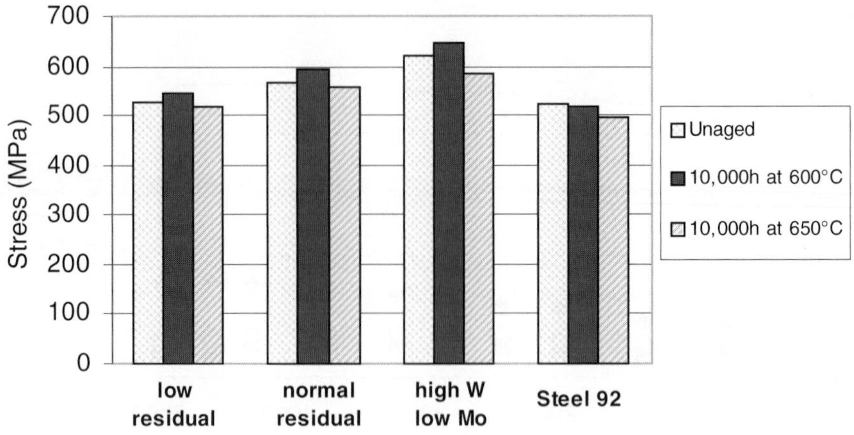

(a)

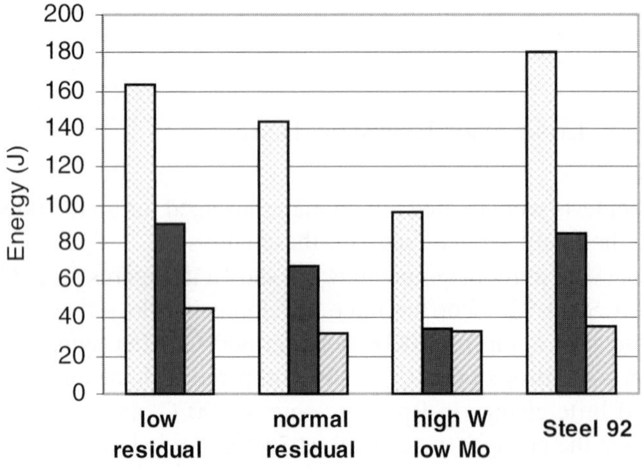

(b)

Fig. 5 (a-b) $Rp_{0.2}$ and Charpy impact toughness at room temperature.

the use of 2.0%W with a corresponding reduction in Mo to 0.5%. The latter produced a significant improvement in rupture strength.

Tungsten and molybdenum improve rupture strength both by solid solution strengthening and by the precipitation of Laves phase, $Fe_2(Mo,W)$. The long term stability at temperature of Laves phase is enhanced at higher concentrations of tungsten.[8] Rupture values available

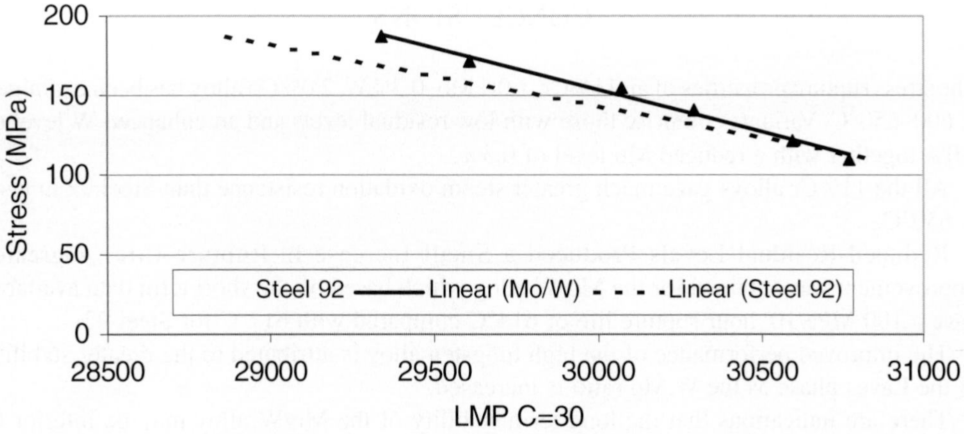

Fig. 6 Comparison of rupture data for the Mo/W alloy with Steel 92.

to date are short term, the maximum rupture life for the Mo/W alloy being 8000 hours, although longer term tests are in progress. The rupture data for the Mo/W alloy, and Steel 92 data taken from the ECCC data sheet,[4] have been fitted using a Larsen-Miller Parameter (LMP), Figure 6. Both data sets fitted well for a C value of 30.

At low LMP values the Mo/W alloy gave better performance than P92, however the rupture stress decreased slightly faster as the LMP values increased. The two crossed at an LMP value of about 30800. Based on these data the 100 MPa, 10^5 hours rupture life occurs at 618 and 614°C for the Steel 92 and Mo/W alloy respectively. (N.B. The value obtained using the more complex relationship in the ECCC datasheet for P92 is 617°C.[4]) Hence the performance of the Mo/W alloy is similar to that of Steel 92 for temperatures up to about 615°C. The steeper gradient of the Mo/W line gives reason to be cautious about the long term rupture properties. The data in the present paper and from other sources indicate that the use of cobalt in conjunction with molybdenum and tungsten leads to reduced long term microstructural stability. This is almost certainly due to the influence of cobalt in reducing the solubility of molybdenum and tungsten, thus accelerating precipitation kinetics of Laves phase.[11] The long term stability of these 12%Cr steels is also limited by the formation of Z-phase, (Nb,Cr)N.[12]

Thus, based on the short term data in the current work the Mo/W alloy produces similar rupture performance to Steel 92 at temperatures up to about 615°C for lives up to about 10^5 hours. However the higher chromium content in the Mo/W alloy gives a significant improvement in steam oxidation resistance, Figure 4.

A small improvement in creep rupture performance was obtained by the use of very low residual levels, but the effect was small compared with that of tungsten. The use of lower residual levels, particularly manganese and nickel, increased the tendency to form delta ferrite on heating which could potentially present problems in hot working operations such as tube making.

CONCLUSIONS

The stress rupture properties of an 11%Cr, 1.0%Mo, 0.3%W, 2.0%Co alloy has been examined at 600–650°C. Variants comprise those with low residual levels and an enhanced W level of 2.0% together with a reduced Mo level of 0.5%.

All the 11%Cr alloys gave much greater steam oxidation resistance than Steel 92 in tests at 650°C.

Reduced Residual Levels Produced a Small Increase in Rupture Life. A greater improvement was obtained for the Mo/W alloy which based on the short term data available gave a 100 MPa/10^5 hour rupture life of 614°C compared with 617°C for Steel 92.

The improved performance of the high tungsten alloy is attributed to the greater stability of the Laves phase as the W:Mo ratio is increased.

There are indications that the long term stability of the Mo/W alloy may be inferior to Steel 92 and this is probably due to the presence of cobalt which reduces the solubility of molybdenum and tungsten, thereby accelerating the precipitation kinetics of Laves phase.

ACKNOWLEDGEMENTS

The authors would like to thank the Management Committee of the COST 522 action for their support and national funding bodies for their financial contributions. Assistance in designing the base alloys was received from Technical University of Denmark and Alstom Power, Baden, Switzerland. Steam oxidation testing was carried out by Alstom Power, Mannheim, Germany. Creep rupture testing was carried out by CESI and CSM, Italy, and Siempelkamp, Germany.

REFERENCES

1. ASTM A213/A213M–99A, 'Standard Specification for Seamless Ferritic and Austenitic Alloy-Steel Boiler, Superheater, and Heat-Exchanger Tubes', 1999.
2. ASTM A335/A335M–99, 'Standard Specification for Seamless Ferritic Alloy-Steel Pipe for High-Temperature Service', 1999.
3. Data Package for Grade 911 Ferritic Steel 9%Cr–1%Mo-1%W, Prepared by a Working Group on Behalf of the European COST 501 Programme on Advanced Boiler Materials, 1998.
4. European Creep Collaborative Committee, ECCC Data Sheets, 1999.
5. F. Masuyama: 'New Developments in Steels for Power Generation Boilers', *Pre-Print of the International Conference on Advanced Heat Resistant Steels for Power Generation*, 1998.
6. A. Strang and V. Vodarek: 'Microstructural Stability of Creep Resistant Martensitic 12% Cr Steels', *Proceedings of the International Conference on Microstructural Stability of Creep Resistant Alloys for High Temperature Plant Applications*, A. Strang, ed., The Institute of Materials, 1998, 117–133.

7. V. Foldyna and Z. Kubon: 'Consideration of the Role of Nb, Al and Trace Elements in Creep Resistance and Embrittlement Susceptibility of 9-12%Cr Steel', *Proceedings of the International Conference on Performance of Bolting Materials in High Temperature Plant Applications*, A. Strang, ed., The Institute of Materials, 1995, 175–187.
8. J. Hald and Z. Kubon: 'Thermodynamic Prediction and Experimental Verification of Microstructure of Chromium Steels', *Proceedings of the IX International Symposium Creep Resistant Metallic Materials*, Hradec nad Moravici, 1996.
9. Data Package for NF616 Ferritic Steel (9Cr-0.5Mo-1.8W-Nb-V), Second Edition, Published by Nippon Steel Corporation, 1994.
10. J. Hald: 'Materials Comparisons Between NF616, HCM12A and TB12M – III: Microstructural Stability and Ageing', *Proceedings of the EPRI/National Power Conference on New Steels for Advanced Plant up to 620°C*, E. Metcalfe, ed., EPRI, 1995, 152–173.
11. S. H. Ryu, Jin Yu and B. S. Ku: 'Effects of Alloying Elements on the Creep Rupture Strength of 9-12%Cr Steels', *Proceedings of the Fifth International Charles Parsons Turbine Conference,* A. Strang et al., eds., The Institute of Materials, 2000, 472–484.
12. K. Kimura et al.: 'Microstructural Change and Degradation Behaviour of 9Cr-1Mo-V-Nb Steel in the Long-Term', *Proceedings of the Fifth International Charles Parsons Turbine Conference,* A. Strang et al., eds., The Institute of Materials, 2000, 590–602.

Effect of Boron and Copper on Precipitate Growth and Coarsening in Martensitic Chromium Steels

Ardeshir Golpayegani
Department of Experimental Physics
Chalmers University of Technology
SE-412 96 Göteborg, Sweden

Mats Hättestrand
R & D Centre, Sandvik Materials Technology
SE-811 81 Sandviken, Sweden

Hans-Olof Andrén
Department of Experimental Physics
Chalmers University of Technology
SE-412 96 Göteborg, Sweden

ABSTRACT

The development of the microstructure of steels P92 and P122 during tempering at 750–770°C and ageing or creep at 600 and 650°C for times up to 26,000 hours was investigated using Secondary Ion Mass Spectroscopy, Energy Filtered Transmission Electron Microscopy and Atom Probe Field Ion Microscopy. Boron was found to segregate to austenite grain boundaries during cooling after austenitising, and spread in the microstructure during the first few minutes of tempering, which is on the same time scale as the growth of $M_{23}C_6$. Boron is only incorporated in $M_{23}C_6$ precipitates and the matrix concentration of B is low, less than 4 ppm. Both VN and $M_{23}C_6$ coarsen during creep, and the coarsening rate of $M_{23}C_6$ is accelerated by strain. The presence of B decreases the coarsening rate of $M_{23}C_6$, possibly due to a reduction in interfacial energy. Laves phase grows during ageing or creep over a long period, 10,000 hours. Strain increases the number density of Laves phase precipitates and decreases their size. In copper containing P122, copper precipitates form during tempering and grow during ageing. Their presence also increases the number density of Laves phase precipitates, which form faster to a smaller size.

INTRODUCTION

Martensitic 9–12% chromium steels are used for critical components in steam power plants such as rotors, piping and valve bodies due to their low thermal expansion, a good thermal conductivity and acceptable corrosion properties. A detailed characterisation of the

Table 1 Chemical composition of the investigated steels (wt.%, bal. Fe).

	Cr	Cu	Mn	Ni	Mo	W	V	Nb	C	B	N	Si	P
P122	11.0	0.87	0.56	0.32	0.42	1.94	0.19	0.05	0.11	0.001	0.05	0.02	0.01
P92	8.96	-----	0.46	0.06	0.47	1.84	0.20	0.07	0.11	0.001	0.05	0.04	0.01

microstructure and its development during creep is necessary to understand the changes in creep resistance since creep strength is the critical property for these high temperature materials.

Large efforts have been made during the last two decades to develop new alloys with increased creep strength.[1, 2] The two Japanese grades NF616 and HCM12A, that have been approved by ASME under the designations P92 and P122, respectively, have proved to have creep strength superior to steels commonly in use today and are considered suitable for use at temperatures up to about 620°C. The two steels have been validated and demonstrated within the EPRI RP1403-50 (1990–1995) project.[3] The alloy design of steels P92 and P122 is based on the well-known grade P91 developed at the Oak Ridge National Laboratory, USA. Increased creep strength has been achieved by the addition of tungsten and boron. The main difference between the two steels is that steel P122 contains 0.9 wt.% copper, added to allow for a higher chromium content.

The general microstructure of the two investigated steels[4] consists of tempered lath martensite, with small amounts of primary Nb(C, N) to prevent excessive grain growth during austenitising. Small secondary MX and $M_{23}C_6$ form during tempering at typically 750°C at martensite lath boundaries and inside laths. During creep the secondary precipitates coarsen, and an intermetallic phase, Laves phase, nucleates, grows and subsequently coarsens. It is believed that the most important factor influencing the creep behaviour of 9–12% chromium steels is the behaviour of precipitates present in the structure.[5, 6] In this paper a quantitative investigation on precipitate composition, volume fraction, size distribution and the kinetics of boron transport from prior austenite grain boundaries to carbide precipitates has been presented. Different parts of this work have been previously published.[7–10]

INVESTIGATED MATERIALS AND EXPERIMENTAL METHODS

The steels P92 and P122 were produced and supplied by Nippon Steel Corporation and Sumitomo Metals Industries, respectively, in the form of seamless pipes with 350 mm outer dimension and 50 mm wall thickness.[4] The chemical composition of the two steels is given in Table 1. Heat treatment was performed as follows, P92: austenitising at 1065°C for 2 hours followed by air cooling and tempering at 770°C for 2 hours, P122: austenitising 1050°C/2 hours/AC and tempering 770°C/3 hours. For studying boron segregation, austenitised samples of steel P92 were tempered at 750°C for times between 30 and 120 seconds in a molten lead bath. In order to determine how precipitate size distributions develop during precipitate

Table 2. Ageing/creep conditions for P92 and P122.

	Ageing		Creep	
	600°C (hours)	**650°C (hours)**	**600°C**	**650°C**
P92	1000	1000	342 hours/1% Strain	1840 hours/1% Strain
	3000	3000	10608 hours/2% Strain	4190 hours/2% Strain
	10000	1000		
	26000	26000		
P122	1000	1000		
	10000	10000		
	17860			

growth and coarsening, steels P92 and P122 have also been subjected to ageing or interrupted creep testing at 600 and 650°C at different times, which are given in Table 2.

Energy filtered transmission electron microscopy (EFTEM) was used to quantify size distributions and mean precipitate sizes of different kinds of precipitates in the two steels. Precipitates of type $M_{23}C_6$, MX (primarily VN), Laves phase and Cu were visualised by producing elemental distribution images of the elements chromium, vanadium, tungsten and copper. The microscopy work was done on a Philips CM 200 field emission gun TEM (FEG-TEM) instrument equipped with a Gatan image filter (GIF). Details of the use of EFTEM and image analysis in the present work can be found elsewhere.[8] With the EFTEM technique it is possible to get a good estimation of the average sample thickness in each image field. Therefore a correction was made to compensate measured particle sizes for truncation effects induced by the limited foil thickness.[8] Regarding experimental parameters the recommendations given by Warbichler et al.[11] were closely followed. For each material condition studied, the accuracy in mean particle size measurement is determined by the number of precipitates measured. Typically between 50 and 150 precipitates were measured, giving a relative error in mean size of ±10 to 15% for MX and $M_{23}C_6$, and ±20 to 25% for Laves phase and Cu.

Atom probe field ion microscopy (APFIM) was utilised to measure the composition of the different phases present in the two steels with high spatial resolution. Standard specimen preparation methods and experimental conditions suitable for steel, as described elsewhere,[7] were used. The instrument used in the present investigation is of our own design and is described in references 12 and 13.

Dynamic secondary ion mass spectroscopy (SIMS) was used to study boron segregation. The surfaces of the specimens were mechanically polished before SIMS analysis. Boron images were obtained using a CAMECA IMS 3F instrument. The primary ions used were 12 keV O_2^- with a current of 3 μA. The secondary particle that was used for boron imaging was BO_2^- with the molecular weight of 43.

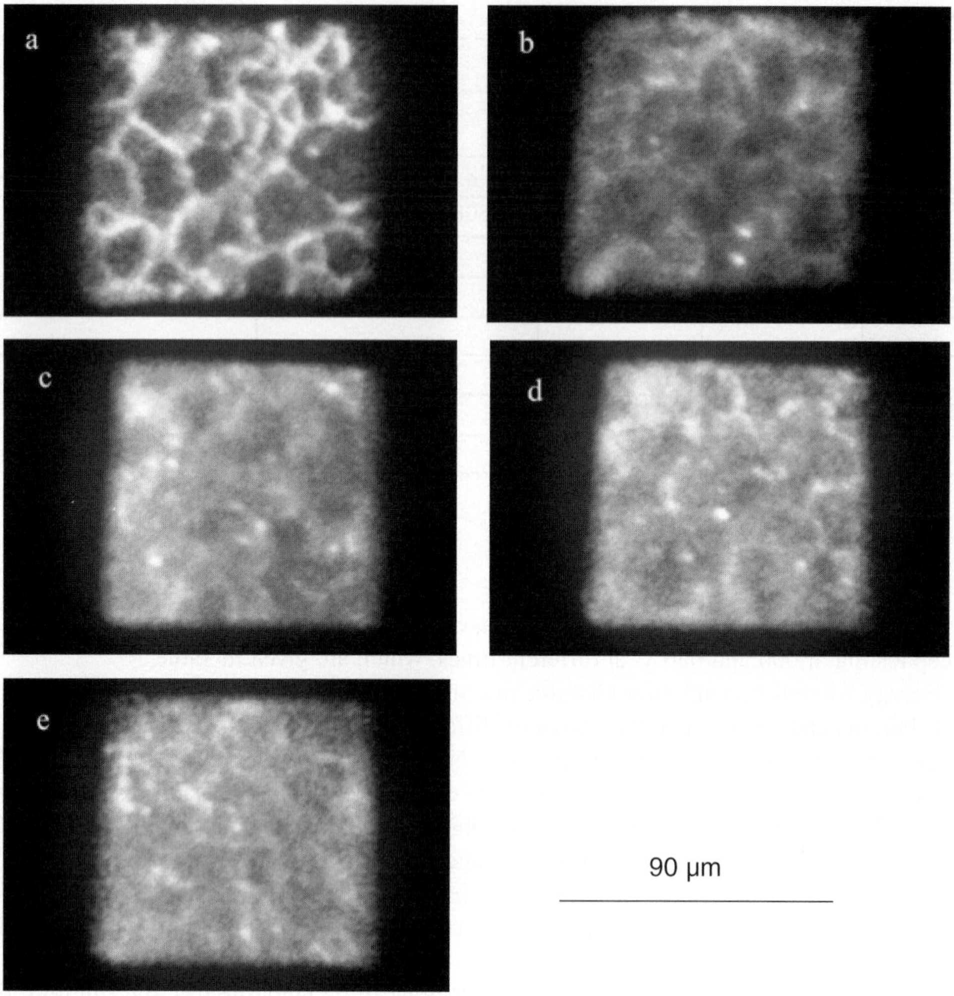

Fig. 1 Boron images of steel P92 (a) austenitised, (b) tempered for 30 seconds, (c) 60 seconds, (d) 90 seconds and (e) 120 seconds.

RESULTS

Boron Redistribution During Tempering

Figure 1 shows SIMS micrographs of the distribution of boron in steel P92 (10 ppm B) after austenitising (1a) and after subsequent tempering for short times at 750°C, from 30 seconds to 2 minutes (1b-e). SIMS is very sensitive to low boron concentrations and gives a clear indication of the boron distribution. Since the spatial resolution is only some 0.5 µm, individual boron-containing precipitates cannot be discerned. It is obvious, that boron is segregated to prior austenite boundaries after austenitising, and that boron diffuses into the martensitic matrix in a time scale of one or two minutes.

Precipitate Growth and Coarsening During Creep and Ageing

Figures 2a-d show EFTEM images of steel P92 aged at 600°C for 1000 hours. Images like these were used to evaluate size distributions of precipitates. The results of particle size measurements in aged or creep-tested material can be found in references.[8–10] The mean equivalent circle diameter of precipitates of type VN and $M_{23}C_6$ in the two steels P92 and P122, as a function of ageing/creep time, is shown in Figures 3–4, respectively. Note that steel P122 was not investigated in the creep tested condition.

Regarding measurements of the size of Laves phase precipitates a direct comparison was made with an alternative technique, field emission gun scanning electron microscopy (FEG-SEM).[14] It was concluded that the EFTEM technique is unsuitable if the particles are too large and/or the number density of particles is too low. Figure 5 shows the mean equivalent circle diameter of Laves phase precipitates in steel P92 and steel P122, respectively, as a function of ageing/creep time. In this case the two techniques EFTEM and FEG-SEM complement each other. The resolving power of EFTEM is needed for small particles, while FEG-SEM provides better statistics for larger particles.

A copper-rich phase was detected in steel P122 in all material conditions. Figure 6 shows the appearance of Cu-rich particles in material aged at 600°C for 10,000 hours. The mean equivalent circle diameter of copper precipitates in steel P122, as a function of ageing time, is shown in Figure 5.

Precipitate Composition and Volume Fraction

The composition of the matrix and different types of precipitates in the two steels P92 and P122 in different ageing conditions was measured with APFIM or, for copper precipitates, with energy dispersive X-ray spectrometry in the TEM (TEM-EDX). A detailed description and the results of these measurements can be found in references.[7] The composition of different types of precipitates present in steel P122 are given in Table 3.

To monitor precipitation of VN, Laves phase and copper the amount of vanadium, tungsten and copper dissolved in the matrix was measured in different material conditions and re-calculated to volume fraction of precipitated phases.[8–10] This type of calculation can be performed if the composition of the precipitated phases is known. Regarding VN precipitates it was found that the amount of vanadium dissolved in the matrix in steel P92 drops during ageing from 0.20 to 0.05 wt.%, which corresponds to a volume fraction of VN of about 0.3%. This volume fraction remains constant during ageing or creep. The precipitated volume fraction of Laves phase and copper as a function of ageing time (at an ageing temperature of 600°C) is shown in Figure 7. In this case precipitation occurs during ageing.

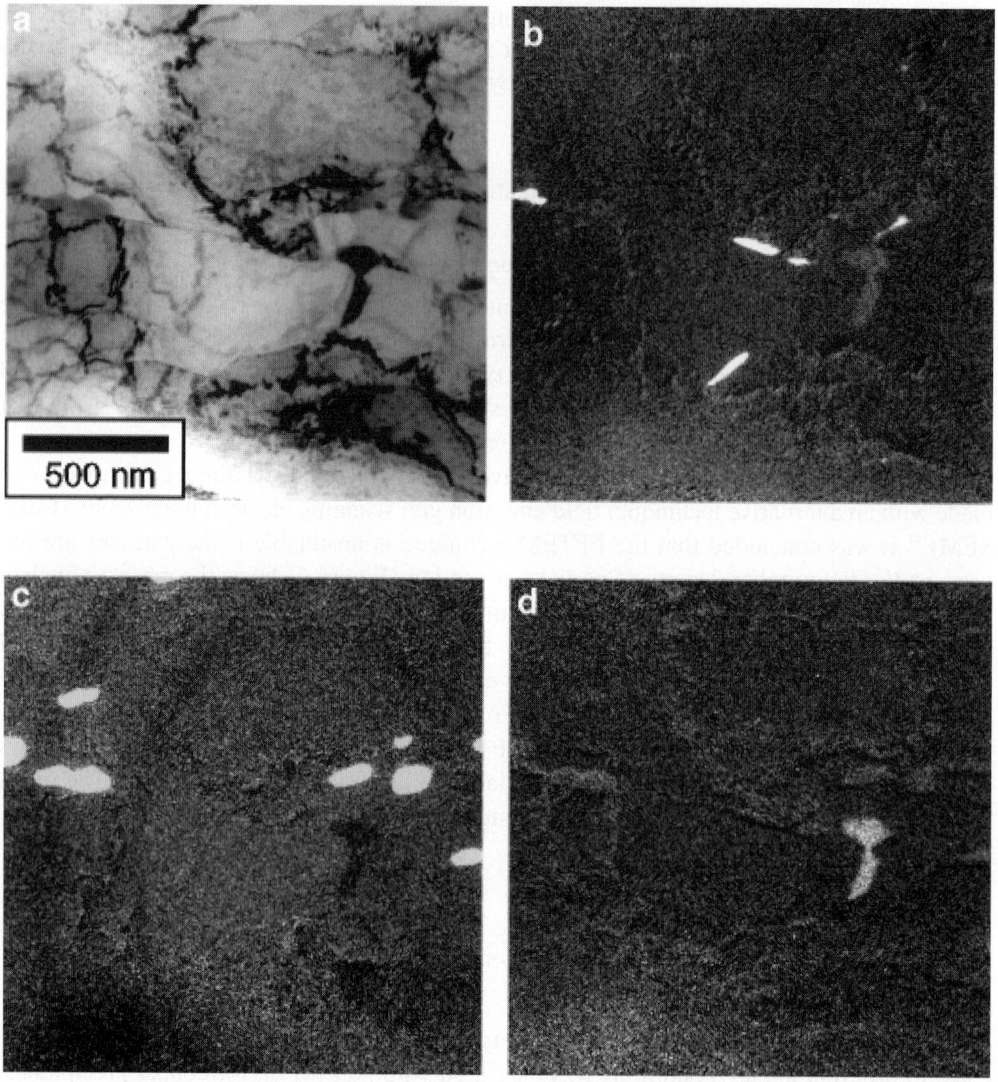

Fig. 2 (a) Bright field image of steel P92 aged at 600°C for 1000 hours, (b) vanadium image showing the distribution of VN precipitates, (c) chromium - $M_{23}C_6$ and (d) tungsten - Laves phase.[8]

In both steel P92 and steel P122 an enrichment of boron was found in $M_{23}C_6$ carbides.[7] The distribution of boron within $M_{23}C_6$ carbides in the two materials is shown in the concentration profiles in Figure 8. As can be seen the boron is evenly distributed inside the carbide in steel P92, while in steel P122 a higher concentration of boron is found close to the carbide/ferrite interface. The concentration profiles also show interfacial segregation of phosphorus.

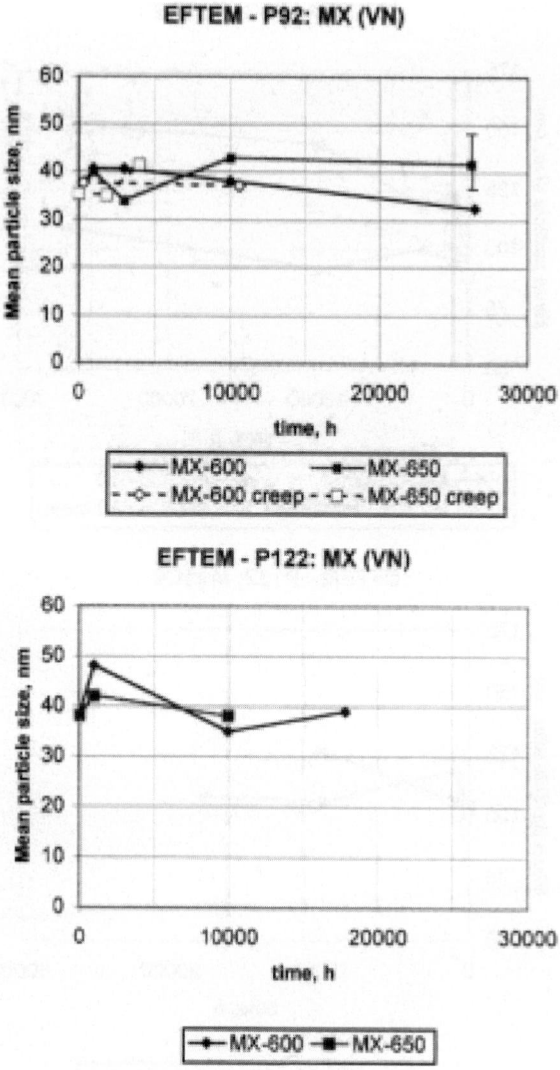

Fig. 3 Mean diameter of VN precipitates in steels P92 (top) and P122 (bottom) as a function of ageing/creep time. A typical error bar is given for one of the measurements.

DISCUSSION

Kinetics of Boron Redistribution During Tempering

The addition of boron to 9–12% chromium steels has been found to be beneficial for creep strength.[5, 15] The mechanism behind this strengthening effect has been difficult to clarify, one reason being that it is very hard to locate the boron in the microstructure. In the present

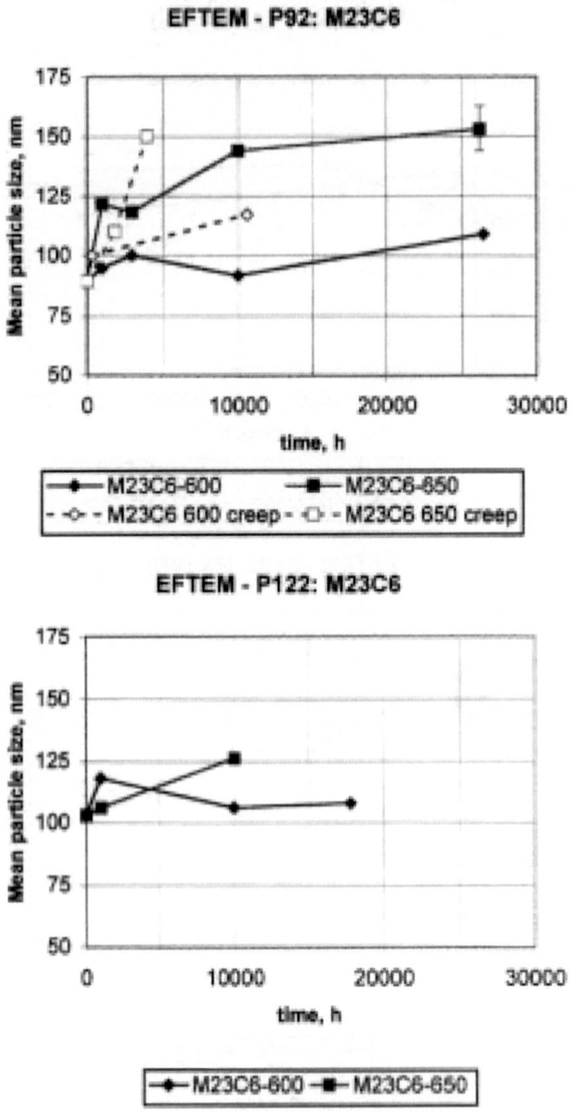

Fig. 4 Mean diameter of $M_{23}C_6$ precipitates in steels P92 (top) and P122 (bottom) as a function of ageing/creep time. A typical error bar is given for one of the measurements.

investigation it has been shown that a large part of the boron added to steel P92 and P122 is incorporated within $M_{23}C_6$ carbides (see Figure 8), which more correctly should be designated $M_{23}(C, B)_6$ in boron containing alloys. Boron enrichment in $M_{23}(C,B)_6$ precipitates was also observed in a number of other boron bearing steels.[16] No boron was detected in any other type of precipitate in these steels, cf. Table 3, and the boron content in the matrix was found

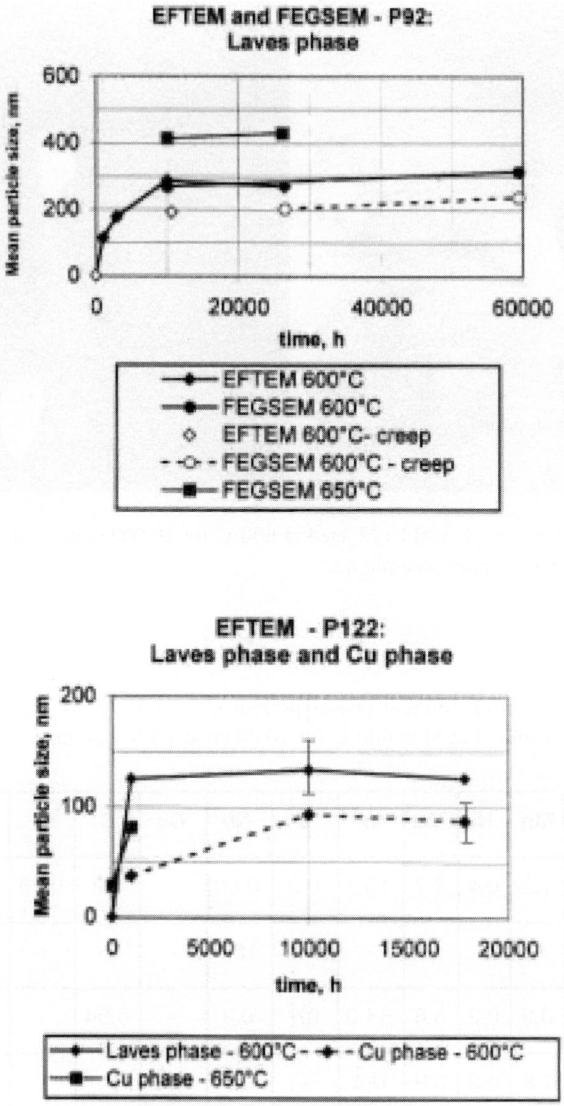

Fig. 5 Mean diameter of precipitates of type Laves phase and copper in steels P92 (top) and P122 (bottom) as a function of ageing/creep time. Cu-rich particles are present only in steel P122. Typical error bars are given for two of the measurements. FEG-SEM data from reference 14.

to be less than 4 ppm by weight[16]. The boron content in $M_{23}(C,B)_6$ was found to depend on the total B/C ratio in the steel, but the B/C ratio in individual precipitates was frequently lower and sometimes higher than this ratio.[16] In most steels the boron content was approximately constant through the $M_{23}(C,B)_6$ precipitate, except for in steel P122, where boron seems to be enriched in an outer part of the precipitate (Figure 8).

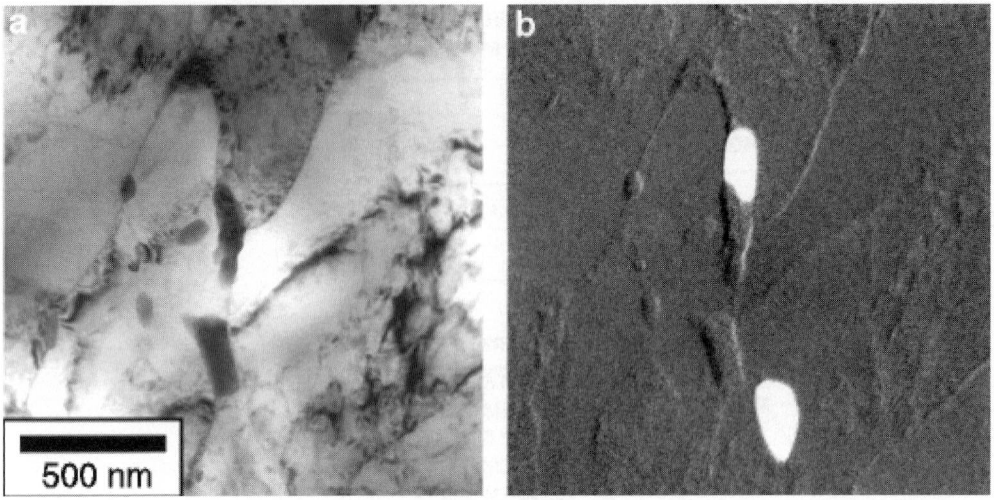

Fig. 6 (a) Bright field image of steel P122 aged at 600°C for 10,000 hours and (b) Copper image showing the distribution of copper precipitates.[8]

Table 3. Chemical composition of different phases present in steel P122. $M_{23}C_6$, Laves phase, Cu and Nb(C,N) were analysed in material aged at 600°C for 10,000 hours VN was analysed in unaged material. All values are in wt.%.[10]

Phase/ Method	Fe	Cr	Mn	Ni	Mo	W	V	Nb	Cu	C	B	N	Si	P
$M_{23}C_6$ APFIM	17.9	58.9	1.2	0.4	2.7	13.5	0.3	0.03	-	5.2	0.03	-	0.01	-
VN APFIM	1.3	15.2	-	-	-	-	57.4	10.6	-	-	-	15.5	-	-
Laves APFIM	28.6	6.4	0.6	0.2	8.6	54.0	0.1	0.3	-	0.04	-	-	1.0	0.20
Cu* EDX	4.7	0.8	1.8	0.2	0.9	0.4	-	0.5	90.8	-	-	-	-	-
Nb(C,N) APFIM	-	-	-	-	-	-	4.0	86.0	-	2.9	-	6.9	-	-

*The TEM-EDX analysis was performed on a thin foil and possibly the amount of iron and chromium is overestimated because of contribution from the ferritic matrix.

The kinetics of $M_{23}C_6$ precipitation during tempering has been studied with TEM and SEM in another steel, KP, which is similar to P92 but with 1%Mo and 1%W.[17] It was found that initially both $M_{23}C_6$ and M_7C_3 nucleate and grow. After 120 seconds, as the diffusion fields of the precipitates begin to overlap, M_7C_3 starts to dissolve, and after 500 seconds the

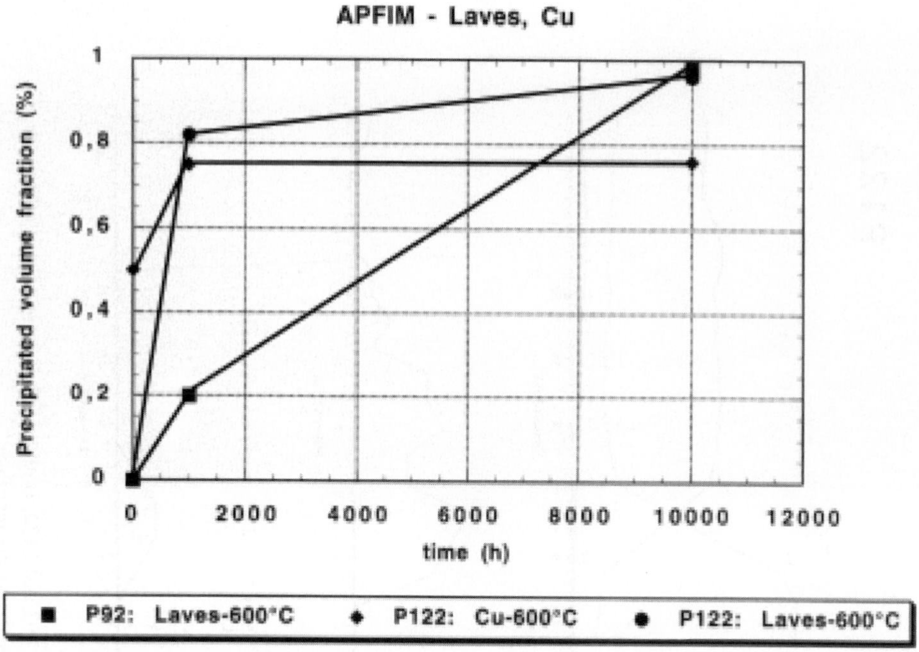

Fig. 7 Precipitated volume fraction of Laves phase and copper in steels P92 and P122 as a function of ageing time. The ageing temperature was 600°C. Cu-rich particles are present only in steel P122.

growth of $M_{23}C_6$ is almost completed and M_7C_3 is almost gone (Figure 9). Thermodynamic and kinetic modelling using the DICTRA software were able to reproduce this result convincingly.[17]

Together with the SIMS observations the following picture emerges: During air cooling after normalising, boron segregates to prior austenite grain boundaries, most likely through the non-equilibrium mechanism described by Karlsson and Nordén for austenitic stainless steels.[18] Non-equilibrium segregation means that boron atoms have been dragged to the boundary by vacancies, not that there exists a binding energy between boron atoms and the boundary. The vacancies being gone after cooling, boron starts to diffuse back into the grain on a time scale of one or two minutes. Together with carbon it then gets incorporated in the growing $M_{23}(C,B)_6$, and, depending on the distance between the prior austenite grain boundary and the precipitate, more or less boron is available during the few minutes that the growth process takes. For precipitates close to or at prior austenite grain boundaries, a higher B/C ratio than in the steel as a whole is possible, as has indeed been observed[16]. The fact that boron diffusion and precipitate growth occur over the same time scale explains the spread in B/C ratio observed by APFIM analysis. It also explains the lack of effect of the presence of boron on the number density of $M_{23}C_6$[19], since nucleation occurs before any boron has arrived to the precipitate (except possibly at the grain boundaries).

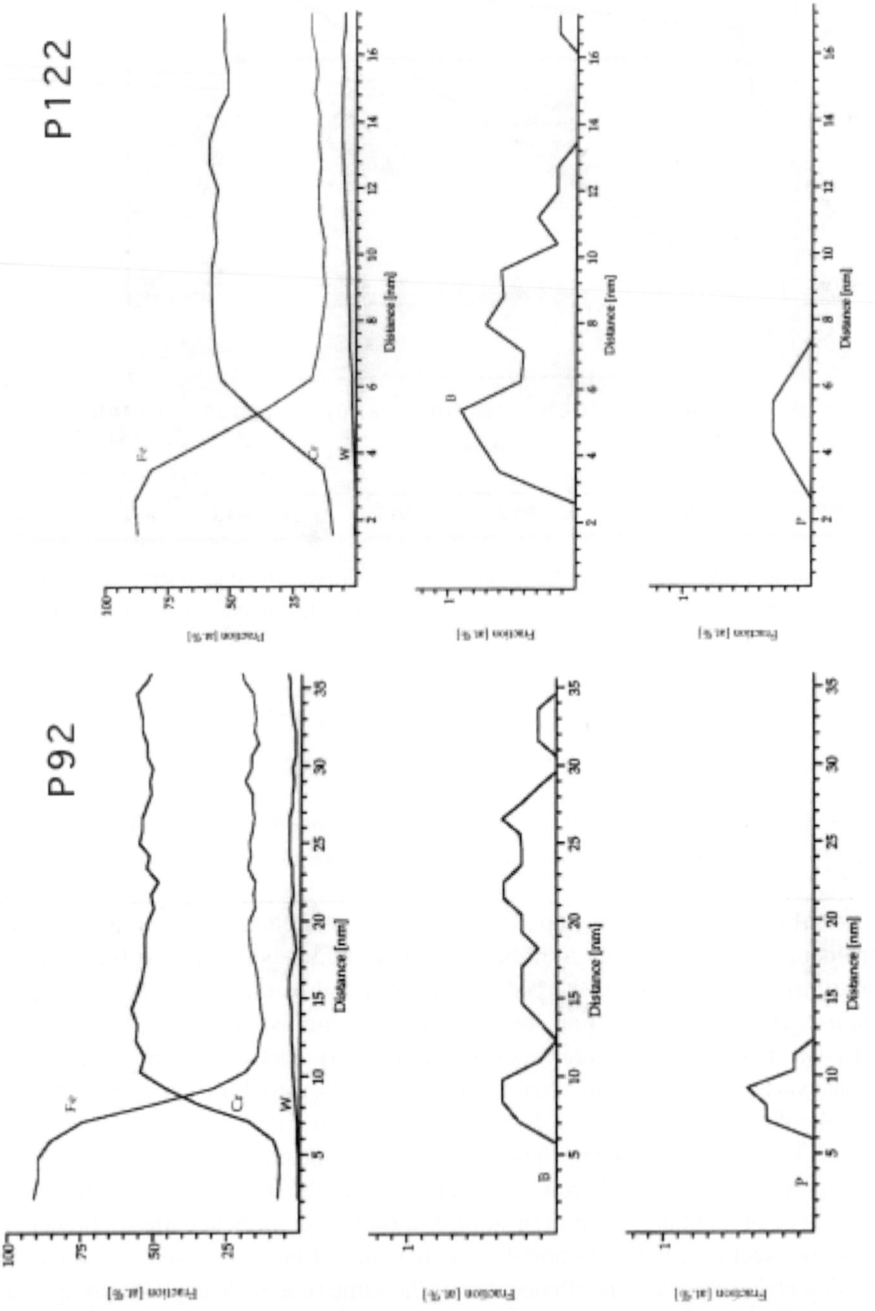

Fig. 8 Concentration profiles through a matrix/carbide interface in steels P92 and P122. Both materials were aged at 600°C for 10,000 hours. Boron is enriched within the carbides and phosphorus segregates to the interface.[7]

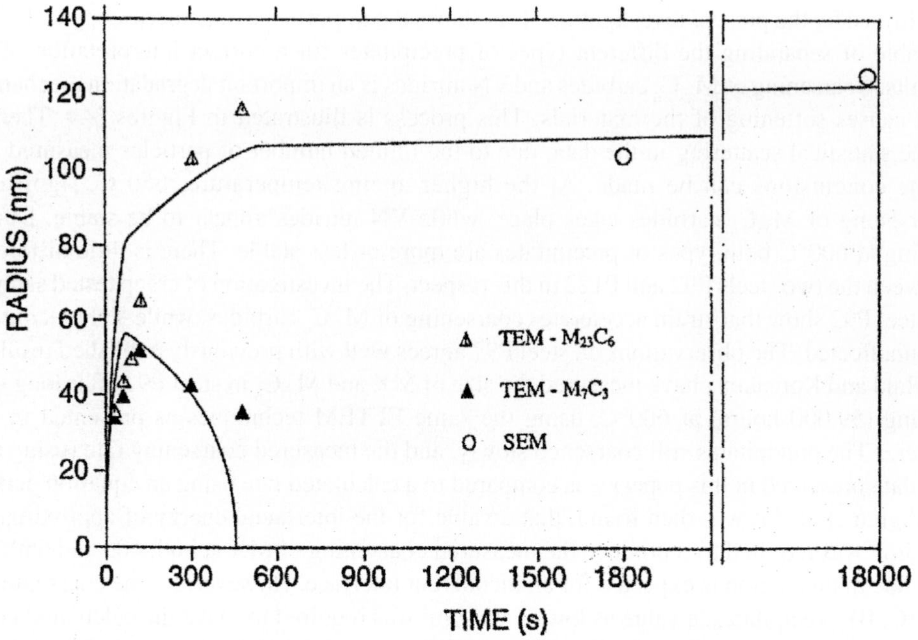

Fig. 9 Growth and coarsening of $M_{23}C_6$ simultaneously with growth and dissolution of M_7C_3 in steel KP. Lines are the results of a growth simulation using the DICTRA software.[17]

In steel P122, boron was found in an outer shell of $M_{23}C_6$ only. One possible explanation for this could be that the few precipitates analysed in this steel by APFIM all happened to lie at the centre of austenite grains. Another possibility could be that precipitate growth is quicker in this steel; its higher chromium content might prevent M_7C_3 precipitation (as suggested by e.g. the Fe-Cr-0.1%C isopleth),[20] which would make the growth of $M_{23}C_6$ quicker since no decrease in growth rate will occur due to the dissolution of M_7C_3. The number density of $M_{23}C_6$ in P122 seems not to be higher than in P92. The two steels have the same carbon content, which means the same volume fraction of $M_{23}C_6$, and they have approximately the same precipitate size i.e. the precipitates have approximately the same number density. This suggests that copper precipitation does not affect the nucleation of $M_{23}C_6$ in this steel.

Kinetics of MX and $M_{23}C_6$ Coarsening

Precipitates of type $M_{23}C_6$ and VN are present in the 'as-received', tempered material. The presence of these particle dispersions is of vital importance for creep strength.[5] During ageing or creep, the volume fraction of $M_{23}C_6$ and VN remains constant. Dynamic precipitation during creep of VN precipitates in steel P92 has been suggested by others,[21] but could not be confirmed by the present investigation. This shows the importance of a microscopy technique capable of separating the different types of precipitates for a correct interpretation of the results. Coarsening of $M_{23}C_6$ carbides and VN nitrides is an important degradation mechanism that causes softening of the materials. This process is illustrated in Figures 3–4. There is some statistical scattering in the data, due to the limited number of particles measured, but some conclusions can be made. At the higher ageing temperature, 650°C, significant coarsening of $M_{23}C_6$ carbides takes place, while VN nitrides appear to be stable. During ageing at 600°C both types of precipitates are more or less stable. There is little difference between the two steels P92 and P122 in this respect. The investigation of creep tested samples of steel P92 show that strain accelerates coarsening of $M_{23}C_6$ carbides, while VN precipitates are unaffected. The observations on steel P92 agrees well with previously published results.[22]

Hald and Korcakova have measured the size of MX and $M_{23}C_6$ in steel P92 after long time ageing (59,000 hours) at 600°C, using the same EFTEM techniques as presented in this paper.[23] The precipitates still coarsened slowly, and the measured coarsening rate (using also the data presented in this paper) was compared to a calculated rate using an equation derived by Ågren et al.[24] It was then found, that a value for the interfacial energy of approximately 0.5 J/m^2 was required to reproduce the measured coarsening of MX at both 600 and 650°C.[23] A value in this region is expected for an incoherent interface. However, for the coarsening of $M_{23}(C, B)_6$ precipitates, a value as low as 0.1 J/m^2 was required to make the calculated curve fit the experimental points at both 600 and 650°C, assuming that the diffusion of Cr, Mo and W controls coarsening. Such a low value is not expected for an incoherent interface, and suggests that boron might be to decrease the interfacial energy between $M_{23}(C,B)_6$ and matrix.[23] It has been found before that the role of boron might be to reduce the coarsening rate of $M_{23}(C,B)_6$ precipitates.[21, 25] Possibly, then, this effect is due to a decrease of the interfacial energy, thus reducing the driving force for coarsening. Another possibility is that the very low boron solubility in the matrix is the rate controlling factor.

Kinetics of Laves Phase Growth and Coarsening

Perhaps the most significant change in the microstructure of steel P92 and P122 occurring during ageing or creep at 600–650°C is the precipitation of intermetallic Laves phase, with a subsequent decrease of the matrix molybdenum and tungsten contents. This is illustrated in Figures 5 and 7. Figure 5 shows the growth of Laves phase particles. The precipitated volume fraction of Laves phase indicated in Figure 7 is merely a reflection of the decreasing matrix tungsten content. The loss of solid solution strengthening from molybdenum and tungsten, caused by the removal of these elements from the matrix, is balanced by a particle strengthening contribution from the Laves phase particles. The magnitude of this strengthening mechanism is related to the number density of Laves phase precipitates. The curves in Figure 5 represent particle growth. When the final size after growth is reached slow particle coarsening occurs. This means that the critical process, which will determine the strengthening effect of Laves phase precipitates, is particle nucleation. Nucleation and growth at 650°C compared to 600°C gives a larger final size and a much lower number density of particles. It is not straightforward to relate particle density to creep strength. Simple Orowan stress calculations performed by Hald[14] indicates a beneficial effect of Laves phase precipitation on creep strength in the case of material used at 600°C. The measurements performed on the copper containing steel P122 show that nucleation of Laves phase is affected by the presence of copper precipitates. It seems that nucleation is enhanced, which gives a higher number density of Laves phase particles.

Copper was added to steel P122 with the intention to suppress the formation of d-ferrite and stabilise the overall constitution of the steel.[26] The present investigation has clarified the redistribution of copper during heat treatment and ageing. Already during tempering at 770°C a large amount of the copper dissolved in the matrix precipitates as small, densely distributed Cu-rich particles. During ageing at 600°C the volume fraction of copper precipitates increases and the matrix copper content stabilises at about 0.1%.[7] The presence of small copper precipitates seems to enhance nucleation of Laves phase particles, giving a finer distribution of this phase. In addition, the copper precipitates themselves should contribute to the creep strength since their smaller size and somewhat smaller volume fraction compared to Laves phase (Figures 5 and 7) means that their number density should be at least as large. The content of copper in P122 has therefore both a direct (Cu precipitates) and an indirect (denser distribution of Laves phase precipitates) beneficial effect on the creep strength of this steel.

CONCLUSIONS

Boron segregates to austenite grain boundaries during cooling after austenitising. During tempering at about 750°C boron diffuses into prior austenite grains and is incorporated in growing $M_{23}C_6$ precipitates. The growth of $M_{23}C_6$ and the diffusion of boron into prior austenite grains occur on the same time scale, a few minutes, explaining the variation in B/C ratio in individual precipitates. No boron was found in any other type of precipitate, and the solubility of B in the matrix at 600°C is less than 4 ppm.

In both steels precipitates of type $M_{23}C_6$ and VN are present in the 'as-received' tempered material. During ageing or creep coarsening takes place and the particles increase in size, while the volume fraction remains constant. VN precipitates are highly stable even at 650°C, while $M_{23}C_6$ precipitates coarsen significantly at this temperature. Strain accelerates the coarsening of $M_{23}C_6$ precipitates. Boron incorporated in $M_{23}C_6$ carbides reduces their coarsening rate, possibly by reducing the interfacial energy of this phase.

During ageing or creep of the two steels precipitation of intermetallic Laves phase takes place. The number density of particles after the growth process is finished after 10,000 hours is considerably lower in material exposed to 650°C compared to 600°C. This contributes to reduction in creep strength at the higher temperature.

In steel P122, which contains 0.9wt.%Cu, precipitation of a Cu-rich phase takes place already during tempering (at 770°C). During ageing the volume fraction of Cu-rich phase increases even further. However, the presence of the copper particles gives a finer distribution and a faster growth time (1,000 hours) of Laves phase precipitates, which should be beneficial for the creep strength.

The most critical processes that degrade the creep strength of the two materials at high temperatures are coarsening of $M_{23}C_6$ carbides and the formation of Laves phase precipitates. Boron is effective in reducing the coarsening rate of $M_{23}C_6$, and copper reduces the size of Laves phase precipitates.

ACKNOWLEDGEMENTS

This work was supported by the Swedish Consortium 'Materials Technology for Thermal Energy Processes' (KME) and by the Research Foundation of VGB, the Technical Association of Large Power Plant Operators, Essen, Germany. Close collaboration with John Ågren, Royal Institute of Technology, Stockholm, and with John Hald and Lea Korcakova of Elsam/Energy E2, Denmark, is gratefully acknowledged, as is collaboration within the COST522 programme Advanced Steam Power Plant.

REFERENCES

1. F. Masuyama: *Advanced Heat Resistant Steels for Power Generation*, R. Viswanathan and J. Nutting, eds., The Institute of Materials, London, 1999, 33–48.
2. M. E. Staubli, K. -H. Mayer, T. -U. Kern and R. W. Vanstone: *Proceedings of 5th International Charles Parsons Turbine Conference*, A. Strang, W. M. Banks, R. D. Conroy, G. M. McColvin, J. C. Neal and S. Simpson, eds., IOM Communications, London, 2000, 98–122.
3. E. Metcalfe and W. T. Bakker: *New Steels for Advanced Plant upto 620°C*, E. Metcalfe, ed., EPRI, Palo Alto, CA, 1995, 1–7.
4. J. Hald: *New Steels for Advanced Plant upto 620°C*, E. Metcalfe, ed., EPRI, Palo Alto, CA, 1995, 152–173.
5. J. Hald and S. Straub: *Materials for Advanced Power Engineering*, J. Lecomte-Beckers, F. Schubert and P. J. Ennis, eds., Forschungszentrum Jülich GmbH, Jülich, 1998, 155–169.
6. A. Strang and V. Vodarek: *Materials for Advanced Power Engineering*, J. Lecomte-Beckers, F. Schubert and P. J. Ennis, eds., Forschungszentrum Jülich GmbH, Jülich, 1998, 603–614.
7. M. Hättestrand, M. Schwind and H-O. Andrén: *Materials Science and Engineering A*, **250**, 1998, 27–36.
8. M. Hättestrand and H-O. Andrén: *Micron*, **32**, 2001, 789–797.
9. M. Hättestrand and H-O. Andrén: *Acta Materialia*, **49**, 2001, 2123–2128.
10. M. Hättestrand and H-O. Andrén: *Materials Sciance and Engineering A*, **318**, 2001, 94–101.
11. P. Warbichler, F. Hofer, P. Hofer and E. Letofsky: *Micron*, **29**, 1998, 63–72.
12. H-O. Andrén and H. Nordén: *Scandinavian Journal of Metallurgy*, **8**, 1979, 147–152.
13. H-O. Andrén: *J. de Phys.*, **47**, 1986, C7 483–488.
14. J. Hald: *Proceedings of 3rd EPRI Conference on Advances in Materials Technology for Fossil Power Plants*, R. Viswanathan, et al., eds., Swansea, 2001, 115–124.
15. V. Foldyna, Z. Kubon, A. Jakobova and V. Vodarek: *Microstructural Development and Stability in High Chromium Ferritic Power Plant Steels*, A. Strang and D. J. Gooch, eds., The Institute of Materials, London, 1997, 73–92.
16. M. Hättestrand and H-O. Andrén: *Materials Sciance and Engineering A*, **270**, 1999, 33–37.

17. M. Hättestrand and A. Bjärbo: *Metall. Mater. Trans. A.,* **32**, 2001, 19–28.
18. L. Karlsson and H. Nordén: *Acta Metallurgica*, **36**, 1988, 13–24.
19. L. M. Lundin, M. Hättestrand and H-O. Andrén: *Proceedings of 5th International Charles Parsons Turbine Conference*, A. Strang, W. M. Banks, R. D. Conroy, G. M. McColvin, J. C. Neal and S. Simpson, eds. IOM Communications, London, 2000, 603–617.
20. R. W. Cahn, P. Haasen, E. J. Kramer: *Material Science and Technology: A Compressive Treatment: Consititition and Properties of Steel,* VCH, Weinheim, Germany, **7**, 1992.
21. D. Henes, S. Straub, T. Sailer, P. Polcik, W. Blum, J. Hald and K.-H. Mayer: *Vortragsveranstaltung 'Langzeitverhalten Warmfester Stähle und Hochtemperaturwerkstoffe'*, VDEh, Düsseldorf, 1997, 42–55.
22. P. J. Ennis, A. Zielinska-Lipiec, O. Watcher and A. Czyrska-Filemonowicz: *Acta Mater.*, **45**, 1997, 4901–4907.
23. J. Hald and L. Korcakova: submitted to *ISIJ International*, 2002.
24. J. Ågren, M. T. Clavaguera-Mora, J. Golczewski, G. Inden, H. Kumar and C. Sigli: *Calphad*, **24**, 2000, 41–54.
25. P. Nowakowski, H. Straube and K. Spiradek: *Materials for Advanced Power Engineering*, J. Lecomte-Beckers, F. Schubert and P. J. Ennis, eds., Forschungszentrum Jülich GmbH, Jülich, 1998, 567–574.
26. A. Iseda, Y. Sawaragiu, S. Kato and F. Masuyama: *Fifth International Conference on Creep of Materials*, ASM International, Materials Park, OH, 1992, 389–397.

The Role of Boron in the 9% Chromium Steels for Steam Power Plants

A. Czyrska-Filemonowicz and K. Bryla
AGH University of Science and Technology
Faculty of Metallurgy and Materials Science
PL-30059 Kraków
Poland

K. Spiradek-Hahn
ARC Seibersdorf research GmbH
Materials and Production Engineering
A-2444 Seibersdorf
Austria

H. Firganek
Institute of Ferrous Metallurgy
Physical Metallurgy Department
PL-44100 Gliwice
Poland

A. Zielinska-Lipiec
AGH University of Science and Technology
Faculty of Metallurgy and Materials Science
PL-30059 Kraków
Poland

P. J. Ennis
Research Centre Jülich
Institute for Materials and Processes in Energy Systems, IWV-2
D-52425 Jülich
Germany

ABSTRACT

The microstructures of the 9% chromium steels P92 (Fe-9Cr-1Mo-1.8W-0.2V-0.06Nb-0.03N-0.1C) and B2 (Fe-9Cr-1.5Mo-0.2V-0.06Nb-.0.015N-0.2C) with boron additions of 30 and 100 ppm, respectively, were investigated by means of optical and transmission electron microscopy in the as received condition and after creep testing at 600°C. Quantitative measurements of the microstructural features of these steels, especially particle size distribution of $M_{23}C_6$, were carried out. Trace autoradiography revealed that boron was incorporated into the $M_{23}C_6$ precipitates in both P92 and

B2 steels. Boron was densely distributed on prior austenite grain boundaries and within martensite laths, and resulted in the stabilisation of $M_{23}(C,B)_6$ precipitates in B2 steel. These growth resistant borocarbides led to significantly improved creep rupture strength of this steel.

INTRODUCTION

Economic and environmental concerns have provided the driving force towards improvements in the thermal efficiency of steam power plants, by increasing the operating temperatures and pressures. In this way, the fuel consumption can be reduced, with a corresponding reduction in CO_2 emissions. However, the more severe operating conditions require the use of new steels with higher creep and steam oxidation resistance. Several new steels have been developed under different international research projects over the last thirty years and have now reached commercialisation. Investigations of the key microstructural features responsible for the strengthening and for the weakening mechanisms that lead to degradation during long-term service tests are necessary for further development of these steels. It has been shown that boron added to high chromium steels enhances creep rupture strength,[1] although the role of this element in 9–12% Cr steels is not completely understood. The $M_{23}C_6$ carbides appear to be more stable in boron-containing steels than in boron-free steels during ageing and creep exposure. It seems that boron retards the growth of $M_{23}C_6$ by forming borocarbides $M_{23}(C, B)_6$, as is the case in austenitic stainless steels.[2] Using atom-probe field-ion microscopy and field emission auger electron spectroscopy methods it was found that, more than half of the boron added was found in the $M_{23}C_6$ precipitates.[3, 4] Unfortunately, it is not possible to use these techniques to reveal the presence of $M_{23}(C,B)_6$ precipitates. In the present work, trace autoradiography has been applied in an attempt to show the boron distribution and to identify borocarbide precipitates.

EXPERIMENTAL PROCEDURES

The chemical compositions and heat treatments of P92, a commercial steel developed by Nippon Steel in Japan under the designation NF616, and B2, a steel developed in the European COST Action 501 Round III, are shown in Tables 1 and 2. Standard uni-axial creep tests were carried out at 600°C and specimens for microstructural examination were selected. The microstructure of the as received and after creep testing materials was investigated using optical metallography and transmission electron microscopy (TEM). Thin foils were used for quantitative determination of microstructural features (dislocation densities, precipitate sizes, sub-grain sizes) and for the phase identification. Details of the TEM procedures and the methods of measurements have already been reported.[5]

In order to show the boron distribution in the microstructure of the investigated steels, the trace autoradiography method was used.[2, 6] Trace autoradiography is a photographic method for recording the distribution of radioactive elements within a specimen. With this method large areas can be examined for boron content in the range 10^{15}–10^{20} B atoms/cm^3. This method involves irradiation of the specimen by thermal neutrons. The irradiated sample is placed in contact with photographic film (foil or plate of organic material), which after

Table 1 Chemical composition of P92 and B2 steels.

Element (wt.) / Steel	C	Si	Mn	Cr	Mo	W	Ni	V	Al	Nb	N	B
P92	0.07	0.02	0.47	9.07	0.46	1.85	0.06	0.19	0.002	0.063	0.043	0.003
B2	0.17	0.08	0.06	9.36	1.55	-	0.12	0.27	0.014	0.060	0.015	0.010

Table 2 Details of test materials, heat treatment of P92 and B2 steels.

Heat Treatment / Steel / Form and Dimensions [mm]	Austenitising	1st Tempering	2nd Tempering
P92 Pipe, ø 300, 40 Wall Thickness	1050°C/1 hour	750°C/1 hour	—
B2 Real Size Rotor, ϕ 840×2400	1100°C/2 hours	590°C/8 hours	700°C/16 hours

exposure and development, yields an autoradiograph. The α tracks from ^{10}B (n, α) ^{7}Li reaction are visible on photographic film and examined with optical microscope after chemical etching in water solution of NaOH and KOH at variable temperatures.

RESULTS AND DISCUSSION

Microstructure and Boron Distribution in the as Received P92 and B2 Steels

STEEL P92

The microstructure of the as received P92 steel consisted of tempered martensite and is shown in Figure 1, the carbides were mainly located on prior austenite grain and martensite lath boundaries. The austenite grain size was measured at ASTM 9, which corresponds to an average grain diameter of 15 μm.

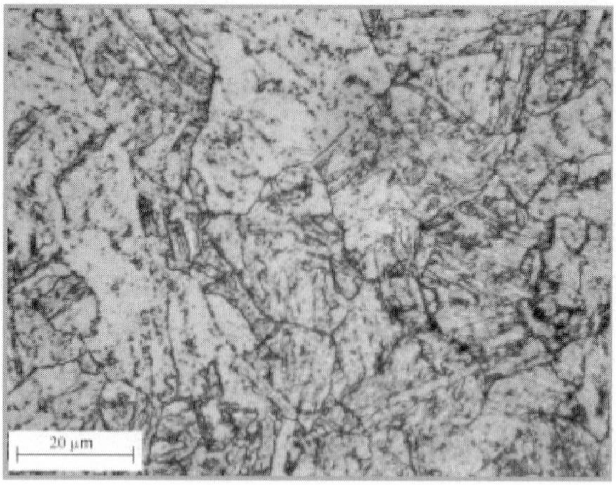

Fig. 1 Microstructure of the as received P92 steel (optical microscopy).

The boron distribution in the as received P92 steel revealed by the trace autoradiography method is shown in Figure 2. The white points on trace autoradiograph correspond to the tracks of boron in a steel microstructure. It should be noted that the tracks on the micrograph arise as a result of ^{10}B (n, α) ^{7}Li nuclear reaction. This is why precipitation sizes are not reflected in the track sizes, which show the amount of reaction products. Boron in the as received P92 steel is predominantly distributed on prior austenite grain boundaries, and hardly visible on sub-grain boundaries. Large and deep tracks on the trace autoradiograph indicate the presence of boron-rich carbides, confirming the presence of $M_{23}(C, B)_6$ borocarbides in P92 steel.

After heat treatment, the microstructure of P92 steel exhibited tempered martensite with a very high dislocation density ($7 \pm 1.0 \times 1014$ m^{-2}) inside the sub-grains. The $M_{23}C_6$ carbides (89 ± 13 nm mean diameter) were precipitated on prior austenite grain boundaries and on sub-grain boundaries (Figure 3a). These carbides consisted of Cr, Fe, Mo, W and C. Fine MX precipitates (globular Nb-rich carbonitrides (Figure 3b), plate like V-rich nitrides and V-wing complexes) were also observed inside sub-grains.[7]

STEEL B2

The microstructure of the as received B2 steel with 100 ppm boron addition consisted of tempered martensite (Figure 4). The prior austenite grain size was ASTM 4-5, average grain diameter of about 90 μm. The very homogenous martensite structure with narrow laths and a few wide laths near grain boundaries were observed. In the as received condition carbides of the $M_{23}C_6$-type were distributed homogeneously on the prior austenite grain boundaries, on the martensite lath boundaries and inside of the martensite laths.[1]

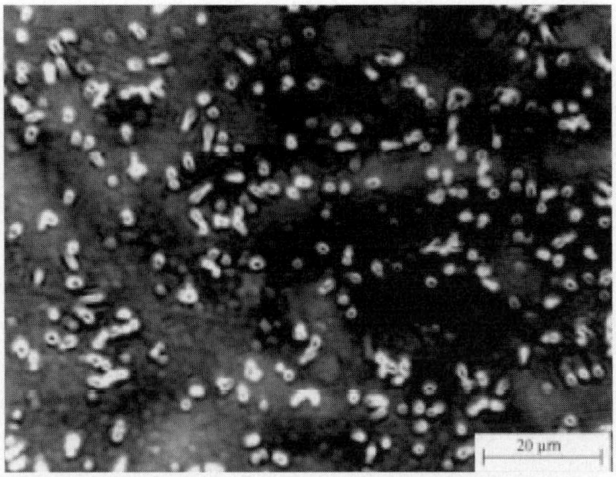

Fig. 2 Boron distribution in the as received P92 steel (trace autoradiograph).

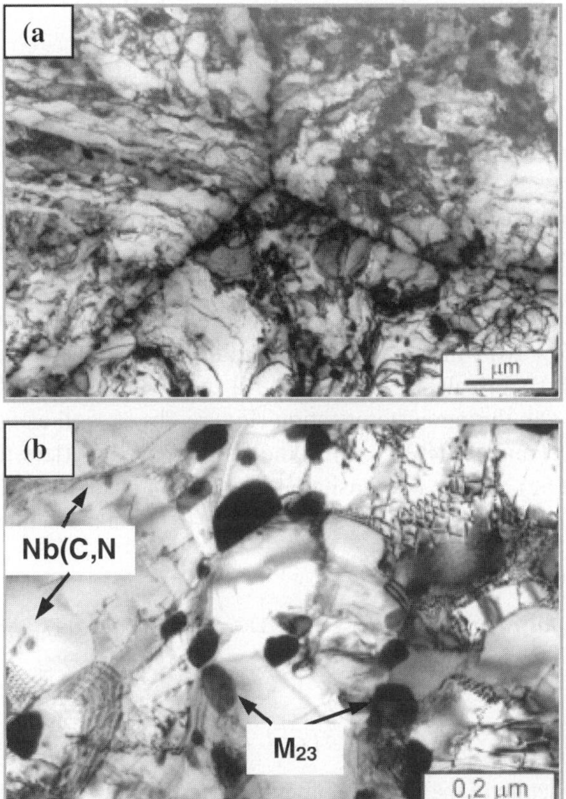

Fig. 3 Microstructure of the as received P92 steel: (a) prior austenite grain boundary and martensite lath (sub-grain) boundaries, (b) $M_{23}C_6$ and Nb(C,N) precipitates (TEM).

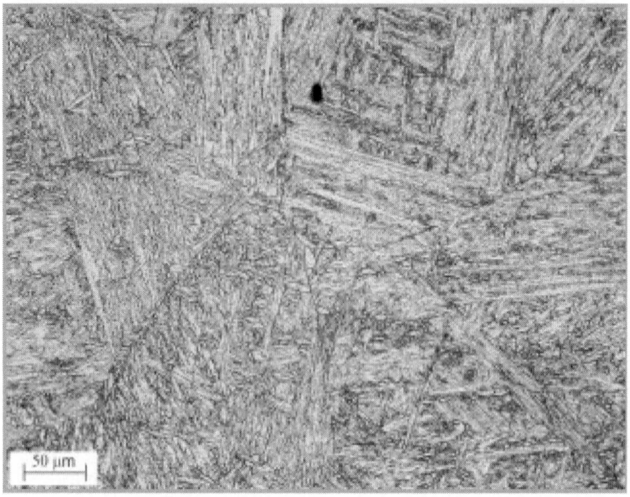

Fig. 4 Microstructure of the as received B2 steel (optical microscopy).

Figure 5 is a trace autoradiograph that shows the boron distribution in as received B2 steel. Boron in this steel was distributed throughout the microstructure. Large and clear tracks on prior austenite grain boundaries are visible and smaller tracks can be observed within the martensite laths. These tracks indicate the presence of $M_{23}(C,B)_6$ in the B2 steel microstructure.

TEM micrographs of the as received B2 steel are shown in Figure 6. A high dislocation density of $4.8 \pm 0.7 \times 1014\ m^{-2}$ was observed within the martensite laths (Figure 6a).

Primary Nb-rich spherical precipitates (0.5-2 μm) which were effective in preventing austenitic grain growth during austenitisation (Figure 6b) as well as very fine MX types of precipitates, V(C, N) and (V, Nb)(C, N) in the form of needles (less than 50 nm in length) were found (Figure 6c). On the martensite laths and prior austenite grain boundaries, $M_{23}C_6$ precipitates (actually $M_{23}(C, B)_6$) of the size of 108 ± 40 nm in mean diameter were observed.

Microstructure and Boron Distribution in Creep Deformed P92 and B2 Steels

Steel P92

The typical microstructure of the P92 steel after creep testing (600°C, 32909 hours, 145 MPa) is shown in Figure 7. The prior austenite grain boundaries and elongated sub-grains are clearly seen and, in comparison with the as received microstructure, the precipitates appear to be larger.

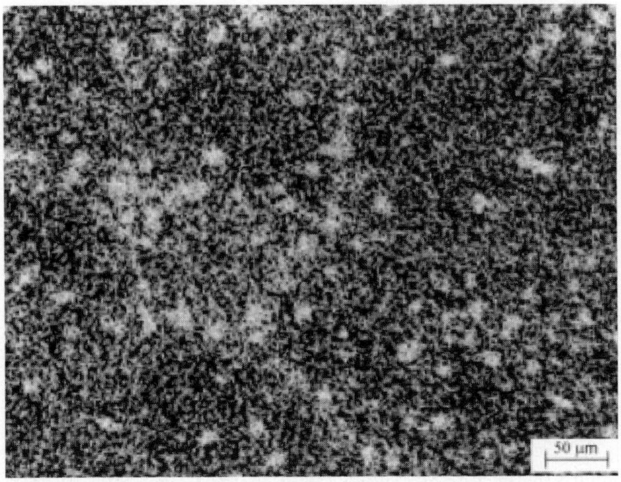

Fig. 5 Boron distribution in the as received B2 steel (trace autoradiography).

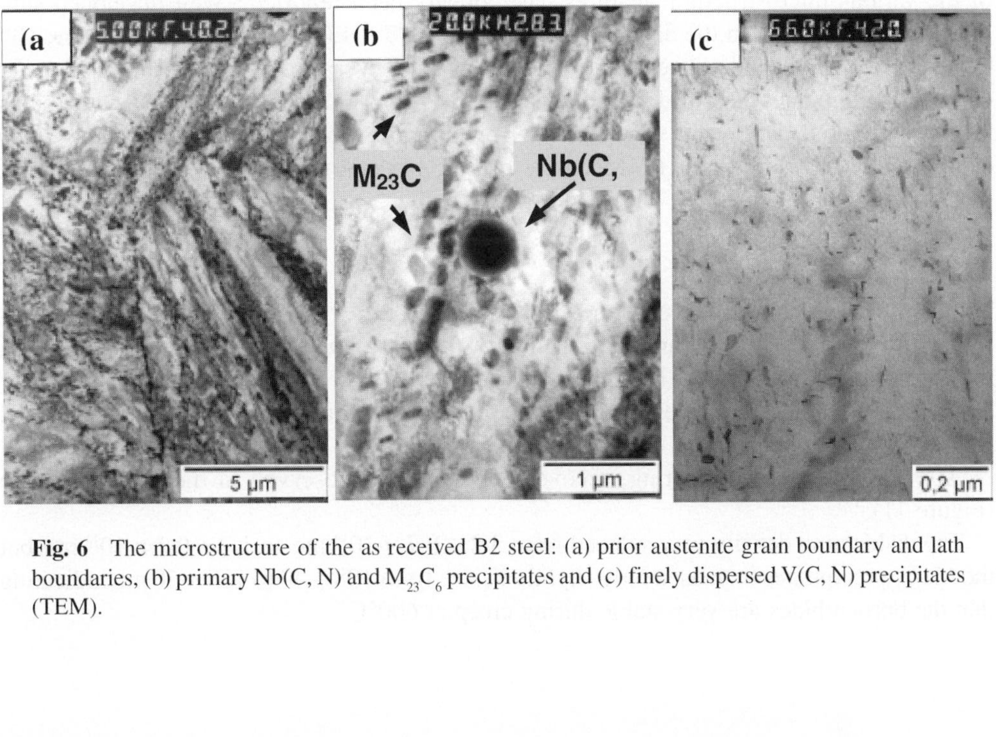

Fig. 6 The microstructure of the as received B2 steel: (a) prior austenite grain boundary and lath boundaries, (b) primary Nb(C, N) and $M_{23}C_6$ precipitates and (c) finely dispersed V(C, N) precipitates (TEM).

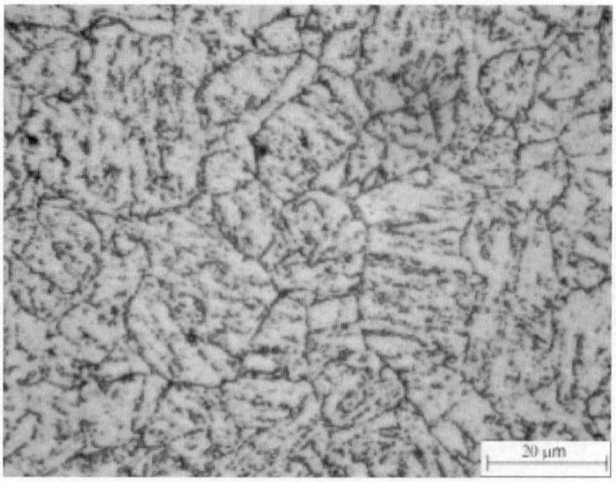

Fig. 7 Microstructure of the P92 steel after creep at 600°C/145 MPa for 32909 hours (optical microscopy).

The boron distribution in creep deformed (600°C, 145 MPa, 32909 hours) P92 steel containing 30 ppm B revealed by trace autoradiography is shown in Figure 8. The very large, deep tracks are the result of coarsening of the $M_{23}(C,B)_6$ precipitates. However as mentioned previously, the track sizes cannot be directly related to the sizes of precipitates.

The extent of the changes in microstructure of P92 after creep testing at 600°C, 145 MPa for 32909 hours is demonstrated by comparing Figures 9 and 3. Quantitative measurements of the various microstructural features have already been reported.[8] Well developed sub-grains and a reduction in the dislocation density (from $7 \pm 1.0 \times 10^{14}$ m^{-2} to $2.3 \pm 0.4 \times 10^{14}$ m^{-2}) in the martensite laths are characteristic effects of recovery processes. The mean diameter of $M_{23}C_6$ precipitates nearly doubled in size (up to 253 ± 30 nm) compared with the as received material, while the very fine MX precipitates were more stable. During creep exposure, the Laves phase $Fe_2(W,Mo)$ was formed, especially close to $M_{23}C_6$ precipitates.

STEEL B2

Typical microstructure of creep deformed B2 steel (600°C/130 MPa for 60458 hours) is shown in Figure 10. In comparison to the microstructure of as received B2 steel (Figure 4), the martensite laths appear to be somewhat wider.

Trace autoradiography investigations to reveal the boron distribution in creep deformed B2 steel are still in progress.

After long-term creep testing, recovery effects were observed in the microstructure (Figure 11).

The dislocation density was reduced from $3.8 \pm 0.7 \times 10^{14}$ m^{-2} to $1.7 \pm 0.7 \times 10^{14}$ m^{-2} but the $M_{23}C_6$ precipitates increased only slightly in size from 108 ± 40 to 125 ± 50 nm indicating that the borocarbides are very stable during creep at 600°C.

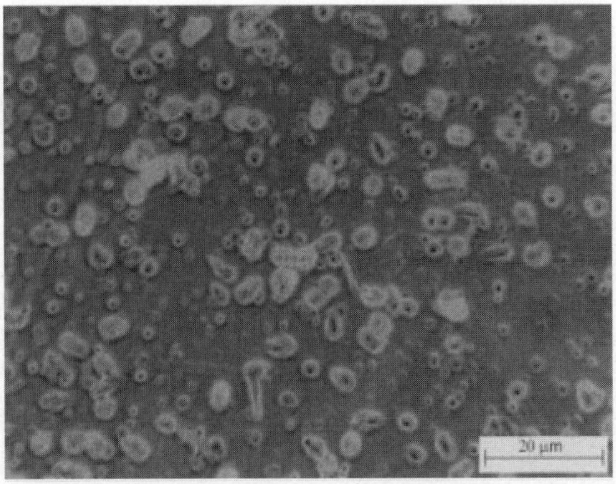

Fig. 8 Boron distribution of the P92 steel after creep at 600°C/145 MPa for 32909 hours (trace autoradiograph).

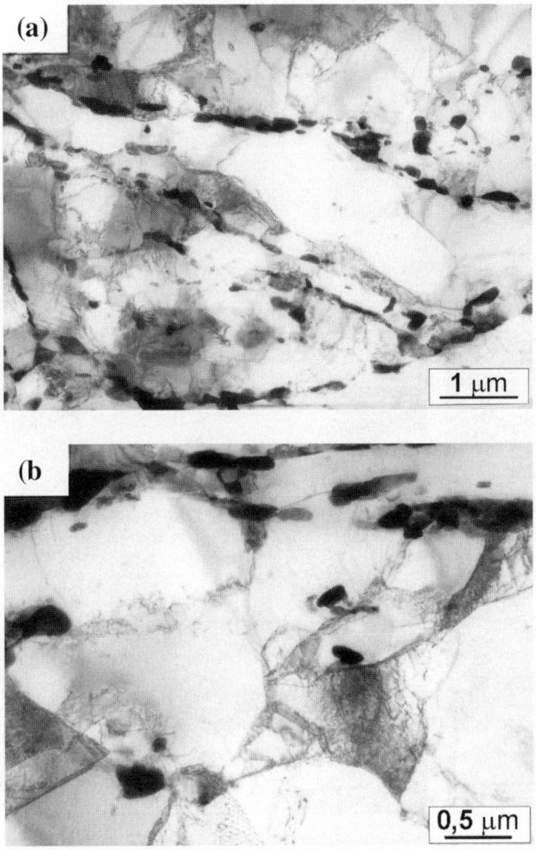

Fig. 9 Microstructure of the P92 steel after creep at 600°C/145 MPa for 32909 hours: (a) prior austenite grain and sub-grain boundaries and (b) recovery effects (TEM).

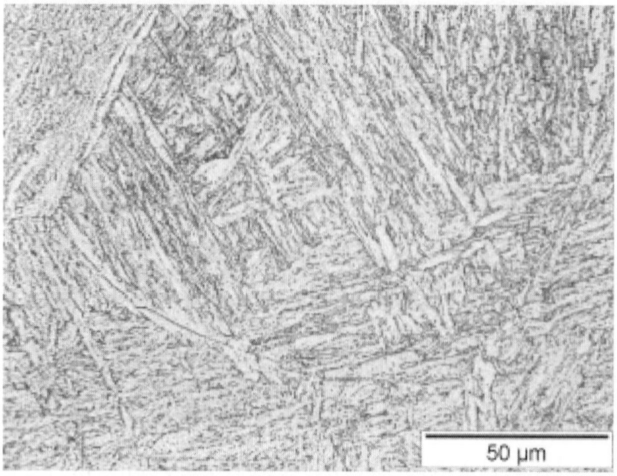

Fig. 10 Microstructure of the B2 steel after creep at 600°C/130 MPa for 60458 hours (optical microscopy).

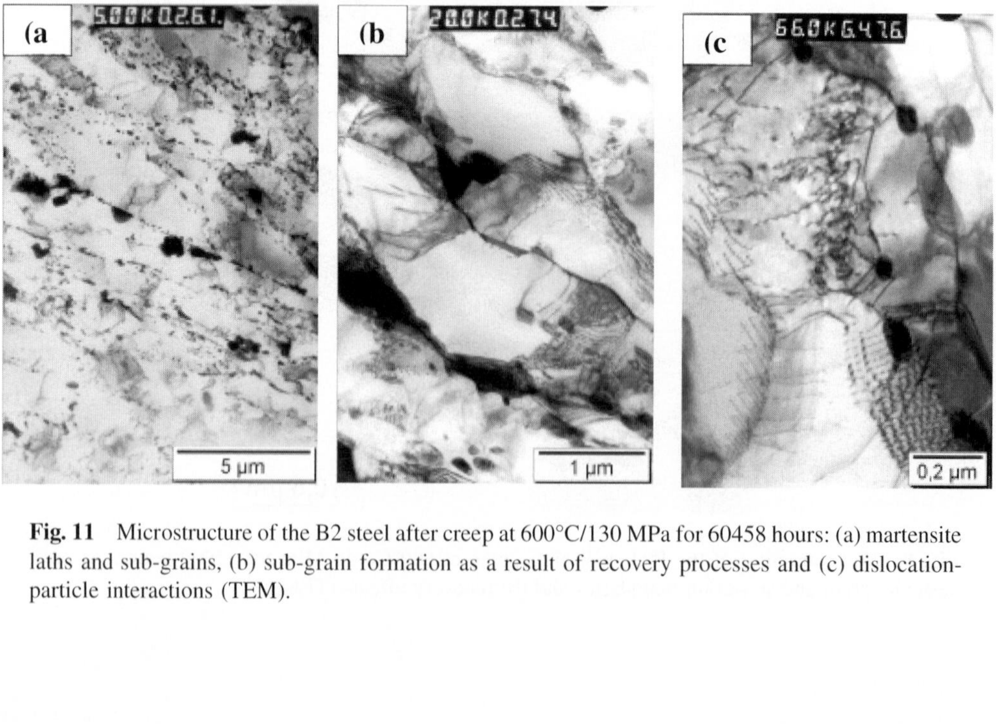

Fig. 11 Microstructure of the B2 steel after creep at 600°C/130 MPa for 60458 hours: (a) martensite laths and sub-grains, (b) sub-grain formation as a result of recovery processes and (c) dislocation-particle interactions (TEM).

This stability during the long-term exposure, attributed to the addition of 100 ppm B and precipitation of boron-rich $M_{23}(C,B)_6$, confirms the earlier observation after shorter exposure durations.[1, 10] Additionally, the very fine MX precipitates are very effective barriers for dislocation movement during creep deformation. The Laves phase (Fe_2Mo) appear in microstructure of B2 steel after several thousands hours of creep test, mainly close to carbides. The development of microstructure during creep deformation up to 60458 hours exposure at 600°C has been described in detail elsewhere.[1, 9, 10]

ROLE OF BORON IN 9% Cr STEELS

The long-term creep results at 600 and 650°C of steel B2 confirmed the beneficial effect of boron on stabilizing the microstructure and significantly improving creep rupture strength this steel.[1, 9, 10] The systematic quantitative analyses of $M_{23}C_6$ carbides in B2 steel indicate their growth resistance during ageing and creep exposure. According with $M_{23}C_6$ particle coarsening expression for Oswald ripening:

$$r^3_{av} \approx K \times t,$$

where: r_{av} - average particle radius, t - time and K - constant; the rate constant K values were calculated for investigated steels[10] (for steel P92: 1.3×10^{-28}, for steel B2: $2.2 - 4.5 \times 10^{-29}$ m^3s^{-1}). These measurements prove the slower growth of these particles in steel B2.

This means that boron in B2 steel retards growth of the $M_{23}C_6$ carbides by forming $M_{23}(C, B)_6$ as it does in austenitic stainless steels.[2] It is known that boron segregates to microstructural defects (e.g. grain- and sub-grain boundaries, dislocations) forming $M_{23}(C, B)_6$ borocarbides in the temperature range of 650–950°C.[2, 4, 12] The investigations of 9–12% steels using atom-probe field-ion microscopy and field emission auger electron spectroscopy has revealed that more than half of the total boron content was incorporated into the $M_{23}C_6$.[3, 4] The basic disadvantage of these techniques is the impossibility of localisation of borocarbides.

The trace autoradiography method has enabled the presence and distribution of boron in $M_{23}C_6$ particles to be confirmed. According to the expectations, boron in steel B2 with 100 ppm B (in the as received condition) is densely distributed on prior austenite grain boundaries and within martensite laths, but in steel P92 with 30 ppm B, mostly on prior austenite grain boundaries (compare Figures 2 and 5).

The stable $M_{23}(C,B)_6$ particles on sub-grain boundaries and/or martensite lath boundaries delay the recovery processes by a pinning effect, which restricts the growth of the sub-grains. This stabilisation of the sub-structure by growth resistant $M_{23}(C,B)_6$ borocarbides considerably improves the creep resistance.[1, 9, 10]

A characteristic feature of the B2 steel was the presence of very small V(C, N) precipitates inside the sub-grains, which act as very effective barriers for dislocation movement during creep deformation causing strengthening of the material. These finely dispersed particles preserve the high dislocation density to be retained during very long-term creep exposure.[1, 9, 10]

CONCLUSIONS

Two 9% chromium steels, P92 (30 ppm B) and B2 (100 ppm) were studied using optical metallography, transmission electron microscopy techniques and the trace autoradiography method. The purpose of the work was to show the boron distribution and identify the role of boron in the as received and creep deformed steels.

The microstructures of both P92 and B2 steels in the as received condition consisted of tempered martensite. The prior austenite grain size of steel B2 was nearly twice that of the P92 steel and there was no distinct evidence of the sub-structure in B2 steel. After creep testing at 600°C, the microstructures of both steels exhibited recovery processes (reduction of dislocation density, formation of sub-grains and their growth, $M_{23}C_6$ particle coarsening). Nevertheless, the recovery processes proceeded more slowly in steel B2.

By means of the trace autoradiography method, the incorporation of boron in $M_{23}C_6$ precipitates was detected. In P92 steel (30 ppm B), boron was distributed preferentially on prior austenite grain boundaries. In B2 steel (100 ppm B), boron is densely incorporated both on prior grain- and laths boundaries and within the martensite laths.

It was confirmed that boron stabilizes the microstructure and effectively retards the recovery processes during long-term creep at 600°C in B2 steel by:

- Precipitation of growth resistant $M_{23}(C,B)_6$ borocarbides,
- Preservation of narrow martensite lath structure and sub-grain boundaries due to pinning by $M_{23}(C, B)_6$ borocarbides,
- Maintenance of a high dislocation density due to the presence of very fine V(C, N) precipitates inside the sub-grains.

ACKNOWLEDGEMENTS

This experimental work was done in co-operation between AGH University of Science and Technology, Research Centre Jülich and ARC Seibersdorf research GmbH in the European COST Actions 501 and 522.

REFERENCES

1. K. Spiradek, R. Bauer and G. Zeiler: 'Microstructural Changes During Creep Deformation of 9% Cr Steel', *Materials for Advanced Power Engineering*, J. Lecomte-Beckers, F. Schubert and P. J. Ennis, eds., Liege, 1994, 251–262.
2. A. Czyrska-Filemonowicz: 'Effect of Boron on the $M_{23}C_6$ Carbides Precipitation in Austenitic Stainless Steels', Ph.D. Thesis, University of Mining and Metallurgy, Kraków, 1976.
3. L. Lundin: 'High Resolution Microanalysis of Creep Resistant 9–12% Cr Steels', Ph.D. Thesis, Chalmers University of Technology and Goeteborg University, Sweden, 1995.

4. M. HÄTTESTRAND and H. O. ANDRÉN: 'Boron Distribution in 9–12% Cr steels', *Materials Science and Engineering*, **A-270**, 1999, 33–37.
5. P. J. ENNIS, A. ZIELINSKA-LIPIEC and A. CZYRSKA-FILEMONOWICZ: 'The Influence of Heat Treatments on the Microstructural Parameters and Mechanical Properties of P92 Steel', *Qualitative Microscopy of High Temperature Materials*, A. Strang and J. Cawley, eds., IOM Communications Ltd., 2002, 191–206.
6. H. FIRGANEK: 'Application of Autoradiographic Methods to the Investigations of Steel and Alloys with Microaddition of Boron, Lithium and Nitrogen', *Proceedings of IV Symposium to the Point Application of Radioisotopic Techniques in Industry, Medicine and Environment Protection*, Warsaw, 1987, 147–158.
7. P. J. ENNIS, A. ZIELINSKA-LIPIEC and A. CZYRSKA-FILEMONOWICZ: 'The Influence of Heat Treatments on the Microstructural Parameters of P92 Steel', *Materials Science and Technology*, **16**, 2000, 1226–1232.
8. P. J. ENNIS, A. ZIELINSKA-LIPIEC and A. CZYRSKA-FILEMONOWICZ: 'Quantitative Microscopy and Creep Strength of 9% Chromium Steels for Advanced Power Stations' *Parsons 2000: Advanced Materials for 21st Century Turbines and Power Plants*, A. Strang et al., eds., IOM Communications Ltd, 2000, 498–507.
9. Ch. STOCKER, K. SPIRADEK and K. BRYLA: 'Microstructural Evolution During Creep and Comparision of the Creep Behaviour of the New Modified 9% Cr Steels with Additions of Co and/or B', *Proceedings of 10th joint International Conference on Creep & Fracture of Engineering Materials and Structures, Part Creep Resistant Metallic Materials*, Prague, 2001, 258–267.
10. K. SPIRADEK-HAHN, P. NOWAKOWSKI and G. ZEILER: 'Boron Added 9% Cr Steels for Forged Components in Advanced Power Plants', *Advances in Material Technology for Fossil Power Plants*, R. Viswanathan, W. T. Bakker and J. D. Parker: EPRI Report 1001462, Institute of Materials, 2001, 165–176.
11. F. ABE, H. OKADA, M. TABUCHI, T. ITAGAKI, K. KIMURA, K. YAMAGUCHI and M. IGARASHI: 'NIMS Efforts in Advanced 9-12% Cr Steels for 650°C USC Boilers', MPA-NIMS Workshop on Advanced 9–12% Cr Steels, MPA Stuttgart, **48**, 2002, 1–13.
12. G. HENRY, PH. MAITREPIERRE, B. MICHAUT and B. THOMAS: 'Kinetics and Morphology of the Intergranular Precipitation $M_{23}(C, B)_6$ Borocarbides in Austenite', *Cent. Doc. Sider. Cir. Inf. Tech.*, **33**(4), 1976, 961–990.

Improvement of Creep Strength by Boron and Nano-Size Nitrides for Tempered Martensitic 9Cr-3W-3Co-VNb Steel at 650°C

F. Abe, T. Horiuchi and M. Taneike
Steel Research Center
National Institute for Materials Science (NIMS)
1-2-1 Sengen, Tsukuba 305-0047
Japan

K. Sawada
Materials Information Technology Station
National Institute for Materials Science (NIMS)
1-2-1 Sengen, Tsukuba 305-0047
Japan

ABSTRACT

In order to improve the long-term creep strength of 9Cr-3W base steel at 650°C, we have been trying to stabilise the martensitic microstructure in the vicinity of prior austenite grain boundaries by the addition of boron and by a dispersion of nano-size MX nitrides. Creep tests were carried out at 650°C for up to about 3×10^4 hours. With a combination of high tungsten of 3% and high boron exceeding 90 ppm, the greater part of the boron added is present as undissolved large particles of tungsten boride after conventional normalising and tempering. The tungsten borides are substantially dissolved by raising the normalising temperature to 1150°C. The dissolution of boron in the matrix promotes an enrichment of boron in the $M_{23}C_6$ carbides, which results in lower minimum creep rate and hence longer time to rupture, especially at low stress conditions. A dispersion of nano-size V- and Nb-nitrides along boundaries as well as in the matrix of 9Cr-3W base steel is achieved by reducing carbon concentration below 0.02%. This gives rise to excellent high-temperature creep strength at 650°C, as shown by approximately two orders of magnitude longer time to rupture than P92.

INTRODUCTION

Tempered martensitic 9-12Cr steels are favoured for high temperature applications, such as boiler and turbine components in power plants, due to their excellent combination of mechanical and oxidation-resistant properties.[1] Suitable values of creep strength coupled with sufficient toughness are achieved by dispersion of fine carbonitride particles and by fine substructures. The microstructure of the steels consists of lath, block and packet substructures containing a high density of dislocations and fine carbonitride particles. The particles usually consist of chromium-rich $M_{23}C_6$ carbides of 100–300 nm in size and of V- and Nb-rich MX carbonitrides of 5–20 nm in size after heat treatments, where M means metallic elements.[2] During long-term exposure at high temperature, intermetallic compounds such as Fe_2(Mo, W) Laves phase can precipitate from supersaturated solid solution. The addition of 9Cr or more in combination with appropriate minor elements, such as silicon and sulphur, is required from the viewpoint of suitable high-temperature oxidation resistance.[3, 4] The analysis of long term creep data in the National Institute for Materials Science (NIMS) Creep Data Sheets suggests that special attention should be paid to stabilize the microstructure in the vicinity of prior austenite grain boundaries for the improvement of the long term creep strength.[5, 6]

Since 1997, NIMS has been conducting a research and development project on advanced ferritic steels for application to large diameter and thick section boiler components such as main steam pipe and header of ultra-supercritical (USC) plant at 650°C and 350 atmospheric pressure.[7] The project involves the development of tungsten strengthened 9Cr steels with long-term creep rupture strength higher than P92 (9Cr-0.5Mo-1.8WVNb steel) and P122 (11Cr-0.4Mo-2WCuVNb steel) at 650°C.

This paper describes the alloy design philosophy for the improvement of creep strength of a tempered martensitic 9Cr-3W base steel by the stabilisation of the martensitic microstructure in the vicinity of prior austenite grain boundaries. The addition of boron, exceeding 0.01 wt.% in combination with minimised nitrogen is shown to be very effective for the stabilisation of the lath martensitic microstructure in the vicinity of grain boundaries, through the stabilisation of $M_{23}C_6$ carbides along lath boundaries. A dispersion of nano-size MX carbonitride particles in combination with the minimised amount of $M_{23}C_6$ carbides also gives rise to excellent high-temperature creep strength at 650°C.

ALLOY DESIGN PHILOSOPHY

In conventional 9-12Cr steels such as P92 and P122, fine MX carbonitride particles are distributed in the matrix within lath as well as along boundaries, while $M_{23}C_6$ carbide particles are distributed only along boundaries after heat treatment.[2] The volume fraction of $M_{23}C_6$ carbide particles in the steels is much larger than that of MX carbonitride particles. On the other hand, both the mean size after heat treatment and the coarsening rate during creep are much larger for $M_{23}C_6$ carbide particles than for MX carbonitride particles. The lower coarsening rate of MX carbonitrides is mainly due to the much lower solubility of V and Nb in the matrix than of Cr. The migration of boundaries, causing the lath or block coarsening, is closely correlated with the onset of acceleration or tertiary creep. Furthermore lath or block coarsening by the migration of boundaries with absorbing excess dislocations is the major process in the acceleration of creep.[8, 9] For the suppression of boundary migration during creep, the refinement of precipitate particles after heat treatment and the reduction of the coarsening rate of precipitate particles for up to long times during creep are required.

We have been trying to stabilize the lath martensitic microstructure in the vicinity of prior austenite grain boundaries for a 9Cr-3W base steel by the addition of boron to stabilize $M_{23}C_6$ carbides at lath boundaries.[10, 11] Boron is known to segregate to grain boundaries.

Table 1 Chemical compositions of 9Cr-3W-3Co-VNb steels with different boron concentration.

C	S	Mn	Cr	W	V	Nb	Co	N	B
0.078	0.31	0.50	8.94	2.94	0.19	0.050	3.03	0.0019	<0.0001>
0.077	0.29	0.51	8.95	2.93	0.19	0.050	3.03	0.0011	0.0048
0.075	0.29	0.50	8.96	2.92	0.19	0.049	3.01	0.0016	0.0092
0.078	0.30	0.51	8.99	2.91	0.19	0.050	3.01	0.0034	0.0139

Table 2 Chemical compositions of 9Cr-3W-3Co-VNb steels with different carbon concentration.

C	Si	Mn	Cr	W	V	Nb	Co	N	B
0.002	0.29	0.51	9.19	2.96	0.20	0.060	3.09	0.049	0.0070
0.018	0.29	0.50	9.16	2.91	0.20	0.058	2.94	0.050	0.0058
0.047	0.30	0.51	9.24	2.90	0.20	0.059	3.07	0.050	0.0063
0.078	0.31	0.51	9.26	2.93	0.20	0.061	3.08	0.049	0.0064
0.120	0.30	0.50	9.27	2.93	0.20	0.058	3.08	0.048	0.0065
0.160	0.30	0.51	9.26	2.94	0.20	0.058	3.06	0.047	0.0061

However, it is also reported that excess addition of boron and nitrogen promotes the formation of boron-nitrides at grain boundaries, which offsets the benefit due to boron and nitrogen.[12] In the present work, the addition of a large amount of boron exceeding 0.01% is examined in combination with no nitrogen addition. On the other hand, to achieve further refinement of precipitate particles after heat treatment, we have been trying to distribute only MX carbonitrides and to eliminate $M_{23}C_6$ carbide particles. It is known that the coarsening rate of V- and Nb-carbides is also very low as low as their nitrides.[13] However, the addition of carbon to a 9Cr steel causes the formation of a large amount of $M_{23}C_6$ carbides rich in Cr. Therefore, it is crucial for 9Cr steels to reduce carbon content to very low amounts, so as to promote the formation of MX carbonitrides as very fine and thermally stable particles for prolonged periods of exposure at elevated temperatures.

EXPERIMENTAL PROCEDURE

The chemical compositions of the steels examined are given in Tables 1 and 2.[11, 14] They were based on Fe-9Cr-3W-3Co-0.2V-0.05Nb (mass%) with different boron, carbon and nitrogen. The steels were prepared by vacuum induction melting to 50 kg ingots. Hot forging

Table 3 Heat treatment conditions of 9Cr-3W-3Co-VNb steels with different boron concentration.

Wt.% B	**Normalising**	**Tempering**
0	1050°C × 1 Hour	790°C × 1 Hour
0.0048	1050°C × 1 Hour	790°C × 1 Hour
0.0092	1050°C × 1 Hour	790°C × 1 Hour
0.0139	1050°C × 1 Hour	800°C × 1 Hour

Table 4 Heat treatment conditions of 9Cr-3W-3Co-VNb steels with different carbon concentration.

Wt.% C	**Normalising**	**Tempering**
0.002	1100°C × 0.5 Hour	800°C × 1 Hour
0.018	1100°C × 0.5 Hour	800°C × 1 Hour
0.047	1100°C × 0.5 Hour	800°C × 1 Hour
0.078	1100°C × 0.5 Hour	800°C × 1 Hour
0.12	1100°C × 0.5 Hour	800°C × 1 Hour
0.16	1100°C × 0.5 Hour	800°C × 1 Hour

and hot rolling were performed to produce plates of 20 mm in thickness for the boron series steels and rods of 20 mm in diameter for the carbon series steels. The plates and rods were normalised and then tempered as given in Tables 3 and 4. Although no nitrogen was added to the boron series steels give in Table 1, residual nitrogen was analysed to be 10-30 ppm. Residual Al was also as low as 30 ppm. The normalising temperature of the 0.0139 boron steel (1080°C (1353 K)) was 30° higher than that of the other boron series steels in Table 1 to obtain the same prior-austenite grain size among the steels (50–60 μm). There was no major difference in Vickers hardness (210–220) among the boron series steels after tempering. No carbon was added to the 0.002°C steel in Table 2. Creep tests were carried out at 650°C (923 K) for up to about 3×10^4 hours under constant load condition, using specimens of 10 mm in gauge diameter and 50 mm in gauge length. In order to estimate the creep strength at 650°C and 10^5 hours by Larson-Miller parameter method, the creep rupture test was also carried out for the very low carbon 0.002°C steel at 700°C (973 K) for up to 4×10^3 hours.

The concentration of boron in precipitates was analysed by field emission scanning Auger spectroscopy (FE-AES, ULVAC-PHI 670xi). FE-AES is quite useful to characterize boron in precipitates because of its high spatial resolution, high sensitivity to light elements and ease to correspond to microstructure. The details of FE-AES measurement are described elsewhere.[11]

NEW STEELS WITH IMPROVED CREEP RUPTURE STRENGTH

The creep rupture data for the 0.0139B and 0.002C steels at 650°C, which are the strongest steels of the boron series (Table 1) and of the carbon series (Table 2), respectively, as will be described later, are shown in Figure 1. In this figure, the results for conventional steels, T91 (9Cr-1Mo-VNb)[15] and P92 (9Cr-0.5Mo-1.8W-VNb)[16] are also shown for comparison. Although only the creep rupture data for up to about 3 x 10^4 h are available at present, the 0.0139B and 0.002C steels exhibit much higher creep rupture strength than T91 and P92 at 650°C. The time to rupture of the 0.002%C steel presented in this study is approximately two orders of magnitude higher than the P92 steel and approximately three orders of magnitude higher than the T91 steel at 650°C and 140 MPa. The creep deformation behaviour and microstructure of the new steels are described as follows.

STABILISATION OF LATH MARTENSITIC MICROSTRUCTURE NEAR PRIOR AUSTENITE GRAIN BOUNDARIES BY BORON ADDITION

Effect of Boron Content on Creep Deformation Behaviour

Figure 2 shows the effect of boron addition on creep rupture strength of 9Cr-3W-3Co-VNb steel at 650°C. At high stresses above 100 MPa, there is substantially no difference in time to rupture between the 0B, 0.0048B and 0.0092B steels, but only the 0.0139B steel exhibits longer time to rupture. With decreasing stress below 100 MPa, the time to rupture significantly increases with increasing boron content. In other words, the degradation in creep rupture strength at long times becomes more significant with decreasing boron content. The present results suggest that the addition of boron scarcely improves the short-term creep rupture strength of 9Cr-3W-3Co-VNb steel at 650°C but that it effectively suppresses the degradation in creep rupture strength at long times.

Figure 3 shows the creep rate versus time curves of the 0B, 0.0092B and 0.0139B steels at 650°C. The creep rate curves consist of a primary or transient creep region, where the creep rate decreases with time, and of a tertiary or acceleration creep region, where the creep rate increases with time after reaching a minimum creep rate. There is substantially no steady-state region. In the transient creep region from the initial stage to a time to reach a minimum creep rate, the creep rate is approximately the same among the steels with different boron content. It should be noted that the transient creep region of the steels with boron continues for up to longer times than that of the steel without boron, resulting in lower minimum creep rate and longer rupture time.

The distribution of boron in the steel is shown in Figure 4, as a function of distance from prior austenite grain boundaries. The results show that boron is enriched in the $M_{23}C_6$ carbides, which is more significant in the vicinity of prior austenite grain boundary. The microstructure observations showed that the $M_{23}C_6$ carbides were distributed mainly along prior austenite grain boundaries and along lath, block and packet boundaries inside the grain.[11] The fine distribution of $M_{23}C_6$ carbides along boundaries was still maintained in the 0.0139B steel during high-temperature exposure. This indicates that boron reduces the rate of Ostwald

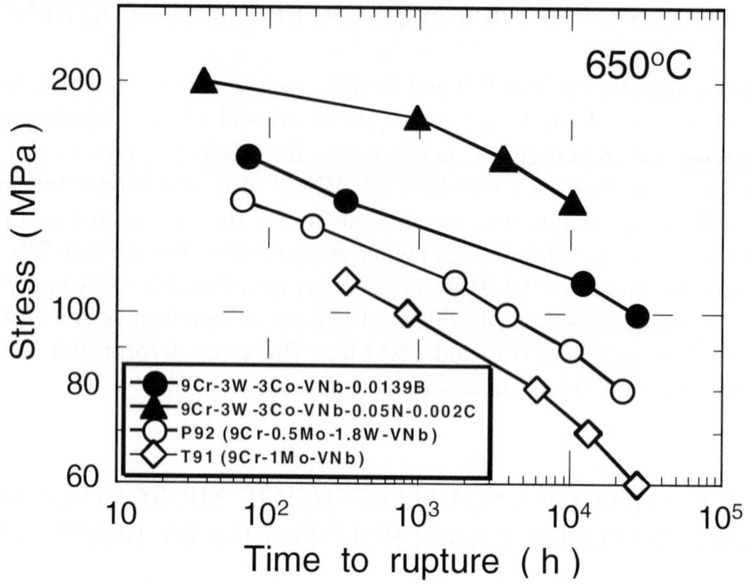

Fig. 1 Creep rupture data for the new steels (solid triangle, the 0.0139B steel in Table 1, solid circle, the 0.002C steel in Table 2) at 650°C, together with those for conventional steels T91 and P92.

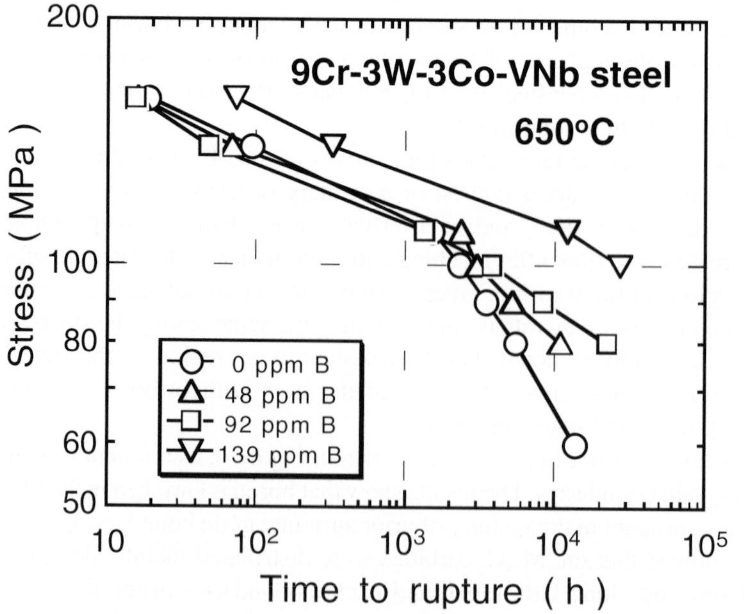

Fig. 2 Effect of boron addition on creep rupture strength of 9Cr-3W-3Co-VNb steel at 650°C.

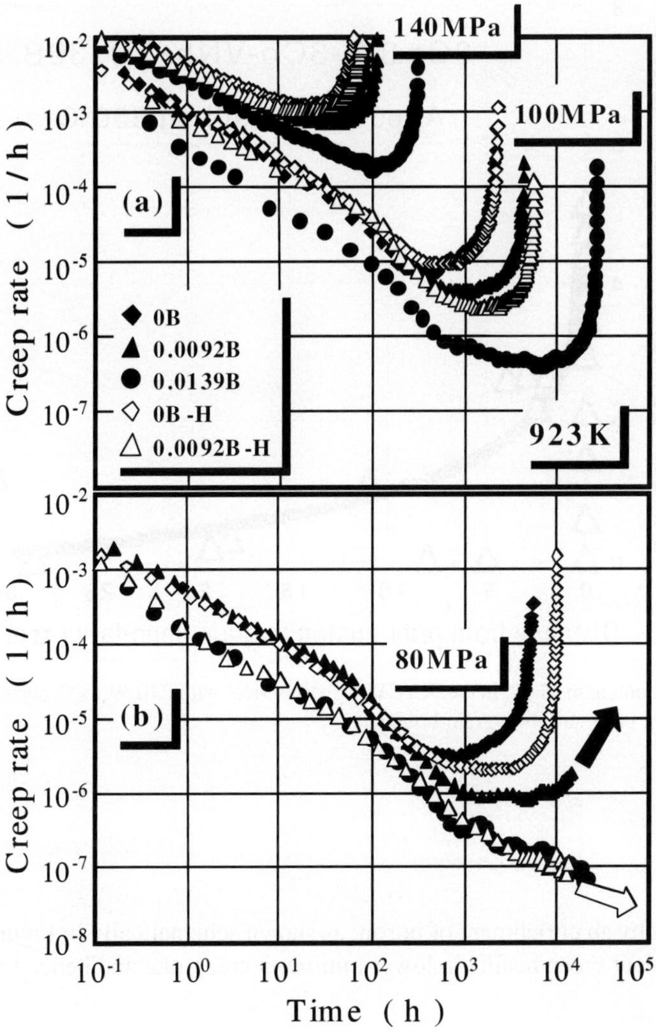

Fig. 3 Creep rate versus time curves of the 9Cr-3W-3CoVNb steels with 0.0139 wt.% boron and without boron at 650°C.

ripening of $M_{23}C_6$ carbides. No evidence was found for the enrichment of boron in the precipitates of Fe_2W Laves phase. An enrichment of boron in $M_{23}C_6$ carbides has been reported by means of Atom Probe Field Ion Microscopy.[17, 18] But they did not specify whether the site of $M_{23}C_6$ in which boron was enriched was located in the vicinity of grain boundaries or not. The present results indicate that the longer duration of transient creep region in the steel containing boron results from the stabilisation of lath martensitic microstructure for up to longer times through the stabilisation of $M_{23}C_6$ carbides in the vicinity of prior austenite

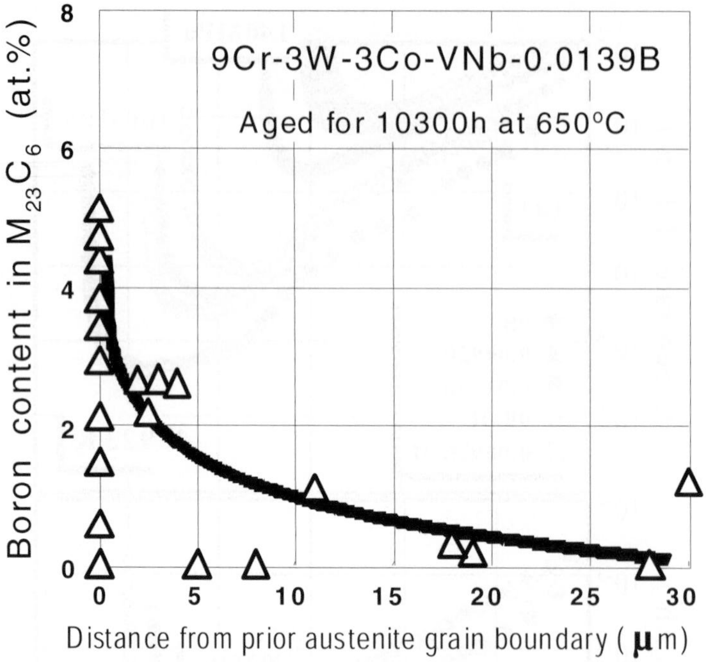

Fig. 4 Boron content in $M_{23}C_6$ in the 9Cr-3W-3CoVNb steel with 0.0139 wt.% boron, as a function of distance from prior austenite grain boundary.

grain boundaries by an enrichment of boron, as shown schematically in Figure 5. The longer duration of transient creep results in lower minimum creep rate and hence longer creep life.

Re-Dissolution of Undissolved Tungsten Borides by High-Temperature Normalising

It should be noted that most of boron added in the 0.0092B and 0.0139B steels have already precipitated as large inclusions of tungsten boride during heat treatment before creep test. Figure 6 shows the microstructure after tempering observed by SEM incorporated in FE-AES and the results of AES analysis of precipitates for the 0.0139B steel. The particles denoted by (b), consisting of mainly W and boron and having a size of over 1 μm, are distributed at random throughout the specimen, which was also typical for the 0.0092B steel. The most probable candidate for the undissolved coarse borides is FeW_2B_2, where parts of Fe and W atoms are substituted with Cr. The present results indicate that the combination of high tungsten and high boron causes a large amount of undissolved tungsten borides after conventional heat treatment. The undissolved tungsten borides consume effective

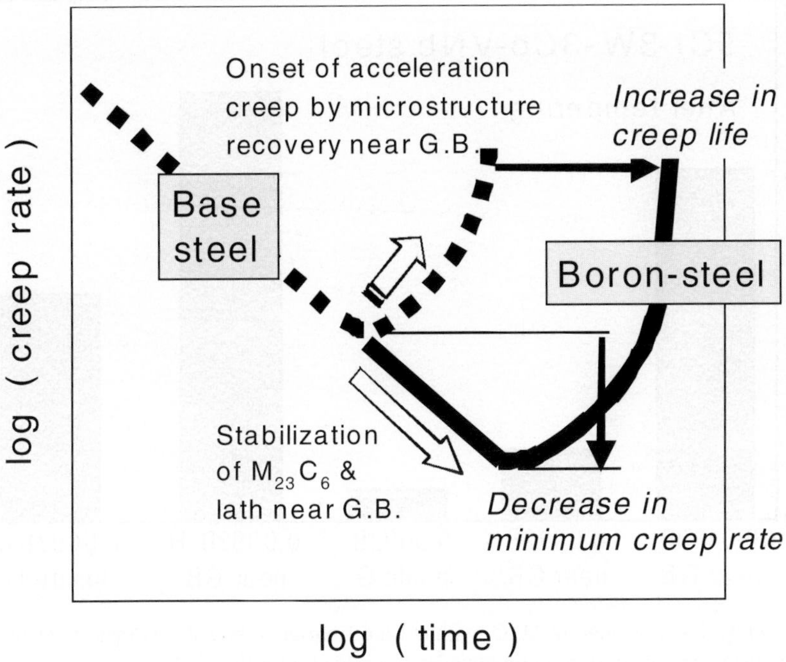

Fig. 5 Schematic creep rate versus time curves, showing decrease in minimum creep rate and increase in creep life by boron addition.

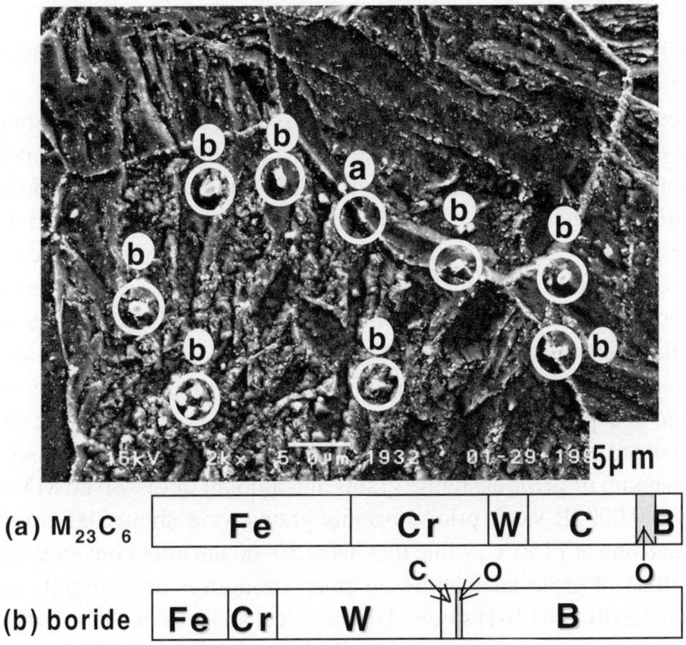

Fig. 6 Analysis of precipitate particles by AES in the 0.0139B steel after tempering.

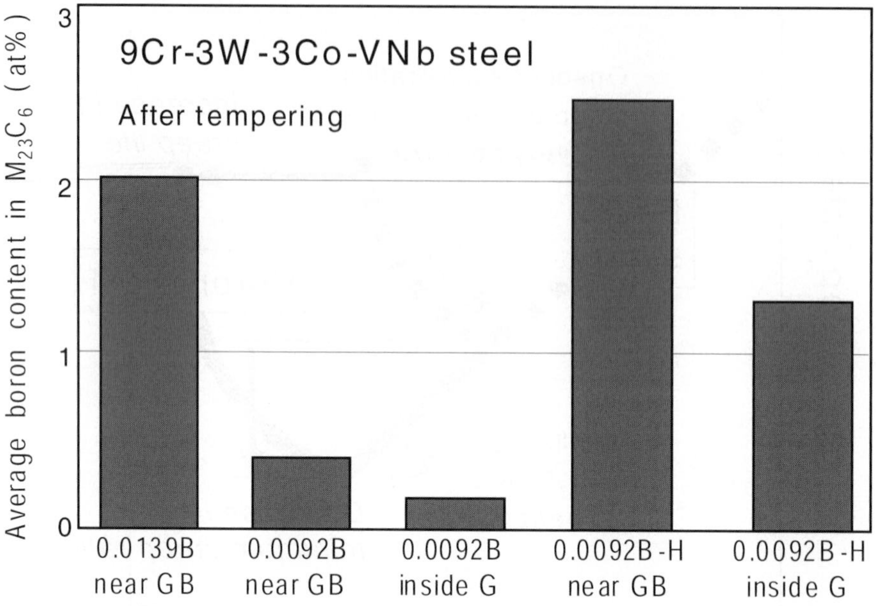

Fig. 7 Average boron content in $M_{23}C_6$ within 2 μm from grain boundaries (near GB) and inside grain more than 2 μm away from grain boundaries (inside G) for the steels.

boron atoms which improves the stability of lath martensitic microstructure by an enrichment in the $M_{23}C_6$ carbides.

The undissolved coarse borides are substantially dissolved by high-temperature normalising at 1150°C.[11] The 0B and 0.0092B steels subjected to the high-temperature normalising are denoted as 0B-H and 0.0092B-H steels, respectively, in Figure 3. Figure 7 shows the average boron concentration in the $M_{23}C_6$ carbides of the 0.0092B and 0.0092B-H steels after tempering, where the $M_{23}C_6$ carbides are roughly divided into 2 groups; those existing within 2 μm from prior-austenite grain boundaries (near GB) and those existing more than 2 μm away from grain boundaries (inside G). The boron content in the $M_{23}C_6$ carbides is significantly increased in the 0.0092B-H steel to about the same extent as the 0.0139B steel. The high-temperature normalising results in lower minimum creep rate and hence longer time to rupture as shown in Figure 3, especially at low stress and long time conditions. Of course, the other factors is also changed by high-temperature normalising in addition to dissolving borides, for example, coarsening of prior-austenite grains and amount of dissolved MX carbonitrides. In both the 0B and 0.0092B steel, prior-austenite grains were about 200 μm after the high-temperature normalising at 1150°C, while they were 50–60 μm after conventional normalising at 1050°C. The effect of grain coarsening on creep strength is only slightly as can be seen from the results on the 0B and 0B-H steels. The chemical analysis of electrolytically extracted residues showed that the amount of undissolved MX carbonitrides in the present steels was very small for both the high-temperature normalising at 1150°C and conventional normalising

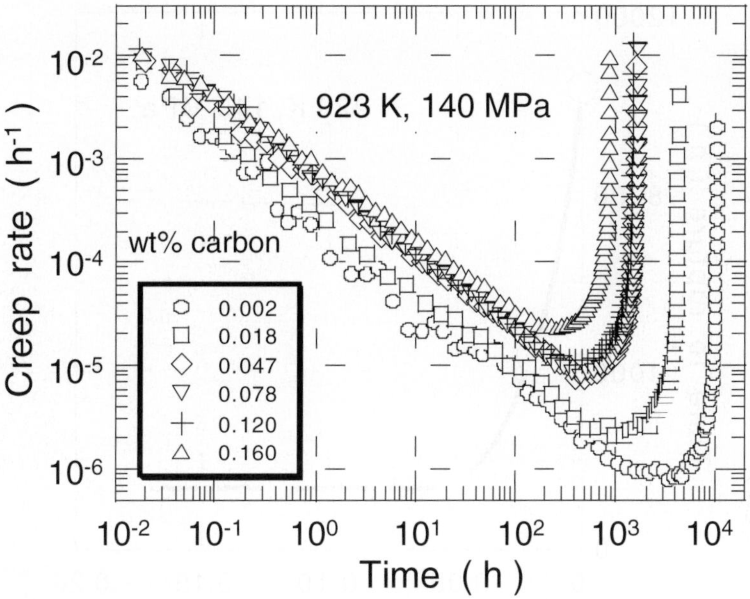

Fig. 8 Creep rate versus time curves of 9Cr-3W-3Co-VNb steel with different carbon concentration at 650°C and 140 MPa.

at 1050°C. This is owing to the very low concentration of nitrogen in the present steels 10–30 ppm. Therefore, the results presented in Figure 3 showing that the effect of high-temperature normalising on creep strength is more significant in the 0.0092B steel than in the 0B steel is mainly caused by re-dissolution of undissolved large tungsten borides. The amount of effectively utilised boron is essential to improve long term creep strength and the high-temperature normalising can increase it.

STABILISATION OF LATH MARTENSITIC MICROSTRUCTURE BY NANO-SIZE MX CARBONITRIDES ALONG BOUNDARIES

Figure 8 shows the creep rate versus time curves of the 9Cr-3W-3Co-VNb steel with different carbon concentrations at 650°C and 140 MPa.[12] The onset of acceleration creep is retarded up to longer times with decreasing carbon concentrations below 0.018%. The retardation of the onset of acceleration creep causes the longer duration of primary creep, which results in lower minimum creep rate and longer time to rupture. The time to rupture is independent of carbon concentration in the higher carbon region above 0.047% but it significantly increases in the lower carbon region below 0.018% as shown in Figure 9, which is also typical in applied tensile stresses within the range of 140–200 MPa. It should be also noted that due to

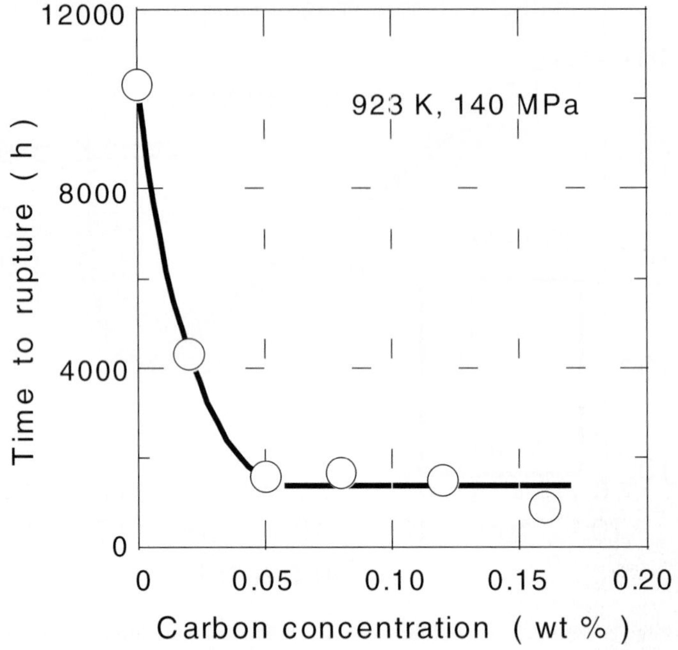

Fig. 9 Carbon concentration dependence of time to rupture for the 9Cr-3W-3Co-VNb steel at 650°C and 140 MPa.

its fine substructures of martensite, the 0.002%C steel exhibits enough toughness values of 100–150 J by Charpy impact test. Enough toughness higher than 50 J is a basic requirement for thick section structural components.

A large number of fine precipitate particles less than 10 nm in size are distributed along prior austenite grain boundaries as well as along lath, block and packet boundaries in the 0.002C steel after tempering, as shown in Figure 10. This is quite different from that in the 0.078C steel, where the large particles of 100 to 300 nm are distributed together with the fine particles. The fine precipitate particles were identified as MX-type carbonitrides and were confirmed via energy dispersive microanalysis to be rich in V and Nb. No Cr-nitrides were detected. The MX carbonitrides were analysed to be substantially V- and Nb-nitrides with a very low carbon concentration of 0.002%. In Figure 10, the large particles in the 0.078C steel were identified as $M_{23}C_6$ carbides rich in Cr. The fine MX carbonitride particles are also distributed in the matrix within lath in addition to along boundaries, while the $M_{23}C_6$ carbide particles are distributed only along boundaries. The amount of $M_{23}C_6$ carbides significantly decreases with decreasing carbon concentration, while the population of fine MX carbonitride particles on boundaries significantly increases. This causes a significant decrease in inter-particle distance and hence a significant increase in pinning force for migrating boundaries.

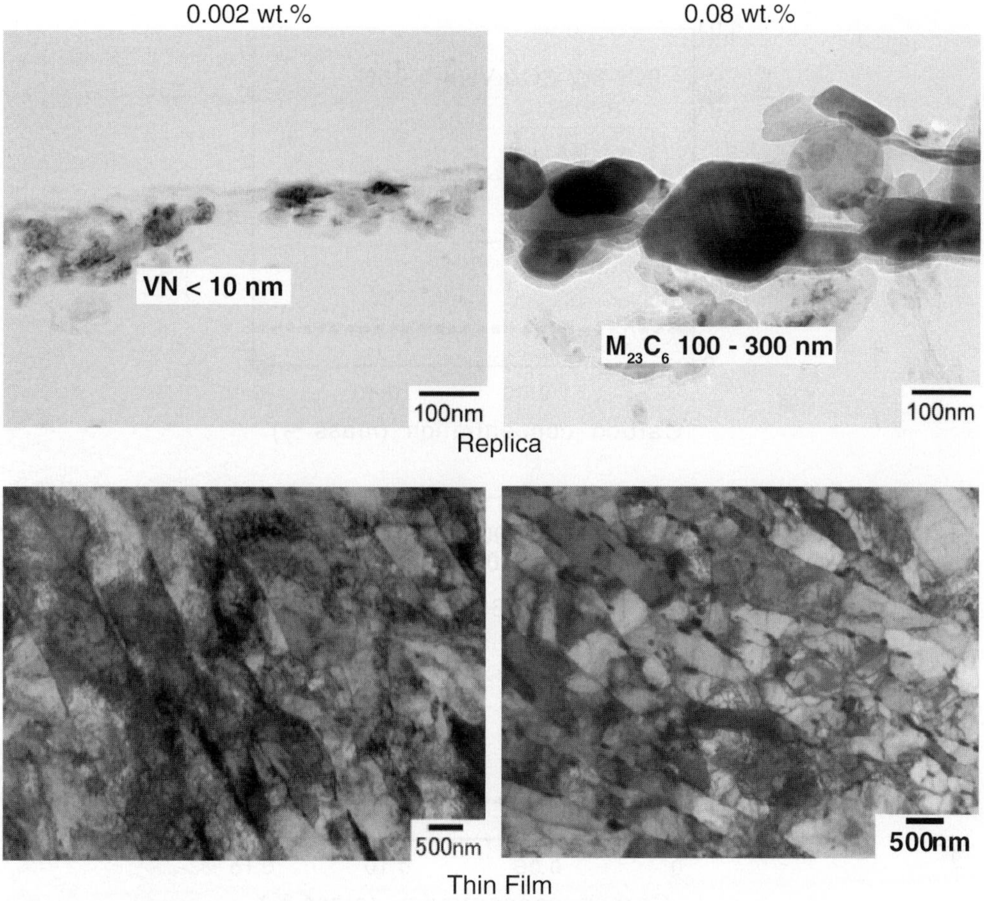

Fig. 10 TEM micrographs of replicas and thin films of the 0.002 and 0.078C steels after tempering.

The mole fraction of equilibrium phases in the 9Cr-3W-3Co-0.2V-0.05Nb-0.05N steel during tempering at 800°C and during creep at 650°C was evaluated using Thermo-calc. This is shown in Figure 11 as a function of carbon concentration. The Thermo-calc evaluation predicts the precipitation of three phases, $M_{23}C_6$ carbides, MX carbonitrides and Fe_2W Laves phase, at 800°C. However, only the two phases of $M_{23}C_6$ carbides and MX carbonitrides were detected in the present investigation after tempering at 800°C for 1 hour. The tempering time of 1 h is too short to precipitate the Fe_2W Laves phase at 800°C. The amount of $M_{23}C_6$ carbides decreases with decreasing carbon concentration, while that of MX carbonitrides is substantially constant in the range of 0-0.15% carbon. The MX carbonitrides are dominant at low carbon concentration below 0.02%. During creep at 650°C, the amount of $M_{23}C_6$ carbides and MX carbonitrides is substantially the same as that at 800°C, while the amount

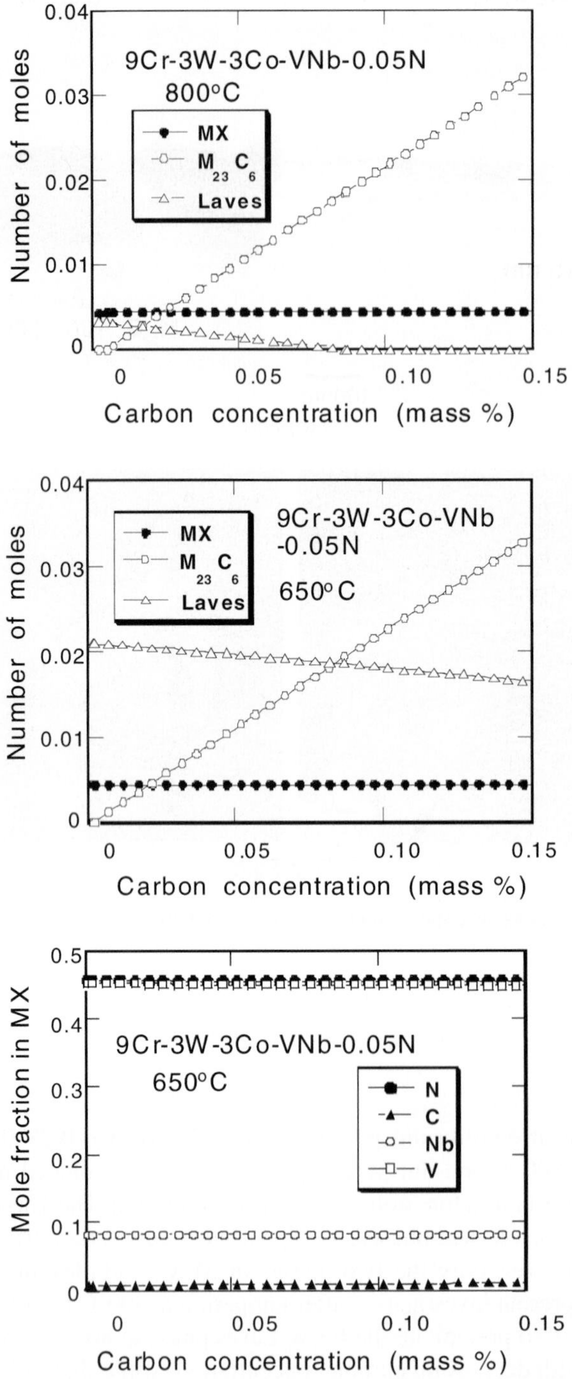

Fig. 11 Thermo-calc evaluation of phases appearing in the 9Cr-3W-3Co-0.2V-0.05Nb-05N steel during tempering at 800°C and during creep at 650°C, as a function of carbon concentration.

of the Fe_2W Laves phase is much larger than that at 800°C. It should be noted that the MX carbonitrides consists of mainly V and N in the range of 0-0.15%C. This suggests that the MX carbonitrides in the present steels are substantially regarded as vanadium nitrides. This is consistent with the experimental results.

The large number of nano-size MX nitride particles linking together along boundaries after heat treatment and the low coarsening rate of MX nitride particles during creep in the specimen with very low carbon of 0.002% suggest that large pinning force for boundary migration is maintained for long times and the onset of tertiary creep is retarded up to long times. This effectively decreases the minimum creep rate and increases the time to rupture. The precipitation of Fe_2W Laves phase also took place during creep at 650°C. The amount of Laves phase precipitated was approximately the same among the present steels with 0.002 to 0.16% carbon. Because the size of Laves phase particles was quite large from 500-700 nm, and the number density was low after creep for 3.6×10^3 hours at 650°C, precipitation strengthening due to the Laves phase is thought to be negligibly small. Furthermore, no evidence was found for the precipitation of detrimental phase of Z-phase, denoted as Cr(Nb,V)N, within the present test conditions. In a conventional 12Cr steel with 0.29Nb, 0.28V and 0.074N, massive particles of Z-phase form at the expense of fine MX carbonitrides and hence the precipitation of Z-phase causes a decrease of precipitation strengthening due to MX carbonitrides.[19] The precipitation of Z-phase is also observed in T91 (9Cr-1MoVNb) after long term creep at 600 and 650°C.[20] The lower concentrations of Z-phase forming elements in the present 0.002C steel (9Cr, 0.05Nb and 0.05N) than in the above 12Cr steel suggest less pronounced formation of Z-phase in the present steel.

Goecmen et al.[21] reported that a dispersion of fine nitrides was achieved in martensitic 9 and 12Cr steels, but they did not provide any creep strength data. Their steels contained high nitrogen concentrations of 0.16 and 0.18% as well as high V concentrations of 0.66 and 0.76%. Consequently, 9Cr steels with high nitrogen concentrations of 0.074 and 0.103% and with low carbon concentration of 0.002% were prepared, and creep tested at 650°C. The results showed that the creep strength was rather lower for the high nitrogen steels with 0.074 and 0.103N than the present steel with 0.05N. The lower creep strength is due to the larger number density of undissolved large nitride particles increases with increasing nitrogen concentration. The undissolved large nitride particles consume fine nitride particles. The MX nitride-forming elements should be completely dissolved by appropriate heat treatments to maximize dispersion strengthening and that the complete dissolution is obtained at relatively low nitrogen concentration of 0.05%.

The creep rupture strength at 650°C and 10^5 hours was estimated by the Larson-Miller parameter method. Using the creep rupture data for up to about 1×10^4 hours at 650°C and those for up to about 4×10^3 hours at 700°C, Figure 12 shows the creep rupture stress for the present 0.002C steel as a function of Larson-Miller parameter, $T (C + \log t)$, where T is the absolute temperature, t is the rupture time and C is a constant. The data for P92[16] are also included in this figure for comparison. In this study, C = 36.11[22] used for P92 was also employed. The creep rupture strength at 650°C and 10^5 hours is estimated to be 100 and 71 MPa for the present 0.002C steel and P92, respectively.

It is concluded that a dispersion of nano-size MX nitride particles is achieved for the 9Cr martensitic steel by reducing carbon concentration to 0.002% and that their low coarsening rate at elevated temperatures gives rise to excellent creep strength up to long times at 650°C,

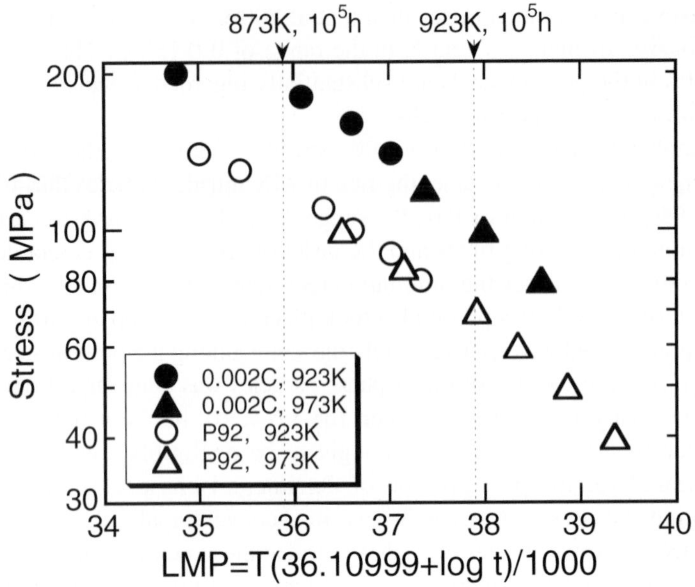

Fig. 12 Larson-Miller plot for the 0.002C steel and ASME-P92.[16]

as shown by approximately two orders of magnitude longer time to rupture than P92. Although the creep rupture strength is higher for the 0.002C steel than for the 0.0139B steel for up to 3×10^4 hours at 650°C as shown in Figure 1, it should be noted that the slope of stress versus time to rupture curves becomes larger for the 0.002C steel than for the 0.0139B steel at long times. Therefore, further stabilisation of martensitic microstructure containing nano-size MX nitrides will be strongly required.

SUMMARY

1. The 9Cr-3W-3Co-VNb steels strengthened by boron addition exceeding 100 ppm and by a dispersion of nano-size MX nitrides exhibit much higher creep rupture strength than conventional steels T91 and P92 at 650°C.
2. The effect of boron is due to the stabilisation of the lath martensitic microstructure for up to long times through the stabilisation of $M_{23}C_6$ carbides in the vicinity of prior austenite grain boundaries by an enrichment of boron in the $M_{23}C_6$ carbides. However, at a combination of high tungsten of 3% and high boron exceeding 90 ppm, the greater part of boron added is present as undissolved large particles of tungsten boride after conventional normalising and tempering. The tungsten borides are substantially dissolved by raising normalising temperature to 1150°C. This increases the concentration of effective boron,

which improves the stability of lath martensitic microstructure, and improves the creep rupture strength at 650°C.

3. A dispersion of nano-size MX nitride particles along boundaries as well as in the matrix is achieved by reducing carbon concentration below 0.02%. The MX nitrides were identified as substantially vanadium nitrides. The dispersion strengthening using nano-size MX nitrides gives rise to excellent high-temperature creep strength at 650°C, as shown by approximately two orders of magnitude longer time to rupture than P92.

REFERENCES

1. W. T. Bakker and B. Nath: *Proceedings of the 3rd EPRI Conference on Advances in Materials Technology for Fossil Power Plants*, The Institute of Materials, Swansea, UK., 2001, 1–4.
2. K. Sawada, K. Kubo and F, Abe: *Mater. Sci. Eng.*, **A319-321**, 2001, 784–787.
3. P. Ennis and W. J. Quadakkers: *Proceedings of the 7th Liege Conference on Materials for Advanced Power Engineering 2002*, F. Schubert, ed., Liege, Belgium, 2002, 1131–1142.
4. H. Kutsumi, T. Itagaki and F. Abe: *Proceedings of the 7th Liege Conference on Materials for Advanced Power Engineering 2002*, F. Schubert, ed., Liege, Belgium, 2002, 1629–1638.
5. H. Kushima, K. Kimura and F. Abe: *Tetsu-to-Hagané*, **85**, 1999, 841–847.
6. F. Abe, T. Horiuchi, M. Taneike, K. Kimura, S. Muneki and H. Okada: *Proceedings of TMS Symposium on Creep Deformation: Fundamentals and Applications*, R. S. Mishra, J. C. Earthman and S. V. Raj, eds., Seattle, USA., 2002, 341–350.
7. F. Abe, H. Okada, S. Wanikawa, M. Tabuchi, T. Itagaki, K. Kimura, K. Yamaguchi and M. Igarashi: *Proceedings of the 7th Liege Conference on Materials for Advanced Power Engineering 2002*, F. Schubert, ed., Liege, Belgium, 2002, 1397–1406.
8. F. Abe, S. Nakazawa, H. Araki and T. Noda: *Metall. Trans.*, **23A**, 1992, 469–477.
9. F. Abe: *Mater. Sci. Eng.*, **A319-321**, 2001, 770–773.
10. T. Horiuchi, M. Igarashi and F. Abe: *Proceedings of the 5th Workshop on the Ultra-Steel*, Tsukuba, Japan, 2001, 176–179.
11. T. Horiuchi, M. Igarashi and F. Abe: ISIJ International, **42**, 2002, S67–S71.
12. K. Hidaka, Y. Fukui, S. Nakamura, R. Kaneko, Y. Tanaka and T. Fujita: *Advanced Heat Resistant Steels for Power Generation*, R. Viswanathan and J. Nutting, eds., The Institute of Materials, London, UK., 1999, 418–429.
13. M. Y. Wey, T. Sakuma and T. Nishizawa: *Trans. Japan Inst. Metals*, **22**, 1981, 733–742.
14. M. Taneike, K. Sawada and F. Abe: *Proceedings of the 7th Liege Conference on Materials for Advanced Power Engineering 2002*, F. Schubert, ed., Liege, Belgium, 2002, 1379–1384.
15. National Institute for Material Science Creep Data Sheet, No.43, 1996.
16. National Institute for Material Science Creep Data Sheet, No.48, 2002.
17. M. Hattestrand and H. O. Andrem: *Mater. Sci. Eng.*, A270, 1999, 33–37.
18. P. Hoffer, M. K. Miller, S. S. Babu, S. A. David and H. Cerjak: *Metall. Mater. Trans. A*, **31A**, 2000, 975.

19. A. Strang and V. Z. Vodarek: *Materials Science and Technology*, **12**, 1996, 552–556.
20. K. Kimura, H. Kushima, F. Abe, K. Suzuki, S. Kumai and A. Satoh: *Proceedings Parsons 2000 Advanced Materials for 21st Century Turbines and Power Plant*, A. Strang, W. M. Banks, R. D. Conroy, G. M. McColvin, J. C. Neal and S. Simpson, eds., The Institute of Materials, London, UK., 2000, 590–602.
21. A. Goecmen, R. Steins, C. Solenthalter, P. J. Uggowitzer and M. O. Speidel: ISIJ International, **36**, 1996,768–776.
22. H. Naoi, H. Mimura, M. Ohgami, H. Morimoto, T. Tanaka and Y. Yazaki: *Proceedings of the EPRI/National Power Conference on New Steels for Advanced Plant up to 620°C*, The Society of Chemical Industry, London, UK., 1995, 8–29.

Innovative Cr-Mo-Co-B Steel Grades for Cast Turbine Components – Preliminary Results of Product Characterisation

S. Cantini and A. Ghidini
Lucchini Sidermeccanica – R & D Department
Lovere (BG), Italy

A. Di Gianfrancesco
Centro Sviluppo Materiali
Rome, Italy

ABSTRACT

The must of cost reduction and CO_2 emission diminution in new steam turbines employed in power generation leads to an improvement of Rankine cycle efficiency. This improvement is possible by increasing temperature and pressure service.

Castings, as well as pipes and forged components employed in the hottest parts of the plant, play a relevant role.

The selection of an appropriate material is likely to be a critical factor in these kinds of applications. Consequently development of new steel grades, with improved creep performances at higher temperatures and improved corrosion resistance, becomes a fundamental aspect of the research.

This paper shows the state of the art of a new Cr-Mo-Co-B steel grade development in the frame of COST522 project, and the preliminary results of metallurgical characterisation on the first industrial cast. The main aspects of designing, simulation and manufacturing are discussed and the first results of creep test on base material and welded joints are presented.

INTRODUCTION

In the past, steam power plants were based on Sub-Critical cycle, with service temperature of about 1000°F (≈ 535°C). As service temperatures were relatively low, GS17CrMoV5-11 ferritic bainitic steel grades and GX22CrMoV12-1 ferritic steel grades were commonly employed. In the 80s, the maturity of the first Super-Critical power plants leaded to the development of 9–12%Cr martensitic steels grades. The basic compositions of 9%Cr-1%Mo have been widely used for many years. As the driving force for new developments is the need to increase the temperature and pressure of the steam circuit, the next generation of the Ultra Super Critical power plant will have service temperature up to 1290°F (≈ 700°C) and very high service pressure.

Table 1 Aim and heat analysis of the first full-scale cast of CB6 steel grade (wt.% *(ppm)).

Elements		C	Mn	Cr	Ni	Mo	Cu	Si	V	Nb	Co
Aim	Min.	0.12	0.10	10.7	0.10	1.40	-	0.20	0.18	0.05	2.80
	Max.	0.14	0.30	11.3	0.20	1.60	0.10	0.30	0.22	0.08	3.20
Heat analysis		0.123	0.17	10.77	0.14	1.48	0.06	0.26	0.20	0.057	2.97
		W	S	P	B	Ti	Al	Sn	As	*Zr	*N
Aim	Min.	-	-	-	0.008	-	-	-	-	-	150
	Max.	-	0.010	0.010	0.015	0.01	0.01	0.01	0.01	-	300
Heat analysis		0.013	0.002	0.008	0.014	0.003	0.007	0.007	0.004	60	238

Traditional materials are no longer employed in new power plants, as they cannot guarantee the 100,000 hours creep rupture strength of 100 MPa for such high temperatures.

Therefore, the European Workgroup COST 522 started a new research called 'Power Generation into 21st century – Advanced steam power plants'.[1]

Within this workgroup, a large number of ferritic-martensitic steel grades have been developed with the aim to reach the target of rupture in creep at 625°C in 100,000 hours with stress of 100 MPa. The best candidate among these new steel grades has been manufactured, as full industrial cast.

MANUFACTURING OF AN INNOVATIVE Cr-Mo-Co-B STEEL GRADE CAST

THE CB6 MATERIAL

In the cast program of COST 522 workgroup, six new trial melts, have been developed, produced in experimental plates and heat treated with the same parameters. The so-called CB6 analysis showed the best performances and so it was selected as best candidate for the production of an industrial cast. CB6 steel grade derives from 9–12%Cr 1% Mo steel families with V, Nb and N (see Table 1).

Molybdenum, in the range of 1–1.5%, improves creep resistance by reducing dislocation mobility. In fact, Mo atoms can directly interact with dislocations or may form precipitates of Mo_2C and Mo_2N that form new phases. In general, the precipitation of new phases can be obtained by appropriate ageing or tempering processes, but it could also be due to in-service stress and temperature.

In order to avoid coalescence phenomena of Mo_2C carbides and Mo_2N nitrides, which lead to a reduction of the precipitation induced hardening, small additions of V (about 0.20%) and Nb (about 0.05%) are introduced. In fact, V and Nb form relatively stable carbides.

Chromium, always present to improve corrosion resistance, in percentage greater than 5% favours Mo_2C carbides and inhibits the formation of complex carbides M_7C_3, $M_{23}C_6$ and

M_6C (where M means an atom but Cr), that are less efficient in hardening and lead to an instable grain boundary.

Small additions of B enable the formation of carbides stabilising grain boundaries and inhibiting $M_{23}C_6$ carbides formation. Al and Ti should be limited in order to prevent AlN and TiN formation at grain boundary, as they could play, in fact, a negative role on microstructure stability.

Cobalt in the range of 2.8 and 3% form Co_2C carbides, inhibits $M_{23}C_6$ carbides formation and increase high temperature hardness and corrosion resistance.

Production and Manufacturing Process

A first full-scale casting, made of 50 tons of CB6 liquid steel grade, was produced:[2] two 4 ton Test-Blocks, one 12 ton flange for case turbine, one 6 ton forging ingot and plates for welding trials have been manufactured. Production process had to be carefully designed in order to achieve the aimed chemical analysis.

CB6 steel grade needs a complex manufacturing process as aimed analysis ranges are very narrow and residuals level is very low (P, Sn, As, Ti < 0.010%, Cu < 0.10%). The most difficult target to achieve was, however, the content of Nb and B, as those elements have an important affinity with O_2 and N_2 and the content of Al and Ti (< 0.010%) being impossible to use Zr or other deoxidisers and denitrifiers.

Component Design Activities

The manufacturing process of components has been designed by both MAGMASoft® and PROCAST® software. The use of two different simulation tools allowed to make a step-by-step comparison of the simulation results.

Foundry processes simulation accurately represents the dynamic, fluid and thermal fields of the casting-mould system, by control volumes approach (Figure 1). Critical zones of castings can be evidenced in the early steps of designing.

The models provide, furthermore, realistic information in order to avoid defects such as porosity, shrinkage and cracks.

In order to validate the predictive model, a test block has been instrumented by 20 thermocouples buried in refractory and in direct contact with steel. Other experimental tests have been carried out in order to evaluate thermal conductivity of different resins and refractories.

Simulation results, verified after pouring and solidification, confirm the great power of the numerical tools tested but also evidence some limits: an accurate measure of chemical and physical materials parameters is needed. Those parameters are typical of the single process/plant and they may be slightly different from default values added in standard tool's libraries.

Only after a correct model calibration on the precise plant / process and material / product, those model can usefully be employed in the whole process design.

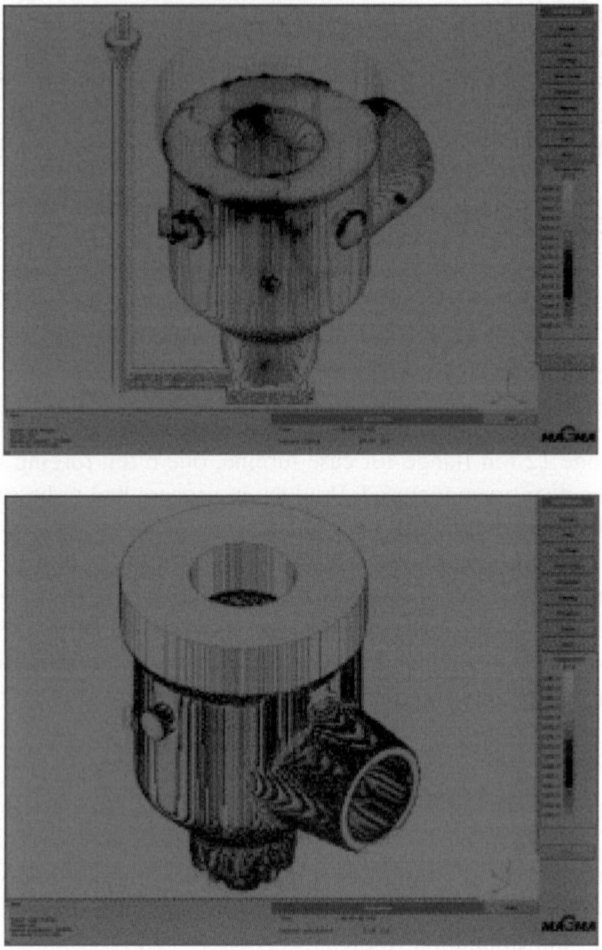

Fig. 1 Foundry process simulation by numerical tools: Turbine case.

MATERIAL CHARACTERISTICS AND PROPERTIES

Preliminary characterisation has been performed on two different plates, obtained in the same mould and cast, in order to define the optimal heat treatment parameters.

The critical transformation temperatures (Table 2) and Continuous Cooling Transformation (CCT) curves have been determined, as well as austenitic grain size and grain size variations with austenitising temperatures and time.

Austenitising temperatures up to 1070°C are enough to solubilize all Mo content and avoid coarsening, but not sufficient to solubilize all the V and Nb content.

Table 2 Critical transformation temperature for new CB6 steel grade.

Critical Transformation Points	Temperature (°C)
Ac1	845
Ac3	950
Ar3	390
Ar1	220

So a compromise has to be found between a good toughness due to a small grain size and a good creep resistance due to the full precipitation of V and Nb, obtainable at higher austenitising temperature.

Moreover, $Cr_{23}C_6$ and Mo_2C are completely solubilised at temperature of 1100°C while VC and NbC solubilize at higher temperatures. To set up the best heat treatment conditions in terms of austenitising and tempering temperatures, the following matrix has been adopted:

Austenitising Temperature (°C): 960–1000–1050–1070–1100

Tempering Temperature (°C): 710–730–750

Mechanical tests, such as tensile testing at room temperature and at 600°C, notch impact toughness and hardness tests have been carried out.

The best compromise between good tensile properties, positively influenced by carbides precipitation, and acceptable impact properties, strongly influenced by grain size, is austenitising at 1100°C. At the same time, in terms of weldability, a higher tempering temperature is preferred. For this reason, and according to mechanical properties (see Figure 2) tempering temperature was set at 730°C.

Tempering process is repeated two times in order to temper the martensitic-bainitic structure and to precipitate carbides, with the first process, and to homogenise carbides, with the second one. After the second tempering, a fully tempered martensitic structure has been obtained (Figure 3).

Figure 4 shows the microstructure by TEM. It is confirmed the fully tempered martensite with high dislocation density in the former martensitic cells with small precipitates inside and some $M_{23}C_6$ on the grain boundary. The tensile properties of the base material as a function of the temperature are shown in Figure 5.

WELDED JOINT MANUFACTURING AND CHARACTERISATION

Two plates, after annealing, austenitising, air-cooling and double tempering, have been used to manufacture a welded joint. The plate dimensions are 700 ×150 × 80 mm with a groove of 50°C (Figure 6).

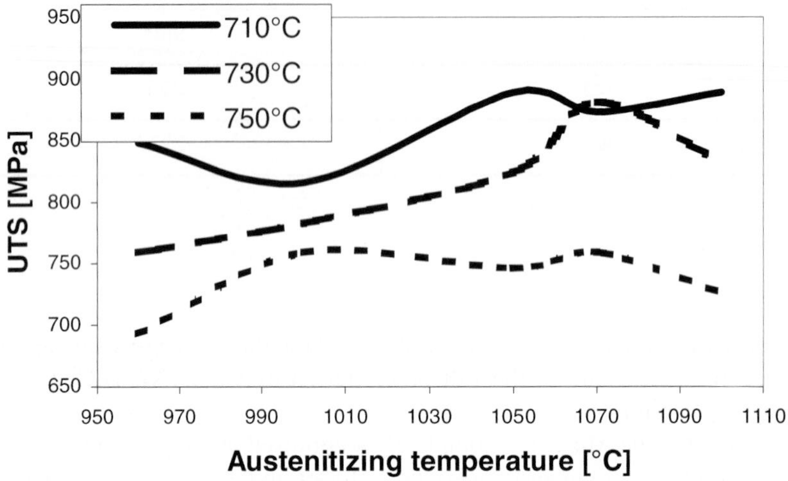

Fig. 2 Choice of the optimal heat treatment.

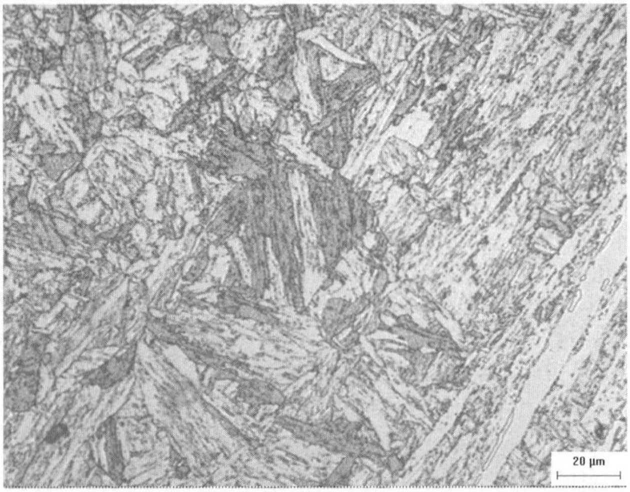

Fig. 3 Microstructure of the base material after the second tempering.

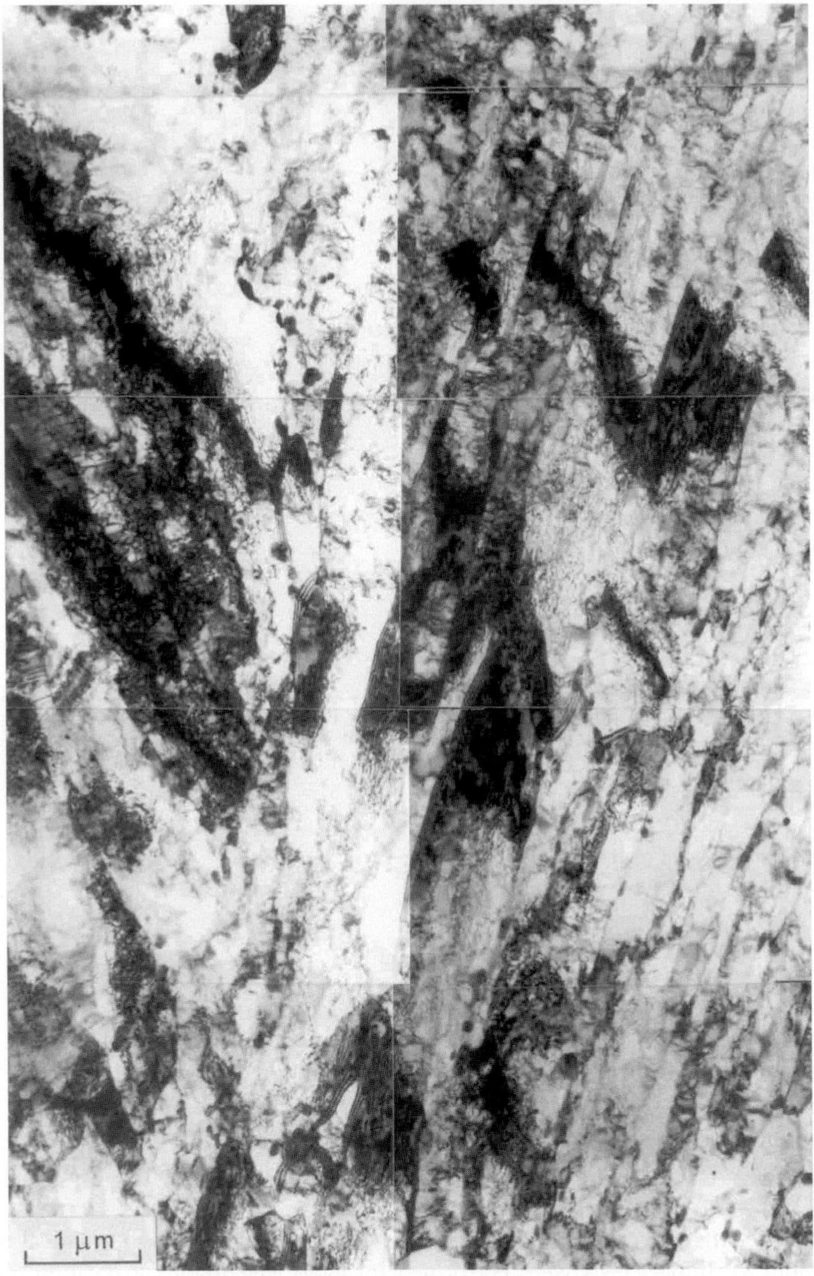

Fig. 4 TEM imagine CB6 structure.

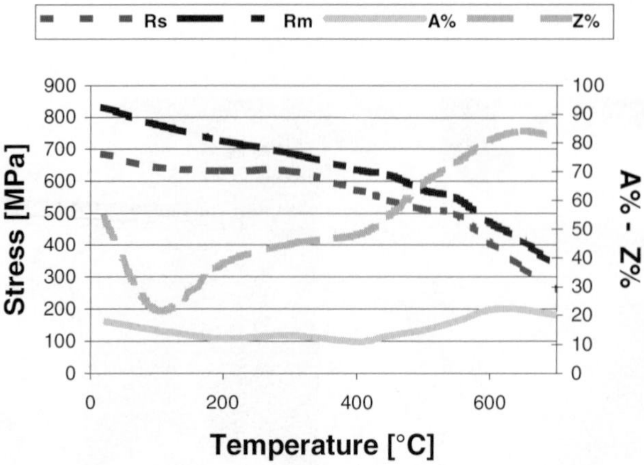

Fig. 5 Tensile properties vs. temperature for CB6 base material.

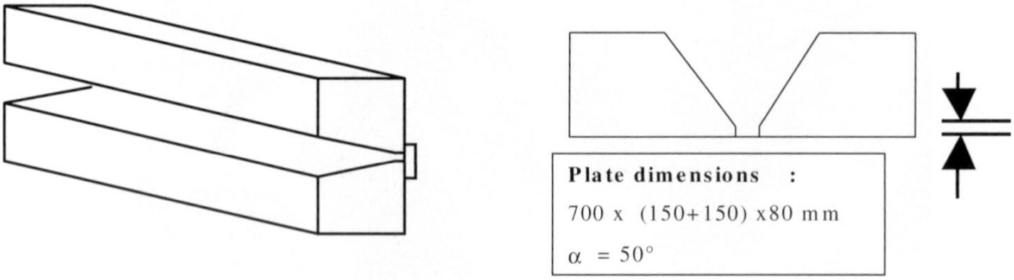

Fig. 6 Draw of weld plates.

A preheating at 200°C before welding has been adopted and the temperature for each intermediate step was 200–300°C. Welding parameters are shown in Table 3.

The microstructure of welded joint after the post weld heat treatment at 730°C for 12 hours is shown in Figure 6, the total time at 730°C is about 30 hours (double tempering plus PWTH).

Several specimens were machined from the welded joint for the mechanical and creep tests (Figure 8). Impact tests were performed for every specimen cut position, results are shown in Table 4.

The hardness profile is shown in Figure 9, it is the possible to observe that the values range between 240 and 320 HV, being the highest values at weld metal and inside heat altered zone (HAZ).

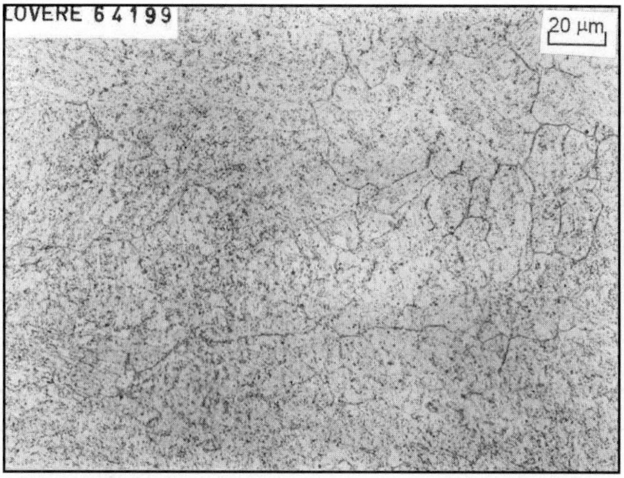

Fig. 7 Microstructure of welded joint (weld metal).

Table 3 Welding parameters.

Beads	Filler Metal		Volt Range	Current Range (A)	Travel Speed 1/cm	Heat Input J/cm
	Trade Designation	Trade Diameter mm.				
1–6	Oerlikon WB5	3.25	21–24	90–130	11–14	8100–17000
7–15	Oerlikon WB5	4.00	23–25	140–160	14–17	11300–17100
16–n	Oerlikon WB5	5.00	24–27	170–210	18–20	12200–18900

Table 4 Impact values at 20°C as function of specimen positions.

Position	Joule (Average on Three Tests)
2A	26
2B	23
3A	29
3B	14.5

Tensile tests at different temperatures on specimens in weld material in upper position (WA) and lower position (WB), were performed and compared to base material characteristics (BM). (Figure 10). It is possible to note that the welded joint is stronger than base material at room temperature, but their behaviours at elevated temperatures are comparable. Also the ductility exhibits the same values of the base material (Figure 11).

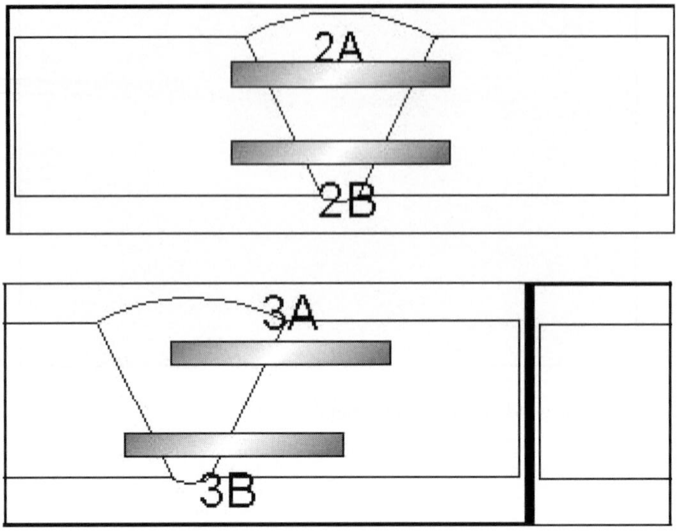

Fig. 8 Position of specimens used for mechanical and creep tests.

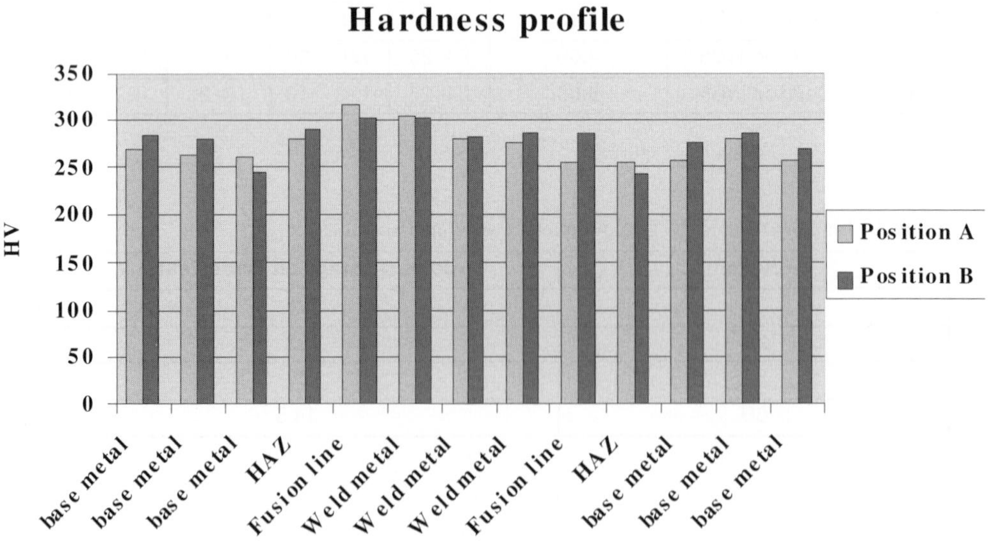

Fig. 9 Hardness gradient of welded joint in upper position (A) and lower position (B).

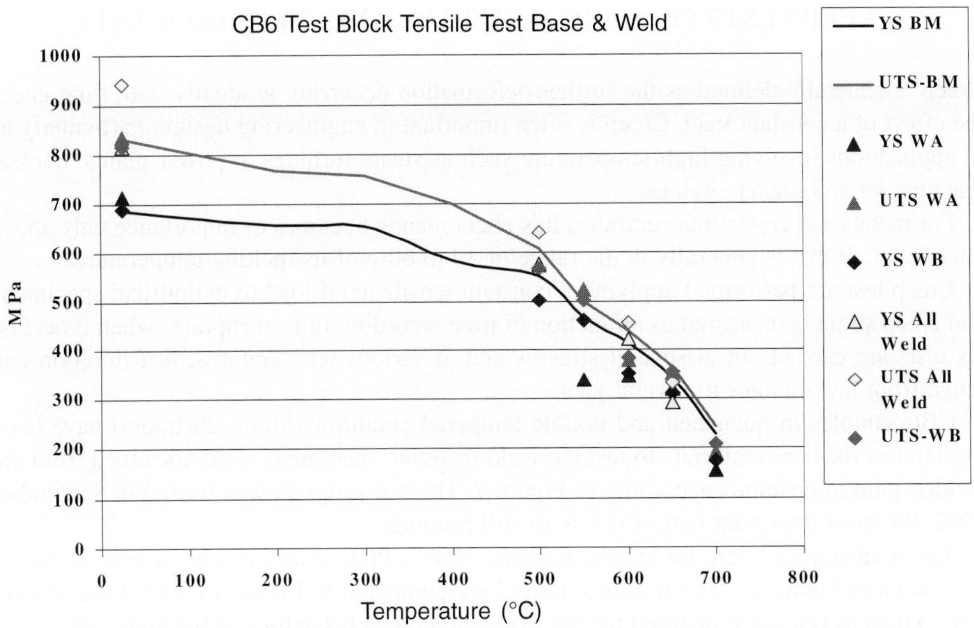

Fig. 10 Tensile properties of welded joint in both positions A (WA) and B (WB) compared to base material (BM) and weld metal (All Weld).

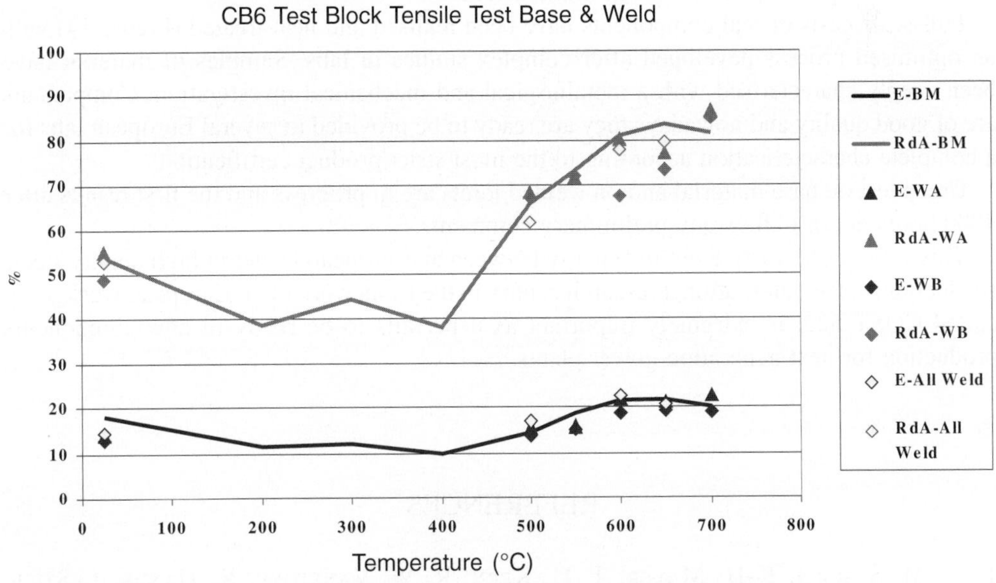

Fig. 11 Ductility of welded joint in both positions A (WA) and B (WB) compared to base material (BM). E% = Elongation, A% = Reduction in Area.

CREEP TESTS ON BASE MATERIAL AND WELDED JOINTS

Creep is generally defined as the further deformation occurring gradually with time under the effect of a constant load. Creep is often important in engineering design, particularly as in applications involving high temperature such as steam turbines in power plants, nuclear reactors, jet and rocket engines.

For metals and crystalline ceramics, this phenomenon becomes of importance only above a temperature that is generally in the range of 30 to 60% of its melting temperature.

Creep test are performed applying a constant tensile axial load to cylindrical specimens and creep strain is measured as a function of time recording time at rupture, when it occurs. As tests are carried out at several stresses and at various temperatures, test duration can range from few minutes to several years.

CB6 samples in quenched and double tempered conditions (total 20 hours) have been used to test the base material. To test the weld material, specimens were machined from the welded joint in positions according to Figure 7. The test program was started in September 2002: the most important part of test is so still running.

Test temperatures were set at 600, 625 and 650°C. Preliminary results of tests at 600°C are shown in Figure 12. The results at 650°C are compared in Figure 13 with those at steel CB2[3] taken as reference material for the evaluation creep behaviour of the new cast.

CONCLUSIONS

A full-scale heat of 50 tons in an innovative steel grade for next generation Ultra Super Critical steam power plant has been manufactured.

Full-scale casts of real components have been realised and heat treated (Figure 14) with an optimised process developed after complex studies in labs. Samples of material have been deeply characterised with a metallurgical and mechanical investigation. Components are of good quality and nowadays they are ready to be provided to several European labs for a complete characterisation according to the most strict product certification.

Creep test on base material and on welded joints are in progress and the first results after 4000 hours doesn't allow any preliminary comment.

This experience, carried out in synergy between an European leader in high quality steels production and an international research centre, in the framework of an European workgroup called COST 522, is extremely important as it permits to be ready in new components production for next generation power plants.

REFERENCES

1. M. Staubli, K-H. Mayer, T. U. Kern, R. W. Vanstone, R. Hanus, J. Stief, and K-H. Schonfeld: *Cost 522 – Power Generation into the 21st Century, Advanced Steam Power Plant: EPRI Conference*, Swansea, 2002.

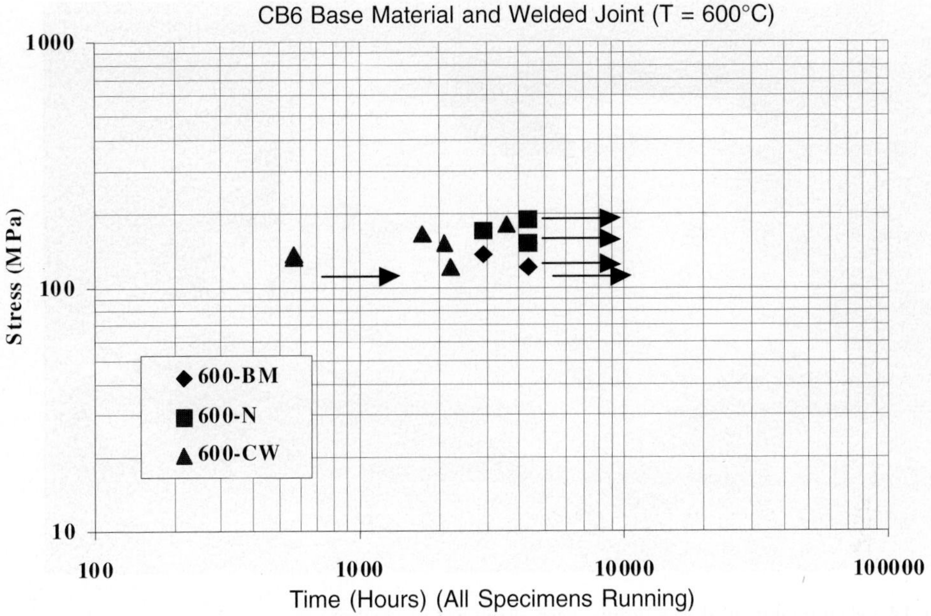

Fig. 12 CB6 base material (BM), notched (N) and welded joint (CW) creep test performed at 600°C (All test are still running).

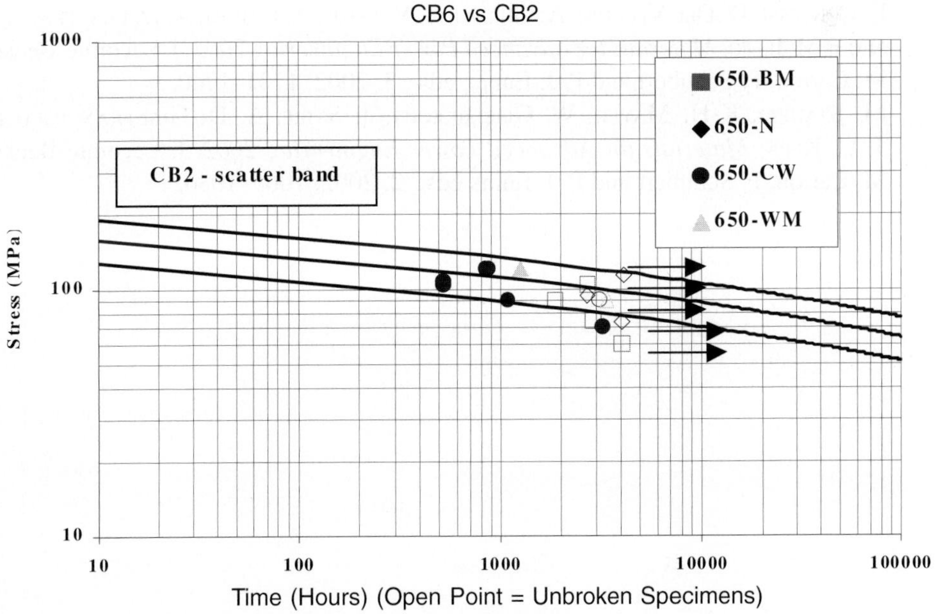

Fig. 13 CB6 vs. CB2 comparison in creep tests at 650°C.

Fig. 14 Heat treatment of cast turbine case in innovative steel grade.

2. E. Contessi, D. Del Vecchio, A. Ghidini, S. Valenti, A. Carosi, A. Di Gianfrancesco and F. M. Ielpo: *Materials for Advanced Power Engineering 2002*, J. Lecomte-Beckers, M. Carton, F. Schubert and P. J. Ennis, eds., **3**, 2002, 1731–1740.
3. M. Staubli, K-H. Mayer, W. Giselbrecht, J. Stief, A. Di Gianfrancesco and T. U. Kern: *Materials for Advanced Power Engineering 2002*, J. Lecomte-Beckers, M. Carton, F. Schubert and P. J. Ennis eds., **2**, 2002, 1065–1080.

ECCC Developments in the Assessment of Creep-Rupture Properties

S. R. HOLDSWORTH
ALSTOM Power
Rugby, UK

G MERCKLING
Istituto Scientifico BREDA
Milan, Italy

INTRODUCTION

The European Creep Collaborative Committee (ECCC) is now in its third phase of activity since its formation in 1992.[1] First focusing on the provision of uniaxial rupture strength values for steels for European product and design standards (1992–96),[2] and then the assessment of weldments (1997–2001), the current emphasis for ECCC working groups is the analysis and application of creep deformation and ductility properties to the assessment of components and multi-axial features. The current initiative is referred to as *ADVANCED-CREEP* (2001–2005) and is supported by the European Commission Thematic Network contract GTC2-2000-33051.

ECCC was originally formed primarily to provide the means for European industry to combine its resources to influence the content of the wave of new European high temperature product and design standards in preparation during the early 1990s. Initially, the principal aims of ECCC were:

- to co-ordinate the generation of creep data throughout Europe,
- to interact with, and supply information to European standards organisations and their technical committees,
- to mutually exchange technical information relating to current and future activities on material developments and
- to develop common rules for creep data generation, collation/exchange and assessment.

The defined objectives are being achieved through the efforts of a number of working groups whose activities are steered by the ECCC Management Committee (Figure 1). The WG3x groups co-ordinate the main creep data generation (testing), collation and assessment activities directed towards influencing the content of European standards/codes (Figure 2). The respective WG3x areas of responsibility are presently:

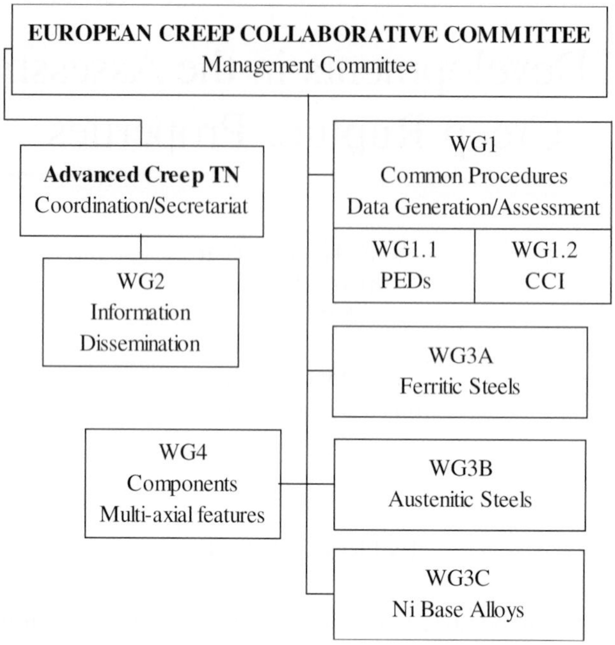

Fig. 1 Current structure of European creep collaborative committee (ECCC).

- WG3A: Low and high alloy ferritic steels,
- WG3B: Low and high alloy austenitic steels and
- WG3C: Nickel base alloys.

Technical support is provided by WG1 and its sub-groups. The main role of WG1 is to provide recommendations on procedures for data generation, collation and assessment to form the basis of common practices followed within ECCC.[3] These recommendations are published and are thereby available to other users. They are acknowledged to have influenced several important standards.[4–6] Two WG1 sub-groups focus specifically on the provision of procedures relating to post service exposed materials (WG1.1) and creep crack initiation data (WG1.2), Figure 1.

At the start of the *Advanced-Creep* phase of activity, a new working group was formed to focus on the assessment of high temperature components and multi-axial features (i.e. WG4). When ECCC was first formed, its main activity was driven by the requirements of alloy producers and plant manufacturers. More recently, the requirements of plant operators have become increasingly important. This is reflected by the creation of WG1.1 in 1997 and in particular now by the formation of WG4 to provide support for plant design and assessment functions (as shown in Figure 2).

The following paper focuses on the present activities of WG1. During the *Advanced-Creep* initiative, the main concerns of this group are the analysis and application of creep deformation and ductility properties to the assessment of components and multi-axial features.

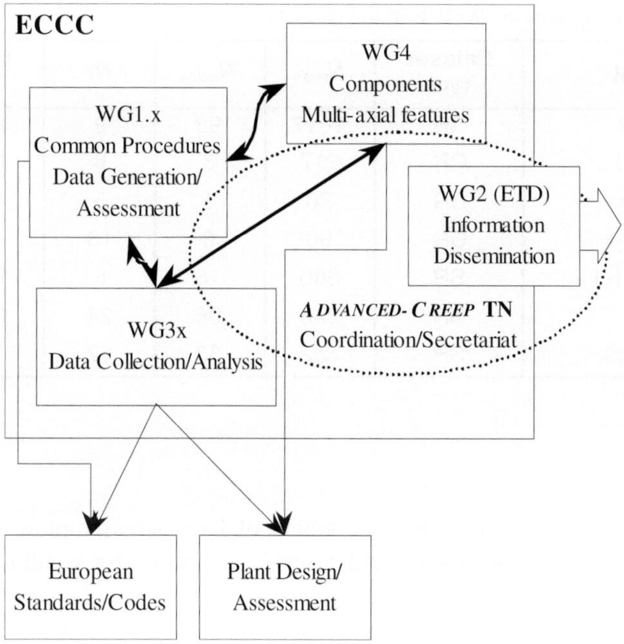

Fig. 2 Relationship between ECCC, the *Advanced-Creep* TN and external organisations.

CREEP-RUPTURE DATA ASSESSMENT

ECCC developments relating to the assessment of stress-rupture data are well documented.[3, 7] Reference has already been made to the requirement during the 1990s for well-substantiated, European industry endorsed creep-rupture strength values for new European high temperature product and design standards. The basis for these was the largest possible acceptable dataset for material of a given specification, e.g. the X10CrMo9-10*(a)* SR dataset in Table 1. The ECCC Recommendation Volumes were created to provide guidance for the generation, collation and assessment of such datasets.[3]

Not least because of the large number of high temperature product standards in preparation at the time, it was necessary to share the workload and the required stress-rupture data assessments were performed by assessors from different countries, adopting various approaches. In order to ensure a level of consistency in the strength values determined in these circumstances, WG1 developed a philosophy which restricted the number of assessment procedures used to those which were well defined and which had been validated in an extensive assessment inter-comparison activity involving the four large SR working datasets referred to in Table 1.[3, 7] Most importantly, the assessment results had to meet the acceptance criteria of a number of post assessment tests (PATs). The ECCC PATs are now widely accepted and form an integral part of modern creep-rupture data assessment procedures, e.g. PD6605, DESA.[4, 8]

Table 1 Summary statistics of WG1 working datasets.

Material	Dataset Type	N_{tests}	N_{heats}	N_T	$t_{u,max}$ kh	t_{tot} kh
10CrMo9-10*(a)*	SR	1017	98	9	141	16607
10CrMo9-10*(b)*	CR	217	8	6	199	4060
10CrMo9-10*(c)*	CR	30	1	5	3	19
X10CrMoVNb9-1	CR	90	6	13	50	842
X19CrMoVNb11-1	SR	360	18	5	128	8147
X2CrNi18-9	SR	843	96	24	111	7549
X5NiCrAlTi31-20	SR	552	33	12	79	7153

Guidance is not restricted to the assessment of large stress rupture datasets, and now encompasses the assessment of smaller datasets,[3, 7] e.g. those for weldments.

CREEP STRAIN ANALYSIS

Background

The current ECCC evaluation of creep strain analysis methods involves a review and evaluation of model equations in common use for representing creep deformation data and an assessment of their effectiveness for various materials and practical applications.

Creep strain $\varepsilon_f(t)$ or $\varepsilon_p(t)$ curves are determined from the results of continuous-measurement or interrupted tests involving the application of a constant load (or stress) to a uniaxial testpiece held at constant temperature (Figure 3a). In continuous-measurement tests, the creep strain, ε_f, is monitored without interruption by means of an extensometer attached to the gauge length of the testpiece. In interrupted tests, the total plastic strain, ε_p, is measured optically at room temperature during planned interruptions ($\varepsilon_p = \varepsilon_i + \varepsilon_f$, Figure 3b). A list of symbols and terms is given in the Nomenclature.

Depending on the nature of the creep model application, the analysis will be of several $\varepsilon_f(t)$ or $\varepsilon_p(t)$ curves determined for a single heat or several heats of the specified material. The creep strain curves may have been determined from a matrix of $t(T, \sigma)$ tests for which T and σ are (*i*) relatively homogeneously distributed or (*ii*) inhomogeneously distributed. Case (*i*) is the ideal situation and generally arises within R & D projects or from well co-ordinated data generation activities. Case (*ii*) is more typical of large multi-national datasets gathered to produce creep strength values for standards.

During the first phase of ECCC activity, WG1 evaluated the applicability of post assessment tests to the results of $R_{p\varepsilon/t/T}$ creep strength assessments.[3e] Such evaluations do not necessarily involve individual $\varepsilon_p(t)$ curve fitting, and typically entail large Case (*ii*) type datasets (e.g. X10CrMo9-10(*b*), Table 1).[9] The prescribed post assessment tests provided an effective

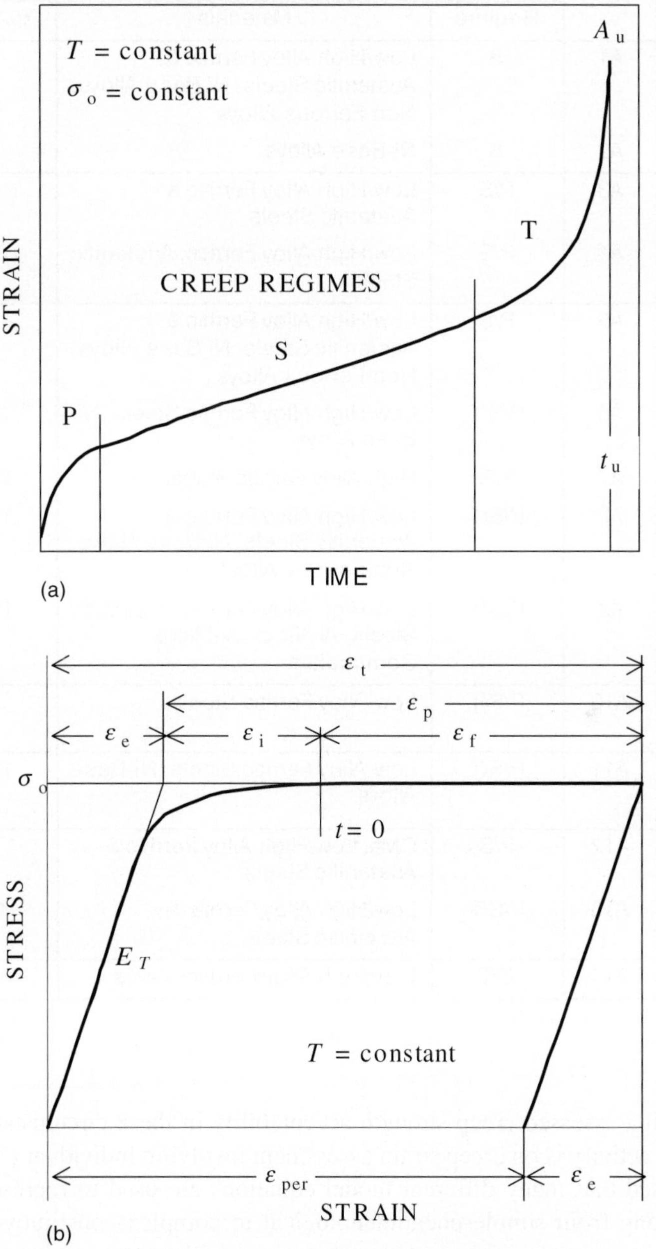

Fig. 3 Schematic diagrams showing (a) primary, secondary and tertiary creep regimes and (b) strains generated during loading of creep test.

Table 2 Range of application of some creep equations with summary of results of 10CrMo9 10(c) creep strain data assessment inter-comparison.

Model Equation	Ref.	Range of Application		Z (10CrMo9-10)	
		Regime	Materials	$t_{0.2\%/\sigma/T}$	$t_{1.0\%/\sigma/T}$
Norton[10]	A1	S	Low/High Alloy Ferritic & Austenitic Steels, Ni Base Alloys, Non-Ferrous Alloys		
Mod-Norton	A2	S	Ni-Base Alloys		
Norton-Bailey	A3	P/S	Low/High Alloy Ferritic & Austenitc Steels		
Bartsch[11]	A4	P/S	Low/High Alloy Ferritic, Austenitic Steels	3	4
Garofalo[12]	A5	P/S	Low/High Alloy Ferritic & Austenitic Steels, Ni Base Alloys, Non-Ferrous Alloys		
Mod-Garofalo[13]	A6	P/S/T	Low/High Alloy Ferritic Steels, Ni Base Alloys	2	2
BJF[14]	A7	P/S	High Alloy Ferritic Steels	15	4
Theta[15]	A8	P/S/T	Low/High Alloy Ferritic & Austenitic Steels, Ni Base Alloys, Non-Ferrous Alloys	17	2
Mod-Theta	A9	P/S/T	Low/High Alloy Ferritic, Austenitic Steels, Al Alloys, Al-Matrix Composites	10	4
Rabotnov-Kachanov[16]	A10	P/S/T	Low Alloy Ferritic Steels		
Dyson & McLean[17]	A11	P/S/T	Low Alloy Ferritic Steels, Ni-Base Alloys	12	3
I. Mech. E[18]	A12	P/S	CMn, Low/High Alloy Ferritic & Austenitic Steels		
Bolton[19]	A13	P/S/T	Low/High Alloy Ferritic & Austenitic Steels	4	13
Omega[20]	A14	S/T	Low/High Alloy Ferritic Steels	468	10

tool for evaluating assessed creep strength acceptability in these circumstances. The focus of current WG1 activity is on creep strain assessment involving individual $\varepsilon_p(t)$ curve fitting.

It is recognised that many different model equations are used to represent creep strain behaviour, ranging from simple-phenomenological to complex-constitutive. A number of model equations commonly used to represent creep strain development in engineering steels are listed in Appendix A. The listing is not exhaustive and simply reflects those expressions most commonly used by organisations currently active in ECCC. Similar creep model equation forms are grouped together in Table 2. For example, the Garofalo, BJF and Theta expressions (eqns A5–A9) share a similar representation of primary creep.

Certain expressions are likely to be better suited for specific materials and analytical applications. For example, the overall primary (P), secondary (S) and tertiary (T) creep strain characteristics of a particular steel (Figure 3a) may not be acceptably modelled by certain creep equation forms. Moreover, some applications only require a knowledge of primary low strain creep behaviour whereas others need a representation of the full creep curve. One purpose of the present evaluation is to identify the respective suitability of commonly used creep equations (e.g. Table 2).

ASSESSMENT INTER-COMPARISON

The latest WG1 creep strain data assessment inter-comparison was based on a Case (*i*) type single-heat, uniformly-distributed creep-rupture (CR) dataset for 10CrMo9-10 ((*c*) in Table 1). The model equations applied by different assessors are evident in Table 2. For most of the analysis approaches adopted, it was first necessary to curve-fit individual $\varepsilon_p(t)$ records, and this activity highlighted the importance of optimising the fitting procedure (e.g. by minimising residual errors). The way in which this is achieved can vary with model equation (and material). For example, good results for the Theta expression (eqn. A8) could be obtained by independently minimising the errors during curve-fitting the primary/secondary and secondary/tertiary regimes of the X10CrMo9-10 $\varepsilon_p(t)$ records.

The results of individual $\varepsilon_p(t)$ curve-fits are combined to provide the master-equation parameters. The effectiveness of the respective master-equations to represent observed $\varepsilon_p(t)$ behaviour was examined in the following way. Plastic strains of 0.2 and 1.0% were selected to represent typical low and high strain industrial requirements. For each assessment, plots of $\log(t^*_{p\varepsilon/\sigma/T})$ versus $\log(t_{p\varepsilon/\sigma/T})$ for the two strain levels were constructed with reference to the following relationships (e.g. Figure 4.):

$$\log(t^*_{p\varepsilon/\sigma/T}) = \log(t_{p\varepsilon/\sigma/T}) \pm \log(2) \quad \text{(dotted lines in Figure 4)} \qquad (1)$$

$$\log(t^*_{p\varepsilon/\sigma/T}) = \log(t_{p\varepsilon/\sigma/T}) \pm 2.5.s_{\text{A-RLT}} = \log(t_{p\varepsilon/\sigma/T}) \pm \log(Z) \quad \text{(chain lines in Figure 4)} \qquad (2)$$

where for a normal distribution, almost 99% of the observed times to specific strain values would be expected to lie within the boundary lines defined by eqn 2.

A perfect prediction of $t_{p\varepsilon/\sigma/T}$ by the master-equation is represented by *Z* equal to zero. Ideally Z is ≤ 2, such that the chain lines in Figure 4 fall on top of (or within) the dotted lines defined by eqn 1. *Z* values of > 4 are unacceptable, whereas values of ≤ 3–4 are marginal but may be regarded as practically acceptable. The *Z* values determined in the present inter-comparison are summarised in Table 2.

The effectiveness of the evaluated models to master-equation predict $t_{p\varepsilon/\sigma/T}$ varied with specific strain value for the 10CrMo9-10(*c*) dataset. For example, the Theta expression is most effective at predicting times to 1% strain and less so for times to 0.2% strain (Figure 4). The modified-Garofalo model is particularly effective for predicting times to low and high strains according to this study (Table 2), although it should be acknowledged that the adopted analysis approach involved an intensive prior individual $\varepsilon_p(t)$ curve fitting procedure. It is of little surprise that the *Z* values for the Omega model predictions of times to the selected strains were poor (Table 2) since this expression was developed to represent tertiary creep behaviour, i.e. $\varepsilon_p > 1\%$.

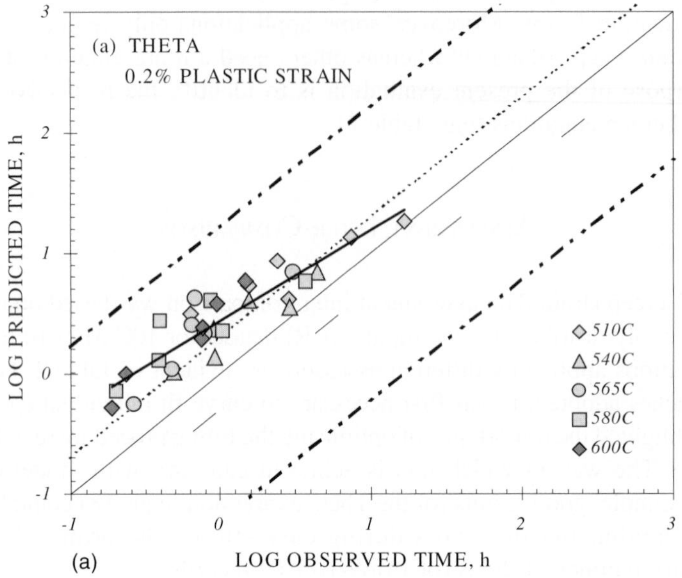

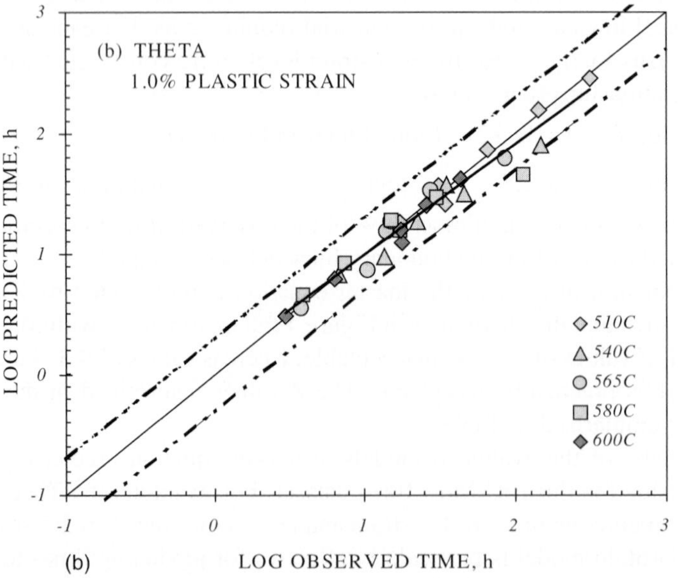

Fig. 4 Example comparisons of predicted and observed times to 0.2 and 1.0% plastic strain for the theta creep equations in the assessment inter-comparison (see text for key to reference lines).

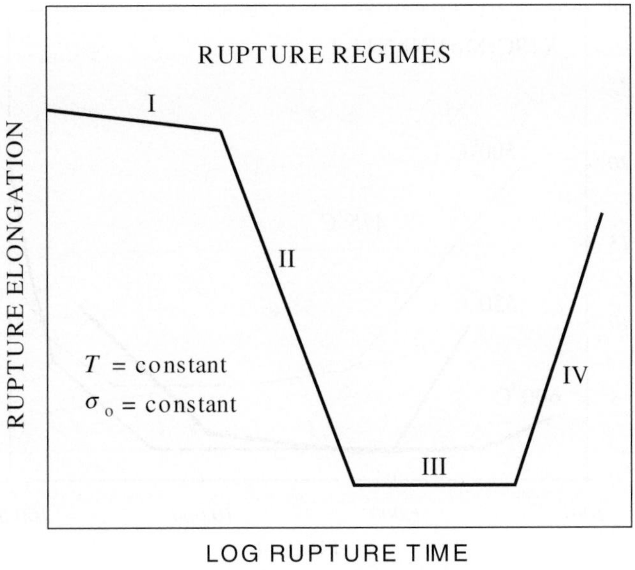

Fig. 5 Schematic representation of regimes of rupture ductility.

Current assessment inter-comparison activity involves a larger Case *(ii)* type multi-source, multi-heat, inhomogeneously-distributed CR dataset for X10CrMoVNb9-1 (Table 1). The heat-to-heat variability in Case (*ii*) type datasets is typically such that the incorporation of a metallurgical characterising parameter into the model equation will be necessary to achieve the same Z values determined for the smaller Case (*i*) type single heat, homogeneously distributed 10CrMo9-10(*c*) dataset.

RUPTURE DUCTILITY

Uniaxial rupture ductility can vary with stress (time) and temperature in a relatively complex way (e.g. Figure 5). The various ductility regimes are associated with distinct rupture mechanisms. For example in ferritic steels, Regime-I involves ductile rupture resulting from the formation of voids typically as a consequence of particle/matrix decohesion. Regime-II is a transition region in which the ductility drops due to the increasing incidence of grain boundary cavitation, but still accompanied by relatively high levels of matrix deformation. In Regime-III, rupture is by the nucleation and subsequent diffusive growth of grain boundary cavities. In Regime-IV, overaging of the microstructure lowers the rate of cavity nucleation and/or growth leading to a progressive recovery of ductility. The mechanisms associated with the identified ductility regimes can differ for different alloy systems. The analytical representation of rupture ductility data has not previously received significant attention.

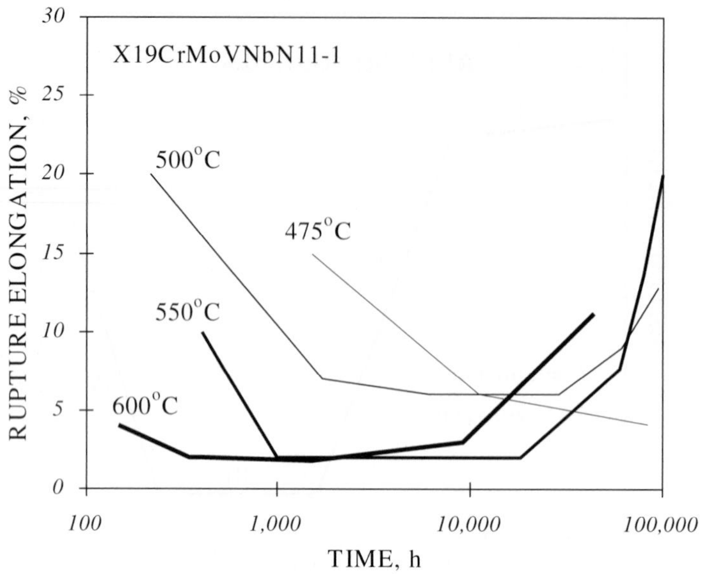

Fig. 6 Minimum $A_u(T)$ rupture elongation profiles for X19CrMoVNbN11-1 steel.

Rupture ductility data assessment comparisons are being conducted on the large stress-rupture working datasets employed previously by WG1[3e, 7] (i.e. for 10CrMo9-10(*a*), X19CrMoVNb11-1, X7CrNi18-9 and X5NiCrAlTi31-20, Table 1. The X19CrMoVNb11-1 dataset is of particular interest in that it exhibits the four ductility regimes shown in Figure 5. Moreover, it is a large multi-source, multi-heat, multi-temperature data collation typical of those used for the determination of creep parameters for standards. The lower data bounds for each temperature are shown in Figure 6. The added complexity due to the multi-heat nature of this dataset is evident in Figure 7.

Initial assessments were performed using the Spindler model, developed to model grain boundary cavity nucleation and growth processes.[21] The form of Eqn. B1 (and all other rupture ductility model equations in Appendix B) is adopted to enable a maximum upper-shelf ductility to be set, which may simply be a temperature dependent tensile ductility function. The Spindler model was specifically developed to relate fracture elongation to strain rate for the determination of creep damage due to secondary loading in terms of ductility exhaustion, e.g. for the assessment of creep-fatigue damage.

The ability of the Spindler model to predict mean rupture ductility with respect to time for the X19CrMoVNb11-1 dataset is shown in Figure 7. A target requirement from $A_u(t)$ assessments is the description of minimum rupture ductility as a function of stress and temperature to long times. The example provided by Figure 7 demonstrates the need for including a metallurgical characterising function in the model form, in particular for the assessment of large inhomogeneously distributed multi-heat datasets.

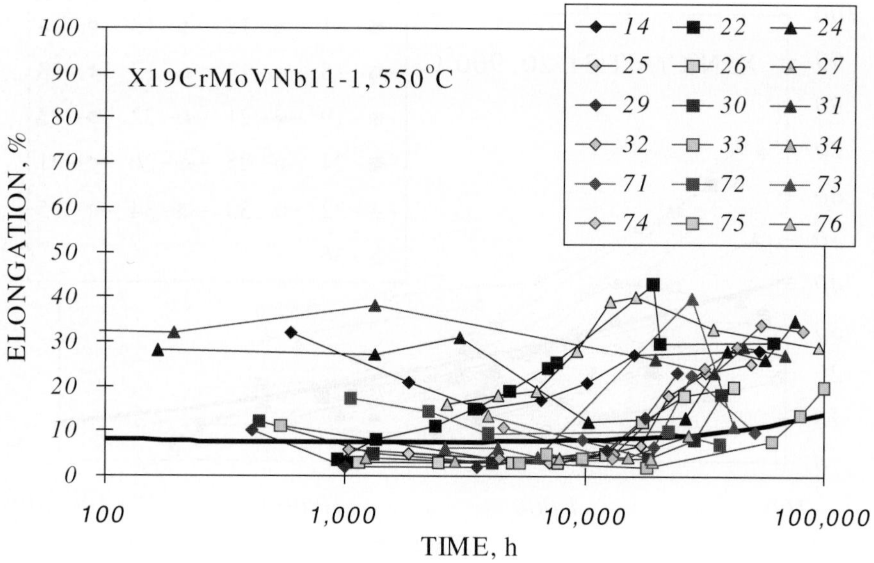

Fig. 7 Comparison of spindler model equation fit (B1, solid line) with X19CrMoVNb11-1 rupture elongation data at 550°C.

The applicability of several simple phenomenological expressions to characterise the large datasets is also being evaluated, i.e. Eqns. B2-B4. An example of the effectiveness of an equation form based on the Soviet model[22] to predict the mean rupture ductility characteristics of X5NiCrAlTi31-20 is shown in Figure 8. A fundamental problem with the analytical prediction of rupture ductility as a function of time is that both parameters are response variables dependent on temperature and stress. The rigorous analytical treatment of rupture ductility behaviour is a challenge which will occupy the resources of WG1 for the remainder of the *Advanced-Creep* work programme.

CONCLUDING REMARKS

ECCC developments in the assessment of creep-rupture properties have been reviewed with particular emphasis on current activities.

A significant effort is being devoted to the formulation of recommendations concerning the assessment of creep strain. There are several model equations available for characterising the primary, secondary and tertiary creep deformation characteristics, ranging in complexity from simple-phenomenological to full-constitutive. The suitability of some of these to specific material classes and analytical applications is reviewed. In those creep strain assessment procedures involving prior individual $\varepsilon(t)$ curve-fitting, best results are obtained by optimising

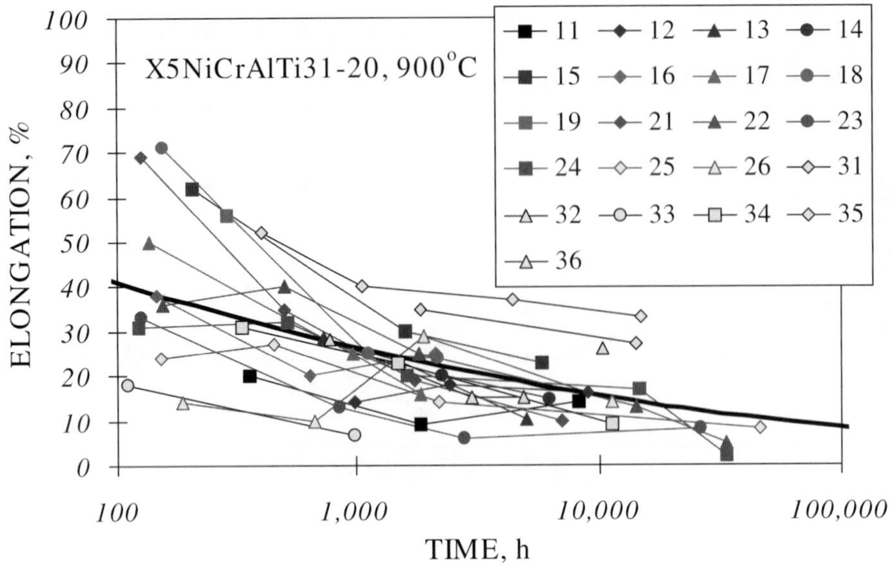

Fig. 8 Comparison of soviet model equation fit (B4, solid line) with X5NiCrAlTi31-20 rupture elongation data at 900°C.

the procedure adopted to fit specific model equations to the deformation characteristics of specific material types. A method of qualifying the effectiveness of a creep strain equation for specific material types and analytical applications is introduced. In most assessments of creep strain characteristics from large, multi-source, multi-heat, multi-temperature datasets, the adoption of a heat-by-heat analysis procedure will be necessary.

The assessment of creep-ductility is complicated by the fact that the data can represent rupture characteristics in up to four different mechanism regimes at main high-temperature material application temperatures. The analytical representation of rupture ductility data has not previously received significant attention. The rigorous analytical description of rupture ductility as a function of time (a second response variable dependent on temperature and stress) is providing a significant ongoing challenge.

In the meantime, the ability of a number of model equations to simply represent the mean rupture characteristics of large datasets of low and high alloy ferritic and austenitic steels is being evaluated. A current development is the incorporation of a metallurgical characterising function into selected model forms as an integral part of defining minimum rupture behaviour, in particular for the assessment of large inhomogeneously distributed multi-heat datasets.

ACKNOWLEDGEMENTS

The authors acknowledge the significant contribution of past and present members of ECCC-WG1 and, specifically to Dr. M. Askins (Innogy), Prof. E. Gariboldi (Politecnico-Milano),

Dr. P. Morris (Corus), Dr. S. Spigarelli (Univ. Ancona), Dr. M. Spindler (British Energy) and Dr. M. Schwienheer (IfW-TUD) for assessment activity referred to in the present paper.

They would also like to acknowledge the support of the EC through the *ADVANCED-CREEP* Thematic Network contract GTC2-2000-33051, and to the co-ordination and secretariat support provided by ETD.

NOMENCLATURE

Several lower case symbols with subscripts are not specifically defined, but are constants associated with the creep strain and ductility equations given in the appendices.

A, A_u	tensile elongation, elongation at rupture
CR	creep-rupture test
E, E_T	elastic modulus, elastic modulus at temperature
ECCC	European Creep Collaborative Committee
m_l, m_1	exponents in spindler equation (B1)
n	Stress Exponent
N_{heats}, N_{tests}, N_T	number of heats, number of tests, number of temperatures
p	time exponent
P	primary (creep regime)
PATs	post assessment tests
Q	activation energy for creep
R	universal gas constant
$R_{p\varepsilon/t/T}$, $R_{u/t/T}$	creep strength and rupture strength for a given time and temperature
$s_{A\text{-}RLT}$	standard deviation of residual log times
S	secondary (creep regime)
SR	stress rupture test
t	time
t_u, $t_{u,\max}$	observed time to rupture, maximum observed time to rupture
t_{tot}	sum of testing times in a given dataset
$t_{p\varepsilon/\sigma/T}$, $t^*_{p\varepsilon/\sigma/T}$	observed and predicted times to given plastic strain
T	temperature
T	tertiary (creep regime)
Z	factor quantifying effectiveness of master creep equation to predict times to specific strains (see Eqn. 2)
ε, ε_e, ε_i	strain, elastic strain, instantaneous plastic strain
ε_f, ε_p, ε_{per}	creep strain, plastic strain, permanent strain
$\dot{\varepsilon}$, $\dot{\varepsilon}_{f,\min}$, $\dot{\varepsilon}_{ave}$	strain rate, minimum creep strain rate, average strain rate
σ, σ_o	stress, initial stress
θ_1, θ_2, θ_3, θ_4	constants in Theta equation (A8)
θ_m	additional constant in modified Theta equation (A9)
Ω	material creep damage susceptibility parameter
ω, $\dot{\omega}$	damage, rate of damage accumulation

APPENDIX A

Creep Strain Equations

A1 Norton[10] $\dot{\varepsilon}_{f,\min} = a_1 \cdot \exp\left(\frac{Q}{R \cdot T}\right) \cdot \sigma^n$

A2 Modified Norton $\dot{\varepsilon}_{f,\min} = b_1 \cdot \exp\left(\frac{Q_B}{R \cdot T}\right) \cdot \sigma^n + c_1 \cdot \exp\left(\frac{Q_C}{R \cdot T}\right) \cdot \sigma^n$

A3 Norton-Bailey $\varepsilon_f = d_1 \cdot \sigma^n \cdot t^p$

A4 Bartsch[11] $\varepsilon_f = e_1 \cdot \exp\left(\frac{Q_1}{R \cdot T}\right) \cdot \sigma \cdot \exp(b_1 \cdot \sigma) \cdot t^p + e_2 \cdot \exp\left(\frac{Q_2}{R \cdot T}\right) \cdot \sigma \cdot \exp(b_2 \cdot \sigma) \cdot t$

A5 Garofalo[12] $\varepsilon_f = \varepsilon_t \cdot [1 - \exp(-b_1 \cdot t)] + \dot{\varepsilon}_{f,\min} \cdot t$

A6 Modified Garofalo[13] $\varepsilon_f = \varepsilon_{f1}\left[1 - \exp\left(-g_1 \cdot \left(\frac{t}{t_{12}}\right)^u\right) + \dot{\varepsilon}_{f,\min} \cdot t + c_{23} \cdot \left(\frac{t}{t_{23}}\right)^f\right]$

A7 BJF[14] $\varepsilon_f = n_1 \cdot [1 - \exp(-\underline{t})]^\beta + n_2 \cdot \underline{t}$ where $\underline{t} = \left(\frac{\sigma}{A_1}\right)^n \cdot \exp\left(\frac{-Q}{R \cdot T}\right)$

A8 Theta[15] $\varepsilon_f = \theta_1 \cdot [1 - \exp(-\theta_2 \cdot t)] + \theta_3 \cdot [\exp(\theta_4 \cdot t) - 1]$ where

$\log(\theta_i) = a_i + b_i \cdot T + c_i \cdot \sigma + d_i \cdot \sigma \cdot T$

A9 Modified Theta $\varepsilon_f = \theta_1 \cdot [1 - \exp(-\theta_2 \cdot t)] + \theta_m \cdot t + \theta_3 \cdot [\exp(\theta_4 \cdot t) - 1]$ where

$\theta_m = A \cdot \sigma^n \cdot \exp\left(\frac{-Q}{R \cdot T}\right)$

A10 Rabotnov-Kachanov[16] $\dot{\varepsilon} = \frac{h_1 \cdot \sigma^n}{(1 - \omega)}$ $\dot{\omega} = \frac{k_1 \cdot \sigma^\nu}{(1 - \omega)^\zeta}$

A11 Dyson & McClean[17] $\dot{\varepsilon}_f = \dot{\varepsilon}'_o \cdot (1 + D_d) \cdot \exp\left(\frac{Q}{R \cdot T}\right) \cdot \sinh\left(\frac{\sigma \cdot (1 - H)}{\sigma_o \cdot (1 - D_p) \cdot (1 - \omega)}\right)$

A12 IMechE[18] $R_{u/t/T} = \left(\frac{a_1 + b_1}{\varepsilon - c_1 \cdot \varepsilon^2}\right) \cdot R_{\varepsilon/t/T} + d_1 + \frac{e_1}{\varepsilon} + \frac{f_1}{\varepsilon^2} - g_1 \cdot \varepsilon^2$

A13 Bolton[19]

$$\varepsilon_f(\sigma) = \varepsilon \cdot \frac{\left(\dfrac{R_{u/t/T}}{R_{\varepsilon/t/T} - 1}\right)}{\left(\dfrac{R_{u/t/T}}{\sigma - 1}\right)}$$

A13 Omega[20]

$$\dot{\varepsilon}_f = \frac{\dot{\varepsilon}_{f,\min}}{\left(1 - \dot{\varepsilon}_{f,\min} \cdot \Omega \cdot t\right)}$$

APPENDIX B

Rupture Ductility Equations

B1 Spindler[21]

$$\ln(A_u) = \mathrm{MIN}\left|\left[\ln(A_1) + \frac{\Delta Q}{R \cdot T} + n_1 \cdot \ln(\dot{\varepsilon}_{\mathrm{av}}) + m_1 \cdot \ln(\sigma)\right], \ln(A(t))\right|$$

B2 Evans & Wilshire[15]

$$\ln(A_u) = \mathrm{MIN}\left|\left[a_1 + b_1 \cdot \sigma + c_1 \cdot T + d_1 \cdot \sigma \cdot T\right], \ln(A(t))\right|$$

B3 Anon

$$\ln(A_u) = \mathrm{MIN}\left|\left[a_1 + b_1 \cdot \ln(\sigma) + c_1/T + d_1 \cdot \ln(\sigma)/T\right], \ln(A(t))\right|$$

B4 Soviet Model[22]

$$\ln(A_u) = \mathrm{MIN}\left|\left[a_1 + b_1 \cdot \log(T) + c_1 \cdot \log(\sigma) + d_1/T + e_1 \cdot \sigma/T\right], \ln(A(t))\right|$$

REFERENCES

1. D. V. Thornton: 'Activities of the European Creep Collaborative Committee', *Proceedings of 5th International Parsons Conference on Advanced Materials for 21st Century Turbines and Power Plant,* A. Strang et al. eds., IOM, Cambridge, London, 2000, 123–128.
2. ECCC Data Sheets, 'Rupture Strength, Creep Strength and Relaxation Strength Values for Carbon-Manganese, Low Alloy Ferritic, High Alloy Ferritic and Austenitic Steels, and High Temperature Bolting Steels/Alloys', D. G. Robertson, ed., ERA Technology, 1999.
3. ECCC Recommendations, 'Creep Data Validation and Assessment Procedures', S. R. Holdsworth, et al. eds., ERA Technology, 2001, *Overview*, **1**, *Terms and Terminology*, **2**, *Data Acceptability Criteria, Data Generation*, **3**, *Data Exchange and Collation*, **4**, *Data Assessment,* **5**.
4. BS PD6605, *Guidance on Methodology for the Assessment of Stress-Rupture Data*, British Standards Institution, 1998.

5. EN 10291, *Metallic Materials, Uniaxial Creep Testing in Tension, Method of Test*, European Norm, 2000.
6. prEN 10319, *Metallic Materials, Tensile Stress Relaxation Testing*, European Norm, Provisional, 2000.
7. S. R. Holdsworth: 'Recent Developments in the Assessment of Creep-Rupture Data', *Proceedings of 8th International Conference on Creep and Fracture of Engineering Materials and Structures*, Tsukuba, JSME, 1999, 1–8.
8. J. Granacher and M. Monsees: 'Assessment Procedure Document for DESA', *Appendix D2 in ECCC Recommendations*, **5**,[3e] 1996.
9. C. K. Bullough and S. R. Holdsworth: 'New Strategies for the Assessment of Long-Term Data from Interrupted Creep Tests', *Proceedings of 3rd International Conference on Engineering Structural Integrity: Life Assessment and Life Extension of Engineering Plant, Structures and Components*, Cambridge, AEA, 1996.
10. F. H. Norton: *The Creep of Steel at High Temperature*, McGraw-Hill, 1929.
11. H. Bartsch: 'A New Creep Equation for Ferritic and Martensitic Steels' *Steel Research*, **66**(9), 1995, 384–388.
12. F. Garofalo: *Fundamentals of Creep and Creep Rupture in Metals*, MacMillan, New York, 1965.
13. J. Granacher, H. Möhlig, M. Schwienheer and C. Berger: 'Creep Equation for High Temperature Materials', *Proceedings of 7th International Conference on Creep and Fatigue at Elevated Temperatures (Creep 7)*, NRIM, Tsukuba, 2001, 609–616.
14. D. I. G. Jones and D. L. Bagley: 'A Renewal Theory of High Temperature Creep and Inelasticity', *Proceedings of Conference on Creep and Fracture: Design and Life Assessment at High Temperature*, London, MEP, 1996, 81–90.
15. R. W. Evans and B. Wilshire: *Creep of Metals and Alloys*, Institute of Metals, 1985.
16. L. M. Kachanov: *Introduction to Continuum Damage Mechanics*, Martinus Nijhoff Publ., 1986.
17. B. F. Dyson and M. McClean: 'Microstructural Evolution and its Effects on the Creep Performance of High Temperature Alloys', *Microstructural Stability of Creep Resistant Alloys for High Temperature Applications*, A. Strang et al. eds., 1998, 371–393.
18. Creep of Steels Working Party, *High Temperature Design Data for Ferritic Pressure Vessel Steels*, Institute of Mechanical Engineering, London, 1983.
19. J. Bolton: 'Design Considerations for High Temperature Bolting', *Proceedings Conference on Performance of Bolting Materials in High Temperature Plant Applications*, York, A. Strang, ed., 1994, 1–14.
20. M. Prager: 'Development of the MPC Omega Method for Life Assessment in the Creep Range', *ASME J. Pressure Vessel Technology*, **117**, 1995, 95–103.
21. M. W. Spindler: 'The Multi-Axial Creep Ductility of Austenitic Stainless Steel', *Fatigue and Fracture of Engineering Materials and Structures*, 2003, submitted for publication.
22. I. I. Trunin, N. G. Golobova and E. A. Loginov: 'New Method of Extrapolation of Creep Test and Long Time Strength Results', *Proceedings of 4th International Symposium on Heat Resistant Metallic Materials*, Mala Fatra, CSSR, 1971, 168.

Prediction of the Long-Term Creep Rupture Properties of 9–12%Cr Power Plant Steels

A. STRANG
Department of Engineering
University of Leicester, UK

V. FOLDYNA
JINPO PLUS, Ostrava
Czech Republic

J. LENERT
Technical University, Ostrava
Czech Republic

V. VODAREK
Vitkovice, Ostrava
Czech Republic

K. H. MAYER
ALSTOM Power (retired), Nuremberg
Germany

ABSTRACT

Microstructural instability has been observed in a number of high chromium tempered martensitic creep resistant power plant steels in the form of sigmoidal inflexions in their long-term creep rupture characteristics. Advanced analytical transmission electron microscopy studies have shown that this behaviour is due to the occurrence of microstructural degradation effects during creep exposure at high temperature in which the steel changes from being initially precipitation strengthened to finally being in a solid solution strengthened condition. This is due to a combination of dissolution and coarsening of precipitates together with recovery effects and the precipitation of more thermodynamically stable phases in the steels. These effects significantly reduce the creep strength of the steels as well as seriously affecting the reliability with which their long-term properties can be predicted. In this paper the relationships between microstructural changes and sigmoidal creep rupture behaviour are discussed for creep exposed 9Cr1Mo(V), 12CrMoVNbN, X19CrMoVNbN 11.1 and TAF650 steels together with the use of parametric extrapolation procedures as means of achieving more reliable predictions of their long-term creep rupture strengths.

INTRODUCTION

During the past decade a new generation of 9–12%Cr creep resistant steels has been developed for boiler and turbine component applications in advanced ultra supercritical power generation plant operating with inlet steam temperatures in the range 600 to 630°C. Whilst these steels have higher creep strengths than the previously used low chromium ferritic power plant steels, there is evidence to show that the high alloy ferritic steels can be microstructurally unstable particularly when used at operating temperatures of 550°C and above.[1–3] This often takes the form of a sigmoidal inflexion in their high temperature creep rupture characteristics where a rapid reduction in creep rupture strength is accompanied by a corresponding increase in creep rupture ductility.[4] Since this phenomenon can occur within the 250,000 hours design life expectancy of modern power generation plant, the extrapolation of short-term creep rupture data on steels prone to this type of behaviour for the purposes of providing design values becomes uncertain.

Sigmoidal creep rupture behaviour was first reported by Bennewitz in 1963, in a study of more than 30 high temperature steels being evaluated at temperatures in the range 450 to 600°C as part of the German National creep testing programme.[5] Although this was a common feature in the majority of the steels studied in this work, some of the alloys did not appear to exhibit this type of behaviour within the durations of the creep rupture data investigated. Despite this, Bennewitz concluded, from the general shape of their creep rupture characteristics, that sigmoidal inflexions would be expected to occur in these steels at creep durations beyond 100,000 hours. Moreover, sigmoidal behaviour was also shown to be dependent on the composition and initial heat treatment of the steel, as well as the temperature of testing. Furthermore, as the testing temperature was increased the sigmoidal inflexions occurred at shorter creep rupture durations and lower test stresses. In addition, sigmoidal behaviour was thought to be associated with precipitation effects occurring during creep exposure and to be particularly sensitive to the creep testing temperature. Whilst Bennewitz's work was pioneering, it is only during the past 10–15 years that the use of modern advanced analytical electron microscopy techniques has enabled the metallurgical factors responsible for this type of behaviour to be identified and understood. Given the lack of today's sophisticated metallographic techniques, Bennewitz's primary conclusion that the sigmoidal creep rupture behaviour of the steels studied was due to the effects of microstructural changes in the steels during long-term high temperature creep exposure has proved to be remarkably perceptive.

Since the publication of Bennewitz's paper sigmoidal creep rupture behaviour has been observed in a number of other creep resistant ferritic steels, with certain of the 9–12%Cr steels being particularly prone to this type of behaviour.[6–15] Furthermore, the use of modern analytical electron microscopy techniques has shown that sigmoidal creep rupture behaviour is due to specific microstructural changes, which occur due to thermal and strain accumulation effects during the creep process.[1] In this process, the material progressively changes from being initially in a high dislocation density precipitation strengthened condition to one, which is fully recovered and finally dependent on solid solution strengthening. This has been clearly demonstrated by studies on a series of creep exposed 12CrMoVNb and 12CrMoV steels by Strang and Vodarek, where, the onset of the sigmoidal inflexion and subsequent rapid reduction in creep rupture strength was shown to be consistent with the dissolution of the fine matrix strengthening M_2X and MX precipitates. In these steels these changes are driven respectively by the precipitation of coarse Z-phase and/or M_6C particles which are thermodynamically more stable than the fine matrix strengthening M_2X and MX precipitates and reduce the creep strength further by scavenging solid solution strengthening elements from the matrix. These changes coupled with significant coarsening of the grain boundary $M_{23}C_6$ precipitates lead to a rapid reduction in dislocation density, recovery and sub-grain formation and severe reductions in the long-term creep rupture strengths of these steels.[1–3, 12, 13]

Experience has also shown that the onset of sigmoidal behaviour in these steels is difficult to predict and as such seriously affects the reliability with which creep rupture data can be extrapolated to provide estimates of their long-term, currently 250,000 hours, creep rupture strengths upon which modern power plant designs are based. This paper considers the effects of microstructural changes on the sigmoidal creep rupture behaviour of a number of 9–12%Cr power plant steels and the use of use of parametric extrapolation for the prediction of their long-term high temperature creep rupture strengths.

MATERIALS

The long-term creep rupture properties and microstructures of a series of 9Cr1Mo(V), 12CrMoVNbN, X19CrMoVNbN 11.1 and TAF650 power plant steels have been evaluated in this study. In the as-received condition, all of these steels exhibited fully tempered martensitic microstructures with $M_{23}C_6$ precipitates at the prior austenite and martensite lath boundaries and fine M_2X and/or MX precipitates within the matrix. In addition TAF650 contained a small amount of Fe_2W Laves phase in the as-received condition. The chemical analyses and heat treatments of the steels investigated are shown in Tables 1 and 2, while typical plots of their creep rupture properties are shown in Figures 1–4.[3, 6, 8, 11]

PARAMETRIC ANALYSES

Analyses of the long-term creep rupture data on all of the steels investigated was carried out using Larson-Miller, Manson-Haferd, Manson-Brown and Manson forms of the general creep rupture parameter equation proposed by Lenert,[16] viz.,

$$\Phi = \frac{\dfrac{\log t_R}{\sigma^q} - \dfrac{\log t_a}{(T-W)^p}}{\left[f(T)-W\right]^r} \qquad (1)$$

where σ is stress, T_R is the time in hours to rupture, T is the temperature in °C, $f(T)$ is a convenient function of temperature and q, Ta, W and r are material constants.

These well known parametric equations can be derived from eqn (1) according to the choice made of the constants shown in Table 3.[16]

Parameter Φ in eqn (1) is a suitable function of the stress represented in this case by a simple polynomial of the form

$$\Phi = A_o + A_1 \log \sigma + A_2 \log^2 \sigma + \text{- - - - - - - - - - -}\ A_m \log^m \sigma \qquad (2)$$

Table 1 Chemical compositions of 9–12%Cr power plant steels in wt.%.

Material	C	Si	Mn	Ni	Cr	Mo	V	Nb	N_{tot}	Others
9Cr1Mo(V)	0.06	0.51	0.43	0.11	9.10	0.94	0.09	-	0.015	
12CrMoVNbN(0.52Ni)	0.16	0.28	0.74	0.52	11.20	0.61	0.28	0.29	0.074	
12CrMoVNbN(0.76Ni)	0.14	0.37	1.00	0.76	11.10	0.57	0.36	0.32	0.062	
12CrMoVNbN(1.15Ni)	0.14	0.13	0.88	1.15	11.74	0.50	0.29	0.30	0.064	
X19CrMoVNbN 11.1	0.20	0.22	0.64	n.a.	10.7	0.62	0.22	0.40	0.064	
TAF650	0.10	0.07	0.55	0.55	10.84	0.14	0.19	0.06	0.016	2.63 W 2.86 Co 0.019 B

Table 2 Heat treatments of 9–12%Cr power plant steels.

Material	Heat Treatment
9Cr1Mo(V)	930°C AC + 2 hours 750°C AC
12CrMoVNbN(0.52Ni)	1150°C AC + 6 hours 650°C AC
12CrMoVNbN(0.76Ni)	1150°C AC + 4 hours 700°C AC
12CrMoVNbN(1.15Ni)	1165°C AC + 4 hours 675°C AC
X19 CrMoVNbN 11.1	1100°C AC + 8 hours 690°C AC
TAF650	1100°C OQ + 2 hours 750°C AC

Table 3 Constants used for specific variations of the lenert general parametric equation.[16]

Procedure	*q*	*r*	*W*	*p*	*f(T)*
Larson-Miller	0	–1	–273.16	0	T
Manson-Haferd	0	1	T_a	0	T
Manson-Brown	0	r	T_a	0	T
Manson	q	r	T_a	0	T

which, when plotted against stress gives the master parametric curve for the material.

Given the sigmoidal nature of the creep rupture data a third order polynomial in logσ was considered to be appropriate for this study, thus eqn (2) becomes

$$\Phi = A_o + A_1 \log\sigma + A_2 \log^2\sigma + A_3 \log^3\sigma \tag{3}$$

Furthermore substituting for Ö, Equation (1) can be rewritten as

$$y(x) = e^{kx}\left\{\Phi(x)\left[f(T) - W\right]^r + \frac{\log t_a}{(T - W)^p}\right\} \tag{4}$$

where $y = \log t_R$, x = logσ and $k = 2.30259q$.

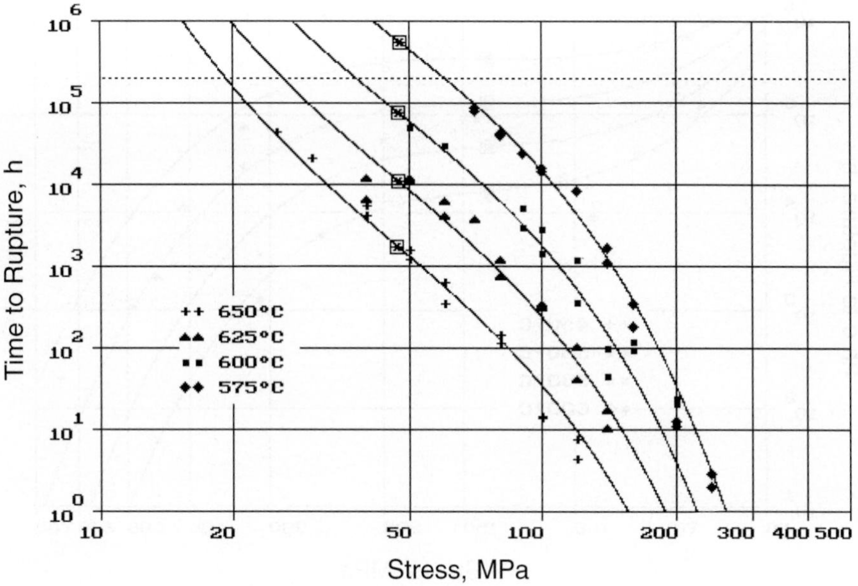

Fig. 1 Manson Parameter Analysis of Creep Rupture Data on 9Cr1Mo(V) Steel.

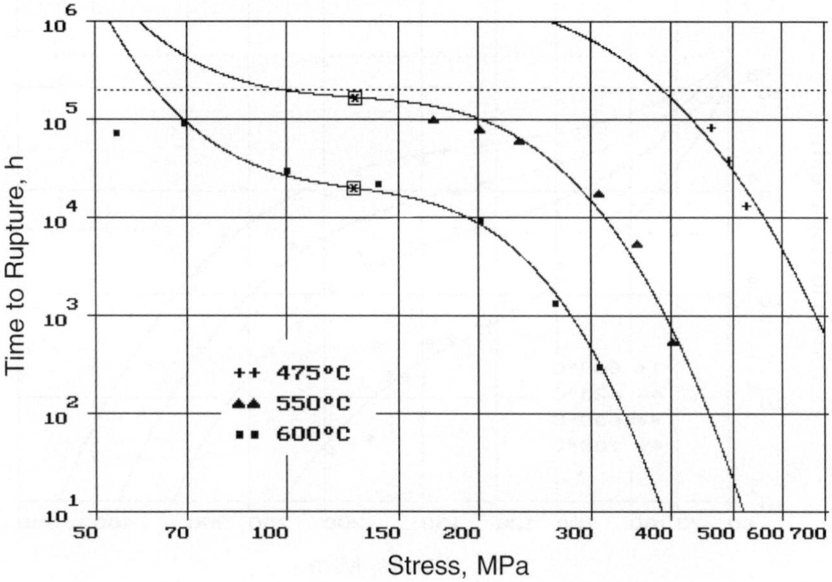

Fig. 2 Manson parameter analysis of creep rupture data on 12CrMoVNbN(0.52Ni) steel.

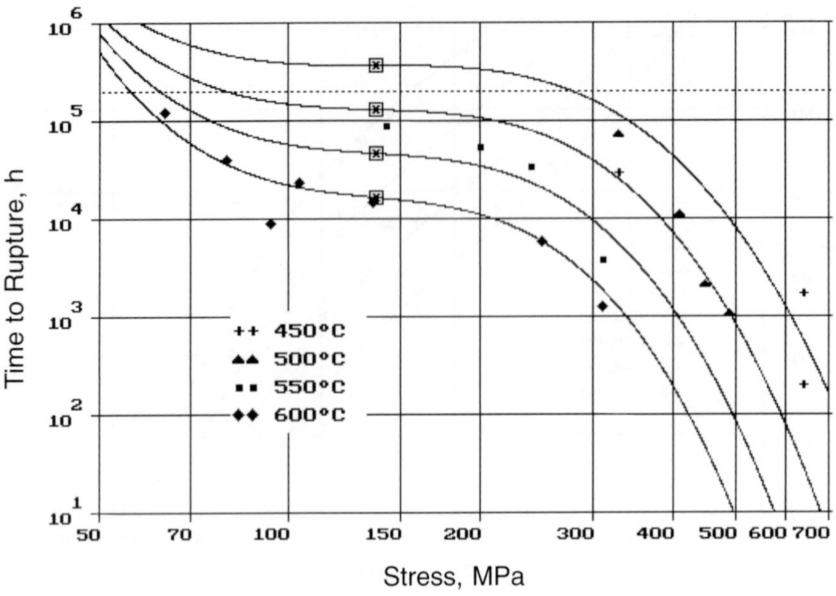

Fig. 3 Manson parameter analysis of creep rupture data on X19CrMoVNbN 11.1 steel.

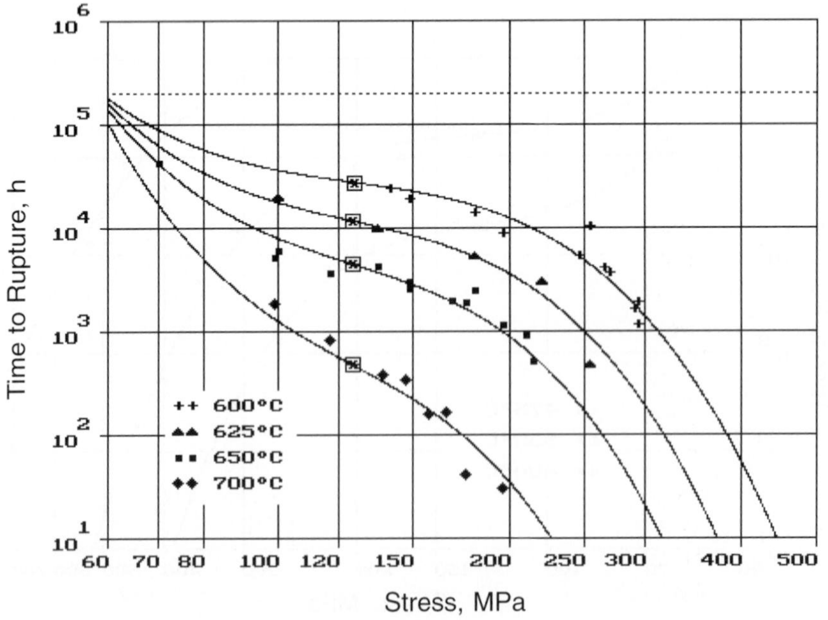

Fig. 4 Manson parameter analysis of creep rupture data on TAF650 steel.

Table 4 Model Functions for various specific parametric equations.

Procedure	Model Function	Parameter	Number of Unknown Constants
Larson-Miller	$\log t_R = \frac{\Phi(x)}{T+273{,}16} + \log t_a$	$\Phi(x) = (\log t_R - \log t_a)(T+273{,}16)$	m + 2
Manson-Haferd	$\log t_R = \Phi(x)(T-T_a) + \log t_a$	$\Phi'(x) = \frac{\log t_R - \log t_a}{T-T_a}$	m + 3
Manson-Brown	$\log t_R = \Phi(x)(T-T_a)^r + \log t_a$	$\Phi'(x) = \frac{\log t_R - \log t_a}{(T-T_a)^r}$	m + 4
Manson	$\log t_R = e^{kx}\left[\Phi(x)(T-T_a)^r + \log t_a\right]$	$\Phi'(x) = \frac{\frac{\log t_R}{e^{kx}} - \log t_a}{(T-T_a)^r}$	m + 5

The parametric Larson-Miller, Manson-Haferd, Manson-Brown and Manson equations can then be formulated as shown in Table 4, from which best-fit equations can be calculated using a least squares approach for the data-sets from each steel.

RESULTS AND DISCUSSION

Analyses of the multi-temperature data set for each steel using the Larson-Miller, Manson-Haferd, Manson-Brown and Manson general parameter equations indicated that the best overall agreement between the calculated and experimental data was obtained using the Manson-Haferd and Manson general equations. From these the best-fit equations derived for each data set were used to calculate the 10,000 and 100,000 hours creep rupture strengths at the different test temperatures for each steel. In addition, the stresses and creep rupture durations corresponding to the sigmoidal inflexion points were calculated for each test temperature. The results of these analyses indicated that there was very little difference between the 10,000 and 100,000 hours creep rupture strengths predicted using the Manson-Haferd and Manson parametric equations. However, estimation of the sigmoidal inflexion points using the Manson-Haferd equation indicated that whilst the durations at which the turning points occurred increased with decreasing test temperatures they did so at a constant stress for each individual steel. This is contrary to Bennewitz's findings, namely that both the stress and duration at which the sigmoidal inflexion occurred decreased with increasing

Table 5 Calculated 10,000 hour creep rupture strengths for 9–12%Cr power plant steels.

Material	Manson Parameter 10,000 Hour Creep Rupture Strengths - MPa								
	450°C	**475°C**	**500°C**	**550°C**	**575°C**	**600°C**	**625°C**	**650°C**	**700°C**
9Cr1Mo(V)					106.1	73.2	48.1	32.5	
12CrMoVNbN (0.52Ni)		561.3		315.9		191.9			
12CrMoVNbN (0.76Ni)		410.9		244.8		112.2			
12CrMoVNbN (1.15Ni)		455.1		245.4		108.9			
X19CrMoVNbN11.1	489.4		382.6	297.6		209.2			
TAF650						213.7	137.0	93.4	73.7

Table 6 Calculated 100,000 hour creep rupture strengths for 9–12%Cr power plant steels.

Material	Manson Parameter 100,000 Hour Creep Rupture Strengths - MPa								
	450°C	**475°C**	**500°C**	**550°C**	**575°C**	**600°C**	**625°C**	**650°C**	**700°C**
9Cr1Mo(V)					69.8	44.2	29.5	21.4	
12CrMoVNbN (0.52Ni)		431.8		201.5		69.1			
12CrMoVNbN (0.76Ni)		343.7		146.1		51.5			
12CrMoVNbN (1.15Ni)		312.6		79.1		59.8			
X19CrMoVNbN 11.1	340.5		210.8	75.3		62.9			
TAF650						67.8	64.3	62.3	60.1

testing temperature.[5] Analyses of the data using the Manson parameter however showed a temperature dependency on both the stress and duration of the sigmoidal inflexion point. On this basis, and since there is little difference in the creep rupture life predictions between the Manson-Haferd and Manson parameter analyses, the remainder of the comparisons between the analyses of the different steels is discussed in terms of the results of the Manson parameter analyses. The results of the Manson parameter analyses on each of the alloys investigated are shown in Tables 5-8 and discussed below.

9Cr1Mo(V) STEEL

Examination of the creep rupture data on the 9Cr1Mo(V) steel indicated that sigmoidal inflexions were only present in the tests conducted at 600°C. However, using the best-fit equations derived from the parametric analysis of the individual data sets, enabled the general

Table 7 Calculated sigmoidal creep rupture inflexion point stresses for 9–12%Cr power plant steels.

Material	Sigmoidal Creep Rupture Inflexion Point Stresses - MPa								
	450°C	475°C	500°C	550°C	575°C	600°C	625°C	650°C	700°C
9Cr1Mo(V)					47.42	47.24	47.07	46.90	
12CrMoVNbN (0.52Ni)		127.82		127.82		127.52			
12CrMoVNbN (0.76Ni)		103.72		102.70		101.93			
12CrMoVNbN (1.15Ni)		129.74		129.10		128.78			
X19CrMoVNbN 11.1	137.26		137.26	136.90		136.90			
TAF650						125.68	125.37	125.37	125.07

Table 8 Calculated sigmoidal creep rupture inflexion point durations for 9–12%Cr power plant steels.

Material	Sigmoidal Creep Rupture Inflexion Point Durations – Hours								
	450°C	475°C	500°C	550°C	575°C	600°C	625°C	650°C	700°C
9Cr1Mo(V)					5.56×10^5	7.50×10^4	1.10×10^4	1.76×10^3	
12CrMoVNbN (0.52Ni)		1.70×10^6		1.67×10^5		2.00×10^4			
12CrMoVNbN (0.76Ni)		6.65×10^6		1.68×10^5		1.21×10^4			
12CrMoVNbN (1.15Ni)		4.14×10^5		5.12×10^4		1.07×10^4			
X19CrMoVNbN11.1	3.66×10^5		1.30×10^5	4.63×10^4		1.64×10^4			
TAF650						2.76×10^4	1.16×10^4	4.43×10^3	4.76×10^2

form of the creep rupture curves for all of the test temperatures in the range 575 to 650°C to be derived and plotted, Figure 1. Furthermore, estimates of the stresses and durations of the inflexion points, as well as the 10,000 and 100,000 hour creep rupture strengths were calculated for each test temperature. Analysis of the data using the Manson general equation indicates that both the stress and duration at which the sigmoidal inflexions occur are temperature sensitive and decrease with increasing testing temperature, as predicted by Bennewitz, Tables 7 and 8.[5] However, the effect of testing temperature on the sigmoidal inflexion stresses for these steels is very small being only between 1–2%, compared with actual values ranging from 160 to 220% for low and high alloy ferritic steels studied by Bennewitz.

In the case of the 9Cr1Mo(V) the estimated inflexion stresses are close to the actual testing stresses and thus enabled a direct comparison to be made between the actual and estimated sigmoidal inflexion durations calculated using the Manson general equation. For example, the calculated sigmoidal inflexion point at 650°C occurs at a stress of approximately 46.9 MPa and duration of 1760 hours. This compares well with an actual creep rupture life of 1613 hours at 50 MPa at the same temperature for this steel. At 625, 600 and 575°C inflexions are predicted at approximately 11,000, 75,000 and 556,000 hours respectively. Whilst the 625 and 600°C predictions compare well with the actual creep rupture data for this steel, that at 575°C is overestimated by a factor of at least three times. This would result in the 250,000 hours creep rupture strength for this temperature being overestimated by more than 60%.

Metallographic examination of the 9Cr1Mo(V) steel indicates that the initial microstructure consists of tempered martensite with $M_{23}C_6$ particles precipitated at the prior austenite and martensite lath boundaries. During creep exposure at 600°C coarsening of the $M_{23}C_6$ precipitates occurs together with precipitation of fine MX particles on dislocations within the sub-grains.[17, 18] Other studies have shown that these particles are isomorphous with VN and would be expected to occur within the whole range of temperatures investigated in this work.[19] In addition it has been observed in the case of the 9Cr1Mo(V) steel that whilst the coarsening rate of the $M_{23}C_6$ precipitates increased with increasing test temperature and creep exposure time the amount of VN present in the matrix decreased.[14, 15]

In the case of the 9Cr1Mo(V) steels coarsening of the $M_{23}C_6$ during creep exposure would lead to progressive reductions in the creep strength with increasing exposure durations. However, further precipitation of fine particles of VN within the matrix tends to counter this by increased precipitation strengthening during creep exposure. At testing temperatures of up to 600°C the reduction in the long-term creep strength is relatively small, since the degree of $M_{23}C_6$ coarsening is small at these temperatures and the precipitation of VN during creep exposure correspondingly large. Furthermore, any sigmoidal inflexion in the creep rupture properties would only be expected to occur after very long exposure durations. At temperatures above 600°C however, significant coarsening of $M_{23}C_6$ occurs leading to reductions in dislocation densities coupled with recovery and sub-grain formation in the microstructure. This in combination with progressive coarsening, and dissolution, of the VN precipitates due to precipitation of coarse M_6X particles leads to significant reductions in creep rupture strength and sigmoidal inflexions being observed at much shorter creep rupture test durations. In addition precipitation of Fe_2Mo Laves phase at all of the testing temperatures has a further deleterious effect on the creep rupture strength of the steels, due to depletion of the matrix in Mo thus lowering the solid solution strengthening of the steel. Vodarek and Strang have reported similar effects in studies conducted on a series of 12CrMoV steels with nickel contents in the range 0.3 to 1.2 wt.%.[13] In these steels increases in the nickel content resulted in accelerated microstructural degradation with more rapid coarsening of the $M_{23}C_6$ precipitates, dissolution of the matrix strengthening MX precipitates, precipitation of coarse M_6X and Fe_2Mo phases and consequent sigmoidal creep rupture behaviour.

12CrMoVNbN STEELS

The parametric analyses was carried out on creep rupture data extending up to 100,000 hours at temperatures of 475, 550 and 600°C on three casts of 12CrMoVNbN steel with nickel contents in the range 0.52 to 1.15 wt.%, Table 1. As in the case of the 9Cr1Mo(V) steel the best fits were obtained using the Manson form of Lenert's general equation. In each instance sigmoidal creep rupture behaviour was observed at 600°C in the steels containing 0.52, 0.76 and 1.15 wt.%Ni at durations of approximately 20,000, 12,000 and 7000 hours respectively. A sigmoidal inflexion was also evident in the 0.76% Ni steel tested at 550°C with strong evidence, based on creep rupture ductility results, that the other two casts were approaching test durations at which sigmoidal inflexions would be expected.[4] As before analyses of the 12CrMoVNbN steels using the Manson general equation indicated that both the inflexion point stresses and durations increase with decreasing test temperatures, Tables 7 and 8. Once more the predicted change in sigmoidal stress with testing temperature was small, namely 4–5%, compared with an actual experimental value of around 150% based on a comparison between the 550 and 600°C data in the 0.76% Ni cast.

Analysis of the creep rupture data from the 0.52% Ni cast using the Manson general parameter indicated a good fit to the experimental data at 600°C with reasonable fits at 550 and 475°C, Figure 2. Similarly good agreement was found between the experimental and calculated values for the 550 and 600°C 10,000 hour creep rupture strengths. However, the corresponding rupture strength at 475°C was about 5% higher than the experimental result. In the case of the 100,000 hour rupture strength there was good agreement between the calculated and experimental values at 600°C. However, the corresponding creep strength estimates were about 17% high at 550°C and 5% low at 475°C.

Good fits were also obtained with the experimental data on the 0.76% Ni cast at 475°C and 600°C. Furthermore the estimated creep rupture strengths for 10,000 hours agreed well with the experimental data for all three test temperatures. However, the fit at 550°C was poor with the parametric curve overshooting the experimental points, especially at durations beyond 20,000 hours. This resulted in the 100,000 creep rupture strength being overestimated by approximately 30% at this temperature. At the other two test temperatures the estimated 100,000 hour creep rupture strengths were in good agreement with the experimental data.

Similar results were obtained in the case of the 1.15% Ni steel although the parametric analysis once more resulted in a significant overestimate of the 100,000 creep rupture strength at 550°C.

Four important points emerge from these observations namely that,

i. Sigmoidal creep rupture behaviour is clearly related to the composition of these steels with increasing nickel contents leading to sigmoidal inflexions occurring at progressively shorter test durations.
ii. The stress and duration at which the sigmoidal inflexion occurs is strongly dependent on the temperature of testing with the inflexion point occurring at longer durations and higher stresses with lower test temperatures.
iii. Extrapolation of creep rupture data on steels prone to this behaviour is very difficult and its success depends on having a multi-temperature data set with sigmoidal inflexions at at least three test temperatures.
iv. The use of multi-temperature data sets which do not meet this requirement will result in gross overestimates of the long -term creep rupture strengths particularly at the lower test temperatures, primarily due to underestimation of the effect of testing temperature on the increase in sigmoidal inflexion stresses at lower temperatures.

Detailed microstructural studies indicated that in the case of the 12CrMoVNbN steels the sigmoidal inflexions were due to two main processes, namely (i) coarsening of the grain boundary $M_{23}C_6$ precipitates with associated reductions of dislocation densities and subsequent subgrain formation in the steels, and (ii) dissolution of fine M_2X and MX precipitates due to the precipitation of Z-phase and M_6X. Through these degradation processes the creep rupture strength of the steel changes from being initially in a precipitation strengthened condition to one dependent on solid solution strengthening. The manifestation of this is the sigmoidal inflexion observed in their creep rupture characteristics. Another aspect observed is that in the precipitation strengthened condition the creep rupture ductility is low and the fractures intergranular while in the solid solution condition the creep rupture ductility is high and the fractures are transgranular. These effects are reflected in the creep rupture ductility curves for these steels where the sigmoidal inflexion in their creep rupture strengths are heralded by an increase in their long-term creep rupture ductilities.[4] Furthermore, the sigmoidal inflexion in the creep rupture properties occurs at progressively shorter test durations with increasing nickel contents and coincides with the corresponding precipitation of Z-phase and M_6X in these steels.

X19CrMoVNbN 11.1

The parametric analyses was carried out on cast 12d of Koenig's creep rupture data on the X19CrMoVNbN 11.1 steel since this cast showed clear evidence of a sigmoidal inflexion in the test results at 600°C, Figure 3.[6] These data extended to durations of approximately 29,000, 73,500, 90,000 and 121,000 hours at 450, 500, 550 and 600°C respectively. Once more the best fit to the data was obtained using the Manson general equation. A close fit was found between the calculated curve and actual creep rupture data at 600°C with this providing good estimates of the creep rupture strengths at both 10,000 and 100,000 hours at this temperature. However although the calculated Manson curve gave a reasonable estimates for the 10,000 hour creep rupture strengths at 550°C and 500°C the fits to the data points were poor with 100,000 hour creep rupture strength being underestimated by over 40% at each of these temperatures, Figure 3. Nevertheless the creep rupture data on this steel shows a clear sigmoidal inflexion at approximately 15,000 hours duration at 600°C together with a probable point of inflexion at around 100,000 hours at 550°C. These observations indicate that X19CrMoVNbN 11.1 is microstructurally unstable at these temperatures and this, together with the observation that Z-phase forms in the steel at these temperatures after exposure for 10 k hours at 600°C[20] suggests that the mechanism of sigmoidal creep rupture behaviour in X19 is similar to that previously reported by Strang and Vodarek in their studies on 12CrMoVNbN steels.[1, 3, 12]

TAF650 STEEL

The creep rupture data on the TAF650 steel determined by Sawada et al.[8] at 600, 650 and 700°C and Mayer at 600, 625 and 650°C extending to maximum durations of 19,000 and 43,000 hours respectively, are shown plotted in Figure 4. A comparison of Sawada et al. and Mayer's data shows that a sigmoidal inflexion is only apparent in Mayer's data at 650°C.

However the close agreement between Sawada et al.'s data with Mayer's curve at the same temperature together with the rapid fall off of rupture strength at the other test temperatures is a strong indication that TAF650 is very prone to sigmoidal creep rupture behaviour in the temperature range 600 to 700°C. Indeed Lenert and Foldyna's analysis of the combined data indicate that the points of inflexion for this steel at 600, 625, 650 and 700°C occur respectively at approximately 27542, 12025, 5100 and 437 hours and an average stress of 125.4 MPa. However, comparisons of the Manson parameter estimated 10,000 hours creep rupture strengths with the actual creep rupture data indicated that the strength was underestimated at 700°C, overestimated at 650°C, but in good agreement with the calculated values at 625 and 600°C. It is therefore concluded that given the limited creep rupture data presently available for this steel, reasonably good estimates for creep rupture strengths are limited to an absolute maximum of 50,000 hours at 625 and 650°C.

Svoboda et al., have recently identified Z-phase and Fe_2W Laves phase in TAF650 after 18,700 hours exposure at 650°C.[11] They have concluded that the sigmoidal creep rupture behaviour observed in this steel at 650°C is due to the of loss of fine matrix MX precipitates due to the precipitation of Z-phase combined with depletion of solid solution matrix strengthening associated with the precipitation of Fe_2W Laves phase.

DISCUSSION

The parametric analyses of the creep rupture data on the four types of 9–12%Cr steels have shown that the best fits are obtained using the Manson-Haferd and Manson general equations with the Manson general equation being preferred on account of its capability of predicting the effect of testing temperature on the sigmoidal inflexion stress. However, the prediction of the sigmoidal creep rupture inflexions and the 10,000 and 100,000 hours creep rupture strengths of the alloys using these derived best-fit equations has proved to be variable. In the case of the sigmoidal creep rupture inflexions the durations in some instances were significantly overestimated, particularly at the lower test temperatures. The net result of this was that the 100,000 hour creep rupture strengths were also overestimated by more than 60%, which is clearly unsatisfactory from the point of view of the safe design of critical power plant components. The unsatisfactory nature of these results is primarily due to the limited nature of the available creep rupture data on these steels. In order to achieve more reliable predictions of the long-term creep rupture strengths of these steels for power plant design lives of 250,000 hours comprehensive multi-temperature creep rupture data are required for each steel in which sigmoidal inflexions exist at least three different test temperatures.

It is now clear that the primary cause of sigmoidal creep rupture behaviour in these steels is due to the effects of microstructural changes in the alloys resulting from the combined effects of thermal and creep strain accumulation during long-term exposure at high temperatures. The susceptibility of steels to sigmoidal creep rupture behaviour is dependent on their chemical compositions and initial heat treatments. The effect of nickel content on the microstructural stability and sigmoidal creep rupture behaviour of the 12CrMoV and 12CrMoVNbN is now well established with nickel contents of > 0.5 wt.% being known to lead to sigmoidal inflexions occurring at shorter test durations and lower testing temperatures.

The use of Cu and Co additions as a palliative to the formation of δ-ferrite in these steels may prove to have similar effects on their microstructural stability in the long-term.

Detailed analytical electron microscopy studies have shown that the mechanism of microstructural degradation during high temperature creep exposure in the 9–12%Cr alloy steels is associated with the initially precipitation strengthened structure being transformed to one dependent on solid solution hardening. In the 9-12CrMoVN steels this is a twofold process in which the coarsening of $M_{23}C_6$ grain boundary precipitates is accompanied by dissolution of the fine matrix MX precipitates due to the precipitation of coarse particles of M_6X. This in conjunction with precipitation of Fe_2Mo Laves phase further weakens the steel by removal of solid solution strengthening elements such as Mo from the matrix. A similar process would be expected in the case of 9-12CrMoWVN steels with Mo and W being depleted from the matrix by precipitation of Fe_2Mo and/or Fe_2W Laves phases.

The mechanism is similar in the case of the 9-12CrMoVNbN steels where the coarsening of the $M_{23}C_6$ precipitates is also accompanied by dissolution of the fine matrix MX precipitates. However the primary driving force in this instance is the precipitation of Z-phase, a complex thermodynamically stable nitride of the form Cr(Nb,V)N. This is often accompanied by precipitation of coarse particles of M_6X and, in certain compositions Laves phases, all of which may lead to further weakening of the matrix through depletion of solid solution strengthening elements such as Mo and /or W.

It has now been clearly demonstrated through these studies that compositional balance is an important factor in achieving long-term microstructural stability and hence high temperature creep rupture strength in this class of alloys. Significant steps towards this are thought to have been achieved in, for example, COST steel B2.[21] However, only time will tell whether higher microstructural stabilities have been realised in the new generation of advanced 9–12%Cr steels.

CONCLUSIONS

Parametric analyses of creep rupture data on a series of creep resistant 9–12%Cr power plant steels has shown that reasonable descriptions of the data can be achieved using Manson-Haferd and Manson general equations. However the despite the paucity of current creep rupture data sets the Manson general equation is preferred due to its potential for modelling the significant effect of temperature on the sigmoidal creep rupture inflexion stress. At present using the best-fit equations derived from the currently available creep rupture data, prediction of the 10,000 and 100,000 hour creep rupture strengths can be overestimated by as much as 60%. For more reliable parametric predictions of creep rupture strengths of these steels for power plant design purposes, further multi-temperature creep rupture data extending to up to 150,000 hours with sigmoidal inflexions at at least three different temperatures are required.

Microstructural studies have shown that sigmoidal creep rupture behaviour in these alloys is due to microstructural degradation during creep exposure resulting in the long-term strength being finally dependent on solid solution strengthening in the alloy. The susceptibility to sigmoidal creep rupture behaviour is closely associated with the composition and initial heat treatment of the steel. Restriction of deleterious elements such as nickel, adjustment of the balance of key alloying elements such as C, Cr, Mo,V, W and N and the addition of B may lead to the development of alloys with greater microstructural stability for advanced long-term, high temperature power plant applications.

REFERENCES

1. A. Strang and V. Vodarek: *Materials Science and Technology*, **12**, 1996, 552.
2. V. Vodarek and A. Strang: *Proceedings of 9th International Symposium on Creep Resistant Metallic Materials*, Prague, 2001, 232.
3. A. Strang and V. Vodarek: *Proceedings of 5th International Charles Parsons Conference (PARSONS 2000)*, Cambridge, 2000, 572.
4. A. Wickens, A. Strang and G. Oakes: *Proceedings of Joint International IMechE Conference on Engineering Aspects of Creep*, Sheffield, 1980, 11.
5. J. H. Bennewitz: *Proceedings of Joint International Conference on Creep*, New York, London, 1969, 69.
6. H. Koenig: *Proceedings of 16 Vortragsveranstaltung – Langzeitvehalten warmfester Stahle und Hochtemperaturwerkstoffe*, Vdeh, Dusseldorf, 1993, 10.
7. K. Kimura, et al.: *Proceedings of 4th International Charles Parsons Conference*, Newcastle upon Tyne, 1997, 257.
8. K. Sawada, et al.: *Materials Science and Engineering*, **A267**, 1999, 19.
9. Y. Hasegawa, et al.: *Key Engineering Materials*, **171–174**, 2000, 427.
10. F. Abe: *Materials Science and Engineering*, **A319–321**, 2001, 770.
11. M. Svoboda, et al.: *Proceedings of International Conference on Materials for Advanced Power Engineering 2002*, Liege, 2002,1521.
12. A. Strang and V. Vodarek: *Proceedings of 7th International Conference on Creep and Fracture of Engineering Materials and Structures*, Univ. of California, Irvine, 1997, 415.
13. V. Vodarek and A. Strang: *Acta Metallurgica Slovaca*, **6**(4), 2000, 388.
14. A. Strang, et al.: *Proceedings of 9th International Symposium on Creep Resistant Metallic Materials*, Hradec nad Moravici, Czech Republic, 1996, 448.
15. A. Strang, et al.: *Proceedings of 4th International Charles Parsons Conference*, Newcastle upon Tyne, 1997, 603.
16. J. Lenert: Trans. VSB – Tech. Univ. of Ostrava, Mechanical Series, **44**, 1226, 37.
17. V. Foldyna: *Steel Research*, **62**, 1991, 453.
18. V. Foldyna, et al.: *Proceedings of 2nd International Conference on Creep and Fracture of Engineering Materials and Structures*, Swansea, 1994, 453.
19. V. Foldyna: *Technike Actuality*, Vitkovic, **1**, 1988.
20. Z. Kubon, et al.: *Proceedings of 5th International Charles Parsons Conference (PARSONS 2000)*, Cambridge, 2000, 485.
21. K. Spiradek-Hahn, et al.: *3rd EPRI Conference on Advances in Material Technology for Fossil Power Plants*, Swansea, 2001, 165.

Long-Term Creep Strength Prediction of High Cr Ferritic Creep Resistant Steels Based on Degradation Mechanisms

K. KIMURA
Creep Group, Materials Information Technology Station
National Institute for Materials Science
2-2-54 Nakameguro, Meguro-ku
Tokyo 153-0061, Japan

H. KUSHIMA and K. SAWADA
Creep Group, Materials Information Technology Station
National Institute for Materials Science
1-2-1 Sengen, Tsukuba-shi
Ibaraki 305-0047, Japan

ABSTRACT

In contrast to the homogeneous recovery in the short-term, inhomogeneous recovery preferentially takes place in the vicinity of prior austenite grain boundaries is important degradation mechanism of high Cr ferritic creep resistant steels in the long-term. Such change in degradation mechanism results in an inflection of stress vs. time to rupture curve and makes a long-term creep strength prediction to be difficult. In order to improve accuracy of long-term creep strength prediction, creep rupture data should be divided into two groups of higher stress short-term region and lower stress long-term one. Change in degradation mechanism and the influence of applied stress on that has been investigated, and the estimation method of the boundary condition between the short-term and the long-term has been discussed. It has been found that the boundary condition corresponds to half of 0.2% proof stress that has been considered to be an elastic limit of the steel. From these results, a new prediction method of long-term creep strength has been proposed on high Cr ferritic creep resistant steels.

INTRODUCTION

From a viewpoint of global environment, improvement of energy efficiency of fossil fired power plant is strongly required, since thermal power plant is one of the main sources of CO_2 emission.[1] It is also beneficial to save energy resources such as oil, coal and natural gas. A lot of efforts have been conducted on research and development of high strength ferritic creep resistant steels that is used for high temperature structural components such as header and main steam pipe.[2] 9Cr-1Mo-V-Nb steel (ASME T91/P91) developed in the early 1980s is improved in creep strength by precipitation strengthening of fine MX carbonitride with addition of vanadium and niobium, and it has already been widely used for high temperature structural components of power plant.[3] Further improvement in creep strength has been attained by replacing a part of molybdenum with tungsten and high strength ferritic creep resistant steels such as 9Cr-0.5Mo-1.8W-V-Nb steel[4], 12Cr-0.4Mo-2W-Cu-V-Nb steel[5] and 9Cr-1Mo-1W-V-Nb steel[6] have been developed in 1990s. These high strength ferritic creep resistant steels have been applied to construction of high temperature structural components of modern thermal power plants. However, some difficulty on long-term creep strength evaluation has been pointed out on such high strength ferritic creep resistant steels.[7] Since relation between stress and time to rupture of those steels shows inflection in the long-term, and it results in large drop in long-term creep strength. In order to evaluate long-term creep strength accurately, it has been proposed that creep rupture data should be divided into two groups of short-term and long-term according to difference in slope of stress vs. time to rupture curve. However, it is not useful for a life prediction from short-term creep rupture data, since boundary condition can not be defined without long-term creep rupture data.

Kimura et al. have investigated on degradation behaviour of 9Cr-1Mo-V-Nb steel (ASME T91) during long-term creep exposure and found that inhomogeneous recovery preferentially takes place in the vicinity of prior austenite grain boundary is a degradation mechanism in the long-term, in contrast to homogeneous recovery in the short-term.[8] It has been also pointed out that such difference in recovery phenomena strongly depend on applied stress and boundary condition between short-term and long-term roughly corresponds to half of 0.2% proof stress.[9] In this study, correlation between applied stress and inflection of stress vs. time to rupture curve has been investigated on several type of ferritic creep resistant steels. A meaning of a half of 0.2% proof stress has been discussed, in order to understand degradation behaviour during creep exposure and improve a long-term creep strength prediction method.

EXPERIMENTAL PROCEDURE

9Cr-1Mo-V-Nb steel (ASME T91),[10] 2.25Cr-1Mo steel (ASTM A542)[11] and 1Cr-1Mo-0.25V turbine rotor steel (ASTM A470-8)[12] were used in the present study. Details of chemical compositions, heat treatment conditions and tensile strength properties are shown in each Creep Data Sheet.[10–12] Microstructure was examined under transmission electron microscope on 9Cr-1Mo-V-Nb steel.

Abrupt loading and unloading test was conducted on 9Cr-1Mo-V-Nb steel at 600 and 650°C, and deformation after abrupt change in load was investigated, in order to discuss an

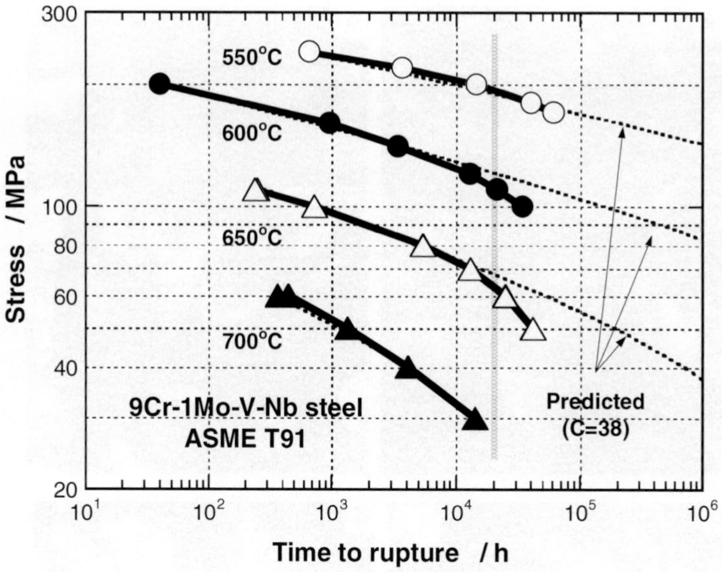

Fig. 1 Stress vs. time to rupture curves of 9Cr-1Mo-V-Nb steel and the creep rupture lives predicted from the short-term data whose rupture lives are less than 20,000 hours by a Larson-Miller parameter with a best fit parameter constant of 38.

elastic limit and plastic deformation at the elevated temperatures. A Larson-Miller parameter was employed for long-term creep strength prediction. New long-term life prediction method has been also discussed by means of Larson-Miller parameter.

RESULTS AND DISCUSSION

Influence of Stress on Recovery

Stress vs. time to rupture curves of 9Cr-1Mo-V-Nb steel tested over a range of temperatures from 550 to 700°C are shown in Figure 1.[10] Predicted creep rupture lives from the short-term data up to 20,000 hours using by a Larson-Miller parameter with a best fit parameter constant of 38 are also shown by the dotted lines in Figure 1. Slope of the curves indicated tendency to increase with decrease in applied stress and increase in temperature. Since such changes in slope significantly take place after about 10,000 hours at 600 and 650°C, creep rupture strength at 100,000 hours at these temperatures are extremely overestimated. Corresponding to changes in slope of the stress vs. time rupture curve, significant decrease in rupture ductility was also observed in the long-term over about 10,000 hours. Moreover, onset strain of accelerating creep stage decreased with decrease in stress.

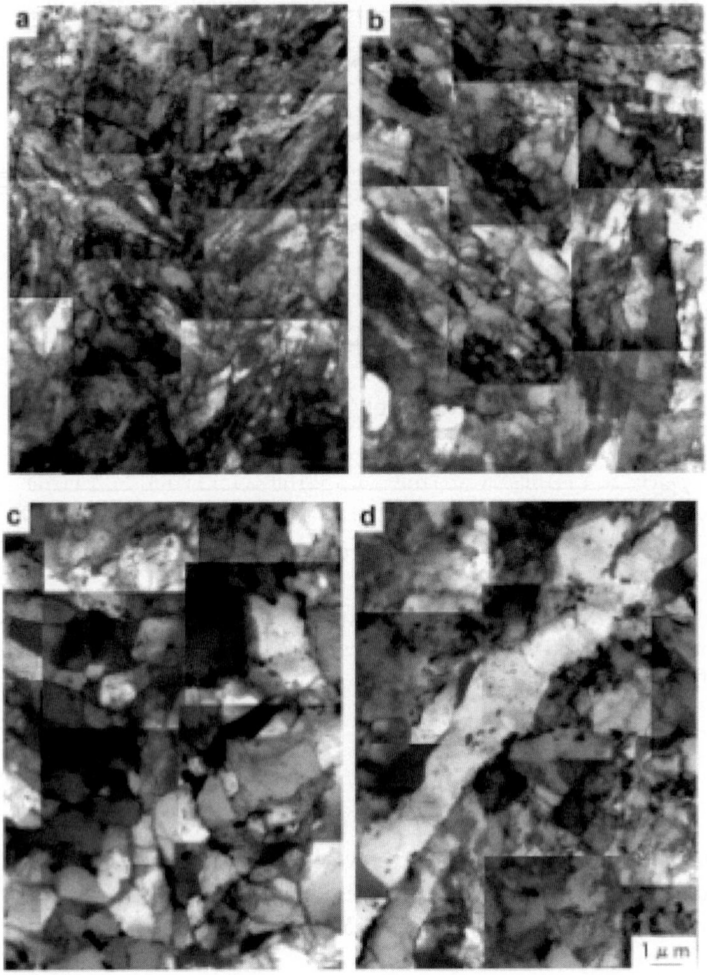

Fig. 2 Bright field TEM images of 9Cr-1Mo-V-Nb steel (a) in the as tempered condition and the specimens creep ruptured at 600°C after (b) 971.2 hours under 160 MPa, (c) 12,858.6 hours under 120 MPa and (d) 34,141.0 hours under 100 MPa.

Bright field TEM images of 9Cr-1Mo-V-Nb steel (a) in the as tempered condition and the specimens creep ruptured at 600°C after (b) 971.2 hours at 160 MPa, (c) 12,858.6 hours at 120 MPa and (d) 34,141.0 hours at 100 MPa are shown in Figure 2. In the as tempered condition (Figure 2a), tempered martensite microstructure was very fine with high dislocation density. Homogeneous progress in recovery of microstructure with increases in lath width and subgrain size and decrease in dislocation density was observed with increase in creep exposure time in the specimen creep ruptured after 971.2 hours (Figure 2b) and 12,858.6 hours (Figure 2c), at 600°C-160 MPa and 600°C-120 MPa, respectively. On the other hand,

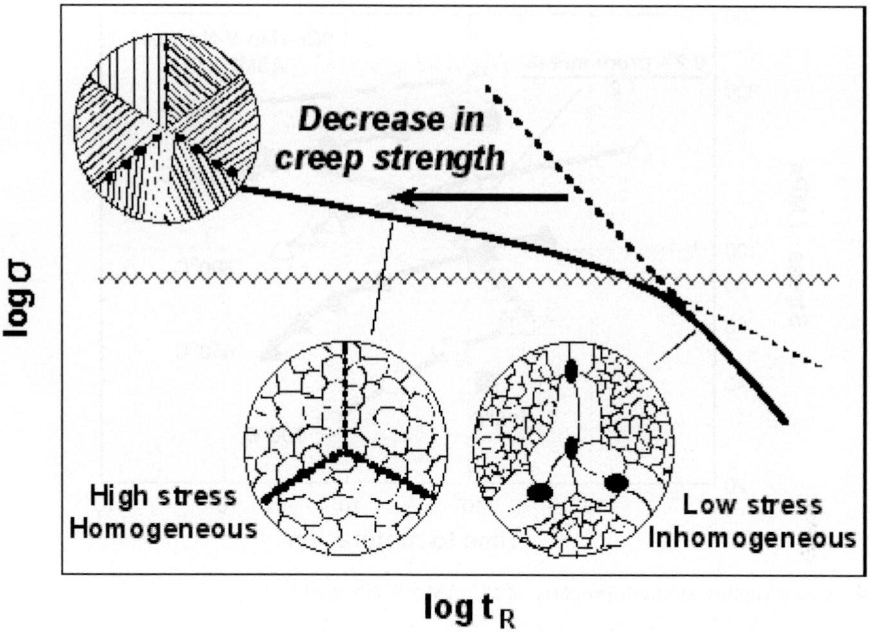

Fig. 3 Schematic illustration on influence of stress on microstructural change during creep exposure.

inhomogeneous progress in recovery was observed in the specimen creep ruptured after 34,141.0 hours at 600°C-100 MPa (Figure 2d). The microstructure within grain was very fine, in contrast to significantly recovered region in the vicinity of prior austenite grain boundary. Although creep exposure time of 34,141.0 hours at 600°C-100 MPa was about three times longer than that of 12,858.6 hours at 600°C-120 MPa, the subgrain size within grain of the former specimen was smaller than that of latter one. Bimodal distribution of subgrain size was clearly observed only in the specimen creep ruptured at 600°C-100 MPa.[13] Inhomogeneous progress in recovery of tempered martensite microstructure was clearly observed in the specimen creep ruptured at low stress, in contrast to homogeneous one in the specimens creep ruptured at higher stresses. Similar phenomenon of preferential recovery in the vicinity of prior austenite grain boundary has been also reported on 1Cr-1Mo-0.25V turbine rotor steel as a degradation mechanism during long-term creep exposure.[14]

Influence of applied stress on progress in recovery during creep exposure is schematically illustrated in Figure 3. Recovery of tempered martensitic microstructure is thought to be promoted in the vicinity of prior austenite grain boundaries through the effects of concentrated internal strain introduced by martensitic transformation, rapid diffusion rate and faster coarsening of grain boundary precipitates, in contrast to those within grain. If the applied stress is high enough to cause plastic deformation, however, recovery within grain is also accelerated and locally recovered region in the vicinity of prior austenite grain boundary is

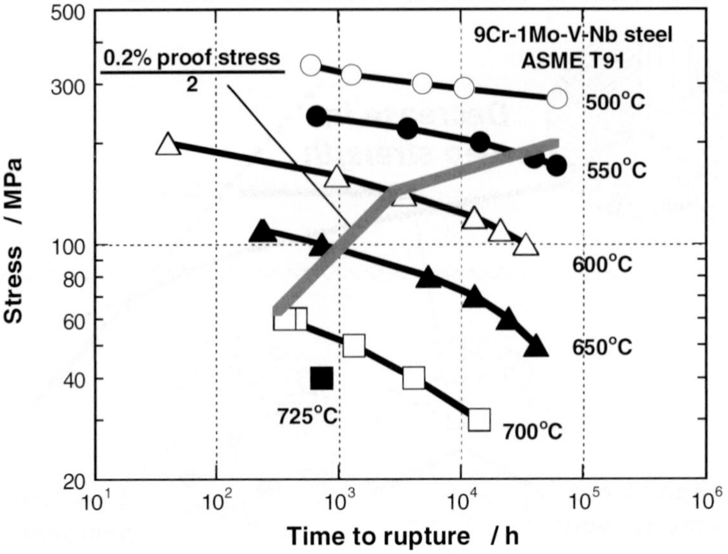

Fig. 4 Creep rupture strength property of 9Cr-1Mo-V-Nb steel.[10]

also easily extended within grain. Consequently, creep strength in the short-term where applied stress is high enough to cause plastic deformation, is significantly lowered. Since such influence of applied stress results in inflection of stress vs. time to rupture curve, stress at the inflection of the curve should correspond to tensile strength property such as elastic limit.

By considering difference in recovery phenomena during creep exposure in the short-term and long-term, it has been supposed that inflection of stress vs. time curve corresponds to tensile strength property, especially elastic limit. Therefore, correlation between shape of stress vs. time to rupture curves and proof stress is investigated on 9Cr-1Mo-V-Nb steel, 2.25Cr-1Mo steel and 1Cr-1Mo-0.25V steel. Microstructure of the 2.25Cr-1Mo steel and 1Cr-1Mo-0.25V steel is tempered martensite and tempered bainite, respectively. Figure 4 shows stress vs. time to rupture curves of 9Cr-1Mo-V-Nb steel, together with half of 0.2% proof stresses at the temperatures.[10] As it has been reported previously, inflection of the curves roughly corresponds to half of 0.2% proof stress at the temperatures.[9] Slopes of the curves in the stress condition lower than half of 0.2% proof stress are larger than those in the stresses higher than that. Stress vs. time to rupture curves of 2.25Cr-1Mo steel and 1Cr-1Mo-0.25V steel are shown in Figures 5 and 6, respectively.[11, 12] Stress conditions of half of 0.2% proof stress are also indicated in the figures. For both steels, stress dependence of creep rupture life is clearly divided into two regions of short-term with small slope of the stress vs. time to rupture curve and long-term with large slope, similar to 9Cr-1Mo-V-Nb steel as shown in Figure 4. From these observations, it has been supposed that half of 0.2%

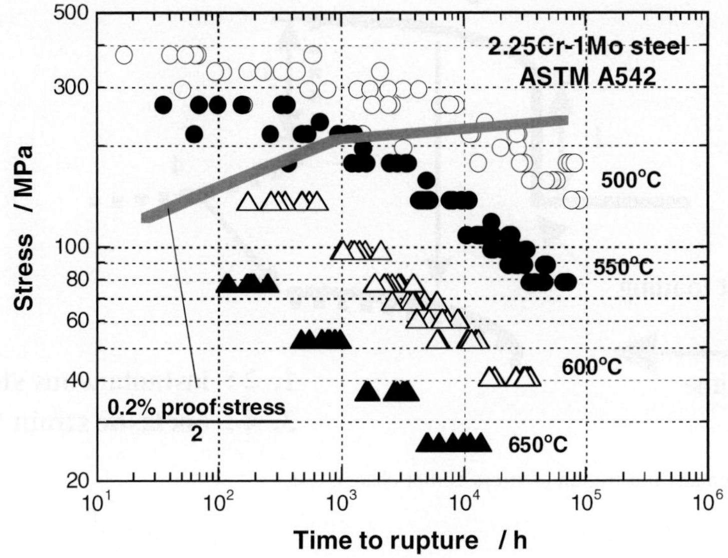

Fig. 5 Creep rupture strength property of quenched and tempered 2.25Cr-1Mo steel.[11]

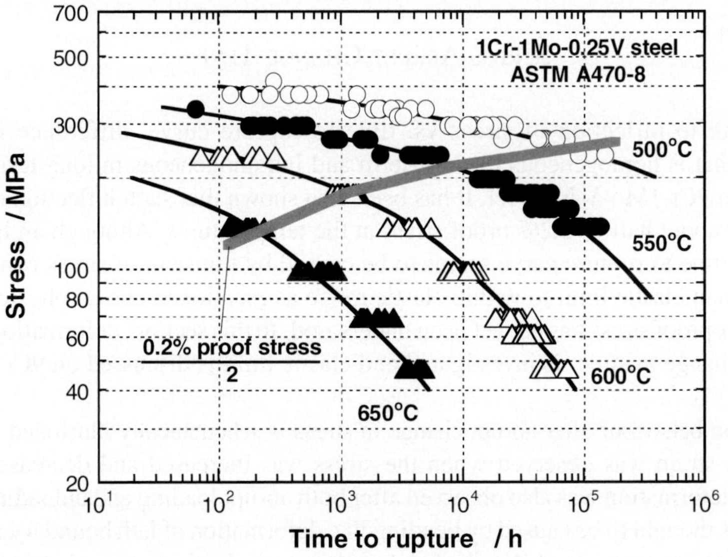

Fig. 6 Creep rupture strength property of 1Cr-1Mo-0.25V turbine rotor steel.[12]

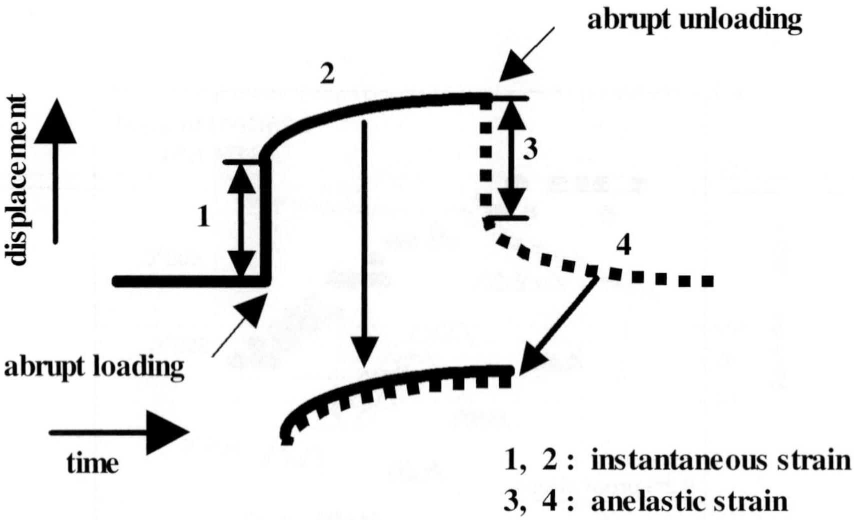

Fig. 7 Schematic illustration of stress abrupt change test.

proof stress is considered to be a boundary condition between short-term and long-term for ferritic creep resistant steels, where stress dependence of creep rupture strength changes.

Stress Abrupt Change Test

Corresponding to inflection of stress vs. time to rupture curve, difference in recovery phenomena, that is homogeneous in short-term and inhomogeneous in long-term, has been pointed out on 9Cr-1Mo-V-Nb steel. It has been also shown that such inflection of the curve takes place at about half of 0.2% proof stress at the temperatures. Although an inflection of the stress vs. time to rupture curve seems to be caused by transient of stress condition from higher level than elastic limit to within elastic range as mentioned above, physical meaning of half of 0.2% proof stress has not yet been understood. In this section, deformation behaviour after abrupt change in stress is investigated and elastic limit is discussed on 9Cr-1Mo-V-Nb steel.

Deformation behaviour after abrupt change in stress is schematically illustrated in Figure 7. Instantaneous strain was observed when the stress was increased and decreased abruptly, and inelastic deformation was also observed after both abrupt loading and unloading. Inelastic deformation is thought to be caused by bending like deformation of lath boundary of tempered martensitic microstructure and detail of deformation mechanism after abrupt loading and abrupt unloading is discussed in the other paper.[15] By comparing a magnitude of inelastic strain after abrupt loading and abrupt unloading, contribution of plastic deformation to a deformation after abrupt loading should be understood.

Table 1 Elastic limit stress obtained by the abrupt loading test and 0.2% proof stress of 9Cr-1Mo-V-Nb steel at 600 and 650°C.[10]

Temperature	Tensile Test		Abrupt Loading Test
	0.2% Proof Stress	1/2 of 0.2% Proof Stress	Elastic Limit Stress
600°C	289 MPa	144.5 MPa	150 MPa
650°C	191 MPa	95.5 MPa	100 MPa

Deformation behaviour of 9Cr-1Mo-V-Nb steel with abrupt loading and abrupt unloading of 140 MPa at 650°C is shown in Figure 8a. Magnitude of time dependent deformation after abrupt loading and abrupt unloading is compared and shown in Figure 8b. Instantaneous strain was observed corresponding to abrupt loading and abrupt unloading of 140 MPa, and followed by time dependent deformation in both changes in stress of loading and unloading (Figure 8a). However, magnitude of time dependent anelastic strain after abrupt loading is much larger than that after abrupt unloading. Difference in magnitude of anelastic strain, that is larger strain after abrupt loading, should be caused by plastic deformation under stress of 140 MPa (Figure 8b). Therefore, elastic limit of the present steel is regarded to be less than 140 MPa at 650°C.

Magnitude of time dependent anelastic strain after abrupt loading and abrupt unloading measured at 600 and 650°C over the range of stresses from 80 to 240 MPa at 600°C and 40 to 200 MPa at 650°C are shown in Figures 9a and 9b, respectively. Magnitude of anelastic strain after abrupt loading was measured after holding of stress for 20 sec. On the other hand, that after abrupt unloading was evaluated by measuring the strain until deformation rate become to be zero. When the loading stress is lower than about 150 MPa at 600°C and 100 MPa at 650°C, magnitude of anelastic strain after abrupt loading and abrupt unloading is almost the same. However, magnitude of anelastic strain after abrupt loading is larger than that after abrupt unloading when the loading stress is higher than those conditions. Consequently, elastic limit of the steel is evaluated to be 150 and 100 MPa at 600 and 650°C, respectively.

Elastic limit stress obtained by the abrupt loading test and 0.2% proof stress of 9Cr-1Mo-V-Nb steel at 600 and 650°C are listed in Table 1.[10] Elastic limit stress evaluated from the abrupt loading test of 150 MPa at 600°C and 100 MPa at 650°C is almost the same as half of 0.2% proof stress of 144.5 MPa at 600°C and 95.5 MPa at 650°C. From these results, it has been supposed that half of 0.2% proof stress is considered to be an elastic limit stress.

Prediction of Long-Term Creep Strength

Stress vs. time to rupture curves of 9Cr-1Mo-V-Nb steel are shown in Figure 10, together with creep rupture lives predicted by Larson-Miller parameter with all the data and best fit

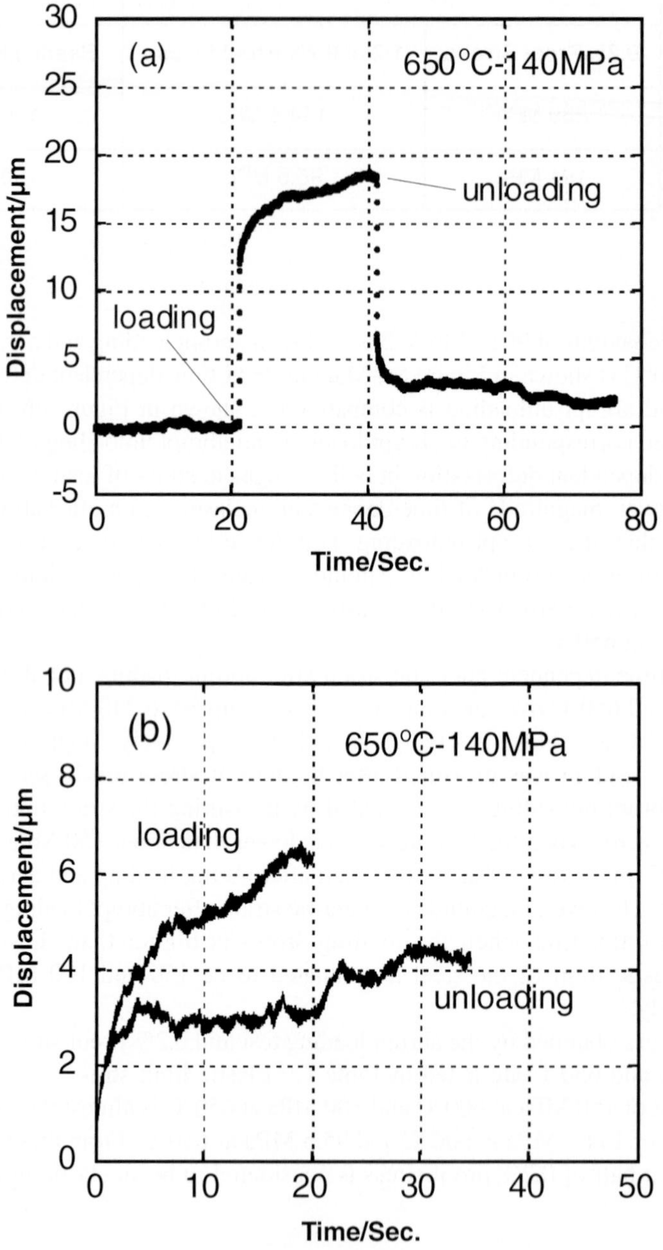

Fig. 8 Deformation behaviour of 9Cr-1Mo-V-Nb steel with abrupt loading and abrupt unloading of 140 MPa at 650°C. Magnitudes of displacement after abrupt loading and abrupt unloading are shown in (b).

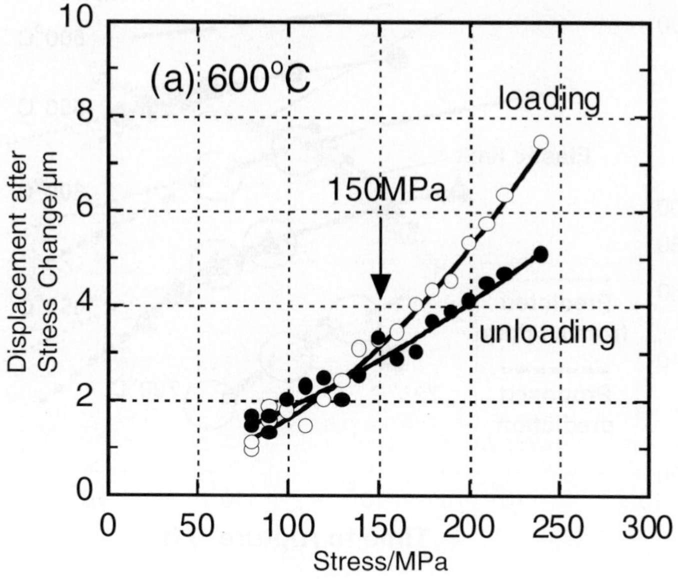

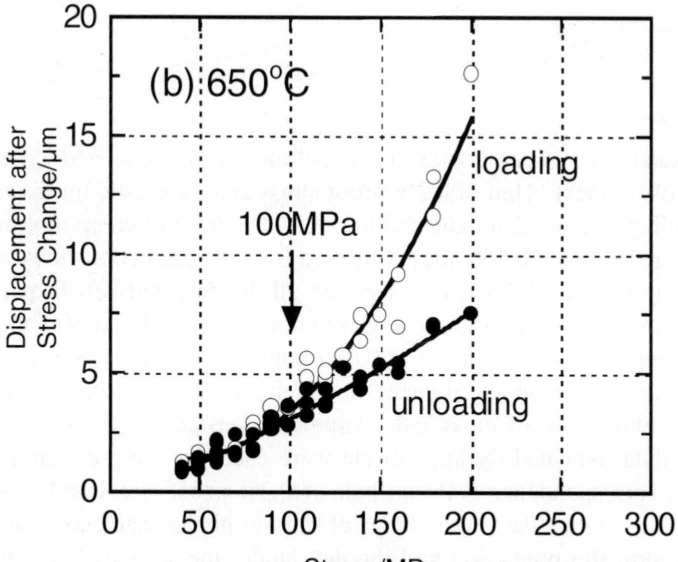

Fig. 9 Stress dependence of displacement after abrupt change in stress of 9Cr-1Mo-V-Nb steel at (a) 600°C and (b) 650°C.

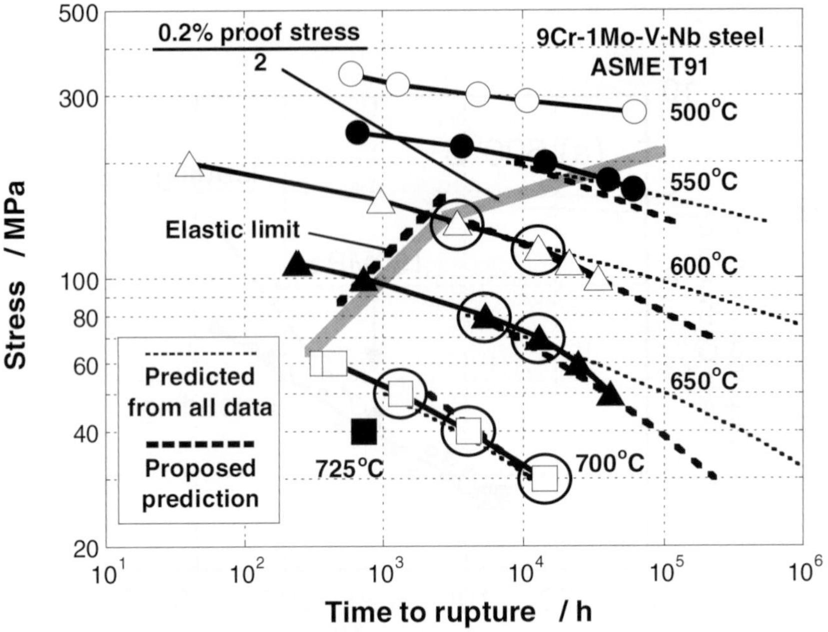

Fig. 10 Long-term creep strength predictions of 9Cr-1Mo-V-Nb steel by a conventional method with all the data and best fit parameter constant (fine dotted lines) and a proposed one with the selected data under stresses lower than half of 0.2% proof stress (indicated by large circles) and a parameter constant of 20 (bold broken lines). Elastic limit obtained by abrupt loading test is also indicated.

parameter constant of 38 (fine dotted lines), and the selected data and a parameter constant of 20 (bold broken lines).[9] Half of 0.2% proof stress and the elastic limit stress evaluated by the abrupt loading test are also indicated in the figure. A lot of creep rupture data including long-term one up to about 60,000 hours was used for a life prediction, however, creep rupture life predicted by Larson-Miller parameter with all the data and best fit parameter constant indicated extremely overestimation at 100,000 hours, especially at 600 and 650°C. On the other hand, creep rupture strength at 100,000 hours was adequately evaluated by Larson-Miller parameter in the proposed manner, in which only the selected data under the stresses lower than half of 0.2% proof stress and a parameter constant of 20 was used. Only the seven creep rupture data indicated by large circle were used for life prediction in the proposed manner. Good correspondence between half of 0.2% proof stress and elastic limit is also recognized. Creep rupture data in the range of stresses higher than elastic limit should not be used for long-term life prediction and the data under the stresses lower than elastic limit should be used. Half of 0.2% proof stress should be referred as an elastic limit stress at the temperatures. It has been concluded, consequently, that long-term creep strength prediction of ferritic creep resistant steels with Larson-Miller parameter should be conducted with the selected creep rupture data under the stresses lower than elastic limit by considering half of 0.2% proof stress at the temperatures.

CONCLUSION

Influence of applied stress on inflection of stress vs. time to rupture curve has been investigated and a meaning of half of 0.2% proof stress has been discussed on ferritic creep resistant steels. The following results are obtained.

1. Corresponding to inflection of stress vs. time to rupture curve, difference in recovery phenomena, that was homogeneous in short-term and inhomogeneous in long-term, was observed.
2. Inflection of stress vs. time to rupture curve took place at the stress condition corresponding to half of 0.2% proof stress at the temperatures.
3. Instantaneous strain was observed corresponding to abrupt loading and abrupt unloading, and followed by time dependent deformation in both changes in stress of loading and unloading.
4. By comparing the magnitude of anelastic strain after abrupt loading and abrupt unloading, elastic limit stress of 9Cr-1Mo-V-Nb steel was evaluated to be 150 MPa at 600°C and 100 MPa at 650°C. These stresses were found to be almost the same as half of 0.2% proof stress at the temperatures.
5. Inflection of stress vs. time to rupture curve should be caused by transient of applied stress from higher level than elastic limit to within elastic range.
6. It has been concluded that long-term creep strength of ferritic creep resistant steels should be predicted by Larson-Miller parameter with the selected creep rupture data under the stresses lower than elastic limit by considering half of 0.2% proof stress at the temperatures.

REFERENCES

1. F. Abe, M. Igarashi, N. Fujitsuna, K. Kimura and S. Muneki: 'Research and Development of Advanced Ferritic Steels for 650°C USC Boilers' *Proceedings of 6th Liege Conference Materials for Advanced Power Engineering 1998*, J. Lecomte-Beckers, et al., eds., Forschungszentrum, Jülich, GmbH, Liege, Belgium, Germany, **5**(1), 1998, 259–268.
2. F. Masuyama: 'History of Power Plants and Progress in Heat Resistant Steels', *ISIJ International* 2001, **41**(6), 2001, 612–625.
3. V. K. Sikka: 'Development of Modified 9Cr-1Mo Steel for Elevated-Temperature Service' *Proceedings of Topical Conference on Ferritic Alloys for Use in Nuclear Energy Technologies,* J. W. Davis and D. J. Michel, eds., TMS-AIME, Warrendale, Pennsylvania, USA, 1984, 317–327.
4. M. Sakakibara, H. Masumoto, T. Ogawa, T. Takahashi and T. Fujita: 'High Strength 9Cr-0.5Mo-1.8W Steel (NF616) Tube for Boilers', *Thermal Nuclear Power*, **38**(8), 1987, 841–850.
5. A. Iseda, A. Natori, Y. Sawaragi, K. Ogawa, F. Masuyama and T. Yokoyama: 'Development of High Strength and High Corrosion Resistance 12%Cr Steel Tubes and Pipe (HCM12A) for Boilers', *Thermal Nuclear Power*, **45**(8), 1994, 900–909.
6. M. Staubli, W. Bendick, J. Orr, F. Deshayes and Ch. Henry: 'European Collaborative

Evaluation of Advanced Boiler Materials', *Proceedings of 6th Liege Conference on Materials for Advanced Power Engineering 1998*, J. Lecomte-Beckers, et al., eds., Forschungszentrum, Jülich, GmbH, Liege, Belgium, Germany, **5**(1), 1998, 87–103.

7. V. Foldyna, Z. Kuboò, A. Jakobová and V. Vodárek: 'Development of Advanced High Chromium Ferritic Steels', *Microstructural Development and Stability in High Chromium Ferritic Power Plant Steels,* A. Strang and D. J. Gooch, eds., The Institute of Materials, 1997, 73–92.
8. K. Kimura, H. Kushima and F. Abe: 'Heterogeneous Changes in Microstructure and Degradation Behaviour of 9Cr-1Mo-V-Nb Steel During Long Term Creep', *Key Engineering Materials*, **171–174**, 2000, 483–490.
9. H. Kushima, K. Kimura and F. Abe: 'Long-Term Creep Strength Prediction of High Cr Ferritic Creep Resistant Steels', *Proceedings of 7th Liege Conference on Materials for Advanced Power Engineering 2002*, J. Lecomte-Beckers, et al., eds., Forschungszentrum, Jülich, GmbH, Liege, Belgium, Germany, **21**(3), 2002, 1581–1590.
10. NRIM Creep Data Sheet, National Research Institute for Metals, (43), 1996.
11. NRIM Creep Data Sheet, National Research Institute for Metals, (36A), 1991.
12. NRIM Creep Data Sheet, National Research institute for Metals, (9B), 1990.
13. K. Kimura, H. Kushima, F. Abe, K. Suzuki, S. Kumai and A. Satoh: 'Microstructural Change and Degradation Behaviour of 9Cr-1Mo-V-Nb Steel in the Long Term Region', *Proceedings of Fifth International Charles Parsons Turbine Conference on Advanced Materials for 21st Century Turbines and Power Plant*, A. Strang, et al., eds., The Institute of Materials, Cambridge, UK, **5**, 2000, 590–602.
14. T. Matsuo, K. Kimura, R. Tanaka and M. Kikuchi: 'Degradation of 1Cr1Mo1/4V Steel at High Temperatures', *Proceedings of Fourth International Conference on Creep and Fracture of Engineering Materials and Structures*, B. Wilshire and R. W. Evans, eds., The Institute of Metals, Swansea, UK, **4**, 1990, 477–486.
15. K. Sawada, F. Abe and K. Kimura: 'Mechanical Response of 9%Cr Heat Resistant Martensitic Steels to Abrupt Stress Loading at High Temperature', *Materials Science and Engineering A*, in press.

Modelling of Precipitation Kinetics and Creep in Power Plant Ferritic Steels

You Fa Yin and Roy G Faulkner
IPTME, Loughborough University
Loughborough, Leicestershire LE11 3TU
UK

ABSTRACT

Microstructure and creep behaviour modelling of power plant steels are important in both predicting component's life and the development of new alloys. In this paper, the authors attempted for the first time to feed the simulated microstructural evolution using the method developed by the authors to a continuum creep damage mechanics (CDM) model to study the creep behaviour of power plant steels. The technique has been applied to a power plant ferritic steel, P92. Reasonably good agreement with creep test results has been observed.

INTRODUCTION

The modelling of microstructural evolution of power plant steels has been an important topic for decades. In recent years, there is increasing interest and development in this topic. Different techniques exist, such as the methods developed by Bhadeshia's group,[1,2] Faulkner's group[3,4] and used in the DICTRA software.[5,6] The authors have recently developed a new approach to this problem using Monte Carlo simulations.[7,8] In such an approach, more information in respect to the evolution of the microstructure of steels, such as particle size distributions and inter-particle spacing, can be obtained in addition to the average size of the precipitates.

Parallel to the development of microstructural evolution modelling, different methods have been developed to predict long term creep behaviour of alloys from short term experimental observations or from using models.[9,10] Recently, Dyson reviewed the application of CDM in materials modelling and component creep life prediction and showed that this method provides a unifying framework for some of the other methods.[11] Different creep damage mechanisms can be integrated into the model, provided that the damage mechanism is understood and the evolution of the particular damage with time is known. Therefore, models of different creep damage mechanisms have to be used. For example, when considering the damage due to particle coarsening, Wagner's model has to be used to derive the evolution of the damage mode with time. The obvious limitation of such an approach is that the Wagner's model is only applicable to intra-granular spherical particles.

Here we report first some of our simulation results on the precipitation kinetics and then the results on CDM modelling by using both the output of simulations and analytical models.

SIMULATION OF PRECIPITATION KINETICS

The details of the simulation of precipitation kinetics used in this study have been reported elsewhere.[7, 8] Here, the essentials are summarised.

The simulation starts with the establishment of the simulation cell, a representation of the material where the phase transformations will be followed. Quench induced chromium segregation to the grain boundaries is then calculated according to the non-equilibrium segregation model developed by Faulkner et al.[12] to give the chromium concentration in grain boundaries of the simulation cell.

The tempering and ageing (or service) time is divided into many small time intervals, Δt. In each time interval, the nucleation and growth or coarsening are considered. The nucleation of precipitates has been considered using classical nucleation theory:

$$I = Z\beta^* \left(\frac{N}{x_\theta} \right) \exp\left(-\frac{\Delta G^*}{kT} \right) \exp\left(-\frac{\tau}{t} \right) \tag{1}$$

where I is the nucleation rate, N is the number of a particular type of atomic site, x_θ is the molar fraction of solute atoms in the nucleus phase, k the Boltzmann constant, T the absolute temperature, ΔG^* the energy required to form the critical nucleus, t the time and τ the incubation time for nucleation. The two coefficients Z and β^* are as follows

$$Z = \frac{V_{\theta a}(\Delta G_V)^2}{8\pi\sqrt{kTK_j\sigma_{\alpha\theta}^3}}, \quad \beta^* = \frac{16\pi\sigma_{\alpha\theta}^2 D x_\alpha L_j}{a^4(\Delta G_V)^2} \tag{2}$$

where $V_{\theta a}$ is the volume occupied by one atom in the nucleus, ΔG_V the free energy change per unit volume of nucleus, $\sigma_{\alpha\theta}$ the interfacial free energy, D the diffusivity, x_α the solute concentration in the matrix, a the lattice parameter, L_j and K_j are defined by

$$V = \frac{4}{3}\pi r^3 K_j, \quad S = 4\pi r^2 L_j \tag{3}$$

In these two equations, V and S are volume and surface area of the nucleus respectively and r the radius or a characteristic length parameter of the particle. Therefore, both K_j and L_j equal unity for spherical particles. For grain boundary precipitates, such as $M_{23}C_6$, they are functions of the contact angle. The incubation time

$\tau = \dfrac{8kT\sigma_{\alpha\theta}a^4}{V_{\theta a}^2(\Delta G_V)^2 D x_\alpha}$ is usually very small, so that the term $\exp\left(-\tau/t\right)$ in eqn (3) can be assumed as unity.

At present, we assume a nucleation is successful when the generated nucleus does not overlap with others. The total number of nuclei accepted in any time interval $t \sim t + \Delta t$ is controlled by eqn (1), i.e.,

$$\Delta N(t) = Z\beta^* \left(\frac{N}{x_\theta}\right) \exp\left(-\frac{\Delta G^*}{kT}\right) \Delta t \tag{4}$$

where $\Delta N(t)$ is the number of nuclei generated in the simulation cell during the time interval. No particular size distribution of the nuclei is introduced here. The nucleation rate decreases with time because the matrix solute concentration decreases due to the formation and growth of precipitates.

The average solute concentration at any time t, $\bar{C}_t$ can be calculated from the initial concentration of solute C_g and the volume fraction of precipitates V_f:

$$\bar{C}_t = 1 - \frac{V_f \rho_\theta N_r}{\rho_\alpha C_g} \tag{5}$$

ρ_α and ρ_θ are the molar density of the matrix and the precipitate phase respectively and N_r is the number of rate controlling atoms in the precipitate molecule. The solute concentration at the surface of each particle also can be determined using:

$$C_r = C_\infty \exp \frac{2\sigma_{\alpha\theta} V_\theta}{RTr} \tag{6}$$

where C_∞ is the equilibrium solute concentration, C_θ the molar volume of the precipitate phase, R the universal gas constant. Thus the concentration gradient at the surface of a particle at time t can be estimated using a mean field approximation.

$$g = \frac{\bar{C}_t - C_r}{\bar{d}} \tag{7}$$

where $\bar{d}$ is the average inter-particle spacing. If g > 0 the particle grows and if g < 0 the particle dissolves. Thus, the increase of the volume of a particle, ΔV, in time interval $t \sim t + \Delta t$ is

$$\Delta V = DSg \frac{\rho_\theta}{C_\theta \rho_\theta - C_r \rho_\alpha} \Delta t \tag{8}$$

CONTINUUM CREEP DAMAGE MECHANICS MODELLING

The continuum creep damage mechanics model in this paper is that due to Dyson.[11] Two types of damage are considered here:

1. Thermally induced damage, including damage due to particle coarsening and due to solute depletion from the matrix,

2. Strain induced damage, including damage due to cavity nucleation and growth and due to multiplication of mobile dislocations. These are briefly reviewed below.

According to Dyson, creep damage due to particle coarsening can be defined as

$$D_P = 1 - \frac{P_0}{P_t} \tag{9}$$

where P_0 and P_t are the inter-particle spacing at the beginning of coarsening and at any time t. Supposing that the coarsening of the particles obeys Livshitz-Wagner equation:

$$r^3 - r_0^3 = Kt \tag{10}$$

where K is a constant determined by diffusivity, interfacial energy, equilibrium solute concentration, molar volume of the precipitate and the temperature, the following relationship can be derived:

$$\dot{D}_P = \frac{k_P}{3}(1 - D_P)^4 \tag{11}$$

where k_P is a parameter determined by K and the initial particle radius. The relationship between k_P and temperature T can be defined using a constant k'_P and an activation energy parameter Q_P:

$$k_P = k_P^{'} \exp\left(-\frac{Q_P}{RT}\right) \tag{12}$$

R is the universal gas constant. It should be noted that eqn (11) only applies to intra-granular spherical particles, such as MX, but is not applicable to $M_{23}C_6$ particles as they are mainly situated at the grain boundaries in high Cr ferritic steels.

The level of damage due to solute depletion from the matrix is defined as

$$D_S = 1 - \frac{\bar{C}_t}{C_0} \tag{13}$$

where C_0 is the initial solute concentration in the matrix and $\bar{C}_t$ is the average solute concentration at time t. The evolution of D_s is described by the following Wert-Zener equation

$$\dot{D}_S = K_S D_S^{1/3}(1 - D_S) \tag{14}$$

where is a material parameter.

The model uses a dimensionless parameter H to model primary creep. H is defined as

$$H = \frac{\sigma_i}{\sigma} \tag{15}$$

where σ is the uniaxial stress and σ_i is an internal back stress generated during stress redistribution within 'hard' regions of the microstructure (particles, sub-grains, etc.) as inelastic strain accumulates. The evolution of is as follows:

$$\dot{H} = \frac{h'}{\sigma}\left(1 - \frac{H}{H^*}\right)\dot{\varepsilon} \tag{16}$$

It can be seen that H can take values from zero to a material dependent maximum H^* ($H^* < 1$). The constant $h' = E\phi$, where E is the Young's modulus and ϕ is the volume fraction of all phases giving rise to the stress redistribution.

The damage parameter for cavity nucleation and growth, D_N is defined as the fraction of grain boundary facets cavitated. When cavities nucleate continuously, the evolution of D_N can be described by the following equation:

$$\dot{D}_N = \frac{k_N}{\varepsilon_{fu}}\dot{\varepsilon} \tag{17}$$

where ε_{fu} is the uniaxial strain at fracture and k_N has an upper limit of $\approx \frac{1}{3}$.

Damage due to the multiplication of mobile dislocations can be described by

$$D_d = 1 - \frac{\rho_0}{\rho_t} \tag{18}$$

where ρ_0 and ρ_t are dislocation density at $t = 0$ and at any time instant, t. The evolution of is as follows:

$$\dot{D}_d = C(1 - D_d)^2\dot{\varepsilon} \tag{19}$$

where C is a material constant.

Considering all the damage mechanisms discussed above, the creep strain rate at any instant is then given by

$$\dot{\varepsilon} = \frac{\dot{\varepsilon}_0}{(1 - D_d)(1 - D_S)}\sinh\left[\frac{\sigma(1 - H)}{\sigma_0(1 - D_P)(1 - D_N)}\right] \tag{20}$$

The two parameters $\dot{\varepsilon}_0$ and are related to temperature

$$\dot{\varepsilon}_0 = \dot{\varepsilon}_0' \exp\left(1 - \frac{Q_{d/j}}{RT}\right) \tag{21}$$

$$\sigma_0 = \sigma_{o,m}\left\{1 - \exp\left[-\frac{\Delta H}{RT_S}\left(\frac{T_S}{T} - 1\right)\right]\right\} \tag{22}$$

where $\dot{\varepsilon}_0'$ is a constant, $Q_{d/j}$ is the combined activation energy for diffusion and jog formation, T_s the solvus temperature when $\sigma_0 = 0$ and ΔH is the enthalpy of solution.

APPLICATION OF THE COMBINED PRECIPITATION/CDM MODELS

The uniqueness of this work lies in taking the precipitation and solute depletion data from the precipitation modelling and inserting it into the CDM models directly. This negates the need to use the analytical eqns (11, 12 and 14).

Table 1 Chemical composition of steel P92 in wt.% (Fe balance).

Cr	Mn	Ni	Mo	W	V	Nb	C	B	N	Si	P
8.96	0.46	0.06	0.47	1.84	0.20	0.07	0.11	0.001	0.05	0.04	0.008

The material studied in this paper is a creep resistant 9% chromium steel used in power plant, namely P92. Many microstructural investigations and relatively long time creep tests have been performed on this steel and thus a relatively good microstructural picture and creep behaviour of the material under different heat treatment conditions has been established. The composition of the steel is shown in Table 1.[13] The steel is austenitised at 1070°C for two hours and air cooled, followed by tempering at 770°C for two hours and then air-cooling. Typical ageing or creep test conditions reported in the literature are 600 and 650°C for up to a few tens of thousands of hours.

We apply our simulation to VN, $M_{23}C_6$ and Laves phase precipitates in this material under the heat treatment conditions reported in the literature, i.e. P92 solution heat treated at 1070°C, tempered at 770°C for two hours and then aged at 600 and 650°C for up to 100,000 hours.

Creep modelling has been carried out using both the Livshitz-Wagner model and our own simulation results. The latter means there is no need to solve eqn (11) as the simulation directly gives the inter-particle spacing at any instant, t. Therefore, eqn (9) is used when we use the simulation results. Similarly, solute depletion due to the formation and growth of Laves Phase is a direct result of the simulation approach. Therefore, eqn (13) was used to model the effects of Laves Phase on creep behaviour, though the analytical model using eqn (14) is also presented here. The parameters used are summarised in Table 2.

RESULTS AND DISCUSSION

Precipitation Kinetics

The precipitation curves of VN and $M_{23}C_6$ particles are shown in Figures 1 and 2 respectively. The lines in these figures represent the number average of equivalent particle radius as a function of time from our simulation, and the symbols are experimental results reported in the literature.[13–15] Some of the experimental observations are made on isothermally aged samples and others are made on creep tested samples. The agreement between the simulation results and the experimental observations is fairly good in both cases of VN and $M_{23}C_6$ particles.

As shown in Figure 1 and 2, particles grow slowly at the beginning of tempering. Then they grow rapidly and reach a semi-stable state before or at the end of tempering ($t = 2$ hours in the figures). For VN (Figure 1), the average size of the particles does not change much until about 10,000 hours or longer depending on the ageing temperature. When aged at 650°C, coarsening sets in earlier than at 600°C and a much more pronounced increase in

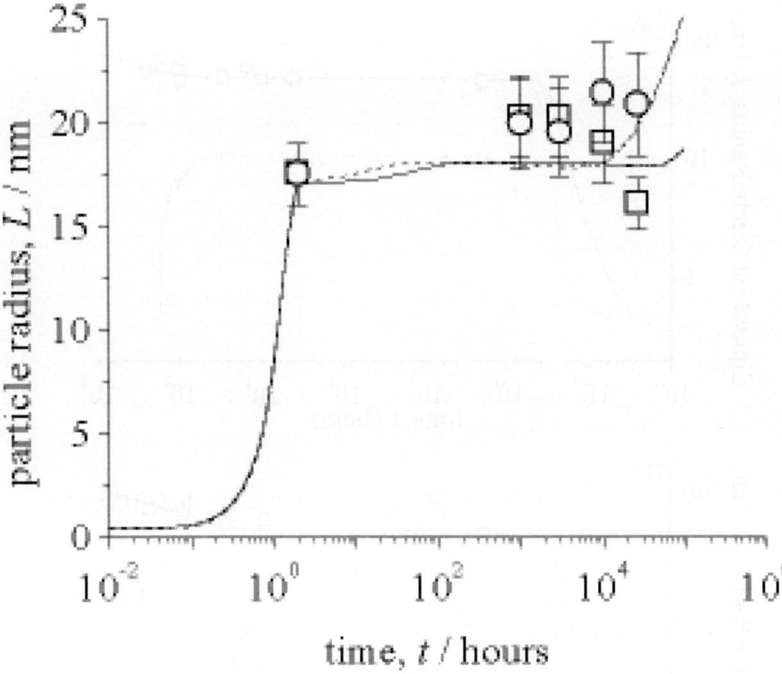

Fig. 1 Precipitation kinetics of VN in P92. Lines: simulated (solid, 600°C, dash 650°C). Symbols: experimental (ref. 13 and 14. Squares, 600°C, Circles, 650°C).

Table 2 Main parameters used in creep modelling.

Parameter	Unit	Value
$\dot{\varepsilon}_0'$	s^{-1}	2400
Q_{dj}	kJ mol^{-1}	270
$\sigma_{o,m}$	MPa	33
$\Delta H/RT_S$	/	1.3
T_S	K	1350
h'	MPa	40000
H^*	/	0.4
ε_{fu}	%	20
k_N	/	0.33
C	/	100
k_P' Q_P	s^{-1} kJ mol^{-1}	3×10^6 300

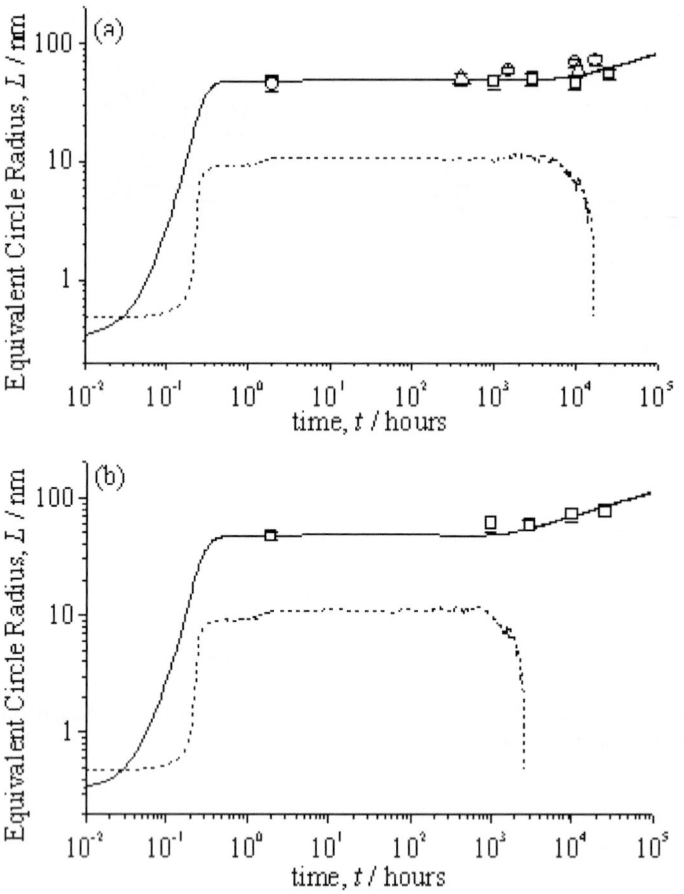

Fig. 2 Precipitation kinetics of $M_{23}C_6$ in P92 at 600°C (a) 650°C and (b) Lines: simulations (solid: inter-granular, dash: intra-granular). Symbols: experimental measurements (squares: ref. 13 and 14, isothermally aged, circles: ref. 15, creep tested, triangles: ref. 13 and 14, creep tested at 160 MPa.

average particle size is found. For $M_{23}C_6$ particles (see Figure 2), similar situations are found for inter-granular particles (solid lines), but a totally different situation here for intra-granular particles. From Figure 2, we can see[1] Intra-granular particles are much smaller than inter-granular particles.[2] After a few thousand hours of ageing, most of the intra-granular particles disappear. This suggests that intra-granular $M_{23}C_6$ particles would have little effect on the long term behaviour of the material.

The inter-particle spacing as a function of ageing time from our simulations is shown in Figures 3 and 4 for VN and $M_{23}C_6$ particles respectively. The inter-particle spacing decreases sharply, corresponding to the sharp increase in the number of precipitates nucleated, at the beginning of tempering. The minimum inter-particle spacing is reached at (VN) or before ($M_{23}C_6$) the end of tempering, indicating that no new particles are nucleating during service.

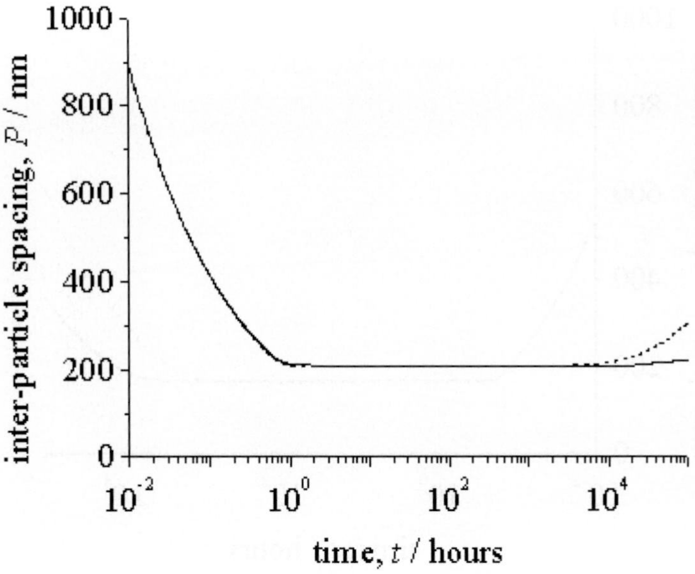

Fig. 3 Simulated inter-particle spacing for VN in P92 at 600°C (solid) and 650°C (dash).

The inter-particle spacing increases sharply after a few thousand to ten thousand hours as a result of coarsening. The effect of higher ageing temperature on coarsening rate is also clearly demonstrated here with a much earlier and sharper increase in inter-particle spacing at 650°C than at 600°C in both cases of VN and $M_{23}C_6$. Comparison of Figures 3 and 4 shows that the increase of inter-particle spacing is much more marked in the case of $M_{23}C_6$ than in the case of VN, indicating that VN is much more stable than $M_{23}C_6$.

CDM MODELLING

CDM modelling was carried out at 650°C using the parameters listed in Table 2. All modelling is performed at constant load with an initial stress of 150 MPa and primary creep is included in all calculations. The situations considered include

1. No damage,
2. Damage due to particle coarsening only using the Livshitz-Wagner model, i.e. eqns (11) and (12) for the damage evolution,
3. Damage due to particle coarsening of $M_{23}C_6$ only and using simulation results, i.e. feeding the evolution of inter-particle spacing with time from the precipitation kinetics simulation to CDM modelling and using eqn (9),
4. Damage due to particle coarsening of VN only from the precipitation kinetics using eqn (9),
5. Damage due to particle coarsening of both VN and $M_{23}C_6$ using the simulation results.

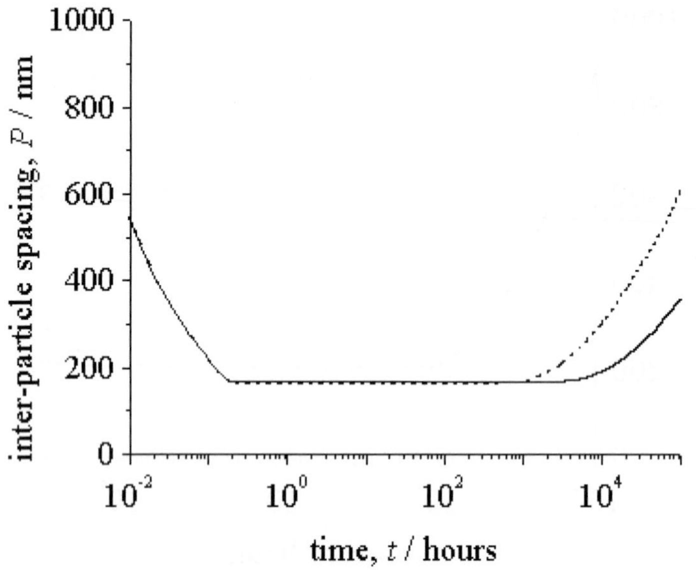

Fig. 4 Simulated inter-particle spacing for $M_{23}C_6$ in P92 at 600°C (solid) and 650°C (dash).

The combined inter-particle spacing as a function of time was worked out from the inter-particle spacing of VN and $M_{23}C_6$ and then eqn (9) was used,

6. Damage due to solid solution depletion from the matrix only using the Wert-Zener eqn, i.e. eqn (14). The equation was solved numerically,
7. Damage due to solid solution depletion of W and Mo from the matrix due to the nucleation and growth of the Laves Phase. The damage parameter as a function of time was calculated from the results of the precipitation kinetics simulation of the Laves Phase, i.e. using eqn (13),
8. Damage due to particle coarsening of both VN and $M_{23}C_6$ particles and due to W and Mo depletion from the matrix due to the formation of Laves Phase, i.e. situations (v) and (vii) were included at the same time,
9. Strain induced damage due to cavity nucleation and growth and due to the multiplication of mobile dislocations were included, as well as the thermally induced damage included in case (viii). The modelled creep curves for all cases are shown in Figure 5.

Figure 5 shows that strain increases slowly for from a few thousand hours to a few tens of thousand hours, and then increases sharply when appreciable damage takes effect. As the figure shows, there is no clear difference between the creep curves with no damage included and with damage due to particle coarsening using the Livshitz-Wagner model This may imply that eqn (11) and (12) with parameters as listed in Table 2 under-estimate the damage caused by the coarsening of intra-granular particles. All the creep damage mechanisms considered speed up the straining of the material considerably. Figure 5(a) shows that creep damage due to the coarsening of $M_{23}C_6$ particles is much greater and increases more steeply

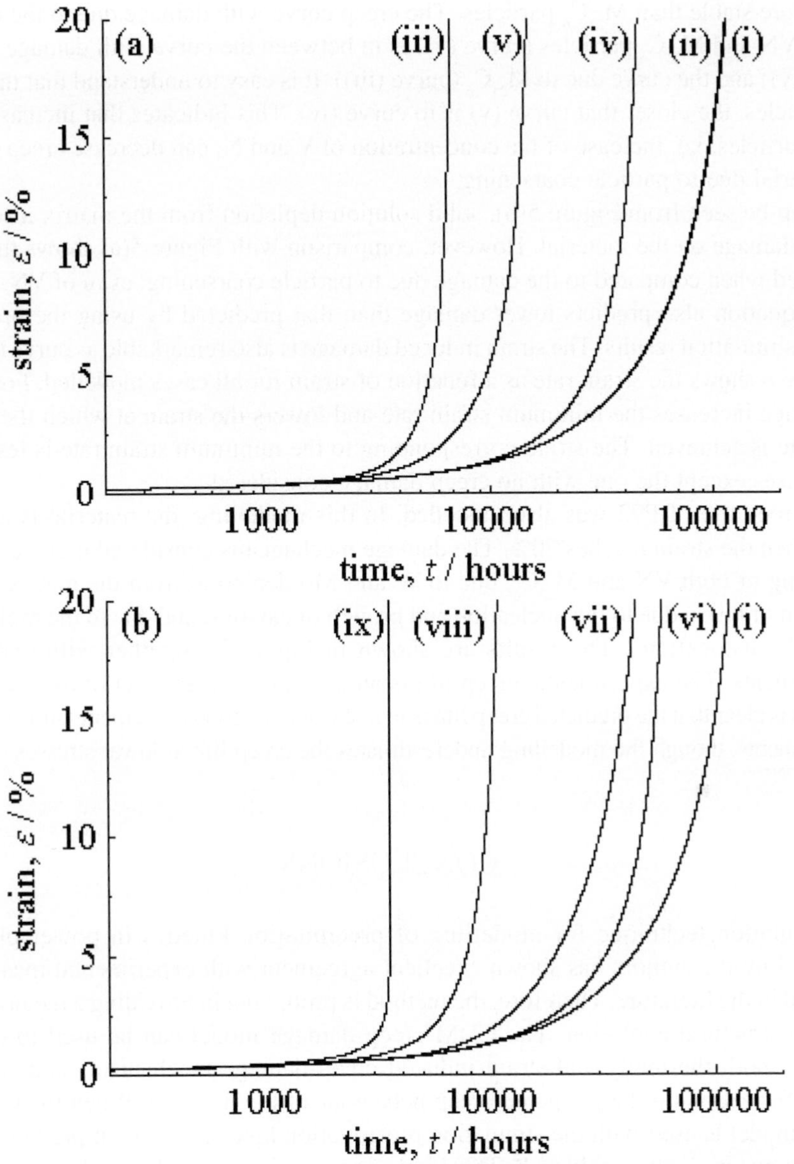

Fig. 5 Modelled creep curves under constant load condition with an initial stress of 150 MPa at 650°C. Primary creep is included in all calculations. (i) with no damage, (ii) with damage due to particle coarsening according to wagner's model, (iii) with damage due to coarsening of $M_{23}C_6$ from simulation, (iv) with damage due to coarsening of VN from simulation, (v) with damage due to coarsening of both VN and $M_{23}C_6$ from simulation, (vi) with damage due to solid solution depletion from matrix according to Wert-Zener Equation, (vii) with damage due to W and Mo depletion due to the formation and growth of Laves Phase from simulation, (viii) with (v) and (vii) considered at the same time, (ix) with strain induced damages due to cavity nucleation and growth and due to multiplication of mobile dislocations and (viii).

than the damage due to coarsening of VN particles. This is because the VN particles are much more stable than $M_{23}C_6$ particles. The creep curve with damage due to the coarsening of both VN and $M_{23}C_6$ particles (curve (v)) is in between the curve with damage due to VN (curve (iv)) and the curve due to $M_{23}C_6$ (curve (iii)). It is easy to understand that the more the VN particles, the closer that curve (v) is to curve (iv). This indicates that increase numbers of VN particles, i.e. increase of the concentration of V and N, can decrease creep damage of the material due to particle coarsening.

As can be seen from Figure 5(b), solid solution depletion from the matrix also can have marked damage on the material. However, comparison with Figure 5(a) shows that it is not so marked when compared to the damage due to particle coarsening, even of VN. The Wert-Zener Equation also predicts lower damage than that predicted by using the precipitation kinetics simulation results. The strain induced damage is also remarkable as curve (ix) shows.

Figure 6 shows the strain rate as a function of strain for all cases modelled. For all cases, the damage increases the minimum strain rate and lowers the strain at which the minimum strain rate is achieved. The strain corresponding to the minimum strain rate is less than 1% for all cases except the one with no creep damage considered.

The creep life of P92 was also modelled. In this modelling, the material is assumed to break when the strain reaches 20%. The damage mechanisms considered include that due to coarsening of both VN and $M_{23}C_6$, due to W and Mo depletion from the matrix due to the formation of laves, due to the nucleation and growth of cavities, and due to the multiplication of mobile dislocations. The results are shown in Figure 7, together with experimental measurements. The experimental creep life is an average of a large set of experimental test results. It is clear that the predicted creep life is in a reasonably good agreement with experimental measurements, though the modelling underestimates the creep life at lower stresses.

CONCLUSIONS

The simulation technique for modelling of precipitation kinetics in power plant steels developed by the authors has shown excellent agreement with experimental measurements published in the literature. Therefore, the method is promising in providing a means to predict the microstructural evolution. The CDM creep damage model can be used to consider a variety of both thermally and strain induced creep damage mechanisms and shows very useful information on the predicted creep behaviour of the material. When the CDM creep damage model is used with the simulation precipitation kinetics results it predicts the creep life of the material reasonably well. This kind of approach can also be used to provide useful information for alloy designers by the effects of different compositions. Further developments are required to take full advantage of this approach.

ACKNOWLEDGEMENTS

We would like to thank the Engineering and Physical Science Research Council (EPSRC) for funding the project (grant number GR/N/13074) and Powergen and Innogy, UK for co-

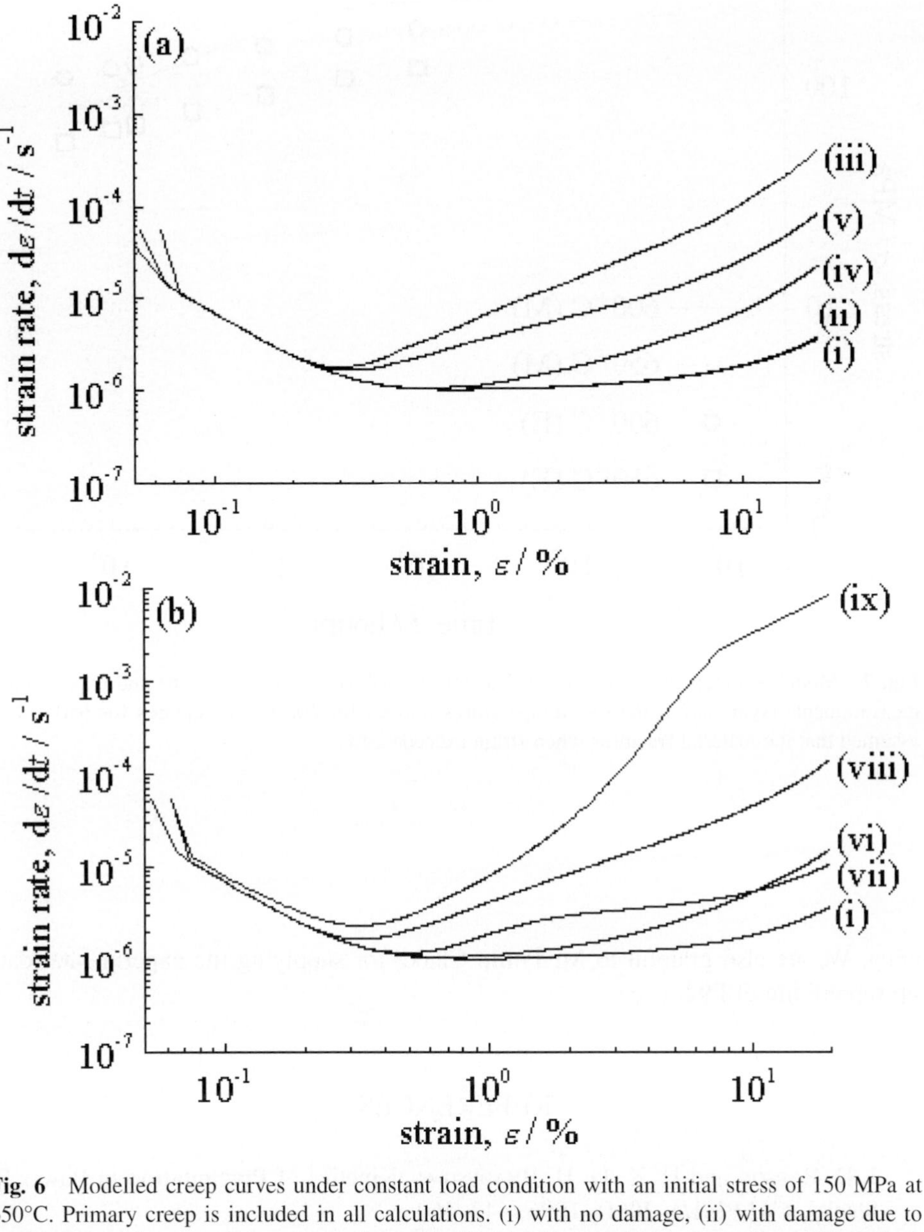

Fig. 6 Modelled creep curves under constant load condition with an initial stress of 150 MPa at 650°C. Primary creep is included in all calculations. (i) with no damage, (ii) with damage due to particle coarsening according to wagner's model, (iii) with damage due to coarsening of $M_{23}C_6$ from simulation, (iv) with damage due to coarsening of VN from simulation, (v) with damage due to coarsening of both VN and $M_{23}C_6$ from simulation, (vi) with damage due to solid solution depletion from matrix according to Wert-Zener Equation, (vii) with damage due to W and Mo depletion due to the formation and growth of Laves Phase from simulation, (viii) with (v) and (vii) considered at the same time, (ix) with strain induced damages due to cavity nucleation and growth and due to multiplication of mobile dislocations and (viii).

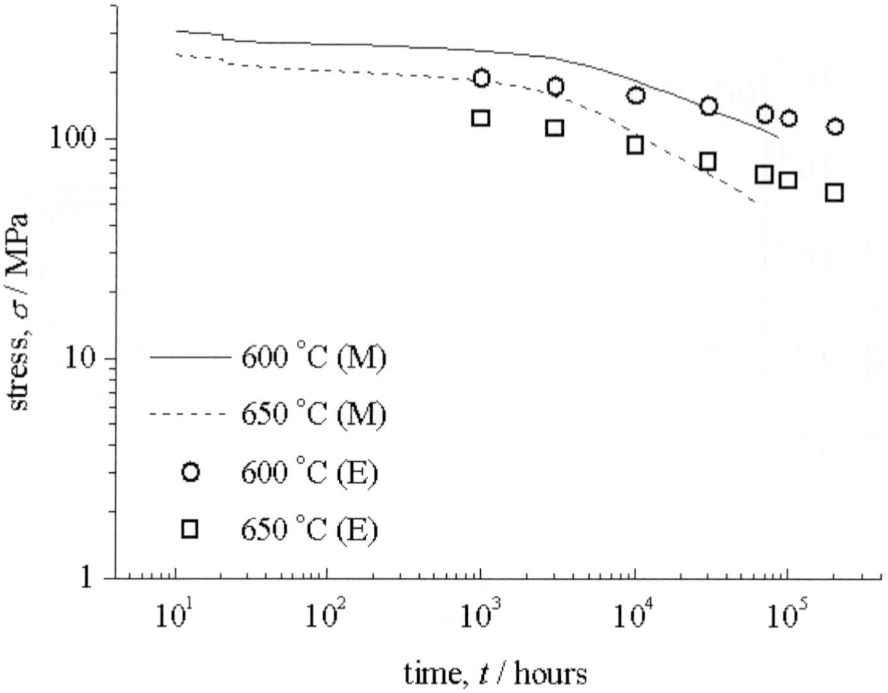

Fig. 7 Modelled creep life (lines) of P92 at 600°C (solid) and 650°C (dash) and experimental measurements (symbols) at the two temperatures (circles for 600°C and squares for 650°). It is assumed that the material fractures when strain exceeds 20%.

funding. We are also grateful to Mr Philip Clarke for supplying the experimental data of creep rupture life of P92.

REFERENCES

1. J. D. Robson and H. K. D. H. Bhadeshia: 'Kinetics of Precipitation in Power Plant Steels', *CALPHAD*, **20**(4), 1996, 447–460.
2. J. D. Robson and H. K. D. H. Bhadeshia: 'Modelling Precipitation Sequences in Power Plant Steels .1. Kinetic Theory', *Materials Science and Technology*, **13**(8), 1997, 631–639.
3. H. Jiang and R. G. Faulkner: 'Modelling of Grain Boundary Segregation, Precipitation and Precipitate-Free Zone of High Strength Aluminium Alloys - I. The Model', *Acta. Mater.*, **44**(5), 1996, 1857–1864.
4. Y. F. Yin and R. G. Faulkner: 'Modelling the Effects of Alloying Elements on

Precipitation in Ferritic Steels', *Materials Science and Engineering A*, **344**, 2003, 92–102.

5. A. BJÄRBO and M. HÄTTESTRAND: 'Complex Carbide Growth, Dissolution and Coarsening in a Modified 12 pct Chromium Steel - an Experimental and Theoretical Study', *Metallurgical and Materials Transactions A*, **32**(1), 2001, 19–27.
6. G. GHOSH and G. B. OLSON: 'Simulation of Paraequilibrium Growth in Multicomponent Systems', *Metallurgical and Materials Transactions A*, **32**(3), 2001, 455–467.
7. Y. F. YIN and R. G. FAULKNER: 'Simulations of Precipitation in Ferritic Steels', *Materials Science and Technology*, **19**, 2003, 91–98.
8. Y. F. YIN and R. G. FAULKNER: 'A New Modelling Approach to Microstructural Evolution in Ferritic Steels', *Power Technology*, J. Lecomte-Beckers et. al., eds., **19**(II), 2002, 1247–1256.
9. B. J. CANE and P. F. APLIN: 'Creep Life Assessment Methods', *Journal of Strain Analysis for Engineering Design*, **29**(3), 1994, 225–232.
10. R. W. EVANS: 'A Constitutive Model for the High-Temperature Creep of Particle-Hardened Alloys Based on the Theta Projection Method', *Proceedings of the Royal Society of London Series A- Mathematical Physical and Engineering Sciences*, **456**(1996), 2000, 835–868.
11. B. DYSON: 'Use of CDM in Materials Modelling and Component Creep Life Prediction', *Journal of Pressure Vessel Technology*, **122**(3), 2000, 281–296.
12. R. G. FAULKNER: 'Segregation to Boundaries and Interfaces in Solids', *International Materials Reviews*, **41**(5), 1996, 198–208.
13. A. GUSTAFSON and M. HATTESTRAND: 'Coarsening of Precipitates in an Advanced Creep Resistant 9% Chromium Steel - Quantitative Microscopy and Simulations', Paper VI in M. Hattestrand, *Precipitation Reactions at High Temperatures in 9-12% Chromium Steels*, Ph.D. Thesis, Goteborg University, Goteborg, Sweden, 2000.
14. M. HATTESTRAND and H. O. ANDREN: 'Influence of Strain on Precipitation Reactions During Creep of an Advanced 9% Chromium Steel', *Acta Materialia*, **49**(12), 2001, 2123–2128.
15. P. J. ENNIS, et al., 'Quantitative Comparison of the Microstructures of High Chromium Steels for Advanced Power Stations', *Microstructural Stability of Creep Resistant Alloys for High Temperature Plant Applications*, A. Strang, et al., eds., IOM, 1998, 135–143.

Critical Stress for Transition of Creep Deformation Behaviour in Virgin and Long-Term Serviced Materials of 2.25Cr-1Mo Steel

T. OHBA, E. BABA and K. KIMURA
National Institute for Materials Science (NIMS)
2-2-54 Nakameguro, Meguro-ku
Tokyo 153-0061
Japan

F. ABE and K. YAGI
National Institute for Materials Science (NIMS)
1-2-1 Sengen, Tsukuba, 305-0047
Japan

I. NONAKA
Ishikawajima-Harima Heavy Industries (IHI)
3-2-16 Toyosu, Koto-ku
Tokyo 135-7633
Japan

ABSTRACT

In order to make clear a critical stress between high and low stress regions in creep and creep rupture properties, proof stress and creep deformation behaviour are investigated for 2.25Cr-1Mo steel at temperatures between 550 and 690°C, using virgin material and long-term serviced material which was already exposed at 577°C for 1.9×10^5 hours in a commercial plant. The 0.2% proof stress, the ultimate tensile strength and the creep rupture strength are lower for the long-term serviced material than for the virgin material. With decreasing stress and increasing test duration, the shape of creep rate versus time curves drastically changes at the critical stress. The critical stress is much higher for the virgin material than for the long-term serviced material and coincides with the 0% offset proof stress, σ_{0P}, which is the stress at the point of departure from a straight line in stress-strain curve in tensile test. The present results suggest whether additional dislocations are introduced or not by loading affects subsequent creep deformation behaviour. It is concluded that the creep and creep rupture data obtained at low stresses below the 0% offset proof stress should be distinguished from ones at high stresses above the σ_{0P} and that only the data at low stresses below the σ_{0P} should be provided for long-term creep life estimation.

INTRODUCTION

High-temperature components used under creep conditions are designed on the base of relevant national design codes, for example, 100,000 hours-creep rupture strength in Japan. On the other hand, the establishment of reliable methods for determining the remaining life has been wished for components being operated for a long time, because there is an economic advantage in using components beyond design life.[1] Therefore, an understanding of long-term creep and creep rupture behaviour is important for the improvement of material reliability.

Recently, long-term creep rupture test data and creep strain data beyond 100,000 hours are available for a number of heat resistant steels and alloys in the National Institute for Materials Science (NIMS) Creep Data Sheets.[2] The analysis of long-term creep data has shown that the creep deformation behaviour of engineering steels and alloys are complicated, especially at low stress and long time conditions, because a variety of microstructural evolution takes place during creep.[2–4] We have recognized that the creep deformation behaviour at low stress and long time conditions is quite different from that at high stress and short time conditions. However, the reason why the creep deformation behaviour changes with decreasing stress and the controlling factors in creep deformation at low stress are still not clear.

The purpose of the present research is to investigate the correlation between the change in creep deformation behaviour with decreasing stress and the proof stress for the virgin and long-term serviced materials of 2.25Cr-1Mo steel. The tensile and creep properties were examined at temperatures between 550 and 690°C after exposure at 577°C for 194,958 hours in a commercial plant and then compared with those in the NIMS Creep Data Sheet for the virgin material. The critical creep stress dividing the high and low stress regions is found to coincide with the 0% offset proof stress in tensile test. The present results suggest whether additional dislocations are introduced or not by loading affects the subsequent creep deformation behaviour.

EXPERIMENTAL PROCEDURE

The materials used were basically two kinds of 2.25Cr-1Mo steel tubes, having a size of 45–50 mm wall diameter and 8–10 mm wall thickness, specified as JIS STBA 24. One of them was the virgin material, the MAF heat[5] being tested in the NIMS Creep Data Sheet Project, the chemical composition of which was 0.10 carbon, 0.23Si, 0.43Mn, 0.011P, 0.009S, 0.043Ni, 2.26Cr, 0.94Mo, 0.07Cu, 0.005Al and 0.008N (mass %). The final heat treatments of the tube were annealing for 20 min. at 930°C for austenitisation followed by cooling to 720°C and annealing for 130 min and then air-cooled to room temperature. Another one had been already exposed at 577°C and 34 MPa for 194,958 hours as a superheater tube in commercial plant. The two materials are called as the virgin and long-term serviced materials in this paper, according to their history. Tensile and creep specimens, having a geometry of 6 mm in diameter and 30 mm in gauge length, were taken longitudinally from the middle of wall thickness of the tubes. Creep tests were carried out at temperatures between 550 and 690°C. For the improvement of the accuracy of creep strain measurement at low stresses, the long-size specimens of 10 mm in diameter and 50 or 150 mm in gauge length were also used in addition to the short-size specimens described above. The long-size specimens were taken from the virgin material of 2.25Cr-1Mo steel plates with 50 mm in thickness. The chemical composition of the plates was 0.12 carbon, 0.22Si, 0.50Mn, 0.003P, 0.002S, 0.18Ni, 2.23Cr, 0.94Mo, 0.13Cu, 0.009Al and 0.0022N. The heat treatments of the plates were annealing for 100 min at 930°C for austenitisation followed by cooling to 730°C and annealing for 105 min and then air-cooled to room temperature. The chemical composition and heat treatments of the plates were substantially the same as those of the tube described above. The increments of gauge spacing were carefully measured basically by a set of two dial gauges with a sensitivity of 1 µm, corresponding to creep strain of 3×10^{-5}, 2×10^{-5} and 7×10^{-6} for the specimens with 30, 50 and 150 mm gauge length, respectively. Tensile tests were carried out at temperatures between room temperature and 650°C at a nominal strain rate of 0.3%/min for up to about 1% proof stress level and of 7.5%/min beyond it. The longitudinal cross section of the specimens was metallographically observed by TEM and the precipitates were identified by x-ray diffraction of electrolytically extracted residues.

RESULTS AND DISCUSSION

Change in Tensile and Creep Rupture Strength by Prolonged Service

The Vickers hardness was measured to be 150 and 128 for the virgin and long-term serviced materials, respectively, at room temperature, indicating that the long-term service at 577^{oC} caused a softening. Both the 0.2% proof stress and the ultimate tensile strength are also decreased by the long-term service at 577°C as shown in Figure 1. The difference in 0.2% proof stress between the two materials is about 100 and 50 MPa at 550 and 650°C, respectively.

Figure 2 shows the relationship between the creep stress and Manson-Haferd time-temperature parameter given by MHP = $(\log t_r - A)/(T_K - B)$, where t_r is the time to rupture, T_K is the absolute temperature and A and B are constants, for the virgin and long-term serviced materials. In this figure, the data at various temperatures are included. At a low temperature of 577°C, the time to rupture of the long-term serviced material is only 0.25– 0.30 of that of the virgin material; the time to rupture of the long-term serviced material is 145.9 hours at 98 MPa, while that of the virgin material is evaluated to be 4350 hours at the same temperature and stress conditions. The difference in time to rupture between the virgin and long-term serviced materials becomes less pronounced with increasing temperature and decreasing stress. At 670°C, the time to rupture of the long-term serviced material is 1651.3 hours at 34 MPa, which is substantially the same as that of the virgin material.

Before creep test, the virgin materials were observed to be two-phase microstructure consisting of ferrite and pearlite.[6] High density of dislocations, fine precipitates of $M_{23}C_6$ and fine needles of Mo_2C were distributed in the ferrite phase, while thin and parallel plates of Fe_3C cementite having several tens nm width were arranged in the pearlite phase. The X-ray diffraction of electrolytically extracted residues showed that extensive diffraction peaks from M_6C appeared in addition to weak ones from $M_{23}C_6$ and Mo_2C in the long-term serviced material. The amount of $M_{23}C_6$ decreased during long-term service at 577°C in favour of the formation of more massive M_6C. It should be also noted that the carbides grew in size and a

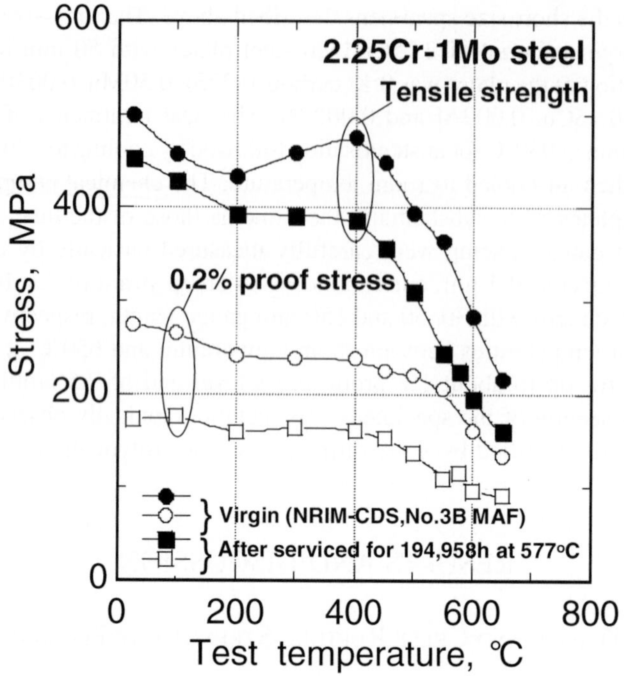

Fig. 1 0.2% proof stress and tensile strength for virgin and long-term serviced materials of 2.25Cr-1Mo steel, as a function of temperature.

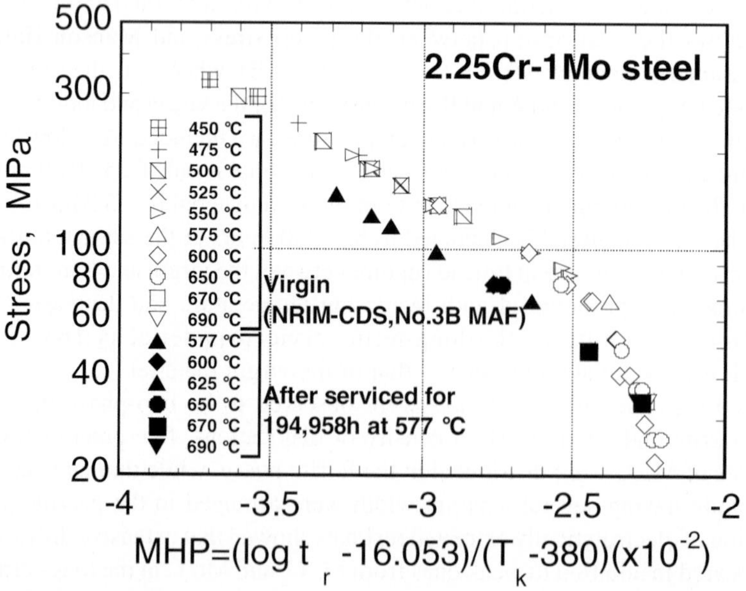

Fig. 2 Relationship between creep rupture stress and Manson-Haferd time-temperature parameter for virgin and long-term serviced materials of 2.25Cr-1Mo steel.

large amount of dislocations present in the virgin material recovered during long-term service. The density of carbides at grain boundaries was lower in the long-term serviced material than in the virgin material. The thin plates of Fe_3C cementite in the pearlite phase were also observed to have coarsened and spheroidised.

The observed large difference in creep rupture strength between the virgin and long-term serviced materials at low temperature and short time conditions results from a large difference in initial microstructure between the two materials before creep tests. The recovery of microstructure becomes more significant also in the virgin material with increasing test temperature and time. This causes a decrease in difference of microstructure and hence creep rupture strength between the virgin and long-term serviced materials at high temperature and long time. The present results shown in Figure 2 are similar as those reported by Kushima et al.[7] Recently, they investigated the effect of initial microstructure on the long-term creep rupture strength for the virgin materials of 2.25Cr-1Mo steel by using three different heats with different initial microstructure. The creep rupture strength differed among the heats depending on initial microstructure at low temperature and short time conditions but the difference became disappeared with increasing temperature and time. The recovery of excess dislocations and the coarsening of carbides occurred during creep and the microstructure after creep test for long times was similar among the three heats. They pointed out that the disappearance of difference in creep rupture strength was due to microstructural evolution during creep.

Microstructural Evolution During Prolonged Service and its Effect on Creep Deformation Behaviour

Figure 3 compares the creep rate versus time curves between the virgin and long-term serviced materials. Because the test temperature for the virgin material (575°C) was only 2° lower than that for the long-term serviced material (577°C), the difference in test temperature would not affect substantially the creep rate. The creep rate curves consist of a primary or transient creep region, where the creep rate decreases with time, and of a tertiary or acceleration creep region, where the creep rate increases with time after reaching a minimum creep rate. In the initial stage of creep less than 0.1 hour, the creep rate is approximately the same between the two materials. This suggests that the difference in initial microstructure scarcely affects the creep rate in the initial stage. The decrease in creep rate with time in the transient region is less significant and the onset of acceleration creep takes place at shorter time in the long-term serviced material than in the virgin material. This results in higher minimum creep rate and hence shorter time to rupture in the long-term serviced material. The minimum creep rate is about 2 orders of magnitude higher in the long-term serviced material than in the virgin material.

In order to clarify the reason why the onset of acceleration creep is promoted in the long-term serviced material, the creep test was interrupted at 1×10^3 hour, which corresponded to a later stage of the transient region as shown by the arrows in Figure 3 and then the microstructure was observed. In the long-term serviced material, dynamic recovery or recrystallisation took place in the vicinity of grain boundaries, resulting in the formation of

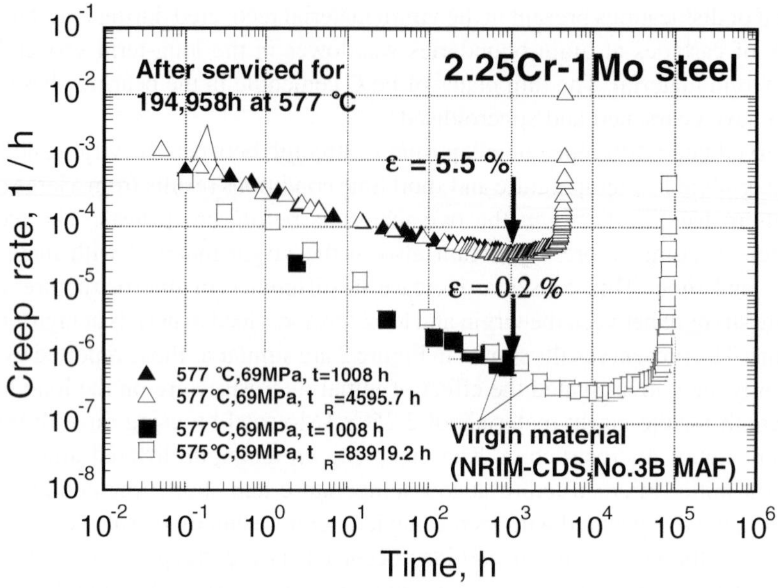

Fig. 3 Creep rate versus time curves for virgin and long-term serviced materials of 2.25Cr-1Mo steel.

subgrains at grain boundaries.[6] The formation of subgrains where the density of dislocations and carbides was very low suggests a decrease of resistance to creep deformation. On the other hand, the TEM observations gave no evidence of the formation of subgrains in the virgin material where high density of dislocations and fine carbides were still maintained for up to 1×10^3 hours during creep. We have revealed for martensitic 9Cr steels that the transient creep is a consequence of the movement and annihilation of excess dislocations which were present after heat treatment or produced by loading and that the acceleration creep is a consequence of gradual loss of creep strength due to the microstructural evolution, such as the agglomeration of precipitates and the recovery of subgrains.[8] The present results suggest that the formation of subgrains near grain boundaries in the long-term serviced material promotes the onset of acceleration creep and that fine distribution of carbides in the matrix and along grain boundaries in the virgin material stabilizes the microstructure and retards the onset of acceleration creep. The promotion of the onset of acceleration creep in the long-term serviced material results in higher minimum creep rate and shorter time to rupture.

Correlation Between Creep Deformation Behaviour in Transient Region and Proof Stress

Figure 4 shows the creep rate versus time curves for the virgin material at 550°C and for the long-term serviced material at 577°C for a wide range of stress. Again, the difference in test

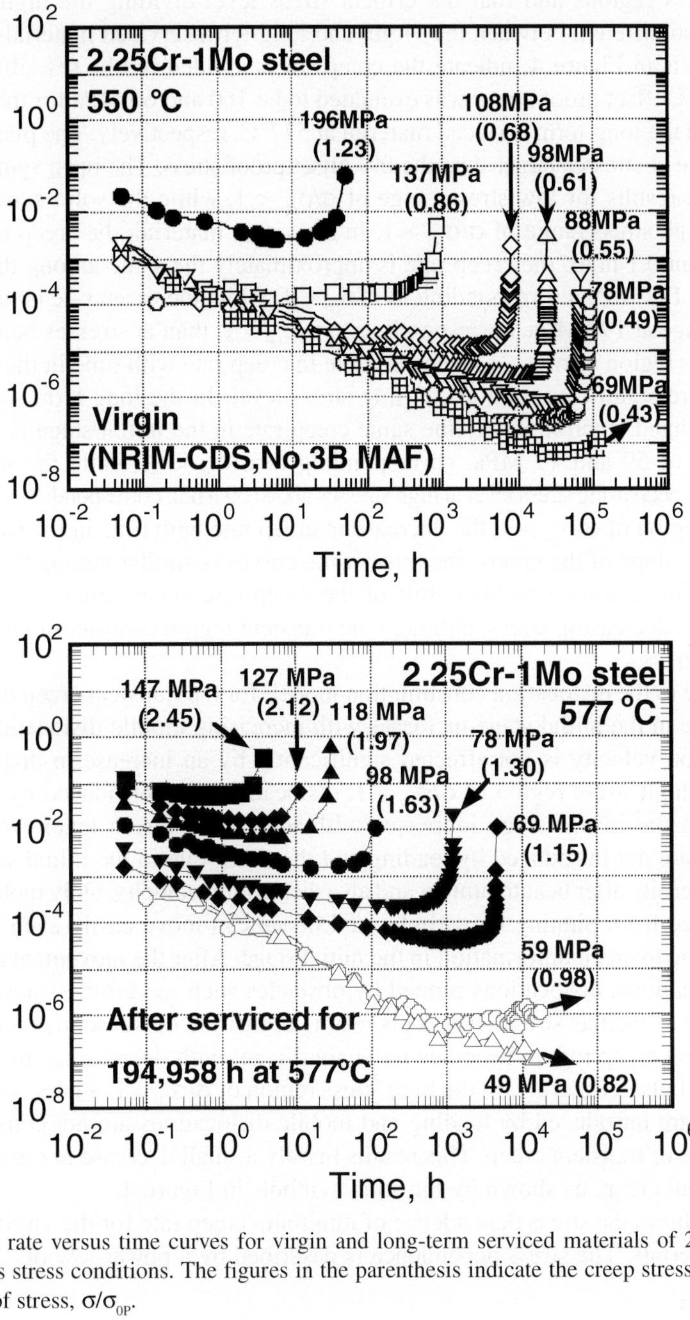

Fig. 4 Creep rate versus time curves for virgin and long-term serviced materials of 2.25Cr-1Mo steel at various stress conditions. The figures in the parenthesis indicate the creep stress divided by 0% offset proof stress, σ/σ_{0P}.

temperature between the two materials would not affect substantially the creep deformation behaviour. It should be noted that the creep deformation behaviour is different between high and low stress regions and that the critical stress level dividing the high and low stress regions is also different between the virgin and long-term serviced materials. The figures in the parenthesis in Figure 4 indicate the creep stress divided by the 0% offset proof stress, σ/σ_{OP}. The 0% offset proof stress was evaluated to be 160 and 60 MPa for the virgin material at 550°C and the long-term serviced material at 577°C, respectively. The plastic deformation can take place at stresses larger than the 0% offset proof stress. The open symbols in Figure 4 represent the results for low stress range of $\sigma/\sigma_{OP} < 1$, while the solid symbols exhibit the results for high stress range of $\sigma/\sigma_{OP} > 1$. In the virgin material, the creep rate in the initial stage less than 0.1 hour, the creep rate is approximately the same among the various stress levels below 137 MPa, corresponding to $\sigma/\sigma_{OP} < 1$, while the creep rate is much larger from the initial stage at 198 MPa, corresponding to $\sigma/\sigma_{OP} > 1$, than at stresses below 137 MPa. In the low stress region of $\sigma/\sigma_{OP} < 1$, the decrease in creep rate with time in the transient region becomes more significant with decreasing stress level. In the long-term serviced material, on the other hand, approximately the same creep rate in the initial stage is also obtained at low stresses of 59 and 49 MPa, corresponding to $\sigma/\sigma_{OP} < 1$, while the initial creep rate decreases with decreasing stress level at high stresses above 69 MPa, corresponding to $\sigma/\sigma_{OP} > 1$. In the high stress region of $\sigma/\sigma_{OP} > 1$, the decrease in creep rate with time in the transient region is small and the slope of the creep rate versus time curves is similar among the different stress conditions. This causes a parallel shift of the creep rate versus time curves to downward direction with decreasing stress, although the transient region continues to longer times with decreasing stress.

According to the dislocation consumption model[8] for the transient creep described before, the creep rate in the initial stage increases with increasing mobile dislocation density when the dislocation velocity is not affected significantly by an increase in dislocation density. Thus, in the high stress region of $\sigma/\sigma_{OP} > 1$, dislocations are introduced by loading and the creep rate in the initial stage increases with increasing stress level. When $\sigma/\sigma_{OP} < 1$, dislocations are not introduced by loading and the creep rate in the initial stage depends on dislocation density after heat treatment and also dislocation velocity. Only mobile dislocations, which are free from pinning by carbides and hence can move easily even if stress is very low, contribute to creep deformation in the initial stage. After the movement and annihilation of free dislocations, dislocations pinned by obstacles such as carbides move depending on pinning force as well as stress level. This results in that the decrease in creep rate with time in the transient creep region becomes more significant with decreasing stress, as shown by the open symbols in Figure 4. In the high stress region of $\sigma/\sigma_{OP} > 1$, a large amount of mobile dislocations are introduced by loading and mobile dislocations are not consumed for up to the later stage of transient creep. This results in only a small decrease in creep rate with time in the transient creep, as shown by the solid symbols in Figure 4.

Figure 5 shows the stress dependence of minimum creep rate for the virgin and long-term serviced materials. The stress dependence is described by a power low of

$$\dot{\varepsilon}_{min} = A\sigma^{n} \qquad (1)$$

where A is a constant and n the stress exponent. The stress exponent n is evaluated to be 12 and 8 for the virgin and long-term serviced materials, respectively. With pure, close-packed metals, the value of n is known to be between 4 and 6 over a wide range of applied stress.[9]

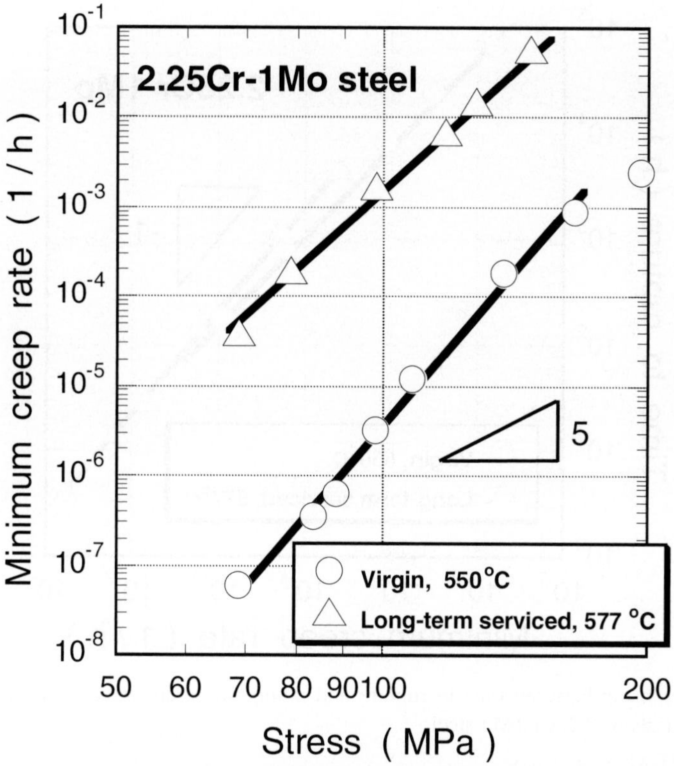

Fig. 5 Stress dependence of minimum creep rate for virgin and long-term serviced materials of 2.25Cr-1Mo steel.

Very large n values up to 40 have been reported by Williams and Wilshire[10] for complex alloys consisting of two or more phases. For both the virgin and long-term serviced materials, the stress exponent n is a little bit larger than 4–6 but it is not extremely large. It should be also noted that the minimum creep rate of the long-term serviced material is much higher than that of the virgin material, as large as about 2 orders of magnitude. Although this is partly due to a little bit higher test temperature for the long-term serviced material, this is mainly due to the difference in stress region between the virgin and long-term serviced materials. As shown in Figure 4, most of the creep test conditions for the long-term serviced material correspond to the high stress region of $\sigma/\sigma_{0P} > 1$, while they correspond to the low stress region of $\sigma/\sigma_{0P} < 1$ for the virgin material. As described previously, the decrease in creep rate with time in the transient region is more significant in the low stress region of $\sigma/\sigma_{0P} < 1$ than in the high stress region of $\sigma/\sigma_{0P} > 1$. This causes much higher minimum creep rate in the long-term serviced material tested in the low stress region than in the virgin material tested in the high stress region.

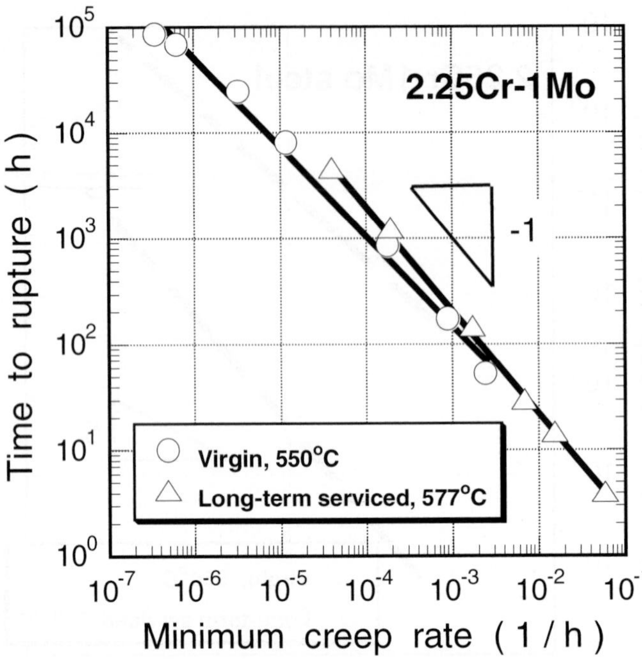

Fig. 6 Relationship between time to rupture and minimum creep rate for virgin and long-term serviced materials of 2.25Cr-1Mo steel.

Correlation Between Creep Deformation Behaviour in Acceleration Region and Proof Stress

Figure 6 represents the relationship between the time to rupture and minimum creep rate for the virgin and long-term serviced materials. The experimental results are described by the Monkman-Grant relationship as

$$t_r = (C/\dot{\varepsilon}_{min})^p \tag{2}$$

where C and p are constants. The exponent p is evaluated to be 0.84 and 1 for the virgin and long-term serviced materials, respectively. The time to rupture versus minimum creep rate relationship is substantially the same between the virgin and long-term serviced materials at short rupture time and high minimum creep rate conditions (196 and 137 MPa for the virgin material). But the time to rupture is shorter in the virgin material than in the long-term serviced material at long time and low minimum creep rate conditions. The difference in the time to rupture versus minimum creep rate relationship between the two materials becomes more significant with decreasing stress and increasing test duration. For the long-term serviced material, all the data in Figure 6 are obtained in the high stress region of $\sigma/\sigma_{OP} > 1$. For the virgin material, the data showing the minimum creep rate larger than about 10^{-3}/h are obtained in the high stress region of $\sigma/\sigma_{OP} > 1$, while those showing the minimum creep rate smaller than about 10^{-3}/hours are obtained in the low stress region of $\sigma/\sigma_{OP} < 1$. Therefore, the present results indicate that the time to rupture versus minimum creep rate relationship is substantially the same between the two materials when the test conditions are in the high stress region of $\sigma/\sigma_{OP} > 1$ for both the two materials. However, in the region where the test conditions are in the low stress region of $\sigma/\sigma_{OP} < 1$ for the virgin material but in the high stress region of $\sigma/\sigma_{OP} > 1$ for the long-term serviced material, the time to rupture of the virgin material is shorter than that of the long-term serviced material even if the minimum creep rate is the same between the two materials. The shorter time to rupture in the virgin material indicates that the increase in creep rate in the acceleration creep region after reaching a minimum creep rate is more significant in the virgin material than in the long-term serviced material.

In order to make clear the reason for the shorter time to rupture in the virgin material at the same minimum creep rate shown in Figure 6, the increase in creep rate by strain in the acceleration region was evaluated based on the results in Figure 7. After reaching a minimum creep rate at low strain, the logarithm of creep rate increases linearly with strain for a wide range of strain in the acceleration creep region. The steep increase in creep rate at strains above 0.20 for the long-term serviced material at 69 MPa is caused by a necking. The linear acceleration of creep rate with strain has been reported for several Cr-Mo steels and 9Cr-W steels.[11–16] Assuming exponential function of strain, the creep rate in the acceleration region is described by[11, 17]

$$\dot{\varepsilon} = \dot{\varepsilon}_0 \exp(n\varepsilon) \exp(m\varepsilon) \exp(d\varepsilon) \tag{3}$$

$$d \ln \dot{\varepsilon} / d\varepsilon = n + m + d \tag{4}$$

where $\dot{\varepsilon}_0$ is the initial creep rate, n the stress exponent for the Norton's law, m the microstructure degradation and d the other parameter associated with damage such as creep voids. Because the microstructure observations gave no evidence of any formation of creep voids, the parameter d in eqs (3) and (4) is neglected and the acceleration of creep rate results from the parameters n and m. The n and m correspond to the acceleration of creep rate by an increase in stress due to a decrease in cross section with strain at constant load test and by a strength loss due to microstructure evolution, respectively. From the slope of the curves in Figure 7, the value of $d \ln \dot{\varepsilon}/d\varepsilon$ is evaluated and shown in Figure 8 as a function of stress. The $d \ln \dot{\varepsilon}/d\varepsilon$ is evaluated to be about 6 for a wide stress range of 69–147 MPa for the long-term serviced material, while it is about 9 for a wide stress range of 137–196 MPa but it increases with decreasing stress below 137 MPa for the virgin material. The value of n is evaluated from the stress dependence of minimum creep rate in Figure 5 to be 12 and 8 for the virgin and long-term serviced materials, respectively. Therefore, the $d \ln \dot{\varepsilon}/d\varepsilon$ is substantially the same as the parameter n for the virgin material at high stresses of 137–196 MPa and for the long-term serviced material at 69–147 MPa, while it is much larger than the value of n and hence a greater part of comes from the parameter m for the virgin material at low stresses below 137 MPa. For the long-term serviced material, the was evaluated only at stresses of 69–147 MPa, corresponding to the high stress region of $\sigma/\sigma_{OP} > 1$, but not at stresses of 49 and 59 MPa, corresponding to the low stress region of $\sigma/\sigma_{OP} < 1$, because of the lack of strain data in the acceleration region. The present results indicate that the $d \ln \dot{\varepsilon}/d\varepsilon$ mainly results from by an increase in stress due to a decrease in cross section with strain for both the virgin and long-term serviced materials in the high stress region of $\sigma/\sigma_{OP} > 1$ but that it results mainly from the microstructure evolution for the virgin material in the low stress region of $\sigma/\sigma_{OP} < 1$.

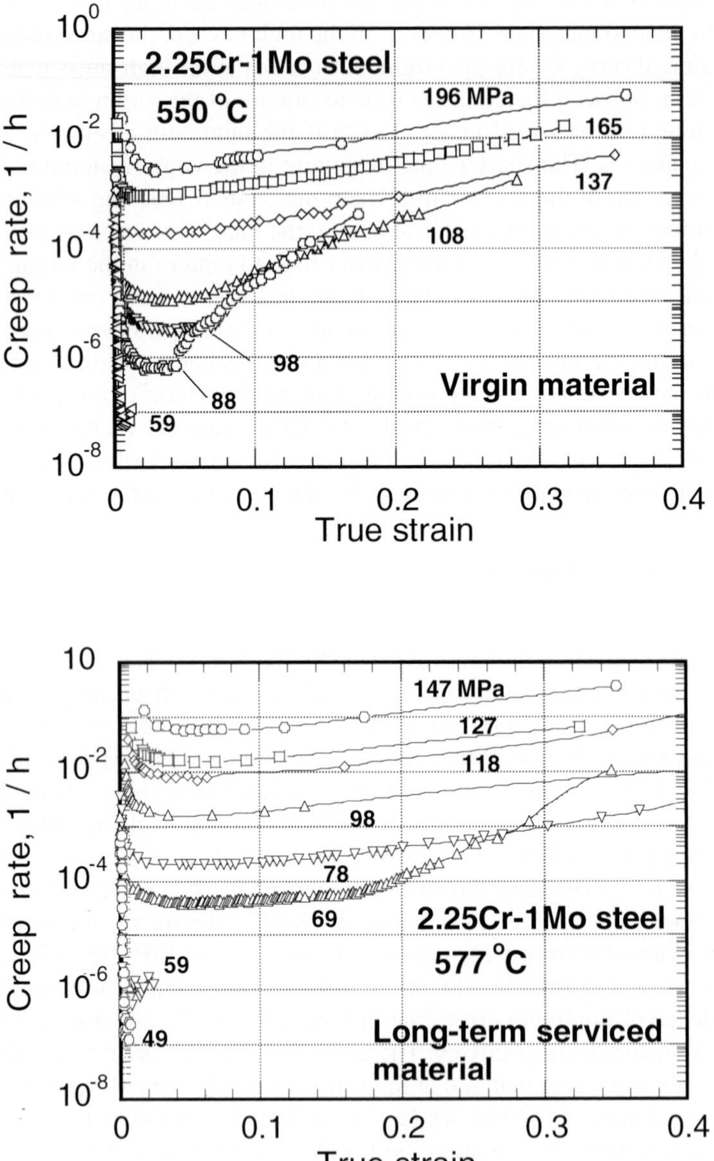

Fig. 7 Creep rate versus true strain curves for virgin and long-term serviced materials of 2.25Cr-1Mo steel at various stress conditions.

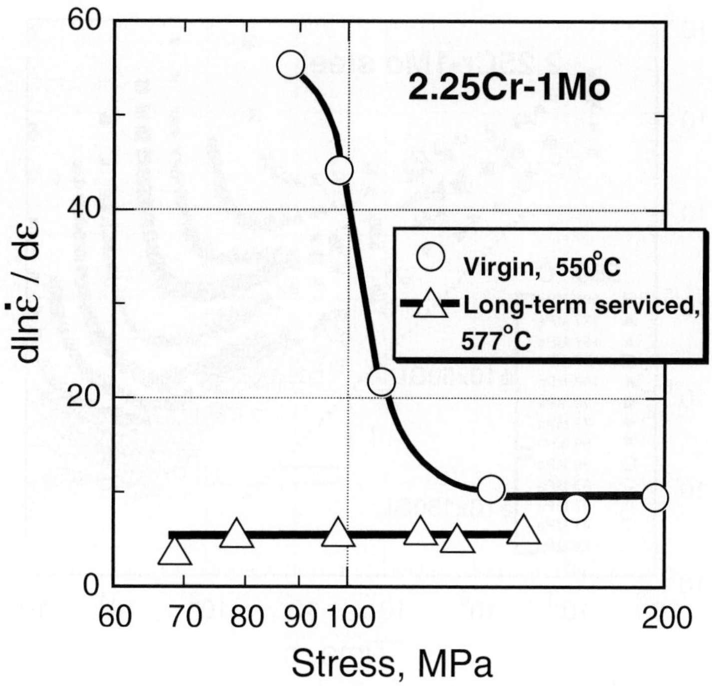

Fig. 8 Increase in creep rate by strain, $d \ln \dot{\varepsilon}/d\varepsilon$, for the virgin and long-term serviced materials of 2.25Cr1Mo steel, as a function of stress.

Creep Deformation Behaviour at Low Stresses

The micro-strains at very low stresses of 60 to 29 MPa have successfully been measured by using the long-size specimens with 150 mm. Figure 9 shows the creep rate versus time curves of the virgin material at low stresses, where the solid symbols were obtained using the specimens of 50 mm gauge length and the open symbols were obtained using the specimens of 150 mm gauge length. The data scattering in the transient creep region is significantly reduced by using long-size specimens. In Figure 9, the slope of creep rate versus time curves in the transient region increases with decreasing stress. The creep rate versus time curves in log-log plot are approximately described by a straight line with a slope of –1 at low stresses below 98 MPa, as shown by the dotted line. The straight line with a slope of –1 corresponds to logarithmic creep described as

$$\log \dot{\varepsilon} = -\log t + D \qquad (5)$$

where D is a constant. We think that the movement and annihilation of excess dislocations, which are present after heat treatment and before creep, are the major process in the transient creep region at low stresses. It should be also noted that small concave feature appears in the same time range of 30 to 100 hours in the transient region. The observed small concave

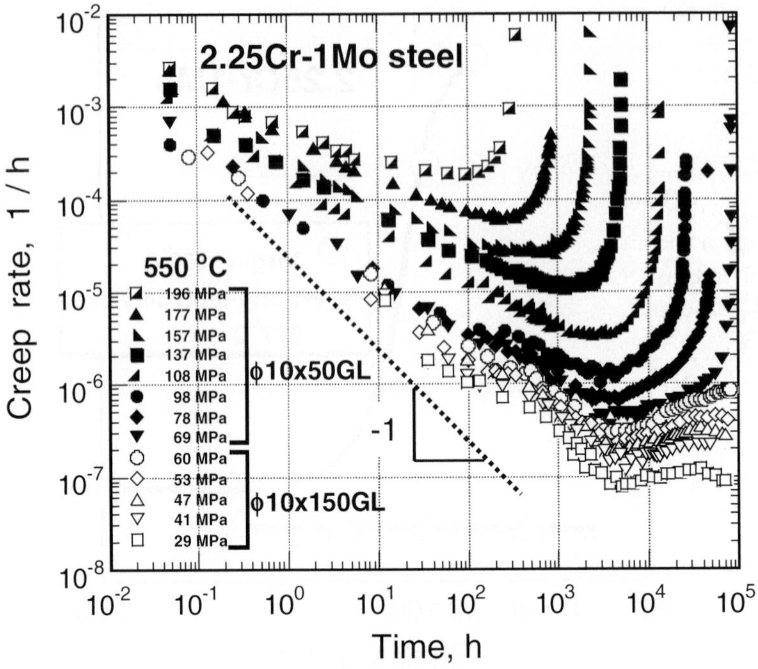

Fig. 9 Creep rate versus time curves of the virgin material of 2.25Cr-1Mo steel at 550°C.

feature may correlate with specific microstructure evolution during creep, although detailed mechanisms are not clear at present.

The present results suggest whether additional dislocations are introduced or not by loading affects the subsequent creep deformation behaviour in both the transient and acceleration regions. Therefore, the creep and creep-rupture data obtained at low stresses below the 0% offset proof stress σ_{OP} should be distinguished from ones at high stresses above the σ_{OP} and only the data at low stresses below the σ_{OP} should be provided for the improvement of reliability of long-term creep life estimation.

SUMMARY

1. The 0.2% proof stress, the tensile strength and the creep rupture strength are lower for the long-term serviced material than for the virgin material. On the other hand, the difference in creep rupture strength between the two materials becomes less pronounced with increasing test temperature and test duration.
2. The microstructure evolution during long-term service at 577°C promotes the formation of subgrains due to dynamic recovery or recrystallisation during subsequent creep test,

which promotes the onset of the acceleration creep region. This results in shorter duration of transient creep region, higher minimum creep rate and hence shorter rupture time for the long-term serviced material than for the virgin material.

3. The creep deformation behaviour correlates with the 0% offset proof stress, σ_{0P}. At high stresses of $\sigma/\sigma_{0P} > 1$, the initial creep rate decreases with decreasing stress and the decrease in creep rate with time in the transient region is small. At low stresses of $\sigma/\sigma_{0P} < 1$, approximately the same creep rate in the initial stage is obtained but the decrease in creep rate with time in the transient region becomes more significant with decreasing stress level.
4. In the acceleration region, the increase in creep rate by strain, $d \ln \dot{\varepsilon}/d\varepsilon$, mainly results from an increase in stress due to a decrease in cross section with strain in the high stress region of $\sigma/\sigma_{0P} > 1$ but that it mainly results from the microstructure evolution in the low stress region of $\sigma/\sigma_{0P} < 1$.
5. The creep and creep rupture data obtained at low stresses below the 0% offset proof stress σ_{0P} should be distinguished from ones at high stresses above the σ_{0P} and only the data at low stresses below the σ_{0P} should be provided for the improvement of long-term creep life estimation.

REFERENCES

1. B. J. Cane and J. A. Williams: *International Materials Reviews*, **32**, 1987, 241–262.
2. F. Abe and K. Yagi: *Proceedings of the 4th International Charles Parsons Turbine Conference*, Newcastle, UK, 1997, 750–765.
3. K. R. Williams and B. Wilshire: *Materials Science and Engineering*, **28**, 1977, 289–296.
4. T. Ohba, O. Kanemaru, K.Yagi and C. Tanaka: *Materials Science Research International*, **3**, 1997, 10–15.
5. National Institute for Materials Science Creep Data Sheet, No.3B, 1986.
6. T. Ohba, E. Baba, F. Abe, H. Irie and I. Nonaka: *Proceedings of the 7th International Conference on Creep and Fatigue at Elevated Temperatures (CREEP7)*, Tsukuba, Japan, 2001, 331–334.
7. H. Kushima, K. Kimura, F. Abe, K. Yagi, H. Irie and K. Maruyama: Tetsu-To-Hagane, **85**, 1999, 848–855.
8. F. Abe, S. Nakazawa, H. Araki and T. Noda: *Metallurgical Transactions*, **23A**, 1992, 469–477.
9. R. W. Evans and B. Wilshire: *Creep of Metals and Alloys*, The Institute of Metals, London, UK, 1985, 69–113.
10. K. R. Williams and B. Wilshire: *Materials Science and Engineering*, **28**, 1977, 289–296.
11. M. Prager: *Pressure Vessel and Piping*, **288**, 1994, 401–421.
12. R. Woo, R. Sandstrom and J. Storesund: *Materials at High Temperatures*, **12**, 1994, 277–283.
13. S. Straub, M. Meier, J. Ostermann and W. Blum: *VGB Kraftwerkstechnik*, **73**, 1993, 646–653.
14. T. Endo and J. Shi: *Proceedings of the Tenth International Conference on the Strength*

of Materials, Sendai, Japan, 1994, 665–668.
15. F. Abe: *Materials Science and Engineering*, **A234–236**, 1997, 1045–1048.
16. K. S. Park, F. Masuyama and T. Endo: *ISIJ International*, **41**, 2001, S86–S90.
17. F. Abe: *Materials Science and Engineering*, **A319–321**, 2001, 770–773.

Materials Development for Boilers and Steam Turbines Operating at 700°C

R. BLUM
ELSAM

R. W. VANSTONE
ALSTOM Power Ltd.

ABSTRACT

In Europe development of boilers and steam turbines for plant operating at temperatures in excess of 700°C is being carried out within the EU supported AD700 project. This collaborative project involves all of the major European power plant manufacturers supported by utilities and research institutes. The first phase of the project began in 1998 and runs until the end of 2003. A second phase began in 2002 and runs until the end of 2005. The objective is the development of all technology necessary for the construction and operation of such plant.

This paper describes the development of the high temperature materials technology essential to underpin the development of the AD700 power plant technology. It describes the factors influencing alloy design and selection and the scope and results of investigations on candidate alloys carried out within the project. It then goes on to describe the programme of full-scale prototype component manufacture which is currently in progress. These prototype components are being characterised through extensive long term testing programmes. The development of joining procedures for these materials is also described.

THE AD700 PROJECT

The overall objective of the AD700 project is to develop and demonstrate a new generation of pulverised coal-fired power plants featuring advanced steam conditions.[1] This will be achieved through the application of nickel-based superalloys to bring the live steam temperature up to about 700°C, resulting in an increase in efficiency from around 47%, representing the current state-of-the-art plant, to around 55%, an increase yielding reductions in fuel consumption and carbon dioxide emission of almost 15%. The plant will have an output within the range of 400–1000 MW, making it suitable for scale generation. The project involves nearly all the major European power plant manufacturers and their material suppliers, the largest European utilities and major research organisations.

The principal innovation which underlies this development is the replacement of iron-based alloys by nickel-based alloys for the highest temperature components. These alloys are already used in the aerospace and gas turbine industries so that the project is to some extent one of technology transfer. However much larger components are required for boilers and especially for steam turbines than are currently produced and there are significant technical challenges to be met to achieve the manufacture of larger components. In addition these components will be required to operate under significantly different conditions of environment, stress and temperature. Therefore demonstration of manufacturing capability and appropriate material characteristics is required. Nickel-based alloys are much more expensive than alloy steels and this aspect is motivating further innovations to minimise the requirement for these alloys. This is being achieved through radical new concepts in plant architecture and turbine construction and also through material developments to maximise the strength of austenitic and ferritic steels.

The first phase of the project began in 1998 and runs until the end of 2003. A second phase began in 2002 and runs until the end of 2005. The objective is the development of all technology necessary for the construction and operation of such plant.

MATERIAL SELECTION

Development and demonstration of appropriate materials and their properties, especially in the case of the nickel based alloys, is critical to establishing the technical feasibility of the new power plant concept. As well as creep strength sufficient for long term operation at these high temperatures (typically 100,000 hours rupture strength of around 100 MPa at the metal temperature is required), materials requirements include corrosion resistance in boiler flue gases and under conditions of steam oxidation, resistance to thermo-mechanical cycling and the ability to be manufactured and welded in thick sections.

Most modern-day nickel based alloys have been developed from a relatively simple Ni20%Cr alloy. In order to achieve the required creep properties, further strengthening through solid solution or dispersion strengthening is necessary. Additions of elements like Mo, W and Co confer solid solution strength. Alloys relying principally on this mechanism, such as Alloys 230 and 617, are used in the solution treated condition and have the advantage of being relatively easy to weld with no requirement for complex post weld heat treatment. However their proof strengths are relatively low and where this property is important additions of Ti and Al to form dispersions of the gamma prime precipitate, conferring high proof strength as well as improved creep strength, can be made. Alloys such as these, for example Alloy 263 and Waspaloy, must be aged after solution treatment to produce the strengthening dispersion and thus post weld heat treatment requirements are more complex. Additions of Ti and Al are limited by the requirement for weldability: where these additions are too high the kinetics of gamma prime precipitation become such that precipitation occurs in the heat affected zone (HAZ) during welding, the reduced ductility resulting from which leads to the potential for HAZ cracking. An alternative approach, to gamma prime strengthening is by alloying with Nb, leading to the more sluggishly precipitated gamma double prime. With regard to corrosion resistance, where flue gas corrosion resistance is required, increased levels of Cr may be necessary. Finally the economic aspects of material selection cannot be neglected so, in order to mitigate the high cost of nickel-based alloys, significant additions of Fe, such as are made in the case of Alloys 718 and 901, may be considered.

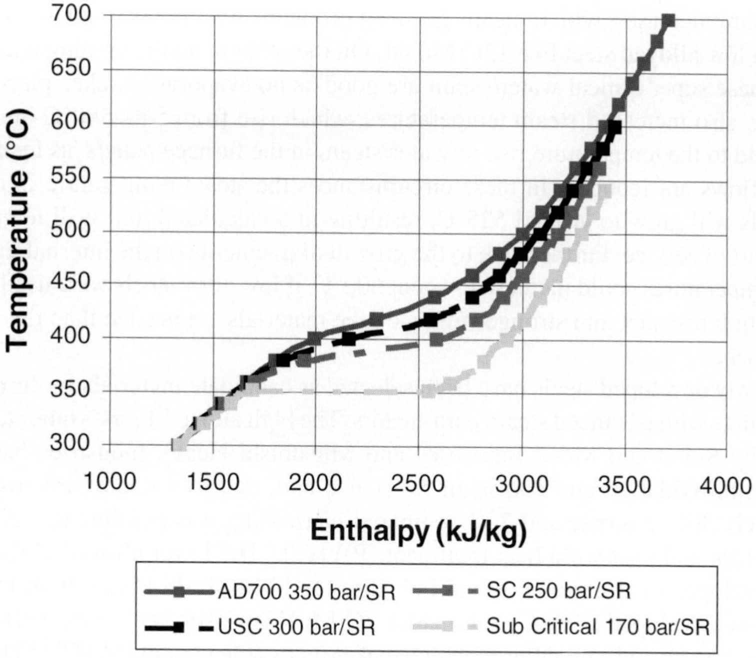

Fig. 1 Water/steam temperature versus enthalpy.

BOILER MATERIALS

In order to realise a 700°C USC boiler, extensive materials development and qualification is necessary, including the use of Ni based superalloys in the most severely exposed components. These developments can be categorized into three groups reflecting the key components of such plant, i.e.:

- Furnace panels
- Superheaters and
- Thick section components and steam lines.

Furnace Panels

The behaviour of water and steam in the furnace panels changes rapidly as steam pressure increases from sub to super critical conditions, and in particular, at constant enthalpy increase (constant heat input) the rise of water/steam temperature also grows rapidly. Therefore, the water/steam temperature at the outlet of the furnace panels of super critical boilers grows constantly, as steam parameters moves towards the advanced conditions of 350 bar and 700°C planned for the AD700 technology.

Typically, the enthalpy at the outlet of the furnace panels would be in the range of 2600–2700 kJ/kg, and Figure 1 indicates how temperature grows from ~375 to ~450°C as steam pressure increases, which means growing problems with the strength of conventional well proven low alloyed steel like 13CrMo 44. On the other hand the cooling characteristics of single phase super critical water/steam are good as no evaporation takes place.

However, also increased steam temperatures, which rise from 540–560°C into the range of 700°C add to the temperature rise of water/steam in the furnace panels, as feed water and live steam flows are reduced. In these circumstances the steam temperature in the furnace outlet panels will grow to around 525°C, resulting in a calculated mid wall temperature of 575°C at start of service. Further, due to the growth of magnetite on the internal tube surface, mid wall temperatures could increase to about 600°C, if low alloy steels were used. Therefore more oxidation resistant and stronger furnace tube materials are needed than the 1Cr 0.5Mo used at present.

Three newly developed steels have been selected as candidate materials for furnace panels in future boilers with advanced steam para-meters. The high alloyed 12%Cr tube steel HCM12 developed by Sumitomo Metal Industries and Mitsubishi Heavy Industries has excellent creep strength, oxidation and corrosion resistance and, due to a duplex microstructure of approximately 30% δ-ferrite and 70% tempered martensite, it is possible to weld this steel without preheat and post weld heat treatment (PWHT).[2] The lower alloyed 2½Cr tube steel HCM2S developed by Sumitomo Metal Industries and Mitsubishi Heavy Industries and the Mannesmann developed 2½%Cr tube steel 7CrMoVTiB1010 both have sufficient high temperature strength and a metallurgy which makes it possible to omit PWHT.[3,4] The chemical composition and mechanical properties for all three steels are given in Figure 2.

Testing of HCM12, HCM2S and 7CrMoVTiB1010 is in progress in Europe to establish practical experience with the handling of these steels. In the furnace panels of an existing subcritical once through boiler, test sections of all three steels have been installed, and service under cycling conditions has been tested for several years. For the HCM2S and 7CrMoVTiB1010 tube materials no problems have been encountered during the production of the test panels or during operation at steam temperatures up to 500°C. For the tube steel HCM12 which has a rather complicated microstructure, lack of sufficient filler material has caused problems with cracks after welding due to embrittlement, and further development of suitable filler material is required before a reliable welding procedure can be established. Nonetheless, test panels of HCM12 have been successfully service exposed with steam temperatures up to 530°C under cycling conditions. Moreover, a full size wood chip fired boiler furnace construction operating with steam temperatures up to 540°C for more than five years demonstrates that HCM12 is a candidate for furnace tubes in future advanced USC boilers.

In connection with the in-plant test and the construction of the wood chip boiler, a computer furnace panel calculation programme has been set up by Elsam to simulate the service exposure and life consumption of a furnace panel in a USC boiler during operation.[5] Calculations first of all demonstrate that at media temperatures higher than around 450°C the temperature rise during operation is strongly dependent on the rate of self oxidation and on the deposition rate of oxides from the feed water. High quality feed water chemistry assuring minimum oxide in feed water is therefore essential if USC plants with advanced steam parameters are to be realised.

Based on these calculations it can be demonstrated that the use of either HCM2S or 7CrMoVTiB1010 is sufficient for media temperatures up to 475°C. For higher media temperatures only HCM12 should be considered due to the need for better oxidation resistance.

Chemical composition, mass %

	C	Cr	Mo	W	Others
1Cr½Mo	0.13	0.9	0.5		
HCM2S	0.06	2.25	0.3	1.6	V,Nb,N,B
7CrMoTiB1010	0.07	2.4	1		V,Ti,N,B
HCM12	0.1	12	1	1	V, Nb

Creep rupture strength

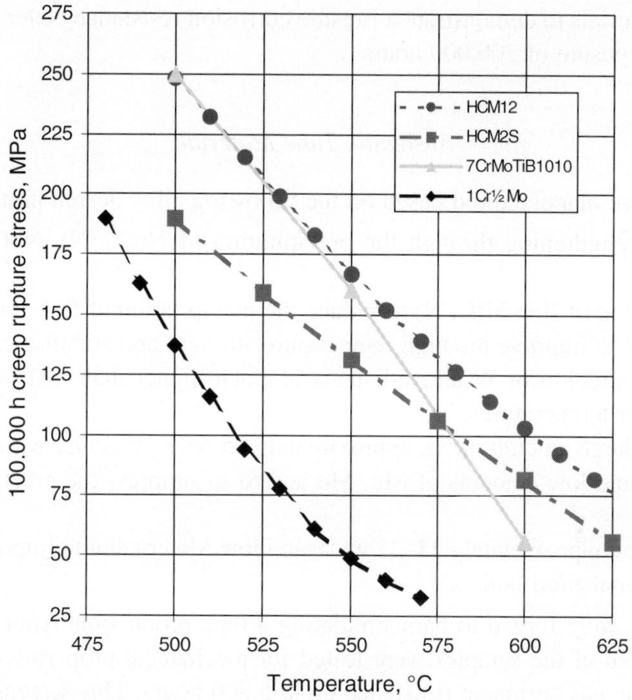

Fig. 2 Materials for furnace panels.

In AD700 fabrication trials and in-plant exposure testing is in progress and the possible use of the nickel alloy INCONEL® Alloy 617 is investigated in parallel as an alternative to HCM12. Of course, alloy 617 has much better mechanical properties and does not generate any internal oxide, and therefore it can be exposed to much higher temperatures than HCM12. The fact that alloy 617 tubes cost roughly 10 times more than HCM12, and call for a far more expensive fabrication, means, however, that the whole boiler economy must be carefully considered before such a choice of material is taken.

SUPERHEATER TUBING

For superheater tubing the aim is to develop an improved austenitic tube material with sufficient strength and flue gas corrosion resistance to operate at steam temperatures around 650°C, and to develop a Ni base superalloy to enable the steam temperatures of 700°C to be achieved. Intensive development work is ongoing in the AD700 project to demonstrate a suitable austenitic tube material with 100,000 hours rupture strength of about 100 MPa at 700°C and a Ni based tube material with 100,000 hours rupture strength of 100 MPa at 750°C - both materials to demonstrate a fireside corrosion resistance better than 2 mm metal loss during an exposure of 200,000 hours.

Austenitic Tube Material

30 trial melts were manufactured based on the following alloy design principles:

- Precipitate strengthening through the precipitation of Nb(C, N), NbCrN and Cu-rich precipitates.
- The precipitation of fine M(C, N) to induce the precipitation of fine and stable $M_{23}C_6$.
- Addition of W to improve the high temperature strength and stabilise carbonitrides. The creep rupture strength of W alloyed steels is much higher than that of the Mo alloyed steels at higher temperatures.
- Addition of a large amount of Ni, approximately 25 wt.%, together with a relatively high amount of N and low amounts of Mn, Mo and Si to suppress the precipitation of sigma phase.
- A Cr content of approximately 23–25 wt.% and low Mo content to improve the corrosion resistance in coal-fired boilers.

All trial melts were forged to bars simulating a tube production. After appropriate heat treatment, fourteen of the samples were tested for mechanical properties as well as steam oxidation and flue gas corrosion resistance up to 5,000 hours. This screening test indicated that four of the fourteen tested trial melts might be able to meet the targets. After another 10,000 hour testing one of the four alloys showing the best properties was selected. The chemical composition and room temperature mechanical properties of this alloy, called Alloy 174, are shown in Tables 1 and 2.

Figure 3 demonstrates the improvements in creep rupture strength compared with some of the best commercial austenitic superheater tube alloys. The extrapolated rupture data of alloy 174 are based on 20,000 hours of testing. Ongoing steam oxidation and flue gas corrosion test demonstrate properties comparable to or slightly better than those obtained for a large variety of 22–25%Cr austenitic superheater steels. So far all targets are fulfilled. Full scale tube production has been successfully demonstrated, and fabricability trials including welding and bending are in progress. Different filler materials have been considered for similar and dissimilar welds. So far Inconel Alloy 625 has been chosen due to its excellent mechanical properties, comparable to Alloy 174.

Long term creep rupture properties and microstructural stability test will complete the characterisation of this super austenite.

Chemical composition, mass%

	Cr	Ni	W	Nb	Cu	Others
Super 304H	18	9		0,4	3	N
NF709	20	25		0,25		Mo, Ti, N
HR3C	25	20		0,4		N
SAVE 25	23	20	1,5	0,4	3	N
Alloy 174	22	25	3,5	0,5	3	N

Creep rupture strength

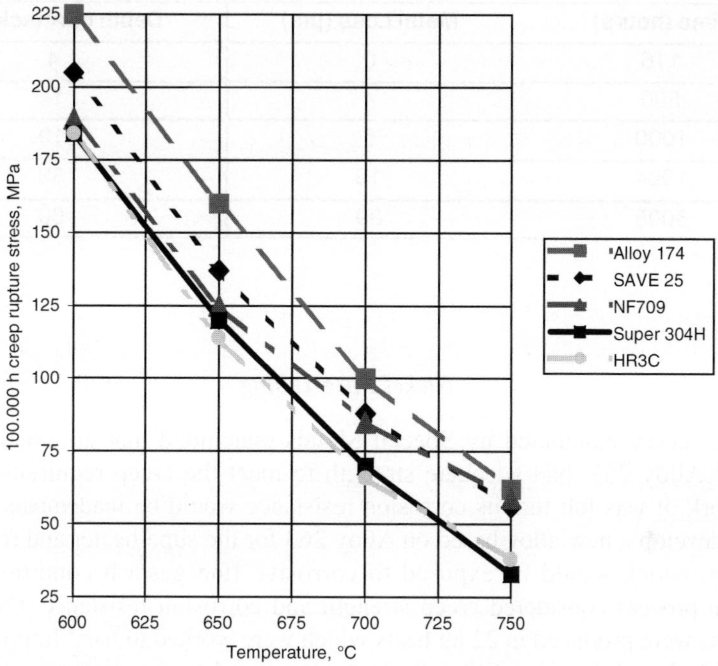

Fig. 3 Materials for superheater tubes.

Table 1 Chemical composition for Alloy 174.

C	Cr	Ni	W	Co	Cu	Nb
0.08	22.2	24.9	3.5	1.5	3	0.49

Table 2 Mechanical properties at room temperature for Alloy 174.

Tensile Strength (MPa)	Yield Strength (MPa)	Elongation (%)	Reduction of Area (%)	Hardness (HV)	Charpy Impact Value (J/cm^2)
785	345	55	70	195	216

Table 3 Composition of INCONEL Alloy 740.

C	Ni	Cr	Mo	Co	Al	Ti	Nb	Mn	Fe	Si
0.03	Bal	25.0	0.5	20.0	0.9	1.8	2.0	0.3	0.7	0.5

Table 4 Penetration results for INCONEL Alloy 740 in coal ash corrosion tests at 700°C.

Time (hours)	Metal Loss (μm)	Depth of Attack (μm)
116	0	4
500	4	14
1000	5	19
1984	16	33
5008	39	60

Nickel Base Tubing

A literature survey conducted by Special Metals concluded that an existing alloy, viz. NIMONIC® Alloy 263, had adequate strength to meet the creep requirement, but, from previous work, it was felt that its corrosion resistance would be inadequate. Hence it was decided to develop a new alloy based on Alloy 263 for the superheater and reheater tubular components, which would be exposed to corrosive flue gas/ash conditions. The alloy optimisation process considered creep strength and corrosion resistance. Over thirty trial compositions were produced in 22 kg heats which were worked to bars. Improvement of the coal ash corrosion resistance of the alloy was developed in a series of coal ash corrosion tests at different temperatures employing samples with a systematic variation in alloy constituents. The tests used an atmosphere of 15 vol.% CO_2 + 10 vol.% H_2O + 1 vol.% SO_2 with balance N_2 and the test specimens were coated with a synthetic coal ash, K_2SO_4 + ($Fe_2O_3/Al_2O_3/SiO_2$) in the ratio 1:1:1.

The resultant optimised chemical composition which was selected is given in Table 3. The new alloy INCONEL® alloy 740, based on alloy 263, is a nickel chromium cobalt alloy which is age hardenable by the precipitation of gamma prime but also benefits from solid solution hardening. Table 4 gives the ash corrosion results for this alloy.

A large creep test matrix is presently underway covering two major test temperatures, 725 and 775°C, with some shorter term tests at 700, 750 and 800°C. Test durations up to 65,000 hours are being targeted, and the results to date are shown in Figure 4.

Ageing trials are underway with target times up to 70,000 hours. Up to now, 1,000, 3,000 and 15,000 hours have been completed at 700, 750 and 800°C. The long term ageing trials are being carried out on material which has been given the initial strengthening/ageing treatment of 4 hours at 800°C. After 1,000 hours at 700°C, the Charpy impact values dropped to some 45% of the values after the initial ageing/strengthening treatment, and further exposure

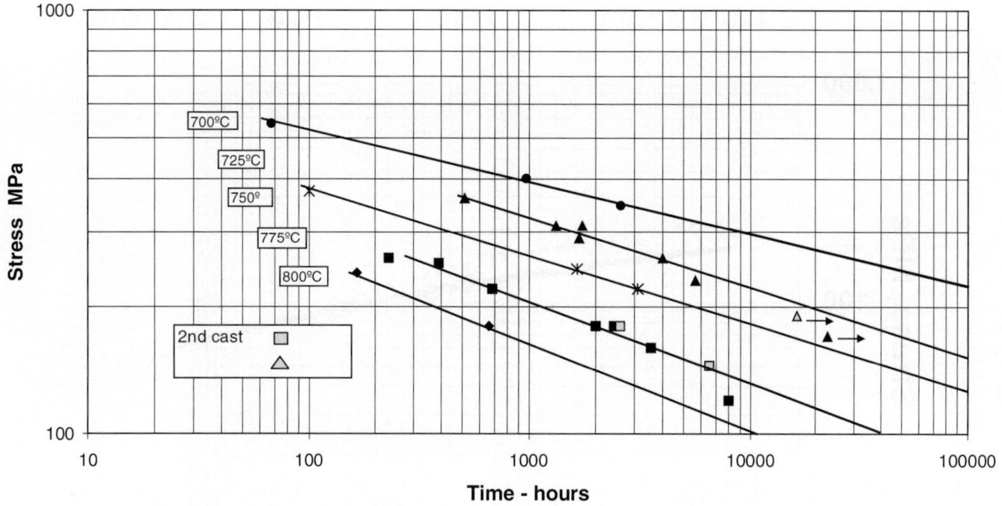

Fig. 4 Inconel alloy 740 creep results.

to 3,000 hours exhibited a reduction to around 27% (of the initial aged/strengthened values). The absolute values after 3,000 hours are around 25 J.

Welding trials have been initiated and have demonstrated that alloy 740 is readily fabricated in the annealed condition. Joining is accomplished with the gas tungsten-arc welding (GTAW) process using both INCONEL Filler Metal 740 and NIMONIC Filler Metal 263. If a high joint strength is required, the deposited weldment may be precipitation hardened. The need for repair of boiler tubes is inevitable and successful welding on aged material has been undertaken although the mechanical test results are not yet available.

A commercial size cast of the alloy has been produced and put through the normal tube production route, and welding trials and fabrication trials as well as long term creep rupture data are planned for these tubes.

Thick Section Components and Steam Lines

For thick section boiler components and steam lines there are two goals for the materials development. An improved ferritic/martensitic 9–12%Cr steel is desirable to expand the present temperature range for ferritic steels up to app. 650°C. A Ni-based superalloy with a 100,000 hour rupture strength of 150 MPa at 700°C is needed to allow construction of outlet headers and main steam lines with acceptable wall thicknesses.

Ferritic Pipe Steels

The task of improving the 9–12%Cr steels further beyond the impressive developments in the last two decades has proved to be very difficult. In the last five years worldwide research

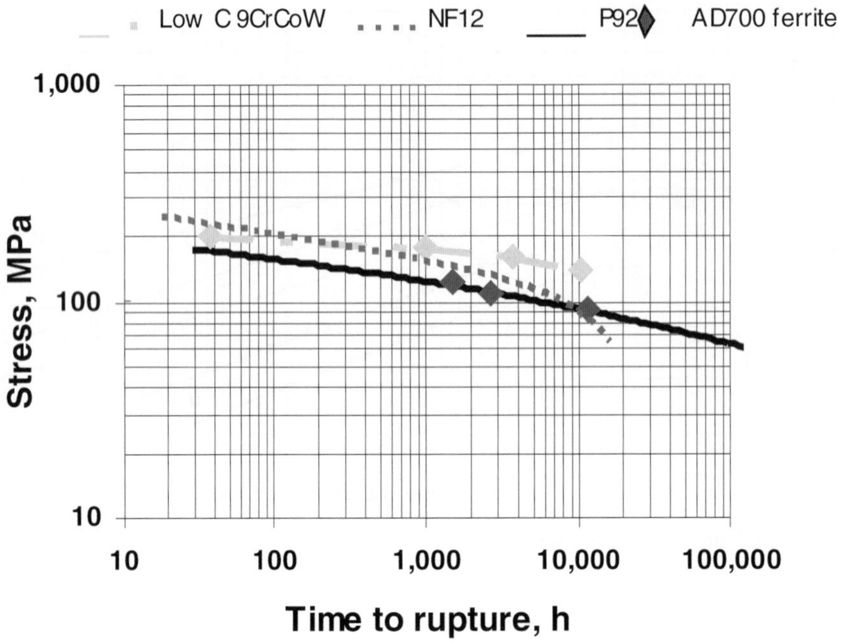

Fig. 5 Ferritic steels 650°C.

has resulted in a large number of new alloys being announced, and from short-term tests they seemed very promising. However, in long-term tests the steels show sigmoidal creep behaviour and so far no ferritic alloy has demonstrated long-term creep strength better than steel P92. In AD700 an attempt was made to improve the creep rupture strength of 9–12% Cr steels. Seven trial melts were manufactured and mechanical properties were obtained up to 12,000 hours. Six of the seven melts turned out to be weaker than P92 and only one melt, a 9%Cr5Co2WVNbN, showed creep rupture data similar to P92. In parallel, tests were made on steel NF12. Short-term data demonstrated a major improvement, but longer term data also showed a dramatic drop in strength for this steel (Figure 5).

Systematic microstructure investigations of new ferritic steels showing sigmoidal creep behaviour have demonstrated that precipitation of the complex Z-phase nitride (Cr(V, Nb)N) takes place in the steels at the expense of the strengthening MX carbonitrides, which decompose. This mechanism is believed to be responsible for the reduction in creep strength, and it seems that high Cr steels are more prone to Z-phase formation than low Cr steels. To be able to improve the strength, a fine-tuning of the composition is needed based on a thorough understanding of recent developments. Recent advances in microstructure characterisation techniques and thermodynamically based microstructure models may prove to be the only way to go further in the development of improved ferritic steels.

A potentially interesting new ferritic steel is the low carbon 9Cr3Co3WVNbN steel currently under development at the NIMS in Japan. This steel is strengthened only by nitrides

Table 5 Composition of Nimonic alloy 263 (Note: Ti + Al : 2.4 – 2.8).

C	Ni	Cr	Mo	Co	Al	Ti	Mn	Fe	Cu	B
0.04	Bal.	19.0	5.6	19.0	0.6	1.9	0.60	0.7	0.2	0.0005
0.08		21.0	6.1	21.0	Max	2.4	Max	Max	Max	Max

and Laves phase precipitates, and the unstable $M_{23}C_6$ carbides are not present. Creep tests up to about 10,000 hours at 650°C show no signs of sigmoidal creep behaviour, see Figure 5. If the low C steels can maintain microstructure stability up to long times, this approach may serve as an important platform for future developments of ferritic steels. The low Cr content will, however, lead to poor steam oxidation resistance, and a surface coating will be needed. At the moment, it does not seem to be possible to obtain high creep strength together with high oxidation resistance at the same time in ferritic steels.

Nickel Base Piping

NIMONIC® Alloy 263 or an improved version of INCONEL® 617 may meet the demands for outlet headers and steam lines at 700°C steam temperature. Long-term creep data and demonstration of fabricability – pipe production, hot bending and welding – are needed before a 700°C power plant can be realised. Alloy 263 is under investigation in the AD700 project, and the improved version of Alloy 617 is investigated by the German national project MARCKO DE2. Alloy 617 pipes were manufactured in late 2002. Commercial scale production of alloy 263 is expected in 2003. The products will be used for qualification of the materials.

The precipitation hardened Nimonic 263 weldable composition, see Table 5, had only limited stress rupture data available as its designed purpose was for industries that do not require 100,000 hour creep data. Therefore, the biggest experimental effort expended on this alloy is in creep testing of commercially available 15 mm diameter bar. Tests to failure along with running tests are shown in Figure 6. These data suggest that the alloy will easily meet the creep criteria. This will enable the design of components with relatively small thicknesses and thereby reduce production costs.

A vital demonstration of the viability of this alloy, in the context of the AD700 programme is the ability to manufacture in thick section and extrusion trials on a recently produced 2 tonne ingot are imminent. It is planned to produce a 300 mm o.d. × 50 mm wall thickness pipe. This trial will be followed by welding trials on the pipe followed by mechanical testing.

In the German MARCKO DE2 project, mechanical testing and welding trials are established on Alloy 617. Both tube and pipe products are available and a review of the existing creep data base, covering creep rupture data for more than 20 heats with testing times above 100,000 hours, resulted in an update of creep strength values for both tube and pipe products. Figure 7 shows the ASME creep rupture data of Alloy 617 and the proposed

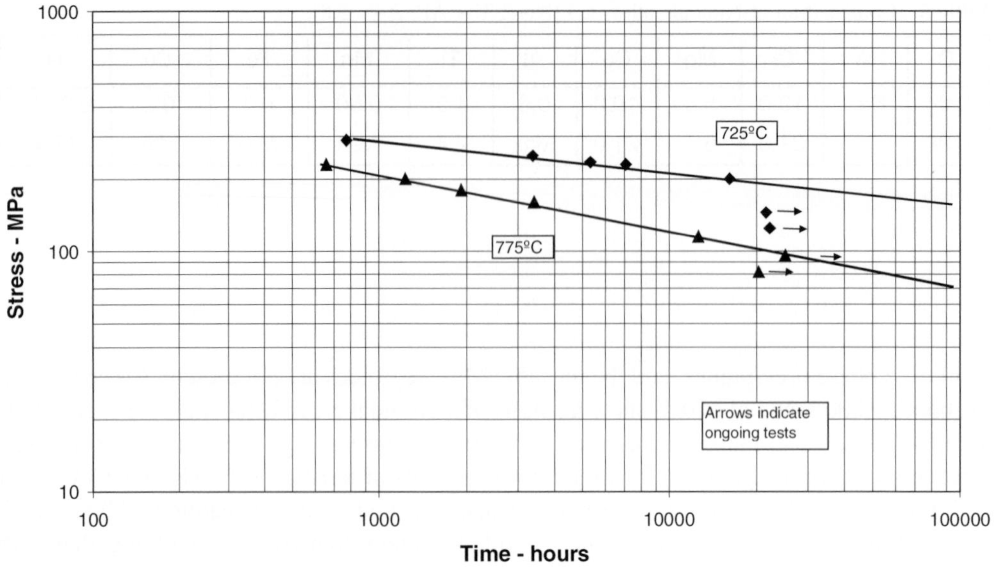

Fig. 6 Nimonic Alloy 263 creep results.

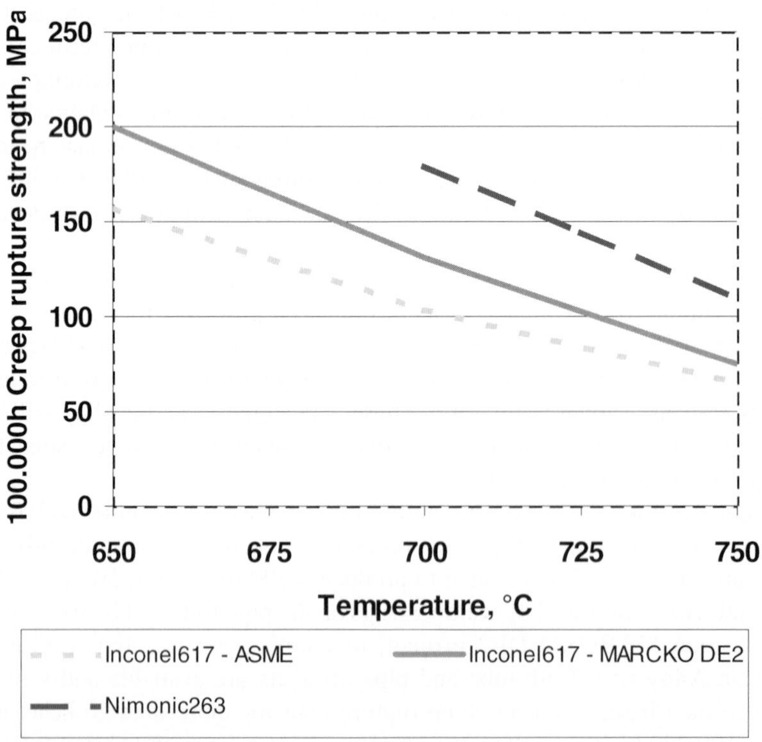

Fig. 7 Super Alloy rupture strength.

revised data obtained in the MARCKO DE2 project as well as the extrapolated creep rupture data of Alloy 263 based on 20,000 hours testing.

TURBINE MATERIALS

In addition to the factors already described, the selection of a first generation of candidate alloys in the AD700 project was influenced by the requirement to produce very large components so that experience of the project participants in producing large forgings in, for example Alloys 706 and 718, and large castings in, for example, Alloy 625 was taken into account. The selection was also influenced by the extent of data already available on the alloy, albeit generally not in the appropriate product form, for example the large body of data on tubes and pipes in Alloy 617.

All these considerations, a review of the literature and of other data available to the project, led to the selection of nine alloys for investigation: Alloys 155, 230, 263, 617, 625, 706, 718, 901 and Waspaloy. In a couple of cases, alloys were selected in more than one heat treatment condition.

Investigation of Candidate Alloys

The main focus of the initial investigation was the identification of the best candidate alloys for the large rotor and casing components. The smaller components such as blading and bolting are equally important, but it was recognised that the product forms required for these components, bar and possibly investments castings, were already available for the gas turbine industry and could be applied to steam turbines with relatively little effort, perhaps limited to additional materials characterisation to establish those materials parameters required for steam turbine design. The nine candidate alloys were produced in various product forms, conventional castings, centri-spun castings, and bar or forgings. One alloy was also investigated in a powder metallurgy form.

A first series of investigations was intended to confirm reported or expected short-term properties and to gain indications of response to ultrasonic testing and welding. These investigations included room temperature and elevated temperature tensile testing, creep testing at 700°C to target durations of 300–3000 hours, and ageing trials at 650 and 700°C. The ageing trials involved impact and tensile testing after exposure for durations of 300, 1000 and 3000 hours. Welding trials were carried out on test pieces in each alloy. The weldments were around 35 mm in thickness and all were made with the same, Alloy 617, consumable. Ultrasonic inspection was carried out on samples of the test materials to establish attenuation properties. Where possible, these investigations were carried out on already available materials but in many cases materials were manufactured specifically for the project.

The mechanical property investigations yielded results which were generally consistent with expected values. In terms of tensile properties castings were generally weaker than wrought products. The properties of centri-spun castings were similar to those of conventional castings. There was some concern that the low proof strength of the castings might lead to poor resistance to thermal cycling. A low cycle fatigue test programme mounted on these

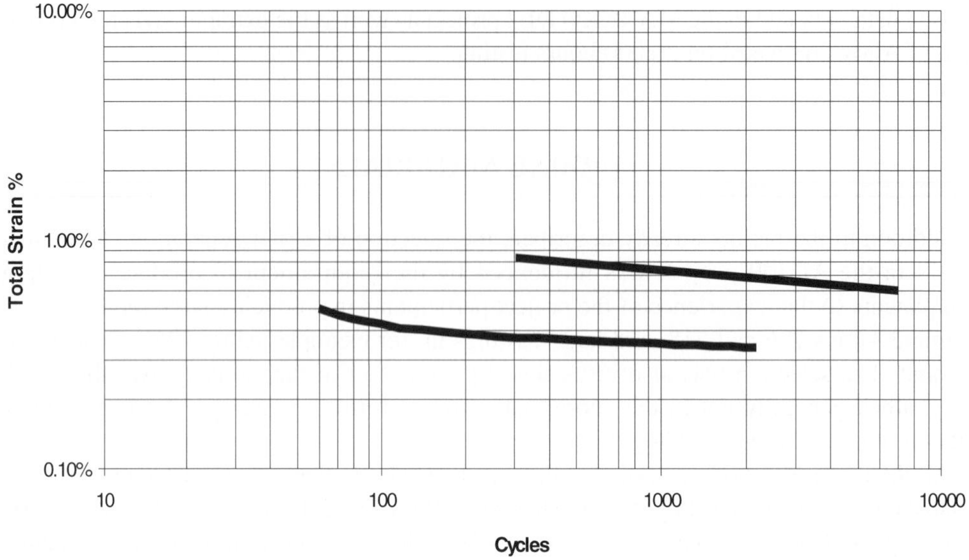

Fig. 8 Upper and lower bounds to low cycle fatigue endurance data obtained from tests on cast Alloys 263, 617 and 625.

alloys has indicated that although relatively low strains lead to endurances of around 100 cycles, the gradient of the endurance curves is shallow so that the strains for endurance of around 5000 cycles are acceptable for steam turbine applications (Figure 8). Nearly all of the alloys responded to long term ageing through increased tensile strength but reduced impact strength. The reductions in impact strength were large enough to be of concern and prompted further investigation. In Alloy 617, values of Charpy-V impact strength were observed to fall to as low as 10 J. However when fracture toughness tests were carried out on this embrittled material, values of fracture toughness in excess of 70 MPa $m^{0.5}$ were measured, values which are considered acceptable for steam turbine applications. The kinetics of embrittlement were very similar in all of the alloys. There was little difference in ageing response at 650 and 700°C and most of the embrittlement appears to be complete after 1000 hours so that the differences between properties measured after ageing for 1000 hours and 3000 hours were relatively small (Figure 9). The creep tests indicated properties generally in line with expectations. However the creep properties of a powder metallurgy product, in Alloy 625, were disappointing. Investigation showed that premature, low ductility creep failures occurred at prior particle boundaries due to the presence of nitrides. However process improvements have been identified which may overcome this problem. One notable success was the development of a modified heat treatment for Alloy 718. The alloy normally undergoes a two-stage ageing treatment involving ageing at around 720°C and then 620°C. An increase in the temperature of these ageing treatments by 30–40°C has resulted in a significant increase in long term creep strength. Figure 10 shows the creep master curve of the modified heat treatment.[6]

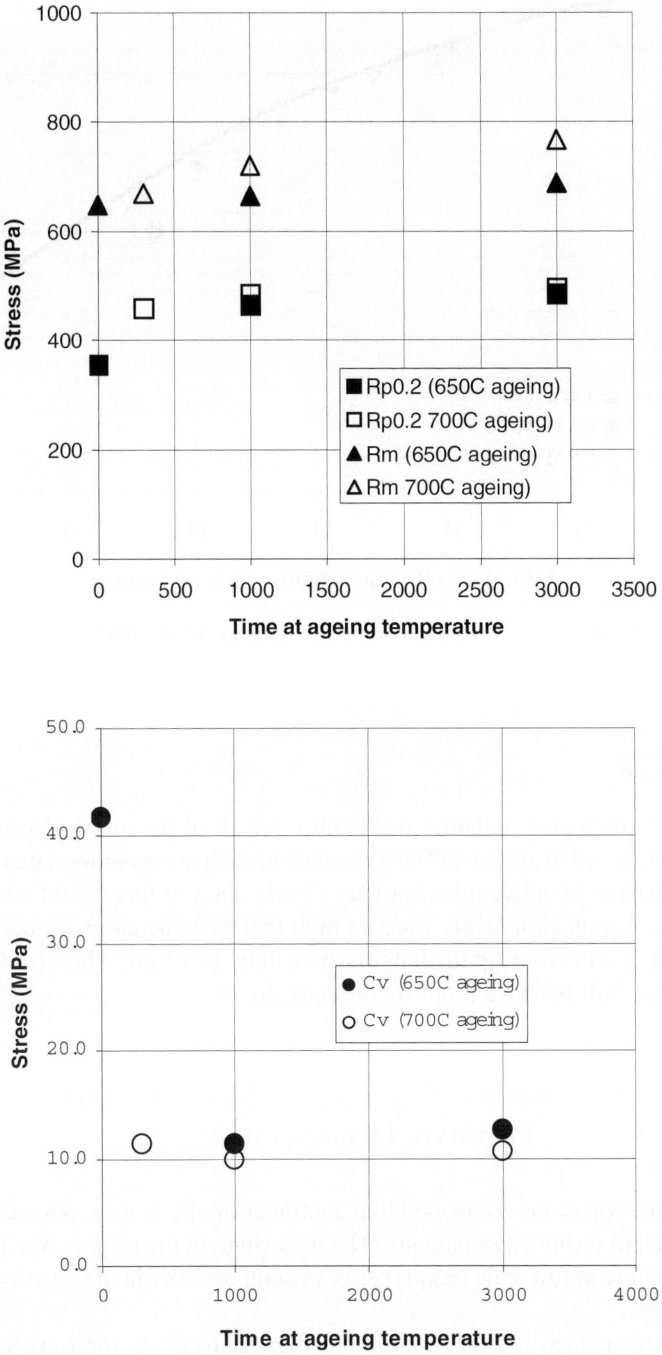

Fig. 9 Kinetics of long term ageing in Alloy 617.

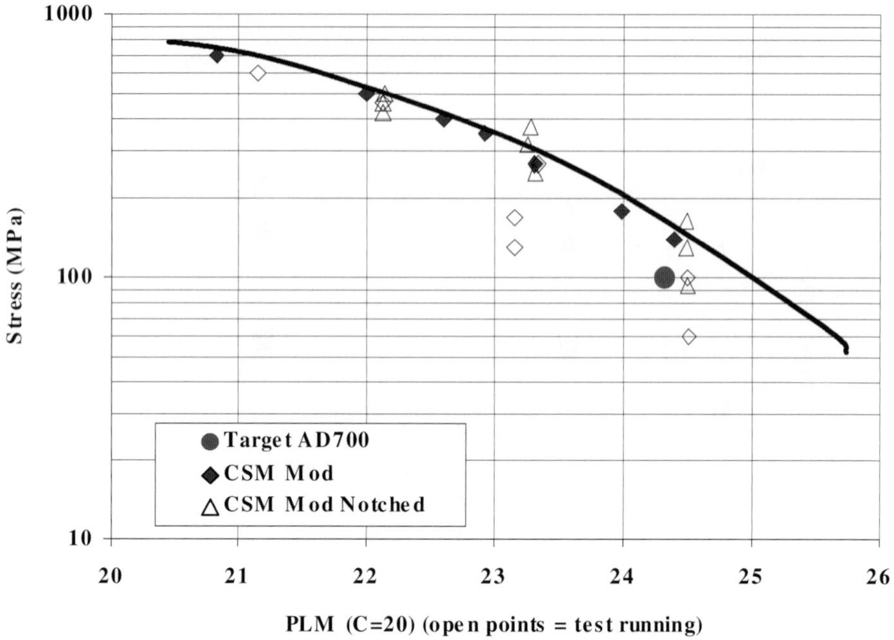

Fig. 10 Larson-Miller master curve of the modified heat treatment specimens.[6]

The welding trials indicated potential weldability for all of the alloys. No heat affected zone cracking was observed in any case. The ultrasonic investigations showed that the wrought alloys could be inspected by ultrasonics but very clearly showed this would not be possible in cast components. Attenuation levels were so high that in some cases no back wall echo was observed even at ultrasonic path lengths less than 100 mm. Therefore volumetric inspection of castings will be carried out by radiography.

Alloy Selection for Full Scale or Model Component Prototype Demonstration

The first round of investigations confirmed that a number of alloys were potentially suitable for application to steam turbine components. The next stage in the project was the selection of alloys for manufacture of full scale prototypes or of components which realistically simulate full scale components.

Even if the properties of castings are poorer than those of forgings, there are strong drivers for their application to valve chest and turbine casings arising from their lower costs and more flexible geometries. Castings require good weldability for the upgrading process arising from defects in the casting itself, but also due to the requirement to join castings to other components. This is especially true in the AD700 concept where the limited size of nickel alloy castings, coupled with the desire to restrict their application to the highest temperature parts of the turbine, means that welded joints in turbine casings, either between two nickel alloy castings, or between nickel alloy castings and steel castings, is likely to be required. The weldability of alloys used in the solution treated condition is generally high due to the relatively low levels of hardening elements such as Ti and Al. Furthermore the absence of phase changes during the welding process means that these nickel alloys require no PWHT to modify the microstructure in the heat affected zone and that they can tolerate any PWHT that is appropriate to the steel base material in dissimilar joints. Taking into account all available date and experience, Alloys 617 and 625 were selected for further casting development.

Very similar factors drove the selection of alloys for forging development. Again limited size and high cost mean that welded constructions will be necessary, again favouring the solution treated alloys. For these reasons Alloys 617 and 625 were also selected for further forging development. However two additional alloys were also selected. It is unlikely that alloy 718 in its standard form has adequate long term creep strength, but, with the heat treatment modification developed in the first round of investigations, it may provide a lower cost alternative to 617 and 625. Therefore this alloy was also selected for continued development. Finally, in case longer term testing showed unexpected reductions in the creep strength of these first three alloys, a fourth alloy, expected to have even higher strength, alloy 263, was also selected.

The initial round of investigations had also revealed that Waspaloy is an excellent alloy for application to smaller, non-welded components such as blading and bolting.

CASTING MANUFACTURE

Model castings (step blocks with sections up to 200 mm) were successfully manufactured in Alloys 617 and 625 (Table 6). Additional blocks were also cast for the manufacture of similar metal joints. Joints 70 mm in depth and 500 mm long were successfully manufactured and inspected using dye penetrant and radiographic techniques. In the case of both alloys, the joints were manufactured using matching consumables. These castings and welded joints were sectioned and are currently being subjected to an extensive and long term testing programme. Investigation of dissimilar joints is planned later in the project.

Within the last year, a full scale valve chest (delivery weight about 3.5 tonnes) has also been manufactured in alloy 617 (Figure 11). Further work is planned on this casting to confirm its properties are similar to those of the castings already under investigation. A second prototype casting is also being planned.

ROTOR FORGING MANUFACTURE

In both of the principal candidate alloys, 617 and 625, full scale HP rotor forgings 700 mm in diameter have been successfully manufactured. Careful selection and control of heat

Fig. 11 617 valve chest casting after removal from mould and shot penning (courtesy of Goodwin Steel Castings).

Table 6 Details of castings for long term property investigation.

617 and 625 casting geometry: Stepped block 300 mm in length with sections of 50, 100 and 200 mm

617 and 625 casting weight: 750 kg

Heat treatment: 617: 1170°C/water quench
625: 1200°C/water quench

Cast analysis:

Alloy	C	Ni	Fe	Cr	Mo	Co	Nb	Ti	Al	Mn	Si	P	S	Others
617	0.06	Bal.	0.6	21.6	9.2	12.2	<0.01	0.53	1.09	0.27	0.34	<0.01	0.001	N, 0.021 Cu, 0.02
625	0.02	Bal.	0.5	21.9	8.4	0.2	3.59	0.3	0.3	0.16	0.18	0.005	0.002	N, 0.033 Cu, 0.08

treatment parameters is necessary to avoid excessive grain growth which leads to poor ultrasonic inspectability. A first attempt to manufacture the 617 forging was unsuccessful but success was achieved on the second attempt after further heat treatment trials and adoption of a modified heat treatment practice. The final result has been sound forgings in a fully

Fig. 12 617 forging during automatic ultrasonic inspection (courtesy of Saarschmiede).

solution treated condition but with minimum detectable defect sizes over a path length of around 1.5 m of better than 3 mm diameter flat bottomed hole equivalent. These forgings have been sectioned and are currently being tested in a long term test programme. More recently, an even larger forging of diameter around 1000 mm suitable for IP rotor forgings, has been manufactured (Figure 12). These forgings will also be investigated to confirm their properties are similar to those already under investigation.

Model forgings have been manufactured in Alloys 718 and 263, heat treatment parameters being controlled to simulate as far as possible the larger forgings manufactured in Alloys 617 and 263. These forgings have also been sectioned for investigation. Details of all forgings investigated in this testing programme are given in Table 7.

Welding development has focused on the two primary candidates and on the requirement for dissimilar joints. Different welding processes and filler metals have been investigated and an optimum solution has been defined. This involves manufacture of the joint using a narrow-gap TIG process and a nickel based filler metal. Manufacture of a full scale joint is planned during the current year to enable a full investigation of weldment properties.

Candidate Materials for Blading, Bolting and Valve Components

As previously stated, the identification of the optimum alloy selection and manufacturing route for blading and bar products such as bolts received a lower priority in the early stages of the programme. It is clear there are many alloys with properties more than sufficient for

Table 7 Details of forgings for long term property investigation.

Alloy	C	Ni	Fe	Cr	Mo	W	Co	Nb	Ti	Al	B	Zr	Mn	Si	P	S	Others
263	0.052	Bal.	0.34	19.5	5.8		19.5		2.21	0.48	<0.005		0.04	0.09	<0.005	<0.002	Cu, <0.10
617	0.060	Bal.	0.35	22.8	9.0	0.02	11.9	0.003	0.48	1.15	0.0002	0.005	0.02	0.05	0.005	0.001	Cu, 0.013, N, 0.004, V, 0.005
625	0.030	Bal.	2.17	21.9	9.1	0.03	0.02	3.7	0.20	0.19	0.004	0.010	0.01	0.02	0.002	0.001	Cu, 0.012, N, 0.006, V, 0.01
718	0.023	54.1	Bal.	18.8	3.0	<0.01	0.04	5.0	0.93	0.46	0.004		0.01	0.09	0.009	0.0006	

Alloy	Dimensions	Heat Treatment (Hold Temp. C/Hold Time, H/Cooling Medium)		
		Solution Treatment	Ageing 1	Ageing 2
263	600 mm diameter × 100 mm	1150/10/Air	800/8/Air	
617	700 mm diameter × 800 mm	1100/3/Water		
625	725 mm diameter × 1230 mm	1100/6/Water		
718	730 mm diameter × 320 mm	1065/1/Air	760/8/Furnace	650/12/Air

Alloy	Rp0.2	Rm	A	Z
	MPa	MPa	%	%
263	615	912	21	18
617	359	772	48	45
625	363	763	57	52
718	1155	1340	19	21

the requirements of steam turbine applications. However a significant decision remains to be made over the manufacturing route for blading. The conventional route involves machining of blades from bar. However as the value of the alloy and the difficulty of machining rises, then the attraction of a near-net shape process grows. Investment castings have been applied very successfully in gas turbines so the application of this route is currently under investigation.

Bolting materials will inevitably be machined from bar and materials with appropriate properties are already available. Selection of the optimum alloy is likely to depend on the particular requirements of a particular steam turbine design.

Wear resistance is an important issue for valve components and hard-facing is applied in conventional steam turbines. The wear performance of conventional as well as more advanced techniques is currently under investigation.

Investigation of Rotor Forging and Casting Model and Full-Scale Prototypes

The materials manufactured as full scale prototypes or as model components for rotor forgings or castings have been sectioned and are currently under investigation. The objective is to provide a basis for full validation of all allowable stresses used in steam turbine design and to investigate all potential failure mechanisms. In the as-received condition tensile, fracture toughness, long term creep (> 30,000 hours), fatigue crack growth, creep crack initiation and growth, low cycle fatigue, creep-fatigue and steam oxidation properties are all being investigated. To investigate the influence of service exposure on properties, tensile, fracture toughness, fatigue crack growth, low cycle fatigue and creep-fatigue tests are being repeated on material aged for 1000 hours at 650°C.

Weldments are being investigated to define cross-weld, heat affected zone and weld metal properties. Cross-weld tensile, long term creep and low cycle fatigue tests are in progress. The heat affected zone fracture toughness, creep crack initiation and growth and fatigue crack growth properties are under investigation. Finally the weld metal properties are being assessed through fracture toughness, creep crack initiation and growth, fatigue crack growth, low cycle fatigue, creep-fatigue and steam oxidation tests.

This programme of testing has been in progress for around 20,000 hours and its results will be presented in future papers. However the results already available have revealed no unexpected behaviours or parameters which would create major difficulties for steam turbine design.

CONCLUSION

Alloys have been identified which meet the requirements for boilers and steam turbines operating at 700–720°C. Good progress is being made in application of these alloys to a series of prototype components to demonstrate the feasibility of manufacture. A comprehensive materials testing programme has been launched to investigate these prototype components and is addressing all critical properties and potential failure mechanisms. Currently the test programme is confirming the expected properties and has identified no technical obstacles to design and manufacture of boilers and steam turbines.

The results of this project have placed the European power generation industry and its supply chain in a powerful position for exploitation of this technology which on its own has a significant potential for mitigation of carbon emissions. When coupled with emerging technologies for carbon capture and sequestration, this technology can also facilitate the transition to zero-emission coal-fired power generation.

ACKNOWLEDGEMENT

The authors wish to acknowledge the efforts of all of their partners in the AD700 project. There are around 50 partners in the two phases of the project and we regret that space does not allow them all to be mentioned in this paper. The authors also acknowledge the financial support of the European Commission through its Framework IV Thermie programme under contract SF/1001/97/DK and through its Framework V Energy programme under contract ENK5-CT2001-00511, and of the Swiss and UK governments for the support provided to participants from their countries.

REFERENCES

1. S. Kjaer, F. Klauke, R. Vanstone, A. Zeijseink, G. Weissinger, P. Kristensen, J. Meier, R. Blum and K. Weighardt: 'The Advanced Supercritical 700°C Pulverised Coal-Fired Power Plant', *Powergen Europe 2001*, Brussels, 2001.
2. A. Iseda, Y. Sawaragi, H. Teranishi, M. Kubota and Y. Hayase: 'Development of New 12%Cr Steel Tubing (HCM12) for Boiler Application', *The Sumitomo Search*, (40), 1989, 41–56.
3. F. Masuyama, T. Yokoyama, A. Sawaragi and A. Iseda: 'Development of Tungsten Strengthened Low Alloy Steel with Improved Weldability', *Materials for Advanced Power Engineering 1994*, Liege, 1994.
4. W. Bendick and M. Ring: 'Stand der Entwicklung neuer Rohrwerkstoffe für den Kraftwerksbau in Deutschland und Europa', *VGB-Konfernz: Werkstoffe und Schweisstechnik im Kraftwerk 1996*, Cottbus, 1996.
5. N. Henriksen, O. H. Larsen and T. Vilhelmsen: 'Lifetime Evaluation of Evaporator Tubes Exposed to Steam Oxidation, Magnetite deposition, High Temperature Corrosion and Creep in Super Critical Boilers', *Power Plant Chemical Technology 1996*, Kolding, Denmark, 1996.
6. A. Finali, A. Di Gianfrancesco, O. Tassa, L. Cipolla and R. Montani: 'Mechanical and Microstructural Qualification of a Prototype IN718 Forged Disk', *Parsons 2003 6th International Charles Parsons Turbine Conference*, Dublin, 2003.

Factors Influencing the Creep Resistance of Martensitic Alloys for Advanced Power Plant Applications

P. D. CLARKE and P. F. MORRIS
Corus R, D & T
Rotherham, UK

N. CARDINAL and M. J. WORRALL
Corus Engineering Steels
Sheffield, UK

ABSTRACT

Over the last 50 years the rising steam temperatures and pressures needed to improve generating efficiency in fossil fuelled stations have placed increasing demands on the creep and oxidation resistance of boiler and turbine materials. These have been met by the development of martensitic steels with higher chromium levels together with additions such as molybdenum and tungsten resulting in alloys such as Type 91 (9%Cr, 1%Mo) and Type 92 (9%Cr, 2%Mo, 0.5%W). Further proposed increases in operating temperature to 650°C and in the longer term to 700°C mean that alloys with superior performance to these are required.

Greater resistance to oxidation can most easily be obtained by further increases in the chromium content. At levels above about 10% however it becomes ever more difficult to obtain a fully martensitic microstructure due to the formation of delta ferrite. This results in loss of strength and reduces the hot ductility which can give problems during hot working operations such as tube making. It can also influence weldability. The higher chromium levels must thus be balanced by austenite stabilising additions such as copper, nickel and cobalt.

As part of the COST 522 programme, work has been carried out to attempt to improve the creep properties of martensitic alloys. The paper reviews briefly the design of creep resistant 9–12%Cr steels and presents initial results on the mechanical properties and creep rupture properties of a martensitic steel composition designed to have creep properties superior to those of Steel 92. For short term creep data, currently for durations up to 8,000 hours, rupture lives in excess of Steel 92 are being achieved for testing at 600–650°C. Testing is continuing and data for durations out to 10,000 hours should be available by the time of the conference.

INTRODUCTION AND BACKGROUND

The move towards higher efficiency power plant has required the development of materials suitable for use under more demanding operating conditions. Over recent years this has led to several international work programmes aimed at developing materials capable of operating at higher steam temperatures and pressures for which improvement of creep strength and oxidation resistance are recognised as essential.

One family of steels that has received considerable attention is the 9–12%Cr martensitic types. These offer greater flexibility of operation for thick section boiler components over more creep/oxidation resistant austenitic steels due to their lower coefficients of thermal expansion and better thermal conductivity. They are also cost effective as a result of the lower alloy content, particularly of expensive additions such as nickel.

Recent developments have included grades such as Steel 92 containing 9%Cr which gives a 100 MPa/10^5 hours rupture life at temperatures approaching 620°C, the highest among commercially available ferritic grades. To achieve enhanced steam oxidation resistance chromium contents above 9% are required. A project in the COST 522 collaborative programme has sought to establish a martensitic 11–12%Cr base composition with better creep strength than Steel 92. The aim of this paper is to briefly describe the design of such an alloy and to present interim test results.

DESIGN OF DEVELOPMENT COMPOSITION

The base composition of 9%Cr-1%Mo has since the 1930's been widely used in coal-fired power plant. This produces a fully martensitic microstructure in air cooled thick walled sections after a conventional austenitising and tempering treatment. It also provides higher creep strength and better oxidation resistance than lower chromium steels, such as the 2¼%Cr-1%Mo grade.

Extensive work in the USA during the mid-1980's led to the development of Steel 91. This involved the addition of Nb and V to the 9%Cr-1%Mo base together with increases in nitrogen and boron producing small, stable precipitates which are important for good creep strength.

More recently the Japanese and Europeans have developed Steel 92 (NF616) and E911 respectively. Both these alloys rely on the addition of tungsten (and reduction of Mo in the case of Steel 92) in relation to the Steel 91 composition to improve creep strength. A further Japanese development was Steel 122 (HCM12A). This contains 11%Cr for superior corrosion resistance and a copper addition for the suppression of delta ferrite which can give toughness and hot workability problems. The compositions of these alloys are shown in Table 1.[1–3]

The temperature for a 100 MPa/10^5 hour creep life for a selection of these alloys is shown in Figure 1.[4] Steel 92 demonstrates the best performance in terms of creep strength and for this reason is recognised as the best of the 9-12%Cr grades currently available.

In order to improve the creep and oxidation performance offered by Steel 92 a development programme was initiated through COST 522. An 11%Cr base was chosen for enhanced oxidation resistance. The following factors were then considered:

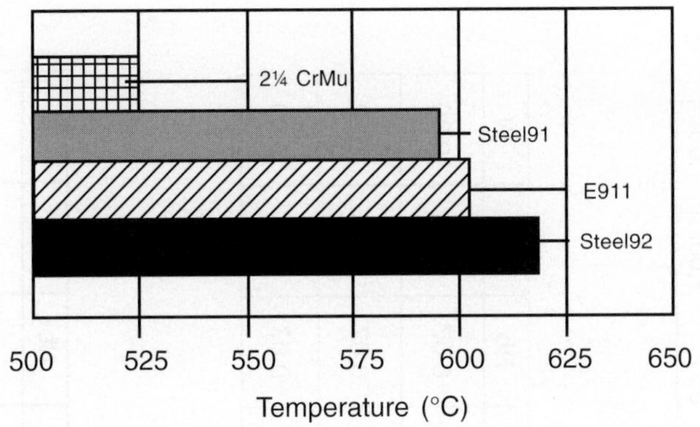

Fig. 1 Maximum temperature in °C for 100,000 hours creep rupture strength of 100 MPa.

- Avoid suppression of the α/γ transition temperature to permit the use of the maximum tempering temperature.[5]
- Obtain microstructural stability through use of low levels of residual elements to reduce coarsening rates of precipitates.[6, 7]
- Use Mo and W to provide rupture strength by solid solution strengthening and Laves phase precipitation.
- Use high W to Mo ratio to optimise the long term stability of Laves phase precipitates.[8]

This led to the design of three development alloys. The first alloy contained lower residual levels (Si, Mn, Ni and Al) than found in conventional 9–12%Cr grades. The level of Mo was set at 1%, W at 0.30% and Co at 2.0% the latter to help suppress delta ferrite. The second alloy is of the same base composition but with Si, Mn, Ni and Al levels more typical of those which can be readily achieved in large scale electric arc steel making. The third variant contained higher W (2.0%) and lower Mo (0.50%) levels than the other two.

The nominal compositions of the development alloys are shown in Table 2.

PRODUCTION OF MATERIAL AND ASSESSMENT PROGRAMME

Experimental Manufacture

A 50 kg air induction melt was produced for each of the development alloys with the aim compositions given in Table 2. Product analyses are shown in Table 3.

Chemical analyses for the alloys were generally close to the respective aims but the levels of Si, Mn, Ni and Al were slightly above the aim in the low residual material. The levels of Si, Mn and Ni were, nevertheless, much lower than found in normal production material and so were considered suitable to study the effect of low residual contents on high temperature properties.

Table 1 Compositions of 9–12%Cr martensite alloys for use in advanced power plant.

Alloy	C	Si	Mn	Cr	Mo	Ni	Al	B	N	Nb	V	W	Cu
Steel 91	0.10	0.35	0.45	9.0	1.00	0.20	0.020	-	0.050	0.08	0.22	-	-
Steel 92	0.10	0.25	0.45	9.0	0.50	0.20	0.020	0.004	0.050	0.07	0.20	1.80	-
E911	0.11	0.30	0.45	9.0	1.00	0.20	0.010	0.003	0.070	0.08	0.22	1.00	-
Steel 122	0.11	0.25	0.35	11.0	0.40	0.25	0.020	0.003	0.070	0.07	0.22	2.00	1.00

Table 2 Nominal compositions of development alloys (wt.%)

Alloy	C	Si	Mn	Cr	Mo	Ni	Al	B	Co	N	Nb	V	W
Low Residual	0.17	0.10	0.05	11.0	1.00	0.09	0.002	0.01	2.00	0.05	0.07	0.26	0.30
Normal Residual	0.15	0.30	0.20	11.0	1.00	0.40	0.005	0.01	2.00	0.05	0.07	0.26	0.30
High W + Low Mo	0.15	0.30	0.20	11.0	0.50	0.40	0.005	0.01	2.00	0.05	0.07	0.26	2.00

Table 3 Chemical analysis of experimental materials (wt.%)

Alloy	C	Si	Mn	Cr	Mo	Ni	Al	B	Co	N	Nb	V	W
Low Residual	0.16	0.15	0.09	11.0	1.04	0.14	0.008	0.009	2.12	0.059	0.065	0.26	0.32
Normal Residual	0.16	0.30	0.22	11.1	1.04	0.43	0.007	0.011	2.16	0.060	0.067	0.27	0.32
High W + Low Mo	0.17	0.35	0.22	11.2	0.50	0.44	0.007	0.009	2.15	0.063	0.068	0.26	2.10

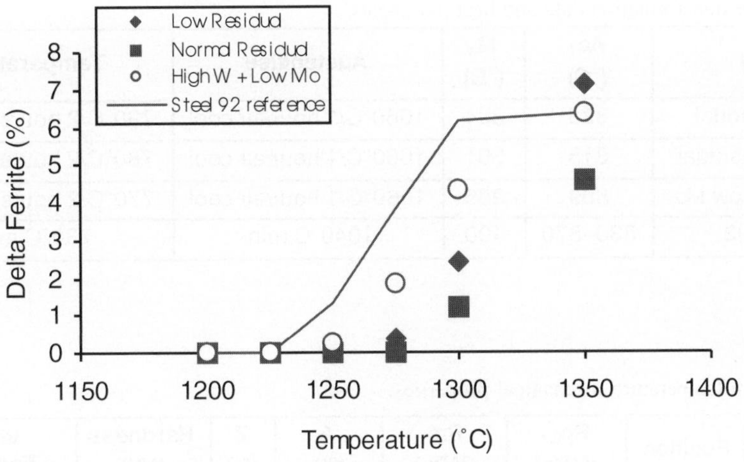

Fig. 2 Delta ferrite as a function of temperature.

The ingots were forged into 50 mm square bar and then rolled to 19 mm diameter round bar for the production of test samples. The final product, both before and after heat treatment, was subjected to ultrasonic and alternating current potential drop test techniques to ensure that only sound material was used for the manufacture of test specimens.

The tendency to form delta ferrite on heating was determined metallographically from 40 mm long bar samples held at selected test temperatures between 1,200–1,350°C for 30 minutes, followed by water quenching. Volume fractions of delta ferrite were then established by point counting. Reference values for Steel 92 were also evaluated using the same technique for an experimental cast made to the composition range shown in Table 1. The area fractions of delta ferrite observed were less than 8% in all cases. Significant levels (>1%) of delta ferrite were observed in the Steel 92 at 1,250°C and above, Figure 2. For the low and normal residual materials >1% delta ferrite was observed at 1,300°C and above and for the Mo/W alloy at 1,275°C and above. Except at 1,350°C, which is well above reheating temperatures for hot working, where delta ferrite was observed, the levels in the experimental alloys were below those found in the Steel 92 for any given temperature.

Heat Treatment

A solution treatment temperature of 1,060°C was selected, being typical of that used for 9–12%Cr martensitic steels.[3, 9] Phase transformation temperatures determined using dilatometry are shown in Table 4. For determination of α/γ transition temperatures (Ac_1's), samples were heated at a rate of 5°C min^{-1}. These were used to allow the maximum tempering temperature to be used without reaustenitisation. The measured Ac_1 temperatures reflected the alloy content ranging from 823°C for the low residual material to 809°C for the Mo/W

Table 4 Phase transformation data and heat treatments.

Alloy	Ac_1 (°C)	M_s (°C)	Austenitise	Temperature
Low Residual	823	324	1060°C/1 hour/air cool	790°C/2 hours/air cool
Normal Residual	815	301	1060°C/1 hour/air cool	780°C/2 hours/air cool
High W + Low Mo	809	285	1060°C/1 hour/air cool	770°C/2 hours/air cool
Steel 92	830–870	400	1040°C min.	730°C min

Table 5 Room temperature mechanical properties.

Alloy	Position	$Rp_{0.2}$ (MPa)	Rm (MPa)	A (%)	Z (%)	Hardness (HV)	Impact Energy (J)
Low Residual	Long.	528	756	22	67	235	163
Normal Residual	Long.	568	786	24	67	258	144
High W + Low Mo	Long.	620	837	22	63	266	96
Steel 91	-	415 min.	585 min.	20 min.	-	265 max.	-
Steel 92	-	440 min.	620 min.	20 min.	-	265 max.	-
Steel 122	-	400 min.	620 min.	20 min.	-	265 max.	-
E911	-	440 min.	620-850	20 min.	-	-	-

alloy. Tempering temperatures about 30°C below these were used in subsequent heat treatments.

For M_s determinations, samples were heated at 200°C min^{-1} to 1,060°C, held for 5 minutes and then cooled at 20°C min^{-1}. These again reflected the alloy content of the steels ranging from 324°C for the low residual alloy to 285°C for the Mo/W alloy. Data for Steel 92 is included in the table for comparison.[9]

Room Temperature Mechanical Properties

Room temperature tensile, hardness and Charpy impact data for samples heat treated according to the cycles in Table 4 are shown in Table 5. As would be expected, the strength increased with alloy content and the Charpy impact toughness decreased. There was no significant difference in ductility between the three development alloys.

Reference data included in the table show that test values for the development alloys exceeded minimum strength and ductility requirements for current 9–12%Cr martensitic grades.[1–3]

Stress Rupture Testing

Stress rupture tests were carried out at 600, 625 and 650°C at stresses designed to give aim lives of 1,000, 3,000, 10,000 and 30,000 hours based on data for Steel 92.[4] At each temperature, tests with aim lives of 1,000 and 3,000 hours have been completed whilst those with projected lives of 10,000 and 30,000 hours are still in progress. The status of the test programme as of March 2003 is shown in Figure 3.

At 600°C and aim durations up to 3,000 hours, the performance of the development alloys was superior to Steel 92 with the Mo/W alloy giving lives up to 4 times those expected for Steel 92. Two tests on each alloy with aim durations up to 30,000 hours had been running for approximately 3,220 hours by the beginning of March 2003.

At 625°C the low and normal residual alloys gave failure lives below those expected for Steel 92. The Mo/W alloy again gave lives well in excess of those expected apart from a test at 125 MPa with an aim life of 10,000 hours which failed after 7,749 hours. Close examination of the failure showed that it was almost certainly premature as it had occurred at a machining defect in the sample. The result has been excluded from Figure 3.

At 650°C the rupture lives given by the development alloys fell below the mean line for Steel 92 although the 1,000 hour aim test for the Mo/W alloy failed after 1,418 hours.

The data clearly demonstrate the beneficial effects of the higher tungsten in the Mo/W alloy. There is also evidence that the use of low residuals does produce an improvement in creep rupture life compared with the higher residual material however, the effect is much smaller than that produced by the tungsten. The limited data are not totally consistent. At 650°C where a direct comparison can only be made for the 110 MPa tests the low residual material failed after 1,244 hours compared with 2,068 hours for the higher residual material.

Steam Oxidation Testing

Steam oxidation tests were carried out on samples from the development alloys together with a sample of commercial Steel 92 in the same furnace. Samples of dimensions 20 × 10 × 3 mm were prepared by spark erosion and tested at 650°C for times up to 5,000 hours. Samples were removed from the furnace and the weight change measured at several points during the test.

Results of the steam oxidation tests are shown in Figure 4. The development alloys with chromium levels of 11% all showed a noticeably greater oxidation resistance than commercial P92 with 9%Cr. After around 5000 hours, weight gains for the development alloys were 0.17 mg cm^{-2} (normal residual), 0.20 mg cm^{-2} (Mo/W) and 0.62 mg cm^{-2} (low residual) which compared to 25.8 mg cm^{-2} for P92.

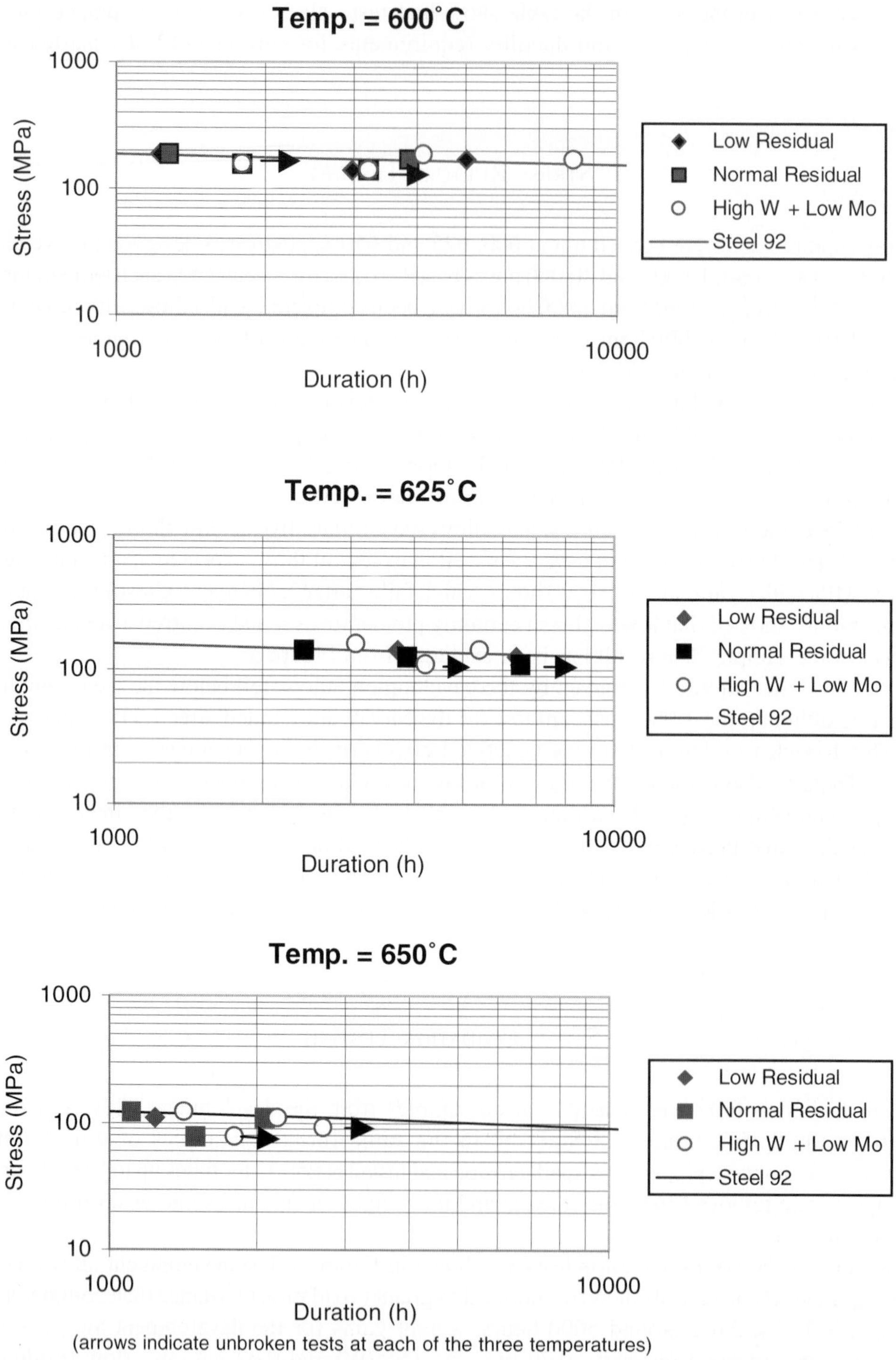

Fig. 3 Creep rupture test values compared with data for steel 92.

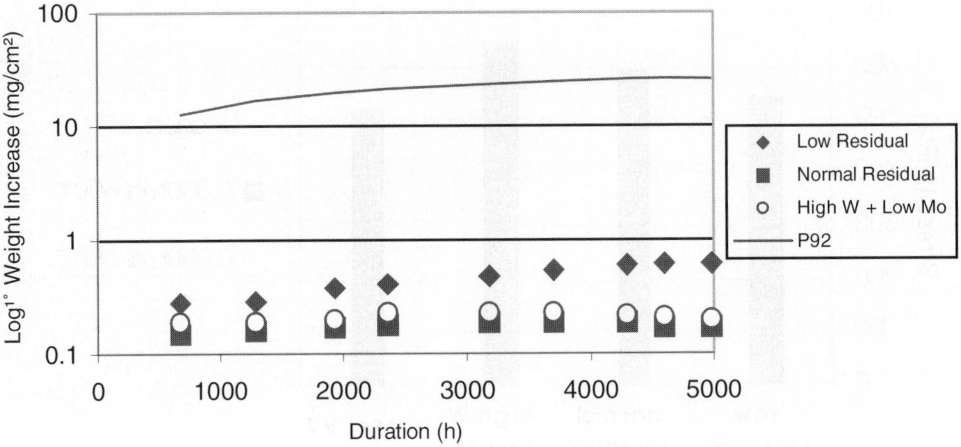

Fig. 4 Steam oxidation test data at 650°C for development alloys and P92.

Long Term Exposure Testing

Tensile and Charpy impact tests were carried out on materials aged for up to 10000 hours at 600 and 650°C covering the likely service range for these alloys.

$Rp_{0.2}$ and Charpy impact data at room temperature are shown in Figures 5a and b together with comparative data for Steel 92.[10] For the three development alloys there was a small increase in the $Rp_{0.2}$ value after aging for 10,000 hours at 600°C. After aging for the same time at 650°C there was a small drop of the $Rp_{0.2}$ value compared with the unaged condition. Data for Steel 92 showed little change in $Rp_{0.2}$ on aging at 600°C, and a small drop after aging at 650°C. In all cases the changes were small, at 35 MPa or below.

All three development alloys and the Steel 92 showed a rapid fall in room temperature Charpy impact energy on aging. The fall was greater after aging at 650°C than at 600°C. After 10,000 hours at 650°C the absorbed energies of the Steel 92 and the development alloys were similar, in the range 32 to 35 J. The absorbed energy of the low residual material was slightly higher, at 46 J.

DISCUSSION

The purpose of the work was to develop a martensitic alloy containing 11-12%Cr with improved oxidation resistance and a creep rupture strength equal to or better than Steel 92 at temperatures up to 625°C. A base alloy with 11%Cr, 1.0Mo, 0.3%W, 2.0%Co was studied. Two approaches were taken to improve rupture strength, the use of low residual levels and

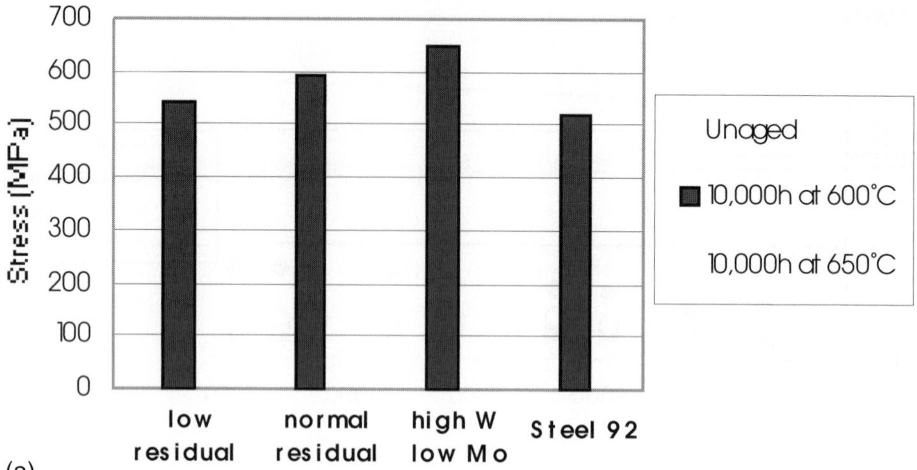

(a)

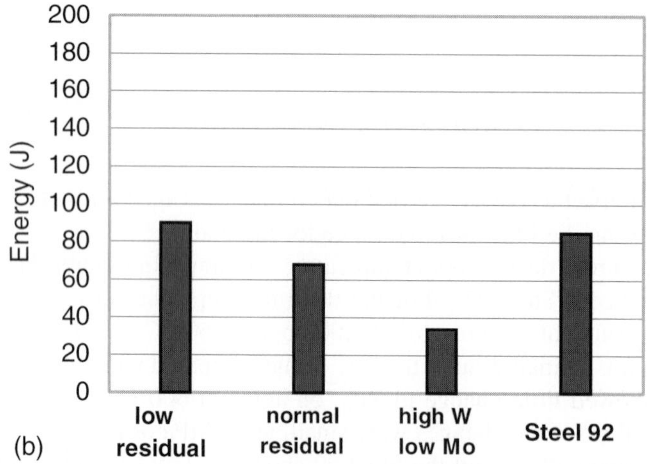

(b)

Fig. 5 (a-b) $Rp_{0.2}$ and Charpy impact toughness at room temperature.

the use of 2.0%W with a corresponding reduction in Mo to 0.5%. The latter produced a significant improvement in rupture strength.

Tungsten and molybdenum improve rupture strength both by solid solution strengthening and by the precipitation of Laves phase, $Fe_2(Mo,W)$. The long term stability at temperature of Laves phase is enhanced at higher concentrations of tungsten.[8] Rupture values available to date are short term, the maximum rupture life for the Mo/W alloy being 8,000 hours,

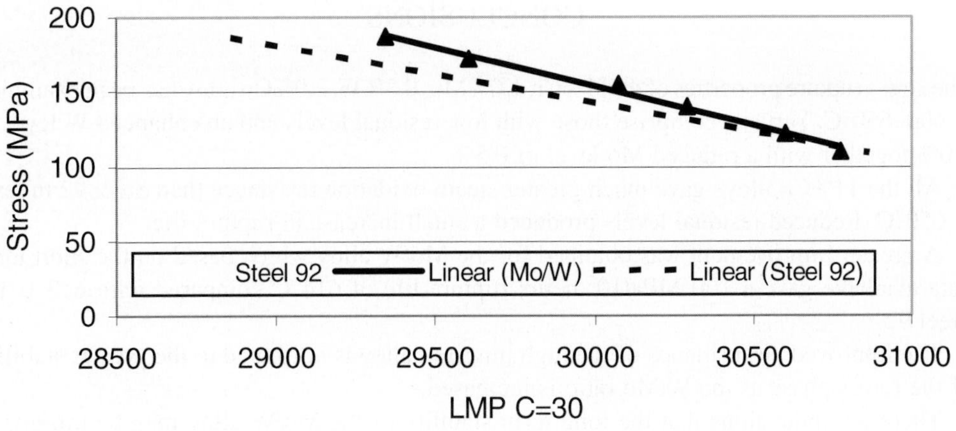

Fig. 6 Comparison of rupture data for the Mo/W alloy with Steel 92.

although longer term tests are in progress. The rupture data for the Mo/W alloy, and Steel 92 data taken from the ECCC data sheet,[4] have been fitted using a Larsen-Miller Parameter (LMP), Figure 6. Both data sets fitted well for a C value of 30.

At low LMP values the Mo/W alloy gave better performance than P92, however the rupture stress decreased slightly faster as the LMP values increased. The two crossed at an LMP value of about 30800. Based on these data the 100 MPa, 10^5 hours rupture life occurs at 618 and 614°C for the Steel 92 and Mo/W alloy respectively. (N.B. The value obtained using the more complex relationship in the ECCC datasheet for P92 is 617°C.). Hence the performance of the Mo/W alloy is similar to that of Steel 92 for temperatures up to about 615°C. The steeper gradient of the Mo/W line gives reason to be cautious about the long term rupture properties. The data in the present paper and from other sources indicate that the use of cobalt in conjunction with molybdenum and tungsten leads to reduced long term microstructural stability. This is almost certainly due to the influence of cobalt in reducing the solubility of molybdenum and tungsten, thus accelerating precipitation kinetics of Laves phase.[11] The long term stability of these 12%Cr steels is also limited by the formation of Z-phase, (Nb,Cr)N.[12]

Thus, based on the short term data in the current work the Mo/W alloy produces similar rupture performance to Steel 92 at temperatures up to about 615°C for lives up to about 10^5 hours. However the higher chromium content in the Mo/W alloy gives a significant improvement in steam oxidation resistance, Figure 4.

A small improvement in creep rupture performance was obtained by the use of very low residual levels, but the effect was small compared with that of tungsten. The use of lower residual levels, particularly manganese and nickel, increased the tendency to form delta ferrite on heating which could potentially present problems in hot working operations such as tube making.

CONCLUSIONS

The stress rupture properties of an 11%Cr, 1.0%Mo, 0.3%W, 2.0%Co alloy has been examined at 600–650°C. Variants comprise those with low residual levels and an enhanced W level of 2.0% together with a reduced Mo level of 0.5%.

All the 11%Cr alloys gave much greater steam oxidation resistance than Steel 92 in tests at 650°C. Reduced residual levels produced a small increase in rupture life.

A greater improvement was obtained for the Mo/W alloy which based on the short term data available gave a 100 MPa/10^5 hours rupture life of 614°C compared with 617°C for Steel 92.

The improved performance of the high tungsten alloy is attributed to the greater stability of the Laves phase as the W:Mo ratio is increased.

There are indications that the long term stability of the Mo/W alloy may be inferior to Steel 92 and this is probably due to the presence of cobalt which reduces the solubility of molybdenum and tungsten, thereby accelerating the precipitation kinetics of Laves phase.

ACKNOWLEDGEMENTS

The authors would like to thank the Management Committee of the COST 522 action for their support and national funding bodies for their financial contributions. Assistance in designing the base alloys was received from Technical University of Denmark and Alstom Power, Baden, Switzerland. Steam oxidation testing was carried out by Alstom Power, Mannheim, Germany. Creep rupture testing was carried out by CESI and CSM, Italy, and Siempelkamp, Germany.

REFERENCES

1. ASTM A213/A213M–99A, 'Standard Specification for Seamless Ferritic and Austenitic Alloy-Steel Boiler, Superheater, and Heat-Exchanger Tubes', 1999.
2. ASTM A335/A335M–99, 'Standard Specification for Seamless Ferritic Alloy-Steel Pipe for High-Temperature Service', 1999.
3. Data Package for Grade 911 Ferritic Steel 9%Cr–1%Mo-1%W, Prepared by a Working Group on Behalf of the European COST 501 Programme on Advanced Boiler Materials, 1998.
4. European Creep Collaborative Committee, ECCC Data Sheets, 1999.
5. F. Masuyama: 'New Developments in Steels for Power Generation Boilers', *Pre-Print of the International Conference on Advanced Heat Resistant Steels for Power Generation*, 1998.
6. A. Strang and V. Vodarek: 'Microstructural Stability of Creep Resistant Martensitic 12%Cr Steels', *Proceedings of the International Conference on Microstructural Stability of Creep Resistant Alloys for High Temperature Plant Applications*, A. Strang, ed., The Institute of Materials, 1998, 117–133.

7. V. Foldyna and Z. Kubon: 'Consideration of the Role of Nb, Al and Trace Elements in Creep Resistance and Embrittlement Susceptibility of 9-12%Cr Steel', *Proceedings of the International Conference on Performance of Bolting Materials in High Temperature Plant Applications*, A. Strang, ed., The Institute of Materials, 1995, 175–187.
8. J. Hald and Z. Kubon: 'Thermodynamic Prediction and Experimental Verification of Microstructure of Chromium Steels', *Proceedings of the IX International Symposium Creep Resistant Metallic Materials*, Hradec and Moravici, 1996.
9. Data Package for NF616 Ferritic Steel (9Cr-0.5Mo-1.8W-Nb-V), Second Edition, Published by Nippon Steel Corporation, 1994.
10. J. Hald: 'Materials Comparisons Between NF616, HCM12A and TB12M – III: Microstructural Stability and Ageing', *Proceedings of the EPRI/National Power Conference on New Steels for Advanced Plant up to 620°C*, E. Metcalfe, ed., EPRI, 1995, 152–173.
11. S. H. Ryu, Jin Yu and B. S. Ku: 'Effects of Alloying Elements on the Creep Rupture Strength of 9-12%Cr Steels', *Proceedings of the Fifth International Charles Parsons Turbine Conference,* A. Strang, et al., eds., The Institute of Materials, 2000, 472–484.
12. K. Kimura et al.: 'Microstructural Change and Degradation Behaviour of 9Cr-1Mo-V-Nb Steel in the Long-Term', *Proceedings of the Fifth International Charles Parsons Turbine Conference,* A. Strang, et al., eds., The Institute of Materials, 2000, 590–602.

An Affordable Creep-Resistant Nickel-Base Alloy for Power Plant

F. TANCRET
Laboratoire Génie des Matériaux
Polytech' Nantes
France

H. K. D. H. BHADESHIA
Department of Materials Science and Metallurgy
University of Cambridge, Pembroke Street
Cambridge CB2 3QZ
UK

INTRODUCTION

Future fossil fuel power plant may in the future operate with steam temperatures as high as 750°C. This would enable an increase in the thermodynamic-cycle efficiency, save fuel and therefore reduce polluting emissions. Nickel-base superalloys are the prime candidates for these relatively high temperatures, but commercially available alloys remain too expensive for large scale power plant applications. With this in mind, we have designed a new affordable nickel-base superalloy, with an expected creep rupture life of 100,000 hours at 750°C under 100 MPa. The design requirements also included forgeability, weldability, corrosion resistance, and microstructural stability over long exposure at service temperature. This procedure resulted in considerable design time and costs savings compared to the usual 'try-and-test' methods.[1–3]

The design procedure included the use of multiparameter non-linear regression techniques (Gaussian processes), coupled with appropriate databases, to model mechanical properties. Phase diagram calculations were used to study microstructural stability and the tendency for chemical segregation during solidification. These methods and alloy-design concepts are briefly reviewed in the first part of this paper.

The new alloy has composition Ni-20Cr-3.5W-2.3Al-2.1Ti-5Fe-0.4Si-0.07C-0.005B (wt.%). It has been fabricated and examined to reveal its microstructure and elevated temperature mechanical properties, including creep data over a period in excess 4200 hours. These experiments are summarised in the second part of this paper.

It is important also to investigate the ability to process the alloy and to optimise fabrication route. Consequently, some aspects of chemical segregation during solidification have been investigated, bearing in mind the requirements for forging and welding operations. The work also includes the modelling and characterisation of the ageing characteristics of the microstructure, which is important for heat treatment optimisation.

MODELLING TECHNIQUES AND ALLOY DESIGN PROCEDURE

The number of variables which affect the performance of an industrial alloy is very large. Given the engineering parameters, this complexity makes it expensive, both in terms of time and money, to achieve the optimum design by experience alone. With multivariate problems it is difficult to conceive of all possible interactions between the variables. Recent papers[4, 5] have demonstrated the power of Gaussian processes used to model precisely such problems. The method involves a non-linear multi-dimensional regression of an output (e.g. a mechanical property) as a function of many inputs (composition, thermomechanical treatments, temperature, etc.). The method works best with a large experimental database covering a wide range of alloy compositions and test conditions. The details are explained elsewhere,[6] but once developed, it can be used to make predictions and the associated uncertainties. The latter include both a noise in the output and the uncertainty with which the model can fit the training data.

Using the Gaussian processes, quantitative models have been produced for nickel-based superalloys, enabling the estimation of the yield stress (*YS*), the ultimate tensile stress (*UTS*), tensile ductility, creep rupture stress (*CRS*), and γ/γ' lattice misfit; all of these as a function of the detailed chemical composition, thermomechanical treatment and test conditions.[1–3, 7]

It is also necessary in the design process to assess the tendency to form undesirable phases (e.g. σ and μ) which might be detrimental to the mechanical properties.[8] Phase diagram and microsegregation simulations were therefore carried out using Thermo-Calc,[9] and a proprietary thermodynamic database developed by Rolls-Royce plc. and Thermotec Ltd.

The engineering requirements include a creep rupture life of 100,000 hours at 750°C under a stress of 100 MPa. The *UTS* to *YS* ratio should be as high as possible, with a minimum of 1.3 at room temperature. The alloy must be easily forgeable and workable, i.e. it must be free of γ' for all temperatures within about 200 K of melting. Furthermore, the quantity of γ' must be less than about 25% in volume in order to ensure weldability[10] and ductility. The alloy must also resist corrosive environments.

An essential priority in this work was to minimise the cost relative to existing alloys. Expensive elements such as Co, Mo, Ta, Nb, Hf and Re... must therefore be avoided. Consequently, the proposed alloy has the following characteristics:

- A high Cr content, typically 20 wt.%, for high temperature corrosion resistance.
- Al and Ti to form γ' for precipitation strengthening.
- W for solid-solution strengthening of both the γ and γ' phases.
- C, to precipitate carbides which limit grain boundary sliding. Its concentration must nevertheless be small enough to avoid the formation of grain boundary carbide films, which are detrimental to creep resistance.[11] In this respect, 0.07 wt.% seemed an appropriate compromise.
- B segregates to boundaries and limits their sliding, at concentrations of ~0.005 wt.%.[9]
- About 5 wt.% of Fe, to reduce cost. Pure Cr is an expensive raw material, so ferrochrome or industrial scrap, which are both cheaper sources of Cr, are used instead.
- Si, at a concentration of about 0.4 wt.% helps deoxidation during melting.

Having fixed the contents of Cr, Fe, C, B and Si, predictions were made with various amounts of W, Al and Ti. The latter were adjusted to find a compromise between a high

Table 1 Nominal and actual chemical composition of the new alloy (in wt.%).

	Cr	W	Al	Ti	Fe	Si	C	B
Nominal	20	3.5	2.3	2.1	5	0.4	0.07	0.005
Actual	20.28	3.308	2.22	1.978	5.025	0.053	0.102	0.0035

creep rupture resistance, a low γ' volume fraction, a low γ/γ' lattice misfit, whilst ensuring the absence of undesirable equilibrium phases at the service temperature (750°C). The final composition arrived at is:

Ni-20Cr-3.5W-2.3Al-2.1Ti-5Fe-0.4Si-0.07C-0.005B, wt.%.

with a heat treatment of 4 hours at 1175°C, 4 hours at 935°C and 24 hours at 760°C, with air cooling in all cases.

Plots of the calculated molar percentages of phases in this alloy are presented in Figure 1, as a function of temperature. The calculated liquidus, solidus, and γ' solvus are important for processing (forging and heat treatment). The expected γ' volume fraction at the service temperature of 750°C is 26 mol.%, which matches the criterion of a weldable superalloy. Figure 1 also indicates the presence of $M_{23}C_6$ carbides at service temperature. These carbides generally precipitate at γ grain boundaries and improve creep resistance. On the bottom scale of Figure 1, it can be seen that 'TiB_2'-type borides are expected to form at high temperatures, and to dissolve at lower temperatures. However, the calculations represent equilibrium between bulk phases and do not take grain boundaries into account. Boron segregates at grain boundaries and this may affect the precipitation of borides, which must be characterised experimentally. Equilibrium also indicates a body-centered-cubic α-Cr phase below 690°C. This phase, even if similarly predicted in some chromium-rich commercial nickel-base superalloys, e.g. Nimonic 81, has not in practice been reported to form. Additionally, the calculated formation temperature range of this α-Cr phase in the designed alloy is below 690°C, much less than the expected service temperature. Finally, and this is one of the most important points, no other undesirable phases are expected to form in the vicinity of the service temperature. Phases like σ, η, or μ are detrimental for high-temperature mechanical properties, in particular for creep resistance.[12]

MICROSTRUCTURE AND PROPERTIES

Two 20 kg ingots of the designed alloy were cast by Special Metals, and rolled into 30 mm diameter cylindrical bars, one of which was investigated in this study. The composition of the melt, from chemical analysis, is given in Table 1. It is rather close to nominal composition except for silicon, but this should not significantly affect the mechanical properties. The strengthening elements, Al, Ti and W are a bit lower than target; carbon and chromium are slightly higher. Samples were cut out from the bar and heat-treated. In what follows, the samples subjected to the designed heat treatment are referred to as being 'fully heat-treated'.

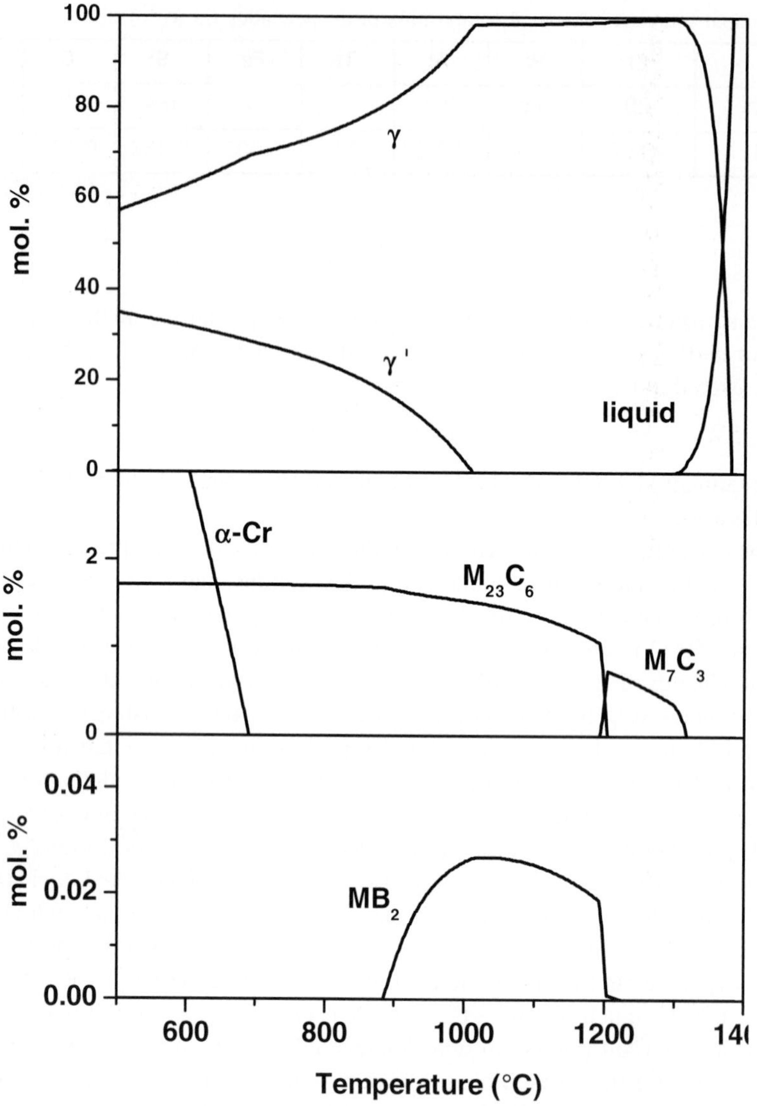

Fig. 1 Calculated equilibrium phase diagram of the designed alloy.

A typical transmission electron microscope (TEM) micrograph of the fully heat-treated alloy is presented in Figure 2. Spherical γ' precipitates of about 100 nm diameter can be seen within the γ grains. The grain boundaries are, as expected, decorated with $M_{23}C_6$ carbides which are there to retard grain boundary sliding.

The compression 0.2% proof stress of the fully heat-treated material was measured on cylindrical specimens of 20 mm length and 8 mm diameter. The results are presented as a

Fig. 2 TEM micrograph of the fully heat-treated alloy.

function of temperature in Figure 3, and compared to the values predicted using the Gaussian processes model. Very good agreement is obtained between the measured and calculated values over the whole temperature range with all measured values within the 65% confidence limits.

Creep tests were performed at 750°C on standard specimens with a gauge length of 25.4 mm and a diameter of 5.64 mm, for stresses of 320, 290, 260, 230 and 200 MPa. The results are plotted in Figure 4, along with the Gaussian processes predictions. There clearly is excellent agreement between the measured and estimated data, giving confidence in the statement that the alloy is likely to meet the intended engineering target of 100,000 hours under 100 MPa at 750°C. The results also validate the design procedure.

PROCESSING ISSUES

Solidification Segregation

In order to process the material safely and efficiently during casting, forging, heat treatment or welding, it is important to ensure that the alloy is able to cope with inevitable chemical segregation induced by non-equilibrium solidification. The segregation can lead to incipient melting which in turn can cause hot cracking during forging and/or welding.

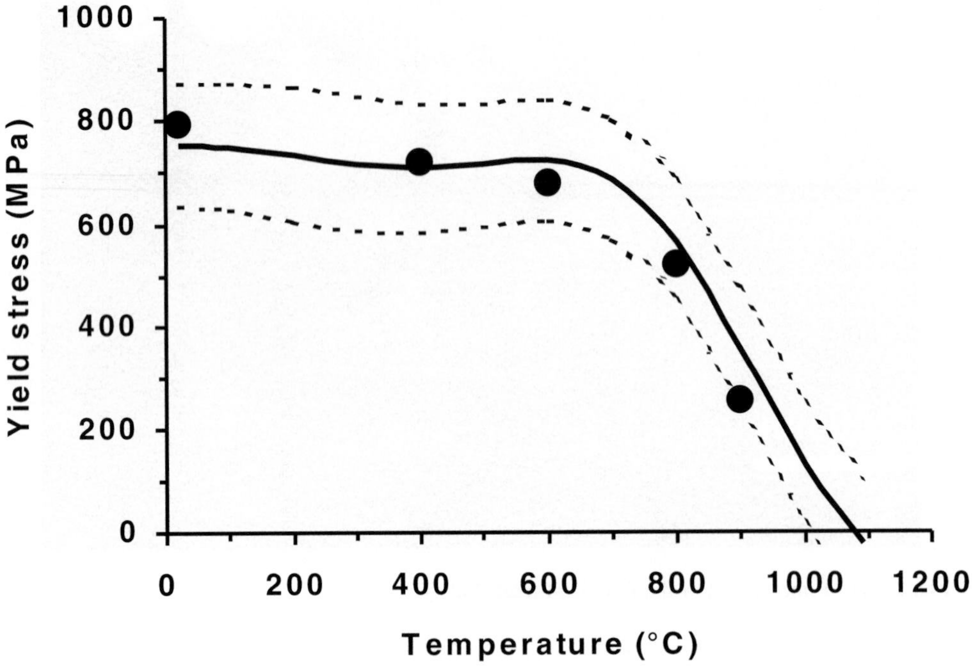

Fig. 3 Evolution of yield stress as a function of temperature. Circles: measurements; solid line: mean Gaussian processes predictions; dotted lines: predicted error bounds.

A special module has been written within the Thermo-Calc software and used to simulate microsegregation during solidification, based on Scheil's approximation. The latter assumes a homogeneous liquid and nil diffusion in any solid. Starting from the liquid, the temperature is reduced in 1 K intervals, and at each step a new liquid-solid equilibrium is calculated, the amount and composition of the liquid being kept as inputs fort the subsequent step. The total amount of solid is the sum of the increments at each step, thus allowing the evolution of the liquid composition to be followed, to predict the composition gradients within dendrites and the primary segregation behaviour. Some of these results have already been presented,[2] but we will focus here on the dendrite concentration profiles. Three simplified dendrite-shapes were investigated: a plate, a cylinder and a sphere. At each computation step, the solidified volume, ΔV, and its composition are calculated. For the plate, cylinder and sphere models, respectively, the increase in dendrite thickness or radius, Δr, can be calculated with:

$$\Delta r_{plate} \propto \Delta V \qquad \Delta r_{cylinder} \propto \frac{\Delta V}{r} \qquad \Delta r_{sphere} \propto \frac{\Delta V}{r^2}$$

This has been related to the computed chemical composition of the solidifying material, and compared to actual concentration profiles in dendrite arms. Measurements have been made using energy dispersive spectroscopy (EDS) within a scanning electron microscope

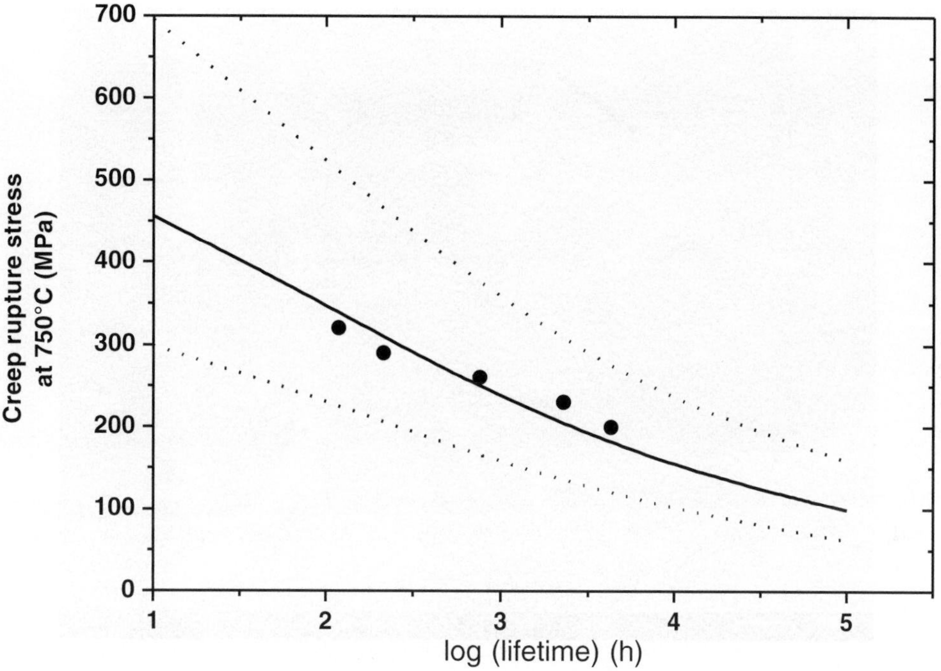

Fig. 4 Creep rupture stress as a function of lifetime. Circles: measurements; solid line: mean Gaussian processes predictions; dotted lines: predicted error bounds.

(SEM). A typical dendrite micrograph is presented in Figure 5, with the scan path from the core to the tip of the dendrite arm. A result is presented in Figure 6 in the case of the titanium concentration profile. There is a rather good agreement between the spherical dendrite model and measurements, which validates the modelling approach.

Precipitation Ageing Kinetics

One major issue to optimise mechanical properties of Ni-base superalloys is the simplest heat treatment necessary to achieve the desired γ' precipitate size, distribution and volume fraction. A three-step heat treatment, including two precipitation ageing treatments, was chosen. It may be possible that identical mechanical properties can be obtained with a single step ageing, but to implement this requires a deeper understanding of γ' precipitation kinetics. In a preliminary investigation we have combined several modelling approaches: a model for diffusion-controlled isothermal growth from a supersaturated solid solution, a thermodynamical simulation software, and a precipitation strengthening model.

First, the material is divided into elementary volumes, each containing one growing precipitate. Before precipitation, the concentration of the diffusive species (Al + Ti) is the

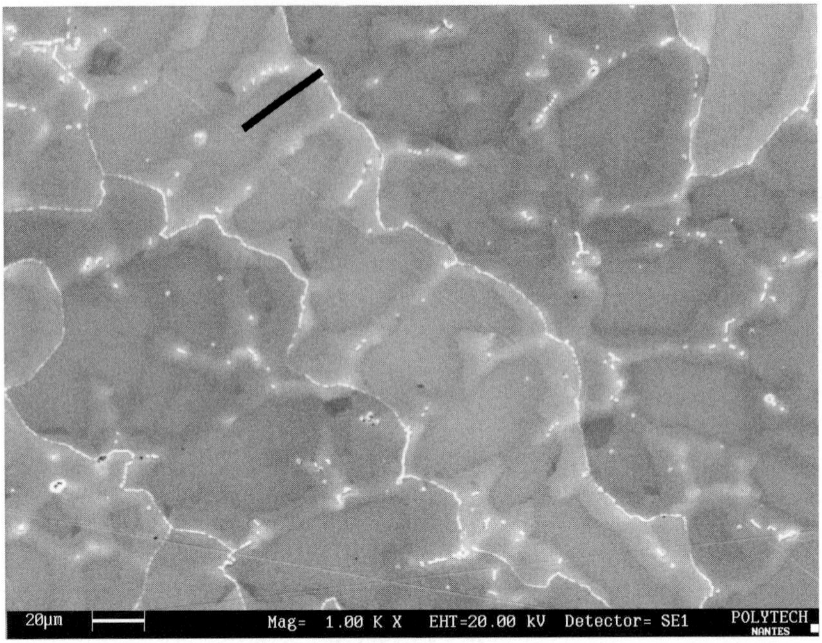

Fig. 5 SEM micrograph of the as-solidified alloy. The black line is the EDS scan path.

average value, C_{av}. It is assumed that local equilibrium exists at the precipitate-matrix interface during growth, the interface concentrations being designated C_{eq}. The elementary volume is then divided into m slices, each of thickness dx. Initial conditions are thus: $C(1) = C_{eq}$ in this first slice, $C = C_{av}$ in all other slices. The flux between slices i and $i + 1$ in the time step dt is computed following Fick's law:

$$dN(i) = -DS\frac{C(i) - C(i+1)}{dx}dt$$

where D is the effective diffusion coefficient and S the slice cross-section. The number dN_p of solute atoms absorbed by the particle in the same time interval is calculated from the concentration gradient at the interface:

$$dN_p = -DS\frac{C(1) - C(2)}{dx}dt$$

The concentration profile at time $t + dt$ in the solid solution, $C(i)_{t+dt}$, is then recalculated by writing that the number of solute atoms in the ith slice is equal to the previous number of atoms in the ith slice ($C(i)_t Sdx$), less those going to the $(i\text{-}1)^{th}$ slice, $dN(i\text{-}1)$, plus the number of atoms coming from the $(i + 1)^{th}$ slice, $dN(i)$, so that:

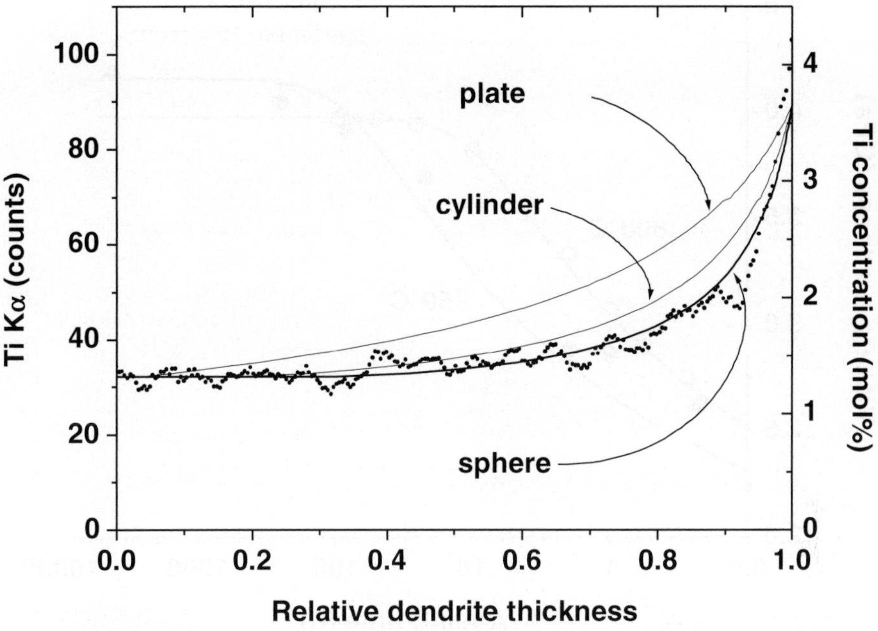

Fig. 6 Comparison between the predicted (right scale) and the EDS measured (left scale) titanium concentration profiles within dendrites, for different dendrite shape models.

$$C(i)_{t+dt} = C(i)_t - \frac{dN(i-1)}{Sdx} + \frac{dN(i)}{Sdx}$$

The total number of Al and Ti atoms incorporated in the particle, N_p, is then the sum of all dN_p calculated during the process. Each of these atoms precipitates one Ni_3(Al, Ti) lattice, which allows an estimate of the γ' volume fraction, V_f, as a function of ageing time. We appreciate that the model here essentially treats the growth problem as a binary approximation, but this may be a good first approximation.

A simple Freidel-type precipitation strengthening model is used to describe the increase in hardness, H, of the material as a function of V_f:

$$H = H_0 + A\sqrt{V_f}$$

This yields a model for the hardness of the material during the precipitation ageing process, with only two adjustable parameters: the effective diffusion coefficient, D, and the proportionality constant A.

The model was then compared to age-hardening measurements made at 750 and 800°C. The material was first made into a single γ phase 4 hours at 1175°C, water quenched, and its Vickers hardness, H_0, measured on a polished surface. Interrupted ageing treatments were then made by heating specimens at 750 or 800°C for various times, followed by water

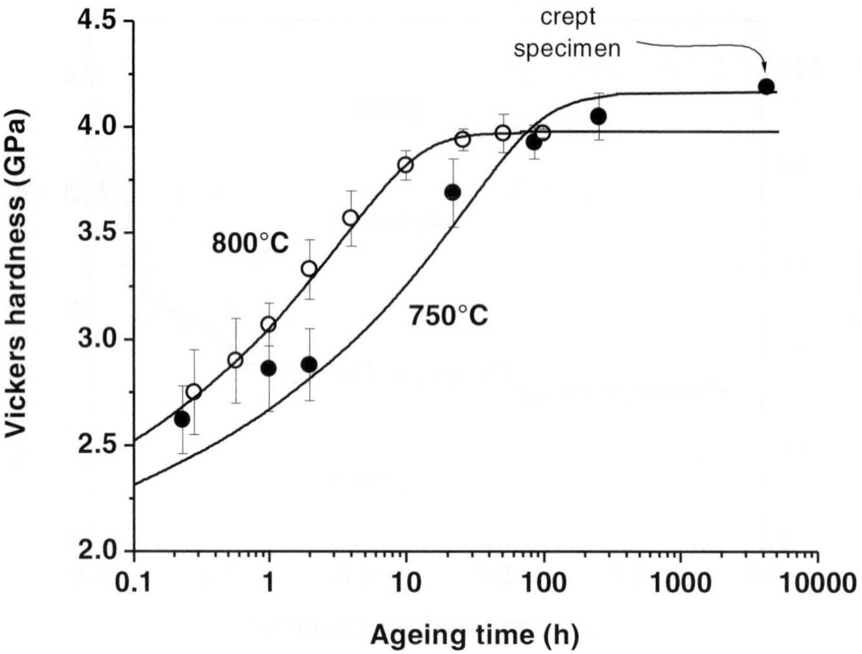

Fig. 7 Evolution of Vickers hardness with ageing time. Symbols: experimental points; lines: model.

quenching. The Vickers hardness was then measured on polished surfaces. Measured hardness data are presented in Figure 7 as a function of ageing time, along with those calculated using the above procedures. The best fit was obtained with $D = 0.025$ nm²/s at 750°C and $D = 0.20$ nm²/s at 800°C, which also allows a rough estimate of the apparent activation energy for diffusion, Q, assuming that:

$$D = D_0 \exp\left(-\frac{Q}{RT}\right)$$

with D_0 a pre-exponential factor and R the gas constant. This yields $Q = 380$ kJ/mol, which is similar to values found in the literature for γ' growth,[13] although measurements at several different temperatures would give better accuracy.

CONCLUSIONS

An affordable, forgeable, weldable, creep-resistant and corrosion-resistant nickel-base superalloy has been designed using Gaussian processes modelling of mechanical properties, and thermodynamical simulations. A designed alloy was then manufactured and tested. Both microstructure and mechanical properties are found to be remarkably consistent with predictions. In particular, creep tests suggest that the alloy will match the engineering requirement of a lifetime of 100,000 hours at 750°C under 100 MPa.

Preliminary modelling and experimental investigations were performed with respect to processing issues. The primary solidification segregation behaviour was successfully modelled using Scheil's approximation, and concentration profiles within dendrites could be predicted, which is important in the forging and welding the alloy. In addition, the precipitation hardening behaviour during ageing was described by an approach combining a diffusion-controlled growth model, a thermodynamical simulation software and a simple strengthening model. This may be helpful in optimising the heat-treatment.

These studies need to be completed by further experimental investigations and model refinements to cover the composition of the interdendritic medium, complex dendrite geometries, aeging at different temperatures, etc.

Further work will also include investigations of the high temperature ductility and forgeability, recrystallisation, welding, etc.

REFERENCES

1. F. Tancret, H. K. D. H. Bhadeshia and D. J. C. MacKay: *Materials Science and Technology*, **19**, 2003, 283–290.
2. F. Tancret and H. K. D. H. Bhadeshia: *Materials Science and Technology*, **19**, 2003, 291–295.
3. F. Tancret, T. Sourmail, M. A. Yescas, R. W. Evans, C. McAleese, L. Singh, T. Smeeton and H. K. D. H. Bhadeshia: *Materials Science and Technology*, **19**, 2003, 296–303.
4. C. A. L. Bailer-Jones, H. K. D. H. Bhadeshia and D. J. C. MacKay: *Materials Science and Technology*, **15**(3), 1999, 287–294.
5. F. Tancret, H. K. D. H. Bhadeshia and D. J. C. MacKay: *ISIJ International*, **39**(10), 1999, 1020–1026.
6. M. N. Gibbs: Ph.D. Thesis, University of Cambridge, UK., 1998.
7. Materials Algorithms Project, http://www.msm.cam.ac.uk/map/
8. T. Cool: Ph.D. Thesis, University of Cambridge, UK., 1996.
9. *Thermo-Calc*, The Royal Institute of Technology, Stockholm, Sweden.
10. K. M. Chang and A. H. Nahm: *Superalloys 718 - Metallurgy and Applications*, E. A. Loria, ed., The Minerals, Metals & Materials Society, 1989, 631–646.
11. A. K. Jena and M. C. Chaturvedi: *Journal of Materials Science*, **19**, 1984, 3121–3139.
12. C. T. Sims: *Journal of Metals*, **18**, 1966, 1119–1130.
13. E. Balicki, A. Raman and R. A. Mirshams: *Metallurgical and Materials Transactions A*, **28**, 1997, 1993–2003.

Mechanical and Microstructural Qualification of A Prototype IN718 Forged Disk

A. FINALI
Società delle Fucine
Terni, Italy

A. DI GIANFRANCESCO, O. TASSA and L. CIPOLLA
Centro Sviluppo Materiali
Rome, Italy

R. MONTANI
Acciaierie Foroni
Gorla Minore
Varese, Italy

ABSTRACT

Higher efficiency in the new fossil fuel fired plants for power generation needs increasing temperatures and pressures service in order to reduce dangerous emissions and guarantee growing outputs. As forged, tube, pipe and cast components used in the hottest part of the plant are severely stressed, the development of new steels and alloys with improved creep performances plays a relevant role.
In particular high and intermediate pressure section rotors of a steam turbine are going to work in much more stressed conditions due to higher temperature and pressure. This paper shows the results obtained for the prototype industrial rotor disk, forged by Società delle Fucine in Italy, using IN718 Alloy produced by AOD + double VAR at Acciaierie Foroni, in the frame of Thermie Advanced (700°C) Pulverised Fuel Power Plant Project. A new heat treatment has been developed by Centro Sviluppo Materiali (CSM) to increase microstructural stability in the range of 650–750°C and improve creep behaviour. The microstructural analysis and the mechanical and creep properties of the prototype rotor qualification are discussed.

INTRODUCTION

The driving forces improving the production of electric power have changed from growth of demand in the 60's to considerations concerning environment and efficiency in the 90's. In 1997 the European Commission approved in the frame of the THERMIE programme the demonstration project *'Advanced ('700°C') PF Power Plant'* proposed by a large group of material and components suppliers of the power industry, R&D centres and leading utilities.

The aim of the project is to develop a new generation of coal fired power plant able to raise up to 375 bar/700°C the steam conditions and to increase electrical efficiency up to 52–55%.[1]

Consequently in the hottest part of the steam cycle it will be no longer possible to use steel components, but new nickel-based superalloys are required; then it is necessary to optimise, test and qualify these materials. In parallel, studies on the design of the large critical components have been performed to optimise the size as function of production limits. At the beginning of the program several existing superalloys have been taken into account for screening tests in order to evaluate the most promising candidates for the component manufacture. The Italian contribution to the program has been focused on the IN 718 super alloy. In this paper the activities carried out to optimise the heat treatment in order to improve the high temperature mechanical behaviour of this alloy are presented and results obtained in terms of microstructural and creep resistance are discussed.

MATERIAL

The material for the screening tests was supplied by Acciaierie Foroni (Italian producer of special steels and superalloys) as a 200 mm hot rolled bar obtained by melting in Electric Arc Furnace, followed by Argon-Oxygen Degassing refining and double Vacuum Arc Remelting. Table 1 shows the chemical analysis of IN718 billet compared with that of the standard UNS7718, the trace elements are the following: Pb < 0.5, Bi < 0.1, Se < 3, Sn < 10 (values in ppm).

HEAT TREATMENT OPTIMISATION

It is well known by literature[2] that the maximum operating temperature for IN718 treated in the standard conditions (UNS7718) is about 650°C. The first activity of the program has been to define a new heat treatment able to improve creep behaviour without reducing tensile, toughness and creep-fatigue properties.

IN718 super alloy is characterised by the precipitation of cuboidal γ' phase (Ni_3Al,Ti) and γ'' phase (Ni_3Nb), acicular and fine, as consequence of the amount of Nb in the matrix. The latter phase especially, during service at high temperature, evolves into the δ phase, coarse and gets acicular, producing a dangerous embrittlement.

A new heat treatment has been developed by CSM. In order to stabilise γ' and γ'' phases, the solution temperature has been increased up to 1065°C, followed by air cooling and two aging stages. Also relatively high aging temperatures have been employed (Table 2).

METALLOGRAPHIC EXAMINATIONS

To evaluate the microstructural stability and the mechanical behaviour after standard and modified heat treatment (HT), aging tests were carried out at 700 and 750°C up to 10,200 hours

Table 1 Chemical analysis of Foroni IN718 billet (wt.%).

	C	Ni	Fe	Cr	Mo	W	Co	Nb	Ti	Al	B	Zr	Mn	Si	P	S
Standard UNS7718	<0.08	50–55	Bal.	17–21	2.8–3.3	/	<1	4.8–5.5	0.7–1.2	0.2–0.8	<0.006	/	<0.35	<0.35	/	<0.015
Foroni Billet	0.03	54	17.5	18.5	3	<0.1	<0.1	5	0.9	0.5	0.004	<0.0005	0.15	0.15	0.008	0.001

Table 2 Heat treatment conditions and mechanical properties.

	Solubilisation (°C/h/Cooling)	First Aging (°C/h/Cooling)	Second Aging (°C/h/Cooling)	YS (Mpa)	UTS (Mpa)	Elong, (%)	Impact Energy (J)	Grain Size (ASTM No.)
Standard	927–1010	720/8/FC	620/8/Air	>1035	>1240	>10		
Foroni Stand.	1000/2/Water	718/8/FC	621/8/Air	1135	1403	20.4	74	5
Foroni Mod.	1065/1/Air	760/8/FC	650/12/Air	1120	1335	27.0	80	4

FC = Furnace Cooling

Table 3 Aging times and temperatures.

Inconel 718	Aging Temperature (°C) Time and (hours)
Standard UNS7718 Heat Treatment	700°C × 3000 hours 700°C × 6000 hours 700°C × 10200 hours 700°C × 3000 hours
CSM Modified Heat Treatment	700°C × 3000 hours 700°C × 3000 hours

Table 4 Larson miller parameter (LMP).

750°C	400 hours	1000 hours	3000 hours	6000 hours
700°C	6000 hours	15,000 hours	60,000 hours	100,000 hours
LMP	23.0	23.53	24.01	24.32

(Table 3). Both tensile strength and toughness decrease after aging (Figure 1). It is important to take into account that aging at 750°C for 3,000 hours, according with Larson Miller Parameter (LMP), is equivalent to 60,000 hours at 700°C (Table 4).

Due to the lack of information about the microstructural evolution of Inconel 718 alloy during aging at high temperature and its influence on mechanical and creep behaviour, a comprehensive study has been carried out by Light Microscopy (LM) and Transmission Electron Microscopy (TEM). A considerable grain growth has been noted in the alloy submitted to modified heat treatment. The grain size measured by LM is 54 μm (ASTM No.5) and 78 μm (ASTM No.4) respectively for the standard and modified treatments. Examples of some microstructures by TEM of as treated materials are shown in Figures 2 and 3. Both materials, according with Electron Dispersion Spectroscopy (EDS) analysis, are δ phase free. It has been observed that the modified heat treatment increases the size of γ'' phase. Figures 4 and 5 show the microstructure of the specimens aged at 750°C for 3000 hours. Also in this case, the modified heat treatment leads to a more stable structure with γ'' phase more uniformely distributed. In the standard treatment, γ'' free zones near the δ phase particles are evident.

It is interesting to observe that near the fracture of the broken creep specimen after 18,261 hours at 700°C and 200 MPa, the δ particles (Figure 6), especially those formed at the grain boundary, are deformed under stress. Usually, δ particles are considered very dangerous because they are brittle and can dramatically reduce the ductility of material.

After 3000 hours at 750°C, the measured precipitate size distribution shows that γ' particles grow and coalesce in similar way in both heat treatments up to 80 nm, γ'' particles grow up to

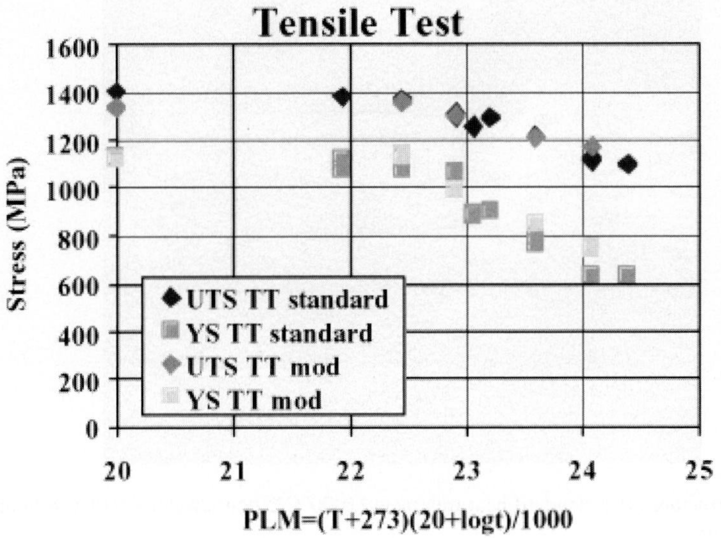

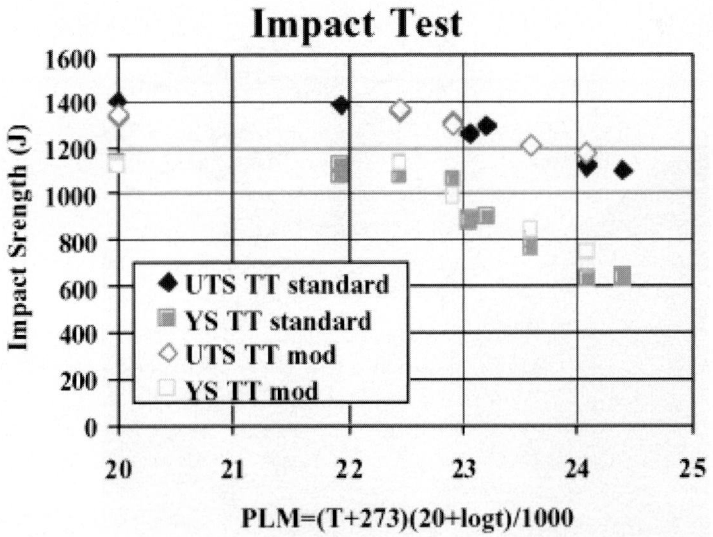

Fig. 1 Effect of aging on mechanical properties.

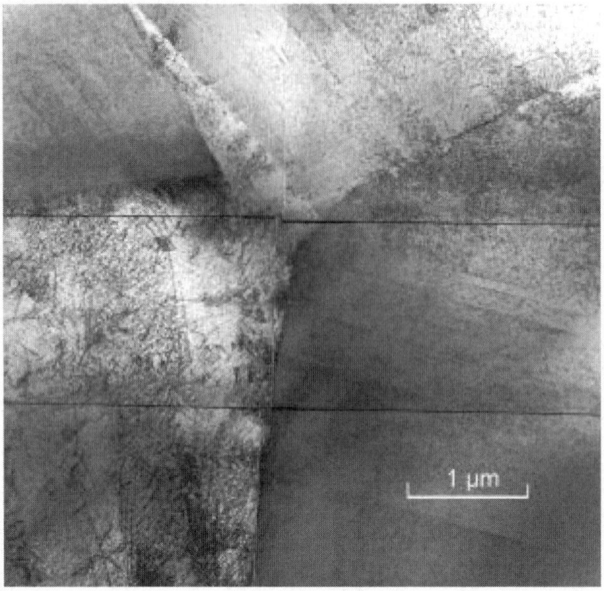

Fig. 2 Microstructure after standard heat treatment (1000°C/2 hours/water – 718°C/8 hours/FC – 621°C/8 hours/air).

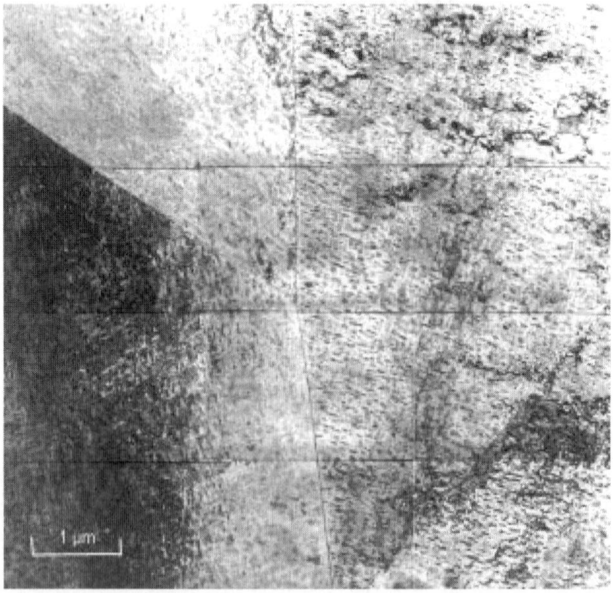

Fig. 3 Microstructure after modified heat treatment (1065°C/1 hour/air – 760°C/8 hours/FC – 650°C/8 hours/air).

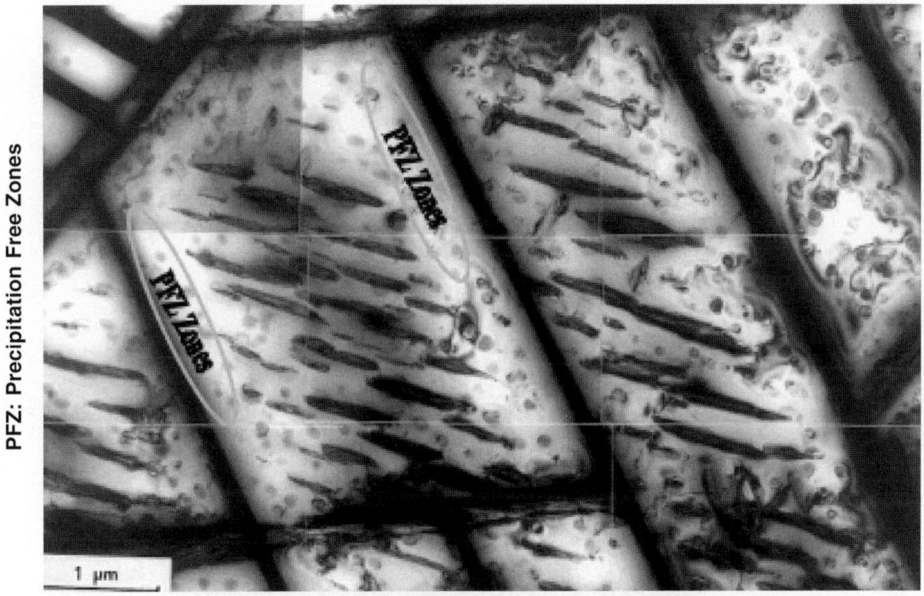

Fig. 4 Standard treated material (1000°C/2 hours/air-718°C/8 hours/FC-621°C/8 hours/air) after 3000 hours at 750°C.

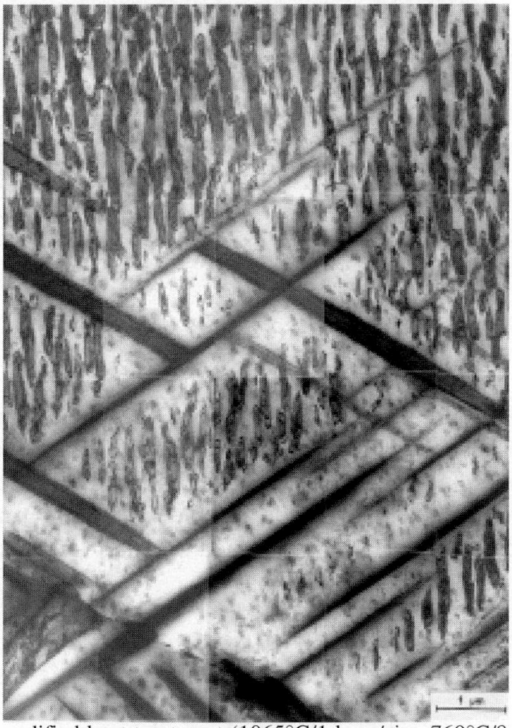

Fig. 5 Specimen after modified heat treatment (1065°C/1 hour/air - 760°C/8 hours/FC – 650C°/8 hours/air) aged 3000 hours at 750°C.

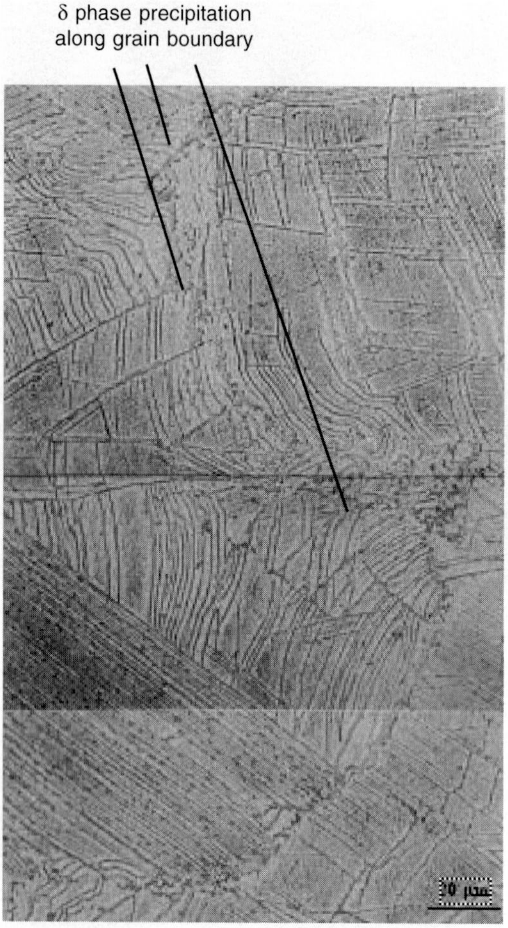

Fig. 6 Deformed grain close to fracture in creep specimen.

900 nm after 5000–10,000 hours and then they decrease to 800 nm (Figures 7 and 8). This phenomena can be explained by the formation of δ phase which precipitates at the expense of γ'' (Figure 9). By means of the modified heat treatment, the nucleation and growth rate of δ phase are decreased. The evolution of the average size of γ', γ'' and δ phases during aging is reported in Table 5 for the material under modified heat treatment condition.

INDUSTRIAL FORGING TRIAL

Following the promising results of the mechanical tests of the screening program and microstructural examinations, an industrial forging trial has been performed. Laboratory

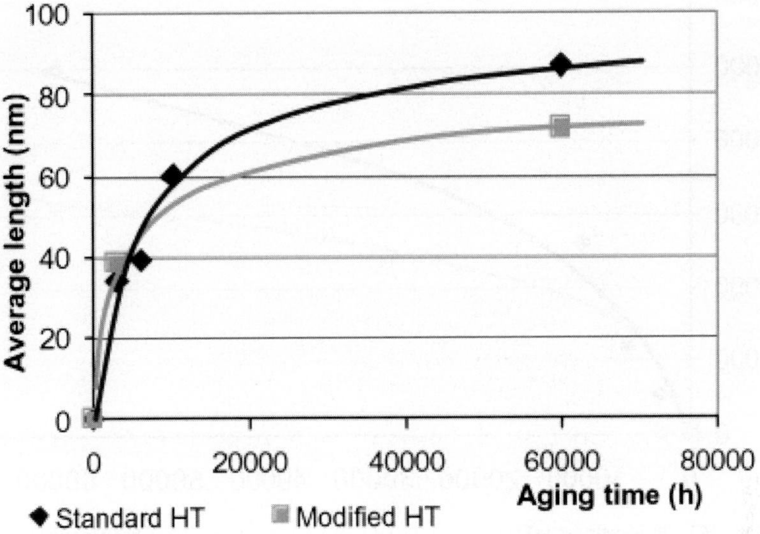

Fig. 7 γ' phase coarsening during 700°C aging tests.

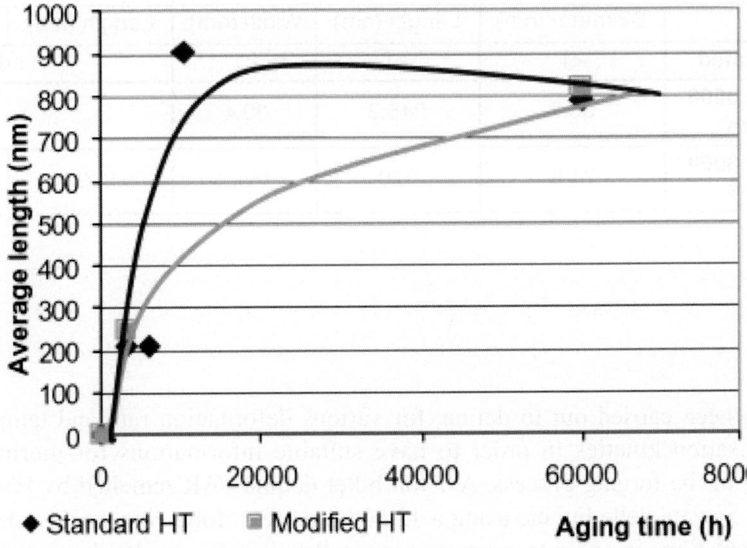

Fig. 8 γ'' phase coarsening during 700°C aging tests.

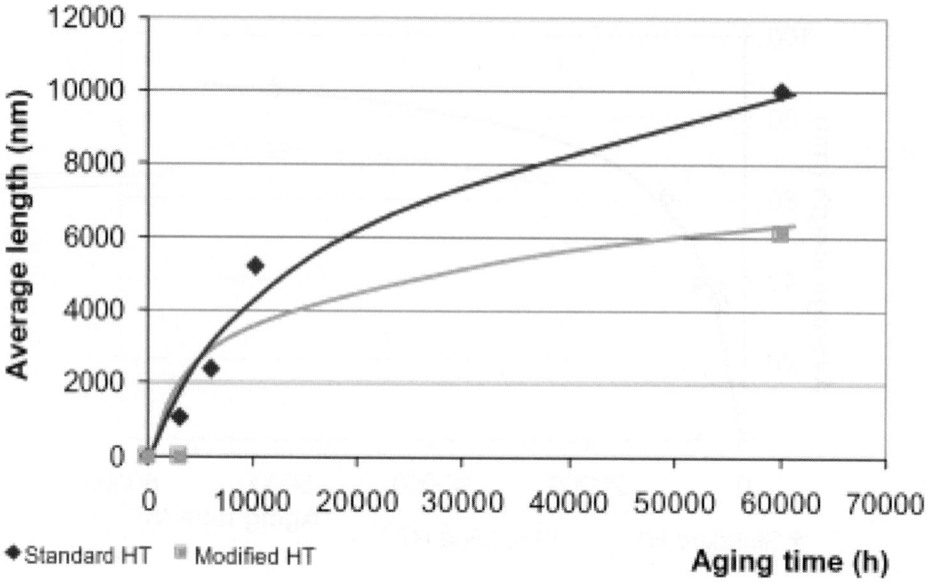

Fig. 9 δ phase coarsening during 700°C aging tests.

Table 5 Average size of the second phases during aging tests for IN718 after modified heat treatment.

	γ'	γ''		δ	
	Diameter (nm)	Length (nm)	Width (nm)	Length (nm)	Width (nm)
As Treated	<1	<1	<1	nd	
700°C x 3000 hours	38.7	245.3	39.4		
750°C x 3000 hours	71.5	820	161.4	6092	167

tests have been carried out to define, for various deformation rate and temperatures, the recrystallisation kinetics in order to have suitable informations for thermomechanical modelling of the forging process. A 1 ton billet double VAR remelted by Foroni has been forged by Società delle Fucine using a 12,600 ton press, following a schedule designed by model simulations in order to promote recrystallisation. Figure 10 shows the billet before forging and the prototype forged disk.

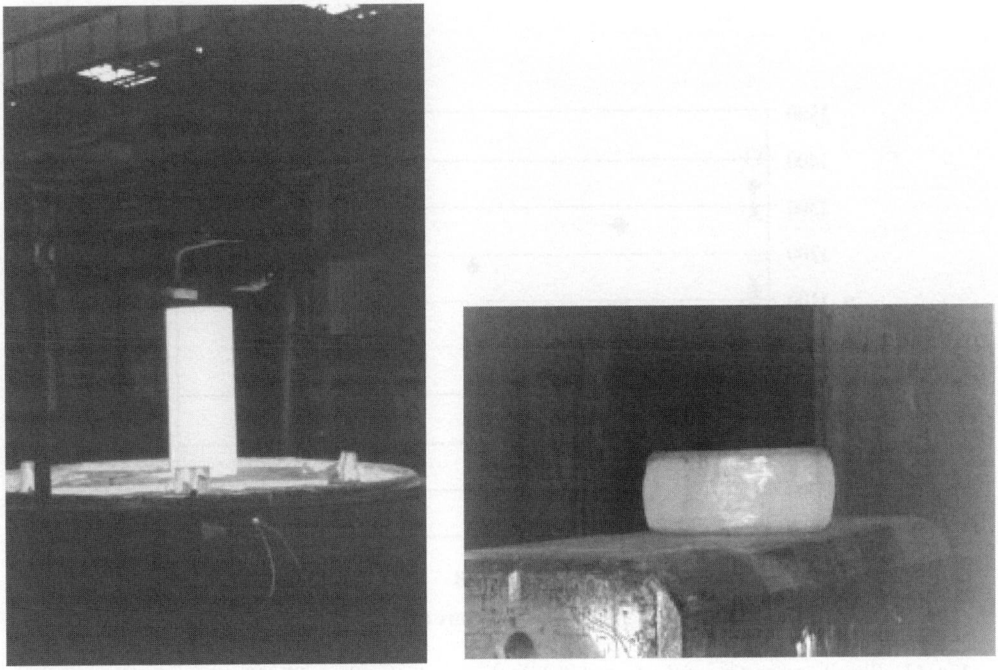

Fig. 10 IN718 billet before and after forging.

MECHANICAL BEHAVIOUR OF PROTOTYPE DISK

The disk was cut and material blocks were heat treated in standard and modified conditions. The trial disk shows a grain size of respectively 80 and 140 μm for standard and modified heat treatments. An increase of 60% in grain size, compared with the hot rolled bar used for the screening tests was achieved. Tensile properties as function of temperature are uniform without relevant differences between radial and axial specimens (Figures 11 and 12). The room temperature properties are in agreement with the standard requirements.

CREEP TESTS

In order to evaluate the creep behaviour of the standard and modified IN718, a long term creep test program was defined. Creep behaviour of standard and modified alloy is compared in Figure 13 with data from literature.[3] The standard treated material creep performance is in perfect agreement with the reference data, while a significant rupture time increase is noted for the new heat treatment, especially at 750°C. The tests on notched specimens showed that IN718 is not sensitive to notches. Creep data have been also included in a master curve using

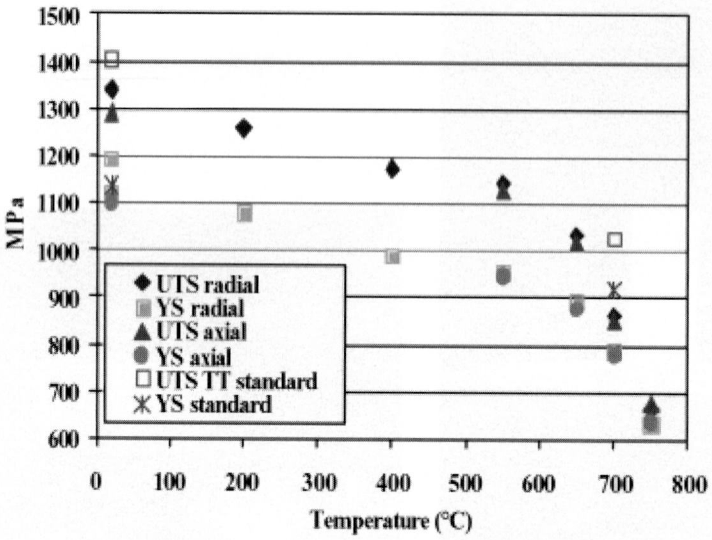

Fig. 11 Tensile properties of the forged disk in the standard and modified heat treatment.

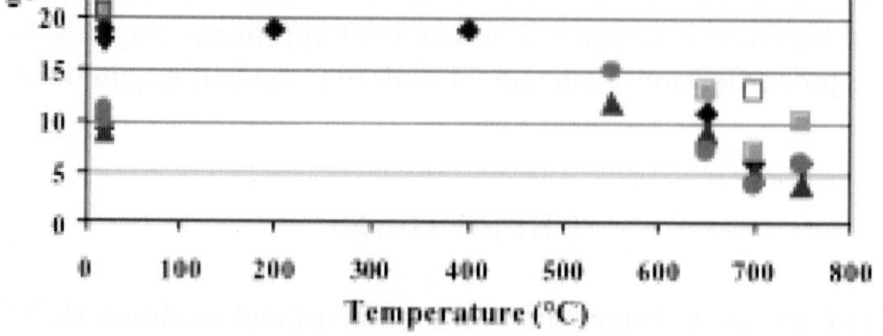

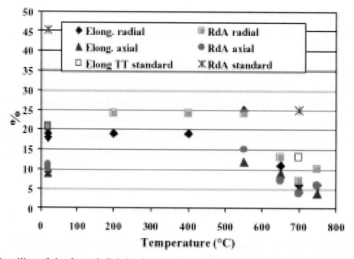

Fig. 12 Ductility of the forged disk in the standard and modified heat treatment.

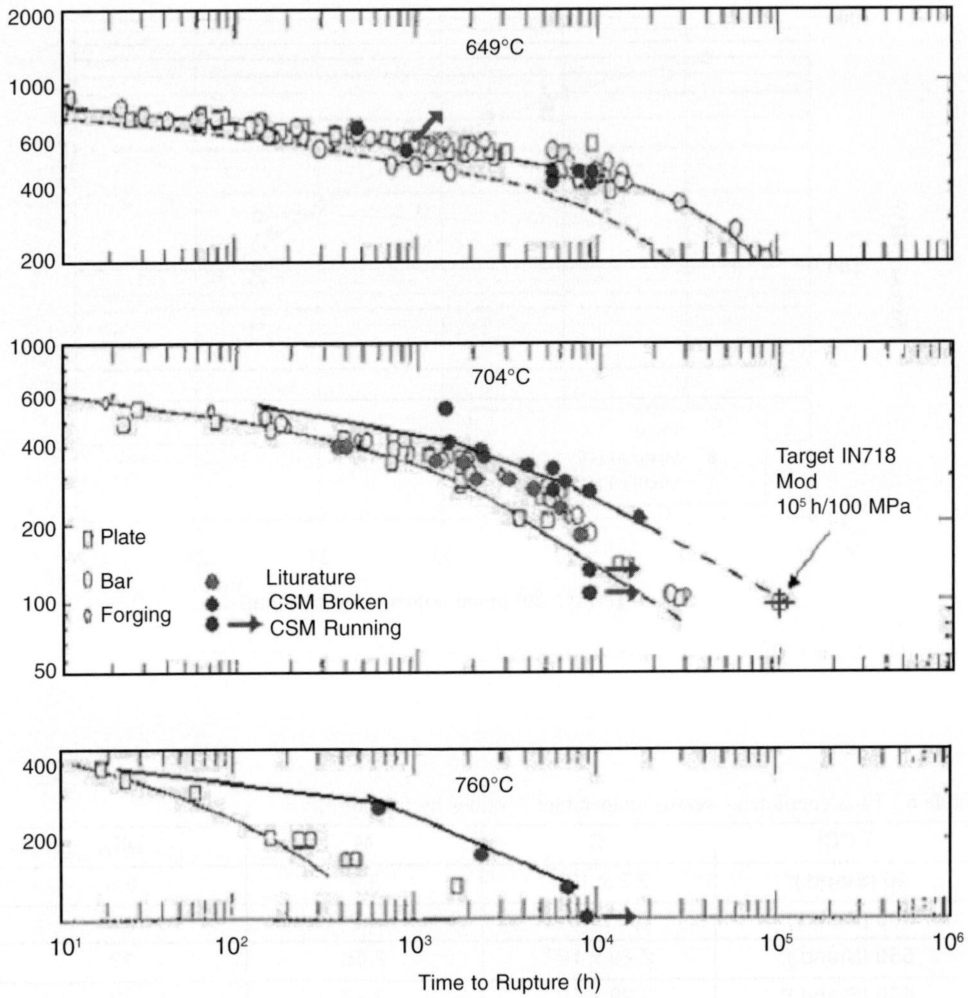

Fig. 13 Comparison of standard and modified heat treatment creep properties.[4]

Larson Miller Parameter: it seems that the 700°C target can be successfully reached (Figure 14).

TOUGHNESS AND CRACK PROPAGATION

Also J integral toughness tests were carried out. The results show that toughness values decrease with increasing temperatures, as expected. Again the behaviour of the material with modified heat treatment is better than that of the standard one (Figure 15).

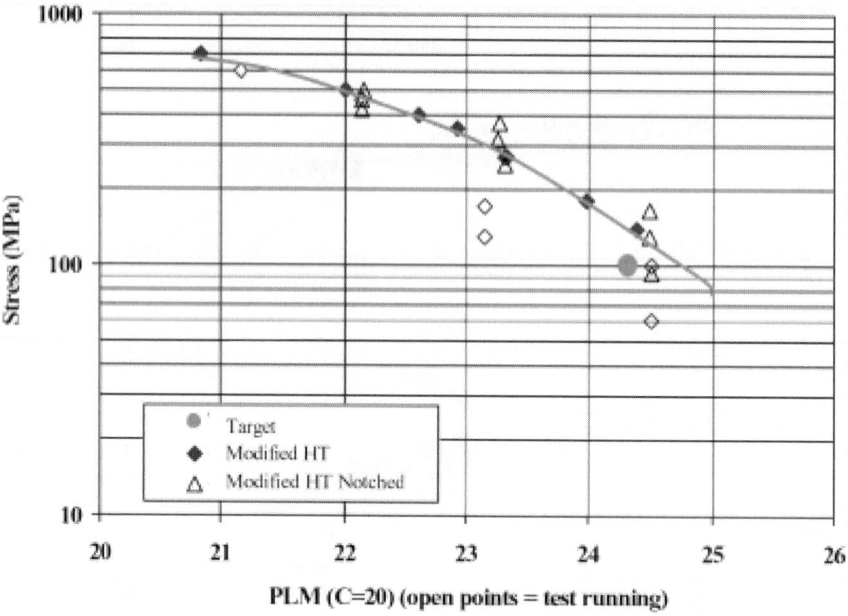

Fig. 14 Larson-Miller master curve of material after modified heat treatment.

Table 6 Paris coefficients versus temperature (*values by literature).[5]

T (°C)	C	M	DK_{th}
20 (Stand.)*	2.2×10^{-12}	3.2	6.8
500 (Stand.)*	5.6×10^{-12}	3.2	8.0
550 (Stand.)*	2.23×10^{-8}	2.66	12
650 (Stand.)*	2.29×10^{-7}	2.17	13
750 (Stand.)*	/	/	17
750 (CSM Mod.)	1.5×10^{-5}	1.71	18

Fatigue Crack Propagation tests have been carried out at 750° with stress ratio R = 0.1 (Figure 16). The typical sigmoidal shape of the curve is quite well described by Paris Law with the following expressions:

$$\frac{da}{dN} = C \times (\Delta K)^m,$$

where C and m are material constants.

The values of the obtained Paris coefficients, m and C, have been compared with those from literature.[4] These values are summarised in Table 6.

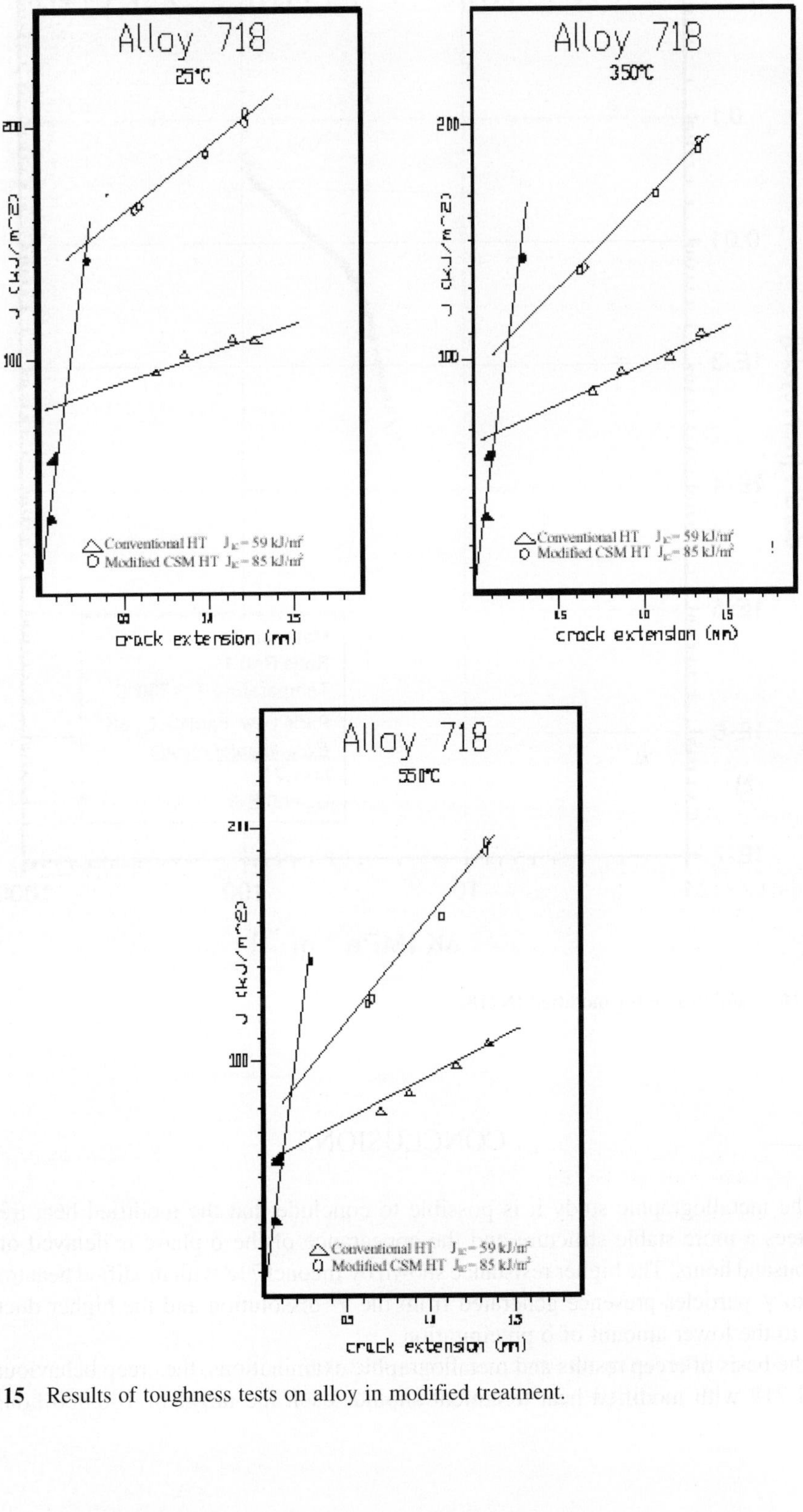

Fig. 15 Results of toughness tests on alloy in modified treatment.

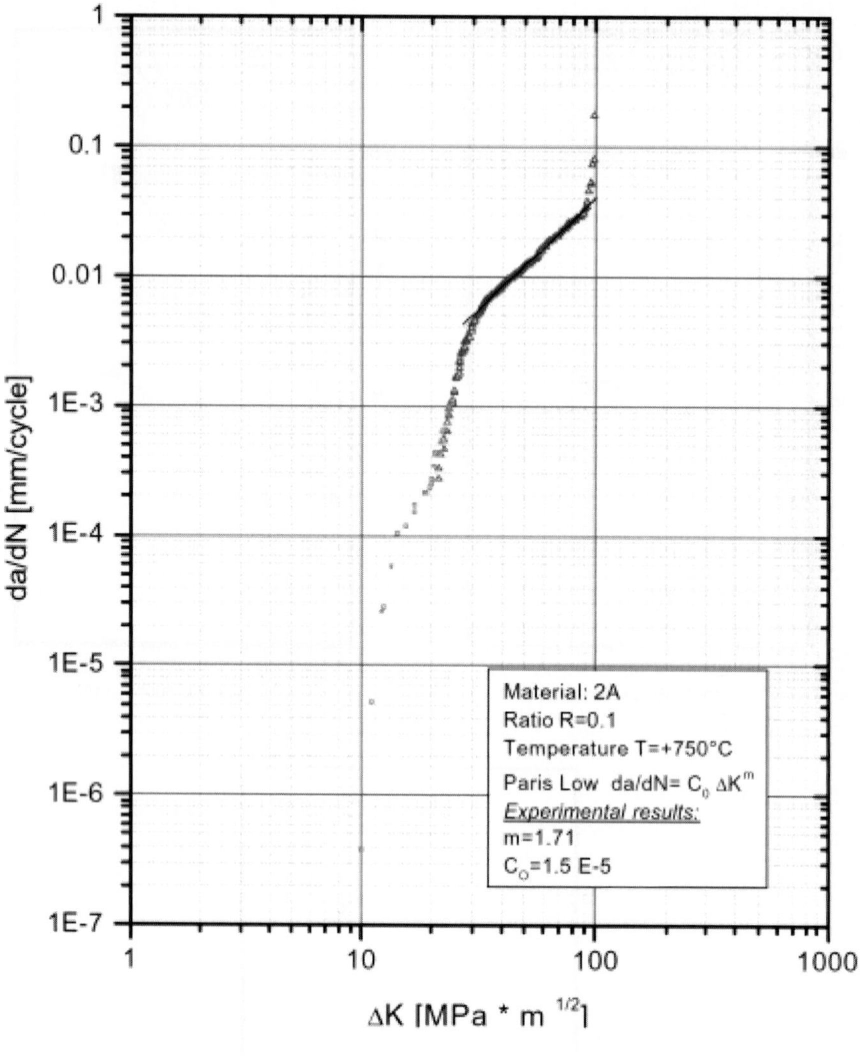

Fig. 16 da/dN curve for modified IN718.

CONCLUSIONS

From the metallographic study it is possible to conclude that the modified heat treatment guarantees a more stable structure and the appearance of the δ phase is delayed of three-four thousand hours. The higher resistance shown by Inconel 718 with modified heat treatment is due to γ' particles presence generated from the γ'' dissolution and the higher ductility is related to the lower amount of δ precipitation.

On the basis of creep results and metallographic examinations, the creep behaviour of the Inconel 718 with modified heat treatment should reach the target of 100,000 hours with

100 MPa at 700°C. According with the results obtained until now, the possible range of time to rupture could be 80,000–120,000 hours. In order to confirm this prediction and better define the slope of the master creep curve, tests are still in progress and other specimens have been loaded.

ACKNOWLEDGEMENTS

Thanks to European Communities for the financial support to the research avtivities

REFERENCES

1. R. Kehlhofer: 'Power Engineering, Status and Trends', *Materials for Power Advanced Engineering*, 1998.
2. Guo, Han and You: 'Creep Crack Growth Behaviour of Alloy 718', *Superalloy 718, 625, 706 and Various Derivates*, Loria, ed., 1998.
3. R. W. Hayes: 'Creep Deformation of Inconel 718 in the 650 to 760°C Temperature Regime', *Superalloy 718, 625, 706 and Various Derivates*, Loria, ed.
4. L. A. James: 'Fatigue Crack Propagation in Alloy 718: A Review', *Superalloy 718 - The Minerals, Metals & Materials Society*, 1989.
5. J. J. Xie, Z. M. Shen and J. Y. Hou: 'Fatigue Crack Growth Behaviour in Alloy Inconel 718 at High Temperature', *Material Science and Engineering*, **89**, 1987, L7–L10.

Creep Modelling of Waspaloy

C. BERGER, A. THOMA and A. SCHOLZ
Institute for Materials Technology (IfW D)
Darmstadt University of Technology

ABSTRACT

The improvement of total efficiency is a need for the establishment of future steam power plants to reduce emissions of greenhouse gases and also to save fossil sources. Consequently, steam inlet temperatures have to be increased. That requires new materials, which bear higher temperatures. In order to achieve a total efficiency of approximately 55%, the steam inlet temperature is increased to 700°C. That means an enhancement of approximately 8% compared with modern power plants. Nickel base alloys have become important for such applications, as ferritic and martensitic steels will not meet such conditions.

This contribution deals with potential materials of the first generation for rotor material in a steam power plant at temperatures of 700°C and above. This essential component has high demands on castability, forgeability and long term stability of the microstructure. Other important aspects to be fulfilled are sufficient creep behaviour and creep crack growth resistance for components with large diameters. All these aspects have been investigated in a joint research program. In a first step special versions of the superalloys Inconel 706, Inconel 617 and Waspaloy are considered.

The determination of creep rupture properties within the applicable temperature range is necessary. For this purpose, creep rupture tests were performed and creep data up to 25,000 hours are considered. This contribution is focussed on the most powerful material Waspaloy with the highest values of creep rupture strength. The investigated material shows an individual behaviour of creep. The creep behaviour is modelled with a modified Garofalo creep equation. This type of equation is a powerful tool for describing long term creep in a wide range of stresses and temperatures. On the basis of this tool recalculation of components under creep and creep crack growth conditions by means of finite element analyses can be conducted successfully.

INTRODUCTION

Earlier steam power plants service conditions were limited to temperatures of about 540°C and pressure of 180 bar. These conditions give a total efficiency of approximately 38%. Modern steam power plants built in the 1990s operate with temperatures just below 600°C and pressure of approximately 250 bar. In future power plants temperatures up to 700°C or higher and a pressure up to 360 bar are of high interest, the aim being an efficiency of approximately 55% is aspired. The use of ferritic steels seems to be limited to temperatures of about 620°C. To bear the required temperature and pressure nickel base alloys are necessary. In a DFG (Deutsche Forschungsgemeinschaft) joint research program nickel base alloys are investigated for the use as rotor materials in steam turbines at temperatures of 700 to 720°C and service durations up to 200,000 hours. In this context castability and forgeability, the long term stability of microstructure, the mechanical long term behaviour and the creep crack growth resistance for large diameters of forged nickel base alloys are important criterions. To study these requirements different partners contribute on this comprehensive research. Investigations on castability performed by ACCESS Aachen. Forgeability is processed by the 'Institut für Bildsame Formgebung' Aachen. Microstructural long term stability is investigated by the 'Institut für Werkstoffe der Energietechnik 2' Juelich. Creep Crack Growth resistance is examined and modelled by the 'Institut für Werkstoffe' Braunschweig. And finally the mechanical behaviour included creep and creep rupture behaviour is investigated by the 'Institut für Werkstoffkunde' Darmstadt.

In addition to the generation and assessment of long term creep and creep rupture data the establishment of creep equations is necessary to describe creep in the whole relevant temperature and stress range of the proposed application.

This paper is focussed on the modelling of creep of Waspaloy. Of future interest are developments combining all efforts and results of the participated partners. As a result of these developments a new alloy called DT750 was identified.

From the large number of existing creep equations the well proved modified Garofalo eqns (1–3) was taken to describe creep of Waspaloy, i.e.

$$\varepsilon_p = \varepsilon_i + \varepsilon_{f1\max} \times H(t) + \dot{\varepsilon}_{p\min} \times t + \varepsilon_{f3} \tag{1}$$

with initial plastic strain e_i, maximum amount of primary creep strain $e_{f1\,\max}$, a time function $H(t)$ of primary creep strain, the minimum creep rate $\dot{\varepsilon}_{p\,\min}$ and the tertiary creep strain e_{f3} (Figure 1).

The established creep equations also support the work of one research partner modelling the creep crack growth resistance.

EXPERIMENTAL

Different types of nickel base alloys were identified in a first step of this DFG-research program. The following alloys are taken as representants of different hardening classes, that are:

- Inconel 706, representing a γ'/γ''-strengthened material,
- Inconel 617, representing a solid solutioned material and finally and
- Waspaloy, representing a γ'-strengthened material.

Alloy Inconel 706[4] was examined in two different heat treatment conditions. The first one has γ' and γ''-phases. The second one additionally decorates the grain boundaries with η-phases. This is due to a special heat treatment with a direct cooling procedure.[5] This heat treatment was originally suggested by Shibata[6] for large components of Inconel 706 as an economic heat treatment procedure. Further details about Inconel 706 and a new developed heat of Alloy 706 indicated as DT706 are presented in this context by Ref. 7.

Inconel 617 is a solid solution strengthened alloy that is characterized by fine $M_{23}C_6$ and primary MC-carbides.[8] Waspaloy contains as well as coherent γ'-precipitations also $M_{23}C_6$-

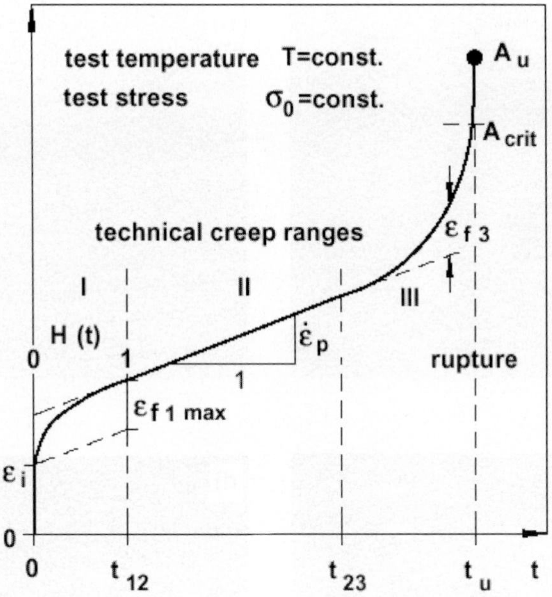

Fig. 1 Linear creep curve and components of equation (1).

precipitations at the grain boundaries in the as-received condition. Two different types of γ'-precipitations are observed. These consist of larger primary γ'-precipitations with sizes of particles of approximately 200 nm and smaller secondary γ'-precipitations. The secondary γ'-precipitations are distributed finely in the matrix.[9]

All materials were examined in two different conditions. These were in an as-received condition after heat treatment and an aged condition after heat treatment and additional ageing of 5000 hours at 750°C (Figure 2). This aged condition should approximate a so called mid-of-life-condition of 100,000 hours at 700°C using a Larson-Miller parameter calculation.

During the ageing procedure on Waspaloy the secondary γ'-precipitations are consumed by growing primary γ'-precipitations. The secondary γ'-precipitations disappear completly after 5000 hours at 750°C. At the same time the volume fraction of the primary γ'-precipitations increases from the as-received condition to about 22–24% in the aged condition.[8]

The creep rupture tests were done according to DIN EN 10 291. There are mainly performed by interrupted tests and partly by an uninterrupted testing procedure. Some of the uninterrupted tests are performed until a certain time or plastic strain is reached and then continued as interrupted tests until rupture.[10] The results of the creep rupture tests are assessed by the usual graphical method that is based on plastic strain-time-diagrams and stress-time-diagrams.[11]

For Waspaloy a creep rupture strength $R_{utT} \approx 260$ MPa for 700°C and 100,000 hours can be derived from long term tests up to 25,000 hours. The course of creep rupture strength R_{utT}

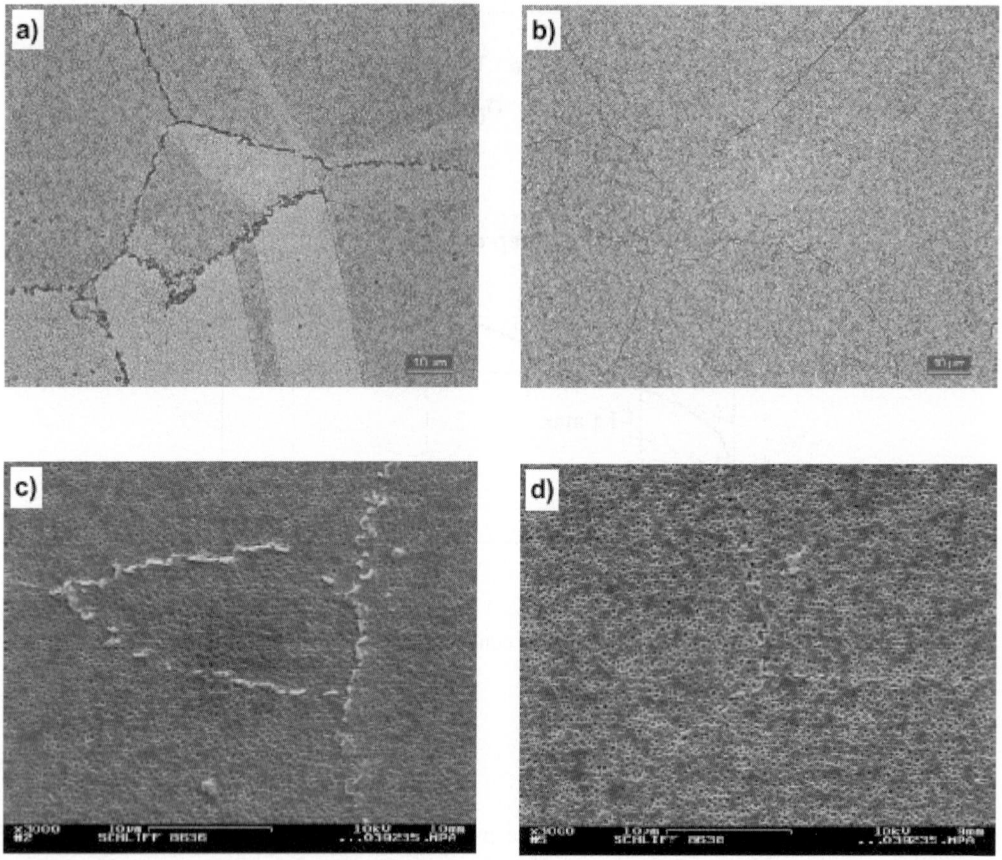

Fig. 2 Microstructure of waspaloy in as-received condition (a, c) and aged condition after additional annealing at 750°C for 5000 hours (b, d).

Table 1 Details of material Waspaloy.

Chemical Composition (mass.%)										
Ni	Cr	Co	Mo	Al	Ti	Fe	Mn	C	Si	Nb
57	19.35	14.0	4.52	1.22	3.13	0.57	0.05	0.03	0.04	0.01
Manufactoring										
Segment of a rotor Ø 165 mm x 70 mm, forged										
Heat Treatment										
1080°C 4 hours/4 K/min to 700°C/air + 850°C 4 hours/air + 760°C 16 hours/air										

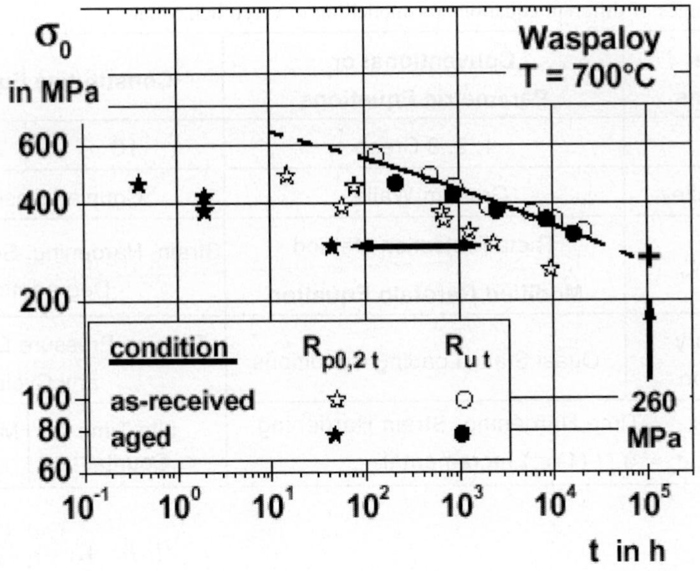

Fig 3 Results of creep rupture strength and 0.2%-creep strain versus time of Waspaloy.

against the temperature T behaves very stable and the ageing treatment shows only small effect on the creep rupture strength $R_{u\,t\,T}$ (Figure 3).

However, the 0.2%-creep strain $R_{p0.2\,t\,T}$ of the as-received and the aged condition differs significant.

Further, it should be mentioned that the rupture elongation A_u observed for Waspaloy is uncritical in both investigated conditions.

MODELLING

There exists a large number of equations to describe the creep behaviour. There are simple equations such as the Norton Bailey equation that is a powerful tool for simple estimations. For higher accuracy parametric equations are available to describe primary, secondary and tertiary creep, e.g. Graham-Walles or the theta-projection-method. The most general solution is obtained by constitutive equations. These equations are designed to describe strain, hardening, softening and degradation (Table 2).

Conventional creep equations are of industrial need to describe creep in the relevant temperature and stress range. For this purpose the modified Garofalo eqn (12) is assumed to be suitable eqn (1). This equation is close to reality and as already mentioned describes primary, secondary and tertiary creep and can be modelled relatively easy.[1–3] They are well adapted to calculate the time dependent stress and strain distribution of components or test pieces which are subjected to quasi-static loading conditions.

Table 2 Examples of different equations for modelling of creep behaviour.

Simple Equations	Conventional or Parametric Equations	Constitutive Equations	
	1, 2, 3 Creep	1D	3D
Norton-Bailey	Graham-Walles	Coupled Design of	
$\varepsilon_p = A\sigma_n t_m$	Theta Projection Method **Modified Garofalo Equation**	Strain, Hardening, Softening and Degradation	
Preliminary Estimation	Quasi Static Loading Conditions	Tension Pressure Demands of any Cycle	
Rules: σ_v, ε_{pv}, Time Hardening, Strain Hardening, $\sigma_0 = \sigma_w / (1+\varepsilon_p)$ Incremental		Life Time Counter	Most General Solution

The experimental basis for the establishment of the creep equation of the type of the modified Garofalo equation for Waspaloy in the as-received condition includes creep data of approx. 25,000 hours at 700°C. Creep tests at 600°C started later but have now exceeded a duration of 10,000 hours. Creep data of more than approximately 15,000 hours at 650 and 750°C are also available.

The parameter identification starts with modelling the initial plastic strain ei, followed by the quasi-static creep range. Finally the parameter identification is completed with defining primary and tertiary creep.

The initial plastic strain ε_i is usually determined from hot tensile tests but additional data points from the creep tests are also considered. With the initial plastic strain ε_i, data points ε_f $(T, \sigma_0, t) = \varepsilon_p (T, \sigma_0, t) - \varepsilon_i (T, \sigma_0)$ could be determined to plot the linear creep curves $\varepsilon_f (t)$.

From these curves the values of minimum creep rate, the maximum amount of primary creep strain $\varepsilon_{f\ 1max}$ as well as the transition times are graphically determined in the following.

A special behaviour is pointed out for Waspaloy in the as-received condition. This special behaviour is illustrated in a plot of a linear creep curve from an uninterrupted creep experiment at 700°C (Figure 4). In this condition two quasi-static creep ranges are observed as already published.[12] The first quasi-static creep range with a creep rate $\dot{\varepsilon}_{p\ min\ 1}$ is identified as part of primary creep $\varepsilon_{f\ 1}$. This is due to the very low strain values observed for $\varepsilon_{f\ 1\ max}$. Compared with the classic creep curve depicted in Figure 1 the difference of the observed individual material behaviour of Waspaloy in the as-received condition can be reduced to an extended primary creep range $\varepsilon_{f\ 1}$ (Table 3).

At the transition time t_{23} the (primary) creep rate $\dot{\varepsilon}_{p\ min\ 1}$ changes to the rate $\dot{\varepsilon}_{p\ min\ 2}$ of the secondary creep range.

The transition times are defined for this as-received condition:

- t_{12} : transition to the primary quasi-static creep rate $\dot{\varepsilon}_{p\ min\ 1}$,
- t_{23} : transition of the quasi-static creep rates $\dot{\varepsilon}_{p\ min\ 1}$ to $\dot{\varepsilon}_{p\ min\ 2}$,
- t_{34} : transition between secondary and tertiary creep range.

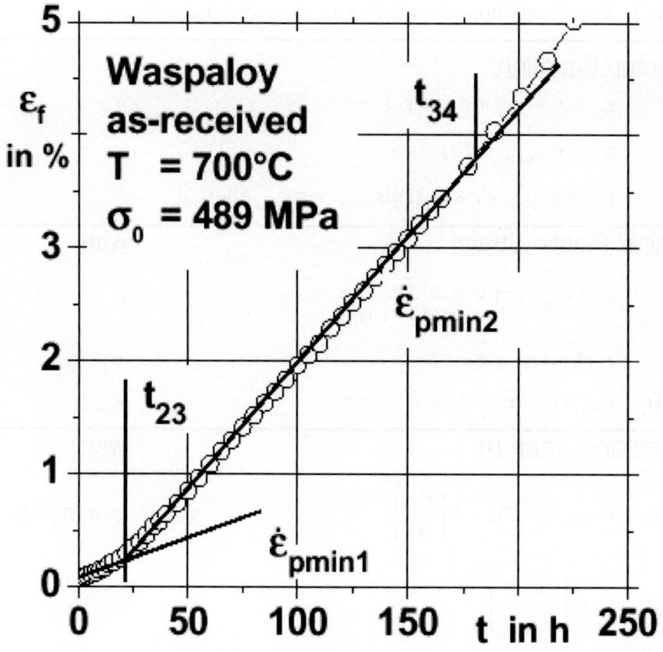

Fig. 4 Linear time-strain-diagram with two quasi-static creep ranges of Waspaloy in as-received condition.

Steps of Modelling As-Received Condition

With the modelled initial plastic strain ei the data points ε_f (T, σ_0, t) were determined as mentioned above and linear creep curves ε_f (t) of all available tests are generated. From these curves the values of the quasi-static creep rates $\dot{\varepsilon}_{p\,min\,1}$ and $\dot{\varepsilon}_{p\,min\,2}$, the maximum amount of primary creep strain $\varepsilon_{f\,1max}$ as well as the transition times can be graphically determined. The graphs of:

- $\varepsilon_{f\,1max}$ against stress σ_0
- $\dot{\varepsilon}_{p\,min\,1}$ against stress σ_0
- $\dot{\varepsilon}_{p\,min\,2}$ against stress σ_0
- transition time t_{12} against $\dot{\varepsilon}_{p\,min\,1}$
- transition time t_{23} against $\dot{\varepsilon}_{p\,min\,1}$
- transition time t_{23} against $\dot{\varepsilon}_{p\,min\,2}$
- transition time t_{34} against $\dot{\varepsilon}_{p\,min\,2}$

were established and checked for consistence in a complex procedure of equibalancing all of these plots. These plots offer the opportunity to check if the determined values for different stresses and temperatures are well correlated. If it is necessary, slightly adjusted values may improve the continuous course of the minimum creep rates $\dot{\varepsilon}_{p\,min\,1}$, $\dot{\varepsilon}_{p\,min\,2}$ and the maximum amount of primary creep strain $\varepsilon_{f\,1\,max}$.

Table 3 Creep equation of waspaloy in as-received condition.

Creep Equation:		
5.	$\varepsilon_p = \varepsilon_i + \varepsilon_{f1} + \varepsilon_{f2} + \varepsilon_{f3}$	
6.	$\varepsilon_{f1} = \varepsilon_{f1\max} \times H(t)$	
7.	$\varepsilon_{f2} = \dot{\varepsilon}_{p\min 1} \times t + H_2(t)[(\dot{\varepsilon}_{p\min 2} - \dot{\varepsilon}_{p\min 1})(t - t_{23})]$	
Initial Plastic Strain:		with:
8.	$\varepsilon_i = \varepsilon_{i\max}(T) \times \frac{\sigma_0^n}{(\sigma_i(T)^n + \sigma_0^n)}$	n
9.	$\sigma_i(T) = \sigma_a + b_\sigma \times T$	σ_a, b_σ
10.	$\varepsilon_{i\max}(T) = \varepsilon_a - b_\varepsilon \times T$	$\varepsilon_a, b_\varepsilon$
Primary Creep (I)		with:
11.	$\varepsilon_{f1\max} = c \times \sigma_0{}^{d} + \left(\frac{\sigma_0}{\sigma_f}\right)^{d_f}$	c, d, σ_f, d_f
12.	$t_{12} = \left(\frac{C_{12}}{\dot{\varepsilon}_{p\min 1}}\right)^{\sigma_{12}}$	C_{12}, α_{12}
13.	$H(t) = 1 - e^{-D \times \left(\frac{t}{t_{12}}\right)^{u}}$	D, u
14.	$\dot{\varepsilon}_{p\min 1} = K_1(T) \times \sigma_0{}^{n}{}_{01} \times e^{a_1 \times \frac{\sigma_0{}^{b_1}}{b_1}}$	n_{01}, a_1, b_1
15.	$K_1(T) = e^{B_1 - Q_1/T}$	B_1, Q_1
16.	$t_{23} = \left(\frac{C_{23}}{\dot{\varepsilon}_{p\min 1}}\right)^{\alpha_{23}}$	C_{23}, α_{23}
Secondary Creep (II)		with:
17.	$\dot{\varepsilon}_{p\min 2} = K_2(T) \times \sigma_0{}^{n}{}_{02} \times e^{a_2 \times \sigma_0{}^{b_2}/b_2}$	n_{02}, a_2, b_2
18.	$K_2(T) = e^{B_2 - Q_2/T}$	B_2, Q_2
19.	$H_2(t) = \frac{1}{2} \times \left(1 + \tanh yp\,(t - t_{23})\right)$	
Tertiary Creep (III)		with:
20.	$t_{34} = \left(\frac{C_{34}}{\dot{\varepsilon}_{p\min 2}}\right)^{\alpha_{34}}$	C_{34}, α_{34}
21.	$\varepsilon_{f3} = C_3 \times \left(\frac{t}{t_{34}}\right)^{f}$	C_3 and f
Validity: (0) $100 \le \sigma_0 \le 600$ MPa and $600 \le T \le 750$°C		
Constants valid for T in K, σ_0 in MPa, t in h, ε in %		

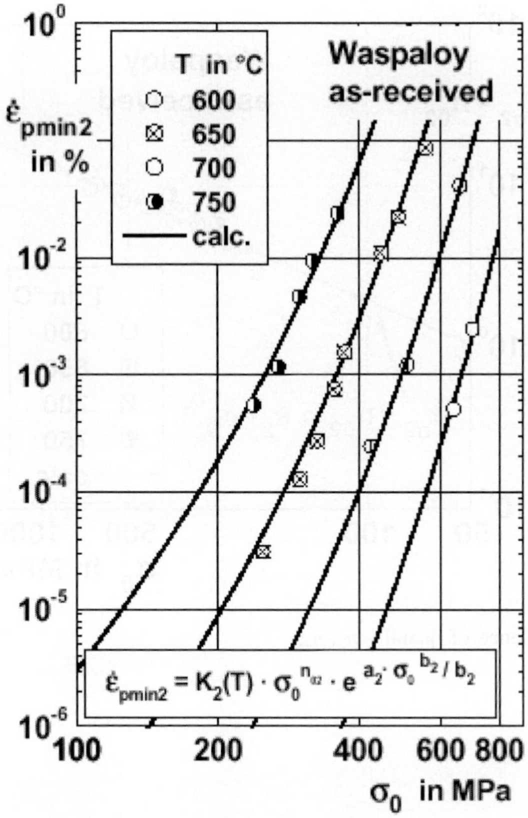

Fig. 5 Minimum creep rate $\varepsilon_{p\,min\,2}$.

On basis of that plots there are performed the modelling of the maximum amount of primary creep strain $\varepsilon_{f\,1max}$ as well as the two quasi-static creep rates $\dot{\varepsilon}_{p\,min\,1}$ of primary creep $\varepsilon_{f\,1}$ and $\dot{\varepsilon}_{p\,min\,2}$ and finally the transition times t_{12}, t_{23}, t_{34}. The determination of a quasi-static creep rate is exemplary described for $\dot{\varepsilon}_{p\,min\,2}$.

From a plot of log $\dot{\varepsilon}_{p\,min\,2}$ against log σ_0 (Figure 5) the dependence of stress σ_0 and temperature T became obvious. The family of curves can be interpreted by an equation of the type.

$$\dot{\varepsilon}_{p\min 2} = K_2(T) \times \sigma_0{}^{n_{02}} \times e^{c \times \sigma_0{}^{c} / d} \tag{2}$$

with constants n_{02}, c, d and the temperature function $K_2(T)$.

In order to model eqn (2) the generalised Norton-exponents.

$$n_{\sigma 2} = \frac{\delta \log \dot{\varepsilon}_{p\min 2}}{\delta \log \sigma_0} \tag{3}$$

were determined from each of the curves at a constant temperature in Figure 5 and were plotted against stress σ_0. From this diagram the coefficient n_{02} was determined as limit of

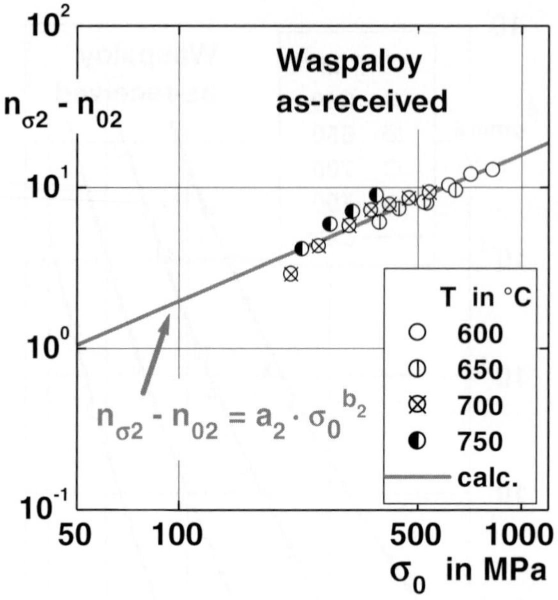

Fig. 6 Stress dependence of quantity $n_{\sigma 2}$-n_{02}.

stress $\sigma_0 \rightarrow 0$. In a next step values of $n_{\sigma 2} - n_{02}$ could be calculated and plotted against stress σ_0 in a double logarithmic diagram (Figure 6).

So the coefficients of an equation

$$n_{\sigma 2} - n_{02} = c \times \sigma_0^{\ d} \tag{4}$$

were determined. The integration of eqn (4) via eqn (3) delivers eqn (2) with an integration constant $K_2(T)$. That constant was modelled with the aid of a simple Arrhenius equation (eqn 18, Table 3).[1, 11] If the minimum creep rate $\dot{\varepsilon}_{p\ min\ 2}$ is recalculated with eqn (2) and the determined coefficients, the measured data points were well interpreted in the range of temperature and stress (Figure 5). The primary quasi-static creep rate $\dot{\varepsilon}_{p\ min\ 1}$ of the as-received condition was determined in analogy.

Generally the principle is respected that all terms of the creep equation that are determined have to be used for the determination of the following constants.

The creep behaviour of the aged condition shows again the well known course of classic creep curves as depicted in Figure 1.

Evaluation of Creep Equation

The grade of the established creep equation of Waspaloy in as-received condition was evaluated by recalculation of the creep data points from the experiments across relevant

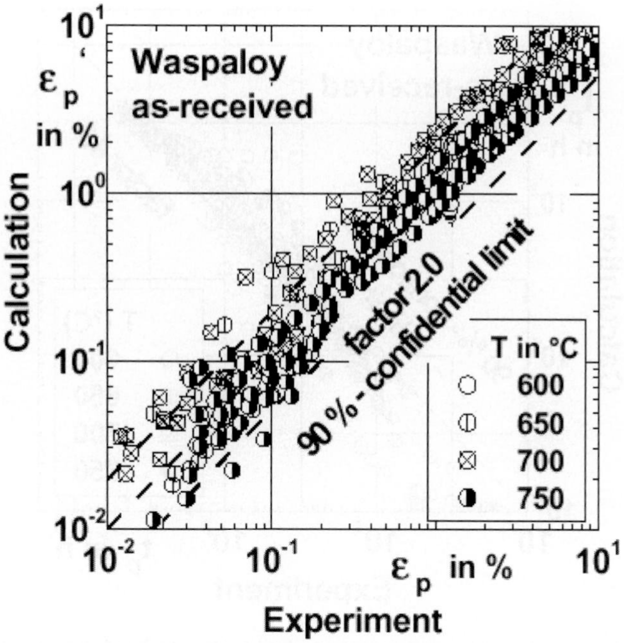

Fig. 7 Strains ε_p from performed creep tests compared with calculated data points with the creep equation for Waspaloy in as-received condition.

ranges of strain, time and temperature. The calculated strain values ep' meets the experimental determined strain values ε_p within an acceptable scatter band (Figure 7).

It also meets the calculated times t_p' the experimental determined times t_p in a slightly wider but also acceptable scatter band (Figure 8). Usually the recalculation of times have greater differences.

SUMMARY AND CONCLUSIONS

In a DFG-research program, nickel-base-alloys for application as rotor-material in future steam turbines were investigated. Creep and creep rupture behaviour of the alloys Inconel 706, Inconel 617, Waspaloy were investigated. All these alloys were examined in an as-received condition and an aged condition. The aged condition was generated by additional annealing at 750°C for a duration of 5000 hours to approximate 100,000 hours at 700°C.

In creep tests on Waspaloy in as-received condition an individual material behaviour can be observed. The classic course of creep curves described with primary, secondary and tertiary creep was slightly modified to describe creep with a typical phenomenological creep equation. The solution was an introduction of a second quasi-static creep range in extension of primary creep just before the usual quasi-static creep rate of secondary creep. With this extension a

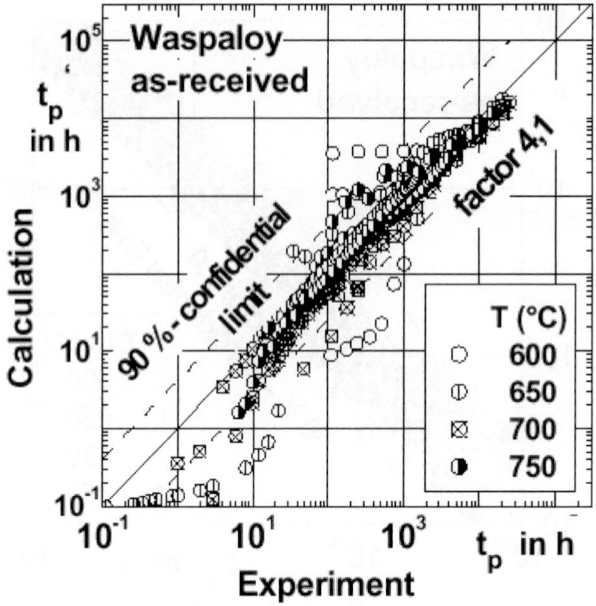

Fig. 8 Times t_p from performed creep tests compared with calculated data points with the creep equation for Waspaloy in as-received condition.

creep equation of the type of the modified Garofalo equation is now available for the description of creep behaviour of this alloy.

This individual material behaviour of Waspaloy disappears after aging at 750°C/5000 hours and the classic course of creep curves re-established. This correlates with observations made by a partner research team. It was reported that a bimodal size distribution of γ'-precipitations were present in the as-received condition. These primary γ'-precipitations were growing while the smaller secondary γ'-precipitations present in the as-received condition disappeared.

The establishment of a creep equation of the modified Garofalo type was performed successfully for Waspaloy in the as-received condition although the necessary requirement of stability of microstrucure was not fulfilled. In the case of the aged condition the modified Garofalo equation in its usual form can be applied.

ACKNOWLEDGEMENTS

Thanks are due to the Deutsche Forschungsgemeinschaft for the promotion of the work.

REFERENCES

1. J. Granacher and T. Preussler: 'Creep of Some Gas Turbine Materials', *Advances in Material Technology for Fossil Power Plants*, R. Viswanathan and R. I. Jaffee, eds., ASM International, Chicago, 1987, 511–518.
2. J. Granacher, M. Monsees, P. Hillenbrand and C. Berger: 'Software for the Assessment and Application of Creep and Creep Rupture Data, *Nuclear Engineering and Design*, **190**, 1999, 273–285.
3. K. H. Kloos, J. Granacher and A. Pfenning: Creep Equations for High Temperature Alloys on the Basis of a Parametric Assessment of Multi-Heat Data, *Steel Research*, **67**(3), 1996.
4. Inconel 706, Brochure from Huntington Alloys Inc., Huntington WV, 1974.
5. S. Müller and J. Rösler: 'On the Creep Crack Growth Behaviour of Inconel 706', *Inter. Congress on Advanced Materials and Processes EUROMAT99*, 1999.
6. T. Shibata, et al.: 'Effect of Cooling Rate from Solution Treatment on Precipitation Behavior and Mech. Properties of Alloy 706', *Superalloys 718, 625 and 706*, E. A. Loria, ed., Pittsburgh, TMS., 1997.
7. J. Rösler, M. Götting, D. Del Genovese, B. Böttger, R. Kopp, M. Wolske, F. Schubert, H. J. Penkalla, A. Thoma, A. Scholz and C. Berger: 'Wrought Ni-Base Alloys for Advanced Gas Turbine Disc and USC Steam Turbine Rotor Applications', *Proceedings of the 7th Liége Conference, Materials for Advanced Power Engineering 2002*, Lecomte-Beckers, Carton, Schubert, Ennis, ed., Liége, 2002, 89–106.
8. H. J. Penkalla, J. Wosik and F. Schubert: 'Microstructure and Structural Stability of Candidate Materials for Turbine Disc Applications beyond 700°C', *Proceedings of 7th Liège Conference*, 2002.
9. H. J. Penkalla, J. Wosik, W. Fischer and F. Schubert: 'Structural Investigations of Candidate Materials for Turbine Disc Applications beyond 700°C', *Superalloys 718, 625, 706 and various Derivatives*, E. A. Loria, ed., Pittsburgh, TMS., 2001, 279–290.
10. J. M. Granacher, M. Oehl and T. Preußler: 'Comparison of Interrupted and Uninterrupted Creep Rupture Tests, Steel Research, **63**, 1992, 39–45.
11. C. Berger, J. Granacher and A. Thoma: 'Creep Rupture Behaviour of Nickel Base Alloys for 700°C-Steam Turbines', *Superalloys 718, 625, 706 and Derivatives*, Pittsburgh, PA, TMS, 2001.
12. A. Thoma, A. Scholz and C. Berger: 'Alloy Design for Ultra High Temperature Steam Turbine Applications: Creep Behaviour and Modelling of Creep', *Proceedings of the 7th Liége Conference, Materials for Advanced Power Engineering 2002*, Lecomte-Beckers, Carton, Schubert, Ennis, ed., Liége, 2002, 1551–1560.

A Study of New Carbon Free Martensitic Alloys with Superior Creep Properties at Elevated Temperatures over 973 K

S. MUNEKI, H. OKUBO, H. OKADA, M. IGARASHI and F. ABE
Heat Resistant Materials Group
Steel Research Centre
*National Institute for Materials Science (**NIMS**)*
1-2-1, Sengen, Tsukuba, Ibaraki, 305-0047
Japan

ABSTRACT

In order to improve the creep strength of the ferritic steel at elevated temperatures over 973 K, a new attempt has been demonstrated using carbon free Fe-Ni martensitic alloys strengthened by Laves phase such as Fe_2Mo Because of the large amount of Ni concentration, the A_{c1} temperature of these alloys was relatively lower than 973 K so that the reverted transformation to austenite occurs within the creep testing temperature range. In this study, creep tests and microstructural observation has been performed on the boron added alloy to clarify the role of reverted austenite and to evaluate the effect of boron on creep properties of the steel. The creep behaviour of alloys is found to be completely different from that of the conventional high-Cr ferritic steels. The alloys exhibit a gradual change in the creep rate with strain both in the transient and acceleration creep regions, and give a larger strain for the minimum creep rate. In these alloys the creep deformation takes place very homogeneously and no heterogeneous creep deformation is enhanced even at low stress levels. The fine precipitates are found to be stable even at 973 K and effectively decrease the minimum creep rate much lower values than that of the conventional high-Cr ferritic steels. Moreover, superior creep resistance is achieved by addition of Mo to the alloy.

INTRODUCTION

High Cr ferritic steels have successfully been used for large diameter and thick section boiler components such as main steam pipe and header in super critical (SC) boilers in fossil-fired power plants.[1] Recent trend towards utilization of clean energy leading to protection of global environment has been accelerating with the application of ultra super critical (USC) boilers. These operate with higher efficiency in power generation than conventional ones and thus release less carbon dioxide etc. into the atmosphere.[2, 3] The USC boiler requires heat resistant materials with improved creep rupture strength at elevated temperatures over 873 K, because of increase in operating temperature and pressure of the steam used. The expected goal of the steam condition for USC boilers is, however, now

Table 1 Chemical composition of alloys used (wt.%).

Alloy	C	Si	Mn	Ni	Cr	Mo	Ti	Al	B	N	Fe
5 Mo	0.001	<0.01	0.49	11.74	4.88	4.89	0.19	0.12	0.0048	0.0006	Bal.
5 Mo-1 Ti	0.002	0.01	0.5	11.81	4.92	4.90	0.94	0.12	0.0046	0.0005	Bal.
10 Mo	0.001	0.01	0.5	11.91	4.93	9.67	0.19	0.13	0.0052	0.0006	Bal.

considered to be 903 K and 30 MPa,[3] which might be the applicable limit of the conventional ferritic steels regarding their creep strength and the steam-oxidation resistance.

Creep deformation of the ferritic steels is controlled by recovery and softening process of the tempered martensite, which strongly depends on the constitution of the microstructure and its change with time.[4, 5]

To achieve long-term stability of the tempered martensite is considered to be the key to increase the creep resistance of the steels at elevated temperatures over 873 K.

Muneki et al. have also noted that the creep behaviour of the Fe-Ni-Co alloys exhibit gradual change in the creep rate with strain both in the transition and acceleration creep regions, and give a larger strain for the minimum creep rate.[6] Furthermore, the addition of Pd and Cr to the carbon free Fe-Ni-Co alloy greatly enhances much the creep resistance at elevated temperatures over 923 K.[7, 8]

In this paper, creep properties over 973 K in the carbon and cobalt free martensitic alloys with 0.005B is discussed.

EXPERIMENTAL PROCEDURE

The chemical compositions of alloys used in this study are shown in Table 1. They were melted as 10 kg-ingots in a vacuum induction furnace. The ingots were pressed and rolled to 16 mm square bars after heating for 1 hour at 1473 K. They were aged for 1 hour between 773 K and 1113 K after solution treatment for 30 min. at 1273 K. To examine the long-term stability of the microstructure of these alloys, specimens were aged for 1 to 6,060 hours at temperatures of 973 K. Creep rupture tests were conducted in the range of temperatures from 923 to 1073 K and in the range of stresses from 60 to 300 MPa using a 6 mm diameter and 30 mm gauge length tensile specimens in the as solution treated condition. The microstructures of the alloys were examined using an optical microscopy and a field emission scanning electron microscopy. Transformation temperatures of A_s, A_f, M_s and M_f were measured by dilatometry by the following procedure. The specimens were induction heated from room temperature to 1273 K heating rate of 10 K/min., kept for 10 min. at 1273 K, and then cooled to room temperature with cooling rate of 10 K/min. Identification of precipitates during the creep test was investigated by XRD analysis from the given crept specimen.

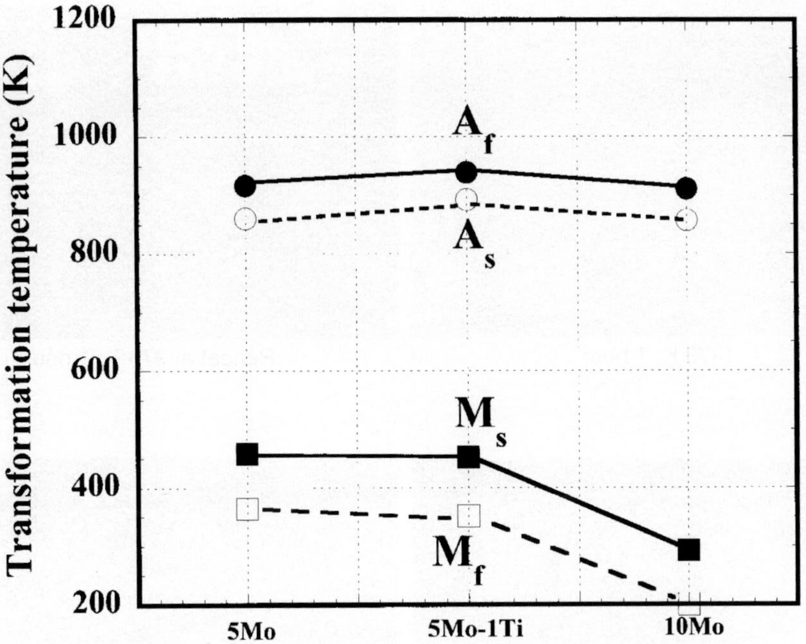

Fig. 1 Change of transformation temperatures by additional elements of the Fe-12Ni-5Cr-Mo-0.005B alloy.

RESULTS AND DISCUSSION

MICROSTRUCTURES DURING CREEP TESTS

Figure 1 shows the effect of additional elements on the transformation temperatures of A_s, A_f, M_s and M_f measured by dilatometry with heating and cooling with the rates of 10 K/min. The A_s and A_f temperatures of the 5Mo and 10Mo alloys were about 860 K and 910 K, respectively. On the other hand, those of the 5Mo-1Ti alloy were identified slightly higher than those of the 5Mo and 10Mo alloys. While, the temperatures of M_s and M_f in the 5Mo and 5Mo-1Ti alloys were about 455 K and 360 K, respectively.

And those of the 10Mo alloy were identified drastically lower than those of other two alloys. As the A_s and A_f are temperatures lower than those of conventional high Cr ferritic heat resistant steels,[6] the microstructure of the matrix of the three alloys are revert austenite during creep tests 973 K in this study.

Figure 2 shows optical micrographs of the Fe-12Ni-5Cr-10Mo-0.005B alloy showing the effect of reheating treatment after solution treatment at 1473 K for 1 hour. As clearly shown in this figure, prior austenite grain size did not changed after reheating treatment at 1173 K.

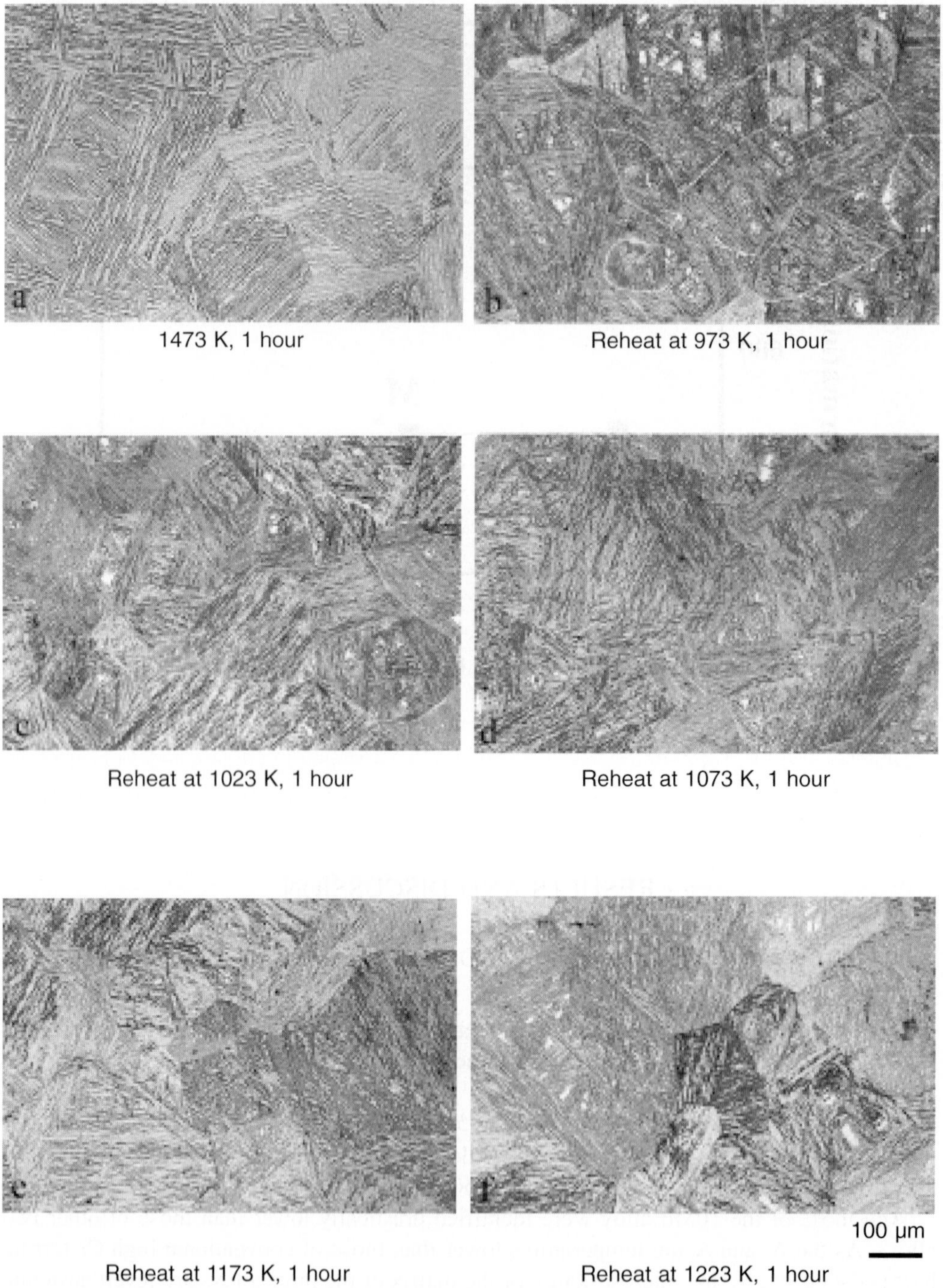

Fig. 2 Optical micrographs of the Fe-12Ni-5Cr-10Mo-0.005B (10Mo) alloy showing the effect of boron addition on the recrystallisation behaviour.

Namely reverted austenite maintained unrecrystallised microstructure even at 1173 K. This is due to a similar effect as retardation of the recovery process and the suppression of recrystallisation by the addition of B into the 18%Ni maraging steel.[9]

Hardness Changes of the Three Alloys

Figures 3a and b show hardness changes of the three alloys after solution treatment at 1273 K for 30 min. They showed higher initial hardness more over HV280. All of the alloys were hardened by the 1 hour aging especially at temperatures of 773 and 823 K, as shown in Figure 3a. Hardness values of the 5Mo-1Ti and the 10Mo alloys were about HV500. Peak hardness of the 5Mo alloy exhibited at about HV400. Hardness of these alloys indicated higher values even at 1073 K than those of their initial values.

Figure 3b shows the time dependence of the hardness change of the three alloys aged at 973 K until 6,060 hours. Hardness of the 5Mo and 5Mo-1Ti alloys gradually increased at about HV320 and HV350 with the time until 100 hours. Hardness of the 10Mo alloys reached about HV400 by 10 hours, and kept higher values at 2,000 hours. All of them were hardened by the aging time and then slightly softened after 1,000 hours aging. However, the hardness of them after 6,060 hours aging was remarkably higher than that of solution treated condition.

Creep Properties at 973 K

Figures 4a and b show creep properties of the three alloys crept at 973 K and 200 MPa. The 5Mo and the 5Mo-1Ti alloys ruptured at very short times, less than 80 hours. Creep resistance was remarkably improved by an increment of Mo contents, as shown the result of the 10Mo alloy. Time to rupture increased to about 2,100 hours. Rupture elongation and reduction in area slightly decreased from the 5Mo alloy at 50.8% and 90.2% to the 10Mo alloy at 32.8% and 78.4%, in respectively.

Creep rate vs. time curves of the specimens crept at 973 K and 200 MPa are shown in Figure 4b. The creep rate of the 5Mo alloy gradually decreased from 6×10^{-2} (1/hour) to 3.3×10^{-3} (1/hour) and then increased and finally ruptured. The deformation mode of other two alloys was similar but their minimum creep rates remarkably decreased. The minimum creep rate decreased with the Ti addition of the 5Mo alloy, and furthermore the rate drastically decreased with the increased Mo addition. The minimum creep rate of the 10Mo alloy was at 1.9×10^{-5} (1/hour), which is about two orders lower than that of the 5Mo alloy.

Creep Properties Above 973 K

Figure 5 shows the stress dependence of the time to rupture at 973 K. Logarithmic time to rupture of the 5Mo alloy increased with decrease of logarithmic stress from about 34 hours at 200 MPa to 952 hours at 140 MPa. Time to rupture at 200 MPa of the 5Mo-1Ti alloy was about 80 hours, which was slightly longer than that of the 5Mo alloy at about 30 hours.

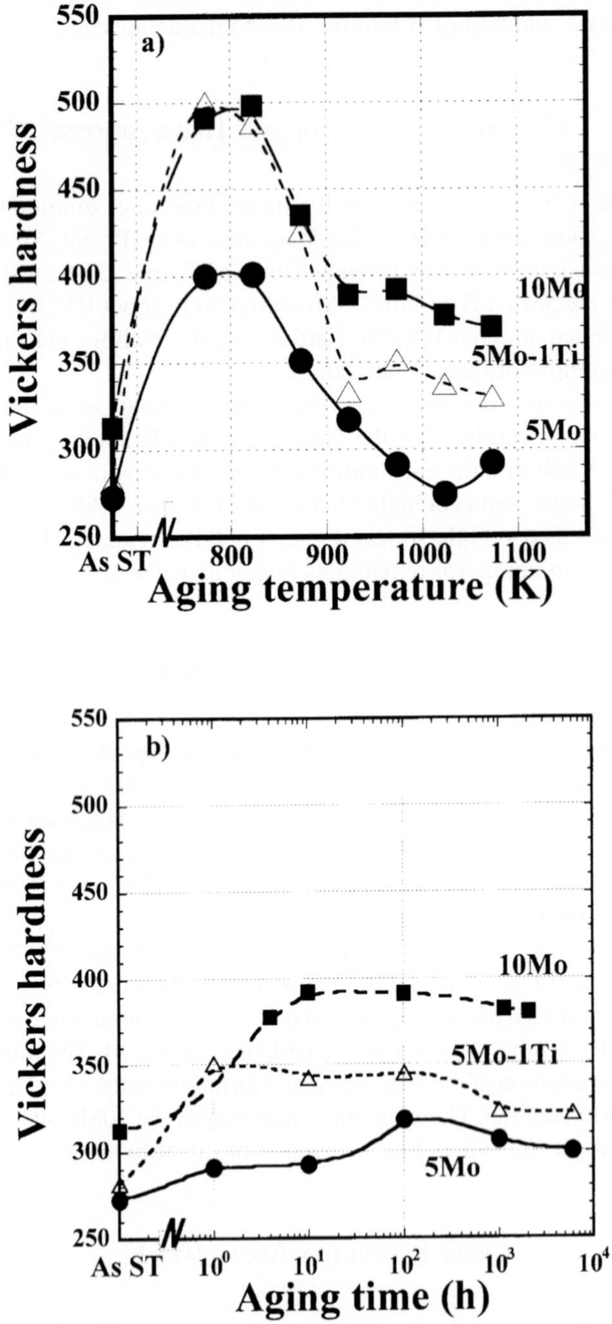

Fig. 3 Hardness change of the three alloys after solution treatment at 1273 K for min. (a) isochronal aging for 1 hour and (b) isothermal aging at 973 K.

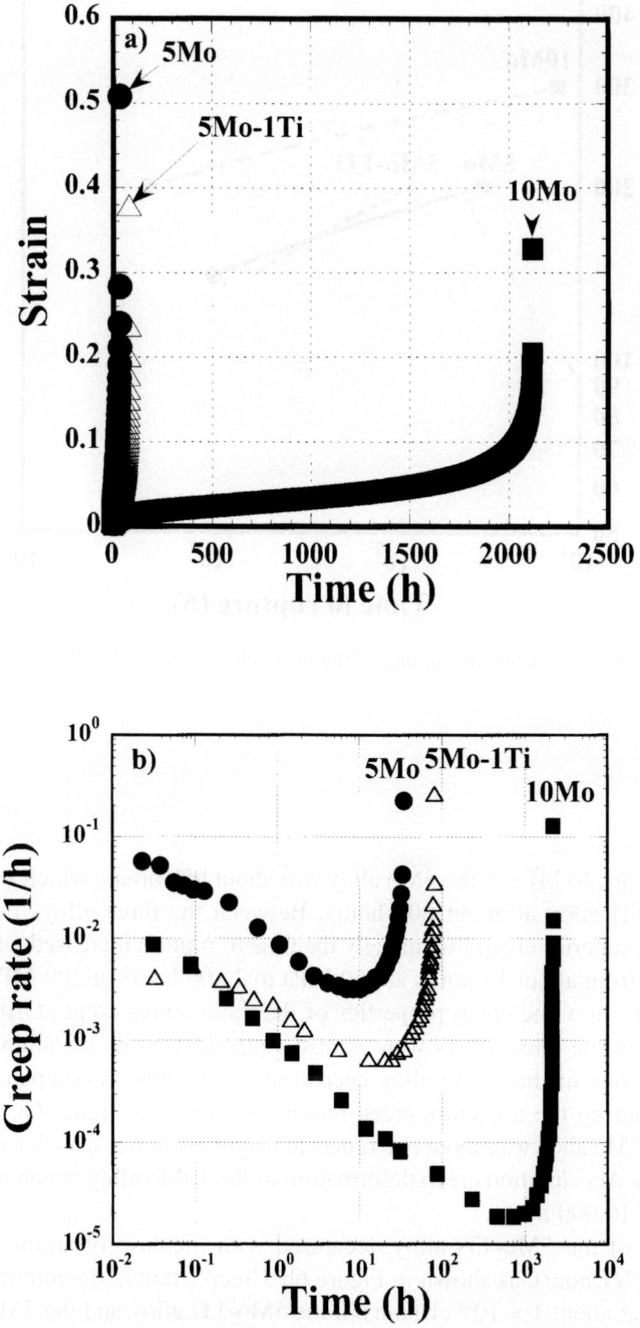

Fig. 4 Creep properties of the three alloys crept at 973 K and 200 MPa, (a) creep strain vs. time curves and (b) creep rate vs. time curves.

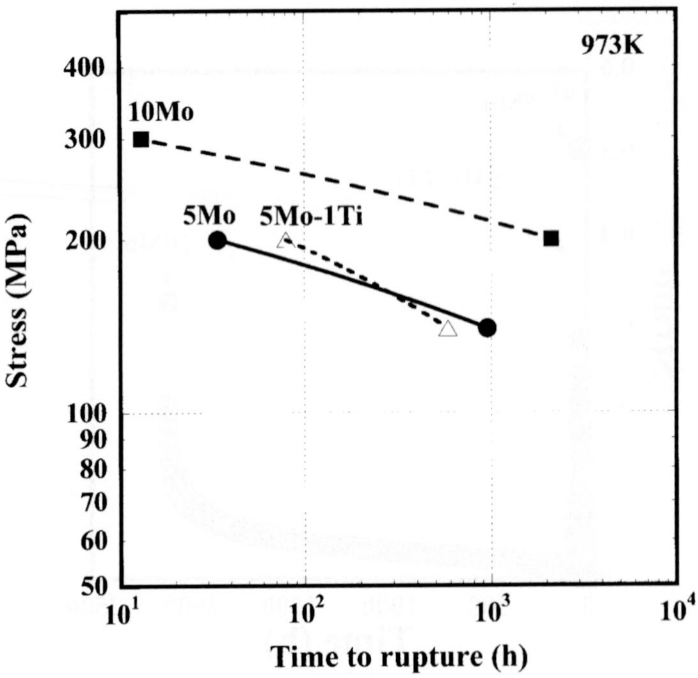

Fig. 5 Stress dependence of time to rupture of the three alloys crept at 973 K.

However, the time at 140 MPa of the 5Mo alloy was about 950 hours, which was longer than that of the 5Mo-1Ti alloy at about 600 hours. Between the three alloys the 10Mo alloy exhibited the most superior creep life, namely the time to rupture increased with decrease of logarithmic stress from about 13 hours at 300 MPa to 2,100 hours at 200 MPa.

Figure 6a and b show the creep properties of the three alloys crept at 1023 K and 100 MPa. Creep rates of the three alloys were almost equivalent at the beginning of the creep test, but the creep rate of the 10Mo alloy decreased remarkably as compared with that of other two alloys during the transition creep region, as shown in Figure 6a. The minimum creep rate of the 10Mo alloy was about two orders in magnitude lower than that of the 5Mo and the 5Mo-1Ti alloys. Acceleration creep deformation of the 10Mo alloy began to pass through the testing time at 10,000 hours.

The creep rate of the 5Mo-1Ti alloy decreased with increase of strain from 1×10^{-2} (1/hour) to 1×10^{-4} (1/hour), as shown in Figure 6b. Creep strain at the minimum creep rate slightly exhibited at about 1×10^{-4} (1/hour) in the 5Mo-1Ti alloy and the 5Mo alloy. Creep strain at the minimum creep rate of the 10Mo alloy extremely small value at about 1×10^{-6} (1/hour). Homogeneous creep deformation of the 10Mo alloy began to start at the very small strain at 0.02. Figure 7a and b show the creep properties of the three alloys crept at 1073 K and 100 MPa. Initial creep rates of the three alloys exhibited similar values at about 1×10^{-2}

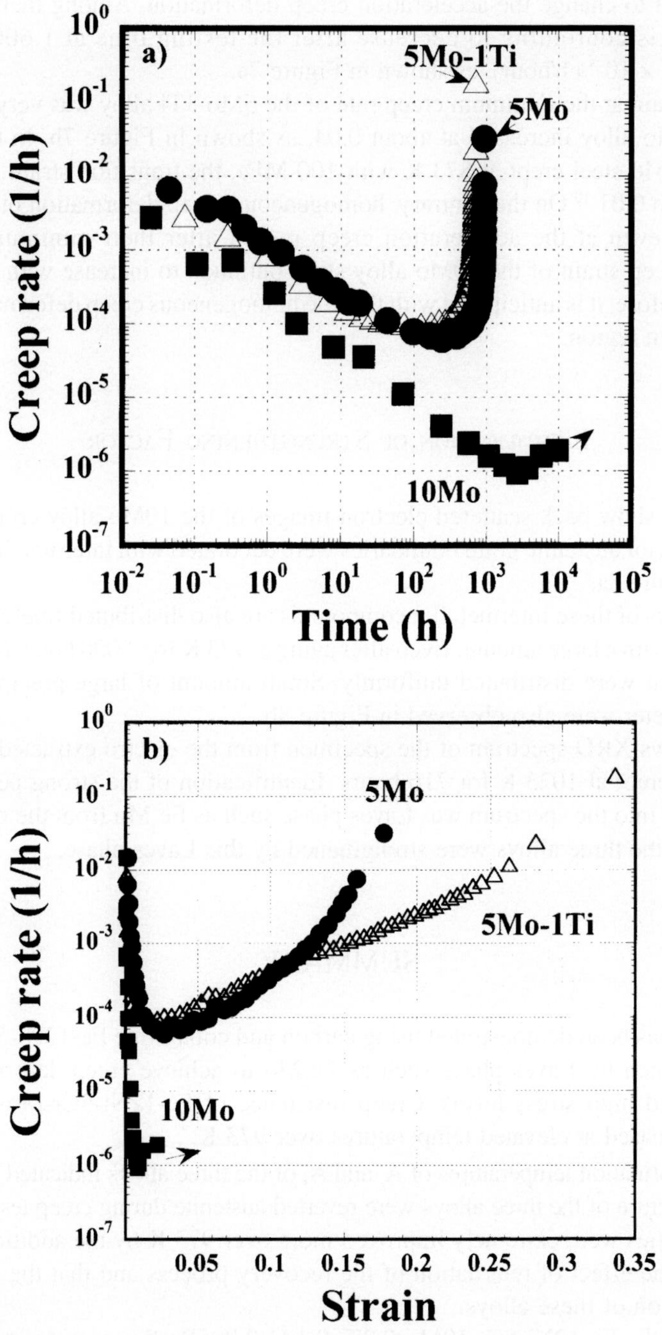

Fig. 6 Creep properties of the three alloys crept at 1023 K and 100 MPa, (a) creep strain vs. time curves and (b) creep rate vs. time curves.

(1/hour). The creep rates of the 5Mo and 5Mo-1Ti alloys gradually decreased with increasing time, and turned to change the acceleration creep deformation. Among them creep rate of the 10Mo alloy is continuing to decrease after the testing time at 1,600 hours at the creep rate of 1×10^{-5} (1/hour), as shown in Figure 7a.

The creep strain at the minimum creep rate of the 5Mo-1Ti alloy was very small, but the strain of the 5Mo alloy increased at about 0.04, as shown in Figure 7b. In the case of the modified 9Cr-1Mo steel crept at 873 K with 100 MPa, the transition strain to acceleration creep is less than 0.01.[10] On the contrary, homogeneous creep deformation of the 5Mo alloy still continued even at the acceleration creep region after their minimum creep rates. Furthermore, creep strain of the 10Mo alloy still continued to increase with decreasing the creep rate. Therefore, it is anticipated with that the homogeneous creep deformation continues to the large strain region.

Identification of Strengthening Factor

Figure 8a and b show back scattered electron images of the 10Mo alloy crept at 973 K for 2,118.6 hours. Prior austenite grain boundaries were decorated with intermetallic compounds, as shown in Figure 8a.

Large amounts of these intermetallic compounds are also distributed finely and uniformly inside the grain with a large amount. Even after aging at 973 K for 2,000 hours fine precipitates less than 100 nm were distributed uniformly. Small amount of large precipitates at about 500 nm in diameter were also observed in Figure 8b.

Figure 9 shows XRD spectrum of the specimen from the electro extracted residue of the 5Mo-1Ti alloy crept at 1023 K for 716 hours. Identification of the strong peaks as marked with solid circle into the spectrum was Laves phase such as Fe_2Mo from the index card. It is considered that the three alloys were strengthened by this Laves phase.

SUMMARY

A new attempt has been demonstrated using carbon and cobalt free Fe-12Ni-5Cr martensitic alloys strengthened by Laves phase such as Fe_2Mo to achieve creep deformation at high temperatures and high stress levels. Creep resistance of Fe-12Ni-5Cr-Mo-0.005B alloys remarkably increased at elevated temperatures over 973 K.

1. As the transformation temperatures of A_s and A_f of the three alloys indicated drastically low, the microstructure of the three alloys were reverted austenite during creep tests over 973 K.
2. Creep properties were extremely improved more over 973 K by the addition of B, which depends on the effect of retardation of the recovery process and that the suppression of recrystallisation of these alloys.
3. Creep life of the Fe-12Ni-5Cr-10Mo-0.2Ti-0.1Al-0.005B alloy was drastically extended from 13 hours at 973 K and 300 MPa to 2,100 hours at 973 K and 200 MPa.
4. Creep testing time at 1023 K and 100 MPa passed over 10,000 hours after the minimum creep rate in the 10Mo alloy.

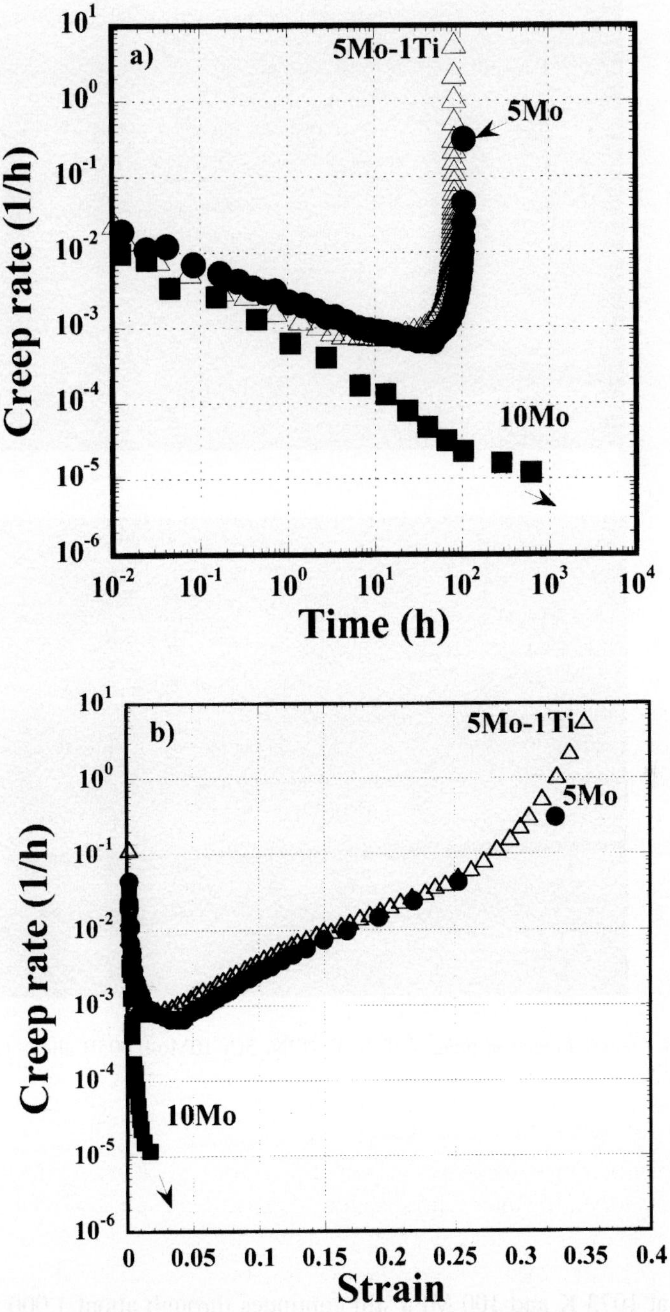

Fig. 7 Creep properties of the three alloys crept at 1073 K and 100 MPa, (a) creep strain vs. time curves and (b) creep rate vs. time curves.

Fig. 8 Back scattered electron images of the Fe-12Ni-5Cr-10Mo-0.005B alloy crept at 973 K for 2118.6 hours.

5. Creep test at 1073 K and 100 MPa still continues through about 1,000 hours before the minimum creep rate.
6. Superior creep resistance in the Fe-12Ni martensitic alloys depends on the distribution a fine and a large amount of precipitates, which is composed of Laves phase such as Fe_2Mo precipitated into the unrecrystallised austenitic matrix during creep tests.

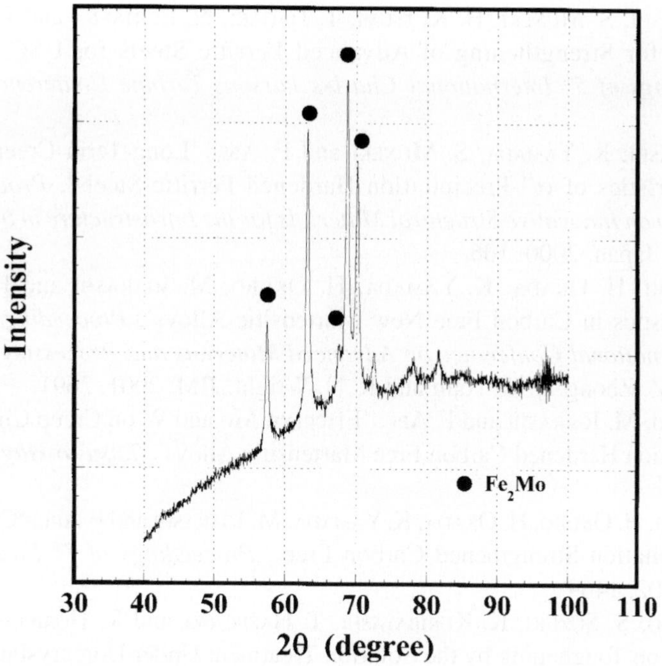

Fig. 9 XRD spectrum of the specimen crept at 1023 K for 716 hours in the 10 Mo alloy.

ACKNOWLEDGEMENT

The authors would like to express their sincere gratitude to Mrs. Wakako Moriiwa, Mrs. Toshiko Ishizuka, Mrs.Hideko Miyazaki and Miss Miyuki Yoshie for their sincere collaboration in experiments.

REFERENCES

1. V. K. Sikka, C. T. Ward and K. C. Phomas: 'Modified 9Cr-1Mo Steel – An Improved Alloy for Steam Generator Application', *ASM International Conference on Production, Fabrication, Properties and Application of Ferritic Steels for High Temperature Applications*, Warren, PA., 1981.
2. F. Masuyama, I. Ishihara, T. Yokoyama and M. Fujita: 'Application of Tungsten Strengthened 9-12% Cr Boiler Steel Pipes', *The Thermal and Nuclear Power*, **46**, 1995, 498.
3. K. Muramatsu: 'Development of Ultra-Super Critical Plant in Japan', *Proceedings of Advanced Heat Resistant Steels for Power Generation*, Miramar Palace, San Sebastian, Spain, 1998, 543.

4. M. Igarashi, S. Muneki, H. Kutsumi, T. Itagaki, N. Fujitsuna and F. Abe: 'A New Concept for Strengthening of Advanced Ferritic Steels for USC Power Plant', *Proceedings of 5th International Charles Parsons Turbine Conference*, Cambridge, 2000.

5. M. Igarashi, K. Yamada, S. Muneki and F. Abe: 'Long-Term Creep Deformation Characteristics of α'' Precipitation Hardened Ferritic Steels', *Proceedings of 4th Workshop on Innovative Structural Materials for the Infrastructure in the 21st Century*, Tsukuba, Japan, 2000, 106.

6. S. Muneki, H. Okada, K. Yamada, H. Okubo, M. Igarashi and F. Abe: 'Creep Chracteristics in Carbon Free New Martensitic Alloys', *Proceedings of 4th Pacific Rim International Conference on Advanced Materials and Processing (PRICM4)*, S. Hanada, Z. Zhong, S. W. Nam and R. N. Wright, JIM, 2001, 2691.

7. S. Muneki, M. Igarashi and F. Abe: 'Effect of Mo and W on Creep Characteristics of Precipitation Hardened Carbon Free Martensitic Alloys', *Tetsu-to-Hagane*, Japan, **88**, 2002, 45.

8. S. Muneki, H. Okubo, H. Okada, K. Yamada, M. Igarashi and F. Abe: 'Creep Properties of Precipitation Strengthened Carbon Free', *Proceedings of 7th Liege Conference*, **21**(3), 2002, 1469.

9. T. Yasuno, S. Suzuki, K. Kuribayashi, T. Hasegawa and R. Horiuchi, 'Effect of B Contents on Toughening by the Solution Treatment Under Unrecrystallised Austenite in 18%Ni Maraging Steel', *Tetsu-to-Hagane*, Japan, **83**, 1997, 671.

10. K. Kimura, H. Kushima and F. Abe: 'Heterogeneous Changes in Microstructure and Degradation Behaviour of 9Cr-1Mo-V-Nb Steel During Long Term Creep', *Key Engineering Materials*, Trans Tech Publications, Switzerland, **171-174**, 2000, 483.

GAS TURBINE TECHNOLOGY

Development and Validation of a New High Load/High Lift Transonic Shrouded HP Gas Turbine Stage

Brian Haller
ALSTOM Power, PO BOX 1
Waterside South, Lincoln
LN5 7FD, UK
brian.haller@power.alstom.com

Gurnam Singh
ALSTOM Power
Newbold Road, Rugby
CV21 2NH, UK
gurnam.singh@power.alstom.com

ABSTRACT

This paper details some of the recent developments that the Company has made to develop and test a new highly loaded high pressure turbine stage design. A new approach was used for the preliminary design of the stage (adapted from the Denton U3 method) and this is described. For the first time the standard empirical loss correlations were abandoned. The targets for the stage are very aggressive. The stage incorporates many innovative features to address aerodynamic and other parasitic losses, including 'Controlled Flow' technology to address secondary flow losses. This new approach is then validated against detailed measurements taken for the stage i.e. the development loop was closed. Studies were made at design and off-design conditions (varying speed and pressure ratio up to higher load conditions). Although detailed interpretation of the unsteady results has not yet been completed, some understanding is drawn from these results.

INTRODUCTION

Worldwide in the gas turbine industry there is a trend towards high load/high lift turbomachinery design with high performance since this can provide the following benefits:

- reduced cost, weight and number of components to cool (due to fewer stages and reduced blade wetted area). A high load HP stage drops the temperature sufficiently to avoid cooling the second stage.

- shorter, stiffer rotor shafts with higher critical speeds and potential to use fewer bearings
- easier maintenance with overhung turbine rotor construction.

All businesses are under intense pressures to increase their profit margins and produce cost effective components and this approach provides a way of meeting this aim.

However success with such an endeavour is critically dependent upon still being able to achieve ultra-high performance which means that the classical 'Smith Chart law' (i.e. decrease of stage efficiency with increased loading) has to be changed. It is believed that this is now achievable using the latest physically-based loss prediction methods and new design technologies.

With such high load/high lift stage designs there are significant aerodynamic challenges since :

- the flow is transonic/supersonic and therefore control of profile/shock/ trailing edge losses is important.
- the high deflections in the blade rows can lead to high secondary losses, particularly in the low aspect ratio HP stages.
- the drive to high lift (high pitch/width) designs can result in high convex curvatures on the blade surfaces which can lead flow over-expansion and strong shocks.
- the high pressure ratios across the stages mean that very careful attention has to be paid to control of leakage flow losses.
- high velocities in the flow path mean that all unnecessary "aerodynamic nasty" /cavity losses have to be avoided.
- unnecessary curvature changes in the transonic flow path have to be avoided.
- there is more potential for interactions, causing unsteady flow /aeromechanical issues.

The objectives of this paper are:

- to show some of the developments the Company has made to develop and test such a high load/high lift HP turbine stage design.
- to show how a sound understanding of the flow physics can be used to derive the optimum design and to present new approaches to address aerodynamic losses.
- to validate the latest 3D steady and unsteady stage viscous methods against the detailed experimental data. The test data are used to 'close the loop' and improve understanding of the flow physics and develop the in-house design methods as shown on Figure 1.
- to show the way unsteady measurements can be used to resolve the rotor flow field, validate CFD, and give an insight into the unsteady flow field downstream of the rotor.

NEW STAGE DESIGN

The new HP stage detailed in this paper is patent applied for and pending (Ref. 1) and has the following overall design characteristics:

Loading $\Delta H/U^2 = 2.0$

Inlet swallowing capacity $\dfrac{\dot{m}\sqrt{T_{01}}}{P_{01}} = 5.0 \times 10^{-4}$

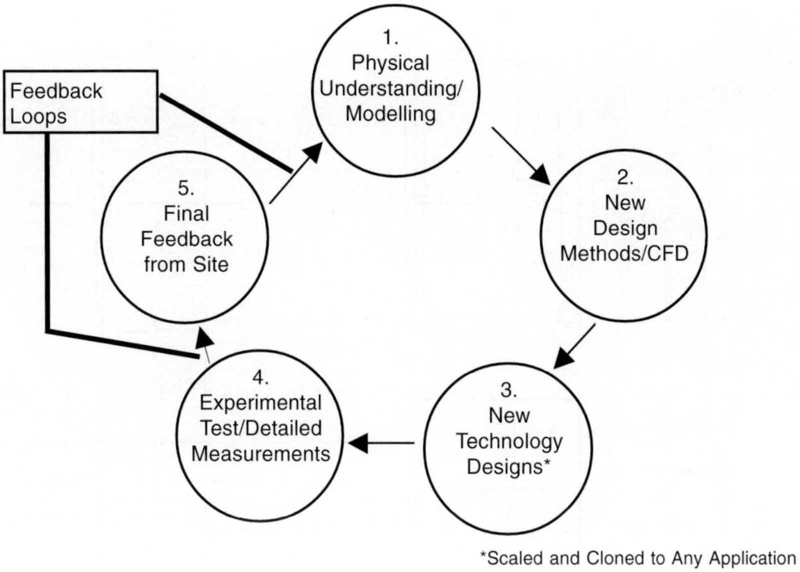

Fig. 1 Development approach.

The design pressure ratio of the stage is around 4:1 but it has the capability to operate efficiently to higher load conditions. The design was targeted at producing an efficiency gain of +4% relative to the current in-service stage.

The methods and philosophies used to design the stage are detailed in the following sections.

Stage Layout and Through Flow Design

The main features of the new design are:

- The low V_a/U to increase the blade heights and reduce the secondary losses whilst keeping the mean diameter the same as the existing design (the V_a/U is 0.409 using the mean of rotor inlet and exit flow conditions).
- 'Controlled Flow design philosophy' (described below).
- a totally conical smooth endwall design was used within the new HP and LP stage designs as past experience shows that this gives the highest performance. Flare was used within the HP NGV as this reduces the flare in the HP rotor and thus reduces endwall losses. The smooth flarepath avoids detrimental endwall curvature effects in the transonic flow.
- to further reduce endwall losses the inter-row axial gap was reduced from around 50% to 30% of the rotor axial width at the root.

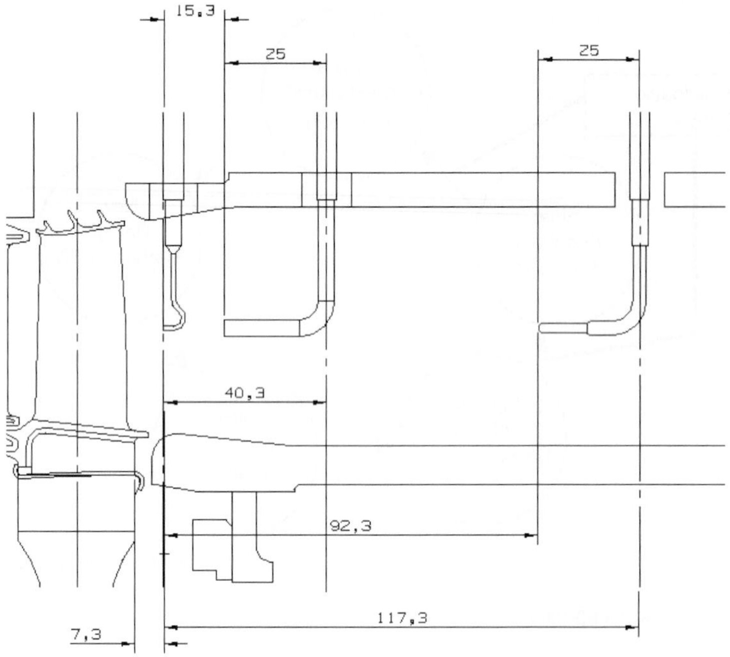

Fig. 2 Layout of the new HP stage in the warm air turbine test facility showing position of downstream traverse planes.

The layout of the new stage installed in the Warm Air Turbine Test Rig (WATR) is shown on Figure 2. There was a 'close-up' traverse plane around half an axial chord downstream of the rotor and two further planes downstream. The measurements in the close-up traverse plane are mainly of interest since :

- this is where the leading edge of the next blade row would be positioned
- more details of the secondary flows are obtained for comparison with the design predictions.

It is well known that a downstream mixing loss occurs for these high load transonic HP stages (around 2% loss in efficiency). Hence the design philosophy is to minimise the length of downstream ducts and suppress the mixing loss by acceleration through the next blade row.

Controlled Flow Design Philosophy

A 'Controlled Flow' philosophy was used for the NGV and rotor (patented ALSTOM steam and gas turbine technology). After 12 years of R & D, exploring many options, this gave the highest performance for ALSTOM steam turbines and was demonstrated by numerous model turbine and site cylinder efficiency measurements to address secondary losses and provide very substantial efficiency gains (Ref. 2).

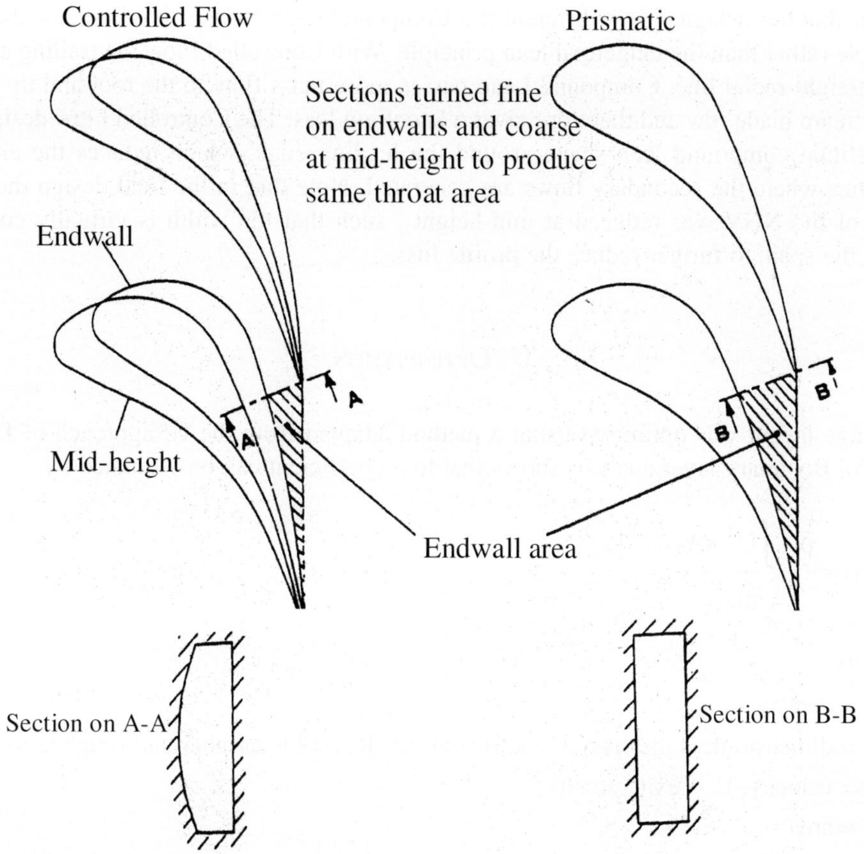

Fig. 3 Controlled Flow design philosophy.

As depicted on Figure 3 Controlled Flow blades are formed by skewing the sections closed towards the endwalls and opening the throat at mid-height to keep the same overall throat area and hence preserve the mean reaction. This philosophy reduces secondary losses on the endwalls in the NGV and rotor due to reducing the areas exposed to high velocity flow downstream of the throat and this can readily be proved by the U3 approach (see later). It also reduces profile losses in the NGV and rotor since the throat widths are larger at mid-height compared to prismatic blades (parabolic curve fitting for loss across the span shows,

$$\text{Mean loss} = \frac{(1\times \text{root value} + 4 \times \text{mean value} + 1 \times \text{tip value})}{6}$$

i.e. the mean section performance dominates). The concept has also been shown to reduce losses in the downstream stage due to the benefit of delivering less mass flow to the root and tip.

In the Controlled Flow design more mass flow is passed through the efficient mid-height sections and less mass flow is passed near the endwalls where the high secondary flow/ vortex losses are generated.

Note that this design is very different to a Compound Lean NGV as it is based on the twist principle rather than the tangential lean principle. With Controlled Flow the trailing edge is on a straight radial line. Compound Lean passes more mass flow to the root and tip of the downstream blade row and therefore gives a knock-on loss. The Controlled Flow design has a beneficial compound lean shape around the leading edge, which reduces the endwall velocities where the secondary flows are generated. Note that in the final design the axial width of the NGV was reduced at mid-height , such that the width is virtually constant across the span, to further reduce the profile loss.

U^3 Optimisation

The stage layout was optimised using a method adapted from the U^3 approach of Denton (Ref. 3). Boundary layer analysis shows that loss coefficient can be estimated by:

$$Loss = \frac{\rho c_d \int U^3 . dA}{\frac{1}{2} \dot{m} U_2^{\,2}}$$

where:

ρ = density

c_d = dimensionless dissipation coefficient =fn(Reynolds number and roughness)

U = velocity, U_2 = exit velocity

A = area

$\dot{m}$ = mass flow rate.

So to reduce loss you have to reduce $\int U^3\, dA$.

This method is used, together with a parametric blade definition, to optimise the blade velocity and turning distributions for minimum stage efficiency debit (profile and secondary loss) - see Ref. 4. It was used to optimise the Va/U for the stage , amount of Controlled Flow across the span, pitch/width and the initial profile designs.

The U^3 method is physically-based rather than being from empirical loss correlations and therefore gives improved insights and trends (e.g. Craig and Cox and Ainley and Mathieson give totally opposite trends for the influence of blade outlet angle on secondary loss!). It also correctly predicts the benefits with aft-loading, optimum passage turning distribution and Controlled Flow as has been demonstrated by previous Company tests. None of the standard empirical loss correlation methods can do this except via "fudge factors" (which break down for new applications).

To derive the optimum design, around 2000 alternative profiles were analysed for each blade row in around 10 second cpu but it would not be possible to practically do this in a realistic timescale using other techniques (plus the spurious losses in Navier Stokes methods due to grid definition, numerics etc. are avoided).

Space precludes a description of all the results from the U^3 method but, as an example, Figure 4 shows that the optimum NGV design has pitch/axial width at the root = 1.6.

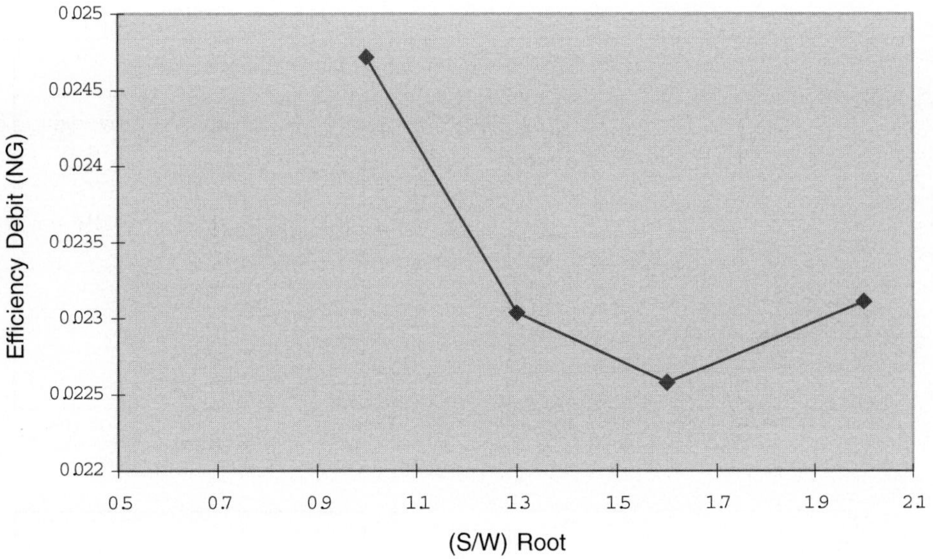

Fig. 4 U^3 optimisation curve for NGV.

A similar study was carried out for the rotor and it was then possible to draw specific conclusions for the optimum NGV and rotor profile designs at root, mean and tip (ie pitch/width, velocity at mid-passage and turning at mid-passage). The profiles were then designed in detail using these values and the performance checked using quasi-3D (forward and inverse) and 3D viscous methods.

3D Analysis

The performance of the stage was then analysed using a number of 3D RANS methods, including the Company 3D-stage viscous time marching method adapted from Dawes (Ref. 5). This code has been extensively validated over 15 years against many 3D-turbine test cases, including Controlled Flow.

Figure 5 shows the predicted isentropic surface Mach number distributions for the NGV and rotor at 10, 50 and 90% height. It can be seen that the peak suction surface Mach numbers are only slightly higher than the outlet Mach numbers and the shock strengths have been minimised. There are no signs of any overspeeds on the incidence-tolerant blade leading edges. It can be seen how Controlled Flow off-loads the sections on the endwalls which reduces the loss. The high lift design of the NGV and rotor are clearly visible.

Figure 6 shows some results from the 3D viscous time marching predictions for the existing and new HP stage designs. It can be seen that the aspect ratio of the new design is much

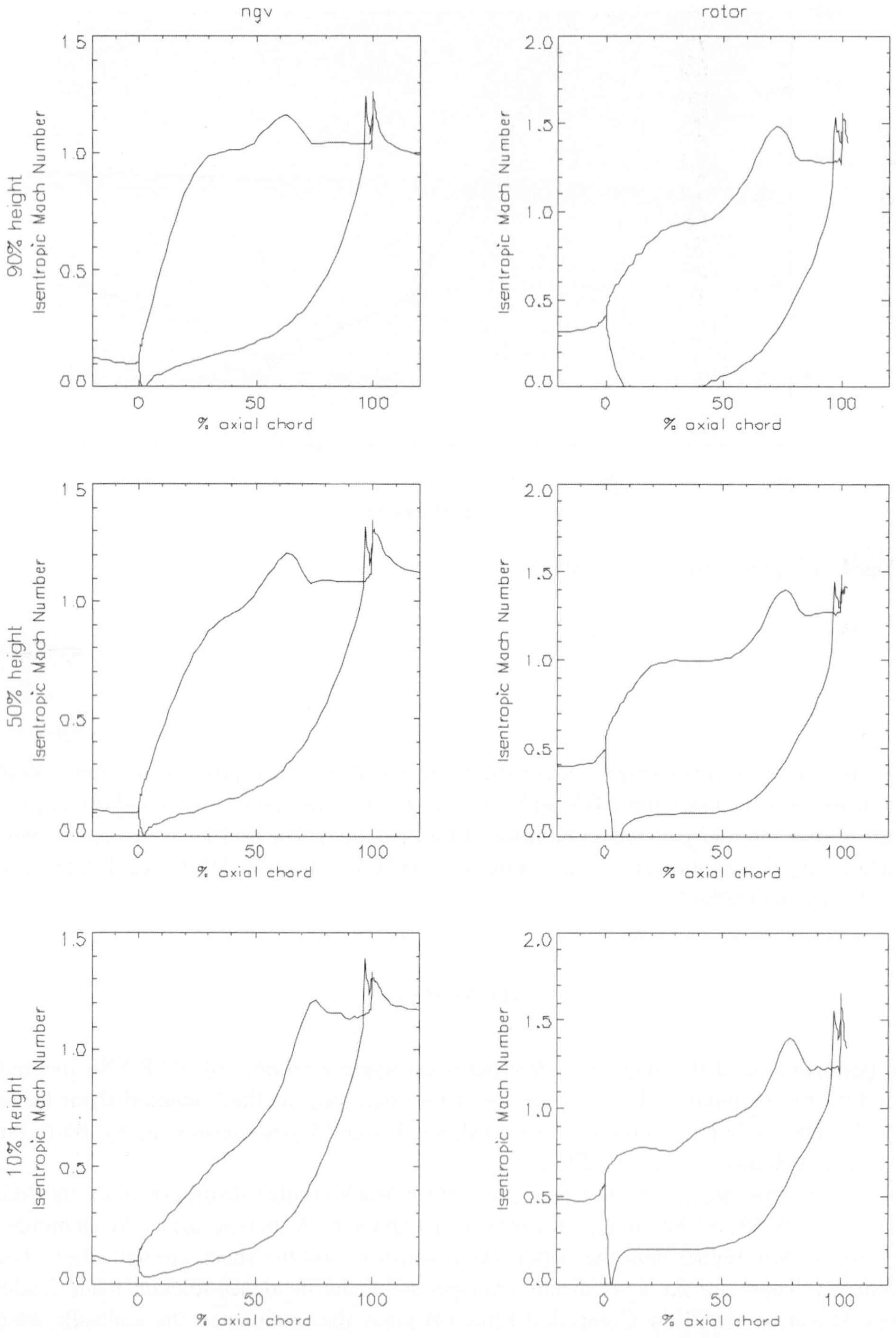

Fig. 5 3D viscous predictions for stage - isentropic mach number distributions on NGV and rotor surfaces.

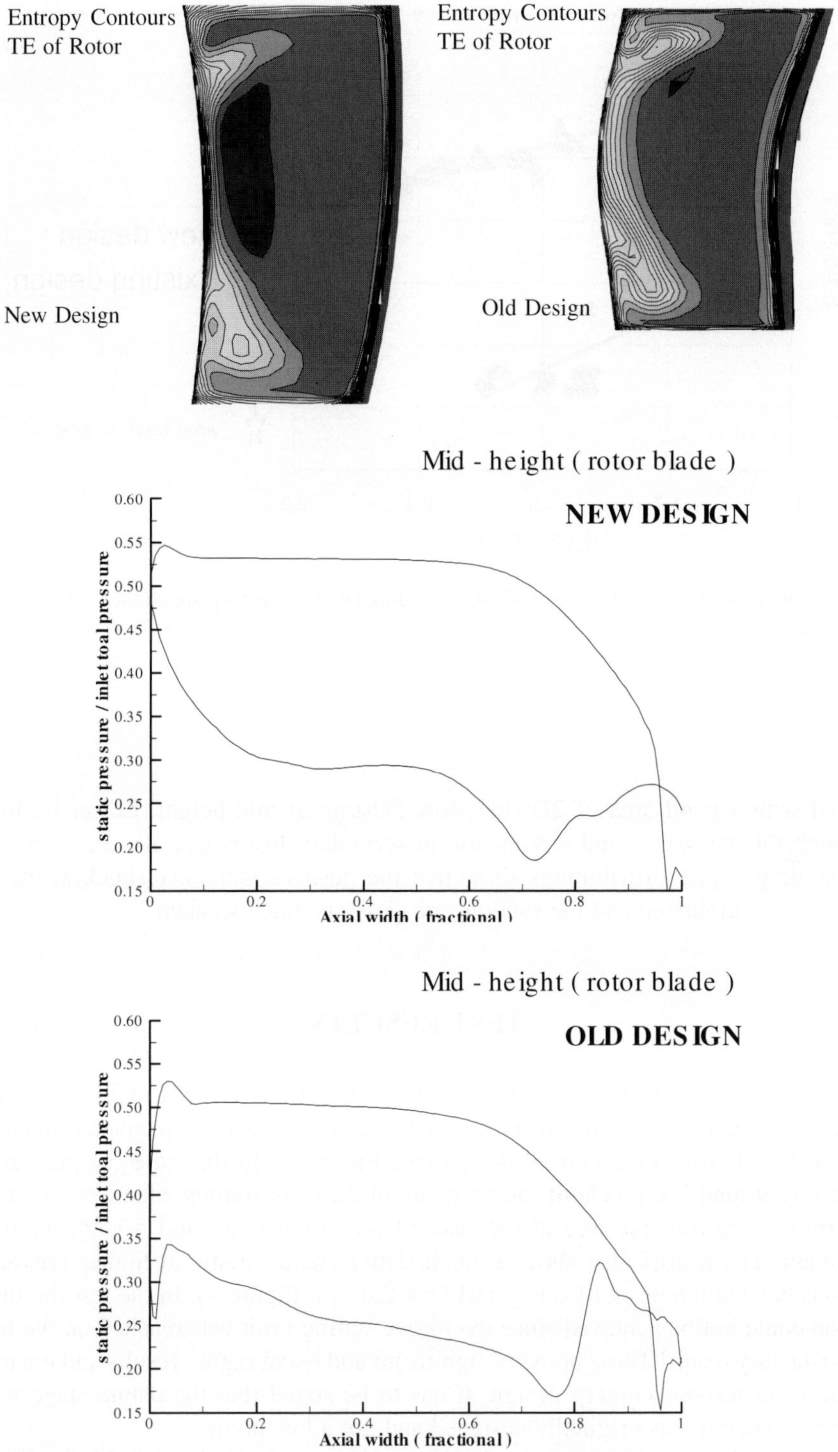

Fig. 6 3D viscous time marching prediction for existing and new HP stage designs.

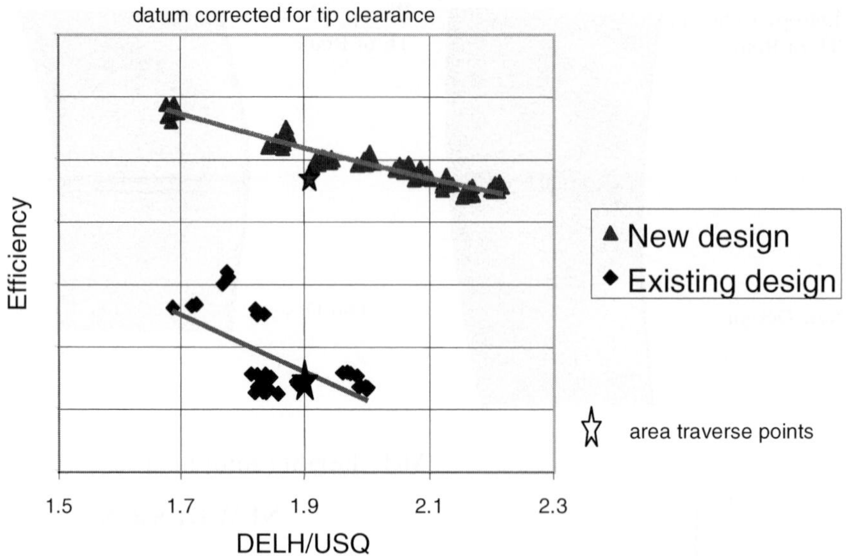

Fig. 7 Variation of stage efficiency with stage loading DH/U^2 (each square on the y-axis represents 2% points).

improved with a good area of 2D flow now existing at mid-height. Lower trailing edge losses with thinner wakes and a reduction in secondary losses can also be seen. The mid height static pressure distributions show that the positive incidence/shock at the leading edge has been eliminated and the suction side shock is much weaker.

TEST RESULTS

The new stage has been tested in a full scale Warm Air Turbine Test Rig at fully representative aerodynamic conditions and has demonstrated a total-to-total mass meaned efficiency gain of over +7% relative to the current design (see Figure 7). In this case the pressure probe traverse was around 2 axial chords downstream of the rotor trailing edge (see Figure 2) and the thermocouple traverse was at the rake plane which is around 5 rotor axial chords downstream. The results also show a much flatter characteristic at higher pressure ratio conditions beyond the design loading DH/U^2 = 2.0 (see Figure 7). In the test the limit load condition could not be achieved since the torque setting limit was reached on the brake - a very satisfactory result! These are very significant and encouraging results and open up new possibilities in turbomachinery design. It has to be stated that the datum stage was fully competitive when it was originally designed and not a low point.

The measured swallowing capacity for the stage agreed very closely with the 3D viscous prediction.

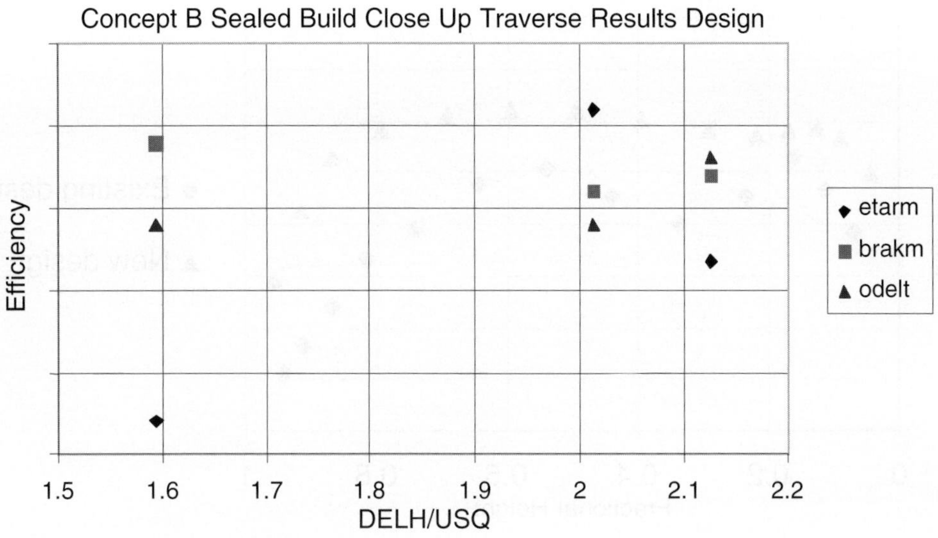

Fig. 8 Stage efficiency established using a variety of methods and plotted against loading (each square on the y-axis represents 0.5%).

The stage efficiency was checked using a number of alternative methods, see Figure 8 (power from brake or measured temperature drop, outlet stagnation pressure for wall statics and continuity, area probe traverse, rakes, use of exhaust plenum thermocouples etc.). However all the methods agreed quite closely for the stage efficiency and this gives a high confidence level.

It has to be stressed that a considerable part of the 7.1% stage efficiency gain relative to the existing design is due to improved sealing and is not all due to blading improvements. The subject of this detailed "efficiency audit" is beyond the scope of this paper but it is nevertheless clear that big advances have been made. It can be stated that the target efficiency gain of 4% has been exceeded.

Of interest, another area traverse was carried out with the thermocouple probe positioned in the plane around 2 rotor axial chords downstream of the stage. This shows more detail of the flow with a very nice flat efficiency distribution for the new design with large improvements over the whole span see Figure 9 . There is a very large reduction in the classical root secondary loss which can be seen on the existing design at around 0.3 fractional height.

'CLOSING THE LOOP' BETWEEN TEST MEASUREMENTS AND PREDICTIONS

Further work has been carried out to 'close-the-loop' by comparing the measurements obtained in the WATR with 3D viscous analyses at design and off-design conditions. The main objectives of this study were :

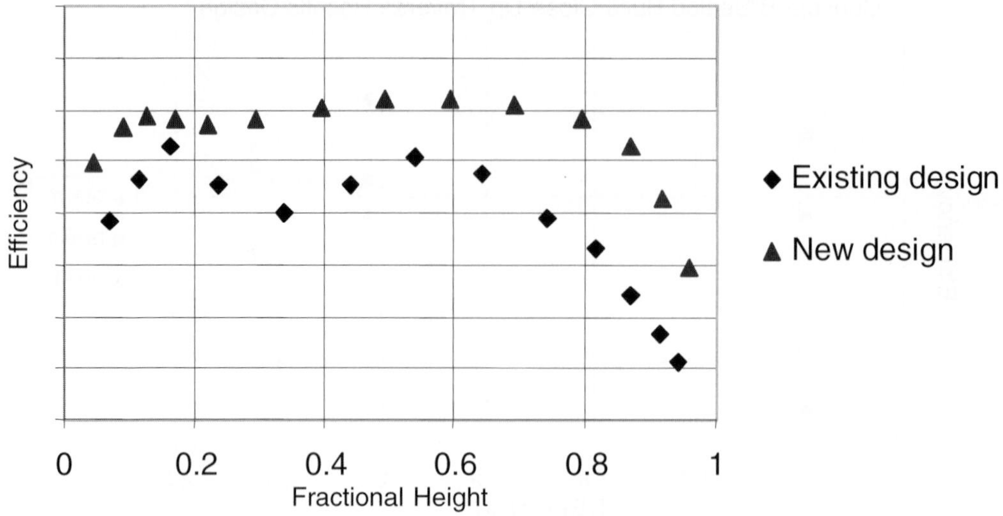

Fig. 9 Measured radial distribution of stage total-to-total efficiency - pressure and thermocouple probes 2 axial chords downstream (each square on the y-axis represents 5%).

- to validate the design methods and approaches
- to determine how the stage was performing relative to the original design intent and if there were any areas of loss which could be reduced further
- to obtain better understanding of the off-design capabilities of the stage (particularly higher loading conditions up to limit load) and the downstream duct mixing loss .

The CFD analysis work is very complimentary with the experimental side and greater insights can be obtained.

Design Conditions

Figures 10, 11 and 12 show the comparisons at exit from the stage in the close-up plane for total-to-total efficiency, absolute swirl angle and the specific mass flow distribution. In general the agreements were very close and satisfactory for practical engineering design processes. It can be seen that the tip leakage flow must be included in the predictions otherwise it has the wrong direction at the tip. The measured stage efficiency distribution bore a remarkable resemblance with the original prediction which was done 'blind' several years before the test.

On the plots some explanation is required for the leakage flow modelling:

- 'Tip flow excluded' - no tip leakage flow
- 'Tip flow extracted/injected' - simplistic model with leakage flow bled off upstream of the rotor and re-injected downstream with realistic estimates for injection angles. The leakage flow = 1.7% of stage flow which corresponds to a radial tip clearance of 0.5 mm.

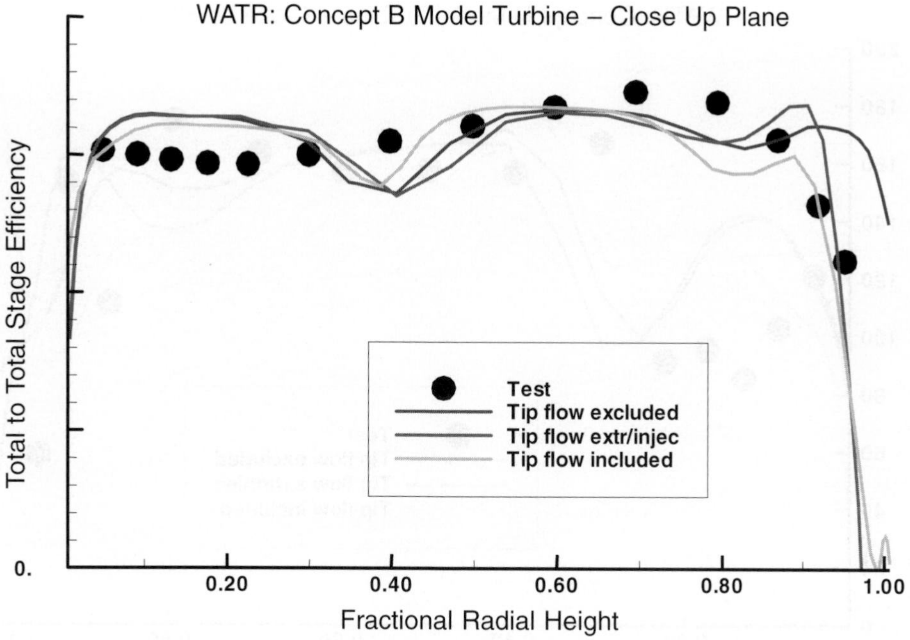

Fig. 10 Measured radial distribution of stage total-to-total efficiency compared to staged analysis (design point, close-up traverse plane).

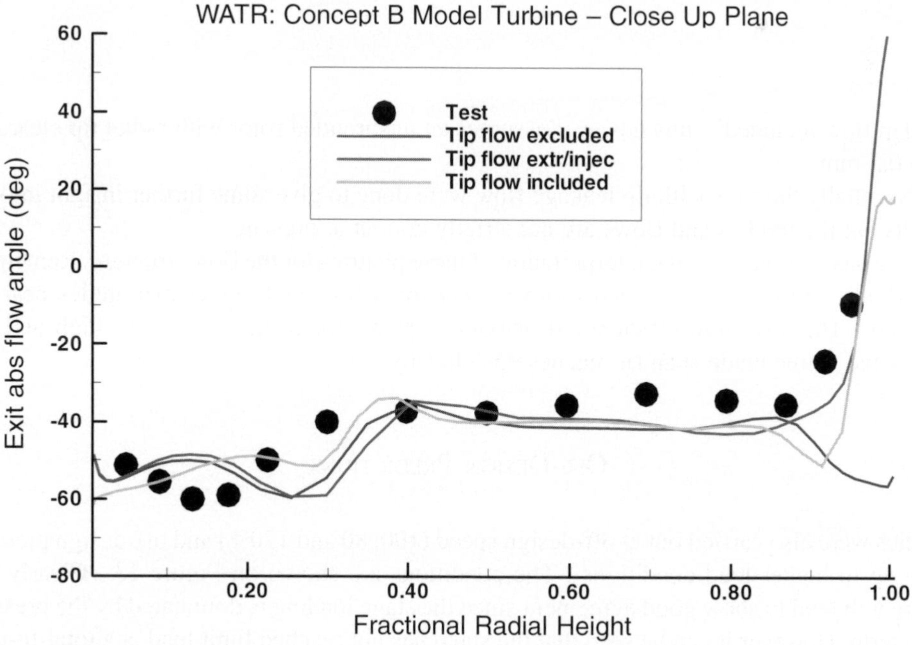

Fig. 11 Swirl angle comparison.

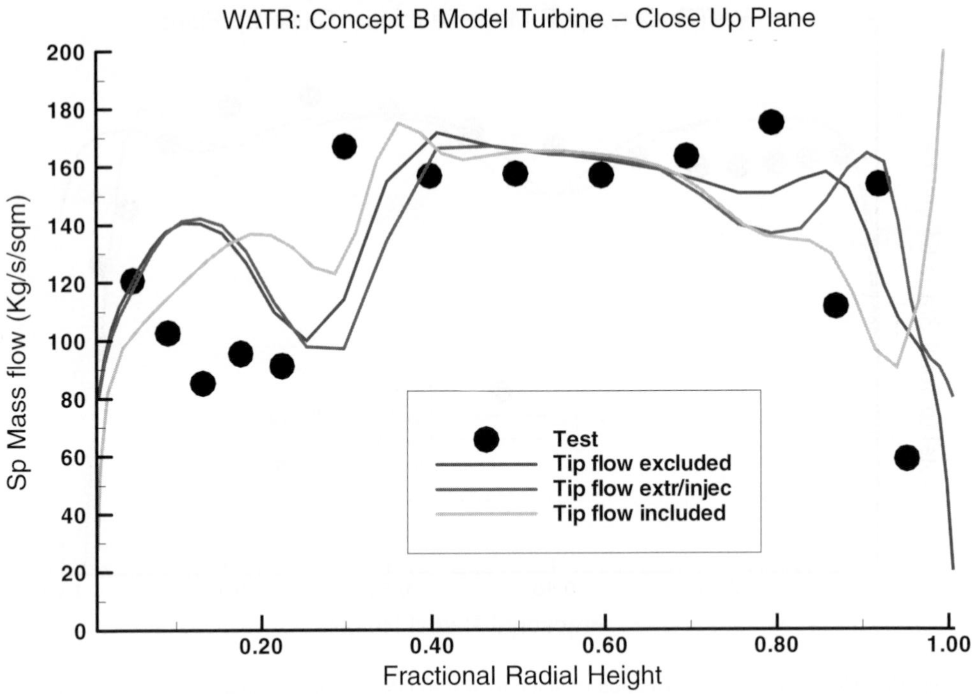

Fig. 12 Specific mass flow comparison.

- 'Tip flow included' - this however assumes an unshrouded rotor with radial tip clearance = 0.5 mm.

Essentially the runs with tip leakage flow were done to give some further insight into the results but the models and flows are not strictly correct at present.

Care has to be taken in the interpretation of these pictures for the flow structure. Remember that Controlled Flow is designed to give lower mass flow and higher exit angles near the endwalls. The measured efficiency distribution speaks for itself and is very high and flat across the whole blade span (it reaches 95% locally).

Off-Design Predictions

Studies were also carried out at off-design speed (100, 80 and 120%) and off-design pressure ratio up to higher load conditions . The conditions are shown on Figure 13 . Clearly this figure will tend to show good agreement since the stage loading is dominated by the pressure ratio term. However it can be seen that the stage has not reached limit load at a total-to-total pressure ratio of 5.2.

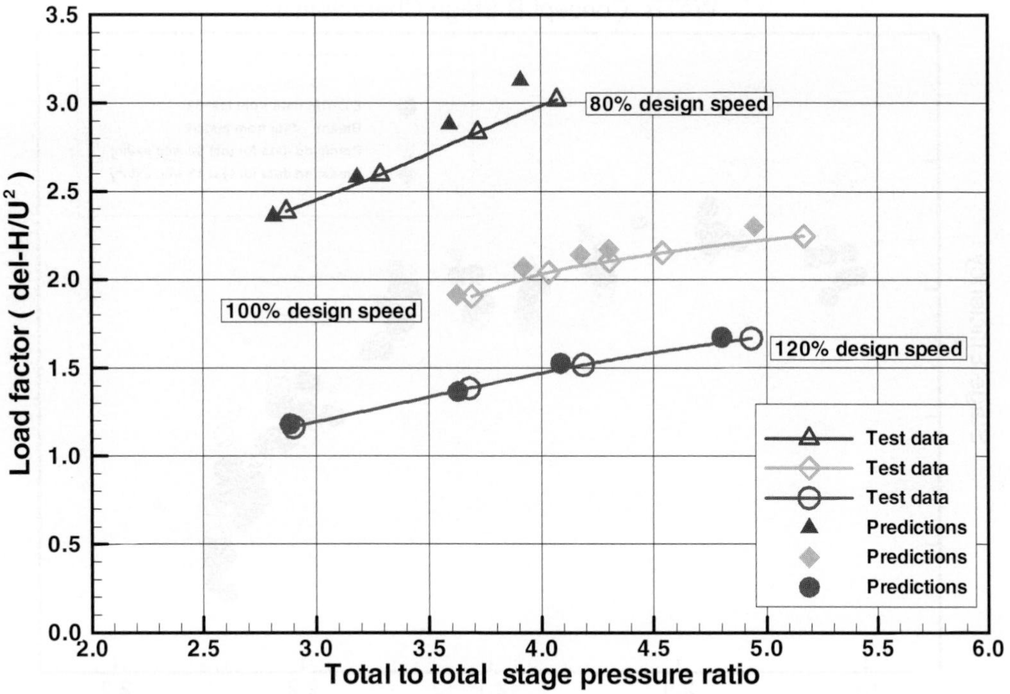

Fig. 13 Summary of stage operating characteristics.

At design speed the variation of stage efficiency with loading is given on Figure14. The agreement with the Staged predictions is very close. The stage efficiency is very flat up to the design loading of 2.0 but drops off above a loading of 2.2.

UNSTEADY MEASUREMENTS

Detailed unsteady probe traverse measurements were carried out at exit from the stage by QinetiQ and these are summarised in Ref. 6.

The results are most easy to understand when they are presented as a space-time diagram. For each radial height the data are plotted with the horizontal axis giving position relative to the NGV and the vertical axis representing time. Features locked to the rotor are seen as diagonal bands across the plot (bottom left to top right); whereas features locked to the NGV are seen as vertical bands. Figure 15 shows the space-time diagram for stagnation pressure. This clearly shows the pair of shockwaves as high pressure bands running parallel to the line indicating the rotor passing event. Both these shockwaves can be seen to be modulated by the upstream blade row although the magnitude of this variation is dependant on the probe

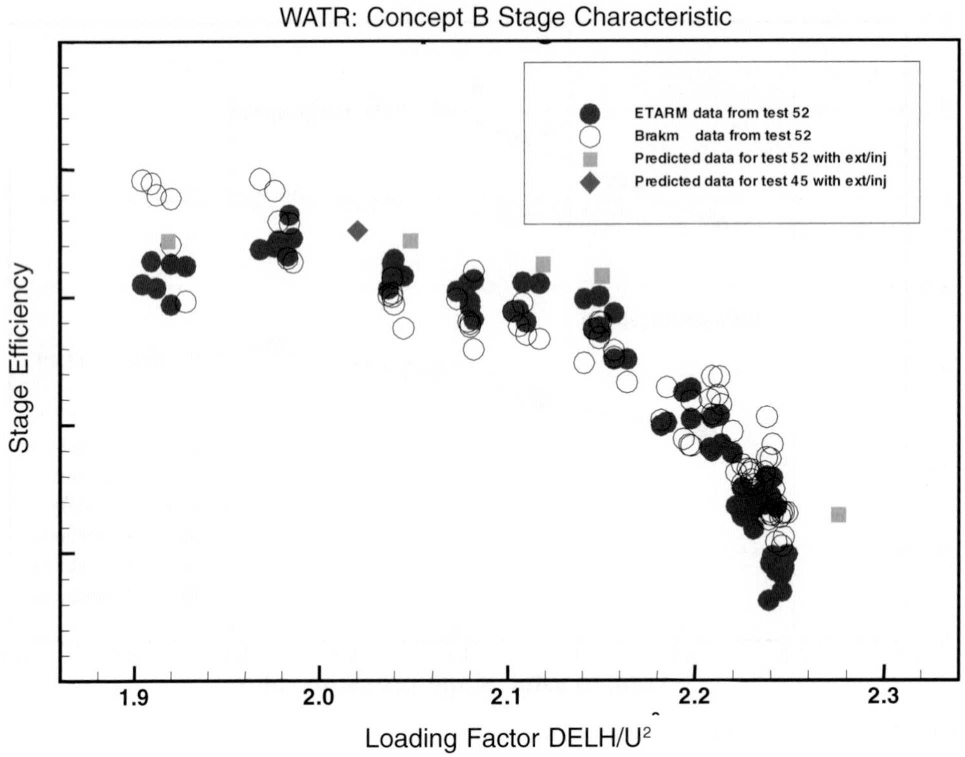

Fig. 14 Variation of efficiency with loading at design speed.

height. When the datum blading was tested (Ref. 7) strong vertical banding could be seen close to the endwalls which was thought to be caused by the NGV secondary flow being driven towards the endwalls as it passed through the rotor. The migration of the secondary flows towards the endwalls has also been observed by other researchers (Ref. 8) With this blading this banding cannot be seen. However, when the random unsteadiness is plotted (Figure 16) clear vertical bands can be seen away from the endwalls. Random unsteadiness is a measure of the turbulence in the flow and high areas are normally associated with wakes and secondary flow regions. This suggests that the NGV wakes and secondary flows are being driven out into the passage by the Controlled Flow geometry. It is thought that the corresponding banding that would be expected in stagnation pressure is 'masked' as it is relatively weak compared to the more dominant rotor shock structure at these heights. It may also be due to the reduced strength of the secondary flows.

It should be stressed that the Controlled Flow design philosophy is intended to pass more flow in the efficient middle part of the blade and less flow near the endwalls, so these measurements are not surprising. Note however that the secondary losses are generated on the endwalls and may be convected to other areas.

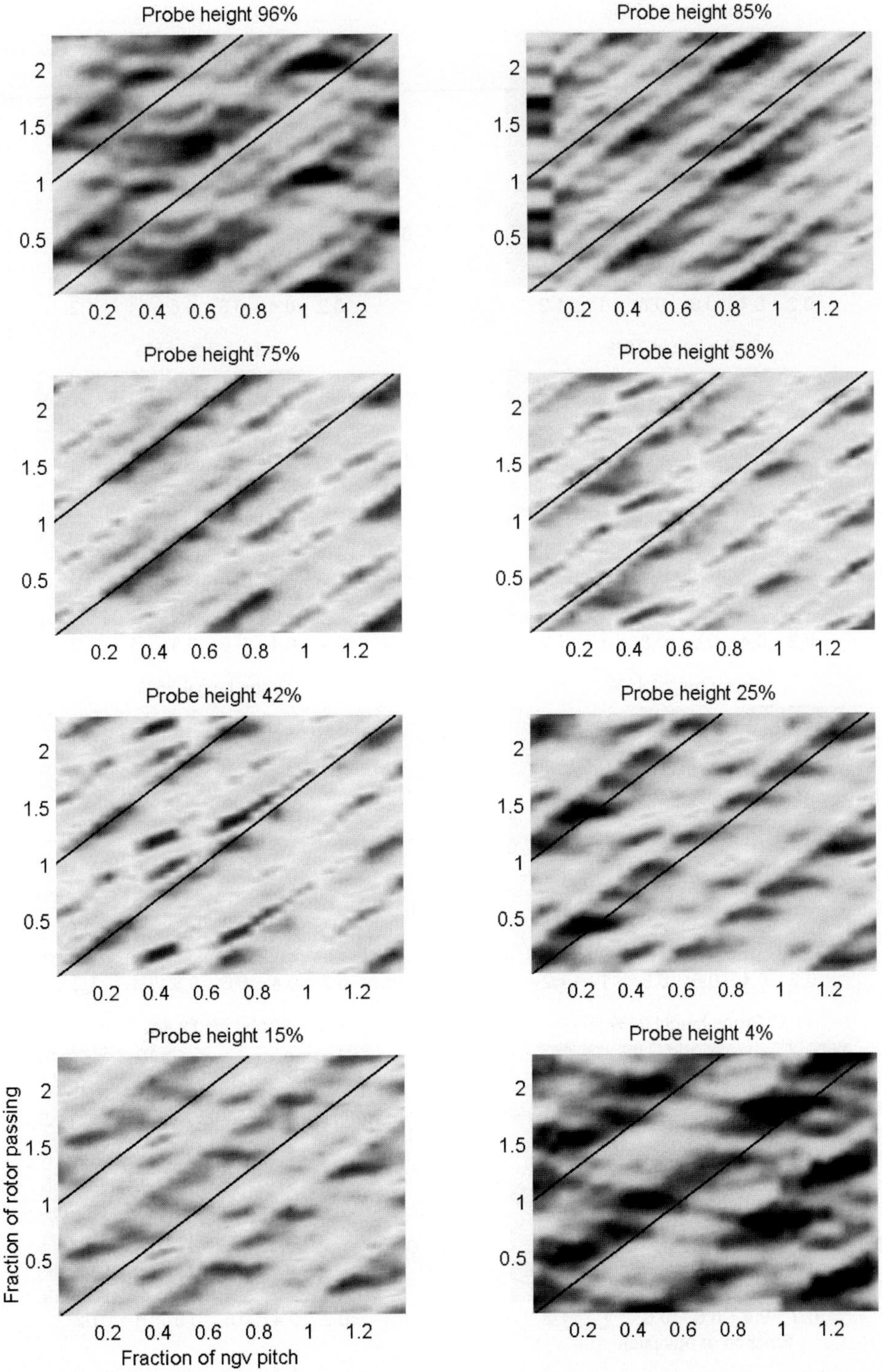

Fig. 15 Space-time diagram showing stagnation pressure measured with the single sensor probe.

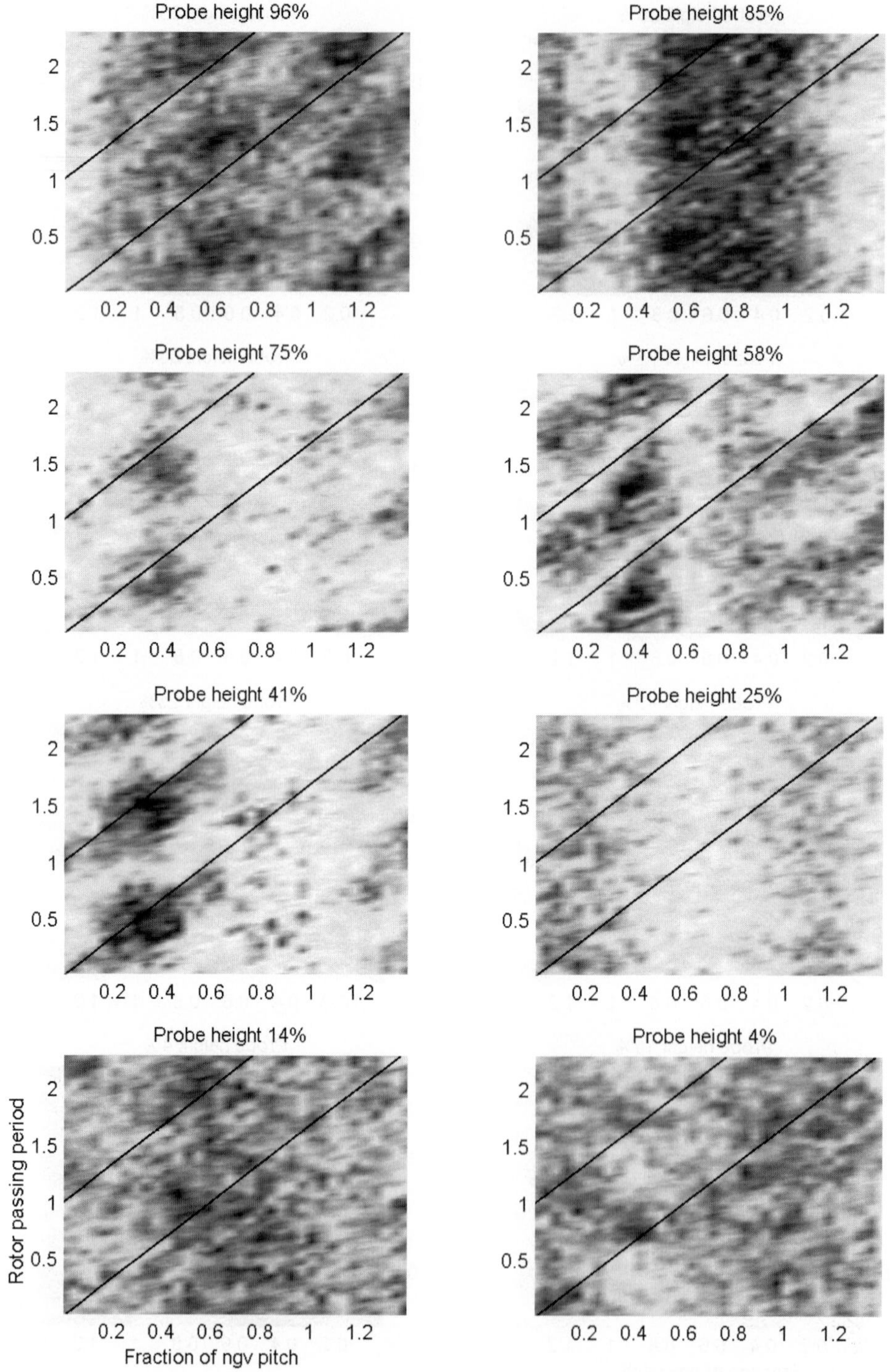

Fig. 16 Space-time diagram for random unsteadiness measured with the single sensor probe.

CONCLUSIONS

A new design philosophy for high pressure turbines, utilising 'Controlled Flow' and 'U3' optimisation has been demonstrated. Very significant and encouraging results have been obtained relative to the existing design in terms of the stage efficiency gain and the improved characteristic at higher loading conditions. The target efficiency gain of 4% was exceeded.

The new design approach has been validated using 3D viscous codes and detailed experimental data.

Unsteady traverse measurements have been used to resolve the rotor flow field giving detail that would not be possible using conventional time averaging techniques.

The unsteady traverse measurements have shown that the Controlled Flow drives the NGV secondary flows away from the endwalls and towards mid-passage. This is opposed to observations from conventional designs. It is surmised that this could be very beneficial in control of the overall temperature distortion (OTDF) at turbine inlet and keeping the hot core flow away from the endwall surfaces and rotor shroud which are difficult areas to cool. However this remains to be proven by the follow-on work.

The unsteady data has revealed some of the effects of blade row interaction but to fully understand these effects further interpretation is being carried out combined with unsteady CFD.

ACKNOWLEDGEMENTS

The authors wish to thank DTI CARAD for funding this work and the contributions from QinetiQ (unsteady measurements) . The off-design analysis was covered by the EU DAIGTS programme.

REFERENCES

1. B. R. Haller: British Patent Application No. 0104002.1 'Controlled Flow Turbine Stage', (number 2359341), 2001.
2. B. R. Haller: 'Application of 3D Computational Fluid Dynamics Methods to Provide Enhanced Efficiency Blading for the 21st Century', Award for Best Paper at 4th International Charles Parsons Turbine Conference, 1997.
3. J. D. Denton: 'Loss Mechanisms in Turbomachines', ASME 93-GT-435.
4. B. R. Haller: VKI Lecture Series on 'Secondary and Tip Clearance Flows in Axial Turbines', *Full 3D Turbine Blade Design*, 1997.
5. G. Singh and B. R. Haller: 'Development of Three-Dimensional Stage Viscous Time Marching Method for Optimisation of Short Height Stages'. *First European Conference on 'Turbomachinery - Fluid Dynamic and Thermodynamic Aspects'*, Erlangen, 1995, 157–180.

6. B. R. Haller and S. J. Anderson: 'Development of New High Load/High Lift Transonic Shrouded HP Gas Turbine Stage Design - A New Approach for Turbomachinery', ASME paper GT-2002-30363.
7. S. J. Anderson and B. Pendery: 'Steady and Unsteady Traverse Measurements on a Transonic Industrial H.P. Turbine Stage', *4th European Conference on Turbomachinery*, (AST-CST-084/01).
8. R. J. Millar, R. W. Moss and Ainsworth: 'Time Resolved Vane-Rotor Interaction in a High Pressure Turbine Stage', submitted to the Journal of Turbomachinery.

Manufacture and Spin Test Evaluation of Bimetallic Gas Turbine Components

M. B. Henderson
ALSTOM Power Technology Centre, Leicester, UK

H. Burlet
CEA, Grenoble, France

G. McColvin
Seimens DDI Turbomachinery Ltd., Lincoln, UK

J. C. Garcia
Turbomeca, Bordes, France

S. Peteves
EC-JRC, Petten, The Netherlands

J. Montagnon and G. Raisson
Aubert & Duval, Les Ancizes, France

S. Davies
Bodycote HIP, Chesterfield, UK

I. M. Wilcock
QinetiQ, Farnborough, UK

P. Janschek
Thyssen-Krupp, Remscheid, Germany

ABSTRACT

To overcome the limitation in temperature capability of currently available materials and achieve higher performance levels for gas turbines, there is an increasing interest in the manufacture of dual alloy discs or bladed discs (bliscs) using a high strength hub/disc region bonded to a creep resistant rim/blade. One promising manufacturing method is diffusion bonding in either the solid state or using a Powder Metallurgical (PM) route under Hot Isostatic Pressure (HIP), which can be augmented, if necessary, using isothermal forging. The manufacture, design and spin test assessment of a number of full-scale discs has been conducted during Phase 2 of a Framework IV European collaborative project (MANDATE) aimed at developing and validating the means of manufacture for bimetallic components. Phase 1 consisted of defining the processing routes (HIP and forging conditions and heat treatment procedures) and assessing the effectiveness of the diffusion bonding process by means of mechanical testing and microstructural examination. In parallel, NDE methods based on ultrasonic inspection were developed. The conclusion from Phase 1 was that the results were sufficiently promising to proceed with Phase 2 of the project and the results of the manufacturing and validation programme are described in the present paper.

INTRODUCTION

Gas turbine engines feature as key components of the most efficient forms of advanced propulsion and power generation technology available. For the foreseeable future all of the major gas turbine engine manufacturers are looking to target improvements in efficiency and lower emissions, whilst offering improved performance capability levels to the operators. Efficiency and emissions targets will be achieved by raising the pressure ratios and turbine inlet temperatures leading to more arduous service conditions on critical turbine components in the hot gas path. This must be achieved without detriment to the reliability, availability and maintainability (RAM) of the engines in operation.[1]

To overcome the limitation in temperature capability of currently available materials and achieve higher performance levels for gas turbines, there is an increasing interest in the manufacture of dual alloy discs or bladed discs (bliscs) using a high strength hub/disc region bonded to a creep resistant rim/blade.[2, 3] This concept enables the different mechanical property requirements for the hub and the rim sections of a disc to be reconciled in a single disc structure. One promising manufacturing method is diffusion bonding in either the solid state or using a Powder Metallurgical (PM) route under Hot Isostatic Pressure (HIP), which can be followed by an isothermal forging consolidation procedure, if necessary. The main barrier to successful application of dual alloy turbine disc technology has been the lack of a suitable, cost-effective manufacturing method.

Recently, a Brite EuRam Framework IV European collaborative project (MANDATE) has aimed at developing and validating cost-effective manufacturing and application technologies for dual alloy gas turbine discs and bliscs (Integrally Bladed Disc). Phase 1 consisted of defining the processing routes, including the HIP and forging conditions and subsequent heat treatment procedures and assessing the effectiveness of the diffusion bonding process by means of mechanical testing and microstructural examination. In parallel, a number of component geometry specific, ultrasonic immersion NDE testing techniques have been developed to inspect the interface of the bonded components. The detection of small planar flaws with a minimum size of 1 mm² has been shown to be possible. The conclusion from

Phase 1 was that the results were sufficiently promising to proceed with Phase 2 of the project to design, manufacture and spin test a number of full-scale, dual alloy discs and bliscs.

MATERIALS REQUIREMENTS AND PROCESSES

Four hub-rim combinations comprising of powder metallurgy (PM), cast and wrought (C & W) and cast polycrystal and single crystal (SC) superalloy products have been selected for evaluation:

- Fine grain (FG) PM U720Li hub bonded to a coarse grain (CG) C & W U720Li ring.
- Fine grain C & W U720Li hub bonded to a coarse grain C & W U720Li ring.
- Fine grain PM U720Li hub bonded to a coarse grain IN738LC ring.
- Fine grain C & W U720Li hub bonded to a coarse grain IN738LC ring.
- Fine grain PM U720Li hub bonded to a SC MC2 dummy blades.

Both high carbon (HC) and low carbon (LC) variants of U720Li have been used to manufacture the outer rings via conventional cast & wrought methods using a supersolvus heat treatment schedule. The major methods of manufacture were identified diffusion bonding under Hot Isostatic Pressure (HIPping) augmented, if necessary, by isothermal upset forging of the bonded bi-metallic cylinders, or hub-blade junctions, to form a dual alloy component. A series of work packages have been conducted to specify the materials needed, target properties and component requirements, as well as manufacture a series of model discs and plates. Development of NDE and microstructural examination methods and the mechanical property characterisation of the constituent segments of a dual alloy model disc (hub, joint and rim) were key activities in qualifying the methods used.[4]

The target components were identified as follows:

- Dual alloy turbine disc for power generation, providing power and efficiency gains by enabling increased disc diameters and higher temperature capability (+50°C).
- Radial turbine wheel (blisc) for helicopter engines and auxiliary power units (APU's), enabling larger wheels to be designed and higher thermal efficiencies (+150°C).

Figure 1 shows a monolithic example of the power generation target component currently manufactured using conventional cast and wrought processing methods. Both ALSTOM and Turbomeca have specified the technical and mechanical requirements for the components and the requisite quality of the selected materials to achieve a range of target conditions (temperatures and stress profiles, blade fixtures and cooling geometries, target lifetimes) and duty cycles anticipated for future engine designs.

SPIN DISC DESIGN

The key objectives were to validate the manufacturing route and test the integrity of the bonded region within a dual alloy disc/blisc. Finite element analyses have been performed to define the loading conditions for the spin tests and, in conjunction with the lifing database, to predict the location of failure after a targeted number of elapsed cycles.

Fig. 1 Example of a monolithic industrial GT rotor stage.

Dual Alloy Spin Disc

A number of disc geometries and features were studied using fully elastic, finite element stress analyses, linked to the forge modelling work conducted by the disc manufacturers that provided boundary conditions as to the location of the bondline and overall achievable dimensions for a dual alloy disc. A 2-D axisymmetric FE model with triangular CAX6 elements was used to optimise the disc profile. Figure 2 shows the predicted radial elastic stress distribution pattern generated within a monomaterial U720Li disc spinning at 20,000 rpm (Emod = 200 GPa, density = 8.0 gm/cm^3, Poisson's ratio = 0.3). The hoop stress was balanced to ensure that the disc would not burst during cyclic loading. It was concluded that the most suitable design was a plane diaphragm disc with a circumferential under-rim stress-concentration feature designed to be aligned radially and coincident with the bondline location, predicted by the finite element forging simulations to be at a diameter of 260 mm. The external diameter was fixed at 363 mm with a central bore of 50 mm diameter. The profile of the hub and drive arm regions was optimised to reduce stress levels locally.

Following agreement of the overall dimensions with the component manufacturers and the spin testers, a rotor dynamics optimisation study was performed to finalise the design of the spin disc and drive arm adaptor.

Bladed Wheel

A Finite Element Analysis of the TURBOMECA full-scale model blisc comprising of a U720PM hub and 13 single crystal dummy MC2 blades has been conducted. A simple 'paddle wheel'

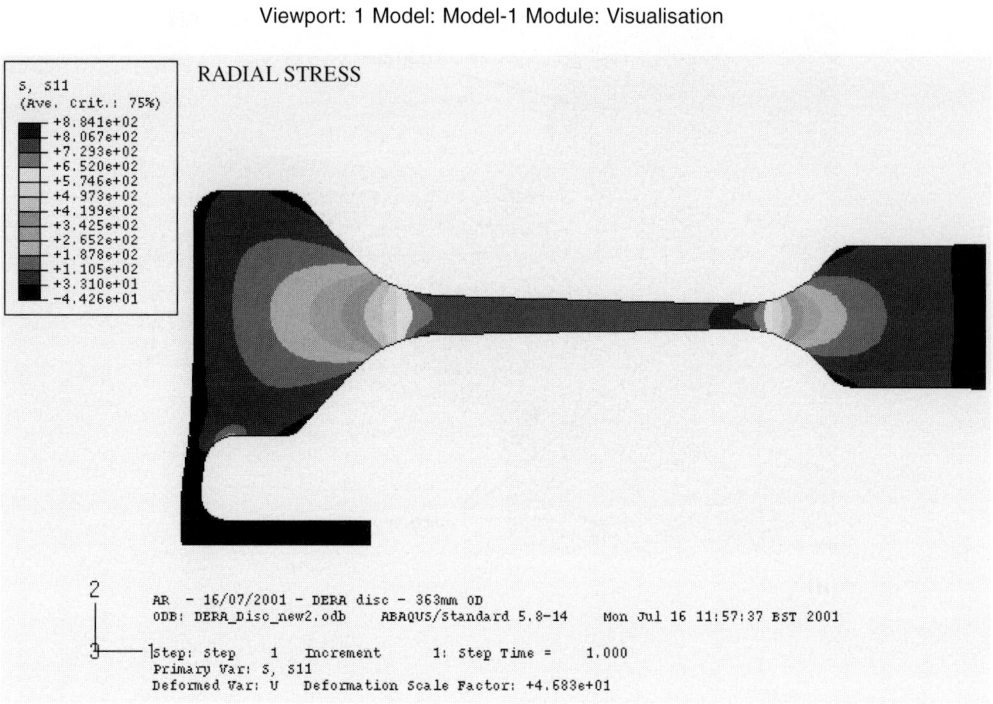

Fig. 2 Predicted elastic radial stress distribution for MANDATE spin disc at 20,000 rpm.

design was selected to ease the disc manufacture and as a consequence the U720PM-MC2 materials junction was more heavily loaded in comparison with what would be expected for a real component.

Both Linear Elastic and Elasto-Plastic Finite Element Analyses were performed using ANSYS Version 5.5. A range of 2D and 3D geometries have been studied to assess the performance of the various models. A uniform temperature distribution of 700°C and a speed of 50,000 RPM were chosen for the analyses. An example of the detailed meshing around the U720PM-MC2 bondline is shown in Figure 3.

Preliminary elastic analyses were able to establish the models and mesh configurations effectively and refine the blade fillet radius and design of the bore holes. The elasto-plastic analysis found a significant stress shakedown around the blade fillet region (Figure 3b), with little difference at peak load between the two materials models used (multi-isotropic and multi-kinematic hardening). Significant differences were found, however, in the predicted residual stress levels after unloading. The critical area for the component was confirmed as the blade fillet at the junction of the two materials and the crack initiation life was estimated to be between 1,000 and 10,000 cycles for a rotational speed of 40,000 RPM.

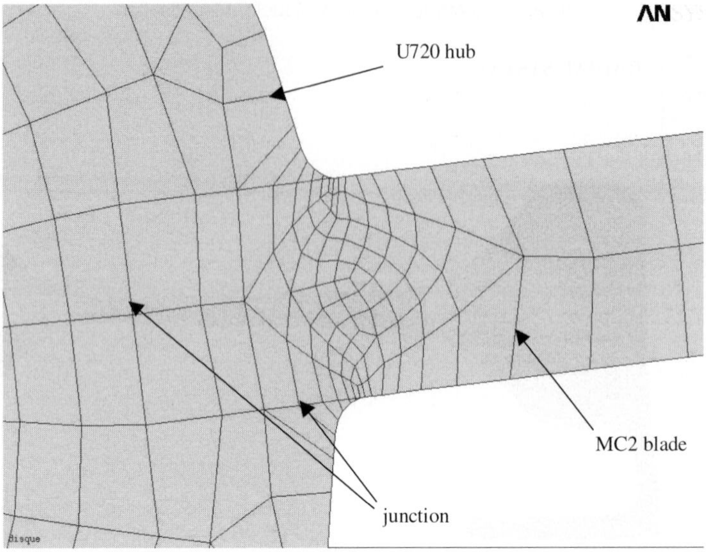

(a)

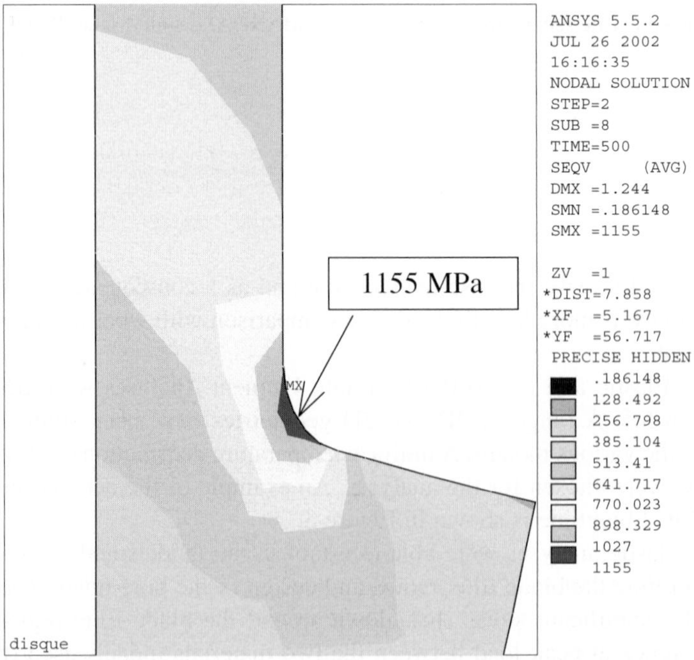

(b)

Fig. 3 (a) Detail on the blade/hub junction and (b) 2D elasto-plastic von-Mises stress distribution for the multi-isotropic hardening model.

DUAL ALLOY DISC FABRICATION

A total of twelve full-scale discs/bliscs (FSDs) were successfully manufactured from combinations of the different grades of U720Li (fine and coarse grained), IN738LC and MC2 materials. Three as-HIPped U720PM-MC2 bliscs, one as-HIPped and eight HIPped plus forged full-scale U720FG-U720CG discs were produced. These have been heat treated and machined as required to provide specimens for life assessment database tests, NDT and microstructural characterisation and finish machining of two HIPped plus forged dual alloy discs (FSD3: U720PM-U720HC and FSD5: U720CW-U720LC) and one blisc (FSD9: U720PM-MC2) to produce spin discs for validation testing.

The process conditions developed during the programme are often proprietary to the component suppliers, however, as much information as possible has been provided in the following summary of the full-scale disc manufacturing programme:

FSD3 (U720PM-U720HC Full-Scale Disc)

This disc was manufactured by means of pre-HIP assembly of a supersolvus heat treated U720HC outer ring for the rim, within a specifically designed can to support the U720PM powder hub material prior to consolidation. Following welding of the can and de-gassing according to conditions determined during the development phase, the assembly was HIPped and isothermally forged at 1100°C, before solution annealing at 1100°C for 4 hours followed by gas fan quenching. Figure 4 shows a similar dual alloy disc (FSD2) in the as-forged condition prior to being heat treated and cut-up into blanks for mechanical property evaluation. Examination of the machined specimen blanks and microstructural sections, as shown in Figure 5, indicated an excellent bond quality and excellent agreement between the actual and predicted (due to FE forge simulations) location of the interface between the two materials. FSD3 was machined to provide a rectilinear profile for NDT examination prior to rough profile machining, final ageing according to HT1 (650°C/24 hours/air cooled, 760°C/16 hours/air cooled) and finish machining to the spin disc design. No evidence of significant defects or debonds were detected during NDT inspection of FSD3.

FSD5 (U720CW-U720LC Full-Scale Disc)

This disc was manufactured using a solid-solid processing route that required careful pre-cleaning and accurate assembly procedures, identified during the development phase of the programme. The pre-HIP assembly and final forged part are shown in Figure 6. Following assembly, de-gassing and HIPping the disc was isothermally forged at 1100°C and solution annealed prior to machining to a rectilinear profile suitable for NDT inspection. After final heat treatment according to HT1, the disc was finish machined to form a spin disc.

Some evidence of near surface debonding at the interface was observed during NDT inspection of FSD5. This region was removed during subsequent machining.

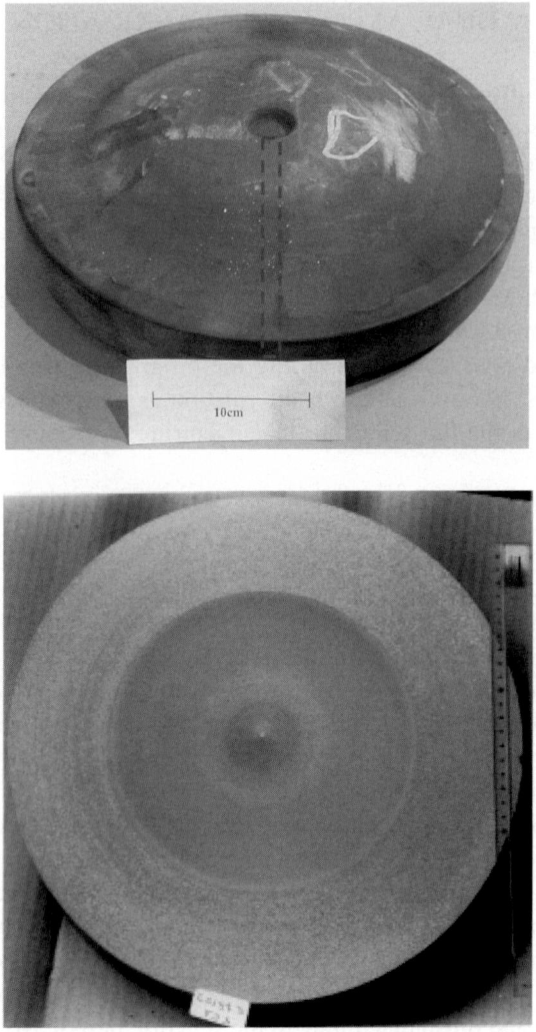

Fig. 4 FSD2 in the as-forged condition (top) and following surface machining and etching to reveal the grain structure (bottom).

FSD9 (U720PM-MC2 Full-Scale Disc)

The U720PM-MC2 blisc design was comprised of a powder hub attached to 13 radially aligned single crystal dummy blades. The assembly and positioning of these parts prior to HIPping according to the design drawings and the stable retainment of their relative positions during processing was critical to the success of the planned mechanical and spin test evaluation studies. Three dual alloy bliscs were manufactured via pre-HIP assembly, followed by canning

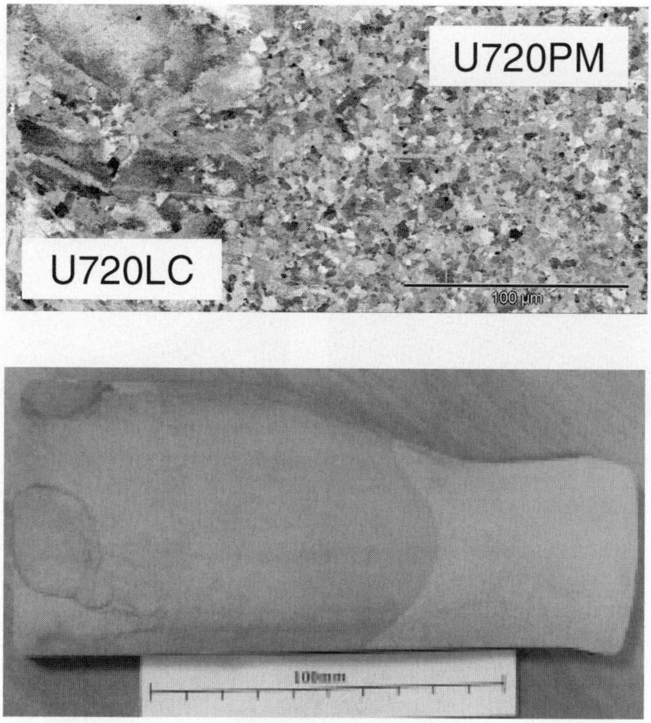

Fig. 5 Cross-sections from FSD2 showing the quality and location of the interface.

and de-gassing and HIPping at 1100°C, prior to solution annealing and heat treatment according to HT1. Figure 7 shows FSD9 after HIPping has been completed.

FSD7 & FSD12 (U720PM/IN738LC AND U720C & W/IN738LC)

A number of specially designed IN738LC cast rings, as shown in Figure 8, were manufactured with a grain size in the as-cast condition of ASTM 2–3. Following post-casting HIP treatment and machining, the rings were assembled with the U720PM (FSD7) and U720CW (FSD12) hubs. After being HIPped these were then isothermally forged at 1100°C (Figure 9) and subsequently solution annealed. Forging of the U720FG-IN738LC combinations proved to be a difficult operation, requiring slow strain rates and accurate temperature control throughout the process, though a number of small surface cracks were evident at the outer parts of the IN738LC rims (see Figure 9). Final ageing according to HT1 was conducted for the discs to be cut-up for mechanical property evaluation and microstructural analysis. The remaining discs have been left in the as-solutioned condition in preparation for further machining, NDT and eventual spin test validation.

Fig. 6 FSD5 encapsulated prior to HIPping (left) and after forging (right).

Fig. 7 Full scale blisc FSD9 after HIPping.

SPIN TEST VALIDATION

To evaluate the cyclic properties of the full-scale dual alloy discs and verify the integrity of the concept a spin testing programme has been conducted. The dual alloy discs selected for spin test evaluation were FSD3 (fine-grained U720PM hub bonded to a coarse-grained C & W U720HC rim) and FSD9 (fine-grained U720PM hub bonded to single crystal MC2 blades). The built and balanced assemblies are shown in Figure 10.

Fig. 8 IN738LC rings after HIPping to remove porosity (left), FSD12 loaded in the HIP vessel (right).

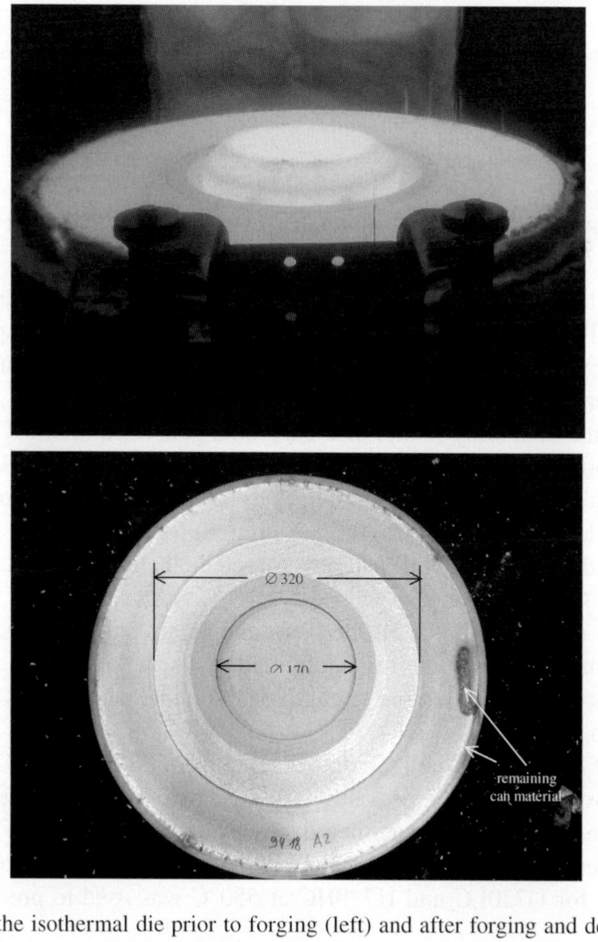

Fig. 9 FSD7 in the isothermal die prior to forging (left) and after forging and decanning (right).

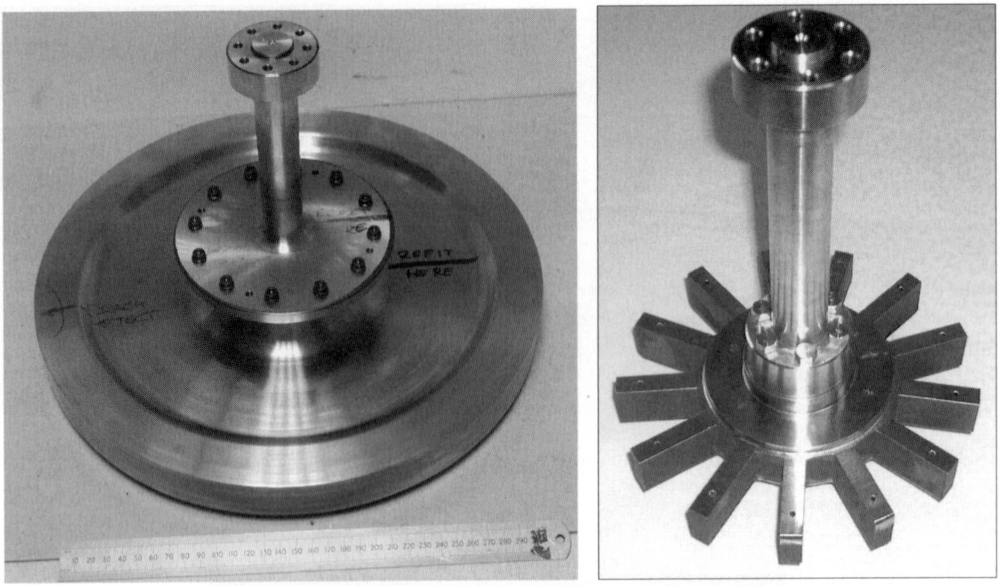

Fig. 10 Fully assembled and balanced FSD3 (left) and FSD9 (right) prior to testing.

Determination of the Spin Disc Test Conditions for FSD3

The lifetime calculation and determination of the spin test conditions were conducted using model disc and full-scale disc lifing data derived from hub, rim and joint specimens extracted from FSD2 and FSD4. A review of the development phase, model dual alloy disc LCF properties has been provided elsewhere.[4]

Trendlines for the dual alloy model disc LCF data generated at 450°C (fine grained hub region), 550°C (joint region) and 650°C (coarse grained rim region) are shown in Figure 11 in terms of the peak stress at half life vs. cycles to failure. Three bands of data are seen:

- The first band describes the highest fatigue strength for the U720PM tested at 450°C.
- The second band describes the HIPped plus forged dual alloy joint and rim data generated at 550 and 650°C, respectively.
- The third band describes the as-HIPped rim and joint data, which show lower fatigue strengths in comparison with the forged dual alloy discs.

These data indicate that the fatigue strength of the joint specimens is determined by the strength of the rim.

Also shown in Figure 11 are the LCF data derived from FSD2 and FSD4 to determine the spin test conditions for FSD3. The peak stress power law fatigue curve fitted to the HIP plus forged FSD rim and joint 550°C specimen data was used to predict the fatigue life for the most highly stressed region of the disc. A Neuber ($\sigma \times \varepsilon$ = constant) approach and the cyclic stress-strain curve for U720LC and U720HC at 550°C was used to predict the shakedown stress level.

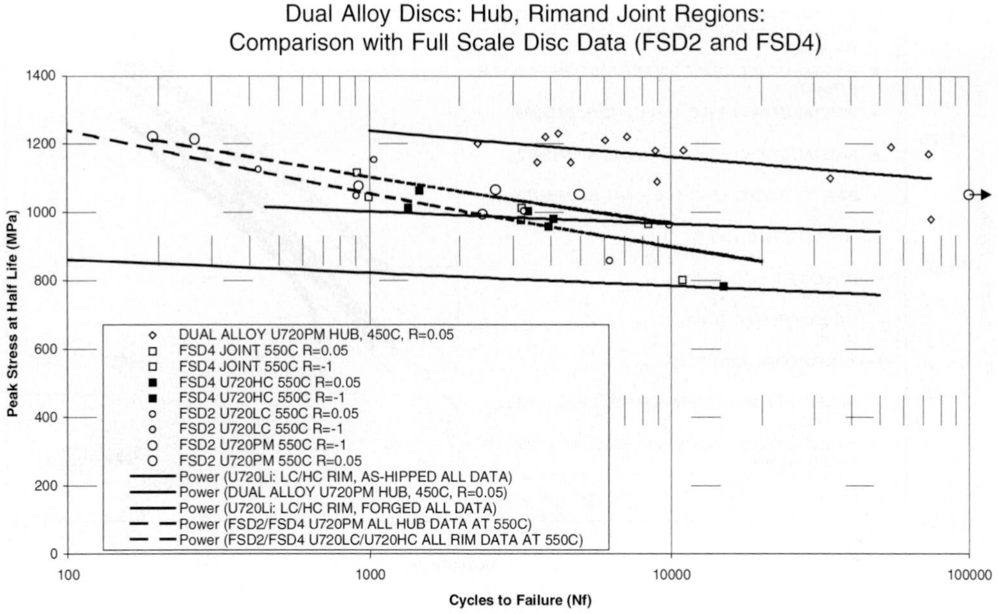

Fig. 11 Summary of the FSD2 and FSD4 hub, joint and rim strain controlled LCF data at 550°C, Rε = 0.05 or –1, 6%/min. ramp rate, in comparison with the corresponding dual alloy model disc trendlines (hub at 450°C, joint at 550°C, rim at 650°C).

Table 1 Summary of spin test conditions for FSD#3.

Temperature (°C)	Trapezoidal Cycle	Speed (rpm)	Peak Elastic Stress (MPa)	Shakedown Stress (MPa)	Predicted Life (Cycles)
550	35:1:35:1s	19,925	900	880	9,680 to 13,160

A series of fully elastic, finite element simulations of the spin disc was conducted at rotational speeds between 20,000 and 25,000 rpm (E-modulus = 195.67 GPa at 550°C) to determine the spin test conditions, which are summarised in Table 1.

Crack growth data generated at 550°C, R = 0.1, frequency = 15 Hz, presented in Figure 12, were used to evaluate the available crack propagation life for the different sections of the dual alloy disc. An estimated fatigue crack growth curve for the joint region, based on a x2 debit factor in measured crack growth rate, was used to account for the spin test 35:1:35:1s trapesoidal cycle. Subsequent FCG testing using a 28:5:28:1 cycle found this estimate to be reasonable. The Paris equation parameters were used in conjunction with the K-calibration due to Pickard[5] to assess the crack propagation risk for the hub, rim and joint regions of the spin disc.

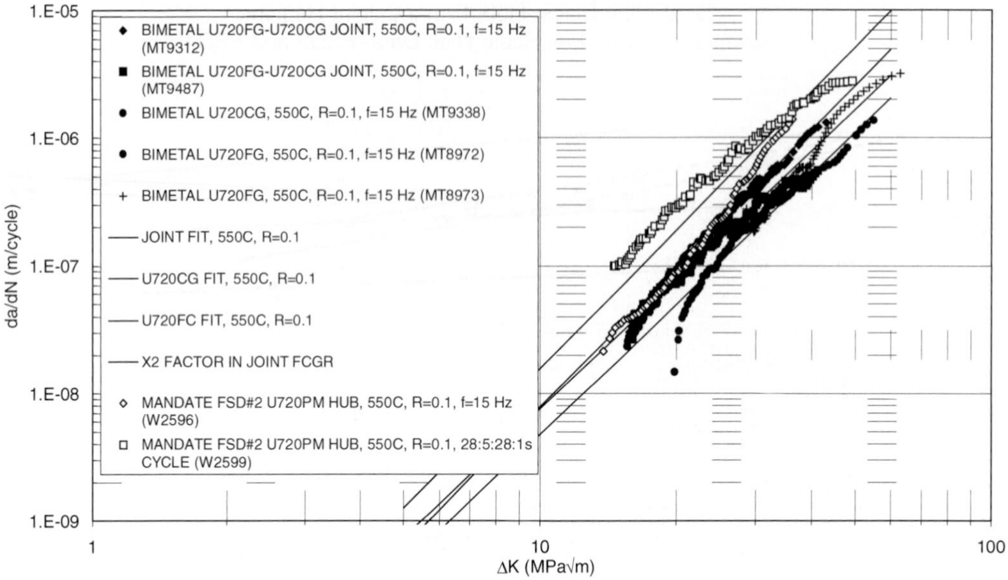

Fig. 12 Constant load amplitude fatigue crack growth curves generated at 550°C, R = 0.1 for Bimetal joint, Fine Grain and Coarse Grain specimens in comparison with Mandate FSD#2 hub specimen crack growth data.

Spin Testing Procedure and Results

For both the dual alloy disc (FSD3) and blisc (FSD9), the balanced assemblies were enclosed within specially designed furnaces capable of achieving a uniform temperature distribution and monitored using six thermocouples placed above and below the disc (see Figure 13). To ensure a constant heat flux and reduce windage losses, air was evacuated from the rig chamber and the pressure maintained at approximately 1 mbar during cyclic testing. Prior to test commencement, both the disc and the blisc were inspected using fluorescent dye-penetrant (FPI) to provide a baseline for any subsequent inspections.

FSD3 (U720PM-U720HC Dual Alloy Disc)

A number of FPI indications were detected close to the joint region, prior to testing, as a consequence of rough machining marks caused by stutter of the cutting tool as it crossed the bondline. The furnace controller was set to ensure a disc bond area temperature of 550°C and the disc was cycled between 100 and 19,925 rpm using a trapesoidal 35-1-35-1 second waveform. Non-destructive inspection of the disc was carried out at regular intervals during the course of the test to identify the presence of cracks. The first definite indication of a

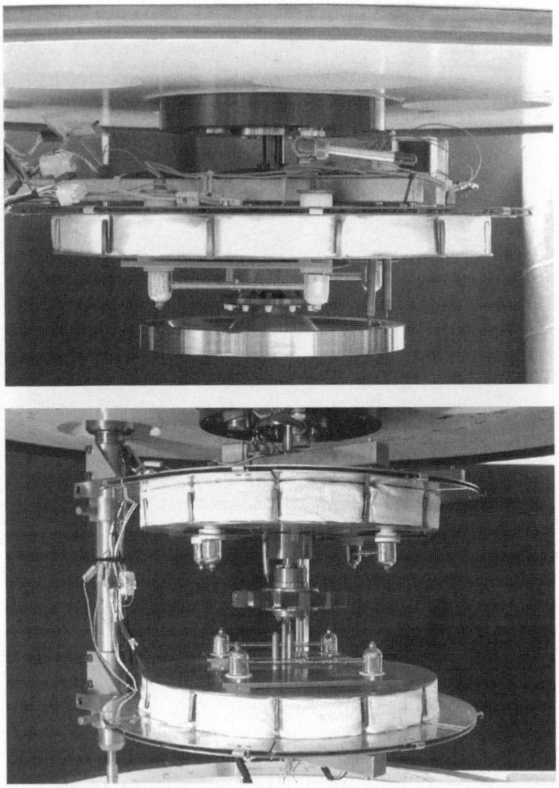

Fig. 13 FSD3 (top) and FSD9 (bottom) mounted on the spin rig.

small crack on the underside of the disc just in from the rim (labelled U1) was found after 3500 cycles. Acetate replicas were unable to definitely identify the crack due the surface roughness. A second small crack (labelled T1) was identified in the rim on the top surface of the disc after 6750 cycles. Due to the difficulty of providing accurate acetate measurements, surface crack length estimates were derived from the FPI measurements.

The disc burst catastrophically after a total of 9602 cycles. The disc fragments were retrieved and pieced back together as shown in Figure 14. The disc can be seen to have failed at a location just inside the rim at crack U1 and then 'unzipped' in a circumferential manner along the bondline. The cracks U1 and T1 were extracted from the disc segments and examined using optical microscopy and a Cambridge 360 scanning electron microscope (SEM).

FSD9: (U720PM-MC2 BLISC)

NDT inspection did not reveal any pre-test surface defects. Once the test temperature of 700°C had stabilised, the blisc was cycled between 100 and 40,000 rpm with a 19:1:19:1

Fig. 14 Re-assembled portions of FSD3 after the disc burst.

seconds trapesoidal cycle. After 109 cycles were completed the test was stopped due to problems with the spin rig drive belt. A further 14 cycles (total of 123 cycles) were completed at a much slower ramp rate (58:1:58:1 seconds cycle), however, due to further drive belt problems the test was discontinued. After strip down, the blisc was FPI'ed and this revealed that two of the MC2 ‘blades’ were cracked at a location remote from the peak stress location at the U720PM-MC2 joint fillet radius. Further analyses of these observations are proceeding.

FSD3 Failure Investigation

Examination of location U1 (Figure 15 and Figure 16) revealed a distinctive thumbnail shaped fatigue crack. The overall crack depth was measured as 3.85 mm, beyond which the failure was dominated by static modes, indicating rapid crack growth, prior to unzipping along the bondline and subsequent disc burst. SEM inspection found a mixed transgranular and intergranular propagation mode and away from the fatigue crack region, a series of approximately parallel machining marks that indicated the location of the bondline (turning marks on the inner bore of the ring). Due to damage inflicted during the burst, it was not possible to identify the initiation site, however, the crack was found to be propagating within the fine grained hub material. Sectioning and polishing of the fracture surface, as shown in Figure 17, found that crack initiation most probably occurred within the coarse-grained rim material close to the bondline. Crack propagation was along the plane of maximum tensile stress within the hub material, but the crack did not follow the bondline until shortly before final failure occurred.

Fig. 15 Crack U1 (as indicated) on the hub segment of FSD3.

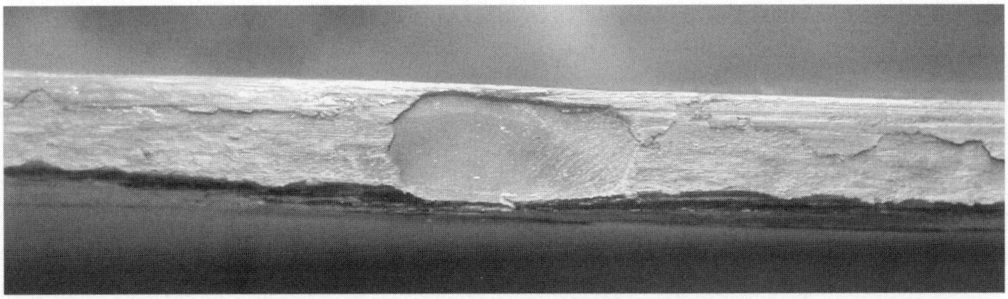

Fig. 16 Higher magnification image of Crack U1 on the hub segment.

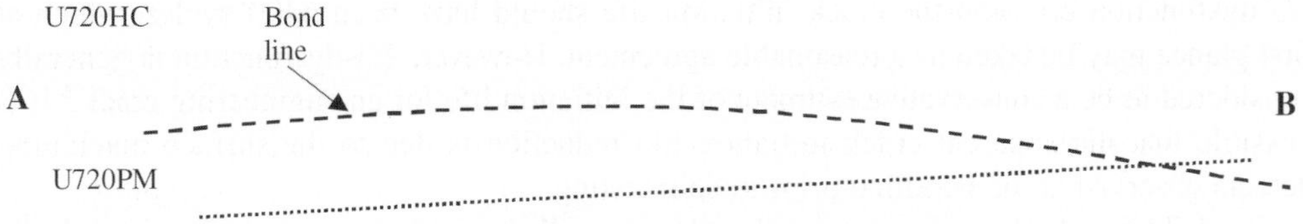

Fig. 17 Crack U1 Rim Piece - FEGSEM image through the fracture surface.

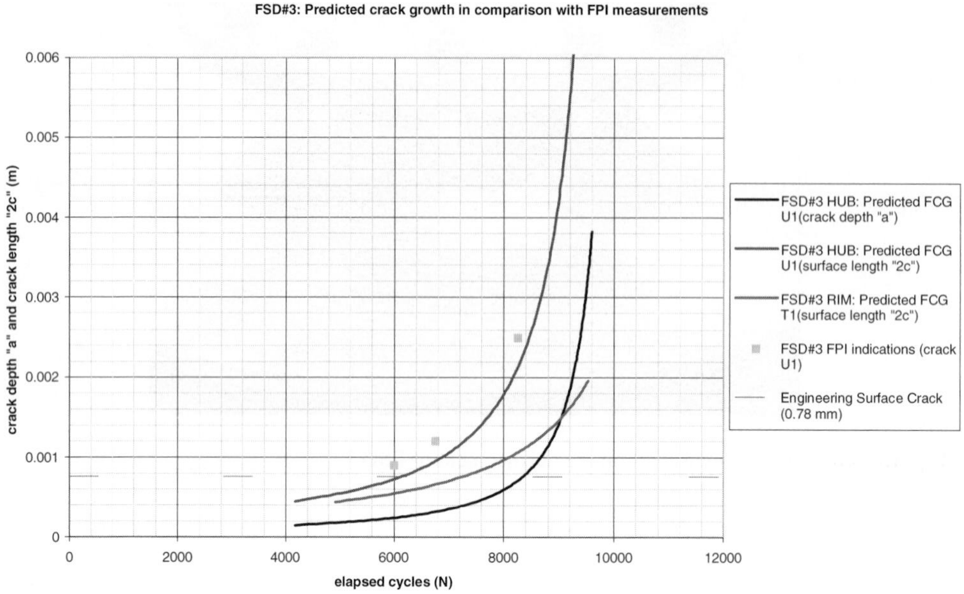

Fig. 18 Comparison of the surface crack length measurements (Crack U1 FPI indications) with the predicted crack growth behaviour for crack U1 and T1.

It was not possible to conduct striation spacing counting from the fracture surface nor make any positive identification of beachmarks. However, a comparison of the estimated surface crack lengths from the FPI indications with the predicted crack growth behaviour is shown in Figure 18. The crack growth prediction was generated using the Pickard solution for a thumbnail shaped crack[5] in conjunction with a back calculation method from the spin test end-point (crack depth = 3.85 mm, dysfunction cycles = 9600). A x2 debit factor was applied to the Paris equation parameters for the hub material data. An aspect ratio of $a/c = 1.5$ (where a = crack depth and c = half surface crack length) has been estimated and applied to the predicted crack depth data. This provides an estimate of the surface crack length as a function of elapsed cycles, as shown in Figure 18. The FPI indications are generally upper bound, however a reasonably good correlation is found between the predicted surface crack length and the estimated FPI crack dimensions.

On this basis the crack initiation life for the spin test to an 'engineering crack' (0.78 mm surface length) was found to be 6000 cycles. According to the generally adopted 2/3-dysfunction criterion the crack initiation life should have been 6400 cycles, which at first glance may be taken as a reasonable agreement. However, 2/3-dysfunction is generally considered to be a conservative estimate of the initiation life for an engineering crack.[6] It is possible that the apparent crack initiation life reduction is due to the surface machining defects observed at the bondline prior to spin testing.

Crack T1 was broken open to reveal a 'thumb-nail' shaped fatigue crack growing wholly in the rim material just above the bondline. This crack measured ~1 mm deep and ~2 mm surface length and revealed a number of localised initiation sites at the surface that did not appear to be associated with machining marks. The fracture surface was distinctly different to that seen in crack U1, as a faceted, predominantly transgranular mode (Stage 1 fatigue) was observed. A FCG back calculation from the spin test end point for crack T1 (a/c = 1) using a x2 debit factor applied to the rim FCG data is also shown in Figure 18. This predicts the crack initiation life (for a 0.78 mm surface engineering crack) for T1 to be ~7200 cycles. This can probably be taken as the true initiation life in the absence of machining damage at the bondline.

CONCLUSIONS

1. The full-scale disc production programme has been completed successfully with manufacture, machining and heat treatment of a number of full-scale dual alloy discs and bliscs in either the as-HIPped or HIPped plus forged condition. The manufacturing routes and heat treatment conditions have been established for each of the dual alloy combinations studied.
2. Two cyclic spin tests have been performed successfully within the project: U720PM-U720HC full-scale dual alloy disc at 550°C and a maximum speed of 19,925 rpm and a U720PM-MC2 full-scale blisc at 700°C to a maximum speed of 40,000 rpm.
3. For the dual alloy disc, catastrophic failure (burst) occurred after approximately 9600 cycles due to low cycle fatigue close to the bondline region. For the U720PM-MC2 blisc approximately 120 cycles were experienced before a test machine problem led to interruption of the test.
4. The experimental cyclic lifetime for the full-scale dual alloy disc agreed well with the predicted life derived from the FE simulations and test data generated within the project.

The project has demonstrated the technical feasibility of manufacturing dual alloy discs and bliscs. Further developments are required to enhance the quality and improve the reliability of the manufacturing processes and additional tests are necessary to better define the lifing methodologies before full-scale engine testing can be undertaken.

ACKNOWLEDGMENTS

The authors acknowledge financial support from the European Community industrial and technology programme, project No BE97-4650 and permission to publish this work from ALSTOM POWER and Turbomeca. Support from the other partner organisations involved in the programme is also gratefully acknowledged.

REFERENCES

1. M. B. Henderson, J. Hannis, G. McColvin and G. Ogle: 'Materials Issues for the Design of Industrial Gas Turbines', *Proceedings of Conference on Advanced Materials*

and Processes for Gas Turbines, G. E. Fuchs. et al., eds., UEF, Copper Mountain, USA., 2002.
2. D. P. Mourer, E. Raymond, S. Ganesh and J. M. Hyzak: 'Dual Alloy Disc Development', *Proceedings of the 8th International Symposium on Superalloys*, TMS, Seven Springs, Champion, USA., 1996, 637–645.
3. Y. Bienvenu, et al.: 'A Study of Fabrication Through Powder Metallurgy of Nickel Based Superalloys Hybrid Structural Parts', *Hot Isostatic Pressing 1993*, Elsevier Science B.V., 1994.
4. H. Burlet, G. McColvin, M. B. Henderson, J. C. Garcia, S. Peteves, H. Klingelhoeffer, M. Maldini, U. E. Klotz, T. Luthi and I. M. Wilcock: 'Microstructural and Mechanical Performance Assessment of Diffusion Bonded Bimetallic Model Discs', *Proceedings of 6th International Charles Parsons Turbine Conference*, Dublin, 2003.
5. A. C. Pickard: 'The Application of 3-Dimensional Finite Element Methods to Fracture Mechanics and Fatigue Life Prediction', Engineering Materials Advisory Service Ltd, Warley, UK., 1986, 93.
6. G. Harrison and M. B. Henderson: 'Lifing Strategies for High Temperature Fracture Critical Components', *Proceedings of Conference on Life Assessment of Hot Section Gas Turbine Components*, R. Townsend, et al., eds., Heriot Watt University, Edinburgh, UK., 1999, 11–34.

Advanced High Temperature Turbine Seals Materials and Designs

W. SMARSLY and N. ZHENG
MTU Aero Engines

E. VIVO
Fiat Avio

D. SPORER
Neomet

M. TUFFS
Rolls-Royce

K. SCHREIBER
Rolls-Royce Deutschland

B. DEFER and C. LANGALDE-BOMBA
Ecole Centrale de Lyon

O. ANDERSEN and H. GOEHLER
IFAM Dresden

N. SIMMS
University Cranefield

G. MCCOLVIN
Alstom Power

INTRODUCTION

Advanced turbine seal materials and designs are under development to achieve higher temperature capability, extended lifetime and reliability than the state of the art technology. Cooling air consumption, inspection cycles interval and repair costs of aero engines will be reduced.

Following results of R&D activities of a European project consortium will be presented:

- Development of design concepts for advanced turbine seals,
- Definition of criteria and requirements of the outer air seals,
- Study the damage mechanisms of the today used outer air seals,
- Assessments of the relevant properties of candidate materials and structures against the advanced design criteria and requirements and
- Lifetime prediction concepts.

Turbine outer air seals have two basic functions: to seal the cavity over the rotating blade and to hold the previous vane. Other functions are to reduce the cavity between vanes and blades and to provide a seal, which can rub in.

In order to define a first version of design criteria and requirements for materials and processes for advanced turbine seals which can be tested in the large scale integrated platform vehicles CLEAN (MTU - SNECMA concept), ANTLE (RR-ITP concept) or any other relevant aero engine or industrial gas turbine, the design, manufacturing and service requirements of outer air seals have been identified by MTU in co-operation with RR, Fiat and Alstom Power.

In general the outer air seal must be able to survive the hot gas environment and resist oxidation and corrosion. Above all, however, they must react in a safe manner when in contact with the rotating blade by easily giving way without damaging the blade tip.

CRITERIA FOR ADVANCED TURBINE SEALS DESIGN CONCEPTS

Criteria for Antle

The RR-ITP concept ANTLE (Affordable Near Term Low Emmissions) and its challenges are shown in Figure 1.[1]

The relevant part for the ADSEALS project is the high and intermediate pressure turbine (Figure 2). For this part outer air seals have to be developed which meet the advanced temperature and lifetime requirements.

In general based on 2D and 3D thermal and stress calculations the potential available candidate materials for high and intermediate pressure turbine outer air seals has to meet the following requirements:

- the peak temperature for the active seal part will be close to 1400°C for the high pressure turbine and above 1200°C for the intermediate pressure turbine.

Therefore the most relevant property for the material choice is the oxidation resistance.

According to the oxidation resistance, among the different classes of available materials, ceramic materials are the most suitable materials for the active seal part. For the abradability, a low strength relatively brittle material should be preferred. For the active seal part of high and intermediate pressure turbine outer air seal the structures has to meet the following structural requirements:

- suitable for rotor movements of radial depths up to 0.5 mm,
- smooth surface towards the gas path in order to minimise the disturbance of the laminar gas flow,
- dense structure to towards the backplate in order to minimise lateral leakage in the structure itself,

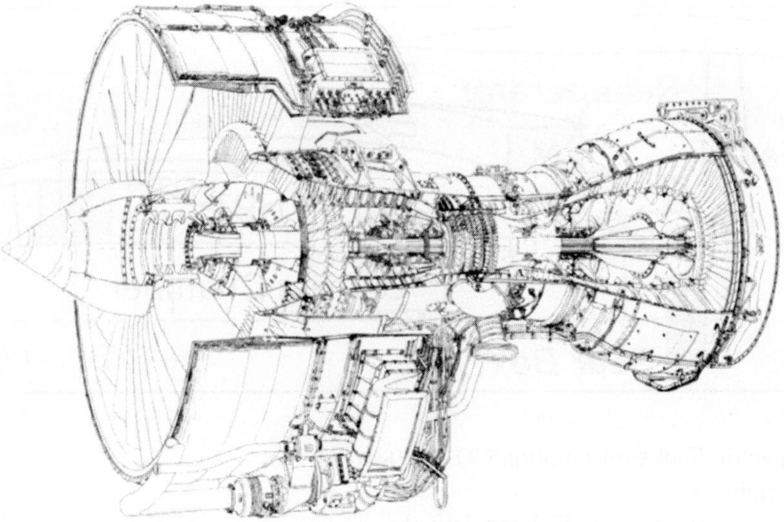

- Time to market reduction –30%
- Cost of ownership reduction –20%
- Life cycle cost reduction –30%
- Reliability improvement +60%
- CO_2 reduction –12%
- No_x reduction –60%

Fig. 1 Advanced aero engine concept: ANTLE- Trend 3 shaft aero engine.

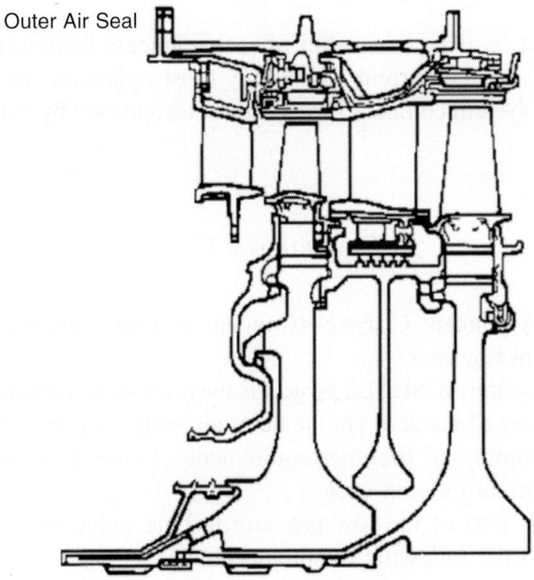

Fig. 2 High pressure turbine outer air seal.

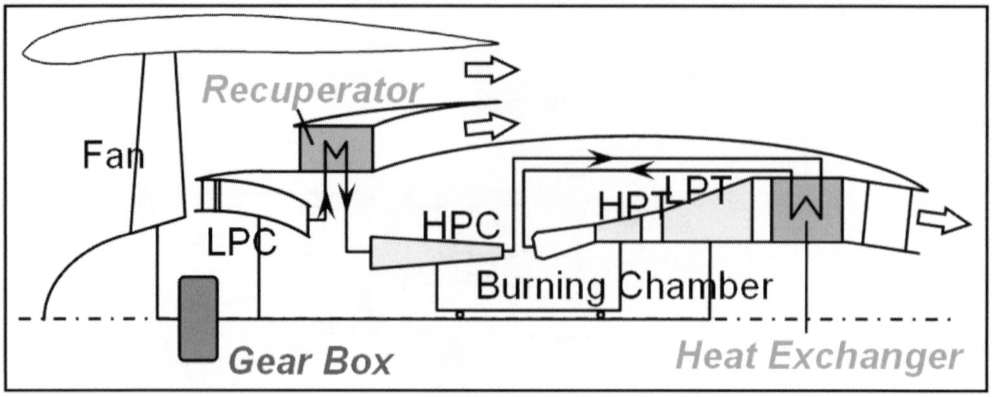

- Specific Fuel Consumption/CO_2 –(30–40%)
- Weight –15%
- Noise ≤ 10 dB (–30 dB Cumulative)
- NO_x Emission < 85% Compared to ICAO 95

Fig. 3 MTU – SNECMA concept CLEAN.

- mechanical stability to resist temperature and pressure gradients and
- abradability for unshrouded rotors: the blade tip should not being damaged when rubbing in.

The unit costs and costs of ownership of seal segments including the repair capability have to be reduced over the current technology (cost reductions of 20 – 30% have to be achieved for the seal system compared with current technology by extending the life time of outer air seals).

Criteria for Clean

The MTU-SNECMA concept CLEAN (Concept of Low Emissions and Noise) and its challenges is shown in Figure 3.[2]

The relevant part for the ADSEALS project is the high speed rotating low-pressure turbine (Figure 4). For this part of the aero engine outer air seals have to be developed which meet the advanced temperature and lifetime requirements. Figure 5 shows in principle the gas temperature for one flight mission cycle.

For the active seal part of the low pressure turbine outer air seal, potential available structures has to meet the following requirements:

- suitable for rotor movements of axial width is about 10–15 mm,
- suitable for rotor movements of radial depths is about 3–5 mm,

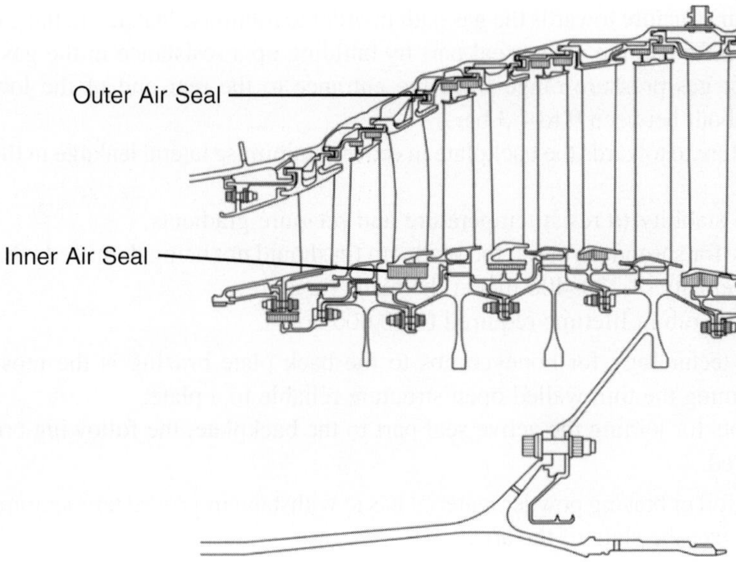

Fig. 4 Low-pressure turbine honeycomb seals.

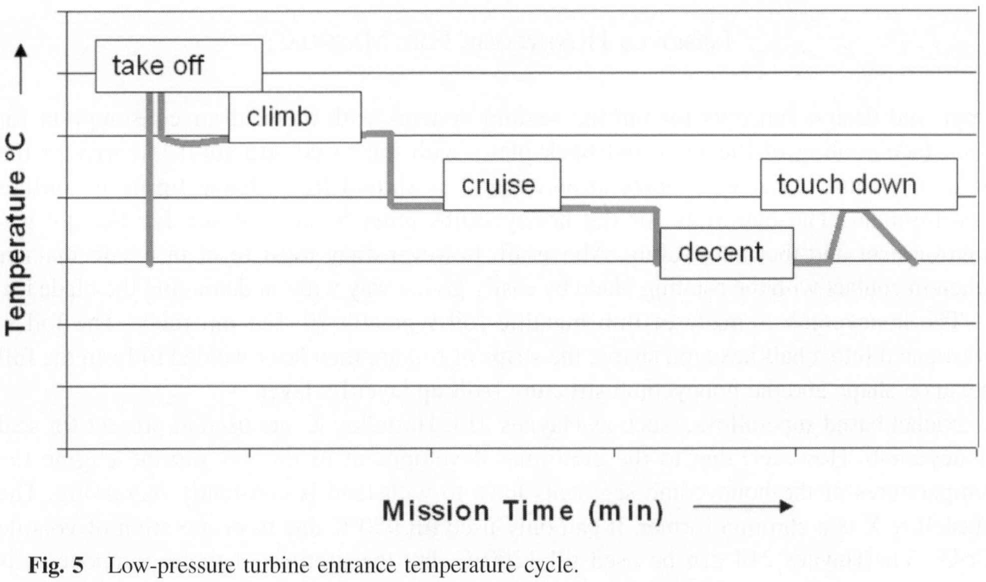

Fig. 5 Low-pressure turbine entrance temperature cycle.

- open cellular structure towards the gas path in order to minimise leakage in the gap between the blade tip fin and the active seal part by building up a resistance in the gas flow path (the exhaust gas pressure range from the entrance to the rear end of the low pressure turbine is about between 9 to 1.3 bar),
- dense structure to towards the backplate in order to minimise lateral leakage in the structure itself,
- mechanical stability to resist temperature and pressure gradients,
- abradability for shrouded rotors: the blade tip fin should not being damaged when rubbing in, with a velocity of 250–400 m/s at the blade tip and
- oxidation and rub in lifetime required to 20,000 hours.

As joining technology for honeycombs to the back plate brazing is the most effective process for joining the thin walled open structure reliable to a plate.

The concepts for joining the active seal part to the backplate, the following criteria have to be considered:

- the brasing foil or brasing powder material has to withstand improved temperature - lifetime conditions over the state of the art ;
- the brasing temperature has to be higher than the service temperature of the seal but not as high that the required properties of the active seal part and the back plate material will be degraded.

If the advanced seal material will meet the required life time, as an estimation for the required target costs for low pressure turbine outer air seals (considering a cost reduction target of 30% over the state of the art and an extended life time of 100% over current technology) a factor of 1.4 could be accepted.

DESIGN CONCEPTS FOR ADVANCED LOW PRESSURE TURBINES

Improved Honeycomb Foil Materials

Improved design concepts for turbine sealing systems with reduced air consumption for back-face cooling of the liner and back plate, with increased structural integrity of the abradable, bonding coat or braze alloy as well as shroud fin or blade tip needs further development. The materials for the honeycombs must be able to survive the hot gas environment and thermal cycling. Above all, however, they must react in a safe manner when in contact with the rotating blade by easily giving way without damaging the blade tip.

The honeycomb is made of thin metallic foil typically 70–125 μm thick. The foil is corrugated into a half hexagon shape; the strips of foil are then laser welded to form the full hexagon shape and the honeycomb structure built up layer by layer.

Nickel based superalloys,[5] such as Haynes 214, Hastelloy X, are used at present for seal honeycomb. However, due to the continual development of the gas turbine engine the temperatures of the honeycomb segments have to withstand is constantly increasing. The Hastelloy X is a chromia former. It can only used till 950°C due to evaporation of volatile Cr_2O_3. The Haynes 214 can be used till 1200°C, but the lifetime of these components is

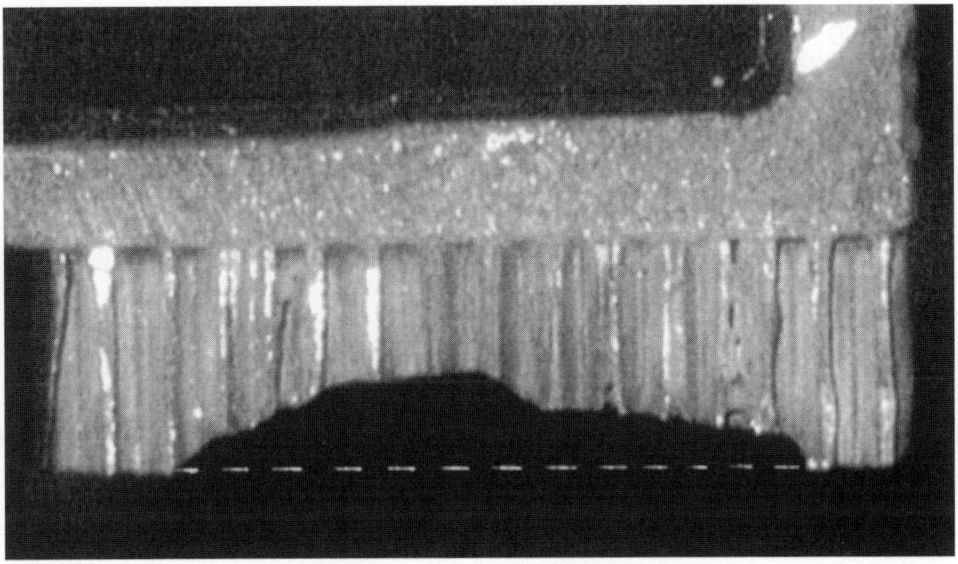

Fig. 6 Damaged nickel alloy honeycomb seal due to oxidation embrittlement and abrasion.

drastically reduced due to internal oxidation of the foil material, because Aluminium diffuse very slowly in the nickel-based alloy.

A solution would be to increase the Al content in alloy matrix with aluminising to increase the oxidation limited life. But the high Al contents can deteriorate the mechanical properties of the alloy, especially ductility flexible as before.

Another way is filling with oxidation resistance materials to reduce the surface that increase the oxidation resistance. But its wear resistance will increase and as the result the abrasion of the turbine blade tip.

A more effective method to increase the oxidation resistance of the honeycombs would be to look for alternative foil materials, e.g. FeAlCr - alloys, like PM2000,[4] Aluchrom YHF,[3] PM2Hf[4] and Kantahl AF. These alloys are aluminium oxide formers with increased oxidation resistance over nickel alloys for temperature up to 1200°C. The aluminium in the iron alloys diffuse faster than in nickel alloys. As the result for the iron alloys a self healing oxidation layer can be formed, whereas for the nickel-alloys, internal oxidation can occur (Figure 6).

HOLLOW SPHERE STRUCTURES

Hollow sphere structures[7, 8] are a new class of lightweight materials within the family of cellular materials. They can have a wide range of properties, which can be achieved by varying the base metal alloy, the sphere size and the structure. The honeycomb structures are superior in structural weight and abradability over filled honeycombs, hollow sphere and fibre mat structures.

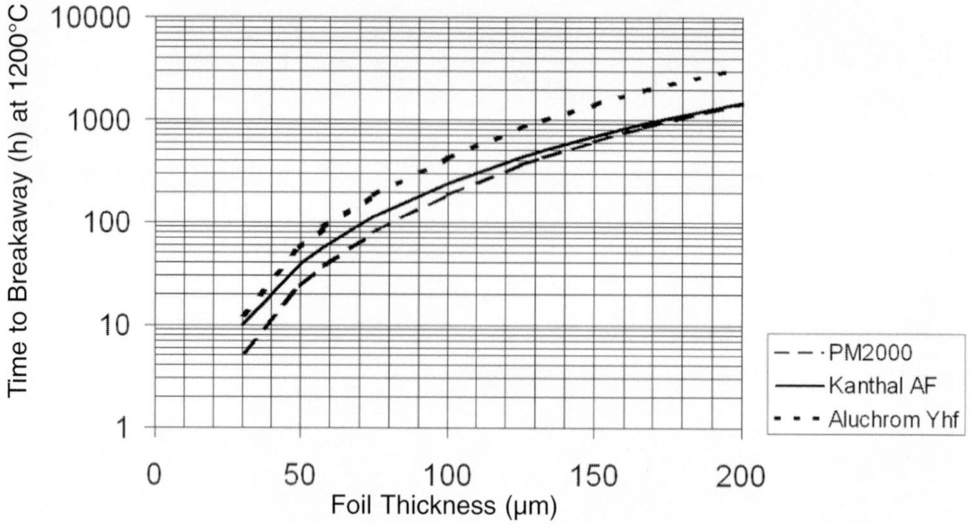

Fig. 7 Oxidation resistance for different advanced FeAlCr honeycomb alloys.

But honeycomb structures have less oxidation resistance compared to the other considered structures because of the larger open surface, which is in contact with hot exhaust gas. If the hollow sphere structure is cut in a way that an open half sphere structure appears towards exhaust gas channel of the turbine, than the abradability is like that of honeycomb structures.

The hollow sphere structure can be brazed on the back plate like honeycombs or potentially by sintering procedures.

DESIGN CONCEPTS FOR ADVANCED HIGH PRESSURE TURBINE SEALS

Porous Ceramic Coatings

The abradable solutions for high pressure turbine seals are based on a series of parallel rails that are machined into the backing plate. The rails are to be filled or overfilled with a porous ceramic coating deposited via thermal spraying. During engine running the blade fin can then rub into the coated rails cutting a path to reduce over tip leakage in subsequent engine running. This solution is shown in Figure 7.

The two principal choices for the ceramic coating were alumina and zirconia, both these materials are considered to be capable of operating at 1200°C in this type of application. It

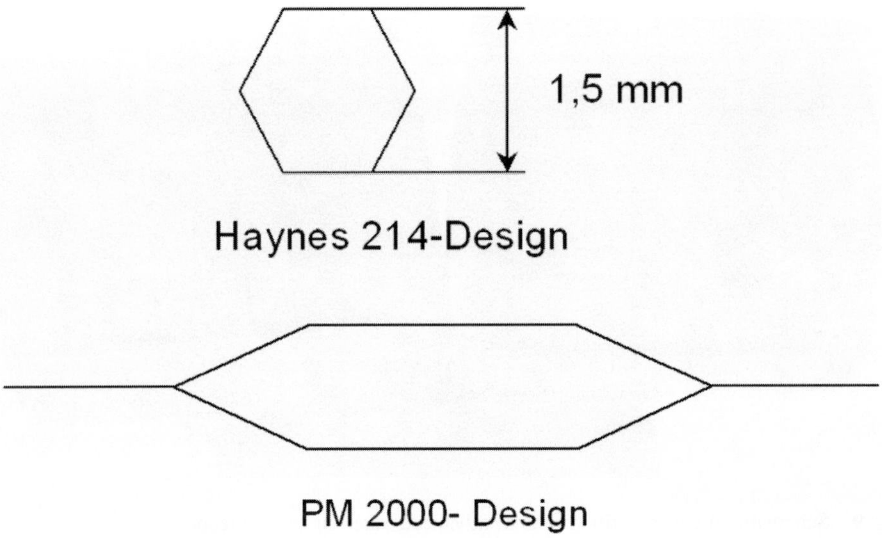

Fig. 8 Design limits for different honeycomb materials.

was decided to proceed with the manufacture of test specimens with rails filled with both alumina and zirconia as the different physical properties may result in a differing performance. Various solutions are under con sideration for the more arduous application of abradable liner systems that will have the potential to act as part of an unshrouded sealing system. In such an application peak component temperatures are expected to be approximately 1400°C.

SUMEGRID

Sulzer Metco developed an innovative combination of investment casting and plasma spraying to produce turbine outer air seals. The grid structure and the back plate are an integrated part made by investment casting of a Ni/Co based alloy. A porous ceramic layer, e.g. zirconia oxide, is sprayed on the top of the grid walls in order to produce an abradable layer (Figure 8).

This concept has the advantage of superior oxidation resistance of the zirconia oxide and the superior abradability of the porous ceramic layer on the top of a honeycomb structure. The ceramic layer also acts as a thermal barrier coating. In addition the metallic back plate and grid could be cooled by air to keep the component temperatures below the materials temperature limits.

Fig. 9 Schematic of ceramic filled rails as advanced outer air seal concept.

MODELL FOR THE OXIDATION AND RUB IN LIFE OF TURBINE OUTER AIR SEALS

The analysis of damages in real aero engines show that oxidation makes the seals honeycomb brittle, so that its life is limited by the oxidation.[11] The reason for this effect is the scale-forming element, Al, in the alloy matrix is consumed. If the remaining Al% is decreased beneath a critical concentration, a breakaway oxidation occurs. Alumina-forming alloys are excellent oxidation-resistant materials at high temperatures. Their resistance relies upon establishment of a stable, slow-growing, and adherent α-alumina. But at lower temperatures and/or in the early stages of oxidation, the metastable oxides γ, δ and θ-Al_2O_3 grow. Once the aluminium content in bulk alloy is too low to reform external oxide, an internal oxidation occurs.

A rapid depletion of aluminium accompanies this internal oxidation process, exacerbating the onset of breakaway corrosion. For pre-oxidation segments, a stable and adherent α-Al_2O_3 scale has been formed over the whole honeycomb after 1 hour at 1090°C in air. This scale protects the base alloy from being oxidised.

But during the services, some 0.5–1% surface scale would be repetitively worn off by rotor tip rub-in (Figures 9 and 10). Once a given protective oxide layer worn off, a fresh alloy would start to oxidise, then an unstable mixed oxide scale would be formed at this position. Based on above analyse the lifetime of OAS segment can be limited by oxidation. Growth and wearing off of the protective alumina scales after long service times to a depletion of aluminium in the alloys, eventually resulting in breakaway oxidation.

This lifetime limited can be predicted using a model, which has been developed by KFA Forschungszentrum Jülich.[6]

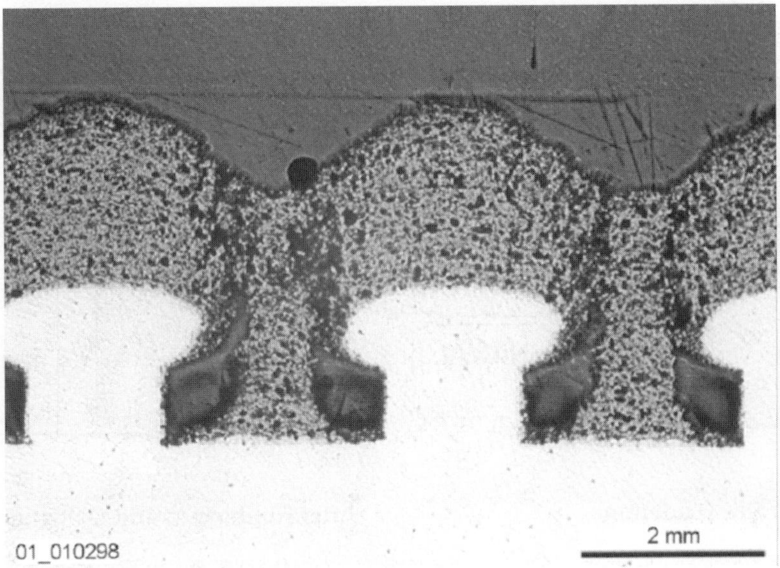

Fig. 10 Plasma sprayed ceramic on the top of the metallic grid.

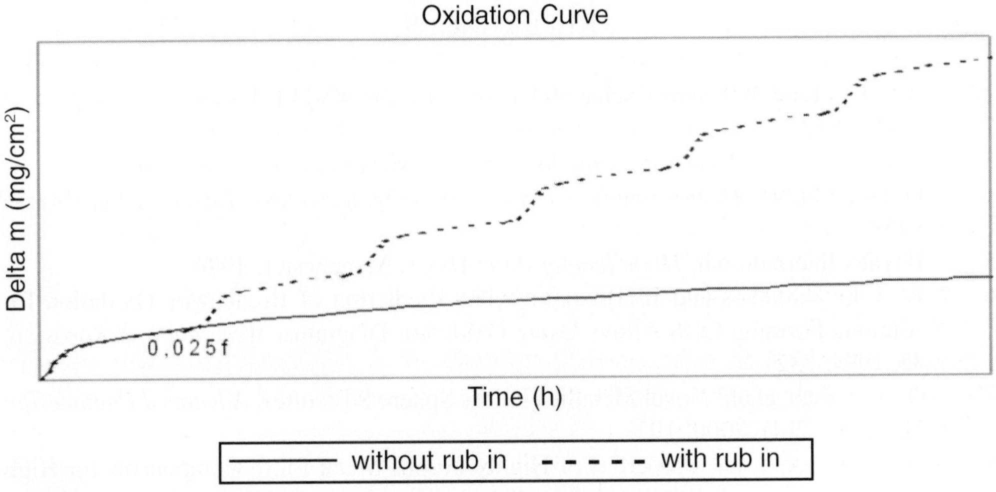

Fig. 11 Comparing the oxidation curves with and without rub-in.

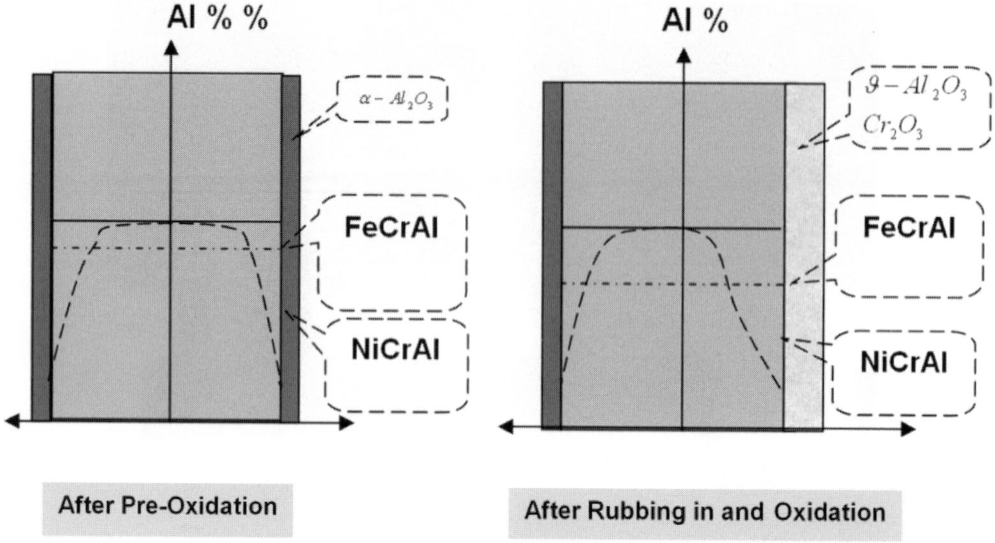

Fig. 12 Al concentration decrease in FeCrAl- and Ni alloys.

The calculation shows that the mechanical damage of a small part of the oxide scale has a great influence on the lifetime of alumina-forming alloys. The FeCrAl - alloy Aluchrom VHF has much better re-healing ability than Ni-based alloy Haynes 214 and, if rubbing-in of tip destroys the oxidation layer (Figure 9).

REFERENCES

1. RR plc, Trend 500 Aero Engine and Targets for the ANTLE Project.
2. MTU/SNECMA, Technology Platform CLEAN and Targets for the CLEAN Project.
3. VDM Krupp, Materials Data for Metallic Materials Used for Honeycombs.
4. Plansee GmbH, *Dispersion-Strengthened High-Temperature Materials*, Lechbruck, 1998.
5. Haynes International, *High-Temperature Alloys*, Manchester, 1990.
6. W. J. Quadakkers and K. Bongartz: The Prediction of Breakaway Oxidation for Alumina Forming ODS Alloys Using Oxidation Diagrams, *Werkstoff und Korosion*, **45**, 1994, 232.
7. O. Andersen, et al.: Novel Metallic Hollow Sphere Structures, *Advanced Engineering Materials*, **2**(4), 2000, 192.
8. O. Andersen, et al.: Properties of Highly Porous Metal Fibre Components for High Temperature Applications, *Proceedings of International Conference on Powder Metallurgy*, **3**, 1998, 33.

Development and Optimisation of Advanced Sealing Systems for Industrial Gas Turbines

IAN W. BOSTON, J. HURTER and G. MCCOLVIN*
Alstom Power

M. G. GEE and G. ALDRICH SMITH
National Physical Laboratory

P. AMOS
Advanced Products Co Inc

ABSTRACT

The drive for ever lower emission levels within large industrial gas turbines coupled with the need for an ever increasing requirement on component life, leaves the combustion design engineer and metallurgist with a difficult compromise. The lower emissions inevitably bring with them increased HCF loading and vibration within an extremely aggressive thermal and pressure loaded environment. Thus wear becomes the key element defining engine life. This report picks up from the baseline tribological testing carried out by NPL and shows how the new material combinations were applied into new seal designs and validated with component specific test facilities. The result of this work was the application of some of the most sophisticated yet cost-effective sealing designs to be found within a modern gas turbine combustor.

INTRODUCTION

The development of the market for gas turbine powered electrical generation stations progressed at an emphatic rate throughout the 90's, with lifetime and engine cycle time requirements increasing almost by a daily rate. The main result of this is that components within the gas path (compressor blades, combustor and turbine blades) suffer by way of cycle life availability. In particular with the combustor there is a need to optimise the air distribution and application to maintain an efficient cycle and avoid over application of cooling air. The combustor is rendered only as good as the sealing system employed to monitor and control this airflow. The combustor is also the area within a low emission engine that is exposed to the greatest vibration loading, a natural byproduct of this very low emission level running (in the order of 1:100th of that seen in modern aero engines).

The design of compressor and turbine blades is now extremely mature and well understood, so again it is the development of the combustor and engine sealing systems that now offers

*Now with Demag Delaval Industrial Turbomachinery Ltd.

the greatest opportunity to set products apart from their competition. It is for this reason that Alstom initiated the advanced combustor sealing and material development programme. This was aimed a taking a broad look at optimising the substrate contact surface, with the evaluation of various high temperature wear coatings, the definition of sealing designs and preloads and the application of new seal materials and philosophies. Design constraints imposed by the incumbent system forced the seal design engineering for the initial engine application.

It should also be noted that often only a small percentage of technology programmes bear fruit in the form of engine application. Such was the success of this programme that the entire engine fleet of GT24 and GT26 large gas turbines are currently being retro fitted with the seals and substrate processing generated from this work.

PROCESS ENGINEERING

Picking up from the initial broad based evaluation of materials and coatings carried out by NPL,[1] the materials for integration into the seals were selected, along with a definitive selection of coatings for the incumbent hardware that form the substrates and wear contact surfaces. The candidate seal materials needed to be incorporated into designs that would deliver the required performance. Manufacturing processes had to be developed to produce the desired seal forms in a reliable and cost-effective manner with consistent high quality. The resultant designs then needed to be qualified in terms of life due to wear using trial seal components in a series of dedicated tribology test rigs. It is important here to remember that wear is the predominant factor effecting seal and system life.

Operating temperatures associated with the hot gas side of the seal are only likely to increase through the natural development life of the gas turbine. It was therefore clear that the metallurgical function of the heat shield and wear contact for the seal would have to be separated from that of the spring pre-load of the seal. Thus the seal was designed as a metal composite welded construction as shown in Figures 1 and 2.

RIG DEFINITION AND TESTING

The rig testing strategy was defined to separate the elements and functions of wear and to gain an understanding as to which one was the most prevalent, the seal and rig programme was very much treated as a system. The rigs that were defined for this phase of testing are listed in Table 1. The test matrix of materials selected from the results of the associated NPL project are listed in Table 2.

VIBRATION – IMPACT TEST RIG

This rig was built to study the wear brought about between two components as they vibrated and impacted against each other with a ‘hammering’ action without sliding. The rig shown

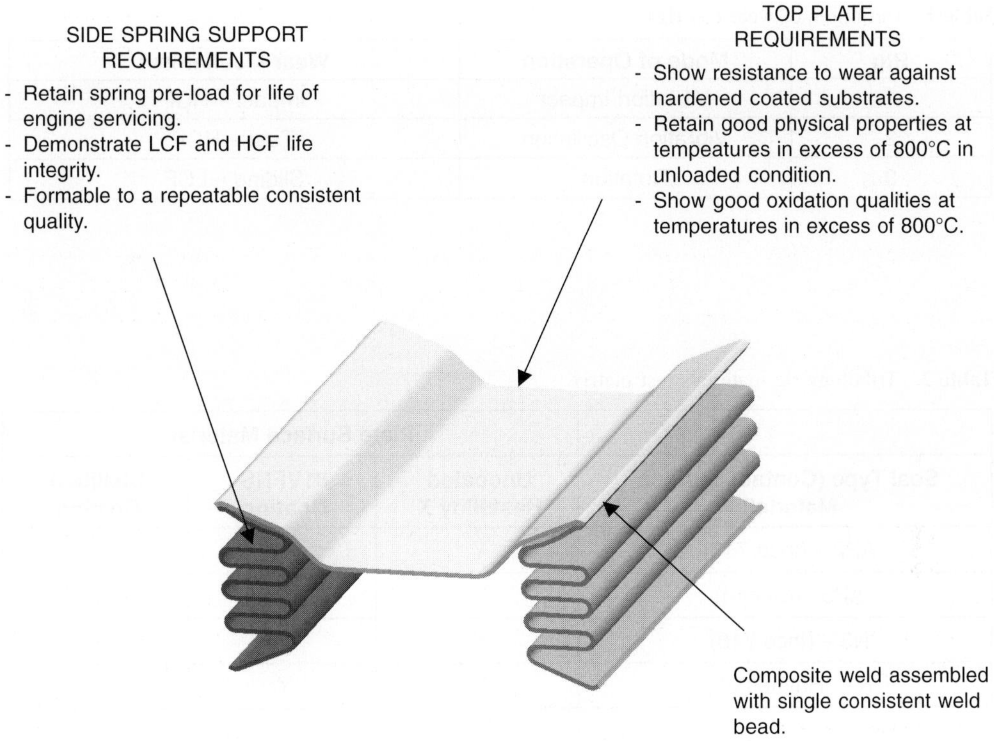
SIDE SPRING SUPPORT REQUIREMENTS
- Retain spring pre-load for life of engine servicing.
- Demonstrate LCF and HCF life integrity.
- Formable to a repeatable consistent quality.
TOP PLATE REQUIREMENTS
- Show resistance to wear against hardened coated substrates.
- Retain good physical properties at tempeatures in excess of 800°C in unloaded condition.
- Show good oxidation qualities at temperatures in excess of 800°C.
Composite weld assembled with single consistent weld bead.

Fig. 1 Seal profile definition.

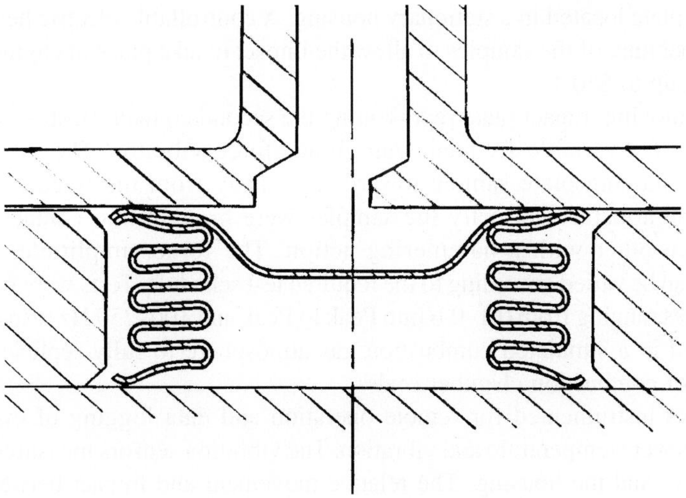

Fig. 2 Seal profile shown within constrained engine standard envelope.

Table 1 Definition of wear test rigs.

Rig	Mode of Operation	Wear Mechanism
1	Vibration Impact	Impact – HCF
2	Vibration Oscillation	Tear – HCF
3	Rotation	Sliding – LCF

Table 2 Tribology rig materials test matrix.

	Plate Surface Material		
Seal Type (Contact Surface Material)	Uncoated Hastelloy X	81VFNS Coating	Stellite 6 Coating
AP2 – (Inco 718)	Y	Y	Y
AP3 –(Ultimet)	Y	Y	Y
N3 – (Inco 718)	Y	Y	Y
K – (Haynes 25)	Y	Y	Y

in Figures 3 and 4 used short length samples of the real engine seal hardware mounted in a vertical fixed plate located in a stationary housing. A controllable electric heater was used to raise the temperature of the samples to allow the impact to take place at engine representative temperatures, up to 550°C.

A vertical moving impact plate representing the second contact substrate was attached to the armature of a variable frequency/amplitude/force vibrator. The vibrator operation oscillated the moving plate sample towards and away from the fixed plate sample. By positioning the armature correctly the samples were brought into contact such that they impacted each other with a hammering action. The force, amplitude, frequency and temperature can be varied according to the required test schedule. Tests were run at amplitudes and frequencies ranging from 0.1–0.6 mm Peak to Peak and 100–250 Hz respectively. Testing was conducted in a simulated combustion gas atmosphere to fully replicate the operating environment of combustion chamber seals.

The rig was instrumented for remote operation and data logging of events. The main measurements were temperature and vibration. The vibration sensors measured the movement of the armature and the housing. The relative movement and impact between the samples was obtained from these measurements. The required test duration and frequency determined the number of impact cycles.

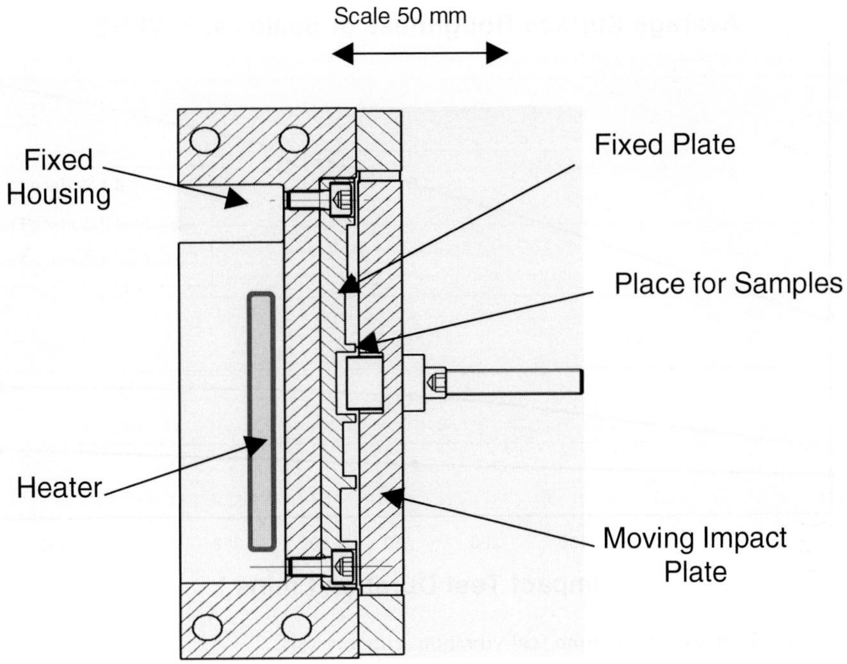

Fig. 3 Schematic of the vibration - impact rig.

Fig. 4 Open view of the rig after testing.

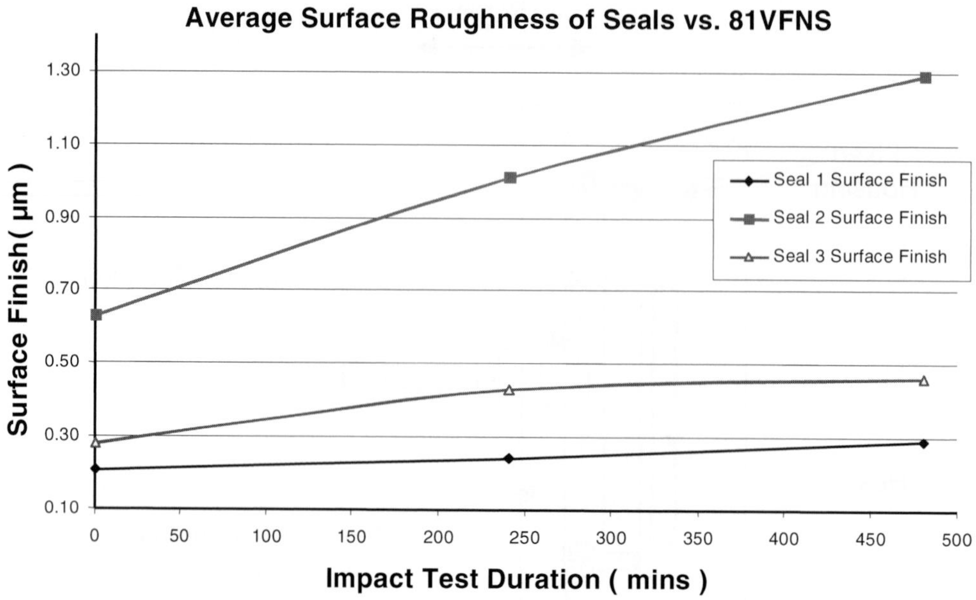

Fig. 5 Example of test results from seal vibration – impact tests.

RESULTS

An example of the results, Figure 5, shows the change in surface roughness of the test sample with time. Changes in roughness were used to provide an indication of the wear taking place as the contacting surfaces hammered against each other. Metallographic analysis was carried out on both the seal samples and the substrates to document the wear characteristic and any material transfer. With this first phase of testing, the Cobalt based coating, Stellite 6, began to show clear benefits to the more traditional nickel-chrome–chromium carbide (81VFNS) coatings in terms of the integrity of the coating.

VIBRATION–OCSILLATION TEST RIG

This rig was built to study the wear brought about by two components as they slide over or rub against each other at high frequency without hammering action taking place. The frequency and amplitude was kept small to replicate a ‘tearing’ effect that may differ in sliding wear from a more conventional long stroke sliding wear mechanism. Here, it should be noted that on investigation to the main driving mechanisms within a gas turbine it was concluded that these small amplitude ‘tear’ modes were the prevailing elements catalysing progressive wear.

The rig, shown in Figures 6 and 7, has a horizontal sliding plate representing the substrate component attached to the armature of a variable frequency/amplitude/force vibrator. Vibrator

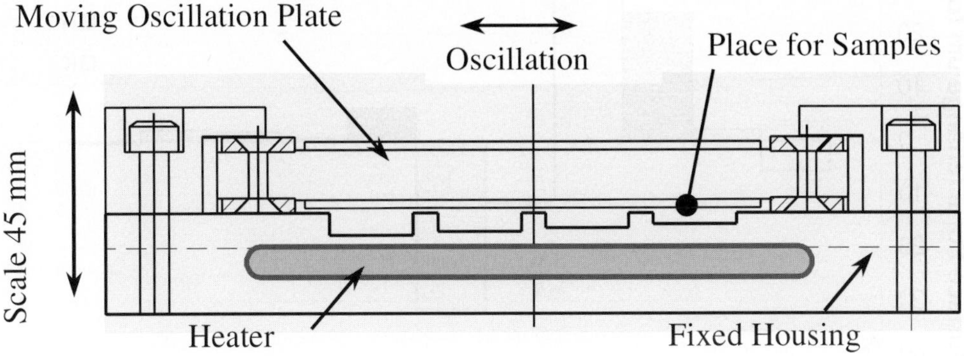

Fig. 6 Schematic section through oscillation – tear rig.

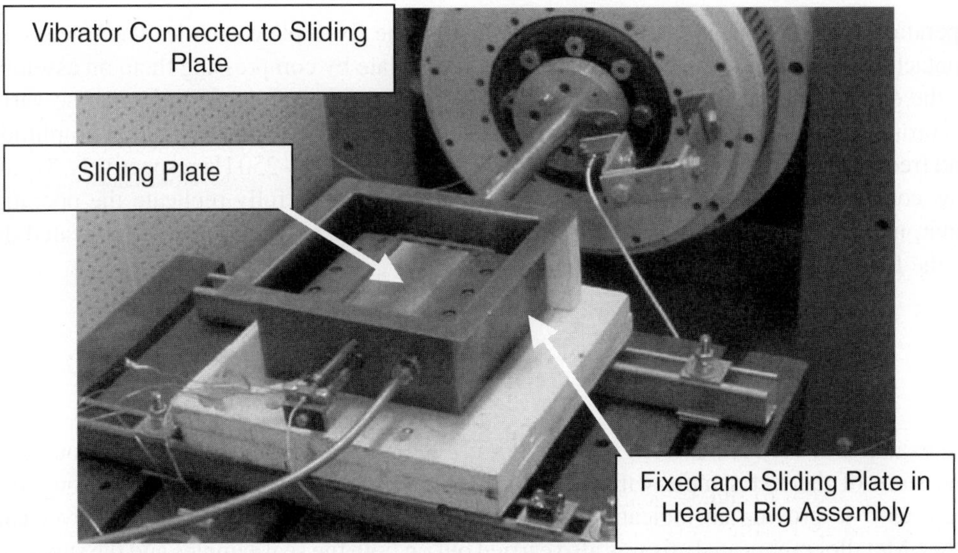

Fig. 7 Open view of oscillation – tear rig.

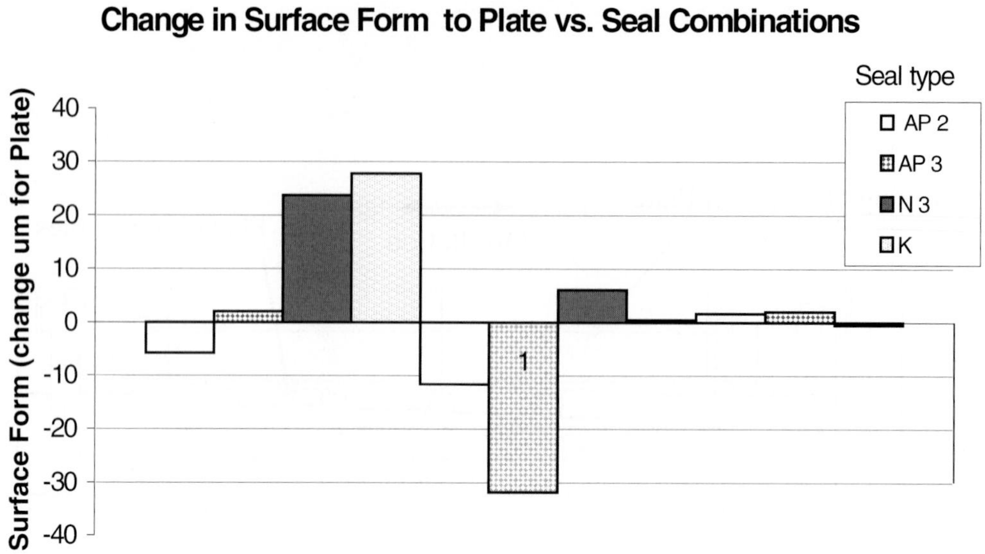

Fig. 8 Example test results.

operation oscillates the sliding plate sample across the fixed plate sample. The true engine contact load was applied to the samples in the fixed plate by compressing them on assembly to the engine fitted height. The force, amplitude, frequency and temperature can be varied according to the required test schedule. As for the Impact rig, tests were run at amplitudes and frequencies ranging from 0.1–0.6 mm Peak to Peak and 100–250 Hz respectively. Testing was conducted in a simulated combustion gas atmosphere to fully replicate the operating environment of combustion chamber seals. Instrumentation and data logging replicated that of the Impact test rig.

RESULTS

An example of the results from this test, Figure 8, shows the change in surface roughness and form of the test sample with time. As with the Impact testing, changes in roughness were used to provide an indication of the wear occurring as the surfaces slide over each other. Metallographic analysis was also carried out on both the seal samples and the substrates to document the wear characteristic and any material transfer. With this phase of testing the relative coefficients of friction were shown to have a big impact on the style of wear taking place. The coated substrates had a lower heat build-up from this tear testing than the uncoated Hastelloy X substrate. This test also showed the interrelationship of the seal style and material, Ultimet beginning to show clear benefits with contacting both a rough hard coating and the uncoated Hastelloy X, imparting wear to have a smoothing effect on the 81VFNS.

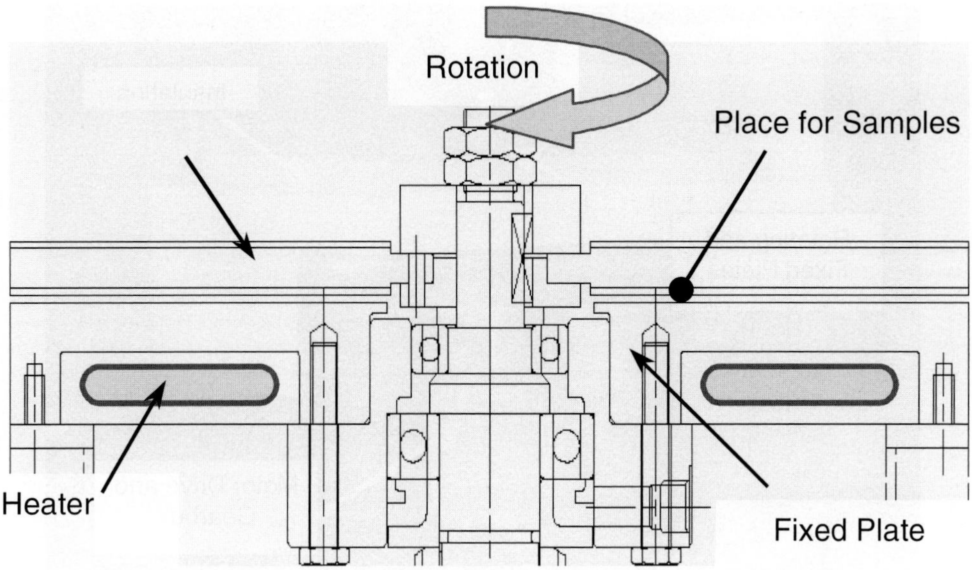

Fig. 9 Rotation – slide rig.

ROTATION TEST RIG

This rig was built to study the wear brought about between two components as they slowly and continuously rub against each other. Seal systems within engines are prone to ageing and creep. Hence as the seals fitted life increases, the contact load reduces and the wear contact area increases by the effect of wear on the seal. This rig had seal sections fitted in slots of varying depths to cover the whole range of engine fitted envelopes and was intended to investigate progressive the wear behaviour.

In this rig shown in Figures 9 and 10, the production engine seal sample length was supported in a horizontal fixed plate located in a stationary housing containing a controllable electric heater. An electric motor applied unidirectional drive to a shaft that penetrated the fixed plate and supported a horizontal rotating disc representing the second component of the wear couple. As the drive shaft rotated the upper disc progressive wear was applied to the samples located in the fixed plate. The heater allowed the wear testing at temperatures up to 350°C. The temperature of this rig was restricted relative to the other rigs by the requirement of a bearing pack at the heart of the rig. This effect was taken into account during the analysis of the results, incorporating data from the NPL programme.

RESULTS

An example of the results from this test, Figure 11, shows the change in the surface roughness and form of the samples with time. Comparing the substrates, the chromium carbide (81VFNS)

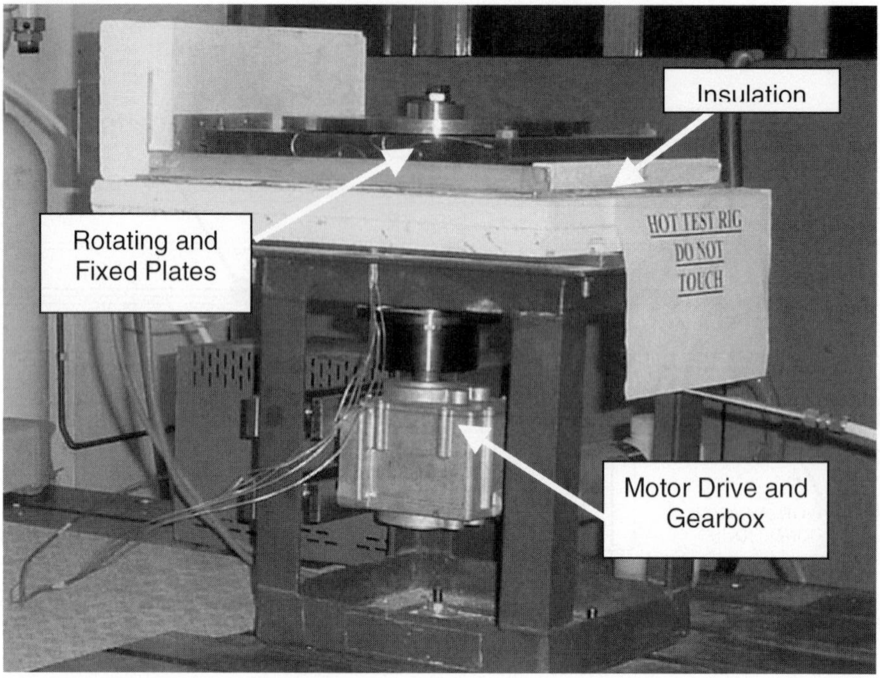

Fig. 10 Side view of rotation – slide rig.

coating has undergone a variable amount of smoothing- polishing dependant on the hardness of the contact seal, whereas the uncoated Hastelloy X was being progressively worn away. The cobalt-based Stellite 6 coating showed the least wear. There was no opportunity for significant heat build up on the substrates as the contact period was too short by nature of the sliding motion.

DISCUSSION

The mechanisms of wear within an Industrial Gas Turbine combustion sealing system are extremely complex and it is accepted that this is open to influences beyond the scope of even this still very thorough testing strategy. The first point to be addressed with this testing programme was to confirm if we had the same relationship between our relative chosen seal contact materials and that of our chosen substrates. To this end the results were extremely encouraging, in particular with the qualification of Ultimet as the chosen material for our seal plate. This material consistently showed that it had superior wear resistance to that of the alternative Inco 718 and Haynes 25 materials, at the engine operating temperatures, in line with the baseline NPL testing.

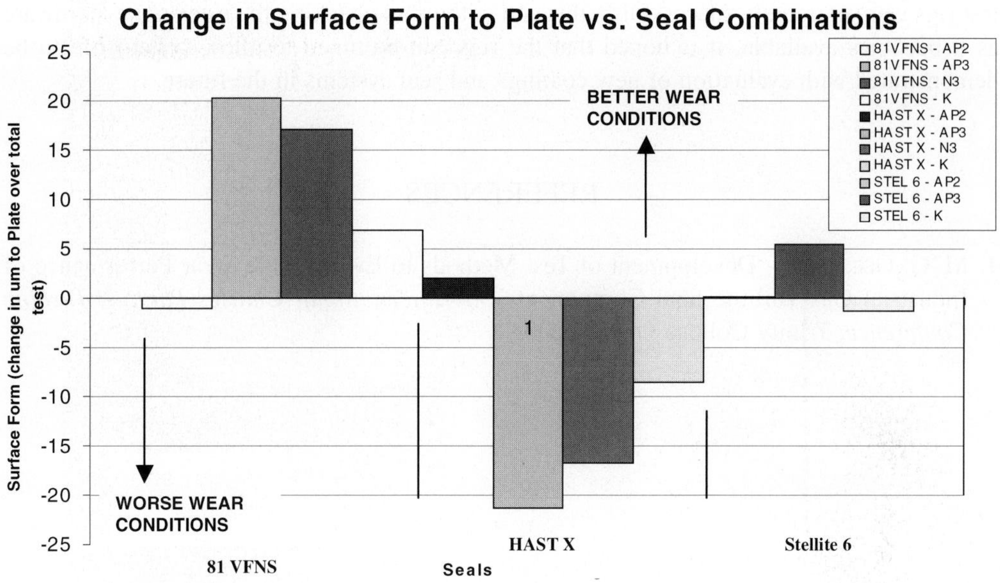

Fig. 11 Wear data.

The benefits of the cobalt-based coatings were also clearly demonstrated time and again with all our testing. However, these coatings tend to command a premium cost and the life of such a coating is not yet established within this environment. The HCF rig test results showed that the chromium carbide coating had undergone an initial running-in phase, resulting in a smoothing of the coating surface, as shown on the surface form charts. Subsequently, the wear was then less progressive and hence may result in a large increase in system life.

To understand more of the implications of system life, the metalographic analysis of the material transfer must be applied and this is ongoing and will be developed, as data from engine run hardware becomes more and more available. In time we may find that there are work hardening effects on the Ultimet that vary significantly relative to the movement mode and may result in material transfer.

The 'burning' of the Hast X on the oscillation test rig was very interesting as it suggests that the hard-face coating may provide benefits to wear principally by a reduction in the coefficient of friction rather than by creating a toughened rubbing surface. Again here we are looking forward to seeing the real engine hardware that is now out in the field.

CONCLUSION

A test system has been established that can begin to replicate and understand the wear mechanisms within an industrial gas turbine combustor sealing system. Principally the HCF test rigs were extremely successful in this and with cross checking the true engine hardware as it becomes available, it is hoped that the rigs can be tuned to allow system life to be demonstrated with evaluation of new coatings and seal systems in the future.

REFERENCES

1. M. G. Gee, et al.: 'Development of Test Methods to Evaluate the Wear Performance of Industrial Gas Turbine Seal Components', *6th International Charles Parsons Turbine Conference*, Trinity College Dublin, 2003.

Defect Reduction in Large Superalloy Castings by Liquid-Tin Assisted Solidification

A. J. ELLIOTT and T. M. POLLOCK
University of Michigan, USA

S. TIN
University of Cambridge
Rolls-Royce University Technology Centre
Department of Materials Science and Metallurgy
Pembroke Street, Cambridge, CB2 3QZ, UK

W. T. KING, S. -C. HUANG and M. F. X. GIGLIOTTI
General Electric Company, USA

ABSTRACT

A large, stepped cross-section casting has been directionally solidified using conventional Bridgman radiation cooling and liquid metal cooling (LMC) techniques. A direct comparison of the two processes using identical moulds and casting parameters indicates considerably enhanced thermal gradients and refined microstructure with the LMC process. High withdrawal rate LMC castings show further improvement compared to the conventional process. The LMC process eliminated freckle-type defects at all withdrawal rates and reduced casting defects at high withdrawal rates. The benefits of LMC process are discussed with regard to process advantage considerations for industrial gas turbine applications.

INTRODUCTION

Production of columnar-grain (CG) and single crystal (SX) nickel-base superalloy aero-engine blades is commonplace today. Improvements to alloy composition, blade design, and coatings continually improve the properties of these materials. However, the basic approach to casting has remained relatively unchanged for some time. Recently, the drive for higher operating temperatures and greater efficiency in industrial gas turbines (IGT) has fuelled interest in extending the higher temperature mechanical property benefits of CG and SX structure to IGT castings.[1] Unfortunately, the Bridgman casting method that is typically used for directional solidification (DS) of aero-engine blades,[2] is not readily scaled up to the more physically massive IGT castings. Radiation cooling is the dominant mode of heat

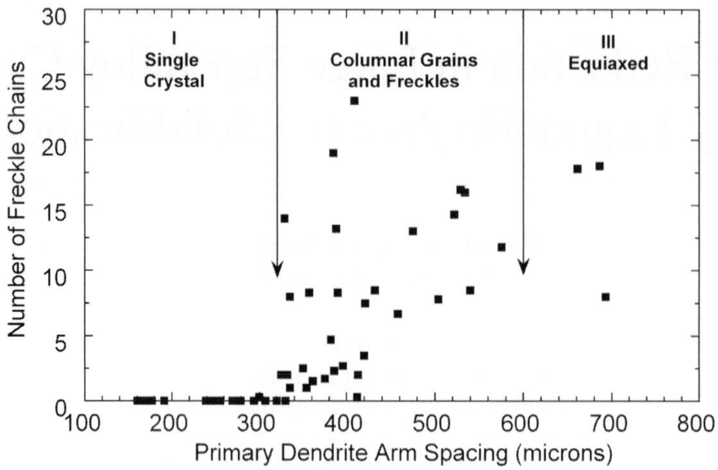

Fig. 1 The dependence of freckle chain formation on primary dendrite arm spacing for experimental single crystal alloys.[9]

transfer in the Bridgman process, except for the first few centimetres above the copper chill plate, where the starter for the process is located.[3–5] Due to the greater quantity of heat that must be extracted, thermal gradients are quite low when casting large IGT blades using the Bridgman process, requiring very slow withdrawal rates. The low gradient and slow withdrawal rate combination shifts the Bridgman process into a high defect occurrence process window for large castings, resulting in poor casting yields.[5–7] Freckle-type defects, which are small chains of equiaxed grains that form due to convective instabilities in the mushy-zone of the solidifying casting, are a common defect in DS castings.[8, 9] Freckle occurrence has been linked to the presence of large primary dendrite arm spacing,[9] which is directly related to thermal gradient and withdrawal rate and consequently the cooling rate during solidification,[10–13] Figure 1.

Maintaining a high thermal gradient at the S-L interface and fast withdrawal rate is critical to reducing defect occurrence in large IGT castings. To achieve these conditions, previously investigated high gradient processes have recently been re-examined and new methods are being developed. The foremost methods include liquid metal cooling (LMC), fluidised-bed quenching (FBQ), and gas cooling casting (GCC), which are all modifications of the Bridgman process. The LMC process, described by Giamei and Tschinkel in the 1970s, was originally developed to produce the high gradient casting conditions necessary for planar front solidification of DS eutectic alloys.[14, 15] LMC cooling is achieved by immersing the casting and mould in a container of low melting temperature metal as it is withdrawn from the hot section of the furnace. FBQ, in which the casting is withdrawn into a fluidised bed, was also considered for production of DS eutectics in the 1980s and is again being examined.[16, 17] GCC, which uses high velocity inert gas to cool the casting as it is withdrawn, has been recently investigated for producing IGT castings.[5]

The LMC process using aluminium as the cooling medium has been utilized in the former Soviet Union for the regular production of aero-engine blades,[6, 18, 19] and is a proven process for smaller aircraft-engine castings. Recent investigations of the process with regard to large castings show promising results.[4, 6, 20–22] LMC gradients are generally described as at least double the conventional Bridgman process,[6, 15, 22] however, there have not been detailed studies that directly measure the benefit of the LMC process in terms of cooling rates, microstructure, and defect reduction. The aim of this research is to provide a direct comparison of the Bridgman and LMC process using tin as the cooling medium. Considering the results of this experimental program, the benefits of LMC over the conventional process and the effects of cooling media contamination are discussed.

EXPERIMENTAL EQUIPMENT & PROCEDURES

A unique furnace with the capability for solidification of Bridgman and LMC-Sn castings was employed to make a direct comparison of the two processes. The furnace is capable of casting 5 kg of Ni-base superalloy into 300 mm tall moulds on a 150 mm diameter chill plate. Castings of both types were made using identical furnace settings at a withdrawal rate of 2.5 mm/min for the comparison. Additional LMC castings were made at withdrawal rates up to 8.5 mm/min to evaluate the capability and possible benefits of combining liquid metal cooling with faster withdrawal rates.

The design of the furnace allowed identical investment moulds to be used for both Bridgman and LMC castings. Simplified castings were designed to have cross-sections representative of large columnar-grained industrial gas turbine blades. The bottommost "tip" section had a cross-section of 95 mm by 51 mm and was 76 mm in height. The middle "airfoil" portion of the casting had a 95 mm by 9.5 mm cross-section and was 76 mm in height. Due to design restrictions, including a large 'root' portion was not feasible; however, a 38 mm by 95 mm cross-section portion with 38 mm height was located on top of the airfoil portion to provide constraint to the airfoil and an intermediate cross-section for analysis. Thermocouples with exposed junctions were inserted through the mould sidewall to a depth of 13 mm inside the casting, in the middle of each casting section. Thermal data was used to determine thermal gradients, cooling rates, and to locate the position of the solid-liquid interface.

A columnar grain variant of René N4,[23] a first generation single crystal superalloy, with minor additions of B and C was selected for this investigation. All castings were made using ingots from a single master heat with nominal composition: Ni-9.7Cr-8.0Co-6.0W -4.7Ta-4.2Al-3.5Ti-0.4Nb-0.15Hf-0.07C-0.008B (wt.%). The liquidus and solidus temperatures of the alloy were extracted from differential thermal analysis heating curves for gradient and cooling rate calculations.

After processing, the cast articles were cleaned with a light abrasive and macro-etched using a solution of aqueous $FeCl_3$, HCl and HNO_3.[24] The macro-etched castings were characterised for surface casting defects and subsequently sectioned, mounted, and polished for microstructural analysis. A microetchant of CH_3COOH, H_2O, HNO_3, and HF in the volume ratio 33:33:33:1 was used to reveal the microstructure. Average primary dendrite arm spacing (PDAS) and volume fraction eutectic were determined from several samples for each casting portion.

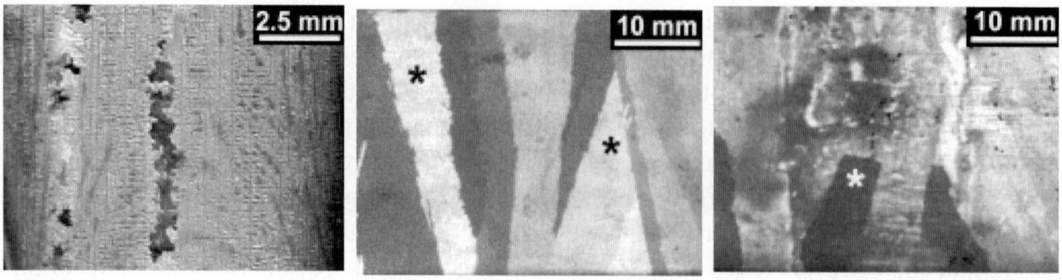

(a) (b) (c)

Fig. 2 Casting defects in columnar grain superalloy castings: (a) freckle-type defects, (b) misaligned grains and (c) planar discontinuity.

RESULTS

Casting Defects

Figure 2 shows three common casting defects in large columnar-grain IGT castings: freckles; misaligned grains; a planar discontinuity. Freckle chains are perhaps the most problematic and difficult to suppress of these defects.[25] Freckle chains form due to convective instabilities in the mushy zone of the solidifying casting, which may fragment growing dendrite arms and result in the formation of small equiaxed grains. The low gradients and slow withdrawal rates necessary for Bridgman solidification of large IGT castings, results in large dendrite arm spacings and tall mushy zone heights. Misaligned grains, which are similar to high angle boundaries in single crystal castings, are detrimental to creep and fatigue properties of CG castings.[26] In this study, a misaligned grain was defined as a single grain being oriented more than 15° off the withdrawal axis or as two grains converging or diverging at greater than 20°. The third major grain-related defect in CG castings is a planar discontinuity, appearing as a nearly horizontal grain boundary indicating the abrupt end of a grain. Grain boundaries transverse to the growth direction are very detrimental to the mechanical properties of the DS castings.[1, 27]

The number and type of surface casting defects are summarised in Figure 3. The LMC castings withdrawn at intermediate withdrawal rates clearly had the lowest defect occurrence. More importantly, LMC processing eliminated the occurrence of freckle-type defects at all withdrawal rates, while several were observed in an identical casting made by the conventional Bridgman process.

Misaligned grains were the most numerous defect observed in this investigation. About half of all the misaligned grains occurred near the bottom of a casting and were quickly outgrown by straighter grains as typically occurs in the sort out zone of columnar grain castings.[13] The LMC castings withdrawn at 5.1, 6.8 and 8.5 mm/min each had only one misaligned grain near the bottom of the casting. The Bridgman and LMC castings withdrawn at 2.5 mm/min both had a higher occurrence of misaligned grains near the bottom of the casting. Faster withdrawal rates may have assisted the competitive growth of straighter grain

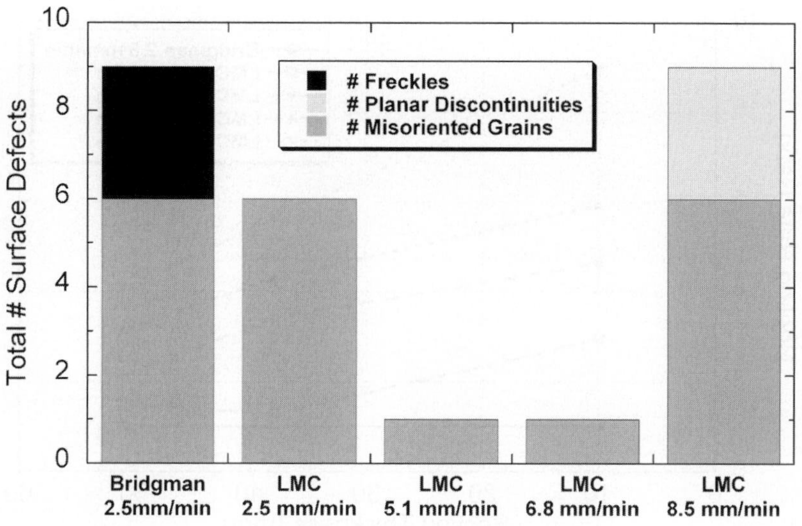

Fig. 3 The number and type of surface casting defects.

orientations in the sort out zones of these castings.[11] The LMC casting withdrawn at 8.5 mm/min was the only casting with misaligned grains in the middle portion of the casting. Additionally, while other castings had grains that were either fairly straight or oriented slightly away from the vertical centreline of the casting, the grains of the 8.5 mm/min LMC casting were oriented towards the centreline of the casting. The orientation of the grains can be an indication of concave solid-liquid interface curvature,[7, 28] which is influenced by the proximity of the S-L interface to the liquid metal cooling bath. The 8.5 mm/min LMC casting was also the only casting to display planar discontinuity defects.

Thermal Gradients & Cooling Rates at The Solidification Front

Thermal gradients calculated from experimental data between the liquidus (1345°C) and solidus (1300°C) temperatures are listed in Table 1. Gradients of LMC castings were almost double Bridgman gradients in the 10 mm thickness section at withdrawal rates up to 6.8 mm/min. At a withdrawal rate of 8.5 mm/min, there was a noticeable increase in gradient in the thinnest section. In the 38 mm thickness section, the LMC casting withdrawn at 2.5 mm/min had a lower than expected gradient because the S-L interface shifted upward into the mould heater of the furnace. However, thermal gradients in the thicker casting sections were generally greater with LMC processing.

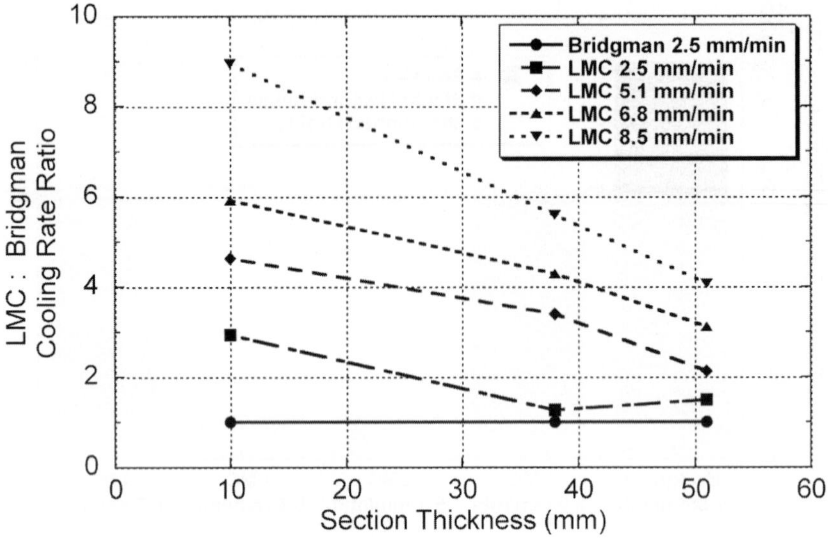

Fig. 4 LMC: Bridgman cooling rate ratio comparison. Ratio calculated using the relation PDAS $\propto G^{-1/2} * R^{-1/4}$.

Table 1 Thermal gradients from each cross-section of experimental castings (°C/cm). Data is missing where thermocouples failed.

Section Thickness (mm)	Bridgman 2.5 mm/min	LMC 2.5 mm/min	LMC 5.1 mm/min	LMC 6.8 mm/min	LMC 8.5 mm/min
10	42	74	-	78	94
38	37	32	-	39	42
51	23	48	29	35	34

The ratios of LMC to Bridgman cooling rates are plotted in Figure 4 as a function of section thickness. Relative cooling rates were estimated from primary dendrite arm spacing measurements using the relationship PDAS $\propto G^{-1/2} * R^{-1/4}$.[10] The benefit of liquid metal cooling is emphasised by this plot showing significant improvements not only in thin sections, but also in the large cross-section portions of the casting.

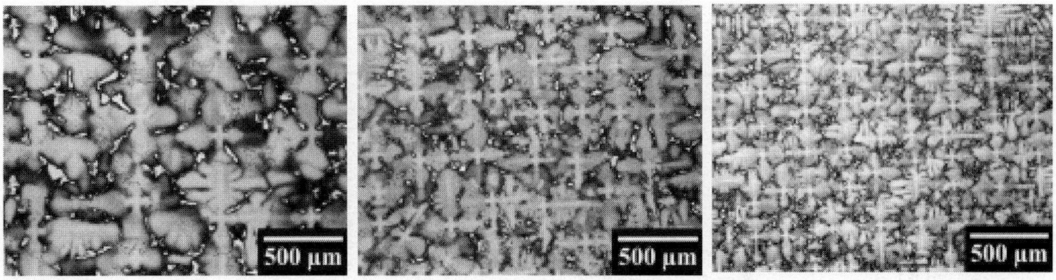

(a) (b) (c)

Fig. 5 Variation in primary dendrite arm spacing and dendrite morphology with casting method and withdrawal rate for section thickness of 10 mm, viewed normal to the [001] growth direction. Casting and withdrawal rate: (a) Bridgman 2.5 mm/min, (b) LMC 2.5 mm/min and (c) LMC 6.8 mm/min.

Cast Microstructure

Microstructural analysis revealed a qualitatively similar appearance of all casting sections. Micrographs of the dendritic structure in sections normal to the [001] growth direction from similar locations of the Bridgman and LMC castings withdrawn at 2.5 mm/min as well as the LMC casting withdrawn at 6.8 mm/min vary significantly in scale, Figure 5. A major reduction in PDAS from 540 to 315 µm was observed when changing from Bridgman to LMC casting methods at the same withdrawal rate in the 10 mm thickness section. At an increased withdrawal rate of 6.8 mm/min, which was optimal for reducing macroscopic casting defect occurrence, the average PDAS was further reduced to 285 µm in the same section.

The broad distribution of cooling rates resulted in significant variation of primary dendrite arm spacings. The average PDAS for each casting sections is plotted in Figure 6 as a function of section thickness. At every section thickness the use of liquid tin coolant provided a marked improvement on PDAS. Increasing withdrawal rate further reduced PDAS. Similar trends were apparent in the PDAS, thermal gradient, and cooling rate data. LMC casting sections generally exhibited higher gradients, faster cooling rates, and smaller PDAS than comparable sections from the Bridgman casting. The 38 mm thickness section of the LMC casting withdrawn at 2.5 mm/min was an exception. It had the largest PDAS of all LMC castings and did not fit the trend exhibited by the other LMC castings. This section also had the lowest gradient of all sections and this can be attributed to the S-L interface shifting out of the cooling media up into the mould heater. Considering all the results for the 38 mm thickness section, it is interesting to note that despite the LMC casting having a lower gradient than the Bridgman casting, the LMC casting still had a significantly smaller PDAS. This may suggest some contribution of radial heat extraction to the solidification process.

Well defined primary dendrites were easily identified in all sections. There were differences in dendrite morphology as a function of cooling condition, with extensive secondary and tertiary dendrite arm growth in many sections. Carbides were also visible in all sections at

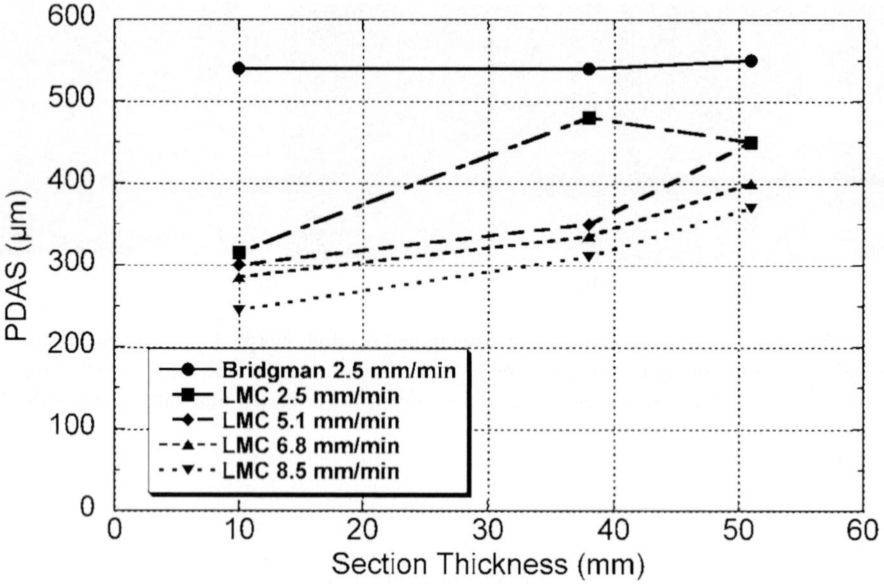

Fig. 6 The variation of primary dendrite arm spacing (PDAS) as a function of section thickness for each casting method and withdrawal rate.

all withdrawal rates, though their morphology varied. Slow cooling rate sections with large PDAS exhibited mainly script-type carbides. Smaller blocky carbides became more prevalent as PDAS decreased. The average volume fraction eutectic did not have a significant correlation with PDAS, cooling rate, or casting method. The eutectic pool size and distribution did correlate with PDAS. Large PDAS sections had fewer, but larger eutectic pools, whereas small PDAS sections had more numerous, but smaller eutectic regions. The finer distribution of eutectic areas in small PDAS sections may be quite beneficial to the solutioning response of liquid metal cooled castings.

Lateral dendrite growth was observed in the upper half of the 10 mm thickness section of the LMC casting withdrawn at 8.5 mm/min, Figure 7. Process data indicates that the S-L interface was located 1.3 to 2.4 cm below the surface of the cooling bath in this experiment. This created a strong radial component of heat extraction in the mushy zone. Radial dendrite growth was more prevalent near the centreline of the casting, apparently due to the accompanying concave nature of the S-L interface that developed due to the fast withdrawal rate and immersion of the S-L interface in the tin. The increased number of misaligned grains in the middle portion of this casting, presence of planar discontinuities, and radial dendrite growth indicate that a withdrawal rate of 8.5 mm/min is near the limit for this particular furnace, casting design, and process conditions.

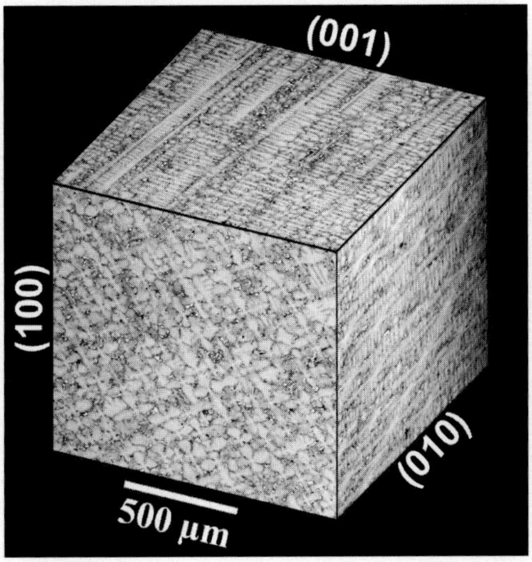

Fig. 7 3D composite micrograph showing lateral dendrite growth in the LMC 8.5 mm/min casting near the top of the 10 mm thickness section. Withdrawal perpendicular to (001) face.

DISCUSSION

Experimental results demonstrate that the LMC process is substantially beneficial in combining higher gradients with the capability for faster withdrawal rates. Thermal gradients of the order of two times higher compared to the Bridgman process and stable withdrawal rates of approximately three times faster, result in considerably improved cooling rates compared to the conventional Bridgman process. These improvements should provide the process capabilities necessary to cast large DS parts for industrial gas turbines reliably.

In this investigation casting defects typical of the Bridgman process were eliminated with the LMC process under optimal conditions. Importantly, it was observed that the LMC process eliminated the presence of freckles, which are a primary concern in large DS castings produced by the Bridgman process.[25] Further, fast cooling rates have been shown to produce refined microstructures which are superior to conventionally cast material.[21,22,29] High gradient casting produces γ' precipitates that are generally more homogenously distributed, smaller, and more square in appearance in the as-cast condition, which may provide enhanced response to heat treatment.[22] Carbide size and morphology have also been related to cooling rate.[21,30] Limited work has also shown somewhat improved creep behaviour of high gradient cast material.[30,31]

A major concern for adopting the LMC process is possible coolant contamination if a mould cracked during withdrawal. Contamination is generally considered less of a concern when using aluminium as the cooling medium because it is an alloying element in most nickel-base superalloys. Tin, however, is generally considered a tramp element because of its chemical similarity to lead and bismuth, which are deleterious to the mechanical properties

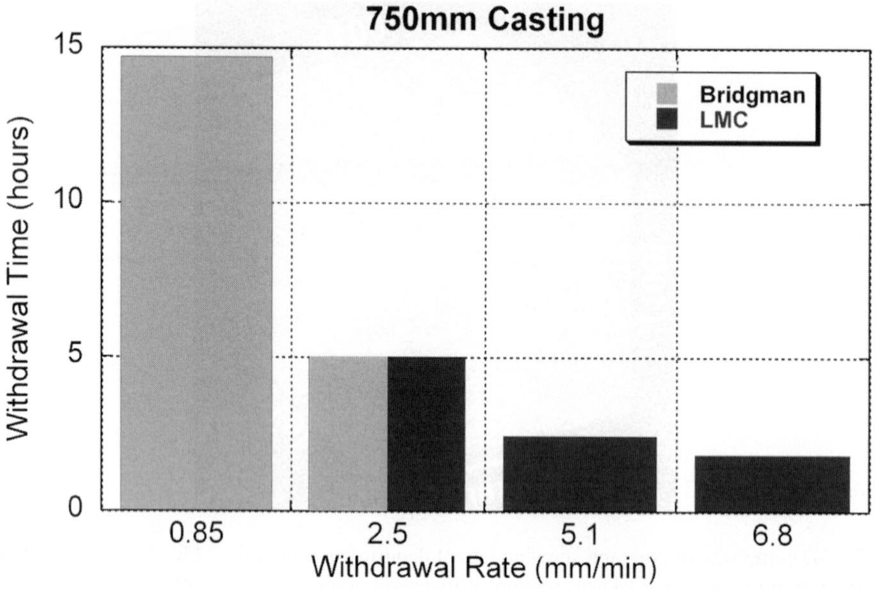

Fig. 8 Withdrawal time of a 750 mm casting using different cooling methods and withdrawal rates.

of nickel-base superalloys.[32–34] In these experiments, when the casting came in contact with liquid tin it resulted in the formation of a uniform thin layer that was easily removed by hand with a light abrasive. In contrast, due to the higher melting temperature of aluminium, reaction with the casting is a much greater concern than when using tin.[22] Limited investigations of the impact of tin contamination on the mechanical properties of nickel-base superalloys indicate no detrimental effects of small amounts of tin. Several investigations noted minimal or no effect on stress-rupture or tensile properties at concentrations approaching 1 wt.%.[33, 34] A more recent study noted no effect of tin on creep strength at concentrations up to 1040 ppm or low cycle fatigue properties up to 4150 ppm.[22]

Other implications of the LMC process for improved manufacture of large DS/SX components merit further analysis. Improved microstructure is not merely an indicator of defect suppression, but also has implications for improved homogenisation and shorter heat treatment cycles. The capability for faster withdrawal rates provides the ability to drastically reduce casting process time. As Figure 8 illustrates, withdrawal times for a 750 mm IGT casting could be between 5 and 15 hours using conventional radiation cooling and withdrawal rates between 0.85 and 2.5 mm/min, which are necessary to maintain oriented structure. This time could be reduced to less than two hours with the LMC process and a withdrawal rate of 6.8 mm/min, which produced the optimal casting in this investigation. The nature of a conductive fluid coolant may also reduce problems resulting from tilted solid-liquid interfaces of cluster moulds that occur with the radiation cooling process. LMC may eliminate this complication and even provide the capability for denser mould clusters.

CONCLUSIONS

1. The LMC process using tin as a cooling medium produced significantly improved gradients with respect to a directly comparable Bridgman process with a fixed casting configuration.
2. The capability for faster withdrawal rates combined with higher gradients provided substantial process advantages for large cross-section castings using LMC.
3. The LMC process was effective at suppressing casting defects including the elimination of freckle-type defects and reduction of casting defects at high withdrawal rates.

ACKNOWLEDGEMENTS

The authors would like to acknowledge the assistance of and useful discussions with Q. Feng, T. K. Nandy, B. Tryon, and C. J. Torbet. The support provided by Tom Van Vranken and PCC Airfoils, Inc. is greatly appreciated. The funding provided by the General Electric Company and the National Science Foundation grant number DMR-0127689 are also gratefully acknowledged.

REFERENCES

1. D. C. Pratt: 'Industrial Casting of Superalloys', *Materials Science and Technology*, **2**(5), 1986, 426–435.
2. R. W. Smashey: 'Apparatus and Method for Directional Solidification', U.S. Patent No. 3,897,815, 1975.
3. C. H. Lund and J. Hockin: 'Investment Casting', *The Superalloys*, C. T. Sims and W. C. Hagel, eds., John Wiley & Sons, 1972, 403–425.
4. A. Kermanpur, M. Rappaz, N. Varahram and P. Davami: 'Thermal and Grain-Structure Simulation in a Land-Based Turbine Blade Directionally Solidified with the Liquid Metal Cooling Process', *Metallurgical Transactions B*, **31B**(6), 2000, 1293–1304.
5. M. Konter, E. Kats and N. Hofmann: 'A Novel Casting Process for Single Crystal Gas Turbine Components', *Superalloys 2000*, T. M. Pollock, R. D. Kissinger, R. R. Bowman, K. A. Green, M. McLean, S.L. Olson and J. J. Schirra, eds., TMS, 2000, 189–200.
6. F. Hugo, U. Betz, J. Ren, S. -C. Huang, J. A. Bondarenko and V. Gerasimov: 'Casting of Directionally Solidified and Single Crystal Components Using Liquid Metal Cooling (LMC)', *International Symposium on Liquid Metal Processing and Casting*, AVS, 1999, 16–30.
7. N. D'Souza, M. G. Ardakani, M. McLean and B. A. Shollock: 'Directional and Single-Crystal Solidification of Ni-Base Superalloys: Part I. The Role of Curved Isotherms on Grain Selection', *Metallurgical Transactions A*, **31A**(11), 2000, 2877–2886.
8. S. M. Copley, A. F. Giamei, S. M. Johnson and M. F. Hornbecker: 'Origin of Freckles in Unidirectionally Solidified Castings', *Metallurgical Transactions*, **1**(8), 1970, 2193–2204.

9. T. M. Pollock and W. H. Murphy: 'The Breakdown of Single-Crystal Solidification in High Refractory Nickel-Base Alloys', *Metallurgical Transactions A*, **27A**(4), 1996, 1081–1094.
10. J. D. Hunt: 'Cellular and Primary Dendrite Arm Spacings', *Solidification and Casting of Metals*, J. D. Hunt, ed., The Metals Society, 1979, 3–9.
11. M. McLean: 'Directionally Solidified Materials for High Temperature Service', *Directionally Solidified Materials for High Temperature Service*, The Metals Society, 1983, 11–54 and 107–149.
12. G. K. Bouse and J. R. Mihalisin: 'Metallurgy of Investment Cast Superalloy Components', *Superalloys, Supercomposites and Superceramics*, J. K. Tien and T. Claufield, eds., Academic Press, 1989, 99–148.
13. W. Kurz and D.J. Fisher: *Fundamentals of Solidification*, 4th rev. ed., Trans Tech Publications, 1998.
14. J. G. Tschinkel, A. F. Giamei and B. H. Kearn: 'Apparatus for Casting of Directionally Solidified Articles', U.S. Patent No. 3,763,926, 1973.
15. A. F. Giamei and J. G. Tschinkel: 'Liquid Metal Cooling: A New Solidification Technique', *Metallurgical Transactions A*, **7A**(9), 1976, 1427–1434.
16. Y. G. Nakagawa, Y. Ohotomo and Y. Saiga: 'Heat Treatment, Microstructure, and Creep Strength of γ/γ'-a Eutectic Directionally Solidified by Fluidized Bed Quenching', *Superalloys 1980*, J. K. Tien, S. T. Wlodek, H. I. Morrow, M. Gell and G. E. Mauer, eds., American Society for Metals, 1980, 267–274.
17. L. D. Graham and B. L. Rauguth: 'System of Casting a Metal Article Using a Fluidized Bed', U.S. Patent No. 6,443,213, 2002.
18. V. V. Gerasimov: 'Method of Casting Parts with Directed and Mono-Crystal Structure', U.S.S.R. Patent No. 2836278/22–02, 1981.
19. G. B. Stroganov, A. V. Logunov, V. V. Gerasimov and E. L. Kats: 'High-Rate Directed Crystallization of Heat Resistant Alloys', *Casting Production (in Russian)*, **12**, 1983, 20–22.
20. T. J. Fitzgerald and R. F. Singer: 'An Analytical Model for Optimal Directional Solidification Using Liquid Metal Cooling', *Metallurgical Transactions A*, **28A**(6), 1997, 1377–1383.
21. J. Grossman, J. Preuhs, W. Esser and R.F. Singer: 'Investment Casting of High Performance Turbine Blades by Liquid Metal Cooling - A Step Forward Towards Industrial Scale Manufacturing', *International Symposium on Liquid Metal Processing and Casting*, AVS, 1999, 31–40.
22. A. Lohmüller, W. Esser, J. Gossmann, M. Hördler, J. Preuhs and R. F. Singer: 'Improved Quality and Economics of Investment Castings by Liquid Metal Cooling–The Selection of Cooling Media', *Superalloys 2000*, T. M. Pollock, R. D. Kissinger, R. R. Bowman, K. A. Green, M. McLean, S. L. Olson and J. J. Schirra, eds., TMS, 2000, 181–188.
23. E. W. Ross and K. S. O'Hara: 'Rene N4: A First Generation Single Crystal Turbine Airfoil Alloy with Improved Oxidation Resistance, Low Angle Boundary Strength and Superior Long Time Rupture Strength', *Superalloys 1996,* R. D. Kissinger, D. J. Deye, D. L. Anton, A. D. Cetel, M. V. Nathal, T. M. Pollock and D. A. Woodford, eds., TMS, 1996, 19–25.

24. G. Petzow: *Metallographic Etching*, ASM, 1978.
25. B. B. Seth: 'Superalloys - The Utility Gas Turbine Perspective', *Superalloys 2000*, T. M. Pollock, R. D. Kissinger, R. R. Bowman, K. A. Green, M. McLean, S. L. Olson, and J. J. Schirra, eds., TMS, 2000, 3–16.
26. T. M. Pollock, W. H. Murphy, E. H. Goldman, D. L. Uram and J. S. Tu: 'Grain Defect Formation During Directional Solidification of Nickel Base Single Crystals', *Superalloys 1992*, S. D. Antolovich, R. W. Stusrud, R. A. MacKay, D. L. Anton, T. Khan, R. D. Kissinger and D. L. Klarstrom, eds., TMS, 1992, 125–134.
27. F. L. VerSnyder and M. E. Shank: 'The Development of Columnar Grain and Single Crystal High Temperature Materials Through Directional Solidification', *Materials Science and Engineering*, **6**, 1970, 213–247.
28. R. N. Grugel and Y. Zhou: 'Primary Dendrite Spacing and the Effect of Off-Axis Heat Flow', *Metallurgical Transaction A*, **20A**(5), 1989, 969–973.
29. P. N. Quested and M. McLean: 'Solidification Morphologies in Directionally Solidified Superalloys', *Materials Science and Engineering*, **65**(1), 1984, 171–180.
30. P. N. Quested and M. McLean:'Effect of Variations in Temperature Gradient and Solidification Rate on Microstructure and Creep Behaviour of IN 738LC', *Solidification Technology in the Foundry and Cast House*, The Metals Society, 1983, 586–591.
31. P. N. Quested, P. J. Henderson, K. Menzies and M. McLean: 'Mechanical Behaviour of Ni-Cr-Base Superalloys in Forms and Conditions Representative of Service', Report No. NPL Report DMA(A)94, COST 50, 1984.
32. R. T. Holt and W. Wallace: 'Impurities and Trace Elements in Ni-Base Superalloys', *International Metallurgical Reviews*, **21**, 1976, 1–24.
33. D. R. Wood and R. M. Cook: 'Effects of Trace Elements on Creep Properties of Nickel-Base Superalloys', *Metallurgia*, **67**, 1963, 109–117.
34. R. L. Dreshfield, W. A. Johnson and G. A. Maurer: 'Effects of Tin on Microstructure and Mechanical Behavior of Inconel 718', Report No. TM-86866, NASA, 1984.

Some Observations on Superalloy Cast Alloy Chemistry & Implications for Alloy Design

G. Durber
Western Australian Specialty Alloys Pty Ltd
2-4 Hopewell Street
Canning Vale, Western
Australia

ABSTRACT

The balance between precipitation hardening and solid solution elements in superalloys is shown to follow a set relationship. Providing this is maintained in any new alloy formulation, then it should be possible to predict the strength of the new formulation. Alloy stability can be determined in advance by examining its location on a crystallographic map.

OBSERVATIONS ON ALLOY CHEMISTRY

Following the discovery in the 1930s of the Ni-Cr-Al-Ti precipitation hardening system, with its excellent high temperature strength and oxidation resistance, the performance of this system was rapidly improved through additions of Co, Mo and W, etc. to create the wrought and cast Ni-base superalloys we are familiar with today. The high temperature strength of these alloys and the rush into applications ensured that manufacturing and in-service problems were encountered in the early stages of their development.

In the mid 1960s, calculation methods capable of predicting the formation of deleterious electron compound phases in superalloys were developed to limit the occurrence of σ in Ni-base superalloys. A comprehensive review of the calculation methods has been presented in Superalloys II.[1] As a consequence, cast superalloys have continued to evolve in a controlled manner.

The temperature capability data shown in Table 1 (alloy composition v phase composition v temperature performance) has been extracted from Superalloys II.[2] Any property v chemistry data deserves close inspection. Traditionally, this type of data has been processed using linear regression analysis to help formulate new chemistries with predicted properties. However, there are valid objections to this methodology and more powerful techniques now exist, e.g. neural network analysis.[3] This technique has the capacity to accommodate unpredicted interactions in order to reproduce observed outcomes. However, providing that linear regression is used in conjunction with some method of phase stability determination and that any new alloy formulation remains within the elemental ranges contained in the database, then the major objections to this approach are overcome. Confidence in the predicted (linear regression) outcome can be further improved by adhering to some subtle alloying combinations that exist in these alloys. These are examined further here.

While the wt.% compositions of the cast superalloys listed in Table 1 (typically comprising 12+ alloying elements and therefore not fully listed here) may initially appear very complex, they do adhere to some basic compositional guidelines to maintain a stable two-phase structure. If these constraints are imposed on any new alloy chemistry, then there is every reason to expect a linear regression analysis to produce a reliable outcome. These guidelines are:

- Ni content generally around 60 wt.% & at.% with just a few exceptions (notably the higher Cr alloys).
- Al + Ti + Nb = 6–8 wt.% (except very high Cr alloys). Now all of this is assumed to partition completely to the precipitation hardening phase, $Ni_3(Al, Ti, Nb) = \gamma'$. Ta also partitions to this phase as would be expected from its chemical and atomic size similarities to Nb, but there is much circumstantial evidence (demonstrated later in some of the chemistry correlations produced) that this partitioning is assumed to be a long way below 100%. The partial partitioning of other elements (e.g. Co, Cr and Mo) to γ' is considered small enough to be ignored both here and in alloy formulation.

  Very significantly, the narrow wt.% range of Al + Ti + Nb converts to a wide range of γ' content ≈ 38–63 at.% (38–68 at.% if Ta is included), demonstrating how effectively precipitation hardening has been developed in these alloys, particularly over the last 20 years. It appears that the physical limits of packing densities are now being approached, e.g. the volume fraction of spheres in an fcc (close packed) array is 0.74. Actually achieving this physical limit would be undesirable, as precipitates would start to coalesce. An upper limit may well have already been reached.
- Solid solution strengtheners Co + Mo + W + Ta + Re generally make up around 20–45 at.% of the matrix following γ' precipitation. The total moves towards the higher end in the newer alloys. Now 25% concentration of solute atoms in a solvent matrix of similar atomic size is ideal for superlattice ordering, particularly where all the elements are transition metals. This level is generally exceeded here, though short range ordering still occurs. Superlattices add to solid solution strengthening through the creation of order/disorder boundaries, which force pairing of dislocations during creep deformation.
- Cr content is adjusted in order to preserve phase stability. A consequence of this is the emergence of clear interdependence between Cr & Al and Cr & (Co + Mo + W + Ta + Re), Figure 1. Correlation factors for both are greater than 0.90.
- Hf is added to DS alloys for grain strengthening at up to 1.5 wt.%
- Residual levels of C, B & Zr for grain boundary strengthening in equi-axed alloys.

Table 1 shows that the levels of hardener and solid solution elements in superalloys were little different in the 1950s from what they are today, but yet the temperature capability of these alloys has advanced considerably. Some of this advancement has happened through development of casting technology, i.e. equi-axed through DS to SC, which has provided strength enhancement through selected crystal orientation. It has also permitted a reduction/elimination of the grain boundary strengthening elements with consequent lifting of solidus temperature. This in turn has led to some re-balancing of the solution strengthening elements. Mo & Ti contents have diminished over the years while W and Ta have tended to increase. The adoption of Ta was probably held back by its relatively high raw material cost, but the performance advantage is now established. The ratio of Al/(Al + Ti + Nb) has also increased,

Table 1 Changes in alloy composition over four decades.

	Approx. Year	Alloy Composition				Matrix & Phase Composition			Temp for 1000 h at 200 MPa
		Cr	Al	Al + Ti + Nb	Mo + W + Ta	Al/(Al + Ti + Nb) and Al/(Al + Ti + Nb + Ta)	γ and γ' inc Ta	Co, Mo, Re, Ta, W	
		wt.% at.%	wt.% at.%	wt.% at.%	wt.% at.%	Atomic Ratio	at.%	Matrix at.%	°C
U500	1956	19.0 17.0	3.0 6.2	6.0 9.7	4.0 2.3	0.64	38.8	31.6	840
R77	1960	14.6 15.5	4.3 8.8	7.7 12.7	4.2 2.4	0.70	50.6	32.4	870
B1900	1965	8.0 8.7	6.0 12.6	7.0 13.8	10.3 4.9	0.91 0.83	59.2 60.5	35.5	900
IN738	1967	16.0 17.5	3.4 7.1	7.5 11.6	6.0 2.3	0.61 0.58	47.8 48.4	19.7	875
R80	1967	14.0 15.4	3.0 6.4	8.0 12.3	8.0 3.6	0.52	49.2	25.3	895
GTD111	1970	14.0 15.6	3.1 6.6	7.9 12.4	8.6 3.2	0.53 0.50	52.6 53.5	26.3	
IN939	1972	22.5 24.5	1.9 4.0	6.6 9.0	3.4 1.1	0.44 0.42	37.2 37.6	30.7	860
IN792	1972	12.5 14.0	3.4 7.4	7.2 12.0	9.9 3.7	0.61 0.55	51.8 53.1	26.2	905
GTD222	1974	22.5 24.7	1.2 2.6	4.3 5.8	3.0 0.9	0.44 0.42	24.3 24.6	25.5	
MM247	1976	8.3 9.5	5.5 12.2	6.5 13.4	13.7 4.7	0.91 0.84	56.7 57.6	34.2	925
R125		9.0 10.3	4.8 10.6	7.3 13.6	12.8 4.8	0.77 0.71	58.5 59.8	35.8	
SRR99	1982	8.5 9.6	5.5 12.0	7.7 14.7	12.3 4.0	0.82 0.77	61.5 62.5	23.3	
PWA1480	1982	10.0 11.7	5.0 11.2	6.4 12.9	16.1 5.4	0.86 0.66	63.7 67.7	30.0	940
PWA1483	1986	12.2 13.8	3.5 7.6	7.6 12.6	11.0 4.1	0.61 0.54	55.3 57.0	28.8	
PWA1484	1988	5.0 6.0	5.7 13.0	5.7 13.0	16.6+ 3Re 6.2 + 1Re	1.00 0.81	61.0 64.]	45.5	
CMSX-4	1988	6.4 7.5	5.7 12.8	6.7 13.3	13.5+ 3Re 4.7 + 1Re	0.91 0.79	63.2 65.4	41.9	
RR3000	1995	2.3 2.7	5.8 13.4	6.1 13.8	14.2+ 6Re 5.0 + 2Re	0.97 0.80	63.8 66.8	29.6	

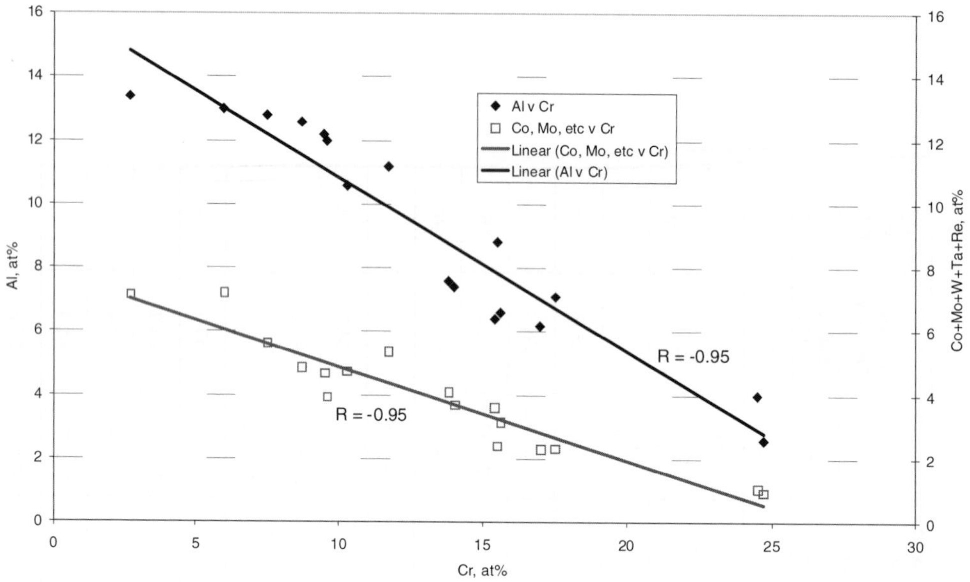

Fig. 1 Al and solid solution content v Cr for the alloys listed in Table 1.

creating an increase in the atomic fraction of γ' for a constant wt.% total of (Al + Ti + Nb). This also confers some oxidation resistance from a higher level of Al as Cr declines.

And combining the two dependencies that are shown in Figure 1, we find:

$$\frac{(\mathrm{Al}-1)}{2} \approx \mathrm{Mo} + \mathrm{W} + \mathrm{Ta} + \mathrm{Re,\ at.\%} \qquad (1)$$

The correlation factor for this relationship is remarkably high, see Figure 2, given the ranges for the individual elements. The two sides of equation 1 balance generally to within 1½ at.%.

It is also observed, Figure 3, for alloys containing 16 wt.% Cr and less:

$$(\mathrm{Ti} + \mathrm{Nb})\ \mathrm{at.\%} \approx 10 - 0.7\mathrm{Al\ at.\%}\ (\text{generally to within } \pm 0.5) \qquad (2)$$

Providing that the chemistry for any new alloy formulation does not deviate significantly from either of these equations, then its calculated strength behaviour should remain reliable.

BALANCING FOR STABILITY

High alloying content brings reduced workability and increased tendency to hot cracking. Precipitation of deleterious second phases, leading to dramatic loss of strength, was seen in

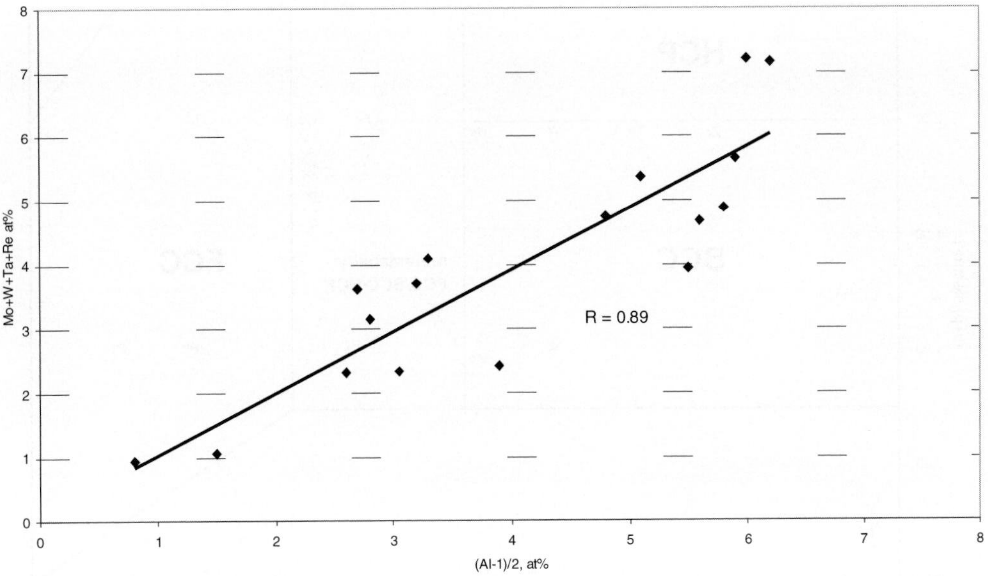

Fig. 2 Observed relationship between Al and solid solution content for the alloys listed in Table 1.

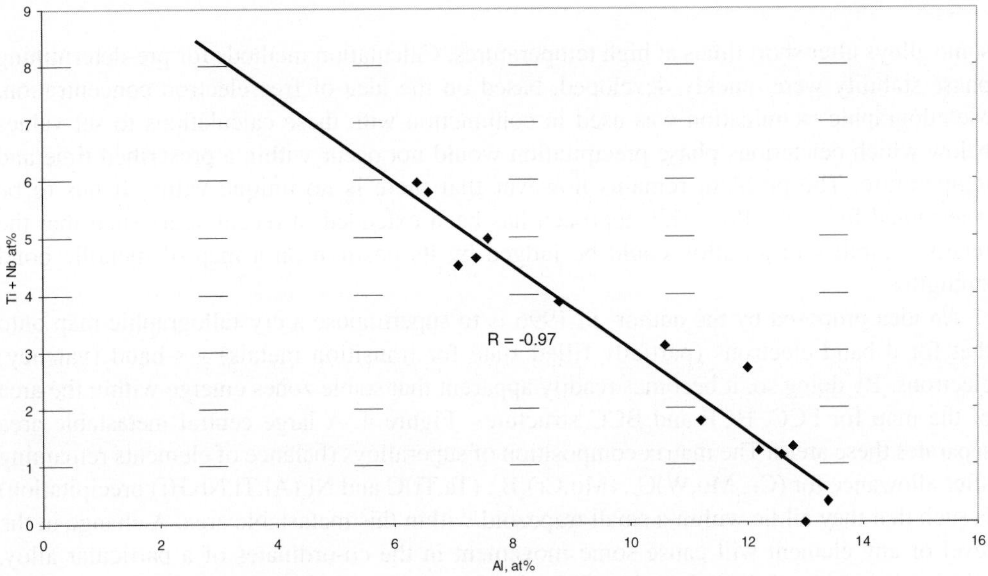

Fig. 3 (Ti + Nb) v Al dependency for those alloys listed in Table 1 with < 16 wt.%Cr.

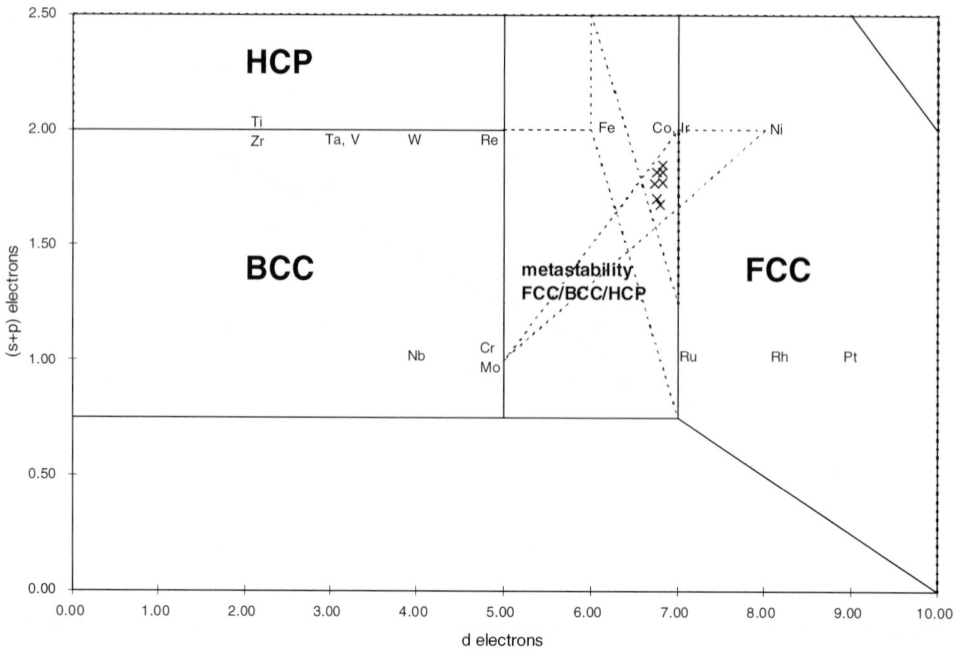

Fig. 4 Crystallographic map superimposed on d v s electrons showing also the locations of various metals and superalloys (x).

some alloys after short times at high temperatures. Calculation methods for pre-determining phase stability were quickly developed, based on the idea of free electron concentration. Metallographic examination was used in conjunction with these calculations to set values below which deleterious phase precipitation would not occur within a prescribed time and temperature. The problem remains however that there is no unique value. It has to be determined for each alloy. This approach has been extended in recent years such that the relative stability of an alloy could be judged by its position on a map of metallic bond strengths.

An idea proposed by the author[4] in 1996 is to superimpose a crystallographic map onto that for d-band electrons (partially filled state for transition metals) v s-band (valency) electrons. By doing so, it becomes readily apparent that stable zones emerge within the area of the map for FCC, HCP and BCC structures, Figure 4. A large central metastable area separates these areas. The matrix composition of superalloys (balance of elements remaining after allowance for $(Cr_{21}Mo,W)C_6$, $(Mo,Cr)_3B_2$, (Ta,Ti)C and Ni_3(Al,Ti,Nb.Hf) precipitation) is such that they all lie within a small trapezoid within this metastable area. A change in the level of any element will cause some movement in the co-ordinates of a particular alloy, pushing it outside of the trapezoid and further into the metastable area or back towards the FCC zone.

All of the cast superalloys listed in Table 1 lie within the trapezoid of Figure 4. Even the wrought superalloy IN718 which has 18% Fe and 18% Cr and Hast-X with 22%Cr and 9%Mo are contained within this region. Another wrought superalloy, Nimonic115 (8.5% Al + Ti, 14.5%Cr), happens to lie just outside the trapezoid (moving towards the Cr position). Now Nimonic115 alloy was one of the first superalloys that suffered from severe structural instability in use.[5] It had to be modified to reduce (not eliminate) this tendency.

DESIGNING FOR STRENGTH

A major key to opening up alloy design is an ability to link stability with strength. There is a mass of data in the public domain on the stress-rupture strengths of individual superalloys, and their strengths are conveniently compared by the use of Larson-Miller plots. It is shown here that the stress-rupture life of any alloy can be defined from alloy chemistry, applied stress and temperature:

Given that the primary and tertiary stages of creep are short compared to the time in steady state creep for normal operation of superalloys,

$$\text{then } \dot{\varepsilon} \propto \frac{1}{t} \tag{3}$$

$$\text{and therefore } \frac{1}{t} \propto \sigma^n \cdot \exp\left(\frac{-Q}{RT}\right) \tag{4}$$

It follows that for constant σ, $\log(t) \propto \frac{1}{t}$

Now the LM curve is a plot of $\log(\sigma)$ vs. P, where $P = T[20 + \log(t)]$ (5)

And for constant P (and hence constant σ), $\log(t) \propto 1/T$

Thus equations 4 and 5 show the same dependency of time upon temperature and it is therefore legitimate to combine the two. This is necessary in order to understand the interdependence of Q & T:

$$P = T\left[20 + \log\left(\frac{a}{\sigma^n} \cdot \exp\left(\frac{-Q}{RT}\right)\right)\right] \tag{6}$$

and for constant P and σ, $Q = K - bT$ (7)

Fitting equation 6 to the established L-M curve for IN738LC produces the result shown in Figure 5, where:

$$t(\text{hours}) = \frac{2{,}100{,}000}{\left[\sigma^n \exp\left(\frac{-Q}{RT}\right)\right]} \tag{8}$$

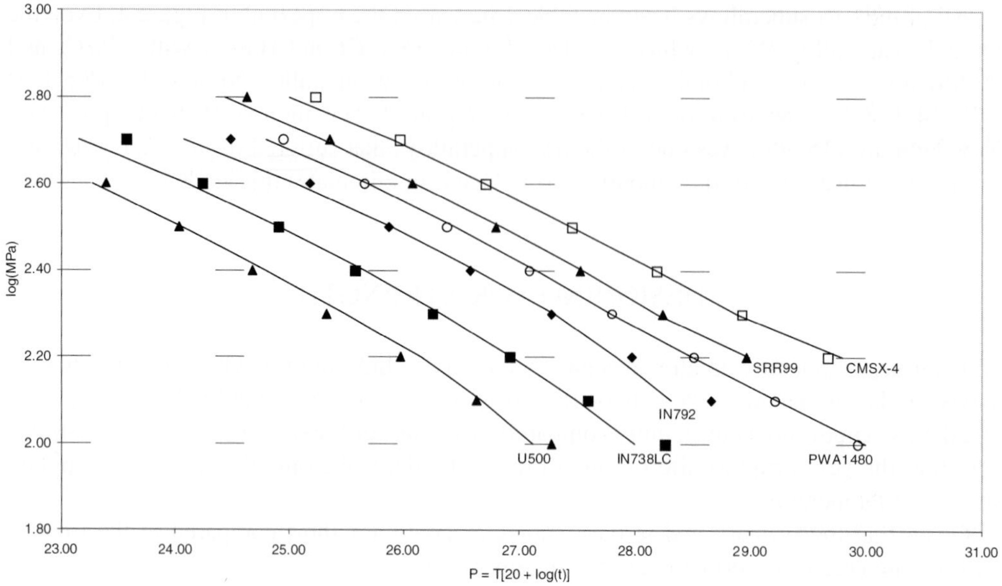

Fig. 5 Observed (solid lines) vs. calculated (points) Larson Miller curves for various alloys.

Table 2 Combined chemistry and structure factors for superalloys.

U500	IN939	R77	IN738LC	R80	B1900
0.965	0.933	0.998	1.000	1.022	1.028
IN792	MM247	MM247 (DS)	PWA1480 (SC)	SRR99 (SC)	CMSX-4 (SC)
1.034	1.047	1.051	1.059	1.076	1.102

where,

$$Q = 793{,}000 - 500T\left(\frac{\text{J}}{\text{mole}}\right)$$

$$\text{stress exponent}, n = \frac{6710}{T}$$

$R = 8.3$ J/K/mole

σ is measured in MPa.

The calculated curves are moved along the x-axis (and rotated slightly about the y-axis) to fit the other alloys by applying a combined chemistry + structure factor (Ź), the values of which are shown in Table 2:

Table 3 Element weighting.

C	Al	B	Co	Cr	Hf	Mo	Nb
–3.00	0.85	–3.00	0.08	–0.13	0.10	0.39	0.99
Re	**Ta**	**Ti**	**V**	**W**	**Zr**	**Ni**	**DS/SC**
0.42	0.30	0.81	0.09	0.33	–4.00	-0.03	0.30

Table 4 IN738LC standard and modified chemistry.

C	Al	B	Co	Cr	Mo	Nb	Ta	Ti	W	Zr	Ź	N_{v3B}
0.095	3.55	.01	8.3	16.0	1.7	0.8	1.7	3.45	2.6	0.03	1.000	2.36
0.11	3.3	.01	7.0	15.0	2.0	1.0	4.2	3.2	3.0	0.03	1.022	2.36

Table 5 IN792 standard and modified chemistry.

C	Al	B	Co	Cr	Mo	Ta	Ti	W	Zr	Ź
0.085	3.5	0.015	9.0	12.5	1.9	4.1	3.9	4.1	0.017	1.034
0.10	4.0	0.015	7.5	11.9	2.0	5.0	3.6	4.5	0.017	1.046

$$P \times 10^3 = \acute{Z} \cdot T\left[20 + \log\left\{\frac{2{,}100{,}000}{\left[\sigma^{\left(\frac{6710}{T}\right)} \cdot \exp\left(-\left(\frac{793{,}000 - 500T}{RT}\right)\right)\right]}\right\}\right] \tag{9}$$

$$\text{For each alloy, } \acute{Z} = 0.021 \times \Sigma\left(element_i \times \check{n}_i\right) + \left\{\frac{DS}{SC}\right\} + 0.9 \tag{10}$$

The values of n for each element (and DS/SC structure) are shown in Table 3. These values have been determined from linear regression of $\acute{Z}$ against alloy wt.% chemistries and structure, with a correlation of better than 0.99.

From equation 10 and the factors shown in Table 3, it becomes possible to predict changes in alloy strengths brought about by adjustment in alloy chemistry. Furthermore, this can be done without compromising alloy stability if the matrix composition remains well within the trapezoid of Figure 4.

Examining a high Cr alloy such as IN738LC, it quickly becomes clear that it is very difficult to increase its $\acute{Z}$ factor without decreasing the Cr content, if the matrix composition is to remain within the trapezoid. However, just a small decrease of Cr permits sufficient adjustment of Mo, Ta, W, etc. to produce a significant increase in $\acute{Z}$, Table 4, equivalent to a 25°C improvement in operational capability.

The modified chemistry falls within the observed tolerance bands of equations 1 and 2. The *Nv* number and the location of the matrix composition within the trapezoid are unchanged in this example, so stability should not be compromised.

A similar exercise on IN792 produces the following: The increased temperature capability is predicted in this example to be 14°C, while the matrix composition remains within the trapezoid and alloy chemistry conforms to the requirements of equations 1 and 2.

CONCLUSION

The likely stability and strength of new alloy chemistries can be quickly determined using the crystallographic map in conjunction with equations 9 and 10.

REFERENCES

1. C. T. Sims: 'Prediction of Phase Composition', *Superalloys II*, C. T. Sims, N. S. Stoloff and W. C. Hagel, eds., John Wiley & Sons, 1987, 217–239.
2. '1000 hours Rupture Strengths', *Superalloys II*, C. T. Sims, N. S. Stoloff and W. C. Hagel, eds., John Wiley & Sons, 1987, 586.
3. H. K. D. H. Bhadeshia: 'Neural Networks in Materials Science', *ISIJ International*, **39**(10), 966–979.
4. G. Durber: 'Strength & Stability Considerations in Alloy Formulation', *Superalloys 1996*, R. D. Kissinger, D. J. Deye, D. L. Anton, et al. eds., TMS, 1996, 111–115.
5. *The Nimonic Alloys*, W. Betteridge and J. Heslop, eds., Arnold, 1974, 15.

Solidification and Phase Stability of Ru-Bearing Ni-Base Superalloys

A. C. Yeh and S. Tin
Materials Science and Metallurgy Department
University of Cambridge
Cambridge, UK

ABSTRACT

Knowledge of the solidification characteristics associated with the unidirectional solidification of advanced Ni-base superalloys is required to understand the mechanical response of these unique materials at elevated temperatures. Although refractory elements, such as Re, W and Mo, are added to Ni-base superalloys to enhance the high temperature creep resistance, these elements also tend to promote the formation of intermetallic Topologically-Close-Packed (TCP) phases that eventually degrade the mechanical properties. Additions of the platinum group metal, ruthenium, to experimental alloys have hindered the precipitation of these deleterious intermetallic phases. In the present investigation, the as-cast and solution-treated microstructures were carefully studied. Electron-Probe Micro-Analysis was performed to study the elemental segregation behaviour. High temperature thermal exposures were used to investigate the microstructural stability. Differential Thermal Analysis was carried out to examine the phases and related thermal properties of the alloys. Results from these analyses will be discussed with respect to the influence of Ru additions on phase stability and partitioning during solidification.

INTRODUCTION

Ni-base superalloys are complex multi-component alloys engineered for structural applications at elevated temperatures. To further enhance high temperature mechanical properties, advanced superalloys contain significant amount of refractory elements, such as Re, W, Ta and Mo.[1–3] These alloying additions serve as potent solid solution strengtheners and alter the lattice misfit between the γ and γ' phases to enhance high temperature creep resistance.[4, 5] The amount of refractory elements allowed in Ni-base superalloys is limited by the tendency to form intermetallic topologically close-packed phases (TCPs), such as μ, P, R, or σ. In addition to serving as crack initiation sites, these brittle refractory-rich phases remove the potent solid-solution strengthening elements from the matrix as precipitation progresses. Although most TCP phases form parallel to {111} planes of the FCC nickel matrix and are compositionally similar, the various phases can be distinguished based on their distinct crystalline structures, Table 1.

Table 1 Summary of the basic crystallography of the σ, μ, P and R phases.

Phase	Typical Example	Unit Cell	Atoms Per Cell**	a (Å)*	b (Å)*	c (Å)*
σ	$Cr_{46}Fe_{54}$	Tetragonal	30	8.8	-	4.544
μ	Mo_6Co_7	Rhombohedral	39	4.762	-	25.61
P	$Cr_{18}Mo_{42}Ni_{40}$	Orthorhombic	56	9.07	16.98	4.752
R	$Cr_{18}Mo_{31}Co_{51}$	Rhombohedral	159	10.9	-	19.54

* the lattice constants of μ and R phases are defined on hexagonal axes.

** based on primary rhombohedral cell.

In high refractory content Ni-base superalloys, TCP formation is most often related to the compositional heterogeneities associated with solute partitioning during solidification of the microstructure. By reducing the degree of segregation within the microstructure either during solidification or through high temperature solutioning treatments, the tendency of the alloy to form TCP phases can be minimised.

The present study investigates the effect of the platinum group metal, Ru, on solute partitioning during solidification and its implications on the long term stability of Ni-base single crystal superalloys. Results from two nominally similar experimental alloys with and without Ru additions are compared and the implications on high temperature mechanical properties are discussed.

EXPERIMENTS

Materials and Solution Treatment

The aim of the current study was to quantify the influence of Ru on segregation during solidification and high temperature microstructural stability. Two experimental alloys provided by Rolls Royce plc for this research are designated RR2100 (no Ru) and RR2101 (2 wt.% Ru). The nominal compositions of these alloys are listed in Table 2. Single crystal bars approximately 12 mm in diameter of the experimental alloys were directionally solidified at Rolls Royce and supplied to the University of Cambridge. The microstructures of both the as-cast and solution-treated specimens were investigated. The solution heat treatment consisted of a twenty hour exposure at ~1,365°C.

Optical and SEM Metallography

The dendritic microstructures of the as-cast and solution-treated specimens of RR2100 and RR2101 were observed under both the optical microscope and scanning electron microscope

Table 2 Chemical compositions of RR2100 and RR2101 (wt.%).

Alloy	Co	Cr	W	Re	Ru	Al	Ta	Hf	Ni
RR2100	12	2. 5	9	6.4	0	6	5.5	0.15	Balance
RR2101	12	2. 5	9	6.4	2	6	5.5	0.15	Balance

(SEM). Samples were prepared using standard metallographic techniques, using a final polish of colloidal silica. The general purpose etching solution for the single crystal superalloy - nimonic (80 ml HCL, 20 ml H_2O, 2 ml HNO_3 and 16 grams $FeCl_3$) was used to reveal the microstructures. Volume fraction estimation of phases contained within the microstructures was performed using KS300 software.

Electron Probe Micro-Analysis

Electron Microprobe Analysis (EPMA) was used to assess the degree of microsegregation present in both of the as-cast and solution-treated microstructures. A Cameca SX50 electron probe microanalyser was used; it was equipped with thallium acid phthalate (TAP) (for Al, Ta, Re), lithium fluoride (LiF) (for Cr, Co, Ni) and pentaerythritol (PET) (for W) diffracting crystals. The scanned area was 1,024 × 1,024 μm with a step size of 4 μm, and x-ray counts were measured at each point for each element by WDS. The beam condition was 20 kV accelerating voltage, 100 nA-beam current and a dwell time of 0.4 seconds per pixel. The amount of Hf was too small to be detected accurately and was not analysed. The qualitative results were converted into elemental concentration maps for each element.

Quantitative analysis of microsegregation was also carried out taking compositional point measurements in the same area with a grid of 21 points by 21 points. The dwell time was increased to 75 seconds per point in order to obtain quantitative composition measurements. Combining these point analyses with a Scheil analysis, the characteristic partitioning behaviour of the alloying elements in these experimental alloys was assessed.

Differential Thermal Analysis

In Differential Thermal Analysis method, the sample is submitted to linear heating, and the rate of heat flow in the sample is proportional to the instantaneous specific heat, which is measured continuously. The platinum resistance thermometer and the heating unit built into the sample holders serves as a primary temperature control of the system. The secondary temperature control system maintains the temperature difference between the two holders at zero by means of a current, which is measured. Solutioned specimens of RR2100 and RR2101 were analysed in the DTA unit. Transitions corresponding to the liquidus, solidus and γ' solvus temperature were identified from the heating curves.

High Temperature Phase Stability

Microstructural stability is critical for Ni-base superalloys designed for high temperature applications. To minimise effects associated with the chemical heterogeneity of the as-cast structure, a solutioning treatment was performed on alloys RR2100 and RR2101. Following solutioning at ~1,365°C, the specimens were capsulated in quartz tubes with argon gas and then subjected to instability tests ranging from 900 to 1,200°C. Specimens were metallographically examined at intermittent intervals ranging from 25 to 1,000 hours. The onset of TCP formation within these specimens was used to assess the stability of these alloys at elevated temperature. From these experiments, an isothermal time-temperature-transformation (TTT) curve was constructed.

RESULTS AND ANALYSIS

As-Cast and Solution-Treated Microstructures

Microstructures of both as-cast and solution-treated RR2100 and RR2101 specimens were studied under the optical microscope and the SEM. The dendritic as-cast microstructures of RR2100 and RR2101 are shown in Figure 1, respectively. Both alloys reveal a severely segregated dendritic microstructure with RR2100 containing a significant volume fraction of γ-γ' eutectic within the inter-dendritic regions. With primary dendrite arm spacings on the order of 370 μm, the volume fraction eutectic in RR2100 was measured to be approximately 15%. Although RR2101 is nominally similar to RR2100, the as-cast microstructure of RR2101 reveals a significant reduction in eutectic area (8%). Despite solidification under constant processing conditions, the addition of 2 wt.% Ru to alloy RR2101 resulted in fewer pools of γ-γ' eutectic forming within the inter-dendritic regions.

After the solution heat-treatment at ~1,365°C for 20 hours, a small degree of residual segregation was still evident in the microstructures of both RR2100 and RR2101. A small volume fraction of TCP phase precipitates was observed within the microstructure of alloy RR2100 after the solution-heat treatment, Figure 2a. The large pools of γ-γ' eutectic observed in the as-cast microstructure are absent in the solutioned microstructure. The location of the TCP phases in RR2100 was confined to the dendrite cores within the γ-γ' microstructure. No TCP phases were observed in the Ru-bearing alloy, RR2101, Figure 2b. The microstructure of the solution heat-treated sample remained stable. Although Ru changes the microstructure in terms of the reducing the volume fraction of γ-γ' eutectic, the overall volume fraction of gamma prime after the solution-heat treatment was measured to be ~70% in both alloys.

Analysis of Microsegregation

The extent of microsegregation in the as-cast specimens of RR2100 and RR2101 was quantified before and after the solution heat-treatment. Figures 3 and 4 show qualitative

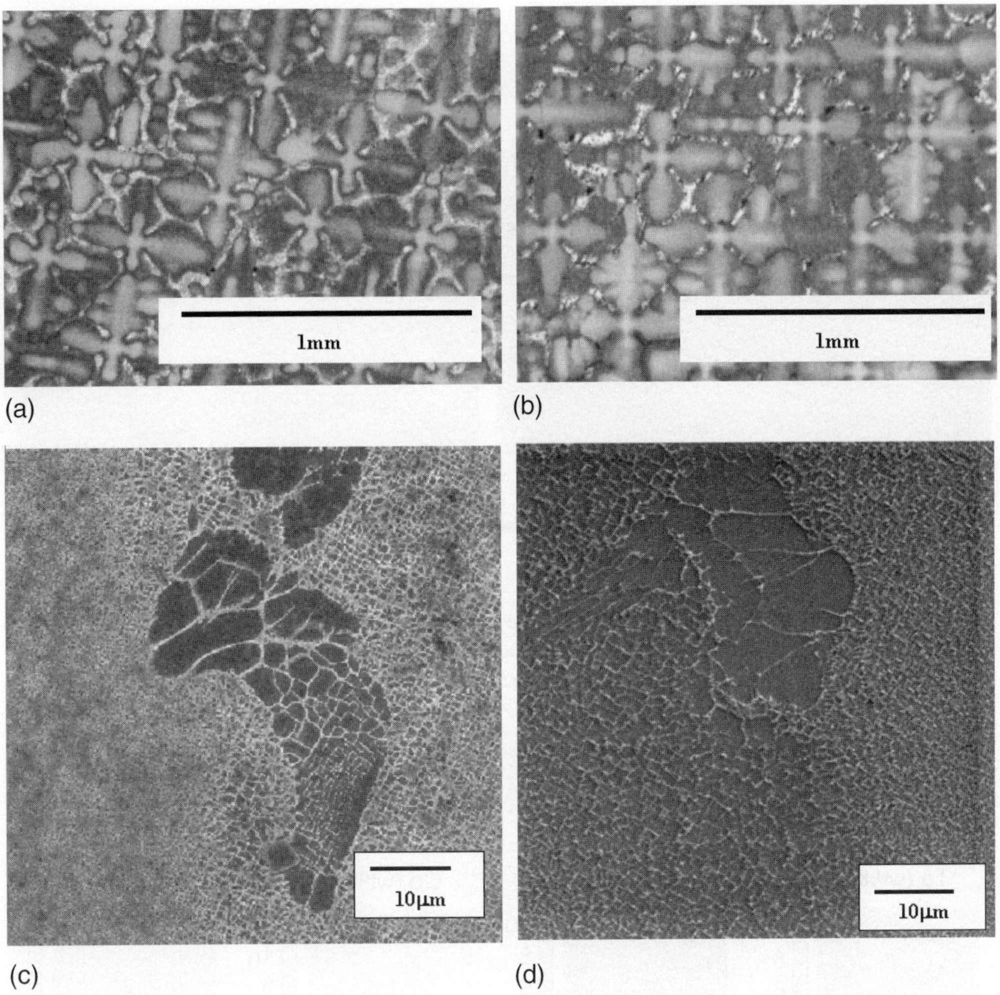

Fig. 1 The as-cast microstructures were observed under optical and SEM. The inter-dendritic regions show γ precipitated in γ'. (a) An optical image of RR2100, (b) An optical image of RR2101, (c) A SEI image of RR2100 and (d) A SEI image of RR2101.

EPMA concentration maps for as-cast and solution-treated RR2101 respectively. With the exception of Ru, the corresponding EPMA concentration maps for RR2100 are identical to those seen in Figures 3 and 4. These results indicate that the characteristic partitioning behaviour of Re, Co, Cr and W causes these elements to accumulate in regions surrounding the dendrite core, while Ni, Al and Ta segregate preferentially to the inter-dendritic regions. Interestingly, Ru is the only element that exhibits essentially no segregation in the as-cast microstructure. After the solution-treatment, a small degree of residual segregation was still

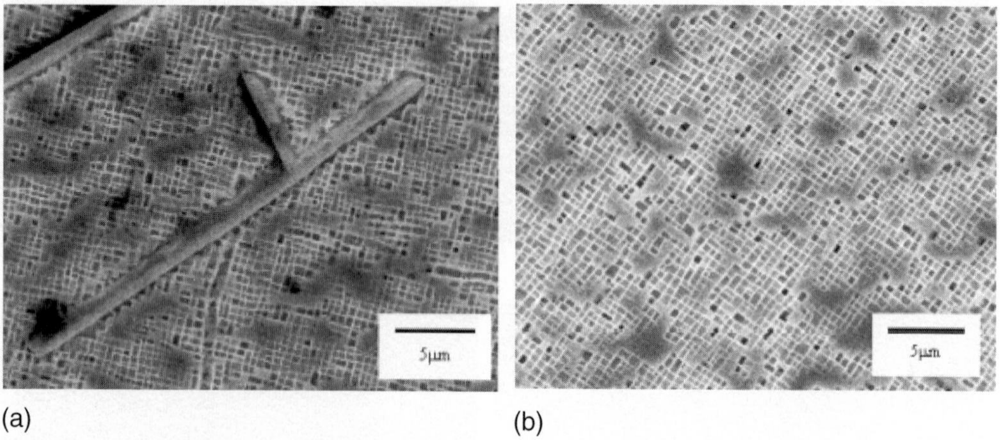

Fig. 2 (a) Microstructure of RR2100 after the solution heat-treatment. Small platelets of TCPs were observed in regions near dendrite cores and (b) Microstructure of RR2101 after the solution heat-treatment. Fine gamma prime precipitates are distributed within the gamma matrix.

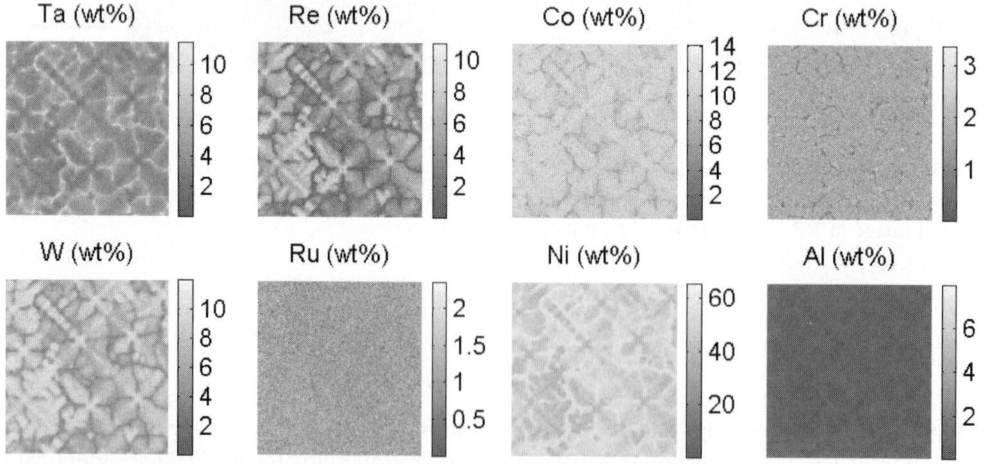

Fig. 3 Qualitative concentration maps for elements in as-cast RR2101. Ru does not exhibit preferential partitioning between dendrite core and inter-dendritic regions.

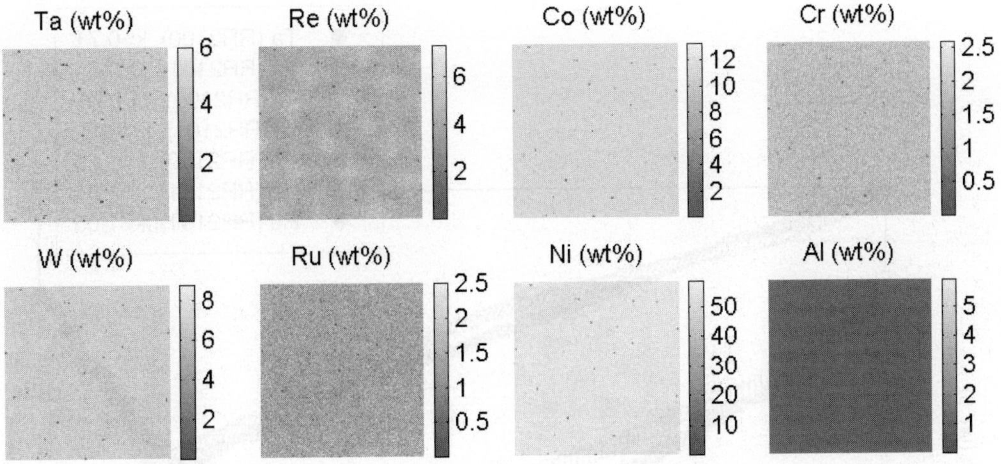

Fig. 4 Qualitative concentration maps for elements in solution-treated RR2101. Residual segregation is observed.

evident within the microstructure. Due to the inherently low diffusivity of Re in Ni, compositional variations still exist within the specimens. In RR2100, the characteristic partitioning behaviour of the alloying elements is identified to the observed pattern in RR2101. Residual Re levels within the dendrite core can be detected in the concentration map after solution-treatment, and results are consistent with the microstructural observations regarding TCP phase precipitation within the dendrite cores after solution heat treatment.

Segregation mapping using point analyses of the two alloys was performed to quantify the differences in partitioning associated with the addition of 2.0 wt.% Ru. Results from the quantitative point mapping analyses for alloy RR2100 and RR2101 is shown in Figure 5. The experimental data for each of the alloying elements in RR2101 were ranked in ascending or descending order on a zero to one volume fraction scale according to their characteristic segregation behaviour (qualitative concentration maps). The elemental distributions were then fitted with the Scheil equation to determine the partition coefficients for the various alloying elements.[6]

$$C_s = kC_0(1-f_s)^{(k-1)} \quad (1)$$

The Scheil analyses assumes that no diffusion in solid and perfect mixing in liquid occurs during solidification. Where C_s is the element concentration, k is the partition coefficient, which is defined by the ratio of solid concentration over liquid concentration, C_0 is the mean composition and f_s is the volume fraction of solid. The partition coefficient associated with each element can be calculated from the gradient of the best-fit line for eqn 1. To ensure the accuracy of those EPMA results, nominal compositions were within good agreement with the mean compositions measured from EPMA. The segregation analyses reveal that in both RR2100 and RR2101, W, Ta and Re exhibit the largest degree of segregation. Other elements,

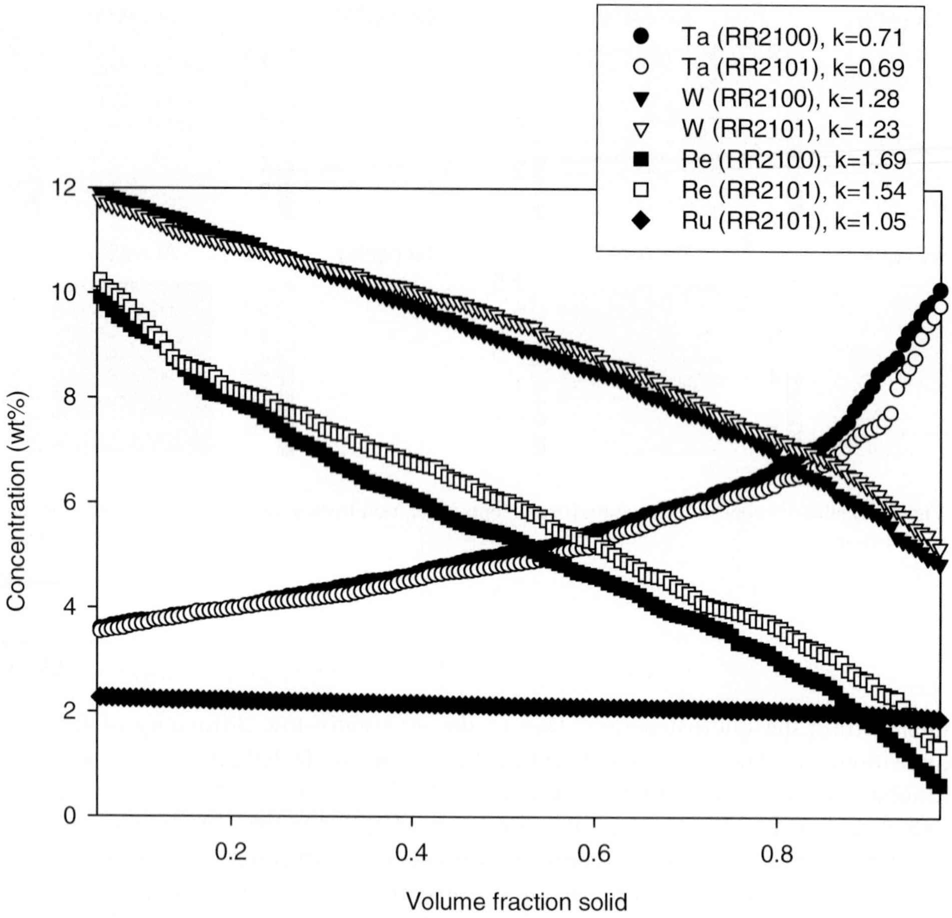

Fig. 5 Comparison between Ta, Re, W and Ru distributions in the as-cast condition for RR2100 and RR2101. The estimated partitioning coefficient (k) from the best-fit Scheil curve are indicated.

including Ru, partition weakly to either the dendrite core or inter-dendritic regions. Interestingly, the 2 wt.% Ru addition to RR2101 appears to influence the characteristic partitioning behaviour of Re. Compared to RR2100, a smaller degree of Re partitioning was observed in RR2101. Comparison of the partition coefficients for RR2100 and RR2101 in Figure 5 shows that only Re segregation was significantly affected by the Ru addition.

EPMA results from the solution-treated samples are shown in Figure 6. The thermal exposure at ~1365°C for 20 hours eliminates a high degree of chemical heterogeneity when compared to the as-cast condition. A small degree of residual segregation, however, still exists within the microstructure of both alloys. In alloy RR2100, Ta and Re levels still range from 4.9 to 5.9 wt.% and 6.5 to 4.5 wt.%, respectively. For alloy RR2101, which exhibited a

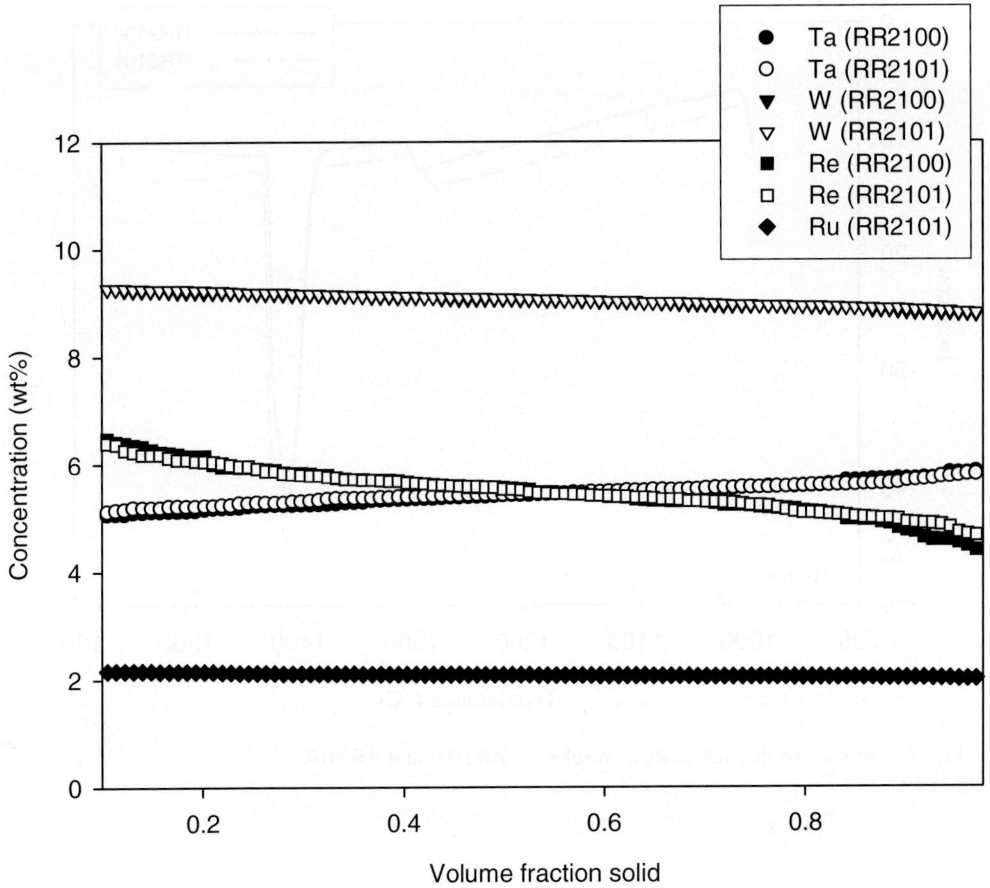

Fig. 6 Measured distribution of Re, Ta, W and Ru in RR2100 and RR2101 after the solutioning heat-treatment.

lower initial degree of segregation in the as-cast state, Ta and Re levels ranged from 5.1 to 5.8 wt.% and 6.4 to 4.6 wt.%, respectively, after the same solution heat-treatment.

Analysis of Thermal Properties

Results from the DTA experiments on RR2100 and RR2101 are shown in Figure 7. The gamma prime solvus of RR2100 is found to be 1,299°C, solidus is 1,387°C and liquidus is 1,425°C. For RR2101, the gamma prime solvus is 1,287°C, solidus is 1,389°C and liquidus

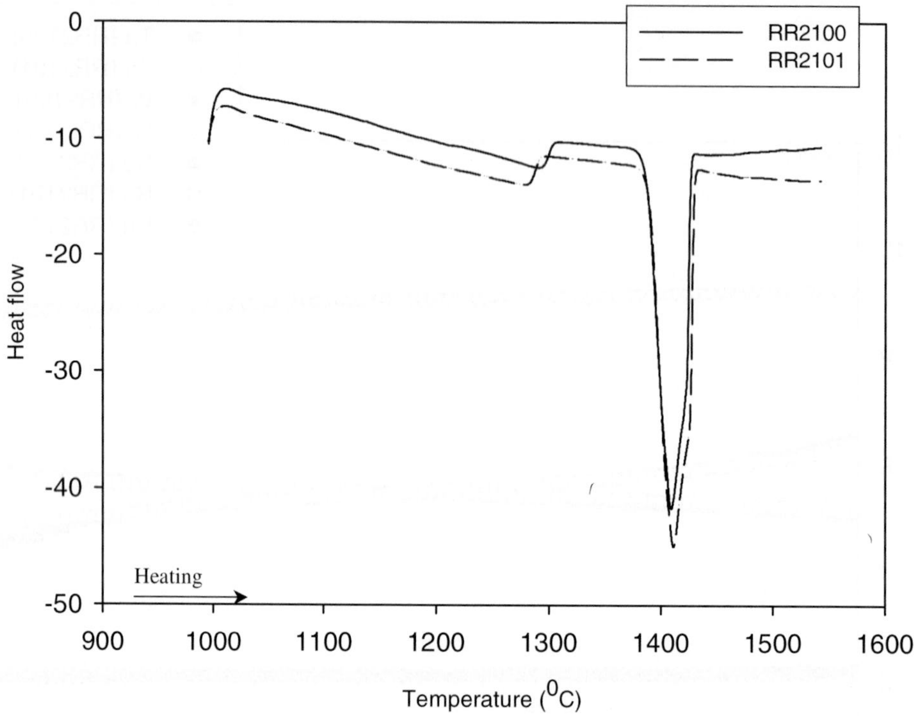

Fig. 7 Differential thermal analysis results for RR2100 and RR2101.

is 1428°C. Compared to RR2100, the 2 wt.% Ru addition to RR2101 modestly lowers the γ' solvus temperature by approximately 12°C while having little influence on the solidus and liquidus temperatures.

The Instability of Alloys

Results from the instability tests performed on these alloys are summarised in Figure 8. The curves on the Time-Temperature plot denote the onset of TCP phase precipitation in RR2100 and RR2101. Based on metallographic analyses of specimens exposed for various times at temperatures ranging from 900 to 1,200°C, RR2100 remains stable for 1,000 hours below 950°C. Precipitation of TCP phases in RR2100 occurs after 100 hours at 1,000°C. In Figure 9, the TCP phases appear as bright phases in backscatter imaging under the SEM due to its high content of refractory elements. With the addition of 2 wt.% Ru, the stability of the alloy increased dramatically as RR2101 is stable for 1,000 hours at 1,050°C. In Figure 10, TEM studies reveal the TCP precipitates in both RR2100 and RR2101 to have similar plate-like geometries. These geometries are characteristic of P-type TCP phase.[7] The measured chemical

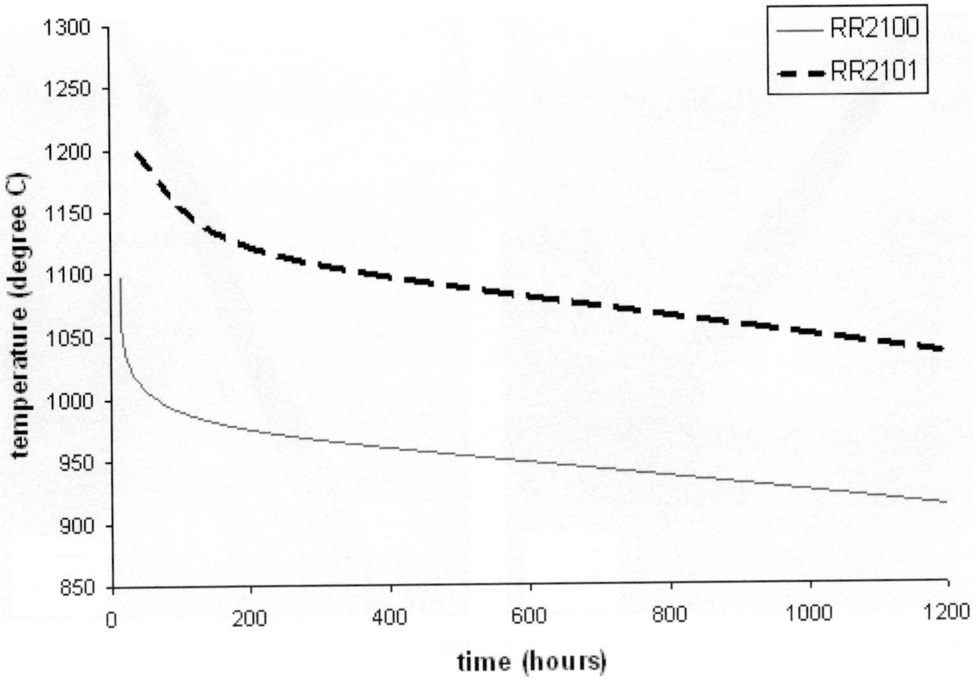

Fig. 8 TTT curves for TCP formation in RR2100 and RR2101.

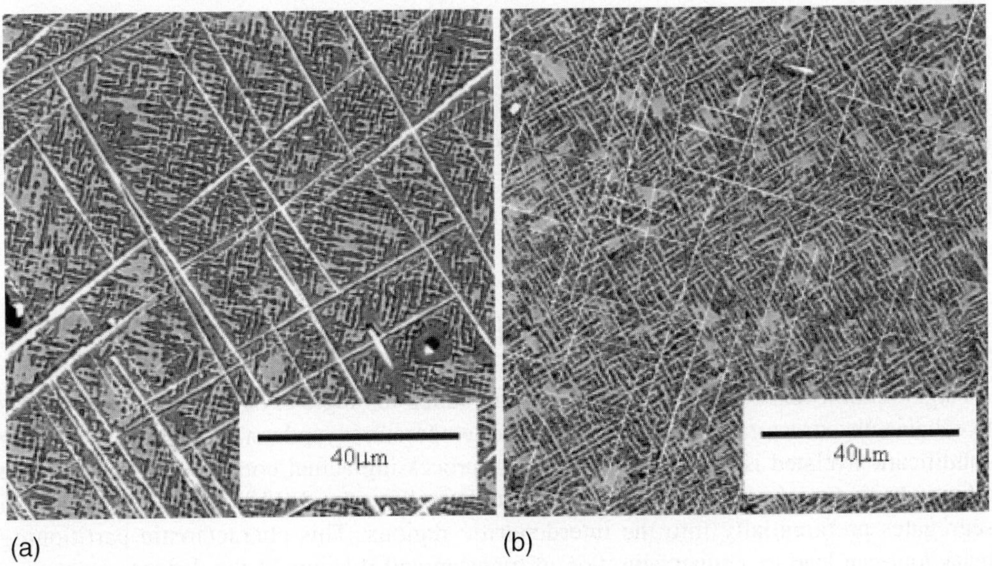

Fig. 9 This shows a backscattered image of the microstructure after isothermal aging at 1,100°C for 1,000 hours. The TCP phases appear as bright phases with needle-like morphologies. (a) RR2100 and (b) RR2101.

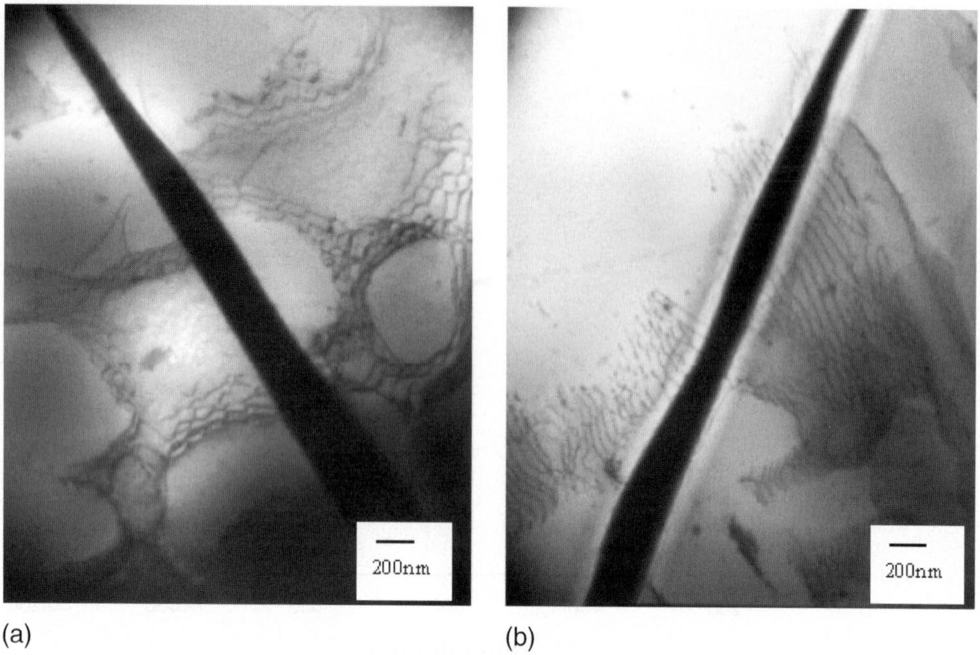

Fig. 10 (a) A bright field TEM image of TCP precipitate (dark phase across the image) in RR2100 after 1,100°C and 1,000 hours exposure time and (b) A bright field TEM image of TCP (dark phase across the image) precipitate in RR2101 after 1,100°C and 1,000 hours exposure time.

compositions of TCP precipitates in both RR2100 and RR2101 are very similar, and suggest that common TCP phases form in both alloys. Further work is currently being carried out to characterise the TCP phases present in these alloys.

DISCUSSION

As significant levels of refractory elements are added to single crystal Ni-base superalloys to enhance the structural properties at elevated temperatures, understanding and controlling solidification related issues become critical in processing actual components. In advanced alloys, rhenium and tungsten tend to segregate heavily to the dendrite core while tantalum segregates preferentially into the interdendritic regions. This characteristic partitioning behaviour can lead to a supersaturation of tungsten and rhenium at the dendrite core upon solidification.[8] The resulting supersaturation of these slowly diffusing elements subsequently provides the primary driving force for precipitation of TCP phases. Despite lengthy solutioning treatments, a substantial degree of residual segregation still exists within the dendritic microstructure, Figure 6, and promotes the formation of these detrimental phases. Ru additions to Ni-base superalloys appear to stabilise the microstructure and delay the onset of TCP precipitation at elevated temperatures.

One possible mechanism by which Ru additions enhance phase stability can be related to the degree of segregation during unidirectional solidification. The driving force for precipitation of TCP phases can be minimised by controlling degree of tungsten, rhenium and tantalum segregation and preventing the dendrite cores from becoming supersaturated with refractory solute.[9, 10] As shown in Figure 5, the extent to which Re, W and Ta are segregated in the as-cast dendritic microstructure is measurable lower in the Ru-bearing alloy, RR2101. During solution heat-treatment, the compositional difference in RR2100 is sufficient to result in the formation of TCP phases, while the microstructure of RR2101 remains stable. Following the solutioning treatment, the degree of residual segregation in RR2100 and RR2101 are similar when measured using the Scheil analysis. Due differences in the as-cast conditions, there exists a slight difference in the distribution of the heavily segregated and slowly diffusing elements, Re and Ta, in the homogenised specimens. It is possible that the improved resistance to TCP formation exhibited by RR2101 is a consequence of this slight difference.

In single crystal alloys, precipitations of TCP phases are often inhibited by a nucleation barrier. Without grain boundaries or other interfaces that can enhance the kinetics of the transformation, a considerable decrease in free energy is required to initiate the onset of TCP formation.[11] In RR2100, existing TCP phases were present in the solutioned microstructure, Figure 2. The presence of these initial precipitates may have possibly enhanced the kinetics of the phase instabilities by providing a large number of nucleation sites at the dendrite cores. It is also possible that the Ru additions to RR2100 influenced the solid-solid partitioning of the constituent elements. Other investigations have attributed compositional changes in the γ matrix and γ' precipitates to Ru additions.[12–14] The thermodynamic factors influencing phase instabilities may also be changing as a result of the Ru addition. Although these preliminary results suggest that Ru additions can be beneficial with respect to enhancing high temperature properties by improving phase stability, an improved understanding of the mechanisms by which Ru alters these properties is required. Such investigations are currently in progress.

CONCLUSIONS

The following statements can be drawn from analysis of results presented in this report:

1. Ru additions influence partitioning of rhenium during solidification. Although the addition of 2.0 wt.% Ru to the experimental alloy did not effect the solidus and liquidus temperatures, the γ' solvus temperature was decreased.
2. The Ru-bearing alloy was significantly more resistant to the onset of TCP formation at elevated temperatures.

The preliminary results and analysis presented in this study suggest that Ru additions are amenable for optimising the high temperature properties of Ni-base superalloys.

ACKNOWLEDGEMENT

The author would like to give special thanks to Dr S. Tin, Dr C. Rae, Prof. S. Reed, Dr R. Broomfield, and Mr A. Ofori. The author gratefully acknowledges the financial support provided by the Cambridge Overseas Trust and Rolls Royce plc.

REFERENCE

1. W. T. Loomis, J. W. Freeman and D. L. Sponseller: 'The Influence of Molybdenum on the γ' Phase in Experimental Nickel-Base Superalloys', *Metallurgical Transactions,* **3**, 1972–1989.
2. A. F. Giamei and D. L. Anton: 'Rhenium Additions to a Ni-Base Superalloy: Effects on Microstructure', *Metallurgical Transactions A,* **16A**, 1985–1997.
3. M. S. A. Karunaratne, P. Carter and R. C. Reed: 'Interdiffusion in the Face-Centred Cubic Phase of the Ni-Re, Ni-Ta and Ni-W Systems Between 900 and 1300°C', *Materials Science and Engineering,* **A281**, 2000, 229–233.
4. H. Murakami, T. Yamagata, H. Harada and M. Yamazaki: 'The Influence of Co on Creep Deformation Anisotropy in Ni-Base Single Crystal Superalloys at Intermediate Temperatures', *Materials Science and Engineering,* **A223**, 1997, 54–58.
5. R. C. Reed, N. Matan, D. C. Cox, M. A. Rist and C. M. F. Rae: 'Creep of CMSX-4 Superalloy Single Crystals: Effects of Rafting at High Temperature', *Acta Materialia,* **47**, 1999, 3367–3381.
6. M. N. Gungor: 'A Statistical Significant Experimental Technique for Investigating Microsegregation in Cast Alloys', *Metallurgical Transactions A*, **20A**, 1989, 2529–2533.
7. M. S. A. Karunaratne, C. M. F. Rae and R. C. Reed: 'On the Microstructural Instability of an Experimental Nickel Based Single Crystal Superalloy', *Metallurgical Transactions A*, **32A**, 2001, 2409.
8. Madeleine Durand-Charre: *The Microstructure of Superalloy*, Gordon and Breach Science Publishers, 1997.
9. T. Grosdidier, A. Hazotte and A. Simon: 'About Chemical Heterogeneities and Gamma Prime Precipitate in Single Crystal Nickel Base Superalloys', *High Temperature Materials for Power Engineering* 1990, Kluwer, Dordrecht, Netherlands.
10. C. M. F. Rae and R. C. Reed: 'The Precipitation of Topologically Close-Packed Phases in Rhenium-Containing Superalloys', *Acta Materialia*, **49**, 2001, 4113–4125.
11. R. Darolia, D. F. Lahrman and R. D. Field: 'Formation of Topologically Closed Packed Phases in Nickel Base Single Crystal Superalloys', *Superalloys 1988*, 1988, 255–264.
12. General Electric Company, *European Patent Application*, Publication Number: 0663 462 A1.
13. H. Murakami, T. Honma, Y. Koizumi and H. Harada: 'Distribution of Platinum Group Metals in Ni-Base Single Crystal Superalloys', *Superalloys 2000*, 2000, 747.
14. Yunrong Zheng, Xiaoping Wang, Jianxin Dong, Yafang Han, H. Murakami and H. Harada: 'Effects of Ru Addition on Cast Nickel Base Superalloy with Low Content of Cr and High Content of W', *Superalloys 2000*, 2000, 305.

Improved Single Crystal Superalloys

KEN HARRIS and JACQUELINE B. WAHL
Cannon-Muskegon Corporation
SPS Technologies, Inc.
Muskegon, Michigan, USA

ABSTRACT

Modern turbine engine performance and life cycle requirements demand single crystal (SX) superalloy turbine airfoil, seal and combustor components. The benefits of single crystal components include superior mechanical and physical properties, along with retention of these properties for thin-walled components. The variety of applications, from TBC-coated first stage blades and vanes which require a fully solutioned microstructure to achieve superior creep strength, oxidation and coating performance to complex multi-vane segments which necessitate more generous grain acceptance criteria to obtain reasonable yield and manufacturing cost, demand SX alloys tailored to the unique requirements of these applications. Improved single crystal superalloys have been developed to respond to these needs.

CMSX-4® [SLS] (Super Low Sulphur) [La + Y] alloy addresses the demand for enhanced oxidation and coating performance, including TBC, while retaining excellent high temperature mechanical properties. Through melt desulphurisation techniques, master melt sulphur content is minimized to < 1 ppm providing more degrees of freedom for SX foundry ceramics when using highly reactive La + Y additions. The add levels of the rare earth element additions (La & Y) can be reduced to provide greater control and to preclude excessive retention in thicker sections and consequential incipient melting problems during solution heat treatment along with adverse effects on fatigue properties.

CMSX-486® a grain boundary strengthened single crystal superalloy, has been developed to allow cost-effective manufacturing for complex SX vane segment castings. This alloy enables generous grain specifications through optimised additions of boron, carbon, hafnium and zirconium and is designed for use as-cast, which eliminates solution heat treatment recrystallisation issues.

Implementation of improved SX superalloys can extend the application of SX components in turbine engines, realizing the potential benefits in strength and design, while reducing manufacturing cost.

INTRODUCTION

Traditional gas turbine hot section alloy development has focused on higher temperature capability to achieve improved efficiency. Consequently, aero and industrial gas turbine engine blades and vanes have progressed from equiax to directionally solidified (DS) to single crystal (SX) castings and alloys, with a commensurate increase in temperature capability (Figure 1). The benefits of single crystal cast components for gas turbine engines have been well documented: single crystal alloys offer improved fatigue, creep-rupture, oxidation and coating properties, resulting in superior turbine engine performance and

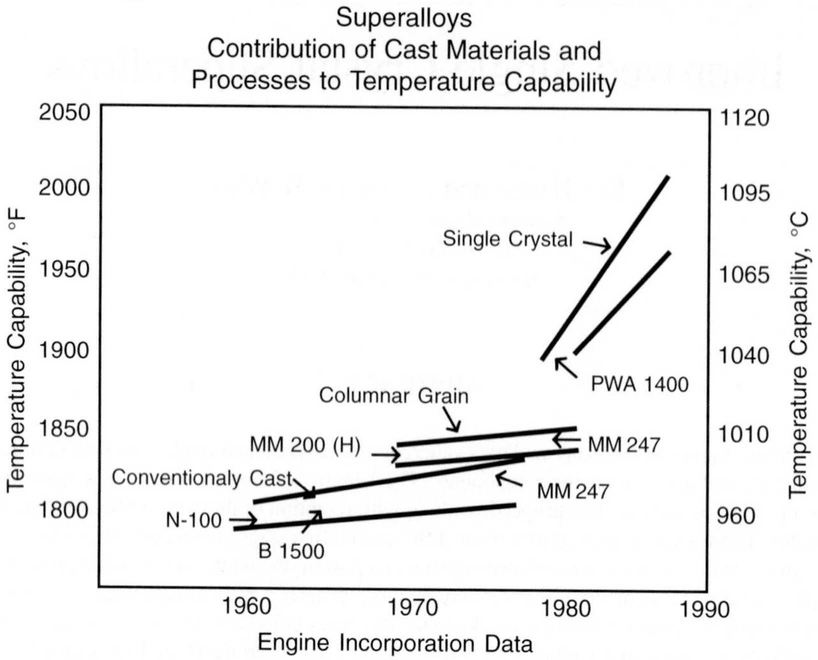

Fig. 1 Increase in temperature capability due to materials and process improvements.

durability.[1] In addition, single crystal alloys retain a higher fraction of their thick section rupture life as wall thickness is reduced (Figure 2).

Due to this attractive combination of properties and performance, the application of single crystal casting technology has extended to a variety of components with sometimes divergent requirements. For example, complex cooled, thin-walled first stage turbine blades are among the most demanding SX castings due to the need for excellent high temperature properties, superior oxidation and coating performance, including the use of prime reliant ceramic thermal barrier coatings (TBC). In contrast, SX multiple vane segments require generous grain inspection criteria to obtain advanced technology castings with a high manufacturing yield. Throughout the industry, affordability is key to successful introduction of new technology or applications. In this paper, the development of two diverse single crystal superalloys which respond to these specific needs will be discussed.

CMSX-4® [SLS] [LA + Y] ALLOY

CMSX-4 is a second generation, Re-bearing nickel-base single crystal superalloy, which has been successfully used in numerous aero and industrial gas turbine applications since 1991. The nominal chemistry is provided in Table 1. CMSX-4 alloy provides an attractive

Table 1 Nominal alloy chemistry (wt.% or ppm).

Alloy	C	B	Al	Co	Cr	Hf	Mo	Ni	Re	Ta	Ti	W	Zr
CMSX-4	--	--	5.6	9	6.5	.1	.6	Bal	3	6.5	1.0	6	--
CM 247 LC®	.07	.015	5.6	9.3	8	1.4	.5	Bal.	--	3.2	.7	9.5	.010
CMSX-3	--	--	5.6	4.8	8	.1	.6	Bal.	--	6.3	1.0	8.0	--
CM 186 LC	.07	.015	5.7	9.3	6	1.4	.5	Bal.	3	3.4	.7	8.4	.005
CMSX-681	.09	.015	5.7	9.3	5	1.4	.5	Bal.	3	6.0	.1	8.4	.005
CMSX-486	.07	.015	5.7	9.3	5	1.0	.7	Bal.	3	4.5	.7	8.6	.005

Table 2 CMSX-4 (SLS) heats – critical chemistry (ppm).

Heat	S (vs. NIST Std. #861)	S (GDMS)	(N)	(O)	Master Heat Reactive Additions
5V0062	.8	.4	1	2	
5V0066	.7	.5	1	2	
5V0077	.4	.4	1	2	Pre-alloyed with 300 ppm La and 300 ppm Y
5V0079	.6	.3	2	2	
5V0080	.6	.6	2	2	
5V0086	.6	.4	2	2	Pre-alloyed with 190 ppm La and 180 ppm Y
5V0087	.7	.3	1	3	Pre-alloyed with 120 ppm La and 130 ppm Y
5V0091	.7	.4	1	2	Pre-alloyed with 243 ppm La and 249 ppm Y
5V0092	.6	.2	1	2	Pre-alloyed with 568 ppm La and 536 ppm Y
5V0093	.6	.3	2	1	Pre-alloyed with 426 ppm La and 391 ppm Y
5V0106	.8	.2	2	2	Pre-alloyed with 222 ppm La and 200 ppm Y

combination of high strength for creep-rupture, mechanical and thermal fatigue, good phase stability and oxidation, hot corrosion and coating performance.[2] Close to 5 million pounds (650 heats) of this alloy has been manufactured to date.

CMSX-4 [La + Y] alloy was introduced to meet the ever-increasing performance requirements for hot section turbine blades and vanes. This development demonstrated that the oxidation behaviour of bare CMSX-4 alloy, with sulphur (S) content #2 ppm, could be dramatically improved by the addition of lanthanum (La) and yttrium (Y) (Figure 3) while maintaining full CMSX-4 alloy mechanical properties.[3] This is important not only for bare oxidation resistance in locations such as the blade tip of shroudless blades or uncoated internal air cooling passages, but also for coating/TBC adherence and performance, particularly included prime reliant predictable life.

The newest improved version of the alloy, CMSX-4 [SLS] [La + Y] incorporates melt desulphurisation techniques, to bring the sulphur content down to < 1 ppm. This has been accomplished with excellent alloy cleanliness, represented by 1-2 ppm oxygen obtained over multiple heats (Table 2). The significance of this SX alloy melting and desulphurisation technology is that the lower sulphur content enables lower levels of La + Y reactive element additions, to produce the same superior oxidation properties, including coating and TBC life.

La + Y additions are typically made on the casting furnace to enable the additions to be tailored to the casting parameters and configuration; however, heats can (and have been) pre-alloyed. The highly reactive elements (La and Y) tie up the residual sulphur (<1 ppm) in the alloy as stable La + Y sulphides, preventing the migration/diffusion of S to the free surface. Sulphur segregation to the free surface has been shown to destroy the strong Van der

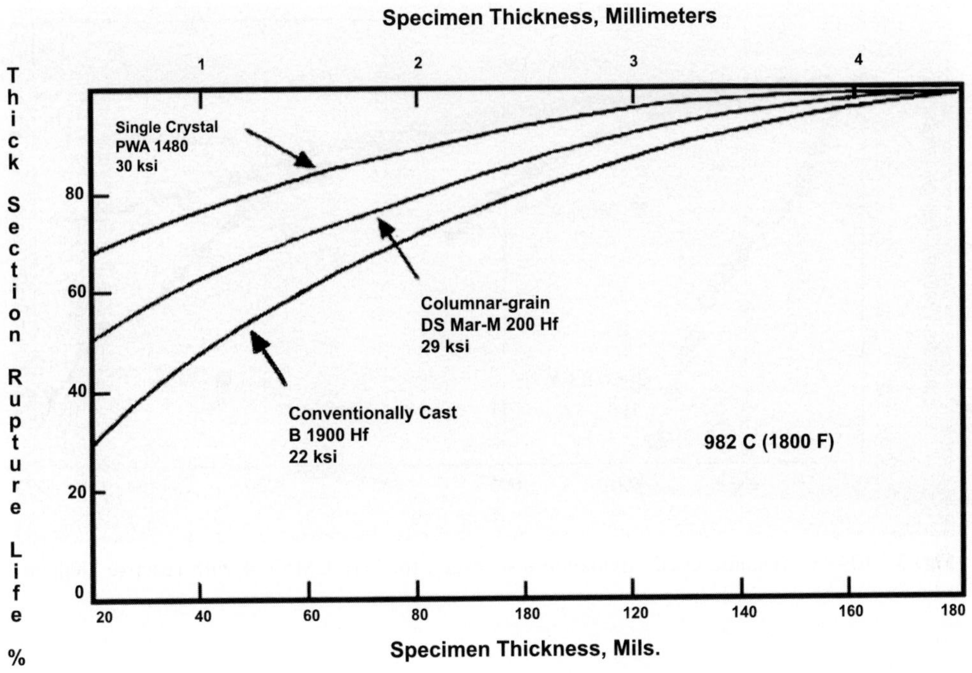

Fig. 2 Rupture life as a function of specimen thickness showing benefit of SX material over DS and equiax components (Courtesy PWA).

Waals bond between the α alumina scale layer and the base alloy/coating/bond coat, resulting in premature coating/TBC spallation.[3] The benefit of reactive element additions has demonstrated a substantial improvement in EB-PVD TBC life (Figure 4).

The ability to obtain superior oxidation performance with lower reactive element additions may permit the use of conventional foundry ceramics, providing manufacturing flexibility and reduced costs. Numerous SX casting trials have demonstrated successful use of standard (Si-containing) foundry ceramics, with excellent grain yield and control of the La + Y content on complex cored SX airfoil castings. Lower reactive element additions with less variation in retained [La + Y] content will also minimize the incidence of incipient melting which can occur due to excess retention in the thicker root sections from higher reactive element additions. It should be noted that, in addition to being more costly, less reactive core materials (such as alumina) are much more difficult to remove following casting, resulting in long, expensive autoclave processing.

Consequently, the development of melt desulphurisation techniques used to produce CMSX-4 [SLS] [La + Y] alloy offers the opportunity to obtain the superior properties of one of the most advanced commercial single crystal superalloys with reduced manufacturing costs due to both less expensive foundry ceramics and less time consuming post-cast processing to remove the very inert core materials.

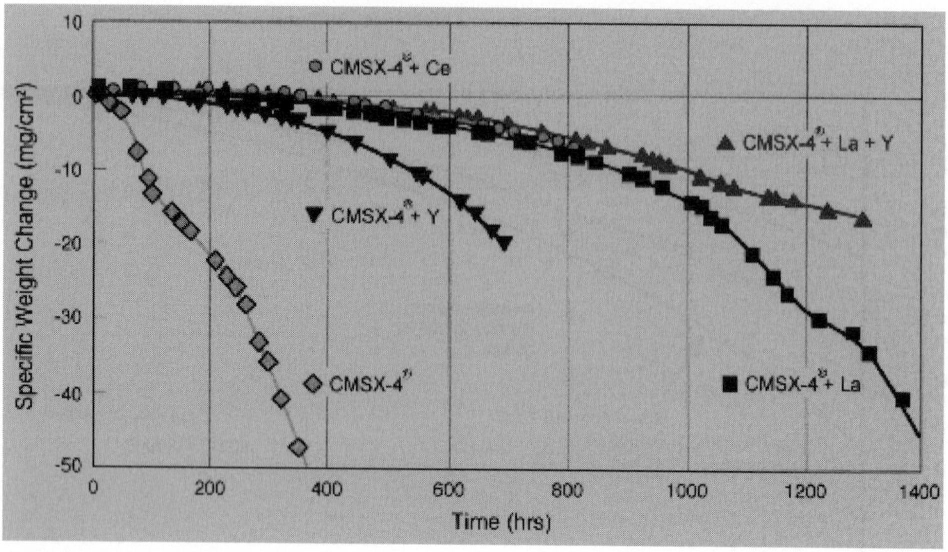

Fig. 3 1093°C dynamic cyclic oxidation test results for bare CMSX-4 with reactive element additions.

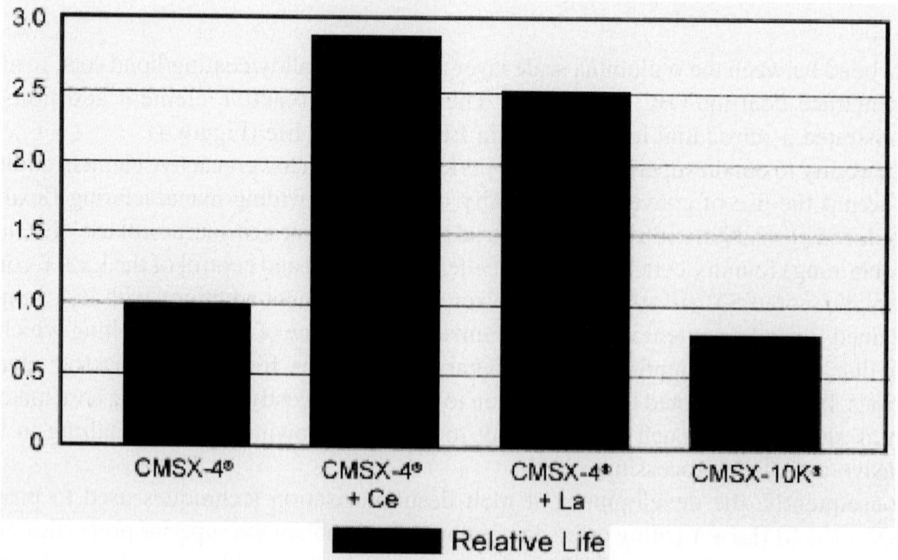

Fig. 4 Reactive element effects on EB-PVD TBC Life 1093°C (2000°F)/10 hours thermal exposure cycles.

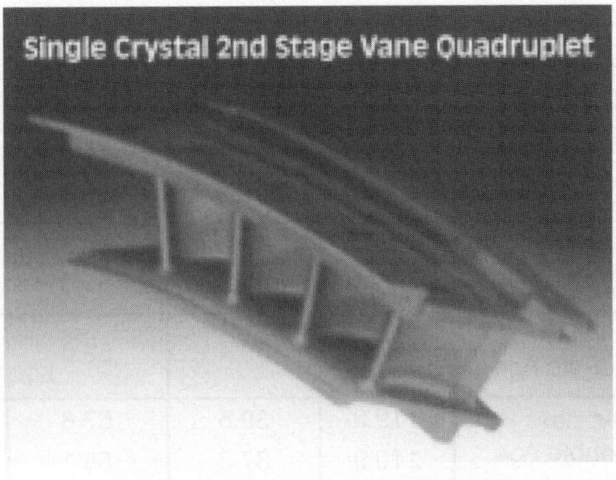

Fig. 5 SX CM 186 LC vane segment casting from RR AE 3007 A1 engine.

CMSX-486® GRAIN BOUNDARY STRENGTHENED ALLOY

Single crystal superalloy vanes have demonstrated excellent performance and durability compared to equiax (polycrystalline) vanes.[4] The concept of single crystal casting an alloy with full grain boundary strengthening elements to obtain SX cast components which could accommodate a generous grain defect specification (improving yield and reducing cost) was first explored in a collaborative program with Rolls-Royce Corporation using the second generation, Re-bearing DS alloy CM 186 LC®. This work demonstrated significant retention of mechanical properties for low angle and high angle boundaries (LAB/HAB) exceeding 20° misorientation.[5] The laboratory testing justified evaluation of relaxed grain inspection criteria, culminating in successful implementation in the 2nd stage vane segment for the RR AE 3007 A1 engine flown on regional jets, such as the Embraer 145 (Figure 5).

This successful application also identified the opportunity for an optimised grain boundary strengthened (GBS) SX alloy, with improved creep-rupture strength over SX CM 186 LC while retaining the lower cost manufacturing achieved through generous grain inspection criteria and minimal heat treatment. Initial alloy development consisted of detailed studies of CMSX-486 alloy chemistry variations, along with evaluation of an alternative composition designated CMSX®-681 alloy. The chemistry of CMSX-486 alloy (Table 1) was established following evaluation of the effects of chemistry variations on creep strength and mechanical property retention across LAB/HAB defects.

Conventional 6.35 mm (1/4") diameter near net shape single crystal test bars were cast at Rolls-Royce Corporation (SCFO) from two heats of CMSX-486 alloy for comparative creep-rupture testing. Early testing established the optimum balance of creep properties are attained in the as-cast + double aged condition, particularly for aero turbine vane applications where 1093°C (2000°F) metal temperatures are critical (Table 3). This is advantageous from a manufacturing standpoint as it reduces the post-cast processing requirements and precludes

Table 3 Typical CMSX-486 creep-rupture properties.

Test Condition	Heat Treat Condition	Life, Hrs.	Time to 1% Creep	Time to 2% Creep	Elong., %	Red. in Area, %
982°C/248 MPa	Partial Sol'n + Double Age	168.1 172.0	51.7 56.4	74.8 80.9	39.7 35.4	47.0 45.1
	As-cast + Double Age	143.0 138.3	48.0 42.9	66.3 61.0	35.7 46.1	48.1 47.0
1038°C/172 MPa	Partial Sol'n + Double Age	114.3	39.4	59.8	28.4	52.5
	As-cast + Double Age	119.2 110.9	39.5 37.3	57.8 56.1	41.7 16.1	49.2 17.2
1093°C/83 MPa	Partial Sol'n + Double Age	472.0 474.2	218.7 145.8	315.9 289.1	33.9 35.2	36.1 43.4
	As-cast + Double Age	643.9 673.9	357.7 360.2	462.1 495.5	33.0 25.4	37.0 40.0

Partial Solution: 1 hr./1238°C (2260°F) + 1 hr./1243°C (2270°F) + 1 hr./1249°C (2280°F) AC/GFC.
Double Age: 4 hrs./1975°F (1080°C) AC/GFC + 20 hrs./1600°F (871°C) AC.

Table 4 Comparison of stress-rupture life, hours.

Alloy	Heat Treat Condition	982°C/248 MPa	1038°C/172 MPa	1093°C/83 MPa
DS CM 247 LC	98% Soln. + Double Age	43	35	161
CMSX-3	98% Soln. + Double Age	80	104	1020
SX CM 186 LC	As-cast + Double Age	100	85	460
CMSX-681	As-cast + Double Age	113	--	528
CMSX-486	As-cast + Double Age	141	115	659

Double Age: 4 hours/1080°C (1975°F) AC/GFC + 20 hours/871°C (1600°F) AC.

the formation of solution heat treatment-induced recrystallisation (RX). Both of these factors have a positive impact on cost of manufacturing.

Consistent with the original development goals, longitudinal creep-rupture properties of CMSX-486 alloy demonstrated the intended improvement in rupture life compared to SX CM 186 LC alloy across a range of temperature/stress conditions (Table 4). In addition, CMSX-486 alloy is stronger than the first generation SX alloy CMSX-3® at 982°C (1800°F)

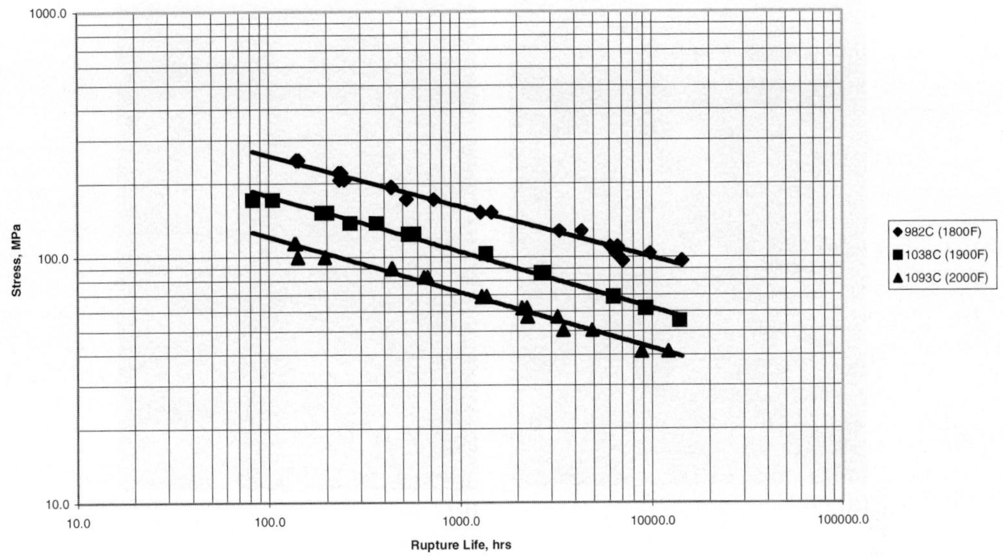

Fig. 6 CMSX-486 alloy stress-rupture at 982 EC, 1038 EC & 1093 EC.

and comparable at 1038°C (2000°F). The alternate composition CMSX-681 alloy showed lower stress-rupture life compared to CMSX-486 alloy at both the 982°C (1800°F) and 1093°C (2000°F) test conditions, possibly due to less favourable γ/γ' mismatch at high temperatures. The higher Ti content in CMSX-486 alloy compared to CMSX-681 will expand the lattice of the γ' probably giving a smaller γ/γ' misfit. There also is significantly higher Ta content in CMSX-681 (6.0%) compared to CMSX-486 alloy at 4.5%. As a result of the less favourable test results, the alternate CMSX-681 alloy chemistry was not pursued further in favour of the CMSX-486 alloy.

The results of long term creep-rupture property testing for CMSX-486 alloy using DL-10 test bars are shown in Figure 6, extending beyond 10,000 hours at each test temperature. It is apparent the desired linear relationship exists out to these time periods. This indicates adequate alloy phase stability in terms of both γ' and TCP phase formation. Optical and SEM microstructures of post-test stress-rupture specimens confirm the stability of CMSX-486 alloy (Figure 7). Some precipitation of acicular looking TCP phases (Re, W and Cr rich) are apparent, though not of sufficient volume fraction to de-alloy the material or nucleate creep cracking. Even so, the Cr content of CMSX-486 alloy has been slightly reduced in recent heats to further inhibit TCP phase formation during long term, stressed high temperature exposure.

To evaluate the potential accommodation for grain defects, bicrystalline slabs were cast using 'seeding' techniques to determine the mechanical properties of CMSX-486 alloy across LAB/HAB grain defects (Figure 8). Creep-rupture testing was undertaken over a matrix of temperature and stress conditions. This data indicates retention of baseline (defect-free)

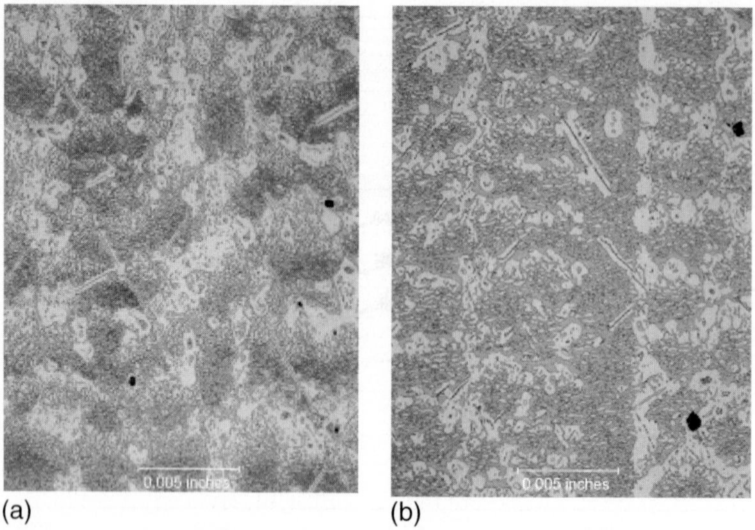

Fig. 7 CMSX-486 alloy post creep-rupture test specimens following (a) 9298 hours at 1038EC/62 MPa (1900EF/9 ksi) and (b) 8805 hours at 1093 EC/41 MPa (2000 EF/6 ksi).

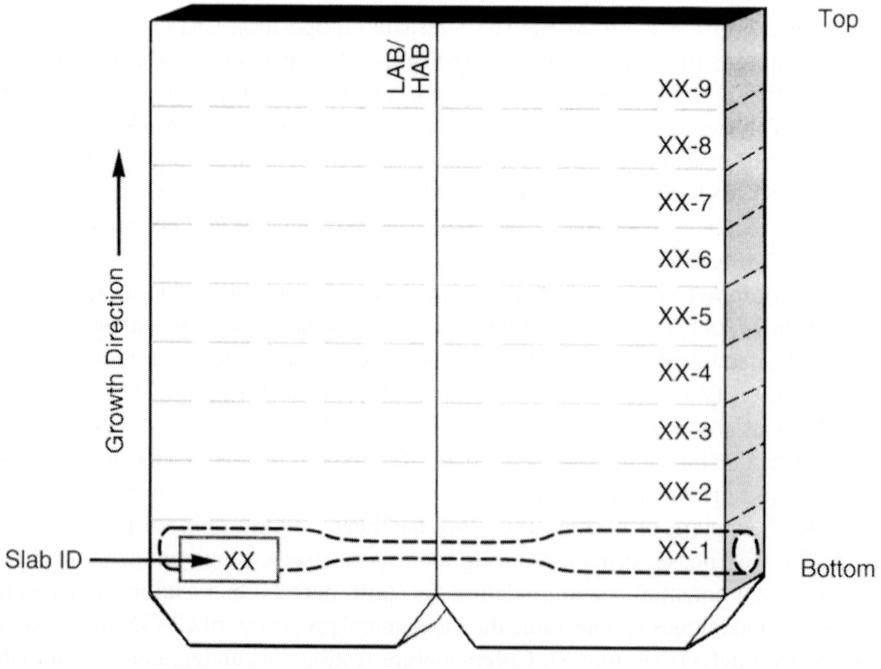

Fig. 8 Sketch of bi-crystal slab and specimen orientation.

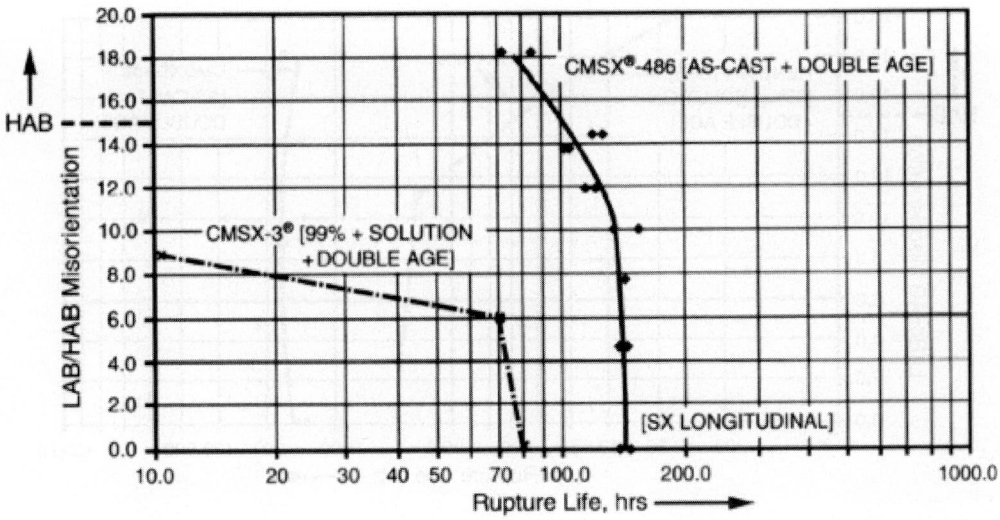

Fig. 9 CMSX-486 & CMSX-3 rupture life vs. LAB/HAB misorientation (982 EC/248 MPa [1800 EF/36.0 ksi]).

properties exceeding 15° degrees misorientation. Examples of rupture life plots vs. LAB/HAB misorientation are shown in Figures 9–11. In comparison to alternative SX alloys, the benefits of GBS alloy CMSX-486 are significant. At 982°C (1800°F), CMSX-486 alloy demonstrates a distinct advantage over CMSX-3 alloy, which contains no GBS elements (Figure 9). Although Rene' N4 (an SX alloy containing C & B) shows accommodation to 10° misorientation, indicating some tolerance for grain defects, CMSX-486 alloy with full grain boundary strengthening elements (C, B, Hf and Zr) shows no reduction in properties to 18° misorientation (Figure 10). As would be expected, the greatest fall-off in LAB/HAB properties occurs at 1093 EC (2000 EF) (Figure 11) due to degradation of grain boundary boride and carbide microstructure and Hf partitioning at this temperature over time. There is an expected decrease in stress-rupture ductility for LABs > 10 E. However, at 982 EC/207 MPa (1800 EF/30.0 ksi) test conditions with a HAB of 16 E, the rupture reduction in area (RA) for CMSX-486 alloy is approximately 8%, whereas René N4 is 0%. Actual grain defect acceptance criteria will be based on component-specific design.

Castability has been evaluated at multiple casting sources for a variety of configurations. A mould of RRC AE 3007 A1 engine HP2 vane segments (Figure 5) was cast at Rolls-Royce [SCFO] using the same SX casting parameters as the production alloy, CM 186 LC. The casting yield and quality results were assessed by SCFO as good and quite similar to the production alloy. Additionally three moulds of outer turbine seal segments were SX cast at a second casting source using a stable alumina prime coat as part of the shell composition. The data show excellent casting yield and quality results. It should be noted that with the relatively high Hf, B & C content for a SX alloy (more similar to a DS alloy), CMSX-486

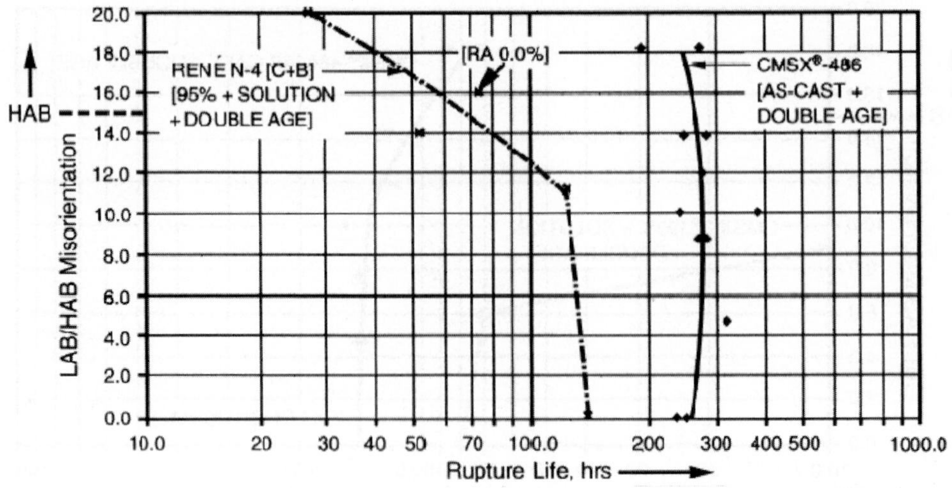

Fig. 10 CMSX-486 & Rene' N4 rupture life vs. LAB/HAB misorientation (982EC/207 MPa [1800EF/30.0 ksi]).

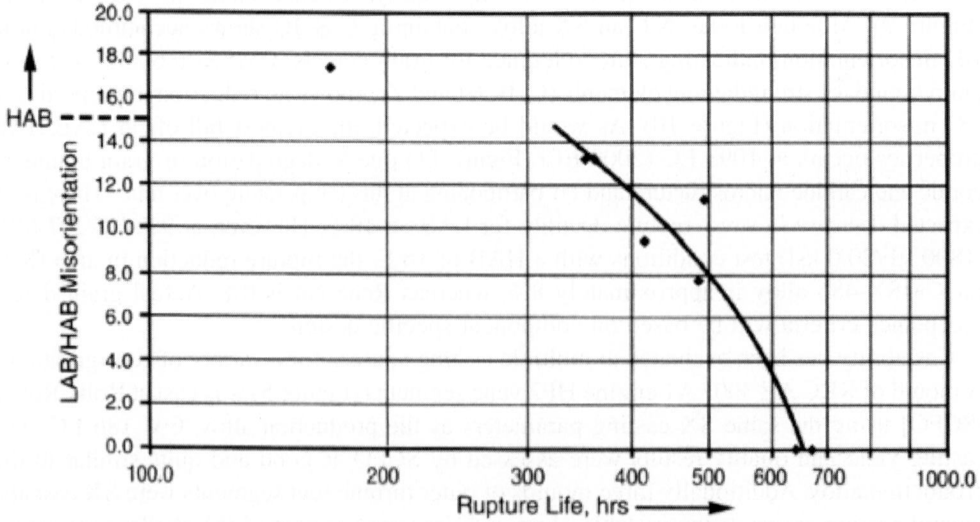

Fig. 11 CMSX-486 rupture life vs. LAB/HAB misorientation (1093 EC/83 MPa [2000 EF/12.0 ksi]).

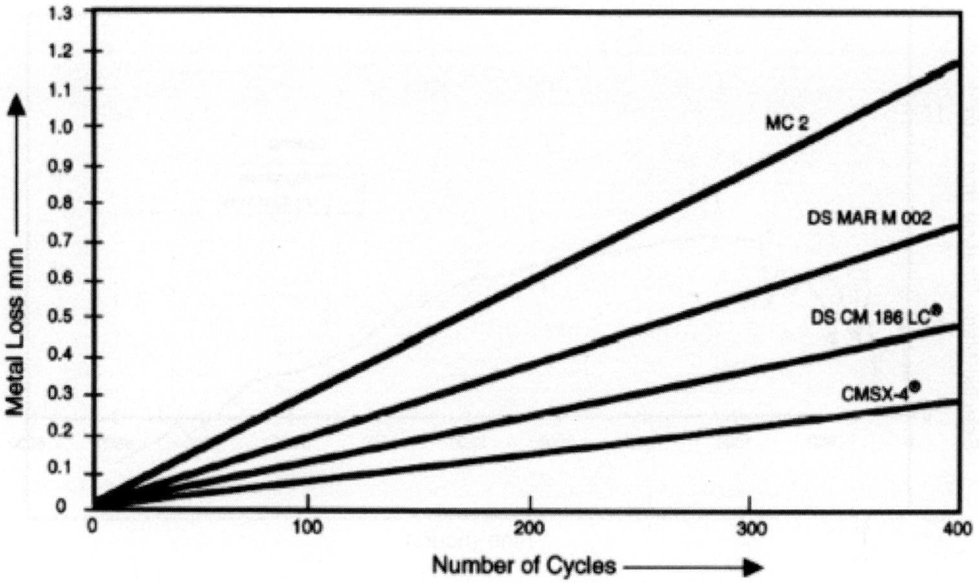

Fig. 12 Burner cyclic bare oxidation 1100 EC, 15 min. cycles, 0.25 ppm NaCl, mach 0.7 average data [courtesy RR].

may well require different casting parameters than those developed for CMSX-4 alloy, for example. A stable alumina prime coat approach is also advisable due to the 1.2% Hf content of the alloy.

DS columnar grain CM 186 LC alloy has been extensively burner rig tested for bare and coated oxidation and hot corrosion (sulphidation) as shown in Figures 12–14.[6] It is expected CMSX-486 alloy will show comparable performance, due to the similarities in alloy chemistry (Table 1).

SUMMARY

The attractive performance and properties of SX alloys have produced many new and challenging applications. However in addition to performance, affordability has become a major consideration for the introduction and expansion of new technology. Improved superalloys help to fit and fill these niches in a cost-effective manner, extending the application of SX casting technology.

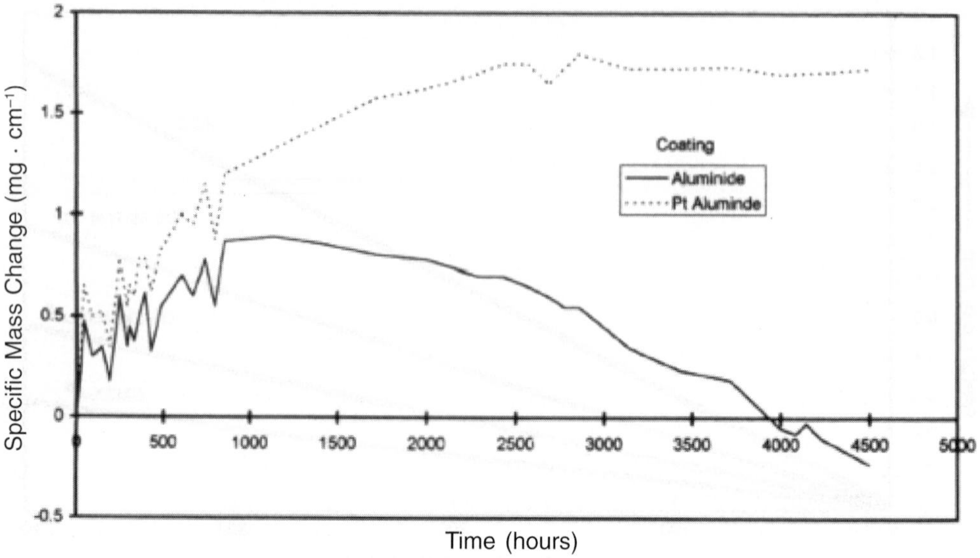

Fig. 13 Burner rig dynamic, cyclic oxidation of coated DS CM 186 LC alloy 1038EC, mach 0.45, JP-5 fuel, one cycle per hour [courtesy RR corp].

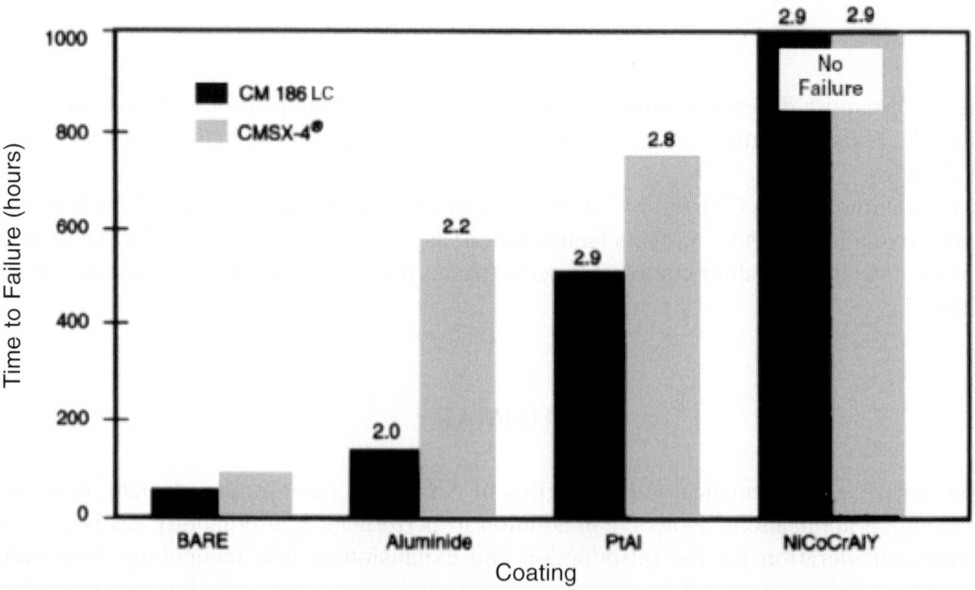

Fig. 14 Bare and coated CMSX-4 alloy and CM 186 LC alloy in type 1 hot corrosion testing (accelerated hot corrosion test 899 EC, 1% S in fuel, 10 ppm salt) [courtesy RR corp].

REFERENCES

1. P. S. Burkholder, M. C. Thomas, D. J. Frasier, J. R. Whetstone, (Allison) K. Harris, G. L. Erickson, S. L. Sikkenga and J. M. Eridon (CM): 'Allison Engine Testing CMSX-4 Single Crystal Blades and Vanes', *Proceedings of IOM Third International Charles Parsons Turbine Conference*, Newcastle Upon Tyne, UK., 1995.
2. M. C. Thomas, R. C. Helmink, D. J. Frasier, J. R. Whetstone, (Allison) K. Harris, G. L. Erickson, S. L. Sikkenga and J. M. Eridon (CM): 'Allison Manufacturing, Property and Turbine Engine Performance of CMSX-4® Single Crystal Airfoils', *Presented at the COST 501 Conference: Materials for Advanced Power Engineering 1994*", Liège, Belgium, 1994.
3. D. A. Ford, K. P. L. Fullagar, H. K. Bhangu (Rolls-Royce plc), M.C. Thomas, P.S. Burkholder, P.S. Korinko (RR Allison), K. Harris and J.B. Wahl (CM): 'Improved Performance Rhenium Containing Single Crystal Alloy Turbine Blades Utilizing PPM Levels of the Highly Reactive Elements Lanthanum and Yttrium', *ASME Journal of Engineering for Gas Turbines and Power*, **121**, 1999, 138–143.
4. P. S. Burkholder, M. C. Thomas, D. J. Frasier, J. R. Whetstone, (Allison) K. Harris, G. L. Erickson, S. L. Sikkenga and J. M. Eridon (CM): 'Allison Engine Testing CMSX-4 Single Crystal Blades and Vanes', *Proceedings of IOM Third International Charles Parsons Turbine Conference*, Newcastle Upon Tyne, UK., 1995.
5. K. Harris and J.B. Wahl: 'New Superalloy Concepts for Single Crystal Turbine Vanes and Blades', *Proceedings of IOM Fifth International Charles Parsons Turbine Conference*, Cambridge, UK., 2000.
6. P. S. Korinko, M. J. Barber and M. Thomas: 'Coating Characterisation and Evaluation of Directionally Solidified CM 186 LC and Single Crystal CMSX-4', Presented at ASME TURBO EXPO '96, Birmingham, UK., 1996.

The Effect of δ Precipitates on the Recrystallisation Behaviour of IN718

R. P. Guest and S. Tin
Department of Materials Science and Metallurgy
University of Cambridge, Pembroke Street
Cambridge, CB23QZ, UK

Experiments have been carried out in order to identify the dynamic and metadynamic recrystallisation of IN718 at 980°C and at a strain rate of $1s^{-1}$. Experiments conducted on both heat treated and as received samples have enabled the influence of δ precipitates on recrystallisation to be analysed. After compression to a true strain of 0.8, the samples had dynamically recrystallised to a volume fraction of approximately 5%. All samples were seen to have fully recrystallised after a 20 second post deformation hold at 980°C. The recrystallisation kinetics of the samples void of δ are shown to be faster than those containing δ. Grain growth, however, is seen to be slower in the samples without δ than those containing δ, although after approximately 1 minute at 980°C the δ particles pin the grain boundaries and stop grain growth.

INTRODUCTION

The recrystallisation behaviour of superalloys, and in particular IN718, has been the subject of research for many years. During thermomechanical processes, such as cogging and forging, microstructural refinement occurs due to the recrystallisation of new grains within the deformed matrix. This reduction in grain size results in enhanced fatigue crack growth resistance, and higher proof/yield strengths, two properties of paramount importance when producing turbine discs.

Recrystallisation in metallic alloys can progress as a result of two distinct mechanisms. One mechanism proposes that grains of low dislocation density nucleate and grow at the expense of highly strained grains with a higher dislocation density. In this case, the material is cleared of dislocations by the movement of high-angle grain boundaries; where high angle boundaries are defined as boundaries with a misorientation of more than 10°.[1] The second mechanism proposes that a network of sub-grains form due to the short-range migration of dislocations. As these subgrains grow and coalesce, the misorientation between them increases until large angle boundaries are produced, creating a recrystallised structure. This second type of recrystallisation is termed in-situ recrystallisation.[2]

Recrystallisation can occur during deformation, termed dynamic recrystallisation, or can occur after deformation whilst the material is still hot, designated metadynamic recrystallisation. Although the underlying mechanisms are the same for both types, the processes resulting in dynamic recrystallisation are complicated by the continual introduction of dislocations due to the ongoing deformation.

Table 1 Nominal composition of the IN718 used in the experiments (wt.%).

C	Si	Cr	Ni	Co	Fe	Mo	Nb+Ta	Ti	Al	B	Cu
0.033	0.08	17.74	53.80	0.19	Bal.	2.95	5.27	1.010	0.46	0.0029	0.02

The presence of precipitates in IN718 plays an important role during hot deformation. The presence of secondary particles will affect flow properties, grain growth, and recrystallisation kinetics. Before forging, billets typically undergo a furnace soak for a number of hours, designed primarily to heat the workpiece to a uniform temperature. During this soak, however, δ phase will almost always precipitate from the matrix. In addition, conventional forging temperatures for IN718 range from 850 to 1100°C, either side of the δ solvus temperature of approximately 1020°C. It is important, therefore, to analyse the recrystallisation kinetics of material both containing precipitates and free from precipitates.

The flow stress response of IN718 during hot deformation has often been correlated with recrystallisation kinetics. Studies have been carried out on specimens ranging from 5.6 mm diameter and 8.4 mm length[3] to those 90 mm in length.[4] Few of the existing investigations report similar results, although most authors report flow-softening during hot deformation. The occurrence of this characteristic response to deformation has been attributed to dynamic recrystallisation at both low,[5] and high strain rates.[6–9] Other investigations have suggested that softening is due to dynamic recovery below 975°C,[3] or at strain rates of $0.1 s^{-1}$ or less.[8] In addition to dynamic recrystallisation and recovery, flow softening by means of adiabatic heating may also occur during hot deformation.[3, 10–11]

The aim of this research is, therefore, to distinguish between the dynamic recrystallisation, metadynamic recrystallisation, and grain growth during the deformation of IN718. Detailed metallographic analysis of deformed specimens with and without δ precipitates has been performed. Results from these analyses are presented and the implications of findings are discussed.

EXPERIMENTAL METHODS

Cylindrical samples of height 12 mm and diameter 8 mm were machined by electrical discharge at a fixed radius from a forged and rolled IN718 billet of diameter 96 mm supplied by Firth Rixson Forging Ltd. The nominal composition of the material is shown in Table 1.

In order to investigate the influence of δ particles, compression tests were carried out on both forged and rolled samples and heat treated samples. Designed to initiate δ precipitation, the heat treatment consisted of 3 hours at 1000°C followed by an air quench.

Samples were compressed in a Gleeble 1500 thermomechanical testing machine equipped with a set of tungsten carbide dies of length 50 mm and diameter 20 mm. Tests were carried out at a temperature of 980°C, and at a strain rate of $1 s^{-1}$ to a true strain of 0.8. A thermocouple was welded to the mid-span of each test piece to provide a feedback loop for accurate temperature control. Graphite paper was used as lubrication between the specimen and dies, and to ensure a good electrical connection through the interface. Specimens were heated to

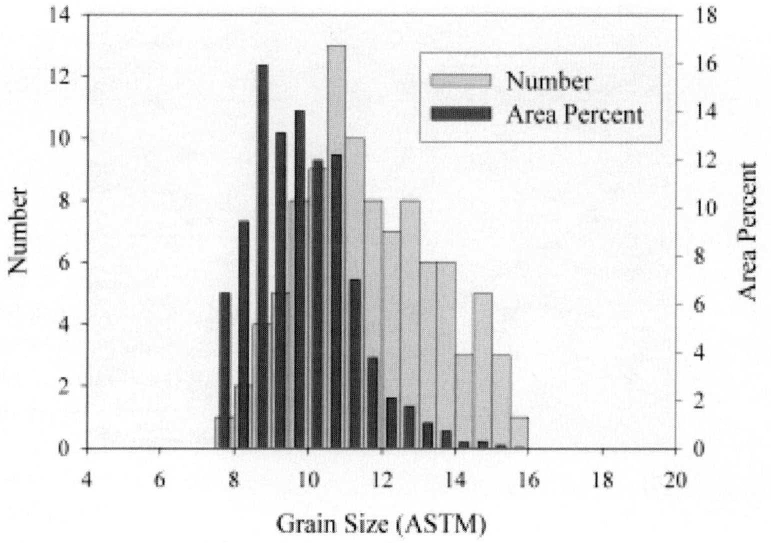

Fig. 1 The grain size distribution of the as-received material.

the test temperature at a fixed rate over a 2 minute period, and then held at 980°C for 1 minute. The samples were compressed and then held at the test temperature for times ranging between 0 seconds and 600 seconds. After the hold time, the samples were immediately water quenched whilst still held in the dies to minimise microstructural changes during cooling of the specimen.

Samples were cut longitudinally parallel to the applied stress direction, mounted, and then ground and polished using standard metallographic techniques. The samples were electrolytically etched using a 5% sulphuric acid solution with a voltage of 4.5 V and a current of 1 A.

Pictures of the centre point of each sample were taken using a Leica DM LM optical microscope. These pictures were analysed using KS 300 software to produce histograms of the grain sizes. Approximately 1000 grains were counted for each sample. The volume fraction recrystallised in each sample was calculated from the optical micrographs using a digital image analysis program.

RESULTS

Figure 1 shows the grain size distribution of the as-received, precipitate free material. Although the modal grain size is ASTM 11, the grain size occupying the highest fraction of the total area is approximately ASTM 9. This apparent shift in grain size is simply due to grains of lower ASTM grain size taking up more area than those with larger ASTM numbers.

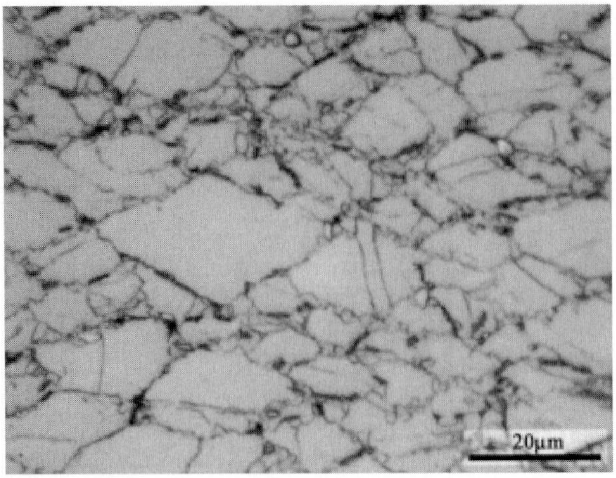

(a)

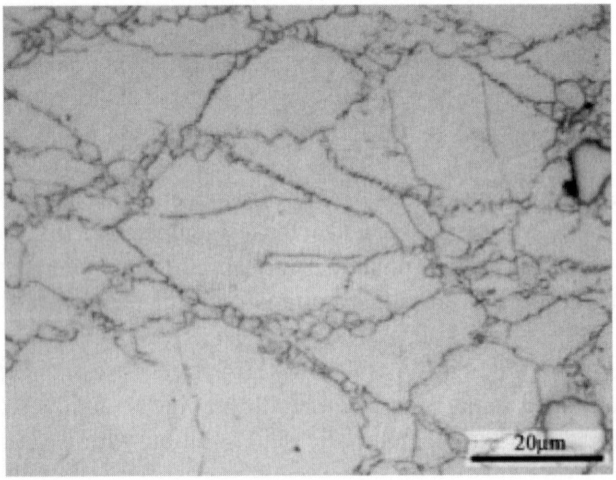

(b)

Fig. 2 Microstructures of (a) heat treated and (b) billet material immediately quenched after deformation at 980°C.

Figure 2 shows optical micrographs of the heat treated and billet material immediately quenched after deformation. The corresponding grain size distributions are shown in Figure 3. Both samples contain recrystallised grains within the microstructure. It should be noted, however, that the recrystallised grains in the billet sample (Figure 2b) were measured to be larger than those in the heat treated sample (Figure 2a). This effect can also be seen solely in the heat treated material as the recrystallised grains are larger on grain boundaries

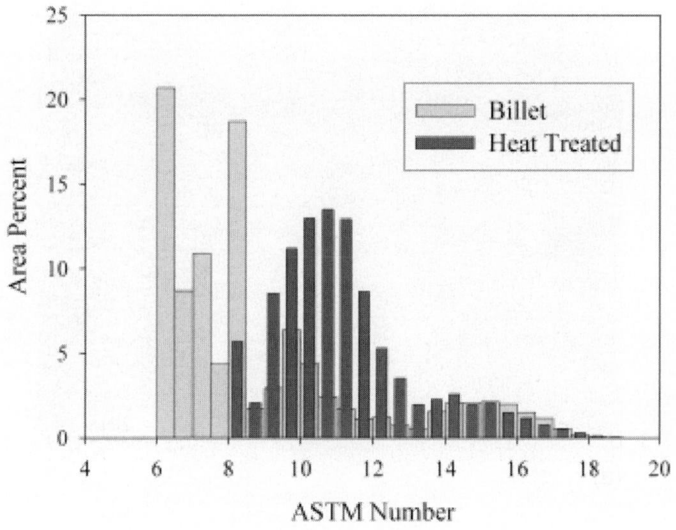

Fig. 3 The grain size distribution in samples of heat treated and non heat treated material after an immediate quench from 980°C.

without any δ precipitates. At regions surrounding the δ precipitates, the recrystallised grains are noticeably smaller. The heat treated material shows an average grain size only slightly smaller (ASTM 10-11) than that of the original billet material (ASTM 9-10), whereas the deformed billet material contains some extremely large grains (ASTM 6-7).

Figures 4 and 5 show the microstructures observed after a 5 second hold after deformation, and the grain size distribution calculated respectively. The heat treated sample has not fully recrystallised, showing a bimodal grain distribution. The deformed billet material, however, shows an almost fully recrystallised structure with an average grain size of ASTM 11.5. After the 5 second hold, the billet material contains a higher number of recrystallised grains than the heat treated material.

After a 20 second post-deformation hold, both billet and heat treated samples have fully recrystallised (Figure 6). The recrystallised grains in the billet sample, however, appear not to have grown to any significant degree whilst the sample has been fully recrystallised.

Figures 7 and 8 show the microstructure after a 30 second post deformation hold period. After 30 seconds, there are no differences in the grain size distributions between the samples, and the microstructures are almost identical, apart from the presence of δ phase in the heat treated specimen. During the next 30 seconds the structures coarsen, although it should be noted that the heat treated sample coarsens to a greater extent, with approximately 33% of the area of the heat treated sample covered with grains of size ASTM 7.5 or larger (Figure 9). Grain growth stops, however, in the heat treated sample soon after a 1 minute hold time, whereas grain growth continues in the billet material, resulting in a grain size of ASTM 5-6 (shown in Figures 10 and 11).

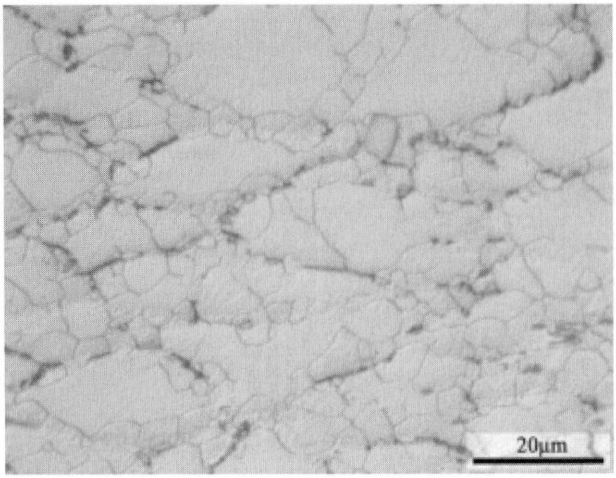

(a)

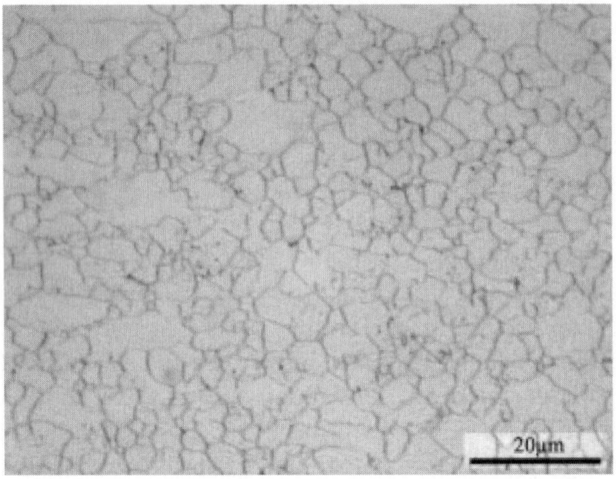

(b)

Fig. 4 Microstructures of (a) heat treated and (b) billet material quenched after a 5 second hold at 980°C following deformation at 980°C.

DISCUSSION

A major problem faced by many researchers is the task of isolating metadynamic recrystallisation from dynamic recrystallisation. This problem arises due to the fact that the characteristic flow stress response is not sufficient evidence to prove or disprove the existence of dynamic recrystallisation. Moreover, the time taken between the end of deformation and the quench is often too long to prevent the rapid microstructural changes from occurring.

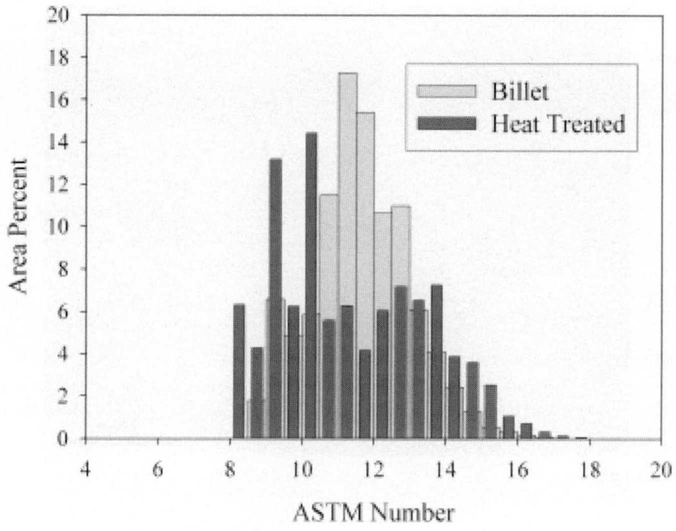

Fig. 5 The grain size distribution in samples of heat treated and non heat treated material after a 5 second hold at 980°C following deformation.

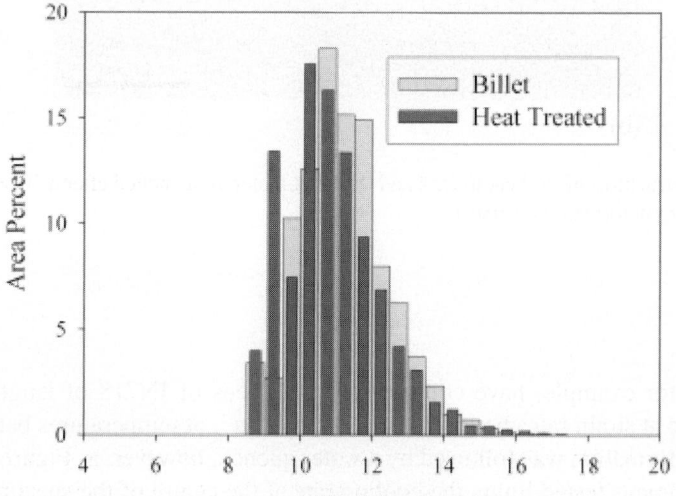

Fig. 6 The grain size distribution in samples of heat treated and non heat treated material after a 20 second hold at 980°C following deformation.

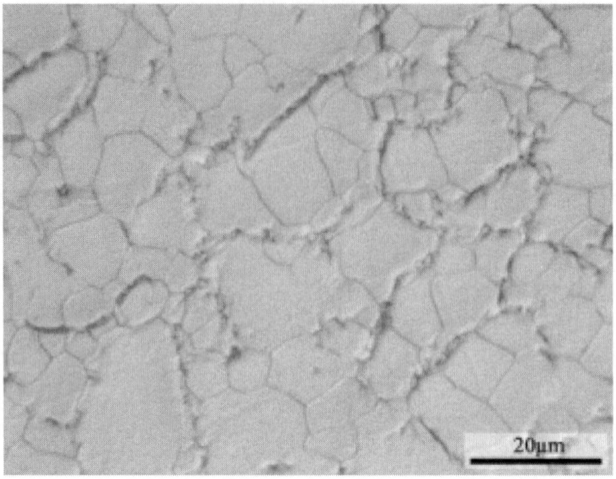

(a)

(b)

Fig. 7 Microstructures of (a) heat treated and (b) billet material quenched after a 30 second hold at 980°C following deformation at 980°C.

Zhang et al.,[12] for example, have compressed specimens of IN718 of length 30 mm and diameter 20 mm at strain rates between 0.001 s^{-1} and 1 s^{-1}, at temperatures between 960 and 1040°C. The deformation was followed by a water quench, however, as Figure 12 shows, the size of the specimens tested limits the cooling rate at the centre of the specimen. Assuming that the specimen was quenched whilst still in the jaws of the machine, the centre of the sample would stay at the deformation temperature for almost 2 seconds, and would still be at 600°C after 5 seconds when immersed in water. The cooling curve during the quench for

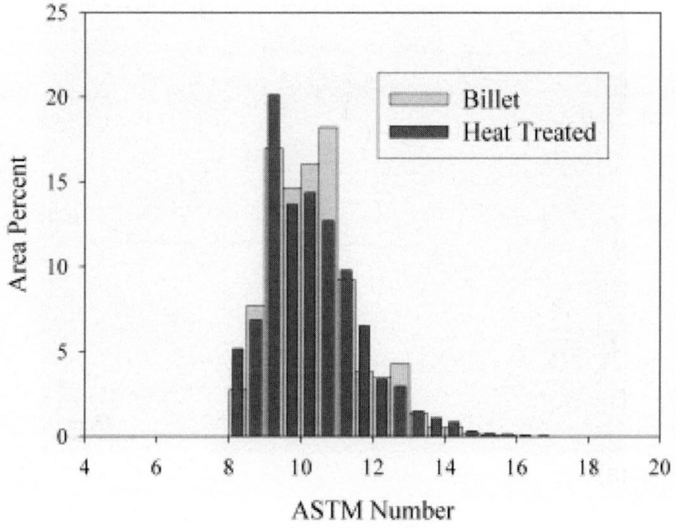

Fig. 8 The grain size distribution in samples of heat treated and non heat treated material after a 30 second hold at 980°C following deformation.

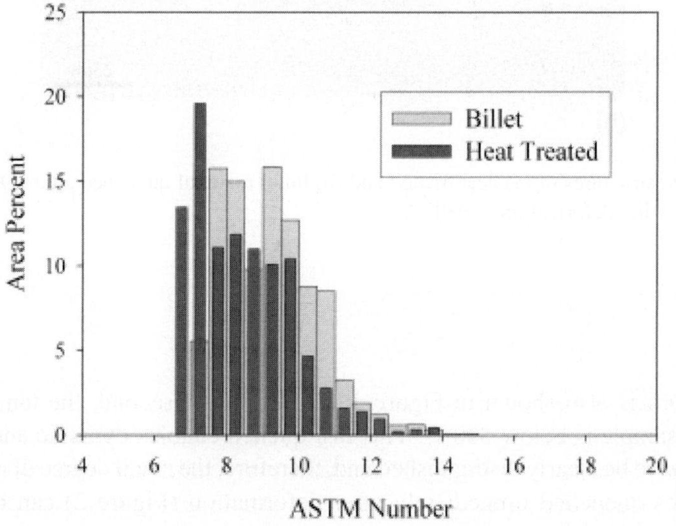

Fig. 9 The grain size distribution in samples of heat treated and non heat treated material after a 1 minute hold at 980°C following deformation.

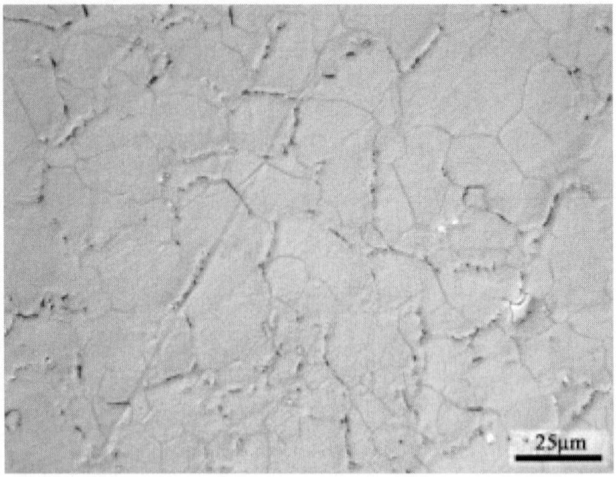

(a)

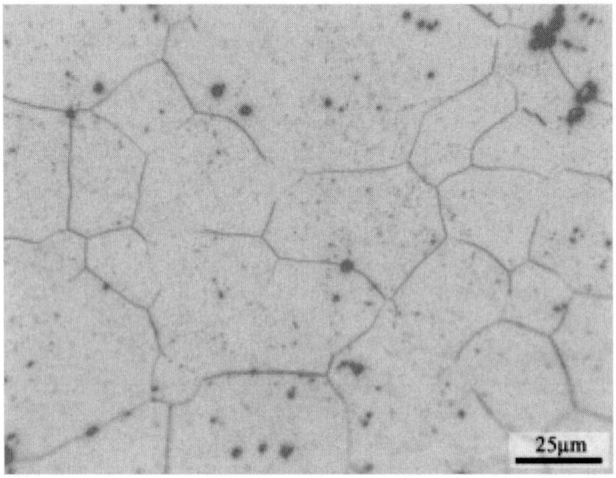

(b)

Fig. 10 Microstructures of (a) heat treated and (b) billet material quenched after a 10 minute hold at 980°C following deformation at 980°C.

the current work is also shown in Figure 12. After only 1 second, the temperature in the middle of the sample is below 800°C. This fast quench enables dynamic and metadynamic recrystallisation to be clearly distinguished and, therefore, the small degree of recrystallisation seen in samples quenched immediately after deformation (Figure 2) can be attributed to dynamic recrystallisation.

Figure 2 shows that considerable grain growth has occurred during the heat-up and soak at 980°C prior to deformation. This is not unexpected since grain growth in IN718 is extremely

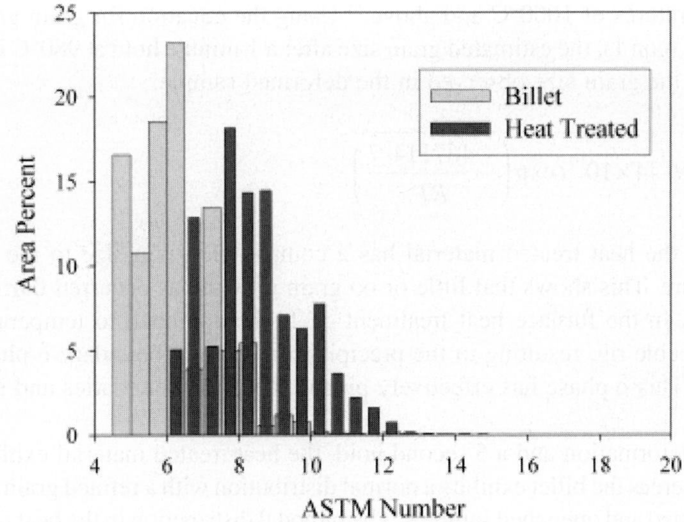

Fig. 11 The grain size distribution in samples of heat treated and non heat treated material after a 10 minute hold at 980°C following deformation.

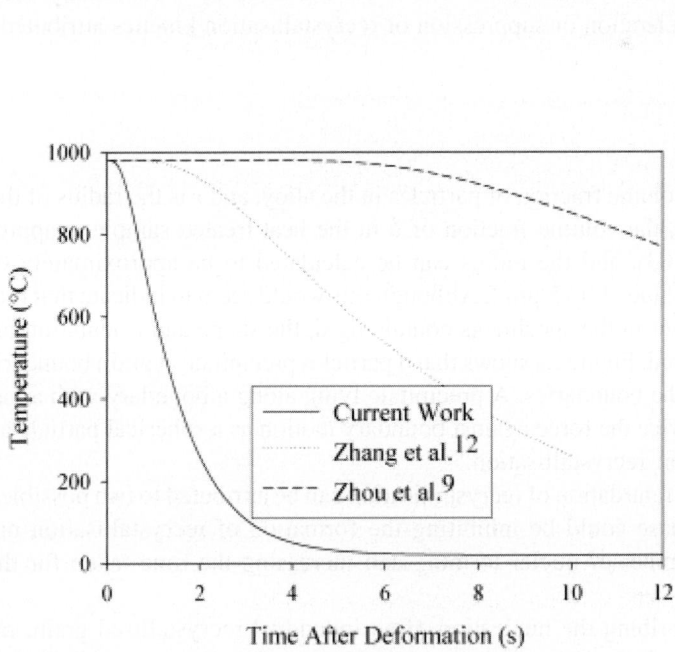

Fig. 12 Calculated cooling curves for the current work and selected other studies.

rapid at temperatures of 1000°C and above.[13] Using the equation for grain growth reported by Huang et al. (eqn 1), the estimated grain size after a 1 minute hold at 980°C is ASTM 6.8,[13] approximately the grain size observed in the deformed sample.

$$d = \sqrt{d_0^{\,2} + 9 \cdot 44 \times 10^{19} t \exp\left(\frac{-467114 \cdot 7}{RT} \right)} \tag{1}$$

In contrast, the heat treated material has a comparable grain size to the original billet starting structure. This shows that little or no grain growth has occurred during the 3 hours heat treatment. In the furnace heat treatment cycle the approach to temperature is slower than in the Gleeble rig, resulting in the precipitation of grain boundary δ phase during the heating cycle. This δ phase has effectively pinned the grain boundaries and inhibited grain growth.

Following deformation and a 5 second hold, the heat treated material exhibits a bimodal distribution whereas the billet exhibits a normal distribution with a refined grain size compared to the as-deformed and quenched samples. The bimodal distribution in the heat treated material is accounted for by a partial recrystallisation of the microstructure. This is in contrast to the billet material where the microstructure has fully recrystallised. After deformation and a 20 second hold both the billet and heat treated material have fully recrystallised, with a mean grain size of ASTM 11.

From the above it is evident that recrystallisation is impeded in the heat treated material.

Secondary phase particles can accelerate, slow, or even inhibit recrystallisation.[1] In alloys systems such as IN718, the presence of particles larger than 1 μm which are widely spaced tend to accelerate recrystallisation.[1] The following criterion is commonly used to delineate between the acceleration or suppression of recrystallisation kinetics attributed to secondary phases.

$$\frac{F_v}{r} > 0 \cdot 2\ \mu m^{-1} \tag{2}$$

where F_v is the volume fraction of particles in the alloy, and r is the radius of the particles. In the present case, the volume fraction of δ in the heat treated sample is approximately 3% (a fraction of 0.03), and the radius can be calculated to be approximately 0.2 μm.[14] This would lead to a value of 0.15 μm^{-1}. Although this would seem to indicate that recrystallisation would occur faster in the specimens containing d, the shape and orientation of the particles must be considered. Figure 2a shows that d particles precipitate at grain boundaries, extending as plates along the boundaries. A precipitate lying along a boundary with an aspect ratio of 4:1 will apply twice the force against boundary motion as a spherical particle and, therefore, potentially inhibit recrystallisation.[15]

The observed retardation of recrystallisation can be attributed to two possible mechanisms. Firstly, the δ phase could be inhibiting the formation of recrystallisation nuclei, thereby reducing the number of nuclei forming and increasing the time taken for these nuclei to reach a specific size.

A model describing the nucleation of an individual recrystallised grain, rather than the recrystallisation of an entire system, has been proposed by Davies et al.[16] This 'ledge model', assumes a stepped grain boundary in a deformed structure, shown in Figure 13a, where both

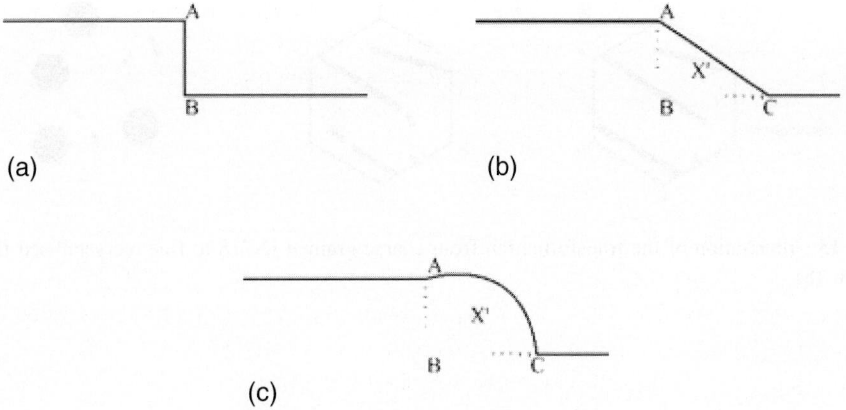

Fig. 13 Illustration of the ledge model for the initiation of recrystallisation (adapted from 17).

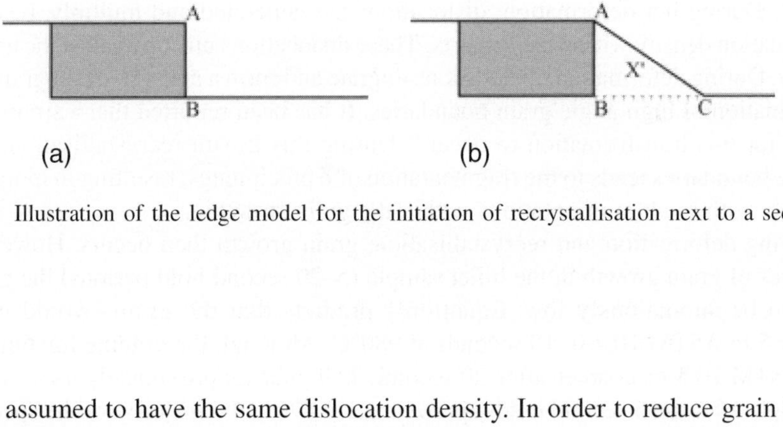

Fig. 14 Illustration of the ledge model for the initiation of recrystallisation next to a secondary particle.

grains are assumed to have the same dislocation density. In order to reduce grain boundary area, line AB migrates to AC, leaving behind two low angle grain boundaries (AB and BC), and a dislocation free region, X'. This boundary migration is thermodynamically favourable as the length of high-angle boundary has decreased from AB + BC to AC. The formation of the two low-angle grain boundaries does not require enough energy to make the migration unfavourable. The difference in dislocation density now acts as a driving force for recrystallisation, and the line AC can 'bulge' into the unrecrystallised grain. If the same process is repeated with a precipitate (shown in Figure 14), line AB cannot simply migrate to form line AC. It must remain in place separating the precipitate from the matrix, therefore, making the transformation thermodynamically unfavourable and inhibiting recrystallisation.

Conversely, a second mechanism has been proposed whereby the δ phase acts as a nucleation site for recrystallisation, and the density of nuclei may be many times greater

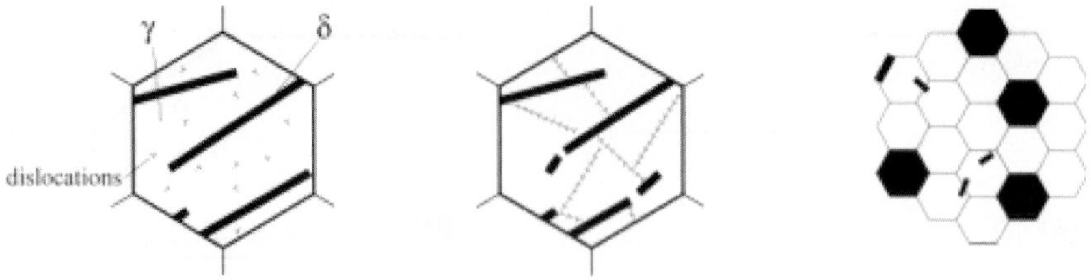

Fig. 15 Illustration of the transformation from coarse grained IN718 to fine recrystallised IN718 (from 18).

around δ particles than on precipitate free grain boundaries. This increase in nuclei density would result in increased competition between recrystallised grains, and a transition from grain boundary migration due to recrystallisation to migration due to grain boundary curvature (grain growth). Boundary migration due to curvature is much slower than that due to dislocation density.

Valitov et al.[18] have proposed a model describing the formation of a large number of recrystallised grains between broken up δ particles. This is shown diagrammatically in Figure 15. During hot deformation, dislocations are generated and multiply, resulting in a high dislocation density within the γ matrix. These dislocations pile up against the undeformed δ particles. During deformation, dislocations migrate and form a network of subgrains, leading to the formation of high angle grain boundaries. It has been reported that a strain of 50% is sufficient for this transformation to occur.[18] During this in-situ recrystallisation, sliding at interphase boundaries leads to the fragmentation of δ precipitates, resulting in more spherical precipitates with random orientations within the grain structure.

Following deformation and recrystallisation, grain growth then occurs. However, in the early stages of grain growth in the billet sample (5–20 second hold periods) the growth rate appears to be anomalously low. Equation 1 predicts that the grains would grow from ASTM 11.5 to ASTM 10.4 in 15 seconds at 980°C. Although the volume fraction of grains of size ASTM 10.5 or coarser after 20 seconds is higher (approximately 29%) than in the billet sample after the 5 second hold (approximately 19%), the majority of the specimen still contains grains in the range ASTM 10.5-12 (44% after 5 seconds and 48% after 20 seconds). This discrepancy can be attributed to one of two factors. Firstly, although the micrograph showing the 5 second hold after deformation for the billet (Figure 4b) shows what seems to be an almost fully recrystallised structure; the structure may not actually be fully recrystallised. A state may exist where the recrystallised grains are the same size as the unrecrystallised ones. The system will, therefore, not progress from recrystallisation to grain growth after a 5 second hold as predicted, but will continue recrystallising. Secondly, and more likely, as the Huang model is an empirically based model, a time step of 15 seconds may well be too short for experimental error not to affect the results substantially.

As demonstrated by the grain distributions for the 30 second hold time, and the 1 minute post deformation hold (Figures 8 and 9 respectively), grain coarsening is more pronounced in the heat treated samples. This accelerated grain growth can potentially be attributed to the increased boundary mobility due to the δ precipitates decreasing the level of niobium from the γ matrix.

For hold times longer than 1 minute, grains in the heat treated samples have been pinned by the δ precipitates and have not grown larger than ASTM 6, with the majority of grains still at approximately ASTM 8 (the grain size prior to deformation). A volume fraction of 4% δ will stop grain growth, and at 985°C only 2.5% is required to dramatically decrease grain growth.[19] After a 10 minute hold, the billet sample, however, without any particles present to inhibit grain growth has coarsened to a grain size of between ASTM 4.5 and ASTM 6.5, very close to that predicted by Equation 1 during the 595 seconds since sample fully recrystallised.

These experiments show that within the microstructure, post-deformation exposures to high temperatures do not play a major role in controlling the grain size. Although the grain size present just after the completion of recrystallisation will not be retained, the structure will not be comprised of grains bigger than those of the initial microstructure. If, however, δ precipitation could be initiated during the first 20 seconds after deformation (at 980°C and strain rate of $1s^{-1}$) the fine recrystallised structure could be retained. Further experiments are being carried out in order to investigate the effect of temperature and strain rate on the recrystallisation time.

CONCLUSIONS

The quench rate at the interior of specimens is extremely important when analysing the recrystallisation kinetics during and after hot deformation. With the current sample dimensions and quench procedure, cooling rates of 500°C/s for the first second after deformation have been observed. Short hold times after deformation have enabled the speed of metadynamic recrystallisation to be analysed.

Samples containing δ precipitates have been shown to exhibit slower recrystallisation, and to a certain degree accelerated grain growth when compared to samples void of precipitates. The accelerated grain growth exhibited by the δ containing samples stops once the original (pre-recrystallised) grain size is attained due to pinning of grain boundaries by the precipitates.

Results from this study are consistent with the grain growth relationship for IN718 proposed by Huang et al.[13] at long times. At shorter times, however, the recrystallisation kinetics play a major role in governing microstructural changes.

ACKNOWLEDGEMENTS

Financial support for this work in the form of an industrial CASE Award between EPSRC and Firth Rixson plc is gratefully acknowledged. The authors would also like to thank Professor D. J. Fray for the provision of research facilities and Dr A. Partridge at Firth Rixson Forgings Ltd. for the provision of materials.

REFERENCES

1. F. J. Humphreys and M. Hatherly: *Recrystallisation and Related Annealing Phenomena*, Pergamon, 1995.
2. H. P. Stüwe: 'Do Metals Recrystallise During Hot Working?', *Deformation Under Hot Working Conditions*, C. M. Sellars and W. J. McG. Tegart, eds., The Iron and Steel Institute, 1968, 1–6.
3. A. A. Guimaraes and J. J. Jonas: 'Recrystallisation and Aging Effects Associated with the High Temperature Deformation of Waspaloy and Inconel 718', *Metallurgical and Materials Transactions A*, **12**, 1981, 1655–1666.
4. W. Horvath, W. Zechner, J. Tockner, M. Berchthaler, G. Weber and E. A. Werner: 'The Effectiveness of Direct Aging on Inconel 718 Forgings Produced at High Strain Rates as Obtained on a Screw Press', *Superalloys 718, 625, 706 and Various Derivatives*, TMS, Warrendale, 2001, 223–228.
5. S. C. Medeiros, Y. V. R. K. Prasad, W. G. Frazier and R. Srinivasan: 'Microstructural Modelling of Metadynamic Recrystallisation in Hot Working of IN 718 Superalloy' *Materials Science and Engineering A*, **293**, 2000, 198–207.
6. S. -H. Cho, K. -B. Kang and J. J. Jonas: 'The Dynamic, Static and Metadynamic Recrystallisation of a Nb-Microalloyed Steel', *ISIJ International*, **41**(1), 2001, 63–69.
7. N. K. Park, I. S. Kim, Y. S. Na and J. T. Yeom: 'Hot Forging of a Nickel Base Superalloy', *Journal of Materials Processing Technology*, **111**, 2001, 98–102.
8. C. A. Dandre, R. W. Evans, R. C. Reed and S. M. Roberts: 'Exploring the Microstructural Evolution of Inconel 718 During Ingot Breakdown: Optimising the Cogging Process', *Compass – International Conference*, 1999, 89–96.
9. L. X. Zhou and T. N. Baker: 'Effects of Strain Rate and Temperature on Deformation Behaviour on IN718 During High Temperature Deformation', *Materials Science and Engineering A*, **177**, 1994, 1–9.
10. C. A. Dandre, S. M. Roberts, R. W. Evans and R. C. Reed: 'Microstructural Evolution of Inconel 718 During Ingot Breakdown: Process Modelling and Validation', *Materials Science and Technology*, **16**(1), 2000, 14–25.
11. M. C. Mataya, E. E. Nilsson and G. Krauss: 'Comparison of Single and Multiple Pass Compression Tests Used to Simulate Microstructural Evolution During Hot Working of Alloys 718 and 304L', *Superalloys 718, 625, 706 and Various Derivatives*, TMS, Warrendale, 1994, 331–343.
12. J. M. Zhang, Z. Y. Gao, J. Y. Zhuang and Z. Y. Zhong: 'Mathematical Modelling of the Hot-Deformation Behaviour of Superalloy IN718', *Metallurgical and Materials Transactions A*, **30**, 1999, 2701–2712.
13. D. Huang, W. T. Wu, D. Lambert and S. L. Semiatin: 'Computer Simulation of Microstructure Evolution During Hot Forging of Waspaloy and Nickel Alloy 718', *Microstructure Modelling and Prediction During Thermomechanical Processing*, R. Srinivasan, ed., TMS, Warrendale, 2001, 137–146.
14. G. Muralidharan and R. G. Thompson: 'Effect of Second Phase Precipitation on Limiting Grain Growth in Alloy 718', *Scripta Materialia*, **36**(7), 1997, 755–761.

15. W. -B. Li and K. E. Easterling: 'The Influence of Particle Shape on Zener Drag', *Acta Metallurgica et Materialia*, **38**(6), 1990, 1045–1052.
16. P. W. Davies, A. P. Greenough and B. Wilshire: 'The Ledge Theory of Recrystallisation in Polycrystalline Metals', *Philosophical Magazine*, **6**, 1961, 795–799.
17. J. W. Christian: *The Theory of Transformations in Metals and Alloys*, Pergamon, 2002, 837–851.
18. V. A. Valitov, B. B. Bewlay, O. A. Kaibyshev, Sh. Kh. Mukhtarov, C. U. Hardwicke and M. F. X. Gigliotti: 'Production of Large Scale Microcrystalline Forgings for Roll Foring of Axially Symmetric Alloy 718 Components' *Superalloys 718, 625, 706 and Various Derivatives*, TMS, Warrendale, 2001, 301–311.
19. Y. Desvallées, M. Bouzidi, F. Bois and N. Beaude: 'Delta Phase in Inconel 718: Mechanical Properties and Forging Process Requirements' *Superalloys 718, 625, 706 and Various Derivatives*, TMS, Warrendale, 1994, 281–291.

Microstructural Effects of Boron Additions on Powder Processed Ni-Base Superalloys

R. J. Mitchell and S. Tin

Rolls-Royce University Technology Centre for Nickel-Base Superalloys
Department of Materials Science and Metallurgy
University of Cambridge
Pembroke Street, Cambridge
CB2 3QZ, UK

ABSTRACT

Continuing requirements for increased fuel efficiency in modern gas turbines have led to significant increases in turbine entry temperature (TET). Ultimately, both engine power to weight ratio and efficiency increase with operating temperature. Recent efforts have focused on the physical and mechanical properties of powder processed Ni-base superalloys for disc applications. Powder processing of alloys, such as Udimet 720 Li, René 88DT and RR1000 minimises processing defects associated with segregation during solidification. Consequently, the high temperature strength of these alloys can be enhanced through the accommodation of elevated levels of refractory elements. In order for these alloys to meet the requirements for turbine disc applications, the microstructures of these alloys must be carefully controlled during processing. Studies have shown that minor additions of C, B, Zr and Hf, for example can significantly influence the microstructural response of these powder processed materials at elevated temperature. The effect of boron on the microstructure of two nominally identical experimental alloys was investigated. Mechanisms by which boron influences the microstructure are discussed.

INTRODUCTION

The in-service requirements of turbine discs, create a unique range of parameters that need to be met for successful operation. Turbine discs are subjected to a centrifugal force associated with their rotation and furthermore experience a temperature gradient, in both the axial (through thickness) and radial (bore to rim) directions. In order to meet these wide-ranging requirements, processing routes are designed to yield an engineered microstructure. To maximise the yield and fatigue properties, it is desirable to have a fine grain size at the disc bore, where the stress experienced is highest. At the rim of the disc, where metal temperatures are highest, a coarse grained structure with few grain boundaries is preferred to optimise creep resistance.

The dominant mechanism responsible for high temperature strength in nickel-base superalloys, is dislocation pinning associated with large volume fractions of ordered $L1_2$ precipitates, γ'. These coherent precipitates, based on the intermetallic $Ni_3(Al, Ti)$ system, form in various morphologies as a direct result of the heat treatment and quench rate experienced.[1]

To optimise the microstructure for in-service conditions, it is typical for nickel-base superalloys to undergo a two stage heat treatment aimed at optimising the γ' size, morphology and distribution. Many studies[2–4] have investigated the relationships that exist between heat treatments and microstructure, with particular attention being paid to the γ' solvus temperature in order to provide the best trade off between mechanical properties. Control of grain growth during thermomechanical processing is particularly important when processing these alloys at, or above, the γ' solvus temperature. Below the γ' solvus temperature, primary grains of γ' inhibit the motion of grain boundaries within the microstructure. Exceeding the γ' solvus temperature dissolves the primary γ' (typically 0.5–2 μm) and enables non-uniform growth of grains.[4] The increase of subsolvus solutioning temperature, decreases the volume fraction of primary γ'. Upon cooling of partially solutioned specimens, the secondary and tertiary γ' size increase to maintain the overall volume fraction. When the solution stage is performed above the γ' solvus, (such as in Rene'88DT) the Zener pinning effect associated with the grains of primary g¢ is lost and grain growth is restricted by carbides and borides at the grain boundaries.[5, 6] Since the high-temperature strength of these materials is largely dependent on maximising the volume fraction of secondary and tertiary γ', supersolvus heat-treatments are desirable.

The rate of cooling from the solution temperature, will initially determine the size, morphology and distribution of γ' precipitates. If the quench rate from the solution temperature is sufficiently high (>130°C/min), then a high density of fine spherical γ' will be precipitated, referred to as secondary γ'. If the quench rate is less severe then both secondary and tertiary γ' form. Coarsening of tertiary γ' occurs after subsequent ageing treatments. However, to minimise the residual stresses associated with quenching large components such as discs, the thermal history of these materials are carefully controlled.

MATERIALS AND METHODS

The work detailed here is based upon two experimental alloys, UC01 and UC02. These alloys are nominally identical in composition with the exception of a slight increase in boron for alloy UC02. The principal aim of this work is to try to ascertain the effect of boron additions on grain growth around the γ' solvus temperature.

Both of the alloys studied were produced via the powder processing route before being extruded into 31 mm diameter rods. The powder products were supplied by Special Metals Inc. and extruded by Wyman-Gordon Ltd. The virgin stock was vacuum induction melted (VIM'd) before being argon atomized (AA) and sieved. The resulting material was placed in mild steel cans with a diameter of 73 mm. This was reduced, with a 5.5:1 area ratio reduction at 83 mms^{-1} and 1110°C in an inert atmosphere, followed by an air cool.

A Netzsch DSC404C High Temperature Differential Scanning Calorimeter was used to ascertain the key solvus temperatures, principally γ' and carbide/boride phases. For each of

Table 1 Composition of RR10007 and experimental disc alloys.[8]

Element (wt.%)	RR1000	Alloy UC01	Alloy UC02
Ni	bal.	bal.	bal.
Cr	15.0	15.0	15.0
Co	18.5	17.0	17.0
Mo	5.0	4.0	4.0
Al	3.0	3.1	3.1
Ti	3.6	4.4	4.4
Ta	2.0	2.5	2.5
Hf	0.5	-	-
Zr	0.06	0.06	0.06
C	0.03	0.045	0.045
B	0.02	0.02	0.035

the alloys a disc of 5.25 mm diameter and 1 mm thickness, with a mass of approximately 200–300 mg was prepared. These dimensions were chosen to best match the internal dimensions of the alumina specimen holders used. In addition, this specimen geometry allowed a Sapphire calibration standard of similar mass to be used. Initially the machine was run with the calibration standard and a series of pure metals, of known response, for the chosen heating rate (10ºC/min) and gas flow rate (50 ml/min argon). The process was then repeated for both alloys with the specimens weighing 200–300 mg, using the above parameters over the temperature range 0–1500ºC. The solvus temperatures were evaluated using Netzsch Proteus software.

Initial studies concentrated on investigating the microstructural response to temperature and time in the as-extruded condition. This work was undertaken by exposing small samples of the alloys (typically 5 mm cube), for one hour, in calibrated Carbolite chest furnaces over the temperature range 1100 to 1200°C in 10°C intervals. Following exposure at temperature, the specimens were removed from the furnace and cooled under forced air convection. Specimens were then prepared using standard metallographic techniques. In addition, further studies were undertaken for durations of 2, 4, 8, 16 and 24 hours at three temperatures to encompass the sub, near (defined as within 5°C of the γ' solvus temperature) and supersolvus conditions.

Of prime importance in these studies were the grain size (measured using the Heyn intercept method) and the size and volume fraction of primary g¢. It is also important to take account of variations that exist in grain size across a typical cross-section and to this end measurements were also made of the maximum and minimum grain sizes, from each series of observations. Imaging was carried out on both a JEOL 820 SEM and JEOL 6340 FEGSEM.

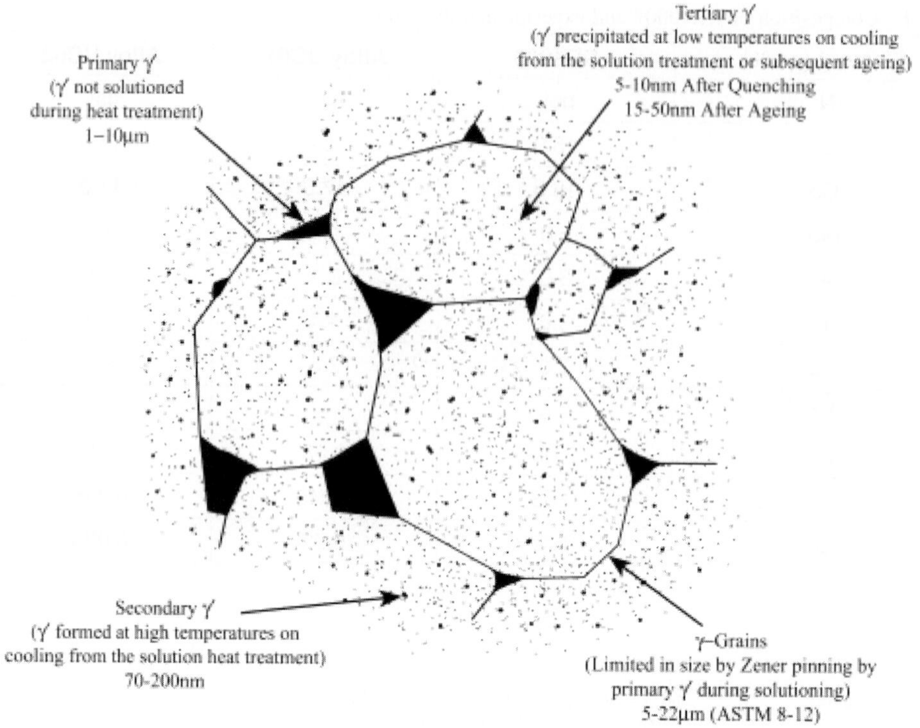

Fig. 1 The major components of a heat-treated polycrystalline superalloy.[4]

RESULTS

In order to accurately measure the very fine grain sizes associated with powder processed material, electron backscatter diffraction (EBSD) was used. This was coupled with orientation imaging microscopy (OIM) to provide accurate information on grain size as well as providing information on crystallographic orientation. Observation of the as-extruded alloys, reveals a fully recrystallized microstructure. A fine grain size (typically 3–7 μm) with blocky primary γ' situated on the grain boundaries and fine intragranular secondary γ' is typical of both alloys, Figure 2. There exists a random distribution of carbide and boride phases visible across the cross-sectioned microstructure. In the as-extruded condition primary γ' sizes are in the range 0.75–1.25 μm for both alloys. The as-extruded microstructure and hardness (~ 420 VHN) were found to be uniform throughout the extrusion length and cross-section.

The grain growth response for both of the experimental alloys is illustrated for sub and supersolvus conditions in Figure 3. Both experimental alloys display a substantial increase in the γ' solvus temperature over the RR1000 alloy. The solvus temperature in these two alloys was determined from the DSC analyses. The difference in boron between these two alloys did not influence the solidus, liquidus or γ' solvus temperature. In both alloys the

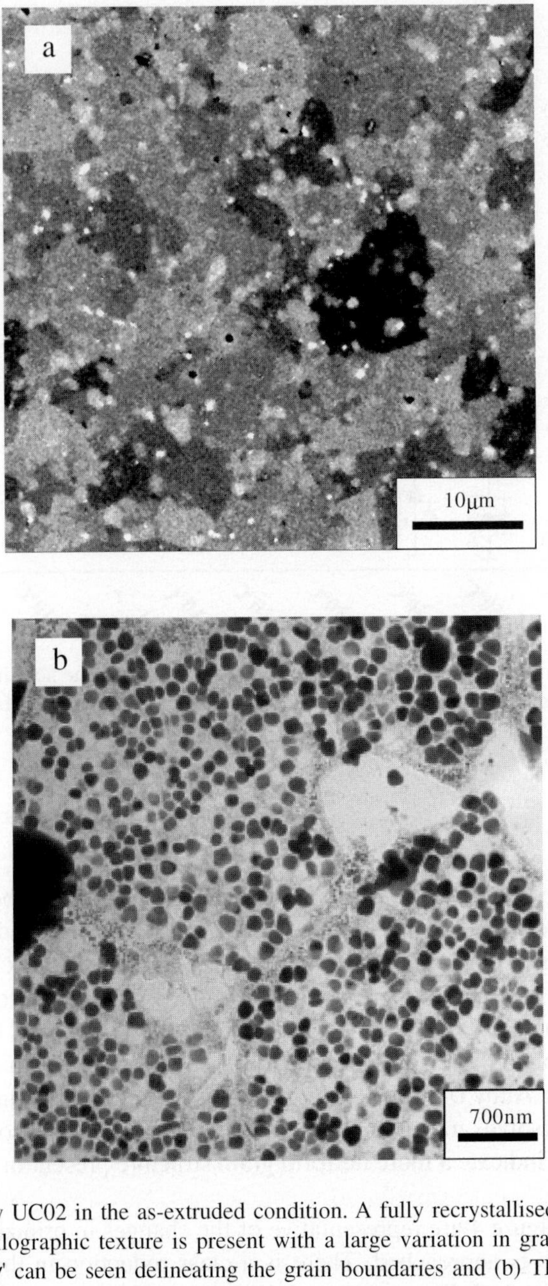

Fig. 2 (a) alloy UC02 in the as-extruded condition. A fully recrystallised microstructure with no preferred crystallographic texture is present with a large variation in grain size about a mean of 5 μm. Primary γ' can be seen delineating the grain boundaries and (b) TEM micrograph showing distribution of γ' in as received alloy UC01. There are three types of γ' precipitates present. The large primary γ' present on the grain boundaries are seen as dense black areas or outlines. There is a uniform distribution of secondary γ' (approximately 70–120 nm) and fine tertiary γ' (1–15 nm), visible on the grain boundaries.

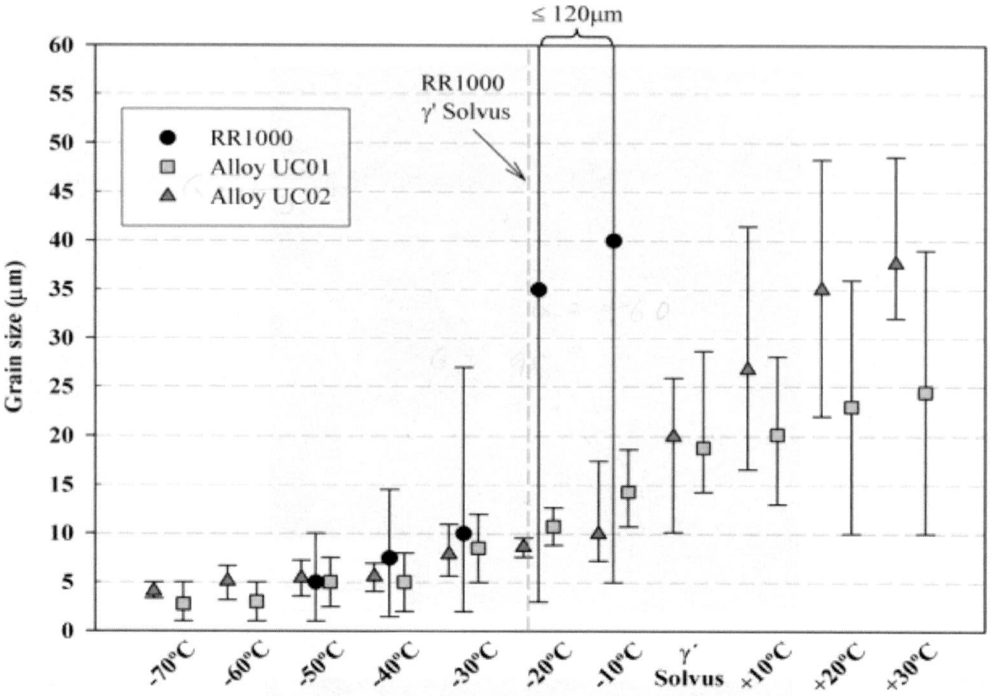

Fig. 3 Effect of solution heat treatment temperature on grain size for a one hour hold, followed by a forced air-cool. For comparison, data on RR1000 established by Manning[9] has been included.

boride/carbide solvus temperature was found to be approximately 35°C above the g¢ solvus temperature. Furthermore, it is clear that both the experimental alloys provide a more uniform grain structure in the supersolvus heat treated state, when compared to RR1000 where grain sizes up to 120 μm were observed for specimens of similar geometry.[9] Alloy UC02 displays similar characteristics to RR1000 with uniform grain growth up to about 10ºC below the γ' solvus temperature. Through the γ' solvus temperature, rapid grain growth occurs before the mean grain size stabilises. Alloy UC01 by comparison, displays a linear grain growth response throughout the sub-supersolvus transitional range. The range of errors associated with grain size measurements, also indicate a more uniform grain structure present in the experimental alloys.

The micrographs in Figure 4 are representative of the changes in primary γ' occurring as the γ' solvus temperature is approached. There is a clear reduction in the overall volume fraction of primary γ', Figure 5. The mean particle radius, however, has increased over the as-extruded alloy. The γ' matrix has become saturated with dissolved primary γ'. At 20ºC below the γ' solvus, the primary γ' that remains is clearly located on the grain boundaries. At temperatures closer to the γ' solvus there is very little primary γ' left on the grain boundaries and there is corresponding grain growth associated with the removal of this pinning force.

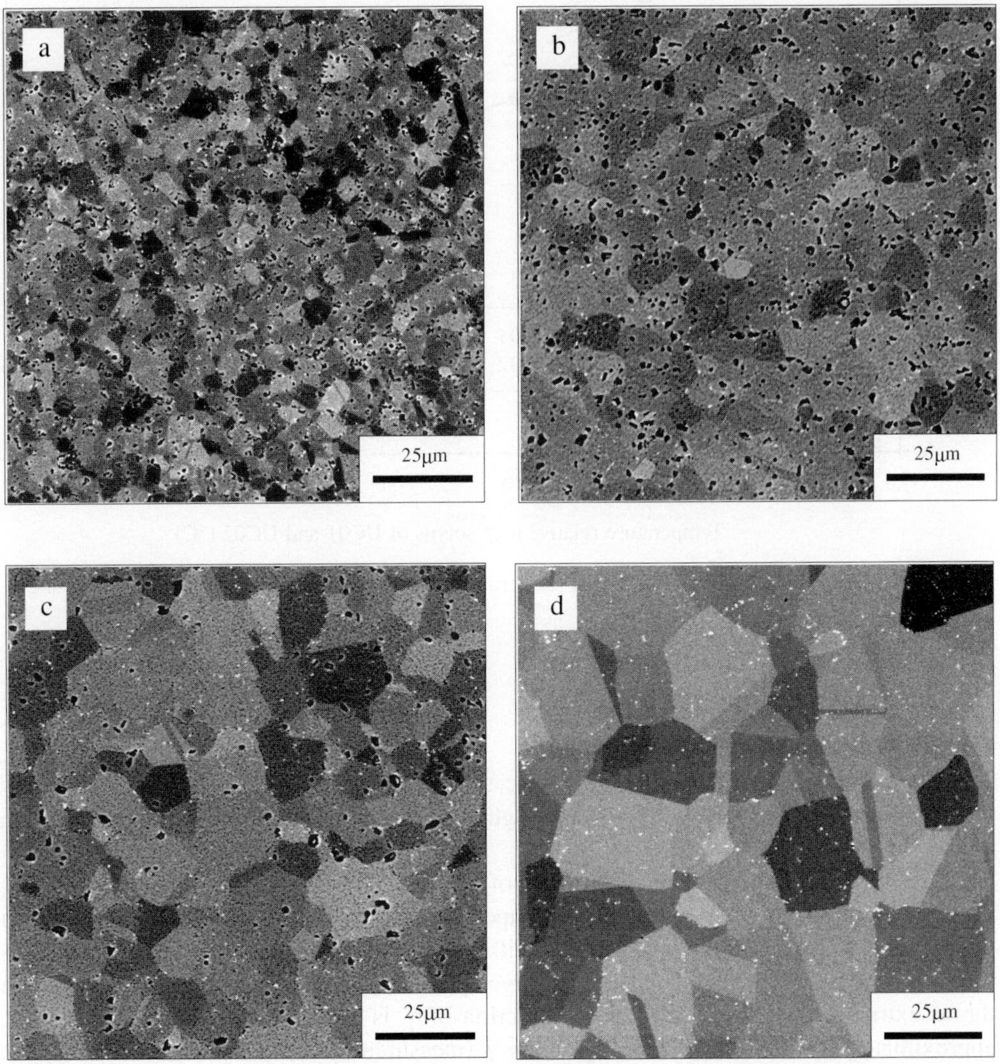

Fig. 4 These four micrographs show the change in microstructure with solutioning temperature for alloy UC01. The temperatures relative to the γ' solvus are (a) –40°C, (b) –20°C, (c) –10°C and (d) +30°C. In the subsolvus micrographs (a-c) the primary γ' has been etched away and it is clear how this phase delineates the grain boundaries in these alloys. Above the γ' solvus grain boundaries are partially pinned by the presence of carbide/boride (white) phases. A clear reduction in the amount of primary γ' is visible with increased solutioning temperature, associated with an increase in the average grain size.

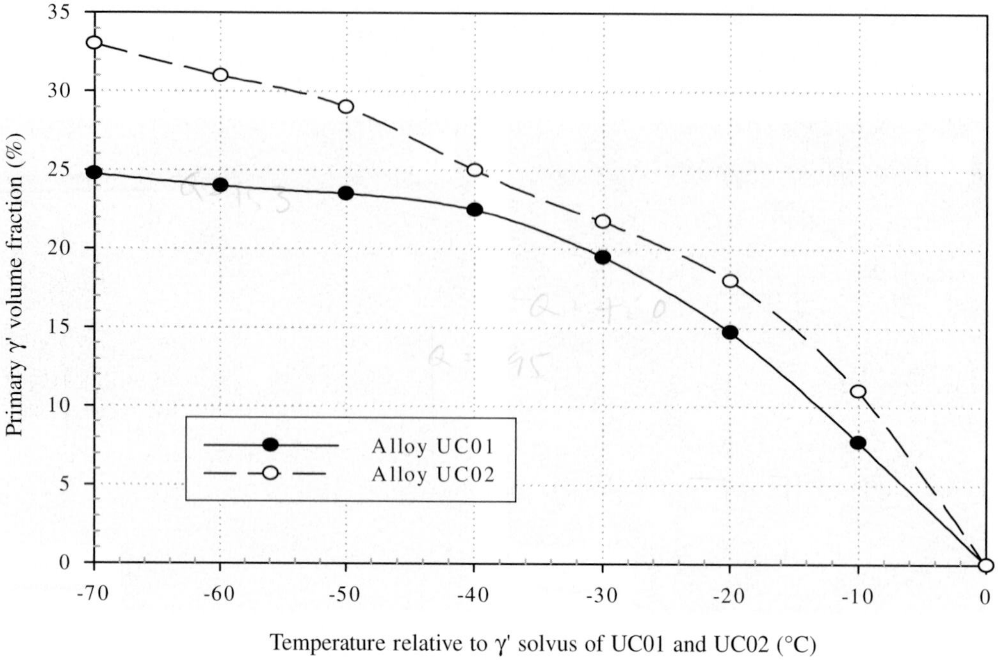

Fig. 5 Effect of solution heat treatment temperature on volume fraction of primary γ' for a one hour hold, followed by a forced air-cool.

Throughout the subsolvus temperature regime, the morphology of the grain boundary primary γ' remains unchanged for both alloys, although coarsening is seen for all temperatures up to the solvus.

Figure 5 shows how the volume fraction of primary γ', present on the grain boundaries, decreases for both alloys as the γ' solvus temperature is approached. A noticeable change in the primary γ' volume fraction also exists within these alloys. Alloy UC02, which contains 0.015 wt.% more boron than UC01, has a larger γ' volume fraction at lower temperatures. In the as-extruded state, the total volume fraction of γ' is greater than 50%. Of this total, approximately 25–35% is present as primary γ' (measured by manual point count method). The remaining γ' is a combination of secondary and tertiary γ'. The rate of dissolution of primary γ' with increasing temperature is very similar for both experimental alloys.

Closer examination of the primary γ' delineating the grain boundaries in both of the experimental alloys, reveals that carbide/boride phases appear to act as initiators for γ' nucleation, as shown in Figure 6. These phases exist in both of the alloys studied, and are randomly distributed throughout the matrix and boundary regions. When located on the grain boundaries they are always found in association with primary g¢, yet where these phases exist within the grains, they are isolated.

To try to understand the effect of boron on the observed grain growth response, values for activation energy Q, were determined by use of the classical grain growth equation:

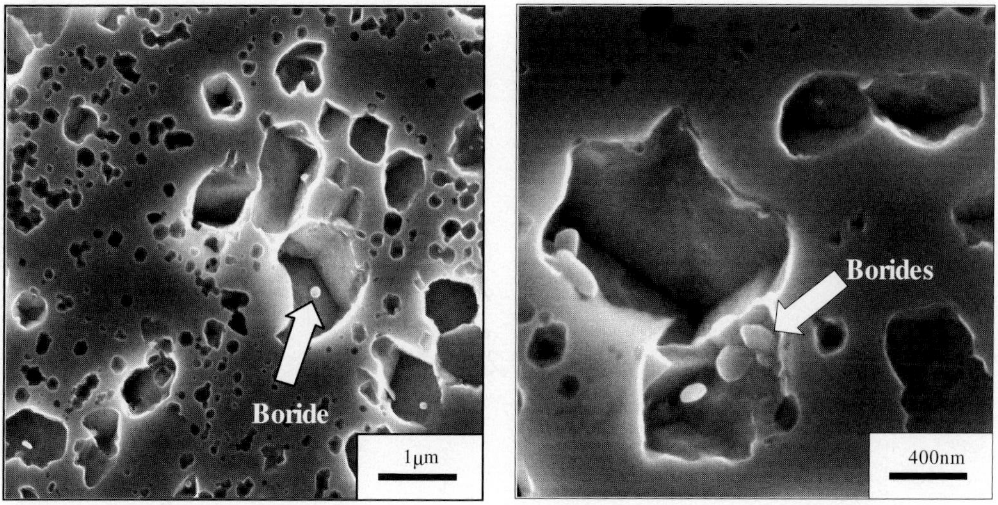

Fig. 6 Primary γ' nucleation in alloy UC02 is often found to occur on grain boundary phases. X-ray diffraction on extracted particles has revealed these to be predominantly M3B2 borides with some MC carbides.

$$D^2 - D_0^2 = kt \exp\left(\frac{-Q}{RT}\right) \tag{1}$$

where D is the average grain size at time t, D_0 is the average grain size at time $t = 0$, k is a constant of proportionality, t is time, R is the universal gas constant and T is the temperature in Kelvin. This equation relies heavily on empirically derived values for k in each alloy system and is therefore limited, because it takes no account of the effect of grain boundary pinning by γ' and/or carbide phases. Nevertheless, it provides information on activation energies, which can potentially be linked to compositional differences. Furthermore, the temperature dependence of the grain size permits the equation to hold in both the sub and supersolvus regimes, by the adjustment of Q.

Figure 7 shows how the activation energy Q, varies for both alloys in the sub and supersolvus states. This slope of this plot equates to $Q/2.3R$. The activation energy for alloy UC01 is greatest in the subsolvus temperature range and an inflection point is visible at the solvus temperature. The change in gradient can be ascribed to the dissolution of primary γ'. The gradient is lower for alloy UC02 in the subsolvus temperature range than for alloy UC01, however this trend is reversed above the solvus temperature. Further tests revealed that the gradients observed for both alloys and for both sub and supersolvus conditions, remained consistent for hold times of up to 24 hours.

The results observed for alloy UC01 and UC02 show that activation energy is linked to grain growth. In the subsolvus temperature range where grain growth in alloy UC01 is marginally higher, Q is correspondingly greater than that found in alloy UC02. At supersolvus

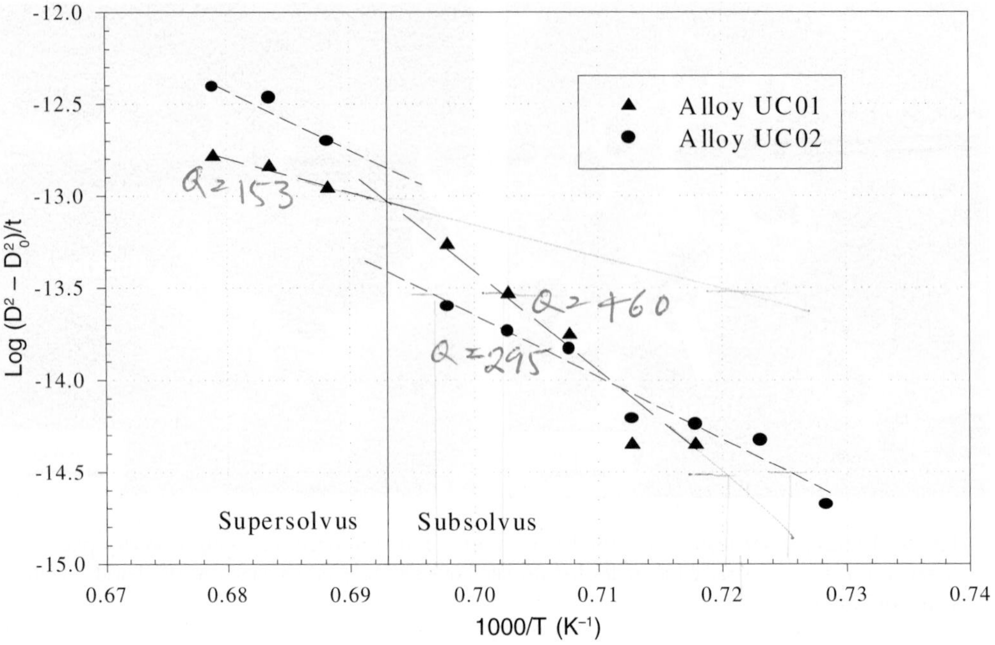

Fig. 7 Determination of activation energy, Q for alloy UC01 and UC02 in both the sub and supersolvus condition.

temperatures, where grain growth is highest in alloy UC02, this trend is reversed. The change in activation energy corresponds to the γ' solvus temperature in both alloys. This change is marked in alloy UC01, but considerably less so in alloy UC02. In the subsolvus region the large difference in activation energy that exists, is most likely attributable to the difference in morphology and size of primary γ' as will be discussed.

DISCUSSION

The only difference between the compositions of the two alloys studied, was the amount of boron present; 0.02 wt.% compared to 0.035 wt.%. Therefore it is highly likely that the mechanisms influencing grain growth in these alloys are being effected by the boron additions. The effect of boron upon both the mechanical properties and microchemistry in Ni-base superalloy systems has been extensively studied.[10–12] Boron is considered to strengthen alloys by one or more of the following mechanisms:

1. reduced grain boundary diffusion,
2. reduced grain boundary surface energy,
3. increased grain boundary cohesion and
4. changing the morphology of γ' and/or $M_{23}C_6$ phases.[13]

The effect of each of these mechanisms on grain growth in alloy UC01 and UC02 are discussed.

Due to the large atomic size difference between boron and nickel, boron tends to segregate on the grain boundaries. The mean boron composition of enriched grain boundaries can be ten times that of the matrix.[14, 15] Boron has a low solubility level above which it forms intergranular M3B2 type borides. These boride phases have been shown to have little effect on improving mechanical properties,[11, 13] however, due to their locations within the microstructure they are likely to have a strong influence on grain development, γ' nucleation and growth.[10, 16, 17] Observation of grain boundaries in both of the experimental alloys, revealed a higher volume fraction of boride phases present in alloy UC02. This combined with the larger volume fraction of primary γ' in this alloy, suggest that borides serve as nucleation sites for precipitation of primary γ' grains.

Segregation of boron on the grain boundaries of Ni-base superalloys has also been reported to form a monolayer, the thickness of which has been estimated to be within the range 0.35–0.5 nm.[13, 18] The thickness of the grain boundary layer gives an indication of the solubility of boron. If it is assumed constant for any given alloy, then the thickness of monolayer developed is a function of the total grain boundary area, which is directly related to the grain size. Work by Garosshen et al. proposed a relationship between boron concentration, monolayer formation and grain size. Additions of 0.01 wt.% boron result in a 0.35 nm thick monolayer and correspond to a grain size of 5 mm. Garosshen et al. assume a thickness of the layer and that there is no segregation of boron within the γ and γ' phases, both of which are likely to be inaccurate. Both alloy UC01 and UC02 have boron additions in excess of this theoretical amount, and it appears that their relative concentrations may straddle the boron solubility limit.

The very fine grain sizes exhibited by both of the alloys in this study suggest a large surface area. If alloy UC01 has boron present to an extent where its solubility level is not reached, then there will be a lack of boride phases present on the grain boundary. The greater amount of boron in alloy UC02 would exceed the solubility requirements for developing a fixed thickness of monolayer film, with any additional boron going to form M_3B_2 borides, resident on the grain boundaries. Due to the effect of boron upon both the grain boundary diffusivity and its propensity to reduce the grain boundary surface energy, a reduction in the grain growth rate might be expected. The inverse behaviour is observed in the case of alloy UC02. Previous work has found that high boron concentrations act to co-segregate nickel as well as chromium and molybdenum.[16, 18] It is likely, therefore that the increased boron concentration at the grain boundaries in alloy UC02 is responsible for the co-segregation of nickel, which results in the formation of larger primary γ' (0.5 μm after extrusion compared to 0.15 μm for alloy UC01). These larger primary γ' particles have been shown to have lower nickel concentrations than nominal values and are considered to act as easy channels for diffusion of Al/Ni during solutioning treatments. This in turn could explain the relative values obtained for the activation energies for the two alloys below the γ' solvus.

This work has provided evidence for a link between boron additions, grain growth response and activation energy. The addition of 0.015 wt.% boron results in a change in the grain growth response through the γ' solvus region. Additional boron also results in the formation of boride phases, which reside principally on the grain boundaries. These in turn act as nucleators for primary γ'. The grain growth response has been linked to changes in activation energy both above and below the γ' solvus.

CONCLUSIONS

- Boron additions to experimental powder processed Ni-base superalloys result in a change in the grain growth response throughout the γ' solvus temperature.
- Grains of primary γ' are nucleated at borides/carbides along the grain boundary.
- Pinning of grain boundaries due to the presence of borides is not the dominant mechanism by which boron additions restrict grain growth.

ACKNOWLEDGEMENTS

Financial support for this work in the form of an Industrial CASE Award between EPSRC and Rolls-Royce plc. is gratefully acknowledged. The authors would also like to thank Professor D. J. Fray for the provision of research facilities and Dr M. C. Hardy and Dr R. C. Helmink at Rolls-Royce plc. for the provision of materials.

REFERENCES

1. P. R. Bhowal, E. F. Wright and E. L. Raymond: 'Effects of Cooling Rate and Gamma-Prime Morphology on Creep and Stress-Rupture Properties of a Powder Metallurgy Superalloy', *Metallurgical Transactions A*, **21A**, 1990, 1709–1717.
2. F. Torster et al.: 'Influence of Grain Size and Heat Treatment on the Microstructure and Mechanical Properties of the Nickel-Base Superalloy U720LI', *Materials Science and Engineering A*, 1997, 234–236 and 189–192.
3. T. B. Gibbons and B. E. Hopkins: 'The Influence of Grain Size and Certain Precipitate Parameters on the Creep Properties of Ni-Cr-Base Alloys', *Metal Science Journal*, **5**, 1971, 233–240.
4. M. P. Jackson and R. C. Reed: 'Heat Treatment of UDIMET 720Li: The Effect of Microstructure on Properties', *Materials Science and Engineering A*, **A259**, 1999, 85–97.
5. R. D. Kissinger: 'Cooling Path Dependent Behaviour of a Supersolvus Heat Treated Nickel Base Superalloy', *TMS Superalloys*, Warrendale, 1996, 687–695.
6. W. P. Rehrer, D. R. Muzyka and G. B. Heydt: 'Solution Treatment and Al + Ti Effects on the Structure and Tensile Properties of Waspaloy', *Journal of Metals*, 1970, 32–38.
7. S. J. Hessell et al.: 'US Patent 5,897,718', 1999.
8. A. J. Manning, D. Knowles and C. J. Small: 'European Patent Application EP 1 193 321 A1', 2001

9. A. J. MANNING: 'Development of a Polycrystalline Ni Base Superalloy for Gas Turbine Disc Application', Ph.D. Thesis, Rolls-Royce University Technology Centre, University of Cambridge, 1999.
10. R. F. DECKER and J. W. FREEMAN: 'The Mechanism of Beneficial Effects of Boron and Zirconium in Creep Properties of a Complex Heat-Resistant Alloy', *Transactions of the Metallurgical Society of AIME*, **218**, 1960, 277–285.
11. R. T. HOLT and W. WALLACE: 'Impurities and Trace Elements in Nickel-Base Superalloys', *International Metals Reviews*, **21**(203), 1976, 1–24.
12. S. FLOREEN and J. M. DAVIDSON: 'The Effects of B and Zr on the Creep and Fatigue Crack Growth Behaviour of a Ni-base Superalloy', *Metallurgical Transactions A*, **14A**, 1983, 895–901.
13. T. J. GAROSSHEN, T. D. TILLMAN and G. P. MCCARTHY: 'Effects of B, C and Zr on the Structure and Properties of a P/M Nickel Base Superalloy', *Metallurgical Transactions A*, **18A**, 1987, 69–77.
14. A. BUCHON, A. MENAND and D. BLAVETTE: 'Phase Composition and Grain-Boundary Segregation in Nickel Base Superalloy: A Preliminary TEM-APFIM Investigation', *Surface Science*, **246**, 1991, 218–224.
15. L. LETELLIER et al.: 'Grain Boundary Segregation in Nickel Base Superalloys Astroloy: An Atom-Probe Study', *Applied Surface Science*, **67**, 1993, 305–310.
16. Y. L. CHIU and A. H. W. NGAN: 'Effects of Boron on the Toughness of γ-γ' Nickel Aluminium Superalloys', *Scripta Materialia*, **40**(1), 1999, 27–32.
17. S. S. BRENNER and H. MING-JIAN: 'Grain Boundary Segregation of Carbon and Boron on Ni_3Al + B/C', *Scripta Metallurgica et Materialia*, **24**, 1990, 667–670.
18. D. BLAVETTE et al.: 'Atomic-scale APFIM and TEM Investigation of Grain Boundary Microchemistry in Astroloy Nickel Base Superalloys', *Acta Materialia*, **44**(12), 1996, 4995–5005.

Microstructural and Mechanical Performance Assessment of Diffusion Bonded Bimetallic Model Discs

H. Burlet
CEA, Grenoble, France

G. McColvin and M. B. Henderson
ALSTOM Power, Leicester, UK

J. C. Garcia
TURBOMECA, Bordes, France

S. Peteves
EC-JRC, Petten, Netherlands

H. Klingelhoeffer
BAM, Berlin, Germany

M. Maldini
CNR-TEMPE, Milan, Italy

U. E. Klotz and T. Luthi
EMPA, Duebendorf, Switzerland

I. M. Wilcock
Qinetiq, Farnborough, UK

ABSTRACT

An alternative method to manufacture dual property discs and blisks has been developed using bonding of PM and solid fatigue resistant hub materials onto creep resistant rim/blade materials. Different HIP cycles, forging operations and heat treatments have been investigated within this project. Microstructural examinations and mechanical testing such as tensile, LCF and creep tests have been performed to qualify the joints and find the best process conditions.

INTRODUCTION

In order to achieve higher efficiency of gas turbines, a research project to develop bimetallic turbine discs or bliscs (bladed discs) has been carried out by a consortium of European partners. Dual property discs enable high temperature creep life in the rim section of the disc or in the blades, and high low cycle fatigue performance in the hub section. Different Ni-base superalloys are thus selected for their mechanical properties and must be joined together. The program includes bonding trials from different hub and rim materials, forging trials on bonded discs, optimisation of the heat treatment, microstructural investigations and mechanical characterisation.

EXPERIMENTAL PROCEDURE

Two hub materials, a powder metallurgy and a fine grain cast and wrought Udimet 720 have been considered in this program. They are referred to as U720PM and U720CW in the paper. The U720PM material was delivered in the form of powder whereas the U720CW was supplied in the form of a forged bar. Both materials were U720Li grades.

Three rim materials have been used: a coarse grain cast and wrought U720Li called U720CG, a cast IN738LC and a single crystal MC2. Both U720CG and IN738LC were supplied in the form of rings. U720CG exhibits a grain size varying in the range ASTM -3/3 whereas the IN738LC displays very long and oriented grains. MC2 was delivered as plates and bars with main orientation <001>.

Manufacturing Process

Dual alloy discs and plates have been manufactured by hot isostatic pressing (HIP). The interest of such a process has already been demonstrated in the past.[1–4] In the case of a powder metallurgy hub, the powder was consolidated during the bonding cycle. One major challenge associated with diffusion bonding of superalloys is the risk of oxide formation along the interface. Special attention has been taken to properly clean the surfaces to be joined. HIP trials have been performed at Bodycote and Aubert & Duval Tecphy.

After the HIPping process, subsequent forging operations were performed on some discs to investigate the consequence of such a consolidation procedure on the mechanical properties of the joint. Forging trials have been performed at Thyssen and Aubert & Duval.

Finally heat treatments have been applied to obtain the desired material properties in both the hub and rim sections. The optimum heat treatments for both materials to be bonded are different, hence a compromise had to be chosen. The solution treatment was restricted by the fine grain material (U720PM or U720CW). This requires a subsolvus treatment to limit the grain size, followed by gas quenching (approximate cooling rate: 70°C/min) to retain the maximum γ' in solution. It is well known that higher cooling rates are required to obtain high strength superalloys but this would generate residual stresses within the dual alloy discs and

Table 1 Experimental parameters.

Rim Material	Hub Material	HIP Condition	Forging Operation	Heat Treatment
MC2	U720PM	1100°C/140 MPa/3 hrs.	no	HT1 & HT3
IN738LC	U720PM	1100°C/140 MPa/3 hrs.	yes	HT1 & HT2
	U720CW	1100°C/140 MPa/4 hrs.	yes	HT1 & HT2
U720CG	U720PM	1100°C/140 MPa/3 hrs.	yes	HT1
	U720CW	1100°C/140 MPa/4 hrs.	yes	HT1

Differences between HT1, HT2 and HT3 are related to the ageing step. HT1 is optimum of U720PM, HT2 optimum for IN738LC and HT3 optimum for MC2.

thus a risk of failure at the interface. Different ageing treatments have been investigated to find the best compromise.

The experimental parameters that have been tested are summarised in Table 1. The dimensions of the U720CG (respectively IN738LC) rims were 190 mm (respect. 150 mm) outside diameter by 86 mm (respect. 90 mm) inside diameter by 61 mm (respect 100 mm) thickness.

Control of the Joint Integrity

Bimetallic samples containing planar artificial flaws simulating debonding areas have been produced to verify the possibility of NDE-methods to detect potential flaws . They were examined using ultrasonic immersion techniques with focused probes. Very high signal-to-noise ratios were obtained. It was proven that flaws of 1 mm² were easily detected whereas flaws of 0.5 mm² were hardly seen.

The bonded discs and plates were sectioned at the end of the whole process to conduct macrostructural, microstructural and scanning electron microscopy (SEM) evaluations and thus determine the quality of the bonds produced.

Mechanical Properties of Both Parent Materials and Joints

Tensile, low cycle fatigue (LCF), high cycle fatigue (HCF) and creep rupture tests were conducted on all the materials in the bonded conditions and also on bimaterial specimens. The joint was located perpendicular to the loading axis of the specimen. In the case of specimen machined from forged discs, the interface was curved and it was difficult to locate it accurately. The so-called 'strain' values obtained from an extensometer placed around the joint are average values obtained from a deformation gradient and must be used very carefully.

MICROSTRUCTURAL EXAMINATION OF THE DISCS AND BLISCS

U720/U720 Combination

Representative U720PM/U720CG bond lines in the as-HIPped and forged states are shown in Figures 1 and 2. As can be seen both processes resulted in bonds free of precipitates. No oxide particles have been detected. The forging operation has generated recrystallisation in the rim material similar to a near necklace structure.

The solid/solid combination U720CW/U720CG exhibits similar results. A recrystallised band of about 50 μm can be seen on the U720CG side of the interface as shown in Figure 3. Figure 4 depicts a typical necklace structure in the more deformed areas of the U720CG rim section.

U720/IN738 Combination

A typical bondline between U720CW hub and IN738LC rim materials is shown in Figure 5. An interdiffusion zone of about 10 μm is clearly visible in the rim side of the interface. It consists of fine recrystallised grains. A semi-continuous line of particles formed along the initial interface. STEM-EDS analysis revealed the presence of Ti, Ta, Nb, Mo and W in these precipitates. It is suggested that this precipitation is the consequence of the recrystallisation in the IN738LC leading to a rejection of these elements towards the interface.[5]

U720/MC2 Combination

A typical U720PM/MC2 bondline in the as-HIPped state is shown in Figure 6. Close to the interface, fine precipitates are visible in a band of 1 μm wide in the MC2 side. Beyond this area, MC2 shows modification of the γ' morphology: rafting or spheroidal shape according to the geometry of the bonded part. The usual cuboidal microstructure of MC2 is present beyond this region. The presence of the evolved microstructure is probably related to stresses generated during the HIP and heat treatment of the bimaterial plates. Some recrystallised zones were observed in the MC2 very close to the interface.

MECHANICAL PROPERTIES OF THE HUB MATERIALS

Tensile and fatigue property data were collected at room temperature and 550°C for both hub materials (U720PM and U720CW) in the as-HIPped or forged states in the fully heat treated condition.

The results of the tensile tests for heat treatment HT1 are summarised in Figure 7. It can be seen that the strength of the PM material is higher than that of the cast and wrought grade, which is due to its inherent finer grain structure. The tensile properties of U720PM did not

Fig. 1 Microstructure of the U720PM hub (right) and U720CG rim (left) after HIP and HT1 process route.

Fig. 2 Microstructure of the U720PM hub (left) and U720CG rim (right) after HIP, forged and HT1 process route.

show significant variation between the different forging conditions investigated. The ductility of U720PM varies between 15 to 25% at both temperatures.

The forging operation lead to a significant decrease of the strength of U720CW. This is probably related to an evolution of the fine γ' precipitates in the material during the preheating

Fig. 3 Microstructure of the U720CW hub (left) and U720CG rim (right) after HIP and HT1 process route.

Fig. 4 Microstructure of the U720CG rim after HIP, forged and HT1 process route.

of the discs before the forging itself. The ductility of U720CW was lower in the forged state compared to the as-HIPped state.

The tensile properties of U720PM were not significantly modified by heat treatment HT2, whereas HT3 had a negative effect: the ductility falling to below 10% at room temperature.

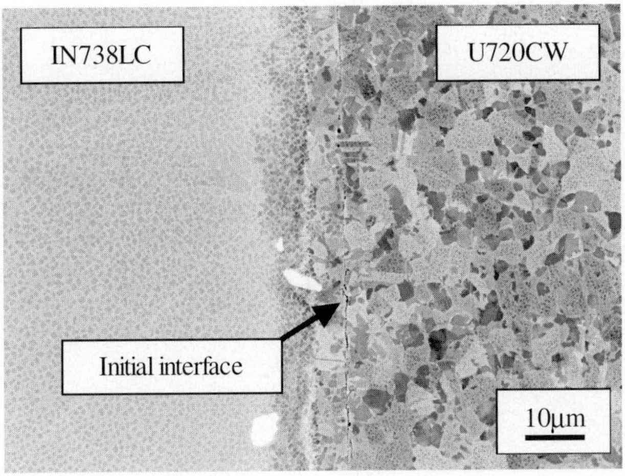

Fig. 5 Microstructure of the U720CW hub IN738LC rim after HIP and HT1 process route.

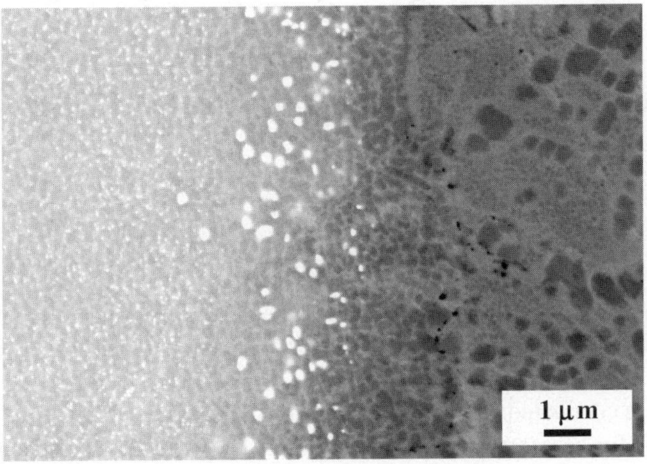

Fig. 6 Microstructure of the U720PM hub (right) MC2 (left) after HIP and HT3 process route.

It is worth noting that there is significant scatter in the tensile properties of the various materials. This is related to their sensitivity to the cooling rate after the solution treatment. In this development project it was known that this cooling rate could vary from one disc to another. Also the local cooling rate inside the disc can vary with thickness. These variations

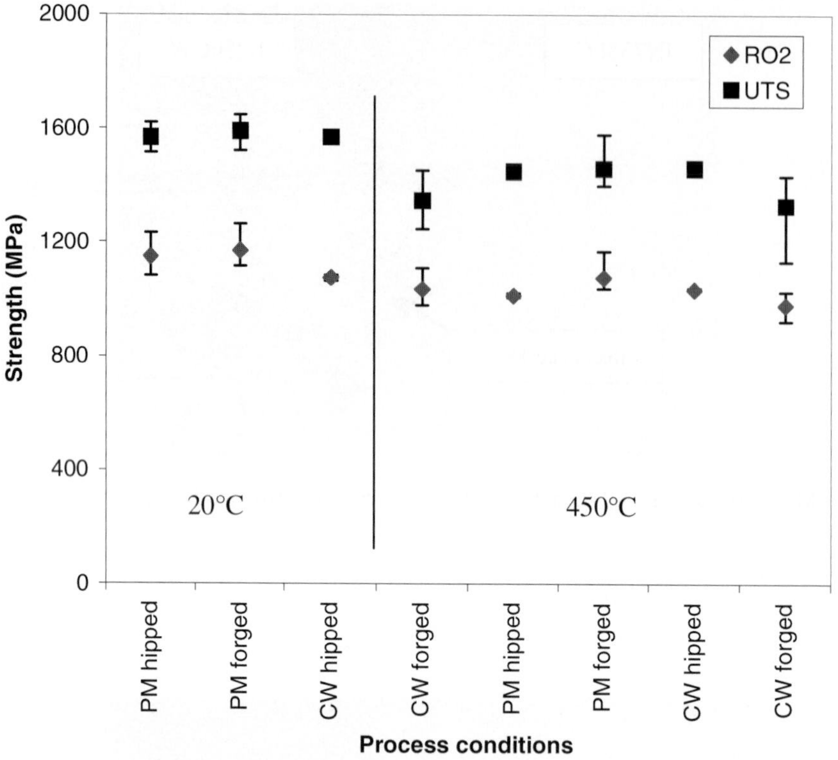

Fig. 7 Tensile properties of U720 hub materials using different process conditions.

generate differences in the microstructure of the materials especially in the γ' size and distribution which are known to greatly influence the mechanical properties of the Ni-base superalloys.

LCF tests were conducted at 450°C and with a strain ratio R = 0.05. The fatigue properties were comparable for heat treatments HT1 and HT2. The comparison of the powder and solid hub materials is illustrated in Figure 8. It can be seen that the cast and wrought grade gives slightly lower resistance than the powder metallurgy grade. This difference is more important when considering the stress/cycles to rupture curves. There were no significant differences between the as-HIPped and forged materials.

MECHANICAL PROPERTIES OF THE RIM MATERIALS

Tensile tests were conducted at room temperature and 650°C for the three considered rim materials U720CG, IN738 and MC2 in orientation <001>. The results are summarised in

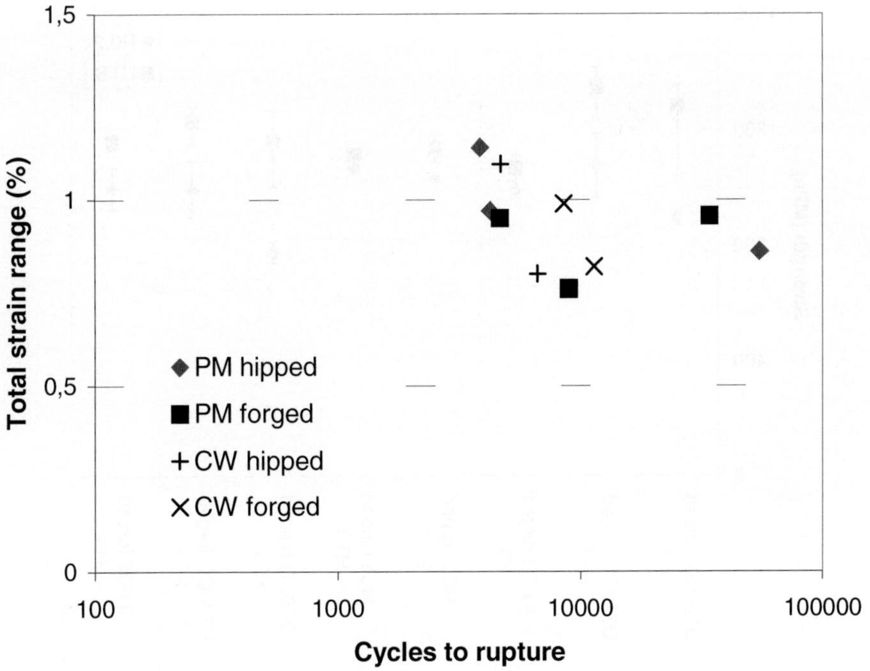

Fig. 8 Fatigue life curves of U720 hub materials at 450°C.

Figure 9. The forging operation seems to increase slightly the strength of Udimet 720. This is probably due to the beginning of recrystallisation observed in the rim section of the discs.

Very few results were available on IN738LC due to the poor quality of the as-received cast rings used for this study. The only interesting result is that heat treatment HT1 which is different from the optimum treatment of this material did not alter its tensile properties. The same conclusion applied for MC2, HT1 giving comparable tensile properties to HT3.

LCF tests were conducted on U720CG material at 650°C and with a strain ratio of 0.05. The forged grade exhibits higher fatigue resistance than the as-HIPped grade (Figure 10), which is consistent with its higher tensile strength.

Creep tests were conducted at 650 and 700°C on U720CG. The stress rupture properties measured on the rim section of the discs were comparable to the properties obtained on U720Li forged bars subjected to similar heat treatments. This means that there was no loss of creep resistance in the disc configuration resulting from the HIP and forge processes used in this study. Moreover the necklace structure seen in the forged discs did not modify significantly the creep properties of the superalloy.

The creep rupture properties measured on MC2 subjected to HT1 treatment gave acceptable values with respect to the blisk application.

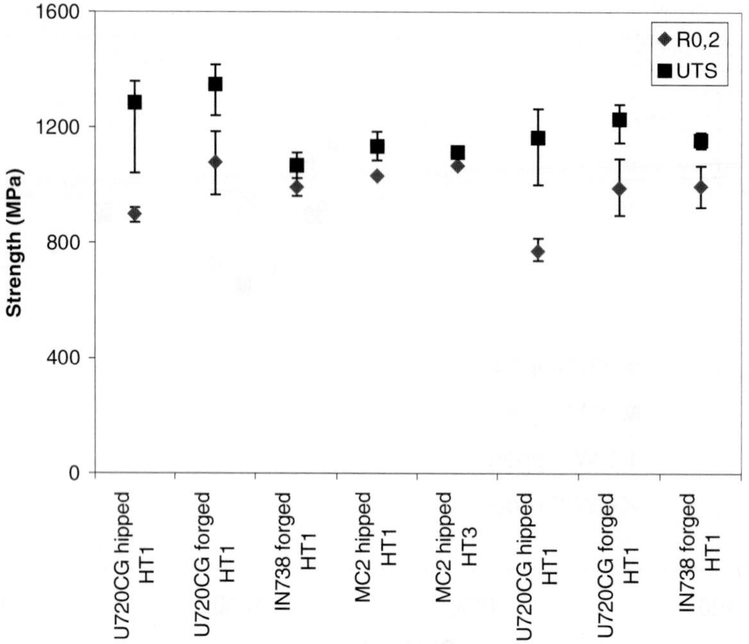

Fig. 9 Tensile properties of rim materials using different process conditions.

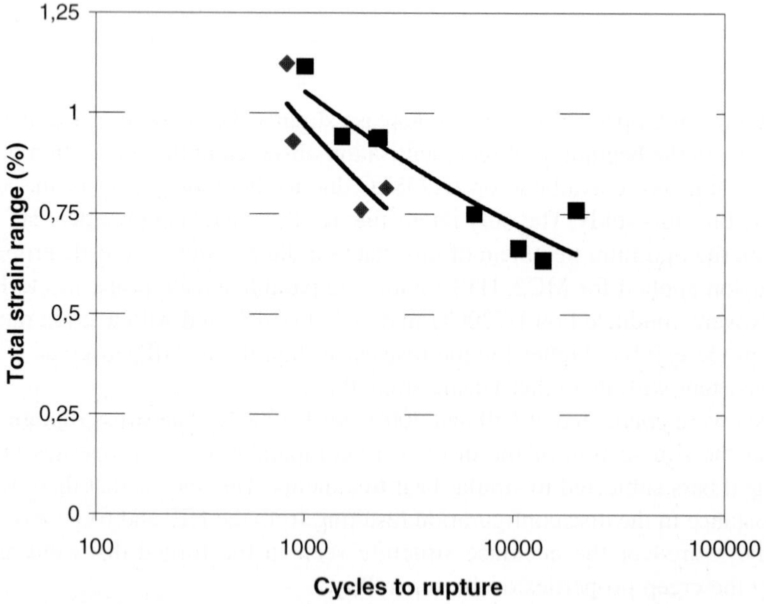

Fig. 10 Fatigue life curves of U720CG at 650°C.

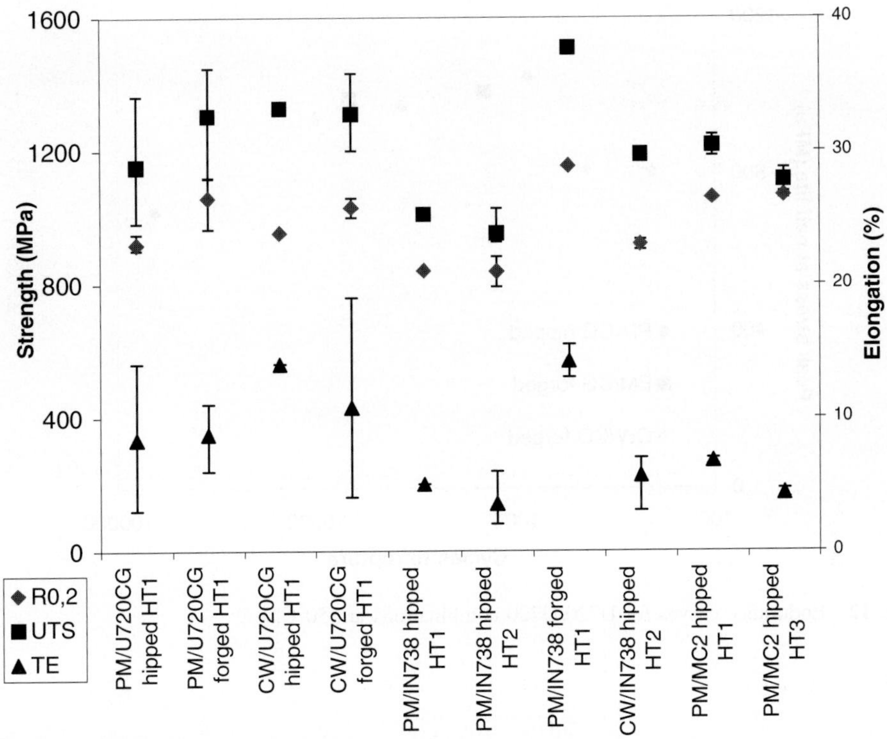

Fig. 11 Tensile properties of the various combinations at room temperature.

MECHANICAL PROPERTIES OF THE COMBINATIONS

Tensile tests have been conducted at room temperature and 550°C on bimaterial specimens machined from the discs and plates with the interface oriented perpendicular to the loading axis. The tensile properties at room temperature for the different combinations are plotted in Figure 11. The 'strain' values reported indicate some average values defined as the ratio of the displacement over the gauge length.

Some plastic deformation occurred in all the specimens. In most specimens, failure occurred in the rim side of the specimen far from the interface. In some of them failure occurred along the interface but with a quite high ductility. The comparison of the various combinations gives the following indications:

- The solid/solid bonds were weaker than the powder/solid one. In some instances failure occurred at the interface during the machining of the specimens. This could be related to the greater difficulty of cleaning and mating up the solid to solid interface and the resultant presence of defects and inclusions,
- The U720/IN738 bond was weaker than the U720/U720 joint. This is probably related to the presence of precipitates along the initial interface,

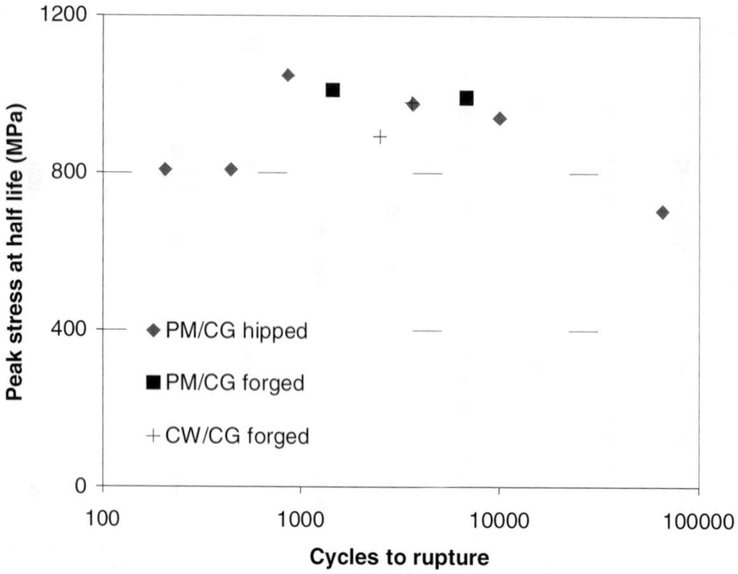

Fig. 12 Endurance curves for U720/U720 combinations at 550°C.

- Some U720PM/MC2 specimens failed very close to the interface in the recrystallised area present on the MC2 side,
- If the bond in the as-HIPped state is weak, the forging operation does not improve its strength. It often lead to some debonding occurring during the upsetting of the discs and
- The tensile properties of U720/U720 combinations are comparable in the as-HIPped and in the forged states.

Some creep tests have been conducted at 550°C on jointed samples. The tests were interrupted after a few hundred hours. No creep damage was detected around the interface, even in the near necklace microstructure in the rim side of the joint specimens.

LCF tests were conducted at 550°C on bimaterial specimens machined from the U720/U720 discs. The tests were carried out under imposed strain range and with a strain ratio R = 0.05. The comparison between the as-HIPped and forged states is not straight forward since the interface is curved in the forged disc and thus the strain distribution around the interface is different compared to a straight bond for a same 'imposed strain range'. The results are thus expressed in 'peak stress at half life' versus the number of cycles to rupture (Figure 12). Considering the U720PM/U720CG results, the results which are derived from 2 discs may be separated within 2 curves: one curve for each disc. This means that significant variation exists between the discs processed under nominally the same conditions. This is related to the control of the cooling rate from solution treatment, which drastically affects the mechanical properties of these superalloys. There was no significant differences in fatigue life between the as-HIPped and forged grades. In all tests, the cracks initiated within the rim material. No damage was detected along the interface itself.

CONCLUSIONS

During this project, the ability to HIP bond turbine discs and blisks with a good quality has been demonstrated. The general conclusions resulting from the microstructural analysis and the mechanical characterisation are as follows:

- The solid/solid joints were weaker than the powder/solid ones due to the greater difficulty of achieving good cleaning and mating up of the parts,
- An optimised heat treatment has been found for each combination,
- An interdiffusion zone was generated in the U720/IN738 combination associated with the formation of fine precipitates along the interface which makes these materials more difficult to bond,
- The microstructure of MC2 was locally modified probably due to the stresses generated during the HIP and heat treatments. Recrystallised areas were observed close to the interface, failure sometimes occurred in these areas,
- If the HIP bond is not good quality, the forging operation cannot improve it and will often generate cracks along the interface and
- The forging operation did not modify the properties of the powder material, but lowered the strength of the fine grain cast and wrought U720, and locally increased the strength of the rim material where recrystallisation occurred.

These results were sufficiently promising to proceed with the design, manufacture and spin testing of full scale bimetallic discs and blisks. This work is presented in another paper of this conference.[6]

ACKNOWLEDGEMENTS

Financial support from the European Community industrial and technology programme, project No BE97-4650, is gratefully acknowledged. One of the authors (IMW) would also like to acknowledge the UK DTI (Aeronautics Research Programme - formally CARAD) for part funding this work.

REFERENCES

1. H. Burlet, L. Briottet, R. Couturier, S. Gallet, G. McColvin, G. Raisson, V. Stamos, J. Olschewski, F. Vogel and T. Kobble: 'Fatigue and Creep Properties of Bimetallic Alloys Produced by Diffusion Bonding', *Advances in Mechanical Behaviour, Plasticity and Damage*, D. Miannay, P. Costa, D. François and A. Pineau, eds., Elsevier, 2000, 127–132.
2. R. Larker, J. Ockborn and B. Selling: 'Diffusion Bonding of CMSX-4 to Udimet 720 Using PVD-Coated Interfaces and HIP', *Journal of Engineering for Gas Turbines and Power*, **121**, 1999, 489–493.
3. T. M. Simpson, A. R. Price, P. F. Browning and M. D. Fitzpatrick: 'HIP Bonding of Multiple alloys for Advanced Disk Applications', *Advanced Technologies for*

Superalloys Affordability, K. M. Chang, S. K. Srivastava, D. U. Furrer and K. R. Bain, eds., Minerals, Metals and Materials Society/AIME , 2000, 277–286.

4. A. R. Price, T. Tom and M. D. Fitzpatrick: 'Sprayformed Materials for Disc Applications', *Gas Turbine Materials Technology*, ASM International, 1998, 45–53.
5. U. E. Klotz, I. M. Wilcock, M. B. Henderson and S. Davies: 'Interdiffusion in HIPped bimetallic Turbine Discs', *Proceedings of 15th International Conference on Electron Microscopy ICEM-15*, Durban, South Africa, 2002, 693–694.
6. M. Henderson, G. Mc Colvin, H. Burlet, J.C. Garcia, S. Peteves, J. Montagnon, G. Raisson, S. Davies and I.M. Wilcock: 'Manufacture and Spin Test Evaluation of Bimetallic Gas Turbine Components', Proceedings of the 6th International Charles Parsons Turbine Conference, Dublin, 2003.

Creep/Fatigue Behaviour of P/M Nickel Disc Alloys

I. M. Wilcock, D. G. Cole and J. W. Brooks
QinetiQ Ltd., Cody Technology Park
Farnborough, Hants
GU14 0LX, UK

M. C. Hardy
Rolls-Royce plc., Derby
DE24 8BJ, UK

ABSTRACT

Demand for turbine engines with greater efficiency, hence increased service temperatures and stresses, have led to the requirement for alloys with good mechanical properties under such conditions. Turbine discs are subjected to a combination of steady state and cyclic loading during operation that may give rise to creep and/or fatigue failure during service. As a consequence materials for turbine disc applications require high temperature strength, microstructural stability and resistance to fatigue crack propagation.

The aim of this work was to establish the relationship between creep and fatigue in high strength disc alloys. Mechanical tests have been conducted on a variety of powder metallurgy (P/M) nickel disc alloys including Alloy 720Li. Fatigue crack growth, low cycle fatigue and creep behaviour have been investigated at temperatures relevant to current and future turbine disc operating conditions.

Subsequent to testing, extensive fractography and microstructural characterisation has been performed. Fractography has allowed the identification of the specific failure mechanisms occurring and causing failure in the mechanical tests which has been further supported by the examination of cross-sections of the specimens.

The results of the mechanical testing have been related to the observations made of the fracture surfaces and microstructure and this has improved the understanding of the mechanisms responsible for failure. The results from this work have provided detailed insight into the crack propagation behaviour within advanced nickel disc alloys, which provides a sound basis for the development of advanced component lifing methodologies.

INTRODUCTION

The constant drive by gas turbine engine users for increased efficiency and power is leading to the development of engines that operate with higher overall pressure ratios and higher compressor discharge temperatures. The increase in temperature and rotational speed mean

Table 1 Nominal chemical composition of alloy 720Li and RR1000 (wt.%).

	Co	Cr	Mo	Ta	Ti	Al
720Li	14.57	15.92	2.98	-	5.18	2.44
RR1000	14–19	14.35–15.15	4.25–5.25	1.35–2.15	3.45–4.15	2.85–3.15

	Hf	B	C	W	Fe	Zr	Ni
720Li	-	0.015	0.015	1.35	0.08	0.042	Bal.
RR1000	0–1	0.01–0.025	0.012–0.033	-	-	0.05–0.07	Bal.

that advanced, high strength materials are required for fracture critical components such as the high pressure turbine (HPT) disc. These higher temperatures also lead to a potential shift in the operating failure mechanisms. Advanced nickel disc alloys are not affected greatly by creep at peak operating temperatures of 600°C, where fatigue is far more of a problem. However, at temperatures in excess of 700°C, creep will play a more important role.

The interaction of creep and fatigue damage mechanisms, along with microstructural and environmental factors, have been investigated in the past in older nickel disc alloys such as AP1.[1, 2] There has also more recently been work on modern alloys.[3–6] Previous work has shown that exposure to oxygen environment can have a significant effect on the fracture mechanisms, with both resistance to fatigue crack growth and low cycle fatigue life improved in the absence of oxygen. These previous studies also noted that there was a change in the appearance of the fracture surfaces as the temperature increased. This suggested that some change in mechanism may have been occurring. An understanding of the performance of these materials and their behaviour is clearly important in the safe operation of gas turbines.

The current work has concentrated on the nickel-based superalloys Alloy 720Li and RR1000, both are solid solution and precipitation strengthened and have been produced by powder metallurgy. A large grain variant of Alloy 720Li (denoted LG) was produced by heat treatment to evaluate the role of grain boundary area and grain size on the mechanical performance of this alloy. The LG microstructure had a corresponding reduced volume fraction of primary gamma prime (γ') and increase in volume fraction of secondary γ'. A comparison of the mechanical properties of the three alloys is presented and related to the microstructural failure mechanisms in operation at the temperatures considered.

EXPERIMENTAL PROCEDURE

Materials

Two variants of the nickel disc alloy Alloy 720Li and RR1000 have been investigated in the current study. The compositions and heat treatments are given in Tables 1 and 2 respectively. A sample of each material was mounted and polished to allow grain size measurements to

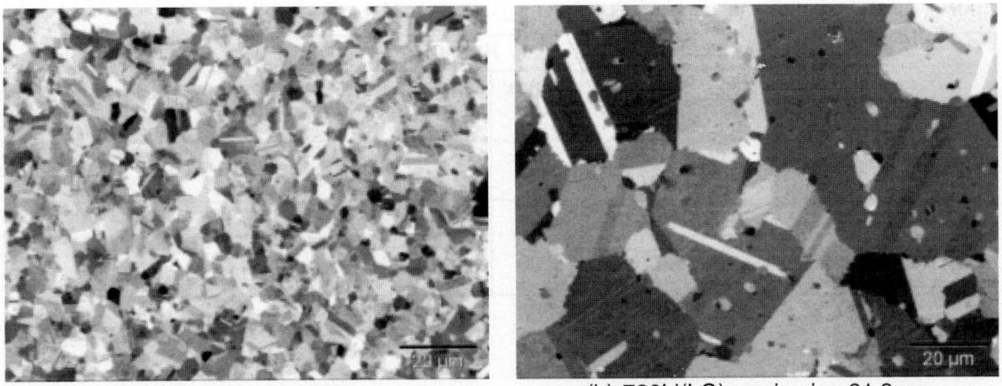

(a) 720Li, grain size 3.8 µm (b) 720Li(LG), grain size 21.3 µm

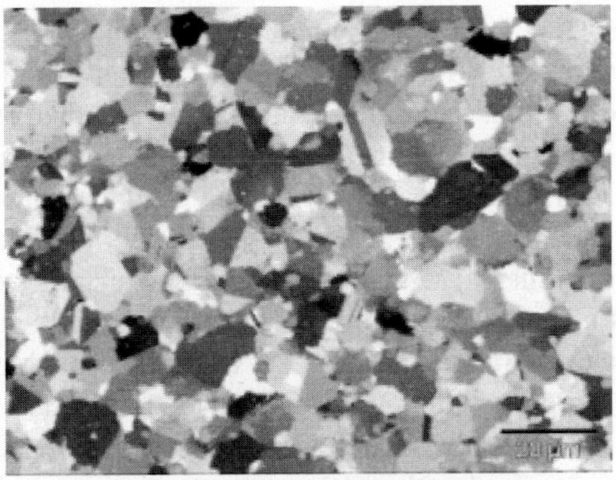

(c) RR1000 grain size 6.9 µm

Fig. 1 FEGSEM (backscattered electron) images showing the microstructures of the three alloys.

be obtained from field emission gun scanning electron microscope (FEG-SEM) images (via line intercept method).

The 720Li material was taken from a disc which had been isothermally forged from a hot isostatically pressed (HIPed) and extruded powder billet. A backscattered electron image of the microstructure can be seen in Figure 1a. This material has a uniformly distributed grain structure with an average grain size of 3.8 µm. The large grain variant of 720Li was produced via a higher solution heat treatment to give the larger grain size. Figure 1b shows the backscattered electron FEGSEM image and the average grain size was measured as 21.3 µm.

The RR1000 material was cut from a disc which had been forged at 1125°C and then heat treated (See Table 2). The microstructure of this material can be seen in Figure 1c and the grain size was measured as 6.9 µm.

Table 2 Heat treatment schedules for the investigated Nickel disc alloys.

	Solution Treatment			Primary Age			Secondary Age		
	Temp/ °C	Time/ Hrs.	Cooling	Temp/ °C	Time/ Hrs.	Cooling	Temp/ °C	Time/ Hrs.	Cooling
720Li	1105	4	Oil Quench	650	24	Air Cool	760	16	Air Cool
720Li(LG)	1135	4	Air Cool	650	24	Air Cool	760	16	Air Cool
RR1000	1120	4	Fan Air Cool	760	16	Air Cool	-	-	-

Mechanical Testing

Fatigue crack growth (FCG) testing was conducted on compact tension specimens in air using a trapezoidal waveform, a R-ratio of 0.1, and at constant load. Tests were performed at 650 and 725°C, and the dwell at maximum load was set at either 1s or 20s. The tests were carried out as outlined in the ASTM standard (E 647-95) using the pulsed dc potential drop method. A spreadsheet was created to convert the raw data measured and logged during the test into FCG curves. The spreadsheet uses the final crack length and potential difference (p.d.) to fix the calibration curves[7] which convert the raw data in voltages to crack lengths and stress intensity factor ranges, in accordance with the ASTM standard.

Strain-controlled low cycle fatigue (LCF) tests were performed on solid specimens in air at 650 and 700°C. Full experimental details are provided elsewhere.[8] Tests were conducted at three different strain ranges (0.7, 0.9 and 1.2%), and R-ratio of either $R_\varepsilon = 0$ or $R_\varepsilon = -1$. A trapezoidal waveform was used with a strain rate of either 0.1 or 0.5% s^{-1} and a 1 second dwell at maximum and minimum strain levels. Tests were stopped when the specimen was about to break (taken as a 20% drop off in load from stabilised loop load), or manually (runout) when it was decided the specimen would not fail.

Creep tests have been conducted on cylindrical specimens in air. Tests were conducted under either constant load or constant stress, at a temperature of either 650 or 725°C. The majority were constant stress tests, but a number of short term tests were conducted under constant load on the basis that very little necking occurs at high stresses (low rupture strains). A number of tensile tests have also been conducted at 650°C.

Fractography

Fracture surfaces from a selection of fatigue crack growth and low cycle fatigue tests conducted in the previous programme, along with some creep specimens, have been examined using a scanning electron microscope (SEM). To compliment the fractography work, a number of fracture surfaces have been sectioned longitudinally (in the direction of crack growth) and mounted and polished so the crack path and adjacent surface could be examined.

RESULTS

Mechanical Testing

Figure 2 shows the growth out curves for the three materials investigated. At all the test conditions the large grained variant of 720Li displays an improved resistance to FCG than the as-received 720Li. RR1000 was shown to have lower FCG rate at both temperatures, in comparison with either 720Li material. For comparison purposes, Figure 2a includes FCG data for Waspaloy (grain size ~50 mm) tested at equivalent conditions and this is seen to have very similar properties to RR1000.[9]

The LCF results are presented in Table 3 and Figure 3 and it can be seen that at both the investigated temperatures the large grain variant of 720Li appeared to offer longer life. When the initial part of the first LCF loop for each material at equivalent test conditions is compared (Figure 4), the LG material can be seen to have a lower strength than the conventional 720Li. In addition, the half life loops show that the LG material shakes down to a lower stress. For an equivalent strain range, the LG specimens experienced a lower stress for the majority of the test duration compared to the conventional 720Li material; this is consistent with their longer lives. The tensile results further support this fact with the 0.2 % proof stress of the LG being lower than the conventional 720Li (Table 4).

Creep tests on all materials produced time – strain curves with little primary or secondary creep; tertiary creep dominated the tests. Figure 5 shows the curves for the Alloy 720Li material tests. The stress rupture versus time data are plotted in Figure 6. The limited evidence in the graph suggests that at 650°C, there is no significant difference in the creep properties of any of the three alloys. At the higher temperature of 725°C, conventional 720Li had noticeably shorter lives than the other materials highlighting its inferior creep properties. It should be noted that the 720Li(LG) results are comparable to RR1000 at 725°C. It is also noted that in general, for the higher stress tests, a lower ductility was exhibited as shown in Figure 7.

Fractography

The purpose of this exercise was to determine the nature of the mechanism by which the specimen had failed under its specific test conditions. Fatigue and crack growth specimens were seen to fail from mechanisms varying from transgranular (the crack growing through grains – generally a flat featureless surface morphology) to intergranular (crack propagates along grain boundaries). In an attempt to facilitate the comparison between failure mechanisms operating at different conditions, a five-point scale has been devised with a typical fracture surface identified for each point. An example of an SEM image for each of the 5 types is shown in Figure 8 (type 1 being fully transgranular to type 5 which is fully intergranular). Table 5 provides a description of the scale along with the results from examinations of all the test specimen fracture surfaces. This allows comparison of the fracture surfaces from different tests (i.e. FCG, LCF and creep) in a more quantitative manner.

Examples of the fracture path from cross-sections can be seen in Figure 9 and 10. A section from a typical type 2 failure is shown in Figure 9. The grain structure is clearly seen

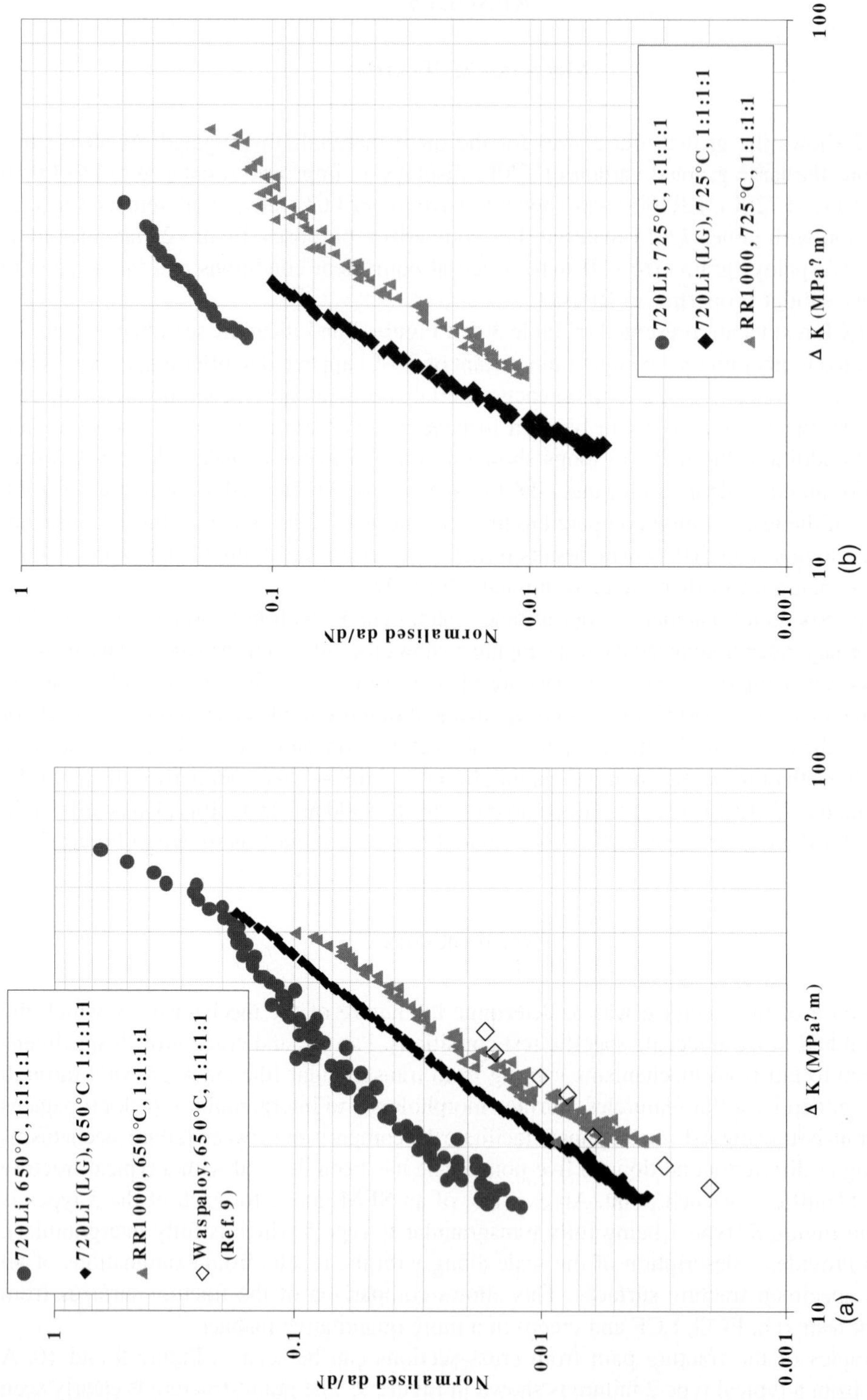

Fig. 2 Fatigue crack growth plots for 720Li and RR1000 at (a) 650°C and (b) 725°C with a 0.25 Hz waveform.

Table 3 Low cycle fatigue results on 720Li and 720Li(LG).

Material	Temperature (°C)	R_ε-ratio	Strain rate (%sec^{-1})	Cycles to Failure, N_f	$\Delta\varepsilon_T$ (%)	Comments
720Li	650	0	0.1	739	1.174	
720Li	650	0	0.1	2116	0.872	
720Li	650	0	0.5	4415	0.920	
720Li	650	0	0.1	3740	0.676	
720Li	650	-1	0.5	1437	1.222	
720Li	650	-1	0.5	29807	0.833	runout
720Li	650	-1	0.5	71968	0.720	runout
720Li	700	0	0.5	566	1.222	
720Li	700	0	0.5	644	1.221	
720Li	700	0	0.1	1183	0.884	
720Li	700	0	0.1	1629	0.784	
720Li	700	0	0.1	4204	0.687	
720Li	700	0	0.5	23620	0.721	
720Li	700	-1	0.1	3738	0.886	
720Li	700	-1	0.5	61028	0.717	

Material	Temperature (°C)	R_ε-ratio	Strain rate (%sec^{-1})	Cycles to Failure, N_f	$\Delta\varepsilon_T$ (%)	Comments
720Li(LG)	650	0	0.5	1349	1.220	
720Li(LG)	650	0	0.5	3461	0.926	
720Li(LG)	650	0	0.5	43950	0.730	
720Li(LG)	650	-1	0.5	5895	0.916	
720Li(LG)	650	-1	0.5	70000	0.717	runout
720Li(LG)	700	0	0.5	1082	1.213	
720Li(LG)	700	0	0.5	4012	0.910	
720Li(LG)	700	0	0.5	56720	0.727	
720Li(LG)	700	-1	0.5	5588	0.921	
720Li(LG)	700	-1	0.5	69614	0.717	

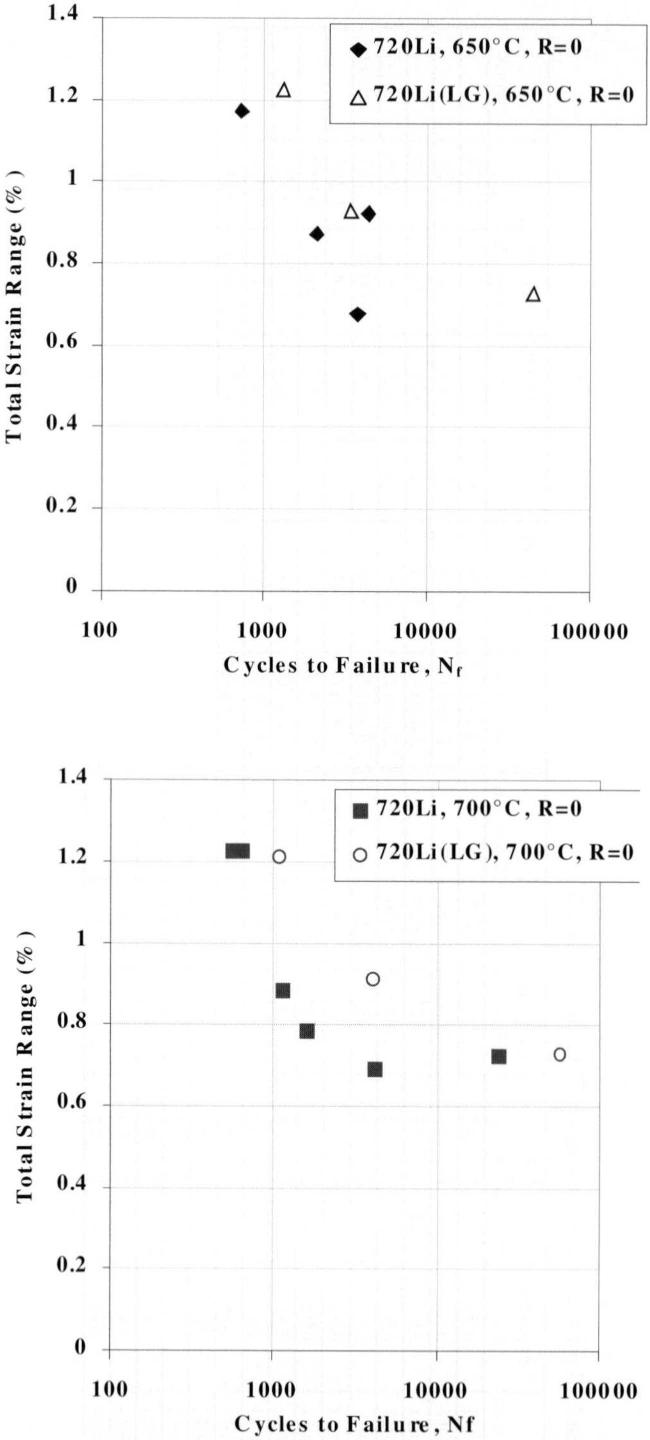

Fig. 3 Plots of low cycle fatigue lives for 720Li and 720Li(LG) at 650 and 725°C.

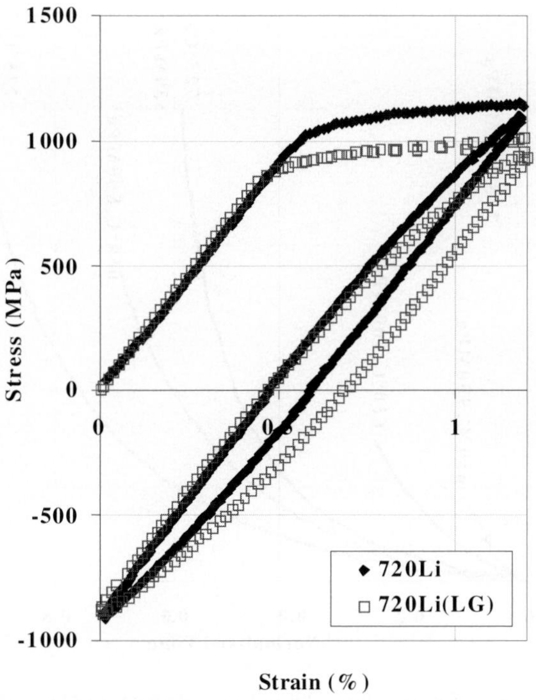

Fig. 4 Initial part of first loop and complete half life loop (cycle number 300) from 720Li and 720Li(LG) both tested at 650°C, $R_{\varepsilon}=0$, $\Delta\varepsilon = 1.2\%$.

Table 4 Tensile data for the 3 alloys at 650°C.

Material	720Li	720Li (LG)	RR1000
Elongation (%)	25	25	27
Youngs Modulus (Gpa)	179	182	177
0.2% Yield Strength (Mpa)	1050	984	1025
UTS (Mpa)	1322	1392	1435

and the crack follows a relatively flat path, implying it is growing through grains. The image shows a small degree of secondary cracking which penetrates less than one grain down into the specimen. Compared to Figure 10 (which shows a type 4 failure – the crack clearly growing round the grains), significant secondary cracking was observed several grain diameters down from the fracture surface.

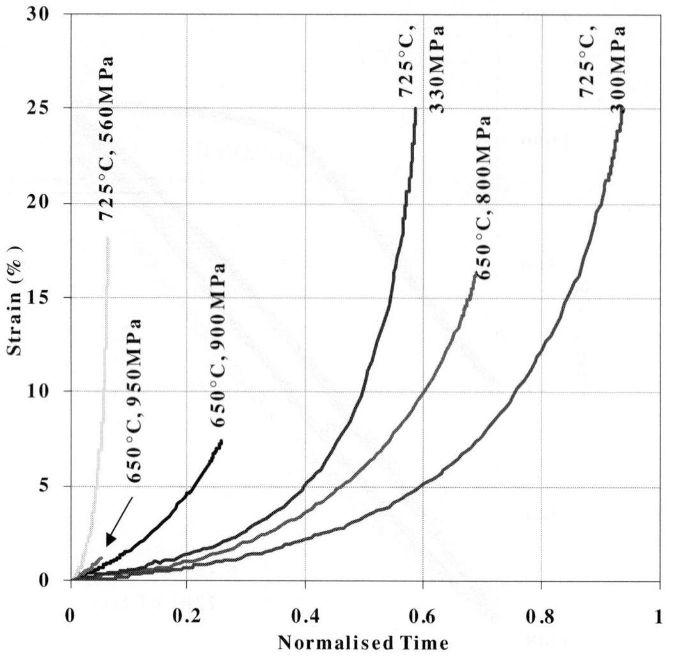

Fig. 5 Time – strain curves for the Alloy 720Li creep specimens tested at 650 and 725°C.

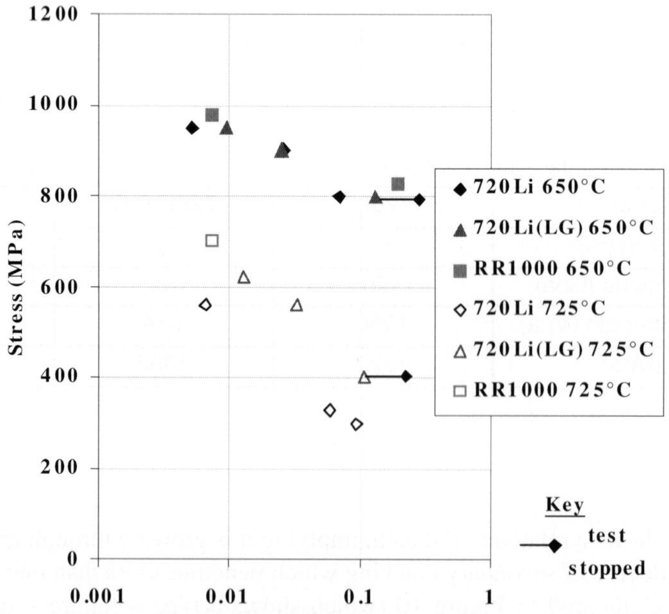

Fig. 6 Creep results for the three alloys tested at 650 and 725°C.

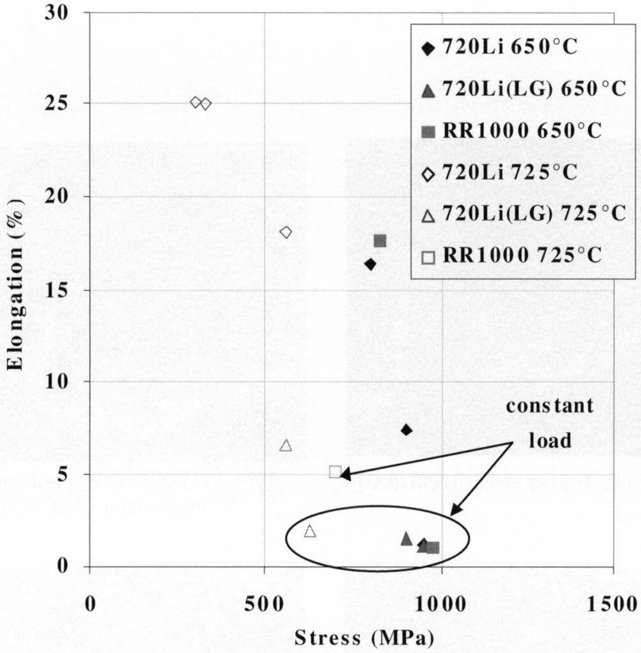

Fig. 7 Elongation of the creep tests identifying those tested under constant load.

Table 5 Fractography failure mechanism scale for FCG, LCF and creep testing (N.B. All LCF tests shown in this table on 720Li were 0.1% sec^{-1} and 720Li(LG) were 0.5% sec^{-1}).

Fatigue Crack Growth (in air)		
720Li	1:1:1:1	1:20:1:1
650°C	4	5
725°C	5	5
720Li (LG)	1:1:1:1	1:20:1:1
650°C	2	3
725°C	4	5
RR1000	1:1:1:1	1:20:1:1
650°C	2	4
725°C	4	5

Creep (in air)	
720Li	5
720Li (LG)	5
RR1000	5

Low Cycle Fatigue (in air)		
720Li	Δε (%)	Mode
650°C	0.9	2
650°C	1.2	3
700°C	0.7	5
700°C	0.8	5
700°C	0.9	5
720Li (LG)	Δε (%)	Mode
650°C	0.7	2
650°C	0.9	2

Scale

1 Fully Transgranular
2 Predominantly Transgranular
3 Mixed Mode
4 Predominantly Intergranular
5 Fully Intergranular

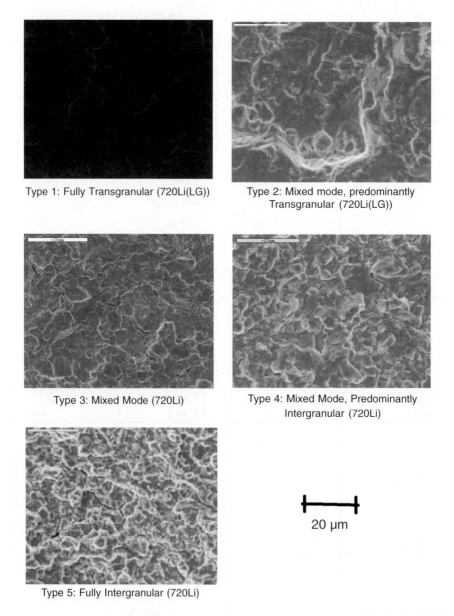

Type 1: Fully Transgranular (720Li(LG))

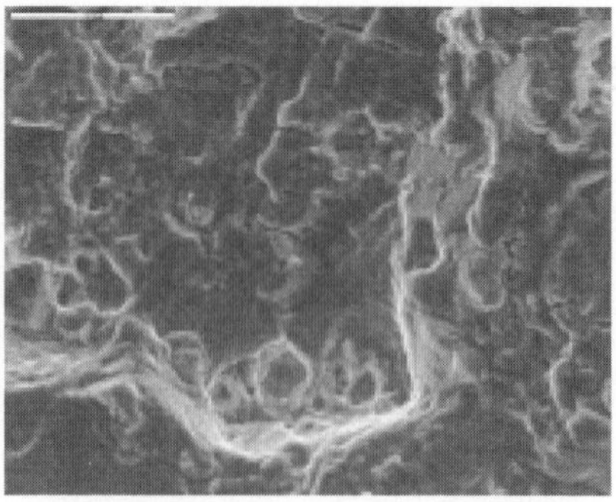

Type 2: Mixed mode, predominantly Transgranular (720Li(LG))

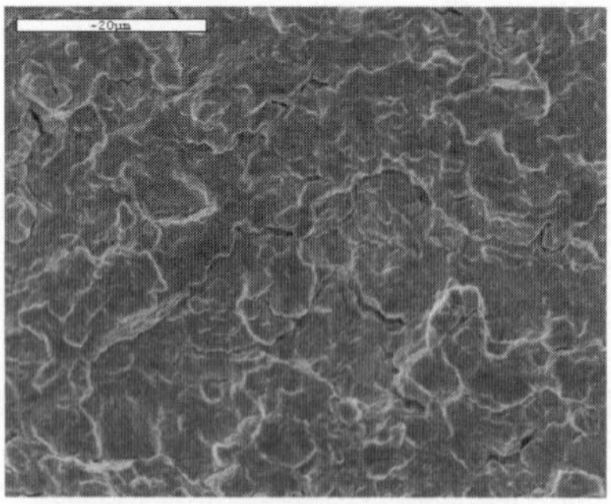

Type 3: Mixed Mode (720Li)

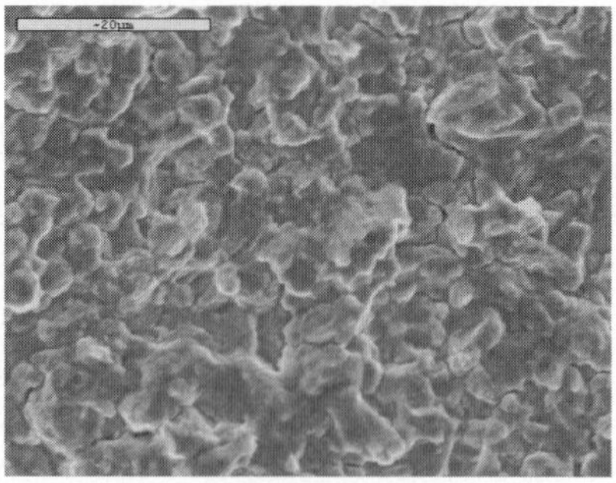

Type 4: Mixed Mode, Predominantly Intergranular (720Li)

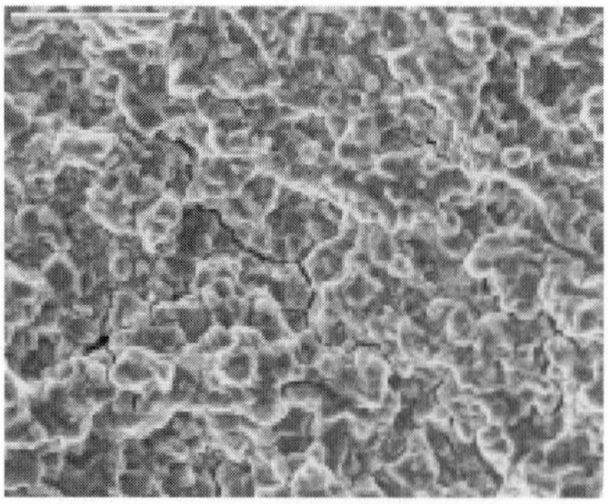

Type 5: Fully Intergranular (720Li)

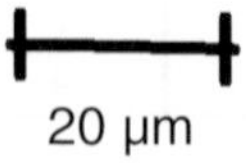

Fig. 8 Examples of fracture surfaces showing the transition from transgranular to intergranular failure.

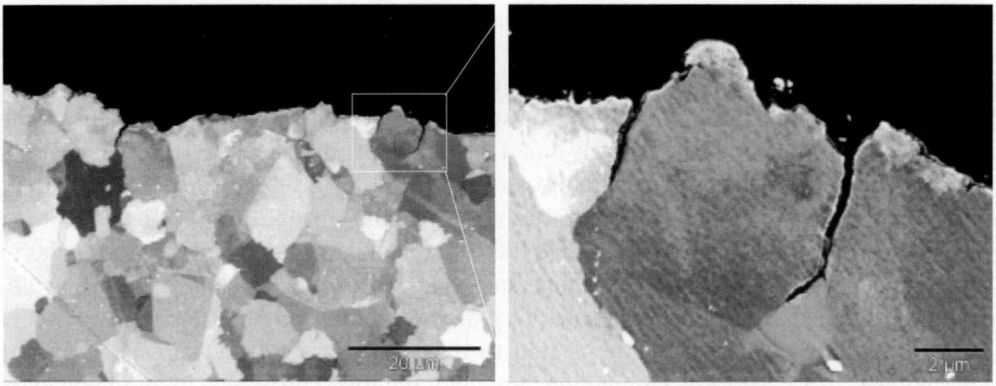

Fig. 9 Cross-section of a type 2 (FCG test on RR1000 at 650°C and 1:1:1:1 waveform).

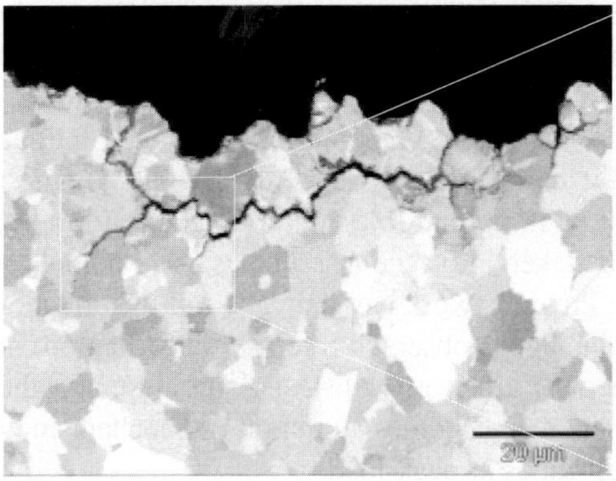

Fig. 10 Cross-section of a type 5 (FCG test on RR1000 at 725°C and 1:20:1:1 waveform).

DISCUSSION

It was apparent from the fatigue and crack growth test results that the LG material demonstrated improved performance over the conventional 720Li material. This is considered to be associated with the large grains giving a more arduous path for the crack to propagate along, leading to the LG variant being more likely to fail in a transgranular manner. Another reason for the improved LCF lives, as previously discussed, is the lower strength of the LG material. However, it must be noted that this result would have been reversed if the tests had been conducted under load control. In general, strain controlled LCF testing is more representative of notched features and stress concentrations, whereas load controlled tests characterise plain features such as bores and diaphragms.

Facets and striations are indicative of transgranular crack growth under conditions which fatigue alone operates. However, it was found that as temperature and dwell were increased, the propensity for an intergranular failure also increased for all three alloys (see Table 5). Comparison of the FCG fracture surfaces was revealing in that all 720Li failed in an intergranular manner, whereas LG and RR1000 exhibited a transition from transgranular to intergranular failure with increasing temperature and dwell time. This may be attributed to the greater grain size of LG and the superior grain boundary strength of RR1000.

The 720Li(LG) material was expected to have an improved creep resistance over conventional 720Li due to its large grains providing improved resistance to grain boundary sliding. The dominant tertiary stage in the creep tests indicated the importance of grain boundary sliding which is considered to require shear displacement from the movement of grains and leads to grain boundary fracture. The improved creep properties of a larger grain variant of as-HIPed 720Li has also been seen previously.[10] This work attributed the improved creep properties seen at 750°C of a coarse grained material to be due to the controlling deformation mechanism being grain boundary sliding. In the current work, RR1000, despite it's relatively fine grain size (approximately equivalent to the 720Li material), displayed an equivalent creep performance to the LG material. This indicates that factors other than grain size are affecting the creep failure mechanisms, such as grain boundary characteristics and precipitate distributions.

Comparison of all test fractography indicates that high temperature, long dwell fatigue tests have surfaces similar to those observed for the creep tests. This may initially suggest that creep mechanisms are playing an increasingly important role on the fatigue test failure. However, the effect of environment must also be taken into account, such as grain boundary oxidation embrittlement ahead of the crack tip. Previous work has shown that if the effect of environment is removed (vacuum) then creep damage is observed (creep void formation) in high temperature (725°C) long dwell (20 second) crack growth tests.[3] In summary, the comparison of the fracture surfaces suggests that creep becomes an important mechanism in fatigue testing at higher temperatures and dwell times at peak loads.

CONCLUSIONS

- Comparisons of the mechanical properties of two fine grain, high strength nickel disc alloys, 720Li and RR1000, have been made.
- Small quantities of a large grain size variant of 720Li were produced in the laboratory by solution treatment at a temperature that was just below the gamma prime solvus temperature. However, it is not considered practical to apply this near solvus heat treatment to a full scale disc forging as it is likely to produce isolated areas of unusual grain growth.
- It was found that the large grain size variant of 720Li and fine grain RR1000 both displayed superior high temperature properties over fine grain 720Li in terms of fatigue crack growth, strain fatigue and creep.

- The improvement in the resistance to crack growth and creep that was shown by the large grain variant of 720Li was considered to have resulted from the increase in grain size. RR1000 achieves equivalent or higher levels of these properties with a fine grain size, which indicates the contribution from other factors such as grain boundary characteristics and precipitate distributions.
- Analysis of specimen fracture surfaces indicated that creep was playing an increasingly important role in strain fatigue tests at higher temperatures and longer dwell times. The effect of environment was not quantified but was also deemed important to the failure mechanism.
- The greater understanding of the failure mechanisms operating will improve material selection and the application of more accurate lifing techniques.

ACNOWLEDGEMENTS

The authors would like to thank MOD who funded this work through the ARP Superpackage 4. Thanks also to Bill Mitten and Singh Ubhi at QinetiQ Farnborough for conducting the creep tests and operating the FEGSEM respectively.

REFERENCES

1. M. R. Winstone, K. M. Nikbin and G. A. Webster: 'Modes of Failure Under Creep-Fatigue Loading of a Nickel-Based Superalloy', *Journal of Materials Science*, **20**, 1985, 2471–2476.
2. J. E. King: 'Effects of Grain Size and Microstructure on Threshold Values and Near Threshold Crack Growth in Powder-Formed Ni-Base Superalloy', *Metal Science*, **16**, 1982, 345–355.
3. N. J. Hide, M. B. Henderson and P. A. S. Reed: 'Effects of Grain and Precipitate Size Variation on Creep-Fatigue Behaviour of Udimet 720Li in Both Air and Vacuum', *Proceedings of Superalloys 2000*, T. M. Pollock, et al., eds., TMS, 2000, 495–503.
4. G. Onofrio, G. A. Osinkolu and M. Marchionni: 'Fatigue Crack Growth of UDIMET 720Li Superalloy at Elevated Temperature', *International Journal of Fatigue*, **23**, 2001, 887–895.
5. M. Marchionni, G. A. Osinkolu and G. Onofrio: 'High Temperature Low Cycle Fatigue Behaviour of UDIMET 720Li Superalloy', *International Journal of Fatigue*, **24**, 2002, 1261–1267.
6. D. W. Hunt, D. K. Skelton and D. M. Knowles: 'Microstructural Stability and Crack Growth Behaviour of a Polycrystalline Nickel-Base Superalloy', *Proceedings of Superalloys 2000*, T. M. Pollock, et al., eds., TMS, 2002, 795–802.
7. M. A. Hicks and A. C. Pickard: 'A Comparison of Theoretical and Experimental Methods of Calibration the Electrical Potential Drop Technique for Crack Length Determination', *International Journal of Fatigue*, **20**, 1982, 91–101.

8. I. M. Wilcock, D. G. Cole, J. W. Brooks and M. B. Henderson: 'Elevated Temperature Cyclic Stress-Strain Behaviour in Nickel Based Superalloys', *Materials at High Temperatures*, **19**(4), 2002, 187–192.
9. G. F. Harrison, L. Grabowski and P. H. Tranter: 'Defects and their Effect of Component Integrity', *Proceedings of the AGARD SMP Workshop 'The Impact of Materials Defects on Engine Structures Integrity'*, 74th AGARD SMP Meeting, Patras, Greece, 1992.
10. S. Dubiez, R. Couturier, L. Guétaz and H. Burlet: 'Creep Behaviour of a Powder Metallurgy Udimet 720 Nickel-Based Superalloy', *Proceedings from Materials for Advanced Power Engineering 2002*, J. Lecomte-Beckers, et al., eds., 2000, I.419–426.

The Characterisation and Constitutive Modelling of As-Cast Single Crystal CM186LC for Industrial Gas Turbine Applications

D. W. Bale
ALSTOM Power Technology Centre
Cambridge Road, Whetstone
Leicester LE8 6LH, UK

M. Toulios
School of Naval Architecture and Marine Engineering
National Technical University of Athens
Athens 157 73, Greece

E.P. Busso
Department of Mechanical Engineering
Imperial College
London SW7 2BX, UK

M. B. Henderson
ALSTOM Power Technology Centre
Cambridge Road, Whetstone
Leicester LE8 6LH, UK

P. Mulvihill
Powergen, Power Technology Centre
Nottingham, UK

ABSTRACT

CM186LC SX is a candidate industrial gas turbine (IGT) blade and vane material as a consequence of its high grain boundary tolerance, good casting yields and low heat treatment costs. A work package within the recent COST 522 action focused on the characterisation of the mechanical properties of as-cast CM186LC SX in the <001>, <011> and <111> crystallographic orientations. The observed low cycle fatigue properties and the effects of orientation, temperature, mean stress and strain rate on the cyclic stress strain and strain life behaviour of CM186LC SX are described. The properties of hollow specimens and the effect of peak tensile and compressive dwell periods on the strain life response is explained. The fracture characteristics of CM186LC SX under cyclic loading conditions are also described. The mechanical behaviour of CM186LC SX is next modelled using a multiscale constitutive approach that accounts for the effects of microstructural features within CM186LC SX. This novel approach is explained, and the effects of eutectic volume fraction and morphology on the mechanical properties of the material are considered.

Table 1 Nominal composition (wt.%) of CM186 LC superalloy.

Ni	Cr	Co	Mo	W	Ta	Re	Al	Ti	Hf	C	B	Zr
Bal.	6	9	0.5	8	3	3	5.7	0.7	1.4	0.07	0.015	0.005

INTRODUCTION

In order to meet stringent emission targets and to improve engine efficiency, the turbine entry temperatures of Industrial gas Turbines (IGT) have steadily increased in recent years.[1] This has necessitated the use of advanced single crystal Nickel based superalloys for first stage turbine blade and vane applications. A candidate alloy for this application is CM186LC SX, a desirable choice owing to the alloys grain boundary and defect tolerance, together with good castability, which results in high casting yields.[1] Each of these beneficial properties are obtained without the need of an expensive solution heat treatment,[1] as is the case in conventional second generation single crystal superalloys, such as CMSX-4.

The current paper introduces the alloy CM186LC SX, explaining recent work to characterise the low cycle fatigue properties and the failure modes determined by examination of the fracture surfaces. Novel approaches to modelling the deformation behaviour of single crystal materials, which account for the contributions of microstructural heterogeneities are also described.

CM186LC SX undergoes a two stage ageing treatment that consists of 4 hours at 1080°C ± 10°C in a vacuum, followed by 20 hours at 870°C ± 5°C. Rapid gas fan quenching in the presence of high purity Argon is utilised after each stage. The nominal alloy chemistry is provided in Table 1.

AS RECEIVED MICROSTRUCTURE OF CM186LC SX

The microstructure of CM186LC SX is made up of two components, regular dendritic regions and eutectic colonies[2] (Figure 1). The dendritic regions consist of a regular distribution of face centred cubic (FCC) cuboidal $Ni_3(Al,Ti)$ γ' particles, which are surrounded by a Nickel based γ matrix. The matrix forms γ channels between the γ' particles, some of which contain spherical cooling γ' particles with a diameter of 20–50 nm. The average γ' particle size was 0.4 μm and the γ' volume fraction was approximately 70%. The interdendritic eutectic colonies exhibit a courser γ/γ' morphology. Microporosity was evident in these regions, as were Tf and Hf rich primary carbides. The volume fraction of the eutectic regions was 20–25%.[1]

THE MECHANICAL PROPERTIES OF CM186 LC SX

Extensive materials testing was conducted within the programme to characterise the mechanical properties of CM186LC SX. The full range of mechanical properties were studied

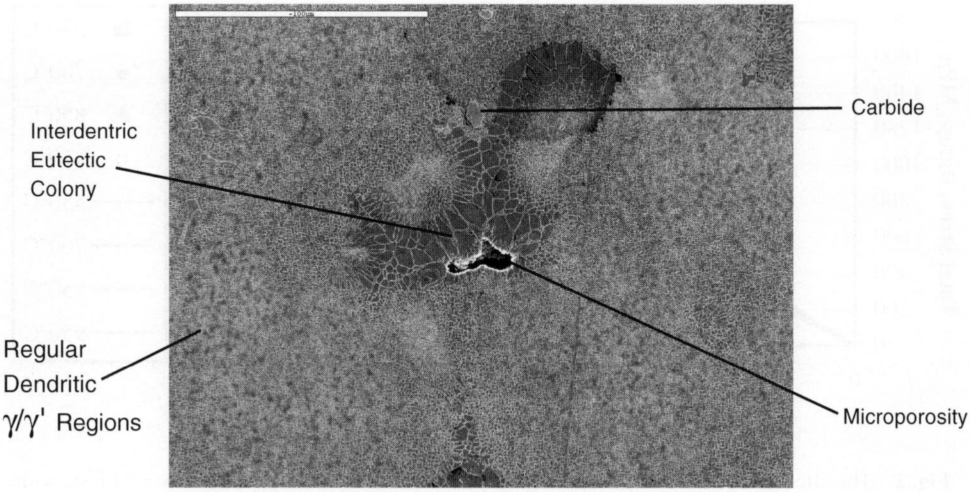

Fig. 1 Micrograph of typical CM186LC SX microstructure.

including monotonic tensile, creep, high cycle fatigue as well as the low cycle fatigue (LCF) properties which are the focus of this paper.[1]

LOW CYCLE FATIGUE PROPERTIES

LCF tests were conducted on solid and cast-to-hollow specimens, intended to simulate the wall thickness present in cooled turbine components. The test programme also considered the effect of crystallographic orientation. Although the properties of the <001> orientation were the primary focus, a limited number of <011> and <111> aligned specimens were also tested. Tests were performed in the temperature range 550 to 950°C and three different strain rates were applied: 0.6, 6 and 60%/minute. Mean stress effects were studied by applying symmetric ($R\varepsilon = -1$) and asymmetric strain controlled loading cycles ($R\varepsilon = 0.5$ and 0.05), where $R\varepsilon = \varepsilon_{min}/\varepsilon_{max}$.

Cyclic Stress Strain Behaviour

The small differences detected between the elastic and plastic portions of the stress strain curves indicated low plastic strain levels as shown in Figure 2. The temperature dependence of the response is strong above 700°C, but relatively weak below this temperature as the results generated at 550°C coincided well with the 700°C data. Limited strain rate effects were observed as the 850°C 6%/minute and 60%/minute strain rate data corresponded well. However, a reduced cyclic strength was observed for the 0.6%/minute tests.

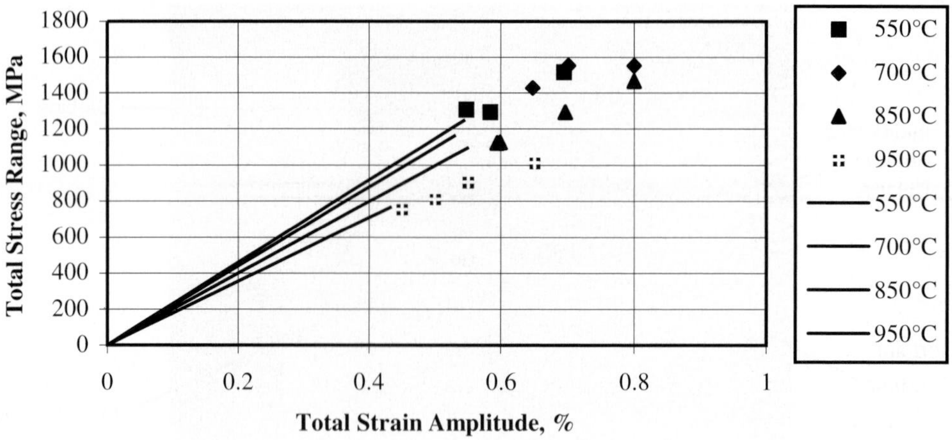

Fig. 2 The effect of temperature on <001> aligned cyclic stress strain curves for $R\varepsilon = -1$ tests with a strain rate of 6%/minute.[1]

The mechanical properties of single crystals are anisotropic with significantly different Young's Modulus values found in each of the three primary crystallographic orientations; <001>, <011> and <111>. The results showed weak orientation dependence,[1] allowing the elastic contribution of the strain life to be normalised to that observed in the <001> direction. This was achieved by modifying the strain range according to equation 1.[1] This method was successfully used to merge the data for each orientation together.

$$\Delta\tilde{\varepsilon}_{<hkl>} = \left[E_{<hkl>}/E_{<001>}\right]\Delta\varepsilon_{<hkl>} \quad (1)$$

where $E_{<hkl>}$ is the Young's Modulus in the specimen and $E_{<001>}$ the Young's modulus in the precise <001> orientation, both measured at the appropriate test temperature.

Strain-Life Behaviour

The <001> orientated specimens consistently gave the greatest number of cycles to failure. The <011> and then the <111> data followed the <001> results. Equation 1 was successfully used to account for the different Young's Modulus values. This technique allowed data for all three orientations to fit a single curve. A plot showing modified strain range versus number of cycles to failure is given in Figure 3. Tests were conducted at 850°C with a strain rate of 6%/minute. Some degree of scatter remains, however all results fall within a 2.5 factor scatter band, indicating that these data are independent of orientation effects, a similar observation was made for CMSX-4.[3]

The effect of mean stress on the strain life response was investigated by comparing results obtained for tests with asymmetric loading cycles ($R\varepsilon = 0.05$ and 0.5) with those found for

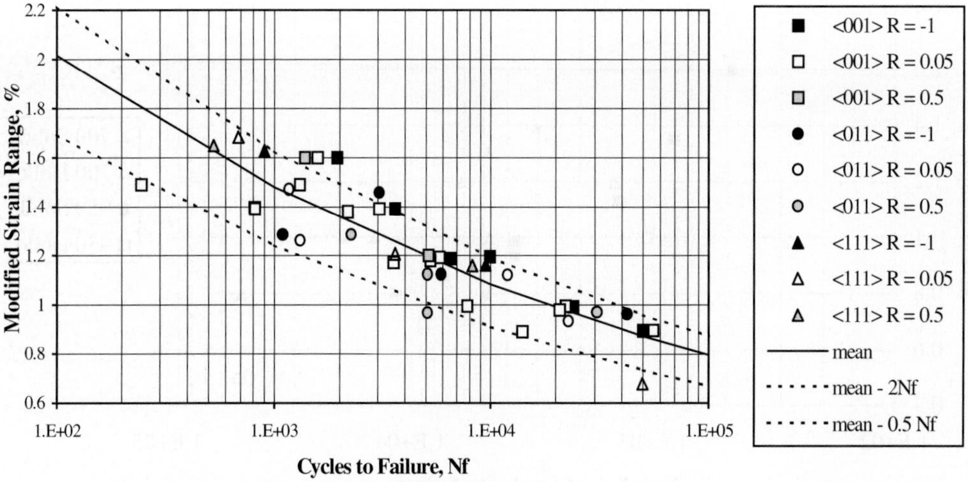

Fig. 3 Modified strain range versus cycles to failure for 850°C, 6%/minute tests.[1]

fully reversing cycles (Rε = –1). Substantial differences in the strain life response for each Rε-ratio was observed at 700°C and below. For tests with higher applied strain ranges and positive Rε-ratios, the mean stress reduced to levels similar to those experienced in tests with a fully reversed loading cycle. Similar life times are therefore observed for all Rε-ratios. For tests with smaller applied strain ranges and positive Rε-ratios, a higher mean stress is maintained throughout the test duration. This resulted in shorter lives for Rε = 0.05 tests, and shorter still for Rε = 0.5 tests.[1]

A mean stress influence was not detected at 850°C and above. At these higher temperatures the mean stress experienced in positive Rε-ratio tests reduces to similar levels experienced in tests with fully reversed loading, even for low applied strain ranges. This behaviour is a result of creep relaxation of peak stress levels during the tests.

Hollow Specimens

In order to simulate the thin walls found in cooled IGT turbine blades a small number of LCF tests were conducted on thin walled tubular specimens. The cast-to-hollow specimens were machined into LCF specimens using surface grinding and central bore honing. Some difficulty in machining concentric bores was experienced, and this problem undoubtedly affected the results, further testing is therefore planned to clarify the results.

In all cases the hollow specimens gave significantly shorter lives than the corresponding tests on solid specimens, particularly at 700 and 950°C where a typical life reduction of half an order of magnitude was observed, as shown in Figure 4. The 700°C tests conducted at

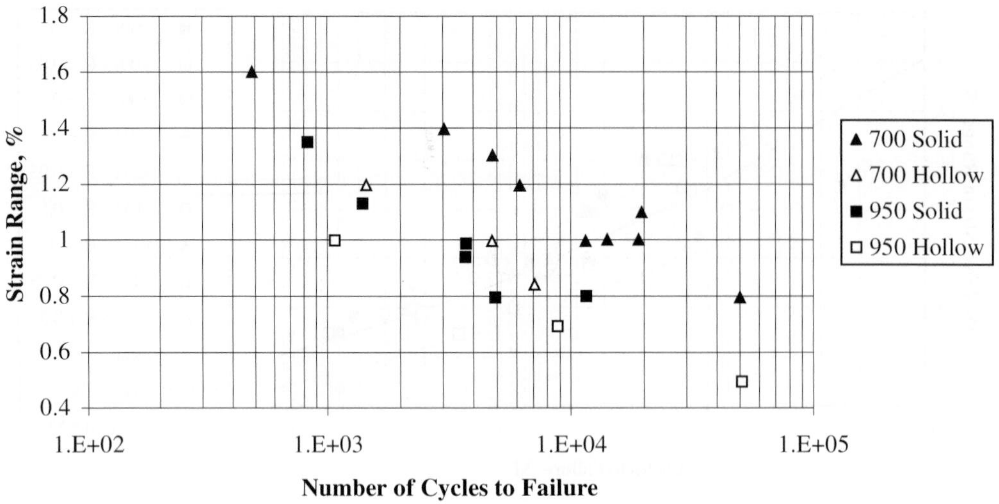

Fig. 4 Strain range versus cycles to failure for solid and hollow specimens at 700 and 950°C ($R\varepsilon = 0.05$, 6%/minute).

low strain ranges gave particularly poor results, a reduction in life of one order of magnitude was observed for the hollow specimens. Interestingly, at high applied strain ranges, the hollow specimen results at 850°C (not shown in Figure 4) fall within the lower bounds of the scatter band surrounding solid specimen results conducted at a similar strain range. However, the 850°C results at lower strain ranges show a reduction in life that is comparable to the results obtained for similar tests at different temperatures. Further work is required to rationalise these discrepancies.

Interaction of Creep and Low Cycle Fatigue Behaviour

A series of tests were conducted at 850 and 950°C with peak tensile and compressive dwell periods ranging from 2 to 60 minutes. Nominal $R\varepsilon$-ratio's of -1 and 0.05 were applied. This provided an understanding of the life reduction associated with short-term dwell periods, the effect of the cycle type (i.e. compressive or tensile dwell) and the effect of increasing dwell time. The short-term (2 min.) compressive and tensile dwell periods caused a significant life reduction when compared with the corresponding continuous LCF test results. These previously reported results[1] showed that at low strain ranges ($\Delta\varepsilon < 1\%$), the corresponding tensile and compressive dwell periods gave similar life times. However, compressive dwell periods caused significantly shorter life times than tensile dwell periods at higher strain ranges.

A few (long-term) tests have been conducted at 950°C, for both $R\varepsilon$-ratios ($R = 0.05$ and $R = -1$), with tensile and compressive dwell period's of 10 to 60 minutes (see Figure 5). It

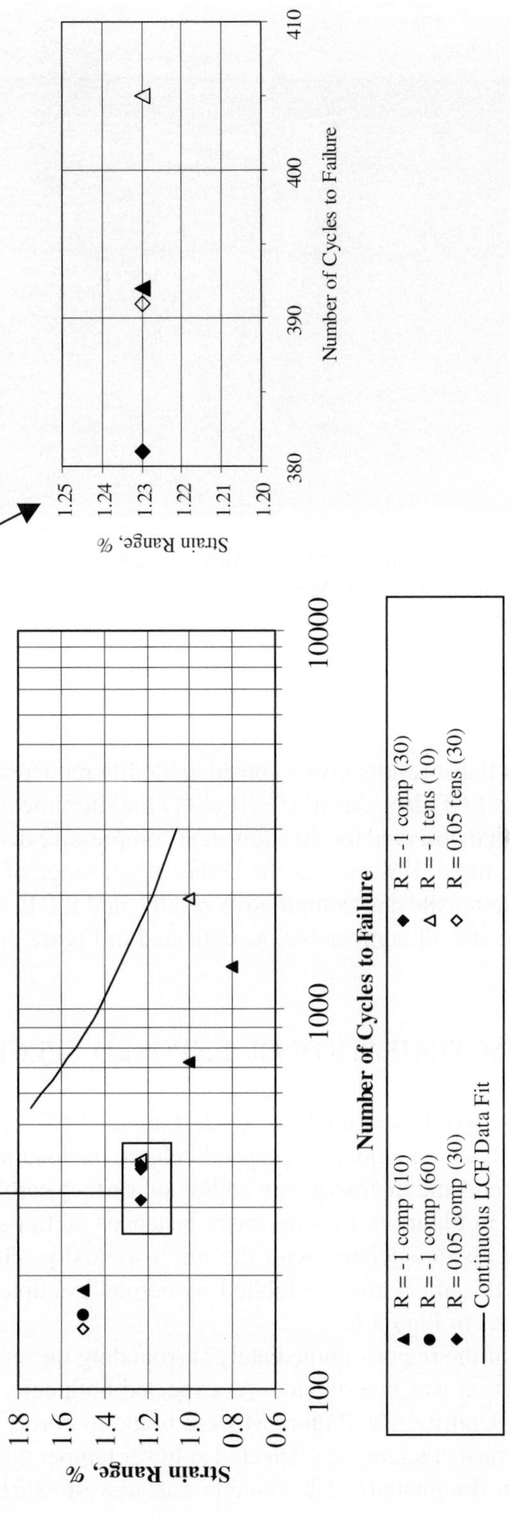

Fig. 5 (a) Strain range versus cycles to failure for continuous LCF fit and LCF tests with peak compressive and tensile dwell periods. (dwell times are given inside the parenthesis within the legend), 950°C 6%/minute) and (b) Dwell tests conducted at $\Delta\varepsilon$1.23%.

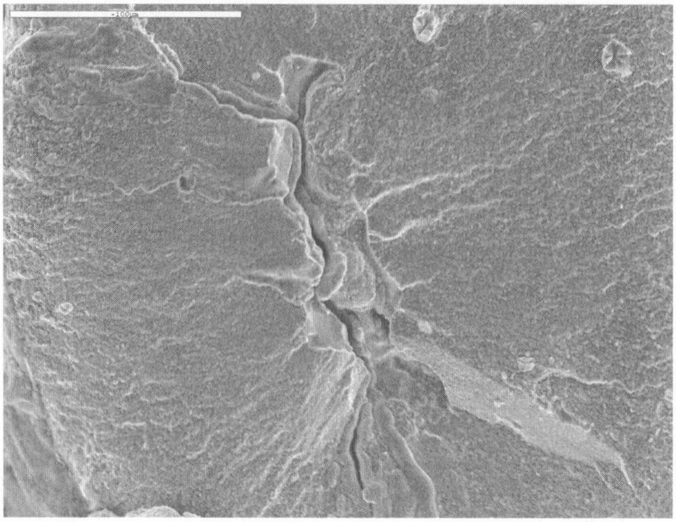

Fig. 6 SEM micrograph of sub-surface microporosity at crack initiation site in <001> CM186LC SX tested at 700°C (6%/min, Rε = 0.05, Δε = 1.2%).[4]

can be seen that all dwell test results give a considerable life reduction when compared with the fit to the continuous LCF data. At Δε 1% (R = –1) the life time of the 10 minute tensile dwell test is over twice that observed for the equivalent compressive dwell test, thus continuing the previously reported trend. However, at the higher strain range of 1.23%, it is noted that the R = 0.05, 30 minute tensile and compressive results, and the R = –1, 10 minute tensile and compressive results are all comparable, as indicated in Figure 5a.

CHARACTERISATION OF EXPOSED SPECIMENS

The fracture surfaces of <001> oriented low cycle fatigue (LCF) specimens tested at 700 and 850°C without dwell were found to lie perpendicular to the loading axis. Cracks initiated from sub surface interdendritic microporosity and/or subsurface carbides, a typical example of the former is shown in Figure 6. In some cases instances of Tantalum and Hafnium rich carbide particles were also associated with the microsporosity. The fracture surfaces of specimens tested at both temperatures exhibited numerous examples of fatigue striations, similar to those observed in Figure 6.[4]

The microstructure of the regions immediately surrounding the subsurface microporosity that caused crack initiation was investigated. As expected both eutectic colonies (Figure 7) and regions of regular dendritic γ/γ' (Figure 8) were found to border micropores.[4]

Some secondary surface cracking was detected at lower temperatures. At 950°C however, surface crack initiation dominated1. The fracture surfaces of <001> specimens tested at

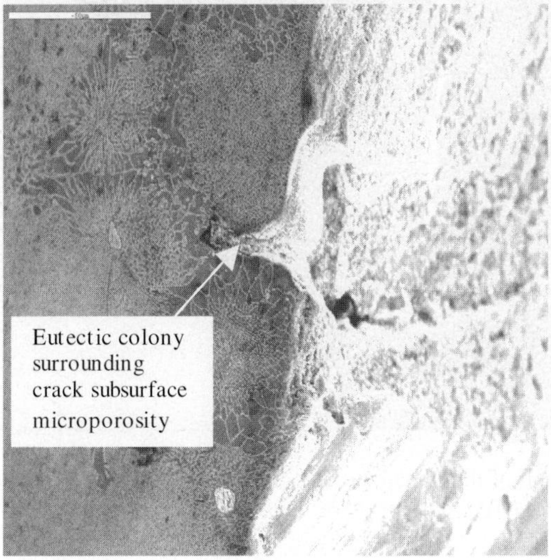

Fig. 7 SEM micrograph of eutectic phase microstructure adjacent to microporosity at crack initiation site of <001> CM186LC SX specimen tested at 850°C (6%/min, $R\varepsilon = 0.05$, $\Delta\varepsilon = 1.0\%$).[4]

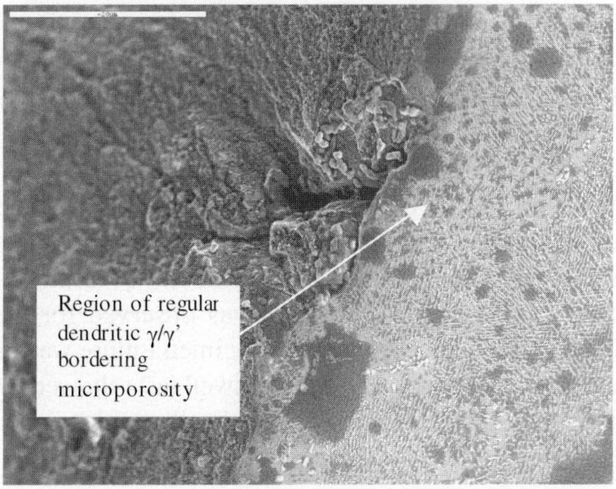

Fig. 8 SEM micrograph of regular γ/γ' microstructure adjacent to microporosity at crack initiation site of <001> CM186LC SX specimen tested at 850°C (6%/min, $R\varepsilon = 0.05$, $\Delta\varepsilon = 1.0\%$).[4]

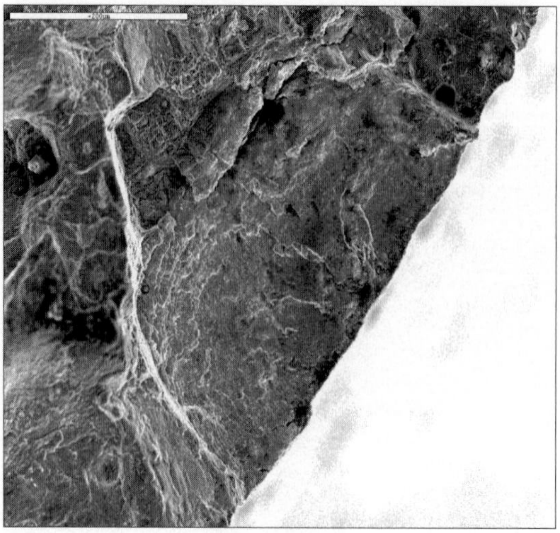

Fig. 9 SEM micrograph of thumbnail surface crack initiation site in <001> CM186LC SX tested at 850°C, (6%/min, $R\varepsilon = 0.05$, $\Delta\varepsilon = 1.0\%$).[4]

950°C were again perpendicular to the loading axis, but cracks were initiating from the surface. The density of surface cracks was much greater than at the lower temperatures and the number present increased as the applied strain range was reduced.[1]

Those specimens dominated by surface cracking exhibited multiple thumb nail initiations sites around the circumference of the fracture surface. In some cases cracks had initiated around the entire circumference of the specimen. A typical example of surface initiation is shown in Figure 9.[4] It was noted that the surface was particular susceptible to cracking at locations where the lines of interdendritic eutectic intercepted the surface, as shown in Figure 10.[4]

A number of LCF plus 2 minute compressive and tensile dwell period specimens were also examined. At 850°C an even distribution of surface crack initiation sites was observed around the circumference of the specimen. Compressive dwell specimens showed a lower surface crack initiation density. Surface cracks were heavily oxidised and aluminium, chromium and cobalt oxides were present. No evidence of γ' rafting was observed at this temperature.[1] A greater crack initiation density was observed for the 950°C LCF with 2 minute tensile and compressive dwell periods. Specimen failure was usually driven by the propagation of a single dominant crack. The tensile dwell tests showed evidence of γ' rafting, however this was not observed for the compressive dwell tests.[1]

MULTI-SCALE CRYSTALLOGRAPHIC CONSTITUTIVE MODELLING

The microstructure of as-cast CM186LC SX is heterogeneous due to the presence of the regular dendritic γ/γ' and the highly irregular γ/γ' eutectic colonies, as described previously.

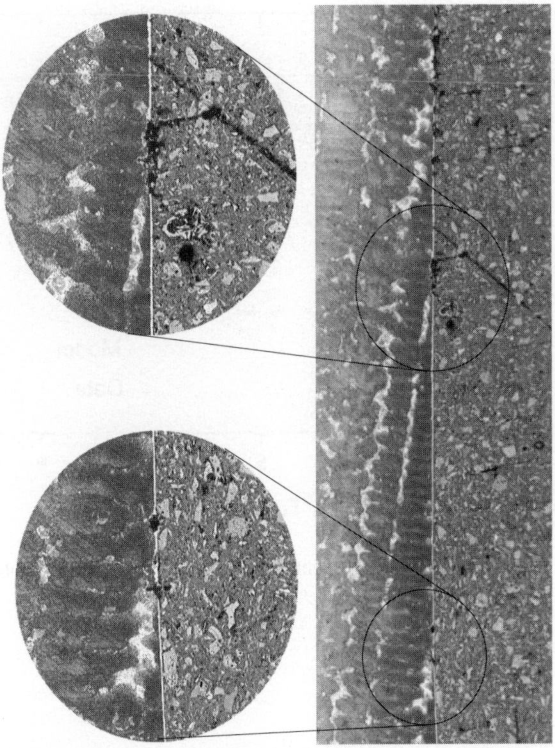

Fig. 10 Optical micrograph of surface cracks which correspond to eutectic regions in <001> CM186LC SX specimen tested at 850°C, (6%/min, $\Delta\varepsilon$ = 1.0%).[4]

A constitutive deformation model has therefore been developed that accounts for this heterogeneity and can be used to quantify its effect on the overall deformation behaviour.

The approach relies upon defining representative volume elements (RVE's) of the microstructure of CM186LC SX in the as-cast condition. The behaviour of the regular dendritic regions is assumed to be that of a eutectic free single crystal superalloy with the same γ' volume fraction. Since the eutectic regions predominately consist of course γ' lamella, they were considered to have the same elasto-visco-plastic behaviour as that of pure Ni_3Al. The behaviour of both regions of the alloy are described by a multi-scale rate dependent crystallographic formulation in terms of material parameters that are explicitly dependent on the characteristics of the γ' precipitate population at the micro scale.[5]

The constitutive formulation was implemented numerically in the finite element (FE) method using a large strain algorithm with an implicit time integration procedure.[6] FE calculations on periodic unit cells of RVE's were performed under 2D generalised plane strain conditions using multi-phase elements. In this approach the constitutive behaviour of each material point within each element is based on the material region to which it belongs. Thus, a single element can incorporate more than one phase.

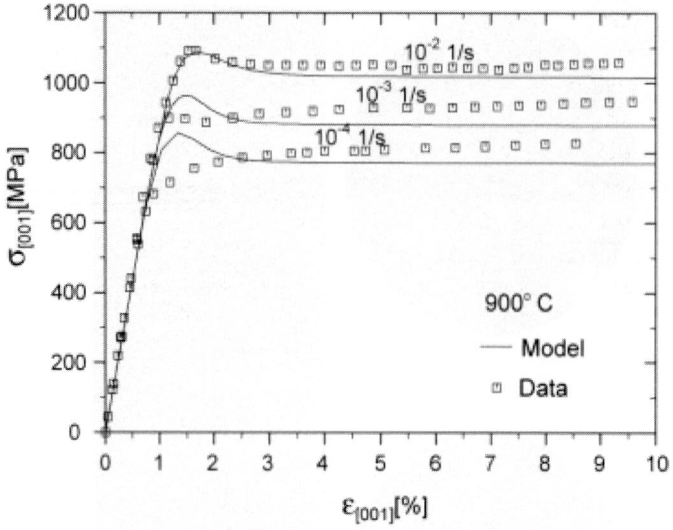

Fig. 12 Comparison between <001> experimental data and the predicted monotonic response given by the RVE, at 900°C and different strain rates.[6]

This work revealed that the smallest RVE size that provided a correct representation of the uniaxial material response to be 650 × 650 μm. The overall material response was found to be strongly dependent on the volume fraction of eutectic, but weakly dependent on the morphology of the eutectic regions.[6]

Finite element simulations of monotonic tensile tests conducted on material oriented in the <001>, <010> and <100> crystallographic directions for different strain rates at 900°C were conducted and compared with monotonic tensile test results. Initially, the model accurately predicted the steady state response,[6] but there were some discrepancies in the elastic regime. Improvements to the estimated elastic modulus values for the eutectic regions lead to a better overall agreement between the model and the test results. A comparison between the simulated and actual test results is given in Figure 12.[5]

Further work will introduce three dimensional RVE's[5] to the model, allowing the nature of the eutectic microstructure along the <001> single crystal growth axis to be considered, leading to a more accurate representation of the deformation behaviour along this axis. Following the model validation, an FE simulation of an IGT blade has been conducted, and the predicted deformation and temperature histories compared with industrial experience.[5]

SUMMARY

The high temperature low cycle fatigue properties of CM186 LC SX have been investigated. The effect of orientation, temperature, strain rate and mean stress on the cyclic stress strain and strain life characteristics of the material have been studied. Tests on hollow specimens were conducted to simulate the thin walls found in turbine blades. Comparison between solid and hollow results tested under comparable conditions showed the hollow properties to give significantly lower cyclic lives, although further work is required to clarify certain discrepancies in the data. The interaction of creep and fatigue loading was also investigated by applying peak tensile and compressive dwell periods to standard low cycle fatigue cycles, a significant life reduction was observed in all cases, and comparable results were obtained for high strain range tests with dwell periods between 10 and 60 minutes. The fracture surfaces of exposed specimens have been studied in order to obtain a good understanding of the specimen failure modes. Finally, a novel multiscale constitutive modelling technique that accounts for the heterogeneity of the microstructure of as-cast CM186LC SX has been developed.

ACKNOWLEDGEMENTS

The authors wish to thank ALSTOM Power for permission to publish this work and the UK Government's Department of Trade and Industry for providing financial support. Support from the other COST 522, Work Package 1.1 partner organisations is also gratefully acknowledged.

REFERENCES

1. D. W. Bale, et al.: 'The Low Cycle Fatigue Behaviour of as Cast Single Crystal CM186LC', *Materials for Advanced Power Engineering 2002, Proceedings of 7th Liege Conference*, J. Lecomte-Beckers, et al. eds., Forschungszentrum, Julich, 2002, 149–158.
2. I. M. Wilcock, et al.: 'The Creep Behaviour of as cast SX CM186LC at Industrial Gas Turbine Operating Conditions', *Materials for Advanced Power Engineering 2002 Proceedings of 7th Liege Conference*, J. Lecomte-Beckers, et al. eds., Forschungszentrum, Julich, 2002, 139–158.
3. C. K. Bullough, M. Toulios, M. Oehl and P. Lukas, 'The Characterisation of the Single Crystal Superalloy CMSX-4 for Industrial Gas Turbine Blading Applications', *Materials for Advanced Engineering 1998*, *Proceedings of 6th Liege Conference*, J. Lecomte-Beckers, et al. eds., Forschungszentrum, Julich, 1998, 861–878.
4. P. Mulvihill, et al.:, 'CM186 LC SX Low Cycle Fatigue Crack Initiation: Fractography and Metallography', Internal Report, Powergen, 2002.
5. G. Regino: 'A Multiscale Constitutive Approach to Model the Mechanical Behaviour of Inhomogeneous Single Crystal Superalloys', Ph.D. Thesis, Department of Mechanical Engineering, Imperial College, London, 2003.
6. G. M. Regino, E. P. Busso, N. P. O'Dowd and D.Allen: 'A Multiscale Constitutive Approach to Model the Mechanical Behaviour of Inhomogeneous Single Crystal Superalloys: Application to as-Cast SX CM186LC', *Materials for Advanced Power Engineering 2002*, *Proceedings of 7th Liege Conference*, J. Lecomte-Beckers, et al., eds., Forschungszentrum, Julich, 2002, 283–291.

Modelling of Creep in Nickel Based Superalloys

A. P. MIODOWNIK
Thermotech Ltd.
Surrey Technology Centre, The Surrey Research Park
Guildford GU2 7YG, UK

X. LI
Sente Software Ltd.
Surrey Technology Centre, The Surrey Research Park
Guildford GU2 7YG, UK

N. SAUNDERS
Thermotech Ltd.
Surrey Technology Centre, The Surrey Research Park
Guildford GU2 7YG, UK

J. -PH. SCHILLE
Sente Software Ltd.
Surrey Technology Centre, The Surrey Research Park
Guildford GU2 7YG, UK

ABSTRACT

A new software programme, JMatPro, is used to obtain steady state creep rates and creep rupture life for multi-component commercial nickel based alloys. A key feature of the programme is that overall properties are obtained through calculation and combination of the properties of individual phases. This includes thermodynamic properties, thermo-physical and physical properties, mechanical properties, and derivative properties such as anti-phase boundary and stacking fault energies. Access to such properties allows the self-consistent calculation of the required input parameters for a standard dislocation creep equation. This is in contrast to many previous attempts that have had to use empirical values for various critical parameters, such as stacking fault energies, anti-phase boundary energies and various elastic moduli, which are all dependent on temperature as well as composition. It is shown that calculations can be made for any desired alloy by entering only the composition of the alloy, the size(s) of γ and/or γ'' (if present), the creep temperature and the applied stress. Good agreement has been obtained between the calculated secondary creep rates/rupture life and the observed results for many commercial nickel based superalloys.

INTRODUCTION

Understanding the factors affecting the performance of Ni-based superalloys is vital to the industrial gas turbine industry and many formulations have been proposed to calculate the secondary creep rates.[1–5] While engineering requirements require the properties of the alloy as a whole to be described, it is the properties of individual phases and microstructural features that play the fundamental role in determining overall properties and which are a necessary input into physically based models. However, values for critical input parameters referring to individual phases are often missing, especially for complex multi-component industrial alloys. In the absence of data related to specific alloys and temperatures, previous treatments have often had to use empirically determined values for important parameters such as the modulus, stacking fault energy or APB energies. This makes it difficult to judge whether a particular approach would still be applicable outside the limited range of alloy composition or temperatures used to justify the initial equations. More importantly, the effect of changing composition and other variables within or outside specification cannot be estimated properly when input parameters are assumed independent of these variables.

The present paper has several objectives. Firstly to show that it is now possible to systematically calculate secondary creep rates and stress rupture life, not only for the wide range of superalloys currently available, but also for new combination of elements that are being suggested for the design of the next generation of alloys. It will be shown that most of the required parameters can be calculated, thus leaving far fewer factors to be empirically determined by comparing theory and experiment. The present approach does not remove the necessity of making experiments, but it does substantially reduce the degree of empiricism inherent in many previous treatments. This can markedly reduce the number of trial experiments when developing new alloys, testing the effect of variations within specification limits, and determining the permissible range of heat-treatments.

CALCULATION METHOD

The present work uses a common formulation (eqn 1) for the secondary creep rate[2] that features both a back stress function and takes the stacking fault energy (γ_{SFE}) explicitly into account.[6] This approach was selected as it contains parameters that have an identifiable physical basis and which can be calculated self-consistently.

$$\dot{\varepsilon} = AD_{eff}\left[\frac{\gamma_{SFE}}{Gb}\right]^{n}\left[\frac{\sigma-\sigma_0}{E}\right]^{m} \tag{1}$$

where $\dot{\varepsilon}$ is the secondary creep rate, A is a structure-dependent parameter, D_{eff} is the effective diffusion coefficient, γ is the stacking fault energy of the matrix, b is the burgers vector, σ is the applied stress, σ_0 is the back stress, with G and E the shear and Young's modulus of the matrix phase at the creep temperature respectively. The back stress σ_o, is calculated following the treatment of Lagneborg and Bergman,[5] setting $\sigma_o = 0.75\sigma$ when $\sigma < \frac{4}{3}\sigma_p$, (where σ_p is the critical back stress from strengthening due to precipitates) and $\sigma_o = \sigma_p$ when $\sigma > \frac{4}{3}\sigma_p$. The exponents n and m exhibit a range of values in the literature, but in this paper have been given fixed values of $n = 3$ and $m = 4$.

A basic feature of JMatPro is that the overall properties of complex materials are calculated by combining the properties of individual phases. It is therefore necessary to start by calculating reliable volume fractions of all the constituent phases over the temperature range of interest. This is achieved for multi-component alloys by combining a suitable thermodynamic database[7, 8] with a fast minimisation engine,[9] following the well-established CALPHAD technique.[10] For the calculation of secondary creep rates in the alloys concerned here, the key phases are γ, γ' and γ''.

The next step is to consider the properties of individual phases, such as the γ_{SFE} for the matrix phase, which is directly required by eqn. 1. Experimental values of γ_{SFE} are usually only available for rather simple alloys and over a limited temperature range, often just room temperature.

JMatPro calculates as many parameters as possible from more basic data held in the programme, in order to minimise the creation of separate databases. While the latter are required for moduli and diffusion calculations, it will be seen that this can be avoided in most other cases. For example, the γ_{SFE} at the creep temperature is calculated from the Gibbs free energy difference between *fcc* and *hcp* structures,[11] which is readily available by using a CALPHAD calculation.

A second example is given by the calculation of the critical back stress σ_p, which is directly related to the strengthening contribution of the γ' and γ'' phases, through a proportionality constant and the volume fractions of γ' and γ''. For the case of γ', this can be calculated using the approach detailed in Saunders et al.[9] For the case of γ'' a slightly different approach is used – in this case a contribution for strain hardening is included due to the large lattice misfit in the major axis of the γ'' precipitates.[12] Fig.1 shows a comparison of calculated and measured[3, 5, 13–15] values for σ_p. The agreement is quite good. Further work is being undertaken to develop a more general approach that is applicable to other strengthening phases.

The various moduli have been calculated using databases for E and Poisson's ratio (ν), such that only the composition of the matrix and temperature are required to obtain G and E[16]. The diffusion coefficient D_{eff} in the above equation is concentration dependent and calculated using the following expressions.[17]

$$D_{eff} = D_o \exp\left(-\frac{Q_{eff}}{RT}\right) \tag{2a}$$

where

$$D_o = \sum_i x_i D_i^0 \tag{2b}$$

and

$$Q_{eff} = \sum_i x_i Q_i \tag{2c}$$

In equations 2b and c, x_i is the mole fraction of element i in the matrix and D_i^0 and Q_i are, respectively, the frequency factor and activation energy for diffusion of element i in the matrix. Although cross terms are neglected, this approach has been previously validated during application to other kinetic phenomena such as TTT curves[9,17] and particle coarsening.[18]

Reference to eqn 1. shows that the only remaining floating parameter is A. If the above treatment is valid, only a single value of A should be required to make viable predictions for secondary creep rates, with the proviso that all the alloys concerned have similar microstructural features. The latter requirement signals that difficulties may arise in achieving a single treatment for both solid solution alloys and alloys containing precipitates. In practice

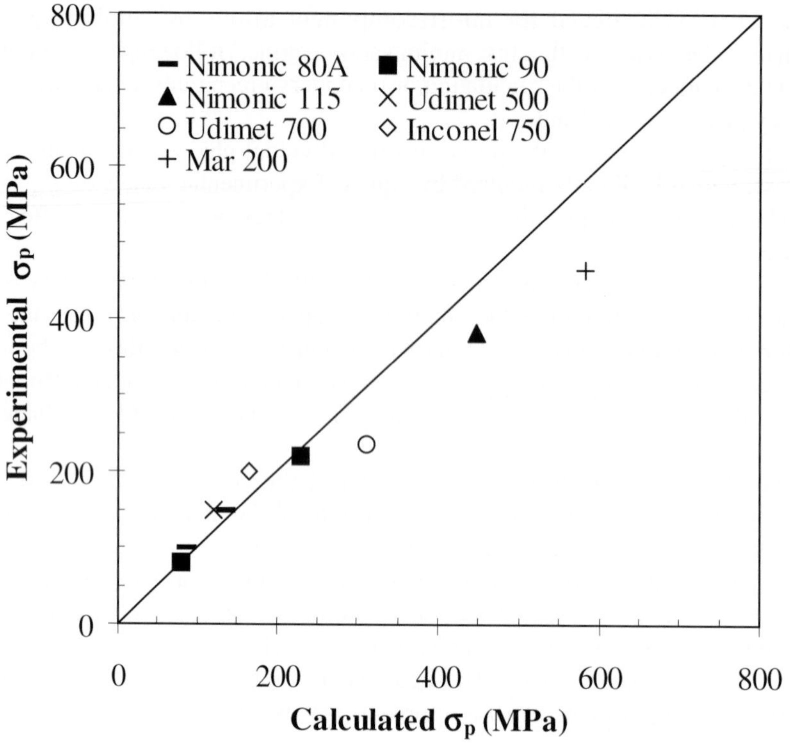

Fig. 1 Comparison between calculated and experimental critical back stress. (note for Nimonic 90 and 80A two results are shown, corresponding to different creep temperatures with subsequently different amounts of γ').

we have found that A is a function of the volume fraction (V_f) of γ' and γ'' and the following equation has been used:

$$A = \frac{10.0^{19.7}}{V_f + 0.005} \tag{3}$$

The need for such an expression arises from the much faster creep rates for solid solution alloys, in comparison to γ' or γ'' hardened alloys. However even after using eqn. 3, the general scatter for solid solution alloys is greater, which suggests that additional variations in the underlying hardening mechanisms need to be addressed.[19]

RESULTS

Figure 2 shows the correlation for a very wide range of polycrystalline alloys calculated using the approach described above. The experimental data for creep rates has been drawn

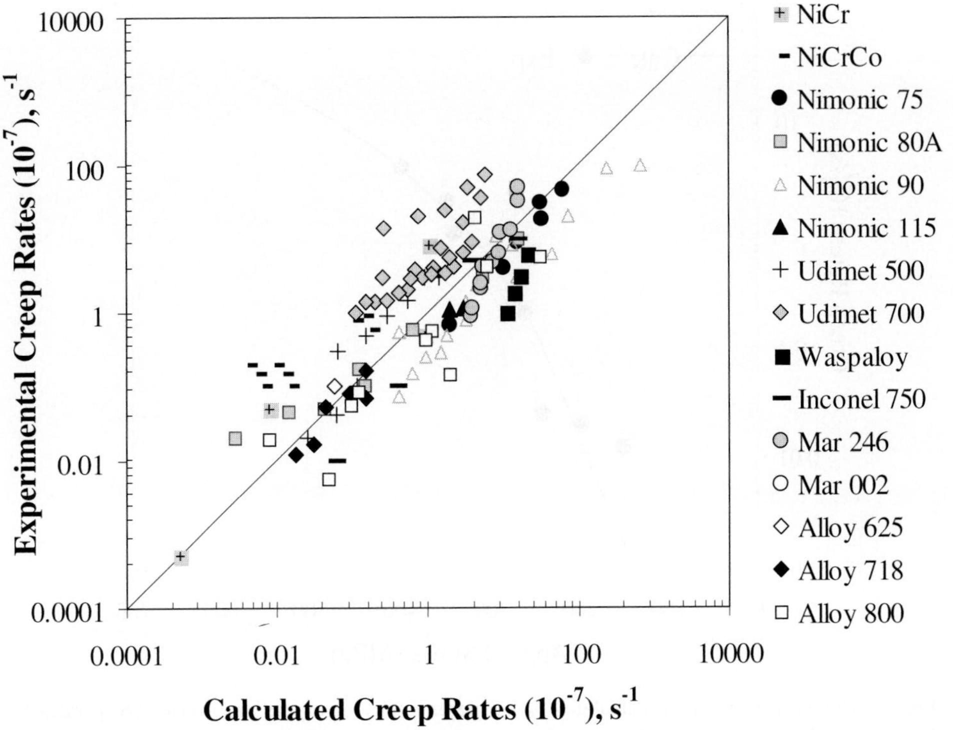

Fig. 2 Comparison between calculated and experimentally observed secondary creep rates for a wide range of Ni-based superalloys.

from a wide range of references.[2, 3, 5, 13, 14, 20–25] Where possible, we have calculated σ_p using information of γ' and γ'' particle sizes reported by the authors, or have used values which can be reasonably estimated from analogous alloys. A specific comparison for the stress dependence of secondary creep rates in Nimonic 80A is given in Figure 3. Curves can be generated to examine the effect of all the other variables in eqn. 1. together with changes in composition and temperature.

The behaviour of single crystal alloys has also been examined, but has not been included in the present paper as this requires a more complicated treatment involving orientation factors and also a slightly different value of A.

Extension to the Calculation of Rupture Strength

As rupture strength is an alternative design criterion in many practical cases, the calculation procedure has been extended to include this property by using the relationship suggested by Davies and Wilshire.[26]

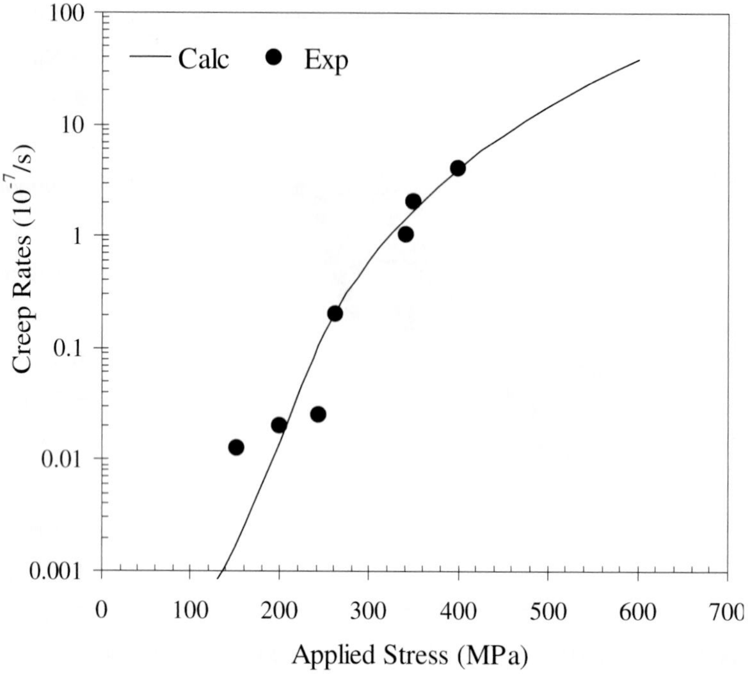

Fig. 3 Comparison between calculated and experimentally observed[13] secondary creep rates for Nimonic 80 A Ni-based superalloy as a function of applied stress.

$$t_r = a\dot{\varepsilon}^b \tag{4}$$

where t_r is the time to rupture, a and b are constants and $\dot{\varepsilon}$ is the secondary creep rate. The correlation between the experimental secondary creep rates and creep rupture life for various Ni based superalloys is given in Figure 4. The experimental data has been drawn from the following references.[23, 27–31]

This useful relationship will, however, only hold if the structure of the alloy remains reasonably stable with time during creep. If the situation refers to short times and relatively low temperatures, then it should still be possible to compare calculation with experiment, even if applied stresses are high, as the microstructure will remain relatively stable.

This has been checked for γ' or γ" hardened disk alloys against the 1000 hrs. rupture strength reported by Sims et al.[32] For the most part comparison is made with the rupture strength given at 760°C by Sims et al.[32] because many disk alloys have a final heat treatment close to 750°C. In this case the amount of γ' and γ" formed after heat treatment will remain stable and coarsening rates sufficiently low to prevent substantial degradation in creep rupture life. Figure 5 shows a comparison of calculated and experimental stress rupture behaviour for a variety of disk alloys.

For two cases, the final heat treatment is at 850°C and we have made the calculation at 870°C for comparison with experimental rupture strength reported at this temperature. For

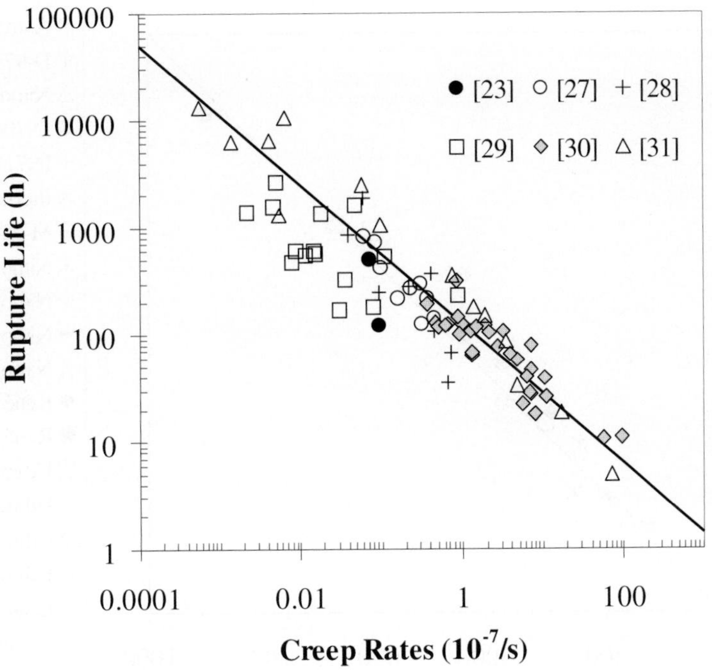

Fig. 4 Creep rupture life versus creep rates.

four alloys the final heat treatment temperature is closer to 650°C and we have therefore used the 650°C rupture stress for comparison. For the case of Udimet 720 only the 650°C rupture strength is given. In this case the amount of γ' at 650° and 750°C is very similar and we have used the strength calculated at 750°C in calculating 650°C rupture strength. For the present, we have not included alloys which exhibit carbide hardening or solid solution hardened alloys, where we have noted that calculated results invariably underestimate rupture strengths. Further work will be undertaken to address this issue.

Where the results presented in standard source books give no information on particle size, we have estimated σ_p in the following way. First, the strength of the solid solution matrix is calculated based on the composition of the matrix at the heat treatment temperature, using a standard grain size of 100 μm. As Sims et al.[32] also report the final 0.2% proof stress after heat treatment, the strengthening contribution of the γ' and γ'' particles can therefore be readily extracted.

DISCUSSION

Considering the relatively simple approach used in the present study, there is really quite remarkable success. It is well understood that the complete modelling of creep is more

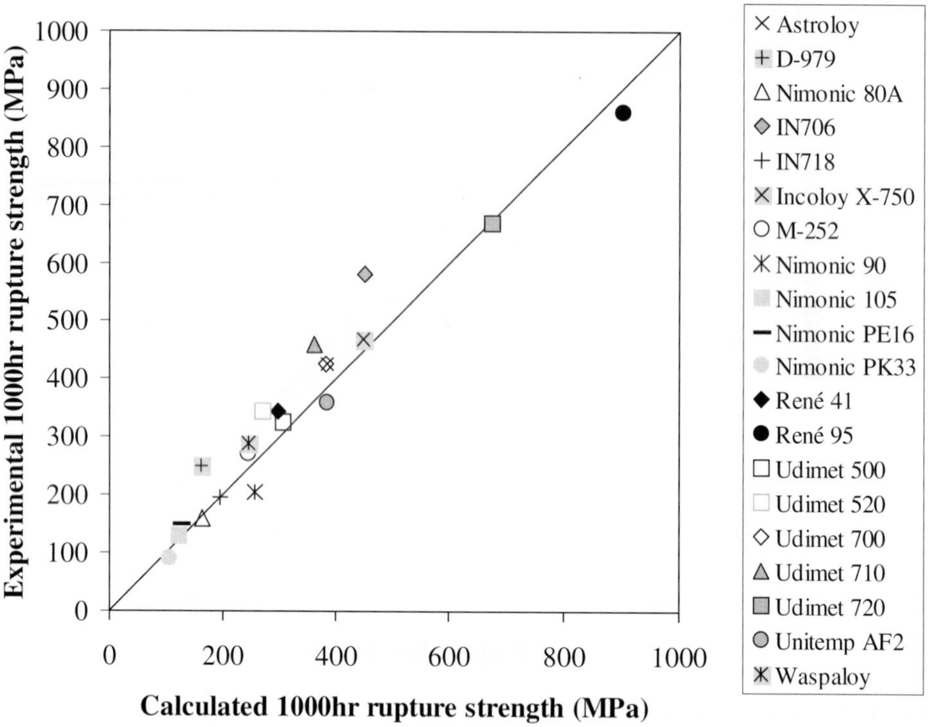

Fig. 5 Comparison between experimental[32] and calculated 1000 hours rupture strength for γ' and γ'' hardened Ni-based superalloys.

complex. For example, a more detailed treatment of microstructural stability should be included because, when creep occurs in the stress range where $\sigma > \frac{4}{3}\sigma_p$ and the temperature is high enough, then the secondary creep rate will vary with time because σ_p will change as γ' and γ'' coarsen. In this case, a coarsening module available within JMatPro[9, 18] allows an estimate of the resulting degradation of properties and will be integrated within the creep module. Also, a better representation of damage accumulation would lead to a more explicit formulation for the tertiary stage of creep and primary creep must be included if a full creep curve is to be calculated.

However, even with its present shortcomings, it is clear that the present approach has some distinct advantages over previous methods. Very few empirical parameters are required, the model has an identifiable physical basis and the calculations are self-consistent. One can therefore expect significant advantages when creep properties need to be extrapolated. Because of the time scales involved, creep experiments are usually performed for periods that are shorter than the real exposure times for industrial gas turbines. Secondly, the approach can be used as part of an alloy design process, where comparative behaviour between various alloys is readily obtained without the need for lengthy experiment.

The model will now be extended so that the shortcomings of the present capability can be overcome and a full creep curve modelled. It is intended to do this by looking at a cumulative function that includes an expression for primary creep as well as a better representation of the effect of tertiary creep. The treatment where there is γ' and/or γ'' coarsening, or where there is a phase transformation simultaneously occurring, needs to be improved. For the latter case, TCP phases may remove slow moving elements from the matrix or metastable γ'' may transform to the stable δ phase. These are further examples where the secondary creep rate will be time dependent and an integration function will need to be applied so that deformation as a function of time is better represented. It should also be possible to include the susceptibility to rafting for single crystals as JMatPro has a capability to accurately calculate γ/γ'' lattice misfit,[33] as well as make modulus calculations.

CONCLUSION

It is possible to calculate steady state creep rates and creep rupture life with the aid of a new software programme, JMatPro. The required input parameters are calculated systematically from the basic properties of individual phases. This substantially reduces the degree of empiricism inherent in many previous treatments and reduces the number of trial experiments when developing new alloys, testing the effect of variations within specification limits and determining the permissible range of heat-treatments. Self-consistent calculations can be made for any desired alloy by entering only the composition, the size of γ' and γ'' particles, the creep temperature and the applied stress. Good agreement has been obtained between the calculated secondary creep rates/rupture life and the observed results for many commercial Ni-based superalloys.

REFERENCES

1. B. Wilshire and H.E. Evans: *Creep of Metals and Alloys*, Institute of Materials, 1985.
2. X. S. Xie, G. L. Chen, P. J. McHugh and J. K. Tien: *Scripta Metallurgica*, **16**, 1982, 483.
3. W. J. Evans and G. F. Harrison: *Metal Science*, **10**, 1976, 307.
4. H. Burt, J. P. Dennison and B. Wilshire: *Metal Science*, **13**, 1979, 295.
5. R. Lagneborg and B. Bergman: *Metal Science*, **10**, 1976, 20.
6. C. R. Barrett and O. D. Sherby: *Trans. Met. Soc. AIME*, **233**, 1965, 1116.
7. N. Saunders: *Superalloys 1996*, R. D. Kissinger et al. eds., TMS, 1996, 101.
8. N. Saunders, M. Fahrmann and C. J. Small: *Superalloys 2000*, K. A. Green, et al., eds., TMS, 2000, 803.
9. N. Saunders, X. Li, A. P. Miodownik and J. -Ph. Schillé: *Materials Design Approaches and Experiences*, J. C. Zhao et al. eds., TMS, 2001, 185.
10. N. Saunders and A. P. Miodownik: *CALPHAD – Calculation of Phase Diagrams*, Pergamon Materials Series vol. 1, R. W. Cahn ed., Elsevier Science, **1**, 1998.
11. A. P. Miodownik: CALPHAD, **2**, 1978, 207.

12. X. Li, N. Saunders and A. P. Miodownik: unpublished research.
13. B. Bergman: *Scand. J. Metallurgy*, **4**, 1975, 97.
14. O. Ajaja, T. E. Howson, S. Purushothaman and J. K. Tien: *Materials Science and Engineering*, **44**, 1980, 165.
15. Z. A. Yang, Y. I. Xiao and C. H. Shih: *Materials Science and Engineering A*, **101**, 1988, 65.
16. X. Li, A. P. Miodownik and N. Saunders: *J. Phase Equilibria*, **22**, 2001, 247.
17. X. Li, A. P. Miodownik and N. Saunders: *Materials Science and Technology*, **18**, 2002, 861.
18. X. Li, N. Saunders and A. P. Miodownik: *Metall. Mater. Trans. A*, **33A**, 2002, 3367.
19. Y. Li and T. G. Langdon: *Creep Behaviour of Advanced Materials for the 21st Century*, R. S. Mishra et al. eds., TMS, 1999, 73.
20. W. J. Evans and G. F. Harrison: *Metal Science*, **13**, 1979, 641.
21. R. W. Lund and W. D. Nix: *Acta Metallurgica.*, **24**, 1976, 469.
22. M. C. Chaturvedi and Y. Han: *Superalloy 718 – Metallurgy and Applications*, E. A. Loria ed., 1989, 489.
23. G. B. Thomas and T. B. Gibbons: *Superalloys 1980*, J. K. Tien et al. eds., TMS, 1980, 699.
24. H. L. Eiselstein and D. J. Tillack: *Superalloys 718, 625 and Various Derivatives*, E. A. Loria ed., TMS, 1991, 1.
25. *Engineering Properties of Alloy 800*, publication 3272A, Henry Wiggin and Co. Ltd.
26. P. W. Davies and B. Wilshire: *Structural Processes in Creep*, A. G. Quarrell ed., Iron and Steel Institute, 1961, 34.
27. K. Harris, G. L. Erickson and R. E. Schwer: *Superalloys 1984*, M. Gell et al. eds., TMS, 1984, 221.
28. F. L. Versnyder and M. E. Shank: *Materials Science and Engineering*, **6**, 1970, 213.
29. A. Ferrari: *Superalloys 1976,* Claitor's Publishing Division, 1976, 201.
30. D. M. Shah and A. Cetel: *Superalloys 1996*, R. D. Kissinger et al. eds., TMS, 1996, 273.
31. G. E. Korth: *JOM*, 2000, 40.
32. C. T. Sims et al. eds.: *Superalloys II,* Wiley & Sons, 1987.
33. N. Saunders, X. Li, A. P. Miodownik and J. -P. Schillé: 'The Modelling of Phase Transformations in the Context of a Generalised Materials Property Capability', *Int. Symp. Computational Phase Transformations*, TMS Annual Meeting, San Diego, 2003.

Low Cycle Fatigue Behaviour of CMSX–4 at Elevated Temperature

C. M. Whan and C. M. F. Rae
University of Cambridge/Rolls-Royce University Technology Centre
Department of Materials Science and Metallurgy
Pembroke Street, Cambridge CB2 3QZ, UK

Turbine blades in aero engines are subject to uniquely demanding and complex operating conditions. The creep component of this deformation has been the subject of detailed research; however, relatively little has been published detailing the microstructural effects of low cycle fatigue (LCF) exposure. In this paper, fracture mechanisms and microstructures of R = 0 LCF tests at temperatures of 750, 850 and 950°C are described. Stresses both above and below the yield stress are considered. At 950°C fatigue cracks initiate at the surface oxide, but at the lower temperatures initiation occurs at internal pores. The microstructures of tests cycled to stresses below the yield point bear some resemblance to creep damage; however, the damage was confined to the γ channels and was less severe than after comparable creep exposure. Cycling to a stress above yield resulted in structures closely resembling tensile deformation. At these stresses, the dislocations were able to shear the γ' precipitates. Knowledge of the effect of stress on LCF behaviour will contribute to our understanding of the deformation in real components around stress concentrators.

INTRODUCTION

Turbine blades in aero engines are subject to uniquely demanding operating conditions. It is common to use single crystal Ni-base superalloys for these applications due to their superior resistance to deformation at high temperatures. Accordingly, the creep properties of single crystal Ni-base superalloys have been the subject of detailed research. However, blades in service are subject to cyclic stress rather than simple monotonic creep. The loading cycle and shape mean that modern blades are subject to complex cyclic stresses and temperature distributions. In order to make some progress in understanding the deformation of blades in service it is important to have a baseline set of data from known stress conditions.

Although considerable work has been done on the effects of creep deformation, there is little published data identifying the microstructural effects of low cycle fatigue (LCF) testing in current single crystal superalloys.[1–5] Maclachlan[3,4] investigated creep, HCF and LCF of CMSX-4 at intermediate stress at 750, 850 and 950°C, R = 0 (where R ratio is defined as

$R = \sigma_{min}/\sigma_{max}$). One stress level was investigated at each temperature. He observed deformation structures at 950°C that were similar to those found in creep, displaying dislocation networks spread homogeneously through the γ channels. LCF structures at 750°C were similar to those seen in high cycle fatigue, with heterogeneous structures including persistent slip bands (PSBs) and stacking faults together with some completely undeformed regions. At 850°C, shear bands were still seen (although to a lesser extent), but there were no stacking faults. Lukas also observed PSBs at intermediate temperatures, between 700 and 760°C.[2] However, as the alloy, the stress levels, and the R ratio are unspecified it is difficult to compare these results with other data. PSBs were also seen by Monier et al,[5] although the alloy had a lower γ' fraction and testing was at R = –1. At the higher temperature of 950°C, Brien and Decamps[1] investigated the superalloy AM1 in a series of conditions under strain control. The emphasis was on R = 0 although some R = –1 tests were undertaken. They found that the microstructures of their R = 0 tests could be divided into two categories, isotropic or anisotropic. Anisotropy arose from either deformation partitioning to particular γ channels, or coarsening of the γ' precipitates into oriented rafts. In this paper the failure mechanism and microstructure of the single-crystal alloy CMSX-4 as a function of stress and temperature are examined. The alloy was tested in LCF with R = 0 at the temperatures 750, 850 and 950°C. Deformation was investigated using transmission electron microscopy, and macroscopic failure analysis using scanning electron microscopy.

Macroscopic failure analysis identifies the initiation point and direction of crack growth, shows the balance between crack propagation and final fast fracture and provides insight into how failure occurred in a particular specimen. The dislocation analysis of these specimens is focused on identifying which dislocation types are present, and where they are located with respect to the γ and γ' phases and interfaces. Changes in precipitate morphology and interface coherency are also examined.

MATERIAL AND EXPERIMENTAL PROCEDURES

Material

CMSX-4 is a nickel-base single crystal superalloy. It is approximately 70% an FCC ordered γ' phase, which precipitates coherently in a matrix of disordered FCC γ. The nominal composition is Ni–6.5Cr–9Co–0.6Mo–6W–6.5Ta–3Re–5.6Al–1.0Ti–0.1Hf (compositions by weight).[6]

Fatigue Tests

Single crystal fatigue test pieces of CMSX-4 were supplied by Rolls-Royce plc and machined by Marsdens. The nominal orientation of all samples considered here is [001]. Specimens were tested at 750, 850 and 950°C at the facilities of Rolls-Royce and Cambridge University. The fatigue tests were load controlled low cycle fatigue (LCLCF) with R = 0. Loading was trapezoidal, with a cyclic frequency of 0.25Hz. The stress levels are summarised in Table 1.

Table 1 Summary of LCF testing conditions.[7]

	Temperature (°C)	750	850	950
Lowest Stress	Stress (MPa)	600	660	480
	Fraction of Yield Stress	0.60	0.73	0.75
	Lifetime (cycles)	100000 (unbroken)	81490	69224
	Lifetime (hours)	111 (unbroken)	90.5	76.9
Highest Stress	Stress (MPa)	1100	880	700
	Fraction of Yield Stress	1.10	0.97	1.09
	Lifetime (cycles)	1829	4992	3323
	Lifetime (hours)	2	5.5	3.7

TEM Observations

Specimens were prepared for TEM investigation by cutting discs from the LCF specimens, perpendicular to the loading axis and at least 5 mm from the fracture surface. These discs were mechanically thinned by abrasion, and then electro-polished using a solution of 10% perchloric acid in ethanol. The polishing conditions were 24 V and –5°C. The foils were then examined in a JEOL 2000 FX TEM operating at 200 kV.

EXPERIMENTAL RESULTS

Fracture Surfaces

The fracture surfaces of the broken specimens from Rolls-Royce plc and Cambridge were examined in Cambridge using a JEOL C820 SEM operating at 20 kV. The fracture surfaces of tests undertaken at 750, 850 and 950°C are presented in Figures 1–3. These figures show that across the temperature range 750–950°C, there is a transition in fracture morphology from pore initiation to surface initiation. Figure 1 shows typical surface-initiated fracture in specimens tested at 950°C. At these elevated temperatures, surface oxidation occurs rapidly. Cracks form within the oxide scale upon cycling and initiate crack growth in the bulk specimen. Crack growth continues until a critical crack size is achieved, and the specimen fails by fast fracture to give a faceted final fracture surface. At 850°C a change occurs and the fracture initiates from internal casting pores. The fracture progression is shown in detail in Figure 2a. Here, cracking has begun at a pore below the surface, and a circular region shows the first stage of crack growth, under vacuum, to the surface. After exposure to the

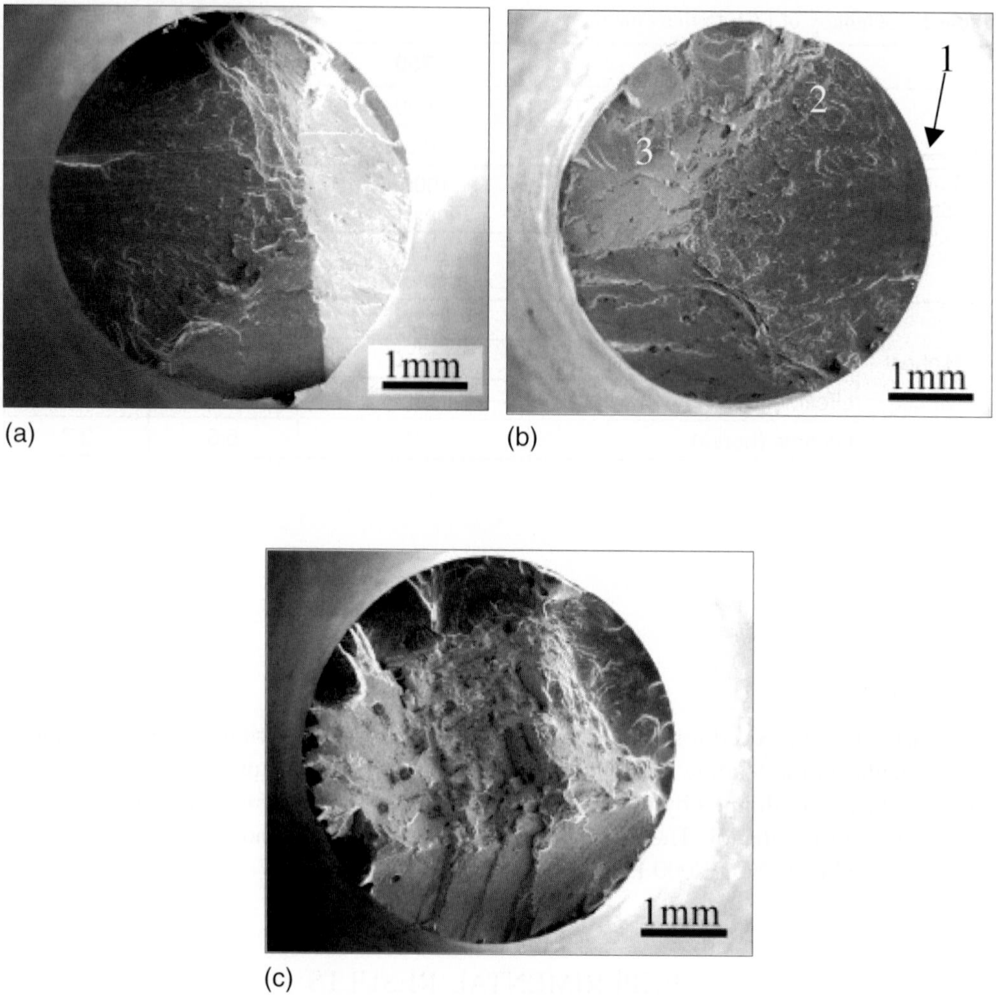

Fig. 1 Fracture surfaces at 950°C at (a) low stress (480 MPa) (b) intermediate stress (c) high stress (700 MPa). Microstructures for a and c are shown in Figures 8 and 9. Features in (b) are highlighted to show the typical progression of fracture in a plain bar due to surface oxide initiated cracking. Here, cracking begins at surface oxide (arrow 1) and subsequently all cracking is under oxidising conditions (2). Final fracture is faceted (3). Figures (a) and (c) demonstrate the same fracture pattern, although it can be seen from (a) that multiple crack initiation is occurring.

environment, oxidation affects the growth causing a change in the appearance of the fracture surface. Figure 3 shows that this fracture mode is also operating at 750°C. The final fracture surface is again faceted. At all temperatures the mechanism remains unchanged with variation in stress, although the fraction of the sample subject to final fast fracture increases with stress.

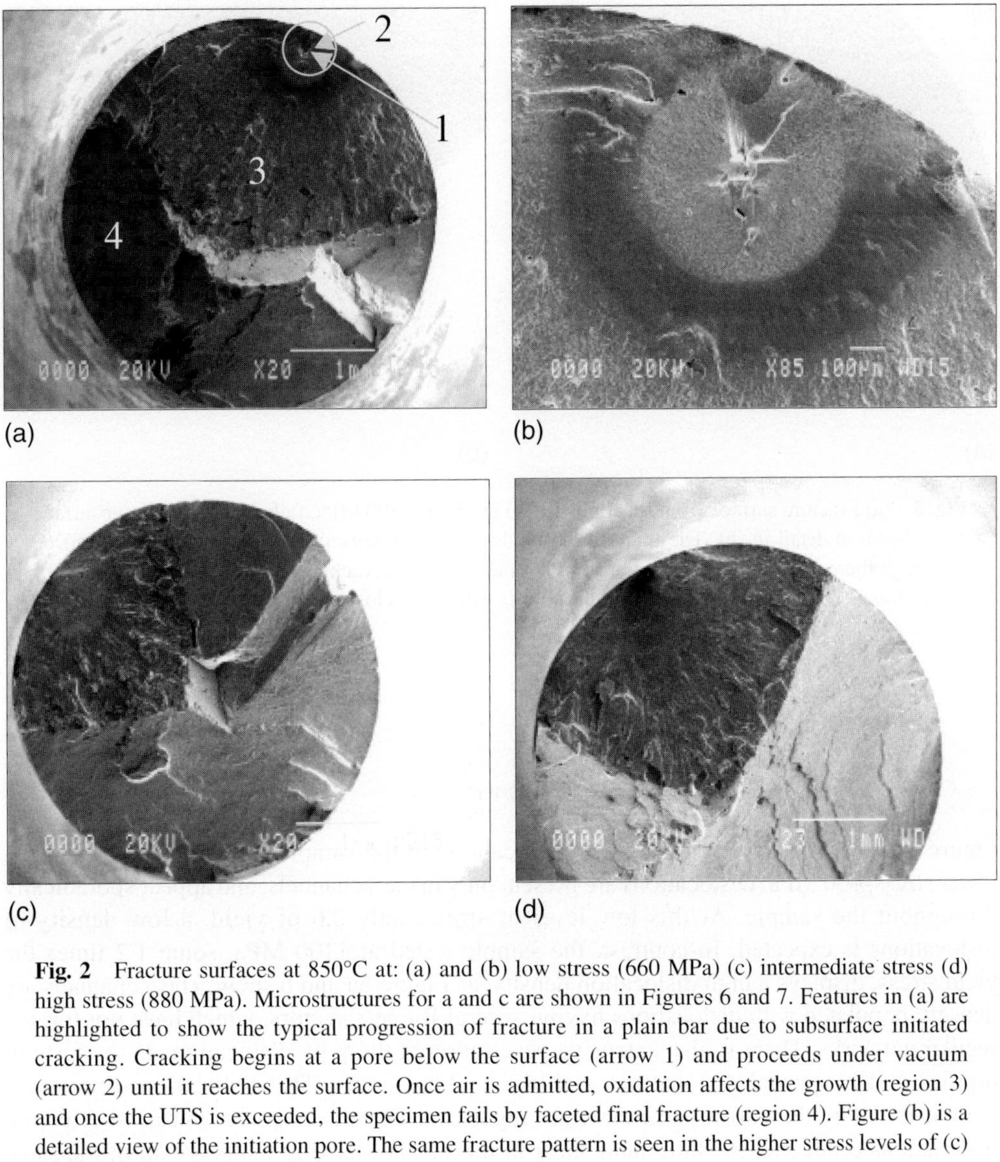

Fig. 2 Fracture surfaces at 850°C at: (a) and (b) low stress (660 MPa) (c) intermediate stress (d) high stress (880 MPa). Microstructures for a and c are shown in Figures 6 and 7. Features in (a) are highlighted to show the typical progression of fracture in a plain bar due to subsurface initiated cracking. Cracking begins at a pore below the surface (arrow 1) and proceeds under vacuum (arrow 2) until it reaches the surface. Once air is admitted, oxidation affects the growth (region 3) and once the UTS is exceeded, the specimen fails by faceted final fracture (region 4). Figure (b) is a detailed view of the initiation pore. The same fracture pattern is seen in the higher stress levels of (c) and (d).

Microstructure

TEM micrographs for high and low stresses at 750, 850 and 950°C are shown in Figures 4-9. The images chosen are representative of the overall microstructures, and the microstructural features are summarised below.

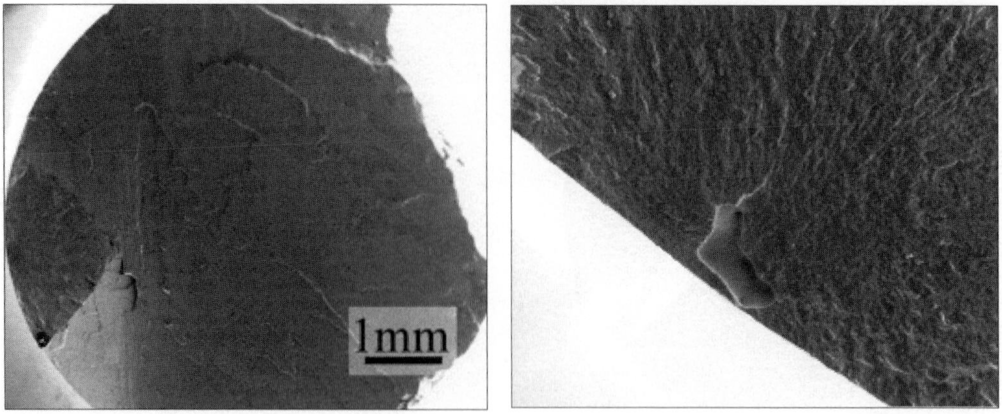

(a) (b)

Fig. 3 (a) Fracture surface from specimen tested at 750°C 1100 MPa, failure initiated at subsurface pore shown in detail in (b). The fracture morphology is essentially the same as that seen at 850°C. Although there is no distinct region of growth under vacuum, cracking has begun at a pore and (b) shows this is simply because the initiation pore is extremely close to the surface.

750°C

Figure 4 shows a very low density of dislocations in the sample tested at 750°C and the lower stress, 600 MPa. Dislocations are present only in the γ channels, and appear sporadically throughout the sample. At this low level of stress, only 0.6 of yield, a low density of dislocations is expected. In contrast, the sample tested at 1100 MPa, some 1.2 times the yield stress, displays a high dislocation density, as Figure 5a and b show. The γ channels are densely populated with dislocations having several Burgers vectors, which have not formed regular networks. There is also extensive penetration of the γ' by dislocations in the form of dipoles and by partial dislocations trailing stacking faults. The dislocation dipoles are demonstrated by the mirror pairs shown in Figure 5b. There is extensive intrusion of partial dislocations producing stacking fault shear in the γ'. The stacking faults are all of the same type, and are on the same plane over the extensive areas observed.

850°C

TEM micrographs from LCF tests conducted at 660 and 880 MPa are shown in Figures 6 and 7 respectively. These micrographs clearly illustrate that dislocation activity is limited to the γ. The low stress sample, tested at 660 MPa, shows tangles of dislocations that are beginning to form irregular networks. After LCF at 880 MPa, the dislocation networks are more developed, but still irregular, and the deformation is more homogeneously spread throughout the γ channels. The dislocation structures after LCF testing at both 660 and 880 MPa

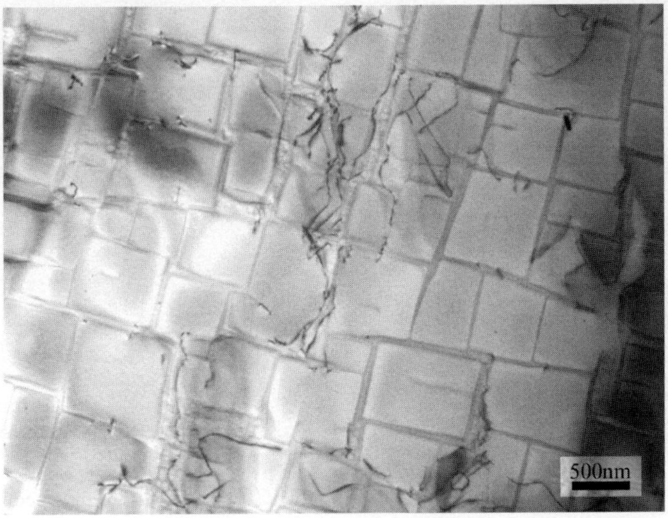

Fig. 4 Transmission electron micrograph of 750°C 600 MPa LCLCF specimen, sectioned perpendicular to tensile [001] direction. Bright field γ = [001]. Very little dislocation debris, most channels completely free of dislocations. No cutting of γ'.

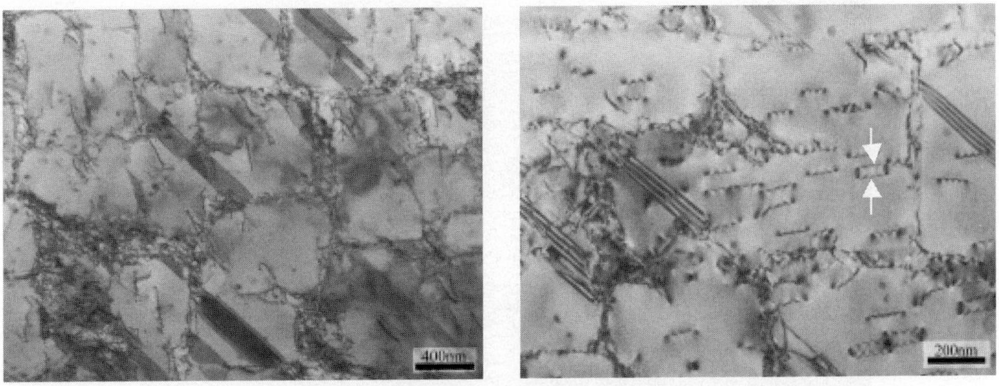

(a) (b)

Fig. 5 Transmission electron micrograph of 750°C 1100 MPa LCLCF specimen, sectioned perpendicular to tensile [001] direction. (a) γ = 220, A high dislocation density can be seen, including stacking faults and dislocation dipoles and (b) γ = 020 Dipoles can clearly be seen as mirror pairs. One such pair is marked with arrows.

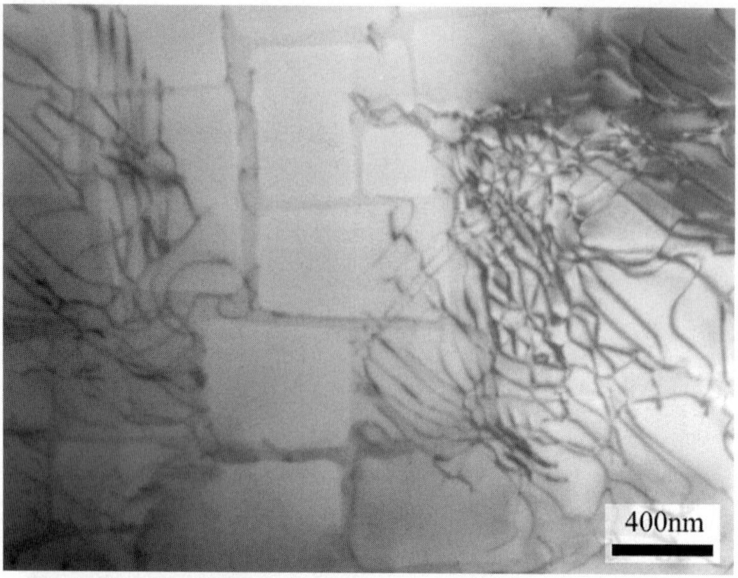

Fig. 6 Transmission electron micrograph of 850°C 660 MPa LCLCF specimen, sectioned perpendicular to tensile [001] direction. Inhomogeneous deformation in the γ, irregular networks beginning to form.

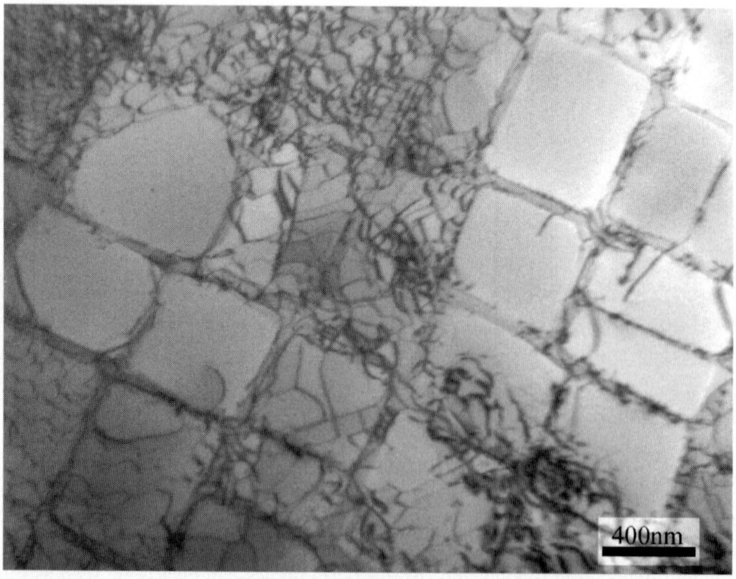

Fig. 7 Transmission electron micrograph of 850°C 880 MPa LCLCF specimen, sectioned perpendicular to tensile [001] direction. Deformation is more homogeneously spread through the γ channels and networks, although still irregular, are more developed than in the low stress sample.

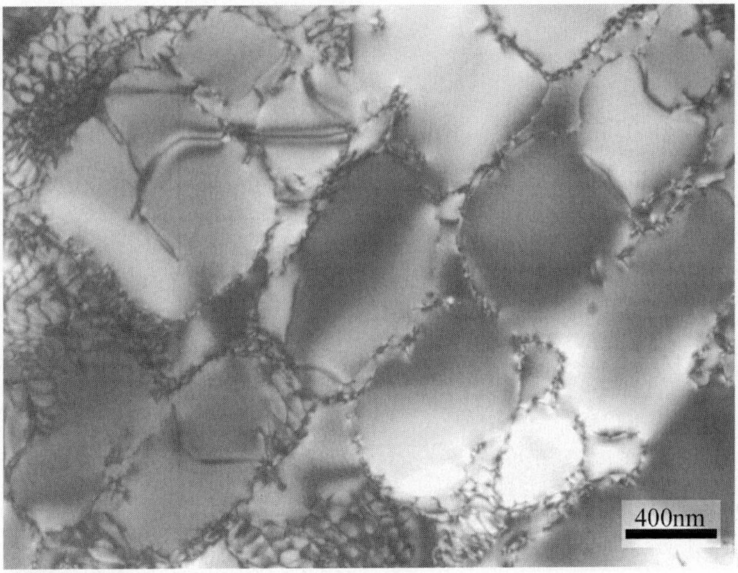

Fig. 8 Transmission electron micrograph of 950°C 480 MPa LCLCF specimen, sectioned perpendicular to tensile [001] direction. Regular networks have formed at the γ/γ' interfaces and significant coarsening of the γ' precipitates can be seen.

show less damage than those seen after creep deformation; following creep deformation at stresses as low as 550 MPa at 850°C stacking faults can be seen in the γ' precipitates.[8]

950°C

At 950°C, the test undertaken at 480 MPa, seen in Figure 8, shows dislocations restricted to the γ channels. The deformation above the yield stress at 700 MPa is shown in Figure 9. Here, some intrusion into the γ', largely by single dislocations, is seen, consistent with stress above the yield stress. In both cases, polygonal networks have formed at the γ/γ' interface, and this provides evidence of slip on at least two systems. Unlike the networks noted above in the 850°C tests, these networks are regular. The network spacing at the lower stress is approximately twice that of the higher stress.

The low stress sample was at high temperature for 77 hours compared to only 4 hours at the higher stress. The low stress sample thus shows signs of rafting, with the precipitates becoming more rounded, and in some instances coalescing. It is the change in precipitate morphology which is largely responsible for the difference in the appearance of the microstructures at the two stresses.

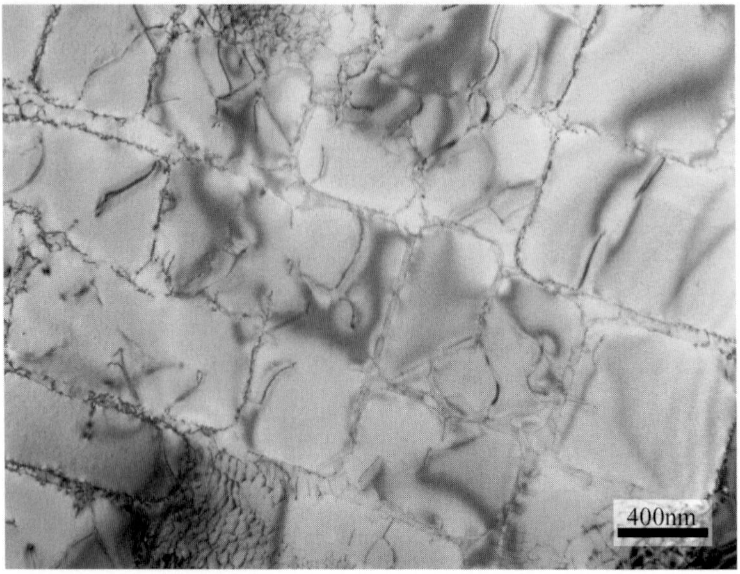

Fig. 9 Transmission electron micrograph of 950°C 700 MPa LCLCF specimen, sectioned perpendicular to tensile [001] direction. Dislocations are mostly restricted to the γ channels, where regular networks have formed. The dislocations are sited on the γ/γ' interfaces. No rafting has occurred.

DISCUSSION

This set of tests has shown that there is a transition in fracture initiation points between 750 and 950°C from pore to surface initiated. At 950°C, the oxide scale develops sufficiently quickly for cracks to initiate in the oxide scale before they would otherwise initiate from internal pores. Below this temperature, oxide scale develops slowly and initiation occurs instead at highly stressed subsurface pores. In this set of tests the initiation point is dependent only on temperature and does not vary with stress. In this testing program, for both high and low stresses at 850°C (and three additional tests at intermediate stresses), the initiation point was invariably a pore. Surface initiation occurred only at 950°C. Maclachlan,[3, 4] however, observed surface initiation at 850°C as well as 950°C, indicating this temperature is likely to be around the transition point of surface/subsurface failure.

The LCF microstructures at lower stresses have been compared with creep microstructures at an equivalent creep exposure. During trapezoidal R = 0 fatigue, only a quarter of the total exposure time is spent at this maximum load. The equivalent creep exposure, therefore, refers to one quarter of the total LCF life.

At 750°C 600 MPa, the LCF test was halted after 111 hours. A crept specimen after equivalent creep exposure would still be in the 'incubation' period prior to primary creep. The observed microstructure after LCF test is reasonably consistent with these very early stages of creep.

The LCF microstructure at 850°C 660 MPa has developed networks in some γ channels, particularly those perpendicular to the tensile axis, but shows little dislocation activity in the γ. This microstructure is much less well developed than that expected from the equivalent creep exposure of 20 hours. This represents approximately one quarter of the total creep life and would produce a crept microstructure well into secondary or steady state creep. Typically the microstructure would have well developed networks at all γ/γ' interfaces and some intrusion into the γ' by stacking faults and dislocation pairs.

At 950°C 480 MPa the creep life has been measured as 22 hrs,[4] very similar to the equivalent creep exposure for the 950°C, 480 MPa LCF test. The microstructure of the LCF test shows regular networks at all γ/γ' interfaces, limited intrusion into the γ', and the early stages of rafting. This dislocation structure is typical of the early stages (<1/3 of total creep life); at failure the creep microstructure showed extensive dislocation pair density in the gamma prime and dense disordered networks at the precipitate interface. However the degree of rafting reflects the total time the LCF test was at temperature rather than the equivalent creep exposure; this is consistent with rafting continuing following the removal of an applied stress once a critical strain is achieved.[9]

Although the stress is not reversed in these tests, the periodic unloading seen during R = 0 fatigue does seem to inhibit the formation of defects such as stacking faults and dislocation pairs which would be a feature of creep tests after equivalent creep exposure. Work by Lukas et al on CMSX-4 at 800°C also suggests that superimposing cyclic stress onto creep retards the creep rate.[10] Although cycling the stress greatly increases the propagation of cracks, and eventually causes fatal fatigue damage, it also seems to cause a reduction in damage by creep.

When the yield stress is exceeded, a key difference is seen in the microstructures as the dislocations enter the γ'. In the 750°C sample, the precipitates are sheared by stacking faults and dipoles, and in the 950°C the shear is mainly by single dislocations. The LCF damage seen here at 750°C is more homogeneous than has been reported previously, with no evidence of persistent slip bands. This can be seen in the high stress tests at 750 and 950°C, 1100 and 700 MPa respectively, Figures 5 and 9. Rather than resembling creep tests, these samples tested above the yield stress display microstructures similar to those seen from tensile testing at similar stress, strain rate and temperature conditions.[11] In the 950°C LCF test and the comparable tensile test at 1000°C, networks have developed in the γ and single dislocations can be seen in the γ'. These single dislocations are not straight and are able to climb rather than being restricted to one slip plane. In both the 750°C above yield LCF test, and the comparable tensile test at 700°C, there are large numbers of stacking faults of a single type. There are also dislocation dipoles present, all of which have the same Burgers vector, consistent with single slip. They could be contributing substantially to the plastic strain by expanding and contracting within individual γ' precipitates in response to the cyclic stress.

CONCLUSIONS

- In CMSX-4 under R = 0 load controlled fatigue, cracks initiate from oxidation at 950°C, and casting porosity at 750 and 850°C.

- There are significant differences in the microstructures of CMSX-4 samples fatigued under load control under varying temperatures and stresses. Significantly:
  - Stacking fault shear has only been observed at 750°C at stresses above yield.
  - Below the yield stress, LCF damage is confined to the γ matrix.
- Dislocation substructures corresponding to LCF at R = 0 above the yield stress resemble those typical of tensile tests.
- Microstructural damage by creep processes in LCF below the yield stress is less than would be expected from comparable creep exposure at the maximum stress. The repeated high stresses seen by specimens in LCF do not produce microstructural damage in the same way as a sustained high stress, suggesting that creep processes are not simply additive under these conditions.
- Different fracture morphologies appear to be the result of surface rather than microstructural effects.

ACKNOWLEDGEMENTS

Financial support and materials for this project from the Cambridge Commonwealth Trust and Rolls-Royce plc are gratefully acknowledged. The authors also thank Professor D. J. Fray for the provision of research facilities.

REFERENCES

1. V. Brien and B. Decamps: 'Low Cycle Fatigue of a Nickel-Based Superalloy at High Temperature: Deformation Microstructures', *Materials Science and Engineering A*, **316**, 2001, 18–31.
2. P. Lukas and L. Kunz: 'Cyclic Slip Localisation and Fatigue Crack Initiation in FCC Single Crystals', *Materials Science and Engineering A*, **A314**, 2001, 75–80.
3. D. W. MacLachlan: 'Creep and Fatigue at High Temperatures in Ni-Base Blade Alloy CMSX-4', Ph.D Thesis, *Materials Science and Metallurgy*, University of Cambridge, 1998.
4. D. W. MacLachlan and D. M. Knowles: 'Fatigue Behaviour and Lifing of Two Single Crystal Superalloys', *Fatigue and Fracture of Engineering Materials and Structures*, **24**, 2001, 503–521.
5. C. Monier et al.: 'Transmission Electron Microscopy Analysis of the Early Stages of Damage in a γ/γ' Nickel-Base Alloy Under Low Cycle Fatigue', *Materials Science and Engineering A*, **188**, 1994, 133–139.
6. J. R. E. Davies: Nickel, Cobalt and their Alloys, Ohio, ASM International, 2000.
7. Rolls-Royce, Fatigue Lifing Test Data, 2002.
8. C. M. F. Rae et al.: 'Primary Creep in Single Crystal Superalloys: Origins, Mechanisms and Effects', *Materials Science and Engineering A*, 2001, In Press.
9. N. Matan et al.: 'On the Kinetics of Rafting in CMSX-4 Superalloy Single Crystals', *Acta Materiala*, **47**(7), 1999, 2031–2045.

10. P. Lukas, L. Kunz and J. Svoboda: 'Retardation of Creep in <001>-Oriented Superalloy CMSX-4 Single Crystals by Superimposed Cyclic Stress', *Materials Science and Engineering A*, 1997, A234–236, 459–462.
11. D. Danciu, H. J. Penkalla and F. Schubert: 'Dislocation Microstructure of CMSX-4 after Tensile Testing with Different Strain Rates at 700 and 1000°C', *Materials for Advanced Power Engineering*, 2002.

Advances in Surface Engineering in Gas Turbines

J. R. Nicholls
Cranfield University, Advanced Materials Department, Building 61
SIMS, Cranfield, Beds MK 43 0AL, UK

ABSTRACT

Surface engineering is now a key materials technology in the design of advanced/future gas turbine engines. This paper focuses on coating systems for hot gas path components, these can vary from low cost aluminide, diffusion coatings to the more exotic, and therefore expensive, thermal barrier coating systems (TBCs).

This paper reviews available coating systems and discuss their relative benefits in terms of performance against manufacturing complexity, and therefore cost. Future trends in the design of environmental and thermal protection coatings are discussed, including the latest advances in modified aluminide coatings, the addition of multiple reactive elements, to overlay coatings diffusion barrier concepts, the design of 'smart' corrosion resistant coatings and the development of structurally modified, low thermal conductivity, thermal barrier coatings.

INTRODUCTION

The drive to improve engine combustion efficiency, while reducing emissions, has meant that the operating temperatures within the turbine section of gas turbine engines, whether aero-, industrial or marine has increased significantly over the last 30 years. For aero engines the mean blade temperatures are around 1050°C, with peak temperatures in excess of 1150°C[1, 2] Under industrial and marine service conditions temperatures are a little lower, 800–950°C, but more severe environments are encountered.[3, 4]

This improved performance has been achieved by novel material design, improved cooling technologies and better manufacturing methods.[2, 3, 7] Thus, uncooled blades have been surpassed by cooled blades which in turn have been replaced by coated, cooled blades. The nickel based superalloy development has similarly progressed from wrought alloys, through cast to single crystal alloys, with latest developments focused on fourth generation single crystal alloys.

As a result the need for coating systems to bestow adequate oxidation and corrosion resistance upon the component became essential, an inevitable result of the reduction in chromium content in the advanced alloys, necessary to achieve the required improvement in mechanical properties to match these higher operating temperatures. Thus, aluminide coatings were first introduced in aero-gas turbine engines in circa 1960 and overlay coatings some 10–12 years later in the early 1970's.[8–12]

Conventionally cast, directionally solidified and first generation single crystal materials could be protected against oxidation and corrosion, to give adequate high pressure turbine blade (HPTB) life using these diffusion aluminide coatings and MCrAlY overlay coatings.[8–12] As the turbine entry temperature (TET) increased, platinum aluminides replaced conventional aluminides[11,12, 20–22] and improved MCrAlY compositions were developed offering better oxidation resistance.[12, 22, 23]

This paper discusses this historical development in diffusion and overlay coating systems to provide environmental protection, to conventionally cast, directionally solidified and the first generation of single crystal superalloys.

Second generation single crystal alloys, though stronger, have proven to be susceptible to hot corrosion, particularly high temperature sulphidation and a wide range of alternative coating strategies have been proposed to address these issues.[12, 22, 24–26]

The latest development in this drive for improved surface protection, high temperatures, and therefore better performance is thermal barrier coatings applied to turbine aerofoil surfaces and other hot gas path components. New developments in structurally modified low thermal barrier coatings will also be reviewed.

COATING DEVELOPMENTS - A HISTORICAL PERSPECTIVE

In general, the development in manufacturing processes for high temperature coatings has paralleled the evolution of gas turbine materials and turbine component design. Early coatings involved surface modification of a component by diffusion, and this was first applied to turbine airfoils in 1957.[27] Such diffusion coatings are still in wide use today. Later, in response to a demand for high performance, modified diffusion coatings and overlay coatings were introduced.[8–12] This occurred in the early to mid 1970's. More recently, the drive for higher turbine entry temperatures has seen the introduction of thermal barrier coatings on nozzle guide vanes and turbine blades, as well as its use within combustor systems.

Diffusion Coating Processes

Diffusion coating processes have been applied for many years to improve environmental resistance of a base alloy by enriching the surface in Cr, Al, or Si. In the 1960's aluminising was first used for the protection of superalloy gas turbine aerofoils.[8, 11, 12, 27] There was renewed interest in the 1970's in siliconising,[29, 30] and later silicon modified aluminides,[31–33] when novel solutions to the low temperature hot corrosion problems were being sought, associated with the higher levels of contaminants in industrial and marine turbine plants burning impure fuels.[4, 19, 30, 33, 34]

Diffusion coatings can be applied to hot gas components using a range of techniques including pack cementation, slurry cementation, over pack CVD* and vapour phase CVD*. Comprehensive reviews of the methods of deposition of diffusion coatings are given in references.[11, 12, 27, 35–39] In pack aluminising, overpack aluminising and vapour phase aluminising the deposition rate and morphology of the coating depends on the aluminium activity in the gas phase, processing time and temperature. Coatings are classified as either 'low activity',

*CVD = Chemical Vapour Deposition

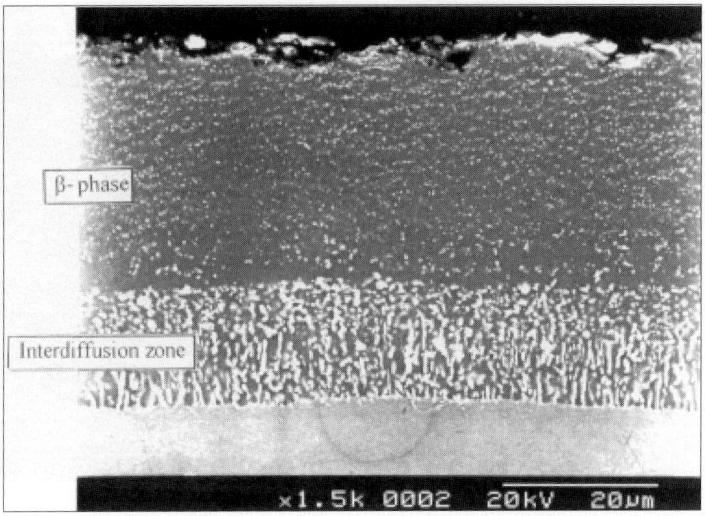

Fig. 1 Fully processed pack aluminide coating (high activity) on IN738.

when outward diffusion of nickel occurs, or 'high activity' when inward diffusion of aluminium occurs. In the latter case a surface layer of Ni_2Al_3 forms and a further heat treatment is required to convert this brittle surface layer to NiAl. This step is usually combined with the heat treatment required to recover substrate properties. Figure 1 illustrates a typical aluminide coating deposited onto a nickel based superalloy using a high activity pack coating process. The coating was deposited using an aluminiding pack contains 2.2%Al at 900°C, and was heat treated for 2 hours at 1120°C, then 24 hours at 845°C. Figure 2 illustrates a similar aluminide coating produced using vapour phase aluminising.

Clearly, the properties of the aluminide coating (or for that matter any diffusion coating) depend upon the process methodology, the substrate composition and the subsequent heat treatment. Typically, aluminide coatings contain in excess of 30 wt.% Al and are deposited to thicknesses between 30–100 μm depending on the type of aluminide formed. They offer satisfactory performance for many aviation, industrial and marine engine applications. Under severe hot corrosion conditions, or at temperature above 1050°C, aluminide coatings offer limited protection. To address these issues modified aluminide coatings and overlay coating (MCrAlY) technologies were developed in the 1970's and development in these two areas continues today to combat the increased demands placed on the modern gas turbine power plants.

Modified Aluminide Coatings

Modified aluminide coatings are fabricated,[11, 12, 27, 30–33, 35–39] either by depositing an interlayer – for example 7 μm of platinum by electroplating or PVD (Physical Vapour Deposition) –

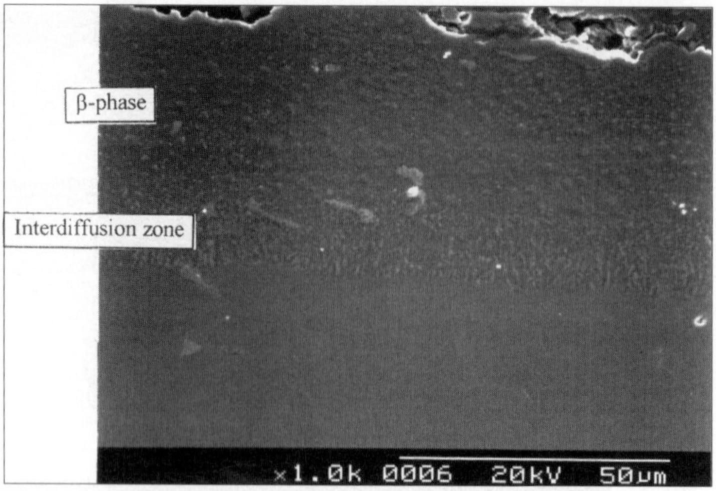

Fig. 2 Vapour phase aluminised coating (low activity) on IN738.

prior to aluminising, when manufacturing a platinum aluminide (Figure 3) or by pretreating the superalloy before aluminising – for example chromising prior to pack aluminising, or by co-depositing elements from a pack, slurry or blending them in the vapour phase–Sermaloy J and Sermaloy 1515 are slurry co-deposited silicon aluminide coatings.[31–33]

Alloying additions that have been considered include Cr, Si and Ta, various rare earths and precious metals, with many of these coatings now commercially available. To improve the high temperature oxidation performance of aluminides a most significant advance was made with the development of platinum modified aluminides. This class of coating is now an accepted industrial standard, out performing conventional aluminides under high temperature oxidation, cyclic oxidation and hot corrosion conditions.[4, 8, 12, 20–22, 27, 34, 38–44] Much work is continuing in this important area both as environmental protection coatings (EPCs) and as bondcoats to thermal barrier coatings (TBCs) as will be reviewed later in this paper.

To combat hot corrosion, additions of chromium[35, 38, 40, 45] and silicon[29–33, 39] were researched in the 70's and 80's. Although, no better than standard aluminides under high temperature oxidation conditions chromium rich diffusion coatings offered improved performance in industrial and marine turbines, burning high sulphur fuels.[40, 45] Additions of silicon were also shown to improve the hot corrosion resistance.[39] At about 10 wt.% silicon addition, silicon modified coatings proved uniquely resistant to Type 1 and 2 hot corrosion,[31–33] if somewhat brittle.[41] This research was the basis for Sermaloy J (Figure 4), a $CrSi_2$ dispersed, β-NiAl diffusion coating.[31–33, 46] Later as engine technologies advanced and directionally solidified or single crystal alloys were more widely adopted, less chromium was available within the alloy to form the $CrSi_2$, (a critical component within Sermaloy J[46]), thus joint research between Sermatech and Rolls Royce lead to the development of Sermaloy 1515, a layered structure

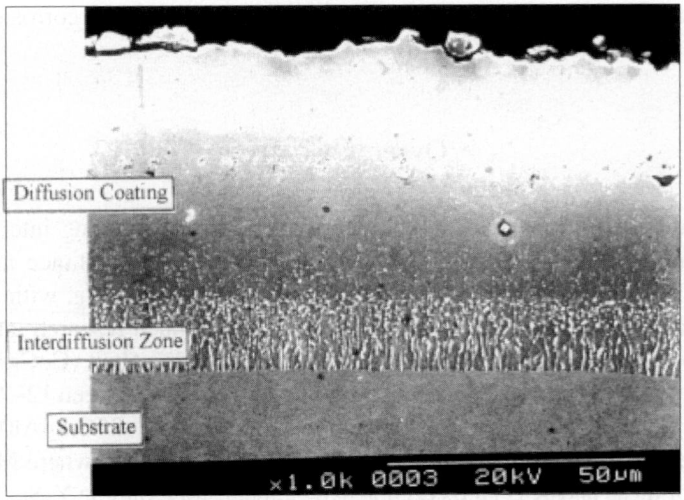

Fig. 3 Platinum aluminide coating (RT22LT).

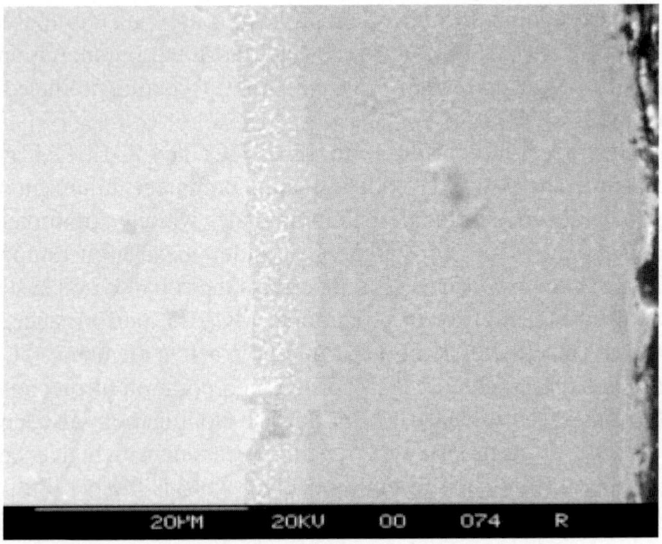

Fig. 4 Silicon aluminide coating (Sermaloy J).

slurry aluminide containing bands of $CrSi_2$ capable of providing hot corrosion resistance to low Cr containing single crystal alloys.[33, 34, 46, 47]

Overlay Coatings

Diffusion coatings, by the nature of their formation, imply a strong interdependence on substrate composition in determining both their corrosion resistance and mechanical properties, hence, the possibility of depositing a more 'ideal' coating, with a good balance between oxidation, corrosion and ductility has stimulated much research interest, since the early 70's. The early MCrAlY coatings were alloys based on cobalt (CoCrAlY) containing chromium additions in the range 20–40%, aluminium additions between 12–20% and yttrium levels around 0.5% with the most successful coating being Co25Cr14Al0.5Y.[9] The most recent coatings are more complex, based on the M-CrAl-X system, where M is Ni, Co, or a combination of these and X is an oxygen active element, for example Y, Si, Ta and Hf[25, 38, 39, 48–51] or a precious metal, for example Pt, Pd, Ru and Re etc.[23, 48, 52–55] The composition of the M-Cr-Al system is selected to give a good balance between corrosion resistance and coating ductility, while the active element addition(s) can enhance oxide scale adhesion and decrease oxidation rates. Current thinking suggests that a combination of active elements is beneficial in reducing coating degradation through their synergistic interaction. Overlay coatings have been deposited using a range of techniques. The earliest production method was electron beam physical vapour deposition (EB-PVD).[9] However, because of the high capital cost in setting up a commercial EB-PVD plant, plasma spray methods have found wide acceptance, particularly the argon shrouded and vacuum plasma spray processes[3, 25, 40, 55–57] and more recently high velocity oxyfuel spraying processes,[25, 58–60] composite electroplating[50, 61, 62] and autocatalytic electroless deposition[26] methods have been used to deposit overlay coating systems. However, coatings produced by EB-PVD processes are still considered the commercial standard against which other process routes are compared. Figure 5 is a micrograph of an EB-PVD CoCrAlY (ATD5B) on MarM002, Figure 6 is an Argon shrouded plasma sprayed NiCoCrAlY and Figure 7 is a NiCoCrAlY coating produced by the composite electroplate method.

Overlay coatings of classic design, with 18–22%Cr and 8–12%Al, generally perform better at higher temperatures where oxidation is the dominant failure mode (above 900°C) reflecting the good adherence of the thin alumina scales which is promoted by the presence of active elements such as yttrium. Generally under these high temperature oxidising conditions NiCrAlY's and NiCoCrAlY's out perform the cobalt based systems.[43, 44] Methods have been investigated to improve the traditional MCrAlY performance. New MCrAlXY alloys have been developed with additions of multiple active elements (Ti, Zr, Hf, Ta and Si for example have been researched)[38, 39, 48–51] or the incorporation of precious metal additions (Pt, Pd, Re and Ru have been studied).[23, 48, 52–55] Surface modifications also have been examined using CVD,[63–65] PVD[48, 49, 66] electroplating[67–71] and re-processing with high energy beams.[38, 51, 72–75] An interesting proposal from such recent processing work is the possibility of depositing a single crystal MCrAlY alloy coating onto a single crystal superalloy.[73, 74] The single crystal, epitaxial coating was produced using laser cladding technology, with controlled solidification of the melt pool.

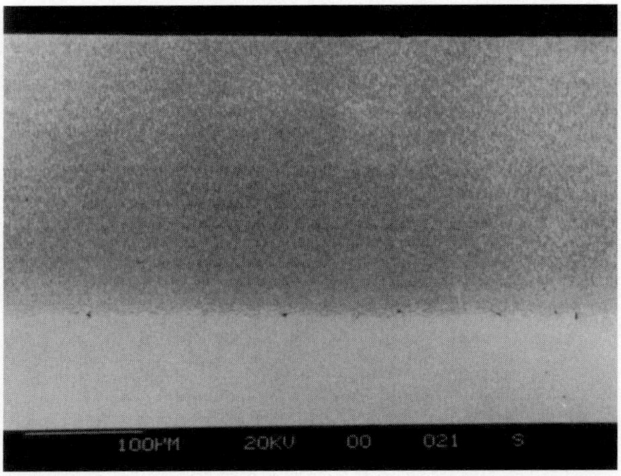

Fig. 5 EB PVD CoCrAlY (ATD 5B) coating on MarM002.

Fig. 6 Argon shrouded plasma spray CoNiCrAlY (LCO22).

Modified Overlay Coatings

Surface modification of MCrAlY overlay coatings by aluminising was first published in 1984, when Restall and Haymen[68] pulse aluminised overlay coating for improve hot corrosion resistance in marine gas turbines.[41] GE29+ and GE34+ are proprietory over aluminised MCrAlY overlay coatings with graded aluminium profiles offering improved performance under high temperature oxidation and hot corrosion studies.

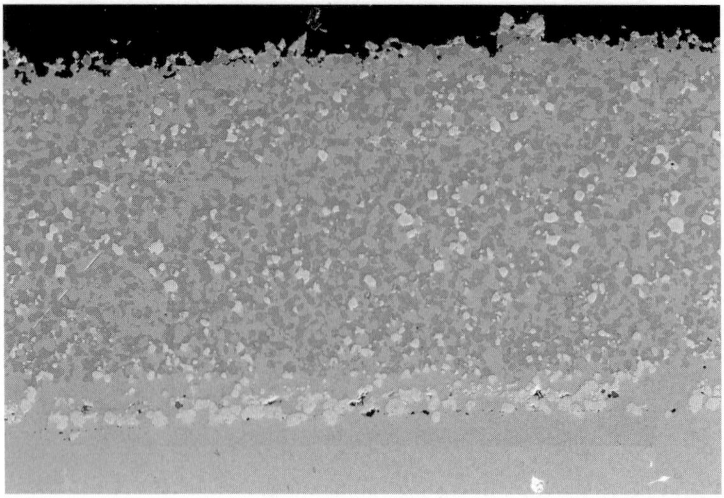

Fig. 7 A NiCoCrAlReY, modified overlay coating produced by the composite electro-plate process.

Recent work has focused on the modification of the surface of MCrAlY alloys with platinum and other precious metals[67–71, 78–80] to improve their performance under high temperature cyclic oxidation conditions.[67–69, 78–80] Such treatments were first proposed in the early 80's as a method of improving the MCrAlY alloy corrosion performance to aggressive deposits from a range of alternate fuel sources.[48] The need for improved cyclic oxidation resistance driven by the pursuit of bondcoats for advanced TBCs has seen recurrent interest in the performance of platinum modified MCrAlY's both as environmental protection coatings and as bondcoats. Figures 8 and 9 illustrate the platinum modified surface microstructures. Platinum modification to the surface of the MCrAlY[67–69, 78] or just to the superalloy without depositing on MCrAlY,[70,71, 79, 80] improves the cyclic oxidation performance. This advantage of the platinum surface treatment relates to the ability this platinum diffused layer has in providing a smoother, stronger more defect free surface. Both strategies are being pursued as TBC bondcoats and will be discussed further in the section reviewing TBC developments.

Smart Overlay Coatings[24, 76, 77]

Smart overlay coatings are functionally gradient coating systems designed to provide high temperature corrosion protection over a wide range of operating conditions. The SMARTCOAT design consists of an MCrAlY base, enriched first in chromium, then aluminium to provide a chemically graded structure. At elevated temperatures, above 900°C, the coating oxidises to produce a protective alumina scale. However, at lower temperatures, this alumina scale does not reform rapidly enough to confer protection under Type II hot corrosion conditions. The SMARTCOAT is therefore designed with an intermediate

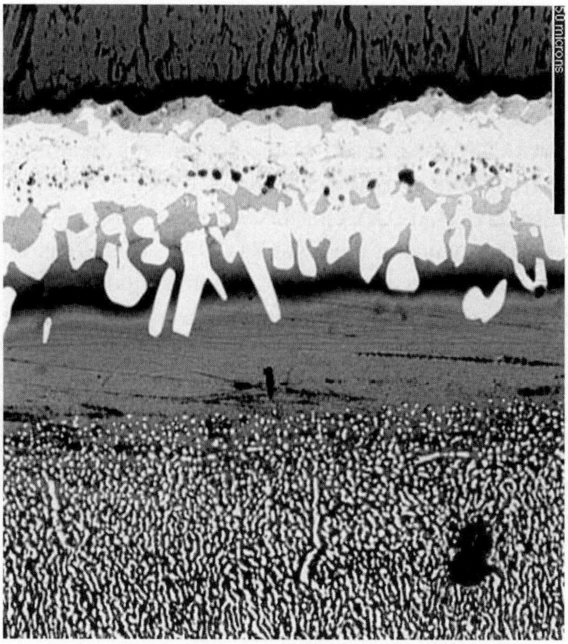

Fig. 8 A platinum diffused, $\gamma + \gamma'$ coating on CMSX4.

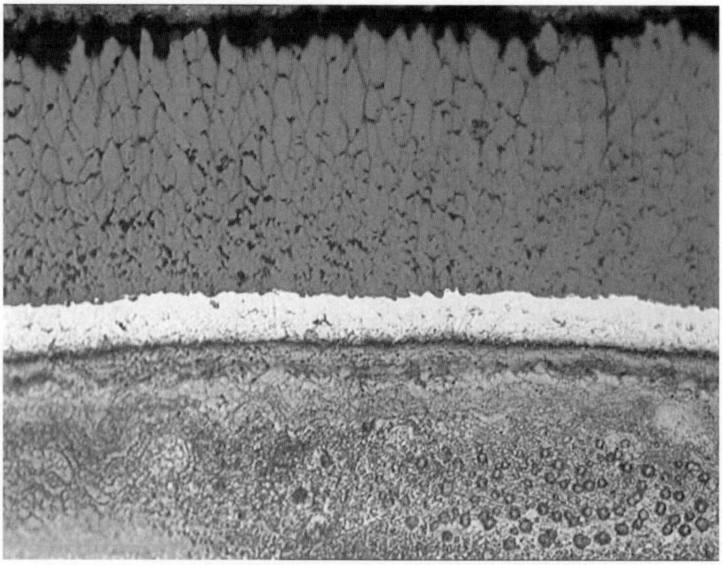

Fig. 9 A platinum diffused NiCoCrAlY overlay coating on CMSX4.

chromium-rich interlayer, which permits the rapid formation of chromia, healing areas of Type II corrosion damage. Figure 10 illustrates the microstructure of SMARTCOAT SmC155, produced by surface treating an air plasma sprayed Amdry 995 basecoat. The three layer microstructure can clearly be seen. The outer layer [A] varies in composition between Ni-15Cr-32Al and Ni-10Cr-21Al, as a result of the aluminising treatment, the interlayer [B] is chromium rich with a composition in the range Ni-60Cr-20Al to Ni-35Cr-40Al, over a basecoat of Amdry 995 (Co-32Ni-21Cr-8Al-0.5Y) for this variant of the SMARTCOAT process.

The hot corrosion performance of the SMARTCOAT structure is illustrated in Figure 11, where SMARTCOAT SmC155 is compared to a conventional platinum aluminide (RT22) and an over aluminised CoNiCrAlY - a similar process and structure to that of GT29+, a General Electric proprietory coating. The hot corrosion test was a 500 hrs. duration salt recoat procedure, with daily replenishment of an 80% Na_2SO_4/20% K_2SO_4 salt mix at equivalent deposition flux of 0.015 mg/cm^2/hrs. The tests were conducted at 700 and 800°C in an air –400 vpm SO_2/SO_3 gaseous environment. At both test temperatures, only the outer aluminide region of the smartcoat was attacked, whereas the platinum aluminide coat was consumed down to the interdiffusion zone at 700°C and complete penetrated at 800°C. For all coatings the substrate was a corrosion resistant superalloy, IN738LC.

Diffusion Barrier Concepts

Even within MCrAlX systems and more advanced overlay coating concepts, diffusion of elements between the substrate and coating can have a major influence on coating performance. Therefore to provide long terms stability it is necessary to develop diffusion barrier coatings that can be used to minimise interdiffusion between the coating and the substrate. Some interdiffusion is of course necessary to provide good adhesion, therefore the diffusion barriers must be tailored to limit the movement of particular elements. Both precious and refractory metals,[23, 48, 81] intermetallics[82] and ceramics[25] have been proposed as diffusion barriers with varying degrees of success. Figure 12 illustrates the deposition of an intermetallic diffusion barrier at the interface of a high Cr containing overlay coating and a superalloy substrate (MarMOO2). The diffusion barrier was developed to limit chromium movement from the coating into the substrate[82] and hence reduce the likelihood of forming sigma phase within the alloy below the coating. Chromium diffusion rates reduced by one to two orders of magnitude over the temperature range 750–1150°C. Ceramics[25] such as TiN/AlN, AlON and ZrO_2 have been investigated as part of the COST 501 programme to limit elemental diffusion. These ceramic layers were deposited using sputter deposition. Each of these ceramic systems were able to reduce interdiffusion for several hundred hours at 1100°C, but ultimately failed locally allowing some interdiffusion. Of these ceramic systems ZrO_2 performed best lasting 586 hours at 1100°C, before cracking in the zirconia layer led to barrier breakdown.

Once the concept of a diffusion barrier is accepted as a method of providing good interface stability, one is no longer constrained in the design of the best overlay coating. No longer is substrate compatibility a requirement in specifying the overlay coating composition. By removing this constraint it should be possible to design overlay coatings with optimised oxidation or corrosion.[82, 83]

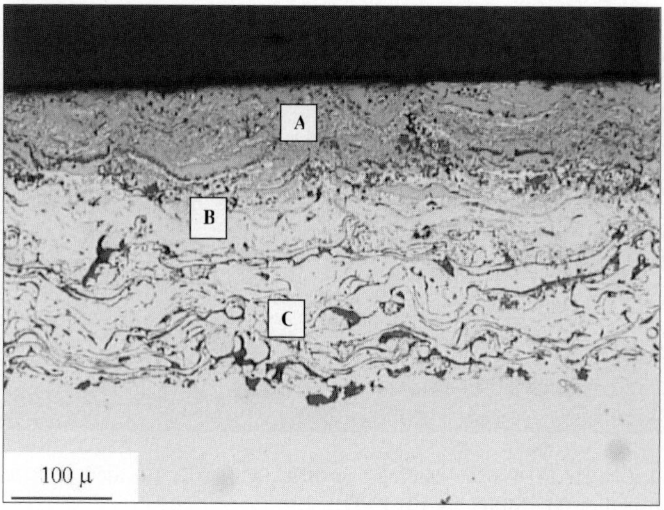

Fig. 10 A micrograph of SMARTCOAT SmC155, showing the three layered microstructure that is Characteristic of SMARTCOAT.[80]

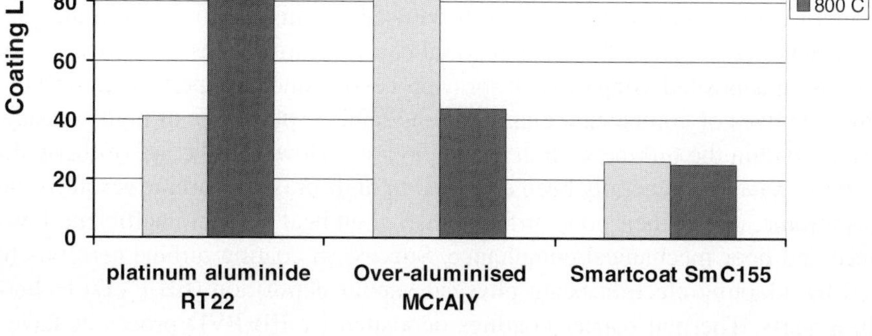

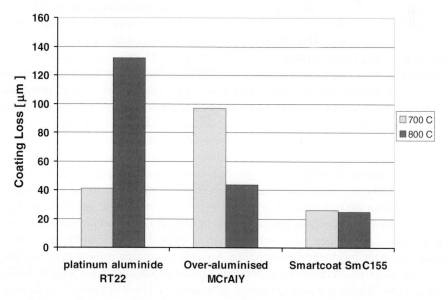

Fig. 11 Hot corrosion performance of SMARTCOAT SmC 155, relative to an over aluminised CoNiCrAlY and a Platinum luminide (RT22).

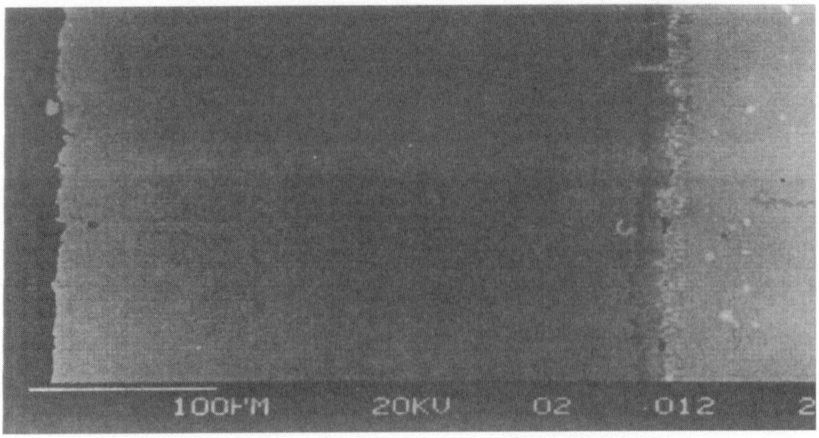

Fig. 12 Graded NiCrAlY overlay coating and diffusion barrier (located at the coating/substrate interface) deposited on MarM002, by EB-PVD.

HISTORIC DEVELOPMENT OF THERMAL BARRIER COATING TECHNOLOGIES

An alternative strategy to the development of environmental protection coatings [EPC], based on corrosion resistant alloys and intermetallics, is to design coating systems that a lower the metal surface temperature. Thermal barrier coatings (TBCs) work in this way and provide the potential to drop metal surface temperatures by up to 150°C, using current technologies in conjunction with component cooling. A 150°C reduction in the component surface temperature would have the effect of reducing oxidation rates by an order of magnitude and would provide further benefits by reducing the components propensity to creep and limiting the extent of interdiffusion between the bondcoat and substrate superalloy.

Thermal barrier coatings (TBCs) have been used for this purpose within the gas turbine engine since the 1970's.[2, 7, 68, 84–91] Early thermal barrier coatings, based on partially stabilised zirconias, were deposited using thermal spray processing and have performed well in service, extending the lives of combustion chambers and annular platforms of high pressure nozzle guide vanes within the turbine section of the engine.[92] However, the use of thermal sprayed TBC systems have not generally been extended to high-pressure turbine aerofoils within the aero-gas turbine, due to their poor surface finish, high heat transfer coefficient, low erosion resistance and poor mechanical compliance. Success in coating turbine aerofoils has been achieved by adopting electron beam physical vapour deposition (EB-PVD) technology to coat these parts. Thermal barrier coatings deposited by EB-PVD processes have a good surface finish,[92] columnar microstructures with high strain compliance[68, 87, 89, 90, 92] and good resistance to erosion[92, 93] and foreign object damage.[94] EB-PVD thermal barrier coatings have been used in production since 1989.[88] Unfortunately, the microstructure which gives the EB-PVD TBC system its high strain compliance and good erosion resistance leads to a relative high thermal conductivity when compared to thermally sprayed coatings.

Early TBCs, manufactured by plasma spraying, were produced by spraying from magnesia or calcia stabilised zirconia.[95] These performed well in service when at operating temperatures below c.a 1000°C. Above this temperature significant diffusion of the magnesium or calcium ions occurred, resulting in precipitates rich in MgO or CaO being formed.[96] This led to an increase in thermal conductivity from 0.8 upto circa 3.5 W/mK associated with the monoclinic phase forming within the ceramic top coat. This rise in the monoclinic content leads to mechanical instability of the coating as a result of the microcracking associated with the martensitic monoclinic/tetragonal phase change on thermal cycling. The limitation of low operating temperature, allied to phase instability, was overcome by the introduction of 8 wt.% PYSZ in the late 1970's a material which is relatively stable for elongated periods at temperatures up to 1500°C with no precipitation of Y_2O_3 from solution.[95, 97] Since their introduction, PYSZ TBCs have performed well in service, significantly increasing the life of components. For example, the application of 7 wt.% PYSZ TBC to combustion can walls reduces the thermal stresses in the can and can result in component lives in excess of 20,000 hours.[92]

Thus, by virtue of its low thermal conductivity, and good thermal stability at temperatures up 1500°C, 8 wt.% yttria partially stabilised zirconia TBCs have become the industrial standard to reduce the heat flux into hot path components within gas turbine engines. Their use allows a high thermal gradient to be subtended across the ceramic thickness without any increase in metal surface temperature, potentially allowing up to 150°C increase in turbine entry temperature (TET).

Although plasma sprayed TBCs have performed well in service on annular surfaces in the engine, their microstructure does yield coatings with the necessary strain compliance, erosion resistance or surface finish required for successful application on blade or nozzle guide vane aerofoils within the aero-gas turbine. As a result their introduction into the high-pressure turbine of such high performance engines has been limited. An example of the structure of the plasma sprayed TBC is shown in Figure 13. In contrast, more recently electron-beam physical vapour deposition (EB-PVD) has been used to deposit TBCs which, because of their columnar microstructure (Figure 14), exhibit very high levels of strain compliance. The columnar microstructure is due to the atomistic nature of the deposition process, with nucleation and growth of the coating controlled by condensation from the vapour phase. In addition to their good compliance, EB-PVD TBC systems also offer other benefits over plasma sprayed TBCs in terms of improved adhesion, surface finish and erosion resistance. The major limitations associated with the EB-PVD microstructures are their high thermal conductivity and the cost of manufacture. Table 1 summarises typical properties of an EB-PVD and plasma sprayed thermal barrier coating, as measured at room temperature.

Bondcoat Oxidation Strain Compliance and Coating Adhesion

It has been recognised that bondcoat oxidation is a major factor affecting the durability of a TBC system.[2, 12, 67–69, 78–80, 84–92, 98–109] Maintaining coating integrity for a TBC is significantly

Fig. 13 Plasma spraying ZrO_2–8 wt.% Y_2O_3 thermal coating (courtesy of Praxair).

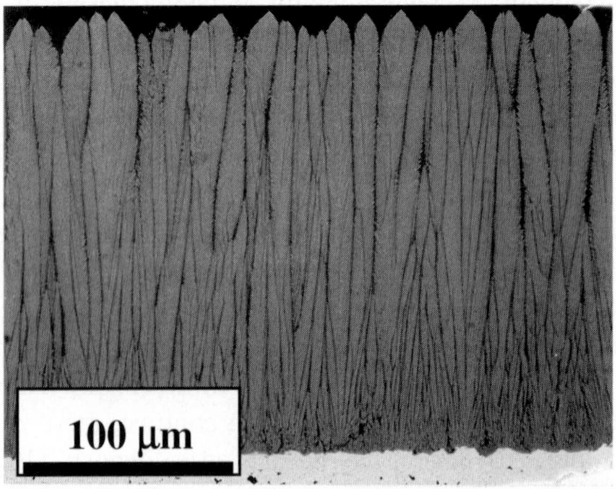

Fig. 14 EB-PVD ZrO_2–8 wt.% Y_2O_3 thermal barrier coating.

Table 1 Properties of TBCs at room temperature.

Property/Characteristic	EB-PVD	Plasma Sprayed
Thermal Conductivity (W/mK)	1.5–1.9	0.8–1.1
Surface Roughness (μm)	1.0	10.0
Adhesive Strength (MPa)	400	20–40
Young's Modulus (GPa)	90	200
Erosion Rate (Normalised to PVD)	1	7

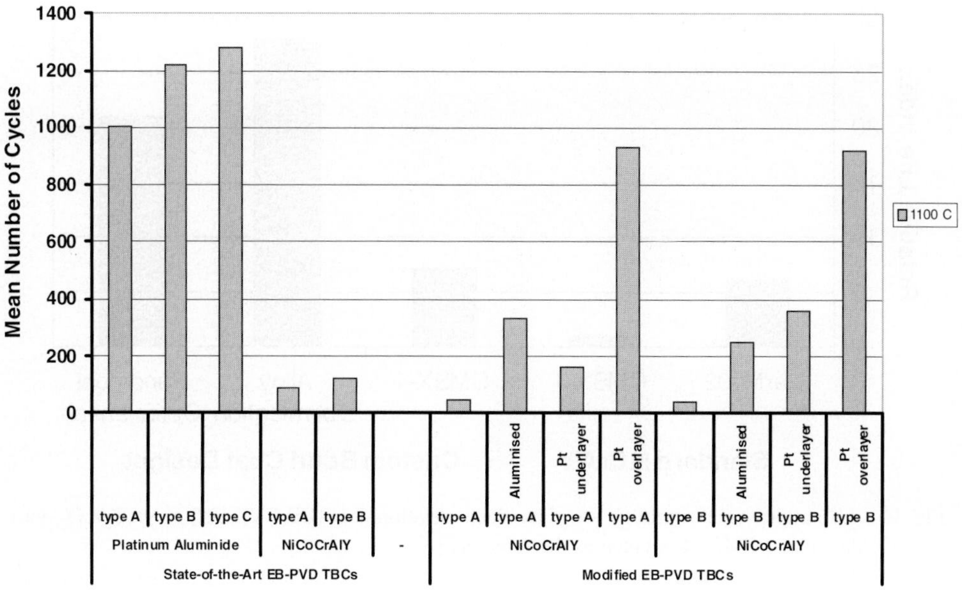

Fig. 15 Cyclic oxidation failure times for various bondcoats with an EB-PVD ZrO_2-8 wt.% Y_2O_3 top coat test at 1100°C (1 hour cycles).

more difficult than for a metallic, environment protection coating due to the relatively large change in properties from the metallic substrate through to the ceramic top coat.

The microstructure of the coating resulting from plasma spraying (Figure 13) differs from that produced by EB-PVD deposition (Figure 14) and thus the coating failure modes are not the same. Bondcoat oxidation is however, a significant factor in both modes of failure.[67–69, 78–80, 84–92, 98–109] The onset of TBC spallation is triggered by some form of microstructural instability at the bondcoat/TGO (thermally grown oxide) or TGO/ceramic interface,[86, 88–91, 106, 108, 109] The nature of this instability and its influence on TBC life is an area of intense research[67–69, 78–80, 84–86, 98–109] which varies from one TBC system to another. However, two major contributions to this failure are the extent of bondcoat oxidation and the strain generated by the thermal mismatch between the ceramic topcoat and the metallic components within the system. Other contributing factors include the strength of the bondcoat, changes in bondcoat microstructure and chemistry, the surface roughness of the component and possible sintering of the ceramic that can modify the overall system compliance. For plasma sprayed systems, failure results from delamination cracking, as a result of in-plane compressive stresses that result in significant out-of-plane tensile stresses. The delamination cracks, within the TBC, are parallel to the interface and near the peaks of the rough bondcoat.[86, 88–91, 106, 108, 109] Thus the rough interface in plasma sprayed coatings, first aids initial adherence through mechanical interlocking, but later drives its delamination due to the generation of out-of-plane stresses as a result of the interaction between bondcoat roughness and oxidation. Thus reducing its final life.

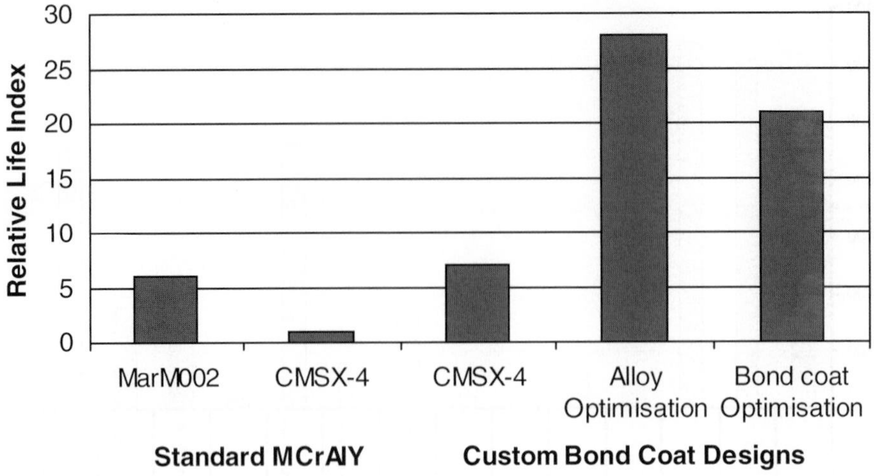

Fig. 16 Life improvement through $\gamma + \gamma'$ bondcoat development [EB-PVD ZrO_2-8 wt.% Y_2O_3 onto NiCoCrAlY coated CMSX4 is taken as the reference].

EB-PVD coatings tend to fail at the alumina scale/bond coat interface, unless substantial bondcoat surface rumpling occurs[68, 69, 78–80, 91, 92, 98–101, 104, 105, 107] whereupon failures may transfer to the TGO/ceramic interface. This failure at the TGO/bondcoat interface results from the progressive reduction in adhesion, as microcracks form and grow.[78, 98–100, 103, 106] Oxidation of the bondcoat is therefore a major driver of the failure of EB-PVD TBCs. The rate of growth of the oxide, provided α-alumina is formed at the ceramic/bondcoat interface, is reasonably constant for all alumina formers (platinum aluminide, MCrAlY, $\gamma + \gamma'$ bondcoat). What differentiates performance is the ability of the TGO to resist spallations as a result of defect formation, microcrack growth and loss of adhesion. Thus critical issues are the ease of void formation at the TGO/bondcoat interface, how readily less protective oxides form beneath the alumina scale and how easily debonding occurs at the TGO/bondcoat interface. This aspect has been extensively studied[78–80, 91, 92, 98–101, 104, 105, 107] and it has been shown by Meier, Pettit and co-workers[68, 69, 78] that the surface treatment of an MCrAlY overlay coating by platinum diffusion acts to bury surface 'grit-line' defects and thus improves TBC adhesion, giving improved TBC lives,

In recent work, Rolls Royce has developed custom designed bondcoats by diffusing platinum into the single crystal superalloy surface[70, 71] - these patented bondcoats have a $\gamma + \gamma'$ microstructure[79, 80] - that have been formulated to prevent or delay these degradation processes. The $\gamma + \gamma'$ bondcoat provide a stronger bondcoat system offer high creep strength thus resisting "rumpling". They also provide a physical barrier and a metallurgical stable microstructure to limit substrate element diffusion.

Figure 16 illustrates such bondcoat development and demonstrates that relative to CMSX4 with a standard NiCoCrAlY bondcoat the new $\gamma + \gamma'$ microstructures provide a 20 fold increase in life for an optimised bondcoat deposited on CMSX4. With further optimisation of the single crystal superalloy (fourth generation alloys) the life can be increased still further.

Low Thermal Conductivity, EB-PVD Thermal Barrier Coatings

As eluded to in the previous sections the first concern of the materials engineer was the thermal stability of the TBC system. Thus, in the early days of TBC research much work was undertaken to determine the best stabilising agent for zirconia. For very high temperature applications (mostly driven by the aero industry) there was broad agreement that 6–8 wt.% yttria partially stabilised zirconia offered the best solution as it showed the highest resistance to spallation under thermal cycling conditions[90] and excellent thermal stability. Later, explanations of this phenomenon identified the formation of the well-known *t*' phase in partially stabilised zirconia[110] and its toughening mechanism.[111] Thus, yttria stabilised zirconia based thermal barrier coatings are viewed as todays 'standard' and will meet the need for thermal protection in the latest generation of high powered, high performance gas turbine engines.

Having addressed many of the mechanical stability issues, the research focus has now changed and is directed towards future engine performance. From Table 1, it is apparent that although the columnar microstructure of the EB-PVD TBC offers improved mechanical properties its thermal conductivity (1.65 W/mK) is significantly higher than its plasma sprayed counterpart (0.8 W/mK). Therein lies the challenge 'How to reduce the thermal conductivity of an EB-PVD TBC to 0.8 W/mK, while still maintaining strain tolerance and thermal stability?'.

Research at Cranfield, under Rolls Royce sponsorship, has examined the influence of tertiary and quaternary Lanthenide dopants on the thermal conductivity of EB-PVD TBCs.[112–114] The aim was to produce partially stabilised zirconia, within the t' phase field, that contained phonon scattering centres to lower the ceramic thermal conductivity. Further, if the dopants also colours the ceramic this could be beneficial in reducing radiative transport. Following this concept, our work on dopant effects concentrating on the role of high atomic mass additions as these should theoretically provide the more effective scattering centres. Dopants evaluated included: NiO (atomic mass 51), Nd_2O_3 (atomic mass 144) and Gd_2O_3 (atomic mass 157) Er_2O_3 (atomic mass 167) and Yb_2O_3 (atomic mass 173). These five dopants were evaluated both as ternary and quaternary additions by manufacturing custom evaporation rods with known dopant levels; selected from 1.0, 2.0, 4.0 and 8.0 mole% addition. The additions were introduced to maximise lattice strains and lattice anharmonicity. The thermal conductivity of each doped TBC was measured using a laser flash technique and the results for various ternary additions are presented in Figure 17, at a 4 mole% addition of tertiary dopant.

Erbia, Gadolinia, Neodymia and Ytterbia all lower the thermal conductivity of zirconia-4.5 mole% yttria (8 wt.% yttria), when added at a 4 mole% level. Nickel oxide, the only divalent oxide addition used, produced only a small decrease in contrast to the work of Tamerin et. al.[115] mole% Erbia reduced the thermal conductivity by 25% and colouring the ceramic pink, Neodymia at 4 mole% lowered the thermal conductivity by 42% and coloured with ceramic blue/lilac. The best system, Gadolinia resulted in a thermal conductivity of 0.88 W/mK, a 47% reduction, achieving thermal conductivities comparable to plasma sprayed

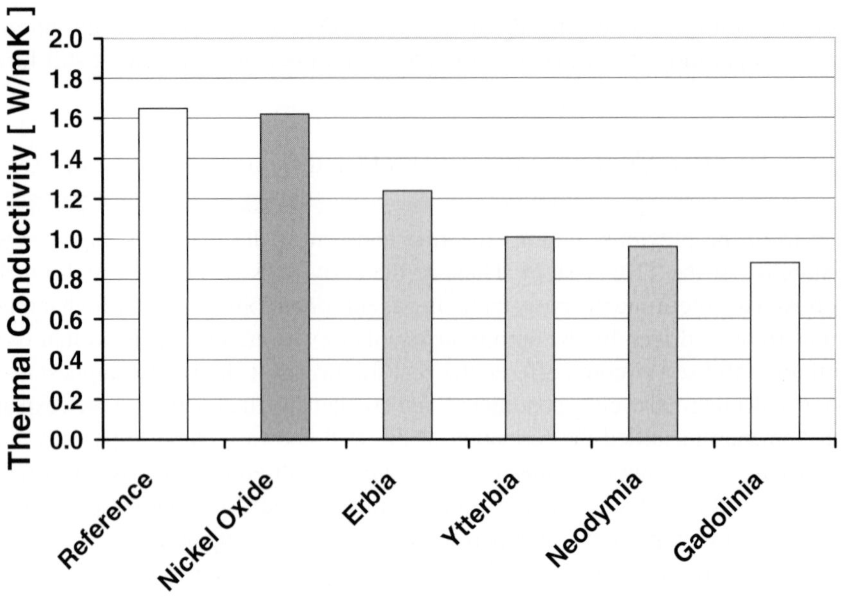

Fig. 17 Thermal conductivities of dopant modified EB-PVD TBCs at 4 mole% addition and 250 μm thickness (data measured at 500°C and corrected to a 250 μm thickness using the method proposed in reference.[113]

TBC, in an EB-PVD system. In all cases the ceramic crystal structure remains t' phase and the modified EB-PVD TBCS have a columnar microstructure.

The foregoing discussion would suggest that to reduce the thermal conductivity of EB-PVD TBCs further one should introduce additional phonon scattering centres into the coating. This is possible by modifying the coating nano-structure. Further, by reducing the mean free path from 250 μm (of the order of the thickness of the coating) to 1-2 μm would cause a reduction in the radiation contribution to the total thermal conductivity.[112] Thus, layering of the ceramic offers a promising route to further lower the thermal conductivity of an EB-PVD TBC, by modifying both the phonon and photon transport within the coatings.

For photon scattering, the periodicity of the layers should be chosen to act as a quarter wavelength filter (between λ and λ/4 of the incident radiation). With radiation wavelengths between 0.3–5.0 μm, the layer periodicity should lie between 0.2–2.0 μm, centred around 0.7 μm.

The method of obtaining such structures in practice uses a glow discharge plasma to vary the density of the ceramic during deposition. Figure 18 illustrates a typical micrograph of the morphological changes that can be introduced. The layers were produced by switching the D.C. bias applied to the substrate between high and low levels during deposition. This has the effect of periodically changing the degree of ion bombardment and thus altering the density of the layers deposited. The measured thermal conductivity for this microstructure is significantly lower than that for a coating produced without ion bombardments. Values

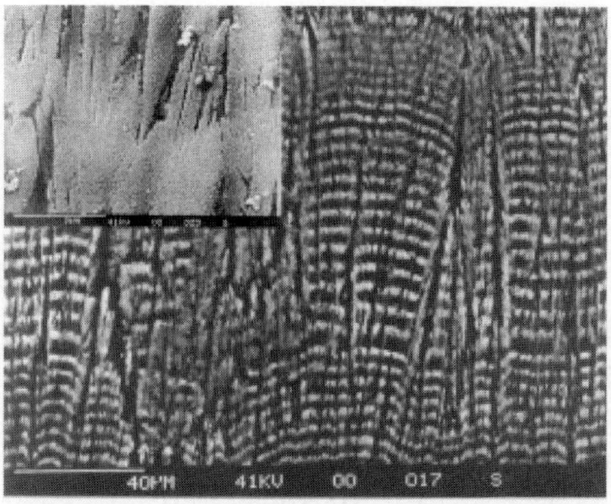

Fig. 18 Layered structured EB-PVD TBC introduced by PAPVD processing of the ceramic. Insert is detail of a coatings boundary.

between 0.95–1.2 W/mK have been achieved[112, 113] for a 2 μm periodic layer structure, as depicted in Figure 18 manufacture with 'standard' zirconia-8 wt.% yttria ceramic. Thus reductions of the order of 37–45% compared to state-of-the-art EB-PVD TBCs are possible for these layered structures. Clearly layering at micron dimensions and the introduction of density change from layer to layer work in combination to reduce the thermal conduction of the coating. The layering periodicity is selected to significantly reduce photon transport, while local changes in layer density act to scatter phonons and thus reduce thermal conduction by lattice vibration as well.

CONCLUDING REMARKS

This paper has reviewed available coating systems to protect hot path components within gas turbines. Trends in the design of environmental and thermal protection coatings have been presented including concepts of smart corrosion resistant coatings, new bondcoat technologies and low thermal conductivity thermal barrier coatings.

To the future, one must look towards custom surface engineering with the substrate and coating system designed to compliment each other such that they provide optimal protection, beit environmental, thermal or a combination of both technologies.

REFERENCES

1. D. S. Rickerby and H. C. Low: *4th European Propulsion Forum*, Bath, UK., 1993.
2. *Coatings for High Temperature Structural Materials*, NMAB, NRC, National Academy Press, Washington DC, 1996.

3. R. Burgel, H. W. Grunling, K. Schneider and U. Werkstoffe: *Korrosion*, **38**, 1987, 549.
4. J. Stringer and R. Viswanathan: 'Gas Turbine Hot-Section Materials and Coatings in Electric Utility Applications' *Proceedings of ASM 1993 Materials Congress Materials Week '93*, Pittsburgh, Pennsylvannia, ASM International, Materials Park, Ohio, USA., 1993, 1–21.
5. D. S. Rickerby: *Turbine Forum*, Nice, France, 2002.
6. J. R. Nicholls and D. S. Rickerby: *3rd International Conference on Recent Advances in Materials, Minerals and the Environment*, RAMM2003, Penang, Malaysia, 2003.
7. D. Driver, D. W. Hall and G. W. Meetham: 'The Development of The Gas Turbine Engine', Applied Science Publishers, London, UK., 1981.
8. G. W. Goward and D. H. Boone: *Oxid. Met.*, **3**, 1971, 475.
9. F. T. Talboom, R. C. Elam and L. W. Wilson: 'Evaluation of Advanced Superalloy Protection Systems', Report CR7813, NASA, Houston, TX., USA., 1970.
10. D. Gupta and D. S. Duvall: *Coatings for Single Crystal Superalloys*, TMS, Warrendale PA., 1984.
11. G. W. Goward: *Surface Coatings and Technology*, **73**, 1998, 108–109.
12. J. R. Nicholls: *JOM.*, **52**, 2000, 28.
13. P. Kofstad: *High Temperature Corrosion*, Elsevier Applied Science, London, UK., 1988.
14. N. Birks and G. H. Meier: *Introduction to the High Temperature Oxidation of Metals*, Edward Arnold, London, UK., 1983.
15. O. Kubaschewski and B. E. Kopkins: *Oxidation of Metals and Alloys*, Butterworths, London, UK., 1967.
16. K. Hauffe: *Oxidation of Metals*, Plenum, New York, USA., 1966.
17. R. A. Rapp and K. S. Goto: *The Hot Corrosion of Metals by Molten Salts*, J. Braunstein et al., Electrochemical Soc., Pennington, NJ., 1981, 81.
18. U. R. Evans: *The Corrosion and Oxidation of Metals*, Matthew Arnold, London, UK., 1960.
19. C. S. Giggins and F. S. Pettit: *Hot Corrosion Degradation of Metals and Alloys*, A Unified Theory-PWA-Report FR-11545, 1979.
20. G. Lehnert and H. Meinhardt: *Surface Treatment*, **1**, 1972, 72.
21. S. R. Levin and R. M. Caves: *J. Electrochem. Soc.*, **120**, 1973, C232.
22. B. A. Pint: *Materials Science Forum*, **397**, 1997, 251–254.
23. N. Czech, F. Schmitz and W. Stamm: *Surface Coatings and Technology*, 1994, 17–21 and 68–69.
24. J. R. Nicholls, N. J. Simms, W. Y. Chan and H. E. Evans: *Surface Coatings and Technology*, **149**, 2002, 236.
25. L. Peichl and D. F. Bettridge: *Materials for Advanced Power Engineering 1994*, **1**, 1994, 717–740.
26. M. P. Bacos, B. Girard, P. Josso and C. Rio: 'MCrAlY Coating by an Electrochemical Route' *Materials for Advanced Power Engineering 2002*, J. Lecomte-Beckers, M. Carton, F. Schubert and P. Ennos, eds., Forschungzentrum, Julich, Germany, **1**, 2002, 429–437.
27. G. W. Goward and L. W. Cannon: *Pack Cementation Coatings for Superalloys, History, Theory and Practice*, ASME Paper 87-GT-50, New York, American Society Mechanical Engineers, 1988.

28. A. H. Sully and E. A. Brandes: *Chromium*, 2nd edn. London, Butterworths, Chap. 7, 1967.
29. P. Felix and E. Erdos: *Werkstoffe U. Korros.*, **23**, 1972, 626.
30. F. Fitzer and J. Schlichting: *High Temperature Corrosion*, NACE-6, 1984, 604–614.
31. F. N. Davis and C. E. Grinell: ASME Pub. 82-GT-244, 1982.
32. M. Van Roode and L. Hsu: *Surface Coatings Technology*, **37**, 1989, 461–481.
33. D. Berry, M. C. Meelu, B. G. McMondie and T. A. Kircherin: ASME Pub. 95-GT359, 1995.
34. J. Stringer and R. Viswanathan: *Advanced Materials and Coatings for Combustion Turbines*, V. P. Swaninathan and N. S. Cheruvu, eds., ASM, 1994, 6.
35. R. Mevrel, C. Duret and R. Pichoir: *Mater. Sci. Technol.*, **2**, 1986, 201.
36. J. R. Nicholls and P. Hancock: *Ind. Corros.*, **5**(4), 1987, 8–18.
37. R. Bianco and R. A. Rapp: *Jour. Electrochem. Soc.*, **140**(4), 1993, 1181–1191.
38. C. Duret, A. Davin, G. Marrijnissen and R. Pichoir: *High Temperature Alloys for Gas Turbines*, R. Brunetaud et al., eds., Dordrecht, D. Reidel Publishing Co., 1982, 53–87.
39. G. W. Goward: *High Temperature Corrosion*, (Conf. Proc. NACE-6), R. A. Rapp, ed., Houston, TX., NACE, 1983, 553–560.
40. M. Malik, G. Morbioli and P. Huber: *High Temperature Alloys for Gas Turbines*, R. Brunetaud et. al., eds., Dordrecht, D. Reidel Publishing Co., 1982, 87–98.
41. J. R. Nicholls, D. J. Stephenson, P. Hancock, M. I. Wood and J. E. Restall: *Proceedings of Workshop on Gas Turbine Materials in a Marine Environment*, Bath UK Ministry of Defence, 1984.
42. J. R. Nicholls and S. R. J. Saunders: *High Temperature Materials for Power Engineering*, E. Bachelet et al., eds., Kluwer Academic Publishers, Dordrecht, 1990, 865–875.
43. A. J. A. Mom: NLR Report MP 81003U, Amsterdam, 1981.
44. R. C. Novak: cited in reference 2, 1994.
45. K. L. Luthra and O. H. LeBlanc: *Materials Science and Engineering*, **88**, 1987, 329.
46. B. G. McMordie and A. Weatherill: *Turbine Forum*, Nice, France, 2002.
47. Sermaloy 1515, U.S. Patent 5,547,700.
48. J. T. Prater, J. W. Patten, D. D. Hayes and R. W. Moss: *Proceedings of Second Conference on Advanced Materials for Alternate Fuels Capable Heat Engines*, J. W. Firbanks and J. Stringer, eds., Report No. 2639SR, 7/29-7/43, Palo Alto, CA EPRI, 1981.
49. J. A. Goebel, C. S. Giggins, M. Krasij and J. Stringer: *Proceedings of Second Conference on Advanced Materials for Alternate Fuel Capable Heat Engines*, J. W. Firbanks and J. Stringer, Report No. 2639SR, 7/1, Palto Alto., CA EPRI, 1981.
50. J. Forster, B. P. Cameron and J. A. Carews: *Trans. Inst. Metal Finish*, **63**, 1985, 115.
51. N. S. Bornstein and J. Smeggil: *Corrosion of Metals Processed by Directed Energy Beams*, Met. Soc. AIME, 1982, 147–158.
52. D. R. Coupland, C. W. Hall and I. R. McGill: *Platinum Metal Review*, **26**(4), 1982, 146–157.
53. I. M. Wolff, L. E. Iorio, T. Rumpt, P. V. T. Scheers and J. H. Potgieter: *Materials Science Engineering*, **A241**, 1998, 264–276.

54. F. Juarez, D. Monceau, D. Tetard, B. Pieraggi and C. Vahlas: *Surf Coating Technol.*, **163**, 2003, 44–49.
55. N. Czech and W. Stamm: *High Temperature Surface Engineering*, J. R. Nicholls and D. S. Rickerby, eds., IOM Communications, London, UK., 2000, 61–65.
56. T. A. Taylor, M. P. Overs, B. J. Gill and R. C. Tucker: *J. Vac. Sci. Tech.*, **3**, 1985, 2526.
57. H. Herman: *Powders for Thermal Spray Technology, Thermal Spray Technology, Powder Science and Technology*, **9**, 1991, 187–199.
58. R. W. Kaufol, A. J. Rotolico, J. Nerz and B. A. Kushner: 'Deposition of Coatings Using a New High Velocity Combustion Spray Gun', *Thermal Spray Research and Applications*, T. F. Bernecki, ed., ASM International, Materials Park, Ohio, USA., 1990, 561–569.
59. L. Russo and M. Dorfman: 'High-Temperature Oxidation of MCrAlY Coatings Produced by HVOF', *Proceedings of the International Thermal Spray Conference*, A. Ohmori, ed., High Temperature Society of Japan, Japan, 1179–1194.
60. E. Lugscheider, C. Herbst and L. Zhao: *High Temperature Surface Engineering*, J. R. Nicholls and D. S. Rickerby, eds., IOM Communications, London, 2000, 67–76.
61. E. C. Kedward: *Metallurgia*, **79**, 1969, 225.
62. F. J. Honey, E. C. Kedward and V. J. Wride: *Vac. Sci. Tech.*, **A4**, 1986, 2593.
63. J. E. Restall and C. Hayman: *Coatings for Heat Engines*, (Workshop Proc.,) R. L. Clarke et al., Washington, DC, US Department of Energy, 1984, 347–357.
64. K. S. Chan, N. S. Cheruvu and G. R. Leverant: *Materials Science Forum*, **369–372**, 2001, 623–630.
65. B. M. Warnes: *Surf. Coat. Technol.*, **163–164**, 2003, 106–111.
66. L. Swadzba, S. R. J. Saunders, M. Hetmanczyk and B. Mendala: *Materials for Advanced Power Engineering*, Forschungszentrum Julich, Germany, 2002.
67. J. R. Nicholls and R. Wing: 'Advanced Coating Systems for Utility Gas Turbines' *Materials for Advanced Power Engineering 2002*, J. Lecomte-Beckers, M. Carton, F. Schubert and P. J. Ennis, eds., Forschungzcentrum Julich, Germany, **1**, 2002, 57–71.
68. M. J. Steiger, N. M. Yanar, M. G. Topping, F. S. Pettit and G. M. Meier: *Z. Metallkund*, **90**, 1999, 1069–1078.
69. N. M. Yanar, G. M. Kim, F. S. Pettit and G. H. Meier: *Turbine Forum*, Nice, France, 2002.
70. D. S. Rickerby and R. G. Wing: 'Thermal Barrier Coating for a Superalloy and a Method of Application Thereof', US Patent 5,942,337, 1999.
71. D. S. Rickerby, S. R. Bell and R. G. Wing: 'Article Including Thermal Barrier Coated Superalloy Substrates', US Patent 5,981,901, 1999.
72. S. R. J. Saunders and J. R. Nicholls: *Thin Solid Films*, **119**, 1984, 247.
73. C. Bezencon, M. Konter, J-D. Wagniere and W. Kurz: 'Microstructural Development in Laser Cladding of a Single Crystal Nickel Based Alloy', *Proceedings of Lasers in Manufacturing, German Scientific Laser Society*, Munich, Germany, 2001, 580–589.
74. C. Bezemcon, J-D. Wagniere, M. Hobel, A. Schrell, M. Konter and W. Kurz: 'Single Crystal Coating of a Single Crystal Turbine Blades by a Laser Cladding Technique' *Materials for Advanced Power Engineering 2002*, J. Lecomte-Beckers, M. Carton,

F. Schubert and P. Ennis, eds., Forschungszentrum Julich, Germany, **1**, 2002, 503–510.

75. G. Muller and D. Strauss: 'Improvement of MCrAlY Properties by Pulsed Electron Beams', *Turbine Forum*, Nice, France, 2002.
76. J. R. Nicholls, N. J. Simms, S. Neseyif, H. E. Evans, C. Ponton and M. J. Taylor: *Electrochem Soc.*, **99–38**, 2002, 305–316.
77. J. R. Nicholls, N. J. Simms, M. Taylor and H. E. Evans: 'Smart Overlay Coatings', *Turbine Forum*, Nice, France, 2002.
78. A. G. Evans, D. R. Mumm, J. W. Hutchinson, G. H. Meier and F. S. Pettit: *Progress in Materials Science*, **46**, 2001, 505.
79. K. Bouhanek, O. A. Adesanya, F. H. Stott, P. Skeldon, D. G. Lees and G. C. Wood: *Materials Science Forum*, **369–372**, 2001, 615–622.
80. O. A. Adesanya, K. Bouhanek, F. H. Stott, P. Skeldon, D. G. Lees and G. C. Wood: *Materials Science Forum*, **369–372**, 2001, 639–646.
81. S. G. Young and G. R. Zellars: *Thin Solid Films*, **53**, 1987, 241–250.
82. J. R. Nicholls, K. J. Lawson, G. Chester, L. H. Yasiri and P. Hancock: *European Research on Materials Substitution*, I. V. Mitchell and H. Nosbuch, eds., Elsevier, London, UK., 1988, 295–307.
83. J. R. Nicholls, K. J. Lawson, L. H. Al-Yasiri and P. Hancock: *Corr. Sci.*, **35**, 1993, 1209.
84. C. H. Liebert et al.: 'Durability of Zirconia Thermal Barrier Coatings on Air Cooled Turbine Blades in Cyclic Jet Engine Operation', Report TMX-3410, NASA, 1976.
85. S. J. Grisaffe: *Thermal Barrier Coatings*, Report TMX-78848, NASA, 1978.
86. R. A. Miller: 'Current Status of TBCs - An Overview', *Surface and Coatings Technology*, **30**, 1987, 1.
87. T. E. Strangman: *Thin Solid Films*, **127**, 1985, 93–95.
88. S. M. Meier and D. K. Gupta: *J. Eng. Gas Turbine and Power*, **116**, 1994, 250.
89. F. O. Soechting: 'A Design Perspective on Thermal Barrier Coatings', *Thermal Barrier Coating Workshop*, NASA-CP-3312, NASA, 1995, 3.
90. S. Bose and J. Demasi-Marcin: 'Thermal Barrier Coating Experience in Gas Turbine Engines at Pratt and Whitney', *Thermal Barrier Coatings Workshop*, NASA-CP-3312, National Aeronautics and Space Administration Lewis Research Centre, Cleveland, Ohio, 1995, 63–77.
91. AGARD Report 823, *Thermal Barrier Coatings*, 1998.
92. P. Morrell and D. S. Rickerby, ibid 20-1, 1998.
93. J. R. Nicholls, Y. Jaslier and D. S. Rickerby: *Materials at High Temperature*, **15**, 1998, 15.
94. J. R. Nicholls, Y. Jaslier and D. S. Rickerby: *Mater. Sci. Forum*, **251**, 1997, 935.
95. J. R. Brandom, R. Taylor and P. Morrell: *Surface and Coatings Technology*, **50**, 1992, 141.
96. R. Brandt: *High Temp. High Press.*, 1979, 13.
97. P. Li, I. W. Chen and J. E. Pennero Hahn: *J. Am. Ceram. Soc.*, **77**, 1994, 118–128, 1281–1288 and 1289–1295.
98. E. A. G. Shilling and D. R. Clarke: *Acta Mater.*, **47**, 1999, 1297–1305.
99. P. K. Wright and A. G. Evans: *Current Opinions Sol. State and Mater. Sci.*, **4**, 1999, 255.

100. D. R. Mumm and A. G. Evans: *Acta. Mater.*, **48**, 2000, 1815–1827.
101. V. K. Tolpygo, D. R. Clarke and K. S. Murphy: *Surface and Coatings Technology*, **124**, 2001, 146–147.
102. G. C. Chang, W. Phuchaeron and R. A. Miller: *Surface and Coatings Technology*, **32**, 1987, 307.
103. S. M. Meier, D. M. Nissley and K. D. Sheffler: *Thermal Barrier Coating Life Prediction Model Development*, Phase II, Report CR-18911, NASA, 1991.
104. B. A. Pint, A. J. Garratt-Reed and L. W. Hobbs: *Mater. High Temp.*, **13**, 1995, 3.
105. J. L. Smialek: *Met. Trans.*, **A22**, 1991, 739.
106. H. E. Evans, A. Strawbridge, R. A. Carolan and C. B. Ponton: *Mater. Sic. Eng.*, **A225**, 1997, 1.
107. A. Atkinson, Á. Selcuk and S. J. Webb: *Oxid. Met.*, **54**, 2000, 371.
108. W. J. Brindley: *Workshop on Thermal Barrier Coatings*, Report CP-3312, NASA, Cleveland, 1995, 189.
109. A. M. Freborg, B. L. Ferguson, W. J. Brindley and G. J. Petrus: *Thermal Barrier Coatings*, AGARD Report 823, 1998, 17-1.
110. R. A. Miller, R. G. Garlick and J. L. Smialek: *Am. Ceram. Soc. Bull.*, **62**, 1983, 1355.
111. L. Lelait, S. Alperine, M. Derrien, Y. Jaslier and R. Mevrel: *Thermal Barrier Coatings*, AGARD Report 823.
112. J. R. Nicholls, K. J. Lawson, D. S. Rickerby and P. Morrell: *Thermal Barrier Coatings*, AGARD Report, 823, 1998, 6-1.
113. J. R. Nicholls, K. J. Lawson, A. Johnstone and D. S. Rickerby: *Mater. Sci. Forum*, **369**, 2001, 595.
114. J. R. Nicholls, K. J. Lawson, A. Johnstone and D. S. Rickerby: *Surface and Coatings Technology*, **383**, 2002, 151–152.
115. Y. A. Tamarin, E. B. Kachanov and S. V. Zherzdev: *Mater. Sci. Forum*, **949**, 1997, 251.

Characterisation of EB-PVD-Thermal Barrier Coatings Containing Lanthanum Zirconate

E. LUGSCHEIDER, K. BOBZIN and K. LACKNER
University of Technology RWTH Aachen
Materials Science Institute, 52056 Aachen
Germany

ABSTRACT

Gas turbine processes are a well developed mechanisms for converting chemical potential energy in form of fuel, to thermal energy and then to mechanical energy for example used in aircraft or generating plants or for generating electric power. Today the most suitable solution to improve the efficiency of gas turbine engines appears to be an increase of higher operating temperatures. The metallic materials used in gas turbine engines have nearly reached their upper limits of thermal stability. In the hottest regions of modern gas turbine engines, metallic materials are used at temperatures above their melting points.[1–3] They can only resist because of air cooling, but as a negative side effect excessive air cooling reduces their efficiency. Therefore, there have been extensive developments of thermal barrier coatings (TBCs). The TBCs have shown great potential in improving the durability and efficiency of gas turbine engines by allowing an increase of the turbine's inlet temperature and by reducing the amount of cooling air at hot-section components. Zirconia offers the properties mentioned before plus a high coefficient of thermal expansion, that is a primary reason for the success as a thermal barrier material on metallic substrates. TBCs have been deposited by several techniques including thermal spraying, sputtering and electron beam physical vapour deposition (EB-PVD). The EB-PVD is the currently preferred technique for demanding applications because of it's special coating structure. Depending on several parameters the ceramic coatings have a columnar grain microstructure that consists of small columns separated by gaps which become smaller inside the coating. These gaps allow the important substrate expansion without the cracking or the spalling of the coat. The main problem of zirconia yttria stabilized thermal barrier coatings (YPSZ) deposited with EB-PVD is a sinter process above 1200°C when the columns loose their property of expansion tolerance due to a closing of their gaps. This process leads to a very fast fatigue of the thermal barrier coatings. For that reason extensive developments of new thermal barrier coatings were needed that allow an increase of turbine inlet temperatures.[4] Investigations on some materials with perovskite, spinelle and pyrochlore structure have shown a great potential of Lanthanum zirconate (pyrochlore) as thermal barrier coating. This ceramic material has a cubic crystal structure. The project introduced today the Lanthanum zirconate coatings were deposited on Ni base Inconel Alloy 600 with a EB-PVD coating device from Leybold. The coatings were characterized by the following analytical methods: x-ray diffraction (XRD), scanning-electron microscopy (SEM), electron dispersive diffraction (EDX), Nanoindentation and thermo-cycling-test.

INTRODUCTION

Gas turbines commonly used in aircraft or generating plants are very often equipped coated first stage blades that have been coated with EB-PVD. The EB-PVD coatings work as a thermal barrier coat (TBC) between the hot gas stream and the cooled metal part of the blades. Due to this, the engine operation can be performed at a higher temperature. Today the TBCs are produced from yttria stabilized zirconia. Beside of the thermal load and an oxidation attack in the first stage of the environment, an erosive attack can also rize to a life limiting factor of the zirconia coating. Not only combustion components, but also particles in the compressed air can increase an erosive attack. In addition the multi-layer system used for gas turbine protection that consists of a considerable number of interfaces, is also a reason for a failure of the coating. The intrinsic stresses of interfaces could be caused by a mismatch in coefficient by thermal expansion, too enhance the gas inlet temperature and the lifetime of the TBC, in this project a lanthanum zirconate coating as TBC was investigated. In a first step, the stochiometric composition of lanthanum zirconate was tried to be realized by preparing segments of lanthanum zirconate powder in a rotating crucible before the deposition run. This method limits the maximum evaporation time to 70 minutes. In a second step the stochiometric composition of lanthanum zirconate was tried to be realized by preparing segments of different powder mixtures of zirconia and lanthanum oxide in a rotating crucible before deposition run. The crystal structure of the deposited EB-PVD-coatings were analysed by x-ray diffraction (XRD). Their microstructure and topography were examined by scanning electron microscopy (SEM). The chemical composition was characterized by electron dispersive diffraction (EDX). The young's modulus and hardness was characterized by nanoindentation. The thermal resistivity were examined by thermal cycling test.

EXPERIMENTAL PROCEDURE

In order to choose a material a number of multiple parameters are of importance. Due to the manufacturing technology (i.e. vaporisation out of one crucible) materials with similar vapour pressures to zirconia should be choosed. In this case the lanthanum zirconate and the lanthanum oxide have much higher pressures as zirconia. This leads to a lower melting point of the two powders compared that of zirconia. Lanthanum oxide shows a complete solubility of zirconia oxide in the melt. With increasing content of lanthanum oxide, opposite to zirconia and decreasing content of zirconia oxide, the liquidus temperature is continuously decreasing. At the composition of 67 mole% lanthanum oxide and 33 mole% zirconia oxide the melting point reaches the stochiometric lanthanum zirconate phase. This phase remains stable from the melting point of 2200°C to room temperature on. The lanthanum zirconate has a cubic crystal structure with a big volume of the unit cell with 1262.51. Below or above this chemical composition metastable phases are possible, like a tetragonal zirconia plus a low fraction of lanthanum zirconate (pyrochlore). Those structures can be produced with PVD. For the production of lanthanum zirconate coatings different powder mixtures with different mixing relations during the deposition process were chosen. In the first case, the crucible was completely filled with stochiometric lanthanum zirconate powder. In the second case, the crucible was filled with different mixtures of zirconia oxide/lanthanum oxide with mixing

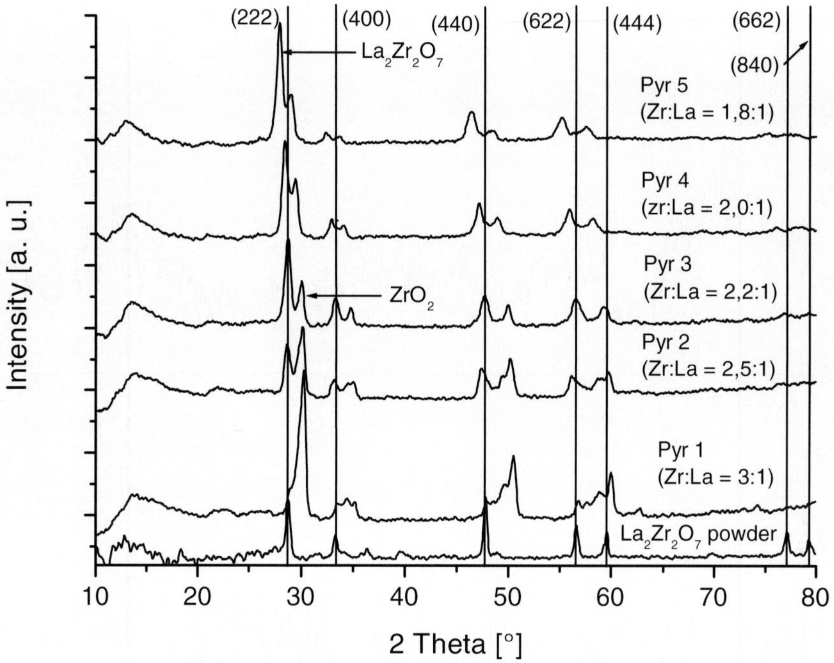

Fig. 1 XRD-illustration of the EB-PVD deposited coatings.

relations of 3 : 1, 2.5 : 1, 2.2 : 1, 2.0 : 1 and 1.8 : 1. Due to that variation of powder the part of lanthanum zirconate structure in the deposited coatings could have been increased. This configuration limits the possible process running time to 70 minutes, as this is the time one crucible rotation needs its lowest mode. As a consequence the thickness of the coat is limited.

RESULTS AND DISCUSSION

To admit, it was not possible to deposit a pure pyrochlore coating from the stochiometric powder out of the crucible. The XRD illustration in Figure 1 allows a qualitative identification of the crystal structure. From coating Pyr 1 to 4 the content of lanthanum zirkonate is increased. The small shift of the strongest peak (222) in Pyr 4 and 5 is due to the content of free lanthanum oxide phase. The coating Pyr 4 has the highest content of pyrochlore. The XRD only does not, allow an appointment of the quantitative amount of lanthanum zirconate in the coatings.

With account of the EDX-linescan analysis, it becomes possible to determine the chemical composition along the thickness of the coating. The EDX image of coating Pyr 4 with the highest content of pyrochlore is shown in Figure 2. The crucible was filled with zirconia oxide/lanthanum oxide with a powder mixture of 2:1 respectively. During the process both

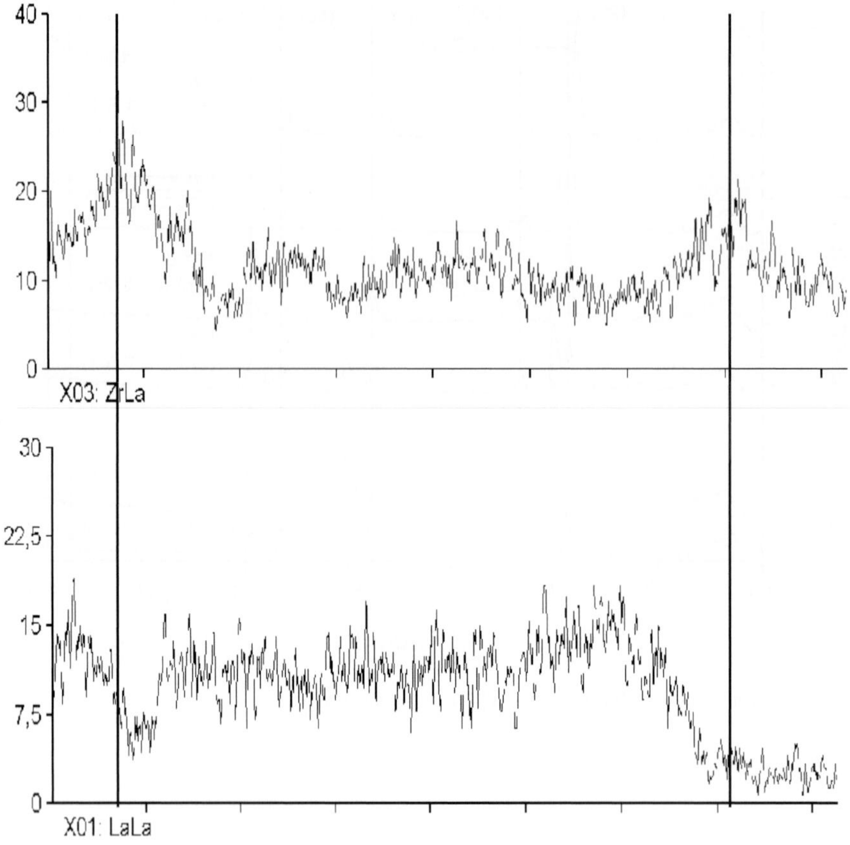

Fig. 2 EDX-Linescan of coating Pyr 4 (coating thickness approx. 43 μm).

powders were vaporised similarilly. For the first two pass of the coating thickness it has been found out that the ratios of both powders were nearly equal. In the last part of the coating thickness the content of lanthanum oxide increased. Thus by changing the chemical composition of coating this way, the content of lanthanum oxide in the powder was too high. The coating thickness measured by EDX-linescan was about 45 μm. The content of lanthanum zirconate for the coating Pyr 4 was calculated with about 90 at-%. During the depositions the substrate temperature conditions were kept constant.

The homologous temperature occurred from the substrate heating temperature and the melting point of lanthanum zirconate is shown in the following equation:

$$\left(\frac{T}{T_M}\right)_{ZrO_2} = \frac{900°C}{2300°C} = 0.39 \quad (1)$$

The SEM images in Figures 3 and 4 show a rough columnar microstructure. In case of the lower melting point of lanthanum oxide, the substrate temperature is sufficient to allow a

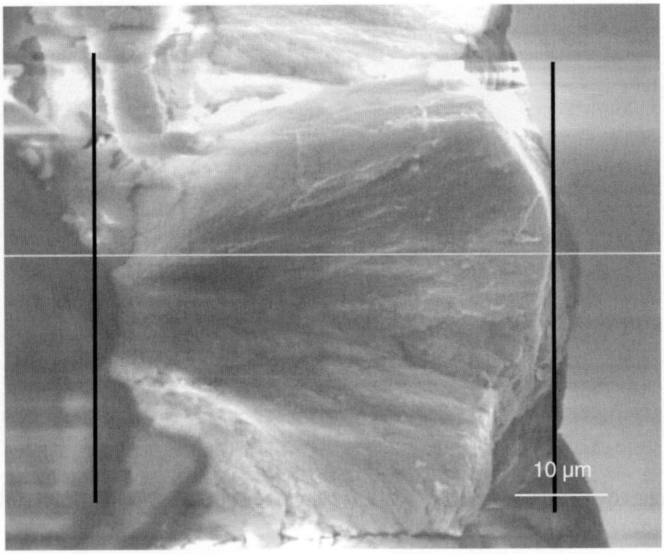

Fig. 3 SEM image of the detected area using the EDX-linescan.

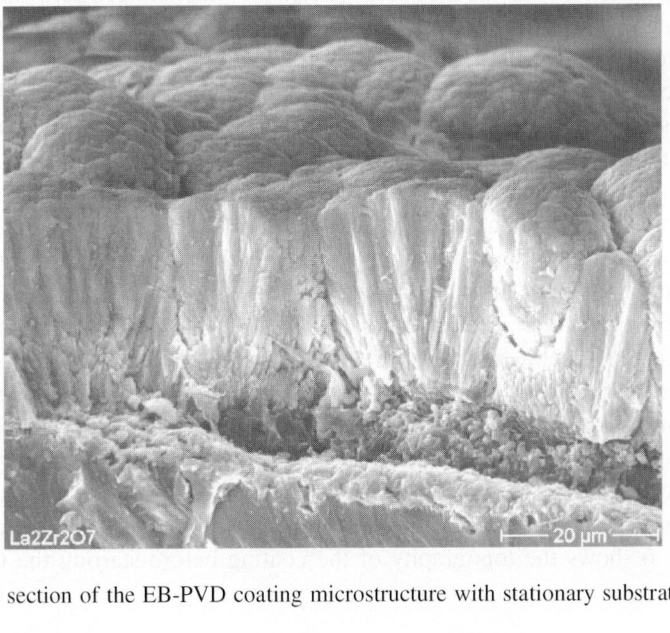

Fig. 4 Cross section of the EB-PVD coating microstructure with stationary substrate.

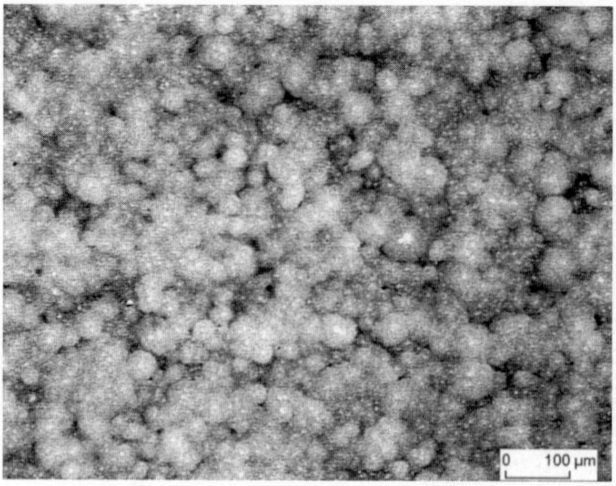

Fig. 5 LM topography image of the EB-PVD pyrochlore coating before starting the thermal cycling test.

surface diffusion. Due to this diffusion bigger grains can arise compared to zirconia coating. The change of a homologous temperature concerning zirconia and the resulting microstructure corresponds with structure zone models like Thornton and Demshichin.[5, 6] The SEM image in Figure 4 confirm the conjecture.

Furthermore the coating was analysed by nanoindentation. The target was an examination of the elastic behaviour of the columns. The results are shown in Figure 5. Examined values were the Young's modulus and the hardness. The measurements were performed on columnar EB-PVD-coating Pyr 4 in in-plane direction.[7] Low values were obtained. Low values for lanthanum zirconate coatings, in comparison to the values of zirconia bulk material (about 210 GPa), were already examined.[8, 9] The Young's modulus is connected with the expansion of the columns. A high Young's modulus determines a decrease in the expansion of the columns. The target is to get a low Young's modulus. The low values for the lanthanum zirconate coating can be described by the same mechanisms as for zirconia case (Young's modulus of about 89 GPa and hardness of about 2.9 GPa). The reason for the lower value is the low adhesion of the zirconia columns to one another. Due to that a final explanation cannot be given on the basis of the presented work.

Thermal cycling tests were performed to evaluate the lifetime of the coatings by exposing the quadratic coated specimens to air for 0.5 hours at 1473 K, and than cooling to 50°C within 5 minutes by the use of forced air cooling with a pressure of 6 bar. The test specimens were inserted into air furnace. The results of the thermal cycling test for lanthanum zirconate promise coating lifetimes lasting more than 500 cycles satisfactory results, as we think. Since a failure of the lanthanum zirconate coating did not occur after an exposure of 500 cycles of, successive thermal cycling test were kept on being performed to examine the coating. Figure 6 shows the topography of the coating before starting the thermal cycling test.

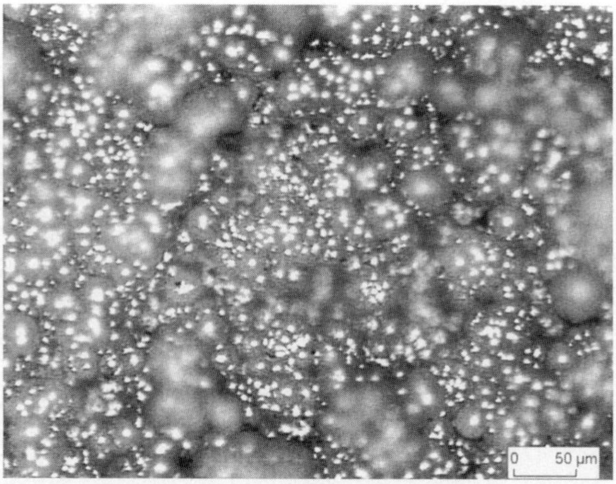

Fig. 6 Topography of the lanthanum zirconate coating Pyr 4 after 350 cycles of thermal cycling test.

Figure 7 shows the topography of the lanthanum zirconate coating after 350 cycles. No failure was observed. The topography of the pyrochlore coating after 550 cycles. It was clearly observed that the coating shows failure in form of big spallations. That means that the oxidation rate of the coating was greatly accelerated under hot-temperature environment. Once the deterioration started, the coated specimen showed abrupt loss in the adhesion strength of its interface between MCrAlY/TBC, which implies a failure of the coating. There is no doubt that the failure of the lanthanum zirconate coating will occur after few thermal cycling tests as long as an YPSZ phase is existing.

CONCLUSION

The presented examinations on the production of stochiometric lanthanum zirconate coatings (pyrochlore) showed the potential of a TBC. A change in the chemical composition can be realized very easily. It was shown that a continuous change in the composition of the source material lead to a continuous change in the chemical composition of the deposited coatings. The coatings exhibited a low Young's modulus necessary for the expansion to relax the intrinsic stresses. The coatings show good resistance to high temperature oxidation. Thermal cycling tests have demonstrated promising results of coating lifetimes lasting more than 500 cycles. In further tests the deposition of stochiometric pyrochlore with the vaporizing material will be realized by two separate powder feeding systems (one for zirconia oxide, one for lanthanum oxide. This way will also allow longer deposition times due to the possibility of a continuous feeding of evaporising material. Future work should also include the deposition of graded coating.

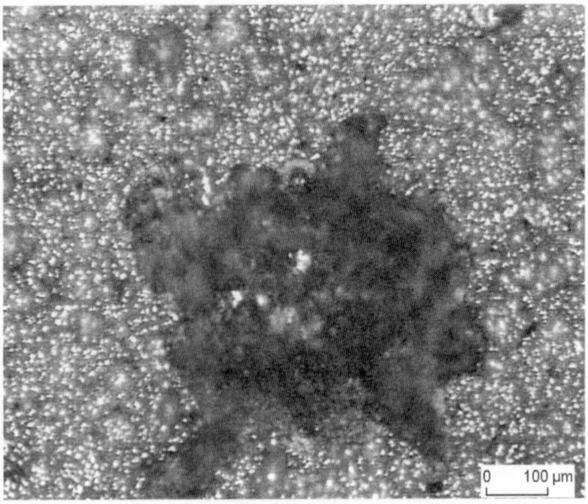

Fig. 7 Topography of the lanthanum zirconate coating Pyr 4 after 550 cycles of thermal cycling test (big spallations could be observed).

ACKNOWLEDGMENT

The authors gratefully acknowledge the financial support of the Arbeitsgemeinschaft industrieller Forschungsvereinigung Otto von Guericke e.V. (AiF).

REFERENCES

1. E. Lugscheider, C. Barimani and G. Doepper: *High Temperature Surface Engineering*, IOM Communications, UK., 2000, 85–94.
2. E. Lugscheider, G. Doepper, H. G. Mayer, A. Seidel and W. Dreier: *High Temperature Surface Engineering*, IOM Communications, UK., 2000, 77–84.
3. F. Jamarani and M. Korotkin: *Surface Coatings and Technology*, **54/55**, 1992 58–63.
4. E. Lugscheider, K. Bobzin and A. Etzkorn: *Proceedings of Life Ass. Hot Sect. Gas Turb. Comp.*, Edingburgh, 1999.
5. J. A. Thornton: *J. Vac. Sci. Tech.,* **A4**(6), 1896, 3059.
6. B. A. Movchan and A. V. Demshichin: *Phys. Metals Metallogr.,* **28**(4), 1969, 83.
7. K. Duan and R. W. Steinbrecher: *Journal of European Ceramics Society*, **18**, 1998, 87–93.
8. C. Funke, R. W. Steinbrecher and F. Schubert: *DLR-Werstoffkolloquium '96*, Köln, 1996, 6.
9. L. Singheiser, R. W. Steinbrecher, W. J. Quadakkers, D. Clemens and B. Siebert: *Mat. Adv. Power Eng.*, Jülich, Germany, 1998, 977.

The Opposite Effect of Re on High Temperature Oxidation Resistance and Hot Corrosion Resistance of Nickel-Based Superalloys

Y. MURATA, MD. MONIRUZZAMAN and M. MORINAGA
Department of Materials Science and Engineering
Graduate School of Engineering
Nagoya University, Furo-cho, Chikusa
Nagoya 464-8603, Japan

R. HASHIZUME
Materials Research Section
Technical Research Center
The Kansai Electric Power Company Inc.
Japan

A. YOSHINARI and Y. FUKUI
Hitachi Research Laboratory
Hitachi Co. Ltd.
Japan

ABSTRACT

In order to elucidate the Re effect on oxidation resistance, high-temperature oxidation tests were conducted with two groups of Ni-based superalloys. One group of alloys was characterized by 10 mol.%Al content (10-Al series). The other group of alloys was characterized by 15 mol.%Al content (15-Al series). Both cyclic and isothermal oxidation tests were carried out at 1373 K for a total time of 720 ks. The oxidation resistance decreased clearly with increasing Re content in the 10-Al series alloys, but did not in the 15-Al series alloys. The alloys having a Re/Al ratio (mol.%) up to 0.1 exhibited very small mass change by oxidation, so the Re/Al ratio is one of the important indications for the design of superalloys having a good oxidation resistance. On the other hand, it is well known that Re is one of the most effective element to improving the hot corrosion resistance of the superalloys. Following the above findings with respect to the Re effects, three Ni-based single crystal superalloys containing Re were proposed with the aid of d-electrons concept.

INTRODUCTION

Despite the recent innovation of coating technology, intrinsic resistance of Ni-based superalloys to high-temperature oxidation and hot corrosion has to be assured in order to prevent the blade of gas turbines from degrading in the case of cracking or spalling of the protective coating materials. In particular, oxidation resistance needs to be improved in response to the general increase in the service temperature. The oxidation resistance of superalloys relies on the protectiveness of Al_2O_3-scales formed on their surfaces during service.

It is well known that Re is unique element which can increase both high-temperature creep strength and the hot corrosion resistance of superalloys.[1] The advanced Ni-based superalloys are characterised by higher Re content; second generation single crystal (SC) superalloys contain about 3 wt.% (about 1 mol.%) Re[2–4] and third generation SC superalloys contain 5 to 6 wt.% (about 1.7 to 2 mol.%) Re.[4–6] So, influence of the Re addition on the oxidation resistance of superalloys raises great concern. Recently we have obtained an indication that Re has a harmful effect on the oxidation resistance of Ni-based superalloys at high temperatures, although the variety of the alloy composition is rather limited.[7] In order to make it more clear, the Re effect on the oxidation resistance has been investigated in this study. Following the above findings with respect to the Re effects, three Ni-based single crystal superalloys containing Re were proposed with the aid of d-electrons concept.

EXPERIMENTAL PROCEDURE

Alloy Preparation

The chemical compositions of the alloys used in this study are shown in Table 1. There are two groups. One group of the experimental alloys is expressed as Ni-8Cr-2Ti-10Al-Re in mol.% (hereafter referred to as A-group), which is characterized by 10 mol.% Al content. The other group is expressed as Ni-8Cr-2Ti-15Al-Re in mol.% (hereafter referred to as B-group), which is characterized by 15 mol.%Al content. The Re content ranges from 0 to 2 mol.% (corresponding to approximately 6 wt.%) with varying 0.25% step in the A-group but 0.5% or 1.0% step in the B-group. The A-group alloys containing 10 mol.% Al are expected to form both Al_2O_3 and Cr_2O_3, whereas the B-group alloys containing 15 mol.% Al are expected to form Al_2O_3 preferentially on the alloy surface during high temperature oxidation.

Following the d-electrons concept,[8, 9] the compositional averages of the two d-electrons parameters, $\overline{Md}$ and $\overline{Bo}$, were calculated since these are known to be related closely to the phase stability as well as the high temperature strength of nickel-based superalloys.[1] The criteria of the $\overline{Md}$ and the $\overline{Bo}$, values for the phase stability and the strength are $\overline{Md} \leq 0.99$ and $\overline{Bo} \leq 0.68$.[1] All the alloys listed in Table 1 were confirmed to satisfy these criteria.

Button ingots of the experimental alloys were prepared by arc-melting under the purified Ar gas atmosphere. Each button ingot was cut into pieces and remelted several times in order to get a compositional homogeneity. The ingots then underwent a solution heat treatment followed by aging heat treatments. The heat treatment conditions were determined on the basis of the experimental data obtained from the differential thermal analysis (DTA). As shown in Table 2, two-steps aging processes were adopted in this experiment.

The heat-treated ingots were cut into small specimens. Those specimens were polished mechanically and were etched in the 1:1 volume% HNO_3 and HCl solution. The

Table 1 Chemical compositions of experimental alloys used, upper in mol% and lower in mass%.

Group	Alloy	Ti	Cr	Ni	Re	Al	Re/Al (mol%)
A-Group	A-1	2.00	8.00	bal.	0.00	10.00	0.000
		1.75	7.60	bal.	0.00	4.93	
	A-2	2.00	8.00	bal.	0.25	10.00	0.025
		1.74	7.55	bal.	0.85	4.90	
	A-3	2.00	8.00	bal.	0.50	10.00	0.050
		1.73	7.51	bal.	1.68	4.87	
	A-4	2.00	8.00	bal.	0.75	10.00	0.075
		1.72	7.47	bal.	2.51	4.84	
	A-5	2.00	8.00	bal.	1.00	10.00	0.100
		1.71	7.42	bal.	3.32	4.81	
	A-6	2.00	8.00	bal.	1.25	10.00	0.125
		1.70	7.38	bal.	4.13	4.79	
	A-7	2.00	8.00	bal.	1.50	10.00	0.150
		1.69	7.34	bal.	4.93	4.76	
	A-8	2.00	8.00	bal.	1.75	10.00	0.175
		1.68	7.30	bal.	5.72	4.73	
	A-9	2.00	8.00	bal.	2.00	10.00	0.200
		1.67	7.26	bal.	6.50	4.71	

Table 2 Conditions for heat treatments.

Alloys	Solution Heat Treatment	Aging Heat Treatment	
		1st Step Aging	2nd Step Aging
A-Group	1633K/14.4 ks, AC	1373K/14.4 ks, AC	1144K/86.4 ks, AC
B-Group	1593K/14.4 ks, AC	1373K/14.4 ks, AC	1144K/86.4 ks, AC

microstructural observation of the etched specimens of all the alloys was made using a scanning electron microscopy (SEM).

Oxidation Tests

The specimens for oxidation tests were prepared by cutting each alloy ingot into 10 × 5 × 1 mm size, and then the surfaces were polished mechanically using emery papers down to #600 grits. Two kinds of oxidation tests were employed in this study. One was the cyclic oxidation test, in which the specimen was kept in a furnace at 1373 K for 72 ks followed by air-cooling in each cycle. This thermal cycle was repeated for 10 times, so the total exposure time at 1373 K was 720 ks. The other was the isothermal oxidation test, in which the specimen was set in a furnace for the exposure to static air at 1373 K for 720 ks followed by air-cooling. However, the mass changes of the isothermal test were much smaller than those of the cyclic one. In fact, the cyclic oxidation is a sever test compared with the isothermal oxidation test, if the test temperature is the same. In the case of nickel based superalloys which show excellent oxidation resistance, cyclic oxidation tests are used commonly in order to evaluate the oxidation resistance of the alloys.[10,11] Then, only the results on the cyclic oxidation test will be shown in this study.

After the oxidation test, specimen surface and the cross-section of the oxidized layer were observed using a SEM operated at 20 kV, and then the EDX analysis was performed in order to get the characteristic X-ray images of the constituents from the oxides. The oxidation products were identified by the EDX analysis and also by the conventional X-ray diffraction technique.

RESULTS

Microstructure

The representative microstructures of A-group alloys and B-group alloys are shown in Figure 1. Here, micrographs shown in (a) and (b) were taken after the solution heat treatment, and those shown in (c) and (d) were taken after the aging heat treatment. Despite the same etching treatment used in these two alloys, their appearance is quite different from each other. Namely, the γ' phase stands out from the γ phase matrix in B-1 alloy and vice versa in A-5 alloy. This is simply due to the difference in the chemical composition between these two alloys. The A-5 alloy nominally contains 1 mol.% (3.3 wt.%) Re and 10 mol.% (4.8 wt.%) Al, whereas the B-1 alloy contains no Re and 15 mol.% (7.6 wt.%) Al. Here, Al is the γ' stabilizing element and Re is the γ stabilizing element, so the γ' phase is more prominent in the B-1 alloy than in the A-5 alloy. Regardless of such a difference in the chemical compositions between them, no eutectic phase was observed in the microstructure shown in (a) and (b). Also, the γ' phase was distributed homogeneously in the γ-phase matrix as shown in (c) and (d). Similar morphology was observed in the other alloys.

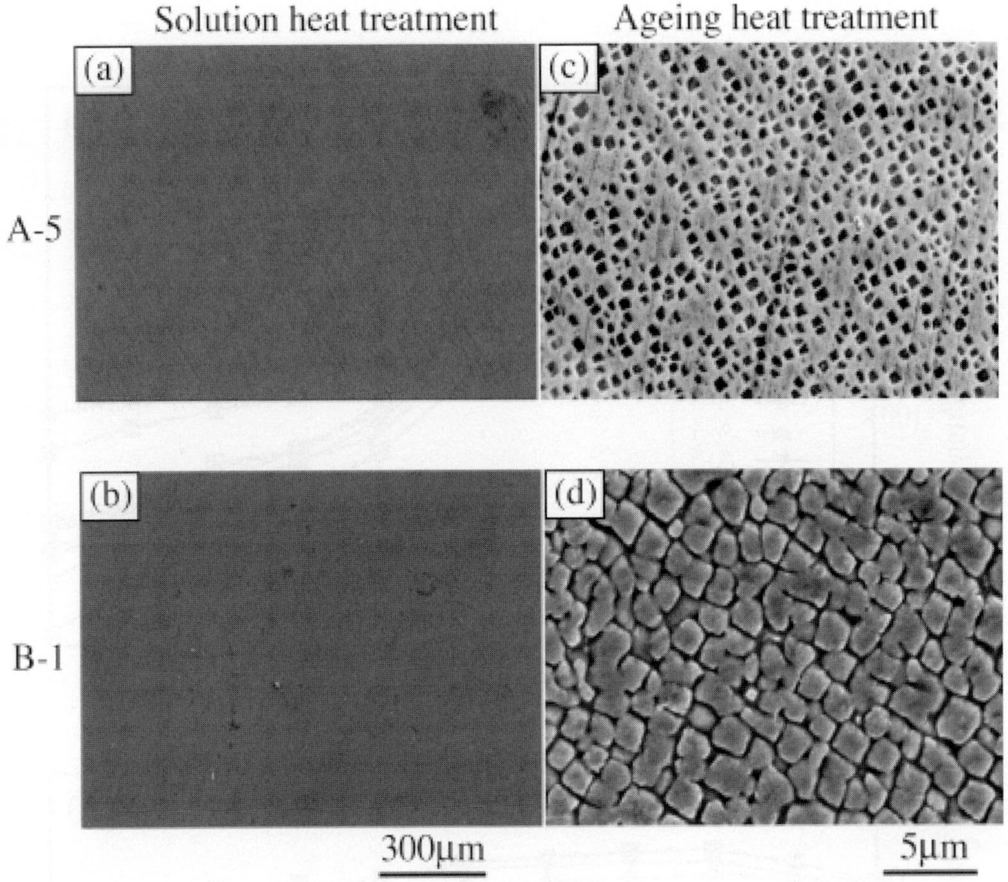

Fig. 1 SEM micrographs taken after solution heat treatment, (a) A-5, (b) B-1 and after aging heat treatment, (c) A-5 and (d) B-1 alloy.

Oxidation Resistance

Cyclic Oxidation Resistance

Figure 2 shows the results of cyclic oxidation tests performed at 1373 K for 10 cycles. As described earlier, the oxidation time in each cycle is 72 ks. The mass change is shown as a function of time in (a) for the A-group alloys and in (b) for B-group alloys. For all the A-group alloys, the mass loss increased gradually with time, although even a little mass gain was observed at the first cycle for A-1 to A-7 alloys. The individual alloys showed a similar trend of losing mass for a definite period of exposure time i.e., a definite number of cycles. For example, the mass loss at any number of cycles was the smallest in the A-1 alloy and the

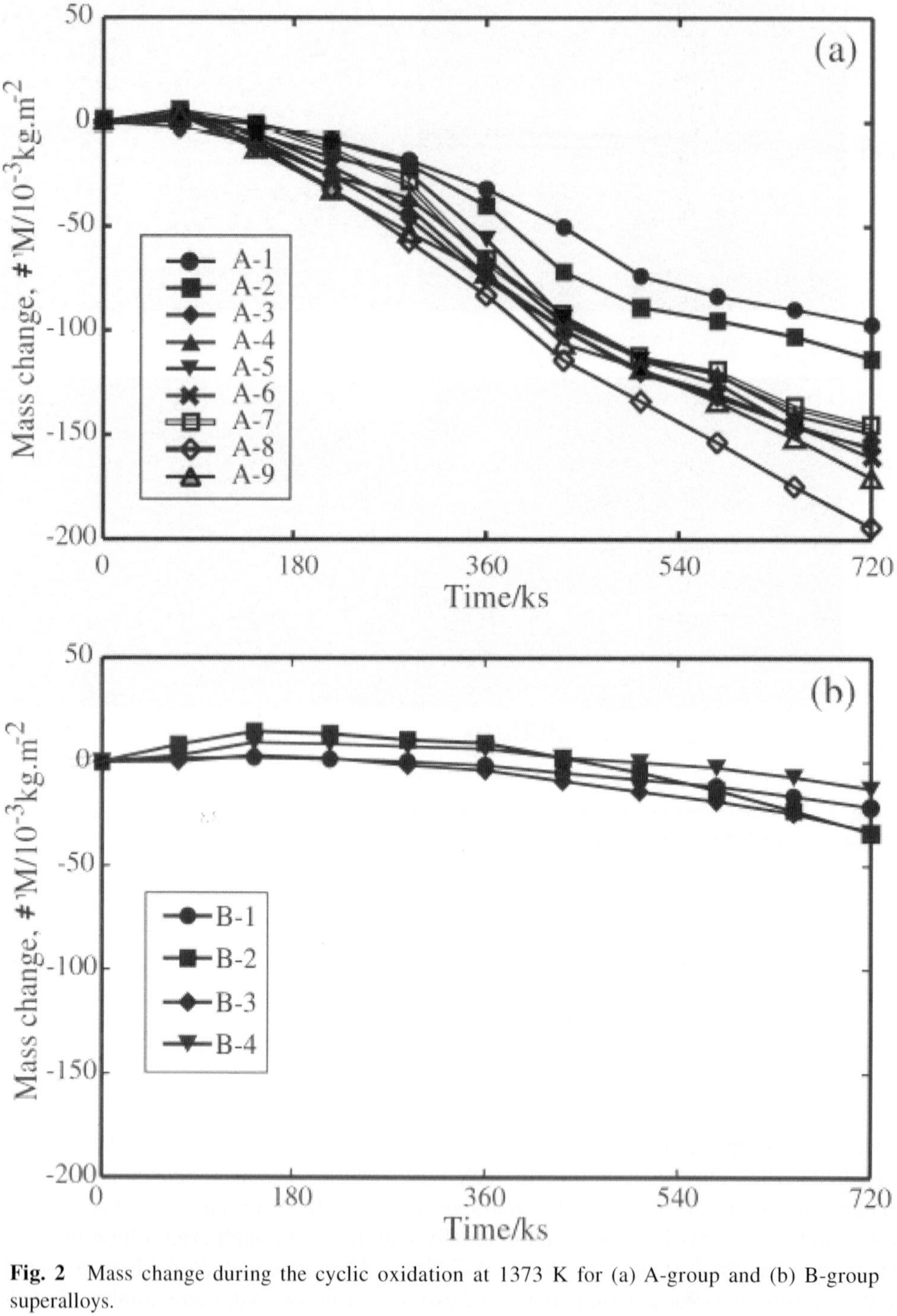

Fig. 2 Mass change during the cyclic oxidation at 1373 K for (a) A-group and (b) B-group superalloys.

largest in the A-8 alloy among the A-group alloys. In the B-group, no significant mass change was observed up to 360 ks (5 cycles) for all the alloys. Namely, a very little mass gain was seen up to 288 ks (4 cycles) and after 360 ks (5 cycles) most of the alloys tended to start losing mass. When compared the two groups with each other, the mass change after 720 ks was much larger in the A-group alloys than in the B-group alloys.

Cyclic Oxidation Products

The SEM image and the corresponding characteristic x-ray images were taken from the cross-section of the specimens after the cyclic oxidation test for 720 ks. The typical results are shown in Figure 3(a) for the Re-free A-1 alloy and in (b) for the A-9 alloy containing the highest Re content (2 mol.% or 6.5 wt.%). In the A-1 alloy, about 17 μm thick oxidized layer was composed of dense NiO in the outer side and dense Al_2O_3 in the inner side. The presence of these oxides on the surface was identified by the X-ray diffraction as explained later. Neither Ti nor Cr concentrated layer was seen in the oxidized layer. On the other hand, the oxidized layer of A-9 alloy shown in Figure 3(b) was about 100 μm thick and it was composed of dense NiO in the outer side and a mixture of porous Cr_2O_3, Al_2O_3 and TiO_2 in the inner side. For the other A-group alloys, the oxidized layer became thicker and porous with increasing Re content in them. Here, in the x-ray image, any Re enriched layer was not observed in oxidation products formed on the Re containing alloys. This is probably due to a high vapour pressure of Re-oxide (Re_2O_7).[12] Similar oxidation products were observed in the B-group alloys, although the Re effect is not so clear as in the A-group alloys. This is due to higher Al content in the B-group alloys than in the A-group alloys, i.e., Al_2O_3 is formed easily in the B-group alloys.

Similar SEM and x-ray images were taken from the specimen surface after the cyclic oxidation test for 720 ks. The results are shown in Figure 4(a) for the A-1 alloy and in (b) for the A-9 alloy. There were localized distributions of Al, Ti, Cr and Ni in the A-1 alloy. For example, the bright area of Al signal corresponds to the dark area in the SEM image. Such morphology was also observed in the A-9 alloy shown in Figure 4(b), but the area fraction of each element was much different between the A-1 and the A-9 alloys. The area of the Al, Ti and Cr-rich region increased but the area of the Ni-rich region decreased in the A-9 alloy, as compared to the respective area in the A-1 alloy. As shown in Figure 3, NiO layer is formed on Al_2O_3 layer, and hence the alloy having a small fraction of the Al-rich region indicates a low degree of the exfoliation of NiO. This corresponds to the case of A-1 alloy showing high oxidation resistance. The oxidized surfaces of the other A-group alloys exhibited the similar morphology but the area fraction was varied depending on the Re content in them.

DISCUSSION

Effect of Re on Cyclic Oxidation Resistance

As shown in Figure 2(a), the minimum mass loss was observed in the Re-free A-1 alloy and the mass loss increased with increasing Re content in the alloy, indicating that Re is an

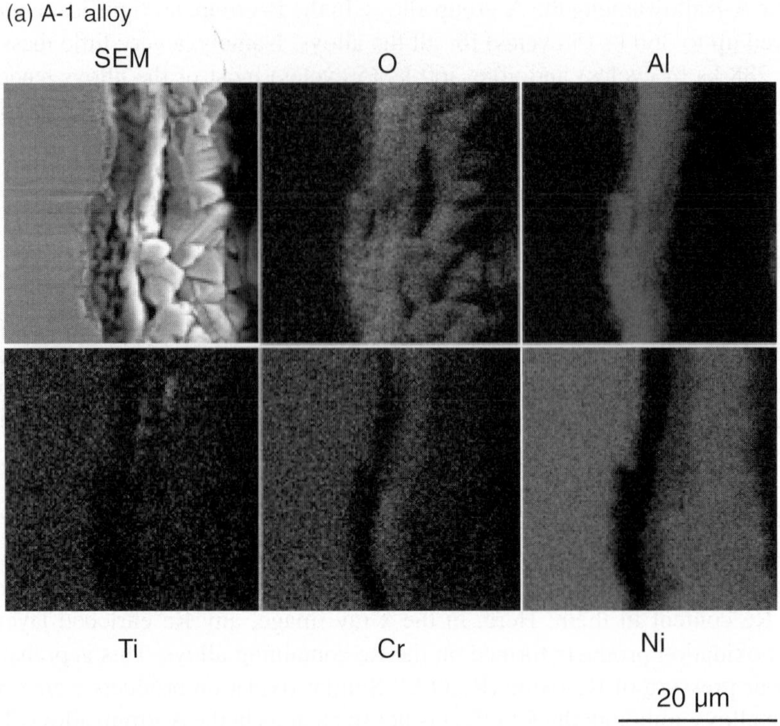

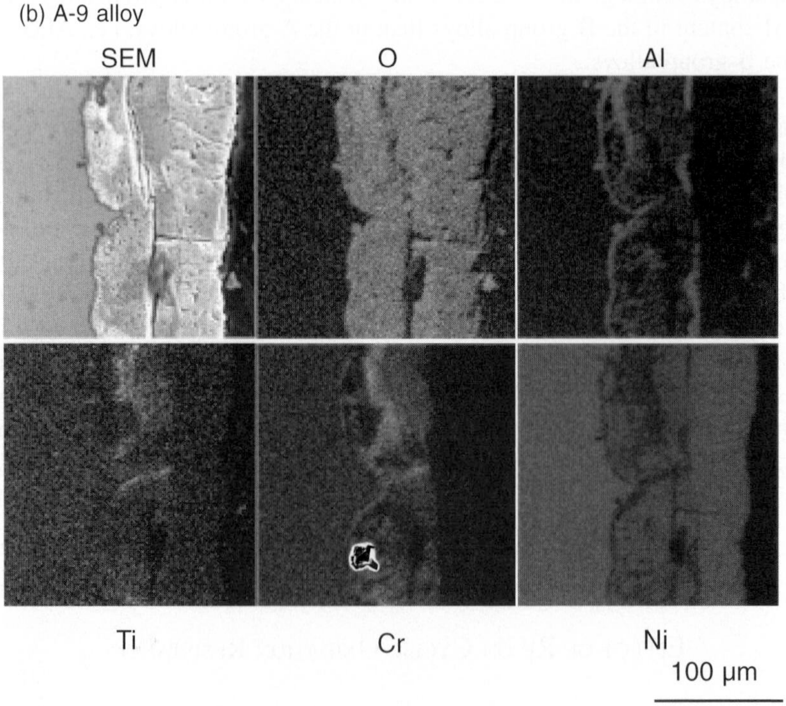

Fig. 3 SEM images and the corresponding characteristic x-ray images taken from the cross section of cyclically oxidized specimens, (a) A-1 and (b) A-9 alloy.

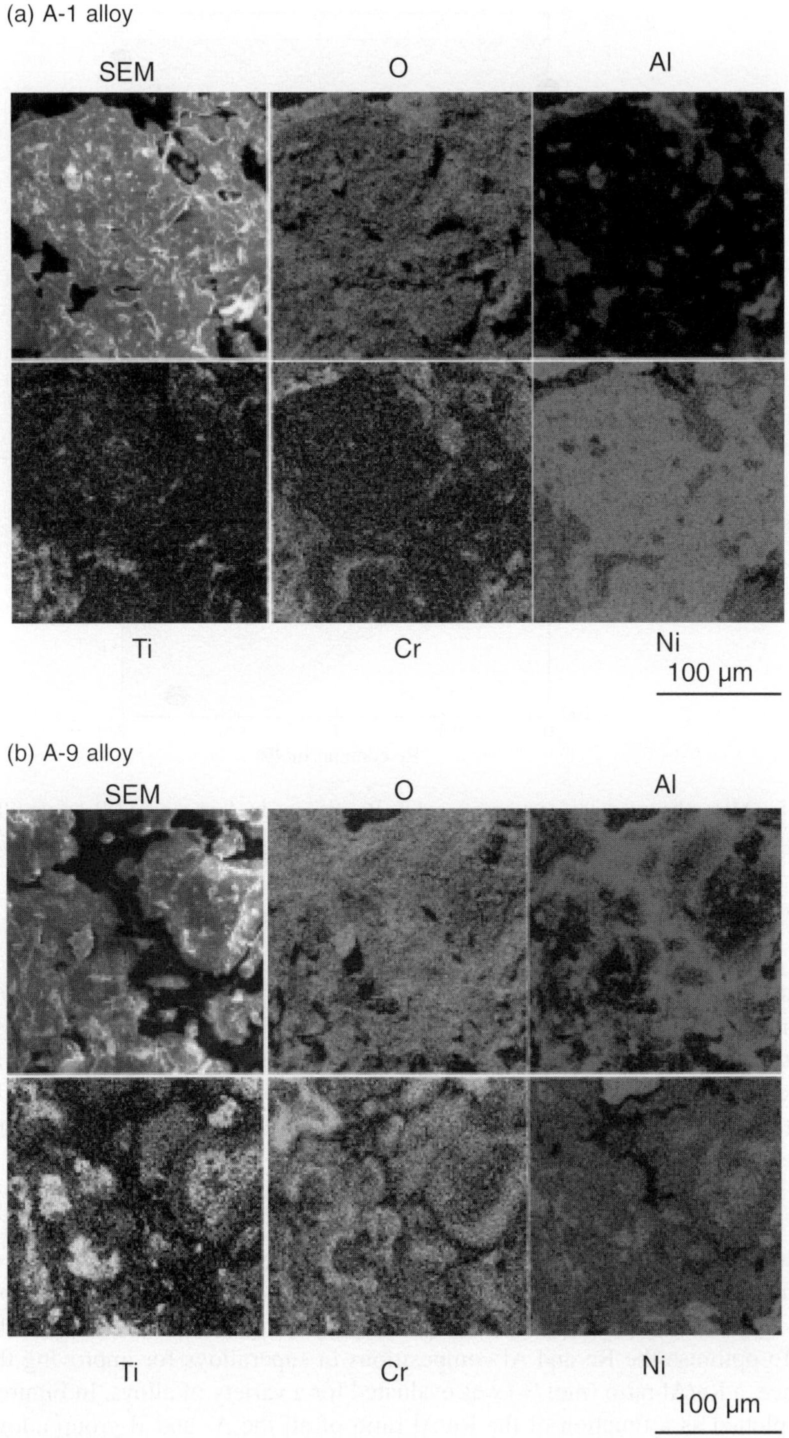

Fig. 4 SEM images and the corresponding characteristic X-ray images taken from the surface of cyclically oxidized specimens, (a) A-1 and (b) A-9 alloy.

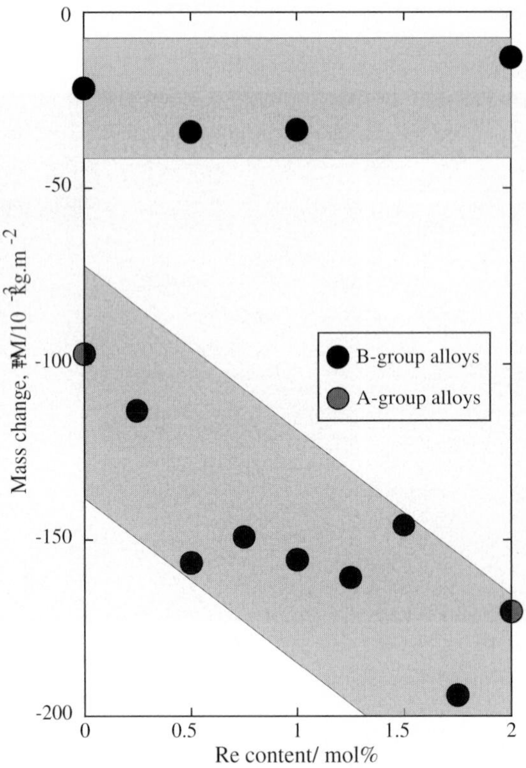

Fig. 5 Mass change as a function of mol.% Re content in A-group and B-group alloys. Mass change was measured after the cyclic oxidation at 1373 K.

element to deteriorate the oxidation resistance. In Figure 5 the mass changes are shown as a function of Re content for the cyclic oxidation test of both the A- and the B-group alloys. The variation of the mass change for the B-group alloys was limited to a given narrow zone regardless of Re content. The range of the mass change was much wider in the A-group alloys than in the B-group alloys. Thus a harmful effect of Re on the oxidation resistance appeared clearly in the A-group alloys.

The difference in the chemical compositions between the A-group and the B-group alloys is only the Al content. So, the smaller mass changes of the B-group alloys than the A-group alloys are due to their higher Al content. Thus it is seen that the deterioration of oxidation resistance due to the Re addition in the 10-Al series alloys (A-group) can be overcome effectively by increasing Al content in the alloys as long as the alloy phase-stability is kept good. To optimise the Re and Al compositions in superalloys for improving the oxidation resistance, a Re/Al ratio (mol.%) was evaluated for a variety of alloys. In Figure 6, the mass loss is plotted as a function of the Re/Al ratio of all the A- and B-group alloys and some other Ni-based alloys listed in Table 3. In spite of the variety of alloying elements in the table, the mass change was found to be small in the alloys having the Re/Al ratio up to 0.1.

Table 3 Chemical compositions of Ni-based superalloys used, upper in mol.% and lower in wt.%.

Alloy	Ti	Cr	Co	Ni	Mo	Hf	Ta	W	Re	Al	Re/Al (mol%)
NHK001	1.29	4.53	12.08	62.42	0.90	0.05	2.33	1.79	1.76	12.85	0.137
	1.00	3.80	11.50	59.16	1.40	0.14	6.80	5.30	5.30	5.60	
NHK002	1.55	5.34	10.99	63.06	0.64	0.04	2.32	2.01	1.46	12.58	0.116
	1.20	4.50	10.50	59.98	1.00	0.12	6.80	6.00	4.40	5.50	
NHK003	1.56	4.77	11.58	62.79	0.52	0.04	2.33	2.03	1.73	12.65	0.137
	1.20	4.00	11.00	59.38	0.80	0.12	6.80	6.00	5.20	5.50	
NHK004	1.17	3.35	14.80	60.69	0.91	0.04	2.41	1.86	1.84	12.93	0.142
	0.90	2.80	14.00	57.18	1.40	0.12	7.00	5.50	5.50	5.60	
NHK005	1.94	4.77	11.77	62.97	0.39	0.05	2.29	1.85	1.80	12.17	0.148
	1.50	4.00	11.20	59.66	0.60	0.14	6.70	5.50	5.40	5.30	
NKH991	2.30	7.06	1.04	71.99	0.77	0.01	2.06	1.90	1.31	11.56	0.113
	1.80	6.00	1.00	69.07	1.20	0.03	6.10	5.70	4.00	5.10	
NKH992	2.33	5.97	9.69	66.67	0.78	0.01	2.08	1.93	1.33	9.21	0.144
	1.80	5.00	9.20	61.92	1.20	0.03	6.05	5.70	4.00	5.10	
NKH993	1.95	4.55	11.62	63.26	0.91	0.05	2.30	1.79	1.80	11.76	0.153
	1.50	3.80	11.00	59.66	1.40	0.14	6.70	5.30	5.40	5.10	
NKH994	1.32	5.09	2.35	72.22	0.72	0.05	2.68	2.02	1.86	11.68	0.159
	1.00	4.20	2.20	66.46	1.90	0.14	7.70	5.90	5.50	5.00	
NKH995	1.71	5.08	5.02	69.85	0.72	0.05	2.19	2.33	1.86	11.19	0.166
	1.30	4.20	4.70	65.16	1.10	0.14	6.30	6.80	5.50	4.80	
NKH996	2.30	9.00	0.00	71.69	0.77	0.00	2.04	2.30	0.70	11.21	0.062
	1.81	7.69	0.00	69.16	1.21	0.00	6.07	6.95	2.14	4.97	
NKH997	2.14	9.06	3.59	68.48	0.50	0.00	2.04	2.53	0.24	11.42	0.021
	1.70	7.80	3.50	66.55	0.80	0.00	6.10	7.70	0.75	5.10	
CMSX-4	1.27	7.60	9.30	63.68	0.38	0.03	2.18	1.98	0.98	12.60	0.078
	1.00	6.50	9.00	61.70	0.60	0.10	6.50	6.00	3.00	5.60	
TMS-75	0.00	3.57	12.58	63.06	1.29	0.04	2.05	2.02	1.65	13.74	0.120
	0.00	3.00	12.00	60.20	2.00	0.10	6.00	6.00	5.00	5.70	

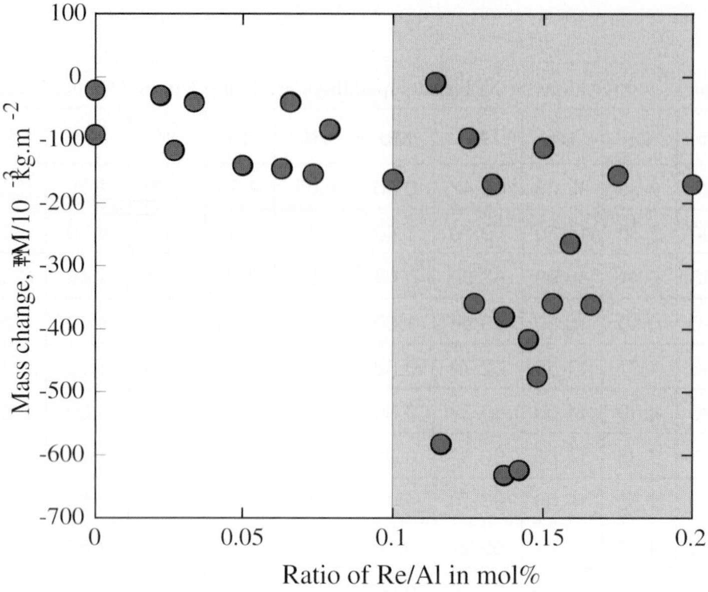

Fig. 6 Mass change as a function of Re/Al ratio (mol.%) of various alloys. Mass change was measured after the cyclic oxidation at 1373 K.

As the ratio increased further from 0.1, the mass change became larger. Thus Re is harmful for the oxidation resistance but its effect is dependent on the Al content. Re is more harmful in the case when the Re/Al ratio is greater than 0.1. Some amount of Re can be introduced into the alloy without hampering the oxidation resistance to any notable extent. However, the total amount of Re + Al is limited by the criteria for $\overline{Md} \leq 0.99$,[1] because these two elements have high Md values, so their excess addition leads to the formation of undesirable phases in the alloy.

Possible Mechanism for Re Effects on Cyclic Oxidation Resistance

It is well known that the formation of a dense oxide layer on the specimen surface is necessary to improve the oxidation resistance of alloys. The dense Al_2O_3 layer is such an effective layer to work as a barrier for further oxidation, resulting in good oxidation resistance.[13] In fact, as shown in Figure 3(a), dense and continuous Al_2O_3 layer was formed on the surface of the A-1 alloy and improved the oxidation resistance. In case of the A-9 alloy, porous Al image was seen in the oxidized layer as shown in Figure 3(b). Re content is only the difference in the chemical composition between the A-1 and the A-9 alloys. Namely, A-1 is Re-free and A-9 contains 2 mol.% (6.5 wt.%) Re. It is thought that Re makes Al_2O_3 porous by the formation of volatile Re_2O_7.[12] In the other experimental alloys, the Al_2O_3 layer became more porous with increasing Re content, indicating that the presence of Re in the alloy limits the formation of dense Al_2O_3 layer on the surface.[12] This is a similar case of alloys containing Mo or W, each of which also forms a volatile oxide, such as MoO_3 and WO_3.

Following the dissociation pressures of these oxides, it is known that Al_2O_3 is formed at first followed by TiO_2, Cr_2O_3 and then NiO. These oxides will appear in the oxide layer in this order from the metal side. Depending on the adhesion of oxide with oxide or oxide with base metal, one oxide or several oxides or all oxides may spall off. For example, only NiO may spall off if the adhesion between NiO and Cr_2O_3 is weak. On the contrary, weak adhesion between Al_2O_3 and Ni-based alloy may promote the entire oxides to be spalled off. In Figure 4(a), the bright area of the SEM image of the A-1 alloy consisted mainly of dense NiO which was supposed to remain adhesive on the surface after the oxidation test. The dark area was composed either or some of Cr_2O_3, TiO_2 and Al_2O_3, depending on the spalling of some oxides. No dense NiO was seen in the dark area of SEM image, so NiO in these areas was supposed to be spalled out. The spalling of only a little NiO resulted in the minimum mass change of the A-1 alloy as shown in Figure 2(a).

On the other hand, a little NiO was observed on the surface of the A-9 alloy, which means that a major part of the outermost NiO was spalled off, resulting in the exposure of the inner Cr_2O_3, TiO_2 and the innermost Al_2O_3 (Figure 4b) and thus the mass loss was large. Thus the adhesion of the protective oxide scales on the specimen surface was strong in the Re-free A-1 alloy but weak in the A-9 alloy.

In the Re containing alloys, Al_2O_3 layer in the oxide products is not dense as is seen in Figure 3(b). This is probably because a volatile oxide, Re_2O_7 is formed 1273 K[12] and it prevents the formation of the dense oxide on the alloy containing Re.

ALLOY DESIGN

On the basis of the findings on the Re effects on the oxidation resistance, three Ni-based single crystal superalloys are designed with the aid of d-electrons concept.[1, 8, 14] The compositional average of the d-electrons parameters, $\overline{Bo}$ and $\overline{Md}$ are known to be related to the phase stability of Ni-based superalloys.[1, 15] In designing the present superalloys, the $\overline{Bo}$ and $\overline{Md}$ values are set as closely as possible to the values of the target region ($\overline{Md}$: 0.985, $\overline{Bo}$: 0.665)[1, 15] where conventionally cast superalloys show maximum high-temperature capability.[1]

On the basis of the Re content higher than the value in the second generation superalloys (i.e., Re > 3 wt.%), the alloys with relatively high contents of total refractory elements (Mo + W + Re) are designed to attain superior creep rupture strength.[13, 16] Also the best resistance in burner-rig test was found in the alloys whose Ta/(W + Mo) compositional ratio was close to unity. Thus hot corrosion resistance in the present study is considered by setting a mol.% Ta/(Mo + W) compositional ratio to be close to unity.[13, 16] Moreover, hot corrosion resistance of the designed alloys is anticipated to be satisfactory due to their high Re content. Good oxidation resistance is predicted by the Re/Al compositional ratio (mol.%) closing to 0.1 as well as the low Ti/Al compositional ratio of the alloys. The chemical compositions of the three designed alloys are listed in Table 4.

Table 4 Chemical compositions of the designed alloys, upper in mol.% and lower in wt.%.

Alloy	Ti	Cr	Co	Ni	Mo	Hf	Ta	W	Re	Al	Md	Bo	Ta/(Mo+W)*	Mo+W+Re*	Ti/Al*	Re/Al*
D-1	1.95	4.55	11.62	bal.	0.91	0.05	2.30	1.79	1.80	11.76	0.983	0.664	0.85	4.507	0.166	0.15
	1.50	3.80	11.00	bal.	1.40	0.14	6.70	5.30	5.40	5.10						
D-2	2.33	5.97	9.69	bal.	0.78	0.01	2.08	1.93	1.33	9.21	0.957	0.665	0.77	4.036	0.254	0.14
	1.80	5.00	9.20	bal.	1.20	0.03	6.05	5.70	4.00	5.10						
D-3	1.50	9.50	2.00	bal.	0.50	0.01	2.00	2.00	1.00	12.30	0.986	0.666	0.80	3.500	0.122	0.08
	1.19	8.18	1.95	bal.	0.79	0.03	5.99	6.09	3.08	5.49						

CONCLUSION

In order to make clear the Re effect on the high temperature oxidation of Ni-based superalloys, oxidation resistance was investigated using a series of experimental alloys, in which the Re content was varied systematically. The results are summarized as follows,

1. Re is an element to deteriorate the oxidation resistance of Ni-based superalloys. The harmful effect of Re appeared clearly in the 10mol.% Al series alloys but did not in the 15 mol.% Al series alloys.
2. Alloys with the Re/Al ratio (mol.%) up to 0.1 showed a very small mass change but alloys with the ratio greater than 0.1 exhibited a large mass change in the oxidation test at 1373 K.
3. The presence of Re in the alloy limits the formation of the dense Al_2O_3 scale on the surface during oxidation, probably by the formation of volatile Re_2O_7 in it.

It is found that Re deteriorates high-temperature oxidation resistance of Ni-based superalloys, although this element is very effective to improving the hot-corrosion resistance of the superalloys.

REFERENCES

1. K. Matsugi, Y. Murata, M. Morinaga and N. Yukawa: 'Realistic Advancement for Nickel-Based Single Crystal Superalloys by the d-Electrons Concept', *Superalloys 1992*, S. D. Antolovich, et al. eds., TMS, Warrendale, PA, 1992, 307–316.
2. A. D. Cetel and D. N. Duhl: 'Second-Generation Nickel-Base Single Crystal Superalloy', *Superalloys 1988*, D. N. Duhl, et al. eds., The Metall. Soc. of AIME., Warrendale, PA, 1988, 235–244.
3. D. L. Erickson and K. Harris: 'DS and SX Superalloys for Industrial Gas Turbines', *Materials for Advanced Power Engineering 1994*, D. Coutsouradis, et al. eds., Kluwer Academic Pub., Dordrecht, 1994, 1055–1074.
4. W. S. Walston, K. S. O'Hara, E. W. Ross, T. M. Pollock and W. H. Murphy: 'Rene N6: Third Generation Single Crystal Superalloy', *Superalloys 1996*, R. D. Kissinger et al. ed., TMS, Warrendale, PA, 1996, 27–34.
5. G. L. Erickson: 'The Development and Application of CMSX-10', 1996, 35–44.
6. Y. Koizumi, T. Kobayashi, T. Yokokawa, T. Kimura, M. Osawa and H. Harada: 'Third Generation Single Crystal Superalloys with Excellent Processability and Phase Stability', *Materials for Advanced Power Engineering 1998*, J. Lecomte-Beckers et al. ed., Forschungszentrum Julich Pub., Julich, 1998, 1089–1098.
7. Md. Moniruzzaman, Y. Murata, M. Morinaga, N. Aoki, T. Hayashida and R. Hashizume: 'Prospect of Nickel-Based Superalloys with Low Cr and High Re Contents', ISIJ International, **42**(9), 2002, 1018–1025.
8. M. Morinaga, N. Yukawa, H. Adachi and H. Ezaki: 'New PHACOMP and its Application to Alloy Design', *Superalloys 1984*, M. Gell, et al. eds., The Metall. Soc. of AIME, Warrendale, PA, 1984, 523–532.
9. M. Morinaga, N. Yukawa, H. Ezaki and H. Adachi: 'Solid Solubilities in Transition-Metal-Based f.c.c. Alloys', *Philosophical Magazine A*, **51**(2), 1985, 223–246.

10. T. M. Simpson and A. R. Price: 'Oxidation Improvements of Low Sulfur Processed Superalloys', *Superalloys 2000*, T. M. Pollock, et al. eds., TMS, Warrendale, PA, 2000, 387–392.
11. C. Sarioglu, C. Stinner, J. R. Blachere, N. Birks, F. S. Pettit and G. H. Meier: 'The Control of Sulfur Content in Nickel-Base Single Crystal Superalloys and its Effects on Cyclic Oxidation Resistance', *Superalloys 1996*, R. D. Kissinger, et al. eds., TMS, Warrendale, PA, 1996, 71–80.
12. C. J. Smithell: *Metals Reference Book*, Butterworth & Co. Ltd., London, **1**, 1967, 266.
13. Y. Murata, R. Hashizume, A. Yoshinari, N. Aoki. M. Morinaga and Y. Fukui: 'Alloying Effects on Surface Stability and Creep Strength of Nickel Based Single Crystal Superalloys containing 12wt.%Cr', *Superalloys 2000*, T. M. Pollock, et al. eds., TMS, Warrendale, PA, 2000, 285–294.
14. M. Morinaga, N. Yukawa and H. Adachi: 'Alloying Effect on the Electronic Structure of $Ni_3Al(\gamma')$', *J. Phys. Soc.*, Japan, **53**(2), 1984, 653–663.
15. Y. Murata, S. Miyazaki, M. Morinaga and R. Hashizume: 'Hot Corrosion Resistant and High Strength Nickel-Based Single Crystal and Directionally-Solidified Superalloys Developed by the d-Electrons Concept', *Superalloys 1996*, R. D. Kissinger, et al. eds., TMS, Warrendale, PA, 1996, 61–70.
16. Y. Murata, Md. Moniruzzaman, A. Yoshinari, M. Morinaga, R. Hashizume and Y. Fukui: 'Surface Stability and Creep Strength of High-Cr Nickel-Based Single Crystal Superalloys', *Advances in Materials Engineering and Technology*, Special Issue, SMP I, vol. 2, M. A. Dorgham, ed., Inderscience Enterprises Ltd., Geneva, Switzerland, 2001, 616–621.

Development of Test Methods to Evaluate the Wear Performance of Industrial Gas Turbine Seal Components

M. G. GEE and G. ALDRICH-SMITH
National Physical Laboratory
Teddington, UK

IAN BOSTON, JONAS HURTER and G. MCCOLVIN
ALSTOM Power, Baden
Switzerland

ABSTRACT

A high temperature wear testing system has been developed at NPL that realistically simulates the conditions that are experienced by combustion chamber seals in the industrial gas turbine. The test system is being used in a programme of work that is being carried out jointly with ALSTOM Power and NPL to evaluate the suitability of candidate materials for combustion chamber seal applications. The aim of the programme is to develop the tools that will enable the performance of materials selected for use in these applications to be assured over the lifetime of the turbines.

BACKGROUND

Because of their potential for high efficiency and flexibility of use, industrial gas turbines are increasingly being used in power generation applications. As part of a development programme to ensure the continued reliable operation of turbines, NPL has been working with ALSTOM Power on the development of wear test methods. These will be used to assess the performance of new materials and designs that will be used to further enhance the operation and reliability of the turbine systems.

The focus of the work reported in this paper is the development of wear test methods that can be used to assess the likely performance of materials for combustion chamber seals. These seals ensure the mechanical and pressure integrity of the combustion chamber during operation. Any loss of the pressure seal would lead to the engine becoming progressively more inefficient and hence major repairs to the engines would be necessary leading to economic losses due to downtime. The wear test systems that have been developed are being used to provide information on the performance of materials under conditions that realistically simulate the conditions that the materials are subjected to in the engine. The results are then used by ALSTOM Power to ensure that the best possible seal design is adopted in their turbines.

The results carried out at NPL are complemented by results of component trials that are being conducted by ALSTOM Power.

There are two main different seals that are being assessed in this paper. At the end of the combustion chamber the membrane seal is an annular component that seals the inner edge of the combustion chamber to the frame of the engine. This is a large component where wear can occur from the relatively slow but large amplitude movements of about 10 mm generated due to differential thermal expansion of the combustion chamber relative to the engine frame. Due to the design of these components the loading on the edge of these components is large.

The other type of seal is the combustion chamber seal, these seal one segment of the combustion chamber to the next. These components are only subject to low loads, and as well as a relatively large overall movement, these components are also subjected to a relatively high frequency low amplitude (0.3 mm) motion due to vibration generated in the engine.

Two different test systems have been designed for the different types of test conditions, but only the low frequency test system will be described in this paper.

EXPERIMENTAL

Test System

The test system that was used was originally used in a project to measure the high temperature wear and friction and ceramics. The test-pieces are sized so that they are push fits in the recesses of the test-piece holding arms. For operation at temperatures below 1000°C, nimonic alloy 90 arms are used; for temperatures between 1000 and 1500°C SiC ceramic arms are used. Test-piece sizes are shown in Figure 1. A schematic of the reciprocating wear test rig is shown in Figure 2.

During testing, the test-pieces and the extremity of the arms are heated to the required temperature by an SiC element split furnace which can be moved away on slides after the test to enable test-pieces to be taken out and replaced ready for the next test. The temperature of the furnace is measured and controlled by a sheathed thermocouple placed near to the samples. The test-piece holding arms are water cooled just outside the furnace to ensure that the rest of the test system is not affected by heat from the furnace.

The lower arm is pivoted so that the lower sample can be pressed into contact with the top sample by the weight placed on the loading pan. The weight is adjusted to give the correct load on the sample, taking account of the lever ratio implicit in the loading system. The vertical load on the sample is calculated from the measured reaction to the loading on the lower arm by the vertical load cell, which is under the lower arm pivot. The wear displacement, or movement of the samples towards each other as wear proceeds is calculated from the vertical displacement measured by the LVDT placed against the lower arm, again taking account of the appropriate lever arm ratio.

The upper arm is guided and constrained by linear bearings to move horizontally. It is driven by a servo actuator through a load cell that measures frictional force. Horizontal

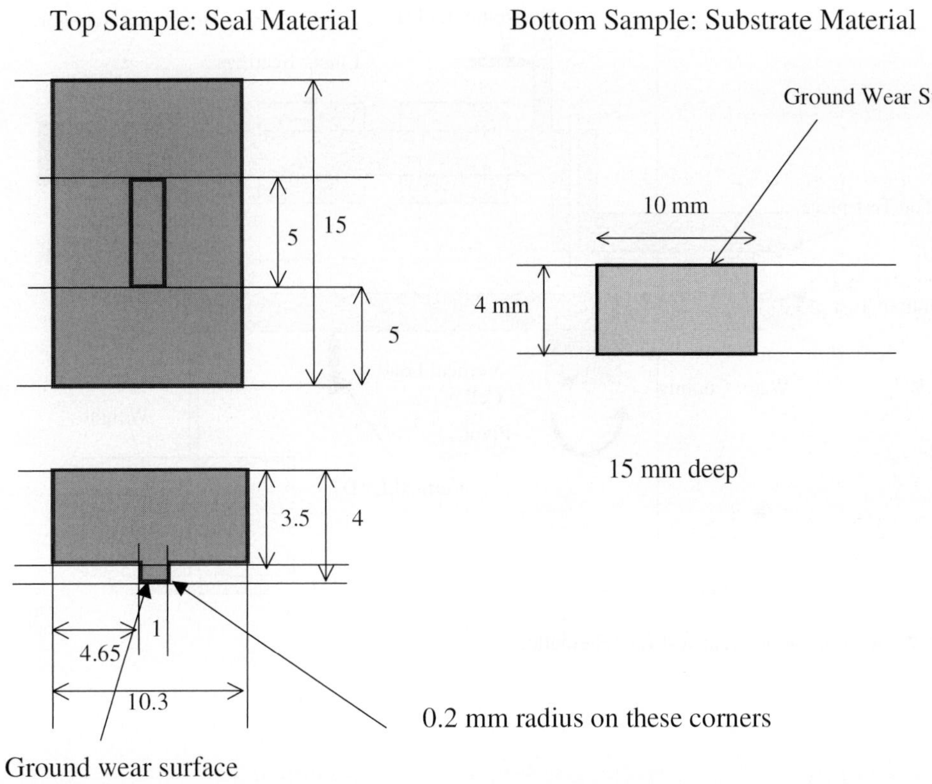

Fig. 1 Test sample geometry.

displacement is measured by another LVDT. Limit switches are set up on the horizontal displacement signal conditioner in such a way that when they are reached the direction of travel is reversed. In this way, a reciprocating motion is established during the test for the two samples.

Test Procedure

The size of the samples was adjusted by careful grinding with SiC paper until they were a push fit in the holding arms. The samples were then ultrasonically cleaned in iso-propanol and hot air dried before being weighed and placed in position in the holding arms. The appropriate dead-load was placed on the balance pan, and the samples were allowed to come into contact. The split furnace was pushed into place over the holding arms and samples, and the gas flow started for the controlled atmosphere.

The furnace was switched on and the correct temperature set on the furnace controller. When the correct temperature was reached (~30 mins), the actuator was started, and the test

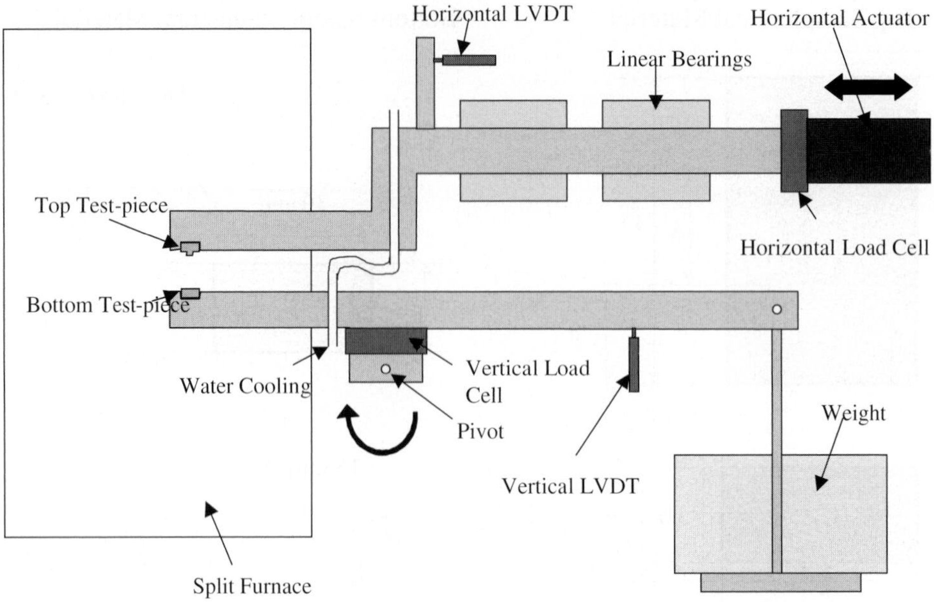

Fig. 2 Reciprocating wear test rig schematic.

allowed to proceed for the required duration. At this point, the furnace was switched off, the samples and holding arms allowed to cool before samples removal. After cleaning the samples were reweighed ready for visual examination and photography using a digital imaging system.

The furnace atmosphere was nitrogen based, nominally containing 15% O_2, 3.4% CO_2 and 1%Ar. Approximately 6.5% water vapour was added to the gas by bubbling through a humidification chamber held at a constant temperature of 37°C.

Tests Performed

Results from an illustrative set of tests are discussed in this paper. The matrix of tests that was carried out and the test conditions that were used is given in Table 1.

All the tests reported here were carried out at a temperature of 550°C with a stroke length of 9 mm, a reciprocating frequency of 0.93 Hz, and a test duration of 1 hour (corresponding to 56 cycles).

The tests marked C in the table simulated the conditions of combuster chamber seals and therefore were tested under an applied load of 125 N. The tests marked M simulated the conditions experienced by membrane seals and were therefore tested at an applied load of 875 N. In all cases the stub sample (with the reduced 5 × 1 mm area of contact) was the candidate seal material, the substrate sample was the material of the opposing surface.

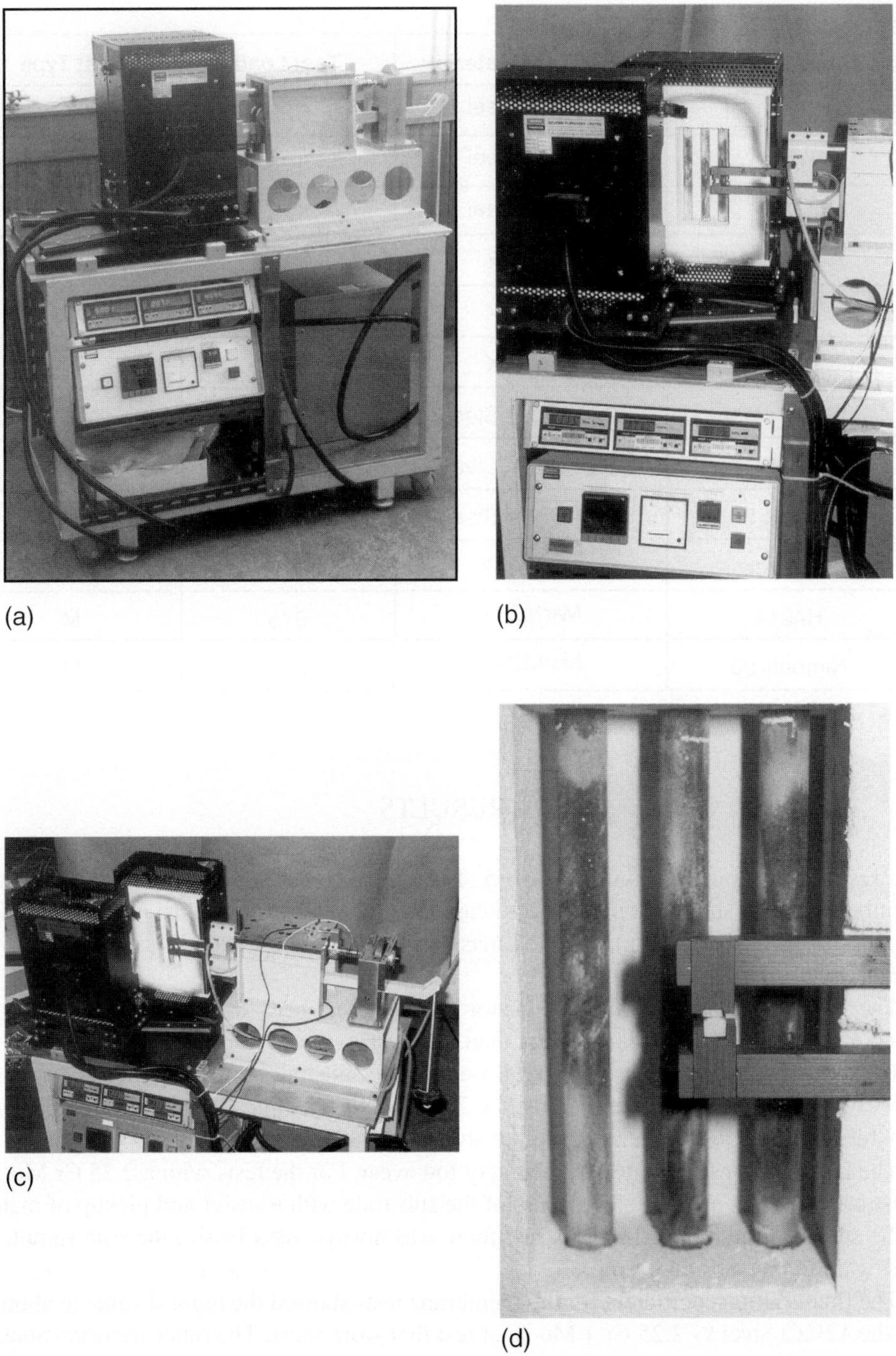

Fig. 3 Various views of low frequency test system, (a) overall, (b) close-up, (c) close-up and (d) sample holders and heating element: note this photograph shows ceramic samples and SiC holders used for very high temperature testing.

Table 1 Matrix of illustrative tests.

Stub Material	Substrate Material	Test Load	Test Type
12%Cr steel	12%Cr steel	125	C
HA25	12%Cr steel	125	C
IN718	12%Cr steel	125	C
St12T	Cast HX	125	C
HA25	Cast HX	125	C
IN718	Cast HX	125	C
12%Cr steel	2.25 Cr Mo Steel	875	M
HA214	2.25 Cr Mo Steel	875	M
Nimonic 90	2.25 Cr Mo Steel	875	M
12%Cr steel	MarM 247	875	M
HA214	MarM 247	875	M
Nimonic 90	MarM 247	875	M

RESULTS

The results are shown in Figures 4 and 5. The mass loss measurements for the combuster chamber seal tests show that in almost all cases the substrate samples lost mass. By contrast, in many cases the stub samples gained mass due to transfer of material from the substrate sample to the stub sample.

The friction measured in the combustor chamber seal tests was slightly lower at about 0.4–0.45 for the 12%Cr steel substrate than for the Haynes 25 substrate at about 0.6.

In the membrane seal tests, very high wear to the substrate took place in the case of the 12%Cr steel stub sample tested against the 2.25 Cr Mo substrate. This was accompanied by transfer of some substrate material to the stub showing as a mass gain.

The rest of the tests resulted in relatively low wear. For the tests with a 2.25 Cr Mo steel substrate there was always loss of mass of the substrate with transfer and pickup of material to the stub. For the MarM 247 substrate there was always mass loss to the stub sample and mass gain to the substrate sample.

The final friction coefficient for the membrane tests showed the highest value at about 0.5 for the 12%Cr steel vs 2.25 Cr 1 Mo steel test that wore most. The other friction values are lower with the lowest values at about 0.3 for the Nimonic 90 stub against 2.25 Cr 1 Mo steel and the HA214 vs MarM 247 sample.

Figure 6 shows photographs of the worn surfaces that compare the appearance of wear surfaces from a test where severe wear took place with samples where only a small amount

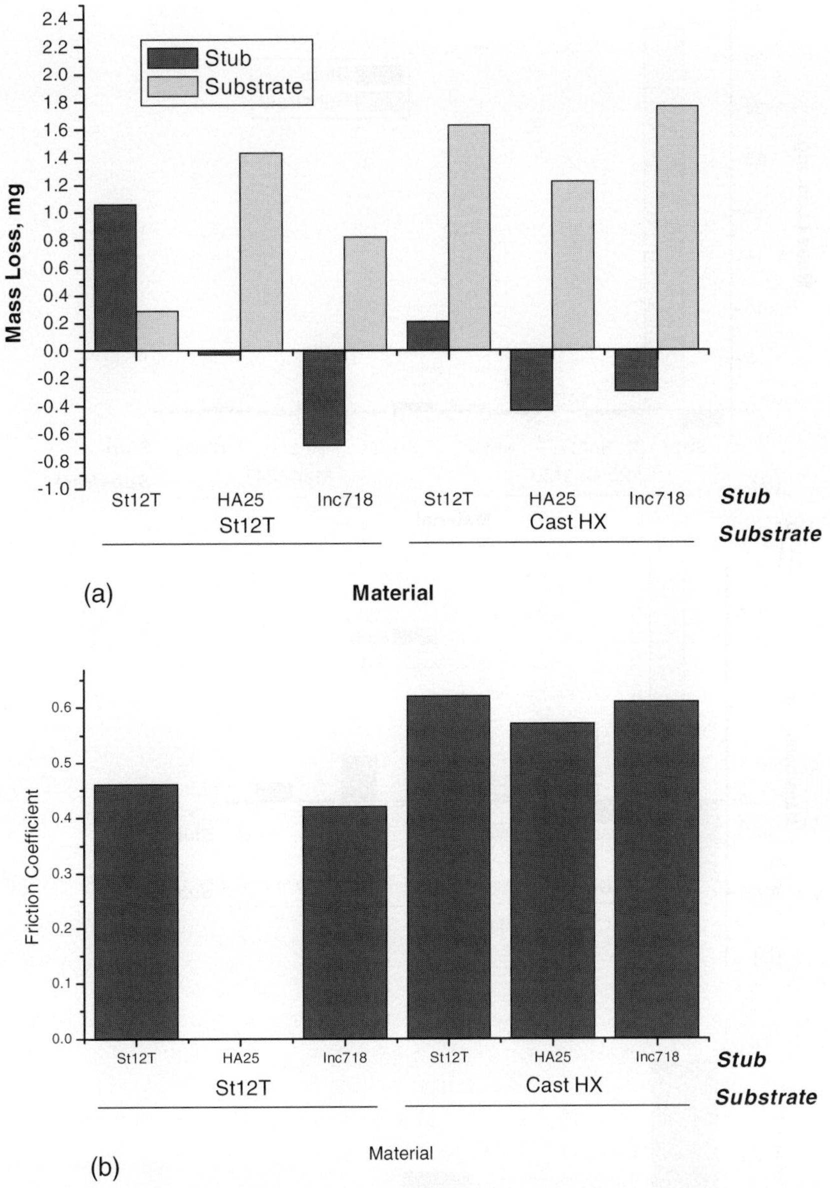

Fig. 4 Results of combustion chamber seal tests, (a) mass loss and (b) final friction coefficient. Note that friction was not measured in the Ha25 vs St12T test.

of wear occurred. In the severe test (membrane test with 12%Cr steel vs 2.25 Cr 1 Mo steel), transferred material can be clearly seen on the surface of the stub. A deep groove has been developed in the surface of the substrate sample with some material pushed out to the front and back edges of the groove.

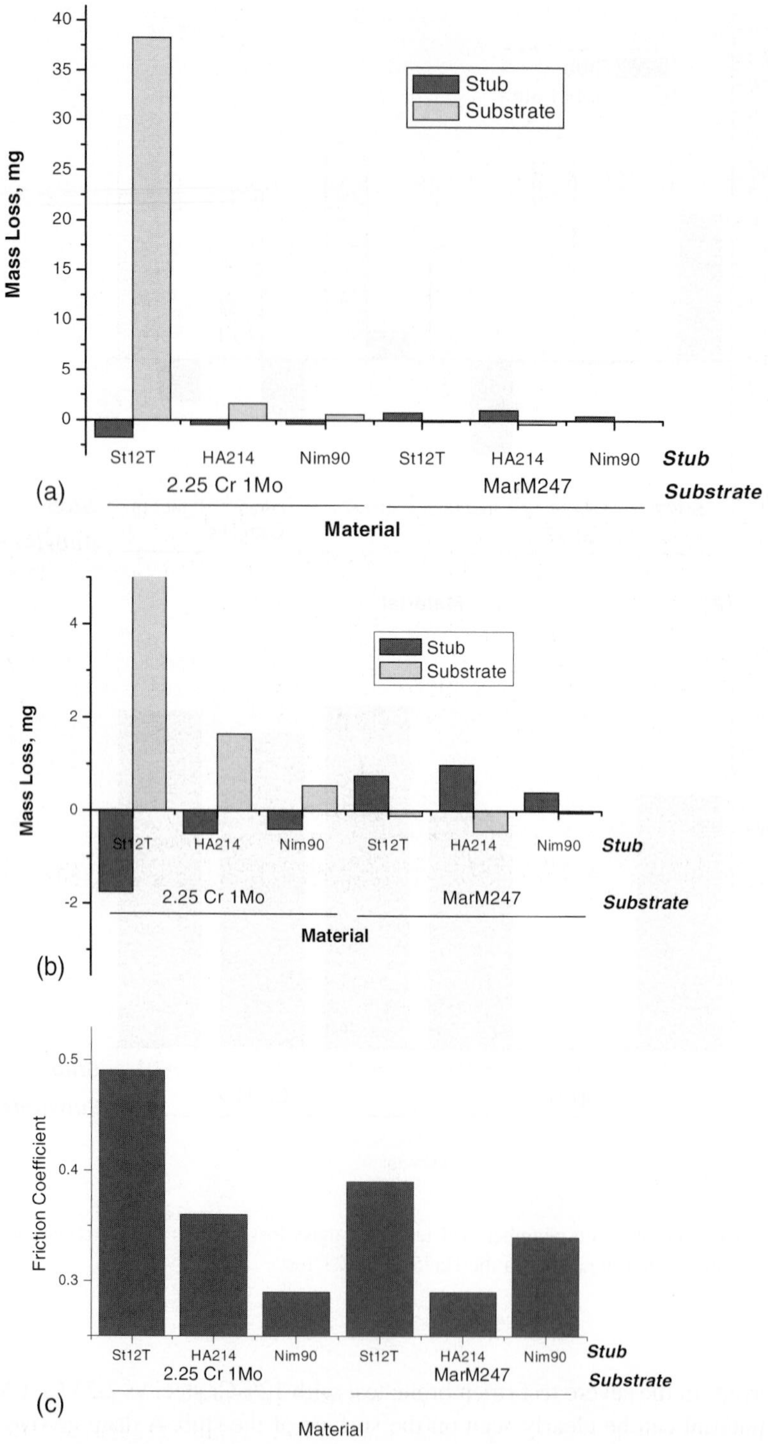

Fig. 5 Results of membrane seal tests (a) mass loss results, (b) mass loss results with expanded vertical scale and (c) final friction results.

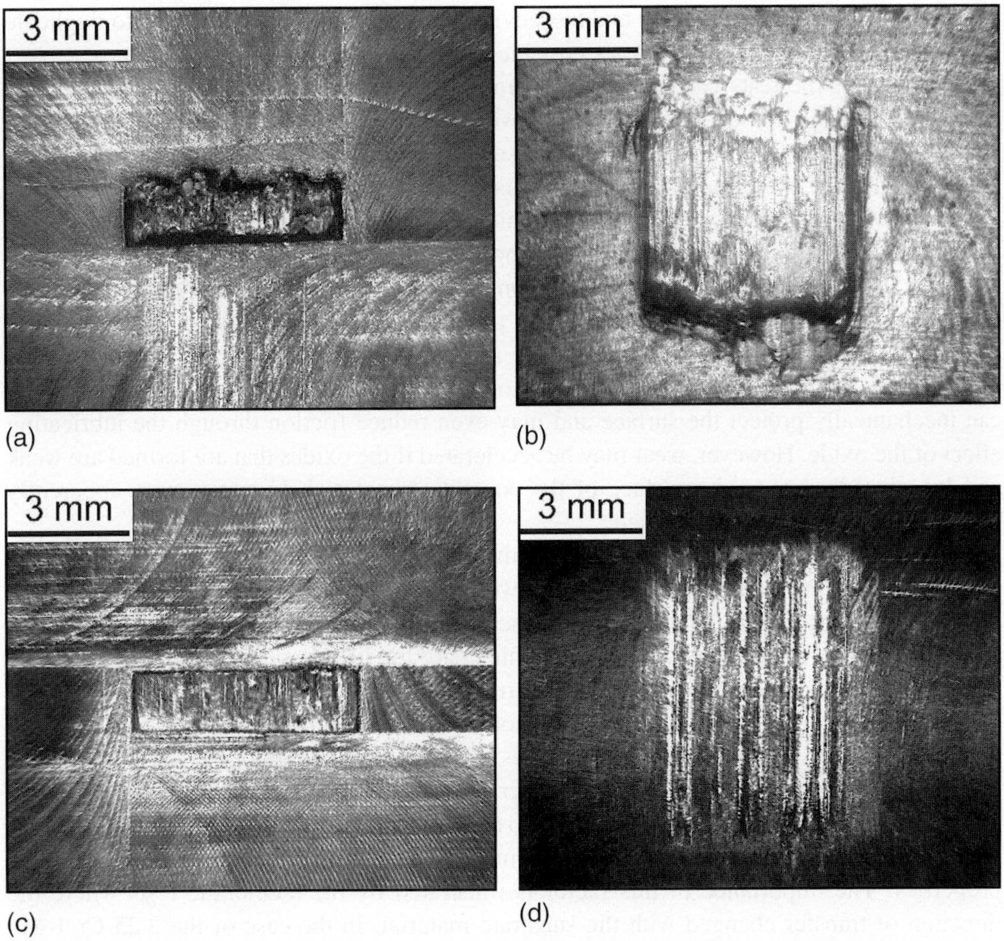

Fig. 6 Photographs of worn surfaces, (a) St12T stub sample from membrane test against 2.25Cr Mo steel substrate, (b) substrate sample from same test, (c) Nimonic 90 stub sample from membrane test against MarM 247 substrate and (d) substrate sample from same test.

In the milder test transfer of material has taken place to the substrate sample. There is, nevertheless, considerable plastic deformation to the surface in the worn area of both these samples.

DISCUSSION

The mechanisms by which wear and friction occur between any two surfaces in sliding contact can be very complex and depend on many of the controlling parameters. In these tests many of these parameters are set to simulate the conditions that are known to occur in the normal operation of the components that are being simulated. Thus the tests load, test speed, test stroke test temperature and test environment have all been controlled to duplicate the service environment as closely as possible.

For some of the tests carried out very dramatic high wear has occurred leaving very significant damage to the test samples. However, it should be noted that there is an inherent acceleration in the tests carried out and presented here. Thus a cycle of testing occurs in about a minute of operation of the test. This is effectively equivalent to a complete start-up and shutdown cycle for a gas turbine. Thus this test would be expected to give considerable wear in much shorter periods than in actual operation.

When the actual mechanisms that can contribute to increasing wear, or in some cases reducing wear are considered, there are many possible effects. These have not yet been fully analysed and only a brief discussion of some aspects can be considered here.

Oxidation can often form a coherent layer on the surface of a contacting wear couple that can mechanically protect the surface and may even reduce friction through the lubricating effect of the oxide. However, wear may be accelerated if the oxides that are formed are weak and loosely adherent to the surface of the contacting material. In some cases apparently coherent oxide films were observed.

More evident in the tests reported here was the severe plastic deformation that was observed in many of the tests, particularly where high wear occurred. In this regard it should be noted that these are elevated temperature tests and the hardness of the materials will be considerably reduced from the room temperature value, enhancing the likelihood of plastic deformation.

Perhaps the most important factor that controls the wear of these materials is the likelihood of transfer of material from one surface to another. This transfer occurs when a strong bond is formed between the two contacting surfaces such that the strength of the adherence of the fragment to the opposing surface is stronger than the cohesive strength of the material. When this occurs the fragment is transferred to the opposing surface. This transfer is governed both by the chemical composition of the two materials, and also by their relative mechanical properties. The importance of this factor is illustrated by the membrane tests where the direction of transfer changed with the substrate material. In the case of the 2.25 Cr 1 Mo substrate the transfer was always from the substrate to the stub, whereas with the MarM247 substrate the transfer was in the opposite direction.

CONCLUSIONS

A test system has been developed that can be used to evaluate the materials used in combustion chamber seals in gas turbine applications. Clear differences in the performance of different materials have been seen that depend on both the chemical composition and mechanical properties of the materials involved.

ACKNOWLEDGEMENTS

The authors would like to acknowledge partial financial support for the test method development from the UK Department of Trade and Industry under the Materials Measurement Programme.

Effect of Thermal Exposure on Microstructural Degradation and Mechanical Properties of Ni-Base Superalloy GTD 111

BAIG GYU CHOI, IN SOO KIM and CHANG YONG JO
High Temperature Materials Group
Korea Institute of Machinery and Materials
South Korea

HAK MIN KIM
Korea Institute of Science & Technology Evaluation and Planning
South Korea

ABSTRACT

Ni-base superalloy GTD 111 is widely used for a blade material in the first stage of land-base gas turbine. In spite of important role of GTD 111 in the high temperature applications, very limited data on microstructure and mechanical properties of the alloy were reported. Microstructural evaluations during thermal exposure and their effects on the creep properties of the material have been investigated. After standard heat treatment, polycrystalline GTD 111 specimens were thermally exposed up to 10000 hours in the temperature range of 871–982°C. Distribution and composition of the existing phases have been analysed both quantitatively and qualitatively using electron microscopes, EDS and image analyser. Growth rate of gamma prime (γ') of the alloy was proportional to the 1/3 power of the exposed time. The activation energy calculated for γ' coarsening was similar to that for self diffusion. MC carbides decomposed into eta(η) + $M_{23}C_6$ during thermal exposure at 871 and 927°C rather than $M_{23}C_6$ + γ'. Precipitation of sigma(σ) phase was observed in the vicinity of MC or $M_{23}C_6$ carbides, especially in the specimen exposed at 871°C. During thermal exposure at 982°C, both precipitation of continuous γ' + $M_{23}C_6$ film along the grain boundary and dissolution of η within γ-γ' eutectic have occurred.

Constant load creep tests with thermally exposed GTD 111 have been conducted to study the effect of thermal exposure and grain boundary structure on creep properties. Various thermal exposure conditions produced two different types of grain boundary structure ($M_{23}C_6$ with γ' film and discontinuous η with $M_{23}C_6$). Size and distribution of γ' were controlled to be similar for both specimens. Formation of intergranular η phase and variation of gamma prime distribution during thermal exposure do not affect creep rupture life and ductility in the present study.

Table 1 Chemical composition of GTD 111 (in wt.%).

Ni	Co	Cr	Al	Ti	Ta	Mo	W	Zr	Fe	C	B
Bal.	9.24	13.86	3.05	4.86	2.91	1.57	3.78	0.008	0.051	0.113	0.013

INTRODUCTION

Ni-base superalloy GTD 111 is widely used for blading material of land-base gas turbine. The alloy is known to have about 20°C creep rupture advantage over another blade material IN738LC, as well as higher low-cycle fatigue strength.[1] GTD 111 is a precipitation strengthening alloy of γ', an intermetallic compound with the general formula of $Ni_3(Al, Ti)$. The alloy is known to have a multi phase microstructure consisting γ matrix, γ' precipitate, γ-γ' eutectic, carbides and some minor phases.[1]

The buckets in current advanced gas turbines are exposed to extremely high temperature and stress which induce significant degradation in microstructure, such as MC carbide decomposition, agglomeration of the γ', formation of minor phases like sigma(σ) and eta(η), and homogenisation of the initial dendrite segregation. Various studies have shown that the changes of microstructure during thermal exposure or service have significant effects on the mechanical properties.[2–4] Therefore, it is worth of studying microstructural evaluations during thermal exposure and their effects on the mechanical properties of the blading material. In spite of important role of GTD 111 in the high temperature performance, very limited data on microstructure and mechanical properties including thermal stability of the alloy were reported.[5–7]

Microstructural evaluations during thermal exposure at three different temperatures up to 10000 hours and their effects on the creep rupture properties of the material have been investigated in the present study. MC carbide decomposition behaviour with thermal exposure and its effect on creep rupture properties also have been discussed.

EXPERIMENTAL PROCEDURE

The chemical composition of the GTD 111 used in the present study is shown in Table 1. The alloy was vacuum melted and cast into rods of 13 mm in diameter. The rods were subjected to standard heat treatment. After standard heat treatment, the specimens were thermally exposed at 871, 927 and 982°C up to 10000 hours. The specimens were mounted, polished and etched in Kalling's reagent to observe the microstructural evolution during thermal exposure by scanning electron microscope(SEM) and optical microscope. The variation of gamma prime distribution was measured quantitatively using image analyser program. Fine particles were observed using transmission electron microscope (TEM). TEM samples were electrochemically polished with a twin jet polisher using the solution of 83% ethanol, 7% glycerol and 10% perchloric acid at –20°C/75V.

Table 2 Results of EDS analysis of γ' for major alloying elements (in wt.%).

	Al	Ti	Cr	Co	Mo	Ta	W	Ni
Dendrite Core	2.98	4.61	9.83	8.52	0.87	2.57	3.29	67.43
Grain Bodnary	3.25	5.54	7.13	7.41	0.77	3.17	2.71	70.03
γ/γ' Eutectic	2.95	8.69	2.48	6.82	-	3.31	1.27	74.48

Constant load creep rupture tests with thermally exposed GTD 111 were conducted to study the effect of thermal exposure and grain boundary structure on creep properties. On the basis of microstructural observation, prior to creep tests the specimens were exposed at 982°C for 1000 and 2000 hours to have different gamma prime distributions with similar grain boundary structure and aged at 927°C for 2000 hours to have different grain boundary structures with similar gamma prime distribution. The temperature during creep test was measured with thermocouples placed on the specimen gage length. The temperature variation of the furnace during creep tests was about ±1°C, and temperature difference within the gage length was also controlled not to exceed ±1°C.

For the study of fracture mode, fracture surface and the longitudinally sectioned specimen whose surface was polished and etched with Kalling's reagent were observed by SEM.

RESULTS AND DISCUSSION

Microstructure of Standard Heat Treated GTD 111

Microstructure of standard heat treated GTD 111 is shown in Figure 1. Primary cube γ' and secondary fine spherical γ' are distributed in γ matrix. Blocky MC carbides with high contents of Ti and Ta were found mainly in interdendritic regions including grain boundaries and γ-γ' eutectic regions. The grain boundaries were covered with discontinuous γ', MC and very fine $M_{23}C_6$ carbides. Blocky and fine platelet phases with high contents of Ti and Ta were found near γ-γ' eutectic after standard heat treatment. Figure 2(d), TEM diffraction pattern of the fine plates, indicates the phase is $\eta(Ni_3Ti)$ which has the crystallographic orientation relationship with γ matrix such as $(111)_\gamma//(0001)_\eta$ and $[0\bar{1}1]//(1\bar{2}10)_\eta$[8]. Very fine particles observed near γ-γ' eutectic could be also identified by selected area diffraction pattern as M_6C or $M_{23}C_6$ as shown in Figure 2.

Difference in contrast came from segregation during solidification could be found by optical microscope as shown in Figure 1. The composition of γ' at each region was analysed by using dispersive x-ray spectroscopy(EDS) in the specimen etched electrochemically as shown in Table 2. Co, Cr and W contents were lower and Ta, Ti, Ni contents were higher in γ' on the grain boundary and γ-γ' eutectic region than those in dendrite core. The result of

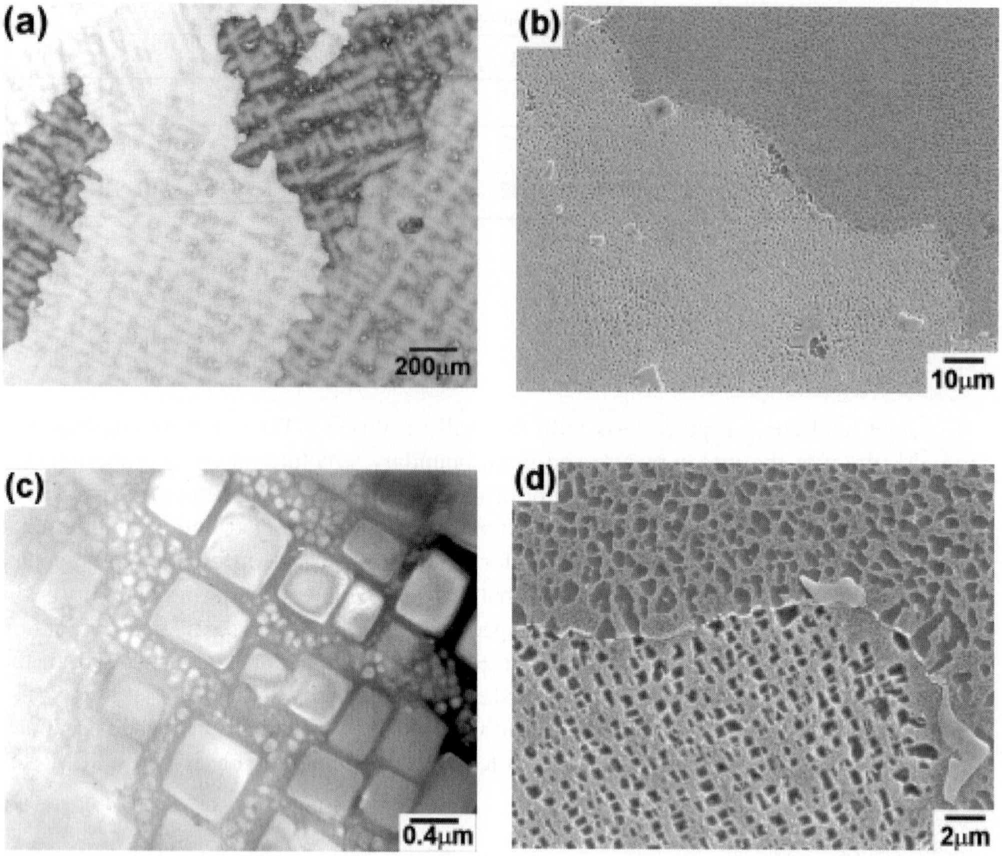

Fig. 1 Microstructure of standard heat treated GTD 111. (a) Optical micrograph, (b) SEM micrograph, (c) TEM dark field image showing cube primary γ' and fine secondary γ' and (d) SEM micrograph showing grain boundary.

Table 3 Results of EDS analysis of each region for major alloying elements (in wt.%).

	Al	Ti	Cr	Co	Mo	Ta	W	Ni
Dendrite Core	2.64	3.48	16.43	10.36	1.96	1.88	6.91	56.00
γ/γ' Eutectic	2.75	8.59	9.62	8.57	1.31	4.41	3.08	61.68

EDS analysis on the specimen etched with Kalling's reagent for 1 second in Table 3 also shows the high contents of Ta and Ti in γ-γ' eutectic area. These findings reflect the fact that the γ phase which grows as a dendrite from the liquid has a relatively high solubility for Cr, Co and W and a low solubility for Ta and Ti.[9]

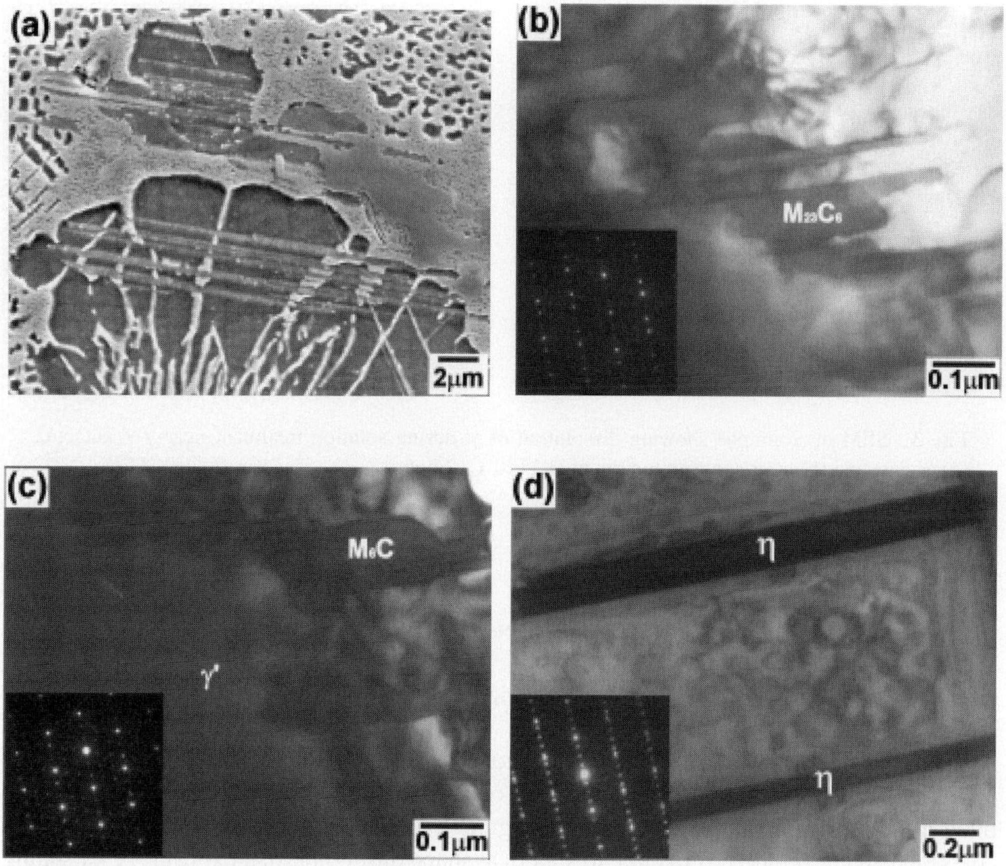

Fig. 2 Fine platelet phase and particles near γ-γ' eutectic. (a) SEM micrograph of eutectic, TEM micrographs and selected area diffraction patterns of (b) $M_{23}C_6$, (c) M_6C and (d) η.

The areas with very fine secondary γ' near γ-γ' eutectic were occasionally found in standard heat treated alloy as shown in Figure 2(a), which were not observed in as-cast condition (Figure 3(a)). The γ' dissolution was found during solution heat treatment at 1120°C as shown in Figure 3(b). This observation indicates that the γ' with high contents of Ti and Ta near eutectic region was dissolved during solution treatment and small γ' precipitated again during aging. The density of fine platelet η phases after standard heat treatment increased compared with that of the as-cast condition. Dissolution of γ' with high contents of Ti and Ta during solution treatment enables the fine platelet η phase to form in the vicinity of eutectic by supply the solute atoms such as Ti and Ta.

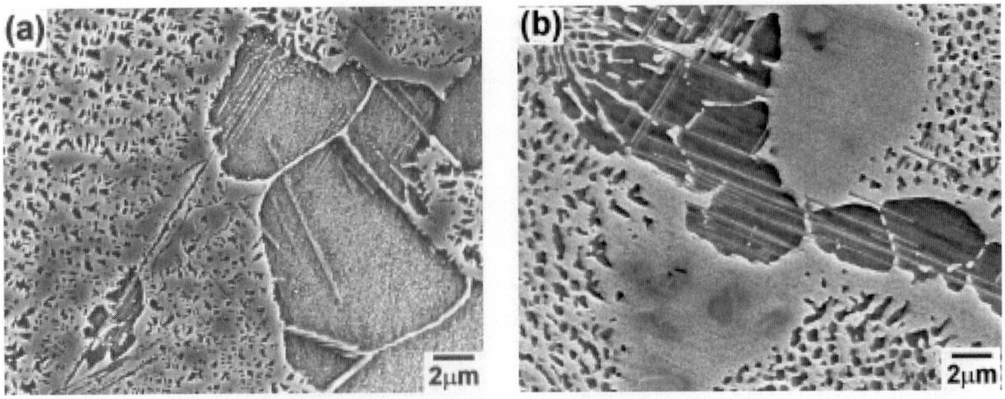

Fig. 3 SEM micrographs showing dissolution of γ' during solution treatment near γ-γ' eutectic. (a) as-cast and (b) after solution treatment at 1120°C, 2 hours.

Microstructural Evaluations During Thermal Exposure

Gamma Prime (γ')

As mentioned above, primary cube γ' and secondary fine spherical γ' distributed in γ matrix after standard heat treatment. The variations of intragranular γ' during thermal exposure at three different temperatures are shown in Figure 4. While fine γ' dissolved, the primary γ' coarsened and developed from cube to spherical shape during thermal exposure. Coarsening rate and morphological change of γ' were fast at high temperature.

Small particles of high density tend to coarsen into larger particles of lower density to reduce total interfacial energy. Assuming volume diffusion is the rate control factor, γ' coarsening during thermal exposure can be given as the following equation,[10]

$$\bar{d}^3 - \bar{d}_o^3 = Kt \tag{1}$$

$$K = \frac{64C_e\gamma_s DV_m^2}{9RT} \tag{2}$$

where d is the average γ' diameter, d_o is the initial average γ' diameter at time 0, t is time, C_e is the equilibrium concentration of γ' solute in γ, γ_s is the interfacial free energy, R is the gas constant, T is the temperature, D is the diffusion coefficient of γ' solutes in γ, V_m is the molar volume of γ'.

A plot of the third power of the mean γ' diameter change versus exposure time at each exposure temperature is shown in Figure 5. The data show good linearity, which indicates γ' coarsening is controlled by volume diffusion.

The diffusion coefficient, D in equation (1) can be given as $D = D_o exp(-Q/RT)$ where Q is the activation energy for diffusion.

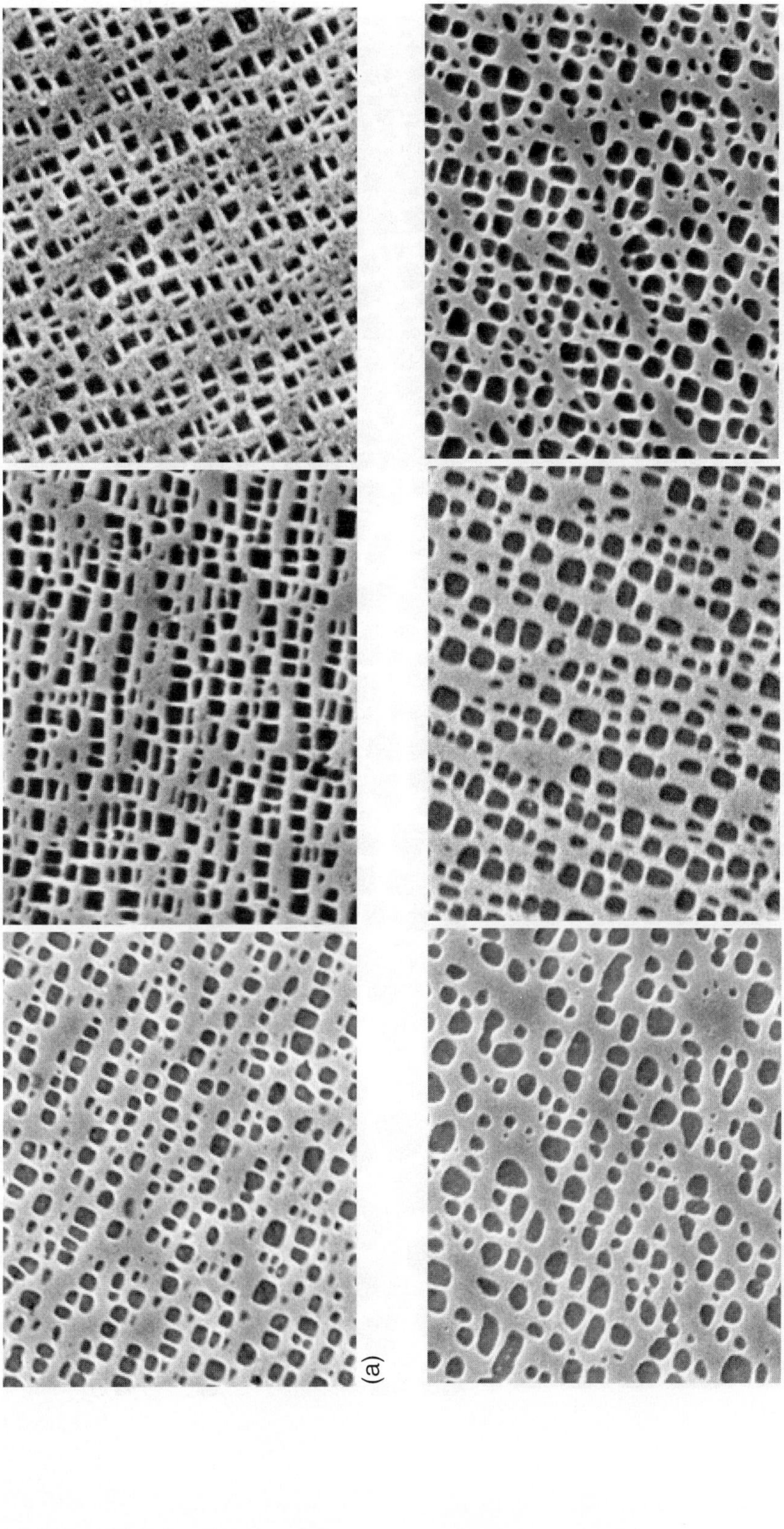

(a)

(b)

Fig. 4 Variation of γ' size and distribution during thermal exposure. (a) 50 hours, (b) 800 hours, (c) 5000 hours and (d) 10000 hours. (c) and (d) in next page.

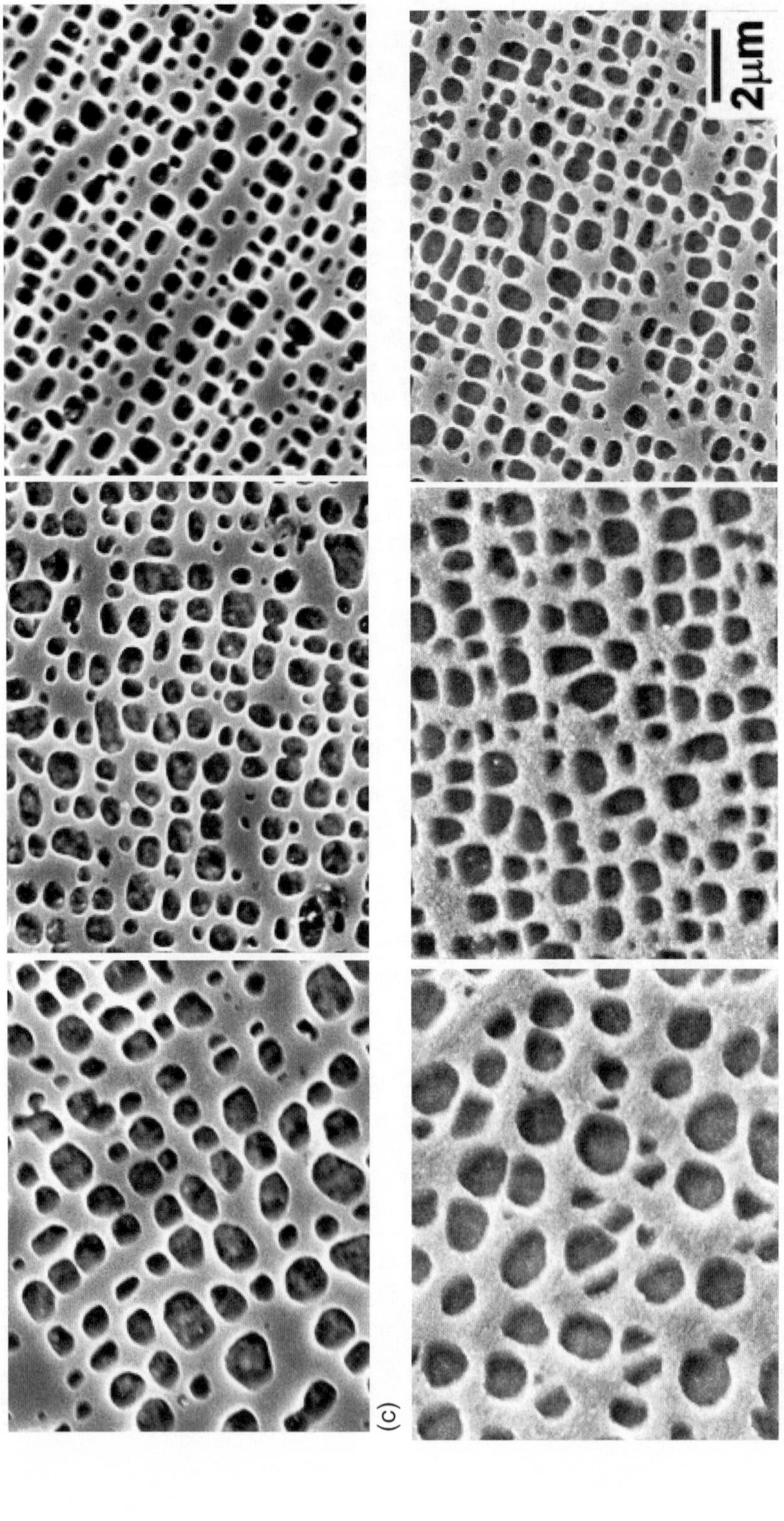

Fig. 4 Contd.

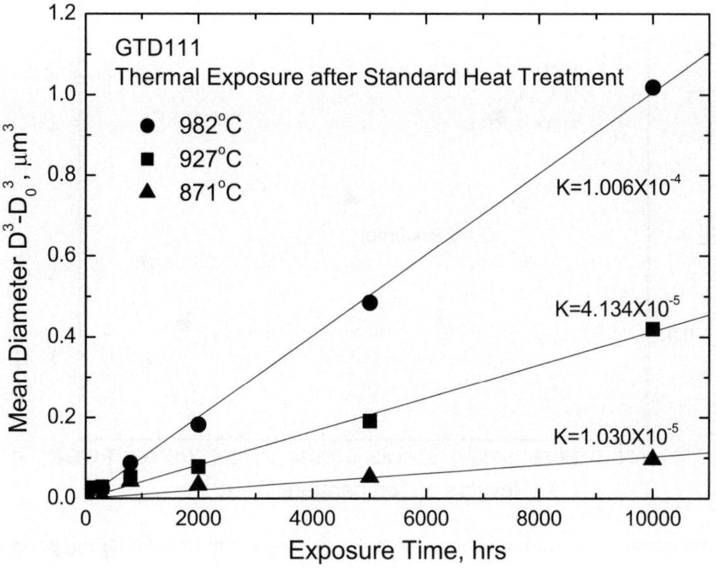

Fig. 5 Variation of γ' size of GTD 111 during thermal exposure.

Therefore, equation (1) can be expressed as follows,

$$ln\ KT = A - Q/RT \tag{3}$$

A plot of *lnKT* versus *1/T* results in a straight line as shown in Figure 6. Activation energy for γ' coarsening can be calculated from the slope of the line, *–Q/R*. In the present study, the value of *Q* is 256 kJ/mol., which is similar to the activation energy calculated in the other report, 259 kJ/mol.[1].

Grain Boundary

Two different carbide, MC and $M_{23}C_6$, were observed on the grain boundary of the standard heat treated alloy. It is known that the MC carbide had a tendency to decompose with thermal exposure into $M_{23}C_6$ plus γ'.[11] In this case, growth of $M_{23}C_6$ carbide on grain boundary makes the alloy locally enriched with γ' forming elements, Al and Ti, allowing the film of γ' to form along the grain boundary.

The decomposition of MC in GTD 111 showed different behaviours at different exposure temperatures. After thermal exposure at 982°C, the grain boundary of GTD 111 was covered with discontinuous $M_{23}C_6$ carbides enveloped by γ' film as shown in Figure 7. However, platelet and/or blocky phase formed from MC after exposure at 927 and 871°C. From the EDS analysis result, Figure 8, the phase could be identified as η phase with high contents of Ti, Ta and Ni. While γ phase looked like platelet phase at the initial stage of formation, it

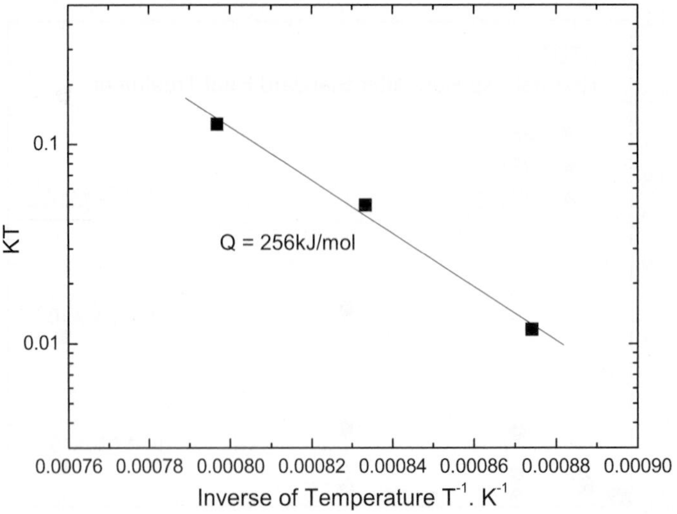

Fig. 6 Determination of activation energy for γ' coarsening using temperature dependence of rate constant K.

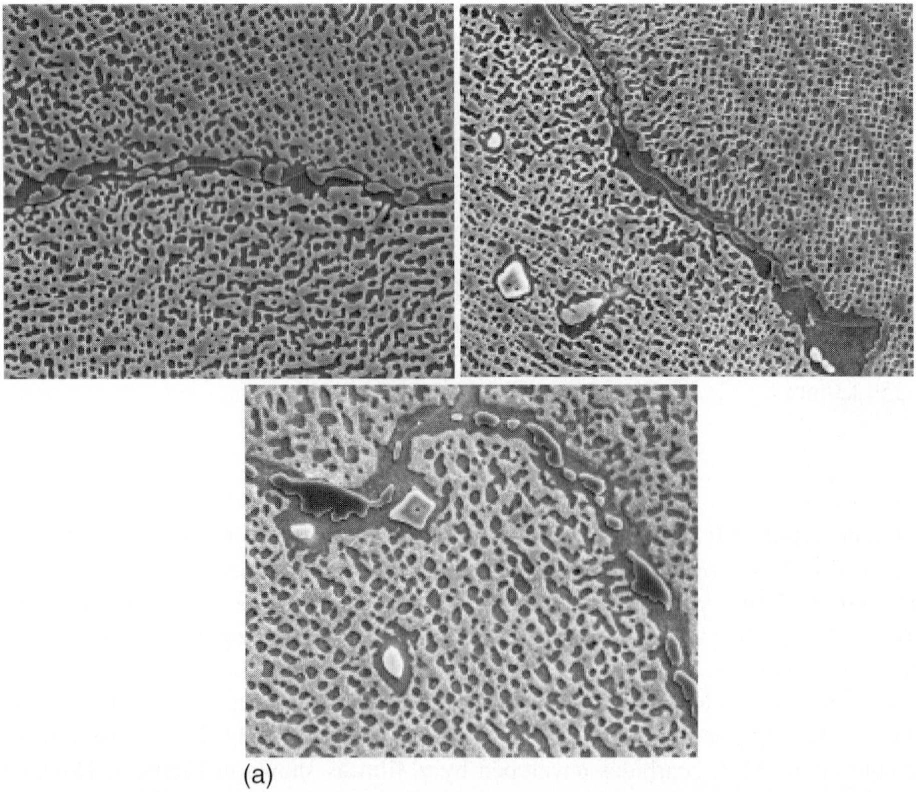

Fig. 7 Evolution of grain boundary structure during thermal exposure at (a) 982°C, (b) 927°C and (c) 871°C. Bright blocky phase is MC, bright acicular phase is σ, dark blocky phase is $M_{23}C_6$, and gray plate and blocky phase are η phase. (b) and (c) in next page.

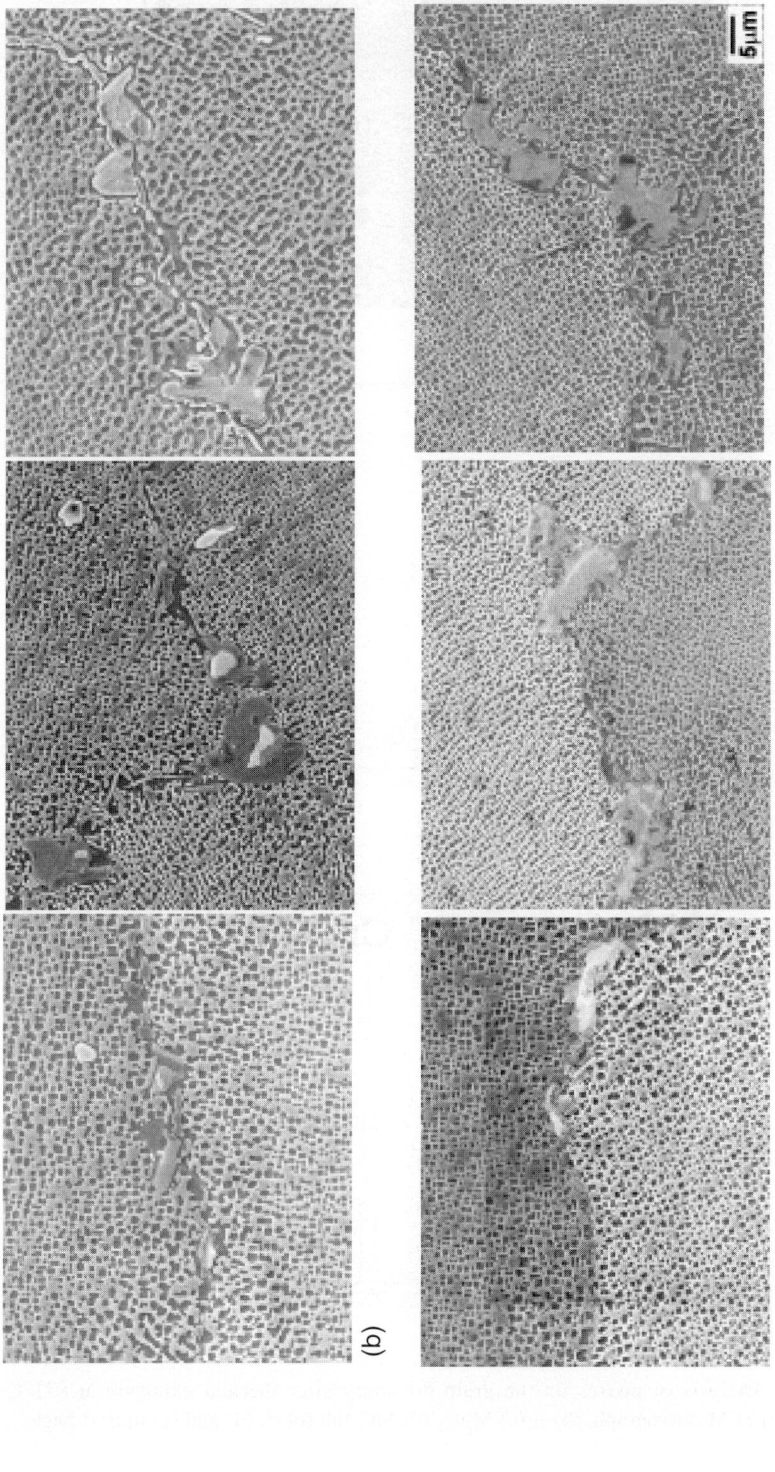

Fig. 7 Contd.

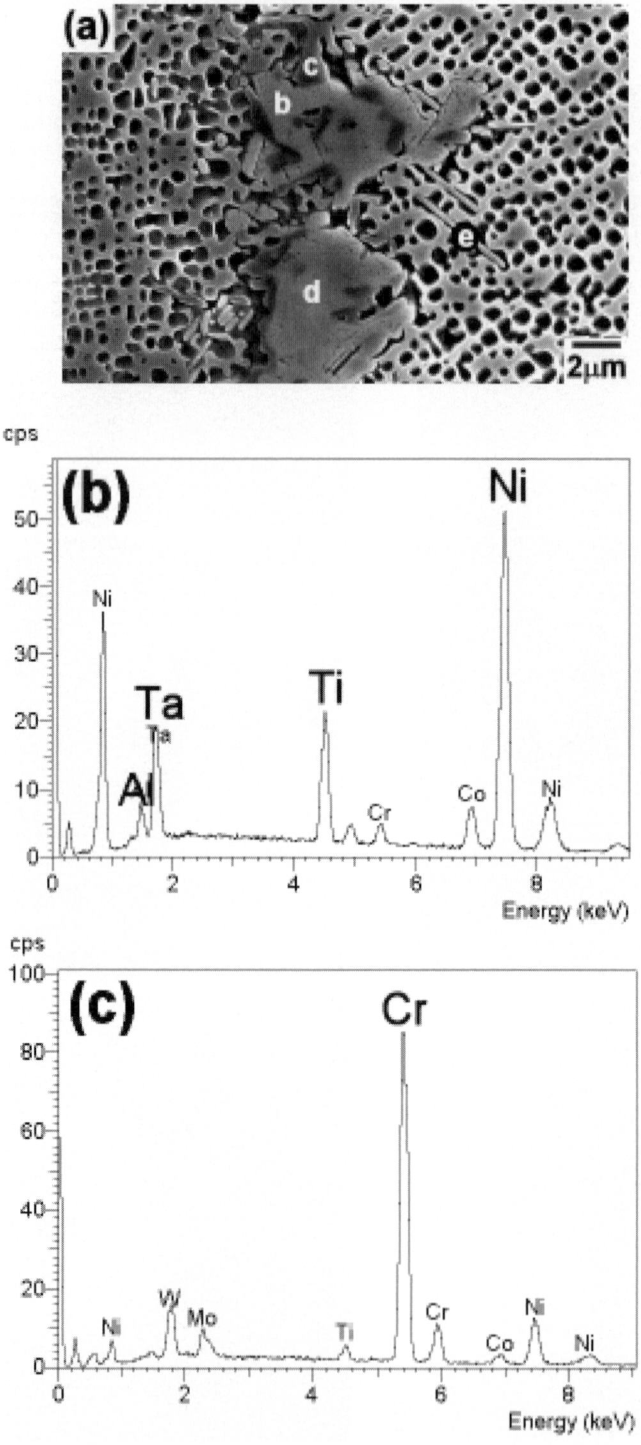

Fig. 8 EDS Analysis of phases on the grain boundary after thermal exposure at 871°C for 5000 hours. (a) SEM micrograph, (b) η (c) $M_{23}C_6$ (d) MC and (e) σ. (d) and (e) in next page

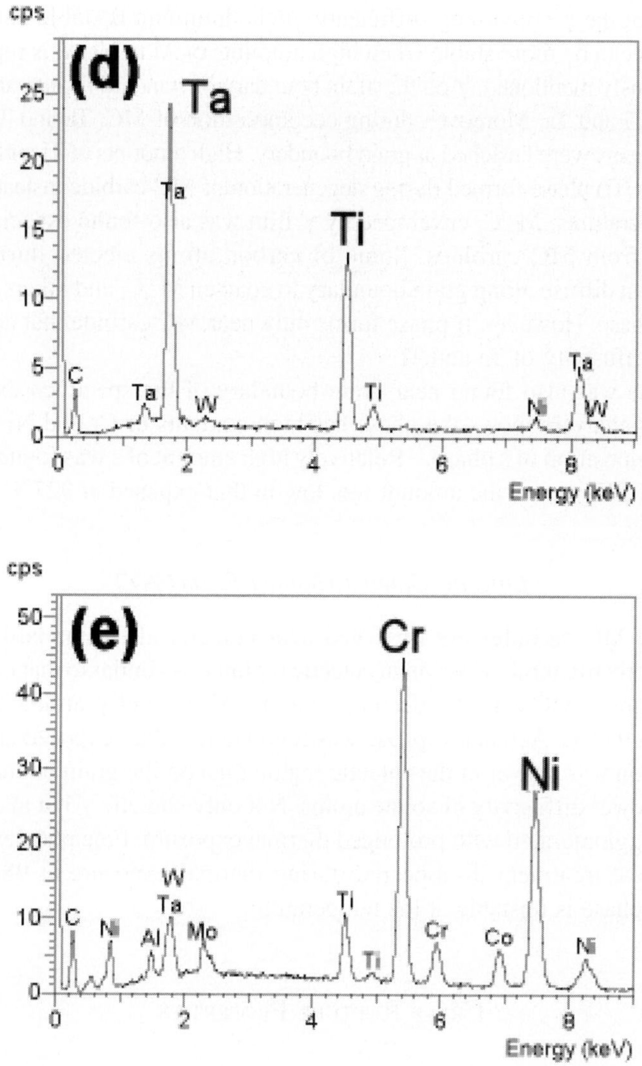

Fig. 8 Contd.

developed into blocky type as exposure time increased. $M_{23}C_6$ with high Cr content formed within η phase was also found. Therefore, the degeneration of MC carbide at these temperatures could given as,

$$\gamma + MC \ \gamma \ \eta + M_{23}C_6$$

Decomposition of MC into η and $M_{23}C_6$ was faster at 927°C than at 871°C due to higher diffusivity of the η forming elements.

It is known that the γ' containing sufficiently high aluminum is stable but another type of Ni_3X compounds can be more stable when high amounts of Al in Ni_3Al is replaced by Ti, Nb or Ta.[11] As previously mentioned, γ' on the grain boundary of standard heat treated GTD 111 had high contents of Ti and Ta. Moreover, during decomposition of MC, Ta and Ti separated from MC then the elements were enriched at grain boundary. High amounts of Ti and Ta made γ' to be unstable and η(Ni_3Ti) phase formed during degeneration of MC carbide instead of γ'.

At these temperatures, $M_{23}C_6$ enveloped by γ' film was also found especially on the grain boundary apart from MC carbides. Some of carbon atoms ejected during MC carbide decomposition can diffuse along grain boundary to coarsen $M_{23}C_6$ and others form new $M_{23}C_6$ carbide near η phase. However, η phase forms only near MC carbide that acts as nucleation site due to low diffusivity of Ta and Ti.

Acicular phase was also found near grain boundary of the specimen exposed at 871°C and 927°C. EDS analysis shows the phase has high contents of Cr and Ni and the result is similar to the composition of s phase.[12] Relatively high amount of s was found in the specimen exposed at 871°C. However, the amount was low in that exposed at 927°C.

Eutectic Gamma-Gamma Prime(γ-γ')

M_6C, $M_{23}C_6$ and MC carbides are observed near eutectic after standard heat treatment. Basically, the microstructural evolution in eutectic region was similar to that on grain boundary as shown in Figure 9. MC carbides decomposed into $M_{23}C_6$ and γ' at 982°C but into η and $M_{23}C_6$ at 871 and 927°C. Acicular σ phase was found in the alloy exposed at 871 and 927°C. However, variation was slower in the eutectic region than on the grain boundary, which may be attributed to lower diffusivity of solute atoms. Not only eutectic γ' but also fine secondary γ' near eutectic agglomerated with prolonged thermal exposure. Fine platelet η phase existed after standard heat treatment disappeared during thermal exposure at 982°C. This result indicates that η phase is unstable at the temperature.

Creep Rupture Properties

The specimens were aged at three different exposure conditions to investigate the effects of grain boundary η phase and coarsening of γ' on creep rupture properties. The γ' size distributions of the specimens exposed at 982°C for 1000 hours, 927°C for 2000 hours and 982°C for 2000 hours are shown in Figure 10. The alloy exposed at 982°C for 1000 hours has similar γ' size distribution to that exposed at 927°C for 2000 hours which has η phase on its grain boundary. The alloy exposed at 982°C for 2000 hours has larger γ' than that exposed for 1000 hours at the same temperature, but both specimens have similar grain boundary structure.

Creep test were conducted at four different test conditions, 927°C/210 MPa, 871°C/320 MPa, 816°C/440 MPa and 760°C/550 MPa. Figure 11 and Table 4 show the results of creep rupture tests.

According to the studies about the effect of η phase on mechanical properties, η can affect mechanical properties, but its effects are dependent on their position and morphology.[14]

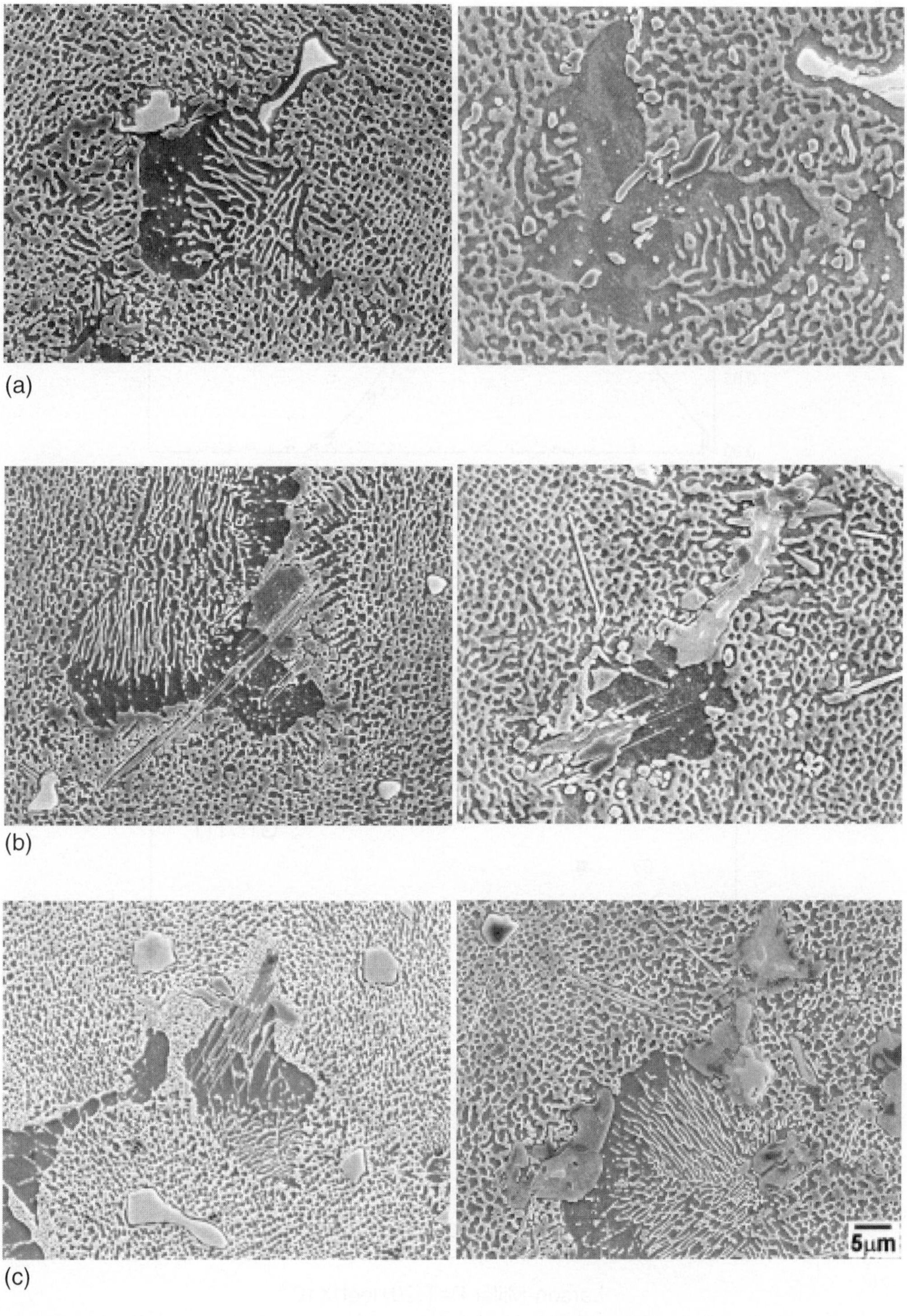

Fig. 9 Evolution of eutectic during thermal exposure at (a) 982°C, (b) 927°C and (c) 871°C .

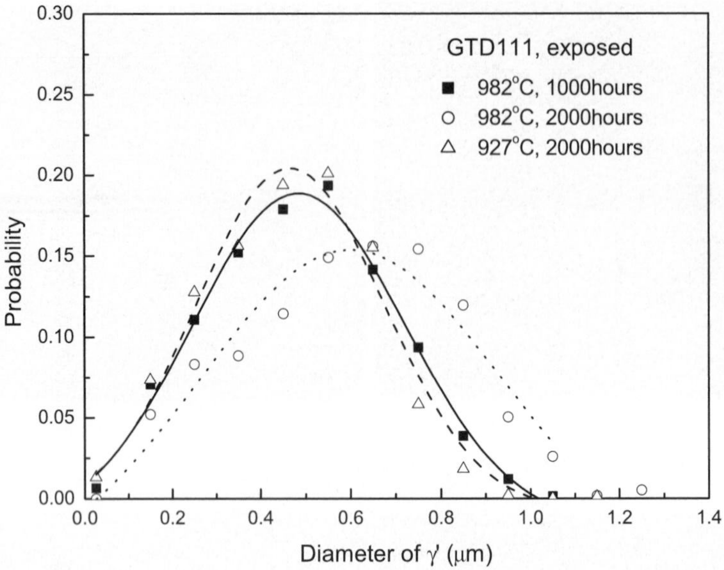

Fig. 10 Distribution of gamma prime (γ') size at three different exposure conditions.

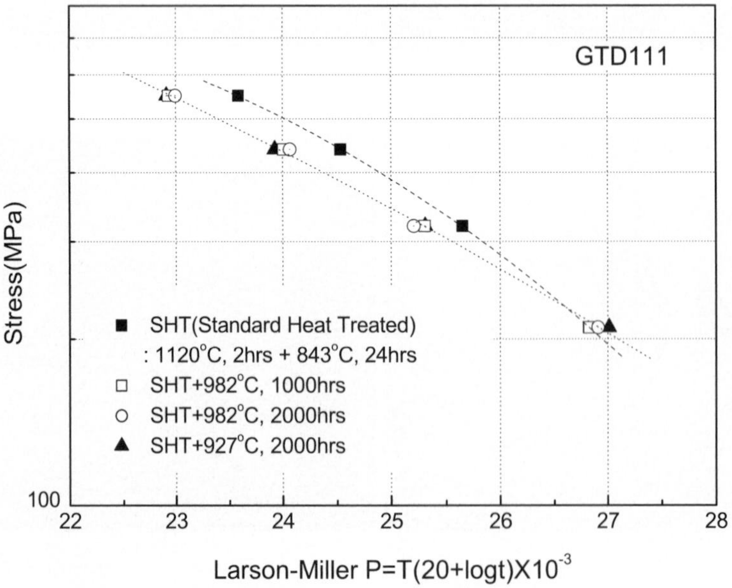

Fig. 11 Larson-Miller plot of GTD 111.

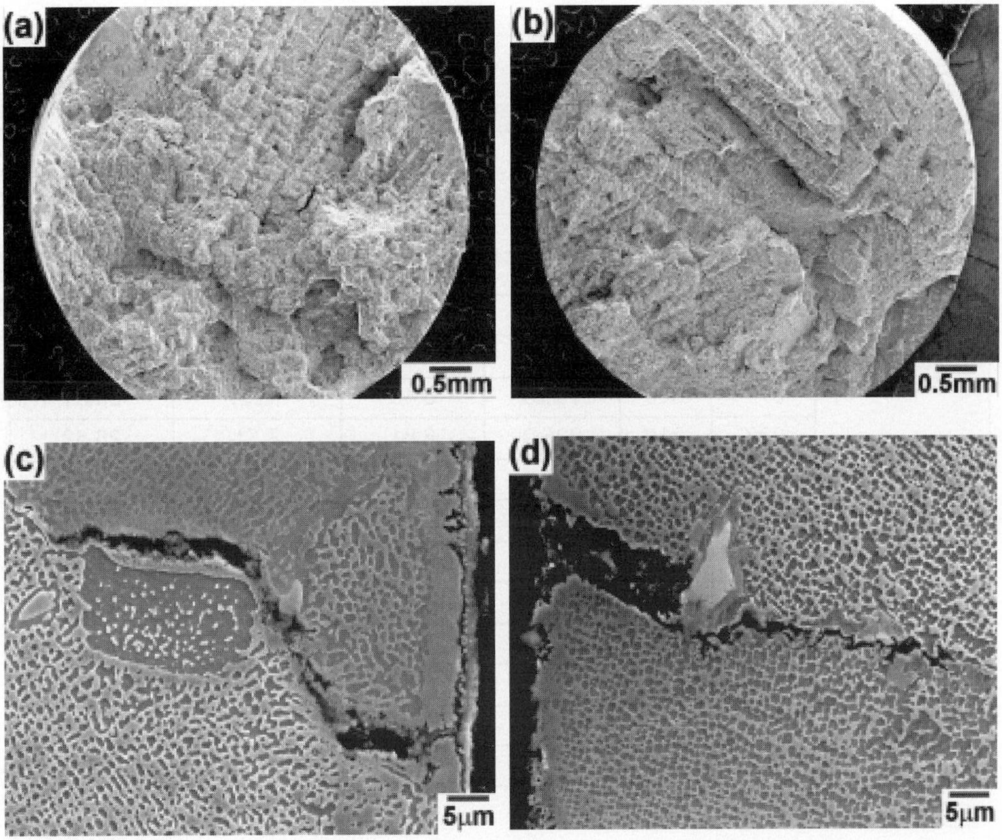

Fig. 12 SEM micrographs showing creep fracture surface (816°C/440 MPa) of the exposed GTD 111 (a) at 982°C, 2000 hours, (b) at 927°C for 2000 hours and secondary cracks on grain boundary of the exposed specimen (c) at 982°C, 2000 hours and (d) thermally exposed at 927°C, 2000 hours.

Regardless of exposed condition, creep rupture lives and elongation of exposed specimens were similar. These indicate that the effects of grain boundary h phase and the size distribution of γ' on creep properties are ignorable in the present study.

Exposed specimens showed relatively short creep rupture lives except the test at 927°C/210 MPa. As the temperature decreased, the deterioration of the creep property due to thermal exposure appeared to be evident. It is known that the fine secondary γ' plays an important role on creep behaviour compared with primary γ' size.[13] Standard heat treated specimen had both primary γ' and secondary fine γ', but secondary γ' dissolved during thermal exposure. At high temperature such as 927°C secondary γ' in standard heat treated specimen was dissolved fast during creep test, therefore it is supposed that its effect on creep property was also diminished at the early stage of creep. Therefore, the creep property was not dependent on exposure condition at 927°C because the effect of primary γ' size on creep strength was not significant in the present study.

Table 4 Results of creep test of GTD 111.

Exposure Conditions	Temp. (°C)	Stress (MPa)	Rupture Time (hours)	Elongation (%)	Reduction of Area (%)	Larson-Miller Parameter
Standard Heat Treated	760	550	682.4	4.64	6.24	23.59
	871	320	263.9	5.00	5.45	25.65
	816	440	334.3	7.24	3.47	24.53
	927	210	223.8	5.43	5.36	26.82
982°C, 1000 hours	760	550	156.6	3.55	8.10	22.93
	816	440	110.3	6.76	7.74	24.00
	871	320	131.9	6.40	10.67	25.31
	927	210	225.0	6.81	7.71	26.82
927°C, 2000 hours	760	550	152.2	5.92	9.73	22.91
	816	440	93.6	5.79	10.23	23.93
	871	320	132.5	5.18	6.39	25.31
	927	210	324.6	8.09	9.96	27.01
982°C, 2000 hours	760	550	182.4	4.87	7.34	23.00
	816	440	125.6	6.59	8.79	24.07
	871	320	108.0	9.61	8.95	25.21
	927	210	264.7	6.93	10.99	26.91

SEM fractographs give the information on fracture mode as shown in Figure 12. Dendrite structure was observed on the fracture surface, which indicates that the failure occurred intergranularly. Secondary cracks along grain boundary were observed on the in all test conditions. These observations reflect that the grain boundary was damaged during creep test. However, the different structure of grain boundary seems to affect neither rupture life nor creep ductility, even though occasional cracked η phases were also observed.

CONCLUSIONS

1. γ' agglomerated during thermal exposure by volume diffusion and the activation energy of the process was calculated as 256 kJ/mol. which is comparable to activation energy for diffusion of Al and Ti in Ni.
2. MC carbide on grain boundary degenerated during exposure in various way with the exposure temperature. They decomposed into $M_{23}C_6$ and γ' film at 982°C but into η and $M_{23}C_6$ at and below 927°C. σ was found most abundantly in the alloy exposed at 871°C.
3. Grain boundary η phase and coarsening of γ' affect neither rupture life nor creep ductility.

REFERENCES

1. J. A. DALEO and J. R. WILSON: 'GTD 111 Alloy Material Study', *Journal of Engineering of Gas Turbines and Power*, **120**(4), 1998, 375–382.
2. H. E. COLLINS: 'The Effect of Thermal Exposure on the Microstructure and Mechanical Properties of Nickel-Base Superalloys', *Metall. Trans.*, **5**(1), 1974, 189–204.
3. M. NAZMY and M. STAUBLI: 'Embrittlement of Several Vickel-Base Alloys after High-Temperature Exposure', *Scripta Metall. Mater.*, **24**, 1990, 135–138.
4. H. E. Collins: 'The Effect of Thermal Exposure on the Mechanical Properties of the Directinally Solidified Superalloy TRW-NASA VIA', *Metall. Trans.*, **6A**(3), 1975, 515–530.
5. S. A. SAJJADI and S. NATECH: 'A High Temperature Deformation Mechanism Map for the high Performance Ni-Base Superalloy GTD-111', *Mater. Sci. Eng.*, **A307**, 2001, 158–164.
6. S. A. SAJJADI, S. NATECH and R. I. L. GUTHRIE, 'Study of microstructure and Mechanical Properties of High Performance Ni-base Superalloy GTD-111', *Mater. Sci, Eng.*, **A325**, 2002, 484–489.
7. S. NATECH and S. A. SAJJADI: 'Dislocation Network Formation during Creep in Ni-base Superalloy GTD-111', *Mater. Sci, Eng.*, **A339**, 2003, 103–108.
8. B. S. RHO and S. W. NAM: 'Fatigue Induced Precipitates at Grain Boundary of Nb-A286 Alloy in High Temperature Low Cycle Fatigue', *Mater. Sci. Eng*, **A291**, 2000, 54–59.
9. M. J. STARINK and R. C. THOMSON: 'The Effect of High Temperature Exposure on Dendritic Segregation in a Conventionally Cast Ni Based Superalloy', *J. Mater. Sci.*, **36**, 2001, 5603–5608.
10. D. J. CHELLMAN and A. J. ARDELL: 'The Carsening of g' Precipitates at Large Volume Fractions', *Acta Met.*, **22**, 1974, 577–588.
11. E. W. ROSS and C. T. SIMS: *Superalloy II*, C. T. SIMS et al. eds., John Wiley & Sons, 1987, 97–134.
12. B. G. CHOI, D. W. JOO, I. S. KIM, J. C. CJANG and C. Y. JO: 'Effect of Thermal Exposure and Rejuvenation Treatment on Microstructure and Stress Rupture Properties of IN738LC', *Korean Journal of Materials Research*, **11**(10), 2001, 915–922.
13. K. KAKEHI: 'Effect of Primary and Secondary Precipitates on Creep Strength of Ni-Base Superalloy Single Crystal', *Mater. Sci. Eng.*, **A278**, 2000, 135–141.
14. G. K. BOUSE: 'Eta and Platelet Phases in Investment Cast Superalloys', *Superalloys 1996*, R. D. Kissinger et al. eds., A Publication of TMS, 1996, 163–172.

Low Cycle Fatigue Lifing of Single Crystal Superalloy CMSX-4

G. L. Drew, D. M. Knowles and C. M. F. Rae
Department of Materials Science and Metallurgy
University of Cambridge
Cambridge, UK

ABSTRACT

The influence of oxidation on crack initiation and propagation under low cycle fatigue has been studied for the single crystal nickel-based superalloy CMSX-4. Specimens tested in air under conditions of 750, 850 and 950°C, over a variety of loading conditions, were examined by scanning electron microscopy. A stress dependent transition in initiation and propagation mechanism was observed above 850°C. Measurements conducted on sectioned specimens have been used to determine the influence of stress on oxide spike growth for cycled specimens. From these observations a model that describes fatigue crack growth as a summation of mechanical and oxidation contributions has been used to compare crack growth and fatigue life test data at 950°C.

INTRODUCTION

The development of single crystal nickel-based superalloys for aerospace turbine blade applications has enabled significant improvements in engine efficiency with increased combustion temperatures. With the elimination of grain boundaries the high temperature low cycle fatigue behaviour has improved typically by an order of magnitude over their polycrystalline predecessors.[1] Despite the pertinence of understanding the mechanisms that ultimately lead to failure of these critical engineering components, detailed studies characterising the mechanical response of alloys such as CMSX-4 are currently lacking.

The in-service operating conditions are very complex, involving transitions in temperature and load under a harsh operating environment. The aim of the present work is to investigate the transitions in the dominating mechanisms of failure with temperature and loading condition under low cycle fatigue such that appropriate models can be applied for estimation of the crack initiation and failure resistance.

MATERIALS AND METHODS

The single crystal superalloy material CMSX-4 has a nominal chemical composition as given in Table 1. For comparison the composition of two other commercially available single

Table 1 Nominal chemical composition (balance Nickel), in wt.%, of CMSX-4.

Alloy	Al	Co	Cr	Mo	Ti	Hf	Re	Ta	W	V
CMSX-4	5.6	9.5	6.5	0.6	1.0	0.1	3.0	6.5	6.4	-
RR2000	5.5	15	10	3	4	-	-	-	-	1
AM1	5.2	6.5	7.7	2.0	1.1	-	-	7.9	-	-

crystal superalloys RR2000 and AM1 are included in the table. Most of the specimens used in this work were supplied by Rolls-Royce plc as tested. Those tested by the author were supplied by Rolls-Royce plc, machined from cast cylindrical bars in the solution treated and aged condition. The solution treatment involves three stages, followed by a two stage ageing heat treatment. The final alloy consists of approximately 70 vol.% cuboidal $L1_2$-type γ' precipitates in a face centred cubic (FCC) γ matrix phase. The average size and distribution of the coherent precipitates is approximately 0.45 μm separated by 0.1 μm channels. The nominal axial orientation of the bars and specimens is [001].

The specimens were tested under strain controlled low cycle fatigue (LCF) at temperatures of 750, 850 and 950°C using radiation heating in air. Either hollow or solid cylindrical specimens were tested, each with 0.25 μm finishes within the gauge length. A variety of strain-ranges with a strain R-ratio of 0, and small number with a strain R-ratio of –1, were examined. As a result specimens were available with a range of loading conditions for each of the three temperatures. The majority of tests were executed until failure, and the lower strain range tests interrupted before failure were run for greater than 100,000 cycles. Included in the set for examination were a several load-controlled specimens, including a series tested at 950°C, stopped at 1/3, 2/3 and 1.0 of life.[2] All tests were conducted at a frequency of 0.25 Hz, using a trapesoidal waveform having a 1 s hold period between each 1 s ramp.

Fractography was performed using depth-of-field and scanning electron microscopy (SEM). SEM was also used for oxide penetration depth measurements on specimen sections mounted and polished to a 1 μm finish. Specimens were prepared with due care to preserve the oxide layer. The longitudinal cuts were made perpendicular to the radial curvature of the specimen to expose the surface features at their radial dimensions. Measurements were recorded under calibrated conditions. An average of 35 measurements, with a minimum 20, was taken per specimen.

RESULTS AND DISCUSSION

The fracture surfaces investigated showed that crack initiation proceeds either from a casting pore within the bulk material or from an oxidised spike or defect at the surface of the specimen. The primary controlling factor, within the range of specimens available for examination, was testing temperature. For specimens tested at 850°C and below, the critical crack was

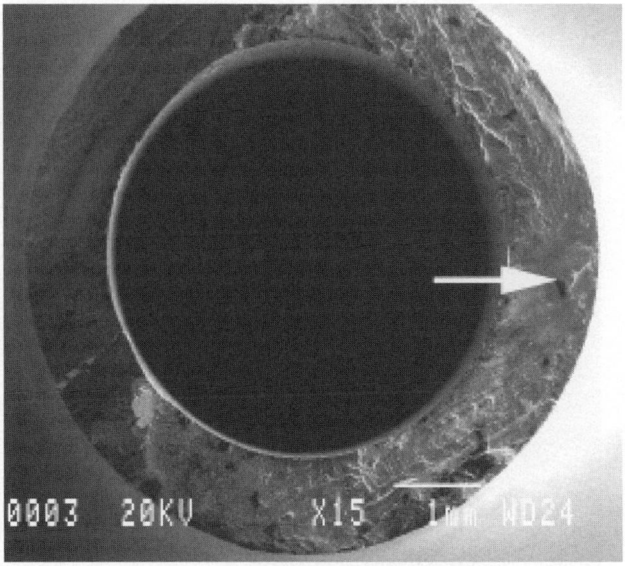

Fig. 1 Fracture surface at low temperature, with critical crack initiation from internal casting pore (inset). (850°C SCLCF).

initiated at an internal casting pore, whereas all specimens tested at 950°C failed via a crack, or cracks, initiated at preferentially oxidised surface sites. The tests at 850°C exhibited both types of initiation site.

Initiation from Casting Pores

Specimens that initiated the critical crack from internal casting pores displayed two types of fracture surface morphology. For the specimens tested at 750 and 850°C the microcrack initially propagated in a radial fashion from the casting pore, resulting in a smooth, penny-shaped surface. At both temperatures this initial growth was perpendicular to the loading axis. This Stage II† crack growth is shown in Figure 1. When the penny-shaped crack reached the specimen surface, and exposed the crack tip to an oxidising environment, one of two behaviours was exhibited. The first was an immediate transition to shear failure, demonstrated by a sharp change in the fracture surface from the penny-shaped region to large facets at near –45° angles. In the second case the crack continued to grow perpendicular to the loading axis for some distance before final shear failure.

† First used by Forsyth[3], the terms stage I and stage II are commonly used to describe the two different modes of fatigue crack propagation that occur in most alloys. Stage I cracking refers to shear on crystallographic slip planes, leading to a zig-zag crack path. Stage II cracking results in a planar (mode I) crack path normal to the stress axis.[4]

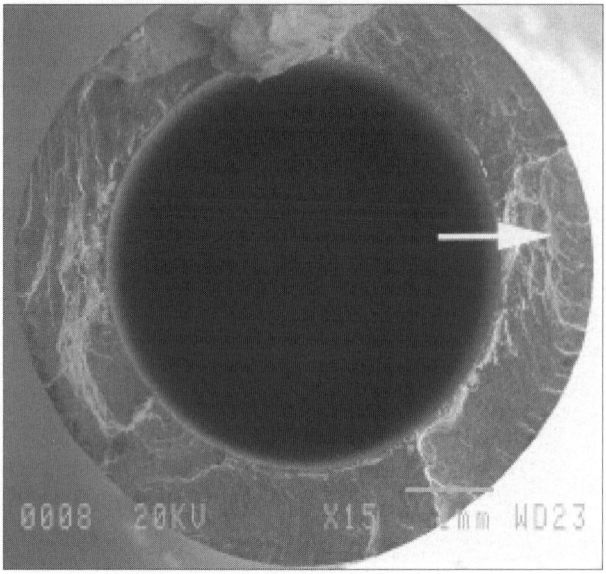

Fig. 2 Fracture surface at high temperature, with critical crack initiation from surface – multiple small sites joined to form the failure crack. (950°C SCLCF).

Initiation from Surface Flaws or Oxide Spikes

The 950°C test specimens initiated critical cracks from the specimen surface. In general multiple sites were initiated, and either one dominated failure or several joined and propagated to failure as one large crack (Figure 2). Crack growth was perpendicular to the load axis, and final failure was by 45° shear of the remaining ligament. The majority of crack growth was non-crystallographic, Stage II type, though in the high stress load-controlled tests there was also a period of crystallographic, Stage III, growth just prior to final shear. As the failure crack grew it deflected to incorporate a number of smaller surface cracks (e.g. Figure 2).

The critical cracks appeared to initiate from preferentially oxidised sites on the specimen surface. In this investigation, the critical crack, or cracks that joined to form the failure crack, initiated from oxide spikes. On several of the fracture surfaces there were also a number of initiated sites that had pores associated with them. The pores were sectioned by the surface of the specimen, and the micro-crack morphology was similar to the other surface cracks. (Figures 3a and b).

The stress and temperature dependence of the crack initiation mechanism for CMSX-4 and a related single crystal alloy, RR2000, is given in Figure 4. The transition for RR2000 is delineated clearly by the data. As a first approximation, the transition for CMSX-4 above 850°C has been drawn to follow the behaviour of RR2000. At least one other mono-crystalline superalloy of similar composition has been shown to exhibit a transition in crack initiation

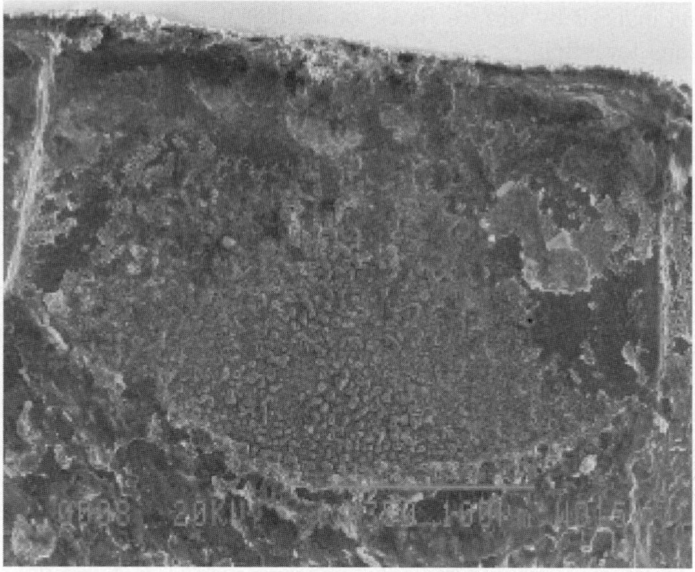

(a)

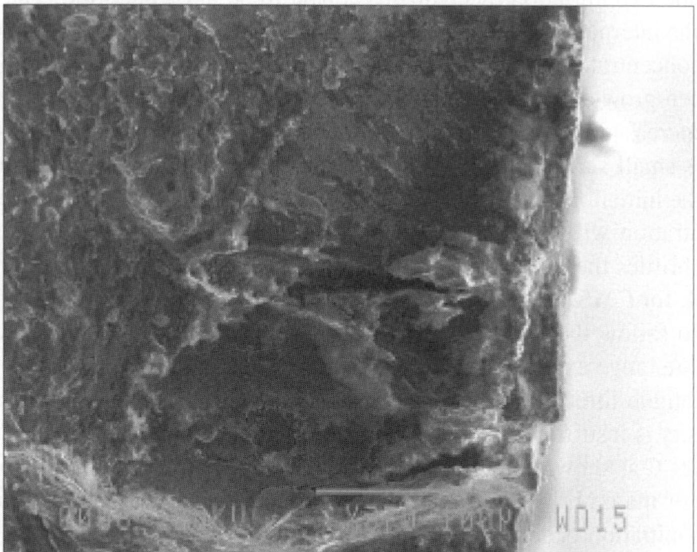

(b)

Fig. 3 Sub-critical surface cracks initiated from (a) an oxide spike and (b) a surface-intersecting casting pore, growing in an oxidative environment. (950°C SCLCF).

characteristics in the vicinity of at 850°C. Fleury and Rémy,[5] working with AM1, have shown that initiation at internal pores prevails for high strain ranges and low temperatures, while crack initiation induced by surface flaws, in particular oxide spikes, is dominant at high temperatures and low strain ranges. In RR2000 the change from pore to surface initiation occurs below 0.9×10^{-2} ($R = -1$) strain at 850°C. A temperature dependent transition zone encompassing 0.3×10^{-2} to 0.9×10^{-2} ($R = -1$) strain at 850°C was indicated for AM1.[5]

The critical crack initiation mechanism for CMSX-4 with respect to temperature is well defined at 750 and 950°C. At 750°C the fatigue limit has been estimated from load-controlled data to lie at a stress immediately below the strain controlled data shown.[6] The upper limit of the 950°C data approaches the yield stress of the material, above which strain controlled tests are no longer informative to this investigation. Load controlled tests beyond the yield stress continue to exhibit oxidation-assisted initiation of surface cracks.[7]

As noted earlier, specimens that failed at 850°C exhibited both types of initiation sites. The critical crack initiated from a single site on a large, irregular internal casting pore, ~100 μm, however smaller oxidised surface initiated sites down to less than 10 μm depth were also recorded. At 850 and 950°C multiple oxide spikes were observed along the gauge of all but one of the sectioned tests, notably including the 950°C load-controlled tests stopped at 1/3 and 2/3 of life. The sizes of the micro-cracks at 1/3 and 2/3 of life are consistent with the evolution of crack depth under LCF at 950°C for AM1.[5] The only exception at 850°C was tested below the fatigue limit, and had a thin cohesive surface oxide similar to that found on the 750°C specimens. No internal initiation sites were found during examination of any of the sectioned specimens.

At 850°C the two initiation mechanisms appear to be in competition. The time to initiate the crack on the internal pore in an inert environment is dependent on the localised plasticity at the stress concentrator. There will be a significant incubation period to initiate the crack, which will then grow at a rate dependant on its size. With higher temperatures the rate of oxidation is increased, and the time to initiate a surface micro-crack decreases. The rate of growth of this small surface crack is dependent both on the rate of oxidation and localised plasticity. If the initiation size of the surface crack is smaller than the internal crack the local stress concentration will be lower. However, the process of oxidation extends the micro-crack and embrittles the substrate ahead of the crack tip, reducing its fracture toughness.

In Figure 4, for CMSX-4 at temperatures up to and including 850°C the transition would be expected to follow the yield stress as a function of temperature, since no specimens in this temperature range exhibited surface initiated critical cracks, down to the fatigue limit at 850°C. The fatigue limit delineates the stress below which the stress field associated with internal porosity is insufficient to initiate a critical crack. Above 850°C for CMSX-4, and at least above 600°C as shown for RR2000, the transition will follow the temperature dependent yield criteria of the embrittled zone of the oxidised spike, since oxidation-assisted critical surface crack initiation occurs before the fatigue limit is reached.

Kinetics of Interdendritic Oxidation

At the lower temperatures and strain ranges oxide penetration was observed as localised regions of greater oxide depth, or 'spikes' (Figure 5). Particularly at the higher temperatures

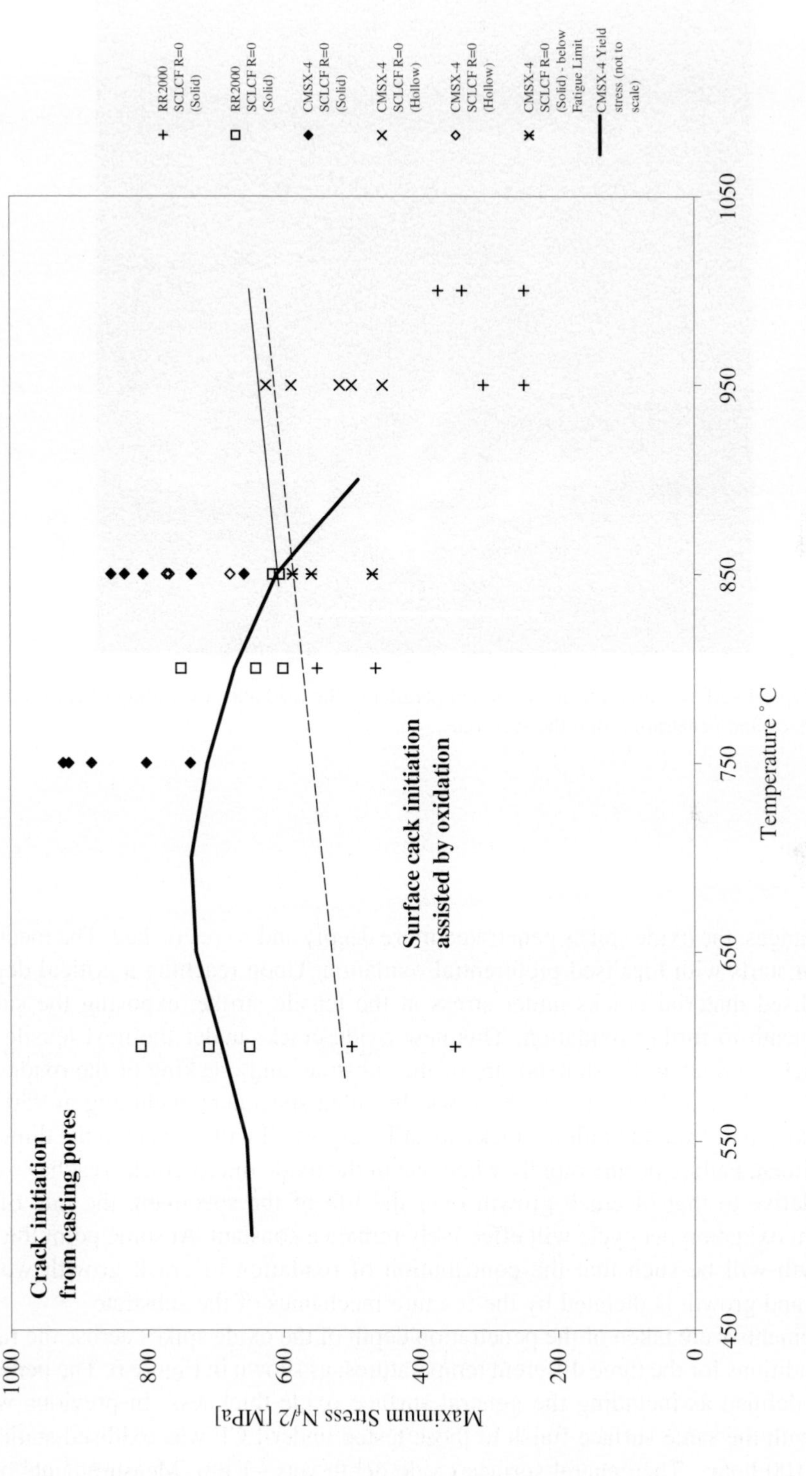

Fig. 4 Stress and temperature dependence of the critical crack initiation mechanism. (NB: RR2000 850 and 950°C data have Rε= −1).

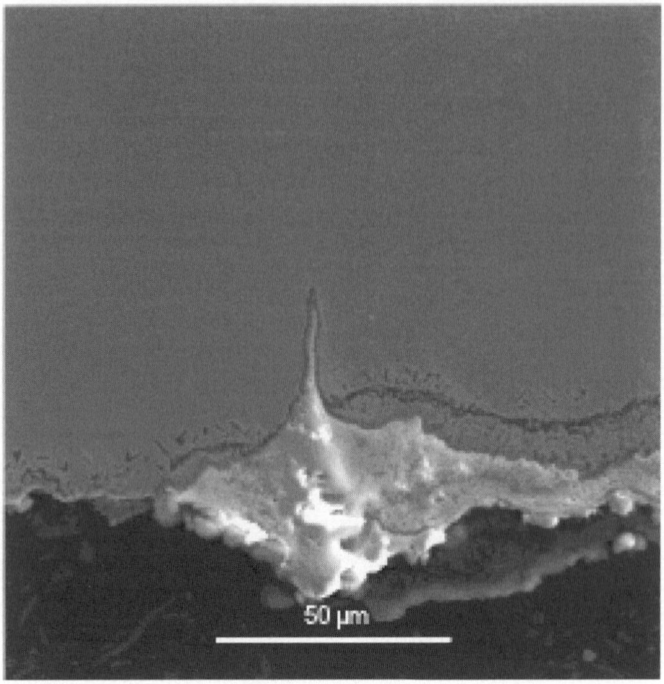

Fig. 5 Typical surface initiation site sectioned parallel to the load axis, exhibiting heavy oxidation at the surface and penetration into the substrate.

and strain ranges, the oxide spikes penetrated more deeply and were cracked. The mechanism of initiation starts with localised preferential oxidation. Upon reaching a critical depth the brittle oxidised material cracks under stress at the tensile stroke, exposing the substrate directly beneath to further oxidation. This new oxide cracks under the next tensile stroke and the crack is initiated. Fresh exposure of the substrate and cracking of the oxide occurs during each cycle and the micro-crack grows. In failed specimens including at 950°C, the micro-cracks (apart from the failure crack) are still very small with respect to the dimensions of the specimen. Failure occurs rapidly when one of the oxide micro-cracks reaches a critical length. Relative to that of crack growth over the life of the specimen, the rate of crack growth from oxidation per cycle will effectively remain a constant. At some point the rate of crack growth will be such that the contribution of oxidation to crack growth would be negligible, and growth is dictated by the fracture mechanics of the substrate.

Measurements were taken of the penetration depth of the oxide spikes across the range of loading conditions for the three different temperatures, as shown in Figure 6. The penetration depth was defined as including the general surface oxide thickness. In previous work[8] a specimen with the same surface finish to those tested under LCF was oxidised statically at 950°C for 100 hours. The general surface oxide depth was ~1 µm. Measurements on non-

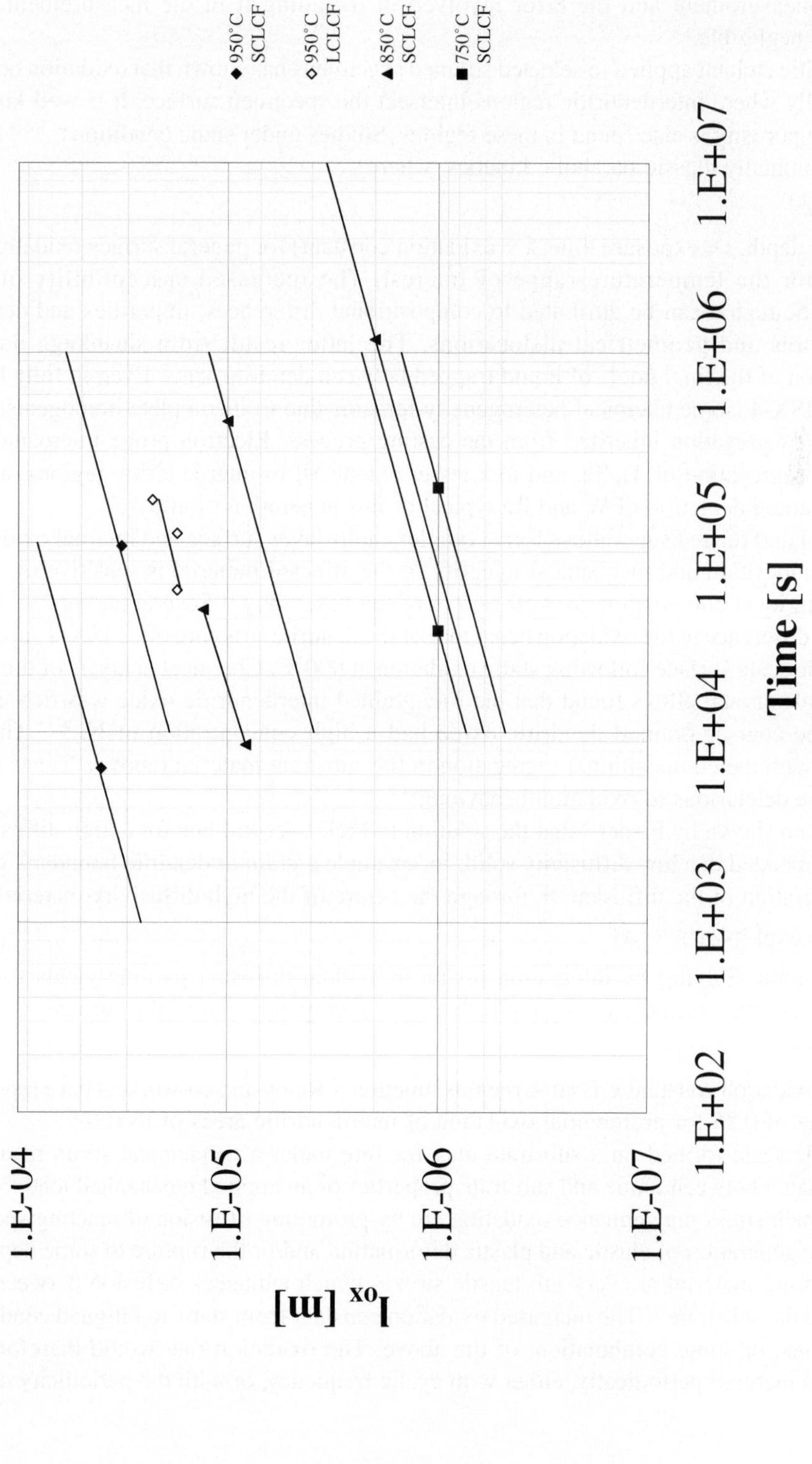

Fig. 6 Variation of the average oxide depth as a function of exposure time for fatigued specimens at 750, 850 and 950°C under a number of strain ranges.

preferentially oxidised areas of fatigued specimens at 950°C gave general surface oxide thickness of the same order. At this temperature, the surface oxide depth is less than 10% of the spike measurement and the error involved in including it in the measurement was considered negligible.

A dendritic etchant applied to selected strained specimens has shown that oxidation occurs preferentially where interdendritic regions intersect the specimen surface. It is well known that casting porosity is also found in these regions. Studies under static conditions[9, 10, 11] have reported nominally classic parabolic kinetics, where

$$l_{ox} = k \cdot t^{0.5}$$

(l_{ox} = oxide depth, t = exposure time, k = oxidation constant) for general surface oxidation of CMSX-4 for the temperature range of interest. The increased susceptibility of the interdendritic region can be attributed to compositional differences, impurities and defects such as pores and geometrical dislocations. The latter result from shrinkage during solidification of the final pools of liquid trapped between dendrite arms. Even in fully heat-treated CMSX-4, some chemical heterogeneity remains due to incomplete homogenisation of the microsegregation inherited from the casting process. Electron probe microanalysis has shown segregation of Ti, Ta, and to a lesser extent Al, to interdendritic regions, along with the marked depletion of W and Re typical of this generation of alloys.[12]

CMSX-4 and related superalloys form complex multi-layer surface and internal oxides in air. The composition and mechanical integrity of the affected material is sensitive to small variations in local composition as well as that of the base alloy.[11] Several authors[10, 11] have observed a difference in the oxidation behaviour of the dendrite structure of CMSX-4 apparent on the oxide scale surface following static oxidation at 900°C. Chemical analysis of this and related single crystal alloys found that the fine grained interdendritic oxide was rich in Ti, whereas the coarser grained dendritic oxide had a high concentration of Ni.[9, 13] This is consistent with the compositional segregation in the substrate material (above). Ti has been shown to be deleterious to oxidation behaviour.[14]

It has been shown by Fisher[15] that the solution to Fick's second law for a high diffusivity material imbedded in a low diffusivity solid, for example a grain or dendrite boundary, gives the concentration of the diffusant, n, through the centre of the high diffusivity material as

$$n(x, t) = \exp[-(4/\pi t)^{0.25} \cdot x]$$

where t is time. Solving by integration for an individual diffusant particle yields for the penetration depth, x (dimensionally equivalent to l_{ox}), perpendicular to the surface

$$x = k \cdot t^{0.25}$$

where the oxidation constant k, is an Arrhenius function.[16] Remy and co-workers have reported an exponent of 0.25 for preferential oxidation of interdendritic areas of IN100.[17]

An oxide scale formed on a substrate may fracture under a mechanical strain resulting from mismatch between oxide and substrate properties or an applied mechanical load.[18] Two ways in which stress may enhance oxidation are by promoting diffusion of reacting species through the generation of elastic and plastic deformation and/or the rupture to some depth, l, of the oxidised material at every nth tensile stroke, which enhances diffusion between the surface and the substrate.[17] The increased oxidation constant from static to fatigued condition is due to one, or some combination, of the above. The oxidation rate would therefore be expected to increase periodically, either with cyclic frequency, or with the periodicity of the oxide fracture event, in the manner of a quasi-functional law as described by e.g. West.[19] The load-controlled low cycle fatigue data at 950°C, interrupted at 1/3, 2/3 and 1.0 of life gave an exponent of 0.22, and oxidation constant $k = 1.49 \times 10^{-6}$. The low cycle fatigue data at 850°C, subject to the same σ_{max}, gave an exponent of 0.28 and oxidation constant $k = 6.25 \times 10^{-7}$. These results suggests that the oxidation kinetics conform to the relationships established for the unstressed state, and are consistent with other investigations on similar alloys.[17, 20]

Since the kinetic law of penetration is the same with as without stress, the assumption is made that oxygen diffusion remains the rate-determining process. Plotting preferential oxide growth under cyclic loading as a function of σ_{max} enables the determination of the stress-assisted oxidation kinetic parameter, k' (Figure 7). The relation is found to follow an exponential law, where the constants A and b are determined at a given temperature.

$$k' = A \cdot \exp(b \cdot \sigma_{max}) (\text{when } \sigma \geq \sigma_0)$$

From a physical basis it has been suggest that this triggering of oxidation kinetics is partly associated with local fracture events of the oxide scale, therefore a threshold corresponding to the fracture properties of the material would be expected. Below a certain applied stress σ_0, the rate of oxidation is independent of stress. [20]

Life Prediction Modelling

Since at 950°C the penetration of oxide spikes and the initiation of surface cracks that lead to failure occur early, the life of the specimen is dominated by micro-crack growth in an oxidising environment. Taking this into account, a formulation similar to the model that has been applied by Chataigner and Remy to the single crystal superalloy AM1[21] has been adopted in order to predict the failure life of the specimens. The interaction between fatigue and oxidation is accounted for assuming elementary crack advance results from both crack opening under fatigue and an additional contribution due to oxidation at the crack tip.

$$\frac{da}{dN_F} = \left(\frac{da}{dN_f}\right) + \left(\frac{da}{dN_{ox}}\right)$$

The fatigue contribution is estimated assuming crack opening occurs only in tension, and is dependent on the crack length, a, maximum tensile stress, σ_{max}, and inelastic strain range, $\Delta\varepsilon_{in}$.[22]

$$\frac{da}{dN_f} = a\Delta\varepsilon_{in}\left[\frac{1}{\cos\left(\frac{\pi\sigma_{max}}{2\sigma_{UTS}}\right)} - 1\right]$$

The contribution from oxidation in each stroke can be estimated from the oxide penetration depth data.

$$\frac{da}{dN_{ox}} = k'.\Delta t^{1/4}$$

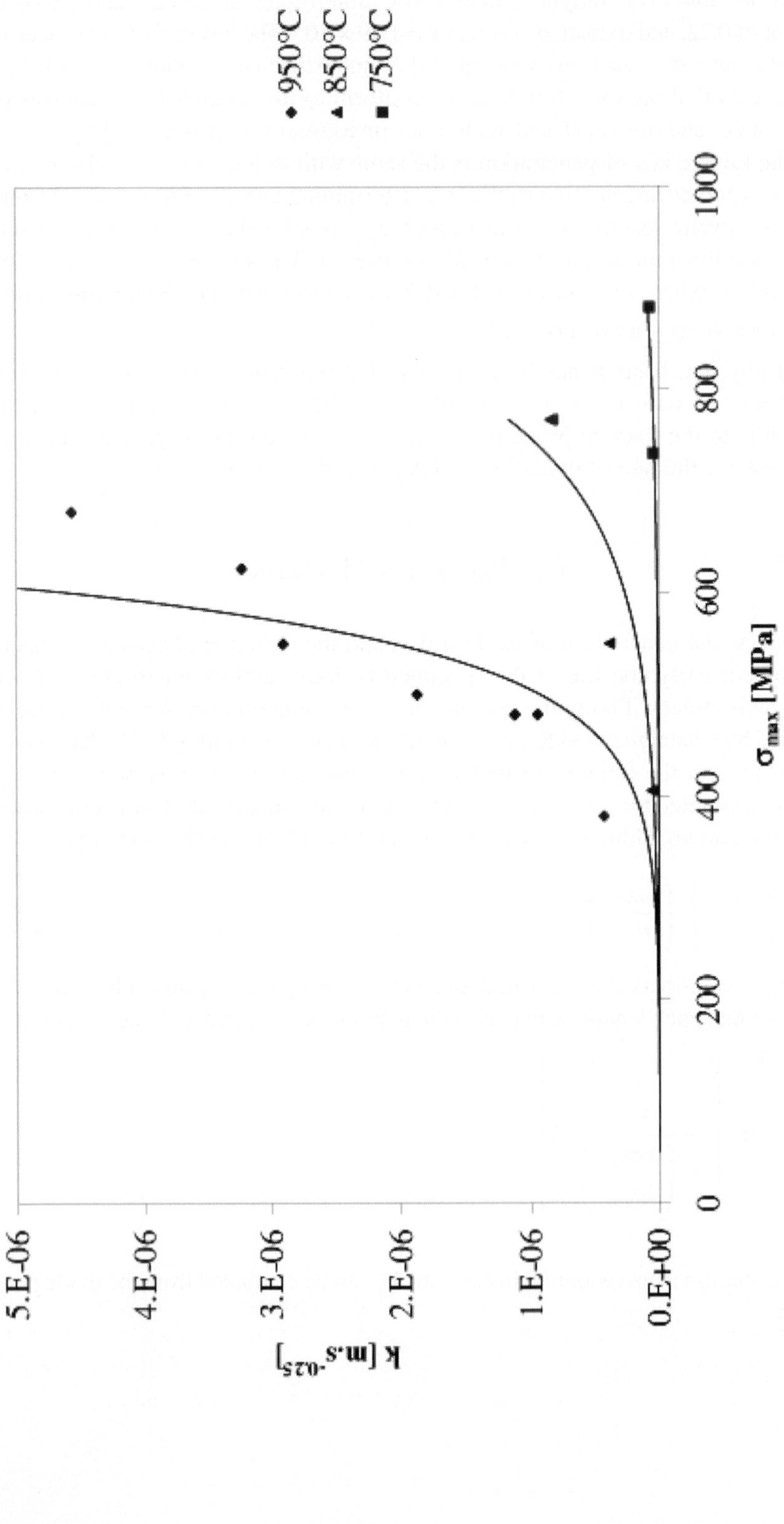

Fig. 7 The influence of maximum tensile stress on the oxidation constant, *k*.

where k' is the oxidation parameter taking into account the exponential-law enhancement of oxide growth by cyclic straining. The formulation of the model assumes crack extension and re-exposure of the substrate in each cycle, therefore da/dN_{ox} is a constant for a given σ_{max}.

The rate of crack growth during LCF and thermo-mechanical fatigue (TMF) has been shown to increase less rapidly as a function of ΔK than for compact tension (CT) tests on CMSX-4.[23] Figure 8 shows a comparison between LCF crack growth data and the predictions from the model with and without the oxidative component. The parameters σ_{max} and $\Delta\varepsilon_{in}$ were determined from a specimen cycled under the same conditions as [22] and not used in the oxidation kinetics study.

It is well known that small crack growth behaviour often does not conform to that measured conventionally from physically large cracks. This is generally accepted as being the result of one or more mechanical, microstructural and environmental factors.[4] In this case, the rate of small crack growth is dominated by oxidation. A study on the single crystal superalloy CMSX-2 correlated the reduction in intrinsic small crack propagation resistance with increasing temperature to the increase in γ' depletion zone ahead of the crack tip.[24] By weakening the material ahead of the crack tip oxidation accelerates the rate of small crack growth. As the crack size increases the contribution from oxidation becomes less significant, and the rate of crack growth rises to follow that of the substrate.

Using this model and the oxidation data measured above, the number of cycles to propagate the crack has been calculated for an independent set of 950°C strain controlled (R = –1) LCF tests. Included in Figure 9 are the results of the calculation omitting the contribution of oxidation to crack growth. The correlation bars represent a factor of 2, and the model is seen to predict the data well.

CONCLUSIONS

- Up to 850°C fatigue failure is controlled by the initiation of a micro-crack from a (anomalously large) internal casting pore. A transition to oxidation assisted surface crack initiation and growth occurs above 850°C such that at 950°C the growth of micro-cracks to failure is dominated by propagation.
- Interdendritic oxidation of CMSX-4 under LCF conditions follows $t^{0.25}$ kinetics, where the influence of loading conditions is taken into account by introducing a function dependent on the maximum tensile stress.
- A model that describes fatigue crack growth as a summation of mechanical and oxidation contributions has compared favourably with crack growth and fatigue life test data at 950°C.

ACKNOWLEDGEMENTS

The financial support for this work was provided by Rolls Royce plc and The Cambridge Commonwealth Trust. The author would also like to thank Professor D. J. Fray for the provision of research facilities and the supporting members of staff at Rolls Royce plc, particularly Dr R. W. Broomfield for supplying the CMSX-4 superalloy.

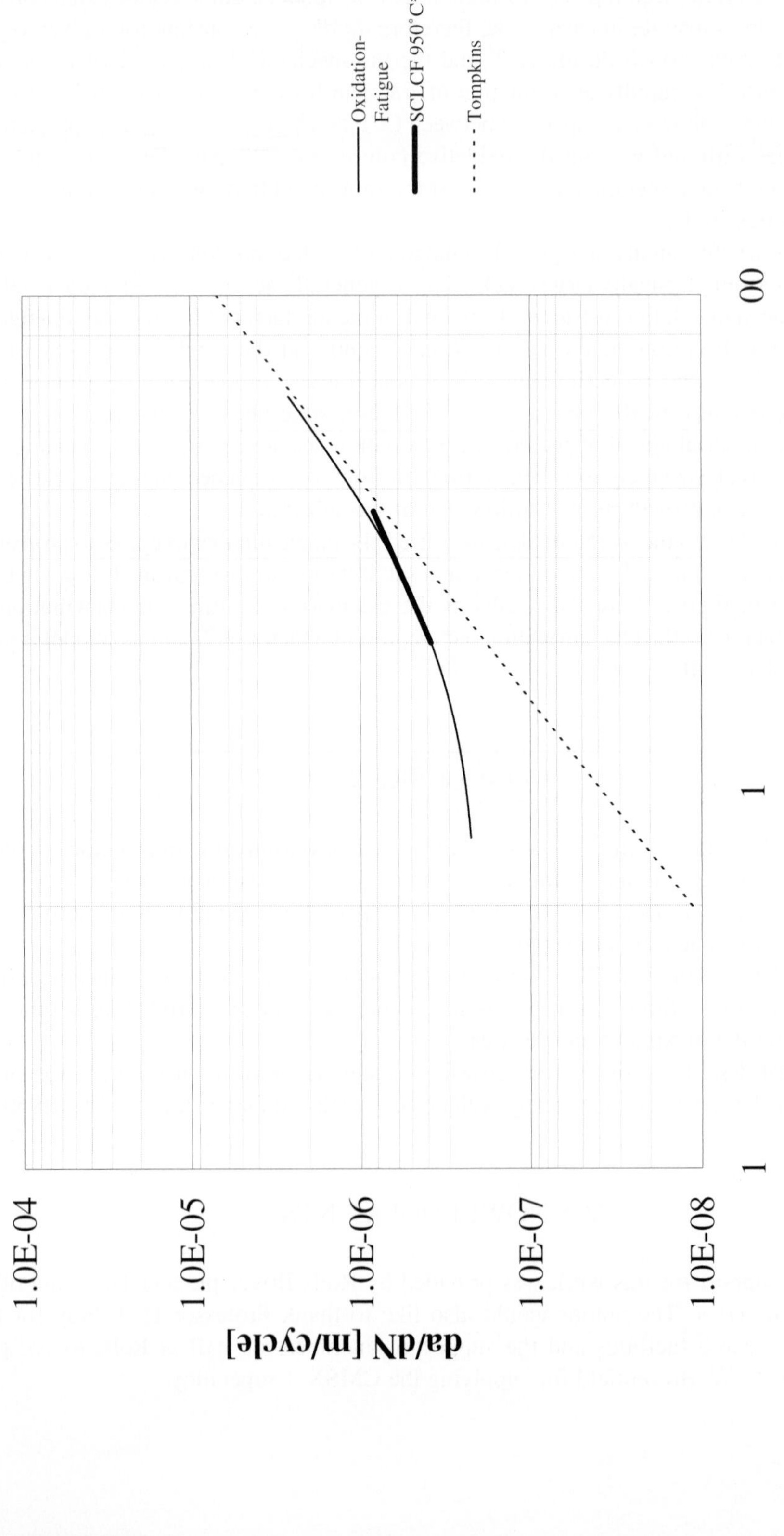

Fig. 8 Comparison of fatigue and oxidation-fatigue crack growth models with LCF crack growth data.[23]

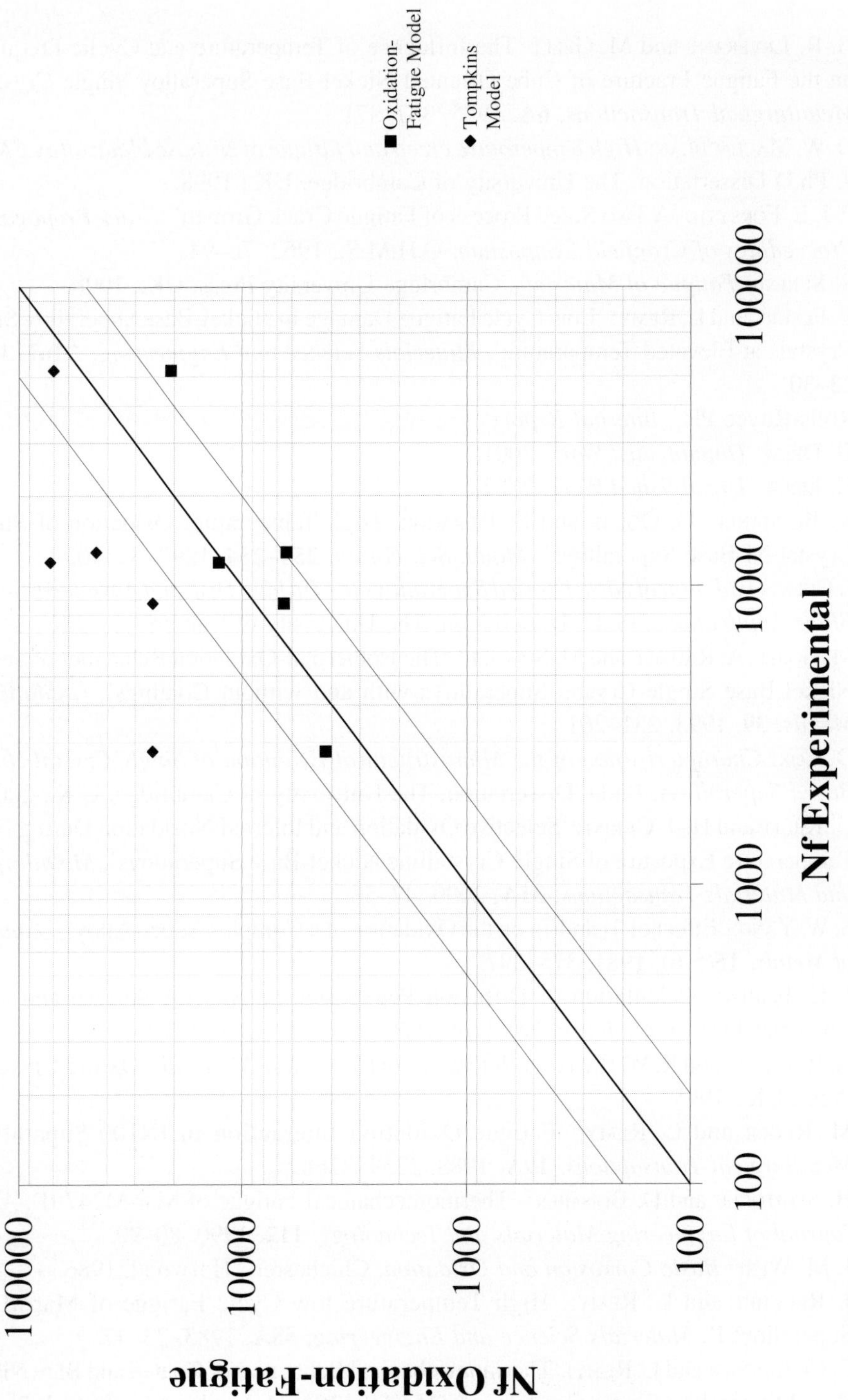

Fig. 9 Comparison of experimental and calculated lives of 950°C SCLCF tests.

REFERENCES

1. G. R. Leverant and M. Gell: 'The Influence of Temperature and Cyclic Frequency on the Fatigue Fracture of Cube Oriented Nickel-Base Superalloy Single Crystals', *Metallurgical Transactions*, **6A**, 1975, 367–371.
2. D. W. MacLachlan: *High temperature creep and fatigue of Ni-base blade alloy CMSX-4,* Ph.D Dissertation, The University of Cambridge, UK., 1998.
3. P. J. E. Forsyth: 'A Two Stage Process of Fatigue Crack Growth', *Crack Propagation: Proceedings of Cranfield Symposium*, O.H.M.S., 1962, 76–94.
4. S. Suresh: *Fatigue of Materials*, Cambridge University Press, UK., 1998.
5. E. Fleury and L. Remy: 'Low Cycle Fatigue Damage in Nickel-Base Superalloy Single Crystals at Elevated Temperature', *Materials Science and Engineering*, **A167**, 1993, 23–30.
6. Rolls-Royce Plc., *Internal Report.*
7. G. Drew: *Unpublished Work*, 2001.
8. G. Drew: *Unpublished Work*, 2002.
9. K. Bouhanek, D, Oquab and B. Pieraggi: 'High Temperature Oxidation of Single-Crystal Ni-Base Superalloys', *Mater. Sci. Forum*, **251–254**, 1997, 33–40.
10. J. Chen: *Oxidation of Nickel-Based Superalloys for Turbine Disk and Advanced Turbine Blade Applications*, Ph.D. Dissertation, The University of Cambridge, U.K., 1996.
11. M. Gobel, A. Rahmel and M. Shültze: 'The Isothermal-Oxidation Behaviour of Several Nickel-Base Single-Crystal Superalloys with and without Coatings', *Oxidation of Metals*, **39**, 1993, 231–261.
12. D. Cox: *Characterisation of the Microstructural Evolution of Single Crystal Nickel-Based Superalloys*, Ph.D. Dissertation, The University of Cambridge, U.K., 2000.
13. U. Krupp and H.-J. Christ: 'Selective Oxidation and Internal Nitridation During High-Temperature Exposure of Single-Crystalline Nickel-Base Superalloys', *Metallurgical and Materials Transactions*, **31A**, 2000, 47–56.
14. S. W. Yang: 'Effect of Ti and Ta on the Oxidation of a Complex Super-Alloy', *Oxidation of Metals,* **15**(5-6), 1981, 375–397.
15. J. C. Fisher: 'Calculation of Diffusion Penetration Curves for Surface and Grain Boundary Diffusion', *Journal of Applied Physics*, **22**(1), 1951, 74–77.
16. A. P. Sutton and R. W. Balluffi: *Interfaces in Crystalline Materials,* Oxford University Press, UK., 1995.
17. M. Reger and L. Remy: 'Fatigue Oxidation Interaction in IN100 Superalloy', *Metallurgical Transactions*, **19A**, 1988, 2259–2268.
18. H. Sehitoglu and D. Boismier: 'Thermomechanical Fatigue of Mar-M247 II', *ASME Journal of Engineering Materials and Technology*, **112**, 1990, 80–90.
19. J. M. West: *Basic Corrosion and Oxidation*, Chichester : Horwood, 1986.
20. J. Reuchet and L, Remy: 'High Temperature low Cycle Fatigue of Mar-M 509 Superalloy: I', *Materials Science and Engineering*, **58A**, 1983, 33–42.
21. E. Chataigner and L. Remy: 'Thermomechanical Behaviour of Coated and Bare Nickel-Based Superalloy Single Crystals, *ASTM STP 1263*, M. Verilli, ed., 1996, 3–26.

22. B. TOMPKINS: 'Fatigue Crack Propagation – An Analysis', *Phil. Mag.*, **18**, 1968, 1041–1066.
23. S. MÜLLER, et al: 'Influence of Load Ration, Orientation and Hold Time on Fatigue Crack Growth of CMSX-4', *Superalloys*, TMS, 2000, 347.
24. M. OKAZAKI, H. YAMADA and S. NOHMI: 'Temperature Dependence of the Intrinsic Small Fatigue Crack Growth Behaviour in Ni-Base Superalloys Based on Measurement of Crack Closure', *Metallurgical Transactions*, **27A**, 1996, 1021–1031.

Microstructural Evolution in Coated Single Crystal Nickel-Based Superalloys

I. Di Martino and R. C. Thomson
Institute of Polymer Technology and Materials Engineering
Loughborough University
Loughborough, Leicestershire
LE11 3TU, UK

ABSTRACT

This work is concerned with the evolution of the microstructure in both the parent material and the coating as a function of temperature and time in some of the latest generations of single crystal nickel-based superalloys. The coating/substrate interactions were studied by investigating the microstructural changes occurring in the interdiffusion zone under various heat treatment conditions. A variety of experimental techniques, including field emission gun scanning electron microscopy, energy dispersive x-ray analysis and elemental mapping, were used for phase characterisation. Thermodynamic equilibrium calculations were performed using a commercial software package, in conjunction with a critically assessed database for nickel-based superalloys. These can then be linked to kinetic models for microstructural evolution which could act as a time-temperature recorder, to allow the prediction of the effective operating temperature of a component and therefore an assessment of its remaining life in service.

INTRODUCTION

The major challenge faced today within the power generation and the aerospace industries is to achieve greater thermodynamic efficiency by increasing turbine inlet temperatures (TIT) and combustion gas pressures in gas turbine engines.[1, 2] Reliability and maintainability are both crucial issues which have to be addressed to reduce overall life cycle costs. Moreover, legislative regulations have to be met to minimise environmental impact by controlling NO_x, SO_x and CO_2 emissions.[2–4] The TIT depends on the maximum temperature which can be safely experienced in the hottest zones of the engines by the hot gas path components, such as blades, combustors and vanes. In such components, nickel-based alloys currently allow for peak metal temperatures of about 900–950°C to be reached within industrial gas turbines, and of over 1100°C in aeroengines.[4, 5]

Market driven requirements for better engine performance are the prime driver towards the development of more advanced superalloys able to withstand increasingly severe operating conditions, both from a thermal and a mechanical point of view, within oxidative and/or corrosive environments. Single crystal nickel-based superalloys are now extensively used for turbine blades, because they show very good high strength/high temperature properties and superior thermal stability when compared to their equiaxed and directionally solidified counterparts.

The protection of hot gas path components, such as turbine blades, against high temperature oxidation and corrosion, or a loss of mechanical properties caused by the thermally activated diffusion of deleterious species into the substrate, is provided by coatings. Both aluminide and MCrAlY coatings are used for aircraft and industrial gas turbine components. Ceramic-based thermal barrier coatings (TBC) are also being increasingly used as external protective layers on top of metallic coatings for their superior high temperature insulating properties.[6]

In addition to extending the component life, when applied on a superalloy substrate that is to be used at very high temperatures, coatings are able to hinder the direct interactions of the substrate with the surrounding environment. They all develop their protective action from the interaction/reaction with the oxygen present in the environment. As a result of this, dense and adherent oxide scales, mainly consisting of Al and Cr-rich oxides, form on the outer surface of the coatings.[1, 7, 8] The presence of such oxides slows down any further oxidative/corrosive attack on the coatings, and helps minimise the diffusion into the parent material of harmful species, such as oxygen and nitrogen. In most commercial coatings, the microstructure is characterised by γ-Ni, β-NiAl which acts as an aluminium sink, and γ'-Ni_3Al. α-Cr, σ-CoCr and other intermetallics can also be present, depending on both the coating and the substrate chemical compositions.[1, 9]

The durability of protective coatings is a crucial issue for industrial as well as aircraft gas turbines since coating life mainly controls the refurbishment and/or replacement of many engine parts, such as blades and vanes. Coating effectiveness tends to decrease as a result of high temperature exposure and thermal cycling which cause Al depletion, reducing the ability of the coating to form a continuous protective oxide layer.[1, 10] Al loss is mainly due to oxidation, spallation and coating/substrate interdiffusion phenomena, which are all related to diffusional transport of Al from the coating towards the coating/oxide interface and into the substrate.[1, 10–12] The latter are responsible for the alteration of the mechanical properties of the coated component due to the formation of new phases, often embrittling, within the so-called interdiffusion affected zone (IDZ) across the substrate/coating interface.[9]

A complete understanding of the diffusion-assisted degradation processes and the microstructural changes taking place within the coating, the IDZ and the substrate is crucial for the development of more accurate life prediction methodologies of coatings and coated superalloy systems in general. A number of diffusion-based oxidation models have been developed.[11–14] The rates of coating/substrate interdiffusion have been quantified and used to predict the operating temperature of engine parts, with the ultimate aim of calibrating finite element heat transfer models for component life and repair calculations.[15] Microstructure methodologies for the creation of remanent-life prediction models in the substrate are usually based on the change of γ' size as a function of time and temperature,[3] but other approaches have been taken, such as monitoring the changes in carbide type and composition in some polycrystalline superalloys and linking such changes to specific time and temperature conditions.[16] If the engine operating temperature is known, it can then be related to microstructural evolution allowing the remaining life of a component to be estimated.[3]

This paper focuses on the microstructural changes observed in a coated single crystal superalloy material as a function of temperature and time. Several investigations were carried out to identify all the microstructural phases and assess the interactions between coating and substrate. Thermodynamic equilibrium calculations were also performed to assist during the phase identification process and to investigate the influence of certain key diffusing elements on the microstructures at different temperatures. Compositional changes occurring within the phases being present were also monitored to assess their potential use as indicators of the materials' thermal history for future development of a microstructure-based life prediction model.

EXPERIMENTAL PROCEDURE

All of the samples investigated in this study consisted of a Cr-rich [001] oriented single crystal nickel-based alloy substrate containing C, coated with a NiCoCrAlY coating, which also contained Ta (< 5.0 wt.%). The substrate material had previously received a commercial high temperature solution heat treatment followed by ageing at a lower temperature, and then was air-cooled.

Three samples were cross-sectioned using electro-discharge machining from cylindrical bars of approximately 13 mm in diameter, on which the coating deposited was approximately 100 μm thick. The three samples were isothermally exposed in still air for 10,000 hours at 850, 950 and 1050°C, here referred to as samples A, B and C respectively. They were then removed from the laboratory furnaces and air-cooled. Metallographic samples were prepared by mounting in an edge-retaining conducting Bakelite, grinding to 2400 grit SiC paper and then polishing on 6 and 1 μm cloths impregnated with diamond paste. They were then electrolytically etched in a 10% phosphoric acid solution at 6 V DC for 5–10 s.

Detailed microstructure characterisation was carried out using a Leo 1530 VP field-emission gun scanning electron microscope (FEGSEM) operated at 20 kV, using both backscattered electron (BSE) and secondary electron (SE) imaging modes. Phase identification was aided by quantitative chemical analyses and elemental mapping performed with an EDAX Pegasus energy dispersive x-ray (EDX) facility attached to the FEGSEM, with an ultra-thin window enabling the detection of light elements. Due to the electron beam interaction volume (~1 μm), in some cases EDX analyses on sub-micron sized particles might have been affected by overlap with surrounding phases. It should also be noted that when elements such as Ta and W were both present within the analysed phase, quantitative results might have also been influenced to a limited extent by overlap of the related elemental energy spectra. In order to minimise any possible errors, quantification of the elements in question was carried out by consideration of their energy spectra for both the L and the M lines.

ThermoCalc Gibbs free-energy minimisation software, version M, with a windows interface, Ettan version 1.00, was used coupled to a specific database, known as Ni-data,[17] to enable the prediction of thermodynamic equilibrium conditions in the multicomponent nickel-based superalloy systems investigated. Yttrium was not accounted for during the calculations.

RESULTS AND DISCUSSION

Thermodynamic Equilibrium Calculations

Al and Cr have generally been regarded to date as the main diffusing elements to be investigated in order to develop models which are able to simulate the degradation of overlay coatings by simultaneous oxidation and coating-substrate interdiffusion in MCrAlY/Ni-based superalloy systems.[10–13, 18] It has been reported that the Al content of the overlay coating and substrate have a greater effect on the coating life compared to Cr. This can be explained through consideration of the diffusion coefficient D_{AlAl} of Al which is weakly affected by the Cr, since its concentration gradient has a small influence on the Al diffusion, i.e. $D_{AlAl} >> D_{AlCr}$.[12]

In this study, thermodynamic equilibrium calculations were carried out on the coating material at 850, 950 and 1050°C to investigate the effects of varying Al and Cr coating content on the predicted equilibrium phases and their relative mass fractions. Diffusion of Al and Cr was simulated by making step-wise changes to their amounts at each temperature, and setting Ni as the balancing element for all modified compositions. It should be noted that such calculations provided no information on the kinetics of diffusion. The results of calculations in which Al and Cr were varied separately are shown in Figure 1.

At 850 and 950°C the equilibrium phases predicted for both sets of modified compositions were γ-Ni, β-NiAl, γ'-Ni_3Al and σ-CoCr. The latter phase was no longer predicted to be present at equilibrium at 1050°C, as shown in Figures 1(e and f), and its proportions were greater at the lower temperature of 850°C. At all of the three temperatures, as the Al content was decreased the weight fraction of γ increased at the expense of β and σ. High Al contents indeed not only result in increased amounts of β but in most cases enhance the tendency for the precipitation of Cr-rich phases.[9] Figures 1(a, c and e) indicate that, as expected, γ became increasingly more stable as the temperature increased, the opposite trend being observed for all the other phases. The β phase was only stable at and above an Al concentration of 6 wt.% at all temperatures. At Al contents greater than 6 wt.%, at 950 and 1050°C, the β phase was predicted to be present in larger amounts than the γ' phase, as shown in figures 1(c and e). As the amount of β phase reduced, with the Al wt.% lowered from 10 to 6, γ' became more stable. However, the γ' phase itself became less stable as the Al content was further diminished from 6 to 2 wt.%.

Figures 1(b, d and f) show that lowering the Cr content caused a decrease in the equilibrium mass fractions of both β and σ, with the opposite effect being observed for γ'. This in turn would suggest that unlike γ', β is not only stabilised by Al but also by Cr. At 1050°C, for which σ phase is not predicted to be thermodynamically stable, the amount of γ decreased consistently as the Cr content was lowered.

Microstructural Characterisation

SEM micrographs are presented in Figures 2–4, which illustrate features of the oxide, coating, IDZ and substrate for samples A, B and C. EDX maps showing elemental distributions in some of these critical areas follow in Figures 5–8. Figure 2(a) shows the microstructure of

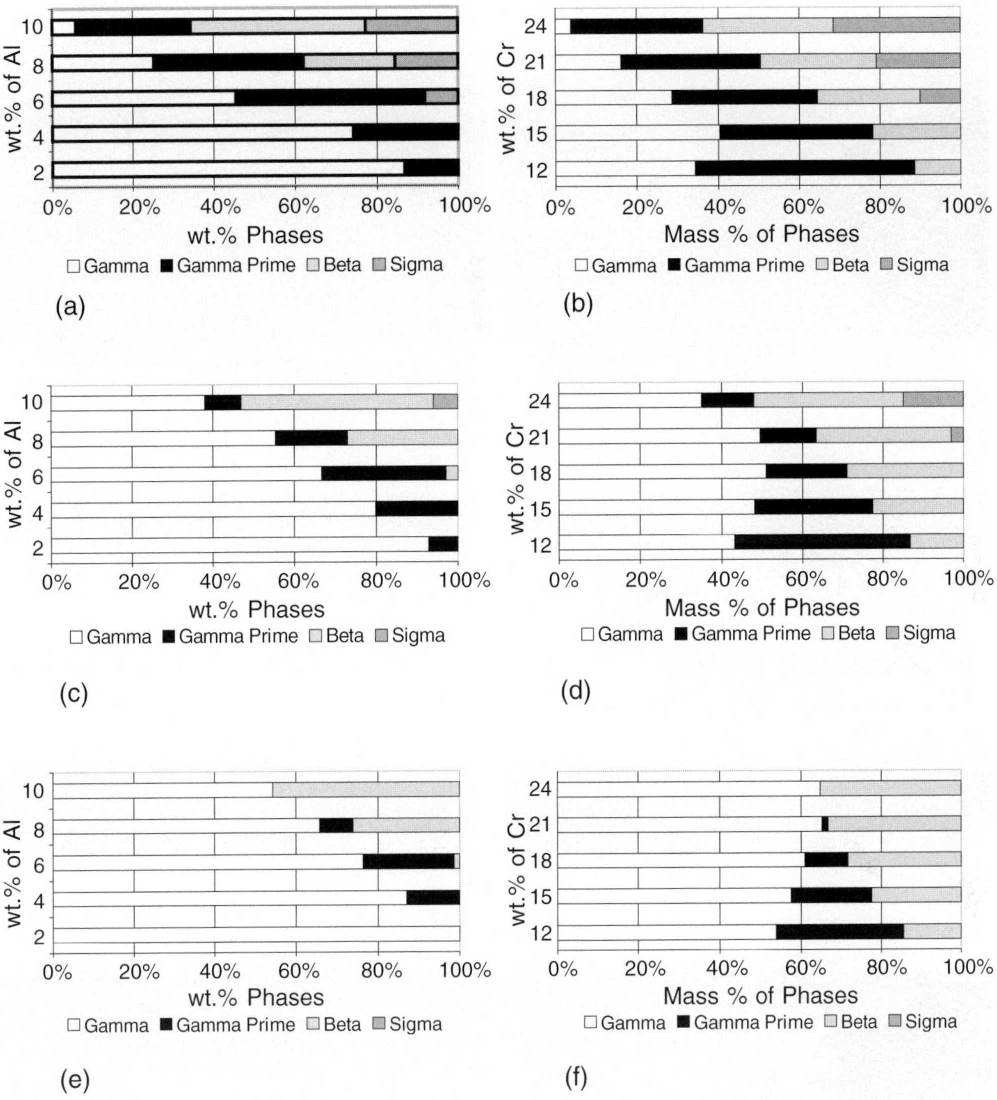

Fig. 1 Results of thermodynamic equilibrium calculations on coating sensitivity to (a) Al at 850°C, (b) Cr at 850°C, (c) Al at 950°C, (d) Cr at 950°C, (e) Al at 1050°C and (f) Cr at 1050°C, with Ni as the balancing element.

sample A (10,000 hours. at 850°C). The external scale, appearing dark in the BSE mode, mainly consisted of Al oxide, presumably Al_2O_3, and had an average thickness of approximately 6–7 μm. Fine precipitates, with a very bright appearance in the BSE mode, and containing up to 70 wt.% of Ta and Y were detected within the scale. Further, Figure 2(a) also shows the

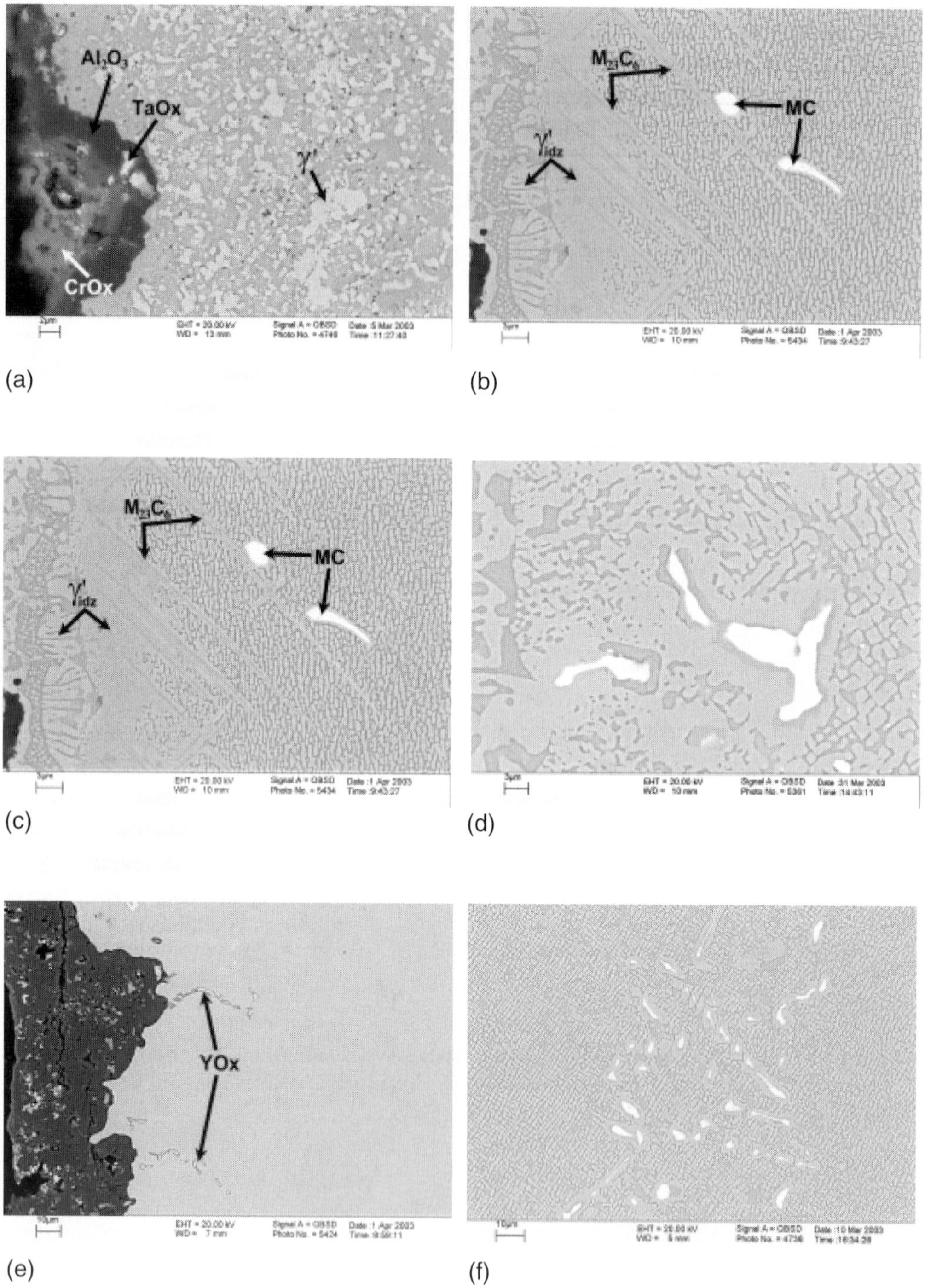

Fig. 2 SEM micrographs showing: (a) oxide (left), coating (right) in sample A, (b) coating (left), substrate (right) in IDZ in sample A, (c) and (d) coating (left), substrate (right) in IDZ in sample B, (e) oxide (left), coating (right) in sample C, and (f) substrate in sample B.

presence of an outer Cr-rich oxide layer, which contained up to 50 wt.% Cr. Its discontinuous and fragmented occurrence within the whole scale suggests that its formation might have arisen after spallation of pre-existing Al_2O_3. Ti-rich oxide particles, which also contained Ta, Al and Y, were observed at the oxide/coating interface, hence indicating that Ti diffusion from the substrate through the whole coating had taken place. In Figure 2(a) it is also possible to observe the presence of some blocky and coarse γ' precipitates (up to 5 μm in size) which were only detected in the region within approximately 20 μm from the oxide/coating interface. In the bulk of the coating most γ' particles were instead 1-2 μm in size. β phase with an average size of about 2 μm was only present within a central coating layer, hence confirming that β depletion had occurred in the vicinity of the coating/oxide interface and also towards the coating/substrate interface.

Figure 2(b) illustrates the microstructure of the IDZ across the coating/substrate interface after 10,000 hrs. exposure at 850°C. The presence of coarse and irregular γ' precipitates, resembling 'fingerprint' shapes, can be seen. Adjacent to these, a continuous γ'-enriched layer was observed, thought to have been caused by the diffusion of Al from the coating into the base alloy. Beneath the coating, γ' exhibited a significant rafting behaviour in the direction parallel to the substrate/coating interface. Such a phenomenon has also been reported elsewhere in literature.[7]

The diffusional transport of Cr from the coating into the base alloy encouraged the formation of Cr-rich phases within the IDZ. Globular particles containing up to 73 wt.% Cr and up to 9 wt.% Co precipitated on the edge of the continuous γ'-enriched layer. Due to their significantly higher Cr content and lower amounts of Co compared to those predicted for σ, these particles were identified as α-Cr, although such a phase was not indicated by the thermodynamic calculations. Cr-rich plate-like phases, with up to 55 wt.% Cr, which also contained W and Mo, were detected with a specific 45° orientation of their longitudinal axis to the orientation of the rafted γ' precipitates. Acicular phases in the vicinity of the coating/substrate interface are usually referred to as s plates, typically regarded as brittle and detrimental to the superalloy mechanical properties.[19] In this case, however, these Cr-rich needle-shaped precipitates were identified as $M_{23}C_6$ carbides, both due to the detection of C peaks during EDX analysis, and their W content (up to 11 wt.%) which was much greater than that predicted for σ. The determination of their crystal structure, in addition to the quantitative chemical analysis, would be required to conclusively identify these phases. Microstructural features similar to those described here have been encountered by previous workers who also proved the Cr-rich plates to be carbides of the type $Cr_{21}Mo_2C_6$.[20] They suggested that N from the combustion atmosphere penetrates into MC carbides located in the vicinity of the IDZ and/or within it, causing a displacement of C. The free C then reacts with Cr and Mo, which are concentrated in this area due to diffusion phenomena.[20] The presence of $M_{23}C_6$ in the IDZ in sample A demonstrated that C diffusion from the substrate towards the coating was significant, and the formation of such carbides was encouraged by the high affinity between C and Cr.[1,9] Figures 5(b) and (c) both relate to the IDZ in sample A and clearly show the presence of Cr and Mo-rich acicular phases growing into the substrate, as well as globular a Cr-rich particles.

Figure 3 shows the microstructure of sample B (10,000 hours at 950°C). The external oxide scale presented similar characteristics to those discussed above for sample A. Although no β phase was present at this temperature, the average oxide thickness was approximately

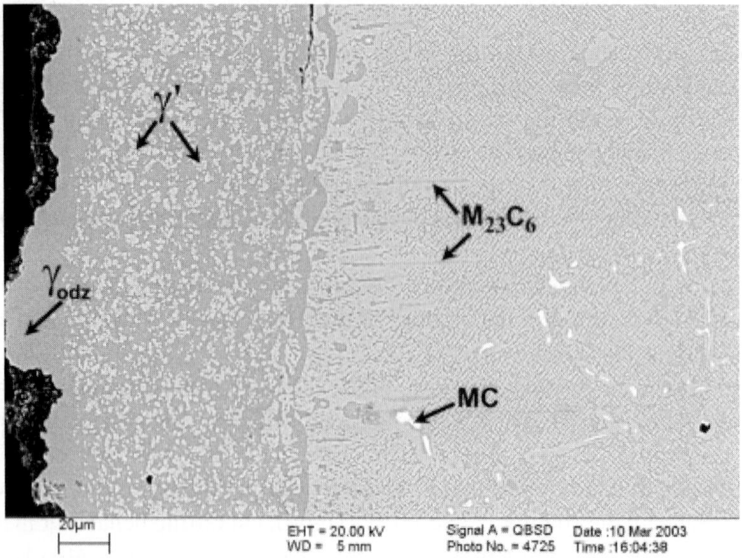

Fig. 3 SEM micrograph showing coating (left), substrate (right) in sample B.

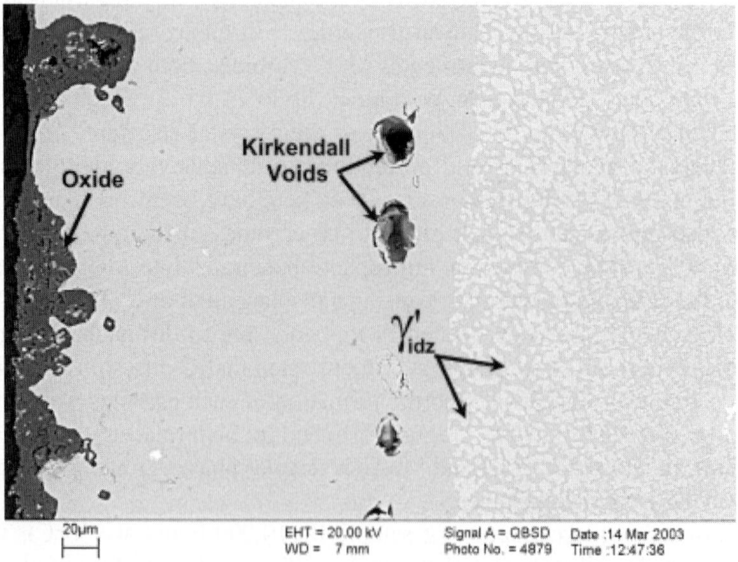

Fig. 4 SEM micrograph showing coating (left), substrate (right) in sample C.

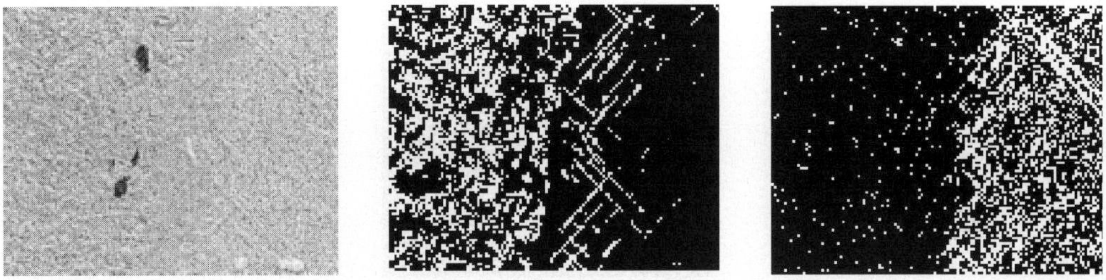

(a) (b) (c)

Fig. 5 (a) SEM micrograph of sample (A) showing coating (left)/substrate (right) interface, (b) Cr and (c) Mo elemental maps of region in (a).

10 μm, hence greater than at 850°C. This observation, coupled with the increased thickness of the oxide, confirms that although β acts as the Al-reservoir in the coating, the Al diffusional transport towards the coating surface after β dissolution is still very significant, hence enabling the oxide to grow further. This explains why in some diffusion-based models the coating life has been related to a critical Al content of the coating, rather than to the volume fraction of the β phase.[18] Particularly evident at this temperature was also the growth of a γ layer adjacent to the oxide, as shown in Figure 3. This is thought to have been a consequence of the outwards Al diffusion and the transport of Cr in the reverse direction.[11] Figures 2(c and d) both show the quite complex microstructure of the IDZ, where MC dissolving to form blocky $M_{23}C_6$ carbides can be seen. Such an observation was confirmed by the EDX elemental mapping investigations carried out across the coating/substrate interface, the results of which are shown in Figures 6(b and c). As observed at 850°C, $M_{23}C_6$ precipitates also occurred as plate-shaped in the IDZ, visible in Figure 3. At this higher temperature, however, such plates were thicker and less elongated than they were at 850°C. The γ' rafting behaviour was still clearly observable beneath the coating/substrate interface, as shown in Figure 2(c). Dissolution of MC to $M_{23}C_6$ could also be seen within the bulk of the parent material, where both phases occurred within a γ' envelope, as shown in Figure 2(f).

At 1050°C the average oxide thickness was approximately 45 μm. Precipitates rich in Ta, Cr, Y and Ti were found within the oxide scale and on its inner edges. Y-rich particles, also containing up to 10 wt.% S, were seen on the oxide/coating interface and also within 40-50 μm from it, hence well inside the bulk of the coating. This suggested that Y had diffused further away from the oxide scale, and it was also able to trap S. The major coating phase at this temperature was γ, since all γ' and β had dissolved. The presence of small Ta-Ti rich carbides was observed in the vicinity of the oxide/coating interface and towards the coating/substrate IDZ. This confirmed that Ta, Ti and C were both able to diffuse from the substrate up to the inner edge of the oxide. The diffusional transport of such elements through the coating is undesirable because Ta can form brittle intermetallic phases at the oxide/ coating interface and Ti promotes the formation of Cr_2O_3, hence preventing the protective action of the Al_2O_3 scale.[9] Figures 7 and 8 give a good indication of the diffusional behaviour of Ta, Y and Ti at 1050°C. In sample C the coating/substrate original interface (OI) could

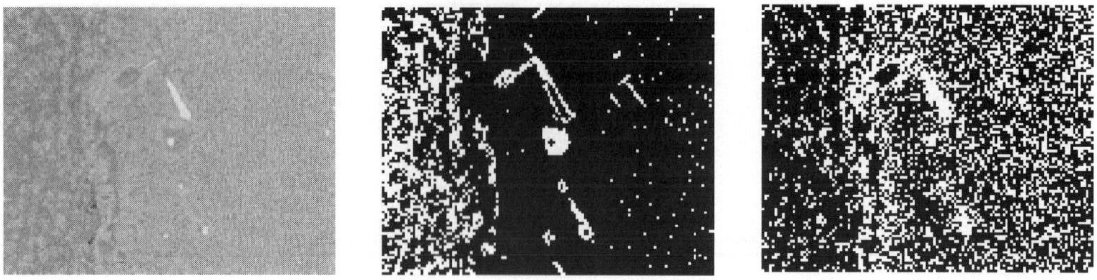

(a) (b) (c)

Fig. 6 (a) SEM micrograph of sample (B) showing coating (left)/substrate (right) interface, (b) Cr and (c) Ta elemental maps of region in (a).

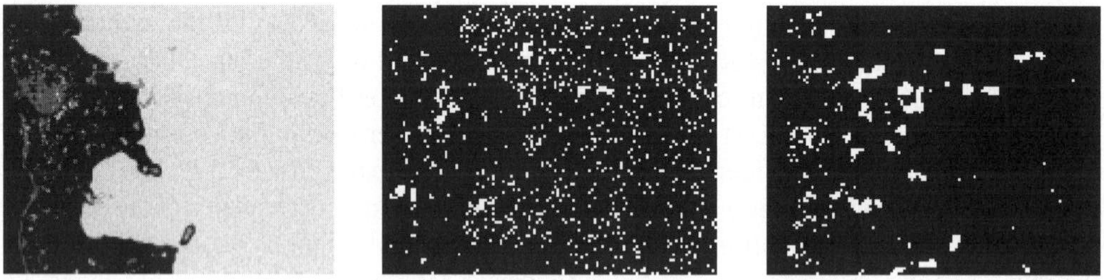

(a) (b) (c)

Fig. 7 (a) SEM micrograph of sample (C) showing oxide (left)/coating (right) interface, (b) Ta and (c) Y elemental maps of region in (a).

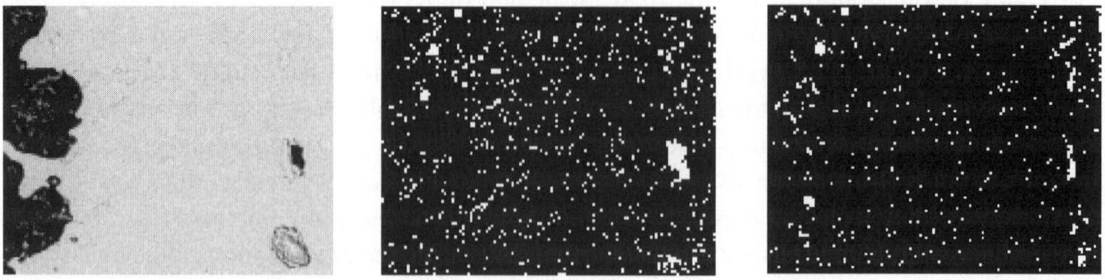

(a) (b) (c)

Fig. 8 (a) SEM micrograph of sample (C) showing oxide (left)/coating (right) interface, (b) Ta and (c) Ti elemental maps of region in (a).

Table 1 EDX results for γ'.

Temp. (°C)	Location	Elemental Composition Ranges (Wt.%)				
		Al	Ti	Ta	Cr	Co
850	γ'_{odz}	6.7–7.7	0.2–0.4	14.3–15.2	5.9–9.4	14.0–16.6
	γ'_{bulk}	7.8–10.6	0.3–0.5	7.0–12.9	6.8–10.5	15.0–18.5
	$\gamma'_{adj\ idz}$	7.0–7.6	0.7–3.0	14.0–14.6	5.1–7.0	13.0–14.8
	γ'_{idz}	7.3–7.6	1.6–2.7	9.2–11.1	3.3–4.1	12.8–13.3
	γ'_{raft}	4.3–5.2	3.3–4.4	8.4–10.0	8.1–9.2	7.1–8.3
	γ'_{sub}	4.2–4.6	5.1–5.2	11.0–11.9	6.4–6.9	6.3–6.6
950	γ'_{odz}	6.2–6.7	1.1–1.5	10.9–13.9	5.0–9.5	11.2–14.6
	γ'_{bulk}	7.0–7.1	1.3–1.4	11.4–11.6	4.9–6.4	11.6–12.9
	$\gamma'_{adj\ idz}$	6.8–7.0	1.4–1.5	13.8–14.1	4.4–4.7	11.7–11.8
	γ'_{idz}	6.7–7.0	1.8–2.4	11.0–12.1	3.5–3.9	11.2–11.6
	γ'_{raft}	5.5–5.7	4.5–4.7	8.8–9.2	3.2–4.5	9.0–9.4
	γ'_{sub}	4.4–4.6	5.0–5.3	11.7–12.0	5.0–5.2	5.9–6.1
1050	γ'_{idz}	4.7–4.8	5.6–5.8	13.1–13.7	7.3–7.8	2.7–2.8
	γ'_{sub}	4.5–5.0	6.0–6.1	11.2–12.9	3.1–3.3	5.9–6.1

odz : within 20-30 μm from oxide/coating interface
bulk : within coating bulk
adj idz : within 15-20 μm from IDZ
idz : within IDZ
raft : within rafting zone
sub : within substrate

clearly be identified by the very significant presence of Kirkendall voids, as shown in Figures 4 and 8. It is indeed reported that Kirkendall porosity generally develops at the interface between the coating and the substrate.[20] In the vicinity of the OI, γ' precipitates characterised by irregular and very coarse morphology were observed, as opposed to those encountered in the substrate. This suggested that agglomeration of this phase might have occurred to a certain extent within the IDZ.

Tables 1 and 2 respectively summarise some of the most relevant results from the EDX analyses performed on γ' and γ in samples A, B and C. The tendency of Al to diffuse towards the oxide/coating interface and towards the substrate was confirmed by these results. Indeed, the Al content in γ'_{odz} was lower than that of γ'_{bulk}, and γ'_{idz} appeared to be enriched in Al when compared to the substrate. The data relating to Cr, particularly those in Table 2, indicated that this element was diffusing towards the parent material, as often assumed in some oxidation diffusion-based models.[12, 18] Both tables also show that important diffusion phenomena and compositional changes in γ and γ' occurred even from these early stages not only for Al and Cr, but most interestingly also for Ta, Ti, W and Mo. Diffusional transport of Ti, Mo and W

Table 2 EDX results for γ.

Temp. (°C)	Location	Elemental Composition Ranges (wt.%)				
		Cr	Co	Ta	W	Mo
850	γ_{odz}	27.7–28.3	33.5–34.8	1.9–2.4	0	0
	γ_{bulk}	22.5–24.0	30.3–30.8	3.9–4.7	0	0.2 – 0.4
950	γ_{odz}	18.2–22.4	22.3–25.0	0.3–0.6	0	1.0 – 1.1
	γ_{bulk}	22.0–22.2	24.8–25.0	4.1–4.4	0.2 – 1.0	0.8 – 0.9
	γ_{idz}	21.1–21.3	22.3–23.1	3.7–3.8	2.4 – 3.0	1.5 – 1.6
	γ_{sub}	14.0–19.2	10.2–12.3	5.1–6.7	4.1 – 5.7	2.4 – 3.0
1050	γ_{bulk}	17.7–18.1	16.4–18.7	6.7–6.9	0 – 2.2	1.4 – 2.0
	γ_{idz}	16.3–17.1	13.5–14.1	6.2–6.6	3.0 – 3.8	1.8 – 2.0
	γ_{sub}	14.0–15.3	9.9–1.6	5.9–7.5	3.5 – 3.8	2.3 – 2.4

odz : within 20-30 μm from oxide/coating interface
bulk : within coating bulk
adj idz : within 15-20 μm from IDZ
idz : within IDZ
raft : within rafting zone
sub : within substrate

appeared more significant as temperature increased. The behaviour of Ta was more complex in that, as previously mentioned, it appeared to be diffusing both towards the region adjacent to the oxide scale and the one immediately preceding the IDZ. All concentration gradients appeared less pronounced at the higher temperatures, since diffusion distances had increased and diffusion profiles extended well into the substrate. This would ultimately lead to elements such as Al and Cr being supplied to the thickening oxide from the substrate, hence resulting in diffusional transport in the reverse direction compared with the early stages of the diffusion process.

CONCLUSIONS

Significant microstructural transformations were observed as a result of isothermal exposure in all of the coated superalloy systems investigated. As expected, both the oxide thickness and the β depleted regions increased with increasing temperature of exposure. Precipitates rich in Y, Ti and Ta were observed in the oxide scale, in addition to alumina and chromia. Chemically-driven diffusion mechanisms involved all major alloying elements, and led to important interdiffusion phenomena across the substrate/coating interface (e.g. γ' rafting, the formation of both Cr-rich plates, growing into the substrate, and globular precipitates). Additionally, MC carbides were observed to transform into $M_{23}C_6$ carbides at the lower temperatures, both in the substrate and IDZ. Thermodynamic equilibrium calculations were able to provide a useful insight into the consequences of diffusional transport for key elements. Changes in chemical compositions of γ and γ' phases moving from the coating into the substrate were also detected and could potentially be used as tracking parameters for microstructure-based life-prediction models. In turn, the accuracy of diffusion-based models could be enhanced by accounting for the diffusional behaviour of elements such as Ta and Ti, which also appear to play a significant role in addition to Al and Cr.

ACKNOWLEDGEMENTS

The authors would like to thank Innogy and Loughborough University for providing funding for this work, and for useful discussions.

REFERENCES

1. C. T. Sims, N. S. Stoloff and W. C. Hagel: *Superalloys II - High Temperature Materials for Aerospace and Industrial Power*, John Wiley & Sons, 1997.
2. J. E. Oakey, D. H. Allen and M. Staubli: 'Power Generation in the 21st Century - The New European COST Action', *Proceedings of the 5th International Charles Parsons Turbine Conference,* A. Strang, W. M. Banks, R. D. Conroy, G. M. McColvin, J. C. Neal and S. Simpson, eds., IOM Communications Ltd., UK., 2000, 642–657.
3. F. C. Price, W. R. Stiles and C. Singh: 'Materials for Marine, Light Industrial and Aeroderivative Gas Turbines', *Proceedings of the 5th International Charles Parsons Turbine Conference,* A. Strang, W. M. Banks, R. D. Conroy, G. M. McColvin, J. C. Neal and S. Simpson, eds., IOM Communications Ltd., UK., 2000, 631–642.
4. M. R. Winstone, A. Partridge and J. W. Brooks: 'The Contribution of Advanced Materials to Future Aeroengines', *Proceedings of the 5th International Charles Parsons Turbine Conference,* A. Strang, W. M. Banks, R. D. Conroy, G. M. McColvin, J. C. Neal and S. Simpson, eds., IOM Communications Ltd., UK., 2000, 779–797.
5. J. T. DeMasi-Marcin and D .K. Gupta: 'Protective Coatings in the Gas Turbine Engine', *Surface and Coatings Technology*, **68**, 1994, 1–9.
6. Y. Itoh and M. Tamura: 'Reaction Diffusion Behaviours for Interface Between Ni-Based Super Alloys and Vacuum Plasma Sprayed MCrAlY Coatings', *Transactions of the ASME,* **121**, 1999, 476–483.
7. A. Sanz, L. Llanes, J. P. Bernadou, M. Anglada and M. B. Rapaccini: 'Influence of the Stress State on the Diffusion Phenomena Across the Interface Between a Protective Coating and a Single-Crystal Superalloy', *Proceedings of Elevated Temperature Coatings: Science and Technology II,* N. B. Dahotre and J. M. Hampikian, eds., TMS, 1996, 373–388.
8. M. M. Morra, R. D. Sisson and R. R. Biederman: 'A Microstructural Study of MCrAlY Coatings', *Surface Engineering,* **85**, 1984, 482–495.

9. W. J. QUADAKKERS: 'High Temperature Coatings - Failure Mechanisms and Future Requirements', *Advanced Coatings for High Temperatures, Turbine Forum*, 2002.
10. K. S. CHAN, N. SASTRY CHERUVU and G. R. LEVERANT: 'Coating Life Prediction for Combustion Turbine Blades', *Journal of Engineering for Gas Turbines and Power-Transactions of the ASME*, **121**(3), 1999, 484–488.
11. J. A. NESBITT, B. H. PILSNER, L. A. CAROL and R. W. HECKEL: 'Cyclic Oxidation Behaviour of $\beta + \gamma$ Overlay Coatings on γ and $\gamma + \gamma'$ Alloys', *Proceedings of the International Symposium on Superalloys*, M. Gell, C. S. Kortovich, R. H. Bricknell, W. B. Kent and J. F. Radavich, eds., TMS, 1984, 699–710.
12. J. A. NESBITT and R. W. HECKEL, 'Modeling Degradation and Failure of Ni-Cr-Al Overlay Coatings', *Thin Solid Films*, **119**, 1984, 281–290.
13. E. Y. LEE, D. M. CHARTIER, R. R. BIEDERMAN and R. D. SISSON: 'Modelling the Microstructural Evolution and Degradation of M-Cr-Al-Y Coatings During High Temperature Oxidation', *Surface and Coatings Technology*, **32**, 1987, 19–39.
14. I. G. WRIGHT, B. A. PINT, L. M. HALL and P. F. TORTORELLI: 'Oxidation Lifetimes: Experimental Results and Modelling', *Proceedings of an EFC Workshop on Lifetime Modelling of High Temperature Corrosion Processes*, M. Scutze, W. J. Quadakkers and J. R. Nicholls, eds., Maney Publishing, UK., 2001, 339–358.
15. K. A. ELLISON, J. A. DALEO and D. H. BOONE: 'Microstructural Evolution of MCrAlY/ Superalloy Interdiffusion Zones', *Proceedings of Life Assessment of Hot Section Gas Turbine Components*, R.Townsend, ed., IOM Communications Ltd., UK., 2000, 311–326.
16. M. J. STARINK and R. C. THOMSON: 'Modelling Microstructural Evolution in Conventionally Cast Ni-based Superalloys During High Temperature Service', *Modelling of Microstructure Evolution in Creep Resistant Materials*, A. Strang and M. McLean ed., IOM Communications Ltd., UK., 1999, 357–371.
17. N. SAUNDERS: 'Phase Diagrams Calculations for Ni-based Superalloys', *Proceedings of Superalloys 1996*, R. Kissinger D. J. Deye, D. L. Anton, A. D. Cetel, M. V. Nathal, T. M. Pollock and D. A. Woodford, eds., 1996, 101–110.
18. J. A. NESBITT: 'COSIM - A Finite-Difference Computer Model to Predict Ternary Concentration Profiles Associated with Oxidation and Interdiffusion of Overlay-Coated Substrates', *NASA Report No. NASA/TM-2000-209271*, 2000.
19. A. STRANG: *Proceedings of the 13th International Congress on Combustion Engines*, CIMAG ed., 1979.
20. L. PEICHL and G. JOHNER, 'High-Temperature Behaviour of Different Coatings in High-Performance Gas Turbines and in Laboratory Tests', *Journal of Vacuum Science and Technology A*, **4**(6), 1986, 2583–2592.

Gamma Titanium Aluminide, TNB

WAYNE E. VOICE
Rolls-Royce, Elt-38
P.O.Box 31, Derby
DE24 8BJ, UK

MICHAEL HENDERSON
Alstom Power, Cambridge Road
Whetstone, Leicester
LE8 6LH

EDWARD F. J. SHELTON
QinetiQ, Cody Technological Park
Farnborough
GU14 0LX, UK

ABSTRACT

Along with other aero-engine manufacturers, Rolls-Royce and Alstom have been developing gamma titanium aluminide (γ-TiAl) for over a decade and it is now time to cash in on this investment. The favourable properties of γ-TiAl are becoming increasingly requisite to future designs and it is necessary to deliver a suitable alloy through an established component processing route and to provide suitable data to give confidence in its use. Because gamma is a new class of engineering material, the amount of information required is considerably more than for conventional alloy development and there has been much to discover about its unique characteristics. The companies have decided to concentrate on a wrought high niobium-containing alloy called TNB. The high strength and high temperature capability of this alloy make it suitable for the initial target applications, HP compressor and LP turbine blades, but impact and defect sensitivity relative to fatigue properties are also key properties and these will be discussed below.

INTRODUCTION

This summarised assessment of the new wrought alloy called TNB was carried out as part of the DTI-CARAD funded programme 'Advanced life assessment methods for γ-titanium aluminide alloys' involving Rolls-Royce, Alstom-Power and QinetiQ. The aim of the project was to increase the UK's competitiveness in the civil aviation gas turbine business by testing and evaluating γ-TiAl (gamma titanium aluminide) alloys for low-pressure turbine and compressor blading applications in aero-engines. This project complements previous work conducted by each of the partners to provide an approach for design and life assessment of γ-TiAl components.

γ-TiAl alloys are a series of intermetallic compounds having a typical composition of titanium and 45–47 at.% aluminium plus additions such as manganese and niobium for ductility and high temperature capability. The beneficial characteristics of these alloys can be summarised as follows:

- Low density (4g/cm^3),
- High stiffness (E =175 GPa at 20°C to 150 GPa at 700°C),
- Density normalised strength similar to cast Ni-based alloys,
- High temperature strength and oxidation resistant to 750°C and
- Low thermal expansion coefficient and high thermal conductivity.

However the alloys are known to possess a number of limiting properties that have continued to restrict their wider application within the gas turbine engine industry. These are perceived as the following:

- Low ductility at low to intermediate temperatures (<2% at room temperature),
- Low fracture toughness (12 MPa √m at 20°C, 25 MPa √m at 650°C) and
- High fatigue crack growth rates leading to poor damage tolerance.

The key benefits offered by γ-TiAl technology for the designers of GT engines are:

- Rotating components at moderate to high temperatures (LPT blades),
- Density reduced centrifugal (CF) loads (reduced disc stresses) and
- High specific stiffness that increases the natural frequencies of compressor and turbine blades.

Mechanical property databases have been previously established for a number of γ-TiAl alloys together with mechanical models to describe their deformation and fracture behaviour. This has demonstrated that it is necessary to assess the defect sensitivity of the alloys and develop methods of damage summation for complex loading situations. The aim of the programme was to address these issues for a range of γ-TiAl alloys and the findings of the most promising alloy, TNB are presented here:

MATERIAL

TNB alloy is a wrought, high niobium-containing alloy (8%Nb, 45%Al and 0.2%C) intended for use in the duplex condition. The extruded material to be investigated was found to have a variable microstructure in the form bands of larger equiaxed g grains drawn out in the working direction amongst a predominantly fine duplex grain structure, see Figure 1. This would be a worst-case situation and, in practice, the additional forging operation would break up the larger grains to give a completely fine grain structure. TNB will be compared to the Ti-45Al-2Nb-2Mn-XD alloy, cast and HIP'ed by Howmet Corporation (XD copyright). Ti-4522XD generally has a near-lamellar grain structure and blocky boride particles but this can vary across the cast section with equi-axed gamma in central regions (HIP closed porosity) and long ribbon, blocky then needle boride particles depending on distance from the surface (cooling rate dependent), Figure 2.

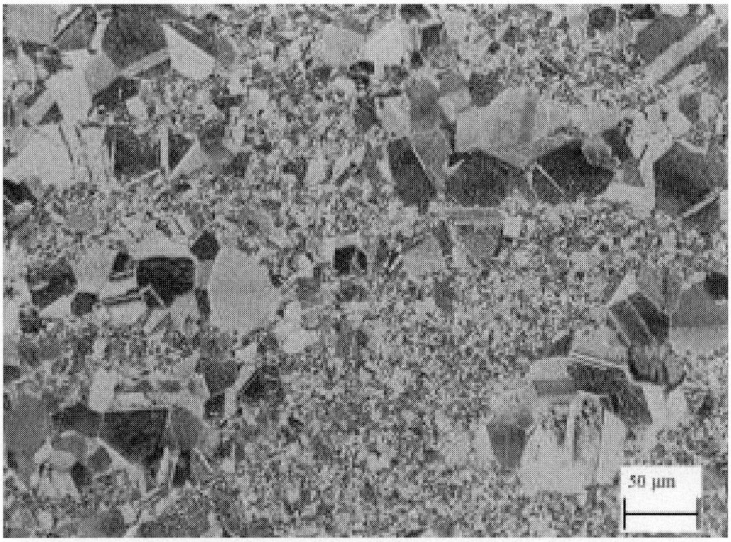

Fig. 1 Extruded TNB showing a variable microstructure with bands of large equiaxed γ grains drawn out in the working direction amongst a predominantly fine duplex grain structure.

DEFECT SENSITIVITY TESTING

This work was aimed at establishing and validating a series of fatigue failure limit diagrams, commonly known as Kitigawa plots, shown schematically in Figure 3,[1] that show fatigue strength as a function of defect or short crack size.

Baseline fatigue data was generated using a specifically designed defect sensitivity test specimen, Figure 4 which contains eight, short machined-in cracks designed to sample across the whole range of grain size and lamella colony orientation with respect to the crack plane. Tests conducted on Ti-4522XD and TNB alloy used specimens having a range of small crack sizes: from 0.05 to 0.25 mm (produced by grinding) and 0.01 to 0.04 mm (produced by electro-discharge machining).

A series of load controlled fatigue tests were conducted on Ti-4522XD and TNB defect specimens at 20°C and elevated temperature at R = 0.1. Incremental fatigue stresses were determined by applying an initial relatively low stress level and if the specimen remained intact after accumulating 10^7 cycles the load was automatically adjusted to provide an increase of 25 MPa. The ultimate fatigue strength was defined as the peak stress level to induce complete rupture of the specimen within a block of 10^7 cycles. This can be justified for γ-TiAl due to its characteristically flat S-N curve. Note that TNB was tested at the 50°C higher temperature of 700°C because of its better temperature capability.

Figure 5 shows a comparison of the fitted Kitigawa plot curves for Ti-4522XD and TNB. The fatigue strength for both alloys at elevated temperature is similar to that at room

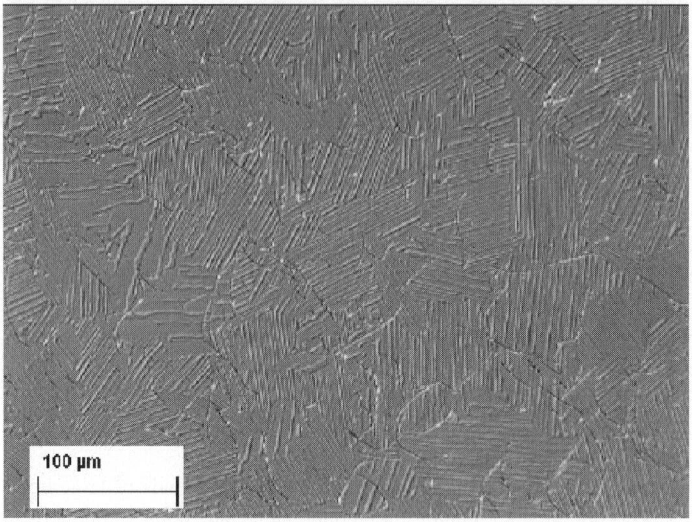

(a)

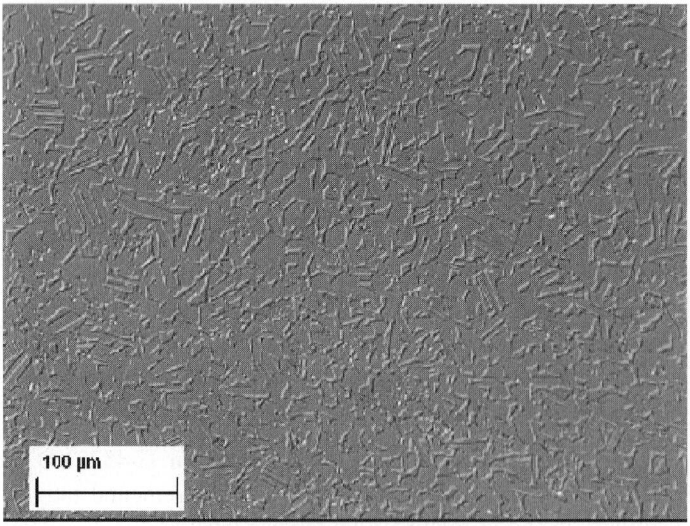

(b)

Fig. 2 Ti-4522XD Variable Microstructure: (a) near lamellar with long boride particles at mid-radius of cylindrical bar and (b) Fully equi-axed region at centre of cylindrical bar due to closing of porosity by HIP'ing.

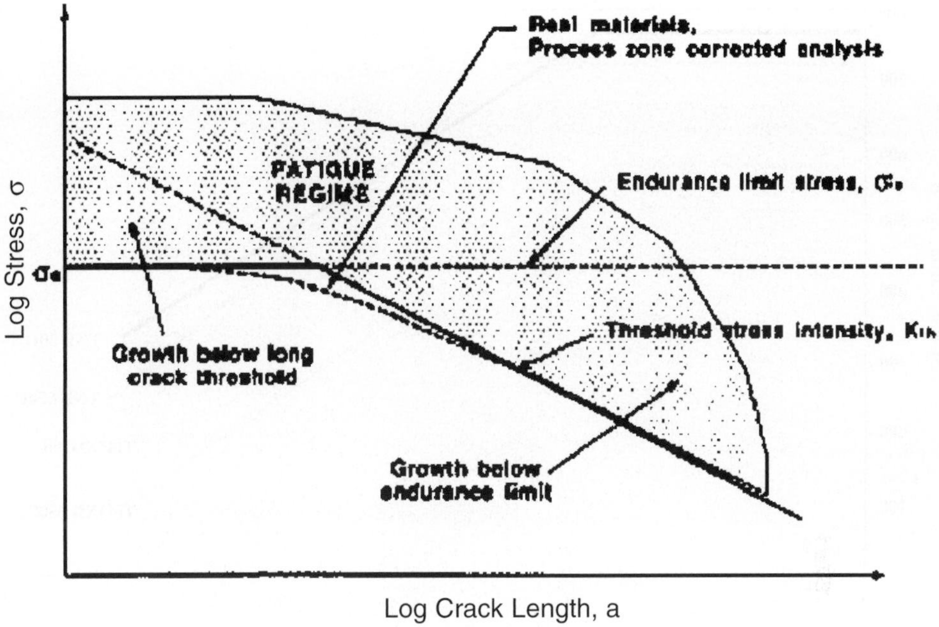

Fig. 3 Schematic representation of the Kitagawa diagram to model fatigue failures in pristine and damaged materials.

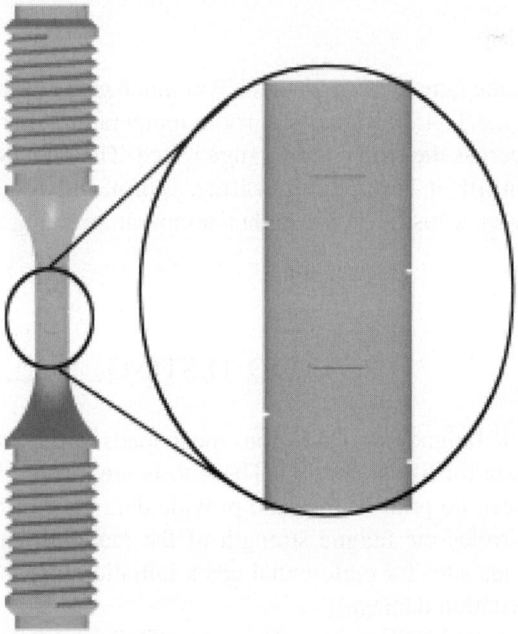

Fig. 4 γ-TiAl defect sensitivity specimen for Kitigawa short crack testing (5 mm gauge dia).

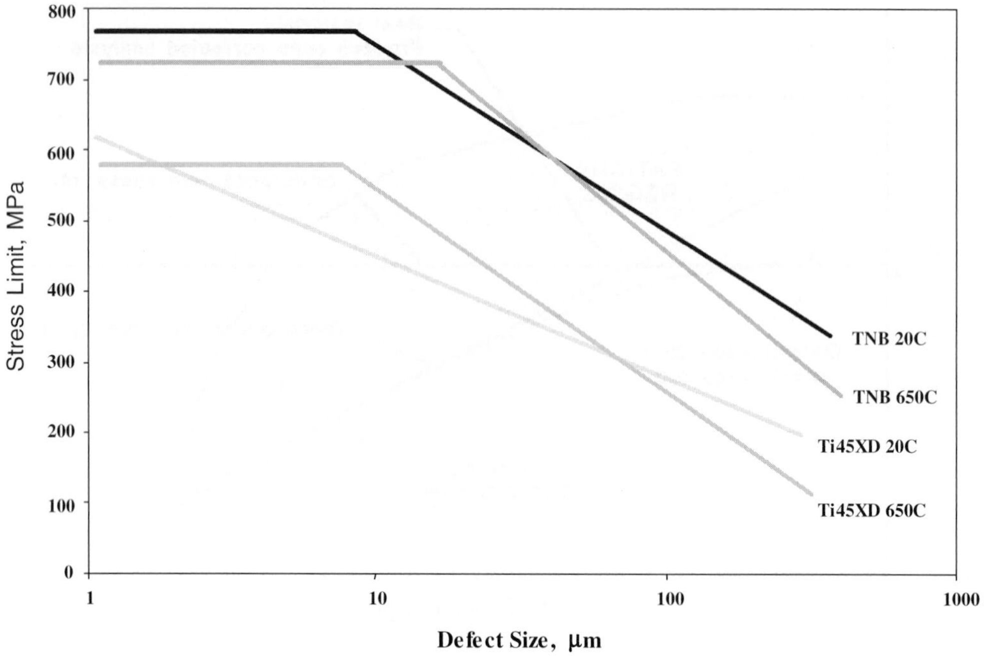

Fig. 5 Comparison of fatigue strengths against defect size for Ti-4522XD and TNB at ambient and elevated temperatures, R = 0.1 (failure <1 × 10^7 cycles).

temperature, however, the fatigue strength of TNB is much greater than Ti-4522XD by about 150 MPa. Apart from the Ti-4522XD tests at room temperature, there is an apparent defect threshold of ~10 µm across the temperature range tested. The effect of defect size is greater on the fatigue strength at room temperature compared to that at 650°C in the Ti-4522XD, probably as a result of the higher temperature being close to the ductile to brittle transition.

PRE-LOAD TESTING

Conventional high cycle fatigue tests (at various mean loads) are used to simulate high cycle fatigue and provide data for life prediction. These tests are invariably performed on virgin specimens that have seen no prior history and provide data for the 'first loading'. However in subsequent major cycles the fatigue strength of the material may be reduced by prior creep, which can provide sites for preferential crack initiation (particularly in materials that are prone to creep cavitation damage).

Sequential creep tests and HCF were performed on TNB alloy specimens; creep exposure was conducted at 700°C over 100 hours to differing degrees of creep strain, i.e. 0.5, 3 and

5%, followed by incremental fatigue testing. The Creep-HCF test results in Figure 6 show little influence of the pre-creep level on subsequent fatigue properties to a level of 3% inelastic strain. Beyond this there was a reduction in creep resistance that probably corresponds to the formation of internal voids. It was also found that pre-LCF exposure at 700°C did not affect HCF relative to the low strain creep. However there does appear to be an underlying effect of thermal exposure at 700°C as the limiting HCF stress was reduced after 24 hours and still further after 150 hours. This is attributed to corrosion pitting which is found on the surface of most test-pieces tested at elevated temperatures, see Figure 7, and may be the result of surface salt deposits from such as finger marks or machining fluids.

These results suggest that the total strain is composed of separate, essentially non-interacting, elastic, plastic and creep parts. In terms of life prediction of combined creep-HCF or combined creep-plasticity histories, the creep, HCF or plasticity damage can be summed separately. Failure occurs when life is exhausted for one of these types of loading. The life prediction methods developed separately to predict creep, fatigue and tensile behaviour may be used efficiently to predict the life of a component subjected to complex load sequences.

HCF VIBRATION TESTING

HCF vibration (flap) testing has provided an in-depth evaluation of the fatigue performance of sub-element specimens of Ti-4522XD, TNB and the nickel alloy IN718 and provided a method of assessing the susceptibility of these materials to foreign object damage (FOD) impact. The simulated blade shape specimen, shown in Figure 8, has been specifically designed to give a direct performance comparison between the different materials. It is designed to fail half way along the leading edge when excited in the second flap mode. The central highest stress region is wide enough for easy targeting with the edge of an impacting 3 mm cube to simulate in-service foreign object damage (FOD). QinetiQ carried out ballistic impacts at impact energies of between 4 and 13 Joules i.e. velocities of up to 300 ms^{-1}. Impacts were based around 0.75 mm deep edge damage as this represents the limit of detection during boroscope inspection. Nominal 0.4 and 2.5 mm damage levels were also selected to give a wider range for study.

The incremental AF (amplitude × frequency) strength levels (for failure in 10^7 cycles) for Ti-4522XD, duplex TNB and IN718 are shown in Figure 9 where an AF value above 1.1 is generally considered to be a good result. Undamaged γ-TiAl specimens show a large scatter in AF strength compared with those for IN718 (AF ranges of 1 compared to 0.2) such that IN718 and the Ti-4522XD have a similar average AF of 1.35, but the scatter of Ti-4522XD makes it unacceptable for use as a compressor blade. This is attributed to the variable microstructure across the HIP'ed cast section particularly in terms of differences in boride particle distribution and proportion of lamellar to equi-axed gamma grains, Figure 2. The TNB alloy gives an average AF strength above 2 with no results below 1.5 so that it actually out performs the currently used IN718. In addition, the scatter in results for TNB should be significantly reduced in the forged product as the banded large grain structure would be refined to increase the usable minimum AF strength.

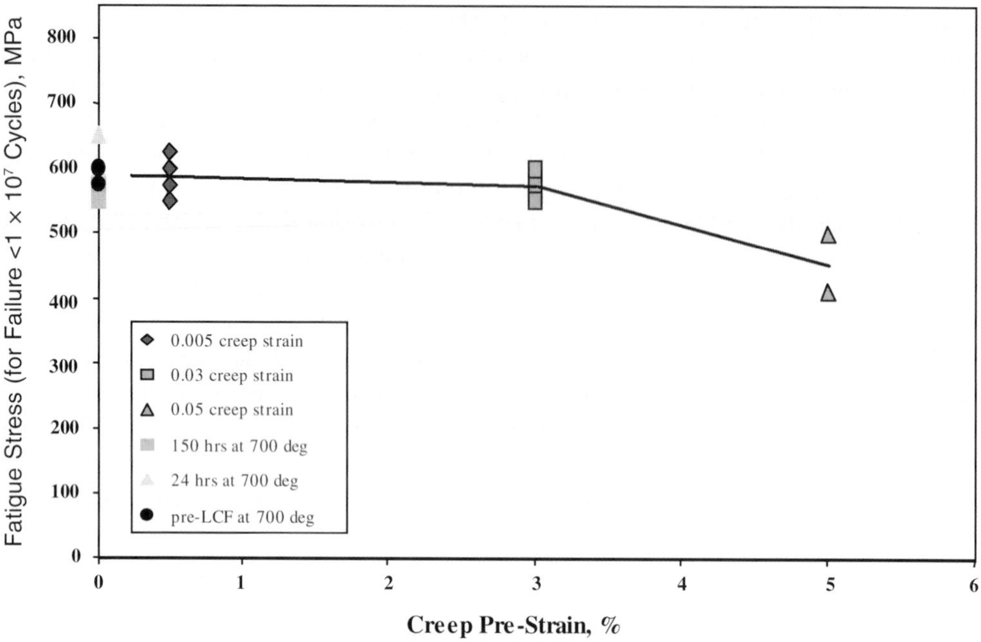

Fig. 6 Pre-Load Testing: Incremental HCF test results for wrought TNB at 700°C shown as a function of pre-creep strain level. Also shown are pre-LCF and thermally exposed results.

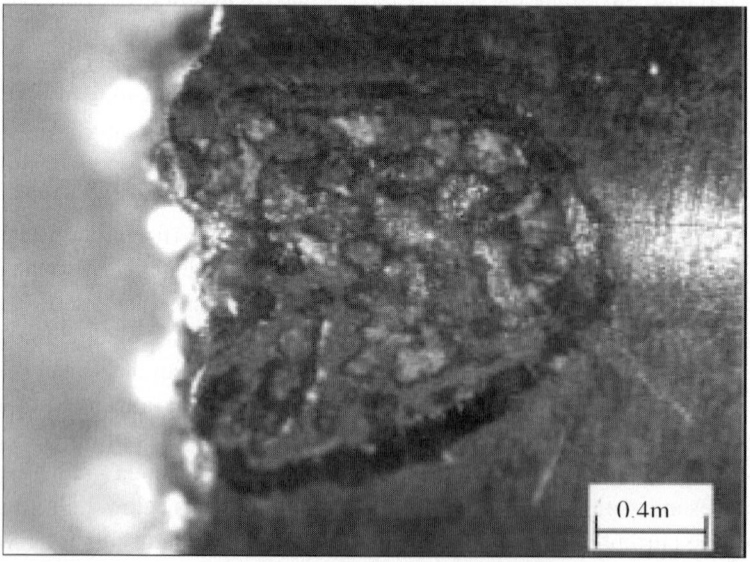

Fig. 7 Corrosion pitting on surface of TNB elevated temperature test-pieces (650°C).

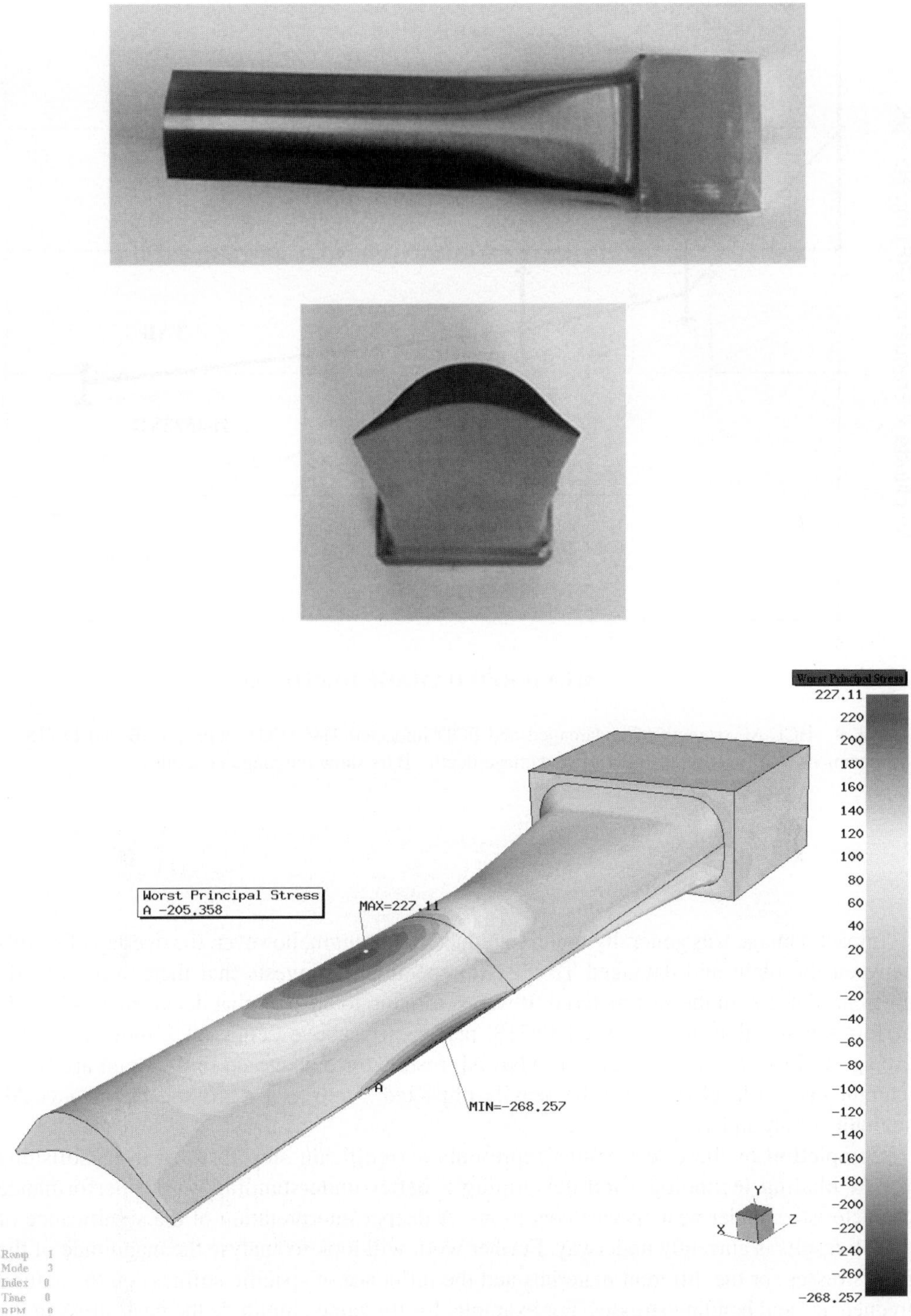

Fig. 8 Specifically designed flap-HCF simulated blade test-piece and stress distribution during second order flap mode.

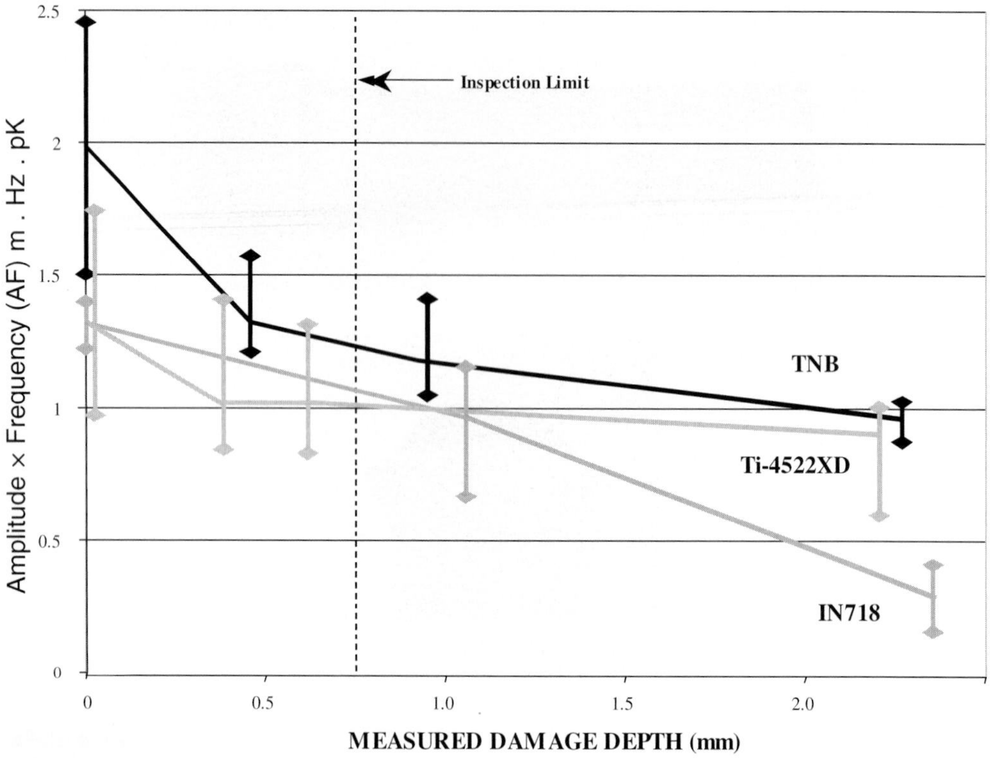

Fig. 9 HCF AF strength of undamaged and FOD impacted Ti4522XD, duplex TNB and IN718 specimens as a function of impact site damage depth. Bars show the range of results.

Impact damage was generally found to reduce AF strength, however, the overlap of results between the plain and damaged Ti-4522XD specimens suggests that there were already inherent 'defects' in the cast material. It was surprising to observe that damaged Ti-4522XD performs better than the damaged IN718, particularly above depths of 2 mm, where the average IN718 AF was just 0.3. The TNB AF results for damage up to 0.75 mm are better than the critical level and even the heavily impacted 2 mm deep TNB specimens gave AF strengths of around 1.

Completion of these test results represents a significant step forward in establishing γ-TiAl blading technology and developing a better understanding of the performance characteristics under near service conditions. A deeper interpretation of the significance of the AF results is currently underway. Further work will look to analyse the magnitude of the peak stresses for the different materials and the influence of specific stiffness on the natural frequencies and bending stresses. For example, for the same amplitude the peak stress in the IN718 specimens is 50% greater than for γ-TiAl, however, the TNB sustains a higher stress at the failure AF and tolerates a much larger displacement.

and the existing blade alloy IN718. The cast Ti-4522XD was shown to be unsuitable because of excessive scatter in properties probably because of variable microstructural features.

REFERENCES

1. H. KITAGAWA and S. TAKAHASHI: 'Applicability of Fracture Mechanics to Very Small Cracks or Cracks in the Early Stage', *Proceedings of the Second International Conference on Mechanical Behaviour of Metals*, American Society of Metals, Boston, 1976, 627–631.

Improving High Temperature Performance in ODS Alloys for Application in Advanced Power Plant

J. RITHERDON and A. R. JONES
Department of Engineering
University of Liverpool
Liverpool, L69 3GH
UK

U. MILLER
Plansee GmbH/Lechbruck
D-86983 Lechbruck
Germany

I. G. WRIGHT
Oak Ridge National Laboratory
Oak Ridge, TN 37831
USA

ABSTRACT

The demand to make power generation ever cleaner, more flexible and efficient is driving the development of advanced materials that can survive increasingly onerous duty in power generating plant. In the USA the DOE funded 'Vision 21' programme is pursuing the design of highly flexible power generation systems capable of utilising diverse mixes of fuels, including coal, gas, oil, and biomass. Boiler tubes and high-temperature heat exchangers are amongst key components where significantly improved in-service properties are being demanded which, in turn, is leading to development of advanced alloy systems combining excellent high-temperature strength and oxidation resistance at competitive cost. In particular, Fe-based and related high temperature ODS alloys with coarse highly anisotropic grain structures show significant potential to offer substantial creep performance and oxidation resistance at temperatures in excess of 1100°C.

The present paper focuses on recent developments designed to improve high-temperature strength, creep resistance and reliability in two such alloys: PM2000 and an ODS-Fe_3Al alloy. Both alloys are mechanically alloyed (MA) materials. Recent results demonstrate the importance of all stages in the powder processing, consolidation and final thermo-mechanical processing of these alloys in determining final microstructure and potential properties. Research on several MA ODS-Fe_3Al alloys has demonstrated how microstructure in consolidated alloys may be affected by processing methods in the earliest stages of production. Specifically, potentially life-limiting defects in service, such as dispersoid inhomogeneities, oxide inclusions and retained fine-grained regions can be traced to controllable aspects of powder handling and production. Advances have been made that promise

alloys with increasingly reproducible quality and with microstructures manipulated to significantly enhance the performance of these ODS alloys.
Particular results discussed include those of a recent European BRITE project aimed at producing elongated grain structures oriented to enhance hoop creep strength in potential PM2000 heat exchanger tubing. The microstructures and properties produced by flow forming are discussed and further options for refinement explored.

INTRODUCTION

Commercial Fe-based mechanically alloyed (MA) ODS alloys such as PM2000, have a composition and microstructure designed to impart creep and oxidation resistance in components operating at temperatures from ~1050 to 1200°C and above. These alloys achieve their creep resistance from a combination of factors including: the dispersion of fine scale (20–50 nm diameter) Y_2O_3 particles introduced during MA which, despite formation of complex oxides involving Al from solid solution (Yttrium Aluminium Garnet, YAG), is highly stable to Ostwald ripening; and the presence of a very coarse, highly textured, high grain aspect ratio (GAR) structure which results from and is sensitive to the alloy thermomechanical processing history.[1–3]

Alloys are, typically, hot consolidated to full density using techniques such as Hot Isostatic Pressing (HIP), extrusion, upsetting or forging following which further hot and cold working (e.g. rolling, drawing etc) is used to produce the final alloy form. Subsequently Fe-based ODS alloys are given a secondary recrystallisation anneal to produce very coarse grain structures for creep resistance. A wide variety of product forms can be achieved, including bar, sheet, wire, tube and foil etc.[4]

Despite the benefits offered by the currently processed range of Fe-based ODS alloys they suffer from a number of performance shortfalls. In particular, the high GAR structures induced for creep resistance lead to anisotropic creep properties, which exhibit maximum creep resistance when the principal creep stress is aligned with the major axis of the grain structures.[5, 6] But the grain structures evolved during secondary recrystallisation of currently available Fe-based ODS alloys strongly align with the principal product forming direction, which means that in a product such as conventionally HIP'd and hot extruded tube the high GAR direction is along the tube axis.[1, 2, 5, 7] Moreover, in the Fe-based ODS alloys this alignment cannot be altered by directional thermal treatments such as zone annealing.[7] So, for Fe-based ODS alloy tubing currently available for high temperature internally pressurised applications, the direction of maximum creep strength is orthogonal to the direction of maximum principal creep stress (the hoop stress). As a result, creep life in the hoop orientation in Fe-based ODS alloy tube may be no better than 20% of that in uniaxially loaded and crept tube.[8] Moreover, pressurised tube burst data for material with current microstructures indicates a creep life (~14,500 hours/1100°C/5.9 MPa pressure) that is ~10% of that likely to be required for tube for application in high temperature heat exchangers (100,000 hours life) for power generation applications.[9]

As well as enhancement of the creep strength of commercially available alloys via microstructural modifications, an additional route to improved high temperature strength is offered by the development of new alloys such as ODS-Fe_3Al. Compared to the ODS-FeCrAl alloys, the prototype ODS-Fe_3Al alloy studied here has inherently superior sulphidation and carburisation resistance and, potentially, may have longer oxidation-limited lifetime.[4] However, the microstructure of the consolidated ODS-Fe_3Al is not yet sufficiently homogeneous and reproducibly defect free for service. In terms of creep resistance, microstructural defects are potentially extremely deleterious since they represent sites for early nucleation of creep cracks and local accumulation of creep damage and form discontinuities in the coarse, elongated secondary microstructure. As there is presently a drive to produce components with the elongated secondary grains orientated favourably in terms of the stresses encountered in service, microstructures with unpredictable grain size, morphology and orientation need to be avoided.[11] The origin of these problems has been traced to various preconsolidation processes within the alloy production route.[12, 13, 14]

A favoured production route for these alloys is MA, during which attrition of the mill components may occur, particularly if high-hardness alloys such as ODS-Fe_3Al are being milled.[13] The resulting fragments of alloy steel become entrained within the powder during subsequent milling and form regions containing none of the strengthening YAG dispersoid found throughout the bulk of the alloy.

The following will cover how improvements can be made to the high temperature performance of ODS alloys by identifying processing quality issues and responding to them to give an alloy with reduced defect concentration from which components with advantageous grain structures may be produced.

MICROSTRUCTURAL DEFECTS AND THEIR ORIGINS

Primary recrystallised samples of a prototype MA ODS-Fe_3Al alloy were annealed at 1275°C for 0, 15, 30, 45 and 60 minutes respectively to provide material at varying stages of secondary recrystallisation. Even after 1 hour annealing, not all of the primary grain structure had been replaced and occasional 'stringers' of fine-grained material aligned in the extrusion direction are retained throughout the alloy.

The basic recrystallisation kinetics were reminiscent of that of the latter half of a typical Johnson-Mehl curve with the proportion of secondary recrystallised material approaching the 100% level asymptotically. It could be argued, therefore, that a small amount of retained primary material is inevitable and that longer annealing cycles should be employed to reduce the amount. However, a reduction in the levels of retained primary material to the negligible levels found in similar commercial alloys would require a significant increase in heat treatment time with potentially deleterious effects, caused by Ostwald ripening, on the size distribution of the YAG dispersion. It is preferable to identify the factors in production processing that promote the retention of regions of primary recrystallised material and remove them.

Figure 1 shows a FEG SEM image of a longitudinal section of notionally fully secondary recrystallised material. A fine-grained 'stringer' is shown which is fully contained within one of the large secondary recrystallised grains. The grains within the 'stringer' exhibited somewhat larger size but similar elongation parallel to the extrusion direction to the grains in primary recrystallised material. The stringers are thought to recrystallise at a lower temperature than the rest of the alloy essentially independent of the wider secondary

Fig. 1 FEGSEM channelling contrast image of a portion of a fine-grained 'stringer' in an ODS-Fe_3Al alloy.

recrystallisation process. The alloy demonstrates an almost bimodal recrystallisation behaviour.

In both TEM and FEG SEM studies, the outer edges of the fine-grained 'stringers' were often decorated with alumina particles or even delineated by stringers of the particles. This coincidence of grain boundaries and particles also occurred *within* the fine-grained regions and at grain boundaries in the secondary recrystallised material as shown in Figure 2. This (mild) pinning of boundaries by alumina particles in the ODS-Fe_3Al was due principally to simple 'Zener' effects derived from the decrease in boundary energy caused by elimination of grain boundary area by pinning particles.[15]

Figure 3 shows a region denuded in YAG dispersoid that was found to span several primary recrystallised grains that were also free of dislocations, whereas typical primary recrystallised material contained both YAG dispersoid and dislocations residual from the extrusion process and trapped by the dispersoid. In those areas of the alloy free of YAG dispersoid, recovery will occur readily and the dislocation concentration will decrease during high temperature consolidation or the early stages of the secondary recrystallisation anneal as seen in Figure 3. The juxtaposition of such areas with areas of higher dislocation density would provide a driving force for recrystallisation. Such areas would not readily be removed by growing, coarse secondary recrystallised grains and would contribute to retained fine-grained areas in the alloy. Following MA, the dispersoid ought to be distributed homogeneously throughout the alloy, but the presence of dispersoid free areas suggested inadequate local powder milling. When powder particles were examined by FEG SEM, regions of chemical inhomogeneity were discovered, some as large as $\approx$100 µm in diameter. These regions are thought to have been fragments of the milling balls or attritor used in the MA process that had become entrained within the alloy powder particles during the MA processing. As such, they would not be expected to contain any YAG dispersoid. Indeed, even in the as-extruded

Fig. 2 FEGSEM channelling contrast image of a secondary grain boundary associated with alumina particles in an ODS-Fe_3Al alloy.

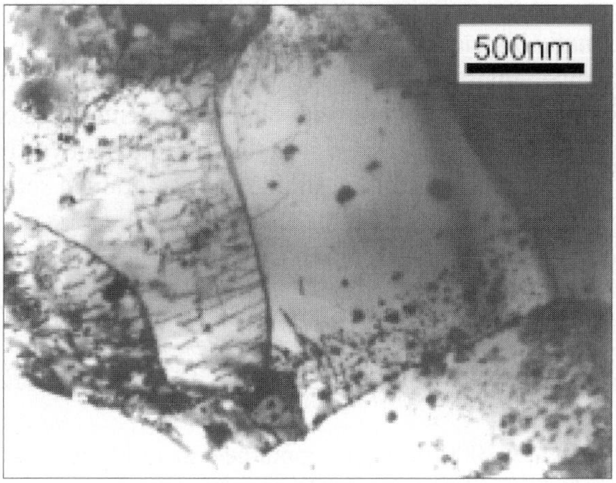

Fig. 3 TEM micrograph showing a region depleted in YAG dispersoid in an ODS-Fe_3Al alloy.

alloy, before any further recrystallisation annealing, stringers of partially recrystallised material may be found, as shown in Figure 4. This effect has been duplicated in PM2000 by the deliberate addition of dispersoid-free Fe powder to the PM2000 powders before consolidation. This resulted in the formation of stringers of recrystallised material which were found after very short annealing times.[16]

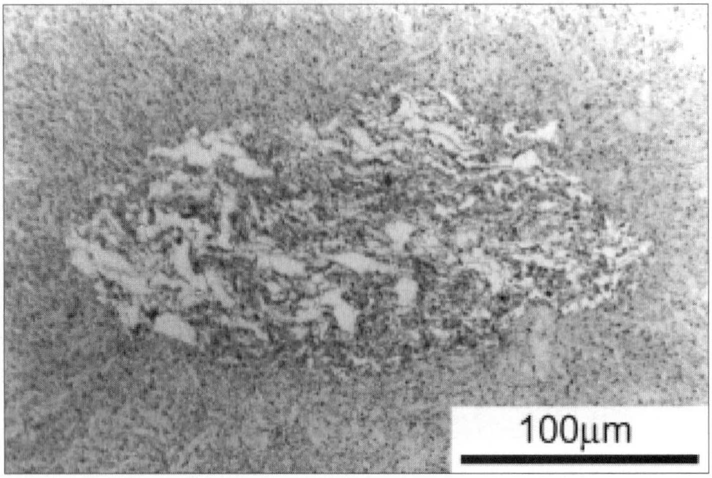

Fig. 4 Transverse section of recrystallising stringer in as-extruded ODS-Fe_3Al.

When compared to commercially available ODS alloys, the consolidated prototype ODS-Fe_3Al contains abnormally high levels of particulate alumina, while the MA powder contained more chemical inhomogeneities. These effects may be due, respectively, to rapid oxidation of the prototype alloy powder and high powder hardness causing greater attrition of milling balls. It seems that these effects are responsible to a large extent for the retention of fine-grains regions within the technologically desirable, coarse-grained secondary recrystallised alloy. Limiting powder oxidation and the entrainment of mill debris into the powder charge would be an excellent first step towards unhindered secondary recrystallisation.

EFFECT OF PROCESS MODIFICATIONS ON POWDER OXIDATION

As-MA ODS-Fe_3Al and PM2000 powders contained in mild steel cans were annealed at 1000°C for 1 hour under a vacuum of 10^{-3} mbar to simulate the types of environment likely to be experienced during commercial consolidation processing. After annealing, the compressive strength of the resulting powder was measured. This was repeated using powders that had been either uniaxially pressed, purged with hydrogen during evacuation or both pressed and purged. The strength data thus collected are shown graphically in Figure 5.

It can be seen that cold compaction of the alloy powders prior to annealing produced a dramatic improvement in sintered strength. Hydrogen purging alone produced the largest enhancement in sintered strength in PM2000 powder. Cold compaction increased sintered strength in ODS-Fe_3Al powder by a factor of ~90 as opposed to the factor of ~3 increase seen with PM2000 powder. It is believed that this was due to more rapid oxidation of ODS-Fe_3Al powder than PM2000 and, therefore, the most effective way to reduce oxidation during

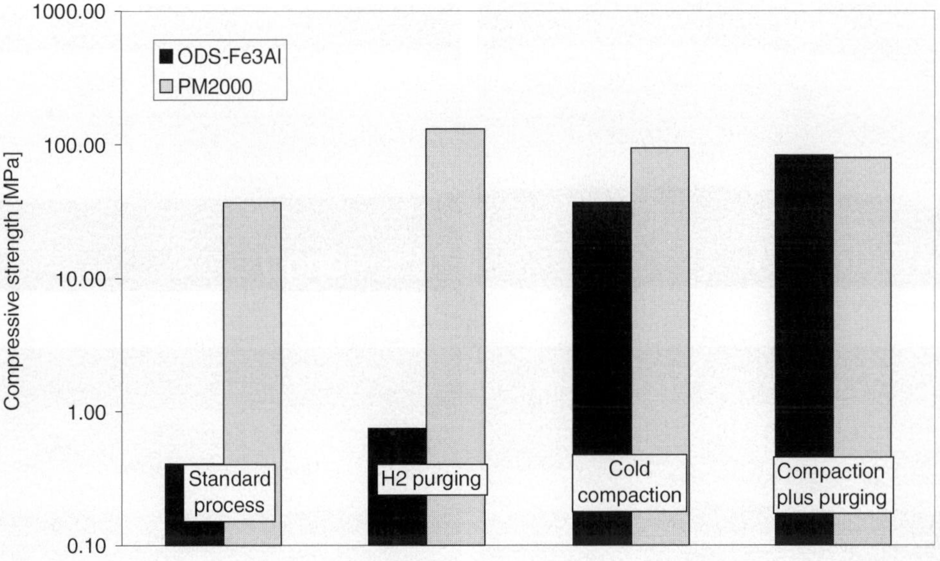

Fig. 5 The effect of process modifications on the sintered strength of ODS-Fe_3Al and PM2000 powders.

consolidation is to minimise the powder free surface area by use of processes such as cold compaction.[14, 17]

The two techniques can be seen to be highly effective, but in an alloy-specific way. It might have been expected, therefore, that the combination of the two techniques would not yield significant further improvements on the application of the individual techniques. In the case of PM2000 powder this was indeed the case, with no obvious additional benefit (within experimental error) gained by combining purging and compaction. However, a further improvement in sinter strength was possible for the ODS-Fe_3Al alloy. A sintered pellet produced by a combination of hydrogen purging and cold compaction had a compressive strength a factor ~200 times higher than that produced by straightforward vacuum annealing. The ODS-Fe_3Al pellet was, in fact, of similar strength to a similarly processed PM2000 pellet. It would appear that a combination of cold compaction and hydrogen purging significantly reduces the oxidation of densely packed ODS-Fe_3Al during simulated consolidation annealing.

IMPROVING HOOP CREEP PERFORMANCE IN ODS ALLOYS

In a recently completed BRITE Euram project the technique of flow forming was used to produce PM2000 alloy tube products with grain structures with improved grain aspect ratio in the hoop orientation.[18] The flow forming techniques applied included water cooled reverse

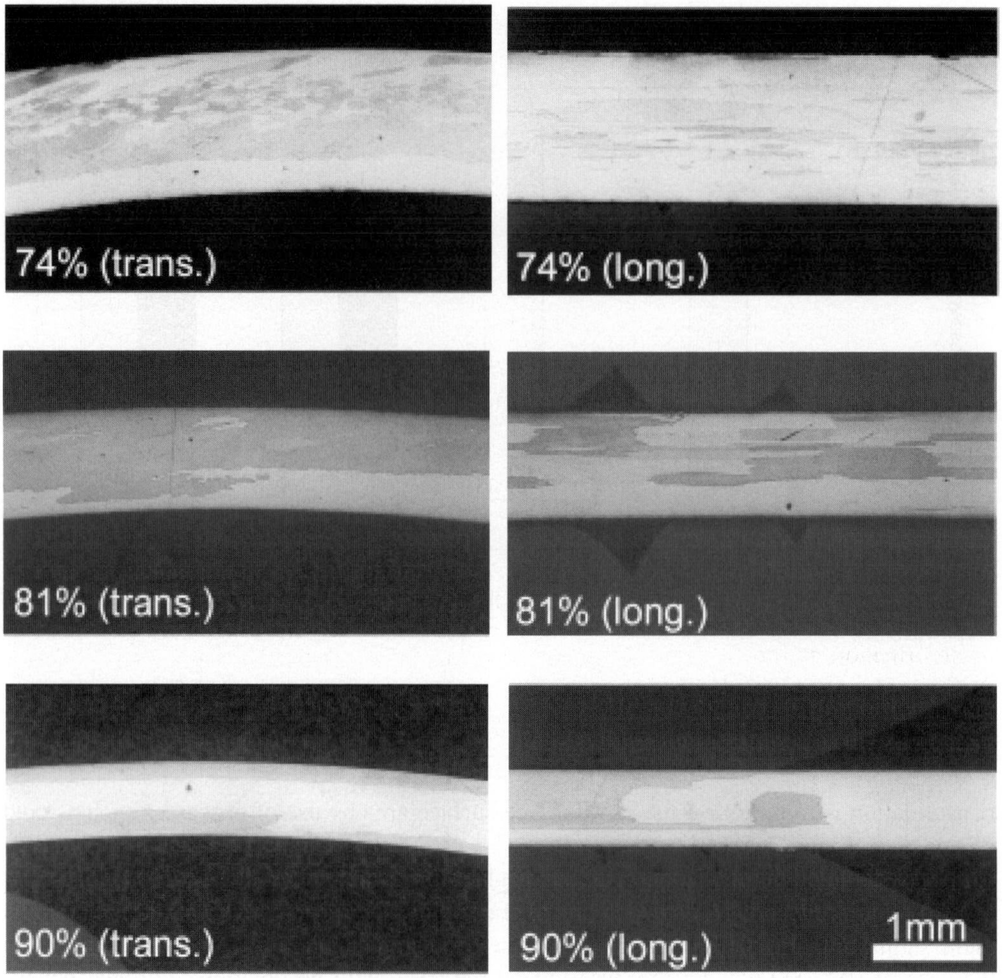

Fig. 6 Optical micrographs of transverse and longitudinal sections of flow-formed PM2000 tubes with varying degrees of deformation, after annealing at 1380°C for 1 hour.

flow forming, where a set of three idle rollers disposed at 120° intervals around a tube preform mounted on a driven mandrel was used in up to 3 passes to produce thin walled tubing products containing total levels of deformation of up to ~90%. The process of flow forming induces a complex pattern of deformation involving both axial and torsional flow in the work piece. Subsequent secondary recrystallisation of these flow formed tubes resulted in evolution of grain structures that were complex and varied in size, aspect ratio and orientation both as a function of position through the tube wall thickness and with the total level of flow forming deformation applied. Figure 6 shows an example of some of the grain structures observed in both longitudinal and transverse sections of tube flow formed to total levels of deformation ranging from 74 to 90% (final wall thicknesses from 1.3 mm down to 0.6 mm, respectively). Essentially, it was found that with total levels of flow forming deformation approaching 90% it was possible to reorient the grain structures within the tube to provide a significantly enhanced hoop orientation GAR compared to extruded and secondary recrystallised tube variants. With lower total levels of applied deformation the flow forming technique produced a pattern of deformation and an associated texture that varied significantly across the wall of the tube produced.[19] As a result, the secondary recrystallisation of such tubes led to grain structures which were finer near the tube outer surface and coarse and aligned with the tube axis, adjacent to the tube inner wall. In all of the tubes produced by flow forming it was found that subsequent secondary recrystallisation led to the formation of a thin (typically ~10 μm thick) characteristic layer of 'blocky' grains at the tube outer surface which followed a helical path identical to that of the set of flow forming rollers and could have grain sizes in the hoop orientation of up to half the tube circumference. An example of this type of grain structure in one of the flow-formed and secondary recrystallised tubes is shown alongside the grain structure of a conventionally extruded and secondary recrystallised tube in Figure 7. Further work on these flow formed variants subsequent to completion of the BRITE Euram project has shown that the formation of these 'blocky' grain structures at the tube outer surface is associated with complex patterns of deformation and local variations in textures that have a periodicity along the tube axis commensurate with the pitch of the set of flow forming rollers.[16]

Results of preliminary assessment of the high temperature hoop creep properties at 1100°C of some of the coarser grained variants produced by appropriate combinations of flow forming and subsequent secondary recrystallisation are shown in Figure 8. The creep data presented include both those from tensile samples extracted in the hoop orientation from tubing flattened at 700°C then secondary recrystallised at temperatures above 1300°C prior to test and of full sections of secondary recrystallised tubing creep tested to failure in air as internally pressurised tube bottles in the laboratory. For context, the plot also contains reference lines which indicate the hoop creep performance expected from standard variants of PM2000 (Pilger) and MA956 tubing and also from PM2000 sheet materials (shaded region). These creep data demonstrate that the application of flow forming followed by secondary recrystallisation enhanced the hoop creep performance of PM2000 alloy. The hoop creep test data obtained at 1100°C from flattened tensile samples extracted from tubing flow formed to 90% deformation before secondary recrystallisation match the creep data from PM2000 sheet. However, 1100°C creep bottle tests performed on the same tubing gave creep results intermediate between standard PM2000 tube and sheet. The creep data from the bottle tests performed on these 90% flow formed and secondary recrystallised tube samples are likely to be a more reliable indicator of the potential for enhanced hoop creep performance in this tube variant than the tensile data, given the larger sample size and more realistic loading conditions represented by pressurised tube testing. However, overall, the data do indicate the clear potential for the application of flow forming techniques to produce PM2000 tubing with microstructures that enable significantly enhanced hoop creep performance compared to current axially extruded variants. Further optimisation of the flow forming technique should enable production of thicker wall PM2000 tube products and enable generation of a performance data base confirming enhanced hoop creep behaviour compared to the standard extruded variants.

Fig. 7 Showing the very different grain structures of PM2000 tubes formed (a) extrusion and (b) extrusion followed by flow forming.

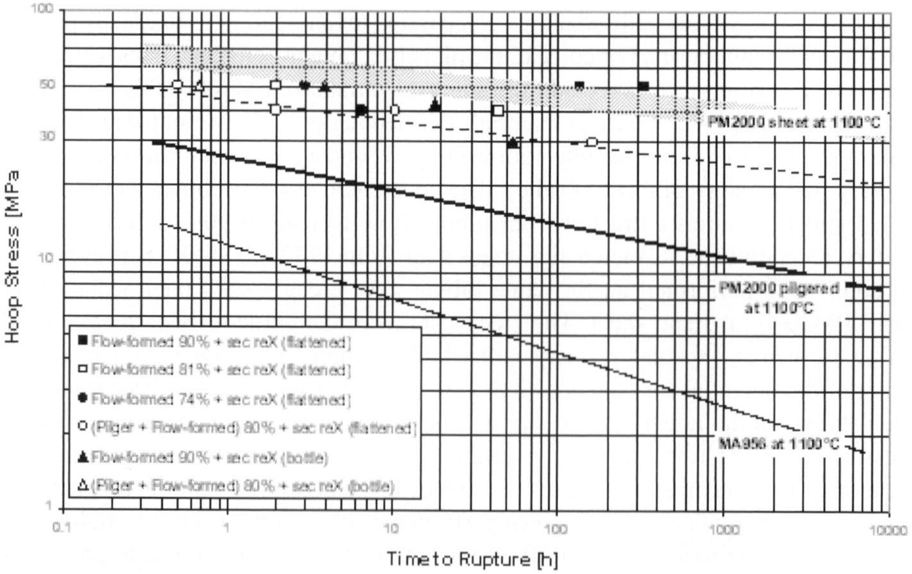

Fig. 8 Hoop creep properties of PM2000 tubes prepared via different routes and in different configurations.

CONCLUSIONS

The advance of the secondary recrystallisation front in ODS-Fe_3Al may be hindered by high concentrations of alumina particles. The mechanism responsible is Zener pinning, which is exacerbated by the planar arrays of alumina present in ODS Fe_3Al.

Areas that are free of the strengthening YAG dispersoid exist in ODS-Fe_3Al. Such areas easily recover and may trigger local recrystallisation that is restricted in extent and interferes with the general process of secondary recrystallisation to coarse grain structures.

Oxidation of ODS-Fe_3Al and PM2000 powders during high temperature annealing processes may be reduced by hydrogen purging of the canned powders, cold compaction of the powders or a combination of these two techniques prior to consolidation.

It is possible to manipulate the grain structure of ODS-FeCrAl tubes by means of flow-forming such that the GAR is much enhanced, compared with standard extruded tube variants, parallel to the hoop direction, the direction of principal stress in pressurised tubes. This has been shown to produce enhanced hoop creep strength in biaxial loading and shows great potential for further improvement.

ACKNOWLEDGEMENTS

This research was sponsored by the Advanced Research Materials (ARM) Programme, U.S. Department of Energy, Office of Fossil Energy under contract DE-AC05-96OR22464 managed by U.T.–Battelle, LLC, and the BRITE Euram III project, 1998-2002, CEC contract BE97-4949.

REFERENCES

1. K. H. Matucha and M. Ruhle: *Metal Powder Reports*, 1993.
2. I. C. Elliott and G. A. J. Hack: *Structural Applications of Mechanical Alloying*, ASM International Conference, Myrtle Beach, Froes and deBarbadillo, eds., 1990, 15.
3. M. J. Bennett, H. Rosemary and J. B. Price: *Heat Resistant Materials*, ASM International, 1991, 95.
4. G. Korb and D. Sporer: *Proceedings Conference High Temperature Materials for Power Engineering*, Liege, Bachelet et al., eds., Kluwer Academic Pub., 1990, 1417.
5. R. Timmins and E. Arzt: *Structural Applications of Mechanical Alloying*, ASM International Conference, Myrtle Beach, Froes and deBarbadillo, eds., 1990, 67.
6. B. Kazimierzak, J. M. Prignon, F. Starr, L. Coheur, D. Coutsouradis and M. Lamberigts: *ibid*, 137.
7. H. K. D. H. Bhadeshia: *Materials Science and Engineering*, **A223**, 1997, 64.
8. B. Kazimierzak: Private Communication.
9. F. Starr, A. R. White and B. Kazimierzak: *Proceedings Conference Materials for Advanced Power Engineering*, Liege, Coutsouradis, et al., eds., Kluwer Academic Pub., 1994, 1393.

10. I. G. Wright, B. A. Pint, P. F. Tortorelli and C. G. McKamey: *Proceedings of 10th Annual Conference on Fossil Energy Materials,* Knoxville, Tennessee, CONF-9605167, ORNL/FMP-96/1, 1996, 359.
11. Y. L. Chen and A. R. Jones: *Recrystallisation - Fundamental Aspects and Relations to Deformation Microstructures*, Risø National Laboratory, Denmark, 2000, 291.
12. A. R. Jones and J. Ritherdon: *Materials at High Temperatures*, **16**(4), 1999, 181.
13. J. Ritherdon and A. R. Jones: *Materials at High Temperatures*, **18**(3), 2001, 17.
14. J. Ritherdon and A. R. Jones: *Proc. 15th Annual Conf. on Fossil Energy Materials*, Knoxville, Tennessee, Judkins et al., eds., Pub. ORNL, 2001.
15. J. Ritherdon, A. R. Jones and I. G. Wright: *Recrystallisation - Fundamental Aspects and Relations to Deformation Microstructures*, Risø National Laboratory, Denmark, 2000, 533.
16. A. R. Jones, J. Ritherdon and D. J. Prior: *Reduction in Defect Content in ODS Alloys,* 17th US DOE Fossil Energy Conference, Baltimore, to be Published, 2003.
17. J. Ritherdon, A. R. Jones and I. G. Wright: *Proc. of the 2001 Powder Metallurgy Congress*, Nice, France, **4**, 2001, 133.
18. BRITE Euram III Project, *Development of Torsional Grain Structures to Improve Biaxial Creep Performance of Fe-Based ODS Alloy Tubing for Biomass Power Plant* (*Graintwist*), CEC contract BE97-4949, 1998-2002.
19. C. Capdevila, Y. L. Chen, N. C. K. Lassen, A. R. Jones and H. K. D. H. Bhadeshia: *Materials Science and Technology*, 2001, 693–699.

Optimisation of ODS-Fe_3Al Alloy Properties for Heat-Exchanger Applications

BIMAL K. KAD and JAMES H. HEATHERINGTON
University of California - San Diego
La Jolla, CA 92093
USA

CLAUDETTE MCKAMEY, IAN WRIGHT, VINOD SIKKA and ROD JUDKINS
Oak Ridge National Laboratory
Oak Ridge, TN 37830
USA

ABSTRACT

A detailed and comprehensive research and development methodology is being prescribed to produce Oxide Dispersion Strengthened (ODS)-Fe_3Al thin walled tubes, using powder extrusion methodologies, for eventual use at operating temperatures of up to 1,000–1,100°C in the power generation industry. Current single step extrusion consolidation methodologies typically yield 8 ft. lengths of 1-3/8" diameter, 1/8" wall thickness, ODS-Fe_3Al tubes. The process parameters for such consolidation methodologies have been prescribed and evaluated as being routinely reproducible. Recrystallisation treatments at 1200°C routinely produce elongated grains (with their long axis parallel to the extrusion axis), typically 200–2,000 μm in diameter, and several millimetres long. The dispersion distribution is unaltered on a micro scale by recrystallisation thermal treatments, but the high aspect ratio grain shape typically obtained limits grain spacing and consequently the hoop creep response. Current efforts are focused on examining the processing dependent longitudinal vs. transverse (hoop) creep anisotropy, and exploring post-extrusion methods to improve hoop creep response in ODS-Fe_3Al alloy tubes. Improving hoop creep in ODS-alloys requires an understanding and manipulating the factors that control grain alignment and recrystallisation behaviour. This represents a critical materials design and development challenge that must be overcome in order to fully exploit the potential of ODS alloys. In this report we examine the mechanisms of hoop creep failure and describe our efforts to improve hoop creep performance.

INTRODUCTION

Oxide Dispersion Strengthened ferritic FeCrAl (MA956, PM2000, ODM751) and the intermetallic Fe_3Al-based alloys are promising materials for high temperature, high pressure, tubing applications, due to their superior corrosion resistance in oxidising, oxidising/

sulphidising, sulphidising, and oxidising/chlorinating environments.[1–4] Such high temperature corroding environments are nominally present in the coal or gas fired boilers and turbines in use in the power generation industry. Currently, hot or warm working of as-cast ingots by rolling, forging or extrusion in the 650–1150°C temperature range is being pursued to produce rod, wire sheet and tube products.[5–7] *A particular 'in service application' anomaly of ferritic and Fe_3Al-based alloys is that the environmental resistance is maintained up to 1200°C, well beyond where such alloys retain sufficient mechanical strength.* Thus, powder metallurgy routes, incorporating oxide dispersions (ODS), are required to provide adequate strength at the higher service temperatures.

The target applications for ODS-Fe_3Al base alloys, in the power generation industry, are thin walled (0.1" thick) tubes, about 1 to 3 inches in diameter, intended to sustain internal pressures (P) of up to 1000 psi at service temperatures of 1000–1200°C. Within the framework of this intended target application, the development of suitable materials containing Y_2O_3 oxide dispersoids, must strive to deliver both a combination of high mechanical strength at temperature, as well as prolonged creep-life in service. Such design requirements are often at odds with each other, as strengthening measures severely limit the as-processed grain size, detrimental to creep life. Thus post-deformation recrystallisation, or zone annealing, processes are necessary to increase the grain size, and possibly modify the grain shape for the anticipated use. This paper describes our microstructure and property optimisation of ODS-Fe_3Al alloy tubes, with a view to improving the high temperature creep response. In particular, we examine thermal-mechanical processing steps to affect and enhance secondary recrystallisation kinetics and abnormal grain growth during post-extrusion processing to create large grains. Such procedures are particularly targeted to improve hoop creep performance at service temperatures and pressure.

EXPERIMENTAL PROCESSING DETAILS

Three separate powder batches of Fe_3Al+0.5wt.%Y_2O_3 composition were milled (labelled as A, B C). Table 1 lists the alloy powder chemistry before and after three separate milling conditions.[9] Two separate analyses (HM, PM) were performed for the starting powders as listed in the table. The powders are labelled as A, B and C in the order of decreasing total interstitial (C + N + O) impurity. Process control agents were used for all milled batches and batch C was milled for a short period. As shown later this interstitial impurity plays a crucial role in the recrystallisation kinetics and the resulting microstructures. The milled powders were encapsulated and vacuum-sealed in annular cans at about 400°C. Sealed cans were then soaked over the 1,000–1,100°C temperature range for 2 hours and directly extruded into tubes via the single step extrusion consolidation over a mandrel. Figure 1 shows a set of tubes in the as extruded and surface finished condition. Further extrusion processing parameters are available elsewhere.[7, 8] The tubes (typically 6–8 ft. length, 1-3/8" diameter and 1/8" wall thickness) are of sound quality and exhibit no cracking or damage after routine machining operations.

Recrystallisation heat treatments were performed on short segments of the extruded tubes in the 1,000–1,300°C temperature range in air. Microstructures were examined using optical, SEM and TEM techniques. High temperature mechanical testing was performed on ASTM

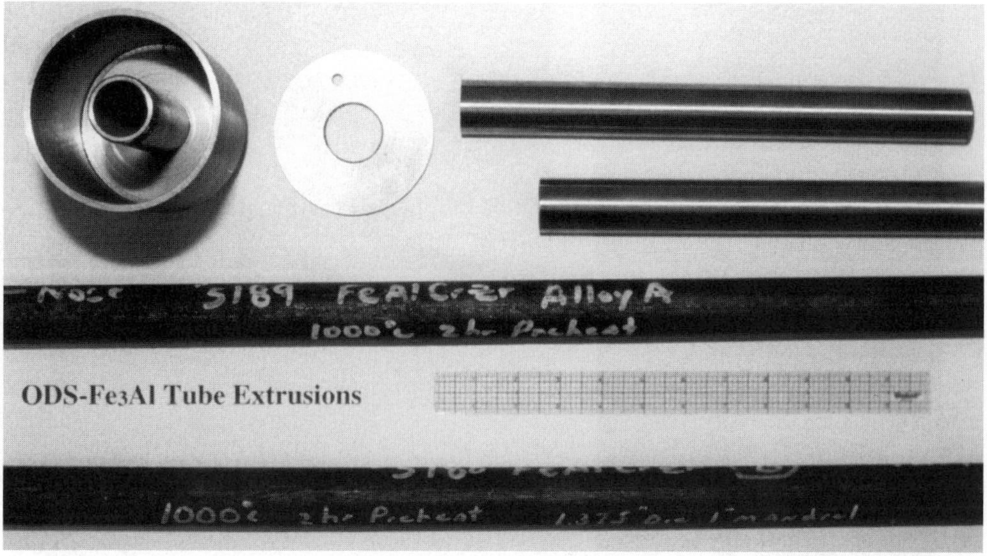

Fig. 1 Assorted ODS-Fe_3Al alloy tubes in the as extruded and surface finished condition produced via an annular can (top left) consolidation methodology.

E-8 miniature samples extracted from the longitudinal and transverse orientations of the tubes. All further testing details are described in the relevant sections below.

RECRYSTALLISATION KINETICS AND GRAIN SIZE

Polished tube cross-section samples were initially examined by optical microscopy. We note that recrystallisation kinetics is affected both by extrusion ratio and the impurity levels in the milled powder batches. Initial recrystallisation treatments were performed on powder consolidations performed at a 9:1 extrusion ratio, which proved very difficult to recrystallise.[9] A marked improvement in recrystallisation kinetics was observed with the increased extrusion ratios. In essence, primary recrystallisation could be initiated (but not necessarily completed) in all tubes extruded at ratios of 16:1 and higher.[7, 8] The nominal recrystallisation treatment is a 1 hour soak at 1,200°C in air. No recrystallisation activity was observed at temperatures below 1,100°C in as-extruded tubes. Figure 2 shows the transverse section microstructures for the A (left) and the C (right) chemistry. We note that for tubes of batch A chemistry, Figure 2a, only a 25% section of the wall thickness (in the interior) exhibits primary recrystallisation. However, tubes of batch C chemistry, Figure 2b, exhibit secondary recrystallisation in a reproducible manner. The recrystallised grains are elongated (with their long axis parallel to the extrusion axis), typically 200–2,000 μm in diameter, and several

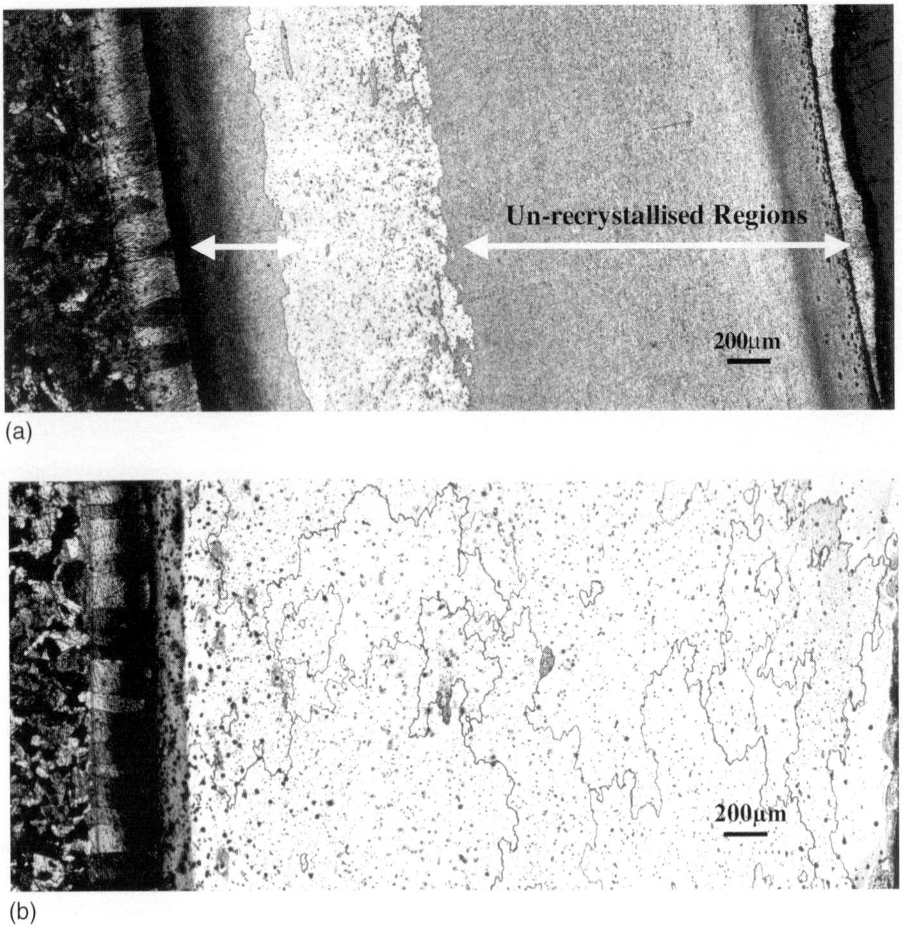

Fig. 2 Recrystallised tube sections of (a) batch A (left) and (b) batch C (right) alloys.

millimetres long. Additional aggressive thermal treatments failed to initiate secondary recrystallisation in the higher impurity (batch A, B) alloys.

MATERIALS CHARACTERISATION

TEM MICROSTRUCTURES

Bright field TEM micrographs of specimens extracted from the heat-treated tubes are shown in Figure 3. The 3 mm TEM discs were extracted from the wall thickness of the tubing such

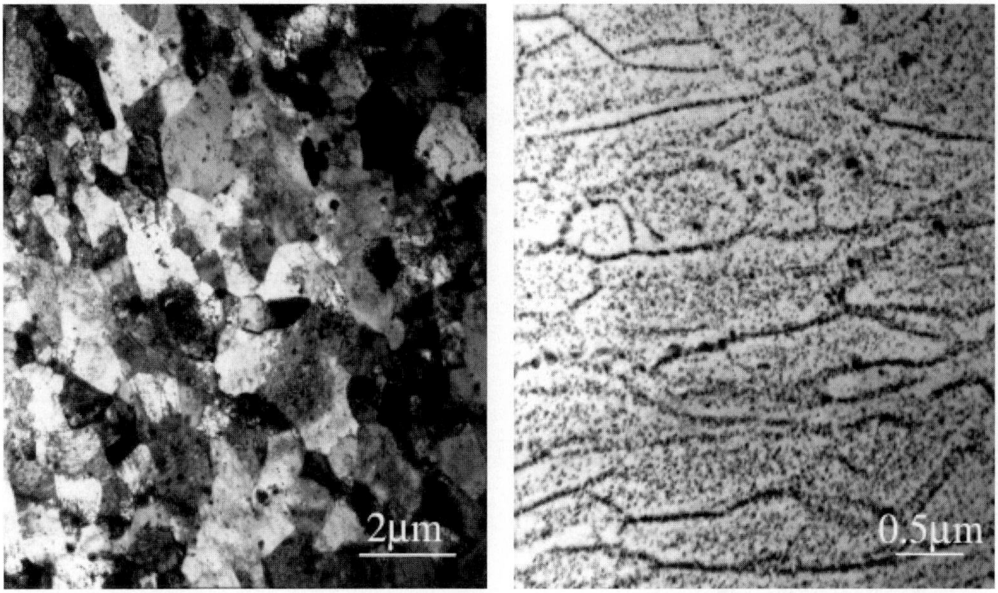

Fig. 3 TEM micrographs of (a) batch A and B batch C specimens extracted from the heat-treated tubes. The viewing direction is along the extrusion axis. Batch A and B exhibit the deformation processed and unrecrystallised {110} texture while batch C exhibits the recrystallised {111} texture.

that foil normal and the extrusion axis are co-incident. With the TEM thin foil perforation expected near the centre of the discs, the microstructures shown below are then representative of the centre of the tube-wall thickness. The high impurity alloys (batch A) exhibit a fine-grained structure, Figure 3a, with a {110} texture.[10, 11] Batch B (not shown here) microstructural evidence is similar to that of batch A. However, batch C exhibits a coarse grain structure with a {111} recrystallised texture,[10] Figure 3b. The precipitate distribution in batch C is rather uniform but also exhibits a cell-type structure on the scale of 1 μm. This cell dimension is consistent with the as-extruded grain size and it is suggested that this particle distribution was originally present on the surface of the milled powders and upon consolidation was incorporated at the as-extruded grain boundaries. The Y_2O_3 precipitates are about 10–20 nm in diameter and their distribution in PMWY-3 alloys is extremely homogenous at about 80–90 nm spacing. This intra-granular distribution is separate from the milling induced impurity that coats the prior particle boundaries. The extent of such impurity pickup is process-dependent.

A magnified view of the batch B sample, Figure 4, illustrates that the grains are effectively pinned by precipitate particles. These precipitates (marked by arrows) tend to be of the order of 0.25 μm, i.e. much larger than the mean Y_2O_3 dispersions of about 10 nm size. Precipitate chemical analyses confirm that they contain negligible amounts of yttrium and instead are rich in aluminium. This is reflected in the relative strengths of the aluminium peak in the matrix and precipitate spectra as illustrated in Figure 4. Looking back to the interstitial

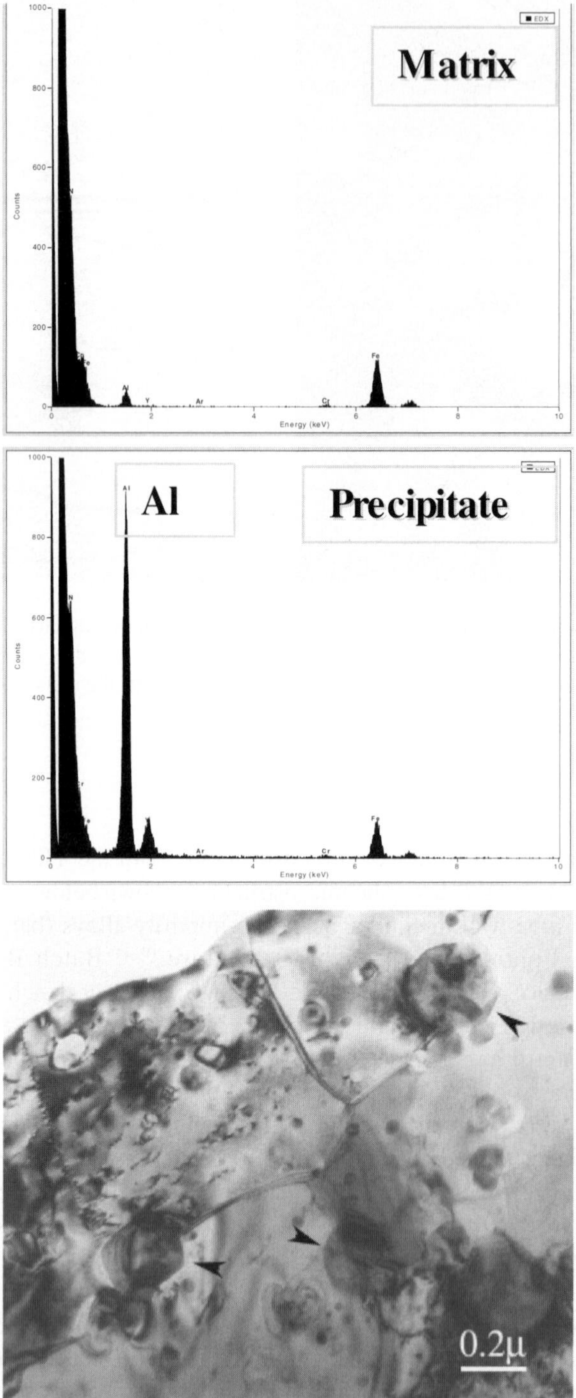

Fig. 4 Impurity pickup at the milling stage results in the formation of large oxides and/ or nitrides of aluminium at the grain boundaries.

Table 1 Chemical analyses of the as-received and milled powder batches.[9]

Sample	Heat Treatment, Air	T°C	Stress	Life, hr	Rate/Day	L-M
#06, C2T	1200/1 + 1300/1 Hour	900	2 Ksi	643		48.17
#20, C2T	1300/1 Hour	900	2 Ksi	2110	7e-5	49.26
#20, C2T	1300/1 Hour	900	3 Ksi	146		46.81
#08, C2T	1200/1 + 1250/10 Hours	900	2 Ksi	1093		48.66
#08, C2T	1200/1 + 1250/10 Hours	900	3 Ksi	93	7e-4	46.37
#05, C2T	1250/10 Hours	1000	1 Ksi	208		51.15
#10, C2T	1200/1 Hour	900	2 Ksi	1325		48.83
#10, C2T	1200/1 Hour	900	3 Ksi	299	4e-5	47.47
#15, C2T	1200/1 Hour	1000	1 Ksi	4315	1e-4	54.04

impurity analysis of Table 1, we note that both batch A and B has a significant level of oxygen, nitrogen pickup depending on the exact milling conditions employed. This impurity is in addition to the oxygen in Y_2O_3 and is interpreted as an overall increase in the precipitate volume fraction via the formation of aluminium oxide and/or aluminium nitride at the prior particle surface.

HIGH TEMPERATURE MECHANICAL PROPERTIES

The ductile to brittle transition temperature (DBTT) for ODS-Fe_3Al alloy tubes was assessed to be of the order of about 800°C i.e., less than the anticipated service temperature range of ODS alloys.[7] High temperature mechanical testing was then limited to the 800–1100°C range (i.e., in the creep regime) for all the three tubes. High temperature tensile and creep properties were evaluated for each of the three alloy chemistries in the longitudinal and transverse orientations. Standard ASTM E-8 miniature specimens are extracted from tube sections for tensile and creep tests. Longitudinal sections are spark machined directly from tubes and transverse sections are machined from flattened (hot pressed at 900°C) tubes. The hot pressing temperature is limited to 900°C to prevent any recrystallisation at this step. Samples are initially heat-treated at 1,200°C for 1 hour. Tensile tests were performed at temperature at a constant strain rate of 3×10^{-3} sec^{-1}.

Tensile Properties

Figures 5a and b show the comparative tensile response of the recrystallised tubes (from the three batches A, B and C) in the longitudinal and transverse orientations. It is surprising that the performance range for the three powder chemistries is rather narrow and belies the vast variations in creep response observed for the three alloys. For the critical transverse loading we note that batch C exhibits the best yield and tensile strength performance over the 800–1,000°C temperature range. No appreciable work hardening is observed in the transverse loading tests particularly at the higher (900–1,000°C) temperatures and it is reasonable to ascribe this behaviour as elastic-perfect plastic material response.

Creep Properties

Longitudinal vs. Transverse Creep Anisotropy

Preliminary creep tests were carried out for the ODS-Fe_3Al as well as the commercially available MA956 alloy tubes. The Fe_3Al samples were subjected to a uniform recrystallisation heat treatment at 1,200°C for 1 hour. Commercial MA956 tubes were recrystallised at 1350°C for 1 hour. Figure 5 shows the longitudinal (L) vs. transverse (T) creep anisotropy for both the Fe_3Al and the Fe-Cr-Al alloys. The MA956 and the ODS-Fe_3Al were tested at 900°C and 1000°C respectively. We note the improved creep response (see arrows) of the Fe_3Al batch C chemistry over the commercial MA956 alloy in both the L and T orientations. The longitudinal and transverse creep response ODS-Fe_3Al powder batches A and B was much inferior to that of batch C. This poor performance stems directly from the milling induced impurities that inhibit primary and secondary recrystallisation. Thus, batch C was selected as the most promising alloy chemistry for further development based on the large recrystallised grain size and superior hoop creep responses.

Transverse Creep Failures in ODS-Fe_3Al Alloys

A mechanistic understanding of hoop creep failures is an important pre-requisite to aid any subsequent microstructural modifications and/or redesign efforts at the system, sub-system or component level. Figure 6a shows a typical transverse creep failure as observed in fully recrystallised ODS-Fe_3Al alloy (batch C) tested at 1,000°C. Significant ductile voids are observed and it is presumed that failure is preceded by void formation and coalescence. A better insight into the origin of voids is obtained by examining the specimen surface immediately below the creep failure. Figure 6b shows creep void formation in crept ODS-Fe_3Al and the origin of such voids is predominantly at the grain boundary. Similar results (not reported here) are also obtained for the MA956 transverse creep tests.

Our program efforts are directed towards exploring metallurgical and microstructural means to enhancing and optimising hoop creep response. Presently variations in recrystallisation heat-treatments and thermo-mechanical processing schemes are being explored for the ODS-Fe_3Al alloys. Table 2 lists the compiled results of various thermal

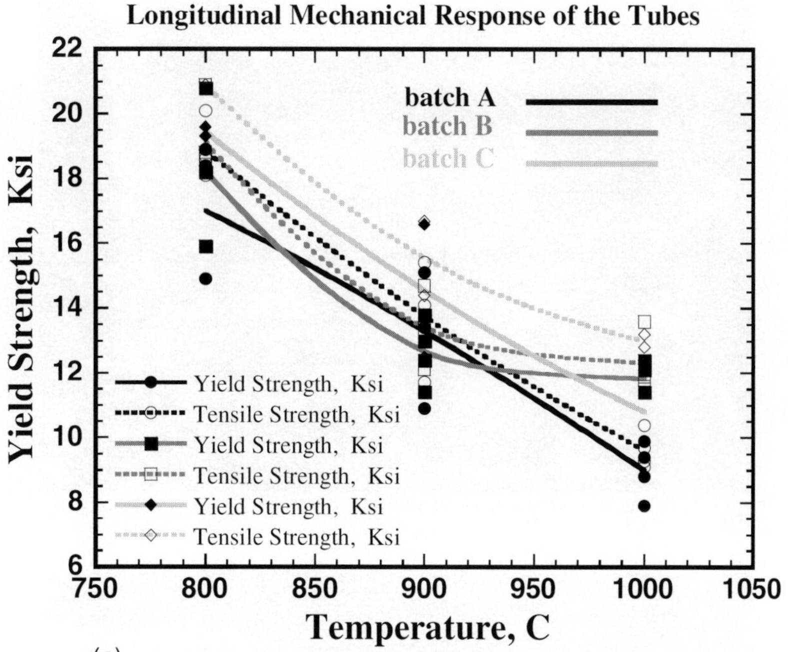

(a)

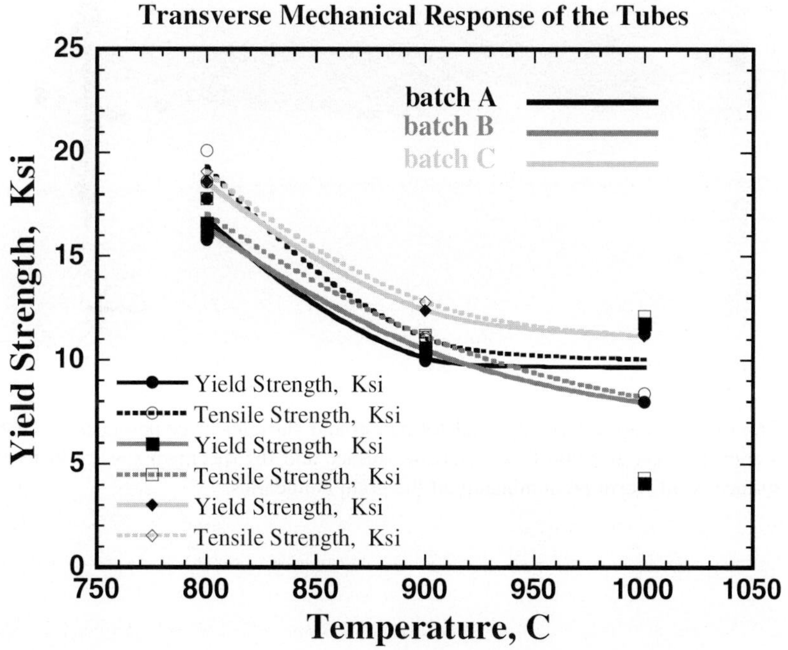

(b)

Fig. 5 High temperature mechanical response of ODS-Fe_3Al alloy tubes (powder batches A, B, and C) in the (a) Longitudinal and (b) Transverse orientations.

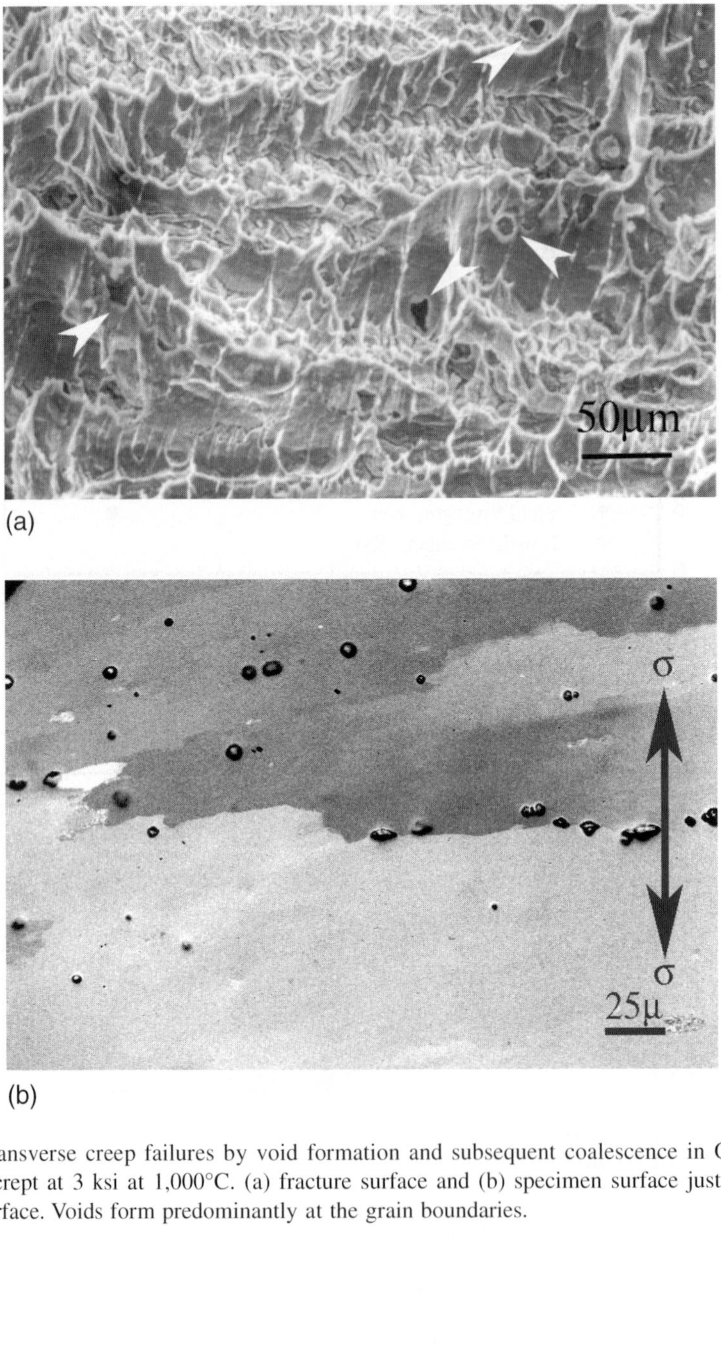

Fig. 6 Transverse creep failures by void formation and subsequent coalescence in ODS-Fe_3Al (batch C) crept at 3 ksi at 1,000°C. (a) fracture surface and (b) specimen surface just below the fracture surface. Voids form predominantly at the grain boundaries.

recrystallisation variations performed for batch C alloys and their ensuing creep response. In each case spark machined samples were heat-treated bare with or without a protective atmosphere. Samples are tested at a constant load and temperature and held for a minimum of 1,000 hours at each test condition before incrementing load. The accumulated test time is

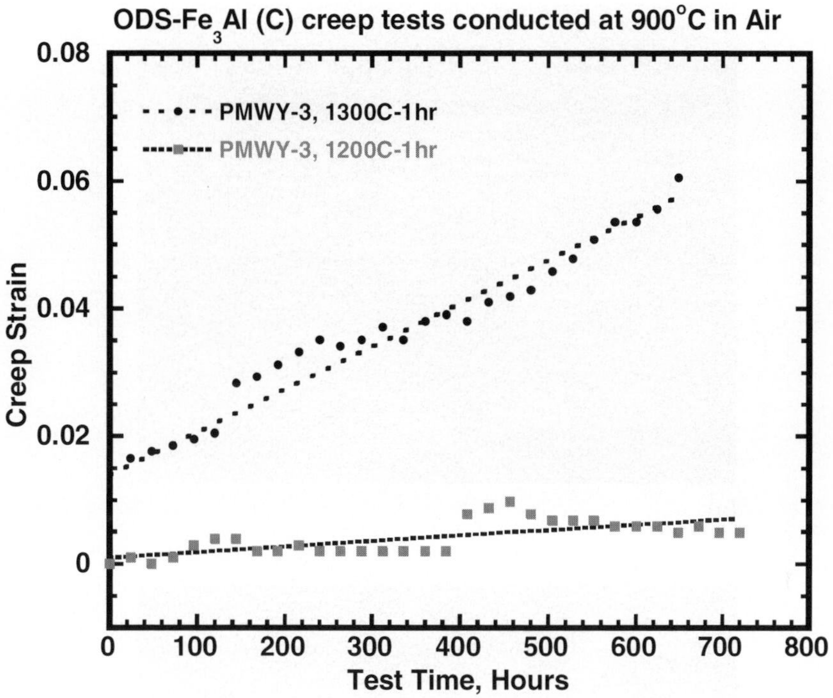

Fig. 7 Comparison of creep rates and strain for different recrystallisation heat-treatments.

recorded for the peak load and all prior load values. The observed creep rate is tabulated in days and L-M is the computed Larsen Miller parameter. For example, Test #10 has 1,325 hours of exposure at 2 ksi and about 299 hours at the 3 ksi increment, and the creep rate is recorded as 4e-5/day at the 3 ksi test condition. It is surmised that a 1,000 hour exposure is sufficient for predicting long-term survivability at that specific test condition on account of the high stress exponent for failure. This is observed in such ODS-alloys particularly in the transverse orientation creep tests.

Aggressive Recrystallisation Heat Treatments

Figure 7 shows a direct comparison of creep response of high purity (batch C) alloys heat-treated at different temperatures in air. The exact data is from Test #6 and #10 (indicated in Table 2) as tested at 2 ksi stress at 900°C. The 1,300°C heat-treatment was conducted by re-heating a test sample previously recrystallised at 1,200°C. It is noteworthy that aggressive thermal exposure does not improve creep-life in ODS-Fe_3Al alloys. A higher creep rate was observed for the 1,300°C treatment. Repeated tests, Table 2, indicate that creep response deteriorates with increasing recrystallisation treatment temperature.

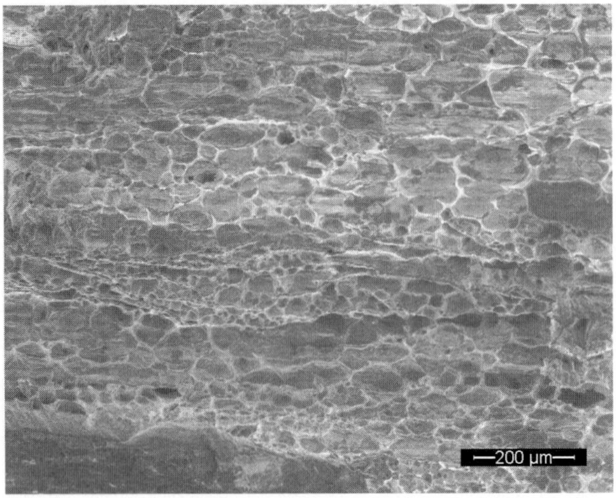

(a)

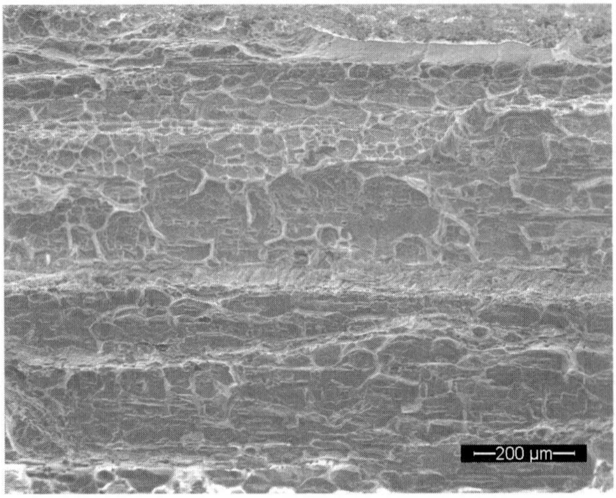

(b)

Fig. 8 Transverse creep failures in ODS-Fe_3Al (batch C) alloy tube samples for a) 1300°C-1 hour, (b) 1200°C-1 hour heat-treatments in air. Failure occurred at 2 Ksi in (a) and at 3 Ksi in (b).

Figure 8 shows a comparison of the transverse creep failures observed for the test #6 and test #10 (as plotted in Figure 7). Note that test #6 failed at 2 ksi whereas test #10 failed at the incremented 3 ksi stress. Both failures appear to have originated in the vicinity of the grain boundary and failed via void formation and coalescence. The ductile lobes have a scale of the order of 60 µm in test #6 (HT; 1300°C-1 hour) compared to about 200 µm for test #10

Table 2 Transverse creep response in high purity (batch C) ODS-Fe_3Al alloys as a function of thermal processing conditions.

Element	As-Received HM	As-Received PM	PMWY-1	PMWY-2	PMWY-3
Fe	Bal.	79.6			
Al	16.3	18.20			
Cr	2.4	2.18			
Zr	20 ppm	26 ppm			
O (Total)	60 ppm	110 ppm	1800 ppm	1900 ppm	1400 ppm
O (in Y_2O_3)			1025 ppm	1053 ppm	1080 ppm
O Balance			775 ppm	847 ppm	320 ppm
O Pickup			665 ppm	737 ppm	210 ppm
N	18 ppm	7 ppm	1264 ppm	145 ppm	88 ppm
N Pickup			1257 ppm	138 ppm	81 ppm
C		24 ppm	667 ppm	360 ppm	303 ppm
C Pickup			643 ppm	336 ppm	279 ppm
H		16 ppm	115 ppm	40 ppm	29 ppm
C + N + O Pickup			2565 ppm	1211 ppm	570 ppm

(HT: 1,200°C-1 hour) Such lobe detail would suggest that significant plastic deformation occurred in the test #6 which is indeed corroborated by the 4% creep strain observed, Figure 7. This observation is in agreement with the high impurity alloys that fail at appreciable strains with significant fine lobe formation. In contrast, high purity batch C samples with improved creep life exhibit limited creep-strain prior to failure, despite their lower over-all creep rates.

Within the context of aggressive thermal treatments it is important to note that commercial ODS-FeCrAl alloys are recrystallised in the 1,300–1,400°C range. Looking back to Figure 5, we noted the improved creep response (see arrows) of the Fe_3Al batch C chemistry over the commercial MA956 alloy in both the L and T orientations. This trend is consistent with the higher recrystallisation temperatures employed for MA956 alloys. This aggressive temperature is in part necessary to overcome the Zener pinning as the Y_2O_3 dispersion volume fraction is about 15% higher in MA956 alloys. Prior studies[12, 13] indicate the formation of subsurface voids when samples are exposed to aggressive environments, and it surmised that similar effects might be operative here. This effect may be particularly detrimental to transverse creep response - which in fact fails via creep void formation and coalescence. Efforts to decrease the recrystallisation temperature in commercial ODS-ferritic and intermetallic Fe_3Al alloys may be a promising avenue of further ODS-alloy development.

SUMMARY AND CONCLUSIONS

High temperature creep response in ODS-Fe_3Al tubes is reported for the longitudinal and transverse orientations. The kinetics of grain growth is affected by interstitial impurity content, which limit the extent of recrystallised regions observed in the tube wall. Consequently, the microstructure exhibiting the best creep response is one of high purity that undergoes complete primary and secondary recrystallisation. The results of property optimisation efforts till date are summarised as follows:

- Powder batch milling appears to be the single most pervasive processing component dominating microstructural and material response. It is suggested that coarse nitrides and oxides of aluminium (formed during milling) inhibit recrystallisation grain growth in high impurity alloys.
- Creep response of the respective powder batches is proportional to the underlying grain structure produced via heat-treatments. Thus ODS Fe_3Al powder batch C with its completely recrystallised tube wall section offers the best creep response.
- Significant deterioration in hoop creep is observed in materials subjected to aggressive recrystallisation temperatures. This observation is consistent with the improved performance of ODS-Fe_3Al alloys over MA956 alloys in equivalent tests performed in our laboratory.

ACKNOWLEDGEMENTS

Research sponsored by the Office of Fossil Energy, Advanced Research Materials Program, U.S. DOE under Contract DE-AC05-00OR22725 with UT-Battelle, LLC.

REFERENCES

1. C. Capdevila and H. K. Bhadeshia: 'Manufacturing and Microstructural Evolution of Mechanically Alloyed Oxide Dispersion Strengthened Superalloys', *Advanced Engineering Materials*, **3**(9), 2001, 647.
2. R. F. Singer and E. Arzt: *High Temperature Alloys for Gas Turbine and Other Applications*, D. Reidel Publishing Co., Liege, 1986.
3. F. Starr: 'Emerging Power Technologies and Oxide Scale Spallation', *Mater. High. Temp.*, **13**(4), 1995, 185–192.
4. B. A. Pint and I. G. Wright: 'Long Term High Temperature Oxidation Behaviour of ODS Ferritics', *Journal of Nuclear Materials*, **307–311**, 2002, 763–768.
5. V. K. Sikka, S. Vishwanathan, and C. G. McKamey: *Structural Intermetallics*, TMS Publication, Warrendale PA, 1993, 483.
6. P. G. Sanders, V. K. Sikka, C. R. Howell and R. H. Baldwin: 'A Processing Method to Reduce the Environmental Effect in Fe_3Al-Based Alloys' *Scripta Metallurgica*, **25**(10), 1991, 2365–2369.
7. B. K. Kad: 'Oxide Dispersion Strengthened Fe_3Al Bases Alloy Tubes', Advanced

Research Materials Report, ORNL/Sub/97-SY009/02, 2001.

8. B. K. Kad, V. K. Sikka and I. G. Wright: 'Oxide Dispersion Strengthened Fe_3Al-Based Alloy Tubes', *14th Annual Conference on Fossil Energy Materials*, Knoxville, TN, 2000, ibid. 'High Temperature Performance of ODS Fe_3Al -Based Alloy Tubes' *15th Annual Conference on Fossil Energy Materials*, Knoxville, TN, 2001.
9. I. G. Wright, B. A. Pint, E. K. Ohriner and P. F. Tortorelli, 'Development of ODS- Fe_3Al Alloys', *Proceedings of 11th Annual Conference on Fossil Energy Materials*, ORNL Report ORNL/FMP-96/1, CONF-9605167, 1996, 359.
10. B. K. Kad, S. E. Schoenfeld, R. J. Asaro, C. G. McKamey and V. K. Sikka: 'Deformation Textures in Fe_3Al Alloys: An Assessment of Dominant Slip System Activity in the 900–1325 K Temperature Range of Hot Working', *Acta Metallurgica*, **45**(4), 1997, 1333–1350.
11. M. Mujahid, C. A. Carter and J. W. Martin: 'Microstructural Study of a Mechanically Alloyed ODS Superalloy', *Journal of Materials Engineering and Performance*, **7**(4), 1998, 524.
12. Y. Iino: 'Effects of High Temperature Air and Vacuum Exposure on Tensile Properties and Fracture of ODS Alloy MA6000', *Materials Science and Engineering*, **A234–236**, 1997, 802–805.
13. A. Czyrska-Filemonowicz, D.Clemens and W.J. Quadakkers: 'The Effect of High Temperature Exposure on the Structure and Oxidation Behaviour of Mechanically Alloyed Ferritic ODS Alloys', *J. of Mater. Proc. Tech.*, **53**, 1995, 93–100.

Development of CMC Combustor Liners for Industrial Gas Turbine Applications

N. Rhodes, M. B. Henderson and S. Norburn
ALSTOM Power Technology Centre
Leicestershire, UK

G. McColvin
Demag Delaval Industrial Turbomachinery Ltd.
Lincoln, UK

ABSTRACT

The use of Ceramic Matrix Composites (CMC's) for combustor liner applications in place of current wrought metallic systems has been targeted as a route for achieving increased firing temperatures for gas turbine engines, thus providing increased thermodynamic efficiency and lower exhaust emissions. The most promising, commercially available candidate CMC materials are SiC/SiC-based fibre composites, however, to achieve the full temperature capability the application of a suitable thermal and environmental protective coating is required. The following paper reviews the work conducted recently within the Brite EuRam Framework V programme 'CINDERS', which is aimed at developing and testing protective coatings suitable for CMC Cerasep G415 combustor liners.
A full characterisation of the CMC, in both the coated and un-coated conditions has been conducted, as well as modelling of the thermal and mechanical properties of the CMC and coatings. Manufacturing and life prediction methods have been developed and integrated into a practical component design 'toolbox' for CMC combustor liners. Several coating systems have been identified for development, characterisation and testing. These have been applied by air plasma spraying and by wet processing methods. Validation assessment has been scheduled by means of rig testing a series of sub-element specimens and combustor liners. NDE techniques have been employed to detect internal defects and damage and to aid service life predictions.

INTRODUCTION

The manufacturers of gas turbines are continually striving to increase the operating temperatures of their engines. This is driven by the need for greater thermal efficiency and reduced emission of harmful exhaust gases. This drive places an ever increasing burden on the materials used in, and the design of, hot gas path components. Historically, materials developments, cooling air design innovations, and thermal barrier coatings have each played a role in increasing the temperature capability of gas turbine engines. However, the use of increasing levels of cooling air from the compressor has placed a penalty, in terms of the efficiency of the engine and the exhaust emissions, which offsets some of the benefits gained.

Therefore, reducing the requirement for cooling air, particularly in combustion chamber design, is seen as key to future engine developments. The introduction of Ceramic Matrix Composite (CMC) materials into hot gas path components such as combustion chamber liners, has long since been identified as a possible route to the achievement of increasing operating temperatures without incurring the penalties associated with increased cooling air use.[1–4]

For the Brite EuRam FPV CINDERS project, a commercially available grade of a SiC fibre reinforced SiC matrix CMC, known as Cerasep G415, has been chosen as the base material. The CMC is made up of layers of 2D, plain weave cloth made from Hi-Nicalon fibres. The matrix is produced by Chemical Vapour Infiltration (CVI) and contains additions that confer some degree of 'self-healing' to the matrix under oxidising conditions. One of the key objectives for this programme is to enhance further the temperature capability of the CMC by the application of a thermal and environmental barrier coating or Thermal Protection System (TPS).[5, 6] Mullite and yttrium silicate were selected as the candidate materials for TPS development. These possess a suitably low thermal conductivity, chemical and microstructural stability up to the operating temperature and a CMC-compatible thermal expansion coefficient that minimises the thermal stresses generated during engine running. The deposition techniques considered for the coatings were the sol-gel process, slurry dipping and air plasma spraying (APS). A thin layer (few microns thick) of elemental silicon has also been considered as a compliant layer between the substrate and the coating to minimise thermal stresses.

EXPERIMENTAL PROGRAMME AND METHODS

Some of the techniques, results and knowledge used to carry out the work in this project are proprietary knowledge, and as such are not detailed in this paper. However, a general description of the salient points and publishable results is given in the following sections.

CMC CHARACTERISATION

Tensile and low-cycle fatigue (LCF) tests were performed on un-coated Cerasep G415 with the fibre architecture orientated at 0/90° to the sample axis. Creep tests on 0/90° samples and a further set of tests using 45/45° samples are on-going. The tensile and LCF work was carried out on an Instron servo-hydraulic machine, using a side loading extensometer with a 25 mm gauge length. Tests were carried out in air at room temperature, 850, 1,100 and 1,300°C. Tensile tests were performed at a strain rate of 0.6%/min, with some tests at 1,100°C repeated at 0.06 and 6%/min. The LCF tests were performed at the same temperatures under load control, using a trapezoidal waveform with a 200 MPa/s loading rate, 1 s tensile and compressive dwell period and a load ratio $R_\sigma = -1$. A further set of tests was carried out at 1,100°C using a 30 s tensile dwell.

Several samples were loaded up to pre-determined stress levels to provide samples for NDE examination. This was intended to provide information on the accumulation of damage within the CMC. Also, some LCF samples that had achieved run-out at 10^5 cycles underwent NDE examination as well. The tensile properties of the composite have been modelled using a bi-linear kinematic material model, which is well established for modelling metal plasticity and also provides a reasonable fit to CMC non-linear behaviour.[7–9] The life assessment methods that are being developed for these materials are based on quantifying and modelling the observed variations in measured mechanical properties, such as the evolution of specimen stiffness and hysteresis loop area during fatigue cycling and damage accumulation.

TPS MEASUREMENTS

Thermal and physical properties of the coatings were measured on freestanding coating material specimens. These measurements include thermal diffusivity by the laser flash method, specific heat capacity by differential scanning calorimetry (DSC) and density measurements using the Archimedes method and by quantitative image analysis. Thermal expansion data was taken from the literature.[10] A number of measurements were made using a push-off technique to ascertain adherence strength.

MICROSTRUCTURAL CHARACTERISATION AND NDE

Microstructural characterisation was carried using optical microscopy, SEM and by non-destructive techniques. The NDE methods used include transmission, pulse-echo, front surface and back surface echo ultrasound techniques, and X-ray imaging. The X-ray imagining system has been coupled with computer aided tomography to generate virtual 2D and 3D images of samples. When used with a synchrotron radiation source, this technique has a spatial resolution of just a few microns. Imaging has been applied to the CMC microstructure, the coating microstructure, the interface between the two, and the accumulation of damage during testing.

DESIGN

The design of the sub-element components and full sized combustor involved some preliminary thermal and stress analyses to qualify basic design concepts. Feasible designs were then subjected to more rigorous thermal and stress analyses to optimise their design and the coating thickness. Manufacturing considerations, regarding the lay-up and impregnation processes of CMC manufacture and the deposition of the coatings made suitable contributions to the design process as well. The integration technology was based on elastic metallic components being used to accommodate thermal and mechanical movements, whilst providing the necessary location of the CMC component with the minimum of induced stress.

The CMC design and coating optimisation were carried out using thermal and stress analyses by finite element methods. Initially, the optimisation of the coating thickness, and to some extent the stress analysis of the CMC was done using estimated property data, or

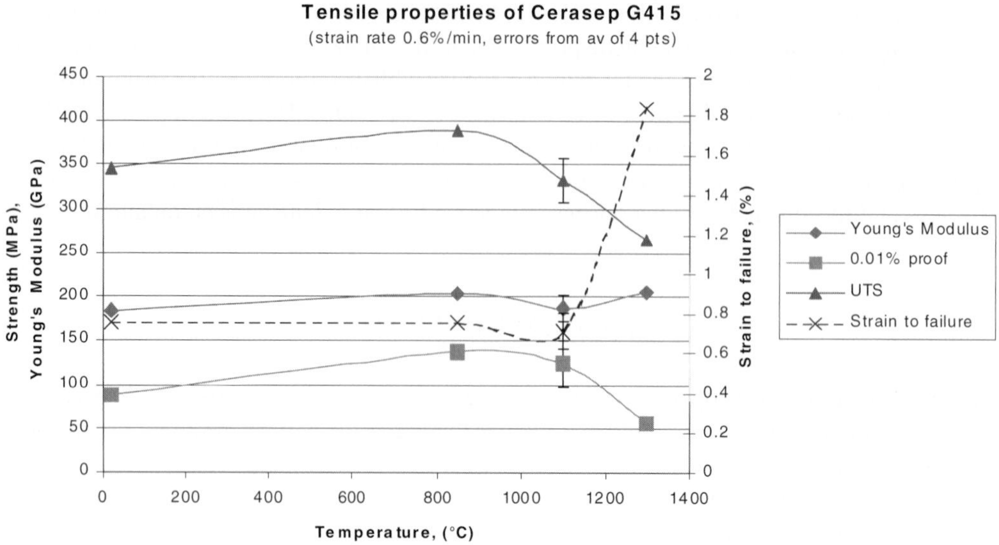

Fig. 1 Temperature sensitivity of the tensile properties of Cerasep G415.

data available for similar materials gathered on previous projects. As the project has progressed, additional data have been generated and the thermal and stress analyses of the combustor sub-elements and full-scale components have been revisited to ensure optimum performance characteristics are achieved.

RESULTS

CMC CHARACTERISATION

The tensile data are presented in Figures 1 and 2. Figure 1 shows the variation in tensile properties with temperature at a strain rate of 0.6%/min, whilst Figure 2 shows the strain rate sensitivity at 1,100°C. The error bars are based on 4 nominally identical tests at 1,100°C. The properties are reasonably constant up to 1,100°C, but above this temperature and at stress levels beyond the yield point (due to matrix cracking) the strength properties are adversely affected. At 1,100°C, the trend of the strain rate sensitivity is that the strength and elongation to failure are both greater at higher strain rates.

Figures 3-5 show selected SEM images of the fracture surfaces of samples tested at the 3 different strain rates. It is worth noting that the samples took approximately 15–20 minutes to cool down from 1,100 to about 800°C, below which oxidation will have virtually stopped. The slow strain rate sample will have been subjected to the test environment for a similar

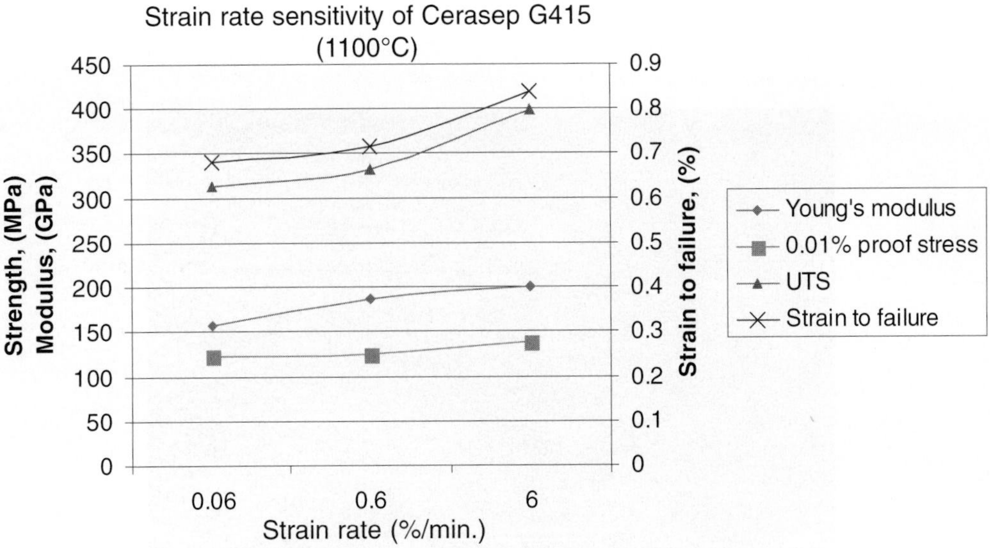

Fig. 2 Strain rate sensitivity of the tensile properties of Cerasep G415.

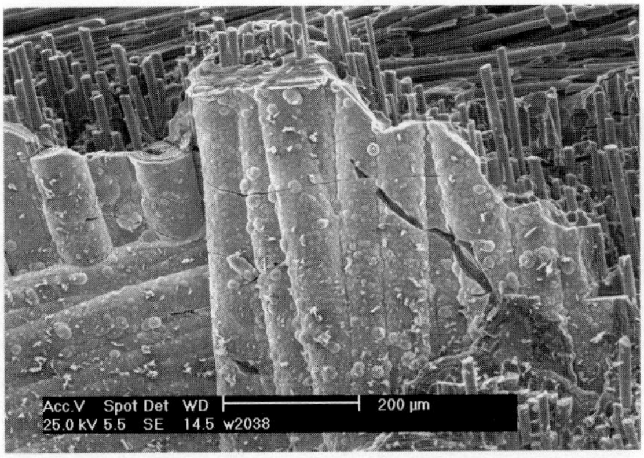

Fig. 3 Fracture surface of Cerasep G415 RT tensile test specimen.

period of time between the onset of matrix cracking and final failure. The fast strain rate sample will have progressed from matrix cracking to failure in a matter of seconds.

A set of S-N curves is shown in Figure 6. The UTS has been used as the N = 1 point. Figures 7 and 8 show examples of the variation in modulus and loop area respectively with elapsed cycles. Data from the same test are used for both plots.

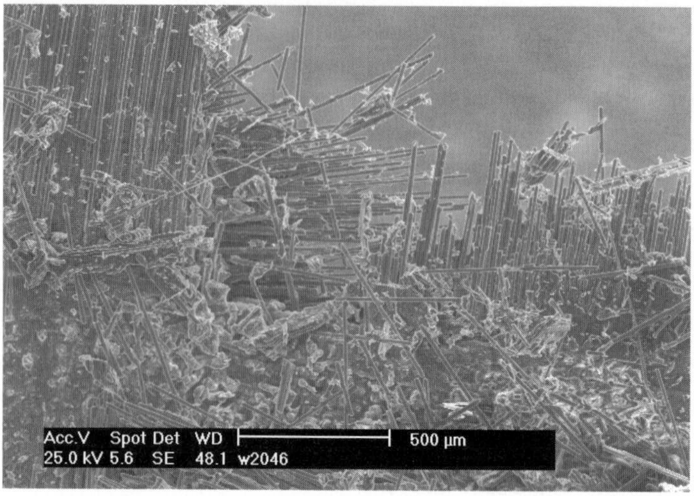

Fig. 4 Fracture surface of Cerasep G415 1100°C fast strain rate sample.

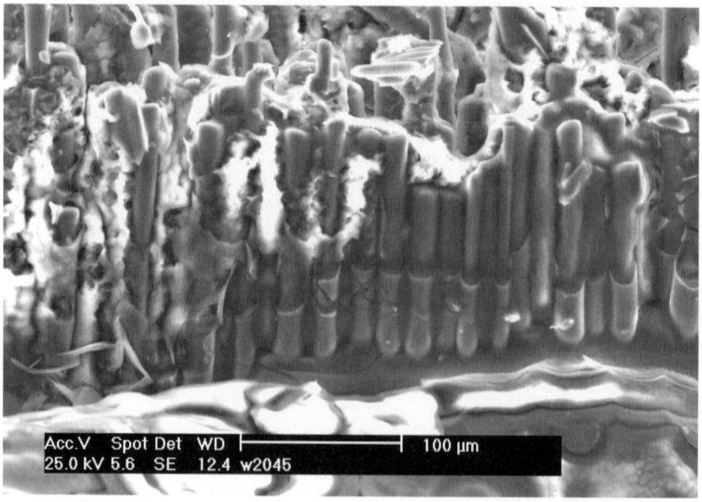

Fig. 5 Fracture surface of Cerasep G415 1100°C slow strain rate sample.

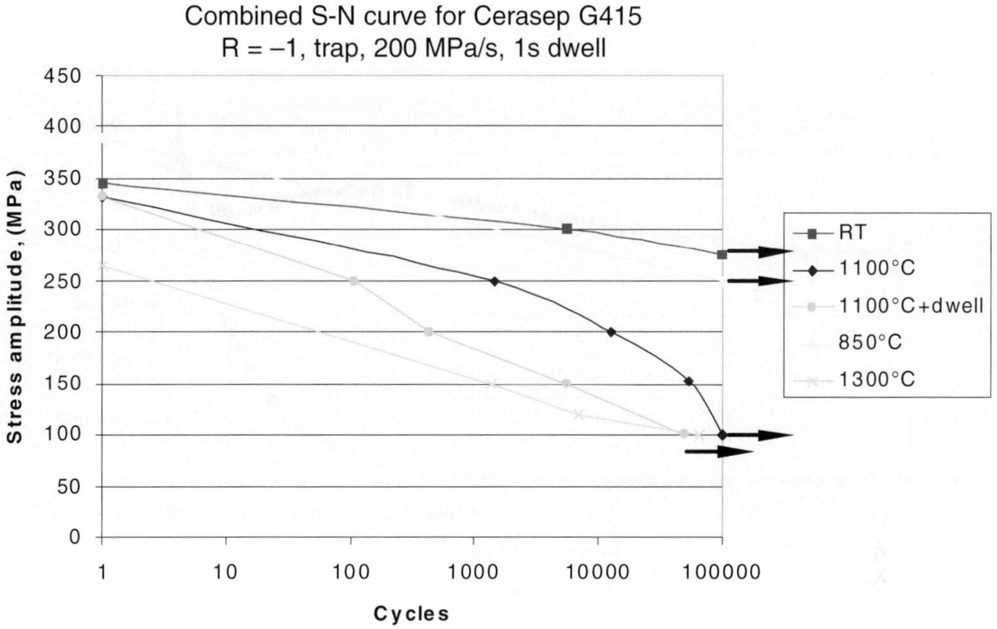

Fig. 6 Load control LCF S-N curves for Cerasep G415.

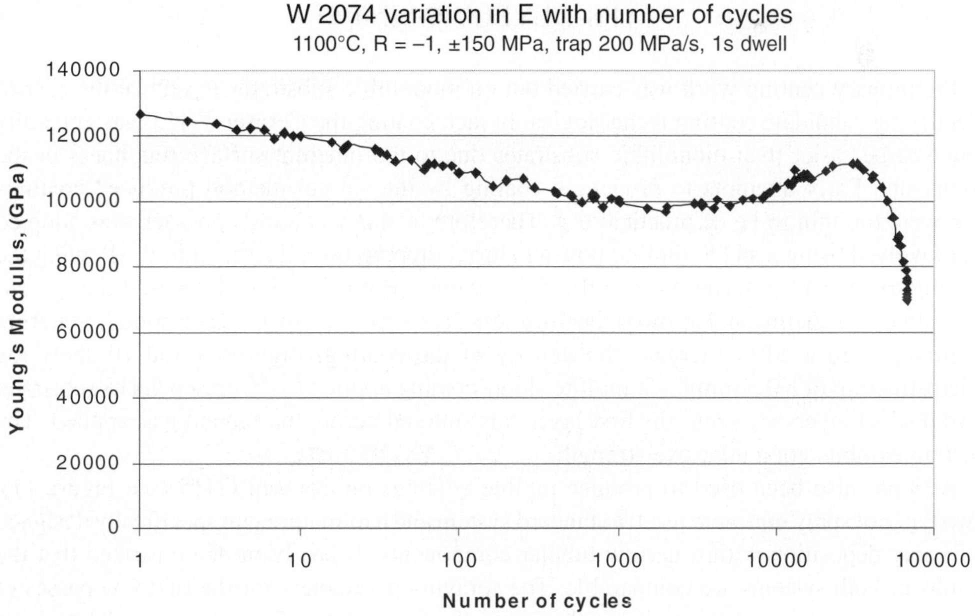

Fig. 7 Variations in modulus with number of fatigue cycles for Cerasep G415 at 1,100°C.

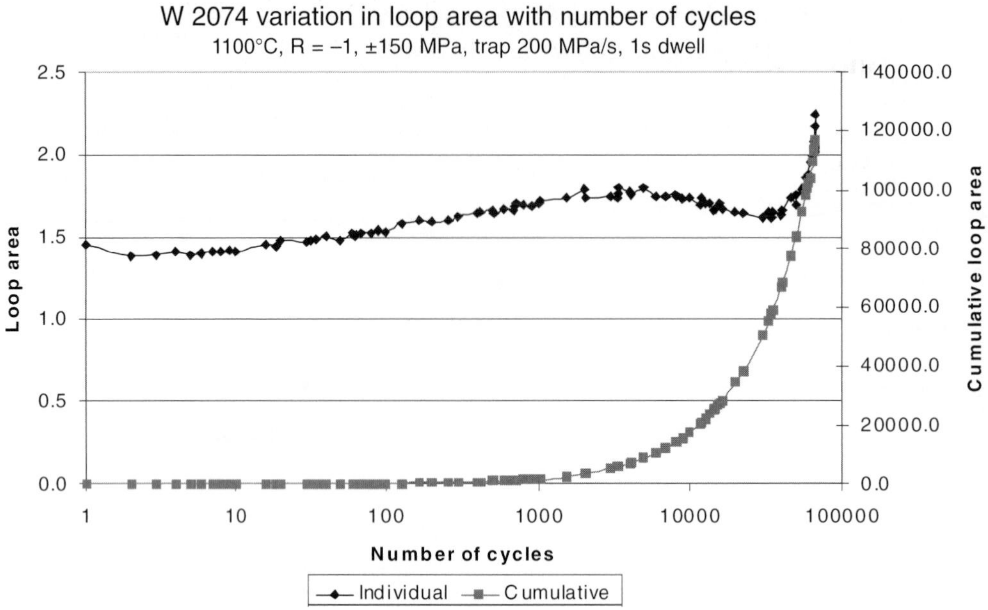

Fig. 8 Example of the variations in hysteresis loop area with number of cycles.

TPS DEVELOPMENT

Preliminary coating work was carried out on monolithic substrates to reduce the cost of identifying candidate coating technologies. In fact, coating the Cerasep G415 was generally found to be easier than monolithic substrates due to the inherent surface roughness of the composite. Early attempts to produce a coating by the sol-gel method produced coatings that were too thin to be of practical use. Therefore, a slurry coating approach was adopted that involved using a stable mullite powder slurry, dipping the substrate into it, allowing the coating to dry and sintering to give the final coating. Binding agents improved the 'green' strength of the coating and allowed the thickness deposited per dip to be controlled. Sintering additives were used to increase the density of the coating. Figures 9 and 10 show the microstructure of a 0.5 mm thick mullite slurry coating applied to a Cerasep G415 substrate. Two distinct layers are seen: the first layer was sintered before the second was applied. The coating exhibits good adhesive strength.

APS has also been used to produce mullite coatings on Cerasep G415 (see Figure 11). Two types of spray gun were used: a standard system and a miniature gun specifically designed for spray deposition within narrow tubular components. It has been demonstrated that the results of both systems are comparable. The spraying parameters for the first few passes of the gun were optimised for mullite adhesion to the Cerasep G415 substrate (analogous to a bond-coat), whilst the remainder were optimised for coating deposition. Coatings up to

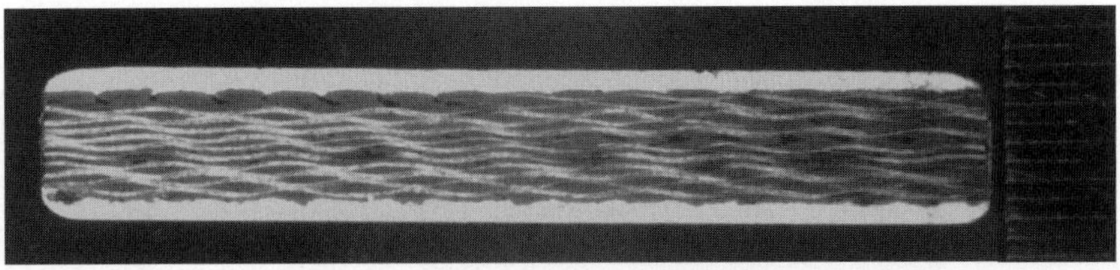

Fig. 9 Example of slurry deposited mullite coating on Cerasep G415.

Fig. 10 Microstructure of slurry deposited mullite coating on Cerasep G415.

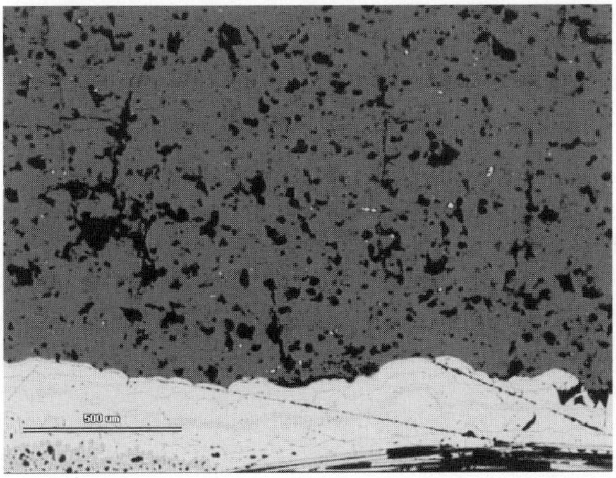

Fig. 11 Microstructure of APS deposited mullite coating on Cerasep G415.

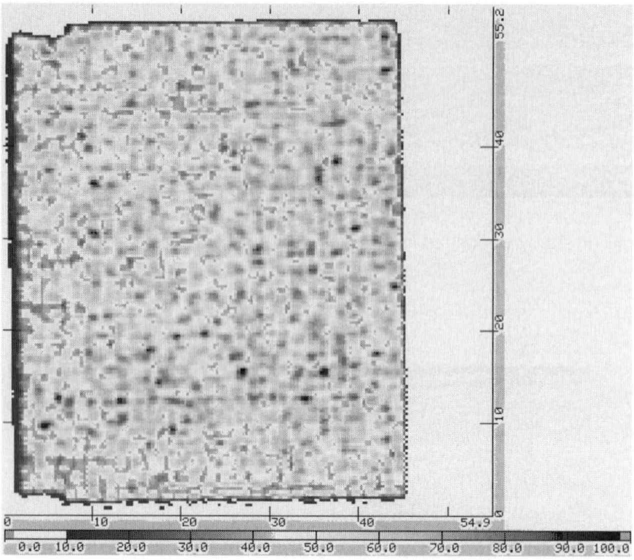

Fig. 12 Transmission ultrasound scan of Cerasep G415 test coupon.

2 mm in thickness have been produced. Figure 11 shows a level of porosity, which has been measured as approximately 11%, and some vertical cracking. It is possible that these may be beneficial in terms of providing some degree of thermal strain tolerance. The adhesion to the substrate looks sound, push-off tests have measured an adherence strength of 36 MPa.

Efforts to produce good quality yttrium silicate coatings have so far been of limited success. The sol-gel route produced only very thin coatings. The use of a slurry application provided thicker coatings that proved difficult to sinter. Further work in this area may include use of the sol-gel material as cement for the larger slurry particles.

NDE

Figure 12 shows an ultrasound pulsed echo image in which the weave structure of the CMC is revealed. Irregularities in the weave structure can be discerned. Figure 13 shows an image of a delamination that was deliberately produced by contaminating the substrate surface with graphite prior to coating application. Such a delamination during production may lead to premature failure of the TPS and thus this type of NDE constitutes a potent quality control and lifing tool for the system.

Figures 14, 15 and 16 show selected virtual sections generated using X-ray computer tomography. Both 2D and 3D sections through the CMC and a 3D section through a coating applied to a monolithic substrate are shown. Clearly resolvable are features such as porosity in both the substrate and coating as well as second phase particles in the coating. Attempts to

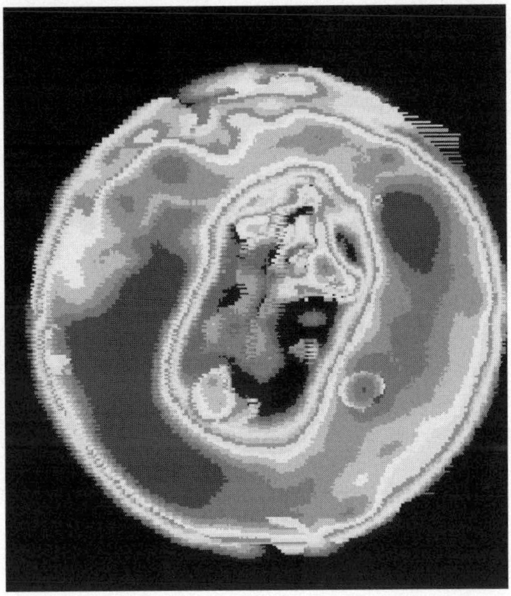

Fig. 13 Pulse-echo ultrasound scan showing deliberate delamination between coating and substrate.

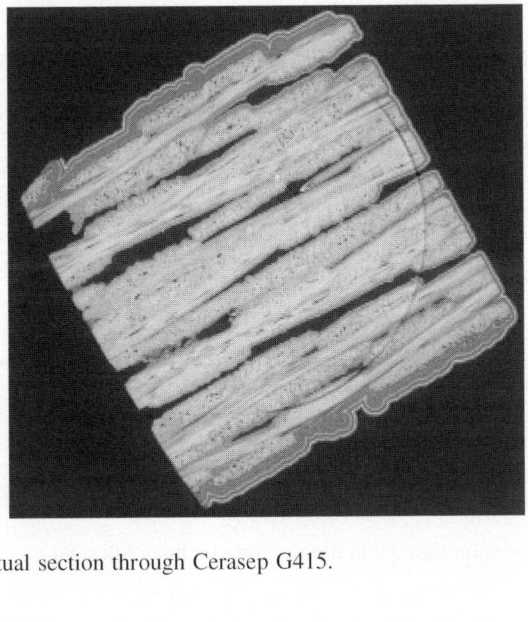

Fig. 14 X-ray CT virtual section through Cerasep G415.

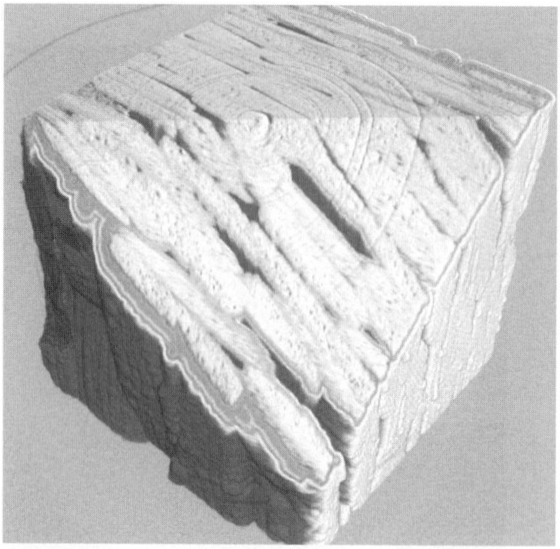

Fig. 15 3D virtual section through Cerasep G415.

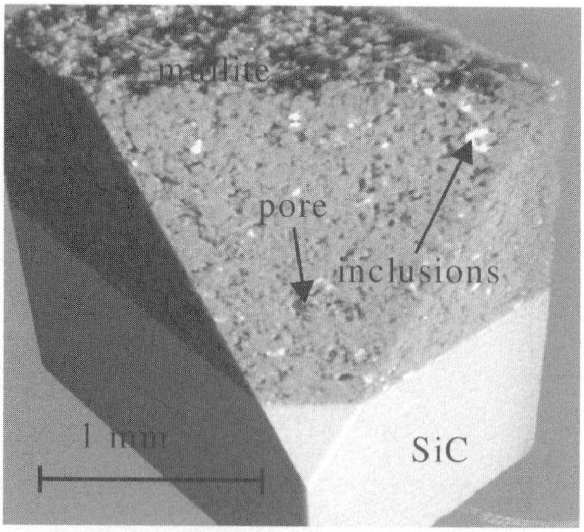

Fig. 16 3D virtual section through mullite coating on monolithic SiC.

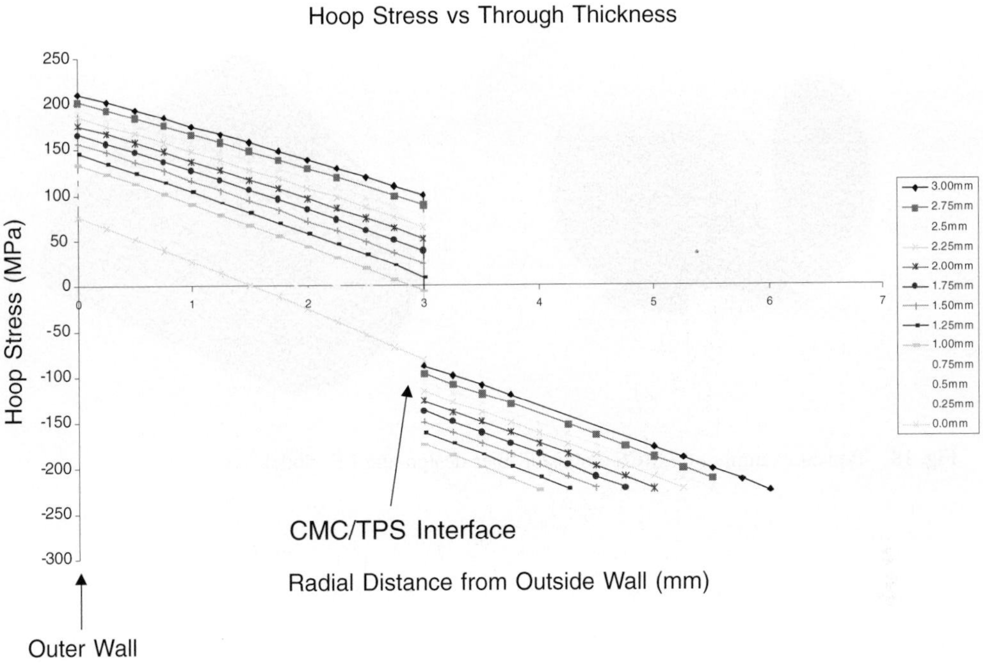

Fig. 17 Stress analysis results for G415 sub-element specimen as a function of position through CMC and APS mullite coating thickness.

visualise the damage to the CMC due to mechanical loading have so far had limited success. This is primarily due to water penetration into the matrix cracks. Further work is proceeding in this area.

DESIGN

Design work for the sub-element specimens and component has been completed. This work consisted of three phases:

i. CMC substrate modelling,
ii. Optimisation of coating thickness (once decisions have been made as to which coatings to apply) and
iii. Integration technology to hold the liner in place in the test rig for the complete design package.

Figure 17 shows the stress analysis results as a function of position through the thickness of the rig test sub-element specimen and as a function of APS mullite coating thickness. Based on these stress analyses, the preliminarily coating thickness was chosen as between 1 and 1.5 mm. Similar results were generated for other coating types.

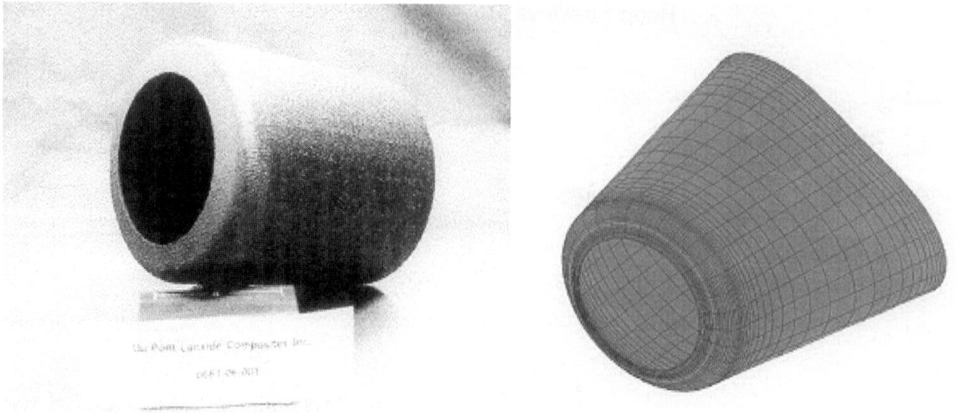

Fig. 18 Typical example of CMC combustor liner design and FE model.

An early example of the target combustor liner design is shown in Figure 18. During the course of the current programme, the design of CMC liners has become simpler to limit distortion of the fibre architecture associated with complex curvatures and to facilitate less convoluted retainment systems. Current methods are based on circumferential spring fixtures and an anti-rotational pinning device capable of accommodating the differential thermal expansion between the component and test rig. Design of the full size component is currently at the stage where the CMC substrate has been fixed to allow manufacture to start. This is necessary considering the long lead-time in delivery of the CVI manufactured CMC parts.

CONCLUSIONS

1. The design, manufacture and integration of CMC combustor liners for gas turbine engines requires a complex balance of manufacturing considerations, materials properties and design calculations. Future implementation of such technology will require a successful demonstration of the materials technology as well as the generation of a 'design toolbox' that consists of a design guide and lifing methods for the component and its integration into the engine. The 'toolbox' forms one of the key deliverables of the Brite EuRam FPV CINDERS project.
2. The current programme has provided much needed mechanical and thermophysical property data for the CMC substrate and candidate coating systems. These data have been applied in the stress modelling and design of a series of sub-element specimens and components. These CMC parts are now under manufacture and will be tested during the remainder of the programme.

3. A number of NDE techniques have been used to characterise the structure and defect distributions within the CMC and coatings. These investigations have been validated using microstructural examination methods.
4. Successful design and application of a thermal protection system (TPS) for SiC/SiC ceramic composites is crucial to the implementation of CMC combustor liner technology.
5. Further considerations that may need to be addressed are the current manufacturing capacity for CMCs and a degree of inertia that surrounds the adoption of CMC's for structural applications.

ACKNOWLEDGEMENTS

The authors would like to gratefully acknowledge the permission of ALSTOM Power for publication of this paper. Many thanks are due to colleagues at ALSTOM and project partners from the Brite EuRam Project CINDERS (ONERA, Flamespray, RIF, SNECMA and Rolls-Royce plc) who have contributed/conducted much of the work reviewed in this paper.

REFERENCES

1. S. R. Levine: 'Ceramics and Ceramic Matrix Composites in Future Aeronautical and Space Systems', *Ceramics and Ceramic-Matrix Composites*, S. R. Levine, ed., ASME, NY, 1992.
2. Van Roode, et al.: 'Ceramic Gas Turbine Development', ASME 1993 GT 309, 1993.
3. J. Shi: 'Ceramic Matrix Composites for Gas Turbine Engines', *Advances in Turbine Materials, Design and Manufacturing*, A. Strang, et al., eds., IOM, 1997.
4. J. Price, et al.: 'Ceramic Stationery Gas Turbine Development Program', ASME 2000-GT-75, 2000.
5. K. L. More, et al.: 'Exposure of Ceramics and Ceramic Matrix Composites in Simulated and Actual Combustor Environments', ASME 99-GT-292, 1999.
6. H. E. Eaton, et al.: 'EBC Protection of SiC/SiC Composites in the Gas Turbine Combustion Environment', ASME 2000-GT-631, 2000.
7. J. Shi: 'Modelling the Tensile Behaviour of a SiC/SiC Composite by Master Curves', ASME 2000-GT-66, 2000.
8. J. Shi: 'Design, Analysis, Fabrication and Testing of a CMC Combustor Can', ASME, 2000-GT-71, 2000.
9. J. Shi: 'Tensile Fatigue and Life Prediction of a SiC/SiC Composite', *Proceedings of ASME TURBO EXPO 2001*, New Orleans, USA., 2001-GT-0463, 2001.
10. J. Burke: 'Ceramics for High Performance Applications', *Proceedings of the Second Army Materials Technology Conference*, MA, USA., 1973, 565.

Single Crystal Braze Repair of SX GT Components

A. SCHNELL and D. GRAF
ALSTOM (Switzerland) Ltd.
Brown Boveri Strasse 7
CH-5400 Baden, Switzerland

ABSTRACT

Economic maintenance of the latest ALSTOM gas turbine generation with single crystal (SX) blading components has required the development of advanced repair processes, such as the braze repair of single crystal components. Due to the high costs of the SX components, and the long lead times for the manufacture of new parts, there exists a necessity for repair techniques that may be used to restore cracked components, without the need to compromise the original mechanical integrity.
The main prerequisite for the implementation of a braze-repair for SX components in a reconditioning process is the guarantee of sufficient mechanical integrity at turbine operation temperatures, including resistance to cyclic effects such as thermal mechanical fatigue (TMF). The high mechanical integrity of the braze joint is achieved by combining suitable crack surface preparation with tailored braze alloys and brazing parameters. Specifically optimised Fluoride Ion Cleaning (FIC) cycles have been employed for complete oxide removal to the crack tip, and braze alloys with strong wetting properties are used to promote the flow of braze alloy for complete crack filling.
The mechanical properties of the braze-repaired SX components must be close to or equivalent to the single crystal material properties of the parent metal in order to be able to apply the brazing process to highly loaded areas of the components.
This paper presents the development of various braze alloys with a high g' content for single crystal repair, as well as the optimisation of brazing heat treatment cycles. This paper also proves the retention of a single crystal microstructure, and demonstrates the test procedures developed to show the required mechanical integrity of brazed repairs for SX gas turbine applications.

INTRODUCTION

Considerable improvements in gas turbine efficiency over recent years have been mainly achieved through increasing the turbine inlet temperature (TIT), in combination with advanced turbine materials such as Nickel-based superalloys in their directionally solidified (DS) and single crystal (SX) form. The ALSTOM GT24/26 Gas Turbines also employ SX materials for the vanes and blades in the High Pressure Turbine (HPT) as well as the first stage blades in the Low Pressure Turbine (LPT). The second row of rotating components in the LPT is made from DS material.

SX turbine blading components are subjected to severe temperature and stress conditions during service, which can cause thermal mechanical fatigue (TMF) cracking to the base material. Such occurrences of base material cracking represent one major factor responsible for the limits imposed on component lifetime. As SX turbine blading components represent a significant portion of the maintenance costs, the installation of reconditioned components - rather than the cost intensive replacement with new parts - is one area of major interest to the power plant industry. The necessity exists for braze repair techniques that may be used to restore SX components, without the need to compromise the original mechanical integrity.

Reconditioning of conventionally cast polycrystalline turbine components has been carried out for many years using advanced forms of well-established joining techniques such as welding and brazing. Recently also the weld repair of worn blade tips made from SX material using Laser welding techniques have been developed and the technology is close to the industrial implementation.[1] The braze repair of TMF cracked SX turbine components with a maintained single crystal microstructure has not yet been reported. The target of the present work was to demonstrate the possibility of high strength repair of cracked SX components with the full retention of the single crystal microstructure.

Braze Repair of Turbine Balding Components

The application of brazing for repair and restoration of cracked and eroded blading components was originally established by the aircraft industry and is today a standard repair technology for land-based gas turbine components. The braze repair techniques currently used in the gas turbine reconditioning industry can be categorised into three applications, depending on the requirements of the damaged components:[2]

- Diffusion Brazing (DB) of narrow cracks or gaps,
- Transient Liquid Phase Bonding (TLP) for joining two separate superalloy parts and
- Wide gap brazing for the repair of wide-open cracks and wall thickness restoration.

The metallurgical stages involved during Diffusion Brazing and Transient Liquid Phase Bonding are similar, and the two terms are often used to describe a brazing process which results in high strength braze joints. It should be noted however, that diffusion brazing is applicable to situations in which the braze alloy cannot be applied as a separate layer between the two joining surfaces. In the case of crack repair by diffusion bonding, the braze alloy melts and flows controlled by capillarity effects in to the gap or crack. For diffusion brazing used as a crack repair process wetting mechanisms play an important role. Therefore, prior to the diffusion brazing process, the ex-service GT components must be subjected to a sequence of mechanical and chemical cleaning processes. In addition to a clean outer component surface, the entire removal of all oxide scales from the crack surfaces is the main prerequisite for the braze alloy to flow into the crack, wetting the surfaces down to the crack tip.

The Fluoride Ion Cleaning (FIC) process

The FIC (Fluoride Ion Cleaning) process has been established as the current standard cleaning method at modern reconditioning facilities. The process subjects the oxidised components to a highly reducing gaseous atmosphere of hydrogen fluoride at elevated temperatures.[3]

This aggressive environment is especially effective in removing deeply embedded oxides in narrow and intricately shaped cracks. The diffusion brazing of cracks has been previously limited to non-critical areas of components as the danger existed that some cracks may only be partially filled and still remain open under the surface. Each type of superalloy forms different oxide scales depending on the chemical composition. Therefore, specifically tailored FIC parameters for each particular superalloy material are required to ensure effective cleaning of all crack surfaces. Extensive FIC parameter studies performed by ALSTOM's reconditioning centre proved that the process is capable of removing deeply embedded oxides from the surfaces of narrow cracks in all currently used superalloys.[4] Effective oxide removal has been proven for alloys such as IN738, IN939, X-45 and MarM247, as well as the SX alloys CMSX-4 and ALSTOM's proprietary SX alloy MK4.

Figure 1 shows a detailed image of an FIC-cleaned crack tip in CMSX-4 material. During the FIC process, Al from the base material at the crack surface is depleted, thereby forming a thin γ'-free layer (~5 μm) adjacent to the crack surface, Figure 1a. The entire crack down to the crack tip has been completely cleaned of all oxides which allows the complete filling with braze alloy during brazing, Figure 1b.

In order to produce high strength braze joints without any detrimental eutectic or brittle phases in the middle of the joint especially long brazing cycles resulting in the isothermal solidification of the joint are required.

HIGH STRENGTH BRAZE REPAIR USING ISOTHERMAL SOLIDIFICATION

It has already be demonstrated in the 1970's that isothermal solidification is the key factor for obtaining high quality brazed joints without the presence of brittle phases at the centre of the joint [5]. During isothermal solidification the melting point depressant (MPD) diffuses from the liquid braze alloy into the parent metal causing the liquid/solid interface to move in the opposite direction to the MPD diffusion. If cooling is initiated before the braze alloy has isothermally solidified, further solidification is driven by the outer temperature gradient. The residual MPD in the remaining liquid is rejected from the growing γ-phase due to very low solubility of the MPD in the γ-phase, so that the middle of joint solidifies in a eutectic manner.

Experiments using MK4 as the base material and the standard braze alloy D15 (Ni-10Co-15Cr-3.5Al-3.5Ta-2.5B) show the significant influence of the brazing time on the microstructure of the braze joint, see Figure 2. The non-isothermally solidified sample shows a microstructure consisting of the brittle ternary eutectic $\gamma + Ni_3B$ + Cr-rich Borides of the type Cr_5B_3 (Figure 2a), whereas with sufficiently long brazing times it is possible to produce fully isothermally solidified joints (Figure 2b).

Isothermal Solidification Time

Since isothermal solidification is time dependent, it is of major economical interest to define the minimum required dwell time at brazing temperature for a certain gap width in any

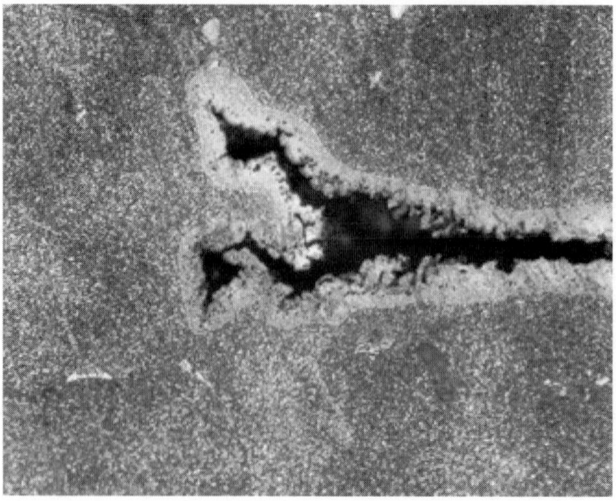

(a)

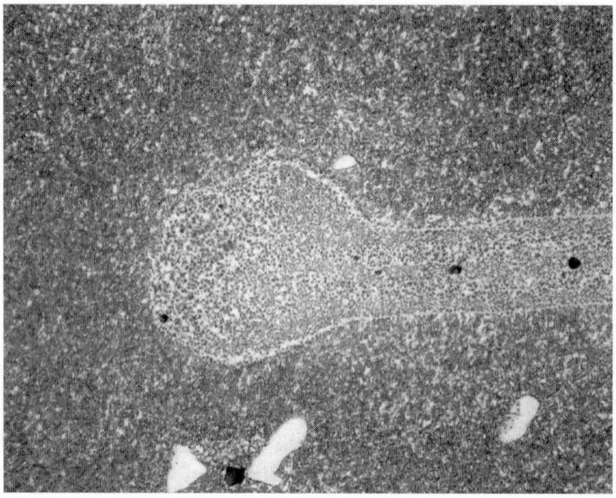

(b)

Fig. 1 (a) FIC cleaned crack-tip in CMSX-4 material prior to brazing and (b) FIC crack-tip in CMSX-4 material. The crack-tip has been completely filled with the braze alloy.

particular parent metal/braze alloy system. Analytical solutions for the required isothermal solidification time have been presented in the literature.[6–8] Considering a moving boundary problem of the solid/liquid interface during brazing, and based on Fick's 2nd law, the isothermal solidification time (t_s) can be expressed as:

$$t_s = \frac{K \times W_o^2}{D_s}\left(\frac{C_F}{C_{aL}}\right)^2$$

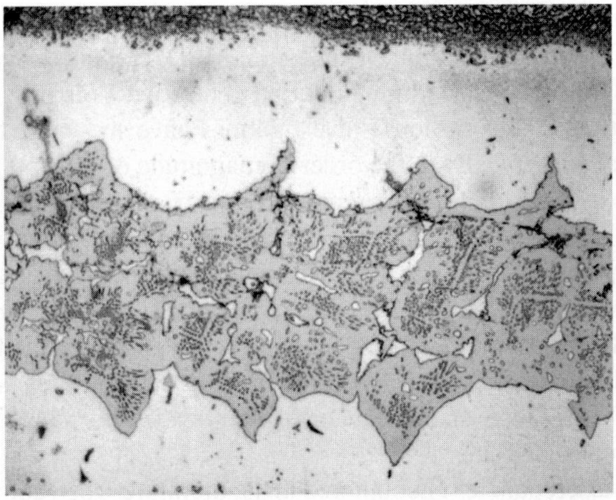

(a)

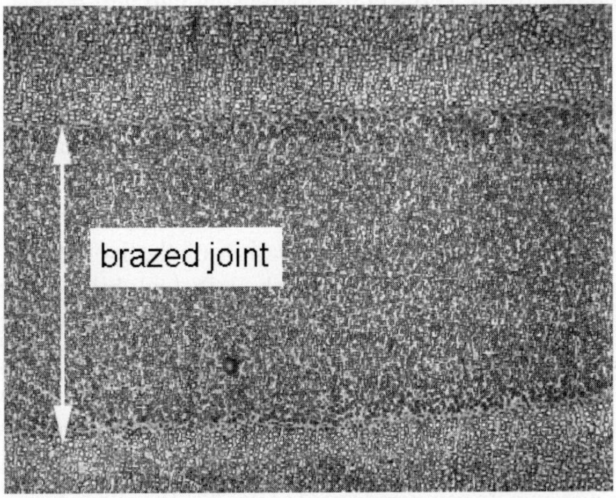

(b)

Fig. 2 Microstructure of brazed joints as a function of the brazing time, (a) Isothermal solidification, interrupted after 3 hours resulting in the formation of a eutectic layer in the middle of the joint, which consists of Ni_3B and Cr-rich borides and (b) Fully isothermally solidified 100 μm wide joint in MK4 material. The aligned microstructure of the γ' particles indicates the single crystalline structure of the braze joint.

where:

t_s is the isothermal solidification time,
C_F is the initial Boron content in the filler metal,
Ca_L is the solubilty of Boron in the base material,
W_0 is the initial gap width,
D_s is the diffusion coefficient of Boron in the base material and
K is a material dependent coefficient.

Based on the above equation the isothermal solidification time (t_s) varies as the square of the initial gap width (W_0) and the initial wt.% Boron (MPD) content. During repair of cracked SX components the initial crack width can not be controlled as is possible during the TLP bonding of new parts. Previous in-house investigations have shown that TMF base material cracks in SX material are typically in the order of magnitude of 80–150 μm, and the severity of cracking must be assessed before the repair process begins.

A further parameter, the boron content in the braze alloy can not be significantly reduced below 2.5 wt.% as a sufficiently high boron content is required to guarantee the flow and wetting properties of the braze alloy. The diffusion velocity of boron into the parent metal and therefore solidification time (t_s) is exponentially influenced by the brazing temperature according to an Arrhenius equation. Raising the brazing temperatures is however limited to values below 1200°C because of the danger of recrystallisation and coarsening of the γ' phase in the SX parent material.

Due to the strong discrepancies encountered in the literature when comparing theoretically and empirically determined isothermal solidification times, a new approach has been formulated, involving the use of a constantly increasing gap width (W_0), and a fixed time for solidification (t_s), in this case, 10 hours. A sample comprising of two MK4 plates has been brazed together (using D15 braze) with a wedge shaped gap of between 0 and 300 μm powder for 10 hours. The maximum gap width at which the joint was completely isothermally brazed, with no residual eutectic phases, has then be measured, see Figure 3. Consideration must also be given to the cooling time from brazing temperature, during which a small amount of further isothermal solidification continues to occur.

The experimental approach allows the estimation of the isothermal solidification time, even without certain unavailable data values such as the solubility concentration (Ca_L) and diffusion coefficient of Boron in the parent material (D_s). The empirical investigation also allows the material dependent coefficient (K) to be estimated. Only little data exists about the diffusivity and the solubility of Boron in the base material at the brazing temperature, therefore a close estimation of the required isothermal solidification time is difficult. In the literature, values for the diffusion of Boron in Nickel have been found to be: Diffusivity D_0 of 6.67×10^{-7} m²/sec and an activation energy Q of 96 300 J/mol.[9, 10] An approximate solubility value of 0.02 wt.% for Boron in the γ-phase has been taken. Using this data and the calibrated material dependent coefficient (K) gained from the wedge sample experiment an approximate estimation of the relationship between isothermal solidification time and the gap width for the MK4/D15 system could be calculated, see Figure 4.

As can be seen from the graph, extended heat treatment times are required for wide gap brazing, such that gaps wider than 200 μm cannot be completely brazed by isothermal solidification within a reasonable brazing time.

Modelling of the Brazing Process Using DICTRA[11]

In parallel to the analytical and experimental approaches, numerical modelling of the isothermal solidification process has been presented in the literature.[12, 13] In-house modelling of the brazing process has been carried out using the DICTRA software tool using ALSTOM proprietary mobility databases. The chosen boundary conditions were as follows: a 100 μm

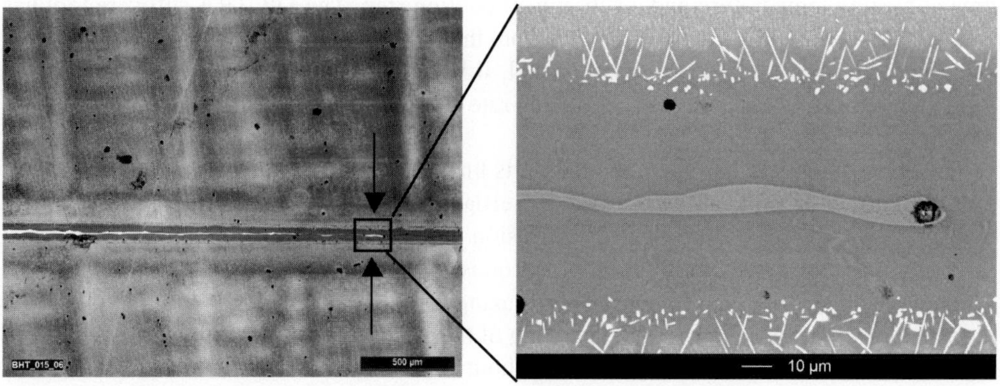

(a) (b)

Fig. 3 (a) Microstructure of the wedge sample etched with BerahaIII in order to visualise the residual eutectic phase and (b) Detail SEM photo of the point at which complete isothermal solidification is achieved.

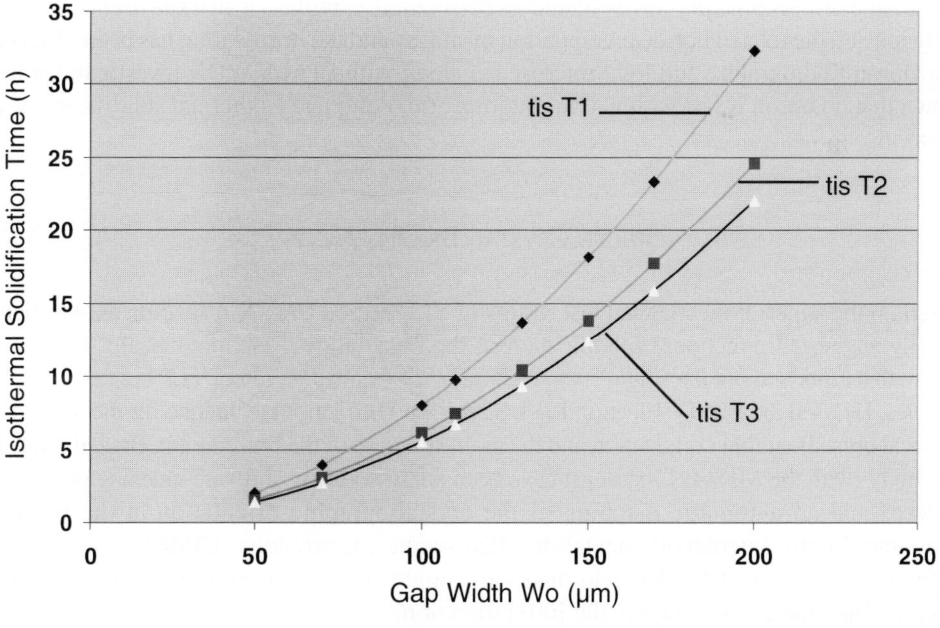

Fig. 4 Calibrated dependency of the isothermal solidification time on the cap width W_0 for the MK4/D15 system.

wide initial gap and a simple brazing cycle consisting of a heating step, an 8 hours dwell time at brazing temperature and a subsequent cooling step. The DICTRA software tool has been used to model the diffusion of boron from the braze alloy into the parent material using simplified mobility data for Boron in Nickel. It was found from the model results that the formation of borides influences the diffusion rate of boron from the filler metal into the base material, see Figure 5.

The observation that Boride precipitation is linked to the diffusion of boron into the base material correlates well with the work of LeBlanc[14] in which he assumed that the in-situ precipitation of borides accelerates the diffusion process. In the equation for the isothermal solidification time, the solubility limit of Boron is extremely low (approximately 0.02 wt.%) which would result in extraordinarily long brazing times. The consideration of in-situ boride precipitation contributes to the understanding of the general overestimation of the calculated isothermal solidification times found throughout the literature.

Although simplified diffusivity values have been used in the DICTRA model, the results show potential that the software tool can be used to simulate any given braze material/braze alloy system. The experimental samples could also be used as a diffusion couple, and can be employed to calculate the diffusivity of boron in the MK4 material using the appropriate diffusion equation.

Boron Diffusion

For all metallographically investigated experimental samples, a marked trend in boron diffusion and the related boride precipitation in and around the brazed joint has been observed. The Boron Radiography and light microscope along with EPMA-WDX investigations have shown that no boron is left within a brazed joint after complete isothermal solidification, see Figure 6.

Single Crystal Brazed Joints

Based on the superior mechanical test results of TLP brazed CMSX-4 specimens, Miglietti[15] already suggested directional solidification of the braze alloy. Nishimoto et al.[16] is the only source that reports about the single crystal nature of the bonded region of TLP bonded CMSX-2 samples. He used the EBSD (Electron Back Scattering Diffraction) technique for the analysis of the local crystallographic orientation and the grain structure of the brazed joint. Brazed joints for this study (with the MK4-D15 system) have been analysed using software indexing of Kikuchi patterns and an automatic mapping of the crystallographic orientation at the EPFL in Lausanne, Centre Interdepartemental de Microscopie Electronique (CIME).

Figure 7 shows a EBSD Kikuchi map for a single crystal brazed joint from a laboratory sample. The blue colour indexes the [001] direction.

The main challenge has been to prove whether complete monocrystalline repair of a TMF crack in a component made from MK4 material is achievable. Single crystal high pressure turbine vanes with TMF cracking from thermal shock testing have been taken and subjected to an entire repair process including FIC cleaning and diffusion brazing processes.

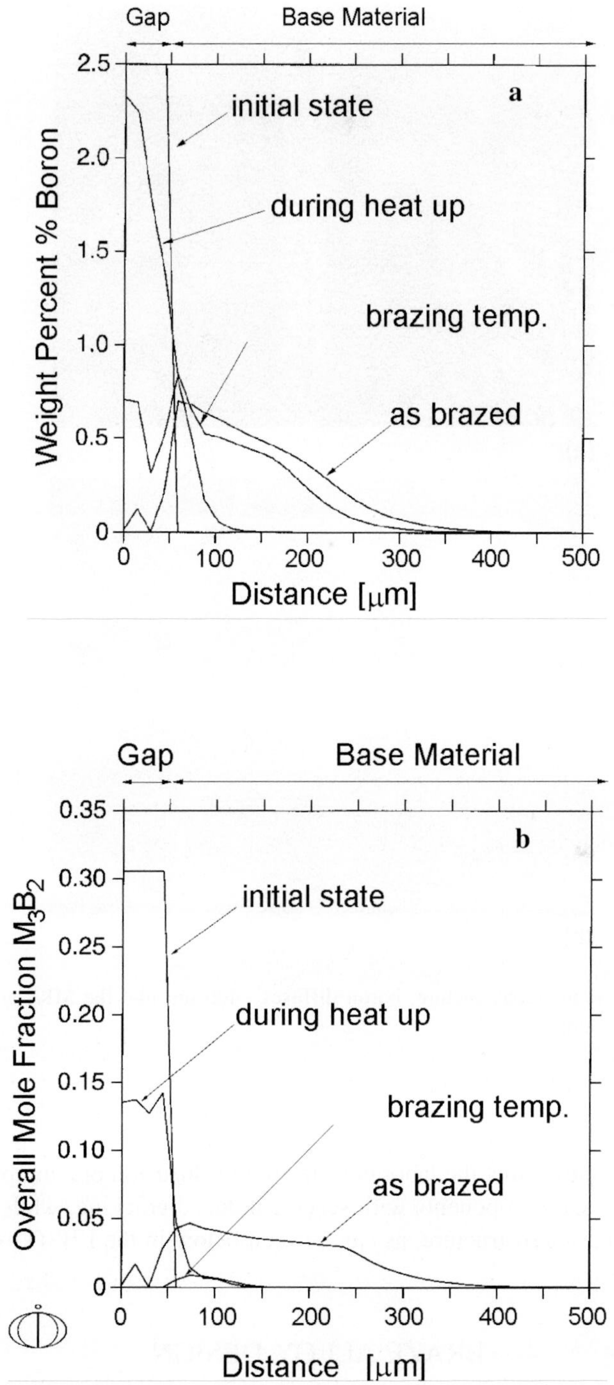

Fig. 5 (a) Boron distribution at different stages during the brazing cycles and (b) Boride formation in the base material due to the diffusion of boron.

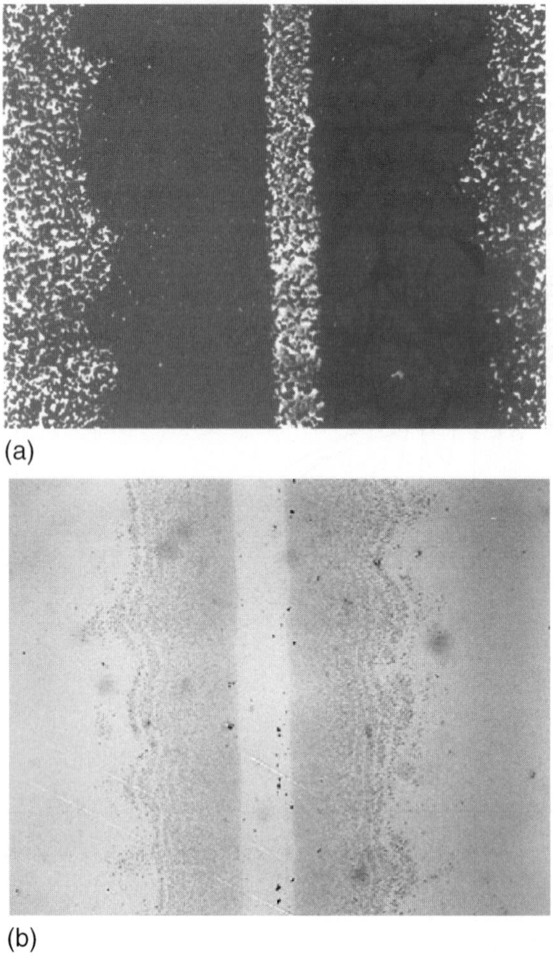

Fig. 6 (a) Boron radiography picture, boron diffuses ~400 μm into the MK4 and (b) Specially eteched MK4/D15 sample.

The results have shown that the laboratory testing of diffusion brazing procedures can be applied to single crystal components with service-induced cracking, while maintaining the desired single crystal microstructure, as can be seen below in the EBSD map, Figure 8.

BRAZE ALLOY DESIGN

The current in-house development activities aim to optimise high strength brazing alloys by the addition of solid solution hardening elements and elements which increase the γ' volume fraction, along with a higher γ' solves temperature which results in improved mechanical

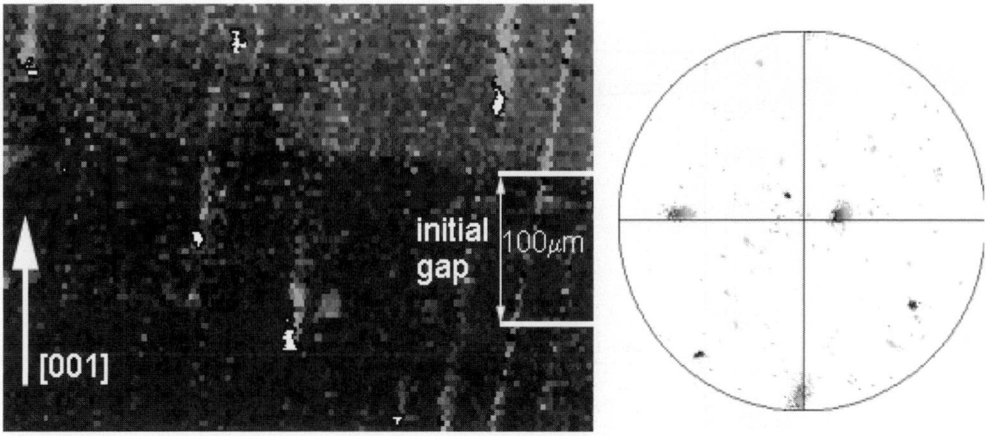

Fig. 7 (a) EBSD Kikuchi map of single crystal brazed joint and (b) (001)-Pole figure, only slight deviations from the 001 direction in the joint.

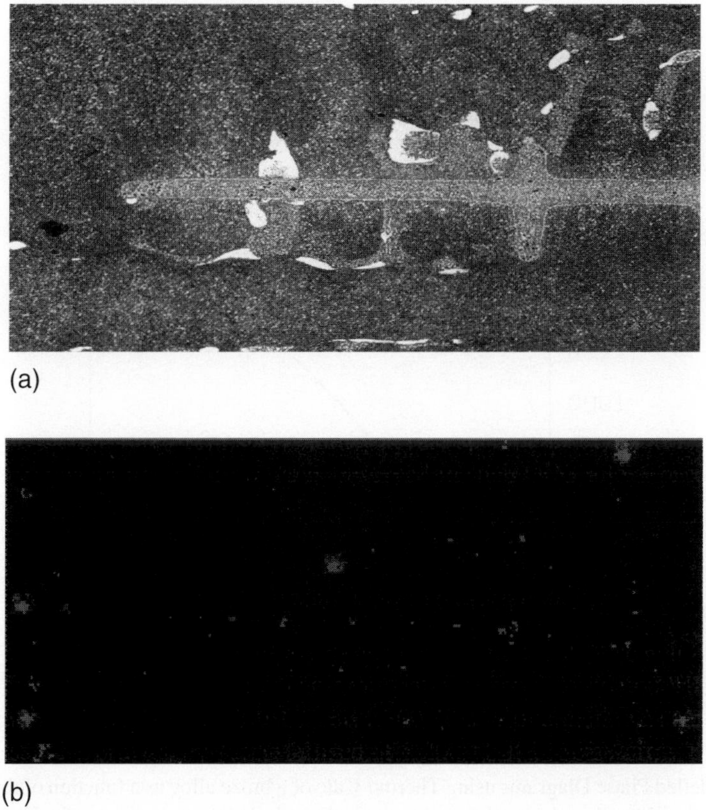

Fig. 8 Microstructure (a) EBSD Kikuchi map and (b) of single crystal brazed crack.

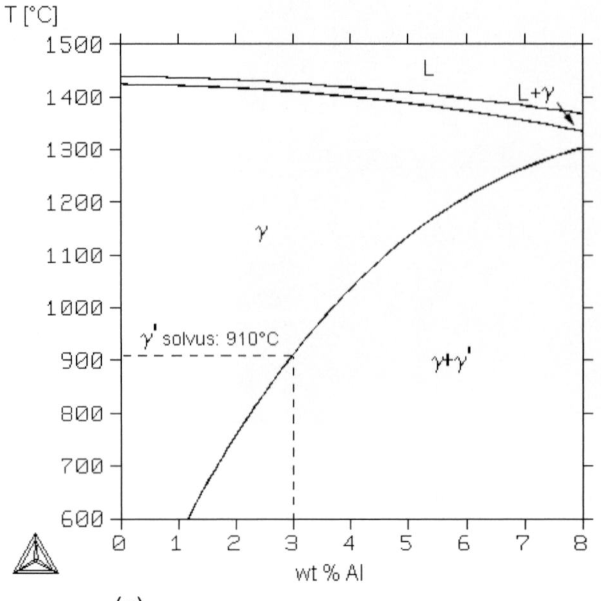

(a)

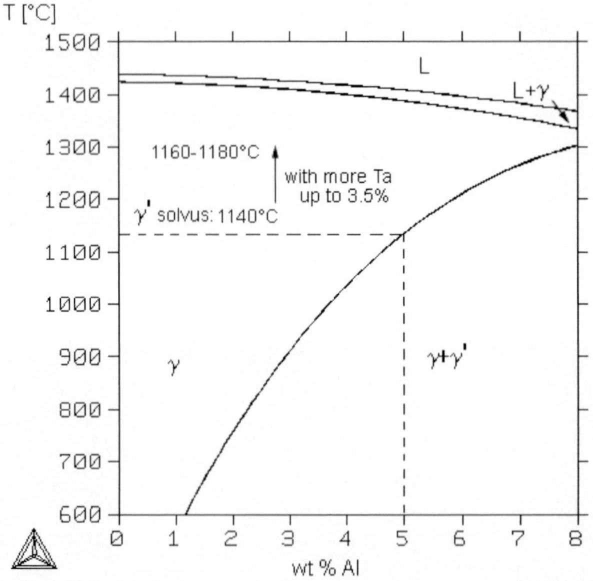

(b)

Fig. 9 Modelled Phase Diagrams using Thermo-Calc of a braze alloy as a function of the Al content. No Boron is considered in the model as the phase diagram shall represent the final composition in the brazed joint after isothermal solidification has been accomplished.

properties at elevated temperatures. This has been modelled by adjusting the elements which control the precipitation and thermal stability of γ' phase. As Thermo-Calc[17, 18] modelling and subsequent practical application of the appropriate alloys has shown, the most notable improvements in γ' properties were found in samples containing higher weight percentages of Aluminium. Four optimised braze alloys based on Thermocalc-Modelling have been produced, and preliminary testing has been carried out showing no adverse effects on the flow behaviour of braze alloys with an increased Al content. The γ' solvus temperature of the proprietary braze alloy lies approximately 150°C higher than that of the D15 braze material, which results in a more stable γ/γ' microstructure at elevated temperatures.

TESTING THE MECHANICAL INTEGRITY

Several approaches to test the mechanical integrity of brazed joints can be found in the literature.[19] Typically test specimens with a butt-brazed joint in the middle of the gauge length are used for LCF (dwell and no dwell) or creep testing. Such laboratory tests are required in order to gain inherent materials properties of the brazed joint for life-time estimations of repaired components. A further important criterion required for the release of braze-repaired components for an engine trial run is the evaluation of the part during simulated transient start/stop conditions. A Thermal Shock (TS) test (which gives an initial, qualitative indication of the service behaviour) has been performed using a cracked and braze-repaired SX-HP vane. A narrow crack at the trailing edge of has the vane has been diffusion brazed and the component was subsequently subjected to Thermal Shock (TS) tests in order to evaluate the quality and integrity of the repair process. The component was not re-coated for the thermal shock testing. During such a TS test the entire component is heated to the average component operating temperature and then rapidly cooled to room temperature. Firstly the braze repaired component was cooled with air, simulating usual shutdown conditions of a turbine. The crack at the trailing edge survived more than 50 air-cooled TS cycles. After the air-cooled TS test series the component has been further cycled with water cooling which generates extremely high TMF loading in the component. The crack at the trailing edge survived more than 50 severe water-cooled TS cycles without any reopening, indicating sufficient mechanical integrity of the brazed component required for an engine trial run.

The thermal shock test procedure has been previously evaluated in-house as a useful tool as it closely represents the severe conditions encountered during start/stop cycles. Nevertheless, the TS test procedure represents only a limited indication of the mechanical integrity of brazed components, and should be combined with the results of laboratory mechanical testing and engine trial runs in order to obtain comprehensive data values for braze repaired components.

CONCLUSIONS

- The FIC cleaning process has been successfully established for MK4 superalloy material as the crack surface cleaning process prior to brazing.

- Full Isothermal solidification is the prerequisite for high strength braze joints.
- The parameters influencing the minimum required isothermal solidification time have been studied and approximate estimations have been given.
- Full retention of the single crystal microstructure in braze-repaired cracks has been achieved and demonstrated.
- Braze alloys containing higher amounts of Al result in braze joints with improved mechanical properties due to a high volume fraction of the g' phase.
- The mechanical integrity of braze repaired SX vanes have been proven by thermal shock testing.

ACKNOWLEDGEMENTS

The authors acknowledge support and contributions of Maxim Konter, Andreas Bögli and Alexander Stankowski (all with ALSTOM Customer Service, Repair Technology).

REFERENCES

1. M. Hoebel, B. Fehrmann and A. Schnell: 'Robot Guided Laser Repair of Single Crystal Turbine Blades', *Power-Gen Europe*, Düsseldorf, 2002.
2. D. W. Gandy, G. Frederick, JT Stover and R. Viswanathan: 'Overview of Hot Section Component Repair Methods', EPRI/DOE Conference Report, Orlando, 2002.
3. W. Miglietti and F. Blum: 'Advantages of Fluoride Ion Cleaning at Sub-Atmospheric Pressure', *Engineering Failure Analysis*, **5**(2), 1998, 149–169.
4. A. Stankowski: 'Advanced Thermochemical Cleaning Procedures for Structural Braze Repair Techniques', Turbo Expo Amsterdam, GT-2002-30535.
5. D. S. Duvall, W. A. Owczarski and D. F. Paulonis: 'TLP Bonding: a New Method for Joining Heat Resistant Alloys', *Welding Journal*, **53**, 1974, 203–214.
6. I. Tuah-Poku, M. Dollar and T.B. Massalski: 'A Study of the Transient Liquid Phase Bonding Process Applied to a Ag/Cu/Ag Sandwich Joint', *Metallurgical Transactions*, **19A**(3), 1988, 675–686.
7. Y. Nakao et al.: 'Theoretical Research on Transient Liquid Insert Metal Diffusion Bonding of Nickel Alloys', *Transactions of Japan Welding Society*, **20**, 1989, 60–65.
8. W. D. MacDonald and T. W. Eagar: 'Isothermal Solidification Kinetics of Diffusion Brazing', *Metallurgical Transactions*, **29A**(1), 1998, 315–325.
9. W. Wang, S. Zhang and X. He: 'Diffusion of Boron in Alloys', *Acta Metallurgica et Materialia* **43**(4), 1994, 1693–1699.
10. R. B. McLellan: 'The Diffusion of Boron in Nickel', *Scripta Metallurgica et Materialia*, **33**(8), 1995, 1265–1267.
11. DICTRA, Version 20, Thermo-Calc AB, Stockholm Technology Park, Björnnäsvägen 21, SE-113 47 Stockholm, Sweden.
12. Y. Zhou and T.H. North: 'Process Modelling and Optimised Parameter Selection During Transient Liquid Phase Bonding', *Z. Metallkd.*, **85**, 1994, 775–780.

13. C. E. Campbell and W. J. Boettinger: 'Transient Liquid-Phase Bonding in the Ni-Al-B System', *Metallurgical Transactions*, **31A**, 2000, 2835–2847.
14. A. LeBlanc and R. Mevrel: 'Diffusion Brazing Study of DS247/Bni-3/Astroloy', *Conference: High Temperature Materials for Power Engineering*, Liege, Belgium, 1990.
15. W. Miglietti et al.: 'Braze Repair of Single Crystal, Ni-Bases Superalloy Aircraft Gas Turbine Engine Components', *DVS, Brazing, High Temperature Brazing and Diffusion Welding*, Aachen, 1998.
16. K. Nishimoto et al.: 'Transient Liquid Phase Bonding of Ni-base Single Crystal Superalloy, CMSX-2', *Welding in the World* , ISIJ International, **35**, 1995, 1298–1306.
17. Thermo-Calc, version M, Thermo-Calc AB, Stockholm Technology Park, Björnnäsvägen 21, SE-113 47 Stockholm, Sweden.
18. N. Saunders: Ni-DATA Database, Thermotech Ltd., Surrey Technology Center, The Surrey Research Park, Guildford, Surrey GU2 5YH, UK.
19. R. W. Broomfield: 'Development of Brazing Techniques for the Joining of Single Crystal Components', *Proceedings of the 5th International Charles Parsons Turbine Conference*, 2000.

Liburdi Powder Metallurgy, Applications for Manufacture and Repair of Gas Turbine Components

R. SPARLING and J. LIBURDI
Liburdi Engineering Ltd.
Dundas, Ontario
Canada

ABSTRACT

The Liburdi Powder Metallurgy (LPM™) joining process has been improved through the addition of new superalloy compositions and the development of hard particle based compositions. Two new compositions based on MarM247 have been developed, tested and applied to hot section gas turbine components. These new compositions show improved microstructures and excellent mechanical properties. In addition to the conventional superalloy LPM™ materials, a hard facing LPM™ material based on wear resistant materials and abrasive compositions incorporating hard particles in an oxidation resistant matrix have also been developed. This paper discusses the mechanical and physical properties of the different materials developed and presents examples of different applications for these materials.

INTRODUCTION

Over the past two decades, many wide gap and modified wide gap processes have been developed and employed in the repair and manufacture of hot section gas turbine components, G.E.'s Advanced diffusion healing ADH,[1] Snecma's rechargement per brasage diffusion RBD, Howmet's effective structural repair ESR,[2] Chromalloy's surface reaction braze SRB,[3] Turbofix, Sermafill,[3] and Liburdi Engineering's LPM™.[4] All of these processes use high and low melting components to limit the amount of melting point depressant and to help bridge large gaps. The patented LPM™ process is the only modified wide gap process capable of filling large cavities resulting from the removal of structural cracks or casting core holes. In particular, the LPM™ process has been used to rebuild components which are missing large amounts of material while maintaining very low levels of melting point depressants. Additionally, the LPM™ process has been extended to applying abrasive and wear resistant materials to change the surface properties of hot section components.[5]

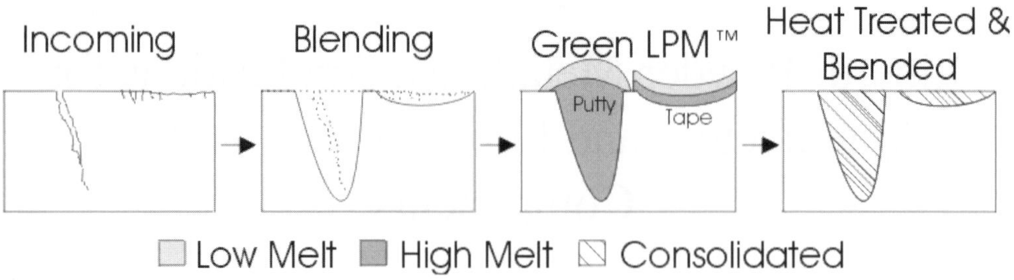

Fig. 1 Sketch showing the steps in an LPM™ repair process.

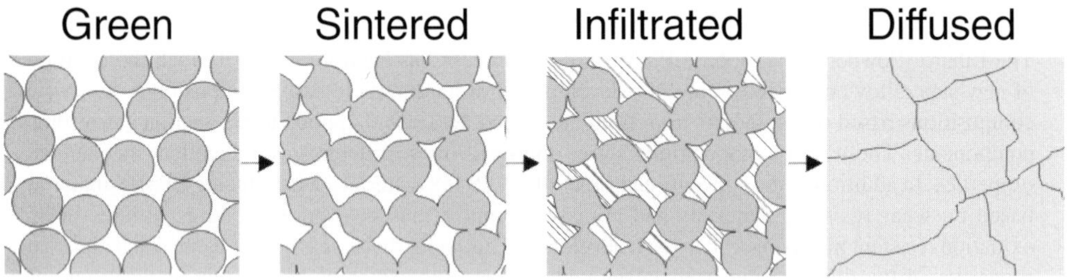

Fig. 2 Sketch, at the microscopic level, showing how the LPM™ is consolidated into a solid.

The LPM™ process is a multi-layer process incorporating high and low melting metal powders to effect sintering and control liquid phase bonding. The powders are mixed with organic binders to produce flexible tapes or clay-like putties. During application of the tape or putty to the new component or repair area, the area is first prepared for LPM™, the low melt material is applied on top of the high melting material and the ratio of the two materials is carefully controlled (Figure 1). The component is then heated slowly in a vacuum furnace to burn off the organic binders and finally heated to and held at a temperature between the melting points of the high and low melt materials. During the heating process the high melting point material sinters to form a rigid structure. Once the melting point of the low melting material is exceeded, the rigid structure is infiltrated by the low melting material at which point liquid phase sintering, transient liquid phase solidification (isothermal solidification) and diffusion bonding occur (Figure 2). The resulting deposits show excellent mechanical strength, creep strength and ductility when compared to conventional wide gap deposits, without the need for lengthy diffusion cycles.

The major advantage of using the LPM™ process is that it does not result in the distortion or cracking commonly associated with welding. This allows for large surface restorations that would otherwise distort or crack the component beyond repairable limits. It also makes it possible to repair cast superalloys components that would crack beyond acceptable limits during welding.

LPM™ has been used in the repair of nozzles, vanes, blades and buckets from G.E. Frame 3, 5, 6B, 7B, 7E, 7EA, 7FA, LM1600 and LM2500, Siemens Westinghouse W191, W251 B8, B10, B12 and W501D5, Rolls Royce T56, RB211, Avon and Spey, Alstom Tornado and TB5400, Dresser Rand DJ270, and others. LPM has also been used in the new manufacture of W501G Row 1 and Row 2 blades and some Solar turbine blades. LPM™ has proven an effective tool in the manufacture and repair of hot section turbine components. The following sections discuss the properties of the newer MarM247 based, abrading, and wear resistant LPM™ compositions and present some typical LPM uses.

SUPERALLOY LPM™

The initial LPM™ composition developed was based on IN738. This composition showed good properties but it was felt that improved properties could be developed using other alloys. A matrix of various high and low melt components was developed and various alloy combinations were evaluated for suitability in the LPM™ process. Both cobalt and nickel based compositions were evaluated. The results showed that compositions based on MarM247 had a very desirable microstructure. Two new MarM247 based compositions were further tested and evaluated; MarM247-3 and MarM247-7.

MICROSTRUCTURE

The microstructure of the original IN738 LPM™ material contained elongated borides. The 100 to 500 μm borides were inter-granular and had length to width ratios of greater than 10:1 (Figure 3). This microstructure resulted in lower ductility but tended to perform well in service.

Both of the newer MarM247 LPM™ compositions have small cuboidal boride particles (Figures 4, 5 and 6). The 10 to 50 μm cuboidal particles are found intra-granularly and have a length to width ratio of about 1:1. As a result, these two new compositions show improved ductility and better fracture properties than the IN738 LPM™ material. The MarM247-7 composition has a slightly higher volume fraction of borides than the MarM247-3, but both have lower boride volume fractions than the IN738 based material. The borides in the MarM274-3 structure tend to be smaller than those in the MarM247-7 structure.

LPM™ materials have been applied to the following alloy systems: GTD111, GTD222, Rene N4, IN738, IN939, IN625, MarM002, MarM247, Rene 80, Rene 125, Rene 142, Hastelloy X, CMSX 4, C1023, U500, U520, U700, U720, 304 stainless steel, 321 stainless steel, 347 stainless steel, ECY768, X40, X45, FSX414 and others. No needle like topologically close packed phases (TCP), microcracks, or abnormal phases were found at the interface between LPM™ and any of the aforementioned alloys. The interface between LPM™ is generally a 50 μm interdiffusion zone that contains a coherent interface with the base alloy (Figure 6).

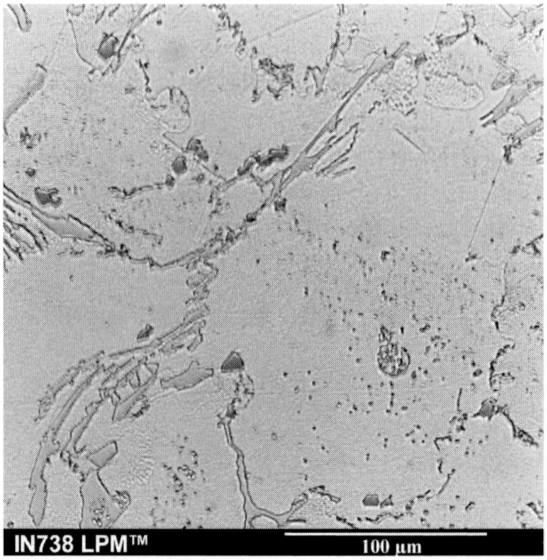

Fig. 3 Optical micrograph of the original IN738 base LPM™ microstructure.

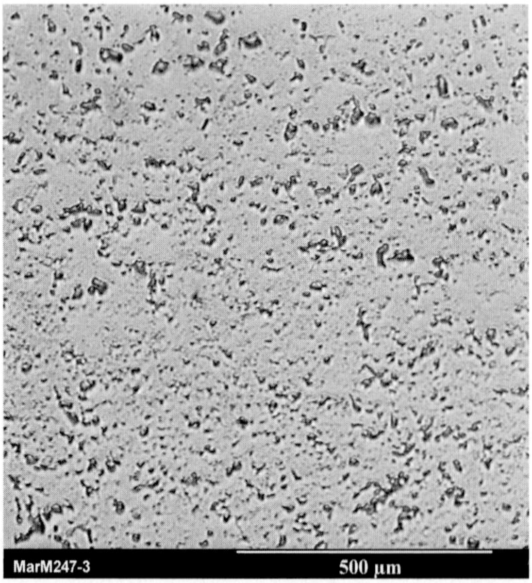

Fig. 4 Optical micrograph of the MarM247-3 LPM™ microstructure. Note the very fine cubic boride particles.

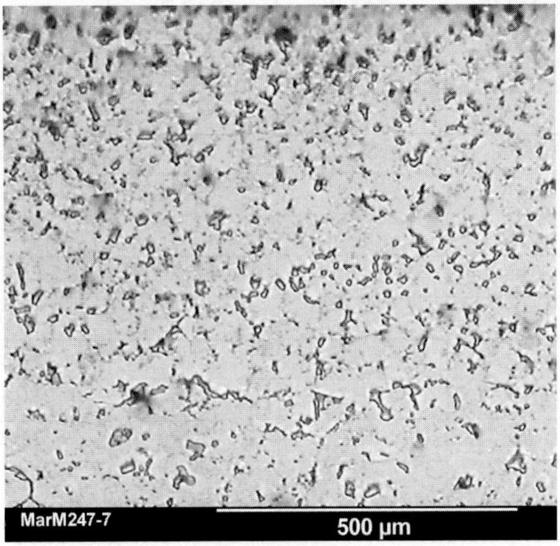

Fig. 5 247-7 LPM™ microstructure with small cubic borides.

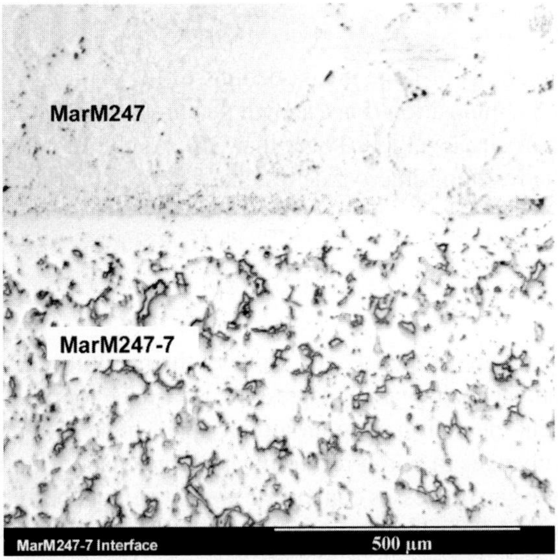

Fig. 6 MarM247-7 LPM™ on a MarM247 substrate. Note the clean interface.

Table 1 Chart showing the compositions of the various LPM™ materials, a typical 50:50 MarM247 wide gap material and the nominal composition of MarM247.

Element	MarM247	247-3	247-7	IN738 LPM™	MarM247 Wide Gap
Ni	59.90	58.13	65.26	63.32	66.95
Cr	8.50	11.95	10.11	15.4	11.00
Co	10.00	13.00	7.52	8.95	9.75
Mo	0.60	0.42	0.45	1.23	0.30
W	10.00	7.00	7.52	1.82	5.00
Ta	3.00	3.00	2.26	1.98	1.50
Hf	1.50	1.05	1.13	0.00	0.75
Al	5.50	3.85	4.14	3.43	2.75
Ti	1.0	0.70	0.75	2.38	0.5
Nb	N/A	N/A	N/A	0.63	N/A
B	0.01	0.90	0.87	0.81	1.5
Atom % B	0.01	4.7	4.5	4.2	7.5

Microscopic examinations of large cross sections of the original IN738 LPM™ and the newer MarM247 based compositions have shown that any pores tend to be smaller than 125 μm in size and total porosity levels are lower than 1%. As the LPM™ material is produced from fine powders, the resulting microstructure has a very fine grain size.

Composition

Due to the multi-layer nature of the LPM™ process, the resulting deposits have a very low weight percent of boron, the melting point depressant used in the low melt component (Table 1). As boron is a very light element having an atomic mass of 11 compared to 59 for nickel, small weight percent changes in boron result in very large changes in the atom percentages. Most conventional wide gap processes contain 1.5 weight percent boron or more. The LPM™ materials contain less than 0.9 weight percent boron due to the deposits containing less than 30% low melting material. The small weight percent and subsequent atomic percent of boron in the LPM™ materials leads to low volume fractions of boride and better ductility than other brazing processes. This also results in higher levels of γ' forming elements and subsequently excellent creep properties as is evident in the relatively large amounts of Al and Ti in the LPM™ material when compared to conventional wide gap materials.

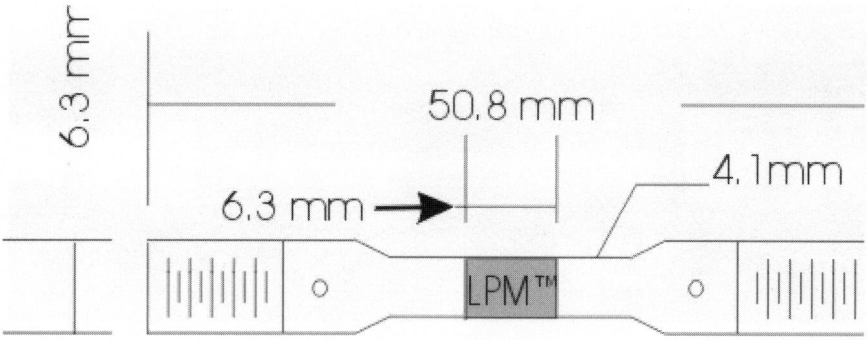

Fig. 7 Sketch of the test bar geometry.

Test Bar Manufacture

Most of the studies done on conventional and modified wide gap processes have concentrated on joint widths not larger than about 1.5 mm. These small joints widths allow for large amounts of boron to diffuse out of the joint into the base material. Often the joints are given long diffusion cycles of up to 24 hours to allow for the diffusion of the boron into the base alloy. Thus, the data generated from these test bars is not representative of the mechanical properties of overlays or relatively thick repair joints formed on actual component repairs. It was felt that for this testing the butt joint should be 6.3 mm wide such that diffusion into the base alloy did not play a role in the joint properties at the centreline of the joint. The test bars with a 6.3 mm joint width are representative of the bulk LPM™ properties and should also test the interfacial strength.

6.3 × 6.3 mm slots were cut down the centre of several 50.8 m × 50.8 mm × 9.5 mm MarM247 blocks. These slots were filled with the MarM247 LPM™ materials and heat treated for not more than 2 hours at the maximum temperature reached during the alloy solution cycle. The blocks were then aged for 2 hours at 1120°C and 16 hours at 843°C to produce the desired γ' precipitate structure. Stress rupture and tensile bars were cut from the blocks such that a 6.3 mm region in the centre of the gauge was LPM™ (Figures 7 and 8). The bars were x-ray and fluorescent penetrant inspected. Test bars with defects larger than 500 μm were rejected.

Tensile Properties

Tensile test bars were tested at a variety of temperatures. Table 2 shows the ultimate tensile strength and reduction in area for each of the test bars. The MarM247-3 material shows slightly better ductility due to the small boride particles and the smaller volume percent of

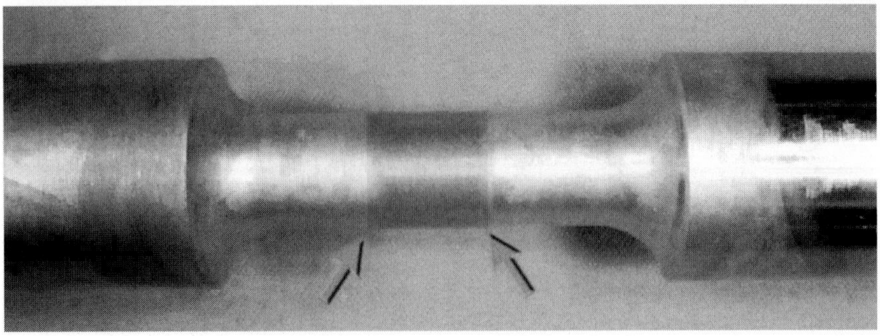

Fig. 8 Photograph of typical test bar in rough machined state.

Table 2 Tensile properties of the MarM247-3 and MarM247-7 LPM™ materials.

Material and Temperature	UTS in MPa	Reduction in Area %
MarM247-7 - 21°C	640	0.8
MarM247-7 - 549°C	680	0.8
MarM247-7 - 760°C	590	0.4
MarM247-7 - 871°C	510	0.6
MarM247-3 - 21°C	770	1.6
MarM247-3 - 649°C	730	1.4
MarM247-3 - 704°C	490	1.5
MarM247-3 - 927°C	340	2.4

borides. The ductility of the MarM247 based LPM™ materials is slightly lower than that for common nickel based superalloys and is considerably higher than the levels for conventional wide gap brazing processes which generally show little or no ductility. None of the bars failed at the interface and all bars failed intergranularly as would be expected. The modulus for both materials was approximately 170 GPa.

Creep Properties

Stress rupture tests were performed using the bars described previously. The tests were conducted to ASTM E139. The stress, time and temperature were varied to generate a broad range of Larson Miller parameter values. The temperature varied from 787 to 982°C and the

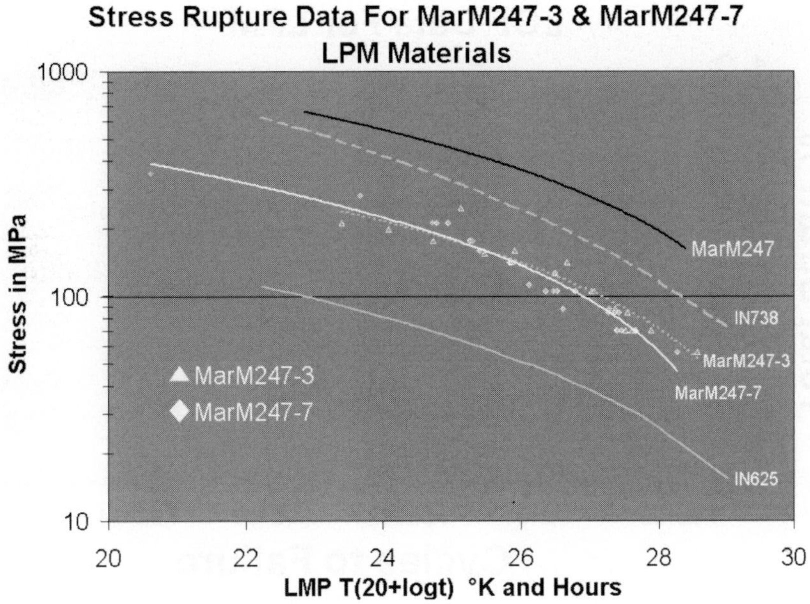

Fig. 9 Graph showing the stress rupture properties of the MarM247-7 and MarM247-3 LPM™ materials.

stress varied from 5 to 35 MPa. The Larson Miller plot is shown in Figure 9. The average reduction in area was 4.2% for the MarM247-7 LPM™ and 7.3% for the MarM247-3 LPM™. These values are similar to those for cast MarM247.

Due to the fine grained nature of LPM™ materials, they tend to have lower creep strengths when compared to large grained cast nickel based superalloys. The LPM™ materials have creep strengths higher than conventional welding alloys such as IN625 for nickel based superalloys and L605 for cobalt based superalloys.

The creep strength compares more closely to fine grained wrought superalloys and cobalt based superalloys. During the studies it was also found that the creep properties of LPM™ varied little with the composition. It has been hypothesised that the creep properties are highly dependent on the grain size. Studies to this effect will be conducted in the future by increasing the grain size through long diffusion cycles.

Low Cycle Fatigue

Low cycle fatigue test bars were produced as described previously and were inspected for surface defects larger than 0.50 mm. The testing was conducted at 850°C using a trapezoidal wave form at an R ratio of –1.0. The specimens were held at the maximum compressive strain for two minutes during each cycle. Failure was defined as the point at which the

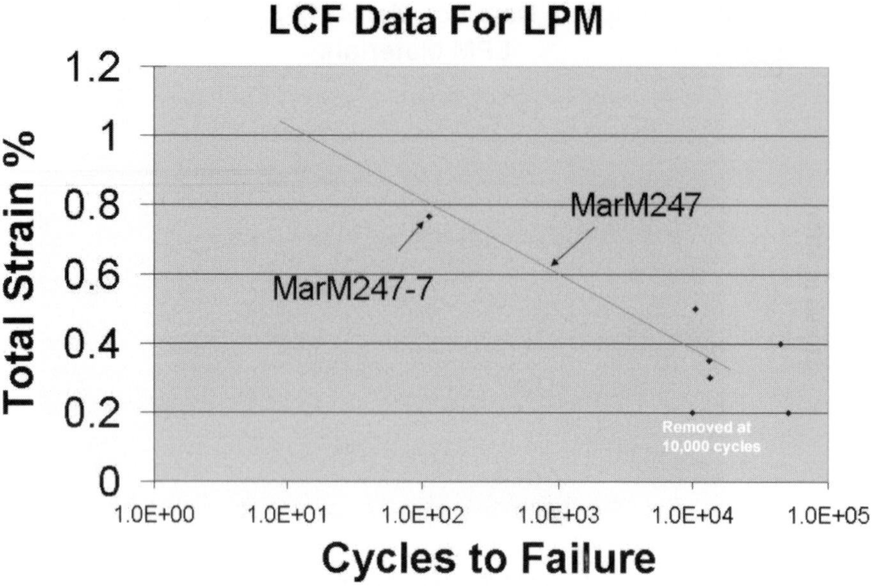

Fig. 10 Graph of LCF data for MarM247-7.

maximum stress amplitude decreased to 50% of the value at 100 cycles. It should be noted that the butt joint in the middle of the test bars produces an effective gauge length of only 6.3 mm and the strain is measured over the total gauge length. During testing the majority of the elongation occurs in the 6.3 mm joint. This means the strain in the joint area is much higher than that measured. The data reported in Figures 10 and 11 are for the total strain over the gauge length of the test bar. Testing indicated the fatigue properties are similar to that of cast MarM247 for MarM247-7 and likely slightly lower properties for the MarM247-3.

SUPERALLOY LPM™ EXAMPLE #1

General Electric MS7001EA Row 1 Nozzles

After the initial 24,000 hour service interval, these FSX 414 cobalt alloy nozzles show extensive cracking especially around the leading edges, trailing edges and between the two airfoils on the outer shroud. The cracking was primarily caused by low cycle thermomechanical fatigue. The cracking at the leading edges was branched and extended through to the internal vane cavities. Some oxidation erosion and associated cracking had also occurred on the outer and inner diameter shrouds between the airfoils. Based on the degree of damage it was felt that conventional weld repairs would either crack or result in excessive distortion of the nozzles. It was decided the LPM™ was the best approach for repair.

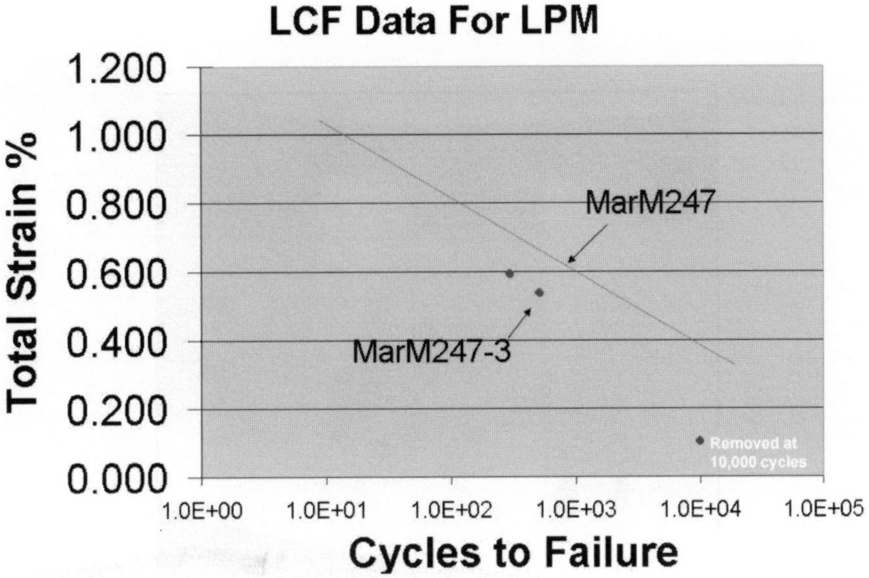

Fig. 11 Graph of LCF data for MarM247-3.

In preparation for LPM™ the cracks were ground out and the eroded surfaces were blended back to unaffected base alloy. This left large holes approximately 25 × 100 mm in size at the leading edges of the airfoils of several nozzles and areas larger than 50 × 50 mm requiring 0.75 mm build ups (Figure 12). The large holes, which are typical of vane repairs where the structural cracks are removed, were filled with LPM™ putty which was sculpted to the original contour of the nozzle. The eroded areas were covered with a 1 mm thick LPM™ tape to restore the dimensions. Small cracks and impact damage were repaired with the LPM™ putty. During repair these areas were successfully repaired using MarM247-7 LPM™.

After the second service interval of approximately 24,000 hours (~48,000 hours total operating time), the nozzles did not show the large thermomechanical fatigue cracks at the leading edges or between airfoils on the outer shroud surface. This suggested the LPM™ had improved the performance of these nozzles due to the improved strength and low cycle fatigue resistance of the LPM™ repair (Figure 13).

After 48,000 hours of service, the nozzles had small cracks on the shrouds, oxidation erosion on the internal and external surfaces and thinned trailing edges. The thinned trailing edges were particularly of concern as the throat area between the airfoils had increased significantly due to material loss during service and repair. This increase in throat area can lead to harmonic imbalance and efficiency loss. As part of the second repair the small cracks and oxidation damage were blended in preparation for LPM™ application and the thinned trailing edges and oxidised areas were thickened by applying LPM™ tape to both surfaces. The cracks were repaired using LPM™ putty. After the second repair the nozzles were again returned to service. The nozzles are expected to return for LPM™ repair at ~72,000 hours in early 2004.

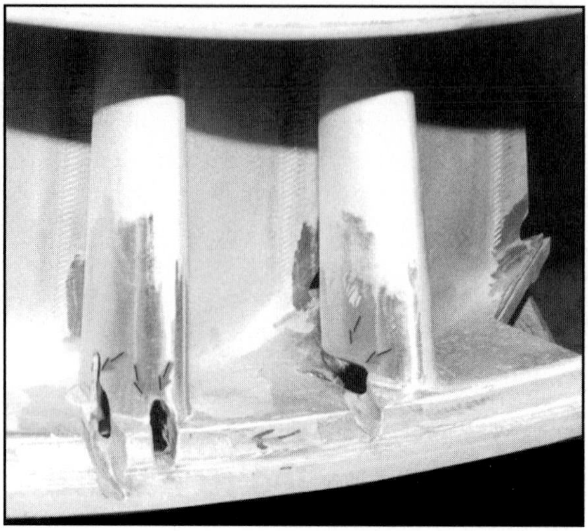

Fig. 12 Photograph showing vane segment prepared for LPM™.

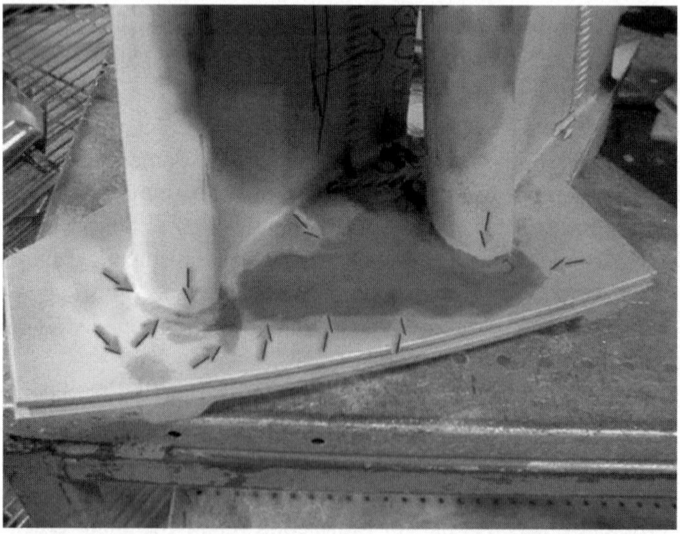

Fig. 13 Photograph of nozzle after 48,000 hours and one repair cycle. The darker areas are LPM™ material.

SUPERALLOY LPM™ EXAMPLE #2

Siemens Westinghouse W501G Row 1 and 2 Core Closure

Liburdi Engineering's LPM™ is specified for closure of the core holes on Siemens Westinghouse 501G row 1 and row 2 directionally solidified MarM247 turbine blades. The casting core holes in the 501G blades are necessary to retain the core during manufacture of the blades (Figure 14). These large core holes are difficult to close via welding due to cracking of the directionally solidified grain boundaries. Conventional wide gap brazing of the 25.4 mm by 6.3 mm casting holes is also not practical. LPM™ was selected to close the blade tips as it has excellent creep properties, very low porosity and can close the casting holes without generating cracks.

LPM™ is applied as putty into the core holes such that each hole is slightly over filled. The blades are then heat treated to consolidate the LPM™. After finishing to contour and machining the tip to nominal dimensions it is not possible to visually determine where the casting holes were (Figure 15). To date over 49 sets of row 1 and row 2 blades have been closed using this process. Figure 16 show a blade tip after more than 8,000 hours of service.

HARD PARTICLE LPM™ MATERIALS

WEAR RESISTANT LPM™

Wear resistant LPM™ (known as LZN™) compositions based on chrome carbide have been developed and used in the manufacture and repair of hot section gas turbine components. These materials are composites of hard particles in an oxidation resistant matrix. The wear resistant LPM™ was designed to replace plasma, high velocity oxyfuel (HVOF) and weld hard facing materials.

The major drawback of plasma and HVOF chrome carbide coatings is their poor mechanical bond to the base alloy. The LZN™ compositions form a metallurgical bond with the base alloy (Figure 17). This metallurgical bond consistently results in failure of the adhesive in the ASTM C633 adhesion test for flame sprayed coatings. The other drawback of plasma spray and HVOF coatings is the necessity of spraying with the line of sight normal to the surface. The LZN can be applied into slots on opposing surfaces and on surfaces which have no direct line of site for a plasma or HVOF gun.

In comparison to weld hard facing, the LZN™ material does not result in base alloy cracking and shows improved hardness. Weld hard facing materials often result in cracking at grain boundaries intersecting the face to be welded. In some components welding results in cracks at or beyond acceptable limits resulting in high levels of deviated or scrapped material. Due to the nature of the LPM™ process LZN™ does not result in cracking of the base alloy. During weld hard facing, the weld pool mixes with the base alloy reducing the hardness of the deposit to be below 50 Rc (Rockwell 'C'). The LZN™ materials have higher volume percents of carbide particles resulting in hardness values of 60 to 65 Rc (Figure 18). The volume percent of carbide is more than 55%. Voids are typically smaller than 25 μm and the porosity levels are below 5%.

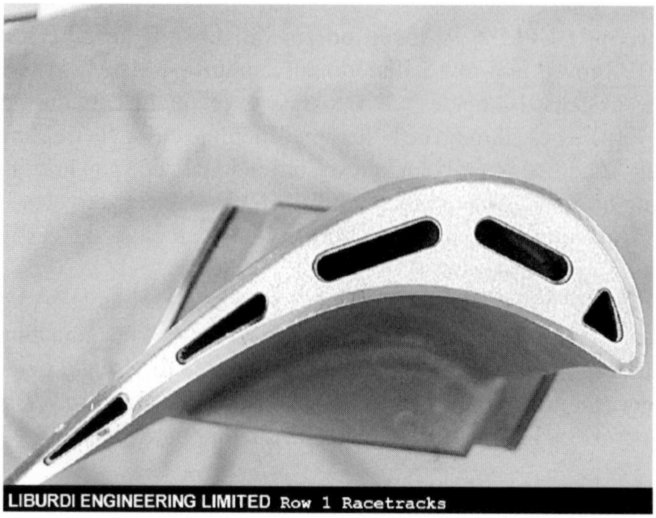

Fig. 14 W501G Row 1 blade prior to LPM™ closure.

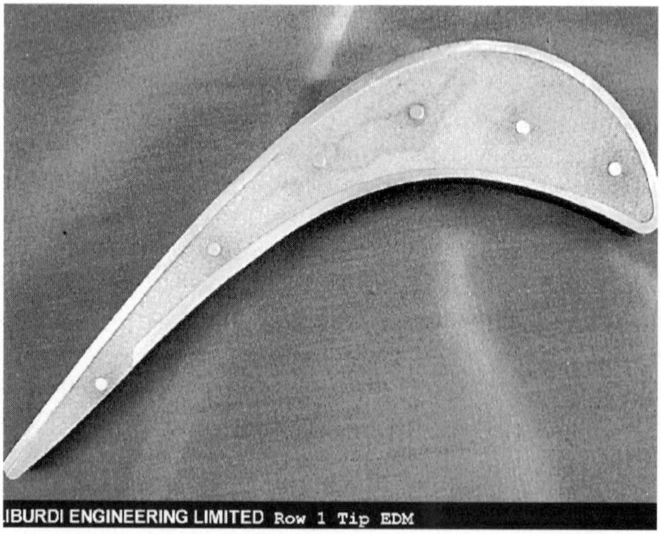

Fig. 15 W501G Row 1 blade tip after LPM closure.

Fig. 16 W501G Row 1 blade tip after 8,000 hours of service.

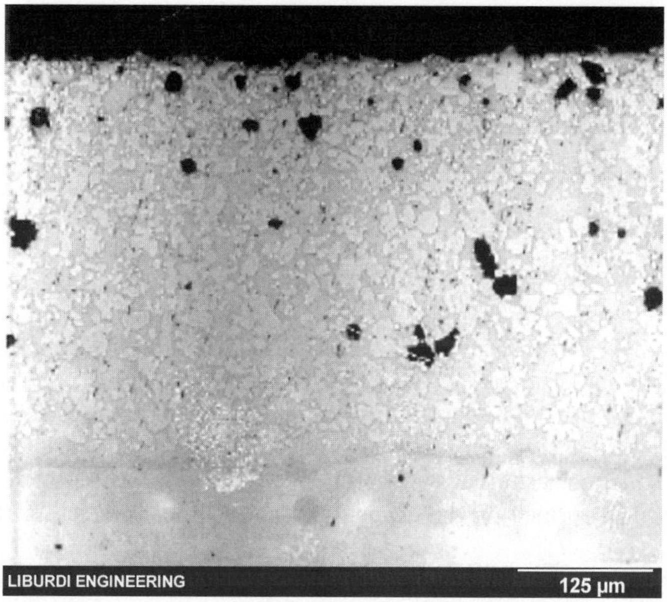

Fig. 17 Optical micrograph showing LZN with a well bonded interface.

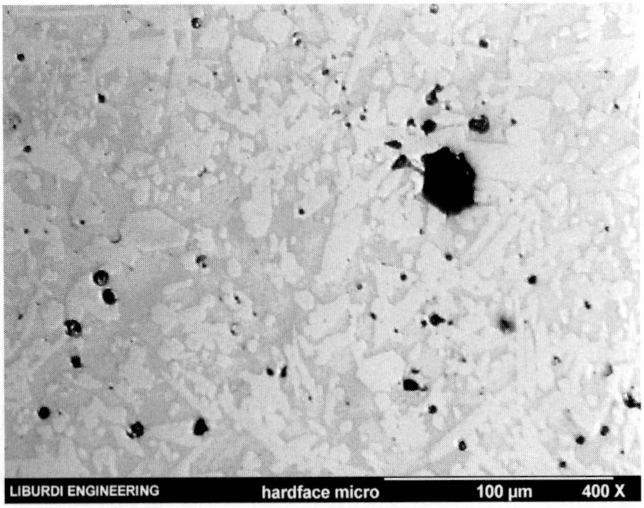

Fig. 18 Optical micrograph of the LZN material.

Table 3 Composition of LZN.

Element	Weight %
Ni	33
Cr	60
Si	4
C	3

Because the LZN™ material can be used for hard-facing of shroud contact faces, transition piece outlets, floating seals, etc., it must be oxidation resistant. The composition of the material is shown in Table 3. The chrome carbide particles are expected to start decomposing above 815°C and it is felt that the oxidation resistant matrix should be stable to temperatures above 980°C. Oxidation testing at 870°C for 1000 hours showed a native chrome oxide scale 2 μm thick had formed on the deposit and no significant attack of the deposit had occurred.

The LZN material has been used to hard face the shroud contact faces on CMSX 4 single crystal turbine blades (Figures 19 and 20). The blades in question had no direct line of site for plasma or HVOF hard facing and the geometry of the contact face did not allow access for a gas tungsten arc weld hard face. As a result LZN was chosen to replace the conventional hard facing techniques. The LZN material is applied to the contact faces as an LPM™ tape

Fig. 19 Photograph showing the unmachined contact face from a CMSX 4 turbine blade.

Fig. 20 Photograph showing the hard facing on the contact faces of the CMSX 4 blade after machining (left- concave, right- convex).

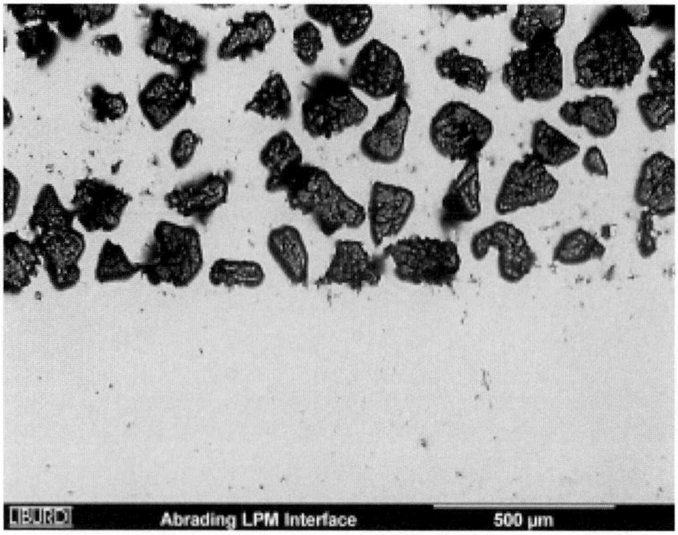

Fig. 21 Optical micrograph showing the abrading LPM™ particles and interface with the base alloy.

material and is given appropriate heat treatments to consolidate it. The LZN material showed no abnormal phases or unusual microstructures at the interface with the CMSX 4 base alloy. The material was machined to contour using conventional aluminium oxide grinding wheels. After machining the contact faces did not show any cracking or spalling.

Abrasive LPM™

Abrasive LPM™ compositions based on large angular hard particles have been developed for use on the tips of unshrouded turbine blades. The tips run against shroud blocks to ensure a tight air seal and subsequently high engine efficiency. Unfortunately the turbine blades are often worn by the shroud blocks as opposed to the shroud blocks being worn by the turbine blades even when running against abradable shroud blocks. The rubbing also results in the removal of any protective coatings from the blade tips. A need was recognised for a harder material which would machine into the shroud blocks, preventing the blade tips from being worn away and prevent the blade tips from oxidizing.

The abrasive LPM™ material is a mixture of large hard particles in a MCrAlY based matrix (Figure 21). This material is made into an LPM™ tape and applied to the blade tip. The tape is contoured to the blade tip in the green state and is given an appropriate heat treatment for the base alloy. After heat treatment the abrading tip is contoured and ground to the appropriate tip height.

CONCLUSIONS

Superalloy LPM™ compositions based on IN738 and MarM247 have proven extremely successful for the repair and manufacture of hot section gas turbine components. These materials show excellent creep, tensile and stress rupture properties for a brazing process. The materials have proven successful during service on a variety of components from different engines.

Hard particle LPM™ compositions used for wear resistance and abrading have also been developed and tested. The wear resistant LZN™ material shows very good hardness and can be used in wear applications where welding, plasma spray and HVOF hard facing methods do not work. The abrading LPM™ material can be used on unshrouded turbine blades where tight tolerances and wear of the tip shrouds into the blade tips is an issue.

ACKNOWLEDGEMENTS

We would like to thank Siemens Westinghouse, Watson Cogeneration and Solar Turbines for their input into the development and use of the LPM™ materials. We would also like to thank the Liburdi Engineering Limited staff for their time and effort in researching and developing the LPM™ materials. In particular Doug Nagy, Paul Lowden and Kevin Wiens should be mentioned for their roles in developing these materials.

REFERENCES

1. W. A. Demo and S. J. Ferrigno: 'Brazing Method Helps Repair Aircraft Gas-Turbine Nozzles', *Advanced Materials and Processes*, **141**(3), 1992, 43–45.
2. R. D. Wustman and J. S. Smith: 'High Strength Diffusion Braze Repairs for Gas Turbine Components', *ASME*, 96-GT-427, 1996.
3. Miglietti, Kearney and Pabon: 'Liquid Phase Diffusion Bond Repair of Siemens V84.2, Row 2 Vanes and Alstom Tornado, 2nd Stage Stator Segments', *ASME Turbo Expo 2001*, ASME 2001-GT-0510, 2001.
4. K. A. Ellison, P. Lowden and J. Liburdi: 'Powder Metallurgy Repair of Turbine Components', *ASME*, 92-GT-312, 1992.
5. I. Nava and D. Nagy: 'Selection of Overlays for Single Crystal Shrouded Turbine Blades', *ASME Turbo Expo 2002*, ASME GT-2002-30661, 2003.

RENEWABLES

Novel Steam Injected Gas Turbine with Condensate Recycling

R. A. Wall
Alstom Power Technology
Whetstone, Leicester, UK

ABSTRACT

The influence of small industrial gas turbine technology on wet condensing cycle design is discussed to illustrate the thermo-economic performance potential for future products.

Steam Injected cycles are well known using existing turbomachinery but are limited in performance (40–45% he) by compromises necessary to preserve compressor stall margin with steam injection into fixed flow capacity turbomachinery. The performance benefits of steam injection include the potential for combined cycle performance at simple cycle cost with flexibility for variable CHP, ultra low nox (< 5 ppm), reduced Carbon emissions by virtue of better efficiency (~17%) and self sufficiency of water supply for steam by virtue of the exhaust condenser.

The basic mechanism for increased net power and efficiency relies on the favourable redistribution of Turbine pressure ratio, between compressor and power turbines, that results from the change in gas properties and increased mass flow in the turbine, relative to the compressor, due to steam injection.

Product potential is examined in terms of performance, aero-mechanical design constraints and system operability requirements for both Single and Twin shaft design. Single Shaft is simpler from an operability point of view but has limited performance (~48% he) due to limited cycle pressure ratio achievable on a single shaft. Twin shaft designs can achieve > 50% he with higher cycle pressure ratio but require variable guide vanes between Compressor and Power turbine to control compressor turbine output if variable steam injection is required for CHP.

INTRODUCTION

Regenerative cycles have shown much promise historically but proven difficult to introduce into a market place dominated by the thermo-economics of large scale gas powered Combined cycles, Coal fired Steam generators and Diesel engines at the smaller power levels.

Steam Injected gas Turbine performance has already been successfully demonstrated by several manufacturers using a range of existing hardware to create variable combined heat & power systems throughout the world in university campuses and food processing plants. These systems depend on the integration of essentially mature technology designed for existent combined cycle powerplant.

However, existing gas turbine hardware, with fixed turbine flow capacity, imposes limitations on performance achievable due to the limited amount of steam that can be introduced without sacrificing too much Compressor stall margin.

Small industrial single and twin shaft design technology is therefore explored to establish the future performance potential for optimised gas turbine design without these restraints.

THERMODYNAMICS

Thermodynamic Process

The wet cycle process scheme consists of a simple cycle gas turbine with heat recovery steam generator, typical of combined cycle or combined heat and power systems, with the option for superheated steam injection into the combustion chamber and capture of water for treatment and recirculation with an exhaust condenser. The degree of superheat may be varied from dry saturated to maximum available depending on requirements for factory process or steam injection. Compressor discharge air and steam are fully mixed prior to combustion in the combustor plenum. The turbine consists of a compressor turbine and power turbine for both single and twin shaft analysis. A basic HRSG is modelled in the standard way with a separate superheater, evaporator and economiser. The condenser is linked to a heat sink for cooling the condenser coolant with either air or water. Water treatment has to be included to ensure suitable reduction of water contamination for gas turbine, Steam generator and Condenser coolant to avoid corrosion and oxidation of hot parts. Recirculation of condensate from the factory process steam depends on the process and may require a separate treatment plant or total loss system for health and safety reasons.

Performance Potential

Wet Cycle Performance potential has been compared over a range of cycle pressure ratios with alternative forms of power production, including Humid Air Turbine (HAT) and Simple cycle, assuming small industrial component efficiencies. Cycle pressure ratio variation has been constrained by typical turbine exit temperature for hot rotating component structural design integrity.

HAT & WET cycles achieve similar efficiencies at ~20 bar but HAT cycle loses specific power at lower cycle pressure ratios due to the method of exhaust heat recuperation. The HAT cycle recovers waste heat using **compressor** discharge air to evaporate water and recuperate exhaust heat. Wet cycle uses **turbine** discharge air, at constant design temperature, to evaporate more water at lower cycle pressure ratios.

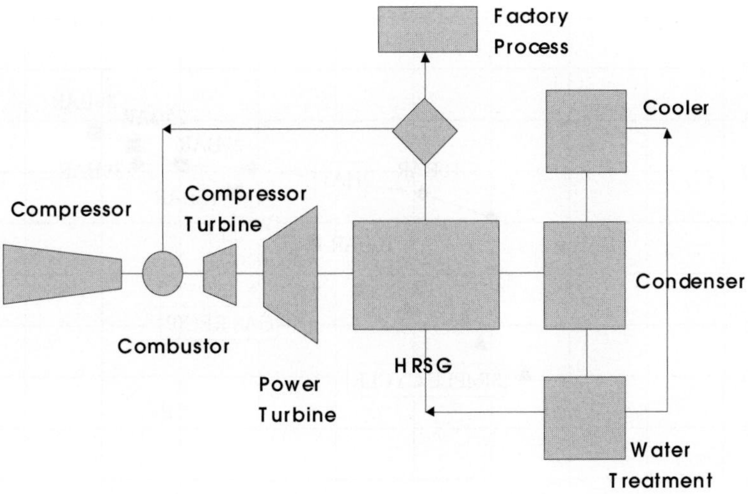

Fig. 1 Process scheme.

The full **wet cycle specific power potential** is approximately twice that of comparable **simple cycle** with similar cycle pressure ratio and approaches combined cycle levels of efficiency (<55% ηe) at 30 bar.

CHP Flexibility

The potential for greater utilisation of hardware is clear when compared to the traditional CHP which typically uses some of the process steam for 'power augmentation' only with little change in electrical efficiency.

This flexibility for variable CHP should prove to be economic for Distributed Power applications which are subject to periodic changes in demand provided the net Cost of Electricity supplied can be less than the Grid supply.

The efficiency, with full steam injection capability, can vary from simple to combined cycle levels depending on the level of steam extraction. Thermal efficiency with full steam extraction, achieves typically 80–90% with Heat/Power ratio of ~2.

Performance Perspective

Parametric variation of cycle pressure ratio shows the potential for variation of efficiency and specific power subject to small industrial component efficiency levels, constant percentage turbine cooling and a 90% effectiveness HRSG Superheater. Turbine exhaust temperature

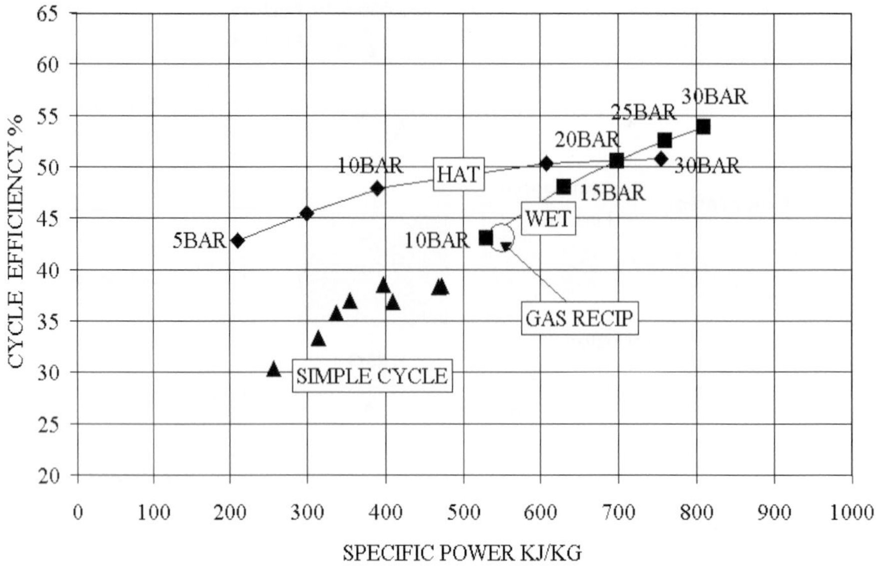

Fig. 2 Cycle performance potential.

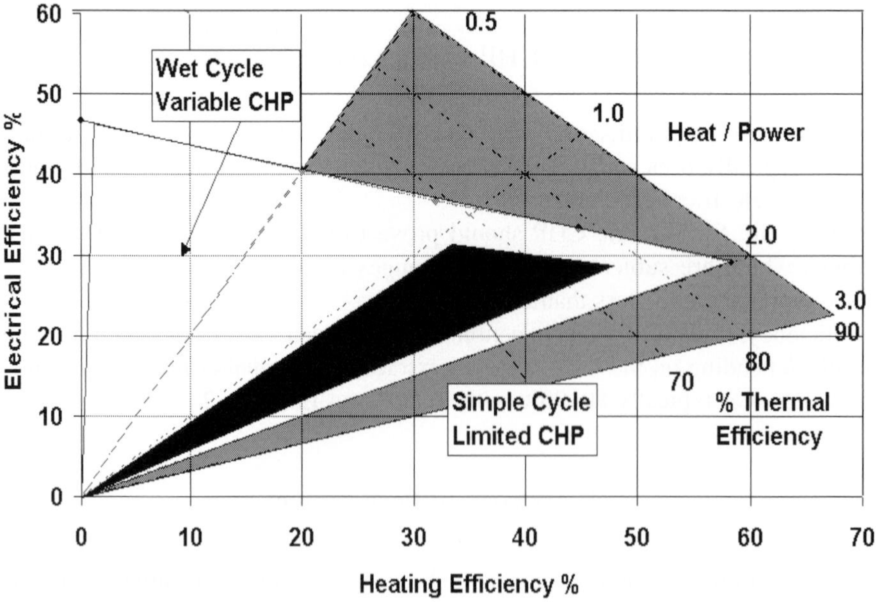

Fig. 3 Variable CHP.

constrains the relationship between cycle pressure ratio and turbine entry temperature for mechanical design integrity of rotating turbine blade support structure.

Ultimately, the exploitation of this performance potential will be subject to the constraints of design technology and system thermo-economics.

Effect of Superheat

Increasing the degree of superheat, that is the difference between the superheater outlet temperature and the saturation temperature of steam(i.e. boiling point) reduces specific power and increases efficiency. Exhaust heat is diverted from the evaporation of water to the superheating of steam such that a higher temperature steam-air mixture enters the combustor to reduce fuel flow for a given firing temperature with less heat lost to the evaporation of water.

Increasing superheat from 30 to 280 K reduces specific power by 20% and increases cycle efficiency by ~1.5% for a 20 bar – 1500 K burner exit temperature cycle.

SYSTEM DESIGN

Single and Twin Shaft Design

Single shaft engine performance is limited by the impact that rotor mechanical design constraints have on the level of cycle pressure ratio that can be achieved by a simply supported rotor bearing system. The last stage power turbine blade root fixture is subjected to blade centrifugal and bending stress at high temperature, typically without cooling, and therefore controls the speed at which the shaft can safely rotate. Increasing the pressure ratio tends to cause an increase in blade root stress due to increased flow area and forces a speed reduction to comply with these design constraints. A limit is reached when rotor dynamic stability is threatened by shaft length required to support the increasing number of compressor and turbine stages. This limit is reached typically for small industrial machines at a cycle pressure ratio between 15 and 20 for conventional blade and disc materials beyond which the rotor has to operate in a supercritical mode.

In the twin shaft engine the compressor shaft speed is independent of power turbine mechanical design and is determined by compressor mechanical design criteria. However it will be shown that the twin shaft design is more complex since it requires a variable stator between compressor and power turbine to control turbine work distribution with variable steam injection.

Compressor

Compressor design stage numbers, which are governed by constant mean stage loading (0.42), exit hub/tip ratio (0.92) and exit mach no (0.295) representative of current design practice,

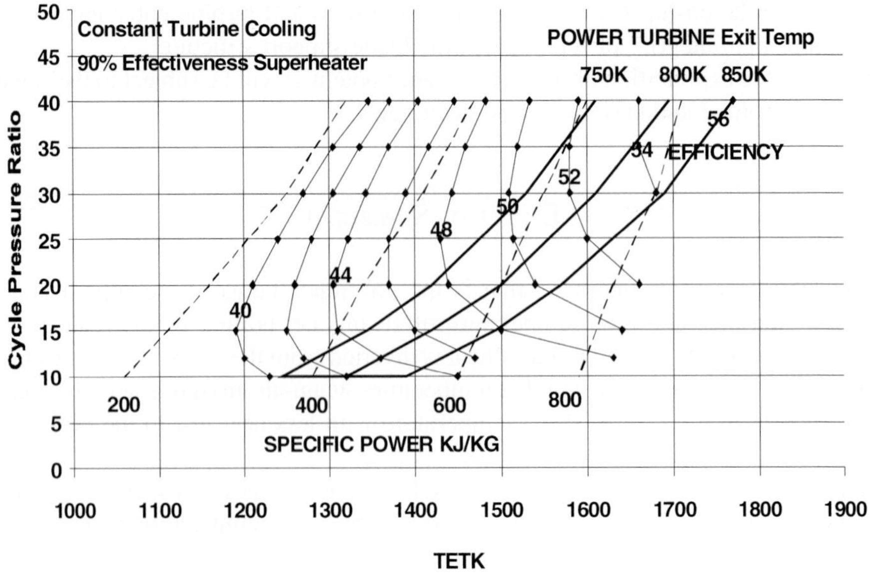

Fig. 4 Superheated wet cycle performance.

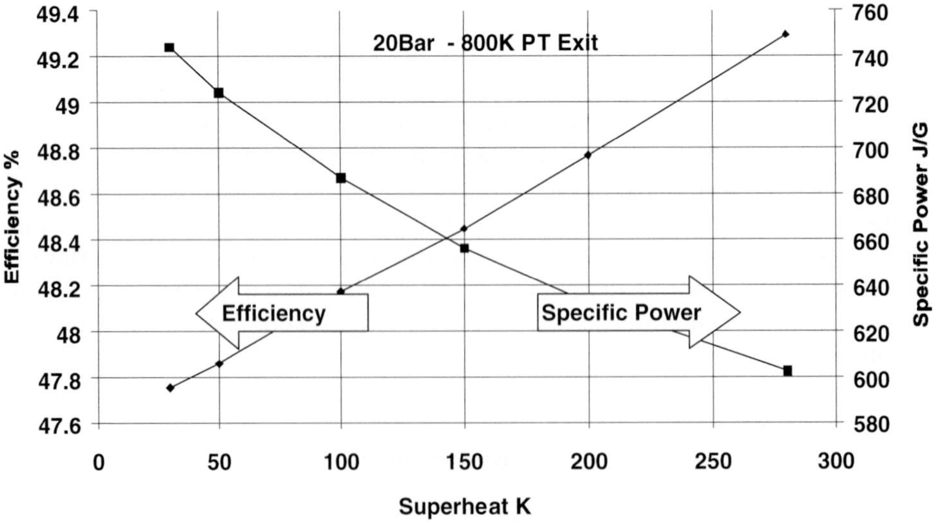

Fig. 5 Effect of superheat on performance..

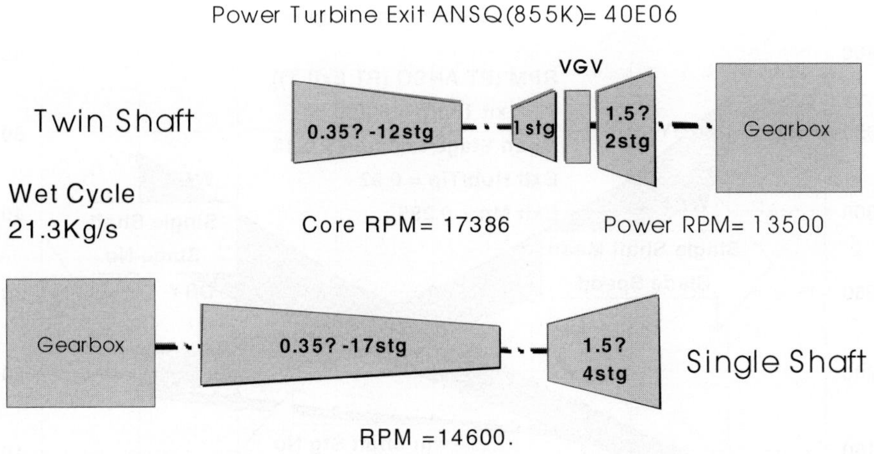

Fig. 6 Twin vs single shaft 15 MW / 21 bar turbomachinery.

rise steeply with cycle pressure ratio for single shaft compared to twin shaft design. Steam injection exaggerates this effect due to extra mass flow of steam in the turbine forcing a further reduction in speed.

Clearly the turbine exit mechanical design constraint has a powerful effect on compressor design configuration but aerodynamic technology requirement is conventional for both Wet and dry design unless the compressor is also humidified.

COMBUSTOR

The combustor design is typical of current design practice for small industrial machines which combine lean premixed burners with supplementary diffusion pilot burner for control of stability and emissions during engine transients. Steam is mixed with compressor discharge air in the burner plenum to provide a homogenous mixture of steam and air prior to combustion.

Combustor cooling can be provided either by the same steam/air mixture or benefit from higher specific heat steam prior to plenum mixing.

The steam/air ratio in the combustion process varies from 20–30%, depending on the degree of superheat, (300–30) K, for 20 bar – 1500 K design.

EFFECT OF STEAM INJECTION ON TURBINE DESIGN

The performance improvement, due to steam injection, depends on the change in gas properties and increase in turbine mass flow relative to the compressor.

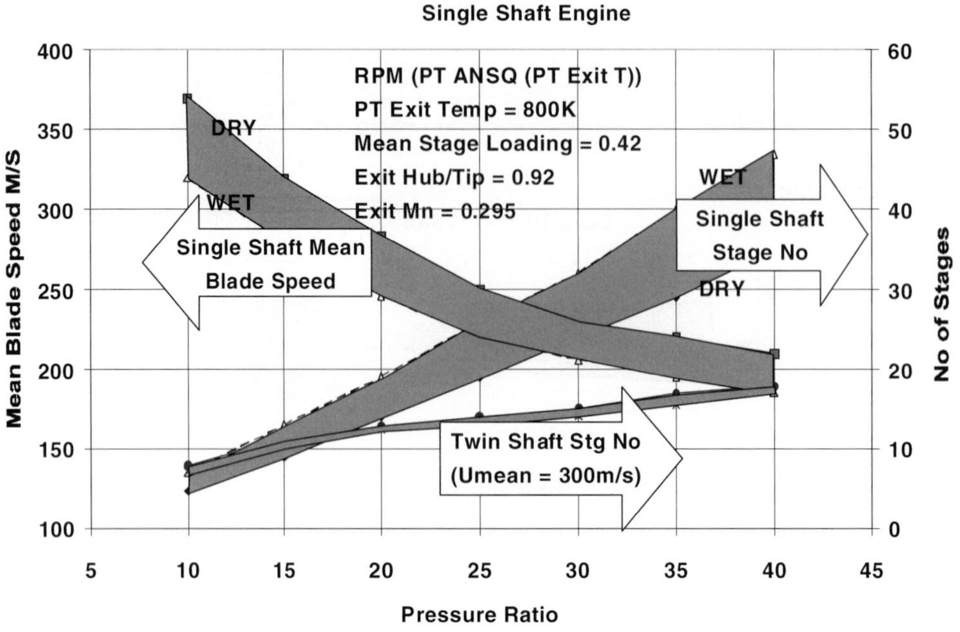

Fig. 7 Wet and dry compressor design.

Gas properties control the relationship between Turbine work and Pressure Ratio and relative mass flow controls the work distribution between compressor and power turbines. The compressor turbine work is reduced by the increase in relative flow for a given compressor power requirement causing a reduction in compressor turbine pressure ratio.

Power Turbine pressure ratio correspondingly increases to compensate for the reduction in compressor turbine pressure ratio to provide a dramatic increase in power turbine work. The chart shows similar magnification of power turbine pressure ratio between Dry(A) and Wet(B) for a range of cycle pressure ratios (15–25).

Turbine Flowpath

A comparison of wet and dry turbine flowpaths reflects the difference in shaft speed resulting from steam injection subject to the turbine blade root fixture mechanical design constraint.

Increased turbine mass flow relative to the compressor, due to steam injection, forces a lower shaft speed for single shaft design and produces a corresponding increase in flow area, mean diameter and stage numbers relative to the dry equivalent. This is likely to produce a small cost increase for the Wet cycle turbine relative to Dry.

Typically, Dry turbine Compressor Turbine (CT) pressure ratio is approximately equal to Power Turbine (PT) pressure ratio whereas Wet CT pressure ratio is half the PT pressure ratio.

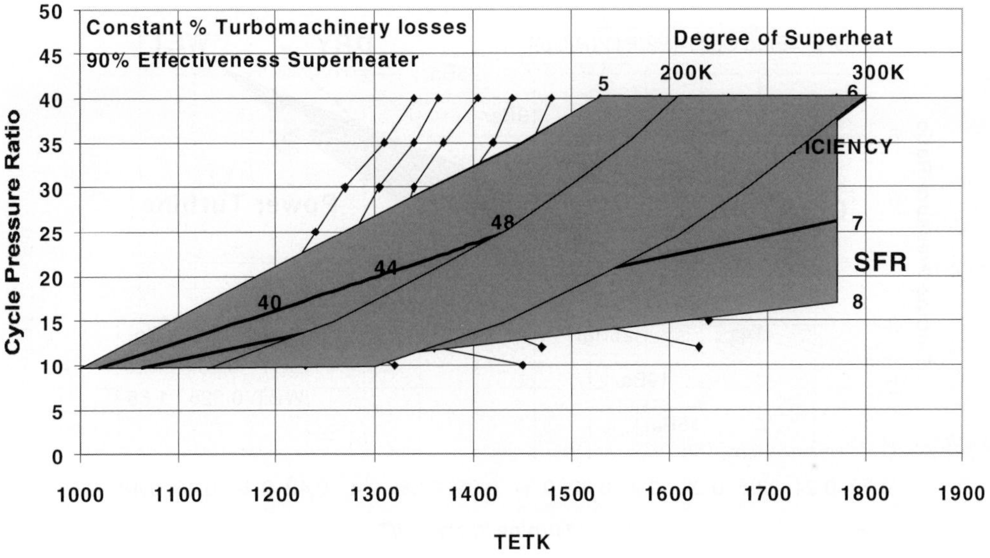

Fig. 8 Superheated wet cycle performance combustion design.

Turbine Cooling

The wet compressor turbine rotor blade cooling flow requirement is increased relative to the dry equivalent by virtue of the associated reduction in compressor turbine work. Reduced stage work increases the average temperature level within the blade while nozzle guide vane requirements remain similar for equal turbine entry temperature.

Rotor blade cooling requirement almost doubles assuming constant cooling efficiency for a single stage compressor turbine design. Some benefit may be obtained by the use of pure steam cooling instead of the steam-air mixture in the combustor plenum.

Maximum Potential

Comparison of wet (20 bar) & dry (25 bar) performance potential, at equal Turbine exhaust temperature and comparable limits of mechanical design feasibility, would suggest that the maximum benefit for Single shaft engines is limited to ~8% ηe with ~50% increase in specific power. It has been assumed that (15–20) stage compressor, designed to conventional stage loading levels, with welded rotor design, represents an arbitrary boundary for subcritical rotor dynamic mechanical feasibility. Improvements in Turbine blade and disc materials, which may facilitate shaft speed increase, would benefit wet & dry design equally.

Twin shaft design is not limited by the single shaft Turbine mechanical constraint but requires variable guide vane flow area modulation between compressor and power turbines if variable steam injection is required for CHP. Increased cycle pressure ratio, afforded by

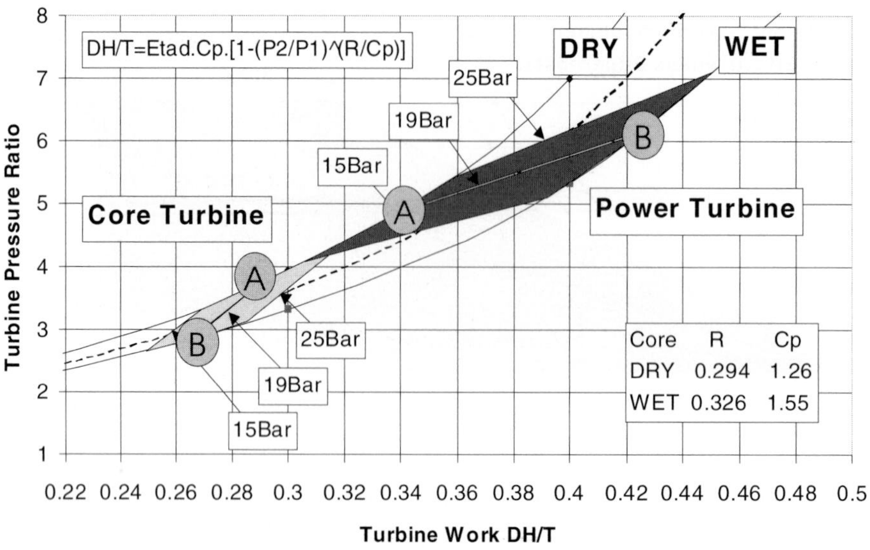

Fig. 9 Effect of steam injection on turbine design.

the twin shaft arrangement, allows cycle efficiency to exceed 50% and specific power to approach double that of typical single shaft, simple cycle machines. However, specific power potential is significantly reduced (35% at 30 bar) if the power turbine entry temperature is limited to (1100–1150) K to avoid cooling the variable turbine nozzle. The complications of variable steam injection can be avoided if auxiliary firing is used to supply steam for variable CHP.

SYSTEM OPERABILITY

Twin vs Single Shaft

Single shaft operation with variation of steam injection does not present an operability problem because output is automatically sacrificed to maintain adequate compressor power by virtue of being attached to the same shaft.

Twin Shaft operation, however, has to be controlled by either Variable Guide vanes situated between Compressor and Power turbines or by the simultaneous injection of steam upstream and downstream of the Compressor Turbine. The latter approach, however, sacrifices some of the potential benefit obtained by injecting all the steam upstream of the compressor turbine. Steam injection between CT & PT has the effect of reducing CT Pressure Ratio by reducing the effective area ratio across the CT. This has the effect of increasing Compressor stall margin in contrast to the reduction of compressor stall margin associated with steam injection upstream of the compressor turbine.

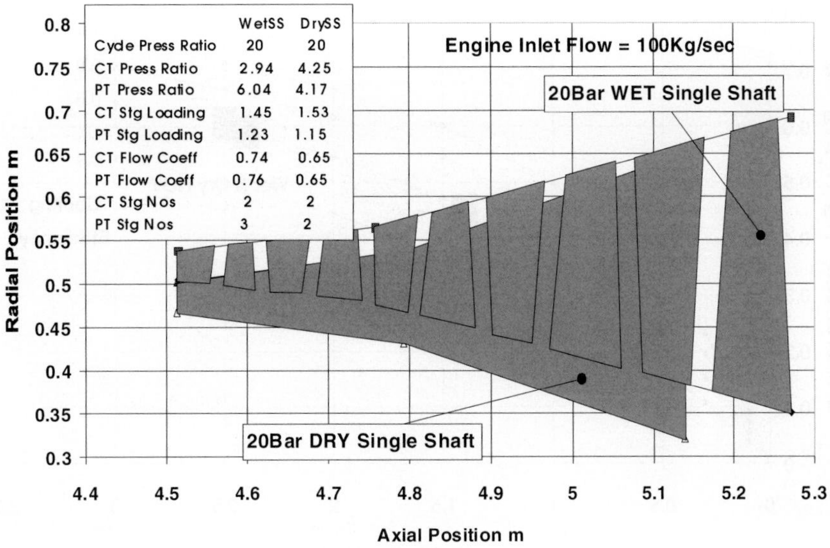

Fig. 10 Wet and dry turbine flowpaths.

Figure compares Twin shaft operability with and without variable geometry and indicates excessive loss of power with steam extraction due to reduction of compressor turbine power without variable geometry.

Compressor

The compressor operating line shifts dramatically with variable steam extraction from ~15% to ~60% Stall Margin equivalence for both single and twin shaft design. This means that significant efficiency may be sacrificed with steam extraction unless compressor characteristics can be modified with the use of variable bleed or guide vanes to minimise the variation of compressor efficiency together with suitable choice of design point.

Compressor variable guide vanes can also offer significant freedom to match variation in demand for power and obtain better operating economics in conjunction with variable steam injection. However it seems likely that supplementary firing in the steam generator would provide better CHP operational flexibility without affecting gas turbine performance.

Combustion

Combustor operability requires control of an extra degree of freedom with the introduction of variable steam injection. The operability strategy would require a phased relationship

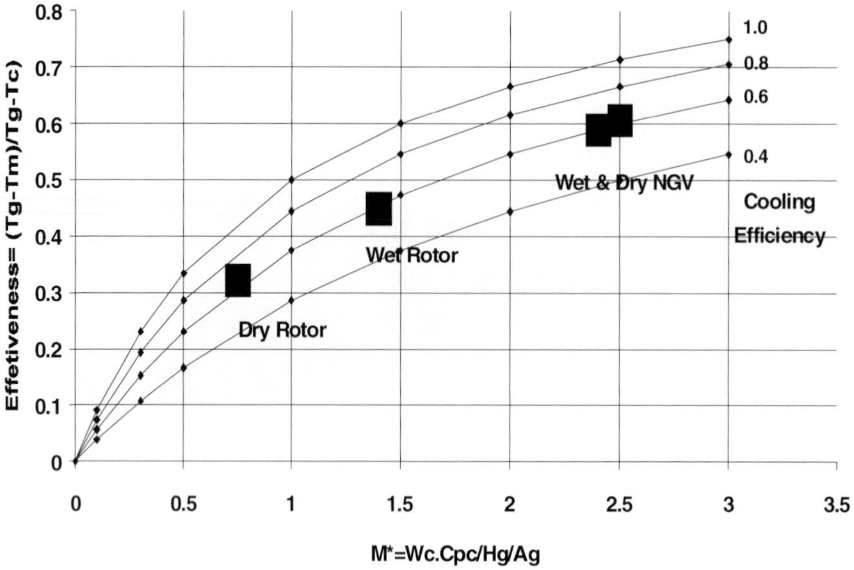

Fig. 11 Cooling effectiveness.

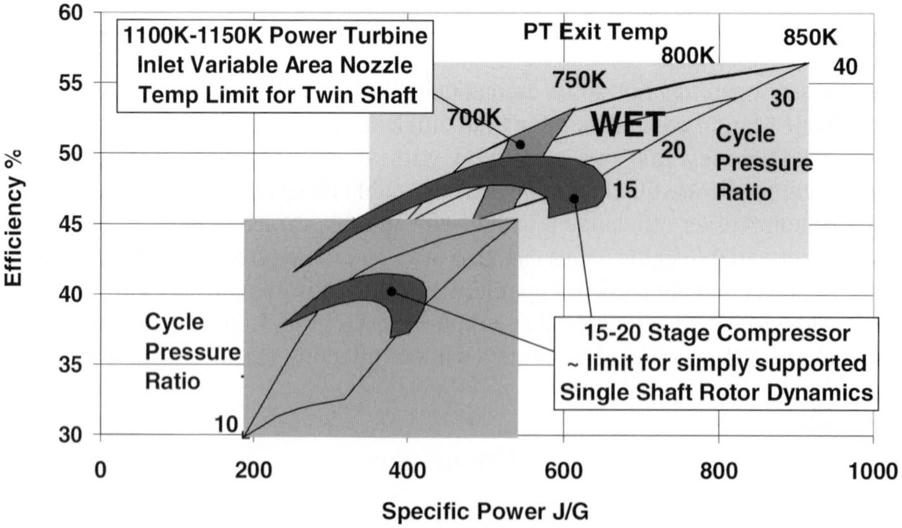

Fig. 12 Wet and dry performance constraints.

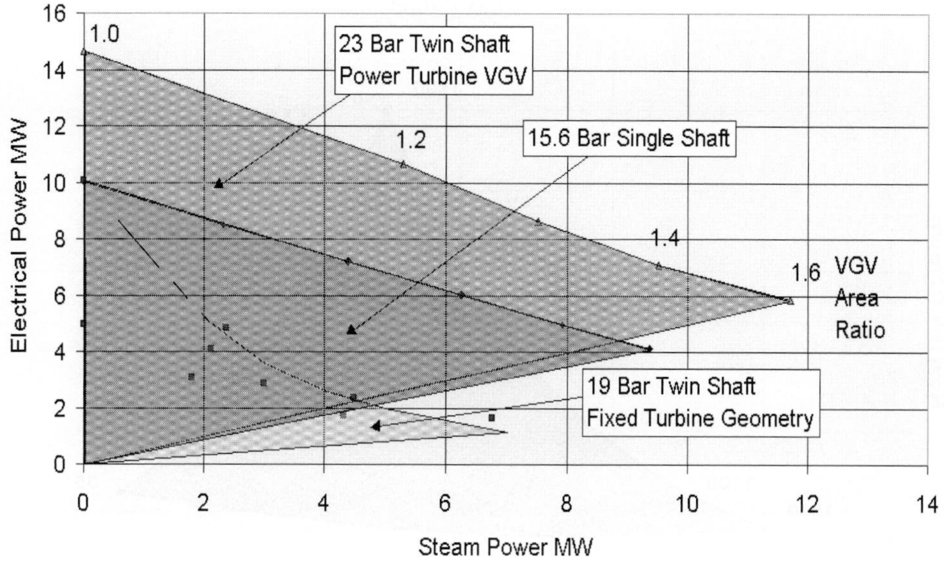

Fig. 13 Derivative engine performance.

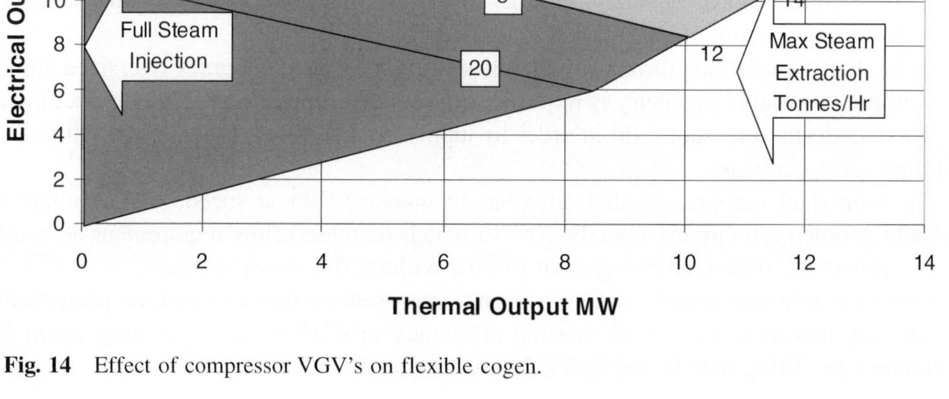

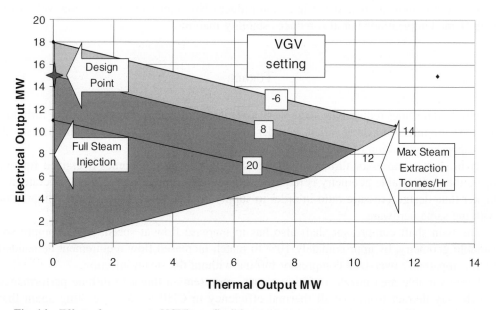

Fig. 14 Effect of compressor VGV's on flexible cogen.

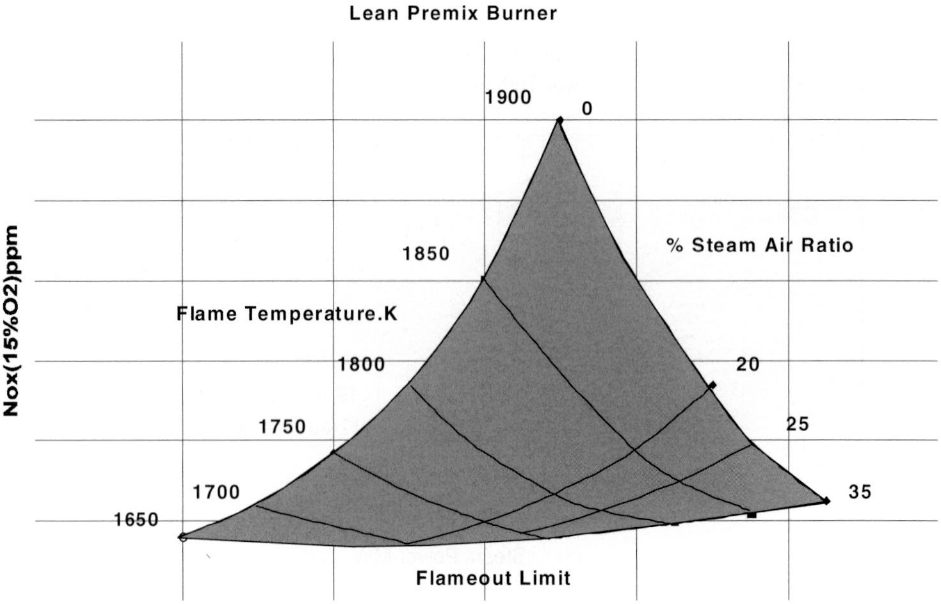

Fig. 15 Combustion with steam addition.

between premixed and diffusion burners to ensure that combustion remains within the margins of flame stability and guaranteed emission levels.

Test results indicate that steam injection reduces Nox emissions to ultra low values for high flame temperatures but also reduces stability margins.

Turbine

Single shaft turbines redistribute stage work associated with compressor and power turbines, as the steam injection rate is varied, in order to satisfy compressor power requirements with corresponding variation of nett output and efficiency.

Twin shaft compressor turbine, with full steam extraction, becomes oversized for dry operation and variable geometry is required to force the compressor turbine to operate at a higher than design pressure ratio in order to supply enough power to drive the compressor with the same hardware.

The twin shaft compressor shaft also has to increase flow at speed, using compressor variable geometry, by approximately 10% to match increased flow requirement demanded by the apparently oversized compressor turbine without the steam injection.

Large variable area nozzle variation clearly represents a threat to turbine performance which may detract from overall thermal efficiency in CHP mode suggesting again that supplementary firing may be the best solution for variable CHP.

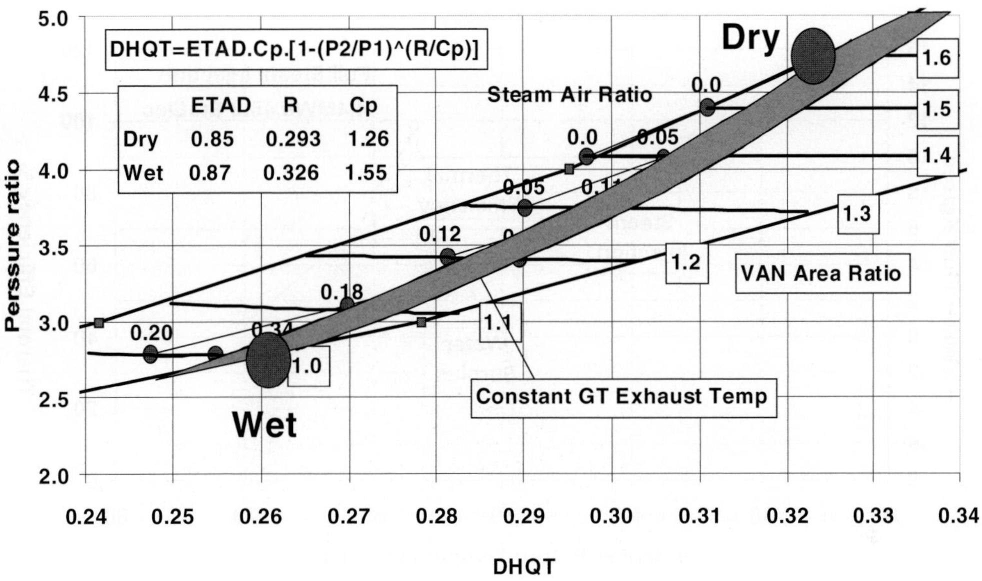

Fig. 16 CT performance with variable steam extraction.

Exhaust Condenser

The exhaust condenser has been considered as a source of heat for district heating as well as a means of capturing water in the exhaust for recirculation and self-sufficiency of water. Recapture of latent heat used for evaporation of water in the steam generator by condensation provides for essentially 100% overall system thermal efficiency. However district heating has been found to be incompatible with recirculation requirements since 100% recirculation requires 30–40°C condenser return temperature and district heating typically requires 70°C return temperature.

Alternative methods of cooling have different cost implications depending on location and might include the use of convection cooling with air, reservoir water or sea water.

Steam Generator

Steam generator or HRSG access to the gas turbine flowpath means that rapid load transients can distort the transient steam air ratio entering the combustion chamber due to either stored energy in the steam generator during load shed or delay in steam generation due to thermal lag. This may influence flame stability or emissions for a short time during transition between conditions and requires an extra degree of freedom in the controls.

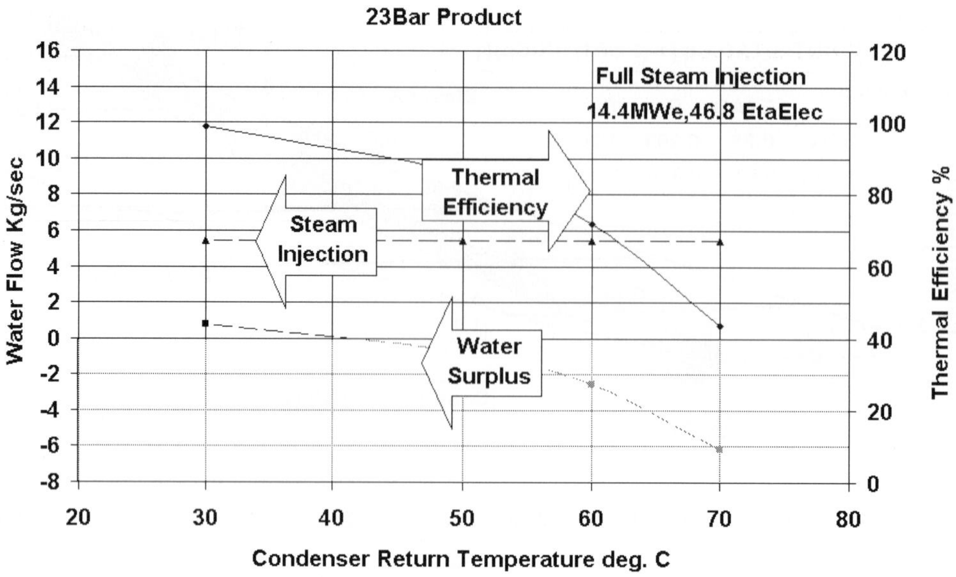

Fig. 17 Condenser design with district heating.

The steam generator also needs protection from water accumulation which may produce shock loading when suddenly boiled by hot exhaust. There may be some merit in providing a steam generator bypass for independent control of steam generator temperature transients.

The once through steam generator (OTSG) may be better suited to such transient performance since OTSG is smaller and simpler and can respond to CHP transients more effectively since it has 1/8 water content of a standard HRSG and can operate dry without the need for bypass ducting. Control is simpler since feed water flow directly controls steam out temperature since there is no barrier between economiser, evaporator and superheater. OTSG is safer since the system can be run dry for shutdown in readiness for restart. The high nickel alloys Incoloy 800 and 825 are used to protect gas and water side from corrosion. The OTSG simplicity may also be more suited, than the recuperator, to achieving microturbine performance for distributed power.

Water Treatment

Injection of contaminated steam is known to threaten the mechanical integrity of all hot part materials due to oxidation, corrosion and hydrogen embrittlement. High standards of condensate treatment, compatible with those already used by gas powered combined cycle systems, is required to protect both HRSG and the gas turbine hot parts.

This should fall within existing experience but may require some new technology development of materials designed to withstand continuous steam injection.

Alternative liquid fuels are more likely to be used for standby mode only, with total water loss, to avoid the expense of more rigorous water treatment plant capable of cleaning water contaminated by much dirtier products of combustion attributable to liquid fuel contaminants.

Total loss system may also be required for health and safety reasons in certain food processing plant to prevent contaminated water entering the steam path.

THERMO ECONOMICS

Market Performance

Wet cycle performance has been shown to reach combined cycle levels of efficiency using small industrial component design technology for twin shaft design, with high cycle pressure ratio, at small to medium power levels. Clearly the application of this approach to larger scale turbomachinery could increase this potential at higher power using aeroderivative component performance levels reflected in power-efficiency trends for all types of cycles.

The additional benefit of variable CHP depends on the practicalities of either cooled variable turbine geometry or auxiliary firing for twin shaft design or reduced performance achievable with single shaft design shown in section 3.7.

Turnkey Cost

Wet cycle specific cost approaches simple cycle levels for derivative small industrial engines due to the increase in specific power caused by steam injection and cycle pressure ratio balancing the cost increase associated with steam generation equipment.

Published data shows that turnkey cost increase associated with steam generation equipment for regenerative cycles is ~(40 to 70)% relative to simple cycle at (10–20) MW electrical output. The chart in section 3.7 shows that powerplant specific power can be doubled with a combination of steam injection and increasing cycle pressure ratio from 20 to 30.

Wet cycle cost prediction results from changes to regenerative cycle steam generating equipment and gas turbine costs due to steam injection. These costs, which are affected by gas turbine specific power (cycle pressure ratio) and increased turbine relative exit flow due to steam injection, place the wet cycle within the simple cycle cost bandwidth.

CONCLUSIONS

- Combined cycle efficiency is achievable at simple cycle cost, using existing powerplant design technology, by injection of steam into twin shaft turbomachinery with cycle pressure ratio greater than or equal to 30.
- Flexibility for variable CHP compromises power generation performance without auxiliary firing. Single shaft design limits cycle pressure ratio due to rotor dynamics. Twin shaft design limits firing temperature to avoid cooling turbine variable geometry.

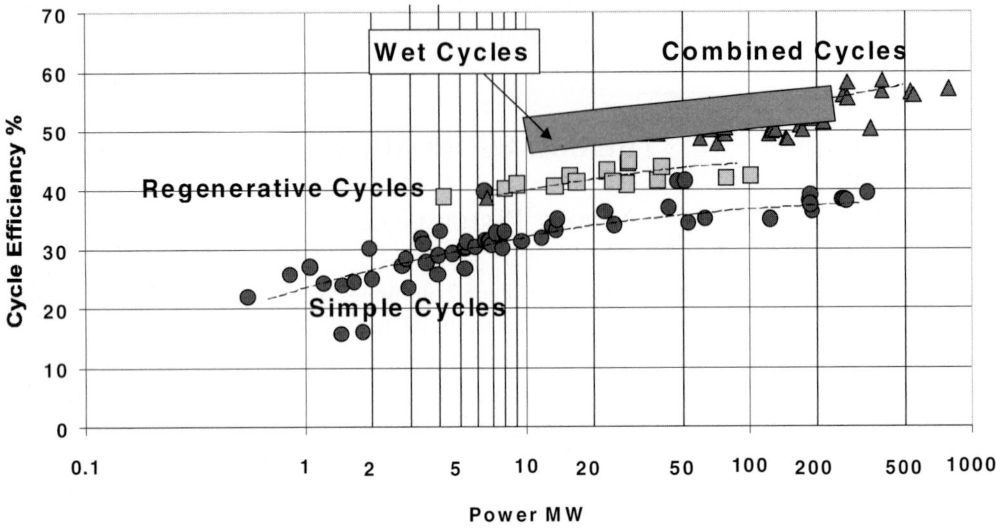

Fig. 18 Thermodynamic performance potential.

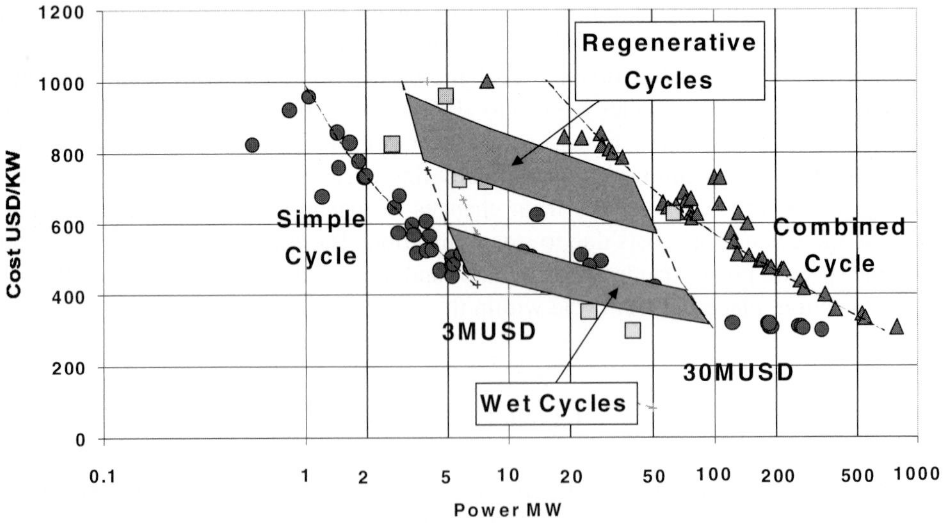

Fig. 19 Turnkey costs.

- Performance competitive for distributed power. Matches grid generator efficiency without the grid. Adaptable to variable demand for heat and power. Environmentally friendly with ultra low NO_x and CO_2. Self sufficient in water supply.

ADDENDUM

In terms of materials, this novel cycle does raise some interesting issues. The condenser will need to be constructed of materials that will be capable of withstanding hot wet conditions, in which the condensate is likely to be somewhat acidic, due to the presence of CO_2, NO_x, and in coastal applications the presence of chlorides. Nevertheless water treatment, should in some respects be easier, as the feedwater will be free from calcium and magnesium scales. The water treatment would, of course, need to be of the all volatile/zero solids type so as to minimise the need for boiler blowdowns and loss of water.

However the condensate will be aerated and, as noted earlier acidic, so that special precautions may be needed to prevent corrosion of deaerator and piping systems. This will be especially critical if the HRSG is of the compact type, as iron compounds in feed water can deposit out in the boiler section, resulting in reduced rates of heat transfer and increased pressure drops. Essentially all of these are engineering problems which can be overcome by using the right equipment and water treatment processes

The higher volume of steam/water vapour in the hot section of the gas turbine may need investigation, since steam could increase oxidation rates, although the use of steam cooled transition ducts and blades in advanced gas turbines suggests that there is no great problem. An issue which does need investigation is the effect of steam on the life of thermal barrier coatings for high temperature blades. Star, in a private communication, has suggested that high steam levels might increase the sintering of TBCs or might increase the oxidation rate of the bond coat.

The wet cycle would be capable of using steam at temperatures up to 873 K, which would be slightly higher than the steam temperatures in the most advanced CCGT plants. On the other hand the steam pressures in the wet cycle are quite modest compared to superheater pressures in modern HRSGs which are around 120 bar, so that conventional materials of construction such as the X20 grades, containing twelve percent chromium should be adequate as there is no great demand for extremely high creep strength.

ACKNOWLEDGEMENTS

Author wishes to thank ALSTOM Power for permission to publish and to acknowledge contributions made by colleagues within ALSTOM and the Swedish EVGT program.

REFERENCES

1. D. Y. Cheng: 'Parallel-Compound Dual Fluid Heat Engine' US Patent No 3978661, 1976.
2. D. Y. Cheng: 'Advanced Regenerative Parallel Compound Dual Fluid Heat Engine', US Patent No 5170622, 1992.
3. Vladimir V. Lupandin, Victor I. Romanoff, Villy A. Krivutsa and Victor Lupandin: 'Design, Development and Testing of a Gas Turbine Steam Injection and Water Recovery System', ASME Paper 2001-GT-0111.
4. R. Digumarthi and Ch-N. Chang: 'Cheng Cycle Implementation on a Small Gas Turbine Engine', *Journal of Engineering for Turbines and Power.* **106**(84-GT-150), 699–702.
5. D. Hein and K. Kwanka: 'Cheng Cycle Cogeneration for a Significantly Varying Demand for Heat and Power', *Proceedings of the Symposium on Energy Engineering in the 21st Century*, Hong Kong.
6. D. Hein, K. Kwanka and M. Nixdorf: 'Strategy for a Cost Optimised Operation Flexible Cogeneration Gas Turbine Plant', ASME 200-GT-0299, 2000.

Candidate Materials for a Modular High Temperature Reactor Power Plant

P. J. Ennis
Research Centre Juelich
IWV-2, 52425 Juelich
Germany

ABSTRACT

The properties of the various high temperature alloys that are being considered for the core and gas turbine components are reviewed in the light of the requirements. Past experience from the German HTR materials programmes and the recent developments in the areas of turbine blade alloys and high chromium steels are discussed. It is shown that there are appropriate alloys now available for the design of the components. The particular problem of the impure helium service environment is described and means for ameliorating the effects of the environment on the materials properties are indicated.

BACKGROUND

The first gas cooled reactors to achieve commercial success were the Magnox reactors, 26 units of which have formed the basis for nuclear electricity production in the UK over the past 40 years. This reactor type is graphite moderated, CO_2 cooled and uses natural uranium as a fuel.[1] Because of oxidation problems with the mild steel components exposed in CO_2, the operating temperature was eventually set at a conservative value of 360°C. A logical development was the Advanced Gas-cooled Reactor (AGR), the aim being to increase the operating temperature to around 550°C for higher thermal efficiency,[2] the higher component temperatures required the use of more expensive high chromium steels.

The next step was a further, significant increase in operating temperature to provide thermal efficiencies similar to those of fossil-fired power stations (around 39%). The restrictions on maximum operating temperature imposed by using CO_2 as coolant were removed by the application of helium as the coolant gas. Temperatures up to 950°C could then be considered. This reactor type became known as the high temperature reactor (HTR, in the USA and Japan HTGR). Experimental and prototype HTRs were constructed in the UK (the European Dragon Reactor), the USA (Peach Bottom and Fort St Vrain), in Japan (VHTR) and in Germany (AVR and THTR). The general characteristics of the HTR are a reactor core made of graphite, and fuel elements in the form of small particles of uranium oxide coated with carbon and silicon carbide. Two core geometries were possible; a prismatic design in which the fuel is contained in channels machined in graphite blocks and a pebble-bed design in which the fuel consisting of coated particles embedded in graphite spheres about 60 mm in diameter are loaded into in a graphite chamber and the coolant flows through the gaps between the spheres. It would have been possible in the 1990s to begin construction of commercial HTR plant, but the changed political climate in the wake of the Chernobyl disaster effectively meant the end of the HTR development work, at least in the USA and Europe. The HTR materials development work world-wide was reported in three special volumes of Nuclear Technology in 1984.[3] The German materials development programme culminated in the publication of a draft design rule document.[4, 5]

Recently, however, there has been revived interest in the HTR system, led mainly by the Republic of South Africa[6] and China. The principle interest is in a small modular reactor system coupled with a helium turbine, which drives a generator.[7, 8] In Europe, a task group has been set up in the Fifth Framework Programme to consider HTR technology. The materials programme for both metallic and ceramic components has been described by Buckthorpe et al.[9, 10]

STRUCTURAL MATERIALS FOR MODULAR HTR

In this section, the metallic components that are regarded as the most critical for safe and reliable operation of a modular HTR with a helium turbine will be briefly discussed. The basis for the discussion is the work carried out in Germany in the 1970s and 1980s for the 'High Temperature Reactor with Helium Turbine Project' (known as the HHT Project) and the 'Nuclear Process Heat Project' (PNP). Although this system was considered for large-scale electricity generation, many of the materials investigated are of interest for the current modular HTR investigations. The operating temperatures considered for HHT and PNP were in the range 750 to 950°C, several high temperature alloys that were commercially available at the time were intensively investigated and for selected alloys design rules were drafted.[4]

ENVIRONMENTAL CONSIDERATIONS

Table 1 lists the alloys that were investigated in the German materials programmes. A significant aspect of materials behaviour was the interaction between the materials and the helium coolant atmosphere. Helium is, of course an inert gas, so reactions involving helium cannot occur. However, the coolant will contain impurities, due to out-gassing of components and small leakages. The main impurity present is water, which may react with the graphite to form CO, H_2 and, under irradiation conditions in the core, CH_4. Although impurities will be present only at microbar levels (Table 2 shows the impurity level ranges used in the German HTR projects), the structural alloys can be affected, the main process being carbon transfer between metal and environment, an effect well known from fast breeder reactor technology. It has been shown[11, 12] that the reaction kinetics play a very important role and the transfer of carbon between metal and environment depends on the flow rates of the gas. At low flow rates, the decomposition of CH_4 at the metal surface is the dominant effect and at the low oxygen potentials, prevailing, protective oxide scales do not form and the metal is

Table 1 Principal alloys investigated in the german HTR materials projects.

Application	Alloy	Composition, wt.%
Turbine Blade	NIMONIC 80A	Ni-0.04C-20Cr-2.5Ti-1.5Al
	IN713LC	Ni-0.07C-12Cr-6Al-5Mo-2Nb1Ti
	TZM	Mo-0.5Ti-0.1Zr
Hot Gas Duct	INCONEL 617	Ni-0.08C-22Cr-12Co-9Mo-1Al
Heat Exchanger Tube	HASTELLOY X	Ni-0.08C-20Cr-18Fe-9Mo
Header	INCOLOY 800H	Fe-0.08C-32Ni-20Cr
Control Rod Cladding	16Cr16Ni Stainless Steel	Fe-0.06C-16Cr-16Ni2Mo-1Nb

Table 2 Compositions of typical simulated HTR helium coolant used in materials testing.

Type	Composition in μbar, Balance He					
	H_2	CH_4	CO	CO_2	H_2O	N_2
HHT Germany	50	5	50	5	5	5
PNP Germany	500	20	15	5	1.5	<5
JAERI B Japan	200	5	100	2	1	-
HTGR USA	500	50	50	-	<0.5	-

carburised. At fast flow rates, typical for reactor operation, the interaction is dominated by H_2O, again in the absence of protective oxide scales, the carbon within the metal is oxidised and so decarburisation occurs. Both these possibilities need to be considered for the operating conditions and the materials applied. The metal-environment interactions can be reduced to acceptable levels if the impurity levels are closely controlled. For example, decarburisation of the materials can be avoided by ensuring that the CO level lies within a critical range so that the equilibrium for reaction, C in metal + $H_2O \leftrightarrow CO + H_2$, lies to the left.

The materials issues concerning out-of-core components will be discussed. The only metallic components that could be envisaged in the core are the control rod cladding tubes. Here the effect of irradiation on the mechanical properties and swelling are of importance, but alternative materials, such as carbon fibre reinforced materials, may offer better solutions. Control rod materials will not be further discussed.

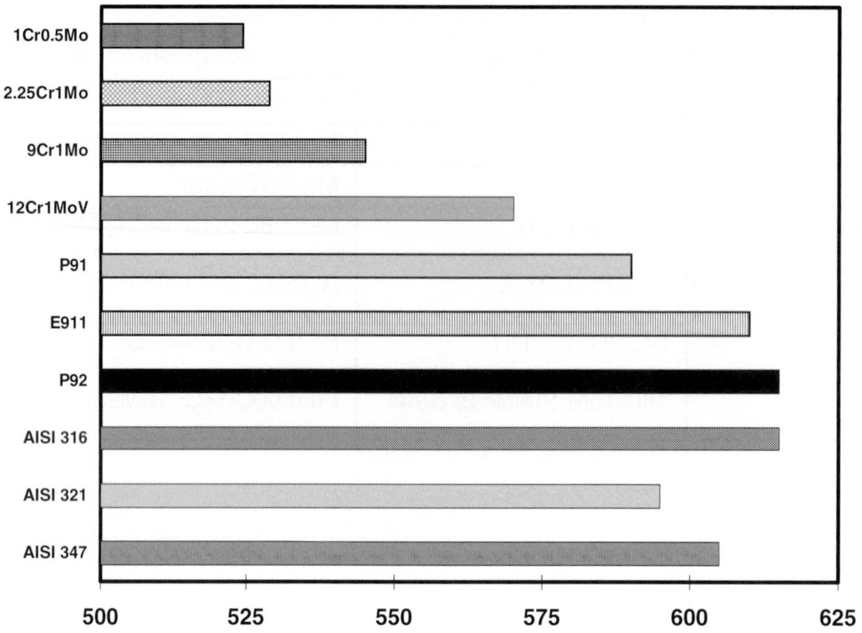

Fig. 1 Maximum operating temperature for power station steels based on 100,000 hours stress rupture strength of 100 MPa.

MATERIALS FOR THE REACTOR PRESSURE VESSEL (RPV)

There are a number of well-qualified, intensively researched RPV steels from light water reactor technology. These steels are certainly candidates for the RPV of a modular HTR. However, in the last twenty years, improved 9% chromium steels (P91, E911 and P92) have become available, which offer substantially increased stress rupture strength than the low alloy RPV steels, and therefore may provide larger safety margins in design and decrease the need for any external cooling. Figure 1 compares the stress rupture strengths of power station steels on the basis of the temperature at which the 100,000 hours stress rupture strength is 100 MPa. The modified 9% Cr steels are now widely used in modern fossil-fuelled power plant operating at steam temperatures up to 600°C.

As far as RPVs are concerned, however, there remain some unsolved problems for the 9% Cr steels. The steels rely on the martensitic transformation to produce the stable, creep resistant sub-structure for long-term service. At slow cooling rates, as could be the case for thick-walled components, the martensitic transformation is suppressed and the creep strength is lost.[13] Furthermore, the 9% Cr steels are susceptible to creep failure in the heat affected zone. Investigation of these two phenomena is required if the 9% Cr steels are to be used for the RPV to take advantage of higher strength under upset conditions. In addition, there is no information concerning the effect of irradiation and the helium environment on 9% Cr steels.

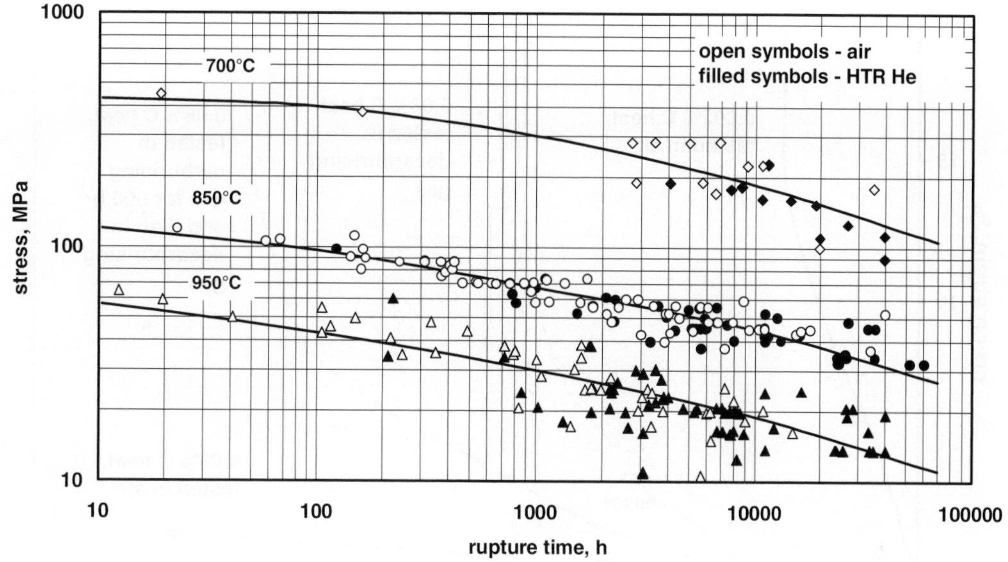

Fig. 2 Stress rupture curves for INCONEL 617 at 700, 850 and 950°C tested in air and in simulated HTR helium atmospheres (carburising).

HOT GAS DUCTS

The principal candidates for this application are wrought, nickel-base alloys, originally developed for aircraft gas turbine combustion chambers, such as INCONEL 617 (Ni22Cr12Co9Mo1Al) and HASTELLOY X (Ni22Cr18Fe9Mo). Extensive characterisation of these alloys was carried out in the German HTR materials programmes. Figure 2 shows a comparison of the stress rupture properties of INCONEL 617 in air and in carburising HTR helium. All the data, for a large number of heats and product forms, form at each temperature a common scatterband. It should be mentioned here that for all the mechanical testing done in the German HTR projects, the experimental data (in the case of creep rupture testing, the time-strain data) has been preserved in the HTM Alloys Databank system developed at JRC Petten,[15] allowing further analyses of the creep behaviour to be carried out. The specimens tested in HTR helium atmospheres exhibited high rates of carbon uptake, the carbon content reaching double the initial value after a few thousand hours testing at 950°C. No appreciable effect on the stress rupture strength was detected. The room temperature ductility was, however, significantly reduced, due to the presence of grain boundary carbide films. The low ductility temperature range extended up to 800°C. From this aspect, carburisation effects must be avoided.

With regard to the effects of decarburisation, the more likely interaction in service because of the high helium flow rates, the creep strength was found to be dramatically reduced.[16] Figure 3 compares creep curves for INCONEL 617 in air, in carburising test gas and in a

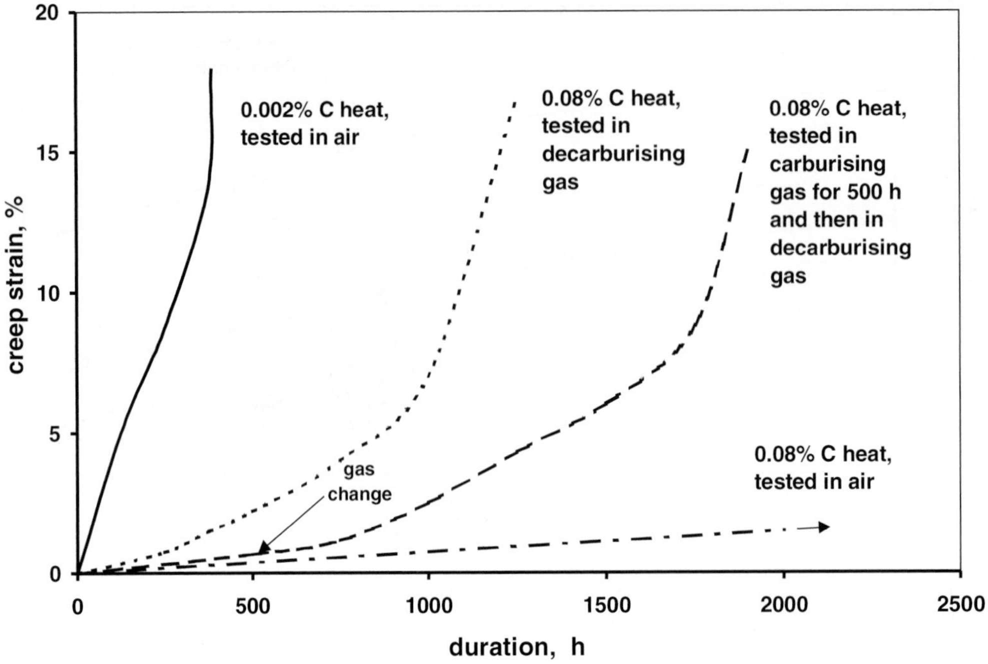

Fig. 3 Creep curves for INCONEL 617 tested at 950°C, 18 MPa: low C (0.002%) heat tested in air, normal C (0.08%) tested in decarburising gas (He-500 μbar H_2–1.5 μbar H_2O), normal C tested initially in carburising gas (He-500 μbar H_2–100 μbar CO-10 μbar CH_4–1.5 0 μbar H_2O) and after 500 hours in decarburising gas normal C tested in air.

decarburising gas. An additional test involved changing the test gas from carburising to decarburising during the test. The reduction of creep strength is clear, illustrating the importance of grain boundary carbides for the creep strength of INCONEL 617. Decarburisation in HTR service must therefore be avoided at all costs.

The high cobalt content of INCONEL 617 could pose a potential problem, although in the German HTR projects, it was found that Co was not incorporated into the oxide scale and so radioactive Co would not therefore enter the gas circuit, even if the oxide spalled off. Notwithstanding, the use of Co-free materials would be more acceptable, for example HASTELLOY XR,[17] a Japanese modification of HASTELLOY X specifically for reactor service, with a stress rupture strength that is a little below that of INCONEL 617.

TURBINE BLADES AND VANES

The combustion temperatures in aircraft and stationary gas turbines have increased over recent years to allow higher efficiencies. The resulting higher metal temperatures have been

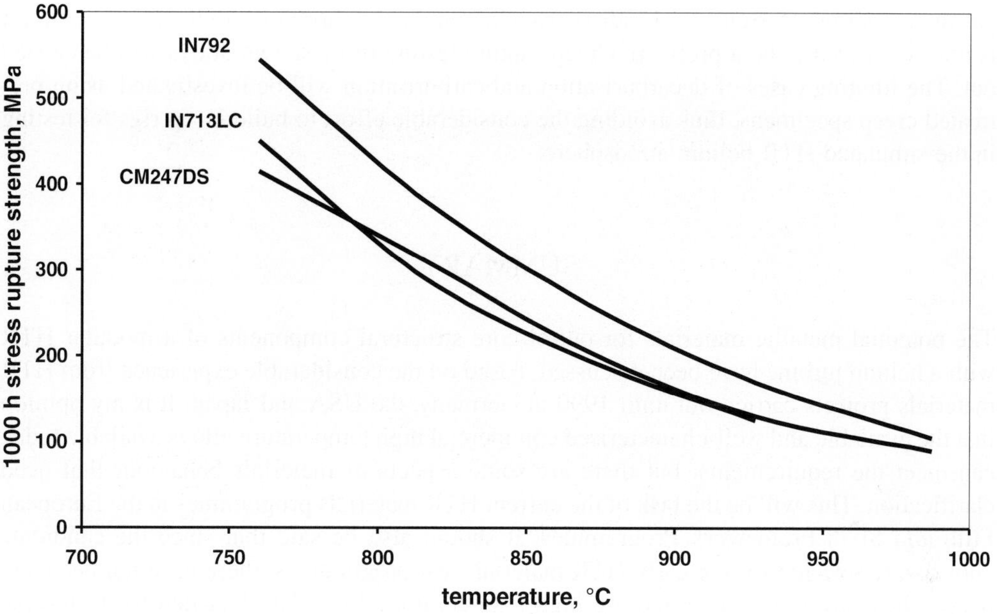

Fig. 4 1,000 hours stress rupture strengths of IN792, CM247Ds and IN713LC.

achieved by alloy development, in particular to introduction of directionally solidified and single-crystal alloys, by complicated blade cooling techniques and by the use of thermal barrier coatings. In this way, metal surface temperatures can be restricted to 950°C, in spite of combustion temperatures reaching 1,400°C. The operating conditions for the HTR helium turbine are in comparison relatively modest. Helium temperatures will be around 850°C. There is therefore no real need to rely on blade cooling or thermal barrier coatings. This will lead to a simpler, more efficient design and established turbine blade materials for which there is a wealth of practical experience may be used.

An interesting material that was tested in the German HHT project was TZM, a molybdenum-based refractory alloy. The application of such a material was possible because of the exceedingly low oxygen potential of the environment. The low temperature ductility of this material was a cause for concern.

One difficult aspect is the in-service inspection of the turbine components. Aircraft engines are regularly inspected for flaws; stationary gas turbines are required to operate for much longer periods between inspections; in the HTR turbine, inspections are made difficult due to irradiation effects.

For the current European HTR materials project, two alloys have been selected for evaluation: IN792 (Ni-0.05C-12.5Cr-9Co-4.5Ti-4W-4Ta-3Al-2Mo), regarded as a chromia scale former and CM247DS (Ni-0.15C-10Co-10W-8Cr-5.5Al-3Ta-1Ti), regarded as an alumina scale former. With both alloys there remains the question of Co but as for the hot gas duct material, the release of radioactive Co-containing particles into the primary circuit is not considered to be a problem. Creep rupture testing of these two alloys will be carried out. The limiting cases of decarburisation and carburisation will be investigated using pre-treated creep specimens, thus avoiding the considerable effort to build creep rigs for testing in the simulated HTR helium atmosphere.

SUMMARY

The potential metallic materials for out-of-core structural components of a modular HTR with a helium turbine have been discussed, based on the considerable experience from HTR materials projects carried out until 1990 in Germany, the USA and Japan. It is my opinion that the available and well-characterised commercial high temperature alloys available today can meet the requirements, but there are some aspects of materials behaviour that need clarification. This will be the task of the current HTR materials programmes in the European Fifth and Sixth Framework Programmes. It should also be said that since the candidate alloys were selected for the early HTR materials test programmes, there have not been any major alloy developments regarding high temperature alloys. The fact that today turbine operating temperatures have increased it is mainly due to the application of cooling and coating technology. Such measures will not be required for the blading of a helium turbine driven by an HTR.

REFERENCES

1. A. Bertini, B. Edmonson, T. Kagayama and D. MacDougall: 'A Review of Operating Experience with UK Designed Magnox Stations', *Gas-Cooled Reactors Today*, British Nuclear Energy Society, London, **1**, 1982, 1–12.
2. R. W. Hall and G. Johnston: 'The Commissioning and Operation of the Advanced Gas-Cooled Reactor in the CEGB', ibid, 23–34.
3. 'Special Issues on High-temperature, Gas-cooled Reactor Materials', *Nuclear Technology*, **66**, 1984.
4. Final Report of German Research Project 'Auslegungskriterien für Hochtemperaturbelastete metallische und keramische Komponenten sowie Spannbeton-Reaktorbehälters zukünftiger HTR-Anlagen', 1988.
5. G. Breitbach, F. Schubert and H. Nickel: *Proceedings of the Workshop on Structural Design Criteria for HTR*', *Jül-Conf-71*, Berichte der Kernforschungsanlage Jülich GmbH, 1989.
6. J. A. de Beer and Z. Olsha: 'Power Generation in Southern Africa', *Materials for Advanced Power Engineering*, J. Lecomte-Beckers, M. Carton, F. Schubert and P. J. Ennis, eds., Forschungszentrum Jülich, Energy Technology series, **21**(3), 2002, 1783–1800.
7. J. Singh and H. Barnart: 'Modular Design Concept for the HTR on the Basis of AVR', *Gas-Cooled Reactors Today*, British Nuclear Energy Society, London, **1**, 1982, 259–264.

8. G. H. Lohnert and H. Reutler: 'The Modular HTR, a New Design of High-Temperature Pebble-Bed Reactor', *Gas-Cooled Reactors Today*, British Nuclear Energy Society, London, **1**, 1982, 265–276.
9. D. Buckthorpe, R. Couturier, B. van der Schaaf, B. Riou, H. Rantala, R. Moormann, A. Buenaventura and B. C. Friedrich: 'High Temperature Reactor (HTR) Materials', *Materials for Advanced Power Engineering*, J. Lecomte-Beckers, M. Carton, F. Schubert and P. J. Ennis, eds., Forschungszentrum Jülich, Energy Technology series, **21**(2), 2002, 759–768.
10. D. Buckthorpe: 'Research and Development Programme on HTR Materials', *Proceedings of ICAPP 03*, Cordoba, Spain, 2003.
11. W. J. Quadakkers and H. Schuster: 'Corrosion of High Temperature Alloys in the Primary Circuit Helium of High Temperature Gas-Cooled Reactors, Theoretical background', *Werkstoffe und Korrosion*, **36**(1) 1985, 141–150.
12. W. J. Quadakkers: 'Corrosion of High Temperature Alloys in the Primary Circuit Helium of High Temperature Gas-Cooled Reactors, Experimental results', *Werkstoffe und Korrosion*, **36**(2), 1985, 335–347.
13. D. F. Lupton and P. J. Ennis: 'Influence of Carburisation on the Ductility of Four High Temperature Alloys', *Res. Mechanica Letters*, **1**, 1981, 245–252.
14. M. Abdel Azim, P. J. Ennis, H. Schuster, F. H. Hammad and H. Nickel: 'The Tensile Properties of Alloys 800H and 617 in the Range 20 to 950°C', *Report of the Research Centre Juelich*, Germany, 1990, 2344.
15. H. H. Over: see website http://matdb.jrc.nl.
16. R. von der Gracht, P. J. Ennis, A. Czyrska-Filemonowicz, H. Schuster and H. Nickel, 'The Creep Behaviour of Ni-22Cr-9Mo-12Co-1Al in Carburising and Decarburising Environments', *Proceedings of International Conference on Creep*, Tokyo, Japan, 1986, 123–128.
17. Y. Kurata, Y. Ogawa and T. Kondo: 'Creep and Rupture Behaviour of a Special Grade of HASTELLOY X in Simulated HTGR Helium', *Nuclear Technology*, 66(2), 1984, 250–259.

The WR-21 – From Concept to Reality

DAVID BRANCH
Chief Engineer WR-21
Rolls-Royce Plc

ABSTRACT

This paper starts by discussing the thermodynamic concepts behind the intercooled and recuperated gas turbine cycle, and the benefits that these offer. It then goes on to show how these thermodynamic principles have been embodied into a practical gas turbine. It continues by covering the design of the WR-21 gas turbine and presents the pedigree of the components that go to make up the WR-21 system.

Following this the development of the system through to production readiness is explained, including the use of rig, model, and full system testing.

The paper concludes by discussing the initial service application of the WR-21, the new Type 45 destroyer for the RN, and the benefits that this system offers the user in this application.

INTRODUCTION

The WR-21 is the world's first modern advanced cycle gas turbine developed specifically for the naval marine market. The main design targets were increased thermal efficiency, reduced signature, reliability, and ease of maintenance, all of which lead to reduced cost of ownership. It has been developed with funding from the US, British, and French Governments, and with an industry team lead by Rolls-Royce of the UK and Northrop Grumman Marine Systems of the USA. In addition major subcontractors are CAE of Canada (engine electronic controller), Honeywell of the USA (intercooler cores), and Ingersoll-Rand Energy Systems (IRES) of the USA (recuperator cores).

The system was developed initially for mechanical drive applications but it is now undergoing trials where it powers an alternator, as part of proving the system for electrical drive use. There are no system hardware changes between the two modes but a different set of control algorithms is used to address the specific requirements of the electric drive operation.

In 2002 the WR-21 was selected to power the latest Royal Navy warship, the Type 45. The first ship of the class, HMS Daring, will commence harbour trials in late 2005, and will enter Royal Naval service in late 2007. The two WR-21s that will power this ship will be production tested this year and will be delivered to the ship in November 2003.

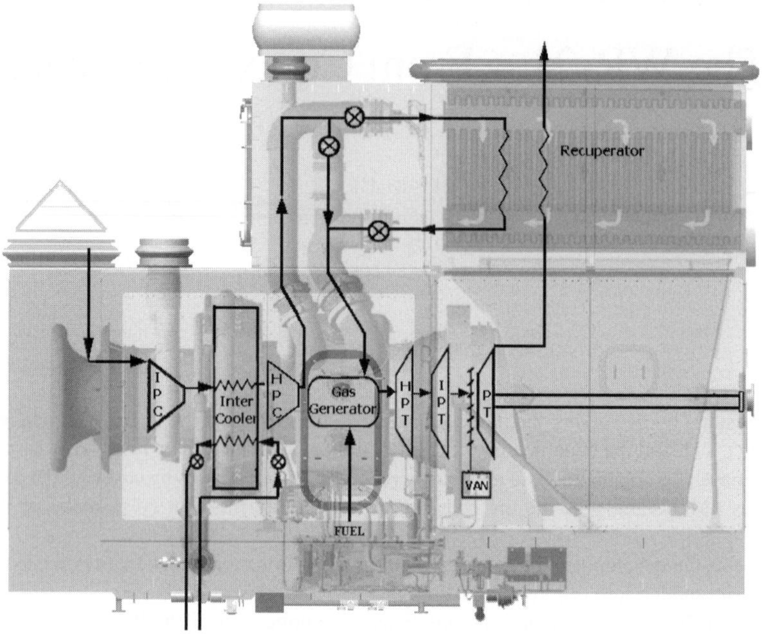

Fig. 1 Advanced cycle schematic overlay on WR-21.

BACKGROUND

The original design brief for the WR-21 came from the US Navy who wanted to have an engine that would give them a significant fuel saving over their current gas turbine powered ships. The target set for WR-21 was to reduce the fuel burn for a typical destroyer annual operating profile by 30% relative to the current usage. The resulting gas turbine also had to have the same footprint as the existing engine. To meet these challenging design targets an advanced cycle was chosen. To meet the reduced fuel burn requirement an intercooled and recuperated cycle was essential. This combination produced a cycle which had a very efficient design point performance but it didn't allow the high levels of part power efficiency required to meet the overall fuel savings when operating a typical destroyer profile, which has large periods of time spent at low and mid powers. To achieve this it was necessary to produce a very flat fuel consumption characteristic from full power down to relatively low powers. To accomplish this a variable geometry free power turbine was chosen. This allowed power turbine aerodynamic capacity to be reduced with reducing power, which results in an almost flat specific fuel consumption characteristic from 100% power to 30% power, resulting in a cycle with a fuel burn characteristic more akin to a diesel cycle than a gas turbine cycle. Figure 1 shows a schematic of the advanced cycle as a schematic overlay on the WR-21 layout and Figure 2 shows the resulting WR-21 specific fuel consumption curve compared to a simple cycle gas turbine. In summary the cycle is; air enters the plenum where it is

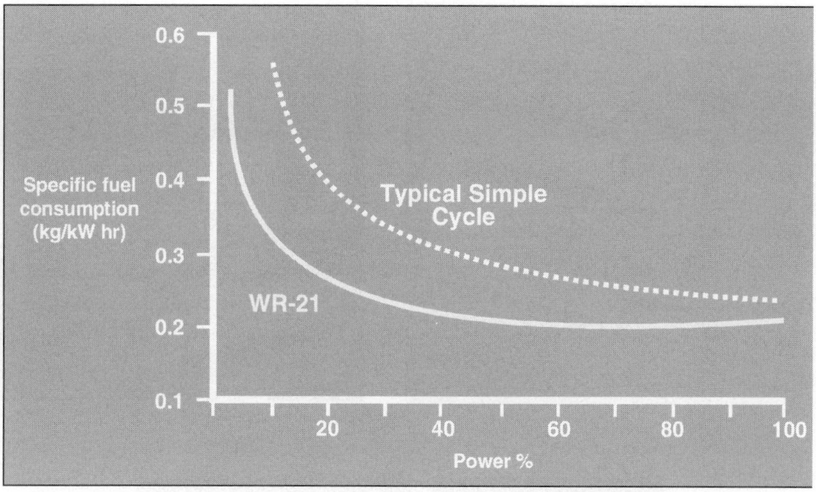

Fig. 2 Advanced cycle SFC compared with simple cycle.

guided into the engine by a radial air intake and continues on into the intermediate-pressure compressor (IPC) and then enters the intercooler (IC).

The compression process is divided approximately 30:70 between the IPC and the high-pressure compressor (HPC), with inter-cooling in between. The air leaving the IPC is cooled by heat rejection, via an on-engine heat exchanger, into a closed glycol/water intermediate loop which itself rejects heat in an off-engine heat exchanger skid. This dual-loop arrangement eliminates all risk of seawater leakage into the engine. From the intercooler, the air goes through the final stages of compression in the HPC. From the HPC air is preheated in an exhaust heat recuperator prior to combustion, thereby reducing the amount of fuel required to reach the desired turbine inlet temperature. After passing through the gas generator turbines, which drive the compressors, gas enters the free power turbine (PT) through a single-stage, variable area nozzle (VAN). After exiting the PT, the gas enters an exhaust collector where it is turned 90 degrees upward, to flow through the recuperator and finally exhaust through the ship's uptake system.

DESIGN

WR-21 Engine System

The WR-21 engine system comprises four main sub-systems; the engine and enclosure module, the freshwater/seawater heat exchanger, the lubrication oil module, and the electronic engine controller. These systems are shown in Figure 3.

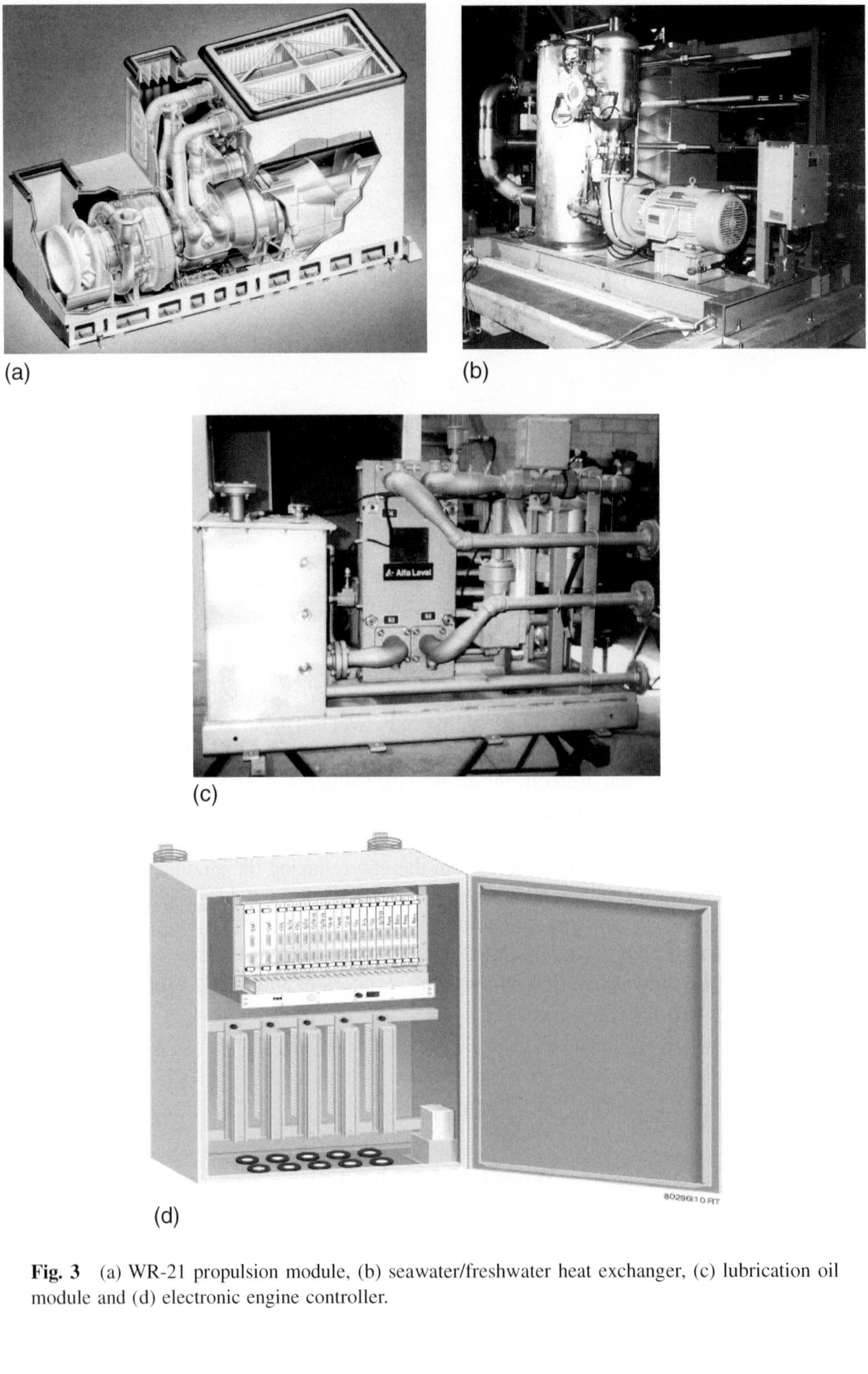

Fig. 3 (a) WR-21 propulsion module, (b) seawater/freshwater heat exchanger, (c) lubrication oil module and (d) electronic engine controller.

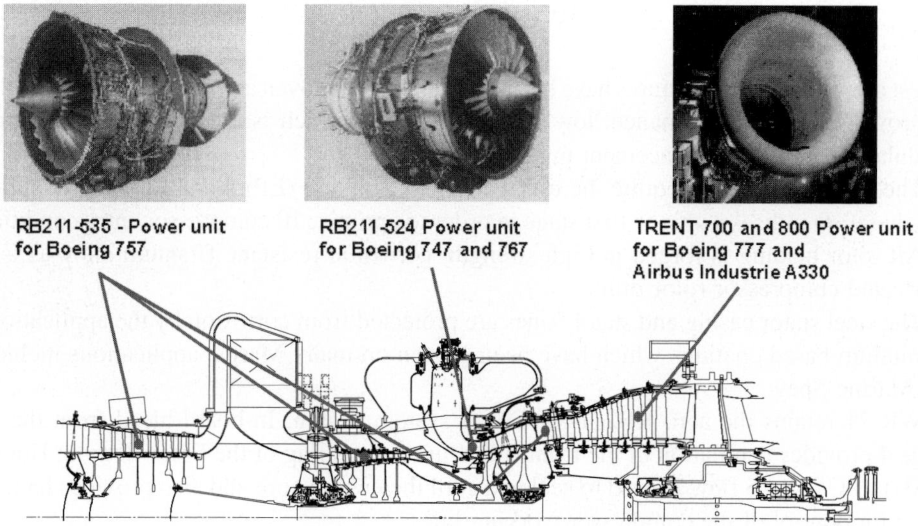

Fig. 4 WR-21 gas turbine aero parentage.

The WR-21 gas turbine is based on components from the latest Rolls-Royce RB211 and Trent family of aero engines. These were chosen to allow commonality of design, with the added benefits of an established reliability database and the option of incorporating future improvements in aero technology with reasonable ease and reduced cost. Figure 4 shows a general arrangement of the WR-21 and the source aero gas turbines.

The WR-21 uses heat exchanger technology for the intercooler and recuperator which is based on the expertise developed by Honeywell and IRES respectively over many years. This has gained much field exposure in their standard product lines.

The enclosure and off-engine skids are designed by Northrop Grumman Marine Systems, utilising system design experience developed over many years successful work as a major US Government Defence Contractor.

WR-21 Detailed Description

The gas generator consists of individually maintainable modules that are joined by curvic couplings allowing all rotating elements to be individually balanced at the factory without need for subsequent balancing following change out. Individual modules are the Intermediate Pressure Compressor (IPC), the Intermediate Case (Intercase and Intercooler), the High Pressure Compressor (HPC), the Delivery Air Manifold (DAM), the Return Air Manifold (RAM), Combustors, High Pressure Turbine (HPT), and the Intermediate Pressure Turbine (IPT), Variable Area Nozzle (VAN) and Power Turbine (PT). In the detailed descriptions of the design that follow each is discussed as a module.

IP COMPRESSOR

Latest aerospace design features have been combined with proven aero compressor technology to provide a high performance, low risk, compressor which is designed to allow simple modular removal and replacement in service.

The WR-21 does not require the upstream Low Pressure (LP) fan used on aero engines, and therefore utilises a unique first stage in order to optimise IP compressor entry conditions.

All rotor blading is forged in high strength, corrosion resistant Titanium alloy as is the fabricated compressor rotor drum.

The steel stator casing and stator vanes are protected from corrosion by the application of aluminium based coatings which have been proven on many Marine applications including the Marine Spey.

WR-21 retains the aero RB211-535E4 air system design. In-board bleed from the IPC stage 4 provides ventilation of the front IPC drum and sealing of the Front Bearing Housing (FBH). IPC stage 6 flow is used to seal and cool the engine core and also provides handling bleed for control of the compressor working line.

An innovative feature of the WR-21 design is the overtip treatment applied to the first stage rotor blade. High Efficiency Casing Treatment Over Rotor (HECTOR) slot technology increases part power surge margin thereby improving specific fuel consumption at low power operating conditions.

INTERCASE

The design of the intercase is unique to the WR-21, but retains many common internal gearbox components from its parent RB211-535 E4 Aero engine.

An increase in length relative to the aero intermediate casing has been introduced in order to accommodate the radially swept gas paths into and from the cores of the intercooler.

The intercase serves two major functions, firstly to deliver air to the intercooler, and return it to the HP compressor system, with minimum aerodynamic loss.

Secondly, the intercase supports the internal gearbox comprising the accessory drive bearings, gears, and radial driveshaft that exits at the bottom of the engine. The majority of the components are common to the RB211-535.

The intercasing comprises a one piece cylindrical precision casting which accommodates the main engine mount attachments. The mounts are capable of withstanding instantaneous shock loads in excess of 25 g in the vertical direction and 12 g fore/aft and athwart ships.

INTERCOOLER

The intercooler system comprises an on-engine water/glycol to air heat exchanger, and an off-engine skid system. The off-engine skid contains a water/glycol to seawater heat exchanger, circulating pump, and reservoir. The on-engine intercooler, shown in Figure 5, consists of ten individual heat exchanger cores contained within a cast casing which provides

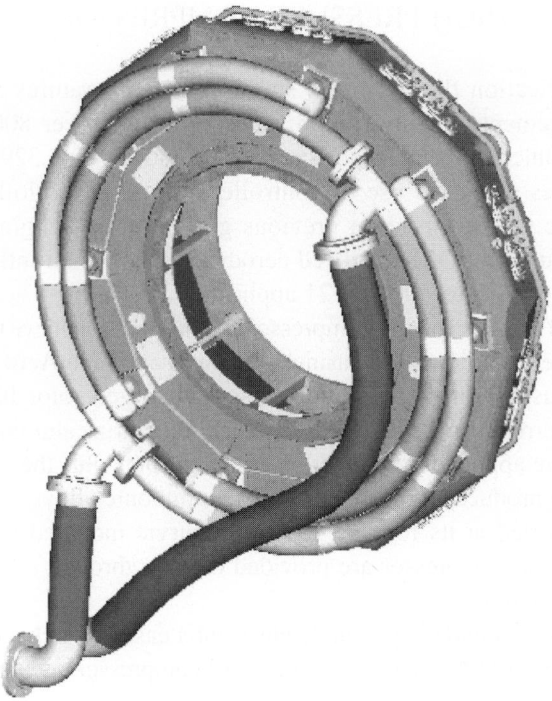

Fig. 5 On engine intercooler.

the aerodynamic flow path for the IP compressor delivery air to the cores, and from the cores back into the HP compressor. The cores are a two pass cross flow design utilising plate fin construction and manufactured from stainless steel with nickel fins.

The intercooler works by reducing the temperature of the air flowing into the HP compressor via heat transfer to the cooling water/glycol. This water/glycol flows in a closed loop and rejects its heat to the seawater in the off-engine skid mounted heat exchanger.

ACCESSORY GEAR DRIVE

The Accessory Drive is mounted to the intercooler and the high pressure compressor by struts and is driven by a radial shaft from the internal gearbox. This radial shaft drives through a bevel gearbox that turns the drive 90° to the direction of the longitudinal axis of the engine. The high speed gearbox drives the oil pump, fuel pump, and oil system breather. The engine starter is also mounted on the high speed gearbox providing torque to the HP spool during starting. All of the accessories are mounted through quick assemble / disconnect fittings on the aft side of the high speed gearbox.

HIGH PRESSURE COMPRESSOR

For the WR-21 application the operating speeds and temperatures are less demanding than the Aero requirements, the maximum speed being just over 8000 rpm versus over 10,000 rpm, and the inlet temperature for standard operation being 329 K versus 790 K.

The blading makes extensive use of controlled diffusion aerofoils in place of older technology parabolic profiles used on previous generations of engines. The increase in efficiency, which results from the improved aerodynamics, was an influential factor in the selection of this compressor for the WR-21 application.

The annulus line of the Aero parent compressor is modified to achieve the flow requirement of the WR-21, but the rotor drum and casings remain as common Aero parts.

The compressor uses a fabricated rotor drum and forged rotor blades which are all manufactured from corrosion resistant titanium. Standard Marine aluminium based corrosion protection coatings are applied to the inner and outer casings plus the stators of stages. The remaining stators are produced in corrosion resistant nimonic alloys.

The rotor is supported at its forward end via a curvic mounted stubshaft, whilst rear support and drive to the compressor are provided directly through a taper bolted flange at the HP turbine disc interface.

The inner casing is surrounded by a single piece outer casing which mounts three handling bleed valves used to vent HPC stage 3 air to control compressor surge margin.

RECUPERATOR DUCTING

A set of high pressure ducts are used to deliver the compressor air to the recuperator. One port and one starboard deliver air from DAM to the inlet manifolds of the port and starboard recuperator modules. The air is then heated in the recuperator by the exhaust gas. It then passes to the outlet ducts and is delivered to the combustors through the RAM. A by-pass leg is fitted to the ducting to allow bypass of the air side of the recuperator in case of battle damage or for routine cleaning.

Valve actuation is hydraulic, controlled by the Electronic Engine Controller and bellows are fitted to accommodate thermal expansion when hot. The ducts are provided with removable insulation pads to facilitate maintenance.

COMBUSTION SYSTEM

The combustion system consists of the combustor, reflex airspray burner (RAB) and discharge nozzle. The general arrangement is shown in Figure 6.

The combustors are mounted from the DAM by a mounting ring with 4 flexible arms. The combustors incorporate chuted ports, which minimise smoke. Extremely low smoke levels of < 5 SAE at combustor exit have been measured leading to zero smoke visibility and negligible rates of recuperator fouling.

Fig. 6 Combustion system.

Cooling of the combustor is achieved by using circumferentially orientated angled effusion cooling holes (AEC). This technology has been developed from the aero Tay combustor and achieves a high cooling film effectiveness.

The combustor is manufactured from Nimonic 86 material which has been chosen for its combination of high strength at temperature and its corrosion resistance. The combustor is further protected by the application of a corrosion resistant CoNiCrAlY bondcoat and a thermal barrier coating system.

Each combustor incorporates a RAB which is welded into the combustor and controls the injection of the fuel-air mixture. It is double-skinned to avoid high end-cap metal temperatures during combustion.

The discharge nozzle turns the hot combustion gases from radial to axial before entry into the HP turbine. It is manufactured in a laminated Nimonic sheet material which contains internal cooling passages achieving a high cooling effectiveness.

HP TURBINE

It is a single stage design which uses the latest aero materials to enhance creep life. Both the rotor blade and shroud segment are cast in single crystal alloy, CMSX4, which has millions of hours of service on civil aero engines. The corrosion resistance of the HP turbine components is further enhanced by the application of silicon aluminide coatings.

The HP Turbine cooling air system uses recuperated return air and HP Compressor delivery air. The recuperated air cools the Nozzle Guide Vane (NGV) and rotor blade and thereby maximises cycle performance. The cooler HP Compressor delivery air is used to cool the outer static structure before entering the IP Turbine NGV.

IP TURBINE

The IP Turbine is derived from the RB211-535E4 aero engine. Commonality with Aero components is high providing cost and reliability benefits. The IP rotor blade is manufactured from a single crystal alloy. The IP NGV is manufactured from Inco 738. Additional corrosion resistance is provided by the application of silicon aluminide coatings to the gas washed surfaces.

The IP rotor blade is produced in single crystal, CMSX4, which has millions of hours of service on civil aero engines. Both the NGV and rotor blade have improved corrosion resistance through the use of silicon aluminide coatings.

Variable Area Nozzle and Inter- Turbine Duct

This module comprises a short inter-turbine duct (ITD) and 54 Variable Area Nozzles (VANs). The ITD provides a transition from the IP Turbine to the Power Turbine and the VAN controls gas entry incidence onto the PT first stage rotor. It is a structural support for the inner bearing ring which carries the VAN assemblies. The inner platform of the ITD strut is extended rearwards to incorporate location for the VANs. To prevent corrosion the ITD strut is cast in Inco 738 and coated with silicon aluminide.

The VANs use the latest aerodynamic methodology and are designed to be both robust and easy to maintain. Each of the VAN assemblies can be removed individually optimising maintainability. A gear ring provides a mechanical drive to all 54 VANs via the VAN levers. Actuation of the gear ring is hydraulic, controlled by the EEC. This method of operation provides low friction, robust, and accurate angular movement and control of the VANs. The gear lever and gear ring are manufactured from cast iron, providing good corrosion resistance and strength, yet economical manufacture.

POWER TURBINE

The Power Turbine is a 5 stage lightweight module, derived from the latest LP turbine used in the Aero Trent 700 and Trent 800. The blading and nozzle guide vane designs closely follow the aero designs and use 3 dimensional orthogonal profiles. Disc stages 2, 3 and 4 are produced from common aero forgings so provide a significant cost benefit.

The rotor blades and NGVs are manufactured in Inco 738 and have a silicon aluminide corrosion protective coating. The discs are manufactured in Inco 718.

The PT casing is a single piece design manufactured from FV535 chrome steel with aluminium based corrosion protection. Casing temperature control is provided by an air duct that directs enclosure ventilation air over the external surface of the casing.

Power Turbine Bearing Support Structure

The design of the Power Turbine Bearing Support Structure (PTBSS) has evolved from experience gained on proven engines such as the Marine Spey.

The module comprises two main assemblies; a spoked support structure and a shaft exchange unit.

The spoked structure is designed to remain in-situ for the life of the ship, while the shaft exchange unit, which contains the bearings, seals and bearing chambers, is designed to be removed with the PTBSS in-situ to allow inspection and replacement of lifed components during service.

The module has rolling element bearings of similar design to the gas generator and uses an oil system that is shared with the gas generator.

Effective bearing chamber sealing is achieved through the use of carbon seals. These seals are a lightly loaded 'circumferential' design which have a long service life in the WR-21 application.

EXHAUST COLLECTOR

The exhaust collector is a welded fabrication of stainless steel which diffuses the exhaust gas prior to recuperator entry. It also contains the tunnel structure through which the high speed coupling shaft passes. The interior of the shaft tunnel is cooled and insulated. The external surface of the collector is insulated to prevent heat transfer to the ventilation cooling air flowing around it.

RECUPERATOR

The recuperator consists of 12 Inco 625 cores arranged in four banks of three, mounted inside a case as shown in Figure 7. This creates a counter flow heat exchanger with cooler compressor delivery air flowing from top to bottom and hotter exhaust gas flowing from bottom to top. Manifolding arrangements are used to create two air inlets and two air outlets to match up to the ducts which carry air to and from the core gas turbine.

The cores are constructed as a series of cells. Each cell consists of an air side passage and a gas side passage. Each is a sandwich of a flat plate with air fins brazed to one side and gas

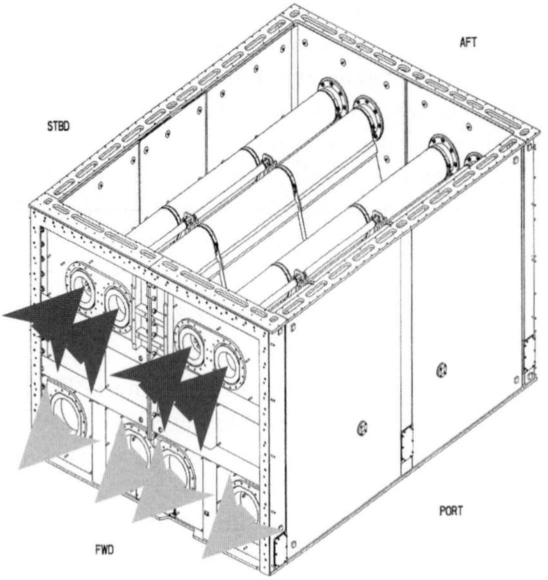

Fig. 7 Recuperator cores and case.

fins brazed to the other. These cells are welded together to form a core. The cores are connected to each other at the air inlet and outlet manifolds using a 'Voss' clamp arrangement. Figure 8 shows a sectional view of a cell.

TEST PROGRAMME

The design process for the WR-21 made much use of model and rig tests. During the project design phase certain risk areas were identified where the analytical technology would be pushed into new areas, or where the consequence of design errors would be so great that the analytical model on its own would be insufficient. In such circumstances rig and model tests were used to confirm the analytical predictions.

For the WR-21 one such area was the intercooler ducting. Here CFD codes were used to determine a candidate ducting design but the consequences of getting this design wrong were high in that the ducting was formed as part of a high value, long lead time cast component. Hence a full size model of the ducting and intercooler cores was manufactured and the geometry tested to determine its performance against targets. This resulted in detailed changes in the duct design that improved its performance, and which could be fed into the manufacture of the final component.

The following summarises the other rig and model tests conducted as part of the WR-21 design together with the design challenges inherent in each.

Fig. 8 Recuperator cell.

AIR INTAKE

The small available space envelope dictated by retrofit requirements required the use of a small intake plenum. The potential of the WR-21 system for installation in many different classes of vessel with differing intake ducting arrangements required the successful intake design to be insensitive to inlet flow pattern. This lead to a series of scale model tests which resulted in the choice of a radial intake design. The detailed design of the plenum was able to be optimised using the model to achieve an intake system which met the stringent requirements made of it.

EXHAUST VOLUTE

The exhaust volute aerodynamic design was a complex design task again for reasons of space. The volute and the diffuser contained within it were required to diffuse the power turbine exit gas flow with minimal pressure losses and then turn the flow through 90 degrees to present a uniform flow field to the recuperator. A water analogy model was used to determine best design for the diffuser/exhaust volute combination to achieve these targets.

The model allowed the easy trialling of various diffuser scarf angles, and detail changes to the volute rear wall to optimise the performance.

COMBUSTION

The design of the combustion system for WR-21 is one of the most difficult design challenges of the whole system. This stems from the wide range of combustion entry conditions that result from the different operating modes of the system. Hence the combustion design makes extensive use of the whole range of rig and model testing. Water analogy models were used to design the manifolds that surround the combustors and to determine the effects of the asymmetric feed of hot air back from the recuperator. Hot rig tests were then used to refine the design of the combustors themselves, and to model conditions not easily achievable on a test engine. The hot rig facility was used all through the engine test programme to allow continued development of the combustors without tying up the complete test system. Potential design improvements were proved using single combustors on the rig and then successful changes were read across into the next engine test.

VARIABLE AREA NOZZLE

As described earlier the design of the operating mechanism for the VANs presented some difficult design challenges. One of the most difficult was to determine the long term life of the bearing arrangement proposed. To ensure that the design was robust enough a rig test was developed to cycle the mechanism in a temperature environment typical of that to be seen in service. This allowed confidence to be gained in the design before committing hardware to a test engine.

IP COMPRESSOR

The IP compressor was based extensively on the aero RB211–535 compressor. However in the aero application the compressor sits behind a swan neck duct; in the WR-21 engine it sits behind a radial intake. To accommodate the different intake flow profile the first stage rotor and stator blades were redesigned. To ensure that the resulting compressor aerodynamic behaviour was well understood a rig test was conducted on the compressor. This measured blade and stator vibrational response as well as aerodynamic performance.

HP/IP TURBINE

Although the HP and IP turbines were based on aero designs they were from two different standards of engine. For this reason a rig test was conducted to ensure that the aerodynamic interactions of the two were understood.

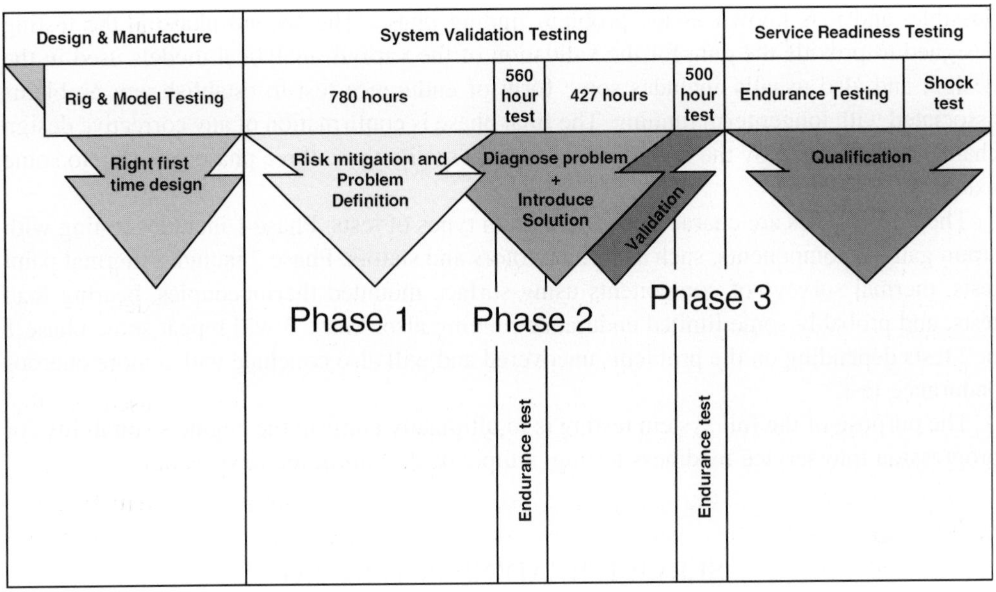

Fig. 9 Test programme phases.

RECUPERATOR SINGLE CORE TESTS

To prove the thermodynamic performance of the recuperator prior to test of the full unit in the engine a single core was instrumented and tested in isolation. This enabled the confirmation of predicted performance as well as gathering of temperature data to be used in the thermal and stress analyses of the full unit.

In addition many materials trials were conducted to establish the necessary material properties, and also to develop the braze processes.

FULL SYSTEM TESTING

The design, analysis, rig, and model testing described above lead ultimately to a design which was carried forward into the system test programme. The purpose of this is to investigate the performance and interaction of all the individual components. These test engines are heavily instrumented to enable the gathering of data on the behaviour of each individual component. The engine is operated across a wide range of test conditions within the limitations of the test facility. The data is then used to validate analytical models which will then be used to cover the whole range of conditions required by the contract specification. The test programme falls into three phases as shown in Figure 9.

The first phase is the conduct of tests which address areas of the design over which there may be some uncertainties. Its purpose is to uncover any problems which exist as soon as possible, and it is known as the problem finding phase. The second phase is the testing designed to provide the data for the validation of the various analytical models used in the design, and also usually includes some form of endurance test to establish any problems associated with longer term running. The final phase is confirmation of any corrective design changes necessitated by the results of phase 1. In reality these three phases overlap to some extent.

The three phases are characterised by certain types of tests. Phase 1 includes testing with strain gauged components, such as turbine rotors and stators. Phase 2 includes thermal paint tests, thermal surveys of components using surface mounted thermocouples, bearing load tests, and probably some limited endurance running also. Phase 3 will repeat some phase 1 or 2 tests depending on the problems uncovered and will also conclude with a more onerous endurance test.

The purpose of the full system testing is to ultimately confirm the engine's suitability for progression into service readiness testing, a topic dealt with in the next section.

SERVICE READINESS TESTING

This testing is the final stage prior to introduction into service. The test replicates the conditions likely to be seen in service operation but uses various techniques to accelerate the effects. In this way the hours run in the test will reflect the types of effects seen in many more hours of operation in service. The typical techniques used to accelerate exposure can be considered under two broad headings; environmental and operational.

ENVIRONMENTAL

A range of sulphur content fuels can be expected to be encountered during an engine's lifetime, dependent on a number of factors, economic, legislative, and geographic. At this time it is believed that the most likely worst sulphur content fuel will contain 1% sulphur by weight. So for the service readiness testing (and in fact for all the development testing conducted) this standard of fuel has been used.

For marine operation salt in air is an obvious hazard. To simulate the effects of this on engine operation and life all the service readiness testing will be with salt added to the intake air at a variety of concentrations.

OPERATIONAL

Typical mechanical drive ships operate their engines at full power for a relatively short period of time, 8% being typical. For the service readiness testing this is increased to approximately 25% of the time giving a dramatic increase in the high temperature exposure of the system.

In typical service operation the factor which tends to determine engine life more than any other is operation in the low to mid power bands where the engine variables (bleed valves and VANs for WR-21) are exercised most frequently. To accelerate the simulation of this type of operation in the service readiness testing a greatly increased number of cycles are run. So the engine may undergo 8–10 such cycles in 6 hours of service readiness testing where it may only see 2 or 3 in service.

FUTURE

The first service application for the WR-21 is in the new UK destroyer, the Type 45. In this application a pair of WR-21s will operate as gas turbine alternators, each driving an Alstom 21 MWe alternator in an integrated package, being close coupled on a common baseplate.

The unique features of the WR-21 offer several advantages to the shipbuilder and the operator. Firstly the extremely good fuel efficiency of the WR-21 means that the ship can easily achieve its design range requirement without having to sacrifice performance. It also brings many associated operational benefits. For a fixed range requirement a WR-21 powered ship does not need to carry as much fuel which allows more space for storage of other materials. So, for example, in an aircraft carrier application more aircraft fuel can be stored, increasing the operational capability of the vessel. Possibilities are also opened up to reduce the necessary Royal Fleet Auxiliary infrastructure if enough of a fleet is WR-21 powered. Secondly the nature of the recuperated cycle means that the system exhaust temperature is significantly cooler than a conventional simple cycle gas turbine. As a result there is not the same need for infrared suppression devices on the ships exhaust ducting. This saves cost and also weight that would be high in the ship's structure where it causes maximum stability problems. It also results in lower exhaust system back pressures which allow the performance of the gas turbine to be better realised in the ship.

The WR-21 is also being actively considered by a number of the world's navies for future ship programmes. This includes the Japanese navy for its future helicopter carrier and destroyer programmes beyond that.

Improving High Temperature Performance of Austenitic Stainless Steels for Advanced Microturbine Recuperators

P. J. Maziasz, R. W. Swindeman, J. P. Shingledecker,
K. L. More, B. A. Pint, E. Lara-Curzio and N. D. Evans
Oak Ridge National Laboratory
Metals and Ceramics Division
Oak Ridge, Tennessee, USA

ABSTRACT

Compact recuperators/heat-exchangers are essential hardware for microturbines, and are also desirable for increased efficiency of small industrial gas turbines. Most commercial recuperators today are made from 347 stainless steel sheet or foil. Larger engine sizes, higher exhaust temperatures and alternate fuels will all require materials with better performance (strength, corrosion resistance) and reliability than typically found in 347 steel, especially as recuperator temperatures approach or exceed 700–750°C. To meet these needs, the Department of Energy (DOE) has sponsored programs at the Oak Ridge National Laboratory (ORNL) to generate data that enables selection of commercial sheet and foil materials, to analyse recuperator components, and to identify or develop more cost effective materials with improved performance and reliability. This paper summarises data on the high temperature creep behaviour of both standard, commercial 347 steel sheet and foils typically used for commercial recuperators and as well as commercial HR 120 foil as a higher performance alternative. This paper also presents some of the initial data on standard 347 steel with modified commercial processing for improved creep resistance. Finally, ORNL efforts on lab-scale heats of 347 steels with modified compositions for improved corrosion and creep resistance are summarised.

INTRODUCTION

In contrast to larger industrial gas turbines, which operate at pressure ratios greater than 10:1, microturbines require recuperation of exhaust heat to achieve desired efficiencies of 30% or more.[1–4] Higher efficiencies mean higher recuperator temperatures, which challenge or exceed the performance of standard 347 stainless steel, which is used in commercial microturbine recuperators today. Since recuperators represent 25–30% of the overall microturbine first-cost, it is also very important to balance the need for higher performance and more durable materials with the need to make them as cost-effective as possible. There is a range of more heat-resistant and corrosion-resistant stainless alloys and superalloys commercially available, but they cost significantly more than 347 stainless steel.

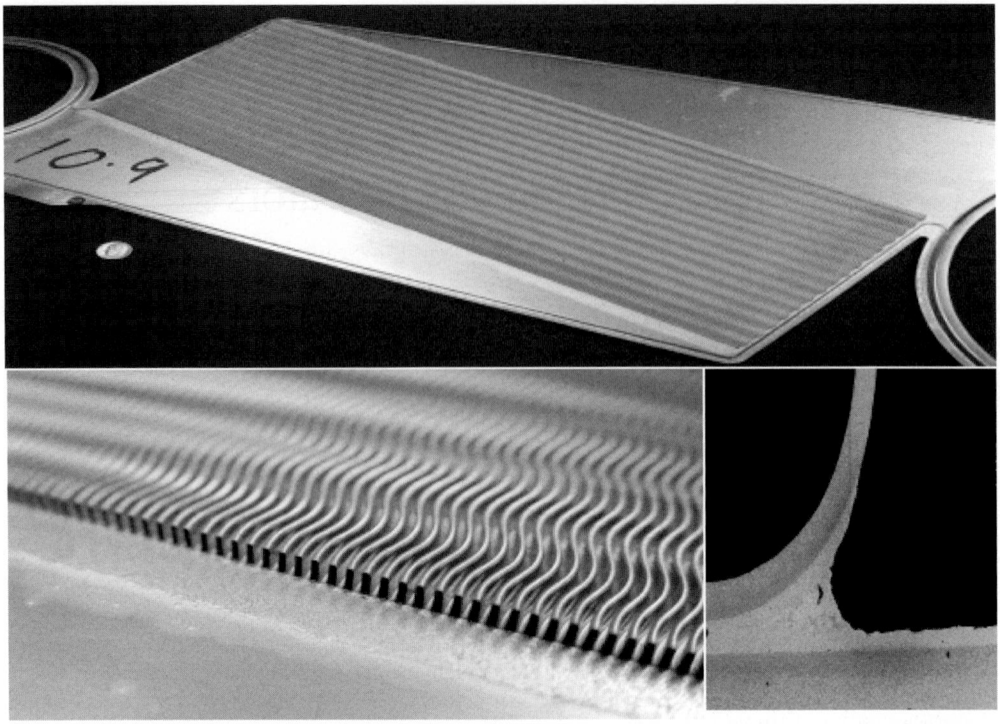

Fig. 1 An air cell from the recuperator of an ingersoll-rand 70 kW power works microturbine showing the folded fins brazed onto the plate at lower (upper) and higher (lower left) magnification, and SEM of a typical Ni-braze alloy joint cross-section (lower right). The fin and the plate are made from 347 stainless steel.

The Department of Energy (DOE) has had an Advanced Microturbine Program[4] for several years, whose goal is to enable design and manufacture of advanced microturbines with efficiencies of 40% or more. Recently, interest in microturbines has expanded to include combined heat and power (CHP) and combined cycle (microturbines and fuel cells) applications, in addition to stand alone distributed power generation applications. Microturbines are attractive because they have very low emissions and they are very fuel flexible (natural gas, flare gas, landfill and sewer gas, and various kinds of biofuels). Two of the main commercial microturbine vendors today in the U.S. are Capstone Turbines, Inc. and Ingersoll Rand Power Systems. Each manufactures its own recuperator, with Capstone producing an annular recuperator with primary surface (PS) welded air-cells,[5] and Ingersoll-Rand making a stack recuperator with brazed plate and fin (BPF) air cells.[6] A typical BPF air cell from Ingersoll Rand is shown in Figure 1. Both the PS and BPF recuperators are compact, counterflow, high efficiency heat exchangers, and they both have some common materials performance needs. To operate reliably at higher turbine exhaust temperatures, they both need materials with better high-temperature strength, creep and aging resistance, as well as better resistance to oxidation in environments containing significant levels of water vapour. However, each kind of recuperator has significantly different manufacturing and operating characteristics, so that there are different paths for upgrading and optimising performance and reliability for each specific technology. Efforts to upgrade the performance of thin foil/ sheet alloys also are directly relevant to heat-exhangers in other applications, including fuel reforming for fuel cells and high-temperature gas-cooled nuclear reactors.

The Oak Ridge National Laboratory (ORNL) has been doing materials research and development for improved high-temperature recuperator performance for several years.[10–14] Initial work focused on developing a systematic data base on lab-scale processed thin foils or sheet of commercial or developmental heat-resistant stainless alloys and superalloys, and included creep testing at 700–800°C and oxidation testing at 650–900°C. ORNL has also recently developed a recuperator test facility based on a modified Capstone 60 kW microturbine[15] for screening and evaluating a wide range of advanced materials.

The first section of this paper summarises the initial effort to screen alloys for improvements in creep resistance at ORNL. The next section of this paper summarises and highlights the most recent ORNL work on: (a) measuring the creep resistance of standard, commercial 347 stainless steel foils and sheet currently used to manufacture recuperators, and of commercial HR 120 as a potential advanced recuperator alloy, (b) efforts to modify the commercial processing of standard 347 steel to improve its creep resistance and (c) to develop new modified 347 stainless steels. The ultimate goal of this program is to facilitate manufacturing of air cells for different kinds of recuperators with upgraded performance and reliability, so that they can be evaluated in engine tests.

INITIAL CREEP TESTING OF HEAT-RESISTANT ALLOYS FOR SELECTION OF RECUPERATOR MATERIALS WITH IMPROVED PERFORMANCE

Commercial grades of the austenitic stainless steels and other alloys were obtained from 1.5–6.5 mm thick production-scale plate stock from a variety of commercial alloy producers. Standard T347 stainless steel was obtained from Allegheny-Ludlum; modified 803 developmental alloys, alloys 740 (formerly thermie-alloy) and 625 were obtained from Special Metals, Inc., alloys HR120, HR214 and HR230 were provided by Haynes International, Inc.; and alloy 602CA was supplied by Krupp VDM. A piece of NF709 stainless steel boiler tubing, obtained from Nippon Steel Corp., was split and flattened for processing into foil. Alloy compositions (some supplied by vendors and other measured by Alstom Power Metallurgical Services Laboratory, Chattanooga, TN for ORNL) are given in Table 1. Plate and thicker sheets were hot-rolled, and the 0.1 mm thick foils were produced by a series of lab-scale cold-rolling and annealing steps at ORNL. Experimental details of processing, specimen preparation and creep-testing of these foils is given elsewhere.[10, 11]

There are several different measures of creep that are important to recuperator performance. These include rupture life, rupture ductility, and time to some level of strain that would significantly distort or affect the original component, which is probably less than 5–10% creep, depending on the specific recuperator type. Most commercial recuperators made from

Table 1 Compositions (wt.%) of heat-resistant austenitic stainless alloys processed into foils at ORNL.

Alloy/Vendor	Fe	Cr	Ni	Mo	Nb	C	Si	Ti	Al	Others
347 Steel (Allegheny-Ludlum)	68.7	18.3	11.2	0.3	0.64	0.03	0.6	0.001	0.003	0.2 Co
Modified 347 Steels (ORNL, mod 2–4)	58	18–19	12.5	0.25	0.4	0.03	0.4	-	-	n.a
NF 709 (Nippon Steel)	51	20.5	25	1.5	0.26	0.067	0.4	-	-	0.16 N
Modified 803 (Special Metals, Developmental)	40	25	35	n.a.	n.a.	0.05	n.a.	n.a.	n.a.	n.a.
Alloy 740 (“Thermie”) (Special Metals)	2.0	24	48	0.5	2.0	0.1	0.5	2.0	0.8	20 Co
Alloy 120 (Haynes International)	33	25	32.3	2.5 max 0.7	0.05	0.6	0.4	0.1	0.1	3 Co max.
Alloy 230 (Haynes International)	3 max.	22	52.7	2	-	0.1	0.4	-	0.3	5 Co max. 14 W + trace La
Alloy 214 (Haynes International)	3.0	16	76.5	-	-	-	-	-	4.5	+ minor Y
Alloy 625 (Special Metals)	3.2	22.2	61.2	9.1	3.6	0.02	0.2	0.23	0.16	-
Alloy 602 CA (Krupp VDM)	9.5	25	63	-	-	0.18	-	0.15	2.2	0.06 Zr

n.a. – Not Available

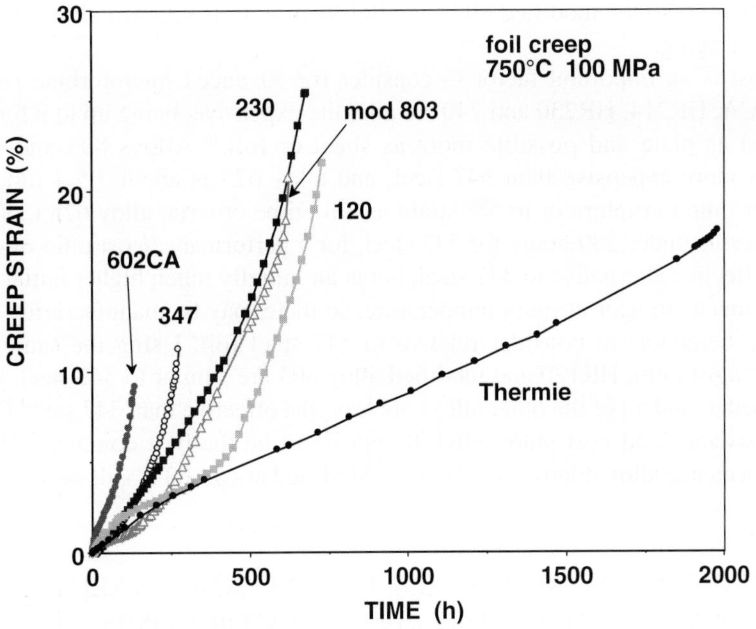

Fig. 2 Plots of creep strain versus time for creep-rupture testing of a range of commercial heat-resistant alloys in air at 750°C and 100 MPa. The alloy designated 'Thermie' has recently been named alloy 740 by Special Metals. All were supplied as thicker plate and processed into lab-scale 0.1 mm thick foils at ORNL.

standard type 347 stainless steel operate at low stresses below 700°C, so that more aggressive creep testing at 100 MPa and 750°C was chosen as an accelerated screening condition for advanced alloys. Creep-rupture data for foil tested in air at that condition for standard 347 steel and various heat-resistant stainless alloys and Ni-based superalloys are shown in Figure 2. Alloys 625 and HR214 are very strong, and have much more creep-rupture resistance than 347 steel (rupture lives of 4,000 hours for alloy 625 and 6,000 hours for HR214, and both have much lower secondary creep-rates), so they are difficult to show on the same graph.[10] Alloy 740 (also formerly known as 'thermie alloy') showed good creep-rupture resistance in the solution-annealed condition, and almost 20% rupture ductility. Alloys HR230, modified 803 and HR120 exhibited creep behaviour between alloy 740 and 347 steel. Clearly, alloys 740, modified 803, HR120 and HR230 all have more creep resistance as foils than 347 stainless steel, whereas alloy 602CA has less. Some of these alloys would have a different ranking in relative strength as plate stock (particularly HR230 would be stronger), which justifies the need for data on sheet and foil performance for such applications. Similar testing of standard alloy 803 foil also shows that it is much less creep resistant than 347 steel.[11] These same alloys were also screened for the effects of aging for 2,500 hours at 750°C on room temperature tensile ductility, which would be a factor for thermal fatigue due to cyclic operation. Here, both the modified 803 and HR120 alloy foils retained almost their original ductility after aging.

Alloy cost is an important factor to consider for advanced microturbine recuperators. Alloys 602CA, HR214, HR230 and 740 are all quite expensive, being up to 9 times the cost of 347 steel as plate and possible more as sheet or foil.[10] Alloys 803 and HR120 are 3–3.5 times more expensive than 347 steel, and alloy 625 is about 3.5-4 times more.[6–10] Using either time to rupture or to 5% strain as a lifetime criteria, alloy 625 (3500 hours to 5% strain versus under 200 hours for 347 steel, for a performance/cost ratio of over 3) is a more cost effective alternative to 347 steel, but at an initially much higher initial cost. Alloy 625 also is much stronger at room temperature, so there may be manufacturing differences (i.e. folding behavior) to consider relative to 347 steel foil. Using the same criteria of performance/cost ratio, HR120 and modified alloy 803 are similar to 347 steel, and HR214 is slightly better, and all of the other alloys are less cost effective than 347 steel. On the basis of creep resistance and cost, alloy HR120 appears to be the best commercially available higher performance alloy relative to 347 steel. Modified alloy 803 is still developmental.[11, 16]

COMMERCIAL MATERIALS TESTING AND EVALUATION FOR MANUFACTURING RECUPERATOR COMPONENTS WITH UPGRADED PERFORMANCE

Work was initiated at ORNL over a year ago on establishing a baseline measure for the creep resistance of the commercial 347 stainless steel foil and sheet stock used to manufacture recuperators. This was the first step in defining the benefit that use of the modified steels or advanced alloys would provide for upgrading the performance of commercial recuperator components. Foils and sheet of several different heats of standard 347 stainless steels that ranged from 0.076 to 0.254 mm in gage thickness were obtained from several different recuperator manufacturers and/or materials suppliers. Commercial 0.09 mm foil of HR120 was also obtained directly from Elgiloy Speciality Metals. Grain sizes ranged from as fine as 2-3 μm to as coarse as 15 μm or more in the various 347 steel foils and sheet specimens, but were coarser in the HR 120 alloy foil. Room temperature tensile properties were consistently similar among the various 347 steel sheet and foils, and the HR 120 alloy foil, with YS ranging from 310–340 MPa, and ductility from 46–65%.

Creep-rupture test data for these various foils and sheet specimens in air at 704°C and 152 MPa are shown in Figure 3 and at 750°C and 100 MPa in Figure 4. There is significant variability in the creep-rupture resistance of 347 steel sheet and foils, with rupture times ranging from 50 to 500 hours and rupture ductilities ranging from 3% to 27%. At 750°C and 100 MPa, rupture times ranged from 50 to 250 hours, and rupture ductility ranged from 8 to 18%. There were some consistent, sensible metallurgical trends for the 347 materials, such as better creep resistance in thicker foils and/or with coarser grain size. However, there also are other factors influencing the creep resistance, because the thinnest gage foil was not the worst performer, nor is the specimen with the coarsest grain size the best. These data clearly indicated that there should be an opportunity to better control the processing and microstructure of standard 347 steel to consistently provide the best properties this alloy has to offer, and then use that to measure the benefit of higher performance alloys like HR 120.

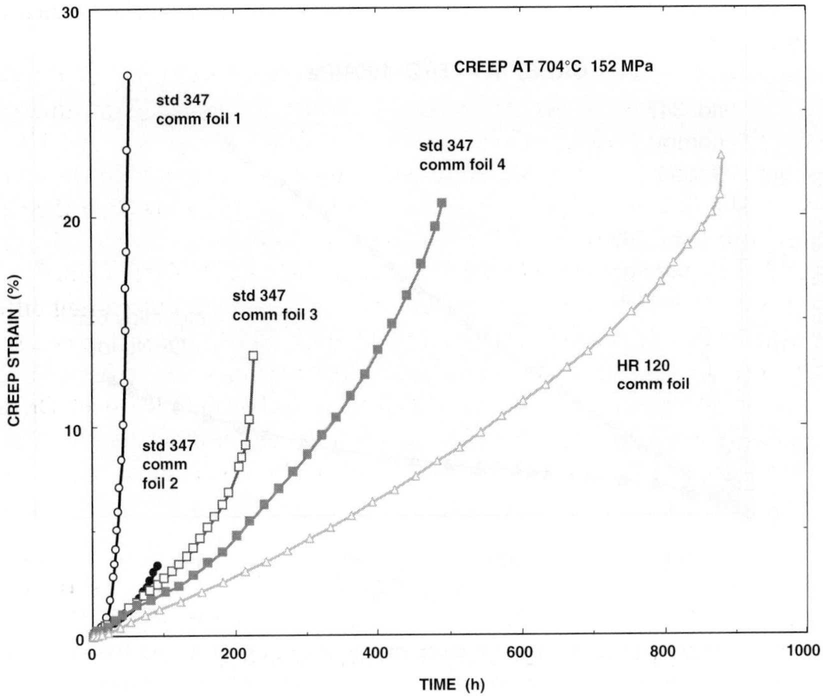

Fig. 3 Plots of creep strain versus time for creep-rupture testing at 704°C and 152 MPa. All specimens were made from standard, commercial 347 foil and sheet stock supplied by recuperator manufacturers or materials suppliers. Commercial foil of HR 120 was supplied by Elgiloy.

Commercial HR120 foil consistently showed much better creep-resistance than standard, commercial 347 steel, lasting about twice as long without rupture as the best heat of 347 steel, and about 15 times longer than the worst 347 steel heat (Figure 3). However, at a 5–10% creep strain limit, the improvement was slightly less, with the HR120 being 33–66% better than the best 347 steel. At 750°C, the HR120 lasted much longer than 347 steel (13 times longer than the best heat), which compares well with alloy 625 (3300 and 4200 hours, respectively), and had a rupture elongation of about 23% (Figure 4). With the 5–10% creep strain limit, the HR120 was 4.5–6 times better than the best heat of 347 steel, but this was still a significantly better performance/cost ratio than was estimated based on the lab-scale processing data shown in Figure 2. While the creep resistance of HR120 (creep rate and time to 5–10% strain) was lower than that of alloy 625 as foil products, these data for commercial HR120 alloy reinforce the previous conclusion that this alloy is a cost effective, higher performance alternative to standard 347 steel for foil applications. However, comparable creep data are needed on standard, commercial alloy 625 foil, and effects of better commercial processing on the creep resistance of 347 steel need to be determined to then enable a complete

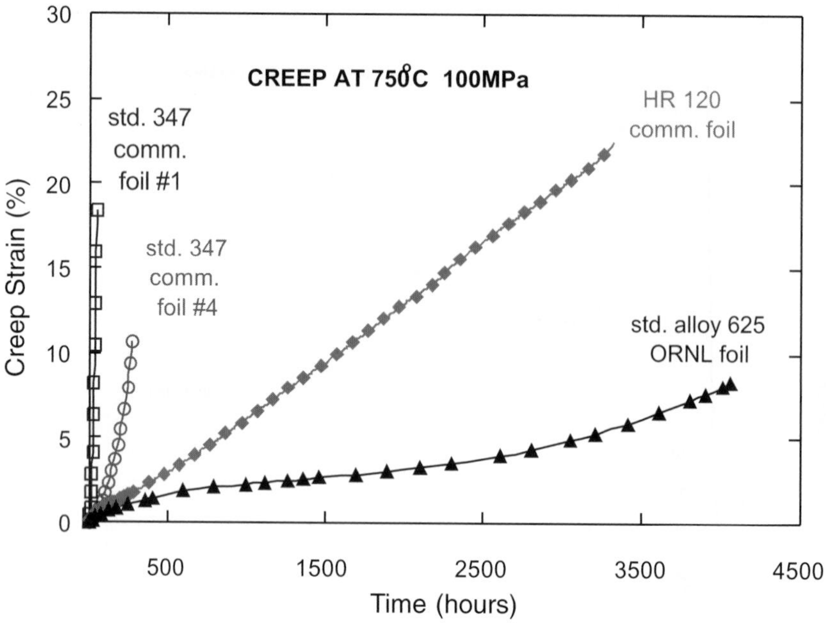

Fig. 4 Plots of creep strain versus time for creep-rupture testing at 750°C and 100 MPa. The 347 steel specimens were from commercial recuperator foil or sheet stock, the commercial foil of HR 120 was supplied by Elgiloy, and foil of the commercial alloy 625 was made into lab-scale foil at ORNL.

cost/benefit analysis to be performed for a specific recuperator. An example of such an analysis showing the factors affecting the choice of 347 steel versus alloy 625 for BPF recuperator performance is given by Kesseli et al.[6]

Analysis of current standard 347 steel recuperator air cells, in both the as manufactured and the engine-exposed conditions, was undertaken to provide baseline information from which to assess approaches for improvement. An example of a fresh braze joint from the middle of a BPF recuperator air cell component is shown in Figure 1. This analysis shows a good, fresh braze joint between the 76 μm 347 steel fin and the 0.254 mm 347 steel sheet plate. The Ni-Cr braze metal is forming nearly a continuous coating on the plate portion of the air cell. This is a different corrosion situation relative to the bare metal foils used to make most other kinds of recuperator, including the PS type. This effort now also includes real-time and accelerated PS air cell testing at the ORNL microturbine testing facility, based on a Capstone 60 kW engine.[15] to better screen relative alloy behavior, measure component or component-relevant properties, and enable testing and evaluating of changes made to upgrade recuperator materials performance.

Finally, a collaborative project was initiated late last year between ORNL and Allegheny-Ludlum Technical Centre (ALTC) to provide commercial coils of standard 347 stainless

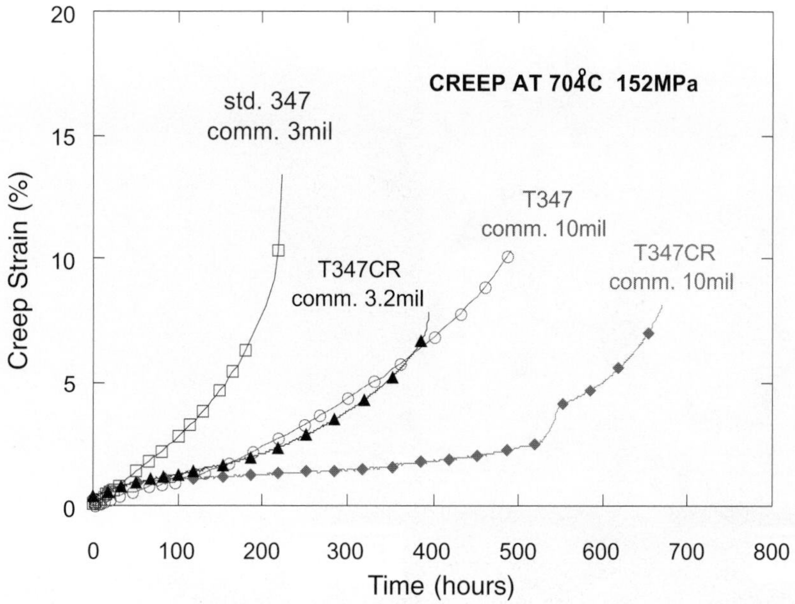

Fig. 5 Plots of creep strain versus time for creep-rupture testing in air at 750°C and 100 MPa. Standard, commercial foil and sheet of 347 steel are compared to similar standard 347 sheet and foil with modified processing conditions designed to change the microstructure to enhance creep resistance (T347CR).

steel foil, processed to have consistently better creep resistance. Ingersoll-Rand Energy Systems has provided specifications for and will take delivery of commercial quantities of the 347 foils and sheets used for manufacturing of their BPF recuperator air cells. Nominal amounts of 0.1 and 0.127 mm foil products also are being made by ORNL and ALTC to extend the study to a wide range of products. Commercial processing parameters have been adjusted to, at a minimum, coarsen the grain size to improve the creep resistance of standard composition T347 steel. The results comparing normal and modified processing for 75–80 μm foils and 0.254 mm sheet of the same commercial heat of T347 steel after creep testing at 704°C and 152 MPa in air are shown in Figure 5. Microstructural analysis of both T347 and T347CR as-processed 0.254 mm sheets to determine grain size (metallography) and intragranular NbC differences (transmission electron microscopy (TEM)) are shown in Figure 6. The creep rupture life of the foil was improved 100% and sheet was improved by 34% by the modified processing and changes in microstructure. The improvements are greater considering a 5% creep strain limit, with 133% for the foil and 90% for sheet. The 0.254 mm sheet clearly shows that the largest change in microstructure due to the change in processing was the elimination of nearly all of the smallest size grains, most likely due to coalescence. Comparing the commercial HR120 foil (Figure 3) to the T347CR (Figure 5) for creep at 704°C, the time to 5% creep strain is now similar, although the former still has a little more

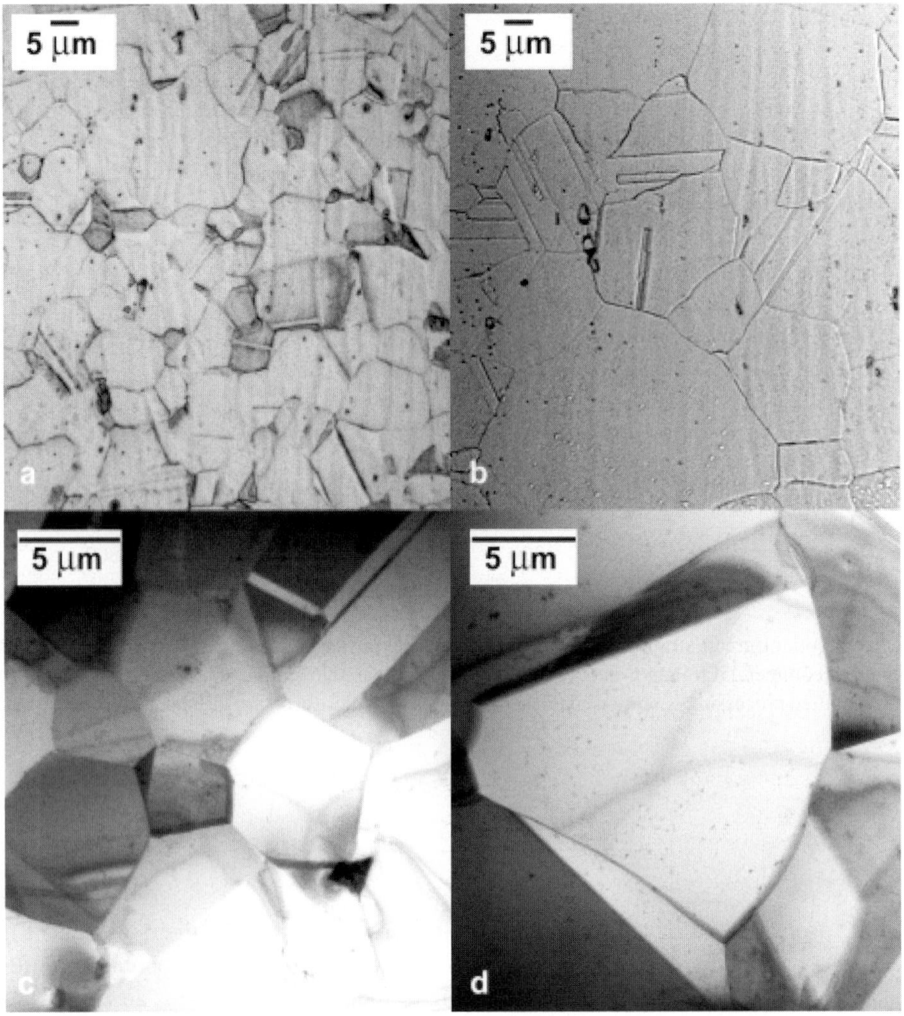

Fig. 6 Microstructure analysis comparing 10 mil sheet comparing T347 (a, c) with standard processing and T347CR (b, d) with modified processing. Metallography is shown in (a) and (b), while TEM is shown in (c) and (d).

than twice the rupture life. These early results clearly indicate that a considerable improvement in the creep resistance of standard composition 347 steel foils and sheet can be made through careful selection of processing conditions.

Clearly an immediate and cost effective improvement in recuperator performance could be made simply from optimising the performance of standard 347 steel. Work continues to creep test these foils and sheet at 750°C and 100 MPa, and to include foils with various thicknesses in this comparison. TEM microstructural analysis also will be completed on the creep tested specimens of the T347CR to determine and understand the role of NbC precipitation in the improvements in creep-rupture resistance.

DEVELOPMENT OF MODIFIED 347 STAINLESS STEELS AND ALLOYS FOR HIGHER TEMPERATURE CAPABILITY

A unique and important aspect of the ORNL work for the last several years has been lab-scale alloy development to modify the composition of 347 stainless steel to significantly improve both the creep-resistance and corrosion-resistance, if possible.[10, 11] While one primary driver for this effort has been the expressed desire of microturbine recuperator makers to provide near-term improvements in performance and durability at the lowest possible cost, another added incentive and opportunity is the recent success achieved by using processing to boost the creep resistance of standard composition 347 steel described above. Several 15 lb heats of 347 stainless steels with modified compositions were melted at ORNL, hot-rolled into plate, then cold-rolled into 0.1 mm thick foils, and finally annealed. The composition ranges of three of the modified 347 steels for the elements common to standard 347 steel are summarised in Table 1, but the several alloying additions that make them different are not disclosed at this time. These modifications and additions were specifically designed to produce 'engineered microstructures', for better aging- and creep-resistance at 700–800°C compared to standard 347 steel, at little or no increase in alloy cost. Data on aging resistance of two of these modified 347 steels (alloys 3 and 4) and oxidation data in 10% water vapour at 800°C have been presented recently,[10, 11] and creep-rupture data for the best alloy (alloy 4) at 750°C and 100 MPa are shown in Figure 7. For comparable lab-scale foil processing, the ORNL mod 347-4 showed over twice the rupture life and 7 times longer time to 5% strain than standard 347 steel. The creep rate of the modified 347-4 compared well with that of alloy 625 until it ruptured. TEM analysis showed that abundant, fine intragranular precipitation, including NbC, and resistance to sigma phase formation, are the microstructural differences that cause the improved creep resistance of the modified 347-4 steel, as shown in Figure 8.

The various modified 347 steels also showed improved corrosion resistance compared to standard 347 steel. Accelerated oxidation testing for about 1000 hours at 800°C in 10% water vapour, to simulate the gas turbine exhaust gas. The improved corrosion resistance of the mod. 347-4 steel at 800°C was found to be directly related to a delay in the onset of break-away oxidation and the formation of non-protective surface mixed Fe-Cr oxides and the associated subsurface metal degradation.[7–9, 11, 14]

Corrosion resistance and the exacerbating effects of moisture become dominant limiting factors for 347 steel at test temperatures of 650 to 700°C, which is more relevant to current microturbine recuperator operating conditions.[11, 12, 14] Oxidation test data at 650°C in 10% water vapour for up to 6000 hours are given in Figure 9. The new modified 347 steels all show much better resistance to oxidation in water vapour at 650°C than standard 347 steel. These limited, initial test data also clearly show that the new modified 347 steels have oxidation resistance that is comparable to the excellent behavior normally found in alloys 625, modified 803, HR120, and NF709, all of which have substantially more Cr and Ni.

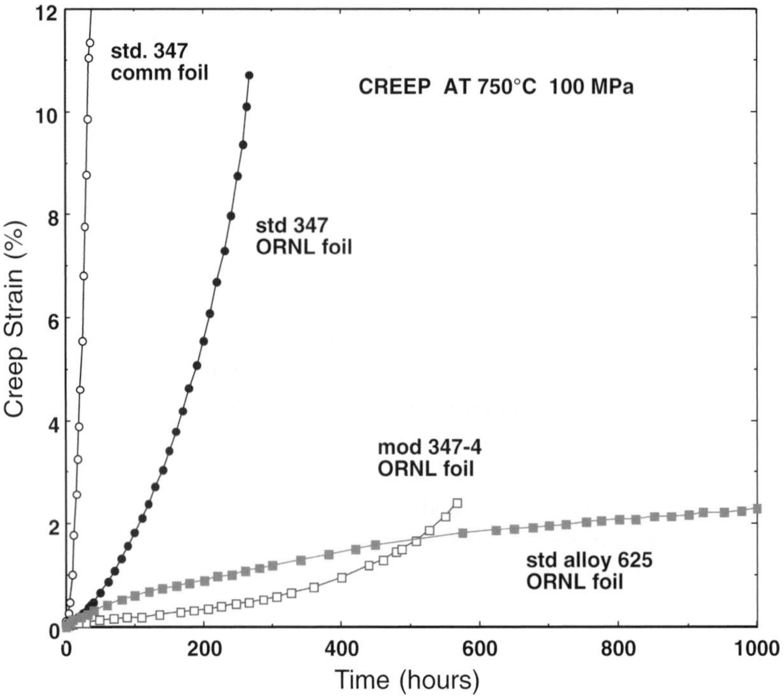

Fig. 7 Plots of creep strain versus time for creep-rupture testing of a commercial standard 347 foil #1, and ORNL lab-scale processing of standard 347, standard 625 and an ORNL lab-heat of a modified 347 stainless steel (347-4). All are 0.1 mm thick foils creep tested in air at 750°C and 100 MPa.

While these preliminary results do not mean that the modified 347 steels will not eventually exhibit the breakaway oxidation in water vapour, the delay in such severe oxidation attack does provide a substantial added benefit of these new alloys for near-term recuperator applications, particularly when combined with their creep resistance. The next step will be to make 1 or 2 larger commercial heats of the best modified 347 steels, so that high quality commercial foil can be made for recuperator component manufacturing trials.

Detailed microstructural analysis of a cross-section of standard 347 foil coupon tested at 700°C and 10% water vapour to oxidation behavior near the surface is also shown in Figure 10. The lower magnification back-scattered SEM image shows directly the heavy surface deposits of iron-rich oxide and the corresponding deep subsurface oxide penetration that significantly reduces the remaining thickness of unaffected metal that are typical of 347 steel suffering breakaway oxidation attack (Figure 10a). The higher magnification SEM image in Figure 10b, made by mapping with the characteristic CrKα x-ray peak in a region with less surface attack, clearly shows the thin chromia layer that remains underneath the Fe-rich oxide deposit, and narrow Cr-depleted grain boundaries just below the chromia scale.

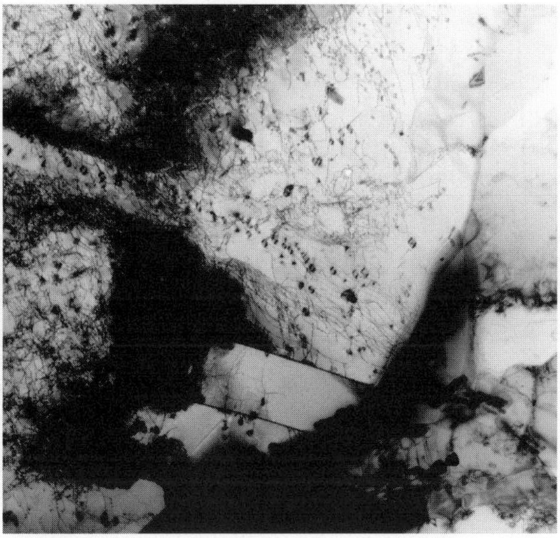

(a)

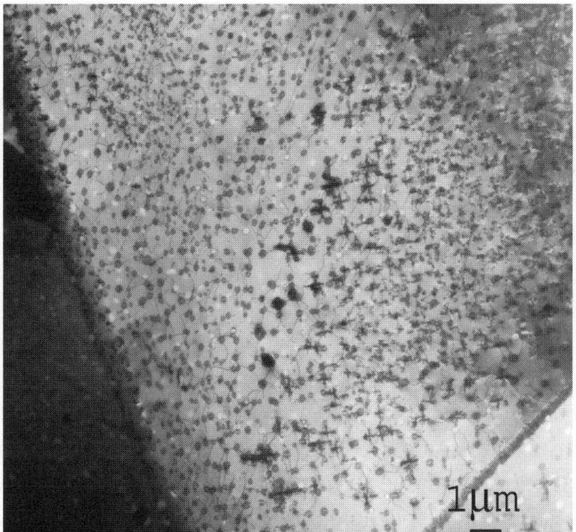

(b)

Fig. 8 TEM analysis of microstructures from the gage portions of the creep specimens tested at 750°C at 100 MPa in air, and whose creep curves are shown in Figure 7. Both are 0.1 mm thick foils processed on a lab-scale at ORNL of (a) standard 347 from commercial plate and (b) modified 347 stainless steel (347-4) from an ORNL lab-scale heat. The obvious difference in intragranular precipitate dispersions is directly responsible for the differences in creep resistance.

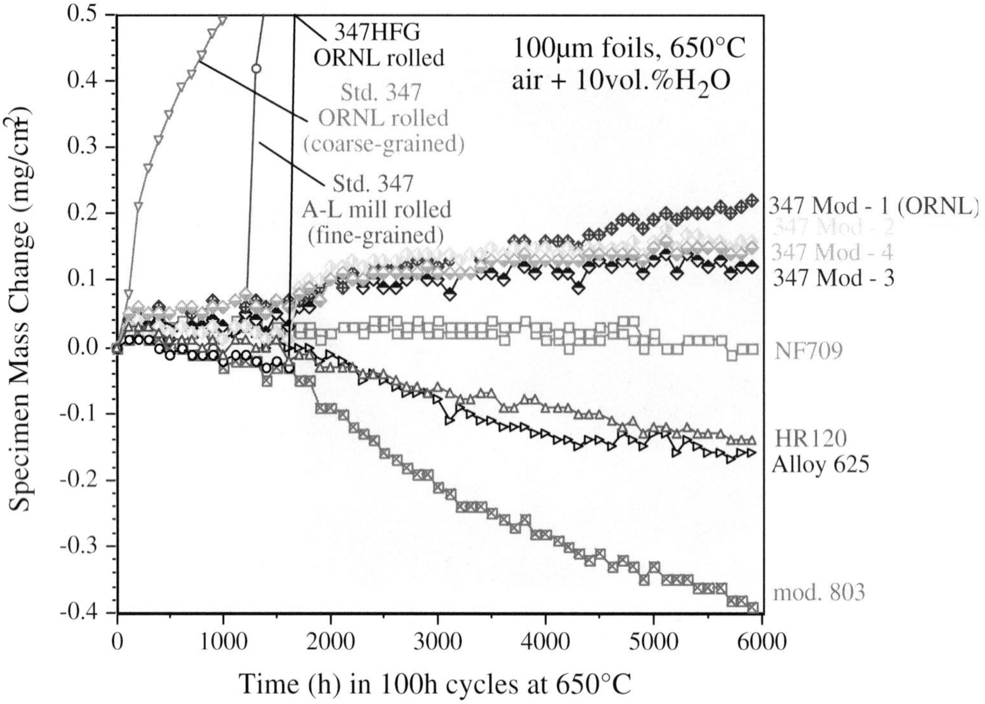

Fig. 9 A plot of mass change as a function of exposure time for specimens tested in air with 10% water vapor at 650°C, with cycling to room temperature every 100 hours to weigh specimens (upper). All specimens are 0.1 mm thick foils of several commercial heats of standard 347 steel, commercial heats of alloys NF709, 625 and HR120, and lab-scale heats of modified 347 steels and modified alloy 803, processed on a lab-scale at ORNL.

This characteristic subsurface structure is caused by the combination of diffusion along the grain boundaries and Cr depletion due to formation of the chromia scale at the surface. The Cr-map image also shows small particles of Cr-rich sigma phase forming along grain boundaries deeper in the foil with much less Cr depletion. Consistently, the subsurface layer with the most Cr depletion along the grain boundaries also shows no formation of Cr-rich sigma phase particles. This characteristic subsurface structure develops rapidly in both recuperator components and lab-test foil coupons, and generally proceeds the formation of the thick Fe-rich surface oxide deposits or break away oxidation. Such microstructural behavior is important in understanding the mechanisms that cause water vapour to accelerate oxidation attack in 347 steel, and in understanding how the advanced stainless steels and alloys resist such attack. Similar microstructural characterisation of the modified 347 steels and foils of the advanced alloys is currently being done.

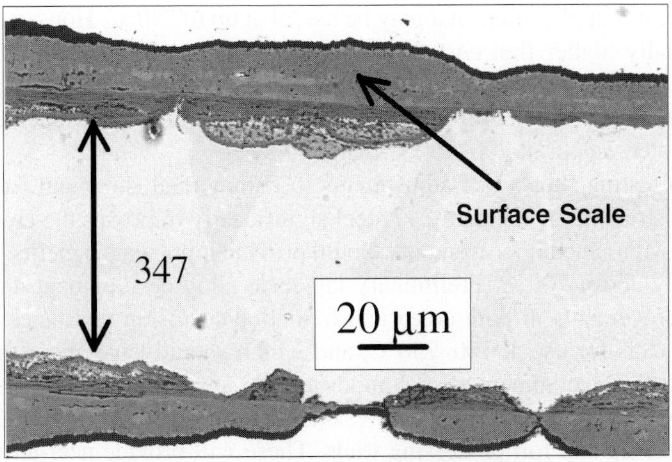

(a)

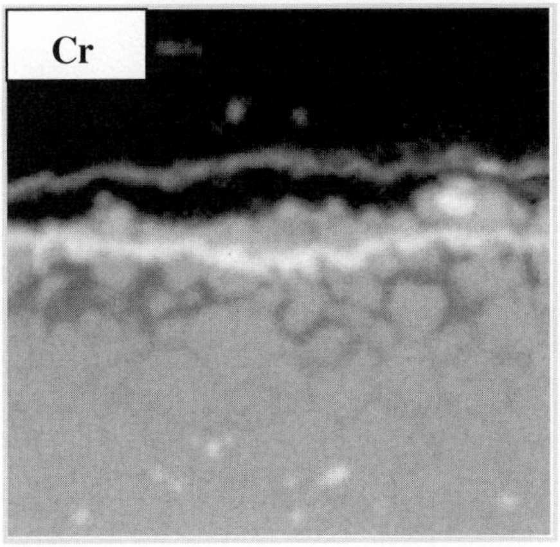

(b)

Fig. 10 The microstructure of a cross-section of standard 347 steel (347 A-L) tested in 10% water vapor at 700°C for 1,000 hours is analysed using backscattered SEM (lower left) and higher magnification X-ray mapping of Cr near the surface. The heavy Fe-rich surface scale and sub-surface attack is typical of break-away oxidation attack in the presence of water vapor.

CONCLUSIONS

A group of cost effective alloys with more aging, creep and oxidation/corrosion resistance than standard 347 austenitic stainless steel has been identified for advanced microturbine recuperator applications. Alloy 625 is a very high performance alternative to 347 stainless steel for recuperator applications that may be useful at up to 750°C. However, alloy 625 also has a substantially higher first cost. HR120 is a commercially available alloy with good creep-resistance at 700–750°C that is a more cost effective performance upgrade relative to standard, commercial 347 steel foil. Both HR120 and 625 alloys have much better oxidation resistance in water vapour at 700–800°C than 347 steel.

Initial creep testing shows that adjustments to commercial sheet and foil processing to tailor the microstructure of standard 347 steel significantly improves the creep resistance at 750°C and 100 MPa. Such improvements should provide immediate benefits for recuperators up to or slightly above 700°C. Preliminary lab-scale alloy development data indicate that substantial improvements in both creep and oxidation/corrosion resistance are possible in modified 347 steels for use at 700–750°C, and which should have about the same cost as standard 347 steel. Development of such modified 347 steels and related austenitic stainless alloys with a better combination of creep-resistance and oxidation resistance should lead to commercial scale-up and foil processing trials. These will provide near-term cost effective recuperator performance upgrades that are not as costly as using HR120 and 625 alloys.

ACKNOWLEDGEMENTS

Thanks to I.G. Wright at ORNL for recommending this work to the conference organizers, and for reviewing this paper. We also thank C. Stinner, D. Shanner and J.A. Salsgiver at the Allegheny-Ludlum Technical Centre for providing standard T347 foil of several thickness and for collaborating to modify commercial-scale processing to make the T347CR foils and sheet. Thanks to Special Metals, Inc. for providing modified 803 alloys, and alloy 740, to Haynes International, Inc. for providing alloys HR120, HR230 and HR214, and to Krupp VDM for providing alloy 602CA. Research was sponsored at ORNL by the U.S. Department of Energy, Assistant Secretary for Energy Efficiency and Renewable Energy, Office of Distributed Energy and Electrical Resources, and by the Division of Materials Sciences and Engineering (ShaRE User Center) under contract DE-AC05-00R22725 with UT-Battelle, LLC, and also supported in part by an appointment to the ORNL Postmaster's Research Participation Program administered by the Oak Ridge Institute for Science and Education and ORNL.

REFERENCES

1. C. F. McDonald: 'Heat Recovery Exchanger Technology for Very Small Gas Turbines', *International Journal of Turbo and Jet Engines*, **13**, 1996, 239–261.
2. M. E. Ward: 'Primary Surface Recuperator Durability and Applications', *Turbomachinery Technology Seminar paper TTS006/395*, Solar Turbines, Inc., San Diego, CA, 1995.
3. K. Takae, H. Furukawa and N. Kimiaki: 'A Preliminary Study of an Inter-Cooled and Recuperative Microgasturbine Below 300 kW', *ASME paper GT-2002-30403*, American Society of Mechanical Engineers, New York, NY, 2002.

4. *Advanced Microturbine Systems – Program Plan for Fiscal Years 2000 – 2006*, Office of Power Technologies, Office of Energy Efficiency and Renewable Energy, U.S. Department of Energy, Washington, D.C. 2000.
5. B. Treece, P. Vessa and R. McKeirnan: 'Microturbine Recuperator Manufacturing and Operating Experience', *ASME paper GT-2002-30404*, Am. Soc. Mech. Engin., New York, NY, 2002.
6. J. Kesseli, T. Wolf, J. Nash and S. Freedman: 'Micro, Industrial and Advanced Gas Turbines Employing Recuperators', *ASME paper GT2003-38938*, American Society of Mechanical Engineers, New York, NY.
7. B. A. Pint and J. M. Rakowski: 'Effects of Water Vapour on the Oxidation Resistance of Stainless Steels', *NACE paper 00259 from Corrosion 2000*, NACE-International, Houston, TX, 2000.
8. H. Nickel, Y. Wouters, M. Thiele and W. J. Quadakkers: 'The Effect of Water Vapro on the Oxidation Behavior of 9%Cr Steels in Simulated Combustion Gases', *Fresenius J. Anal. Chem.*, **361**, 1998, 540–544.
9. H. Asteman, J.-E. Svensson, M. Norell and L. -G. Johansson: 'Influence of Water Vapor and Flow Rate on the High-Temperature Oxidation of 304L, Effect of Chromium Oxide Hydroxide Evaporation', *Oxidation of Metals,* **54**, 2000, 11–26.
10. P. J. Maziasz and R. W. Swindeman: 'Selecting and Developing Advanced Alloys for Creep-Resistance for Microturbine Recuperator Applications', *Journal of Engineering for Gas Turbines and Power* (ASME), **125**, 2003, 51–58.
11. P. J. Maziasz, B. A. Pint, R. W. Swindeman, K. L. More and E. Lara-Curzio: 'Selection, Development and Testing of Stainless Steels and Alloys for High-Temperature Recuperator Applications', *ASME paper GT2003-38762*, Am. Soc. Mech. Engin., New York, NY, 2003.
12. B. A. Pint, R.W. Swindeman, K. L. More and P. F. Tortorelli: 'Materials Selection for High Temperature (750–1000°C) Metallic Recuperators for Improved Efficiency Microturbines', *ASME paper 2001-GT-0445*, American Society of Mechanical Engineers, New York, NY, 2001.
13. B. A. Pint, K. L. More and P. F. Tortorelli: 'The Effect of Water Vapor on Oxidation Performance of Alloys Used in Recuperators', *ASME paper GT-2002-30543*, American Society of Mechanical Engineers, New York, NY, 2002.
14. B. A. Pint and R. Peraldi: "Factors Affecting Corrosion Resistance of Recuperator Alloys," *ASME paper GT2003-38692*, Am. Soc. Mech. Engin., New York, NY, 2003.
15. E. Lara-Curzio, P. J. Maziasz, B. A. Pint, M. Stewart, D. Hamrin, N. Lipovich and D. DeMore: 'Test Facility for Screening and Evaluating Candidate Materials for Advanced Microturbine Recuperators', *ASME paper GT-2002-30581*, American Society of Mechanical Engineers, New York, NY, 2002.
16. M. A. Harper, G. D. Smith, P. J. Maziasz and R. W. Swindeman: 'Materials Selection for High Temperature Metal Recuperators', *ASME paper 2001-GT-0540*, American Society of Mechanical Engineers, New York, NY, 2001.

The Use of Microturbines in Biomass and Waste-to-Energy Projects

D. Robertson and M. Newnham
Bowman Power Systems
Southampton, UK

ABSTRACT

Microturbines are very small high-speed gas turbine engines of a simple radial design. They can be used for power generation, utilising a high speed alternator and inverter technology to configure the power output to the customer's needs. As a result of their scale (power output in the range 30–400 kW), simplicity and multi fuel capability, microturbines are ideally suited for integration with a range of biomass and waste-to-energy processes. Both gasification (via combustion of the producer gas) and combustion of dry biomass/waste (by heat exchange from the flue gases) can be used to supply primary energy to the microturbine. Alternatively, for wet biomass and waste, the microturbine can be fuelled with gas from an anaerobic digester. Bowman is developing microturbine variants for each of these applications. This paper seeks to describe some of these developments.

INTRODUCTION

Bowman Power Systems (BPS) was formed in 1994 to address the growing opportunities in the Distributed Power industry. Since its inception it has been developing the TURBOGEN™ family of small-scale compact power generation systems. These MTG (microturbine generator) systems are based on microturbines (compact gas turbines) and high-speed generator technologies, together with associated power electronics.
Microturbines are very small high speed gas turbine engines of a simple radial design, closer in concept to low cost turbochargers than the more complex axial designs of large industrial gas turbines - which are often derived from aero engines. Conservative operating temperatures eliminate the need to use high cost, high temperature materials and coatings. This, combined with their simplicity, enables low cost volume production to be a realistic goal.

In comparison with power generation systems based on reciprocating engines, the advantages and features of microturbine-based systems include:

- Fuel flexibility (gas, liquid and renewables).
- Low emissions (NO_x and CO both below 10 ppm on natural gas).

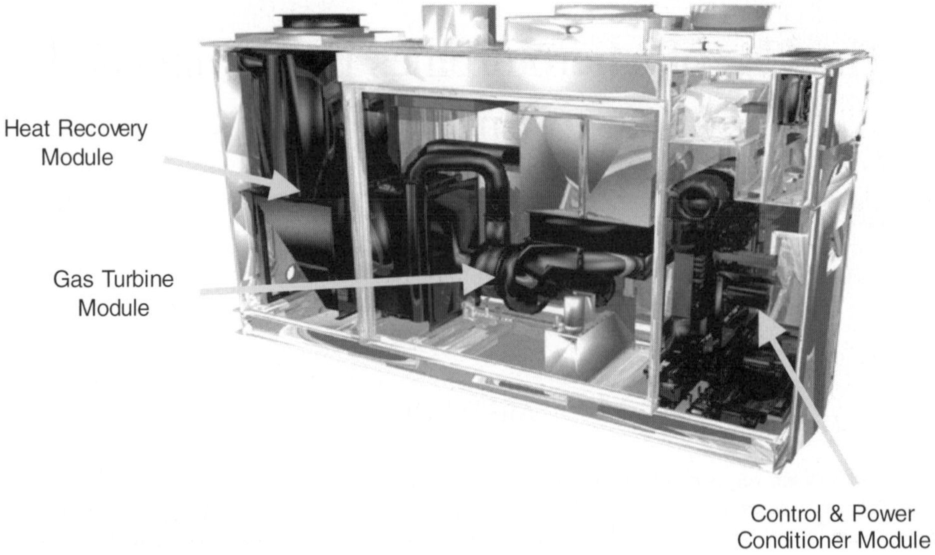

Fig. 1 BPS 80 kWe turbogen CHP unit.

- High grade heat easily recoverable from the exhaust gas stream (particularly useful for process applications and to drive absorption chillers).
- Low noise and vibration.
- Variable heat to power ratio.
- Compact.
- High reliability.
- Low maintenance.

Many of the above features are particularly important for biomass applications, which are frequently located in rural areas. Biomass tends to have relatively low density, both in terms of bulk and in terms of useable energy. This means that transportation costs are relatively high compared to fossil fuels. Consequently, locating the energy conversion process close to a concentrated source of biomass is clearly an advantage. It is also typical for some rural areas to suffer from high electricity prices and/or a weak grid. In this scenario, distributed power becomes particularly attractive.

POWER GENERATION FROM BIOMASS

Biomass is organic matter such as wood, crops, animal and agricultural wastes – the term is usually only used for solid organic matter or liquid slurries. As it has a reasonable (if variable)

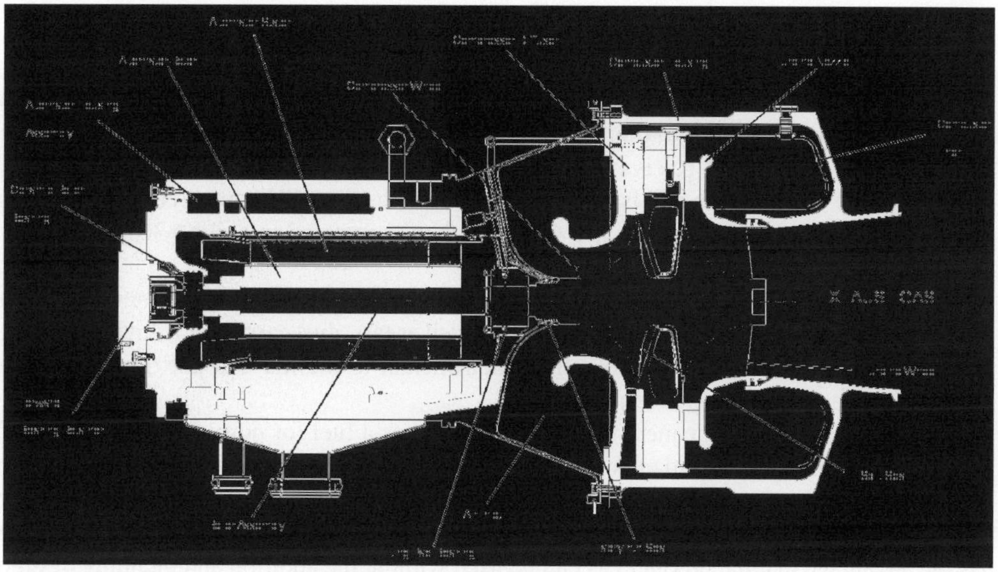

Fig. 2 Cross section of TA80 microturbine generator.

calorific value, biomass has the capacity to be used as a fuel – as people have done for heating and cooking for thousands of years.

For power generation and CHP applications, the main biomass fuels are wood, either grown as an energy crop (e.g. short rotation coppicing) or waste/bi-products of other processes and agricultural wastes such as straw, poultry litter, dung, etc. Closely allied processes include energy-from-waste (also called waste-to-energy) and incineration (municipal solid waste, clinical waste and chemical waste).

In recent years, interest in renewable energy has intensified in response to concerns about the global warming effects associated with increasing levels of carbon dioxide in the atmosphere. When fossil fuels (coal, oil, natural gas) are burnt, carbon dioxide is produced. Whilst this is also the case for biomass, there is no net increase in atmospheric carbon dioxide levels - since the carbon dioxide released was absorbed from the atmosphere during formation of the biomass material. Therefore biomass is considered 'CO_2 neutral'.

Biomass can be converted to electrical and thermal energy with low emissions in various ways. Processes to convert biomass into a form suitable for power generation fall into two main categories – thermal and biological. Thermal processes have generally been developed or adapted from conventional fossil fuel technologies. In most large biomass power generation systems, hot flue gases from a biomass combustor are passed through a boiler to raise steam to drive a steam turbine. However, this technology is not economically viable at small scale. There are two main alternative thermal processes suitable for combining with microturbines for distributed power generation – gasification and indirect firing.

GASIFICATION

Gasification is the total or partial chemical transformation of a fuel under the action of heat in sub-stoichiometric conditions (i.e. insufficient oxygen for complete combustion) to produce a product gas. The product gas is typically a mix of hydrogen, methane, carbon monoxide, carbon dioxide and nitrogen and is combustible, but with a low CV compared to natural gas and other conventional fuels (typically 4.5–6 MJ/Nm^3 compared to around 36 MJ/Nm^3 for natural gas).

There are a number of possible physical arrangements for the gasifier, and tar formation is a potential problem for some arrangements. As well as tars, dust carryover from the gasifier, and moisture and acid gases are potentially detrimental to the life and performance of an engine – hence gas cleaning is required.

Once the gas has been cleaned, it can then be used as a fuel for powering an engine - IC or turbine. The gas can be used either alone or after blending with natural gas or bottle gas or liquid fuel. The handling and management of these blended fuels to make them suitable for use in a reciprocating engine is very expensive, particularly at small scales. The gas turbine has the advantage that the product gas can be used without the addition and blending of high CV fuel. However, the gas has to be compressed, and both the compressor and combustor need to be specially designed to suit the fuel gas flow and heating value.

Bowman Power Systems is working with Rural Generation Limited (RGL), a specialist supplier of biomass gasification and combustion plant, to develop an integrated wood gasification/microturbine CHP unit of 75 kWe output. The gasifier is a fixed bed downdraft arrangement, which has been shown to produce low tar levels and to be a reliable gas generator. After testing of the new combustor and fuel gas compressor, a prototype microturbine CHP unit will be installed at RGL's demonstration site in Londonderry, N. Ireland for endurance testing. This project is part-funded by the Department of Trade and Industry.

INDIRECT FIRED MICROTURBINE

In an indirect fired microturbine system, the hot gas from the biomass combustor passes through the primary side of a gas-to-air heat exchanger. On the secondary side of the heat exchanger, air from the microturbine compressor is heated prior to expansion in the turbine.

Bowman Power Systems is working with Talbott's Heating, a specialist supplier of small-scale commercial and industrial biomass-fuelled heating and cogeneration plant, to develop a ground-breaking new waste-to-energy CHP concept, based on an indirectly-fired microturbine. Using a modified Bowman Power TG50 packaged microturbine-generator unit specially converted for indirect firing, a prototype unit with an output rating of 30 kWe/ 100 kWth has been developed in an 18 month project. Part-funded by the Department of Trade and Industry, the prototype CHP plant has successfully completed proof-of-concept testing, and is currently undergoing extended endurance tests.

Unlike gasification approaches where combustible, but contaminated (i.e. dust and tars) gases are used for direct firing, the indirect system provides a completely clean airflow through the turbine. This eliminates the problems of abrasion, blade fouling and high

Fig. 3 Rural Generation Ltd.'s gasifier demonstration site, Londonderry.

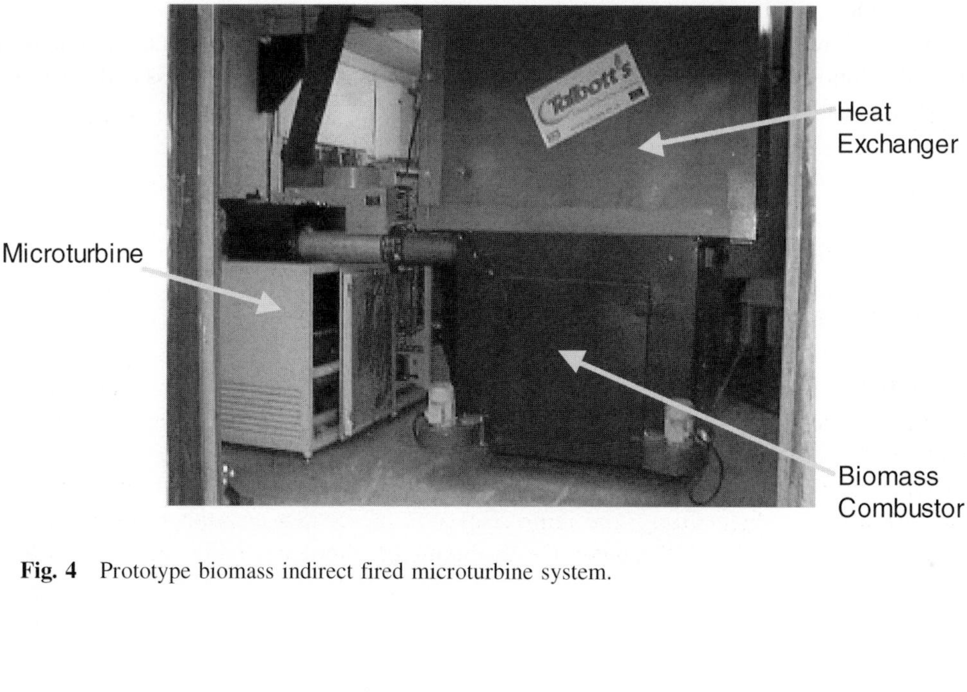

Fig. 4 Prototype biomass indirect fired microturbine system.

mechanical wear normally associated with direct-fired, gas turbine-based biomass systems, while relatively high thermal inertia and low operating temperatures reduce problems due to thermal shock.

ANAEROBIC DIGESTION

Anaerobic digestion is a well-established biological process for treating agricultural, household and industrial residues, and sewage sludge. Anaerobic digestion occurs naturally in swamps and landfill sites (the methane content of landfill gas can be around 50-60%, but reduces as the site becomes exhausted).

Specially-built digesters can also be used to optimise the gas yield. Anaerobic digesters range from small on-farm units of 30 m^3 to centralised anaerobic digesters of in excess of 1000 m^3. This is the preferred route for the utilisation of wet biomass such as farm slurries and green crop wastes – since the very high moisture content reduces the economics of thermal processes. Typically, the gas produced is 50–80% methane, 20–40% carbon dioxide, with trace levels of hydrogen, carbon monoxide, nitrogen, oxygen and hydrogen sulphide. Once the gas has been cleaned, it can be used as a medium-CV fuel suitable for combustion for both heat and power generation. Waste heat from the engine can be redirected into the digester to increase the yield of methane. There are a number of challenges to successful utilisation of digester gas in an engine. These include maintaining the consistency of gas quality, water saturation of the gas, and corrosion from acids components of the gas (e.g. H_2S).

BPS is developing a low cost power generation system for biogas utilisation, under a third Department of Trade and Industry part-funded project. The proposed approach uses modular 60 kWe microturbine units to cover both on-farm applications (singly) and the much larger centralised anaerobic digesters (multiple units paralleled). The heat required by the digester is recovered from the microturbine exhaust by way of an exhaust gas to water heat exchanger.

Bowman Power Systems believes that these three approaches represent a comprehensive range of solutions to the very differing requirements of smaller scale biomass and waste-to-energy applications.

Development and Performance Evaluation of Novel Rotor-Generators for Tidal Current Energy Conversion

J. A. Clarke, A. D. Grant and C. M. Johnstone
Energy Systems Research Unit
Department of Mechanical Engineering
University of Strathclyde
Glasgow, UK

INTRODUCTION

Recent policy developments within the electricity supply industry have favoured the development of renewable technologies, as demonstrated by the recent UK Government's 'Energy White Paper'[1] and the Scottish Executive's 'Securing a Renewable Future'.[2] In both cases, aspirational targets of 20% and 40% respectively have been set for the generation of electricity from renewable sources by 2020. Since the current economically viable renewable sources are stochastic in nature (e.g. wind power), achieving these targets will result in higher levels of vulnerability within the electrical supply network. This, in turn, will increase the levels of control and reserve plant required to prevent supply disruption. In an effort to address this undesirable situation, more predictable renewable energy technologies require to be developed. Tidal current technology has been identified as an important contributor because the energy yield and time of occurrence may be predicted in advance. Further, by arranging for the strategic location of tidal power generation systems at several locations, a continuous base load power supply should be achievable. This latter attribute is important, as sufficient base load supply is crucial to maintaining electrical network integrity. It is for these reasons that the UK Government is increasing its support for the development of tidal energy technology as a medium to long-term energy supply system.

DEVELOPMENT OF TIDAL TECHNOLOGY

Technology for the exploitation of marine currents is still in its infancy, being under development in the UK for only the last two decades. Work to date has shown the main research challenges to be associated with:

- The power capture device (rotor versus oscillating aerofoil),
- Power take-off (hydraulic or mechanical transmission),

- Device structural support (tensioned mooring or rigid structural piling) and
- Connection of the generated power to the supply network.

At the present time two power capture devices are being investigated for commercial development: the oscillating aerofoil driving hydraulic accumulators[3] and a horizontal axis turbine driving a mechanical shaft.[4] The former device is a development of technology and principles emanating from the offshore marine industry, while the latter device is an evolution of the wind turbine in relation to the requirements of the sub-sea environment. In the case of the tidal turbine, although the fundamental fluid dynamic interactions between rotor and stream are the same as in wind energy conversion, there are certain differences that are likely to cause divergences in technological development. Some of these are obvious and will influence materials selection and structural design, e.g. the higher density of the fluid medium and the greater possibilities of surface fouling and corrosion. Some are less obvious and will have a profound impact on the take-up of marine power: the predictable range of the current velocities at a given site, and the relatively low levels of turbulence in the stream.

The structural loading on wind turbines contains a large stochastic element, which arises from a combination of effects. These include wind shear (from the atmospheric boundary layer), misalignment of the rotor with the wind direction, interaction between the rotor and the supporting tower and, most significantly, the presence of turbulence in the approaching wind. This last effect manifests itself as short-term variations in both wind speed and direction. Directional changes caused by large-scale turbulent eddies have a particularly severe effect on dynamic loading. Another factor is the possibility of extreme winds, which requires statistical analysis to determine the 50 or 100 year maxima to be used in structural design calculations. Wind turbines are of course shut down as a matter of routine under storm conditions.

Tidal current turbines will operate in a more predictable environment. Maximum current velocities can be predicted with reasonable accuracy and it should not be necessary to enforce turbine shut down except in an emergency. Dynamic loads may still occur as a result of velocity shear and misalignment, but these are also predictable. Incoming turbulence will generate fluctuating loads, although the range of excursions, particularly in the direction of flow, will be relatively small. Some stochastic inputs will also arise from the effect of storm surges, which may increase current velocities and introduce dynamic loading due to surface wave action. Most of the potential sites are in shallow water, and the rotor blade tips may approach within a few metres of the free surface, where agitation of the water beneath large waves may be significant. This clearly requires systematic investigation, but it may be that the effects are small. Sites will generally be close to land, and the fetches for surface wave development will be limited.

Most recent tidal energy research and development has concentrated on turbines that follow established wind turbine configurations (predominantly of the horizontal-axis type). Deployment of at least two large prototypes is presently under way. However, research continues into determining the operational performance envelope of tidal stream rotors in real conditions.[5]

Eventually, it is probable that the conditions experienced by tidal current turbines will permit precise design solutions, tailored to specific sites. It may also be possible to exploit ideas first suggested for wind energy but rendered impractical for that application by high levels of free stream turbulence. One possibility is the creation of vortices to act as energy concentrators,[6] another is the use of contra-rotating rotors, the subject of this paper.

CONTRA-ROTATING TURBINES

The potential benefits of using a pair of horizontal-axis rotors, turning in close proximity, are the same as originally postulated for wind energy conversion:[7] greater energy capture, higher relative velocities in the power train, minimal reactive torque transferred to the supporting structure and reduced environmental impact.

ENHANCED ENERGY CAPTURE

A contra-rotating turbine has the potential to extract more energy from the flowing stream than a single-rotor device because it eliminates swirl from the downstream wake. Careful blade design is vital to ensure proper interaction between the two rotors. This must be maintained over an acceptably large range of tip speed ratios, during which parts of the aerofoil blades may be operating under stalled conditions. Tidal stream rotors experience lower turbulence levels than their counterparts in wind energy conversion, which makes it more feasible to realise the potential advantages of a contra-rotating turbine.

The use of variable pitch blade geometry would make a practical design easier to achieve, but has thus far been avoided by developers of tidal current turbines in order to reduce complexity. For contra-rotating machine to be competitive, it would be initially desirable to use fixed-geometry rotors.

HIGHER POWER TRAIN VELOCITY

The low rotational speeds of single rotor tidal current turbines result in large step-up ratio gearboxes and bulky multi-pole generators being used to satisfy power quality requirements for electrical network connection. Using the two drive shafts from the contra-rotating rotor to drive the generator's rotor in one direction and the stator in the opposite direction will result in a substantial increase in the drive train's relative velocity. Due to developments in variable speed generators for the wind turbine industry, it may be possible to directly couple the generator drive shafts to the rotor, eliminating the need for a gearbox and lowering the installation mass.

Minimal Reactive Torque

Tidal turbines are slow-turning devices and experience a high reactive torque. The structural implications of eliminating or substantially reducing this torque are much greater than for wind turbines. With a contra-rotating machine, it should be possible under ideal conditions to eliminate the reactive torque. This will enable the deployment of a low cost tensioned mooring system, as opposed to the expensive, rigid structure tower and complex seabed piling/foundations required by single rotor systems. In practice, tidal current turbines will experience a range of approach flow velocities, and it remains to be seen whether low reactive torque can be maintained over the full operating envelope of the device.

Lower Environmental Impact

Marine fouling on and around the turbine should be reduced as a result of the higher relative velocities that will be created by the contra-rotating rotors, making it difficult for marine life to become stationary, deposit and flourish within the immediate environment. The environmental impact of such a device should be less than that of conventional tidal current turbines as the scouring effect of the rotating wake downstream from the rotor will be largely eliminated.

PROJECT PURPOSE

The objective of the project being reported was to develop and test an innovative tidal current turbine consisting of dual, co-axial, contra-rotating rotors directly driving a novel contra-rotating electrical generator. In addition to device performance, the work considered:

- The extremities in tidal flow conditions that the rotor will have to withstand,
- The implications for reduced structural support requirements and
- The reduction in installed cost.

The dual rotor system has a dissimilar number of blades on the upstream and downstream rotors in order to eliminate the power lulls that are produced when the downstream blades are in the shadow of the upstream blades. The research is attempting to establish the optimum blade configuration: e.g. this might comprise 3 blades (120° apart) on the upstream rotor and 4 blades (90° apart) on the downstream rotor. Each rotor drives a shaft, the upstream rotor turning in a clockwise direction and the downstream rotor in an anticlockwise direction. CFD simulations are being undertaken to establish the most effective shaft coupling configuration to enable maximum torque delivery for a range of tidal conditions. The relative rotational speeds between the two shafts will be considerably faster than a conventional turbine shaft, potentially requiring a smaller gearbox to drive the electrical generator. An alternative configuration is also being considered whereby a direct generator drive is used, as is becoming more common on large wind turbines. In addition to establishing and testing a prototype device, the project is also developing a blueprint for a real scale pilot plant and attempting to quantify the potential market.

INITIAL PERFORMANCE APPRAISAL

A major advantages associated with contra-rotating tidal rotors is the reductions in reactive torque (enabling the use of a lower cost mounting structure) and downstream swirl (minimizing environmental impact). In the former case, the reduction of this energy loss path means that the rotor is able to utilise this energy and so improve the capture efficiency compared to a similar sized conventional rotor. (It has long been recognised that turbine performance will improve if attempts are made to minimise swirl, as demonstrated by modern wind turbines, which are designed as high-speed, low-torque machines.) To quantify the potential gain in power production, the power contained in the swirling flow downstream of the rotor plane

Table 1 Horizontal-axis rotor specification.

No. of Blades	**3**	**Aerofoil Section NACA 44 Series**			
Radius at Tip at Root	10 m 1.5 m	Chord at tip at root	1 m 1.6 m	Pitch Angle at Tip at Root	2° 24°
		Chord Variation	Linear	Pitch Variation	Linear
Stream Velocity		1.8 m/s	Fluid Density		1025 kg/m^3

was investigated. Predictions were made for a conventional rotor using standard blade-element theory. The specification of this rotor, as modelled within the present work, is given in Table 1.

Enhanced Energy Capture

Predictions were obtained for the values of the axial and tangential velocity components in the stream at exit from the rotor plane and as a function of radial position along the blade. Distributions of these velocities are shown in Figures 1 and 2, after normalisation against the velocity of the approaching free stream.

In Figure 1 the tip speed ratio is 4 and the predicted power coefficient is 0.401, which is close to optimum for the modelled rotor. In Figure 2, the predicted power coefficient for a tip speed ratio of 2 is 0.099, indicating poor performance. For a tip speed ratio of 4, the blade is close to stall near the root, but otherwise is running at angles of attack between 5 and 7°, as can been seen in Figure 3. In contrast, the results shown in Figure 3 are well into the stall-regulated zone at a tip speed ratio of 2. Here, the entire blade is stalled and the relatively high axial velocities reflect the reduced power extracted by the rotor. The predicted power coefficient is now 0.099.

The tangential velocity components show a different pattern. In the highly stalled case (Figure 2), the swirl behind the rotor is gentle but fairly uniform. In contrast, the swirl pattern in the near-optimum case (Figure 1) more closely approaches a free vortex, reducing from root to tip. Velocity magnitudes are much greater in this case.

The power contained in the rotating, tangential motion in the wake is given by

$$\dot{m}.\frac{u_\theta^2}{2} \qquad (1)$$

where $\dot{m}$ is the mass flow rate and u_θ is the local tangential component of velocity. The power can be summated over the entire flow field for any given situation. This has been done for the simulated rotor for a range of tip speed ratios, and the results are shown in Figure 4. Here the power has been normalised against the power in the approaching free stream. It can be seen that the rotating component never exceeds 1.4% of the free stream value, and under stall regulation it falls well below this value. From this result it appears that the use of a contra-rotating turbine could only marginally increase the power coefficient by about 0.01.

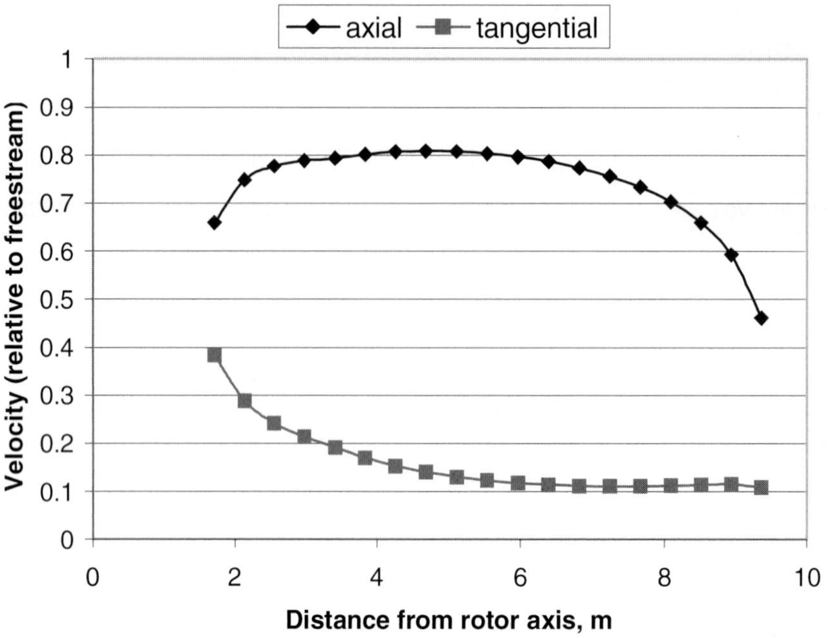

Fig. 1 Axial and tangential velocities along the rotor blade length (tip-speed ratio 4).

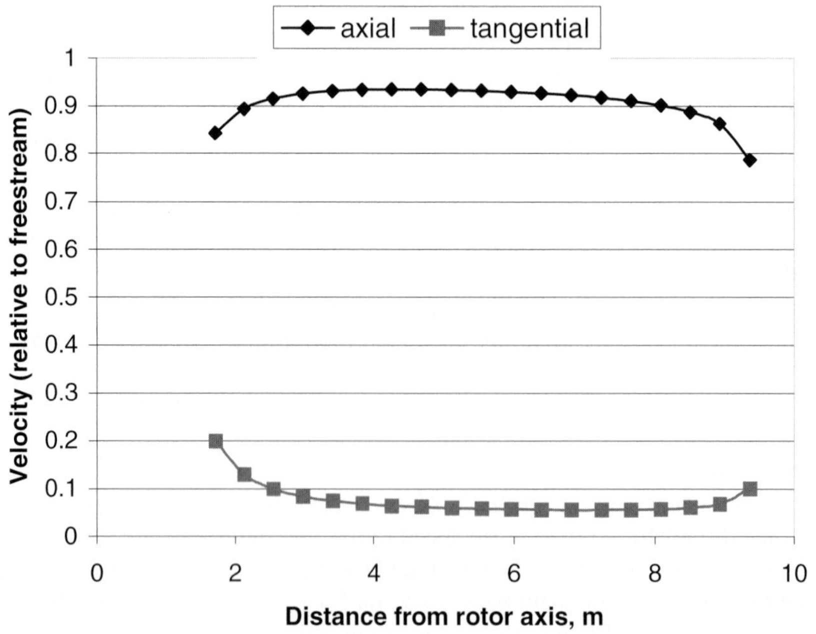

Fig. 2 Axial and tangential velocities along the rotor blade length (tip-speed ratio 2).

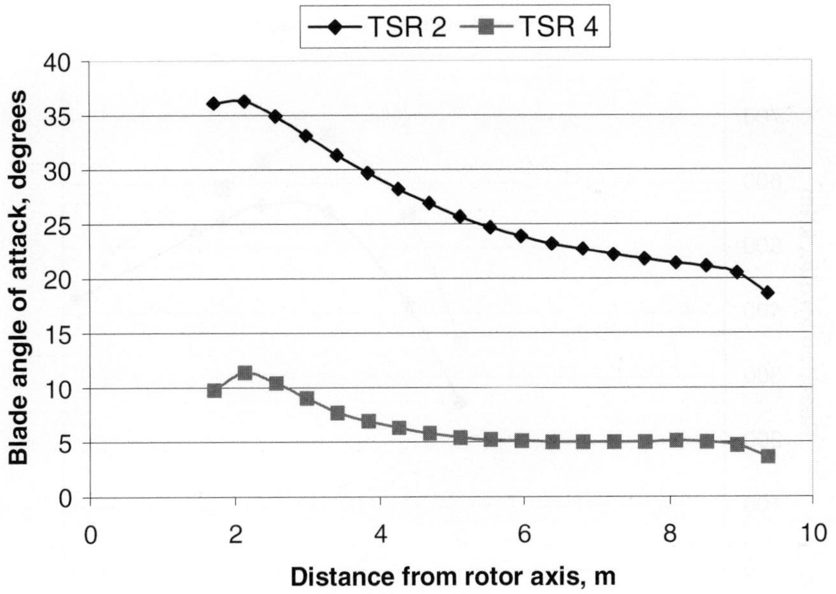

Fig. 3 Blade angle of attack for tip-speed ratios 2 and 4.

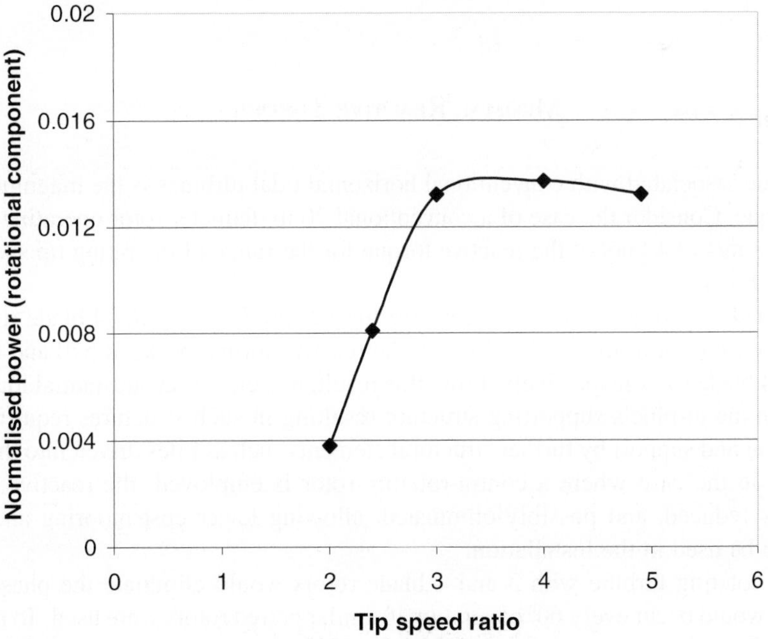

Fig. 4 Normalised power contained within the rotating, tangential motion of the wake.

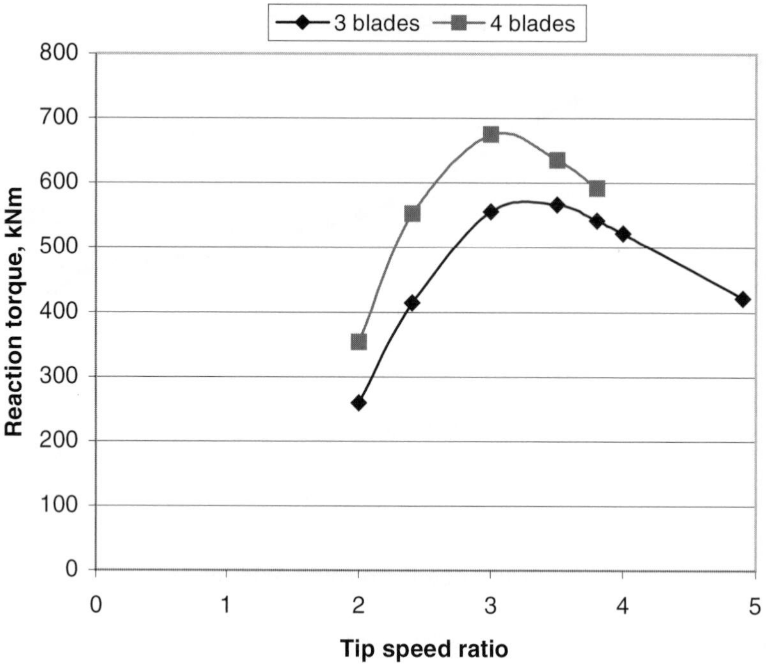

Fig. 5 Reactive torque associated with 3 and 4 bladed, 20 m diameter turbines in a tidal stream of 1.8 m/s.

Minimal Reactive Torque

A major issue associated with conventional horizontal tidal turbines is the magnitude of the reactive torque. Consider the case of a conventional 20 m diameter rotor operating in a tidal stream of 1.8 m/s (3.4 knots); the reactive torque for the range of operating tip speed ratios is plotted in Figure 5.

As expected, the torque (and hence reactive torque) produced by the 4 blade rotor is the greater of the two: for a tip speed ratio of 3, the reactive torque peaks at 570 and 680 kNm for a 3 and 4 blade rotor respectively. From this result, it is clear that substantial loading will be placed on the turbine's supporting structure resulting in such structures requiring heavy reinforcement and support by further structural elements such as piles driven into the seabed. Conversely, in the case where a contra-rotating rotor is employed, the reactive torque is substantially reduced, and possibly eliminated, allowing lower cost mooring and support structures to be used in the installation.

A contra-rotating turbine with 3 and 4 blade rotors would eliminate the phased power pulsing that would occur every 60° of rotation if similar paired rotors were used. To neutralize the impact of reactive torque on the turbine supporting structure, equal and opposite torques must be developed. This implies that the two rotors will operate at different tip speed ratios.

The predictions discussed so far correspond to a fixed geometry rotor, which if operating in a fixed speed mode would limit the power output at high current velocities (low tip speed ratios) by stalling the blades. All marine current rotors constructed to date have adopted this philosophy, but the technology is still in its infancy and variable pitch blades for wind turbines, unknown at the outset, are now the norm on large machines.

However, stall-regulated blades may not be the best option for a contra-rotating turbine. The downstream rotor blades must of course pass through the wakes of those upstream, and if the latter are heavily stalled they will cause a significant degree of disturbance. Where power output regulation is required, pitch control to reduce blade angles of attack towards zero is probably preferable in order to minimise cyclic loads. The downstream rotor, which would be the beneficiary, experiences no such complications and could operate satisfactorily with fixed geometry.

Higher Power Train Velocity

For a conventional single turbine rotor configuration (20 m diameter turbine operating in a tidal stream of 1.8 m/s), the rotor will rotate at approximately 1.7 RPM ($\omega = 0.178$ rad/s). When coupled to a mechanical drive system powering an electrical generator, such a system requires a specialised gearbox with a substantial step-up ratio. If using a conventional two pole-pair generator (1,500 RPM), the hypothetical step-up gearing ratio needs to be in the region of 882:1. If a multi-pole generator is to be used to reduce the engineering challenges associated with the gearbox (>10 pole-pairs reduces the step-up gearing ratio to <176:1), then this introduces complexities associated with rotor speed control–especially where the power generated must comply with the voltage and frequency tolerances permitted under the electricity regulations governing the connection of generation plant to the public electricity supply.[8] In both these cases, the gearboxes are large heavy objects that induce substantive loadings on their structural supports.

The introduction of a contra-rotating rotor should increase the relative shaft speed supplied to a planetary gearbox in the case of a conventional electricity generator. Or, potentially, it could serve as a direct drive to a new and novel contra-rotating electrical generator, thus eliminating the requirements for a gearbox and further minimising the reactive torque. As the shaft speed could increase by as much as a factor of 2 ($\omega < 0.356$ rad/s), the step-up ratio associated with a planetary gearbox would be reduced to approximately 450:1 and 90:1 for a two and ten pole-pair generator respectively. This would considerably reduce the mass to be supported.

Where a contra-rotating generator was employed, the rotor would rotate in one direction and the stator in the opposite direction. This relative increase in generator speed will increase the voltage generated at the terminals as given by where ω is the rotational speed and Φ the magnetic flux. This increase in generator speed and voltage will reduce the number of pole-pairs required and thereby the size/mass of the generator.

$$E_{av} = \frac{2}{\pi}\omega\int_0^{\Pi} \beta l r \cdot d\vartheta = \frac{2}{\pi}\omega\Phi$$

CONCLUSIONS

This paper has reported the interim outcomes from the performance appraisal of a novel contra-rotating tidal current rotor. The work was undertaken to quantify the performance benefits that might be achieved relative to conventional, singular rotor technology. The findings have demonstrated that

- A marginal improvement in rotor performance can be expected due to the capture of the energy normally dissipated in the downstream swirl,
- The reduction in the downstream swirl will minimise seabed scouring,
- A reduction of reactive torque should be possible, enabling a mooring-based supporting system to be employed as opposed to the expensive structural support systems employed with conventional rotor systems,
- The increase in relative shaft speed reduces the sub-sea mass to be supported and the complexity associated with a high step-up ratio gearbox and
- The introduction of variable pitch control on the upstream rotor will optimise the angle of attack for all stages of the tidal stream while ensuring proper interaction with the downstream rotor.

The next stage of the work programme involves additional mathematical modelling of the two-rotor turbine configuration. The expectation is that this will lead to the construction of a prototype for use in empirical research using tow tank testing. The data from such tests will be used to verify the parameter values assumed within the modelling stage. In this way, modelling and laboratory testing will then be used together to identify approaches to the minimisation of the reactive torque over the full operating envelope of the turbine and the development of a new generator system based on direct coupling of the turbine.

REFERENCES

1. UK Government Department of Trade and Industry, 'Our Energy Future – Creating a Low Carbon Economy', DTI, London, UK, URN 03/658, 2003.
2. Scottish Executive, 'Securing a Renewable Future: Scotland's Renewable Energy', Scottish Executive, Edinburgh, UK, 2003.
3. T. Trapp and C. Lomax: 'Developing a Tidal Stream Energy Business', *Proceedings of Oceanology International*, London, UK., M02-013-01.ADT, 2002.
4. P. Fraenkel: 'Marine Current Turbines: A New Clean Method for Power Generation', *European Energy Venture Fair*, Zurich, Switzerland, 2002.
5. *Engineering and Physical Science Research Council*, 'SUPERGEN Wave and Tidal Research Programme', EPSRC, Swindon, UK., 2003.
6. A. D. Grant, C. McGill and M. Thiess: 'Tidal Stream Energy Conversion: Power Augmentation Using Vortices Generated by a Delta Wing', *Proceedings of the 3rd European Wave Energy Conference*, Patras, Greece, 1998.
7. A. F. Stobart: 'Wind Turbine', Patent Number WO 92/12343, 1992.
8. *Electricity Association Engineering Guidance*, 'Recommendations for the Connection of Embedded Generation Plant to the Regional Electricity Companies' Distribution System', London, UK., **59**, 1991.

The Future of Renewable Power in the UK and New Zealand with a Focus on Geothermal Materials Challenges

T. P. LEVI and K. A. LICHTI
Materials Performance Technologies
New Zealand

F. STARR and A. SHIBLI
European Technology Development Ltd.
UK

ABSTRACT

New Zealand and the UK are advanced countries, which share many common features in the generation of power. One significant difference is that New Zealand has no nuclear power and produces around 65% of its electricity from renewable sources, mainly from large-scale hydro. In this respect New Zealand has natural advantages over the UK in the use of renewables. Nevertheless both New Zealand and the UK, because of environmental concerns and worries about the long term future of natural gas reserves, and are now trying to move to the newer greener alternatives, namely electricity produced from wind, biomass, geothermal and wave energy. Hydro power is well developed as is some geothermal and wind. This paper will briefly consider several renewable processes but there is particular emphasis on behaviour of materials in geothermal systems in New Zealand and elsewhere. The importance of long distance energy links in a renewable scenario is also indicated.

NEW ZEALAND AND UK COMPARISONS

New Zealand and the UK are advanced countries, which share many common features in the generation of power. The two countries are of a similar size, with the population split between two separate islands. In the case of the UK these are Great Britain and Northern Ireland, and with New Zealand the North and South Islands. The climates are only superficially similar, with New Zealand being much nearer the equator than the UK. In terms of our hemisphere, New Zealand would stretch from a point in the Atlantic, to the north west of Spain, down to the coast of Morocco, in North Africa. Auckland, the biggest city in New Zealand, is on the exact opposite side of the Globe to Gibraltar. The relatively mild weather in New Zealand stems from it being surrounded by the Pacific, so that the country has a maritime rather than a continental climate. However, the South Island is quite mountainous, with many glaciers and lakes, typical elevations being three times that of the UK. These climatic differences have a major impact on the current and prospective use of renewables.

Population is quite different too, that of the UK is on 60 million, whereas in New Zealand the figure has just reached 4 million. In both countries the population is concentrated in big towns and cities, so that the supply of gas or electricity is, in principle, fairly simple and economic. But most people live in separate houses rather than apartment blocks, contrary to those in Continental Europe. This militates against large scale domestic cogen systems in which a centralised power unit, as well as producing electricity, uses its waste heat to supply hot water for washing and for space heating via hot water or steam pipes laid underneath streets or roads. On a final energy use per head basis, each person in both countries uses about 3 tonnes of oil equivalent.[1,2] Here it is worth pointing out that neither country relies on fuel oil for power production, most of it is used as gasoline or diesel. The UK, in fact, currently exports a net third of its output, even though reserves are on the decline.

The UK and New Zealand both have gas fields and natural gas is now a major player in energy use. New Zealand, until recently, was converting natural gas into gasoline. The manufacture of synthetic gasoline has now been discontinued, although a significant proportion of gas is used to produce fertiliser and methanol, with much of the rest being used for power production. Not much gas is used in New Zealand for domestic heating as the climate is so mild. In the UK there was an embargo for a long period on using 'North Sea Gas' as a power station fuel, gas being earmarked for space heating, cooking and industrial furnaces. The picture has changed with the run down of the coal industry and the rise of the combined cycle gas turbine. Now more electricity is produced in the UK from natural gas than from coal, as the reserves, in fact, are virtually exhausted. Production has declined, fairly uniformly, from a peak of almost 300 million tonnes per annum in 1913 to about 30 million tonnes of coal today.[3] UK economic reserves amount to little more than ten years of output at the current rate. This compares with the UK primary energy needs of about 260 million tonnes of oil equivalent per year.[2] These figures for the UK Coal Industry may surprise some, as these are often quoted in terms of billions of tonnes of coal and hundred of years of reserves.[4] But to get at these will require the robotic mining of small deep seams, whereas the industry has actually concentrated on mechanised systems which can only work where the seams are thick and shallow.[5]

There has been a vast expansion of the gas infrastructure in the UK. Pipelines run between the Continent and the East Coast and there is now a pipeline between the West Coast and Ireland. On a European basis the network is even bigger, having started with the discovery of a huge field at Groningen in the Netherlands in 1959. Western Germany gets much of its gas from Russia and there is pipeline from Algeria, across the Mediterranean, to France. This network has enormous implications for a renewable future, particularly for the UK as pipelines, depleted gas fields, underground storage caverns, and gas holders could be used to transmit and store synthetic natural gas or hydrogen.

The heyday for indigenous gas in both New Zealand and the UK is also coming to an end, and there is some concern about the future. This is especially critical in the UK where the fuel for domestic heating is gas. Hence in conjunction with its use in power generation, over 60% of the primary energy input is from this source. The optimistic view in the UK is that, in the medium term, new fields will continue to be found or we will import gas via undersea pipelines from Norway or Russia. In New Zealand too, there is the expectation that new fields will continue to be discovered. If not, liquefied natural gas might need to be tankered in. However, as much of New Zealand gas is used for electricity generation or methanol production, there is a case for using more coal for power stations and using it to manufacture chemicals. Coal reserves in New Zealand are regarded as healthy, with around 3.5 million tomes being extracted every year, with about half sent for export.

Turning now to the long distance transmission of energy, the **UK National Grid** is much older than the **gas pipeline system** having been started in the mid-thirties. The principal innovation over recent years has been the transmission of electricity via an undersea direct current link. This is now economic over very long distances. There are schemes to supply the British Isles with geothermally generated power from Iceland, a distance of a 1000 kilometres. The situation at present is more prosaic. The south of England gets power across a short 30 kilometre link, using surplus nuclear energy from French PWRs and, very recently, a 500 MW link has been established across the Irish Sea.

Neither of these direct current links are vital to Britain or Northern Ireland. It could be argued, however that a mal-distribution in the UK mainland supply was developing with the concentration of coal fired power stations in the centre of the country, mainly along the River Trent. The situation is now much better with the rise of combined cycle. These plants can be built virtually anywhere since cooling water requirements can be eliminated through the use of air-cooled condensers.[6]

The situation is quite different in New Zealand where the electricity links between the North and South Islands are vital, especially as the use of gas in the domestic and commercial sectors is quite limited, particularly in the South Island. Indeed it can be argued that the link shows what might be needed in a renewable fuel economy. During the colder part of the year, a large amount of power, produced from hydro plants in the South Island, is sent via the link to the North. If hydropower runs low because of a dry season, as has been the case in recent years, power must be supplied from the North Island, as there are no thermal power stations in the South. This has caused some difficulties due to line inadequate capacity.

But New Zealand is significantly nearer a renewable economy than the UK, producing a sizeable fraction of its electrical energy from hydro and geothermal sources. New Zealand has no nuclear plants.[7] In contrast, the UK has no geothermal energy and comparatively little, in percentage terms, hydro plant, needing to produce about a quarter of its electrical power from nuclear energy. However, because of the huge discrepancy in population, the amounts for hydro are not that different; New Zealand's hydro output being about 1.7 million tonnes oil equivalent whereas that of the UK is about 0.4 million tonnes. There is more scope for expansion of hydro in New Zealand than there is in Britain. Nevertheless, hydro faces much the same problem with public opinion, in New Zealand, as does the large scale expansion of nuclear power in the UK.

RENEWABLE ENERGY IN THE EUROPEAN UNION

The world summit on sustainable energy held in Johannesburg in 2002 centred many of the discussions on the issue of renewable energy. It is surprising that the second largest contributor to global electricity production is the 'renewables' accounting for 19% in 2000, ahead of nuclear, gas and oil, although coal was the largest producer with 39%. Interestingly, 17% of the electricity generated from renewable sources came from hydro power plants. Today, technological advances, public awareness of environmental issues and governmental mandates will ensure that the proportion of renewables power will continue to increase.

In the European Union (EU), over the past decade, hydroelectric power consumption has increased by approximately 10% and accounted for almost 5% of the total EU power consumption in 2000. Although other renewables, including geothermal, solar, biomass and wind quadrupled between 1991 and 2000, they still only accounted for 1% of the total EU energy consumption in 2000. It is believed that natural gas and renewable energy will be the two fastest growing fuels over the next 20 years. In 2001 the European Parliament approved a Renewables Directive requiring the EU to double the renewable share of total energy consumption by 2010. To this end European countries have been implementing strategies to increase their use of renewables. The UK hopes to increase the share of electricity generated by renewables from 3% in 2001 to 10% in 2010.[8] In July 2001 the Minister for Industry and Energy made an announcement regarding hydro power that included a 10–20 MW increase in output from each of 30 hydroelectric power stations, which will follow from refurbishment and upgrading. In addition, the British Government is set to invest over the next 3 years approximately $360 million into renewable energy sources. An example of this investment is new wind energy projects that have increased generation capacity by approximately 380 MW and this figure is expected to rise to 800 MW by 2006. The global wind power prospects look encouraging as over the past 20 years the price per kilowatt-hour for wind generation capacity has fallen by a fifth.[9] Italy, for example, relies heavily on oil imports to meet its energy demands and intends to diversify as a top priority. Italy has significant renewable resources such as solar, biomass and geothermal that could be utilized in their goal of doubling the countries energy production from renewable sources by 2012.

Most European countries have been net importers of energy for some time and this will be the situation for the UK quite soon. Only Norway produces a huge surplus, exporting about nine times as much as it uses.[10] Denmark produces a small surplus, but its future seems more assured than that of the UK because of its commitment to highly efficient coal fired stations of the cogen type, plus a very heavy investment in wind power. However at some stage Europe as a whole will need to become more self-reliant. If we assume that the public antipathy to increases in nuclear and hydro power will remain, this will mean an expansion of biomass, wind, wave, tidal, solar thermal and solar photovoltaic energy and, where feasible, geothermal energy. Fortunately, for the UK, by the time that renewables become really important, the European economy is likely to be even more integrated, and energy flows across national boundaries will be even more commonplace than at present.

The geothermal map for Europe as a whole does suggest that geothermal energy can make a contribution. Sadly the prospects for the UK seem limited as for most of the country the ground temperature at a depth of 3 km is below 80°C. Eire is even worse off, it would appear, with temperatures below 60°C.[11] If the UK is a user of geothermal power this will be via the electricity grid. Favoured locations for geothermal energy in Continental Europe would seem to be the west coast of Italy, many parts of what was Yugoslavia, and regions along the Rhine valley, close to Stuttgart in Germany. Here temperatures can exceed 160°C, which, as described in Section 4.1, are seen to be good for power production. In addition

Table 1 Projected electrical output for scenario 2 in year 2050.

Source	Annual Average Gigawatts	Annual Oil Equivalent Million Tonnes
Wind	14.4	10.8
Solar PV	5.0	3.8
Wave	3.75	2.8
Tidal	2.45	1.8
Hydro	0.9	0.7
Energy Crops	10.2	7.7
Agricultural Waste	5.7	4.4
Municipal Waste	1.9	1.4
Total	**44**	**31.6**
2002 Average Demand	**38**	**29**

there are large parts of France, Eastern Spain and Germany where the 3 km depth temperature is between 100–160°C. Here the arguments for power production are more marginal. The prospects for these are improved if combined with cogen schemes or, alternatively, the hot water is used for power station feed heating.

UK FUTURE ENERGY SCENARIOS

The Royal Commission on Environmental Pollution has reviewed various scenarios for the UK in which the aim is to reduce greenhouse gas emissions. Their proceedings were published in June 2000, at about the time of the last Parsons Conference, under the title 'Energy in a Changing Climate'.[12] Four scenarios were considered: Scenario 1 can be described as business almost-as-usual, with the modest attempts to curtail energy demand still resulting in a 30% increase by 2050. The other three scenarios involved massive reductions in energy demand through conservation schemes, etc., with electricity use being cut to about annualised rate of 20–25 GW. Scenario 2 is a projected nuclear free option, involving a major expansion of renewables, plus fossil fuel power stations in which there was **no requirement to sequester carbon dioxide.** Suppression of energy use seems a more realistic approach than sequestration, which itself breeds massive conversion losses. For the renewables sector the split in Scenario 2 is shown in Table 1.

Given that even at the present time, UK **winter peak demand** is about 50% MW more than the average, this means that by 2050 the UK would need up to seventy 400 MW fossil fuel plants just for peak lopping. An issue, which greatly troubled the Royal Commission, is that renewable power is subject to the vagaries of the weather, even discounting the absence of solar energy in the winter. Hence, a very large number of 40 MW standby fossil fuel plants would be needed for those times when renewables simply ceased to operate. Presumably the worst time would be at a period of slack tides, on a dark winter's evening, with widespread zero wind anticyclonic conditions. Nevertheless even in these circumstances, and discounting the fossil plant input, the country would still have a sizeable amount of power to hand. In addition to the energy from biomass and farm wastes, totalling about 15 GW, there would be another 3.5 GW from micro CHP systems. Here it may be argued that the Royal Commission have been unduly conservative, and has also neglected the possibility of importing electrical energy from the Continent or Iceland. Some of this would be using geothermal energy. Done properly, the additional reinforcement of the grid could be an energy saving option in its own right. A recent study from UMIST has shown that if lines are sized to take ten times the normal load, the extra costs over a 20 year period are saved in reduced line losses.[13] The Royal Commission also has overlooked the possibility of house-by-house storage of power using miniature flywheels, these fitting in extremely well domestic cogen schemes.

RENEWABLE ENERGY IN NEW ZEALAND

For the year ending September 2002 renewable energy contributed about 34% of the total primary energy supply in New Zealand of which about 65% was towards electricity generation. Pete Hodgson, the New Zealand Energy Minister, announced the Government target of consumer energy from renewable sources as an additional 30PJ (960 MWh) by 2012.[14, 15] This figure is expected to be attained from development of wind, solar, biomass, hydro and geothermal projects. If for example a growth of between 25–55 PJ (800–1769 MW) of consumer energy can be met by renewable energy this would equate to an additional 19% of renewable source[16] increasing the total renewable production significantly. This strategy is intended to create a sustainable future and assist New Zealand in meeting its international climate change commitments.

WIND

New Zealand, with its prevailing westerly wind, has several suitable sites where speeds average above 36 km/h.[17, 18] The typical capital cost for large turbines attached to the distribution network are NZ$ 1500–2500/kW. Approximately 145 GWh are contributed annually to New Zealand's grid.

HYDRO

Hydro-electricity is a mature and well-understood technology and given the correct circumstances is sufficiently cost effective that it can compete with fossil fuels.

The first public electricity supply in New Zealand, powered by the Reefton hydroelectric plant, started in 1888.[19] Continued development means today large hydro power schemes are the mainstay of the New Zealand electricity supply,[20] producing in the region of 60% of the 39,000 GWh of electricity presently required.[21] Of the many hydro power stations operating in New Zealand at least 10 are between 50 to 70 years old, have the capacity to produce over 600 MW of power and have already produced in excess of 240,000 GWh. Despite their age there have been very few failures and track records suggest that age alone is not a major problem.

Estimate of the capital costs for developing a hydro resource in New Zealand are variable. However, typical current costs are NZ$1,500–NZ$ 8,000/kW. The operational and maintenance costs are in the region of NZ$ 15/kW/year.[22]

Biomass

It is estimated that biomass accounts for at least 15% of total world energy and this figure is likely to be more than doubled in some developing countries. The most recent Danish energy policy sees renewable sources contributing 30% of the total electricity production by 2010 and more than 75% by 2030.[23] Biomass and wind are expected to play the most dominant roles. Overseas co-firing of biomass (e.g. straw and wood) with coal is increasingly common.

The geography and climate of New Zealand provide it an advantage in the biomass energy market. The annual woody biomass residue from forestry alone is estimated to be between 4 and 6 million tonnes. Assuming an energy value of 9 MJ/kg gives 54 PJ this would equate to approximately 10% of New Zealand's total consumer energy demands.[24] In reality, woody biomass contributes significantly less, some estimates put the figure at around 5% of New Zealand's total primary energy demand, and just under 1% of electricity production[25] this is however over optimistic. Landfill gas sites have been successfully commercialised at five locations within New Zealand and have been in operation for some time. In the calendar year 2002 biomass and landfill provided 1.3 PJ of primary energy[26] which equates to less than 0.5%. Within the next decade many advanced and innovative renewable technologies are likely to be commercially viable in New Zealand, some are already being demonstrated and it is believed that several will become competitive with fossil fuels as new and more advanced technology emerges.

The capital costs associated with biomass naturally vary with fuel type and process used. In general it is estimated for landfill, including garbage collection costs, a typical figure in New Zealand is NZ$2,250/kW. Operational and maintenance costs are in the region of NZ$ 70/kW/year.[23]

GEOTHERMAL IN NEW ZEALAND

The Geothermal and Engineering Background

There are four main types of geothermal resources: hydrothermal, geopressurised, hot dry rock and magma. Of these four types currently only hydrothermal resources are commercially exploited. Hydrothermal resources arise when hot water and/or steam is formed in fractured or porous rock in shallow to deep reservoirs (0.1–4.5 km). This is a result of either the intrusion of molten magma into the earth's crust or the deep circulation of water through a friction heated fault or fracture zone. The water in a hydrothermal resource may have been present for a very long period and in some situations the reservoir may begin to run dry after commercial extraction begins. In other cases the water from rain, nearby springs or even the sea will provide a continuous supply, although this too can be depleted. Reinjection of water may be necessary. High temperature resources (180–350°C) are usually heated by molten rock, whilst the low temperature resources (100–180°C) can be produced by either process.[27] In general the high temperature resources are used for power production and the lower temperature resources are used for direct heating, agriculture and low temperature industrial processes as well as tourism. Anticipated developments in hot dry rock and geopressurised reservoir technology should facilitate greater utilisation of geothermal resources in the future.[28] Magma energy production is in concept stage only.

If the exit temperature is in excess of 200°C, it will be possible to produce enough steam directly to drive a steam turbine. At lower temperatures, if a direct system is specified, the mixture of steam and hot water must be flashed off at a lower pressure in a separating vessel, to reduce the amount of water in the steam. Depending on the geothermal heating mechanism, and the source of the water, the steam is likely to be contaminated and this has implications for steam turbine failure mechanisms and materials of construction.

The presence of H_2S, which is common in geothermal steam, could induce hydrogen embrittlement. At Yamagawa in Japan the source of the hydrothermal water was the sea.[29] At Sumikawa deposition on stator blades was occurring due to re-evaporation of contaminated water. This was prevented by water cooling the blades, so as to create a continuous water film on their surfaces (note that the steam temperature was only 151°C.[29] Particulates can give rise to erosion as has been found in a 110 MW turbine in Mexico.[30]

At still lower temperatures an indirect or binary cycle is needed. Here the steam/water mixture is directed into heat exchangers in which an organic liquid such as butane, propane or a fluorocarbon is evaporated.[31] The vapour produced is then used to drive a turbine in a quasi Rankine cycle. Here again, to maximise energy production and improve system controllability, the water phase will be removed from the steam phase and used in the heating of the organic liquid in different parts of the equipment. More recently a Kalina cycle, based on ammonia/water has been used in an Iceland cogen plant.[32]

Historical Development

Electricity was first generated from geothermal water at Larderello in Tuscany, Italy in 1904. Italy is one of the largest producers of geothermal energy in the world, but behind the United States, the Philippines and Mexico. Today Italy has an installed geothermal capacity in excess of 550 MW. Table 2 details some of the global installed generating capacities for 2000.

In New Zealand the high temperature fields are principally located in the Taupo Volcanic Zone. In general the geothermal fields are on the order of 12 square km and spaced about 15 km apart.[20] High temperature geothermal resources have been exploited for energy production since 1958 when the Wairakei plant was commissioned. Secondary uses of produced fluids include agriculture (e.g. drying Lucerne crops at Broadlands, Ohaaki) and aquaculture (pilot prawn farm at Wairakei). Geothermal plant in New Zealand were intended

Table 2 Installed geothermal generating capacities.[32]

Nation	Installed MWe	GWh Generated	% National Capacity	% National Energy GWh
Australia	0.17	0.9		
El Salvador	161	800	15.39	20
Iceland	170	1,138	13.04	14.73
Indonesia	589.5	4,575	3.04	5.12
Italy	785	4,403	1.03	1.68
Japan	546.9	3,532	0.23	0.36
Mexico	755	5,681	2.11	3.16
New Zealand	437	2,268	5.11	6.08
Philippines	1,909	9,181	n/a	21.52
USA	2,228	15,470	0.25	0.4

for base-load and present generation capacity is in the order of 2644 GWh, about 7% of the total electricity demand. This represents about 5% of the world total installed geothermal generating capacity for a country with < 0.1% of the population. Moderate and low temperature geothermal fields are distributed across the North and South Islands.[20]

Costs and Financial Risk Assessment of Geothermal Plant

The capital costs of geothermal power for small, medium and large plants in high, medium and low quality geothermal resources are given in Table 3. The indirect costs vary significantly depending mainly on location, accessibility and infrastructure. In addition annual operational and maintenance costs need to be accounted for and a figure of 18% total cost is a reasonable guideline.[33] Table 4, is for comparative purposes and details the approximate costings for some fossil fired plant as well as other renewables.

Risk assessment is an important part of the development of geothermal fields. International geothermal developments in Pacific Rim countries have shown that using a systematic methodology of exploration and prioritisation success rates of over 80% are achievable. Once prospective sites have been selected then sociological and environmental studies can be performed.

Materials Issues in Geothermal Plant

The selection and use of materials for geothermal energy applications for conventional fluids is well established, whilst research on utilisation of more aggressive fluids continues. Power

Table 3 Direct capital costs (us$/kw installed capacity).[33]

Plant Size	High Quality Resource	Medium Quality Resource	Low Quality Resource
Small plants	Exploration : US$ 400–800	Exploration : US$ 400–1000	Exploration : US$ 400–1000
< 5 MW	Steam field : US$ (200–400)	Steam field : US$ 300–600	Steam field : US$ 500–900
	Power plant : US$1100–1300	Power plant : US$1100–1400	Power plant : US$1100–1800
	Total : US$ (1700–2500)	**Total : US$ 1800–3000**	**Total : US$ 2000–3700**
Medium plants	Exploration : US$ 250–400	Exploration : US$ 425–600	Not normally suitable
5 – 30 MW	Steam field : US$ (180–500)	Steam field : US$ 400–700	
	Power plant : US$ (500–1200)	Power plant : US$ 950–1200	
	Total : US$ (930–2100)	**Total : US$ 1600–2500**	
Large plants	Exploration : US$ 100–200	Exploration : US$ 100–400	Not normally suitable
> 30 MW	Steam field : US$ (200–450)	Steam field : US$ 400–700	
	Power plant : US$ 750–1100	Power plant : US$ 850–1100	
	Total : US$ (1050–1750)	**Total : US$ 1350–2200**	

Table 4 Power station capital, operating and maintenance costs and efficiency at september 2001.[22]

Technology	Size	Capital Cost	O & M cost		Efficiency
			Fixed	Variable	
	MW	NZ$/kW	NZ$/kW	NZc/kWh	%
Conventional pulverised coal with FGD	400	2330	45	0.80	36
Integrated coal gasification combined cycle	400	2840	62	0.19	43
Gas combined cycle	400	856	30	0.12	45
Advanced gas combined cycle	400	1229	28	0.12	49
Combustion turbine	160	706	12	0.02	30
Advanced combustion turbine	120	986	17	0.02	37
Generic distributed generation (base loaded)	2	1297	8	3.56	31
Geothermal	50	3000	93*	(≈6)	(10)
Wind	Variable	2000	28	n/a	(25)
Hydro	Variable	1500–8000	15	n/a	(≈90)

* Includes fixed and variable costs.
( ) Figures in brackets modified for recent experience.

plant developers and operators, in seeking increased efficiencies and cost savings are challenging the limits of conventional fluid technology. The established limits are encapsulated in a set of common 'rules of thumb' that are being forgotten.[36] Key issues and rules are:

1. Corrosion of carbon and low alloy steels is controlled by formation of protective sulphide and oxide corrosion products in near neutral pH fluids. These films are unstable when oxygen is present. Hence, oxygen contamination of geothermal fluids must be avoided.
2. Where stainless steels are selected, to avoid the thick films which are encountered with carbon steels or where erosion can occur, the stainless alloys specified must be resistant to pitting corrosion and chloride and sulphide induced stress corrosion cracking (SCC).
3. H_2S is present in most produced geothermal fluids and hence hydrogen diffuses into the steels. NACE Standard MR0175[37] must be applied so low strength materials with a minimum of cold work are preferred.

Geothermal wells produce either dry stream or two-phase fluid. Dry steam wells produce steam contaminated with non-condensable gases CO_2 and lesser amounts of H_2S and NH_3 and occasionally HCl. Two phase fluids (water and steam) typically contain CO_2, H_2S and NH_3 with acidity being near neutral. If sulphate or HCl acidity is present in two-phase fluids, as in near-volcanic systems, then acidity can be lower.

The first rule is illustrated in Figures 1, 2 and 3. Figures 1 and 2 show potential-pH 'Pourbaix' type diagrams for carbon steel in 180°C geothermal steam where the Total Sulphur

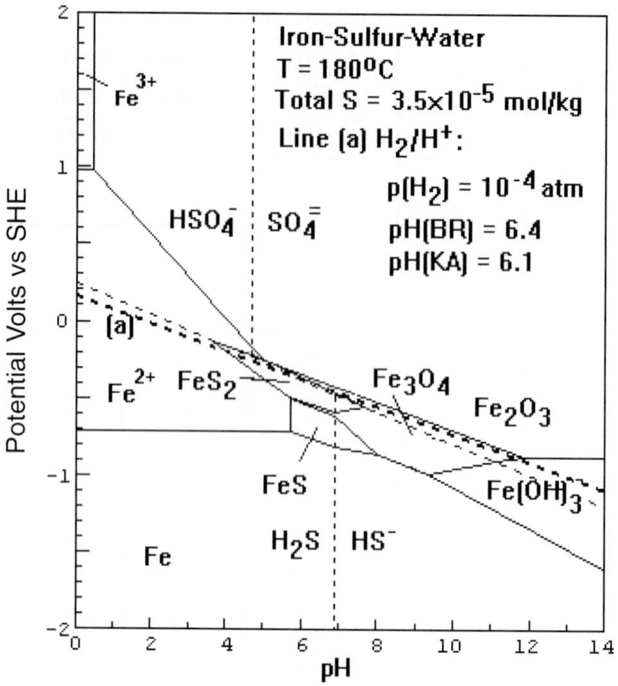

Fig. 1 Potential-pH pourbaix diagram for iron-water-water system with total water = 3.5×10^{-5} mol./kg. Note the sulphide regions meet to displace the magnetite.

concentration varies from 3.5×10^{-5} mol./kg (Figure 1) to 1.5×10^{-5} mol./kg (Figure 2). High temperature pH of geothermal steam, separated from produced waters, is on the order of 5.5 to 7.0.[38] The corrosion products that form first in fluids where the Total Sulphur is greater than 3.5×10^{-5} mol./kg are iron sulphides such as pyrrhotite and troilite. As the layer of corrosion product spreads to fill the entire surface the access of the corrosive fluid to the metal surface is dependent on diffusion processes. The reduced access of sulfur species gives a reduction in Total Sulfur so that magnetite becomes stable next to the metal surface as illustrated in the Pourbaix diagram of Figure 2 and the cross section in Figure 3. These films require near neutral pH and long times, on the order of days and weeks to form in order to block the metal surface.

Figure 4 shows material loss as a function of time measured using on-line probes. The initial high corrosion rate is reduced to low levels in about 10 days of exposure. A consequence of sulfide and oxide film formation is that many geothermal fields show similar corrosion rates as once the films are formed corrosion becomes dependent only on upset conditions. If oxygen is introduced localised corrosion predominates. Figure 4 also illustrates how upset conditions, where air is introduced, can increase corrosion rates.

Operators and developers pressing the bounds of the technology have tended to find the limits of pH empirically, however much research has recently been done to measure these

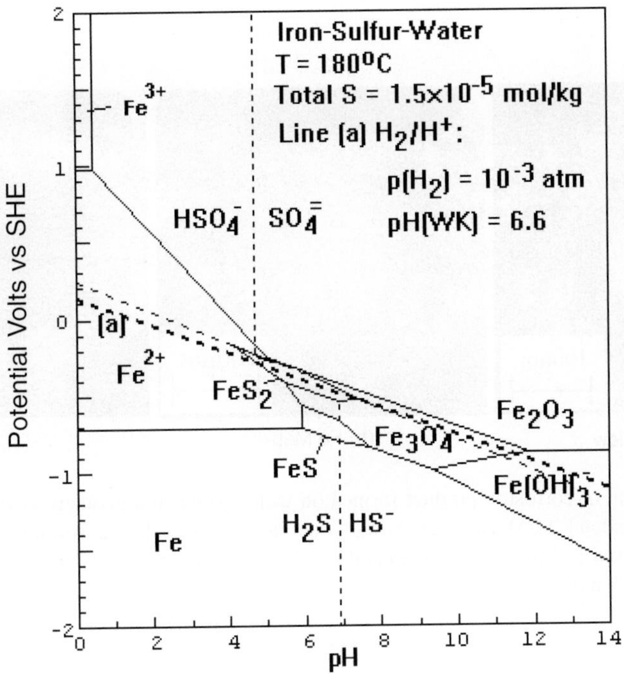

Fig. 2 Potential-pH pourbaix diagram for iron-water-water system with total water = 1.5×10^{-5} mol./kg. Note the sulphide regions are separated and magnetite stability is predicted.

and to develop thermodynamic models as a function of Total Sulphur and pH to describe them. Steam contaminated geothermally produced HCl gas can be utilised if NaOH is injected to neutralise the acidity before the steam condenses. Other acid two-phase geothermal fluids can be used provided the required passive films are developed and maintained, fluids with pH as low as 4.5 can be successfully utilised. Adjustment of lower pH fluids by NaOH injection has been tried but as the effect of stabilising anhydrite and scaling of wells is limiting viability; scale inhibitors are being investigated.[38]

The second rule recognises that if chloride and sulphide are present and pitting corrosion and SCC of stainless steels will occur if the following conditions are encountered:

- Concentration of corrosive species, e.g. chlorides, sulphur, polysulphides and polythionates.
- Moisture or wetness.
- Tensile stress, e.g. residual stress from fabrication or welding.
- Temperature in the range 60 to 180°C.
- Alloy susceptibility.
- Oxygen.

If any one of these were absent or present at very low levels then SCC would not be expected to occur or would take a very long time to initiate. Oxygen is the one parameter that is normally controlled but can be encountered at shutdowns and startups, and on-line at

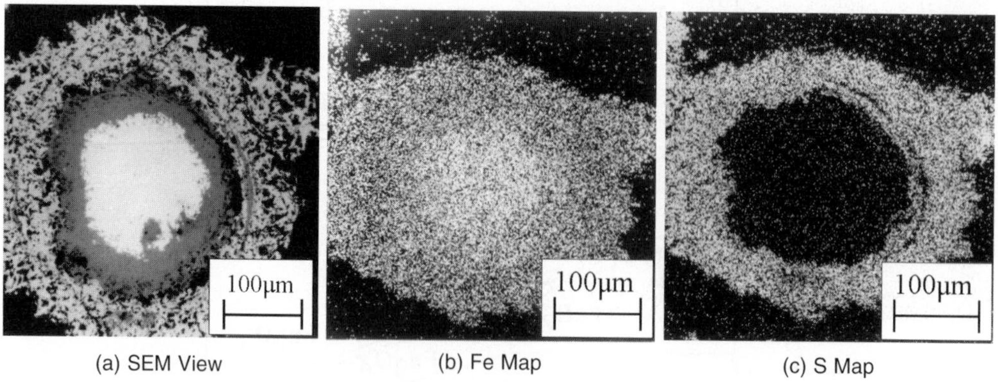

Fig. 3 Illustration of corrosion product formed on weld spatter in a geothermal steam pipeline at Kawerau, New Zealand. SEM view (a) shows a dual layer film while the iron map, (b) shows both layers are iron based and the sulphur map and (c) shows the inner layer is an iron oxide and outer layer is an iron sulphide.

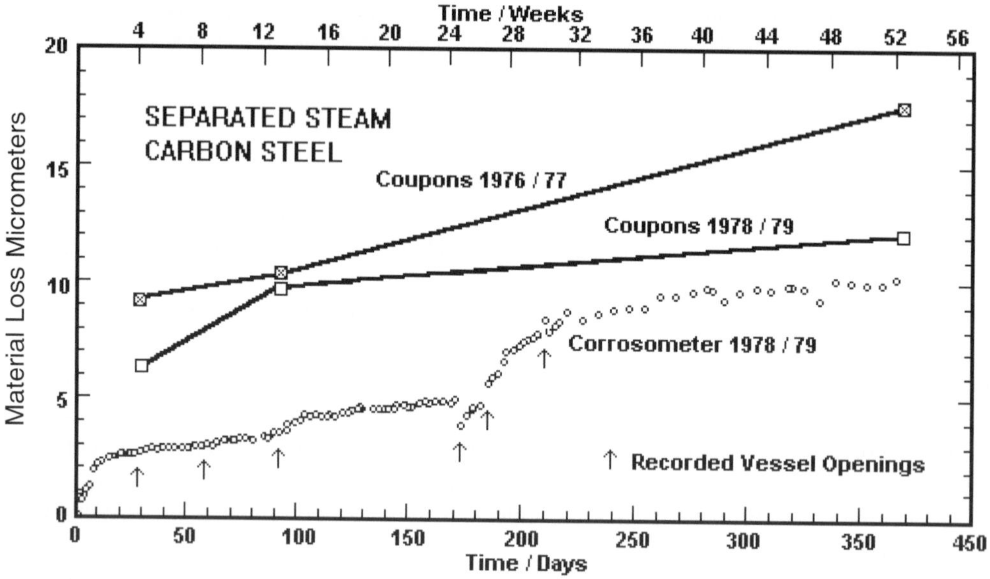

Fig. 4 Illustration of corrosion results from ASTM coupons and on-line electrical resistance corrosometer probes. Note the changes in material loss at times of pressure vessel openings (arrowed). The normally passive films are disrupted by thermal stresses and the local environment becomes acidic when air mixes with the H_2S.

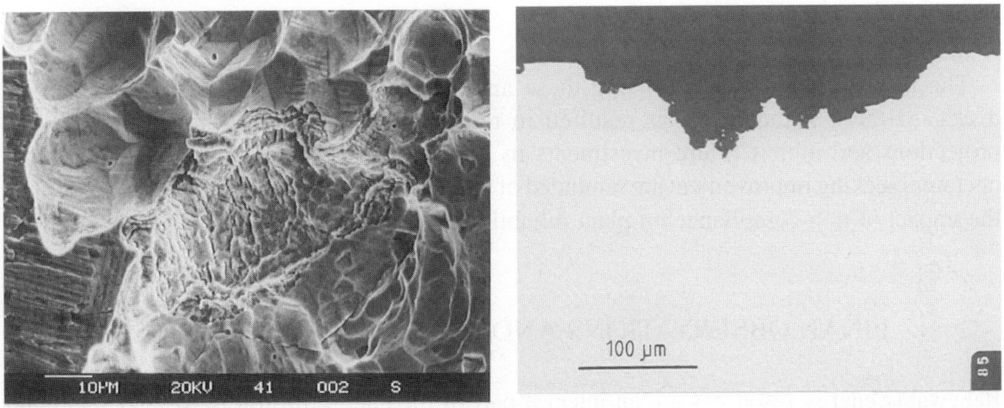

Fig. 5 Performance of sandvik alloy 2RK65 after 63 weeks at 100°C under a drip solution of geothermal steam condensate with 30 mg/kg chloride added.

gland seals and valve stem packings, in gas exhausters and in cooling water circuits. Guidelines for selection of alloys having the required resistance have been developed.[39] Figure 5 illustrates how Sandvik Alloy 2RK65 was susceptible to pitting corrosion under evaporative concentration conditions at 100°C but did not stress crack.[40]

Changing of production conditions can lead to unexpected pitting and Stress Corrosion Cracking failures:

- Water washing of turbines with aerated water has caused pitting of turbine blades and corrosion fatigue.
- Two cycle shifting of geothermal wells has caused pitting on valve stems, galling and seizure.
- Standby conditions has caused SCC of a gas exhauster impellor at a weld repair.

The third rule deals with the preferred selection of low strength materials and conforming to NACE MR0175. Failure to comply with these rules, for all metals and alloys, invariably results in cracking as a result of hydrogen embrittlement. Geothermal environments are reducing and H_2S acts as a catalyst for ingress of hydrogen produced from corrosion reactions.[41] The typical guideline of hardness being less than Rockwell C 22 should only be relaxed for alloys proven by experiment and in-service experience. It can be argued that the levels of H_2S present would not require the Oil and Gas standard to be applied but experience has shown that it should be applied, in all geothermal environments, as the absence of hydrocarbons makes geothermal fluids more aggressive.

The risk of failures is being increased by the need for economies in size and plant fabrication techniques:

- Higher strength turbine blades used to give smaller turbines.
- Rapid welding processes that result in high residual stress in welded pressure vessels.
- Cold working used to fabricate plant where hot forming equipment is not available.

- Avoidance of heat treatment of fabricated components to reduce cost and minimise risk of distortion.

The need to keep re-emphasising these ancient rules is astounding. The tendency to disregard these parameters has resulted in recent failures, which have upset financial projections and inhibit future investments in geothermal energy sources. Developers and operators seeking improvement are reminded of the reasons for these rules and should consider the impact of non-compliance on plant reliability.

FINAL OBSERVATIONS AND CONCLUDING REMARKS

Renewable energy resources are an integral part of the energy profile in New Zealand and can compete with fossil fuels in generating electricity, hydro currently supplying about 60% of the total electricity demand and woody biomass and geothermal supplying about 99% of direct industrial energy used.[42] Of concern for New Zealand is the dependence on hydro and fossil fuels for primary production as environmental factors and diminishing low cost supplies can have dramatic impact on availability and cost.

Although more that NZ$ 1 billion has been invested since 1996 in new generation when the wholesale market began. 27% of the New Zealand's energy needs are being met by stations built in the past seven years, and new generating capacity is required to provide a sustainable energy grid.[43] Some projects under consideration at present are: a plan to build a new power station (either a 192 MW dual-fired station or a 150 MW distillate fired station with the potential to convert to gas in the future.[44] This should be commissioned by June 2004. In the renewables area is projected an expanded wind farm (from 32 to 64 MW), a wind farm to produce between 40 and 80 MW, a hydro station with a 570 MW capacity. However, resource consent and building for hydro projects take a long time and no generation would be expected prior to 2008.[43] Finally, in the distant future it is highly likely that fossil fuel usage will increase. This is hardly surprising as New Zealand has 10 billion tonnes of economic coal reserves.

In the near term, a mothballed, oil-fired power station which has the capacity to generate 250 MWe is being considered for operation.[45] Understandably, until the coal transport infrastructure is improved it would be more economic to import coal to meet winter power shortfalls.[46] It is believed that geothermal development on an international scale will continue to grow steadily in coming years although the rate it still expected to be somewhat linked to the relatively low costs of fossil-fired generation. However, that does not preclude rapid development of wells in individual countries such as New Zealand, Italy, Iceland or South America.

In some respects the UK situation is not too dissimilar to New Zealand. However the large population means that any decision about energy will have major implications for land use, the EU economy as a whole, and the environment. Even now there are big arguments about the visual impact of wind turbines, which of all the renewables promises the most rapid growth in the short term and certainly will be needed if the UK is to meet its 2010 target.

It does seem to the British authors that the issue of trans-border energy flows will be an important factor in ensuring continuity of supply. The proportion of power that comes as a gas or as electricity through a 'grid' is a moot point, but during and after the transition to a renewable economy this will become a critical issue. One feature of gas and direct current electricity systems is that they have tended to be built so that the energy flow is in one direction only. It is one cause of the lack of capacity on the North-South undersea link in New Zealand. The requirement for two-way flows and the need to size the Grid and associated plant to take much more than the near term demand is as much a political as an economic decision. A good sign is that the UK National Grid has firm plans for a two-way 1200 MW link between Northern England and Norway and a similar link between the UK and Holland is also envisaged. Ireland as a whole, particularly with the strong economic growth in Eire, is having to develop a gas and electricity grid, which ultimately will get much of its basic energy from overseas.

Indeed any cost benefit analysis about 'grids' needs to take into account the fact that although types of generating plant come and go, the basic transmission system expands, rather than contracts. Here one can make an historical point that a vestige of the world's oldest energy conversion and transmission system still stands on the former gas works site in Fulham, in London, in the shape of a gas holder built in 1830. Not much is left of the original gas distribution system, however, this was built by screwing lengths of musket barrel together, these being 'surplus to requirements' after the Napoleonic Wars! The current life of UK gas distribution system is 50 years. It is based on the creep life of the material of construction, polyethylene. The life of the high pressure transmission steel pipeline network is fatigue based because of the need to line pack to store gas at night. No doubt corrosion is the long term limit with the electricity grid.

Technical issues, like the above, which govern the life of transmission and distribution networks, always have economic implications. Badly thought out decisions about materials of construction can get some technologies a bad name. As the authors have emphasised, this is important consideration in the development of geothermal resources for power production. Factors, which need to influence materials choice, include the source of the hydrothermal water, temperatures/pressures and the presence of hard particulates. These will vary from site to site. The risk of corrosion by SCC, pitting or erosion can be high, but as this account shows, these problems are easily dealt with. In short, the neglect of well established rules in the design of geothermal systems is foolhardy. Similarly the way that a site is run, so as to minimise oxygen pick up will make a huge difference to the level of corrosion and, in consequence operating and maintenance costs.

Looking again to the future, there are developments taking place with small-scale units in New Zealand that could have enormous implications for power production in Northern Europe. A form of Stirling engine, of about 0.8 kW output, intended for single household cogen systems, is in quantity production at a Christchurch based factory in New Zealand's South Island. What is ironic about this initiative is that because of the absence of a natural gas network in New Zealand, Europe is looked upon as major outlet, although the company has produced hundreds of portable power generation systems for leisure use, even though production only started a year or two back. This New Zealand Stirling is currently being evaluated in a combined central heating boiler/ power generation system by organisations in the UK and The Netherlands. Somewhat further from production is another type of Stirling based on the 'free piston' principle, designed by a US company, and huge effort in England and Japan is now going into its commercialisation One of the authors of this paper was responsible for initiating the work in the UK on these two machines and can attest that Stirling engine based Domestic-Cogen is practical, economic and energy saving.[46]

This type of power unit, although only producing about a kilowatt, could make a huge difference to power production and energy conservation in the EU since very little energy is wasted. Heat in the cooling water or exhaust is picked up by the household central heating system. Given development to a 2 kW size, and assuming that one was installed in every house, there would be no difficulty in meeting a UK peak power demand of around 40 GW, negating the fears of the Royal Commission. We would then have quite literally 'Power from the People' rather than to the people, as is the situation today. Clearly, there are potential competitors to the Stirling engine for this duty. They should be capable of working on any solid and liquid fuel, without the need for chemical processing to convert it into hydrogen, a current drawback of the fuel cell. Hopefully some of these competitors will be turbine based!

So given the right sort of political lead and commercial backing for house-by-house generation, wide spread use of wind, solar, wave and geothermal power, plus massive interlinking of the gas and electricity grids, it would appear that Europe and elsewhere could progress to a genuinely renewable energy economy. This in itself will give a more stable and environmentally safer world. We leave you with these final thoughts!

ACKNOWLEDGEMENTS

The authors would like to thank Andy Bloomer of Century Resources, Wairakei for critically reviewing this paper. The New Zealand Foundation supported the preparation of this manuscript for Research, Science and Technology under contract number CO8X0219.

REFERENCES

1. Energy Data File and Flow Chart, New Zealand.
2. Energy Flow Chart 2001, UK Department of Trade and Industry 2002.
3. A Review of Remaining Reserves at Deep Mines for the Department of Trade and Industry, UK Department of Trade and Industry 2002.
4. BP Statistical Review of World Energy.
5. M. W. Thring: 'The Engineer and Energy', *The Engineers Conscience*, Ipswich Book Company, 1992, 59–130.
6. F. Starr and D. Pierce: 'Air Cooled Condensers for Combined Cycle Plant: The Metallurgical and Fabrication Background', *Parsons 2000*, Strang, et al., eds., IOM 2000, 228–236.
7. 'New and Renewable Energy: Prospects for the 21st Century', UK Department of Trade and Industry, 2000.
8. www.eia.doe.gov
9. www.ewea.org
10. Chart 5.1 'Ratio of Energy Production to Primary Energy in OECD Countries 2000', UK Department of Trade and Industry Energy Sector Indicators, 2003.

11. 'Geothermal Energy: Power from the Heart of the Planet', Understanding Global Issues 4/93.
12. 'Energy in a Changing Climate' UK Royal Commission on Pollution in 2003.
13. G. Strbac and N. Jenkins: 'Network Security of the Future UK Electricity System', Report to PIU UMIST 2001.
14. Media Release EECA 'Energy Supply Sector Enters New Era', 2002.
15. New Zealand's Renewable Energy Target, Energy Efficiency and Conservation Authority and the Ministry for the Environment, 2002.
16. International Energy Agency R&D Wind Annual Report, 2001.
17. Energy Efficiency and Conservation Authority, Energy-Wise Renewables – 3, 1997.
18. www.eeca.co.nz
19. http://environment.about.com
20. www.energyinfonz.co.nz
21. Energy Data File January 2003, Report from the Ministry of Economic Development, 2003.
22. www.med.govt.nz
23. www.ecd.dk
24. www.bioenergy.org.nz
25. 'Energy-Wise News' 2000.
26. Private Communication Graeme Speden, Press Secretary Office of Hon Pete Hodgson, New Zealand Parliament, 2003.
27. www.acre.murdoch.edu.au
28. Renewable Energy Technology Roadmap, ITR 2002/129, Department of Industry, Tourism and Resources, Canberra, Australia, 2002.
29. Y. Nakagawa and S. Saito: 'Geothermal Power Plants in Japan Adopting Recent Technologies', *Proceedings of World Geothermal Congress 2000*, Kyushi, Tohuku, Japan, 2000.
30. Z. Muzur, FZ. Sierra-Espinosa, G. Urquiza-Beltan and J. Kubiak-Senior: 'Erosion of the Rotor Disc in Geothermal Turbine of 110 MW', *Proceedings of 2000 International Joint Power Conference*, Miami Beach, Florida, 2000.
31. ORMAT Web Site for Description of Organic Rankine Cycle for Geothermal Energy, www.ormat.com
32. M. Mirrolli, M. Hjartarson, HA. Mlcak and M. Ralph: 'Testing and Operating Experience of the 2 MW Kalina Cycle Geothermal Power Plant in Husevik', *Iceland OMMI Free Internet Journal on Power Plant: Operation Maintenance and Materials*, **1**(2), www.ommi.co
33. G. W. Hunter: 'The Status of World Geothermal Power Generation 1995–2000', *Proceedings of World Geothermal Congress 2000*, Kyushu-Tohoku, Japan, 2000.
34. www.worldbank.org
35. Private Communication Andy Bloomer, 2003.
36. T. Marshall: 'Geothermal Corrosion', *Metals Australasia*, 1981, 12.
37. NACE Standard MR0175-92, Item No. 53024, 'Sulphide Stress Cracking Resistant Metallic Materials for Oilfield Equipment', 1992.
38. K. A. Lichti, P. T. Wilson and M. E. Inman: 'Corrosivity of Kawerau Geothermal Steam', *Geothermal Resources Council Transactions*, **21**, 1997, 25.

39. K. A. Lichti, C. A. Johnson, P. G. H. McIlhone and P. T. Wilson: 'Corrosion of Iron-Nickel Base and Titanium Alloys in Aerated Geothermal Fluids', *Proceedings of World Geothermal Congress*, Florence, Italy **4**, 1995, 2375.
40. K. A. Lichti, H. Bijnen and P. G. McIlhone: 'Pitting Corrosion and SCC of Some Engineering Alloys in Aggressive Chloride/Sulphide Environments', *Proceedings of Australasian Corrosion Association Conference*, Rotorua, New Zealand, 24, 1984.
41. G. D. McAdams, K. A. Lichti and S. Soylemezoglu: 'Hydrogen in Steel Exposed to Geothermal Fluids', *Geothermics*, **10**(2), 1981, 115.
42. 'Availability and Costs of Renewable Sources of Energy For Generating Electricity and Heat', Report by East Harbour Management Services Ltd. for the Ministry of Economic Development, 2002.
43. The New Zealand Herald, 2003.
44. The New Zealand Herald, 2003.
45. The New Zealand Herald, 2003.
46. The New Zealand Herald, 2003.
47. F. Starr: 'Power from the People', *Ingenia*, Royal Academy of Engineers, (also available on the ETD Ltd web-site www.etd1.co.uk), 2001.

Subject Index

B

G

H

I

M

P

Q

Author Index

P

R

S

T

V

W

Y

Z